EFFICIENT TRANSPORTATION AND PAVEMENT SYSTEMS: CHARACTERIZATION, MECHANISMS, SIMULATION, AND MODELING

PROCEEDINGS OF THE 4TH INTERNATIONAL GULF CONFERENCE ON ROADS, DOHA, QATAR, 10–13 NOVEMBER 2008

Efficient Transportation and Pavement Systems

Characterization, Mechanisms, Simulation, and Modeling

e software mentioned in this book is now available for download on our Web site
: http://www.crcpress.com/e_products/downloads/default.asp

Editors

Imad L. Al-Qadi
University of Illinois at Urbana-Champaign, Urbana, IL, USA

Tarek Sayed
University of British Columbia, Vancouver, BC, Canada
Qatar University, Doha, Qatar

Nasser A. Alnuaimi
Qatar University, Doha, Qatar

Eyad Masad
Texas A&M University at Qatar, Doha, Qatar

CRC Press
Taylor & Francis Group
Boca Raton London New York Leiden

CRC Press is an imprint of the
Taylor & Francis Group, an **informa** business

A BALKEMA BOOK

Cover photo by Chris Miburn

CRC Press/Balkema is an imprint of the Taylor & Francis Group, an informa business

Typeset by Vikatan Publishing Solutions (P) Ltd., Chennai, India
Printed and bound in Great Britain by Cromwell Press Ltd, Towbridge, Wiltshire.

Published by: CRC Press/Balkema
P.O. Box 447, 2300 AK Leiden, The Netherlands
e-mail: Pub.NL@taylorandfrancis.com
www.crcpress.com – www.taylorandfrancis.co.uk – www.balkema.nl

ISBN: 978-0-415-48979-9 (hbk)
ISBN: 978-0-203-88120-0 (ebook)

Efficient Transportation and Pavement Systems – Al-Qadi, Sayed, Alnuaimi & Masad (eds)
© 2009 Taylor & Francis Group, London, ISBN 978-0-415-48979-9

Table of Contents

HMA characteristics and evaluation

Flexible pavement design and assessment

Efficient Transportation and Pavement Systems – Al-Qadi, Sayed, Alnuaimi & Masad (eds)
© 2009 Taylor & Francis Group, London, ISBN 978-0-415-48979-9

Preface

This conference follows the success of the previous three conferences held in Kuwait (2002), United Arab Emirates (2004) and Oman (2006). Upon the completion of the Third Gulf Conference on Roads that was held in Mascut, Oman in 2004, an important decision was made to expand the fourth conference into an international one and to also plan it as a refereed conference. We are so pleased that the conference name and content have changed to reflect this important international expansion.

Today, transportation engineers are always being challenged not only to plan, design and build durable and safe highway systems, optimize the materials used and reduce traveling delay time, but all the while keep a sustainable environment. Hence, this conference is considered timely to share the international experience of academics and professionals with those of the gulf region on transportation systems and pavements. The discussed topics include highway planning, network optimization, traffic congestion and safety, fright transportation, pavement material characterization, analysis, design, and modeling, and pavement mechanics.

This conference represents a major step forward in the gulf series conferences, not only as it becomes international, but also as it becomes a fully refereed conference. All papers were reviewed by at least by two reviewers. Seventy seven papers are included in the proceedings covering recent advances in transportation materials and systems. In addition, the Conference includes six Keynote Speakers, and technical sessions arranged in two parallel tracks. In addition, two workshops on road safety and pavement materials, design and evaluation will be held prior to the conference.

Qatar is very pleased to host the Conference considering the rapid transportation growth in the region and the growing interest in this topic in the Gulf. Papers from 20 different countries were accepted for presentation and publication indicating the international nature of the Conference. The Conference Proceedings provide both a state-of-the-art and state-of-the-practice record on *Traffic Control and Management, Road and Traffic Safety, Highway Design and Transportation Planning, Binder Characteristics and Their Effect on Hot-Mix Asphalt (HMA) and Pavements, HMA Characteristics and Evaluation, Flexible Pavement Design and Assessment, Pavement Distresses and Maintenance, and Infrastructure Management.* In addition papers on *Research, Development, and Standardization* are included.

Finally, we would like to acknowledge the Scientific and Organizing Committees for their continuing input and suggestions throughout the Conference planning and the many individuals who helped review the papers. In addition, we would like to gratefully acknowledge the efforts of Ms. Lina Said for her work on organizing the paper reviews among other tasks and Ms. Zina Al-Azhem for her work on local organization of the conference. The financial support and assistant in logistics by Ministry of Municipality Affairs and Agriculture, Ministry of Interior, Public Works Authority, Urban Planning and Development Authority, and Qatar University in Doha are appreciated. Without the collaboration of all, this conference would never have happened.

November 2008
Doha, Qatar

Keynote presentations

Efficient Transportation and Pavement Systems – Al-Qadi, Sayed, Alnuaimi & Masad (eds)
© 2009 Taylor & Francis Group, London, ISBN 978-0-415-48979-9

Towards long lasting pavements in the gulf

Imad L. Al-Qadi
University of Illinois at Urbana-Champaign, Urbana, IL, USA

Flexible pavements in hot-weather regions are especially susceptible to severe "near-surface" deterioration, including near surface-initiated cracks and permanent deformation (rutting). As a result, creating long-lasting and smooth-riding pavements for these regions is a serious challenge and one of the reasons we have gathered at this conference.

Long-lasting flexible pavements, which combine a rutting-resistant, impermeable, and wear-resistant surface course with a durable base layer, are designed to be structurally sound and to provide more than 30 years of service without major rehabilitations. Although flexible pavement design has gradually evolved from empirical methods to mechanistic–empirical approaches that consider traffic variation, environment, emerging pavement structural configurations, and evolving pavement construction materials, these design improvements have not completely addressed the cracking and rutting issues stated above that are even more pronounced in the Gulf region.

Rutting is a complex phenomenon that depends on various factors such as hot-mix asphalt (HMA) characteristics (aggregate structure, binder stiffness, and air void content), environmental factors (temperature and moisture), loading conditions (vehicle speed, tire load and pressure, and load wandering), and pavement structure design (combination of layer stiffness, interface characteristics, and base support). The instable shear flow in the upper HMA layers results from the complex stress state (normal and shear) induced by moving wheel load. Hence, the rutting-resistant HMA is primarily dependent on its shear resistance and stability.

Therefore, the mechanistic-empirical design of long-lasting flexible pavements in hot regions must consider accurate tire-pavement interaction, material properties, accurate numerical modeling, and full-scale testing. However, current flexible pavement design criteria only accounts for circular loading, inflation pressure equivalent to contact stress, elastic materials, and stationary loading. Unfortunately, these assumptions are inconsistent with realistic loading conditions and may result in erroneous pavement response calculation and pavement performance predictions; especially in hot regions. A linear elastic numerical solution under static loading is questionable at best for predicting pavement responses under moving wheel loads.

The reality of the tire-pavement contact area is that truck tires produce highly non-uniform vertical contact stresses as well as surface transverse and longitudinal tangential stresses. The horizontal, tangential stresses are the result of the inward pressure of tires and the resistance to movement by surface contact. The effect of these stresses is mainly at the surface and shallow depths and is manifested by surface deterioration.

This presentation introduces a well-developed three-dimensional (3D) finite element (FE) model. The FE model incorporates HMA linear viscoelasticity, measured 3D contact stresses, continuous moving load, and implicit dynamic analysis. It has been successfully validated with field measurements. The model's calculated responses, which include shear strains and octahedral shear stresses, have been used to analyze the 3D effect of tire contact stresses on HMA.

The presentation will also address the effects of vehicle accelerating/braking on HMA rutting and/or shoving potential, address the effect of loading on HMA layers' interface and its relationship to slippage and shoving/rutting, and present the application of Accelerated Pavement Testing (APT) to compare the long-term pavement performance between different types of HMA with respect to rutting.

The presentation will show that vertical shear strains near the surface could be responsible for "near surface" cracking and shoving/rutting, while surface tensile strains near the tire-edge rib could be responsible for top-down cracking. The presentation also questions the validity of "bottom-up" fatigue cracking in relatively thick pavements and alternatively suggests that these cracks may instead be initiated at 50–100 mm below the HMA surface.

These new developments in understanding the shear strain/stress distribution and their effect on HMA performance will lead to more appropriate layer stiffness design regarding shoving/rutting in flexible pavements. The end result of these developments will be longer-lasting, durable, and smooth-riding pavements in hot regions which are especially threatened by rutting/shoving and "near surface"-initiated cracking.

Efficient Transportation and Pavement Systems – Al-Qadi, Sayed, Alnuaimi & Masad (eds)
© 2009 Taylor & Francis Group, London, ISBN 978-0-415-48979-9

Pushing the envelope: Transportation system management under competing objectives

Hani S. Mahmassani
Civil and Environmental Engineering, McCormick School of Engineering and Applied Science & Managerial Economics and Decision Sciences, Kellogg School of Management, Northwestern University, IL, USA

Managing the transportation infrastructure entails balancing different objectives that may be in conflict—maximize throughput, maximize reliability and service quality, minimize accidents, minimize vulnerability, minimize environmental degradation and energy use, maximize productivity, and so on. Improvements in one objective sometimes entail degradation in another, requiring difficult trade-offs to be made. This talk will highlight the role of emerging technologies and new operational concepts in the management of transportation networks and delivery of transportation services, in a way that seeks to push the system towards improvements along several objectives simultaneously. The talk will also discuss methodological implications and approaches for both off-line evaluation as well as real-time operational decision-making in this context.

1 BIOGRAPHICAL SKETCH

Dr. Hani S. Mahmassani joined Northwestern University as the William A. Patterson Distinguished Chair in Transportation in September 2007, with joint appointments in the McCormick School of Engineering and the Kellogg School of Management. He was previously the Charles Irish Sr. Chaired Professor and Director of the Maryland Transportation Initiative at the University of Maryland, after 20 years on the faculty at the University of Texas at Austin. He received his PhD from MIT and MSCE from Purdue.

Dr. Mahmassani specializes in multimodal transportation systems analysis, planning and operations, dynamic network modeling and optimization, intelligent transportation systems, system vulnerability and security applications, and freight systems analysis. He has served as Principal Investigator on over 110 research projects funded by national and international agencies, and published over 250 articles in journals and conference proceedings.

Dr. Mahmassani is Editor-in-Chief of *Transportation Science*, and Associate Editor of *Transportation Research C* (*Emerging Technologies*), and the *IEEE Transactions on Intelligent Transportation Systems*. He serves on several NAS and NRC committees and panels, and is a member of steering committee of ITS America Forum for academia and consultants as well as a former Coordinating Board Member, ITS America. He serves as evaluator and advisor to various university research and academic entities, national research institutes, government research programs, and corporate R&D units. He is a consultant to several companies and government agencies in the areas of intelligent transportation systems, transportation network modeling and simulation, policy analysis, strategic systems planning, operations and logistics.

Efficient Transportation and Pavement Systems – Al-Qadi, Sayed, Alnuaimi & Masad (eds)
© 2009 Taylor & Francis Group, London, ISBN 978-0-415-48979-9

Experience on cold in place pavement recycling using the foamed asphalt technique

Andreas Loizos
Laboratory of Highway Engineering, Department of Transportation Planning and Engineering, National Technical University of Athens (NTUA), Athens, Greece

Cold in-place recycling (CIPR) using the foamed asphalt technique offers a promising strategy for the rehabilitation of distressed pavement layers. Due to the high cost of traditional pavement rehabilitation and the environmental benefits associated with the use of recycled pavement materials for the rehabilitation of pavements an increase in the use of recycled materials and techniques has been observed in recent years. Information concerning laboratory testing techniques and related construction aspects are satisfactorily documented in current international literature. However, limited information is available concerning the field performance of foamed treated recycled materials and where available it has been limited to low and medium volume roads.

Due to the lack of field performance data concerning the above mentioned technique the Greek Ministry of Public Works contracted the Laboratory of Highway Engineering of the NTUA, with its experience and study of the aforementioned technique, to undertake a long-term in-situ comprehensive monitoring of pavement performance and data analysis research study. Due to the importance of the Trans European Highway, the investigated highway sections are considered, both for the National and the European database, as an LTPP site.

The field investigation was based primarily on non-destructive tests (NDT), as well, laboratory testing on specimens and cores was also conducted. In-situ collected data were analyzed, in order to estimate the structural condition of the recycled pavement.

The present presentation shares knowledge and experience gained from the above mentioned long-term monitoring study on CIPR using the foamed asphalt technique, concerning pavement and recycled material design, as well as mix design, taking into account current international standards. A critical discussion of the CIPR technique, both for newly built asphalt pavements and rehabilitation of existing pavements, as well as the main findings of the data analysis concerning the conducted field experiment are also presented and discussed.

Keywords: Cold in place recycling, foamed asphalt, NDT

Efficient Transportation and Pavement Systems – Al-Qadi, Sayed, Alnuaimi & Masad (eds)
© 2009 Taylor & Francis Group, London, ISBN 978-0-415-48979-9

Applying fundamental materials engineering to solve pavement infrastructure problems

D.N. Little
Texas A&M University, Texas, USA

Fundamental material and engineering properties of asphalt mixture components can be used to evaluate the performance of asphalt mixtures and to form the baseline by which to understand the fundamental mechanisms of damage. This paper illustrates how two material properties, surface free energy (*SFE*) and molecular properties of the asphalt binder (e.g., shape and size), are used to understand and assist in the performance modeling of the fatigue damage process including the balance between crack propagation during loading and crack healing during rest periods. The paper also explains how *SFE* can be used to select asphalt mixture components (aggregate and asphalt binder) that are the most compatible with each other among the available candidates so that the resulting mixture will have the highest probability to resist moisture damage.

Efficient Transportation and Pavement Systems – Al-Qadi, Sayed, Alnuaimi & Masad (eds)
 ISBN 978-0-415-48979-9

Advances in road safety engineering

Tarek Sayed
Department of Civil Engineering, University of British Columbia, Vancouver, BC, Canada

There has been a serious concern about traffic safety since the start of the automobile age, approximately eleven decades ago. In spite of this concern, traffic safety problems have prevailed over the past century causing enormous economic and social costs. It is commonly accepted that there are many costs associated with vehicular mobility such as air pollution, noise, visual intrusion, collisions, as well as others. However, the economic and social costs associated with road collisions greatly exceed other mobility costs due to the pain, grief, loss of property, injury, and deaths attributed to road collisions. Worldwide, about 1.2 million people are killed in road collisions each year. In the US, more pre-retirement years of life are lost due to road collisions than from the combined effects of the two leading diseases, cancer and heart disease. In Qatar, the road safety problem has persisted over the last decade. Every year, about 250 people are killed in traffic collisions across Qatar. The majority of these fatalities involve young Qataris. Thousands of other collisions that cause property damage occur annually. The cost of these collisions is estimated in billions, and is borne by the health care system, the insurance industry, emergency response services, the legal system, and ordinary people whose lives are shattered.

Road collisions occur due one or a combination of the three components of the road system: the driver, the vehicle and the road environment. Several studies have shown that more than 90% of collisions involve some driver error. This statistic can mislead one to think that the focus of road safety improvement programs should be on the driver. However, driver related road safety improvement programs may not be the most cost effective solution to safety problems. There is a growing appreciation in the road safety community for the need to improve traffic safety through better road engineering. After a long period when the emphasis was on driver education and police enforcement, there has been a relatively recent renewed commitment across the world to examine the role of road engineering in improving safety. Geometric design guides in many countries now present explicit relationships between road design decisions and safety consequences, moving road engineering away from reliance on minimum standards without regard to safety. Road safety audits at the planning and design stage of new projects have also emerged as a new and effective tool to improve safety such that roads are planned and designed with due regard for safety.

The focus of this presentation is on engineering initiatives that can be employed to improve road safety. An overview of traditional road safety improvement programs will be provided along with emerging trends in road safety planning and engineering. These emerging trends include: safety conscious planning, the explicit evaluation of safety in road design, safety audits, the use of risk-based design, and the use of intelligent transportation systems (ITS) technologies.

Efficient Transportation and Pavement Systems – Al-Qadi, Sayed, Alnuaimi & Masad (eds)

Dynamic early merge and dynamic late merge at work zone lane closures

A.E. Radwan
College of Engineering and Computer Science, University of Central Florida, Orlando, FL, USA

R.C. Harb
CATSS University of Central Florida, Orlando, FL, USA

To improve traffic safety and efficiency in work zones, several states in the U.S. introduced and tested the Dynamic Lane Merge System (DLM), an intelligent work zone traffic control system. The dynamic lane merging, can take two forms; dynamic early merge and dynamic late merge. Previous DLM applications demonstrated the strengths and weaknesses of each merging scheme. However, up to date the literature lacks of a comparison between the two aforementioned dynamic merging schemes. This manuscript reviews previous DLM applications and suggests two simplified dynamic merging systems; Simplified Dynamic Early Merging (SDEM) and Simplified Dynamic Late Merging (SDLM) for testing on short term work zones.

Transportation systems

Traffic control and management

Efficient Transportation and Pavement Systems – Al-Qadi, Sayed, Alnuaimi & Masad (eds)
© 2009 Taylor & Francis Group, London, ISBN 978-0-415-48979-9

Towards fully integrated adaptive urban traffic control

Baher Abdulhai
Canada Research Chair in Intelligent Transportation Systems,
University of Toronto, Toronto, Canada

Hossam Abdelgawad
Department of Civil Engineering, University of Toronto, Toronto, Canada

ABSTRACT: Advancements in Intelligent Transportation Systems (ITS), communication and information technologies have the potential to considerably reduce delay and congestion through an array of network-wide traffic control and management strategies. At the University of Toronto's ITS Centre, a comprehensive ITS research and teaching program was initiated in 1998. One of the prime goals of the centre is to advance the state of art in traffic control and management. Over the past decade, researchers developed a set of traffic control and management components with an integrated vision in mind. This set includes computer systems for adaptive freeway incident detection, traffic flow forecasting, adaptive arterial signal control, adaptive freeway and corridor control, and a platform for live traffic data gathering, storing and dissemination. In this paper we give an overview of the vision and the overall system, identify its essential components developed to date, and briefly describe the technical background of each component together with quantitative performance analysis.

1 INTRODUCTION

Although transportation plays a fundamental role in societal fabrics and national economies worldwide, we often do not pay attention until we are stuck in a traffic jam. Congestion, environmental degradation and global warming are all quickly becoming 'household' concerns all over the world.

There are many categories of solutions to traffic problems. The list of potential solutions ranges from better demand management to reduce car dependency to better supply management using cutting edge technology or "Intelligent Transportation Systems" (ITS).

The ITS field has been aggressively growing over the past decade or so. ITS capitalize on emerging computer and communication technologies to better manage the increasingly congested and dynamic transportation networks of today. ITS involves the application of Information Technology (IT) and other advanced methods and techniques to improve the performance of transportation systems and to increase the contribution of these systems to our economic and social well-being. ITS involve a complex interplay between technology, human perception, cognition and behaviour, and social, economic, and political systems. Thus, ITS is inherently multi- and inter-disciplinary in nature.

The "heart" of ITS undoubtedly lies in system's surveillance and control in real time. Integrated system control has the potential to considerably reduce delay and congestion through an array of network-wide traffic control and management strategies. At the University of Toronto's ITS Centre, a comprehensive research and teaching program was initiated in 1998. One of the prime goals of the centre is to advance the state of art in traffic control and management. Over the past decade, we developed a set of traffic control and management components with an integrated vision in mind. This set includes computer systems for adaptive freeway incident detection, traffic flow forecasting, motorist information for border crossing, adaptive arterial signal control, adaptive freeway and corridor control, and a platform for live traffic data gathering, storing and dissemination. In this paper we give an overview of the vision and the overall system, identify its essential components developed to date, and briefly describe the technical background of each component together with quantitative performance analysis. A common thread amongst all of the system's components is a technological core based on Artificial Intelligence (AI). More specifically, three AI technologies were utilized; neural networks, genetic algorithms and reinforcement learning. The motivation is to utilize the capabilities of AI to produce a self-learning perpetually adaptive traffic control systems as described in the paper.

Firstly, the paper briefly highlights the role of advanced traffic management system (ATMS) in traffic control and management. In the following sections, a set of ATMS developed at the ITS center and Testbed at University of Toronto are described. In particular, we highlight The ITS Center and Testbed (ICAT) platform (Foo, B. *et al.*, 2006), Genetic Adaptive Incident Detector-GAID (Roy, P. & Abdulhai, B., 2003), Reinforcement Learning Signals Control (Abdulhai, B. *et al.*, 2003) and Automated Adaptive Traffic Corridor Control (Jacob, C.& Abdulhai, B.,2006). We provide brief technical background and some relevant quantitative measures. Interested readers are advised to refer to the corresponding references for more comprehensive details and case studies.

2 INTEGRATED ATMS OVERVIEW

As our roads and freeways are becoming more and more congested, capacity expansion is becoming less feasible and less desirable from a sustainability perspective. A viable alternative to conventional capacity expansion is to improve the operational efficiency of existing infrastructure through ATMS, a branch of ITS. The goal of ATMS is to improve the overall efficiency of the traffic network, which results in a reduction of traffic delay, pollution and incidents. Traditional traffic management strategies, such as prompt response and clearing of an incident, are already making very significant improvements in the efficiency and safety of the system. However, further improvements can be achieved with a deeper understanding of the underlying traffic flow behaviour, and an advanced traffic management strategy that is tailored to that behaviour. Advanced computer analysis and control algorithms are required to achieve these goals, and a research, training and test facility is necessary to develop these algorithms. (Foo, B., *et al.*, 2006)

ATMS construct a group of independent but interrelated elements comprising a unified whole in order to achieve certain goal(s) while maintaining the system's stability. Figure 1 depicts what

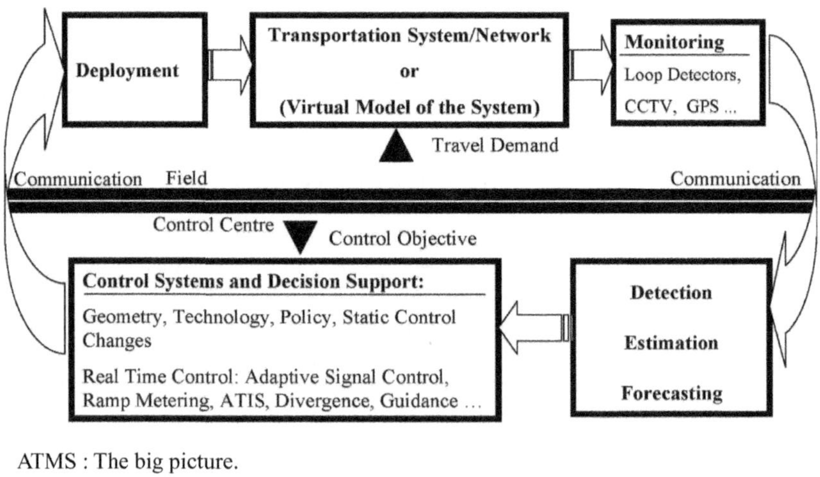

Figure 1. ATMS : The big picture.

we call "ATMS: The Big Picture". As illustrated in the figure, the system is conceptually split into a real world side and a control center side that are connected via wireline and wireless communication technologies. Interacting with the real transportation network is often desirable. However, this may not always be feasible. Instead a virtual replica of the transportation system is approachable through *microsimulation modelling tools and techniques*. In either the real system or in the virtual model, the system's state and functionality are monitored using detectors. Monitoring the transportation system encompasses updating the system's states through loop detectors, closed circuit TVs, GPS, etc. Surveillance information is then *communicated* to the control centre. Surveillance data are used to intelligently *detect* and possibly *forecast* uncertainties that might affect the stability of the transportation system. A *controller* is then called in order to explore a set of actions that would enhance the state of the system. The optimal control actions are then communicated back to control actuators in the field for deployment. This ATMS cycle continues perpetually.

3 THE ICAT PLATFORM

The ITS Centre and Testbed (ICAT) at the University of Toronto was officially opened in November, 2001, and the ICAT was brought online with both the provincial Ministry of Transportation of Ontario (MTO) and municipal City of Toronto through fibre-optic connections in March, 2003. ICAT is a facility that enables the research and development of advanced computer analysis and control algorithms that are required to further improve the operational efficiency of Toronto's traffic network. The ICAT receives live traffic data and video from the MTO and the City of Toronto through the dedicated fibre-optic connections. The ICAT database stores 20-second loop detector data, messages displayed on variable message signs (VMS), operator incident logs and 3-minute snap-shots of camera images. This data can be accessed by practitioners and researchers using the ICAT platform web interface (Foo, B. *et al.*, 2006).

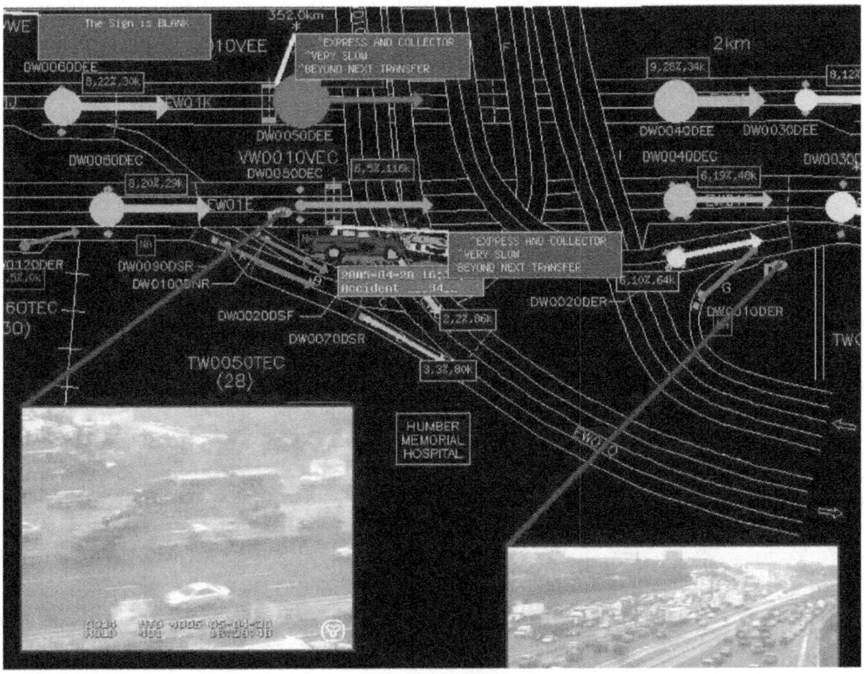

Figure 2. Sample traffic data map (*Source: Foo B. et al., 2006*).

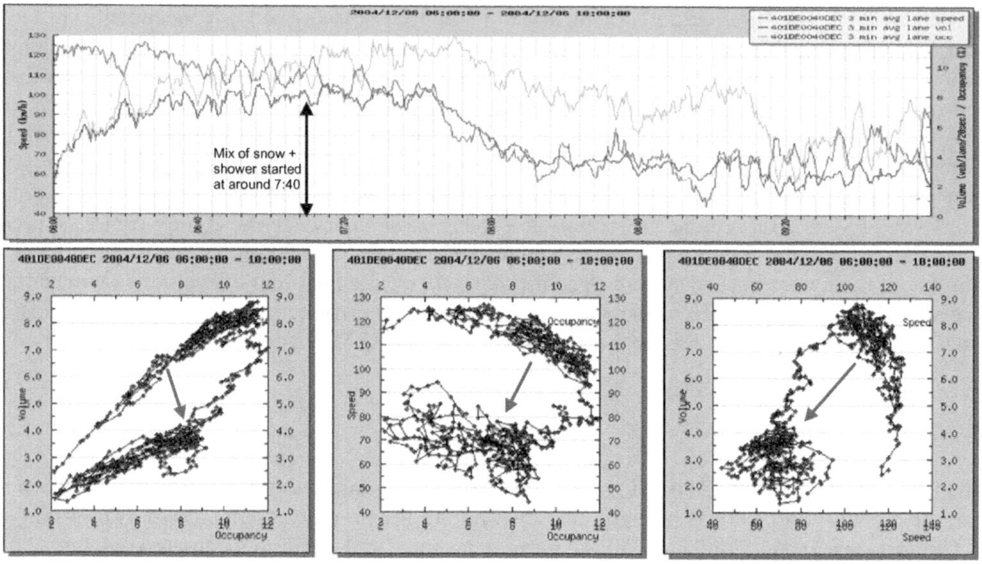

Figure 3. Traffic data plots available on the ICAT website showing the effects of mixed precipitation on traffic (*Source: Foo B. et al., 2006*).

3.1 *ICAT platform web interface*

The ICAT Platform website is a front end human interface for the received traffic data. The website is coded with the help of free software for configuring the layout's content and plotting the data). It has the capability to display (real time or historical) data on a highway schematic map, make time series plots as well as fundamental plots of traffic data, display current and historical VMS messages, and display current and historical incidents logged by the operator. Figure 2 shows a sample of the traffic data map available on the ICAT website. The loop detector data is superimposed on the schematics where all three variables (volume, occupancy and speed) are represented graphically. Figure 3 shows the traffic data plots taken on December 6th, 2004 in the morning. According to Environment Canada (2005), a mixture of rain and snow started at around 7:40 am. A significant change of traffic behaviour can be observed from the three fundamental plots (Foo, B. *et al.*, 2006).

4 INCIDENT DETECTION: A GENETIC ADAPTIVE INCIDENT DETECTION

Successful Freeway Traffic Management Systems (FTMS) and ITS rely on timely detection of any traffic disruptions or occurrence of incidents. Ironically, incident detection remains one of the weaker links in implementing ITS traffic control concepts because of the far from perfect present performance of existing automated incident detection algorithms. The problem is accentuated with the different traffic, geometric and environmental conditions at different locations, resulting in vastly different requirement of detection algorithm parameters at each detector locations.

In the endeavor to develop an incident detector with better obvious features of higher Detection Rate (DR), lower False Alarm Rate (FAR) and low Mean Time to Detection (MTTD) with an added feature of minimized location dependence (better transferability), a new incident detector based on genetically optimized probabilistic neural network has been developed. The detection logic comprises of a genetically optimized Probabilistic Neural Network (PNN) for freeway incident detection (Abdulhai B. & Ritchie S. (1997), Abdulhai B. (1996), Abdulhai B. & Ritchie S. (1995)).

An important advantage of the present GAID (Genetic Adaptive Incident Detection) incident detector is a distinct smoothing parameter for each of its input dimensions (variables). This not only increases the generalization accuracy of the algorithm, but also makes it adapt to new site conditions without human intervention via automated optimization using Genetic Algorithms (GA) (Abdulhai, B., 1996).

4.1 *Genetic adaptive incident detection: Brief description*

PNNs have long established themselves fundamentally as classifiers. The idea behind linking PNN and incident detection started with an attempt to separate the incident 'classes' from normal 'classes'.

GAID is a combination of two subset procedures. The first one is 'GAID-Optimizer' that contains the genetic optimization of the multi- σ PNN. The second one is 'GAID-Detector' which is a neural network classifier that processes incoming loop detector data to detect any probable incident and hence generate an alarm. The GAID-Optimizer contains detailed data and graphs showing the progress of the GA training across the generations. It also contains windows for table and charts showing the changes of relative importance of σ parameters across the runs. A snapshot of the GAID-Optimizer is shown in Figures 4.

GAID-Detector implements modified input feature space rather than the direct volume and occupancy values. Aptly called 'Deviations from the Averages for the Time and Location' (ATL) (Masters, T., 1995) the detector subtracts average of volume and occupancy values for the given time period and freeway location from the present values of volume and occupancy. As the traffic data to be classified become available, and instead of feeding them directly into the detection algorithm, the historical ATLs are first subtracted from the on-line values and only the deviations from these ATLs are fed into the algorithm. The resulting algorithm would be more transferable

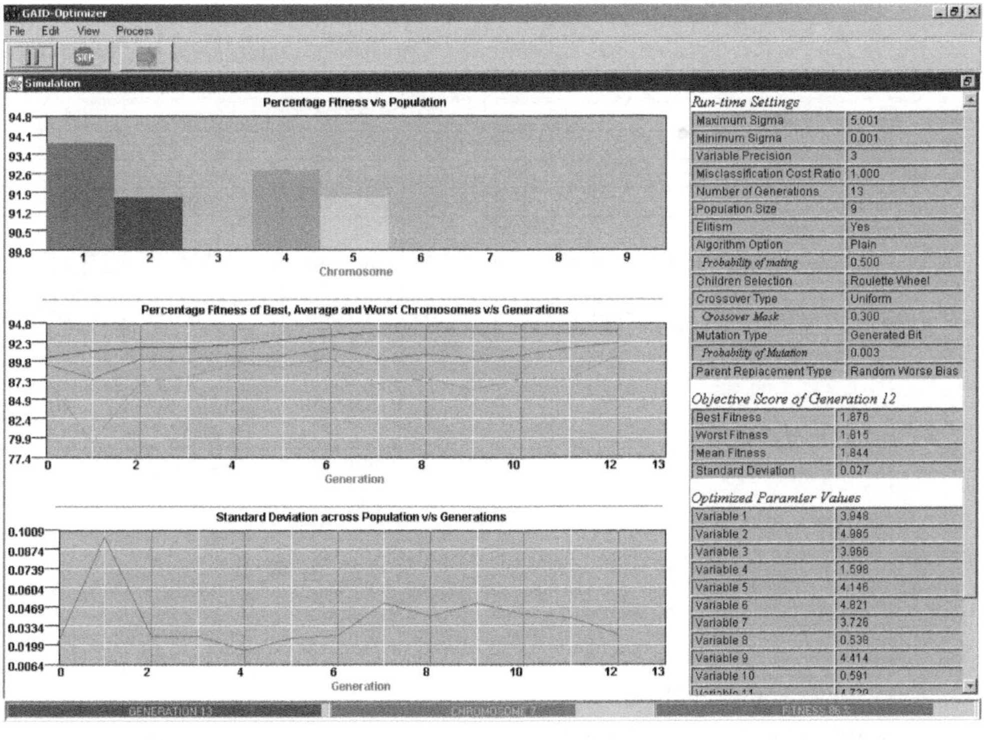

Figure 4. Snapshot of GAID-Optimizer during optimization progress (*Source: Roy P. & Abdulhai B., 2003*).

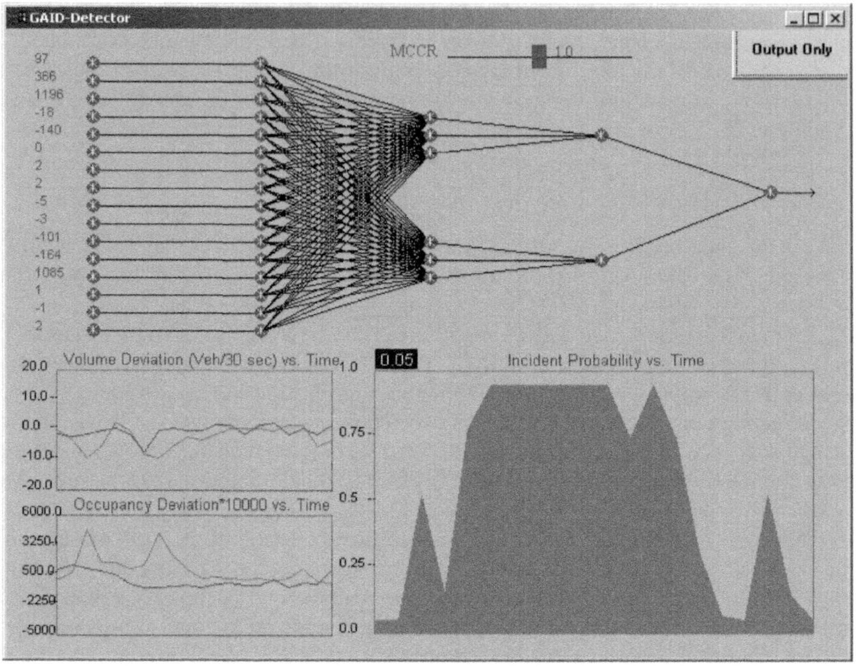

Figure 5. Snapshot of the GAID-Detector (*Source: Roy P. & Abdulhai B., 2003*).

Table 1. Performance comparison of incident detection algorithms (I405N training data and I405N validation data) (*Source: Roy P. & Abdulhai B., 2003*).

Attributes	PNN	GAID	McMaster (no optimal)	McMaster (optimal)
Incident Classification %	73.99	76.91	6.14	64.69
Normal Classification %	90.99	92.00	99.19	69.40
Overall Classification %	85.50	87.12	69.11	67.88
Detection rate %	96.90	97.67	37.98	93.8
False Alarm (/detector/number of normal observation)	0.0173	0.0149	0.0029	0.0354
False Alarm (/detector/hr of non-incident condition)	2.0759	1.7893	0.3486	4.2448
Mean Time to Detection (sec)	86	84	222	84
curvature			-525.92	-400
slope			191.13	200
intercept			-2.26	5
ocrit			0.23	0.27
vcrit			12	24

compared to one classifying absolute volume and occupancy values (Masters, T., 1995). A snapshot of the GAID-Detector is shown in Figure 5.

4.2 *Comparative analysis of GAID*

In addition to the usual Detection Rate (DR), False Alarm Rate (FAR) and Mean Time to Detection (MTTD), the authors suggest three more criteria: percentage correct classification for incident condition, percentage correct classification for normal condition and percentage overall correct classification. GAID is compared to the McMaster incident detection algorithm (Hall, F. *et al.*, 1993) and the original PNN (Abdulhai, B., 2003).

A sample of the comparative results is shown in Table 1. All algorithms were trained with the training data and their performances were evaluated on the validating data. The McMaster optimized parameters are tabulated below the table. Note the significant difference of parameters for the McMaster algorithm and the optimized McMaster algorithm. From Table 1, the original McMaster algorithm fails in the detection and correct incident classification. Even with the optimized parameters, the percentage correct classifications are very low compared with the PNN and GAID. GAID's performance appears the best. GAID has added benefit of being highly transferable. On-site learning capability of GAID in real time is simple, fast and fully automated, conditioned on user simple approval. (Roy, P. & Abdulhai, B., 2003).

5 REINFORCEMENT LEARNING SIGNAL CONTROL

The ability to exert real-time, adaptive control over a transportation process is potentially useful for a variety of ITS services, including control of a system of traffic signals, control of the dispatching of paratransit vehicles, and control of the changeable message displays or other cues in a dynamic route guidance system, to name a few. A key limitation of conventional control systems is a requirement for one or more pre-specified models of the environment. The machine-learning research community, related to the artificial intelligence community, provides us with a variety of methods that might be adapted to transportation control problems. One of these, particularly useful due to its conceptual simplicity, yet impressive in its potential, is reinforcement learning (see Sutton & Barto (1998) or Kaelbling *et al.* (1996) for comprehensive overviews or Bertsekas and Tsitsiklis (1996) for a more rigorous treatment) (Abdulhai, B. *et al.*, 2003).

5.1 *Reinforcement learning and Q-learning*

In its simplest terms, reinforcement learning (RL) involves an agent that wishes to learn how to achieve a goal. It does so by interacting dynamically with its environment, trying different actions in different situation in order to determine the best action or sequence of actions to achieve its goal from any possible given situation. Feedback signals provided by the environment allow the agent to determine to what extent an action actually contributed to achievement of the desired goal. A simple RL algorithm is known as Q-learning. The agent is the entity responsible for interpreting sensory inputs from the environment, choosing actions on the basis of the fused inputs, and learning on the basis of the effects of its actions on the environment. At time t, the Q-learning agent receives from the environment a signal describing its current state s. The state is a group of key variables that together describe those current characteristics of the environment that are relevant to the problem.

Based on its perception of the state $\boldsymbol{s}$, the agent selects an action $\boldsymbol{a}$, from the set of possible actions. This decision depends on the relative value of the various possible actions, or more precisely on the estimated Q-values $\boldsymbol{Q}_{s,a}$, which reflect the value to the agent of undertaking action a while in state s, resulting in a transition to state $\boldsymbol{s'}$, and following a currently optimal policy (sequence of actions) thereafter. At the outset, the agent does not have any values for the Q-estimates and must learn these by randomly exploring alternative actions from each state. A gradual shift is effected from exploration to exploitation of those state/action combinations found to perform well.

As a result of taking action $\boldsymbol{a}$ in state $\boldsymbol{s}$, the agent receives a reinforcement or reward $\boldsymbol{r}_{s,a}$ which depends upon the effect of this action on the agent's environment. The objective of the agent in seeking the optimum policy is to maximize the accumulated reward (or minimize the accumulated penalty) over time. A discount rate may be used to bound the reward, particularly in the case of continuous episodes. The discount rate reflects the higher value of short-term future rewards relative to those in the longer term.

The combination of state $\boldsymbol{s}$, action $\boldsymbol{a}$, and reward $\boldsymbol{r}_{s,a}$ is then used to update the previous estimate of the Q-value $Q_{t-1(s,\,a)}$ recursively according to the following training rule:

$$\delta = \alpha_{s,\,a}(\boldsymbol{r}_{s,a} + \gamma_t \cdot \boldsymbol{MAX}\,(\boldsymbol{Q}_{t-1(s',\,a')}) - \boldsymbol{Q}_{t-1(s,\,a)}) \tag{1}$$

where:

δ = the increment to be added to the previously estimated Q-value, $\boldsymbol{Q}_{t-1(s,\,a)}$ to geat $\boldsymbol{Q}_{t(s,\,a)}$

$\alpha_{s,\,a}$ = a training rate in the interval [0,1]

$\boldsymbol{r}_{s,\,a}$ = the reward received for taking action $\boldsymbol{a}$, while in state $\boldsymbol{s}$

γ_t = a discount rate in the interval [0,1], applied to future rewards

$\boldsymbol{MAX}\,(\boldsymbol{Q}_{t-1(s',\,a')})$ = the previously estimated Q-value following the optimum policy starting in state $\boldsymbol{s'}$

$\boldsymbol{Q}_{t-1(s,\,a)}$ = the previous estimate of the Q-value of taking action $\boldsymbol{a}$ while in state $\boldsymbol{s}$.

This particular training rule is relevant to stochastic environments such as the traffic environment in the case study outlined in the next section.

5.2 *Key elements of the signal control implementation*

The initial test application of Q-learning to the problem of traffic signal control involved a single, isolated intersection. The isolated signal and linear system implementations involve two-phase operation without turning flows. The test-bed consisted of a simulated two-phase signal controlling the intersection of two, two-lane roads. Vehicle arrivals were generated using individual Poisson processes with pre-defined average arrival rates on each of the four approaches. The average rates could be varied over time to represent different peak-period traffic profiles over the two-hour simulated episodes. In practice, the agent would operate continuously. The essential elements of the isolated signal control case study are defined in the next section.

5.2.1 *State, action and reward definitions*

The isolated intersection state information available to the agent included the queue lengths on the four approaches and the elapsed phase time. In addition to local queue lengths, various combinations of upstream and downstream queue lengths and the offset of signal changes controlling upstream and downstream movements are being evaluated as state elements. Since the addition of elements to the state definition dramatically increases the size of the state-space, a balance has to be sought between the benefit of this information and its impact on problem tractability. Sensing of queue lengths would be most effectively achieved using video imaging technology in combination with artificial neural network or other pattern recognition techniques.

The isolated signal agent was operated with a fixed cycle length as context. Each second, between a point ten seconds into the green phase and a point ten seconds from the end of the green phase, arbitrary limits fixed to ensure minimum practical phase lengths, the agent selected an action: either remain with the current signal indication or change it. Action selection consists of a decision, made at the time of the previous phase change, as to when to make the next phase change.

The definition of reward (actually a penalty in this case) is relatively straightforward, being the total delay incurred between successive decision points by vehicles in the queues of the four approaches. The delay in each one-second step, being directly proportional to the queue length, was modified using a power function to encourage approximate balancing of queue lengths.

5.2.2 *Exploration policies*

Convergence of a Q-learning agent on a suitable Q-function, particularly where the process being controlled is stochastic, requires adequate exploration to ensure that all areas of interest across the state space are visited sufficiently often. An ε-greedy policy was tested where the best action is exploited with probability ε and an exploratory action is chosen randomly with probability $1-\varepsilon$. Good results were achieved where the probability of choosing the best action was annealed,

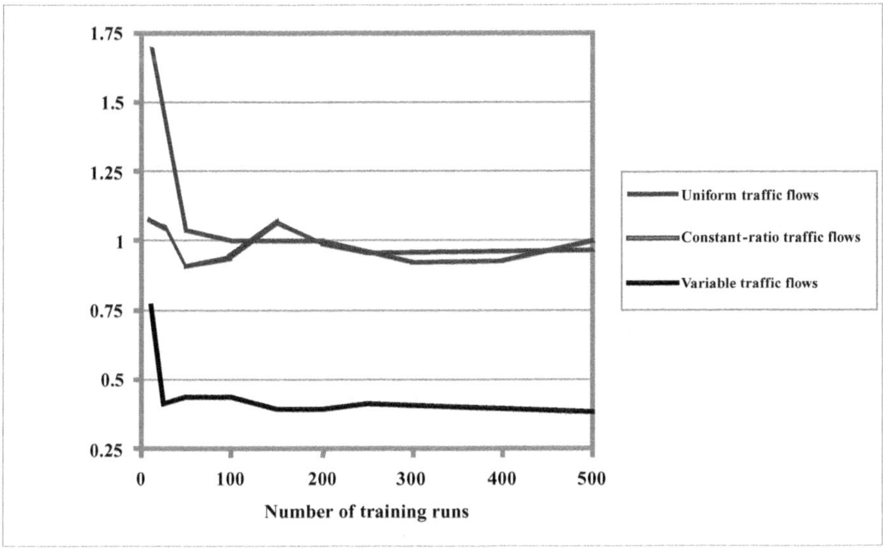

Figure 6. Isolated traffic signal: Average delay per vehicle ratio (Q-Learning/Pre-timed) (*Source: Abdulhai et al., 2003*).

starting with random exploration and increasing to the point where the best action was chosen with probability 0.9, provided that state had been visited at least 35 to 50 times (Abdulhai, B. *et al.*, 2003).

5.3 *Preliminary test results*

The following discussion presents selected results obtained from the isolated signal testbed. In this case, performance was compared with that of a commonly used pre-timed signal controller. Tests were conducted using three different traffic profiles to evaluate the performance of the Q-learning agent under varying conditions. Figure 6 summarizes the results of these tests. The graphs in Figure 6 reflect the average vehicular delay across individual sets of 50 test episodes, typically conducted after each of 10, 25, 50, 100, 150, 200, 250, and 500 training episodes. Each training and testing episode was equivalent to a two-hour peak period involving 144 signal cycles. In accordance with typical practice, the pre-timed signal-phasing plan used as a baseline for comparison utilized constant phase times based on the critical peak-hour flow rates by direction. It is evident from the results the Q-learning agent outperforms pre-timed signals when traffic patterns are highly variable.

6 AUTOMATED ADAPTIVE TRAFFIC CORRIDOR CONTROL

The two most promising control tools for freeway corridors are traffic-responsive ramp metering and/or dynamic traffic diversion, using variable message signs (VMS). Technically, the use of these control methods independently might limit their potential usefulness. Therefore, integrated corridor control using ramp metering and VMS diversion simultaneously can be synergic and beneficial. Administratively, freeways and adjacent arterials often fall under different jurisdictional authorities. Common lack of coordination amongst those authorities due to lack of means for information exchange and/or possible 'institutional grid-lock' could hinder the full potential of technically-possible integrated control. Therefore, fully automating corridor control could alleviate this problem. Motivated by the above, the aim of the research approach is to develop a self-learning adaptive integrated freeway-arterial corridor control for both recurring and non-recurring congestion (Jacob, C. & Abdulhai, B., 2006).

In this study, a simple but powerful reinforcement learning- the Q-learning approach of Watkins (Watkins C., 1989)—is selected to address the stochastic nature of traffic situation. The Q-learning model developed in this study was trained and tested on a microscopic simulation model of key corridors in Toronto using the Paramics micro-simulation suite (Quadstone Ltd., 2000). The study involved testing the methodology in three different applications, namely, freeway control using multiple ramp metering, integrated corridor control with a single ramp and VMS and Integrated corridor control with multiple Ramps and VMS (Jacob, C. & Abdulhai, B., 2006).

6.1 *Elements of the Q-learning control agent*

At any time, the state of the network is based on direct traffic measurements, such as speed, volume and occupancy, which can be obtained from the loop detectors on the road. Depending upon the control methodology, the system states in this study are defined as a certain combination of average speeds on the competing routes, average percentage occupancy on the detector(s), status of VMS and ramp, presence or absence of incidents. The set of actions available to the agent in any particular state are a combination of VMS and ramp signal settings.

The reward can be either positive or negative, based on whether a benefit (e.g., delay reduction) or penalty (e.g., delay) is taken as the outcome of the action. The agent is trained to minimize the total time spent by all vehicles in the agent's network periphery. As described by Papageorgiou (Papageorgiou *et al.*, 2003) this is equivalent to maximizing the time-weighted exit flows from the section. This value is taken as the reward for the agent's action. Each simulation-step will be divided into K time-steps of duration T sec (20 sec.). The number of vehicles crossing the detector placed at the end of the agent's network section during the kth time-step s(k) is obtained from the flow detector placed at that location:

$$\text{Time-weighted outflow} = S = T^2 \sum_{k=0}^{K-1} (K-k)s(k) \tag{2}$$

6.2 *Measurements of effectiveness*

In order to evaluate the feasibility of the approach, the results obtained from the learning agent have been compared with the results obtained from the base-case scenario. Result obtained from simulations without using the learning agent (no-control case) was selected as the base-case. ALINEA control (Papageorgiou *et al.*, 1991) which is considered as one of the most commonly used and effective ramp metering algorithms was selected for comparing the effectiveness of the proposed machine learning based freeway control using multiple ramps. In the test runs, the action selected will be the one with maximum Q-value.

6.3 *Freeway control using multiple ramp metering*

A 16 km stretch of the Queen Elizabeth Way (QEW) in Toronto with seven on-ramps was found to be a suitable section to test the proposed control methodology. Two different models were tried with different state definitions:

States for model 1

In this model, the agent receives the percentage occupancy on the detector just downstream of the ramp as well as the current metering rate in order to represent the state of that portion of the network.

States for model 2

Here, the state of the network is represented by means of four variables: occupancy on the main line just upstream of the ramp; occupancy on the ramp just at the entrance to the main line;

Table 2. Results from the QEW ramp metering (*Source: Jacob C. & Abdulhai B., 2006*).

	Without Control (Base case)	ALINEA (Best case)	Q-Leaning (2 States, Look-up-table)	Q-Leaning (2 States, CMAC)	Q-Leaning (4 States, CMAC)
Mean Travel Time (sec)	572.9	541.5 (5.48%)	544.5 (4.96%)	524.7 (8.41%)	516.1 (9.91%)
Total Vehicle Distance (m)	81999444	84923677 (4.56%)	84319430 (3.82%)	84966304 (4.7%)	86204672 (7.02%)
Mean Speed (kmph)	50.9	56.5 (11.00%)	55.9 (9.82%)	57.9 (13.75%)	59.4 (16.7%)
Stop Time (sec)	2444	1987 (18.70%)	1800.7 (26.32%)	1441 (41.04%)	1517.8 (37.9%)

occupancy on the mainline at the end of the stretch for that particular ramp and the current metering rate at the ramp. The first two states will be a representative of the demand from the main line and demand from the ramp respectively. The third variable reflects the state downstream of the ramp.

Table 2 provides the results obtained from the ramp metering agent approach using a look-up table to store Q-values and with a CMAC based function approximation. Results from ALINEA and a base no-control case are also shown. The percentage improvement obtained with each measurement of effectiveness (MOE), as compared to the no-control case, is shown in brackets.

It shows that Q-learning has performed better than the ALINEA control, except the look-up table approach was slightly inferior. It is clear that the Q-learning with function approximation capabilities performs much better than the state of art control in ALINEA. When more detailed network information is provided using four state variables, the results improve from that of the two-state case, but at the expense of more learning time. Therefore, a compromise should be sought between performance and efficiency.

6.4 *Integrated control of a simple corridor with one ramp and one VMS*

In this, a part of the waterfront corridor from the Greater Toronto area (built in the ITS Centre and Testbed, University of Toronto), comprised of the Gardiner Expressway along with the Lakeshore Boulevard—a parallel signalized artery—is selected. Figure 7 shows the network used in this study. The agent is provided with learning scenarios of incident cases and various incident states are included as state information. This will facilitate the agent to provide effective control on non-recurrent congestion. An incident scenario is defined by (a) the location of the incident, (b) the incident severity, and (c) the estimated duration of the incident. Three cases of incident scenarios are considered for the learning of the agent ;(A): a single-lane-blocking incident on the Gardiner for 10 minutes; (B): double-lane-blocking incident on the Gardiner for 15 minutes; (C): a single-lane-blocking incident on the Gardiner and a double-lane-blocking incident on the Lakeshore for 15 minutes (Jacob, C. & Abdulhai, B., 2006).

The states considered are: average speeds and incident status on both the routes, status of VMS (75% diversion is assumed when the VMS is active) and status of ramp. The set of actions available to the agent in any particular state are a combination of VMS and ramp settings. The algorithm tries to minimize the total time spent by all the vehicles in the entire corridor. Table 3 provides the

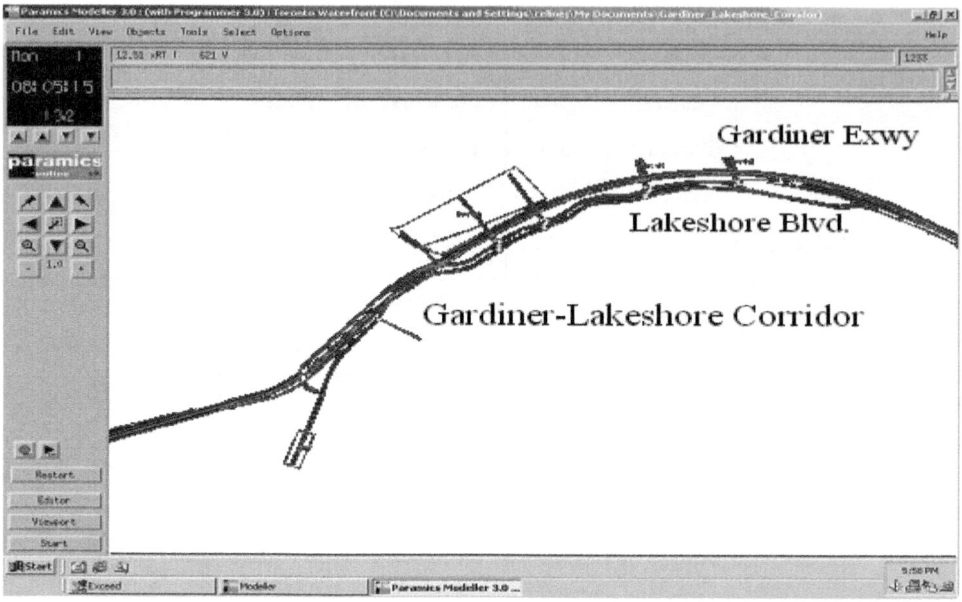

Figure 7. Gardiner/Lakeshore corridor with single VMS and ram (*Source: Jacob C. & Abdulhai B., 2006*).

Table3. Results of integrated control with single VMS and ramp with incident scenarios (*Source: Jacob C. & Abdulhai B., 2006*).

	Case A		**Case B**		**Case C**	
	No Control	**Integrated Q-L**	**No Control**	**Integrated Q-L**	**No Control**	**Integrated Q-L**
Mean Travel Time (sec)	653	562 (13.9%)	826	688.3 (16.7%)	647	568.2 (12.2%)
Total Vehicle Distance*10^{-7} (m)	3.8	4.2 (8.5%)	3.3	3.8 (14.1%)	3.8	4.1 (6.5%)
Mean Speed (kmph)	28.5	33.5 (17.5%)	22.0	26.7 (21.4%)	28.4	33.2 (16.9%)
Stop Time (sec)	3877	3246 (16.3%)	5495	4073 (25.8%)	3832	3234 (15.6%)

result obtained from the integrated control with all the three incident scenarios. The percentage improvement obtained with each MOE, as compared to the no-control case, is shown in brackets.

6.5 *Integrated control with multiple ramps and VMS*

In this case, an extended part of the Gardiner/Lakeshore corridor from Humber River to Don Valley Parkway with two ramps (on and off ramps) and two VMS controls is selected. Figure 8 shows the corridor built in Paramics. The corridor is divided into two segments. The first part is from the start of the corridor for eastbound traffic, up to the location of the second off ramp (at Jameson Avenue) from the Gardiner, and the second part includes the remaining section.

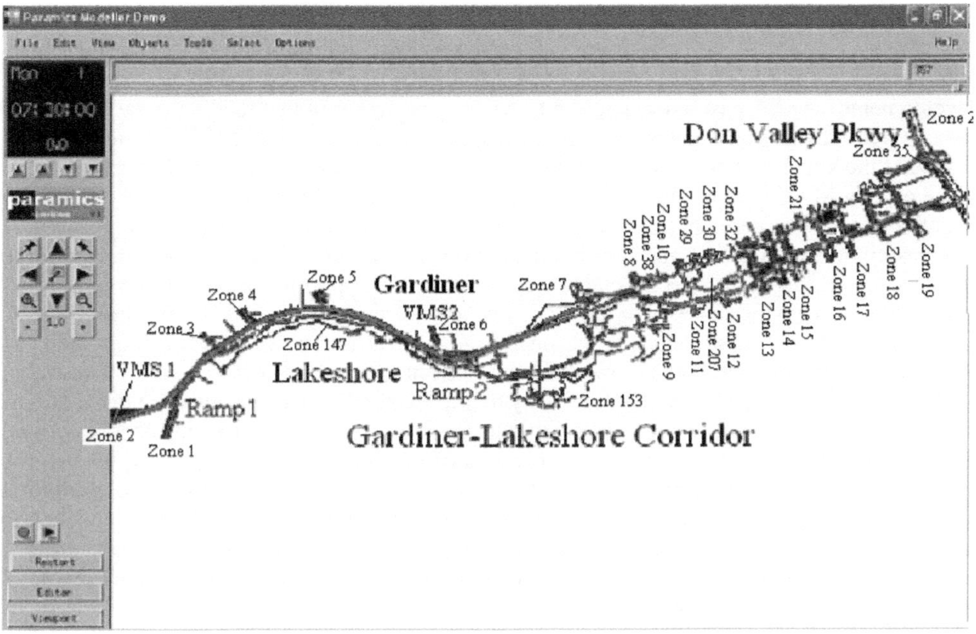

Figure 8. The Gardiner/ Lakeshore corridor with multiple ramps and VMS.

Table 4 Comparison of integrated, semi-integrated and independent Q-learning agents with base case (*Source: Jacob C. & Abdulhai B., 2006*).

Control / MOE	No-Control (Base Case)	Integrated Q-learning	Semi-integrated Q-learning	Independent Q-learning
Mean Travel Time (sec)	984.6	913.5 (7.2%)	930.7 (5.5%)	946.6 (3.9%)
Total Vehicle Distance (m)	123689811	142964811 (15.6%)	140445903 (13.5%)	138550943 (12.0%)
Mean Speed (kmph)	22.2	26.1 (17.6%)	25.1 (13.1%)	24.3 (9.5%)
Stop Time (sec)	7784.2	6198.3 (20.4%)	6675.6 (14.2%)	6780.8 (12.9%)

Each part includes one VMS and one on-ramp, which are to be controlled by a single integrated Q-learning agent. The results of the integrated, semi-integrated, and completely independent models, after it has converged, are presented in Table 4. The percentage improvement obtained with each MOE, as compared to the no-control case, is shown in brackets. The comparison shows that a fully integrated system performs much better than a semi-integrated and independent system. Similarly, the performance of a semi-integrated system is better than that of completely independent system. This shows that the effectiveness of the approach increases with the degree of integration.

7 SUMMARY AND CONCLUSION

In this paper, we presented a description of the ITS Centre and Testbed and some of the key ATMS developments over the past ten years. The ICAT is a research tool that facilitates the research and development of computer analysis and control algorithms for ATMS and ATIS applications.

One of the key developments within the ICAT is Genetic Adaptive Incident Detection which is based on two AI techniques, probabilistic neural network and genetic algorithm. The main benefits of GAID are the increase in the overall generalization accuracy of the detection algorithm, better transferability, and post transferability adaptation via automated genetic re-optimization. It is designed to avoid any developer dependent handcrafting of complex parameters. It accounts for prior probability of incidents, severity of the incidents, as well as unequal costs for misclassification

A number of control strategies based on Reinforcement Learning have been presented. As a branch of the artificial intelligence, reinforcement learning appears to offer significant advantages in ITS where real-time, adaptive control is key to improving effectiveness and efficiency. The ability to learn through dynamic interaction with the environment is seen as a significant benefit relative to control methodologies that rely on pre-specified models of these processes. We presented the application of reinforcement learning to the problem of traffic signal control. Preliminary results from the application of Q-learning to an isolated, two-phase traffic signal are encouraging. In addition, the application of reinforcement learning or more specifically Q-learning is presented for the independent, combined and integrated control of freeway and freeway/arterial traffic corridor using VMS and ramp metering.

REFERENCES

Abdulhai, B., (2003) "ITS-eh! Meet Canada's Flagship Intelligent Transportation Systems Centre and Testbed," *IEEE Intelligent Systems Magazine.*

Canadian Climate Data. Environment Canada (2005) http://www.climate.weatheroffice.ec.gc.ca/climateData/canada_e.html. Accessed January 5, 2005.

Roy, P. & Abduhai, B. (2003) "GAID: Genetic Adaptive Incident Detection for Freeways". Transportation Research Record 1856 pp. 96–105.

Foo, B., Ashton and Abdulhai, B. (2006) "Building an online its research and training facility—the IACT Database and Platform" Intelligent Transportation Systems Conference ITSC, IEEE

Abdulhai, B., Pringle, R. and Karakoulas, G. (2003) "Reinforcement Learning for True Adaptive Traffic Signal Control" ASCE Journal of Transportation Engineering.

Jacob, C. & Abdulhai, B. (2006) "Automated Adaptive Traffic Corridor Control using Reinforcement Learning: Approach and Case Studies" Transportation Research Record 1959 pp. 1–8.

Abdulhai, B. & Ritchie, S. (1997) "Development of a Universally Transferable Freeway Incident Detection Framework", Presented at the Transportation Research Board 76th Annual meeting.

Abdulhai, B., (1996) "A Neuro-Genetic-Based Universally Transferable Freeway Incident Detection Framework", Ph.D. Dissertation, University of California Irvine.

Abdulhai B. & Ritchie S. (1995) "Performance of Artificial Neural Networks for Incident Detection in ITS", ASCE Transportation Congress, pp. 227–238.

Sutton, R., & Barto, A. (1998). Reinforcement Learning—An Introduction. MIT Press, Cambridge, Massachusetts.

Kaelbling L., Littman M. and Moore, A. (1996). "Reinforcement Learning: A Survey." Journal of Artificial Intelligence Research, 4, pp. 237–285.

Bertsekas D., & Tsitsiklis J. (1996). Neuro-Dynamic Programming. Athena Scientific, Belmont, Massachusetts.

Watkins C. (1989). "Learning from Delayed Rewards" Ph.D. Thesis. King's College, University of Cambridge, Cambridge, UK.

Papageorgiou, M., Diakaki, C., Dinopoulou, V., Kotsialos, A., and Wang, Y. (2003) Review of Road Traffic Control Strategies. Proceedings of the IEEE, Vol. 91, No.12.

Papageorgiou, M, Hadj-Salem, and Blosseville, J.M. (1991) ALINEA: A local feedback control law for on-ramp metering. In Transportation Research Record: Journal of the Transportation Research Board, No. 1320, pp. 58–64.

Masters, T. (1995) Advanced Algorithms for Neural Networks. A C++ Sourcebook. John Wiley and Sons, Inc. Toronto.

Modeller and Programmer V3.0 User Guide and Manual (2000). Quadstone Ltd., Edinburgh, UK.

F.L. Hall, Y. Shi and G. Atala (1993) "On-line testing of the McMaster incident detection algorithm under recurrent congestion", Transportation Research Record: Journal of the Transportation Research Board, No. 1394, pp. 1–8.

Efficient Transportation and Pavement Systems – Al-Qadi, Sayed, Alnuaimi & Masad (eds)
© 2009 Taylor & Francis Group, London, ISBN 978-0-415-48979-9

Development and comparison of dynamic transit signal priority strategies

Wael Ekeila
Department of Civil Engineering, The University of British Columbia, Vancouver, BC, Canada

Tarek Sayed
Department of Civil Engineering, The University of British Columbia, Vancouver, BC, Canada

Mohamed El Esawey
Department of Civil Engineering, The University of British Columbia, Vancouver, BC, Canada

ABSTRACT: Transit Signal Priority (TSP) is a popular strategy used to enhance the performance of transit systems by modifying the signal control logic to give transit signal priority through signalized intersections. Conventional TSP strategies used in most cities have been shown to offer significant benefits in minimizing the delays of transit vehicles. However, there have been concerns about some shortcomings of conventional TSP strategies which limited their applications. The main concern is the potential negative impact on the cross street traffic. Another concern is the static nature of conventional TSP strategies and the lack of responsiveness to real-time traffic and transit conditions. This paper describes the development and evaluation of a dynamic Transit Signal Priority control system which has the ability to provide signal priority in response to real-time traffic and transit conditions. The dynamic TSP system consists of three main components: a virtual detection system, a dynamic arrival prediction model, and a dynamic TSP algorithm. The virtual detection system was developed in a VISSIM microsimulation model. Subsequently, several bus arrival prediction models were developed using linear regression and Neuro-fuzzy methods. Techniques such as Kalman and Bayes filters were used to refine the prediction. Finally, a dynamic TSP algorithm was developed to determine what TSP strategy to use and when to apply it. The dynamic TSP system was tested and compared to the conventional TSP system using the microsimulation model. Scenarios with varying simulation parameters and traffic volumes were tested. Results showed that the dynamic TSP system outperformed the conventional one.

1 INTRODUCTION

Transit Management Strategies have become a necessity rather than an option. Faced with the increase in transit demand and the fixed capacity of the existing infrastructure, urban planners and traffic engineers are developing and improving transit facilities and movements. Transit Signal Priority (TSP) has become one of the most popular ITS solutions. With TSP, priority in the traffic network is awarded to person trips rather than vehicle trips.

TSP has been widely used in many cities in North America, Europe, and Asia. Conventional TSP strategies used in most of these cities have been shown to offer significant benefits in minimizing the delays of transit vehicles. However, there have been some concerns about the shortcomings of conventional TSP strategies which limited their applications. The main concern is the potential negative impact on the cross street traffic. Another concern is the static nature of conventional TSP strategies since they assume a fixed arrival time for the detected transit vehicles. A third concern is the lack of responsiveness to real-time traffic and transit conditions.

This research attempts to overcome the shortcomings of the conventional TSP systems by developing a "Dynamic" transit signal priority system. The proposed DTSP system is composed of three components; an Automatic Vehicle Location (AVL) detection system, a transit arrival prediction model, and a priority strategy selection algorithm. Two case studies are presented to test and compare these algorithms. The first case uses a simulated hypothetical four-legged intersection while in the second case study a proposed Light Rail Transit (LRT) corridor in British Columbia is simulated. VISSIM was used in the simulation of the two case studies.

2 PREVIOUS WORK

At a signal, priority can be awarded to a transit vehicle in different ways and methods. How, when, and where to apply transit signal priority is what differentiates between priority concepts. Several concepts are described in Chada and Newland (2002), which included: active and passive priorities, direct and indirect priorities, and conditional and unconditional priorities. They also described several priority design criteria. These TSP design criteria are based on various parameters such as transit vehicle's delay, time headway, occupancy, location in queue, and the time since the last priority was provided.

Several priority strategies and techniques have been developed through a number of case studies and projects carried out in different cities in North America. Baker et al. (2002) listed some of these strategies which included: *Green Extension, Red Truncation, Phase Splitting, Cycle Length Reduction*, *Transit Coordination, Queue Jumps, Phase Insertion*, and *Phase Rotation*.

Because of the importance of real-time information dissemination to transit users and planners, the development of transit arrival prediction models has become an active research area. Transit arrival prediction models are also important for the application of advanced transit control operations such as dynamic TSP. Accuracy is the key issue when applying arrival prediction models in TSP strategies.

Lee et al. (2005) used online microsimulation-based arrival prediction models in developing dynamic TSP systems. The developed method consisted of two major components: an online microsimulation travel time prediction model and a priority operation model. When a transit vehicle is detected, the prediction model is activated to retrieve signal timing information and traffic data from the upstream and the downstream sensors. The developed prediction model was validated using PARAMICS through comparing the predicted bus travel times to the actual travel times. On the other hand, the priority operation model consisted of a library of six priority plans that would be evaluated by the arrival prediction model. After evaluating the travel time of each priority plan, the most appropriate plan with the least travel time would be chosen and sent to the signal controller.

Liu et al. (2003) attempted to integrate adaptive signal control techniques with advanced transit signal priority techniques. The priority awarded to a transit vehicle would be dependent on a weighting factor given to each transit priority call. A bus with priority would be converted into a relevant number of passenger vehicles using the weighting factor. The weighting factor was determined using the traffic demand and queuing conditions of every approach at an intersection and on the lateness of the transit vehicle. The computed weighting factor would be used to recalculate the signal timings and splits keeping into consideration the signal's parameters such as minimum and maximum greens, permissive start and end times, and force-off green times.

Li et al. (2005) developed an adaptive transit signal priority system on an actuated signalized corridor. The proposed system tried to give priority to transit vehicles while minimizing the negative impacts of the cross street traffic and pedestrian safety. The system consisted of: a continuous detection system, a priority request system, a signal control algorithm that adjusted signals to provide priority, and communication links between them. AVL systems were used to track transit vehicles. Then an arrival predictor, that used both historic and real-time data, was used to predict bus arrival time. Two TSP solutions were used in this study, green extension and early green. The algorithm was tested through simulation and a field test. Results showed considerable time savings in terms of transit delays, while maintaining an acceptable delay level for the minor traffic.

3 METHODOLOGY

3.1 *The microsimulation model*

In order to test and compare conventional and dynamic TSP algorithms, a four-leg hypothetical intersection was modeled using VISSIM. Each of the four approaches consists of two thru lanes of 3.5 meters width. In the model, one transit line was modeled with a transit stop upstream of the west bound (Figure 1). The dwelling time was defined as a normal distribution with a mean of 15 seconds and a standard deviation of 3 seconds (Ekeila, 2005). A two-phase signal was used to model the primary signal logic. The signal had a cycle length of 65 seconds with green time duration of 29 seconds for the east-west phase and of 26 seconds for the north-south phase.

Unlike the conventional TSP system which is composed of fixed check-in and check-out detectors, the dynamic TSP system uses AVL for the dynamic detection of transit vehicles. The use of the AVL system as a detection system in dynamic TSP offers several advantages such as:

- Many transit agencies are currently using AVL systems for schedule adherence and management purposes. Hence, there will be almost no additional cost required for a new detection system.
- Unlike fixed detectors where the bus is detected at a certain location of the street, AVL systems track vehicles continuously. Hence they can be considered as a set of fixed virtual detectors.

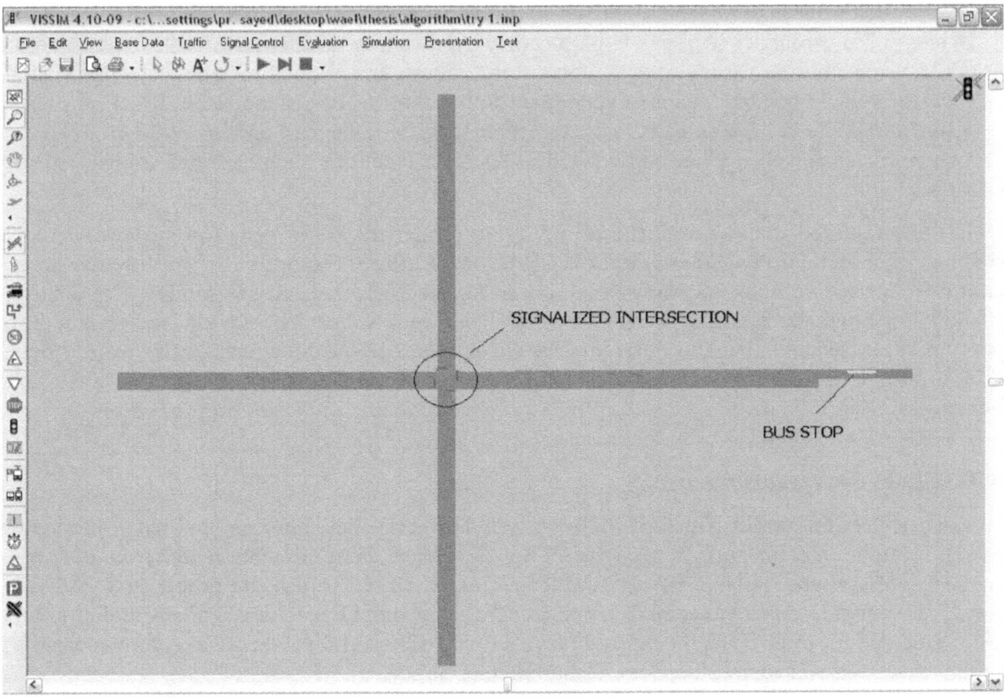

Figure 1. Layout of the simulated intersection.

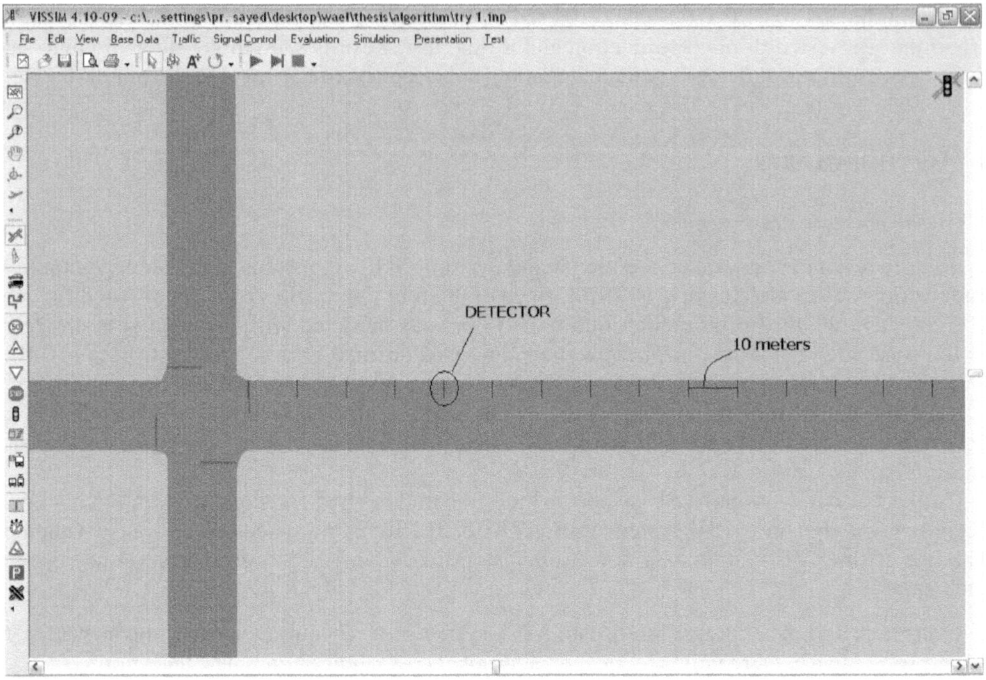

Figure 2. Modeling of AVL detection.

- AVL systems are able to send other types of information that are useful in the application of dynamic TSP strategies (e.g. passenger occupancy, dwelling time, speed...etc).
- AVL systems have a variety of detection criteria. They can provide time-step detection (where a transit vehicle can be detected every specific time interval), distance-step detection (where a transit vehicle can be detected as it travels a specific distance), and event-based detection (where a transit vehicle can be detected at specific locations such as intersections and bus stops).

In the simulated intersection, the only way to model the AVL detection system was as a distance-step detection system since VISSIM does not have the option of continuous detection of a certain vehicle. As shown in Figure (2), the AVL system was modeled as a group of detectors that are located 10 meters away from each other. The length of the detectors was taken as zero so that the detectors would act as a point detector. As the front wheels of the transit vehicle passes over a detectors a single impulse would be sent to the signal controller.

3.2 *The arrival prediction model*

In order to test the performance of the proposed TSP strategies, an accurate arrival prediction model is required. A linear regression model was developed using simulation data from the hypothetical intersection. As well, the same intersection was used to test the proposed TSP strategies. The simulation data included distance, dwelling time, and travel time. Three simulation runs were executed for three different exposure levels (750, 1000, 1200 veh/hr). Travel time measuring points were located every 100 meters for 800 meters (Figure 3).

In each simulation run, travel time measurements were collected for four different busses. In total, 96 data points were used for model development. The arrival prediction model was developed using linear regression. In the model, y presents the predicted travel time which is a function of a vector of independent variables (x_i). The constant b is the y-intercept. This is mathematically expressed as:

$$y = m_1x_1 + m_2x_2 + \cdots + m_nx_n + b \tag{1}$$

The vector of independent variables (x_i) included two variables; DISTANCE and DWELL. DISTANCE is defined as the distance traveled by the transit vehicle to the intersection (meters), while DWELL is defined as the dwelling time of the transit vehicle at a stop (seconds). The developed model is shown in Table (1).

As shown, the model parameters are significant at the 95% confidence level. Moreover, the model shows very good fit to the data with $R^2 = 0.99$. This was, to some extent, expected since the four transit vehicles run through the same environment and traffic conditions. Accordingly, the data obtained from the four buses and the ten measurement points are highly correlated.

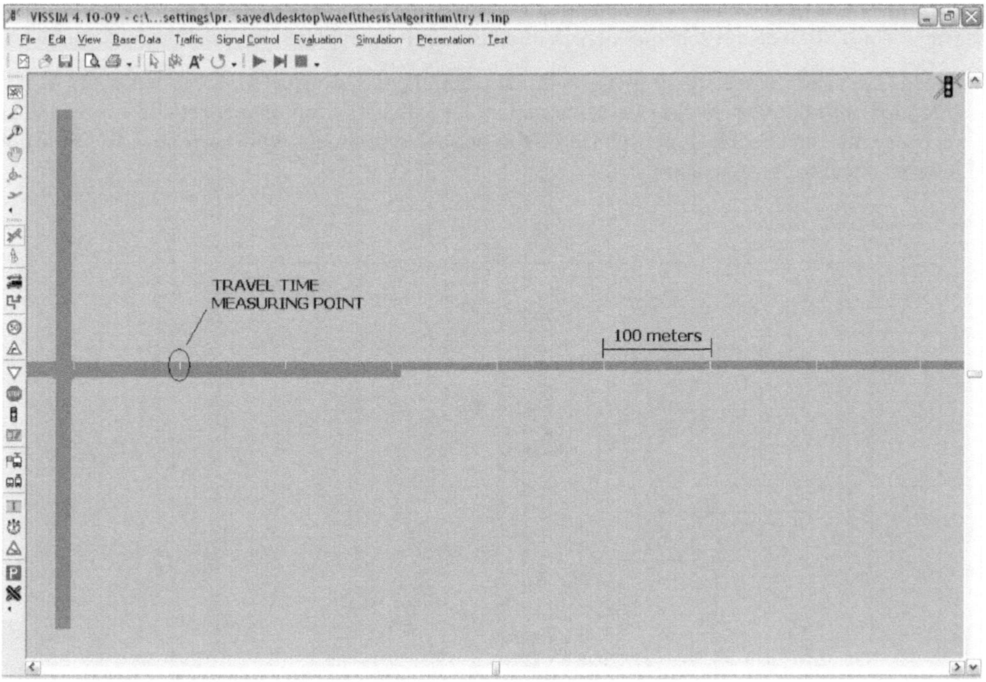

Figure 3. Travel time measurement points.

Table 1. The simulation-based linear prediction model.

Simulation Based Linear Prediction Model	R^2	Variable	Coefficient	T-Statistic
		DWELL	1.479	24.13
	0.99	DISTANCE	0.135	69.46
		CONSTANT	2.274	–

3.3 *Algorithms development*

The third component of this research was to develop an adaptive algorithm that provides transit vehicles with the most suitable transit priority strategy at signalized intersections. This algorithm would make use of the AVL system and the developed transit arrival prediction model. The algorithms were developed on VAP; an optional add-on module of VISSIM for the simulation of programmable signal controls. Five different TSP algorithms were developed; a Classic Algorithm, a Dynamic Algorithm, an Adjusted Dynamic algorithm, a Dynamic-Bayes Algorithm, and a Dynamic-Kalman Algorithm.

3.3.1 *The classic algorithm*

The classic algorithm was designed to compare the performance of the conventional TSP system and the newly proposed dynamic TSP system. Conventional TSP systems are usually composed of a simple detection system and an algorithm that provides priority to transit vehicles when detected. The general procedure of the classic algorithm is presented in Figure (4).

The classic algorithm consists of two components; a detection system and TSP decision scheme. The detection system of the classic TSP system utilized a check-in detector located 50 meters away from the approach stop line, and a check-out detector placed at the stop line.

In the conventional TSP system, the decision would be instantly taken once a transit vehicle passes the check-in detector. If no detection occurred, the algorithm would run the normal signal settings. The TSP decision scheme incorporated two TSP strategies:

- Green Extension: If the signal phase is green when the bus is detected, the green phase will be extended until the bus checks out or the maximum extension limit is reached. To achieve offset recovery, the succeeding green phase of the opposing approach will be reduced with a time interval equal to the extension.

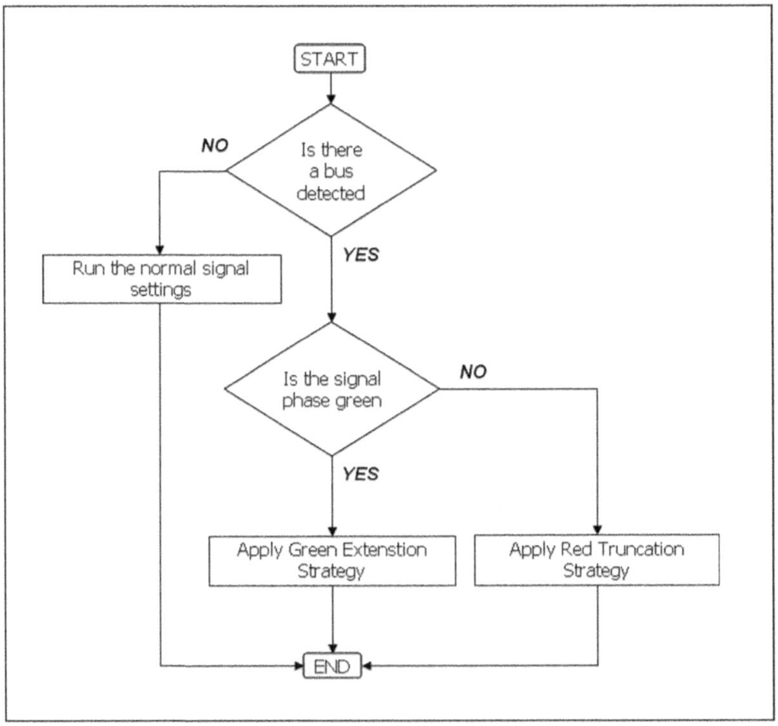

Figure 4. The conventional TSP flowchart.

- Red Truncation: If the signal phase is yellow or red when the bus is detected, the red phase will be truncated with a time interval equal to the maximum extension, taking into consideration the minimum signal timing required for pedestrians. To achieve offset recovery, the green phase will be extended with a time interval equal to the maximum extension.

3.3.2 *The dynamic algorithms*

Four dynamic algorithms were developed (Dynamic, Adjusted Dynamic, Dynamic-Bayes, and Dynamic-Kalman Algorithms). The four algorithms were used to test the performance of the proposed dynamic TSP system. All the algorithms share the main structure. However, they differ in their computational procedures and data assignments. The general procedure of the four dynamic algorithms is presented in Figure (5).

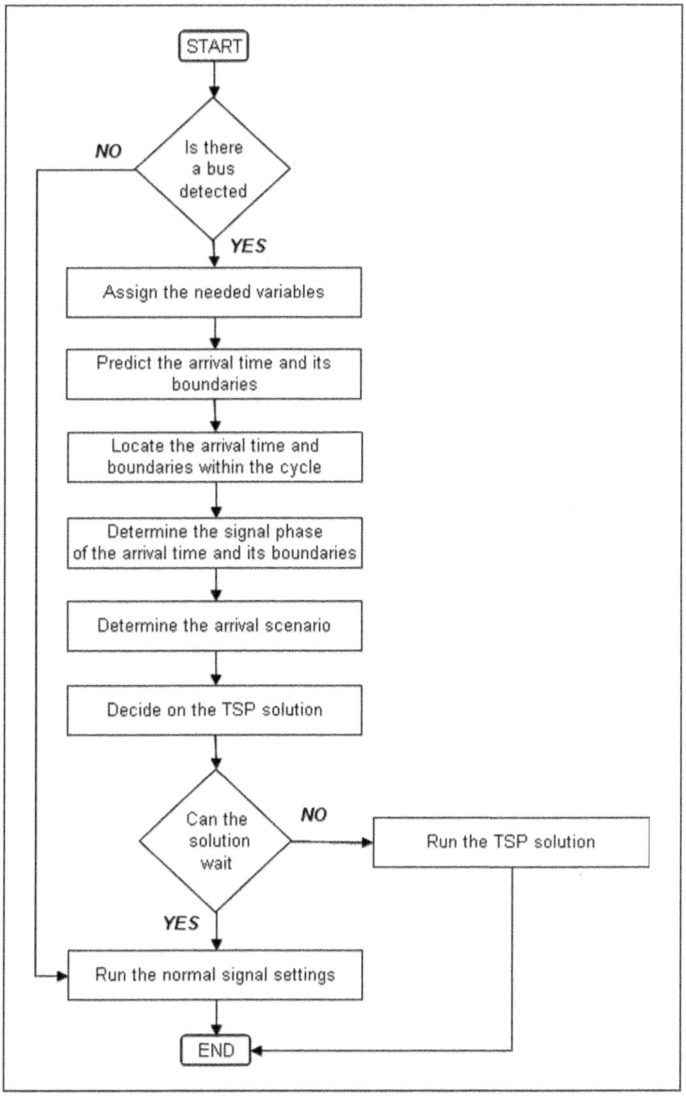

Figure 5. The dynamic TSP flowchart.

As shown in the figure, the algorithms are composed of seven major sequential steps. The next section describes these seven steps and illustrates the differences between the four algorithms.

1. *Transit vehicle detection*
In all the four algorithms, whenever a transit vehicle had been detected, data were collected for the variables defined in the prediction model. In the Dynamic-Bayes algorithm, the travel time of the previous bus, from the detector to the stop line, was retrieved as well. The Bayes algorithm used the previous bus travel time to enhance the model's travel time prediction. In the Dynamic-Kalman Algorithm, the average travel time of all previously simulated busses and their variances were retrieved rather than the travel time of only the previous bus. A Kalman Filter is then applied to fuse the model's prediction and the average travel times and variances of previous buses.

2. *Arrival Time Prediction*
This is the most crucial step in the algorithm, and it differentiates between the Dynamic, Adjusted Dynamic, Dynamic-Bayes, and Dynamic-Kalman algorithms. In the Dynamic and the Adjusted Dynamic algorithms, the prediction model developed earlier was used without any adjustments to predict the transit vehicles' arrival time. Lower and upper boundaries of the arrival time were defined to determine the arrival scenarios in a later stage. The time length between the predicted arrival time and any of its boundaries was defined as the *boundary length*. The lower and upper boundaries, in some cases, would be about one standard deviation from the arrival time. Sayed and Zein (1998), estimated the variance of a linear prediction model as shown in Equation (2). This equation was used in the algorithms to calculate the variance of the prediction models.

$$Var = s_d^{\,2}\left(1+\frac{1}{n}\right) + s_{a1}^{\,2}\left(x_1 i - \bar{x}_1\right)^2 + \cdots\cdots + s_{an}^{\,2}\left(x_n i - \bar{x}_n\right)^2 \tag{2}$$

Var = variance of the prediction model
s_d = standard error in y (the predicted dependent value)
s_a = standard error of an independent variable
n = number of independent variables in the model
x_i = value of the independent variable
$\bar{x}$ = mean value of the independent variable

In the Bayes algorithm, the arrival time prediction incorporated three steps: calculation of the arrival time and its variance from the prediction model, computation of alpha which is defined as the degree of confidence in the model's prediction as in Equation (3), and finally using alpha as Equation (4) to calculate the arrival time that would be used in the algorithm.

$$\alpha = \frac{1}{1+\left(\dfrac{Var_{Model}}{P_{Model}}\right)} \tag{3}$$

$$P_{Bayes} = \alpha P_{Model} + (1-\alpha) P_{Previous} \tag{4}$$

In the Kalman algorithm, the arrival time prediction is calculated using the arrival time and its variance from the prediction model, and the average travel time and variance of all previous simulated busses. This is expressed mathematically by Equation (5).

$$P_{Kalman} = \frac{Var_{Previous}}{Var_{Model} + Var_{Previous}} P_{Model} + \frac{Var_{Model}}{Var_{Model} + Var_{Previous}} P_{Previous} \tag{5}$$

3. *Arrival time allocation*

After predicting the arrival time and the lower and upper boundaries, the locations of the arrival time and its boundaries within the cycle were determined. In all the four algorithms, the arrival time allocation was carried out as follows:

a. The arrival time, in seconds, was added to the current cycle second.
b. The number of cycles left until arrival (N), was calculated.
c. The location of the arrival time within the cycle was determined by subtracting the length of N cycles from the sum of the arrival and current times.
d. The same procedure was repeated for each boundary.

4. *Signal phase determination*

In VAP a signal phase is defined using three parameters; GE (Green End), YE (Yellow End), and RE (Red End) (VISSIM 4.3 User Manual). In a cycle, a signal phase can be arranged in one of three scenarios as shown in Figure (6). Knowing the three parameters GE, YE, and RE the phase scenario was determined for both the arrival time and its boundaries.

5. *Arrival scenario determination*

In the four dynamic algorithms, after knowing the signal phase of the arrival time and its boundaries, the corresponding arrival scenario was determined. As shown in Figure (7), seven different arrival scenarios were defined:

- SCENARIO 1: A scenario where the predicted arrival and its upper boundary are in the green phase, while the lower boundary is in the preceding red phase.
- SCENARIO 2: A scenario where the predicted arrival and its lower and upper boundaries are all in the green phase.

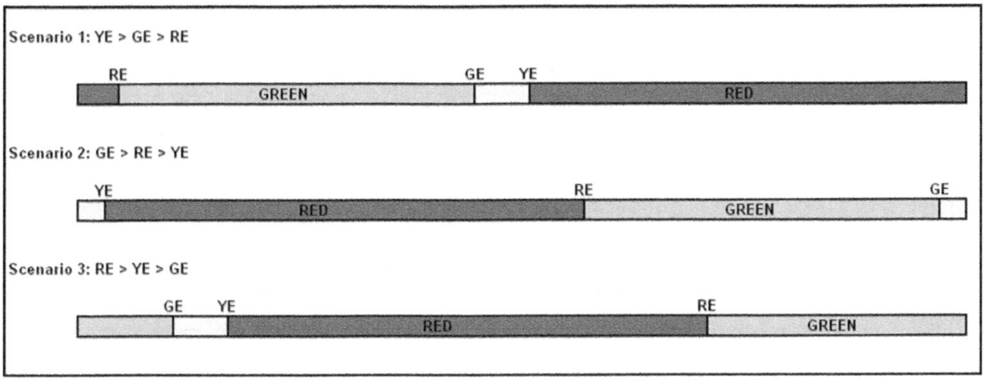

Figure 6. Phase scenarios.

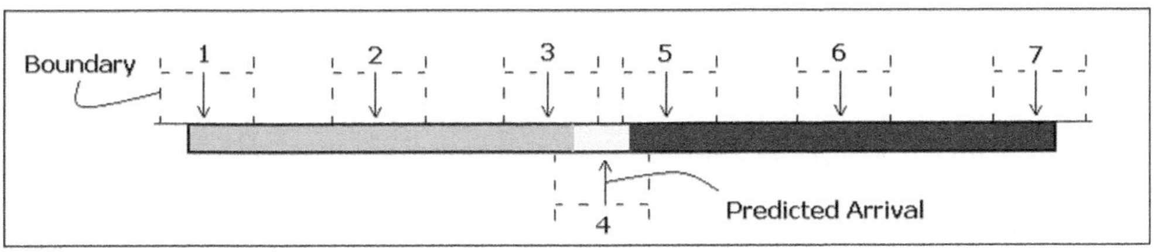

Figure 7. Arrival scenarios.

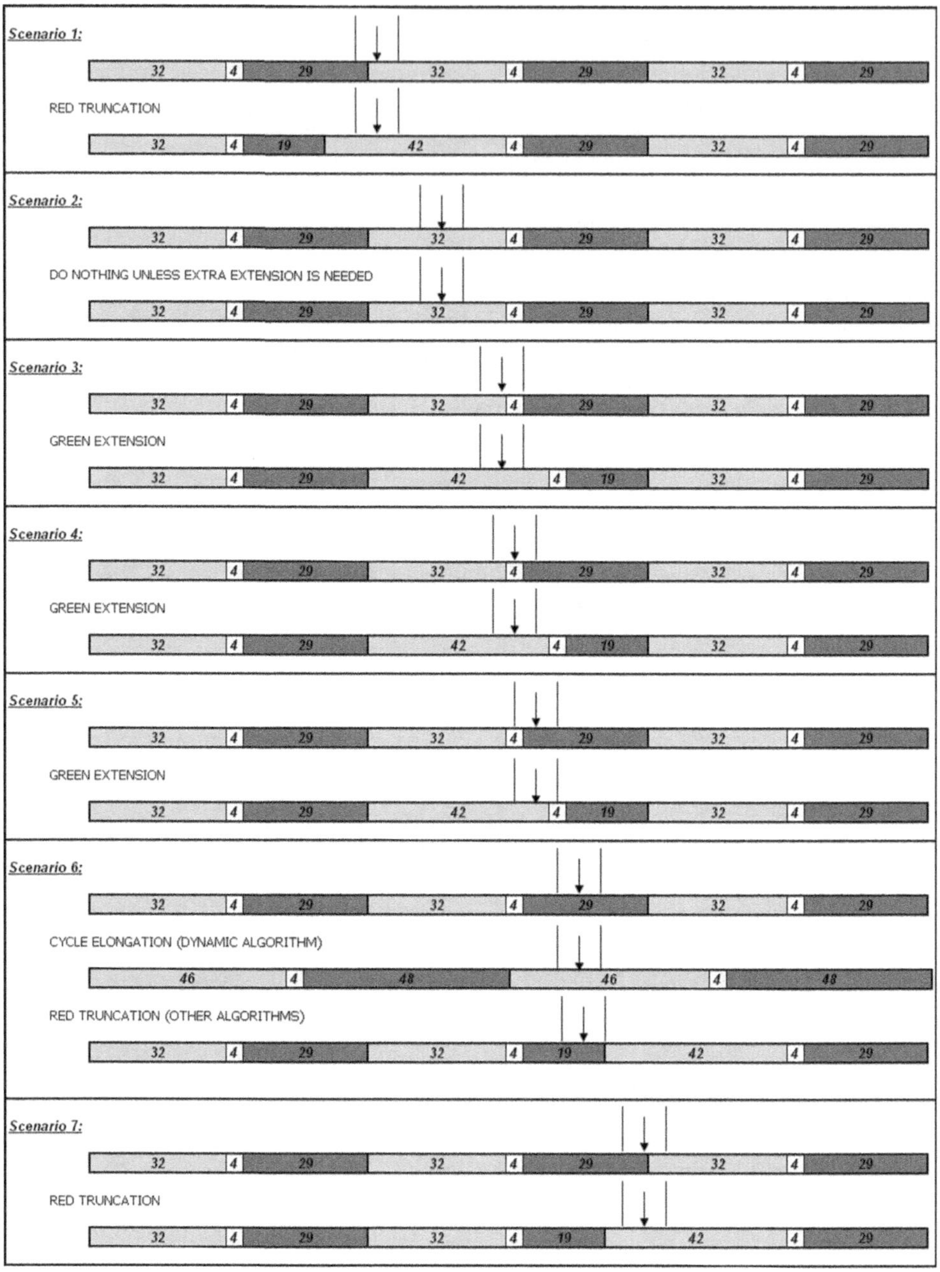

Figure 8. Illustrations of TSP solutions.

- SCENARIO 3: A scenario where the predicted arrival and its lower boundary are in the green phase, while the upper boundary is in the succeeding yellow or red phase.
- SCENARIO 4: A scenario where the predicted arrival is in the yellow phase.
- SCENARIO 5: A scenario where the predicted arrival and its upper boundary are in the red phase, while the lower boundary is in the preceding yellow or green phase.

- SCENARIO 6: A scenario where the predicted arrival and its lower and upper boundaries are all located in the red phase.
- SCENARIO 7: A scenario where the predicted arrival and its lower boundary are in the red phase, while the upper boundary is in the succeeding green phase.

6. *TSP solutions*

The type of solution or TSP strategy chosen was based on the determined arrival scenario. In this research, three TSP solutions were used:

- SOLUTION 1: This solution is based on the Green Extension Strategy. In this strategy, a green extension will be given to the green phase until the bus checks out the stop line or the maximum extension is reached. Offset recovery is applied in this solution where the succeeding opposing green phase will be reduced with a time interval equal to the extension.
- SOLUTION 2: This solution is based on the Red Truncation Strategy. In this strategy, the red phase will be truncated earlier so that when the bus arrives, the signal will be green. Offset recovery is also applied in this solution where the green phase will be elongated with a time interval equal to the red truncation.
- SOLUTION 3: This solution is based on the Cycle Extension Strategy. In this strategy, the cycle will be extended to have a length of one and half of the normal cycle length in order to make sure that the transit vehicle will arrive at a green phase and in order to retain signal coordination. The solution will be executed for two cycles, replacing three cycles of normal cycle length, then the cycle will be reduced again to its normal length.

In each of the four dynamic algorithms and for each scenario, one of the three defined solutions would be used. In order to understand the way that different solutions would work with the scenarios, Figure (8) illustrates the solutions of different scenarios for a period of three normal cycle lengths.

SOLUTION 3 was only used in the Dynamic algorithm in SCENARIO 6. This is the only difference between the Dynamic and Adjusted Dynamic algorithms. This adjustment was made to check the performance of this unpopular strategy on the intersection at different exposure rates. The hypothesis here is that some TSP solutions could outperform others according to flow conditions. Although presented theoretically in many papers, the application of the Cycle Extension Strategy in real life projects is rare.

7. *Taking the decision*

The time at which the decision is taken is one of the most important and dynamic components of the developed algorithms. After running all the preceding steps and determining the most suitable TSP solution, the algorithms checked whether to apply the TSP solution or wait for some more time. If the algorithm would wait, all the preceding steps would be repeated again to check whether

Table 2. Time limits of the seven arrival scenarios.

Scenario	Decision Time Limits
Scenario 1	At the start of the preceding red phase
Scenario 2	N/A
Scenario 3	At the start of the current green phase
Scenario 4	At the start of the preceding green phase
Scenario 5	At the start of the preceding green phase
Scenario 6 (SOLUTION 3)	At the start of the second preceding green phase
Scenario 6 (SOLUTION 1)	At the start of the current red phase
Scenario 7	At the start of the current red phase

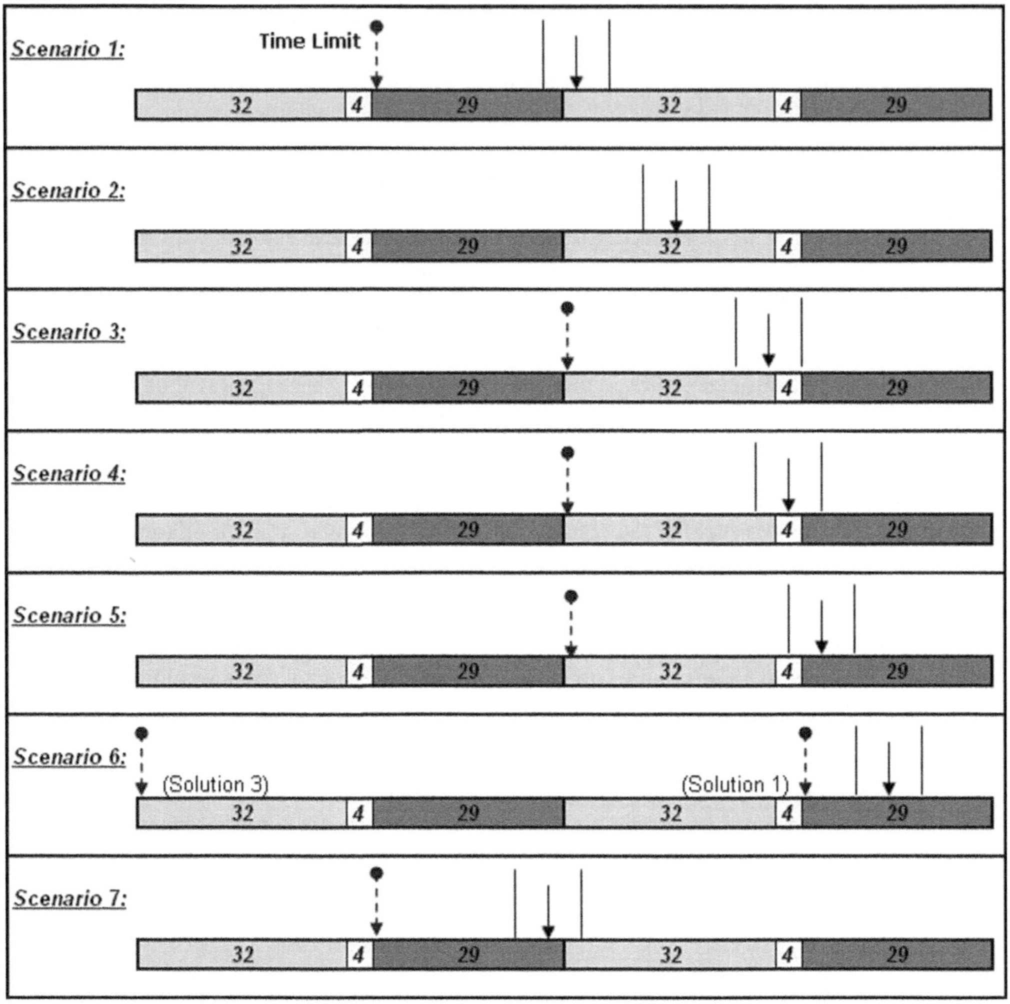

Figure 9. Time limits of the seven arrival scenarios.

the previously determined arrival scenario and TSP solution would still be valid. This component of the dynamic algorithms tries to choose the optimum TSP solution taking into consideration any traffic changes that might cause changes in the arrival scenario. When to take the decision is based on the arrival scenario and on the nature of the chosen TSP solution. The developed decision time limits for the seven arrival scenarios are listed in Table (2) and illustrated in Figure (9).

The dynamic algorithm runs the normal signal settings until a decision is taken. When a decision is taken, none of the previous steps will be executed, except for detection, until the solution ends and offset recovery is achieved. When the next bus arrives, the algorithm will be ready to grant it the priority it needs.

4 RESULTS

In order to test the performance of the different algorithms, 30 different simulation runs were executed. For each of the five algorithms, 3 different simulations were executed with various seed

Table 3. Validation results (Exposure = 750 vehs/hr).

	Classic	*Dynamic*	*Adj Dyn*	*Bayes*	*Kalman*
SUM	214.4	141.2	178.1	196.9	180.8
AVG	5.4	3.5	4.5	4.9	4.5
S. D.	7.0	3.8	5.9	6.5	5.8

Table 4. Validation results (Exposure = 1000 vehs/hr).

	Classic	*Dynamic*	*Adj Dyn*	*Bayes*	*Kalman*
SUM	232.7	229.3	203.1	215.0	218.4
AVG	5.8	5.7	5.1	5.4	5.5
S. D.	6.9	5.5	5.8	6.1	6.0

values. The simulations were executed at two different exposure levels (750 and 1000 veh/hr). The simulation period was set to 2 hours (7200 seconds). In this period, 40 busses crossed the intersection. Comparisons between the five algorithms were based on the total delay of all buses at the intersection. In VISSIM, delay is defined for a vehicle whenever its speed drops below 5 km/hr. Tables (3) and (4) show the sum of delay of all busses, the average delay, and the standard deviation for each algorithm.

Generally, all the developed dynamic algorithms outperformed the classical algorithm. The Dynamic algorithm outperformed the Adjusted Dynamic algorithm at the traffic flow level of 750 veh/hr. However, at the higher traffic flow level of 1000 veh/hr, the Adjusted Dynamic algorithm outperformed the Dynamic one. This supports the hypothesis that some TSP solutions could be more beneficial than others according to traffic flow level. Although Kalman Filter and Bayes technique were used in previous research [(Shalaby and Farhan (2004), Farhan and Shalaby (2002), Wall and Dailey (1999)], and showed improved prediction results. In this research, Kalman and Bayes algorithms outperformed the Dynamic algorithm only at the traffic flow of 1000 veh/hr. However, the Adjusted Dynamic algorithm outperformed them both at the two volume levels tested.

5 SUMMARY

In this research, a Dynamic TSP system was proposed to replace the Conventional TSP system used in most cities. Microsimulation was employed to test and compare the performance of four Dynamic TSP algorithms and a conventional TSP algorithm on a hypothetical intersection. The Dynamic algorithms used an AVL system to collect bus related information. Using an arrival prediction model, the algorithms recursively check the predicted arrival time of the bus until it reaches a time limit when a decision has to be taken. Based on the arrival scenario of the bus, an appropriate TSP solution is chosen and applied to the signal. The results showed that the four Dynamic TSP algorithms outperformed the Conventional TSP one. In addition, using previous buses travel times in the prediction of current bus arrival time showed to be inappropriate for use in TSP applications. The previous bus might have arrived at a green or red phase. Therefore, the travel time of the previous bus does not really help in determining the travel time of the current bus since the arrival phase and the queue length could be different.

A number of research ideas can be further researched to enhance and improve the developed Dynamic TSP system. The dynamic TSP algorithm can be enhanced by adding a larger library of TSP solutions or some newly defined scenarios. In addition, the algorithm can comprise more flexible and effective offset recovery and signal compensation strategies. Finally, a more advanced algorithm can be developed to handle TSP for networks rather than a single intersection or an arterial. In real-life applications of high-frequency transit lines similar to the one used in this study, bunching is a major issue. TSP systems for such contexts have to explicitly include strategies to deal with 2 transit vehicles arriving very close to each other. Also, any practical TSP system should also deal appropriately with 2 transit vehicles arriving at the intersection from two opposite approaches along the same corridor. All these issues will be considered for future investigation.

REFERENCES

1. Ekeila, W. "Dynamic Transit Signal Priority". Unpublished Master of Applied Science, University of British Columbia, 2006.
2. Chada, S. and Newland, R. "Effectiveness of Bus Signal Priority". National Center For Transit Research, University of South Florida, USA, 2002.
3. Baker, R.J. et al. "An Overview of Transit Signal Priority". White Paper, Intelligent Society of America, 2002.
4. Lee, J., Shalaby, A., Greenough, J., Bowie, M. and Hung, S. "Advanced Transit Signal Priority Control Using On-line Micro-simulation-Based Transit Prediction Model". Presented at the 84th Annual Meeting of the TRB, Washington, D.C, 2005.
5. Liu, H., Skabardonis, A. and Zhang, W. "A Dynamic Model for Adaptive Bus Signal Priority". Presented at the 82nd Annual Meeting of the TRB, Washington, D.C, 2003.
6. Li, M., Yin, Y., Zhou, K., Zhang, W., Liu, H., and Tan, C. "A Dynamic Model for Adaptive Bus Signal Priority". Presented at the 84th Annual Meeting of the TRB, Washington, D.C, 2005.
7. Ekeila, W. "Vancouver's Streetcar Micro-simulation Report". University of British Columbia & City of Vancouver, BC, Canada, 2005.
8. Sayed, T. and Zein, S. "Traffic Conflict Standards for Intersections". Transportation Planning and Technology, Vol. 22, pp. 309–323, 1998.
9. VISSIM 4.3 Manual, PTV, Germany, 2007.
10. Shalaby, A. and Farhan, A. "Prediction Model of Bus Arrival and Departure Times Using AVL and APC Data". Journal of Public Transportation, Vol. 7, No. 1, 2004.
11. Farhan, A., Shalaby, A. and Sayed, T. "Bus Travel Time Prediction Using AVL and APC". ASCE 7th International Conference on Applications of Advanced Technology in Transportation, Cambridge, Massachusetts, 2002.
12. Wall, Z. and Dailey, D. "An Algorithm for Predicting the Arrival Time of Mass Transit Vehicles Using Automatic Vehicle Location Data". Presented at the 78th Annual Meeting of the TRB, Washington, D.C, 1999.

Efficient Transportation and Pavement Systems – Al-Qadi, Sayed, Alnuaimi & Masad (eds)
© 2009 Taylor & Francis Group, London, ISBN 978-0-415-48979-9

Optimized strategy for integrated traffic and transit signal control

Jinwoo Lee
Bureau of Intelligent Transportation Systems and Freight Security, Sauder School of Business, University of British Columbia, Vancouver, Canada

Amer Shalaby
Department of Civil Engineering, University of Toronto, Toronto, Canada

Baher Abdulhai
Department of Civil Engineering, University of Toronto, Toronto, Canada

ABSTRACT: This paper proposes a unique and innovative approach to integrate transit signal priority control into a traffic adaptive signal control strategy. The proposed strategy was named OSTRAC (Optimized Strategy for integrated TRAffic and TRAnsit signal Control). The cornerstones of OSTRAC include an online microscopic traffic flow prediction model and a Genetic Algorithm (GA) based traffic signal timing module. A sensitivity analysis was conducted to determine the critical GA parameters. The developed traffic flow model demonstrated reliable prediction results through a test. OSTRAC was evaluated by comparing its performance to three other signal control strategies. The evaluation results revealed that OSTRAC efficiently and effectively reduced delay time of general traffic and also transit vehicles.

1 INTRODUCTION

The demand for integrated signal control of traffic and transit flows through urban arterials has been growing, from both traffic and transit control perspectives. From the perspective of general traffic control, transit vehicles in mixed traffic flows often become the sources of traffic flow disruptions. In addition to the comparatively slower driving speeds of transit vehicles, frequent and regular stops to board and discharge passengers may further interfere with traffic progression. Nevertheless, transit vehicles have often been disregarded by most traffic signal control strategies mainly due to the complexities of the inter-vehicular behaviours between general traffic and transit vehicles. Consequently, traffic signal timing plans programmed to solely accommodate general traffic have frequently been disrupted by transit vehicles.

On the other hand, from the perspective of transit, coordinated traffic signals often force transit vehicles on urban arterials to stop more frequently than general traffic. Due to the nature of transit operations and the slower travel speeds, transit vehicles frequently drop out of the progression band established for general traffic. As a result, a substantial fraction of transit delay time is incurred by traffic signals (1). The method of Transit Signal Priority (TSP) was conceived to alleviate transit signal delay by providing preferential treatments to transit vehicles at signalized intersections. TSP has been proven successful in speeding up transit vehicles along arterial corridors (2, 3). However, TSP control often disrupts the normal traffic signal operation, which is programmed for optimal traffic flow control. This is due to allocating more green times to the approaches with transit routes.

The primary goal of this study is to develop an optimized strategy for integrated traffic and transit signal control at signalized intersections. In the development procedure, a specific focus was dedicated to overcome the limitations of the current traffic signal control and TSP control practices. The proposed strategy was named OSTRAC (Optimized Strategy for integrated TRAffic and TRAnsit signal Control). A detailed description of OSTRAC and its functional components are provided next.

2 OPTIMIZED STRATEGY FOR INTEGRATED TRAFFIC AND TRANSIT SIGNAL CONTROL

Typically, traffic adaptive signal control strategies are configured with two functional components including a traffic flow model that estimates the near future traffic conditions, and a signal timing method that determines timings based on the estimated future traffic conditions. These two functional components cooperatively and iteratively search for the optimal traffic signal timings.

OSTRAC was also developed based on the common framework of traffic adaptive strategies. Figure 1 illustrates the operational procedure of OSTRAC, highlighting its functional components.

As shown in the above figure, the cornerstones of OSTRAC are the microscopic traffic flow prediction model and the Genetic Algorithm (GA) traffic signal timing module. OSTRAC finds optimal traffic signal timings by iteratively evaluating numerous candidate timings, which are selected based on the searching mechanism of GA. This signal timing module contains an optimization function that is composed of a set of performance measures of both general traffic and transit vehicles. This module provides TSP by giving a heavier weight to the bus delay time component in the function. One distinct feature of the proposed signal timing module is that it can disregard giving TSP depending on the estimated traffic conditions of the intersection.

The primary role of the traffic flow model is to provide the estimated fitness of traffic signal timings proposed by the GA signal timing module. Distinct from the conventional traffic flow models, the proposed microscopic prediction model was developed to represent each vehicle as a separate object with its own characteristics. The traffic flow model acquires required information of the incoming traffic through the traffic sensors implemented at different locations in the transportation network.

OSTRAC is basically a real-time signal control strategy. To obtain desirable results, consideration was given to the establishment of an operational scheme of OSTRAC in a real-time

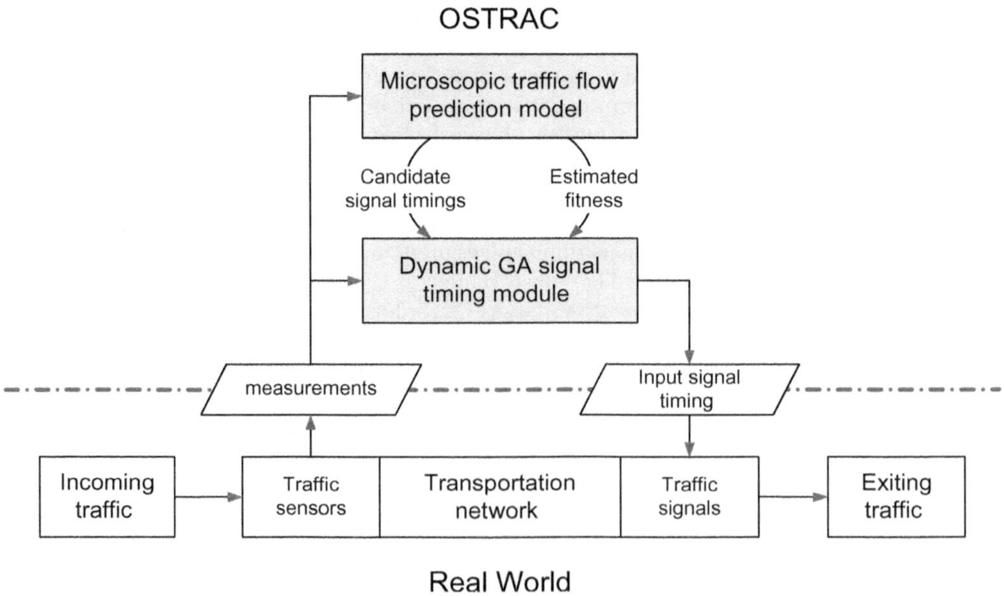

Figure 1. OSTRAC operational framework.

framework. The following section presents the proposed real-time operational scheme, followed by a detailed description of the functional components of OSTRAC.

2.1 *Acyclic real-time signal control scheme*

One evident strength of the traffic adaptive control over pre-timed control is the capability to adjust traffic signal timings in real-time using on-line traffic data. Therefore, the proper design of an operational scheme is key to obtaining desirable outcomes of such a real-time traffic signal control strategy. The fundamental operational scheme of OSTRAC employs the acyclic rolling horizon control approach. The acyclic real-time control scheme does not explicitly consider the traditional cyclic signal control concept, but rather determines the optimal phase switching times during the pre-determined horizon period (4). The term "horizon period" is different from cycle time in that one signal phase may be provided more than once, or not provided at all, in one period. In a rolling horizon manner, only the first small fraction (i.e., rolling fraction) of the computed plan is implemented to take advantage of the latest sensor data. One clear advantage of the acyclic control scheme over the cyclic method is the flexibility in adjusting traffic signal timing to respond to large and rapid changes of traffic demand. On the other hand, the traditional cyclic control scheme has relative benefit over the acyclic scheme in maintaining coordination between adjacent traffic signals.

The proposed scheme was further enhanced to react to transit vehicles. In the absence of transit vehicles, OSTRAC repeatedly executes its control procedure in a certain time interval, usually a few

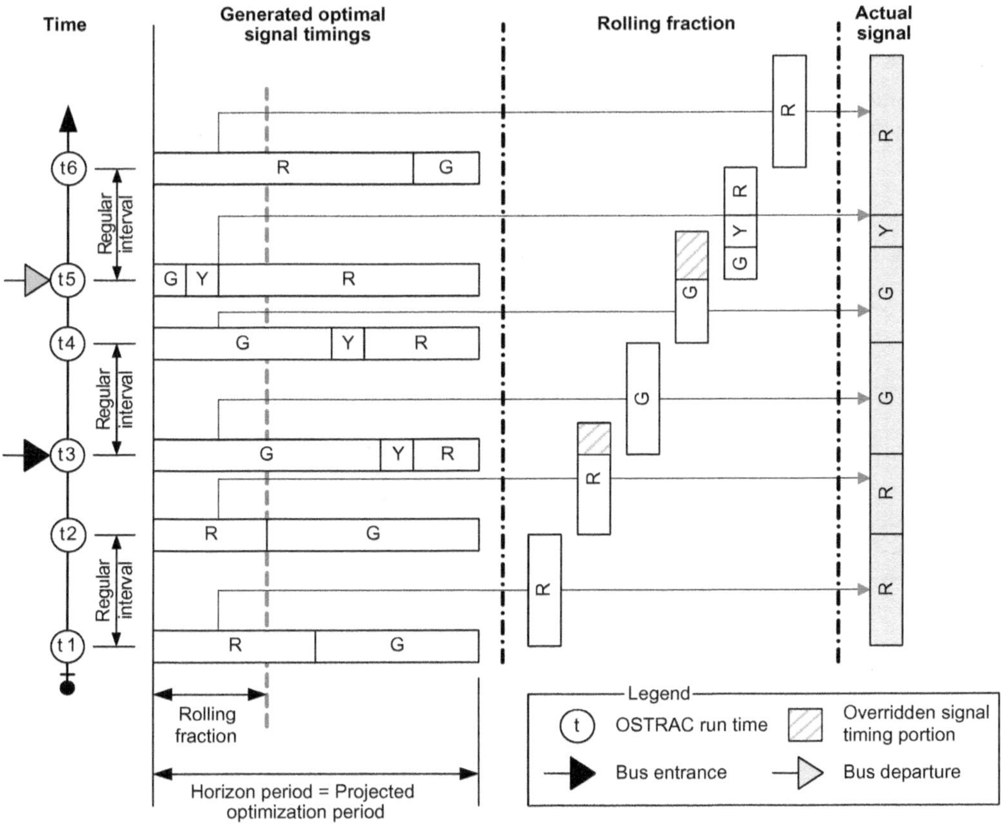

Figure 2. Acyclic real-time signal control scheme of OSTRAC.

seconds. OSTRAC also re-optimizes the traffic signal timing, overriding the existing interval-based operation when transit vehicles either enter or leave the intersection approaches for two reasons.

When a transit vehicle enters into an intersection approach, rather than substituting the existing traffic signal timing plan, as does the conventional TSP control, OSTRAC re-optimizes the traffic signal timing with a higher weight given to the transit vehicle performance measure (i.e., transit vehicle delay) in the optimization function. This immediate response to the transit vehicle enables an early signal timing modification, which indicates a more efficient handling of the approaching transit vehicle with minimal impacts on the existing traffic signal system.

When a transit vehicle exits an intersection approach, OSTRAC immediately re-optimizes the traffic signal timing with adjusted coefficient values, and it is more likely to assign a heavier weight to general traffic performance. A quicker return to the normal signal operation will minimize the potential effects of transit-focused signal control on the general traffic, especially in the non-transit approaches.

Figure 2 demonstrates the OSTRAC operation with a transit vehicle entrance and departure during the operation.

As illustrated in Figure 2, OSTRAC is operated using an interval-based method until a bus arrives at the intersection approach at t_3. At this point, the GA signal timing module re-starts its control procedure to generate a new optimal signal timing plan, which is more likely to be a transit favourable signal timing plan. The last few seconds of the t_2 fraction time is overridden by the new plan generated at t_3. OSTRAC maintains the regular operation interval until this bus leaves the approach. When the bus departs the approach at t_5, OSTRAC re-optimizes the signal timing with a stronger emphasis on the general traffic performance. The following OSTRAC operations keep the regular operation interval unless a new transit vehicle requests prioritized signal control.

2.2 *Experimental network design*

For this study, an arbitrary signalized intersection was modelled in Paramics to be used for the following purposes:

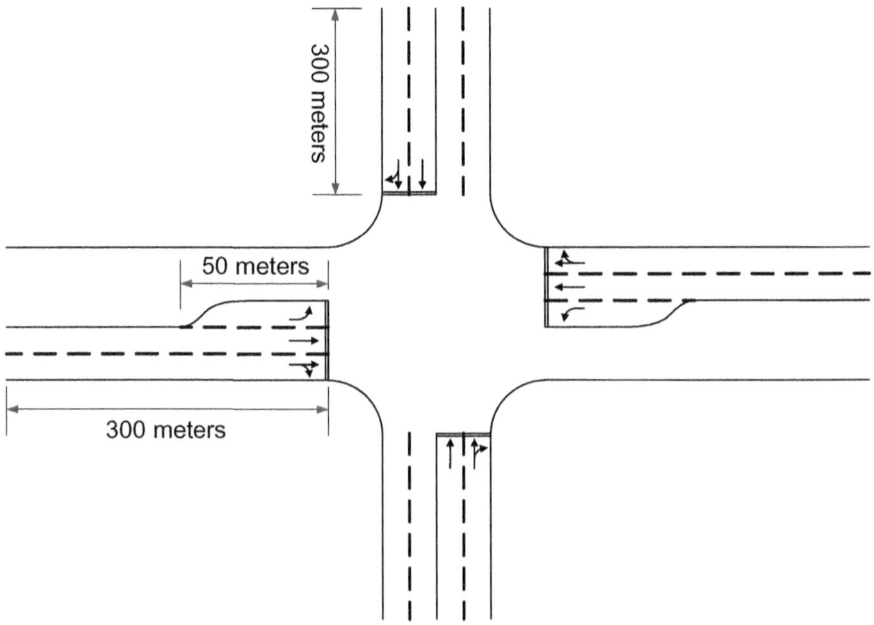

Figure 3. Experimental intersection layout.

- Performing a sensitivity analysis of the GA parameters;
- Validating the microscopic traffic flow prediction model, and;
- Evaluating the performance of OSTRAC.

The modelled intersection is four-legged with three lanes in the east and west approaches and two lanes in the north and south approaches. Figure 3 illustrates the layout of the experimental intersection.

The east and west approaches were designed to have one left-turning exclusive lane, one exclusive through lane, and one shared through and right-turning lane. The north and south approaches have one exclusive through lane and one shared through and right-turning lane. The left turn bays start from 50 meters upstream on the east and west bound approaches. Left turns from the north and the south bounds are not allowed. Each intersection approach is 300-meter long.

2.3 *Dynamic GA signal timing module*

OSTRAC minimizes the optimization function, defined in the signal timing module in order to improve intersection performance. For this study, a Performance Index (PI) function was developed, combining general traffic delay and transit vehicle delay. Delay is a major consideration in the operation of signalized intersections. In general, the operation of traffic signals aims at minimizing either total vehicle delay or average vehicle delay.

The proposed module also accommodates a routine to adjust the relative importance of each performance measures constituting the PI function based on the traffic measurements. For instance, OSTRAC may pursue to minimize signal delay of general traffic when no transit vehicle is present. When a transit vehicle approaches, the signal timing module may assign a higher priority to the transit vehicle delay component in the optimization function. However, if any of the non-bus approaches are in congested conditions, the proposed module also can disallow TSP.

The role of the GA signal timing module is to search for the signal timing plans that optimize the given PI function. In the proposed acyclic signal control scheme, the PI of a signalized intersection is a function of the signal control variables, including the time and sequence of each signal phase. Therefore, for a given optimization horizon h, the goal of the GA optimizer is to determine the control variables that minimize the PI. This can be expressed as follows:

$$\text{Minimize } PI(h) = \alpha\left\{\sum_{v=1}^{V} d_v \Big/ V\right\} + \beta\left\{\frac{\sum d_{bus}}{N}\right\} = f\left(s_k^h, p_k^h\right) \quad \text{for every control horizon } h$$

Subject to the following constraints:

1. $\sum_{1}^{Kh} p_k^h = h_L \quad for\ k = 1, \ldots, K_h$

2. $p_k^h \le g_{\max}(k) \quad for\ k = 1, \ldots, K_h$

3. $g_{\min}(k) \le \begin{Bmatrix} p_k^h + p_{Kh-1}^{h-1} \\ p_k^h \end{Bmatrix} \quad \begin{matrix} if\ k = 1\ and\ s_k^h = s_{Kh-1}^{h-1} \\ else\ if\ k = 2, \ldots, K_h - 1 \end{matrix}$

Where,

- h = optimization horizon
- $PI(h)$ = performance index of horizon h
- α = coefficient of vehicle delay time fraction ($\alpha = 0.2$ in this study)
- V = total number of vehicles in the intersection
- v = vehicle index ($v \in V$)

- d_v = delay time of vehicle v
- β = coefficient of bus delay time fraction (β = 1.0 in this study)
- d_{bus} = total bus delay time
- N = total number of buses eligible to request signal priority
- k = signal phase (i.e., *kth* signal phase)
- K_h = total number of signal phases during *hth* horizon
- S_k^h = index of *kth* signal phase during *hth* horizon
- P_k^h = duration of *kth* signal phase during *hth* horizon
- h_L = length of time horizon
- $g_{max}(k)$ = maximum time for signal phase k
- $g_{min}(k)$ = minimum time for signal phase k

The GA optimizer generates candidate signal timing plans (i.e., chromosomes) to satisfy the three constraints. First, the sum of the signal phase times should be equal to the optimization horizon length. In addition, the GA phase times must be greater than or equal to the pre-defined minimum phase time, and at the same time they must be less than or equal to the maximum phase time.

2.3.1 *Genetic algorithms*

GAs are robust machine learning search methods, which derive their behaviour from a similitude of the evolution processes in nature. GAs have been widely used to solve multi-dimensional optimization problems in numerous application areas since its evolution in 1975 by Holland (5). In GAs, the term gene refers to the parameters that encode a particular representation of some elements of a candidate solution. A Chromosome, which is a full candidate solution (i.e., in our case, it is a traffic signal timing plan) to an optimization problem, is composed of these genes. Each chromosome has a fitness value corresponding to its performance, compared with other members of a population of candidate solutions. In this study, the fitness of one chromosome is the estimated PI values of candidate traffic signal timing plans. Typical GAs follow a sequence of decisions that can be summarized as follows:

Step 1. Problem encoding
Step 2. Random generation of initial population
Step 3. Evaluation of the fitness of each chromosome in the population
Step 4. Selection for reproduction
Step 5. Crossover and mutation
Step 6. Test for stopping criteria. Return the solution, if satisfied; repeat from step 3 onward, if not satisfied

The optimization process of GAs begins by encoding and mapping solutions into chromosomes. An initial population of chromosomes is then generated at random. An initial population of random chromosomes is generated in the next step. At each generation, one population of chromosomes evolves into a new population by replacing some of its members through natural selection methods together with genetic-inspired operations such as mutation and crossover. Reproduction is the process of selecting solutions from the current generation that will be allowed to contribute to the next generation. Since reproduction selects individuals based on their fitness values, chromosomes with less fitness values have a higher probability to be selected in a minimization problem. The crossover operator generates two new offsprings by copying selected bits from two chosen parent chromosomes. Mutation produces small changes to the bit string by choosing bits at random, then changing their values. Crossover is a key operator in the GA, and is used to exchange the main characteristics of parent individuals and pass them on to the children. Mutation is applied after crossover to maintain the genetic diversity of the population and recover possible lost characteristics during crossover. Only small values are adopted for mutation probability to avoid excessive randomness in the search process (6). This routine is repeatedly processed until a terminating condition is met (for example, the desired number of generations is reached).

2.3.2 *Sensitivity analysis of GA parameters*

When applying GAs, there are some critical parameters that should be determined, including population size, number of generations, crossover probability, and mutation probability.

The population size specifies the number of chromosomes in each generation. A higher number generally results in a more robust search, but requires more time to converge. The number of generations indicates the maximum number of evolutions that a GA operation should accomplish. The crossover probability is a percentage specifying what fraction of selected pairs undergo crossover. The mutation rate determines what percentage of genes in the population should be replaced with random values each generation. A high value may disrupt convergence, but a low number may not keep the population diverse enough and lead to premature convergence.

The selection of appropriate GA parameters is crucial for efficient GA operations. A number of previous studies recommended the optimal settings of the GA parameters, including Dejong (7) and Grefenstette (8). However, there are no guidelines available for the selection of GA parameters that are applicable to global optimization problems. Therefore, we conducted a sensitivity analysis for the following GA parameters and values:

- GA selection method: the roulette wheel selection, tournament selection, and rank-based selection method with elitism;
- Population size: 10, 20, and 50;
- Crossover probability: 0.3, 0.5, and 0.8; and
- Mutation probability: 0.01, 0.03, and 0.05.

Using the above parameter values, a total of 81 (i.e., 3^4) scenarios can be defined. In order to find a good parameter set, but also to maintain a reasonable degree of computation effort at the same time, the analysis was performed in three steps.

- The GA selection method was first evaluated using the median values of the other parameters. For instance, each GA selection method was tested with a crossover probability of 0.5, a mutation probability of 0.03, a population size of 20, and a generation number of 50.
- For the selection method that produced the best optimization results, the population size was determined using the median values of the other two parameters. For each population size of 10, 20, or 30, the performance of the GA optimizer was evaluated with a crossover probability of 0.5 and a mutation probability of 0.03.
- With the chosen GA selection method and the population size, all possible combinations of the crossover probabilities and mutation probabilities were evaluated.

The developed testbed intersection was used for the analysis. The GA optimizer was configured with each test parameter set to examine the optimization convergence under identical environments. The objective function was defined to minimize the general traffic delay and the results were gathered from ten runs for each parameter set. The binary encoding method was adopted in the proposed GA module mainly due to its simplicity. More detailed information about the conducted sensitivity analysis is available elsewhere (9). On the basis of the analysis results, the following specifications of the GA-based signal timing module were determined:

- Rank-based selection method with Elitism
- Population size 10
- Number of generations 100
- Mutation probability 0.03
- Crossover probability 0.6

2.4 *Microscopic traffic flow prediction model*

In a previous study by the authors, a microsimulation model was developed for the purpose of transit delay time estimation (10). Within this model, the movement of individual vehicles is governed by a set of pre-defined rules dealing with different driving situations. The established driving rules and their primary roles are summarized below:

- *Initializing rules*
  Define the characteristic of each vehicle including the aggressiveness level, desired driving speed, desired acceleration speed, desired deceleration speed, perception time, desired gap from the leading vehicle;
- *Free flow driving rules*
  Specify the vehicle behaviour in free flow speed. Any car will follow these free flow driving rules, if the headway distance from the leading car is greater than its critical distance. The critical distance of one vehicle is defined as the headway distance where this vehicle would be affected by the behaviour of the leading vehicle;
- *Car following rules*
  Apply when the headway distance of one vehicle is less than the critical distance. The car following rules determine vehicle behaviour, which is whether to maintain, decelerate, or accelerate speed in response to the behaviour of its leading vehicle;
- *Lane changing rules*
  Vehicles following the leading one may attempt to change lanes. The lane changing rules consist of some pre-defined conditions that must be satisfied before lane changing, and the behaviour of vehicles when they are changing lanes or considering lane changing;
- *Traffic signal reaction rules*
  Define the behaviour of vehicles when they are under the effect of traffic signal operation. These rules apply to the vehicles approaching the downstream intersection and override all the other driving rules. For instance, even if one vehicle is in free flow conditions, this car must reduce its speed if the red phase is being provided to the approach, and;
- *Transit vehicle rules*
  Distinct operation of transit vehicles are defined in the transit vehicle rules. These rules primarily describe the behaviour of transit vehicles associated with passenger service.

In this study, the proposed traffic flow model was further enhanced to deal with general traffic in addition to transit vehicles. Also, the performance of the proposed traffic flow model was validated before implementation into OSTRAC. The reliability of the transit vehicle delay prediction was previously evaluated (10). The emphasis was placed on devising a study that would provide indications of the prediction performance of general traffic delay.

The evaluation of the proposed traffic flow model was conducted by comparing the prediction results to those of the Paramics microsimulation software. Assuming the Paramics results as the actual values, the model was calibrated to produce similar prediction results with those of Paramics before the test. The testbed intersection was used for the evaluation. Arbitrary traffic volumes and a traffic signal timing were created at the intersection. The total period was one and a half hours including three sub-periods to represent low, intermediate, and heavy traffic flows. In each sub-period, the OSTRAC prediction model was configured to run 20 times in random intervals. This setting is to ensure that the proposed model can make predictions under various operational environments of the intersection. Figures 4 (a), (b), and (c) graph the average vehicle delay times predicted by the OSTRAC model and the Paramics software in the low, intermediate, and heavy traffic flow periods, respectively.

The prediction results of the OSTRAC model were evaluated using four basic measures including: the mean error (ME), the mean absolute error (MAE), the root mean square error (RMSE), and the mean absolute percent error (MAPE). Table 1 provides the summary of the evaluation results. Note that the actual values are the Paramics results, and the predicted values are the results of the proposed model.

Two conclusions can be drawn from the presented results. First, regardless of the traffic flow levels, the predicted vehicle delay times were not significantly biased, either over-estimated or under-estimated. The obtained ME values were 2.21 seconds, 1.84 seconds, and –1.17 seconds for the low, intermediate, and heavy traffic flow rate, respectively. These results indicate that the proposed model

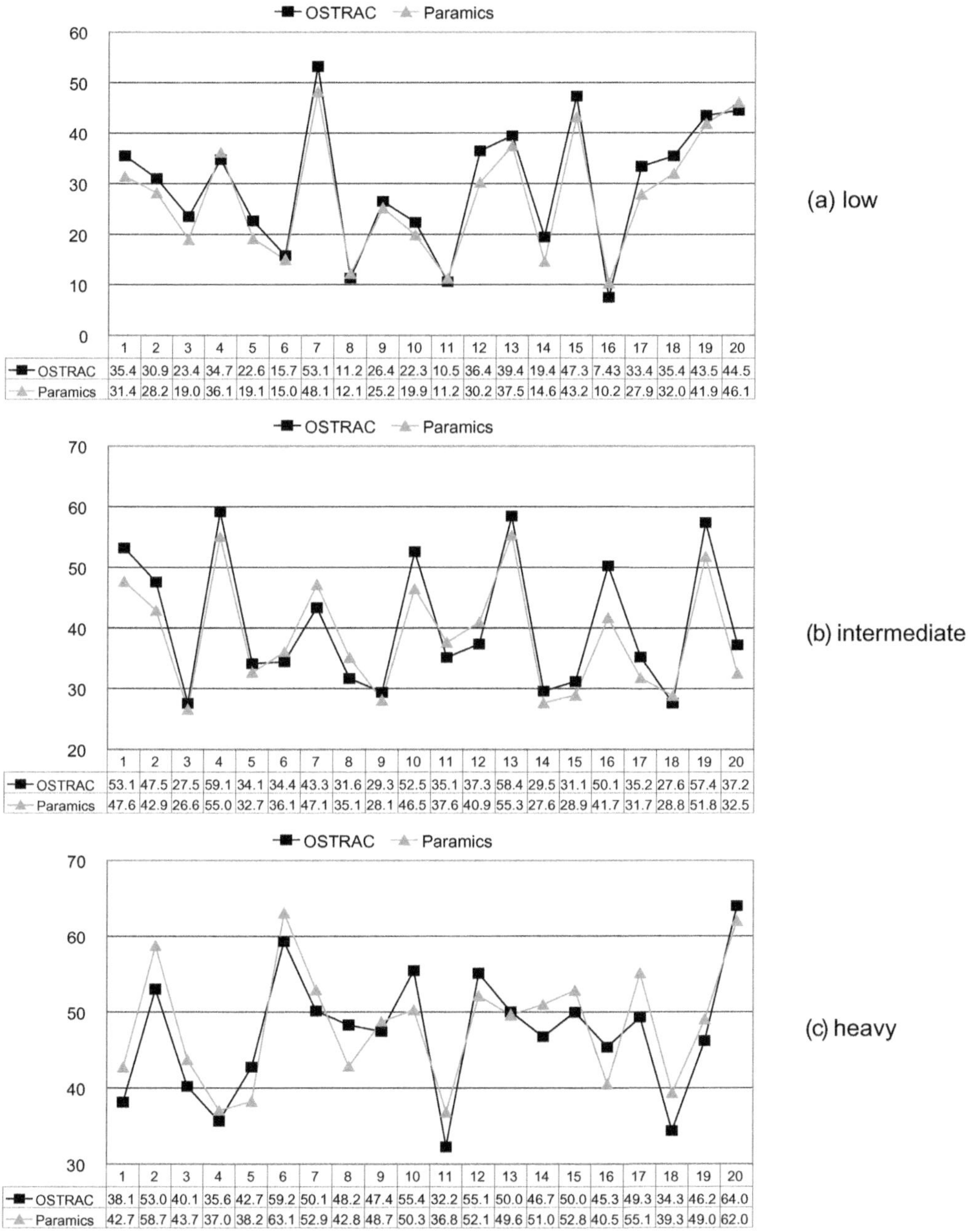

(a) low

	1	2	3	4	5	6	7	8	9	10	11	12	13	14	15	16	17	18	19	20
OSTRAC	35.4	30.9	23.4	34.7	22.6	15.7	53.1	11.2	26.4	22.3	10.5	36.4	39.4	19.4	47.3	7.43	33.4	35.4	43.5	44.5
Paramics	31.4	28.2	19.0	36.1	19.1	15.0	48.1	12.1	25.2	19.9	11.2	30.2	37.5	14.6	43.2	10.2	27.9	32.0	41.9	46.1

(b) intermediate

	1	2	3	4	5	6	7	8	9	10	11	12	13	14	15	16	17	18	19	20
OSTRAC	53.1	47.5	27.5	59.1	34.1	34.4	43.3	31.6	29.3	52.5	35.1	37.3	58.4	29.5	31.1	50.1	35.2	27.6	57.4	37.2
Paramics	47.6	42.9	26.6	55.0	32.7	36.1	47.1	35.1	28.1	46.5	37.6	40.9	55.3	27.6	28.9	41.7	31.7	28.8	51.8	32.5

(c) heavy

	1	2	3	4	5	6	7	8	9	10	11	12	13	14	15	16	17	18	19	20
OSTRAC	38.1	53.0	40.1	35.6	42.7	59.2	50.1	48.2	47.4	55.4	32.2	55.1	50.0	46.7	50.0	45.3	49.3	34.3	46.2	64.0
Paramics	42.7	58.7	43.7	37.0	38.2	63.1	52.9	42.8	48.7	50.3	36.8	52.1	49.6	51.0	52.8	40.5	55.1	39.3	49.0	62.0

Figure 4. Test results of the traffic flow prediction model.

slightly over-estimated vehicle delay in the low and intermediate level scenarios. On the other hand, in the heavy traffic level scenario, the model produced a minor under-estimation of delay time.

Second, based on the calculated MAE, RMSE, and MAPE values, the proposed model provided highly reliable prediction results, which are less than 5 seconds of absolute differences from the actual vehicle delay times. The results also revealed that the MAPE values improved as the intersection

Table 1. Prediction accuracy test results.

	Measures			
Traffic flow level	ME (sec)	MAE (sec)	RMSE (sec)	MAPE
Low	2.21	2.95	3.38	12.35%
Intermediate	1.84	3.48	3.97	8.71%
Heavy	−1.17	3.69	4.00	7.90%

Where,

$$ME = \frac{1}{N}\sum_{n=1}^{N}(p_n - a_n) \quad MAE = \frac{1}{N}\sum_{n=1}^{N}|p_n - a_n|$$

$$RMSE = \sqrt{\frac{1}{N}\sum_{n=1}^{N}(p_n - a_n)^2} \quad MAPE = \frac{1}{N}\sum_{n=1}^{N}\frac{|p_n - a_n|}{a_n}$$

N = number of observations
p_n = predicted value
a_n = actual value

approach became more congested. However, this does not necessarily indicate that the proposed model performed at a substandard level during the low traffic flow period. Since the vehicle delay times are relatively smaller values in the low traffic flow period, even a minor numerical error may cause significant deterioration in the MAPE value when it is expressed as a percentage error.

3 OSTRAC EVALUATION

The purpose of the evaluation is to assess the benefits of OSTRAC *via* simulation analysis. The evaluation was conducted by comparing the OSTRAC performance to three other traffic signal control scenarios as described below:

- Scenario 1: Pre-timed signal control without TSP
- Scenario 2: Pre-timed signal control with a conventional TSP method
- Scenario 3: Adaptive signal control with a conventional TSP method
- Scenario 4: Integrated traffic and transit signal control (OSTRAC)

The pre-timed signal timings for the first and second strategies were obtained using the Synchro software. In order to model a traffic adaptive signal control, OSTRAC was implemented with its optimization function to include only the general traffic delay component. In order to model a conventional TSP operation, the two most commonly used TSP strategies, namely transit phase extension and non-transit phase truncation, were implemented. The second and third scenarios would demonstrate the efficiency of an integrated traffic and transit signal control (i.e., OSTRAC) compared to non-integrated signal control strategies (i.e., pre-timed and traffic adaptive) with conventional TSP.

The scenarios were evaluated under two different traffic demand levels representing typical off-peak and peak period traffic conditions. The total evaluation period extended three hours. This evaluation period was also divided into six 30-minutes sub-periods that were assigned different traffic flow rates aiming at volume-to-capacity (v/c) ratios to be maintained between 0.3 to 0.7 for the off-peak period scenarios, and between 0.5 to 1.0 for the peak period scenario. Table 2 provides the used traffic flow rates.

The traffic signal control at the test intersection was configured to operate three signal phases: the east-west green, the north-south green, and the east-west left turn green so that each phase

Table 2. Input traffic flow rates.

	Analysis period		East			West			North			South		
Off-peak period	Start	End	LT	TH	RT	LT	TH	RT	LT	TH	RT	LT	TH	RT
	10:00	10:30	71	360	48	42	353	41	0	345	69	0	216	49
	10:30	11:00	49	318	39	62	257	39	0	246	54	0	185	65
	11:00	11:30	47	292	31	55	330	34	0	297	40	0	271	53
	11:30	12:00	58	377	66	82	247	43	0	257	64	0	258	33
	12:00	12:30	75	344	42	65	278	52	0	199	55	0	203	61
	12:30	13:00	53	400	56	41	332	61	0	256	90	0	258	65

	Analysis period		East			West			North			South		
Peak period	Start	End	LT	TH	RT	LT	TH	RT	LT	TH	RT	LT	TH	RT
	16:00	16:30	36	469	74	68	321	15	0	178	125	0	339	25
	16:30	17:00	49	521	82	68	343	25	0	199	118	0	365	22
	17:00	17:30	54	556	87	72	389	26	0	213	136	0	422	35
	17:30	18:00	60	606	83	95	441	32	0	250	139	0	466	32
	18:00	18:30	56	498	64	68	420	29	0	198	125	0	430	19
	18:30	19:00	50	476	55	61	398	24	0	167	122	0	411	23

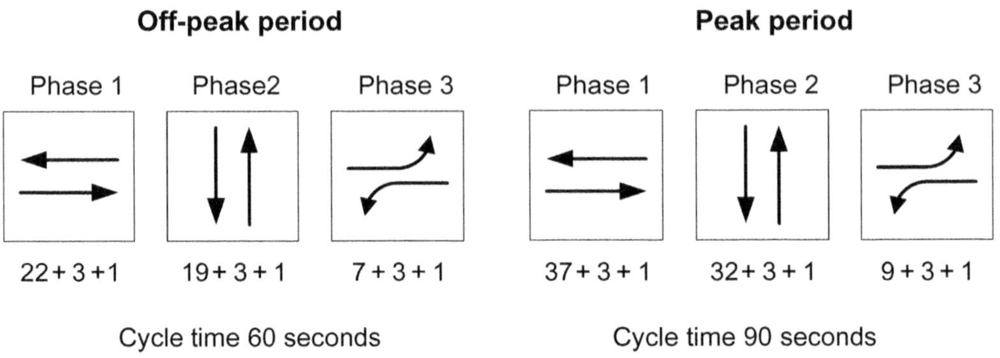

Figure 5. Pre-timed traffic signal timing plans.

is followed by an intergreen period containing 3 seconds of yellow and 2 seconds of all-red. Figure 5 presents the implemented traffic signal timing plans. Note that the average traffic flow rates were used to develop the off-peak period signal timing. The peak period signal timing was calculated using the traffic flow rate of the sub-period with the greatest total vehicle arrival flows during the peak period (i.e., 17:30–18:00).

The test intersection accommodated two-way bus services with a far-sided bus stop in the east and west approaches. The modelled transit route has a high frequency of 12 buses per hour (i.e., 5 minutes of headway). The bus release times from the upstream link were set to values between 4 minutes to 6 minutes for a realistic modelling of the bus operation. Approximately 50% of the buses were scheduled to release later than their schedules. The transit phase extension strategy was designed to provide a maximum of 15 seconds of extensions in 1-sec intervals starting at the end of the scheduled transit green phase. The non-transit phase truncation shortened the second signal phase (i.e., north-south green) by 7 seconds in the off-peak period, and by 10 seconds in the peak period.

OSTRAC determines whether to provide prioritized signal control to buses, when they pass the upstream transit sensor. The actual headway is one of the criteria of the decision. OSTRAC provides prioritized signal control only for the buses that are running behind schedule. Another decision criterion is related to the traffic conditions in the non-bus approaches. OSTRAC continuously monitors the flow rates of incoming traffic into each intersection approach. If more traffic has entered to any non-bus approach than a certain threshold (i.e., 80 vehicles per a 5 minute period), OSTRAC disregards providing TSP by simply considering all the buses in the intersection as general traffic. Note that the selected threshold roughly represent the degree of saturation in the approach is greater than 0.9.

3.1 *Evaluation results*

The evaluation focused on the impact of the OSTRAC operation on the general traffic and buses in terms of their signal delay times. Table 3 provides the summary of the results, classified by the evaluated signal control scenarios. Note that the Table further categorizes the general traffic delay times into traffic movement groups that use different signal phases. The first phase group includes the east-west through and right-turning movements. The second phase group consists of the north-south through and right-turning movements. The east-west left-turning movements compose the third phase. The table also provides the percentage changes in each measure, achieved by the tested signal control strategies with respect to those of the base scenario. Note that a total of 20 simulation runs were conducted for each evaluation scenario. The presented delay times are the averages of those simulation results.

In summary, throughout the off-peak and peak evaluation periods, OSTRAC demonstrated consistent and balanced performance by substantially reducing delay times of both general traffic and

Table 3. Evaluation results.

Evaluation scenarios		Average signal delay times (sec)				
		General traffic				
Traffic demand levels	Control strategy	Phase 1 movements	Phase 2 movements	Phase 3 movements	Total traffic	bus
Off-peak period	Base scenario (Pre-timed without TSP)	26.47	29.16	33.76	28.17	37.23
	Scenario 2 (Pre-timed with a conventional TSP)	25.01	33.05	35.92	29.77	33.48
		(–1.47%)	(+13.34%)	(+6.40%)	(+5.69%)	(–10.07%)
	Scenario 3 (Adaptive control with a conventional TSP)	22.68	27.81	30.92	25.40	33.09
		(–14.32%)	(–4.63%)	(–8.41%)	(–9.84%)	(–11.12%)
	Scenario 4 (OSTRAC)	22.80	25.56	30.67	24.57	31.70
		(–13.93%)	(–12.35%)	(–9.15%)	(–12.79%)	(–14.85%)
Peak period	Base scenario (Pre-timed without TSP)	44.97	49.11	47.01	46.77	56.44
	Scenario 2 (Pre-timed with a conventional TSP)	43.95	56.76	51.13	49.57	50.91
		(–2.27%)	(+15.58%)	(+8.76%)	(+5.99%)	(–9.80%)
	Scenario 3 (Adaptive control with a conventional TSP)	37.71	47.73	43.45	41.51	47.14
		(–16.14%)	(–2.81%)	(–7.57%)	(–9.95%)	(–16.48%)
	Scenario 4 (OSTRAC)	37.40	42.33	42.40	39.71	47.68
		(–16.83%)	(–13.80%)	(–9.80%)	(–15.09%)	(–15.52%)

buses as compared to the base scenario. The adaptive control with a conventional TSP operation was able to reduce considerable delay times of both traffic and buses as compared to the base scenario. However, the evaluation results also revealed that this strategy compromised considerably the performance of the non-bus phase movements (particularly phase 2) as compared to OSTRAC. The pre-timed control with a conventional TSP operation successfully improved the bus operation at the test intersection, but it caused significant delays to the traffic movements using the non-bus signal phases. The findings from the tables are illustrated in Figure 6 for a better understanding of the evaluation results.

As shown, the second scenario involving pre-timed control with a conventional TSP strategy reduced the bus delay time by 10.07% in the off-peak period, and by 9.80% in the peak period compared to the base scenario. With regards to the general traffic delay time, this strategy demonstrated a remarkable contrast between the results of the bus phase movements and the non-bus phase movements. In the off-peak period, this strategy slightly reduced the delay time of the bus phase movements by 1.47%. On the other hand, however, it significantly deteriorated the delay time of the non-bus phase traffic movements. This scenario increased the vehicle delay time of the second phase movements by 13.34%, and of the third phase movements by 6.40%. This is because the implemented active TSP strategies naturally assign more green time to bus signal phase (i.e., phase one) and less green time to non-bus phases (i.e., phase two and phase three). A similar pattern was also found in the peak period results. The tested strategy provided the most efficient signal control for the bus phase movements by reducing their average delay time by 2.27%. The strategy performance then substantially declined. The delay time of the second phase movements increased by 15.58%, while that of the third phase movements increased by 8.76%.

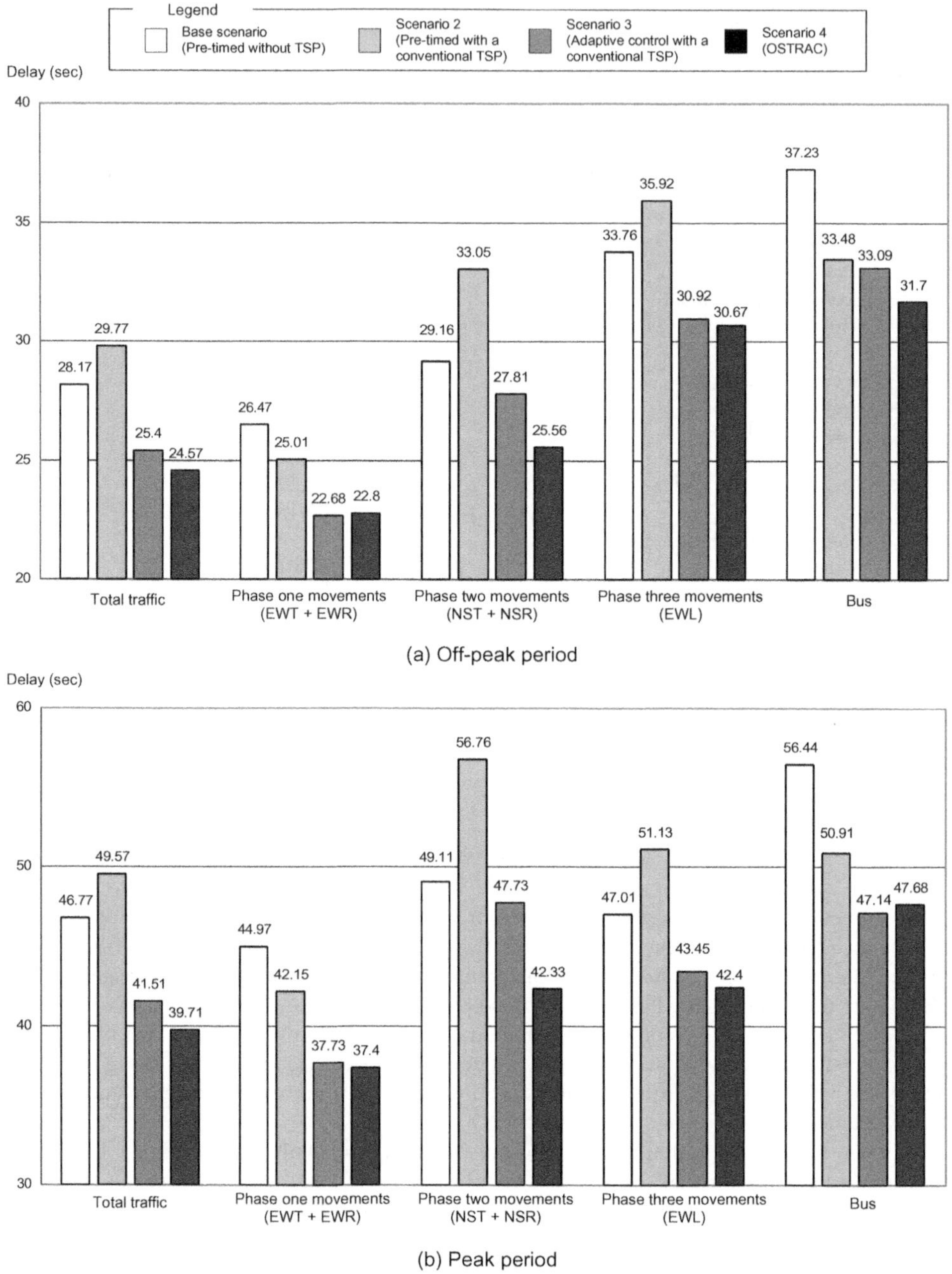

Figure 6. Evaluation results.

The third scenario operated a traffic adaptive signal control with a conventional TSP. This scenario was able to substantially reduce the overall traffic delay as compared to the base case and second scenarios. However, as demonstrated by the previous scenario (i.e., pre-timed with a conventional TSP), the benefit of this strategy was lopsided to the bus phase traffic movements. The average delay time of the first phase traffic movements was reduced by 14.32% in the off-peak period, and by 16.14% in the peak period. However, the performance significantly declined when it dealt with the non-bus phase movements, especially for the second phase movements. The achieved delay time reduction was only 4.63% in the off-peak period, and 2.81% in the peak period. These results can also be partly explained by the fact that conventional TSP operation often disrupts the operation of traffic adaptive signal control.

Another remarkable result was that this scenario substantially improved the bus operation. The reduced bus delay time was 11.12% in the off-peak period, and 16.48% in the peak period. It is worth noting that this achievement is outstanding as compared to the second scenario that operated the same TSP strategies. This result may have been attributed to the traffic adaptive control, which successfully improved the traffic conditions in the bus approaches. In addition to the traffic signal control, bus travel times are also be impacted by the traffic conditions in the bus approach. Therefore, buses in the intersection may have taken an advantage of the substantially reduced delays in the bus approaches.

OSTRAC showed exceptionally efficient operation for the bus phase movements by reducing their average vehicle delay time by 13.93% in the off-peak scenario, and by 16.83% in the peak period scenario. It also performed at a standard level for the phase two movements (i.e., north-south through and right-turning) by reducing their delay times by 12.35%, and 13.80% in the off-peak and peak periods, respectively. For the phase three movements (i.e., east-west left-turning), OSTRAC achieved 9.15% of the average delay time reduction in the off-peak period and 9.80% in the peak period. The bus priority control of OSTRAC is an important factor that promoted successfully vehicle delay time reductions for the bus phase movements. When a bus requests signal priority, OSTRAC searches for the signal timings that are more beneficial to the buses rather than to the overall traffic condition. This process may re-calculate a traffic signal timing in a way to provide more green time to the bus signal phase, which consequently results in relatively more benefit to the bus phase movements.

Another important control objective of OSTRAC was to improve the operation of buses. OSTRAC achieved 14.85% and 15.52% of the bus delay time reductions in the off-peak and peak period, respectively. Note that the achieved bus delay time in the peak period was slightly less than that of the third scenario, which operated in traffic adaptive control with a conventional TSP. In fact, this result is not unexpected given that the OSTRAC operation disregards the bus prioritized control when any of the non-bus approaches are in saturated conditions. This operation pursues to minimize the average traffic delay time of all traffic movements in the intersection rather than giving absolute priority to the buses.

4 CONCLUSIONS

This paper introduced a unique and innovative approach to integrate transit signal priority control into a traffic adaptive signal control strategy. The proposed strategy was named OSTRAC (Optimized Strategy for integrated TRAffic and TRAnsit signal Control). The cornerstones of OSTRAC are the microscopic traffic flow prediction model and the Genetic Algorithm (GA) traffic signal timing module. These two functional components cooperatively and iteratively search for the optimal traffic signal timings. The main task of the traffic flow model is to provide the estimated performance of candidate traffic signal timings that are proposed by the signal timing module. The signal timing module selects the candidate traffic signal timings based on the searching mechanism of GA.

A sensitivity analysis was conducted to determine the critical GA parameters before application into OSTRAC. The performance of the traffic flow prediction model was also validated. Two conclusions were drawn from the results. First, regardless of the traffic flow levels, the predicted vehicle delay times were not significantly biased, either over-estimated or under-estimated. Second, the proposed model provided highly reliable prediction results, which are less than 5 seconds of absolute differences from the actual vehicle delay times based on the calculated MAE, RMSE, and MAPE values.

The performance of OSTRAC was evaluated on a hypothetical intersection within the Paramics software. The evaluation was conducted by comparing the OSTRAC performance to three other traffic signal control strategies including: a pre-timed signal control without TSP, a pre-timed signal control with a conventional TSP method, and a traffic adaptive signal control with a conventional TSP method. The evaluation results can be summarized as below:

- The pre-timed control with a conventional TSP scenario successfully improved the bus operation at the intersection, but it also caused delay time increase of the non-bus phase traffic movements;
- The adaptive control with a conventional TSP scenario was able to substantially reduce the overall traffic delay as well as the bus delay time as compared to the pre-timed control. However, the observed benefit was seriously lopsided to only the bus phase traffic movements, and;
- OSTRAC clearly proved its efficiency in improving both traffic and bus operation through the evaluation. OSTRAC consistently and significantly reduced the delay time of both general traffic and bus as compared to the other three control strategies.

ACKNOWLEDGMENTS

This research was supported by the Natural Sciences and Engineering Research Council.

REFERENCES

1. Evans, H. and Skiles, G. Improving Public Transit through Bus Preemption of Traffic Signals. Traffic Quarterly, Vol. 24, No. 4, 1970, pp. 531–543.
2. Courage, K.G., and Wallace, C.E. Evaluation of Some Bus Priority Strategies on NW 7th Avenue in Miami. *In Transportation Research Record: Journal of the Transportation Research Board*, No. 626, TRB, National Research Council, Washington D.C., 1977, pp. 32–35.
3. Wood, K., and Baker, R.T. Using SCOOT Weightings to Benefits Strategic Routes. Traffic Engineering and Control, 1992, pp. 226–235.
4. Gartner, N.H. OPAC: A Demand Responsive Strategy for Traffic Signal Control. In *Transportation Research Record 906*, TRB, National Research Council, Washington, D.C., 1983, pp. 75–81.
5. Holland, J.H. *Adaptation in Nature and Artificial System*. Ann Arbor, The University of Michigan Press, 1975.
6. Mitchelle, T. Machine Learning. McGraw Hill, 1997.
7. DeJong, K.A. and Spears, W.M. An Analysis of the Interacting Roles of Population Size and Crossover in Genetic Algorithms. Proceeding of the First Workshop Parallel Problem Solving from Nature, Springer-Verlag, Berlin, 1990. pp. 38–47.
8. Grefenstette, J.J. Optimization of Control Parameters for Genetic Algorithms. IEEE Transaction on Systems, Man, and Cybernetics, Vol. SMC-16, No. 1, Jan./Feb. 1986, pp. 122–128.
9. Lee, J. Development of an Optimized Strategy for Integrated Traffic and Transit Signal Control, Unpublished Ph.D. dissertation, University of Toronto, Toronto.
10. Lee, J., A. Shalaby, J. Greenough, M. Bowie, and S. Hung. Advanced Transit Signal Priority Control Using On-line Microsimulation-based Transit Prediction Model. *In Transportation Research Record: Journal of the Transportation Research Board*, No. 1925, TRB, National Research Council, Washington D.C., 2005, pp. 185–194.

Efficient Transportation and Pavement Systems – Al-Qadi, Sayed, Alnuaimi & Masad (eds)
 ISBN 978-0-415-48979-9

Exploring the potential benefits of VII-Enabled dynamic speed control in signalized networks

Ghassan Abu-Lebdeh
Department of Civil Engineering, American University of Sharjah, Sharjah, UAE

Hui Chen
Department of Civil and Environmental Engineering, Michigan State University, East Lansing, MI, USA

Zaher Khatib
Department of Civil Engineering, University of Sharjah, Sharjah, UAE

ABSTRACT: This paper demonstrates how control with dynamic operating speed (not speed limit) can be used to realize different control objectives including improved traffic flow and safety. The analysis shows that using dynamic speed control (DSC) can improve operations of signalized networks by increasing throughput, decreasing delay and number of stops, and reducing speed noise and speed variation thus lowering the likelihood of accidents. It is shown that using dynamic speed control synchronously with dynamic traffic signal control in signalized networks can enable more sustainable and reliable traffic operations thus leading to higher traffic throughput, reduced accident potential, reduced fuel consumption, and lower emissions levels. As physical capacity of urban networks and signal-only improvements are no longer sufficient to cope with current congestion levels, time has come for new generations of advanced control concepts such as DSC. As information availability improves (through initiatives such as vehicle infrastructure integrated (VII) systems) and with the advances in wireless communications, computation, sensor technologies and their deployment in intelligent transportation system settings, designing and implementing advanced control concepts is now both possible and feasible. A dynamic speed control (DSC) system which communicates the near-optimal speed optimized based on real-time data to guide drivers on their best speed selection is one such concept.

1 INTRODUCTION

When the term "dynamic speed" is used it typical refers to dynamic speed limits. However, this paper is about an entirely different type of dynamic speed: it is dynamic operating speed. This is the speed at which drivers are expected to operate, much like the speeds at which airliners operate as they approach a hub airport. Here, the airliners operating speeds may be increased or decreased based on both safety and airport capacity constraints. In a signalized network, speeds on different links would be optimized synchronously with signal control parameters in real time based prevailing conditions to achieve certain operational and safety objectives. Put differently, speed in this case is a decision variable not a constraint as is the case in conventional traffic signal control. This is the *dynamic speed* this paper is about.

Research into possible applications of dynamic speed in signalized networks is currently non existent. But the advent of intelligent transportation systems, largely motivated by the affordability and spread of communication, sensor, and computation technologies gave rise to ideas that otherwise were either impossible or too difficult to implement. At the same time traffic congestion problems are reaching crises levels in many urban areas. In urban areas where a larger portion of the roadway system is signalized thoroughfares, innovations in controlling such facilities assume more urgency. Using dynamic speed as a mean of traffic control has the potential to improve both safety and operation. The vehicle infrastructure integrated systems (VII) make this possible (more on VII later).

Most current research on improving operations in signalized networks has emphasized developing strategies that deal only with signal timing parameters such as cycles, green splits, phases, and offsets. In all existing and under-development control strategies, speed is used as a constraint only. That this does not need to be the case always is a key argument in this paper. The objective of this paper is to conceptually examine the limitations of signal-only approach to controlling signalized networks under different traffic conditions. The paper then explores, at a fundamental level, corresponding remedies using dynamic speed algorithms and analyzes likely benefits of different control algorithms. The paper presents the ideas and the supporting evidence using a theoretical approach and common assumptions on fundamental traffic properties. The paper is not about providing a complete set of algorithms nor is it about quantify the likely benefits.

The remainder of this paper is organized as follows: background material on dynamic speed control is described next followed by a discussion on the deficiencies of signal-only control in signalized networks. Discussion of variants of dynamic speed control algorithms for different traffic flow conditions and different control objectives then follows along with an assessment of potential benefits. The paper ends with concluding remarks and planned future work.

2 BACKGROUND

Per the Manual on Uniform Traffic Control Devices (FHWA, 2003a), posted speed limits are set to assist drivers in choosing a safe travel speed. Factors that shall be taken into considerations when determining appropriate speed limits are: 1) road characteristics such as shoulder condition, grade, alignment, and sight distance; 2) the 85th percentile and pace speed (pace speed is defined as a 10 mph range of speeds that usually takes in 70% of all drivers; 3) roadside development and environment; 4) parking practices and pedestrian activity; and 5) reported crash experience for at least a 12-month period (FHWA, 2003b). All these factors are not time-dependent.

Time-dependent factors such as weather, roadway and traffic conditions, and signal timing do influence what speed is appropriate for both safe travel and stable flow. The idea of dynamic speed control (DSC) is to dynamically calculate near optimal speeds to guide drivers in setting their speeds in response to some or all of these time-dependent factors. This would be the speed that drivers should operate at, not lower and not higher. It is not a speed limit that just should not be exceeded. Variable speed limit systems which also manage speeds have been widely used in the past for work zones (Coleman et al. 1996; TRB, 1998; Lin et al., 2004) and freeway traffic management (Florida DOT, 2003; Paterson, 2003; Hegyi et al., 2003). But those systems deal with speed limits rather than optimal operating speeds.

Managing traffic speed is a good idea both for safety and operational reasons. Researches have indicated that the potential risk of rear-end collisions on freeways can be reduced if traffic flow speeds are properly regulated with variable speed limits, VSL (Coleman et al. 1996, TRB, 1998). Lin et al. (2004) developed a VSL adjustment algorithm to fulfill the objectives of queue reduction and throughput maximization on highway work zones. While various studies and algorithms were developed for freeway applications, research on managing speed in signalized networks is very limited. Chen and Abu-Lebdeh (2006a, 2006b) developed a framework for integrating dynamic speed control and dynamic signal control. That study demonstrated the feasibility of such control and provided preliminary assessment of the operational improvements. However not all potential benefits of using dynamic speed in signalized networks have been explored.

At the heart of dynamic signal control is information, and the vehicle infrastructure integrated (VII) systems initiative provided much impetus for that. Dynamic speed control (DSC) in signalized networks is envisioned to be implemented in an VII system setup with functionalities emulating those found in closed-loop control systems. Vehicles and infrastructure components are envisioned to be able to communicate and exchange information to enhance both traffic safety and operations. As noted by the US Department of Transportation, the "coordinated deployments of communication technology in all vehicles by the automotive industry and on all major U.S. roadways by the transportation public sector are the fundamental building blocks of the VII concept [USDOT, 2008]. In a VII setting, the roadway would be divided into segments. Each segment has a "hot post" that collects data from traffic along that particular segment using Dedicated Short Range Communications (DSRC). Vehicles can communicate with roadside units (RSU), or "hot posts" and within themselves via the hot posts. Hot posts of different roadway segments can communicate with each other. Operationally, a vehicle would transmit anonymous on-board sensor data to an RSU every time it passes one. A vehicle would store time samples of the data it collected between successive RSUs and then transmits these data samples as it passes the RSU. The anonymous data received at an RSU would be sent to an aggregation point from which it is then forwarded to authorized subscribers (e.g., Traffic Operations Centers, DOTs, etc.). Each aggregation point may receive data from several thousand RSUs. All data are organized and ordered by the geographic coordinates from the vehicles and would be available to authorized subscribers. At a higher level, each section of the road "knows" what all other roadway sections are experiencing, which means that vehicles on any given segment can "know" what is happening with traffic on downstream segments. A simplified architecture of data flow is shown in Figure 1 (ITS-JPO, 2005).

In signalized networks, a VII-based DSC system would require a display medium, which could be either variable message signs (VMS), or in vehicle display devices. The display media is directly connected via dedicated communication channels to a traffic management center where real time traffic data from the hot posts and traffic signals are collected, processed, and used in decision making. There, near-optimal speeds are determined. The VMS and/or the in-vehicle devices would be used to display a speed (depending on control objectives and level of information made available to motorists) when vehicle enters a link. If VMSs are used, they would be installed overhead along with traffic signals. The VMSs can be used to provide lower information level where a group of vehicles are instructed to follow a given speed. Such setup would limit the flexibility of the control algorithm but would require less detection, computation and communication resources. Departures, arrivals and queue status would be evaluated at the end of each cycle and only one speed decision is made for vehicles leaving the intersection during the same cycle. On the other hand, if in-vehicle devices are to be used to provide higher level of information whereby speed information can be customized for every individual vehicle (or vehicles in a given platoon; in

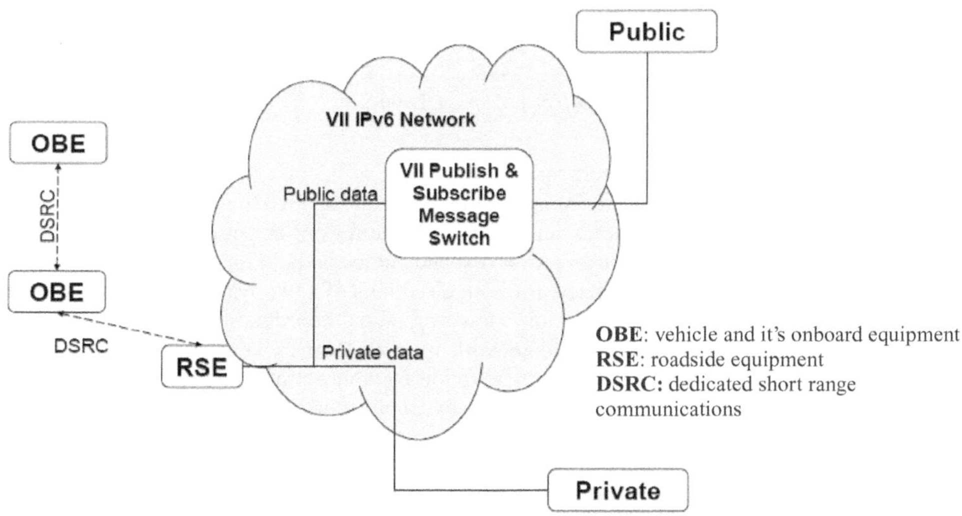

Figure 1. Simplified architecture data flow (*Source: ITS-JPO, 2005*).

this case speeds of vehicles would be additionally constrained to meet given safety thresholds). With this higher level of information more flexible control can be provided but in this case more detection, computation and communication resources will be required. The arriving and departing volumes, and queue status would be evaluated at smaller time steps all depending on the vehicle arrival rate.

3 LIMITATIONS OF SIGNAL ONLY CONTROL AND STUDY MOTIVATION

In this section, the relationship between signal timing and vehicle speeds are discussed along with the limitations of signal-only control algorithms.

3.1 *Difficulties with fixed-speed signal coordination*

Signal coordination is the most common mean to operate signalized networks to achieve specific control objectives such as minimization of delay and/or maximize of capacity utilization. Central to achieving this is the progression of traffic movements between neighboring intersections. Currently, all signal control and coordination algorithms assume constant traveling speed over time for a given road link. This could be the speed limit or some other value determined based on local practices. Because traffic conditions change over time, using a constant speed value over time to calculate the ideal offset is problematic. In some cases, there are long queues waiting at downstream intersections, hence vehicles released from an upstream intersection would not be able to accelerate to the pre-specified fixed speed value before reaching the tail of downstream queue. In other cases, drivers may choose to drive at higher speed than the fixed speed value used in the calculation of the ideal offset if they are not aware of the speed that they are supposed to drive at. Overestimation or underestimation of optimum speeds that vehicles should operate at can easily undermine the effectiveness of signal coordination. Hence, if we can determine the optimum speed and communicate it in real time to drivers then that will help maintain the overall traffic operations at an optimum level. This is the logic behind using DSC.

3.2 *Signal coordination for two-way streets and networks*

For a two-way street, it is assumed that the offset of one direction is defined as offset1 and the offset of the other is defined as offset2. According to Roess et al. (2004), the offsets of these two directions must satisfy:

$$offset\,1 + offset\,2 = nC \tag{1}$$

where n is an integer and C is the cycle length.

Based on equation 1, if the offset of one direction is specified, then the offset of the other direction is automatically set. If the offset is set to the ideal one for one direction, it is not always possible to have ideal offset in the other direction if speed values on both directions are fixed.

Since it is difficult to find ideal offsets for both directions of a two-way street, one may believe that a one-way street system is the solution. However, signal coordination for networks is much more complicated. Even in one-way signal network systems, there still exist network closure problems (Roes et al., 2004). Network closure refers to the fact that setting offsets for one direction on three links automatically determines all offsets between all four signals in any set of four signals of a roadway loop (see Figure 4).

For example, in the grid network shown in Figure 2, if offsets1, offset2 and offset3 are specified, the fourth offset, offset4 is automatically determined.

Offsets determined by the offsets of other directions or other links are called locked-in offsets. If the offsets that are not locked-in are set to ideal offsets, locked-in offsets may not be suitable for

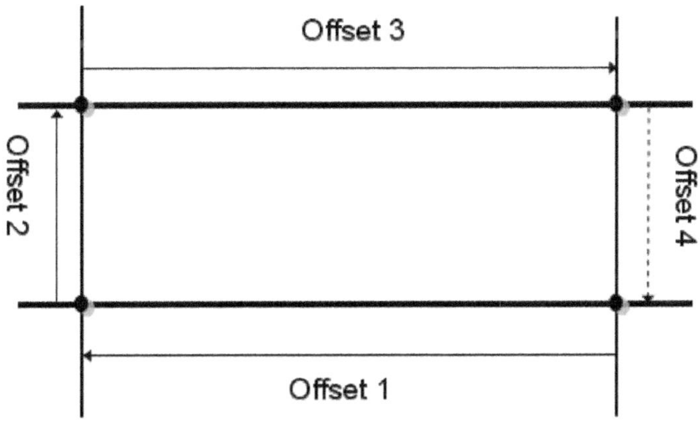

Figure 2. Network closure illustration.

the traffic they are serving. For instance, if offset1, offset2, and offset3 in Figure 2 are set to ideal offsets according to the prevailing traffic condition on the corresponding links, offset4 may not be an ideal offset for the traffic it is serving. In a network with two-way streets, if the offsets for one direction are specified, the offsets for other directions are also locked-in offsets, thus more of the offsets may not be suitable for the traffic they are serving.

3.3 *Inefficient use of green time*

In order to use green time efficiently, not only offsets need to be ideal, but also traffic is expected to pass the signal at or near saturation flow rate. But in reality, especially when traffic demand is lower than capacity, vehicles enter the system at lower than saturation flow rate. When only fixed speeds are used to guide drivers, vehicles may or may not be traveling in platoons and hence may not always maintain saturation headways. In this case more green time will be needed to process the same number of vehicles traveling at longer time headway compared to the case where vehicles are traveling at shorter, saturation headways. Developing a vehicle-based algorithm to specify the speed selection for every vehicle, or group of vehicles that are traveling in a platoon, can regulate traffic flow into a pattern more suitable to the control strategy in place. For example, vehicle-based speed control algorithm can be used to decrease headways between vehicles by systematically setting higher speeds for following vehicle than lead vehicles until required headways are achieved, then following vehicles would be given the same speed as the lead vehicle to maintain that headway. With such control less green time will be needed to process the same number of vehicles hence system capacity can be used more efficiently and delay would be decreased.

The two problems described above not only have negative impact on operations of signalized networks, but also on safety and the environment. These will be discussed in detail in the following sections, and solutions based on dynamic speed control will be presented and the benefits analyzed.

4 DYNAMIC SPEED CONTROL AND ITS POTENTIAL BENEFITS

The potential benefits of using dynamic speed control algorithms are discussed using the two-intersection system shown in Figure 3. To start, assume the signal timings are optimized by existing signal timing algorithms (with fixed speed limit). As discussed in the previous sections, because of changes in traffic conditions over time, network closure conditions, and unbalanced flow on

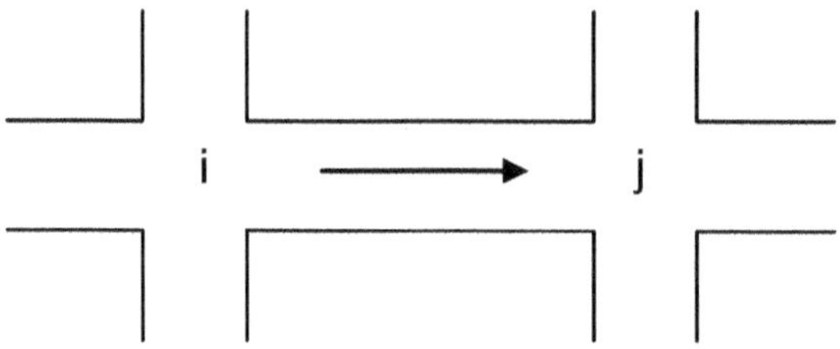

Figure 3. Two intersection system.

different directions, etc., sub-optimal signal timing are inevitable especially in larger urban traffic signalized networks. This section discusses deficiency of signal-only control under different traffic conditions. Dynamic speed control remedies will be proposed based on different control objectives. The proposed dynamic speed algorithm would adjust speed optimally to guide vehicles to travel according to prevailing traffic conditions and the given signal timings.

4.1 *Better signal coordination*

When fixed speed is used in the optimization of signal timing, non-ideal offsets are inevitable because of the change in traffic conditions over time, network closures conditions, and unbalanced flow on different directions. Having a sub-optimal signal coordination plan has negative impacts on the operation of the signalized networks. Two examples of which are illustrated in Figures 4(a) and 5(a).

In Figure 4(a), since the offset between intersections i and j is smaller than the ideal offset, the first vehicle (vehicle #4) released from intersection i during a typical cycle arrives at the downstream intersection j and is stopped du to the red. Under this scenario, the total wasted green is t1. But if dynamic speed control were to be used (in this case vehicles would be guided to travel at higher speeds than the fixed speed limit), then such wasted green could be avoided. The vehicle trajectories of the improved traffic flow due to use of DSC would be as shown in Figure 4(b). In this case, under the same signal timing plan, the system with dynamic speed control produces higher system throughput, lower delay, less number of stops and shorter average travel time.

In the second case, shown in Figure 5(a), the offset between intersections i and j is larger than ideal offset. Vehicle #4 was the first vehicle released from intersection i during a typical cycle and arrived at intersection j before the traffic clears there. The vehicle has to slow down and stop before accelerating back to the desired speed. In this case if the vehicles leaving intersection i were guided to travel at lower speeds, smoother traffic flow would be achieved although system throughput, delay, and travel time may be the same. The smoother traffic flow achieved with Dynamic Speed Control as shown in Figure 5(b) is more favorable from a safety, fuel consumption, and environmental emissions stand points. These will be discussed in more details in a later section.

4.2 *More efficient use of green time*

If vehicles were to travel at uniform saturation headways, green will be wasted only if the offset is different from the ideal one as shown in Figure 4(a). But vehicles typically enter traffic networks at random and larger than saturation headways, as shown in Figure 6(a). This is especially true for lower demand situations. Under such conditions, even though the offset is set to the ideal offset value, green time can still be wasted between the vehicles if all vehicles are guided to travel at the same speed. But a vehicle-based dynamic speed control can potentially prevent that.

When vehicles travel in the system with larger headways, more green time is needed to process the same number of vehicles. In order to use green time more efficiently, dynamic speed could be used to guide the following vehicles to travel faster than the leading vehicle until tight headway are achieved. Figure 6(b) shows the improved traffic flow after a vehicle-based dynamic speed

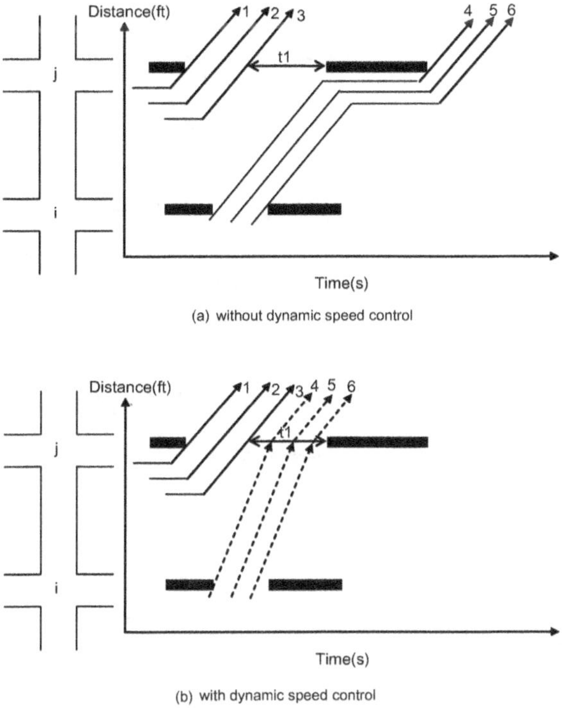

(a) without dynamic speed control

(b) with dynamic speed control

Figure 4. Vehicle trajectories when actual offset is smaller than ideal offset.

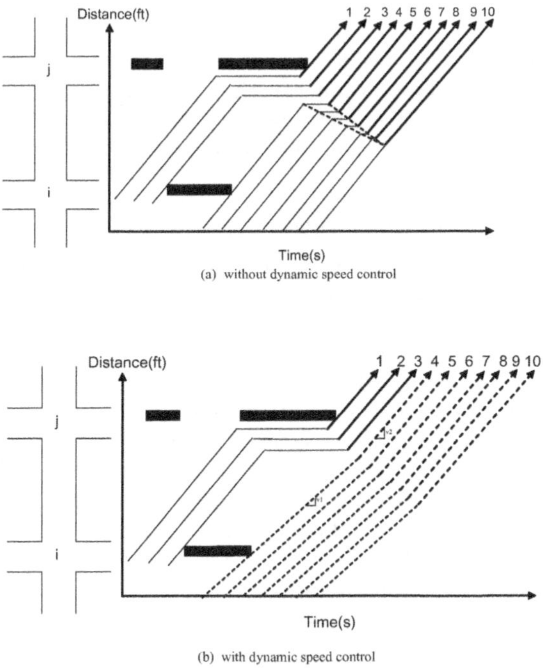

(a) without dynamic speed control

(b) with dynamic speed control

Figure 5. Vehicle trajectories when actual offset is bigger than ideal offset.

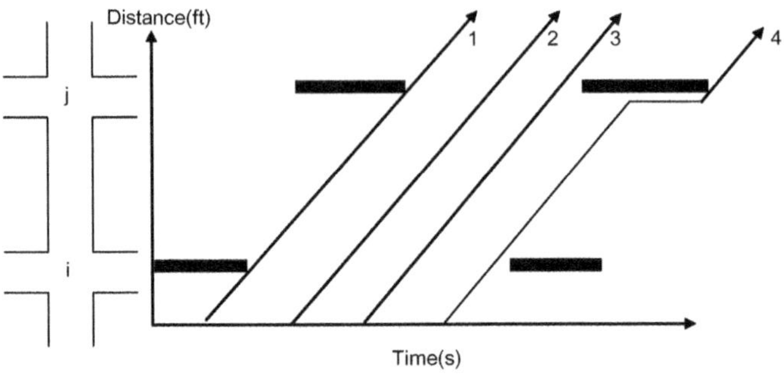

(a) without dynamic speed

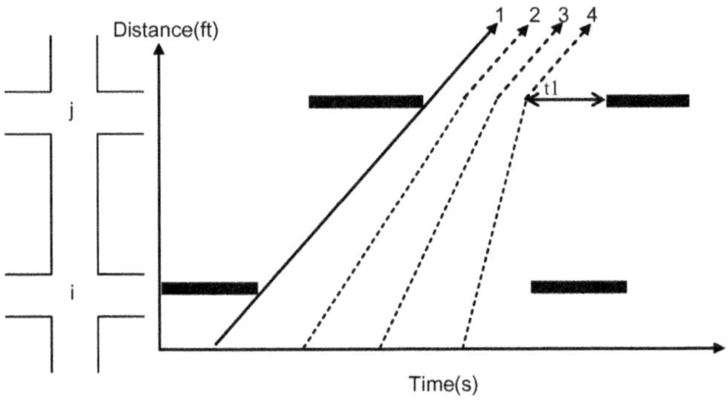

(b) with dynamic speed

Figure 6. Vehicle trajectories when actual offset is bigger than ideal offset.

control is implemented. Green time saved in this case can be used to serve more vehicles if traffic demand is high, or it can be used for other conflicting movements such as left turning traffic from an opposite direction or through and left turning traffic of a crossroad. This is especially valuable when the network is experiencing unbalanced flow in different directions, which is common.

4.3 *Safety and fuel consumption implications*

Vehicle speed noise and speed variation between vehicles have direct impacts on road safety, fuel consumption and vehicular emission. Vehicle speed noise refers to the deviation of a vehicle speed from its desired travel speed. Speed variation refers to the speed difference between vehicles in a vehicle platoon. Thorton and Lyles (1996) concluded that a major factor leading to an accident is not speed itself but the variation of speed. Oh et al. (2005) also proved that speed variation is the most different parameter between normal traffic conditions and traffic conditions that are right before accidents. The results of previous studies of optimal speed profiles prove that fuel consumption is approximately minimized by operating at constant speeds (Schwarzkopf & Leipnik, 1977) and the magnitude of fuel consumption increases as a result of deviations from a constant speed. (Chang & Morlok, 2005). In order to design safe and fuel economic roadway system, speed noise and speed variation need to be minimized. Deploying dynamic speed control in signalized network is one step further towards this objective.

Figure 7 shows the speed profile of vehicle #4 of Figure 3 before and using dynamic speed control.

Under dynamic speed control, vehicles are given guidance to travel at lower speed compared to control with fixed speed. The effect is that under the same signal timing plan, a vehicle may have

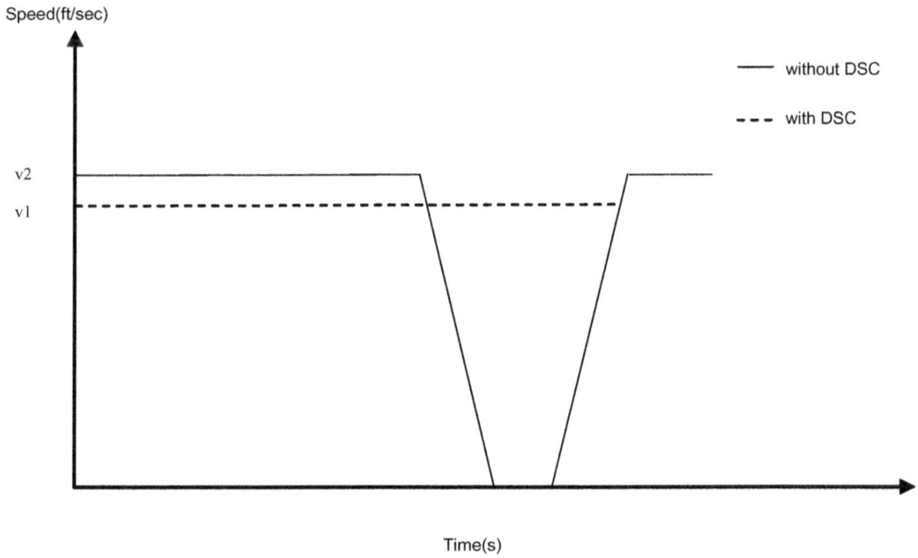

Figure 7. Speed profile of vehicle 4 in figure 3.

the same average speed and travel time, but there are less speed noises. Because vehicle #4 is the leading vehicle released from intersection i within that cycle, less speed noise for vehicle #4 also meant less speed deviation for the following vehicles. This can be seen from Figure 5 (a) and (b). It is clear that the number of vehicles that are impacted by the stopping shockwave is fewer or even none and the intensity of speed noise is less for the following vehicles. Also, because of the reduced acceleration and deceleration period and the smaller number of vehicles affected, smaller speed variation is achieved in the traffic flow under the DSC control.

4.4 *Dynamic speed control provides more flexibility*

Besides the potential benefits that DSC can bring about to operations, safety, and fuel consumption, it also adds flexibility to the control of signalized networks. In many cases, dynamic speed can supplement signal timing design and improve the effectiveness of signal timing control. Consider the case where a signal along a long corridor has been coordinated with other signals for normal traffic conditions. If for any reason there is an overflow and residual queues form on one or more links, the offsets between the signals at the end of these links need to be changed according to prevailing (new) traffic condition in order to maintain good coordination quality. Such change may necessitate changing signal timing for few or even all of the signals along the arterial. Changing in signal timings along the arterial might also impact the signal timing and signal coordination of crossroads. Such transition between different signal timings and coordination plans may induce considerable delays (ITS Decision, 2001). However with VII and DSC in place, good coordination can be achieved by adjusting only the speeds on the impacted link or links. No changes to signal timings would be needed elsewhere. One can argue that using dynamic speed control may in some cases be easier and more cost effective to use than traditional signal-only control algorithms.

5 CONCLUSION

Dynamic speed control is a promising application of vehicle integrated systems (VII). It adds critical flexibilities to the control of signalized networks. This paper analyzed and summarized potential benefits that may be realized from using dynamic speed control in the control of signalized networks. Implementation of dynamic speed control can improve system throughput; decrease delay, number of stops, and travel time; and reduce vehicular emissions and fuel consumption. It is one step further towards designing more reliable and sustainable and reliable signalized network systems; ones that are operationally efficient, safer, more fuel-conserving, and environment-friendly.

REFERENCES

Chang, D. and Morlok, E. Vehicle Speed Profiles to minimize work and fuel consumption. Journal of Transportation Engineering, Vol. 131, No. 3, 2005.

Chen, H. and Abu-Lebdeh, G. 2006. Development of a Framework for an Integrated Dynamic Signal-Dynamic Speed Traffic Management Algorithm for Signalized Networks. *Compendium of papers CD-ROM of 85th Transportation Research Board annual meeting.* Washington DC.

Chen, H., and Abu-Lebdeh, G. 2006. Assessment of Capacity and Flow Improvements of Combined Dynamic Signal Control and Adaptive Speed in Signalized Networks. *Proceedings of the 5th International Symposium on Highway Capacity and Quality of Service*, volume 2, pp. 669–678, Yokohama, Japan.

Coleman, J.A.; Pniati, J.F.; Cotton, R; Parker, M.: Covey, R.; Pena, H.; Graham, D.; Robinson, M., McCuley, J.; Taylor, W.C., Morford, G.. 1996.: FHWA Study Tour for Speed Management and Enforcement Technology. FHWA-PL-96-006.

Committee for Guidance on Setting and Enforcing Speed Limits. 1998. Managing speed, Transportation Research Board, Special Report, 254.

FHWA. 2003a. Manual on Uniform Traffic Control Devices.

FHWA. 2003b. Safety, Speed management Workshop. Federal Highways Administration, U. S. Department of Transportation, Washington, D.C.

Florida DOT. 2003. Florida to implement variable speed limit. Urban Transportation Monitor. V. 17, NO. 11.

Hegyi, A., De Schutter, B and Hellendoorn, J. 2003: Optimal coordination of variable speed limits to suppress shock waves.. *IEEE Transaction on Intelligent Transportation Systems*, volume 6, No. 1, March.

ITS Decision (2001). Traffic Signal Control. http://www.calccit.org/itsdecision/serv_and_tech/Traffic_signal_control/trafficsig_report.html. Last update 11/01/01.

ITS-JPO (Joint Program Office). 2005. VEHICLE INFRASTRUCTURE INTEGRATION(VII)-VII Architecture and Functional Requirements, Version 1.1.

Lin, P-W, Kang, K-P and Chang, G-L. 2004: Exploring the effectiveness of variable speed limit controls on highway work-zone operations. Intelligent Transportation System. 8:1–14.

Oh, J., Oh, C., Ritchie, S. and Chang, M. 2005. Real-time Estimation of Accident Likelihood for Safety Enhancement. *Journal of Transportation Engineering.* Volume 131, NO. 5, pp. 358–363.

Paterson, D. 2003. *Melbourne's Variable Speed Limit System. Proceedings of the 21st ARRB and 11th REAAA conference. Transport: Our highway to a Sustainable Future.*

Roess Roger P., Prassas, Elena S. and Mcshane, William R. 2004. *Traffic Engineering, 3rd* edition. Pearson, Prentice Hall.

Thorton, M and Lyles, R.W. 1996. Freeway speed zones: Safety and compliance issues. *Transportation Research Record* 1635, pp. 18–25.

Schwarzkopf, A.B. and Leipnik, R.B. 1977. Control of highway vehicles for minimum fuel consumption over varying terrain. Transportation Research, 11(4), pp. 279–286.

USDOT. 2008 VII Concept of Operations. http://www.its.dot.gov/vii/vii_concept.htm. Last updated 4/23/2008.

Efficient Transportation and Pavement Systems – Al-Qadi, Sayed, Alnuaimi & Masad (eds)
© 2009 Taylor & Francis Group, London, ISBN 978-0-415-48979-9

A study of driver behaviour with regards to traffic control devices

Dr. Fahad Al-Rukaibi
Civil Engineering Department, Kuwait University, Kuwait

Dr. Mohammed Ahmed Ali
D A Watt Consulting, Calgary, Alberta, Canada

Dr. Ahmad Aljassar & Engr. Lulwa Al-Abdulmuhsen
Civil Engineering Department, Kuwait University, Kuwait

ABSTRACT: This paper presents the preliminary results of a research study being undertaken at Kuwait University to investigate the driver non-compliance with TCDs in Kuwait. Specifically, the study aimed to quantify non-compliance for: STOP signs, YIELD signs, and red signal indication at a selected number of intersection sites in Kuwait. At each selected site, driver non-compliance with TCD was person observed and recorded during both the peak and off peak periods. Data on road inventory was also collected along with the non-compliance information to investigate the effects of road geometry on non-compliance. A number of sites in all the six Governorates in Kuwait were surveyed using pre-tested survey forms. Results of a questionnaire survey of the drivers randomly chosen in selected areas to investigate their traffic law awareness and their self reported compliance with traffic rules are also presented in the paper. A number of recommendations are made to improve the driver compliance with the TCDs in order to enhance the road traffic safety in Kuwait.

1 INTRODUCTION

Kuwait has one of the highest accident fatality rates in the world. In the year 2006, over 450 deaths were reported from over 60,000 traffic crashes; in addition ove-r 2.75 million traffic violations were recorded. The analysis of traffic crashes in Kuwait has indicated driver error as the most prominent cause. In spite of the growing concern over the seriousness of driver non-compliance with traffic control devices (TCDs) in Kuwait, the extent of this problem has not been addressed in recent years.

Kuwait has implemented new, more stringent traffic laws in November of 2001 in order to minimize traffic violations and thereby reduce traffic accidents. To complement the laws, over 70 cameras were installed to record speed and red-light running violations. In spite of the improved enforcement efforts, the traffic safety situation has not improved. In the year 2006, there were over 2.75 million traffic violations reported in Kuwait. Included in this, are over 313,000 Red-light running and 840,000 over-speeding violations recorded by cameras. The number of these kinds of serious violations has increased over the years.

The recent economic boom witnessed due to a myriad of reasons such as a perceived stability in the region after removal of Saddam Hussein in Iraq and increase in oil revenues due to increase in prices has led to a significant increase in the expatriate population and resulting traffic and congestion issues on Kuwait Road Network. The Kuwait Government and authorities have identified the road traffic issues including the road safety as priority areas of action. Research and development are primary to developing plausible solutions to these serious issues. KFAS and Kuwait University have included traffic safety research as areas of high priority. This research project is part of this increased interest and endeavors by Government and Public Authorities in Kuwait to produce solutions arising from increasing traffic and safety issues in Kuwait. This paper presents the results of a preliminary analysis of the data collected in the study.

Specifically, the research project aims at the following objectives to be undertaken during the course of this one year study:

1. To undertake an analysis of the traffic control devices violations records collected from the Ministry of Interior (MOI) in Kuwait.
2. To investigate the driver compliance to stop-controlled intersections, signalized intersections and roundabouts.
3. To study driver behaviour with regards to the TCDs.
4. To determine the relationships of traffic control violations and driver behaviour.
5. Investigate the effect of road geometry on non-compliance of traffic control.

More than 700,000 motor vehicle crashes occur at stop signs each year in the U.S. (NHTSA 1998). About one third of these crashes resulted in injuries and over 3,000 lead to fatalities. Stop sign violations were found to be major cause (70%) of accidents in a study of crashes at stop controlled junctions in 4 U.S. cities. About 12% of these accidents were rear-end crashes (Retting et. al. 2003). Several studies have documented poor compliance at stop signs, characterized by failure to stop or to look adequately for oncoming traffic (McKelie 1986, Pietrucha et. al. 1989). Treat et. al. (1979) reported improper lookout to be the leading cause of crashes, accounting for nearly one fourth of all investigated motor vehicle collisions.

Non-compliance to signal indications at signalized intersections is also a major cause of crashes at intersections. It is estimated that red light running crashes in the 6-year period 1997–2002 in the U.S. produced an average annual estimate of 225,000 crashes, more than 193,000 persons injured, 855 fatal crashes, and 943 deaths (Retting 2004). Red-light running and the associated risk of severe crashes at signalized intersections have been an ongoing concern to many safety professionals. The proportions of crashes that occur due to such driver behavior represent a substantial number of crashes at urban and suburban signalized intersections in Michigan. In the year 2001, crashes related to red-light running in Michigan represented about 28% of the crashes of all severities and 40% of fatal and serious injury crashes occurring within signalized intersections (Schattler and Datta 2004). Roundabouts also witness significant number of vehicle crashes in many countries around the world (Baranowski and Waddell 2003). In the Arabian Gulf significant non-compliance of TCDs was reported by Koushki et. al. (1993).

Stop signs, signal control and roundabouts are used at intersections to control potentially conflicting traffic movements. Transportation engineers attempt to provide "positive guidance"through a combination of devices at intersections to provide information to drivers when they need it. Signs, signals, pavement markings, cones, barricades, and warning lights are designed with dedicated colors, shapes, and sizes based on the different functions they provide. They regulate, guide and warn vehicular and pedestrian traffic about road conditions. Uniformity of design (color, shape, and size) helps drivers to quickly understand the messages of traffic control devices. Consistency is crucial for ensuring driver respect, recognition, and proper timely reaction. When traffic control devices are properly selected and located for good day and night visibility, recognition, and comprehension, driver and pedestrian compliance can ensure safety on the roadway.

2 DATA AND METHOD

The traffic violations and accident statistics were collected from the Ministry of Interior, Kuwait for further analysis. The TCD compliance surveys were undertaken at signalized intersections, roundabout intersections, 4-way stop control intersections and right-in/right-out stop control intersections and signalized intersections.

A total of 10 stop controlled intersections (Table 1 and Table 2), 10 roundabout intersections (Table 3) and six signalized intersection (Table 4) were sampled for direct observation/video sur-

Table 1. Selected right-in/right-out stop controlled intersections.

Junction No	Location	Governorate
T1	Khalid bin Abdulaziz street and Street 101, Mubarak Al-Abdullah Area	Hawalli Governorate
T2	Beirut Street, Hawalli	Hawalli Governorate
T3	Block 10, Street 1, Bayan	Hawalli Governorate
T4	Block 1, Street 1, Al-Oyoon Area	Al-Jahra Governorate
T5	Block 4, Street 43, Al-Khaldeya	Al-Asema
T6	Block 1, Street 10 & Sultan Al-Kulaib, Oudaileya	Al-Asema

Table 2. Selected 4-way stop intersections.

Junction No	Location	Governorate
N1	Street 101 between Block 1 & Block 2, Al-Naeem	Al-Jahra Governorate
N2	Road 4 and Road 6, Shuwaikh Commercial Area	Al-Asema
N3	Block 3 between Street No.35 & Maysalon Street, Qadsiya	Al-Asema
N4	Block 1, Babel Street, Al-Khaldiya	Al-Asema

Table 3. Selected roundabout intersections.

Junction No	Location	Governorate
RA1	Street 1 & 3, Mubarak Al-Abdullah	Hawalli Governorate
RA2	Street 101 & 122, Mubarak Al-Abdullah	Hawalli Governorate
RA3	Qutaiba street & Beirut street, Hawalli	Hawalli Governorate
RA4	Alsaqaby Street & Quraish Street, Al-Nuzha	Al-Asema Governorate
RA5	Al-Nuzha Street & Quraish Street, Al-Nuzha	Al-Asema Governorate
RA6	Road 10 & 14, Bayan	Hawalli Governorate
RA7	Al-Masjed Al-Aqsa Street & Road 20, Bayan	Hawalli Governorate
RA8	Ferdous Street, Khaldeya	Al-Asema Governorate
RA9	Ferdous Street, Khaldeya	Al-Asema Governorate
RA10	Abdulla Al-Hajri & Sultan Al-Kulaib St., Oudaileya	Al-Asema Governorate

veys from different Governorates in Kuwait. Typical survey forms were developed based on the Manual of Transportation Studies. Peak and Non-peak period surveys were conducted by the Civil Engineering Students of Kuwait University.

Two types of data were collected in these surveys; Inventory and site description and compliance data. Site data is expected to indicate any geometric and design deficiency which could be related to potentially high violations at the location. For stop and roundabout junctions the following four categories of driver action are identified:

1. Not stopping
2. Practically Stopped

Table 4. Selected signalized intersections.

Junction No	Location	Governorate
S1	Algous Street and Road No. 280, Al-Qurain Residential Area	Mubarak Al-kaber Governorate
S2	Mosa ben Naser Street and extention of Beirut Street, Hawalli	Hawalli Governorate
S3	Alghaws street and Khalid ben Abdulaziz street, Bayan & Mishref	Hawalli Governorate
S4	Third Ring Road and Damascus Street	Al-Asema Governorate
S5	Al-Rehab and Ashbelya residential areas	Al-Farwaneya Governorate
S6	Al-Siddeq & Al-Salam, South Surra	Al-Farwaneya Governorate

3. Stopped by traffic (forced stop) and
4. Voluntary full stop

Categories 1 and 2 represent total and partial non-compliance while the categories 3 and 4 represent compliance to the TCDs. Red light violations were observed by turn movements for signalized intersections.

Driver behaviour questionnaire survey was undertaken using a pre-tested questionnaire. The survey questions were designed to elicit the following information:

- Compliance with TCDs, traffic signs, road markings and posted speeds
- Driver behaviour with regards to shoulder lane, parking, yellow light and mobile phone use
- Behaviour at stop signs, roundabout yield sign
- Self reported violations and accidents, and
- Personal information

The surveys have been administered in person by the Civil Engineering students of Kuwait University on sampled drivers. The data from the questionnaire was analyzed using SPSS software.

3 RESULTS

3.1 *Traffic violation and accident statistics*

A summary of traffic violations data for the years 2006 is presented in Table 5. A total of 2.76 million violations have occurred in this year. A large number of red light running and over-speed violations were recorded as in previous years. Over 80,000 seatbelt violations also raise cause of safety concern in Kuwait. The rising trend in the number of different types of traffic violations has become an important issue for policy planners in Kuwait in the Ministry of Interior and the other concerned departments of Government of Kuwait. A higher traffic council has been established to address these issues. The Ministry of Interior is trying to make its penalty regime very stringent. Over 25 million Kuwait Dinars were collected in fines and several drivers' licensees were suspended and dozens were imprisoned for serious violations.

The accident statistics for the years 1995–2005 were also collected from the MOI and the results were analyzed to determine the accident and violation rates as shown in Table 6. Number of traffic violations as well as the number of traffic accidents have shown increasing trend. The number of fatalities and serious injuries however been decreasing during the same period.

Table 5. Traffic violations (2006).

Traffic violation	Capital	Hawally	Farwania	Al-Jahra	Al-Ahmadi	Mubarak Al-Kabeer	Others*	Total violations
Drunk driving	1	2	4	0	1	5	0	13
Red-light runnng	2829	10033	1612	2459	1568	3237	291689	313427
Over-speed violations	1202	4445	738	1142	942	9580	822428	840477
Driving in opposite direction	21793	29902	15470	897	4329	2783	13	75187
Driving without license	8514	14355	5796	4800	4481	13981	1	51928
Driving without license plate	2339	3148	1183	1372	1405	2586	6	12039
Seat belt violations	3098	20047	11024	4530	27146	15978	2	81825
Traffic rules violations	6730	4433	2656	790	4334	25270	26	44239
Parking violations	97883	64164	71007	34784	106967	30753	25	405583
Parking in places reserved for handicapped	1817	559	2354	279	2832	518	0	8359
All violations	291541	387773	219868	141862	266384	332887	1117170	2757485

* Traffic violations recorded live by camera

Table 6. Summary of accident and traffic violation statistics (1999–2005).

Description/year	1999	2000	2001	2002	2003	2004	2005
Total population	2,107,195	2,228,363	2,309,102	2,419,928	2,546,684	2,753,656	2,991,189
Total number of drivers	–	–	–	657,798	753,542	842,276	924,785
New car registrations	60,752	64,594	75,131	90,271	112,579	110,384	116,768
Canceled/scrapped cars	15,133	17,539	25,584	17,899	73,974	24,502	27,857
Total number of cars	754,500	801,555	851,091	947,382	954,978	1,042,617	1,134,042
Traffic violations	443,240	664,215	1,459,277	1,740,527	2,384,397	3,043,451	3,045,265
Total number of accidents	26,635	27,696	31,028	37,650	45,376	54,878	56,235
Collision accidents	24,887	26,069	29,277	35,989	43,612	53,123	55,035
Other accidents	1,748	1,627	1,751	1,661	1,764	1,755	1,200
Total injuries	1,743	1,125	1,566	2,249	1,332	824	863
Minor injuries	1,178	736	1,043	1,641	927	584	576
Serious injuries	565	389	523	608	405	240	287
Accident fatalities	333	331	300	315	372	398	451
Average daily violations	1,214	1,820	3,998	4,769	6,533	8,338	8,343
Average daily accidents	73	76	85	103	124	150	154
Serious accidents/day	1.548	1.066	1.433	1.666	1.110	0.658	0.786
Accident fatalities/day	0.912	0.907	0.822	0.863	1.019	1.090	1.236
Violations/1000 vehicles	587	829	1,715	1,837	2,497	2,919	2,685
Accidents/1000 vehicles	35	35	36	40	48	53	50
Injuries/1000 vehicles	75	49	61	64	42	23	25
Accident fatalities/ 1000 vehicles	44	41	35	33	39	38	40
drivers/Violations	–	–	–	2.65	3.16	3.61	3.29
Accidents/driver	–	–	–	0.057	0.060	0.065	0.061

(continued)

Table 6. Summary of accident and traffic violation statistics (1999–2005). (*continued*)

Description/year	1999	2000	2001	2002	2003	2004	2005
Serious injuries and fatalities/1000 driver	–	–	–	1.403	1.031	0.757	0.798
Accidents/100 thousand persons	1,264	1,243	1,344	1,556	1,782	1,993	1,880
Serious injuries and fatalities/100 thousand persons	42.6	32.3	35.6	38.1	30.5	23.2	24.7

3.2 *TCD compliance surveys*

The data collected from direct observations/video surveys were entered into several work sheets in excel for further analysis. Each intersection was analyzed for TCD compliance for each approach wherein a TCD was installed. For the purpose of summarization for this paper all the responses in each type of intersection were summed and percent compliance/non-compliance to the TCD was determined. Table 7 shows the results for 3-way stop controlled intersections, Table 8 shows the summary results for 4-way stop controlled intersections and the results for yield control compliance are presented in Table 8.

A high proportion of drivers were found to be in violation of TCD compliance in all types of intersections as shown in Table 7. Approximately 25% of the drivers did not stop at the right-in/right-out intersections during the morning peak and 21.4% in the afternoon peak. The non-compliance was higher during the non-peak period with over 40% of the drivers not stopping before entering the main road. Full compliance at these intersections was found to be very low (5–10%). Full compliance as well as full-non-compliance was higher in the off peak at these intersections.

Very high degree of non-compliance was observed at the 4-way stop controlled intersections. It should be noted that these intersections are usually located on local streets within the residential neighborhoods or industrial areas. Significantly higher proportion of drivers not complying with stop-control leads is a cause of concern for traffic safety. Over 60% of the drivers were found to be full-non-compliant to the stop-control while a meager 3% drivers were found to be in full-compliance to the stop-control signs at 4-way stop control intersections.

Large numbers of intersections at secondary-secondary and secondary-local road intersections in Kuwait are roundabouts which operate with typical yield control. The surveys aimed to capture the full-voluntary yielding behaviour of drivers, the results as shown in Table 9 indicate an utter lack of respect for the yield control at the roundabout intersections. Approximately 3–5% of the respondents were found to be in full-compliance and 40% in full non-compliance to yield control.

3.3 *Driver behavior questionnaire*

Frequency analysis of the driver behaviour towards the TCDs was undertaken using SPSS software. The sampled drivers consisted of representative driver population known in Kuwait.

The results presented in Table 10 indicate a rather alarming situation with regards to compliance to the TCDs. A significant proportion of drivers interviewed disclosed negligence towards following the traffic rules including the TCDs. The first part of table shows the driver responses to compliance with STOP, YIELD, Traffic light, road marking and speed limit rules. Over 30% showed disregard for STOP, 23% for YIELD and 21% towards the traffic light. This self reported disregard towards these important traffic control devices points to the lack of education in terms of traffic safety among the drivers in Kuwait. Over 26% drivers reported not following the directives of road markings and over 35% respondents did not obey the speed limit in secondary and local roads. This result also points towards focusing on driver education and more stringent speed

Table 7. TCD compliance of right-in/right-out stop controlled intersections.

Driver Behavior	Period	Responses in right turn movement	
		No.	%
Non Stopping	AM peak (7–8 am)	185	**25.9**
Practically Stopped		354	49.6
Stopped By Traffic		140	19.6
Voluntary Full Stop		35	4.9
Total responses		714	100.0
Non Stopping	PM peak (1:30–2:30 pm)	140	**21.4**
Practically Stopped		285	43.6
Stopped By Traffic		188	28.8
Voluntary Full Stop		40	6.1
Total responses		653	100.0
Non Stopping	Off peak (3–4 pm)	299	**40.6**
Practically Stopped		181	24.6
Stopped By Traffic		180	24.5
Voluntary Full Stop		76	10.3
Total responses		736	100.0

Table 8. TCD compliance at 4-way stop-controlled intersections.

		All Approaches							
		Reponses in each turning movement							
		Left		Straight		Right		Total	
Driver Behavior	Period	No.	%	No.	%	No.	%	No.	%
Non Stopping	AM peak (7–8 am)	57	55	147	**56**	117	**77**	321	**62**
Practically Stopped		34	33	39	15	28	19	101	19
Stopped By Traffic		6	6	75	28	3	2	84	16
Voluntary Full Stop		7	7	3	1	3	2	13	3
Total responses		104	100	264	100	151	100	519	100
Non Stopping	PM peak (1:30–2:30 pm)	183	54	280	**67**	171	**69**	634	**63**
Practically Stopped		72	21	51	12	42	17	165	16
Stopped By Traffic		64	19	78	19	18	7	160	16
Voluntary Full Stop		22	6	10	2	18	7	50	5
Total responses		341	100	419	100	249	100	1009	100
Non Stopping	Off peak (3–4 pm)	80	41	193	**65**	120	**64**	393	**58**
Practically Stopped		88	45	47	16	62	33	197	29
Stopped By Traffic		20	10	49	17	2	1	71	10
Voluntary Full Stop		9	5	6	2	3	2	18	3
Total response		197	100	295	100	187	100	679	100

enforcement in the neighborhoods and internal roads in residential districts. Traffic calming and other speed reduction measures shall be considered in improving the neighborhood traffic safety. The second part of Table 10 shows the responses to congestion and parking related attributes. Again, a significant number of surveyed drivers have shown apathy towards parking rules, respect of emergency lanes and disregard to safe driving habits.

Table 9. TCD compliance at roundabout intersections (yield control).

Driver Behavior	Period	All Approaches No.	%
Non Stopping	AM peak (7–8 am)	3647	**44.5**
Practically Stopped		2127	26.0
Stopped By Traffic		2009	24.5
Voluntary Full Stop		408	5.0
Total responses		8191	100.0
Non Stopping	PM peak (1:30–2:30 pm)	3002	**32.3**
Practically Stopped		3130	33.7
Stopped By Traffic		2873	30.9
Voluntary Full Stop		295	3.2
Total responses		9300	100.0
Non Stopping	Off peak (3–4 pm)	3121	**43.4**
Practically Stopped		2098	29.2
Stopped By Traffic		1767	24.6
Voluntary Full Stop		204	2.8
Total responses		7190	100.0

Table 10. Self reported driver behaviour.

Description	% responses					
1. How often do you OBEY the following?	Never	Rarely	Some times	Usually	Nearly all times	All the times
the STOP signs at un-signalized junctions?	4.6	6.7	18.8	19.5	20	30.5
GIVE-WAY signs at roundabouts?	2.9	5.9	14.3	21.4	25.7	29.8
SIGNAL indicators at signalized junctions?	1.9	5	13.5	19	23.7	35.8
Road markings?	4.4	6.9	14.5	24.9	24	25.2
the speed limit of 100 or 120 kph on free-ways?	6.7	8.2	15	21.3	23.3	25.6
the speed limit of 80 or 60 kph on primary and secondary roads?	7.4	10.2	18	23.4	20.6	20.3
the speed limit of 45 or 30 kph on local and access roads?	13.1	11.6	16.4	21	14.4	23.5
2. Please tell us if you do the following while driving...						
I use shoulder/emergency lane to drive when the road is congested?	51.9	18.2	14.9	7	4.1	4
I park on the street when I don't find parking?	44.7	23.6	18.8	6.6	3.8	2.3
I park on the road when I am in a hurry?	51.6	21.5	15.4	6.3	3.6	1.7
I park in the handicapped parking?	66.8	15.7	8	4	3.2	2.3
I pass the signal when it is yellow?	16	17.2	29.4	14.6	12	10.8
I use mobile phone while driving?	9.8	10.9	21.9	19.4	15.7	22.3

Table 11. Self reported compliance to STOP and Give-way signs.

	% responses	
Description	STOP Sign	Give-way
stop all the times	32.6	30.5
slow down all the times	28.5	24.8
stop or slow down when I see traffic	35.5	40.9
I don't respect this sign	3.6	3.8

Table 11 shows the results of driver responses specific to STOP control and YIELD control. Approximately 30% of respondents fully obeyed this controls and a significant proportion of surveyed drivers seem to be ignorant of the serious safety consequences of their non-compliance to the TCDs.

4 CONCLUSIONS

This paper has presented the results of a research study being undertaken at Kuwait University funded by the Kuwait Foundation of Advancement of Sciences (KFAS). The study is aimed to quantify non-compliance for: stop signs, yield signs, and red signal indication at selected intersection sites in Kuwait. The results of the direct observation and video surveys have indicated very high degree of non-compliance at the 4-way stop controlled intersections, right-out STOP control intersections and YIELD controlled roundabout intersections. Significantly higher proportion of drivers not complying with stop-control is a cause of concern for traffic safety. Over 60% of the drivers were found to be full-non-compliant to the stop-control while smaller insignificant 3% drivers were found to be in full-compliance to the stop-control signs at 4-way intersections. Similar results were noted at other types of intersection control. Driver interview surveys undertaken during the survey period at sites near the selected intersections have also indicated lack of knowledge and/or acceptance of these types of controls on the streets. The rising trends in traffic violations and accident as well as road fatalities reinforce the concept of stricter enforcements, improved driver education, tighter driver licensing controls and engineering measures such as 'Traffic Calming' and regular traffic safety audit on Kuwait's Road Network. The indications from the Governmental Authorities in Kuwait are very encouraging. The Government of Kuwait has recently established Higher Traffic Council and has put traffic safety and congestion issues on top priority. This ongoing study will make many recommendations to improve compliance to the TCDs which will go a long way in reducing road traffic accidents and saving lives in Kuwait.

ACKNOWLEDGEMENTS

The authors wish to acknowledge the financial support provided to this research study by the Kuwait Foundation of Advancement of Sciences (KFAS) project number 2007150803. Thanks are also due to the CE-366 class of 2007 for participating in data collection.

REFERENCES

Institute of Transportation Engineers, A manual of Traffic Studies, Chapter 14: Traffic Control Devices Studies and Chapter 15: Compliance with Traffic Control Devices Studies.

G.S. Russel and L. Austin, Danger Compensation Effects of Stop Signs at Intersections, Accident Analysis and Prevention, vol. 15, No. 2, 1983.

W.J. Frith and N.M. Derby, Intersection Control by Stop and Give Way Signs—The Conclusion of Polus, Accident Analysis and Prevention, vol. 19, No. 3, 1987.
FHWA Resource Center, Leaflet on Safety Effects of Traffic Control Devices (TCD)/Low Cost Safety Improvements,
J.M. Stuart, An opinion Survey and Longitudinal Study of Driver Behaviour at Stop Signs, Canadian Journal of Behavioural Sceince vol 18 (1), 1986.
F. Alrukaibi, M. Ali, A. Aljassar, Student Drivers—Traffic Safety Attitudes and Driving Behavior: A Case Study in Kuwait, Transportation Research Record No. 1969, Transportation Research Board.
P.A. Koushki and A.M. Al-Ghadeer, Driver Noncompliance with Traffic Regulations in Rapidly Developing Urban Areas of Saudi Arabia, Transportation Research Record No. 1375, Transportation Research Board.
R.A. Retting, H.B, Weinstein, M.G. Solomon, Analysis of Motor Vehicle Crashes at Stop-Signs in Four US Cities, Journal of Safety Research, vol. 34 (2003).
P.A. Koushki, R.L. Smith and A.M. AL-Ghadeer, Urban Stop Sign Violations in Saudi Arabia, Transportation Quarterly, vol. 46, No. 3, 1992.
V.R. Houten and R.A. Retting, Increasing Motorist Compliance and Caution at Stop Signs, Journal of Applied Behaviour Analysis, vol. 34, No. 2, 2001.
N. Deveaunuse, K.Kim, C. Peek Asa, D. McArthur, and J. Kraus, Driver Compliance with Stop Signs at Pedestrian Crosswalks on a University Campus, Journal of American College Health, vol. 47, No. 7, 1999.
K.L. Schattler and T.K. Datta, Driver Behaviour Characteristics at Urban Signalized Intersections, Transportation Research Record, No. 1862, 2004.
Institute of Transportation Engineers, Improving Intersection Safety Through Design and Operations, CD ROM of Conf. Today's Transportation Challenge: Meeting Our Customer's Expectations, Florida, 2002.
M. Pietrucha, K. Opiela, R. Knoblauch, and K. Cringler, Motorist Compliance with Standard Traffic Control Devices, FHWA-RD-89–103 Report.

Efficient Transportation and Pavement Systems – Al-Qadi, Sayed, Alnuaimi & Masad (eds)
© 2009 Taylor & Francis Group, London, ISBN 978-0-415-48979-9

Integrated traffic information system for Urban/Inter-Urban traffic management

S. Garda
Principal Engineer (WSP)

D. Weston
Senior Engineer (WSP)

ABSTRACT: With the increase in deployment of Intelligent Transport Systems (ITS) within the Gulf States to improve road network management, there is an opportunity for many benefits to be realised by integrating such systems on a local and cross-border level. This paper details the primary benefits for integrating urban and inter-urban traffic management from a road operator's perspective, but there is also a secondary and very crucial benefit that is also available from the same systems, i.e. integrated traffic information services for the traveller. This can add value to the travelling public in order to improve the overall journey experience. This paper describes the work done in this area in Europe and discusses why and how a similar approach can be taken by the Gulf States to tackle congestion, improve road safety and allow efficient road usage.

The paper covers how integration of ITS systems in participating countries in Europe enables the sharing of traffic data via standardised methods. Using standards and principles such as Urban Traffic Management and Control (UTMC), DATEX II and Travel Information Highway (TIH) has meant that a lot of significant work has been undertaken in this area. A similar concept can be adopted for the Gulf countries.

Finally, the paper addresses the types of ITS data sources that provide the information to the road operator for network management and the methods of delivering traffic information to the public.

1 DEMAND FOR INTEGRATED TRAFFIC INFORMATION SYSTEM IN THE MIDDLE EAST

There is an increase in urban population in many of the Gulf States as well as an increase in the movement of vehicles and people between the neighbouring countries. To address the local traffic problems there are activities being undertaken for the deployment of ITS systems to help ease congestion. The challenges being presented to road operators centre around ensuring the roads are safe to use and that congestion is kept to a minimum. For these challenges to be met, good decision making for taking appropriate next steps is required from both the road operators and travellers. An enabler for this decision making process is having an integrated traffic information system in place. For this to occur there is a need to focus on integration of ITS systems so that the maximum benefits can be realised.

An integrated traffic information system that collects and analyses both local and inter-urban traffic data can be extremely useful to those in charge of managing the road network as well as being a useful source of traffic information for the public. This approach helps local and national road authorities meet their objectives of minimising congestion, reducing pollution and making the roads safer.

Figure 1. Traffic congestion on urban roads.

In Europe, significant progress has been made in achieving system integration, largely through the use of common standards and principles such as the Urban Traffic Management and Control (UTMC) standard defined by DfT of the UK, and the Travel Information Highway (TIH) principles defined by the TIH Community in the UK.

2 A PRINCIPLED APPROACH TO INTEGRATED TRAFFIC MANAGEMENT

Informing travellers plays an important part in the efficient management of a transport network. Provision of high quality and timely information can cause a reduction in the amount of travellers entering already congested conditions, with travellers choosing to delay their journey, find an alternative route or instead use public transport. In order to achieve the maximum benefit it is necessary to provide travellers with a complete picture of the information pertaining to their journey. As journeys often involve urban and inter-urban sections the provision of information from the whole network is essential.

Consider the example of a freight vehicle driving from Abu Dhabi to Dubai. With an integrated traffic information system in place, he/she would have access not only to local traffic information in Abu Dhabi and Dubai (provided by the respective local authority), but also to inter-urban traffic information pertaining to the journey to Dubai (provided by a national traffic authority). This would enable freight operators to plan journeys better and react to incidents that may have occurred on the travel route to avoid delays. For this example to work effectively and seamlessly there needs to be collaboration and data sharing among both local and national traffic authorities.

In the UK the Highways Agency National Traffic Control Centre (NTCC) and the many local authorities work together to share data as part of the UTMC framework and TIH community. This 'partnership' provides travellers with information about local roads (urban) as well as the national network (inter-urban) through the use of websites, variable message signs, radio broadcasts, mobile devices (PDAs) and kiosks.

For an integrated traffic information system to be effective there needs to be an agreed set of principles that enforce interoperability and compatibility between both urban and inter-urban systems. An example of this is the TIH set of principles that defines the type and format of data that is exchanged between two or more parties. An agreed set of standards ensures data can be shared between traffic authorities thus enforcing the principle of inter-urban data exchange. This ideology has been trialled in Europe with the exchange of traffic data between the UK and France using a data sharing standard known as DATEX II. DATEX II specifies what data is exchanged, the format and content the data and the method of transmitting the data to third parties across Europe.

Figure 2 below illustrates how different ITS systems such as accident detection, road sensors and satellite communication can be integrated and linked to a centralised Traffic Control Centre (TCC).

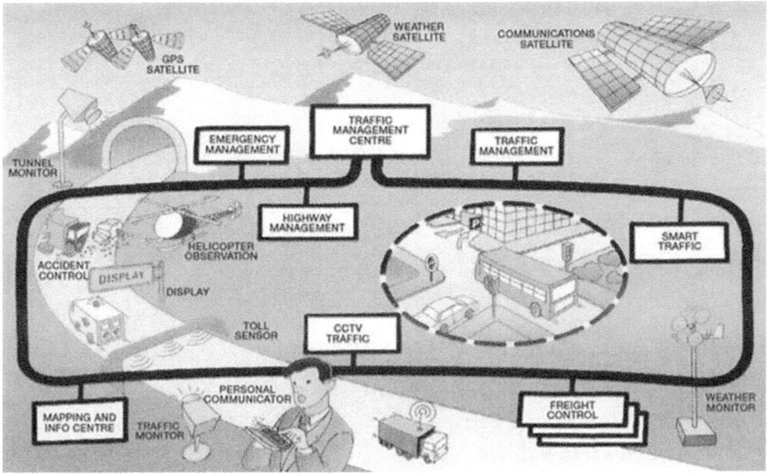

Figure 2. ITS systems integration to deliver traffic information to both traffic managers and travellers.

Numerous TCCs can be linked together to share information using UTMC and DATEX II and subsequently provide the public with both urban and inter-urban traffic information.

3 DELIVERING TRAFFIC INFORMATION TO NETWORK MANAGERS AND THE PUBLIC

An integrated traffic information system relies on accurate, reliable and real-time data from multiple urban and inter-urban sources. These can include automated traffic detection/classification systems such as CCTV, ANPR and incident/roadwork information.

Having processed the incoming traffic data, the local and regional road authorities would have the responsibility of disseminating the relevant information to the travelling public. In the UK there are various roadside and in-car technologies that are used to provide real-time journey information which can be categorised as being either pre-trip (before starting a journey) or in-trip (accessed during a journey or at a break in a journey) sources of information.

3.1 *Pre-trip*

For example, in England the Highways Agency hosts a traffic information website which is based on a mapping interface that displays the locations of incidents, current and planned roadworks and current congestion occurring on the road network. In addition, the website features journey time predictions and VMS sign setting information.

CCTV images of the ‘current state of the roads’ are also displayed on the map for the user to access. This type of website is intended to be used by drivers before they commence their journey so that are able to make informed journey planning decisions.

Other examples include having relevant information delivered via SMS to the mobile device just before a journey is commenced.

3.2 *In-trip*

In-trip systems provide drivers with up-to-date information throughout their journey and can be sub-divided into in-vehicle systems and roadside/information point systems.

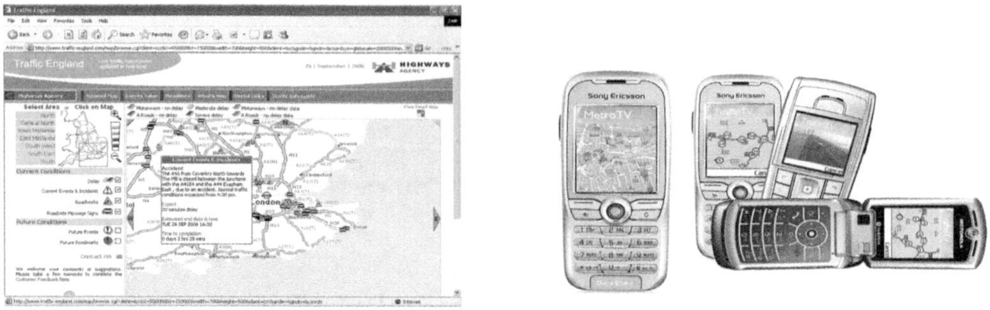

Figure 3. Examples of traffic information services for the public.

The most commonly used in-vehicle source of traffic information is radio news broadcasts that can cover both local and national road networks. Roadside message signs (VMS) provide information to drivers at strategic points throughout the network, i.e. to inform of an accident, bad weather or a temporary diversion route.

Another example is the Highways Agency Information Point (HAIP). This provides drivers with access to the Highways Agency traffic information website to find out the latest road conditions for their route. HAIPs are kiosks that can be installed at motorway service stations, hotels and airports or anywhere that large numbers of drivers convene.

4 BRINGING THE INFORMATION TOGETHER

An example of practically bringing information together to enable both the road operators and travelling public to make informed decisions about the roads and journeys is the STREETWISE project which WSP has been involved with. This is a Euro-Regional project aimed at delivering road users with seamless, reliable and accessible travel information services that uses common data standards and protocols for data exchange. The project covers the Trans European Road Network in the UK and Republic of Ireland. STREETWISE is an example of traffic information services sharing data across urban and inter-urban boundaries and a similar deployment could be adopted in the Gulf States to provide drivers with journey information via the pre-trip and in-trip information sources described earlier. STREETWISE has adopted the integrated traffic information system concept which has led to assisting the travelling public through an integrated urban/inter-urban traffic information service.

Figure 5 above illustrates the concept of having TCCs in the respective Gulf countries that manage the local and urban road network as well as provide the data for travel information services to the local public. These TCCs would use standards such as the UTMC platform for integration of the individual ITS systems in its region. The respective TCCs would then also be able to share their traffic data with TCCs in neighbouring countries using standards such as DATEX II and principles like TIH. By obtaining this cross-border traffic information, the network managers would be able to manage their roads more effectively as they would be aware of vehicle and people movement in and out of their region. This additional traffic data can then also be converted into travel information that is suitable for the public to use for planning their journeys as well as react to unplanned events that have taken place. This type of travel information availability would have great value to road users such as commuters and freight operators as it would allow better planning and reaction to incidents.

Combining the different uses of the same data set together results in overall efficiency in road usage and congestion management.

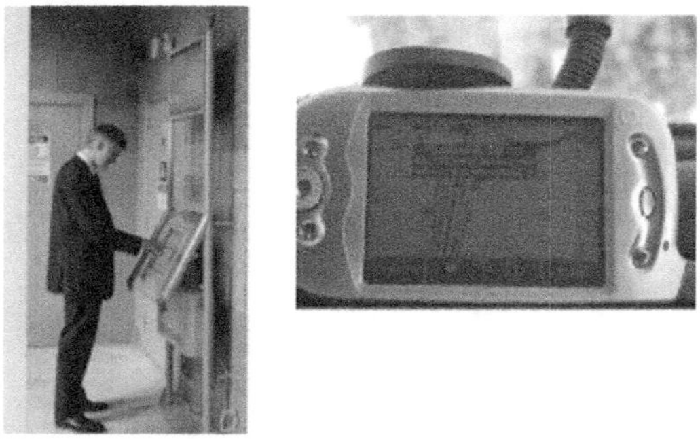

Figure 4. Examples of in-trip traffic information services.

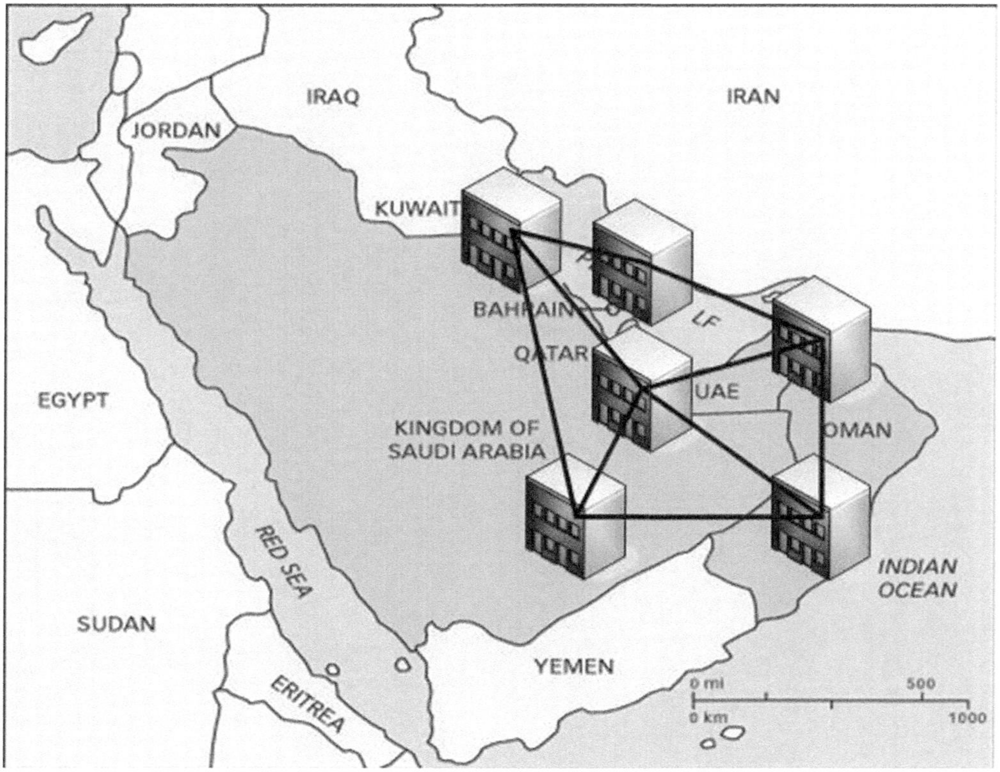

Figure 5. Conceptual illustration of interconnected Traffic Control Centres in the Gulf.

REFERENCES

DATEX II—http://www.datex2.eu/ (website accessed 9 July 2008).

DfT—www.dft.gov.uk (website accessed 9 July 2008).

STREETWISE—http://www.nra.ie/Publications/DownloadableDocumentation/Transportation/file,3659,en.PDF (website accessed 9 July 2008).

TIH—http://www.tih.org.uk/index.php/Main_Page (website accessed 9 July 2008).

UTMC—http://www.dft.gov.uk/pgr/roads/tpm/utmc/ (website accessed 9 July 2008).

Efficient Transportation and Pavement Systems – Al-Qadi, Sayed, Alnuaimi & Masad (eds)
© 2009 Taylor & Francis Group, London, ISBN 978-0-415-48979-9

Using GIS as a tool to integrate land use, traffic analysis, and travel forecasting

N. Hawkins, O. Smadi, X. Chai & H. Isebrands
Center for Transportation Research and Education, Iowa State University, Ames, IA, USA

ABSTRACT: This paper describes an effort to develop and implement a scenario planning tool to assist staff, at a rapidly growing Iowa community, in determining future right-of-way needs within hard to plan for areas. The area under review consisted of the entire southwest quadrant of the city, where successive large development tracks were rapidly changing community dynamics and boundaries. City staff needed more information than provided by the regional model and yet realized the area was too large to be analyzed through sequential traffic impact studies. Key features of the research include an evolution of tools to address staff needs including a spread sheet method which included; integrating GIS roadway network and parcel data into a spreadsheet format that assists with conducting trip generation, distribution, and assignment; and developing tools to export roadway geometric and turning movement data into a popular traffic analysis/simulation program. The paper discusses the limitations experienced in applying this tool particularly as the planning area grew in size and staff wanted to migrate to city-wide use. This effort illustrates an alternative approach to bridging the gap between local and regional planning and in integrating land use planning with GIS and traffic operational analysis.

1 INTRODUCTION

Traffic demand is the critical factor in establishing the size of a roadway. Unfortunately, determining this demand is a significant challenge especially within high growth areas such as in the City of West Des Moines, Iowa. Even with enormous growth and shifts in roadway travel demand, city staff are still charged with the dual responsibility of not only "building it right" but to also having "sized it right" in the first place. Under high growth conditions, there is tremendous motivation to identify and reserve future right-of-way needs in advance of development and soaring land prices.

Several published reports give perspective to the magnitude of potential issues related to traffic forecasting.

- A recent international study presented the results from the first statistically significant study of traffic forecasts in transportation infrastructure projects. For 50% of road projects, the difference between actual and forecasted traffic is more than ±20%; for 25% of road projects, the difference is larger than ±40%. Highly inaccurate traffic forecasts combined with large standard deviations translate into large financial and economic risks (1).
- An examination of the regional forecasting model for Hennepin County Minnesota found that forecasted volumes were not accurate enough to correctly assign the number of lanes needed (accuracy rates of 54% for local streets, 55% for county roads, and 67% for the state trunk highways) (2).
- A study for the Kansas DOT found a poor correlation between what was forecasted in 1970's and what actually happened after the horizon year had been reached in terms of socioeconomic and demographic data, which are the major inputs used by travel demand models to forecast future traffic volumes on roadway links (3).

Businesses are looking for the "best opportunities" and site planning can change on a continuous basis until construction is complete. This often leaves agencies scrambling to assess impacts, needed infrastructure, and funding. Within the City of West Des Moines, the success of large scale development has had a ripple effect in attracting successive projects of all sizes and complexities. The efforts of City staff to identify and reserve needed roadway right-of-way, through diligent review and use of the scenario planning tool described below, has no doubt saved the city millions of dollars in right-of-way costs, business disruption, and motorists delay, frustration, and safety.

Forecasting future traffic is a key input into the planning, design, maintenance, and overall budgeting processes for the City. Forecasts which are either too high or low can have extremely negative impacts particularly in areas of accelerated growth.

Estimates of future travel demand are typically done either at a big picture regional level or at a very detailed driveway level. City staff felt that planning for large development impacts along entire corridors were not well served by either of the above two methods. In an effort to identify right-of-way needs along corridors more effectively, staff supported the Center for Transportation Research and Education (CTRE) in a research effort to try and create, and apply, a planning tool which would bridge the gap between regional and driveway level planning.

In cooperation with City staff, CTRE developed a spreadsheet based analysis tool which can be described as working between the regional model and traffic impact approaches, see Figure 1. This new tool allows the City to quickly analyze the traffic impact on West Des Moines roadways from various "what if" land use scenarios and the impact that each development has on the transportation network. It helps guide the growth of the ultimate street layout by estimating future traffic levels and identifying future congestion areas. This information can, in turn, be used to determine right-of-way needs for a future roadway—how many lanes, potential traffic signal locations, and so on—and help alleviate traffic problems before they become realities.

Another issue faced by staff was that the modeling process is not transparent for users, especially for those not who are not skilled users. The TransCAD program itself is considered expensive, with the majority of costs in the staffing, training, and data management required. Planning at the local level is more project-specific and on an as-needed basis. Manipulating the regional model to provide local roadway design details requires hand smoothing and the opportunity to introduce error in the results complicates running multiple scenarios. Staff felt that a smaller or sub-area scenario planning tool, based upon land use as opposed to socioeconomics, would provide at least an alternative approach to projecting future vehicle volume demand.

A wide variety of travel demand models are in use. A listing is provided by the FHWA's Traffic Analysis Toolbox, Volume II (4). Available travel demand modeling tools include:

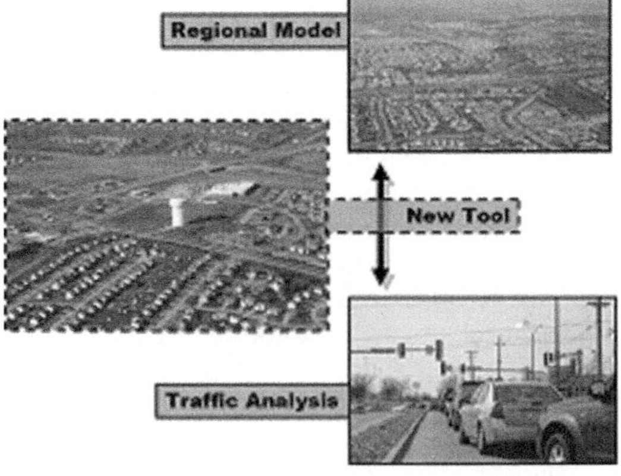

Figure 1. New tool illustration.

- b-Node Model: http://mctrans.ce.ufl.edu/store/description.asp?itemID = 482
- CUBE/MINUTP: www.citilabs.com/minutp/index.html
- CUBE/TP+/Viper: www.citilabs.com/viper/index.html
- CUBE/TRANPLAN (Transportation Planning): www.citilabs.com/tranplan/index.html
- CUBE/TRIPS (Transport Improvement Planning System): http://citilabs.com/trips/index.html
- EMME/2: www.inro.ca/products/e2_products.html
- IDAS: http://idas.camsys.com
- MicroTRIMS: http://mctrans.ce.ufl.edu/store/description.asp?itemID = 483
- QRS II (Quick Response System II): http://my.execpc.com/~ajh/index.html
- SATURN: http://mctrans.ce.ufl.edu/store/description.asp?itemID = 157
- TModel: www.tmodel.com
- TransCAD: www.caliper.com/tcovu.htm
- TRANSIMS (Transportation Analysis Simulation System): http://transims.tsasa.lanl.gov

There are three prominent software tools available to assist in conducting traffic impact analysis: TRAFFIX, TEAPAC, and WinTASS (5). These programs assist at the driveway and local roadway level, but the focus of this research fell somewhere between this driveway level and the regional level.

The new tool was based upon more specific West Des Moines input information than included within the regional model. As an example, the tool includes smaller traffic analysis zones which are comprised of multiple parcels and uses a denser ultimate roadway network that was created by staff. The new tool provides a glimpse of travel demand as based upon full build out of the area as opposed to artificial limits of 2025 or 2030 as used in the regional model. This analysis of local conditions under a parcel by parcel but yet full-build scenario allows staff to make informed decisions through analysis of lots of potential land use development scenarios.

2 COMMUNITY GROWTH

The City of West Des Moines has experienced significant commercial and residential growth over the last five years. This growth is a result of large-scale development in the southwest quadrant of the city. A new regional mall added over two million square feet of retail, and across the street from this, Wells Fargo constructed an office complex with an ultimate build-out to accommodate 13,000 employees. Table 1 illustrates this growth through commercial construction valuations by year (6). As shown, commercial construction in 2004 alone added over $287 million in property valuation. For 2006, the commercial uses are further defined, with retail representing over half the new uses.

3 SPREADSHEET TOOL DEVELOPMENT AND APPLICATION

This section summarizes the features and application of the scenario planning tool as developed and applied within the City of West Des Moines.

Table 1. West Des Moines commercial construction valuation (2002–2006).

Year	Added Valuation	
2002	$	15,174,732
2003	$	146,511,627
2004	$	287,631,919
2005	$	86,878,860
2006	$	91,226,617

2006 By Use	
Retail	51%
Office	27%
Industrial	22%

Purpose. To provide an enhanced planning analysis tool to assist in forecasting traffic demands within the western growth area of the City of West Des Moines; to use this tool while working with city staff to update all land uses and travel distributions within the study area and provide staff with the ability to make roadway geometric assessments, given the rapidly changing scenarios and conditions.

Scope. The study area is shown within the dashed lines of Figure 2. This massive planning area covers over 30,000 acres, 7,000 sub-parcels, 331 intersections, and 1,992 roadway links.

Spreadsheet Tool Features. This research developed a software tool that takes land uses and a roadway network from a GIS format and generates a spreadsheet tool that allows trip distribution and assignment. The distribution and assignment effort can be done manually, by shortest path, or through creation of multiple paths. The generated intersection volumes and roadway network are then processed for input into a commercially available traffic analysis program (Synchro) in which traffic signal timing, queue lengths, delay, and other performance features can be fully evaluated. As a final step, the outputs from the operational analysis can be brought back into the GIS environment by color or other thematic legends. A full user's manual has been prepared for using the tool and for converting operational analysis information back into the GIS format. These documents were not included within this report, but are available online at Iowa State University's Center for Transportation Research and Education project website (http://www.ctre.iastate.edu/research/detail.cfm?projectID = 1103299779), along with the developed software.

The research effort required a significant effort by staff to organize all existing and future parcel and roadway information into a GIS format. Figure 3 shows how the GIS format allows a simple click on any parcel to reveal descriptive (parcel) and calculated information (a.m. and p.m. peak-hour trips).

Figure 4 shows how parcel trips are distributed away from the centroid (user-selected percentages in any direction). As shown, this would be for either a.m. or p.m. trips leaving the parcel.

Figure 5 shows how trips are assigned to the roadways using one of several methods. The tool allows the user to choose between shortest path, multiple shortest path, or complete manual assignment. The example shows trips leaving the centroid of the parcel to the tip of the arrow (specific access

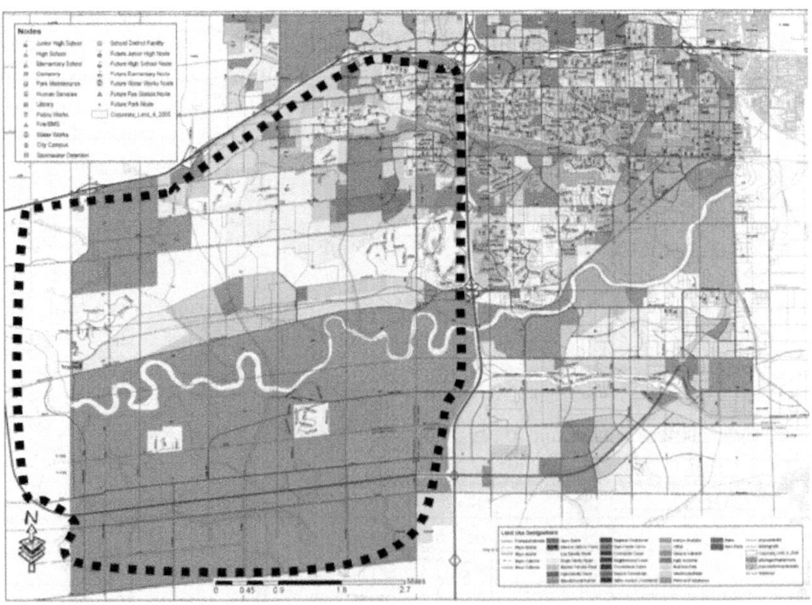

Figure 2. City of West Des Moines study area.

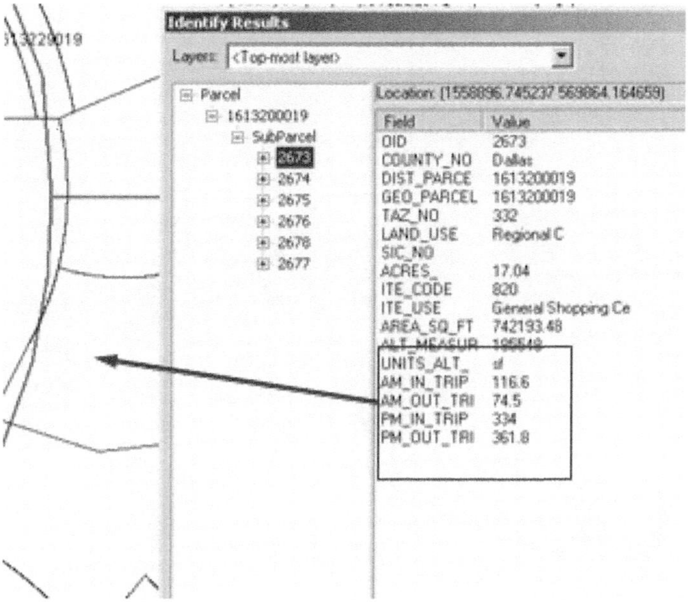

Figure 3. Parcel information in GIS.

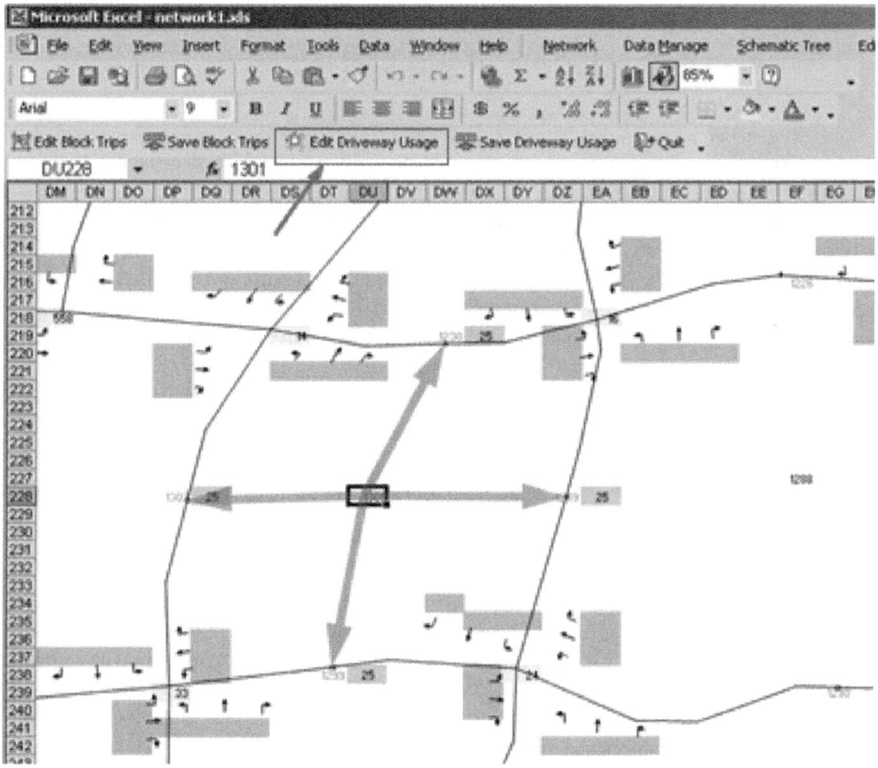

Figure 4. Trip distribution (spreadsheet tool).

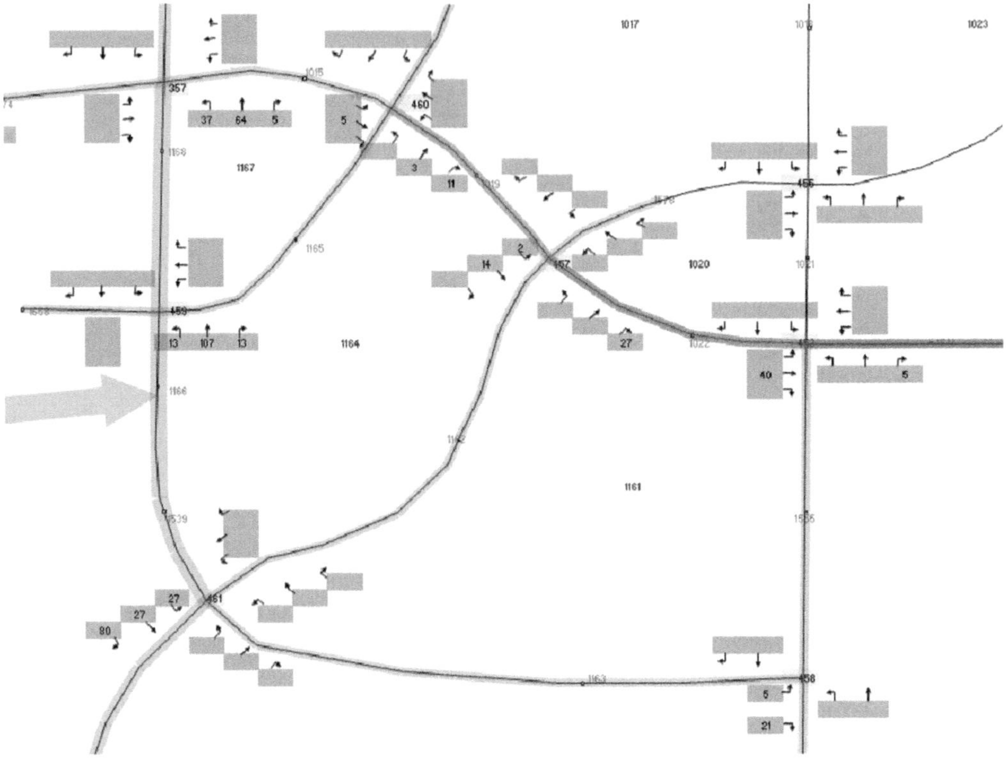

Figure 5. Trip assignment (spreadsheet tool).

point on the roadway shown). This example also shows how the tool shades the roadway with a red color, of which the width and shade provide feedback on the volume of vehicles using specific routes (for this particular movement out of the parcel at this particular driveway).

Figure 6 shows how the information looks as it is automatically converted into a traffic analysis program (Synchro). All intersections can then be evaluated in terms of meeting capacity demands, turning lane storage, backups, and other measures of effectiveness.

Results. Figure 7 shows how the traffic analysis information can be brought back into the GIS format to view performance information at any scale.

The scenario planning tool is an incredible resource for the City of West Des Moines and has been directly credited with saving substantial dollars in establishing right-of-way needs ahead of development. Land uses within the area are in a constant state of change, and maintenance of the land use database is a bonus for both land use and traffic planning.

4 ADJUSTING FROM LARGE AREA TO CITY-WIDE PLANNING

As the spreadsheet tool was applied to larger and larger study areas, limitations to this approach became clear. The City of West Des Moines is not on an island and the influence of regional traffic needed to be accounted for to arrive at meaningful results. Accordingly, CTRE is working with staff to modify the traffic impact planning approach. This revised approach uses the strengths of the regional model but still includes customized local land-use and roadway network information. This approach supports city-wide planning needs and simplifies staff's ability to conduct city-wide scenario planning efforts.

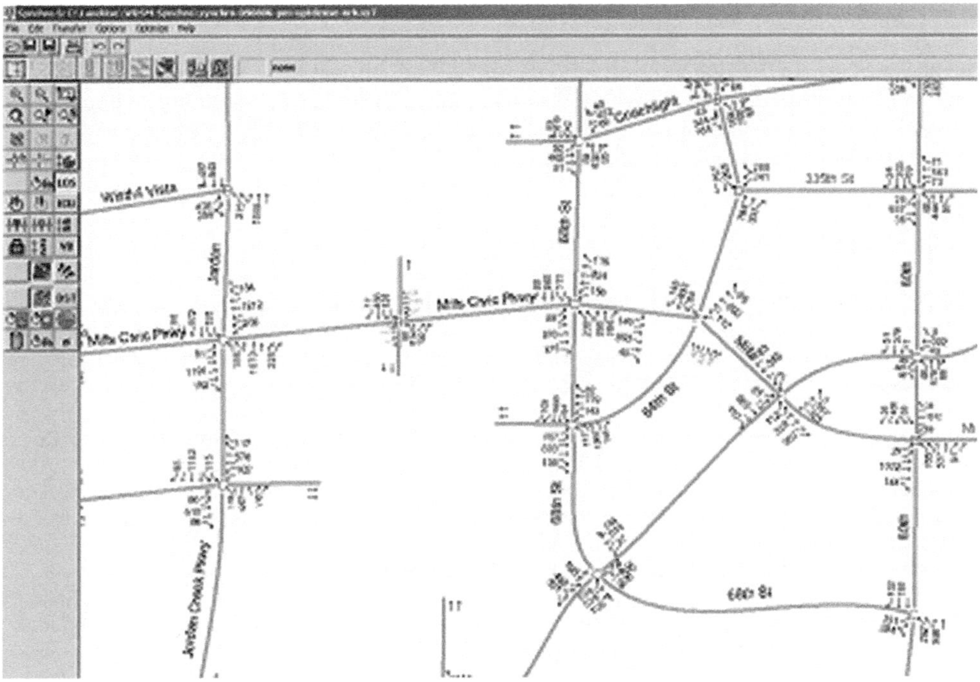

Figure 6. Traffic analysis view.

Figure 7. Intersection LOS in GIS format.

5 SUMMARY

The above efforts to develop and refine tools for staff are critical in adequately planning for and supporting transportation growth within the community. The dynamic conditions, within the City of West Des Moines, will continue to necessitate the need for innovation, analysis, and communication of traffic impacts for the foreseeable future.

REFERENCES

1. Flyvbjerg, Bent, Mette Skamris Holm, and Søren L. Buhl. "Inaccuracy in Traffic Forecasts." Transport Reviews, vol. 26, no. 1, January 2006, 1–24.
2. Hennepin County Transportation Systems Plan. http://wwwa.co.hennepin.mn.us/files/HCInternet/Static%20Files/103190389Chapter_05_Traffic_Forecast_Meth.pdf July 19th, 2000. Chapter 5 pages 1–14. Accessed 6/1/2008.
3. Eustace, D. Development of Multiple Growth Strategies for Use in Developing Traffic Forecasts: A Robustness Approach. Report K-TRAN: KSU-01-3. Topeka, KS: Kansas Department of Transportation, 2005.
4. Jeannotte, K. Traffic Analysis Toolbox Volume II: Decision Support Methodology for Selecting Traffic Analysis Tools. Report FHWA-HRT-04-039. Washington, DC: Federal Highway Administration, Office of Operations, 2004.
5. Rathbone, D. "Traffic Impact Analysis Software." The Urban Transportation Monitor, 5 August 2005, 10–12.
6. City of West Des Moines, Iowa. "Development Retrospective 2006." 2007. 4–5. http://www.wdm-ia.com/userdocs/forms/Development_Retrospective_06.pdf.

Efficient Transportation and Pavement Systems – Al-Qadi, Sayed, Alnuaimi & Masad (eds)
© 2009 Taylor & Francis Group, London, ISBN 978-0-415-48979-9

Validation of the VT-Meso vehicle fuel consumption and emission model

Huanyu Yue & Hesham Rakha
Virginia Tech, Charles Via, Jr. Department of Civil and Environmental Engineering, Blacksburg, VA, USA

ABSTRACT: The VT-Meso model estimates light-duty vehicle fuel consumption and emission rates on a link-by-link basis based on three independent variables: average speed, number of vehicle stops per unit distance, and average stop duration. The model uses these variables to construct synthetic drive cycles for each roadway segment and then predicts average fuel consumption and emission rates for four modes: decelerating, idling, accelerating, and cruising. This paper validates the VT-Meso model estimations against microscopic second-by-second energy and emission estimates using the INTEGRATION software for different simulation scenarios. The exercise demonstrates that the VT-Meso model successfully predicts fuel consumption rates both in terms of absolute and relative values. Alternatively, the vehicle emission rates were consistent in terms of trends across the various simulation scenarios. The results demonstrate that such a modeling approach may be sufficient for the environmental evaluation of alternative traffic operational projects.

1 INTRODUCTION

Macroscopic vehicle fuel consumption and emission models are currently the primary tools for evaluating the regional environmental impacts of transportation projects. In typical applications, a transportation planning model is first used to determine the average speed and total vehicle-miles of travel for the network or facility being considered. Then, an emission model such as MOBILE6 (EPA, 2002) or EMFAC (CARB, 1991) is used to compute the average fuel consumption and emission rates for the facility.

Macroscopic models produce single fuel consumption and emission rates for each average speed input. These models assume that all vehicles pollute similarly for the average speed and vehicle-miles traveled considered and that variations in driver behavior can be neglected (An et al., 1997). This presents a problem when the drive cycles encountered in the field differ from those assumed within the models because estimated emission rates may not correspond to actual emissions. A particular problem occurs when comparing drive cycles to identical average speeds because identical emission rates are then estimated for all cycles despite differences in the second-by-second speed profiles.

There are two general approaches to overcome this limitation of current state-of-the-art procedures. One is a microscopic approach, and the other is a mesoscopic approach. The VT-Meso model was developed to estimate light-duty vehicle (LDV) fuel consumption and emission rates on a link-by-link basis based on average speed, number of vehicle stops per unit distance, and average stop duration. The model uses these variables to construct synthetic drive cycles with four operation modes: deceleration, idling, acceleration, and cruising. Within each mode of operation, fuel consumption and emission rates are determined using relationships derived from instantaneous microscopic energy and emission models. The model allows the user to calibrate typical deceleration and acceleration rates.

The objective of this paper is to validate the proposed VT-Meso model estimations against microscopic energy and emission estimates using simulated data. In a previous effort to validate the model, VT-Meso model estimations were compared against Environmental Protection Agency (EPA) field data (Rakha *et al.*, 2006). The model was demonstrated to successfully predict absolute fuel consumption and HC, CO, CO_2, and NO_X emission rates for the 14 EPA drive cycles. This paper investigates the ability of the VT-Meso model to correctly predict changes in vehicle fuel consumption and emission rates across different drive cycle scenarios using the INTEGRATION microscopic simulation package.

The remainder of this paper is divided into four sections. The first section presents an overview of the INTEGRATION modeling framework. The second section provides an overview of the VT-Meso model structure. The third section describes the network construction and scenario development exercises. The fourth section presents the model's evaluation and results. Finally, the conclusions of the study and recommendations for further research are presented.

2 STATE-OF-THE-ART ENERGY AND EMISSION MODELS

This section reviews the state-of-the-art energy and emission models in North America to put the proposed research effort within perspective. Emission models can be categorized as macroscopic, mesoscopic, and microscopic models. Macroscopic models use average aggregate network parameters to estimate network-wide energy consumption and emission rates. Mesoscopic models use average aggregate link parameters to estimate energy consumption and emission rates on a link-by-link basis. Alternatively, microscopic models estimate instantaneous vehicle fuel consumption and emission rates that are aggregated to compute network-wide emissions. Some of the widely used microscopic & mesoscopic models are described briefly in this section.

The CMEM model is a power demand-based emission model that was developed at the University of California, Riverside (Barth et al. 2000). The model estimates LDV and LDT emissions as a function of the vehicle's operating mode. The development of the CMEM model involved extensive data collection for both engine-out and tailpipe emissions of over 300 vehicles, including more than 30 high emitters. These data were measured at a second-by-second level of resolution on three driving cycles, namely: the federal test procedure (FTP), US06, and the modal emission cycle (MEC). The model is based on a simple parameterized physical approach that decomposes the entire emission process into components corresponding to the physical phenomena associated with vehicle operation and emission production.

The EPA's Office of Transportation and Air Quality (OTAQ) is developing a modeling system termed the **MO**tor **V**ehicle **E**mission **S**imulator (MOVES) (Koupal et al. 2002). This new system estimates emissions for on-road and non-road sources, covers a broad range of pollutants, and allows multiple scale analysis (fine-scale to national inventory estimation). Currently MOVES is under development and a final version of MOVES will be released later in 2007. The macroscale on-road inventory implementations (national default and domain specific) use estimates of vehicle miles traveled (VMT) as a starting point for generating total activity for most emission processes. The mesoscale level of analysis proposed to be supported by MOVES requires that the user supply sufficient information to calculate bottom-up emissions at the link/zone level for running exhaust, start, brake wear, and tire wear processes. The most likely source of this information would be output from a traditional travel demand model. The microscale level of analysis proposed within MOVES is based on necessary input for the dispersion model CAL3QHC, which is EPA's recommended model for intersection analysis.

The Mobile Emission Assessment System for Urban and Regional Evaluation (MEASURE) model was developed to estimate hourly transportation facility-level automobile exhaust emissions using geographical vehicle registration data, accurate digital road data, travel demand forecasts, and zone-based socioeconomic data (Bachman et al. 2000). This model, which was developed for use with a GIS database, predicts emissions based on knowledge of typical amounts of time a vehicle spends accelerating, decelerating, cruising, and idling. MEASURE uses empirical statistical distributions of vehicle activity by facility type. The modules are based on statistical analysis of real-world data, and are integrated on an hour of the day basis. Fleet composition and vehicle activity in terms of modal variables are generated for the principal network and for the local roads. Modal variable distributions are defined as a function of the road type, level of service, and other Highway Capacity Manual (HCM) parameters. Congestion level is estimated from traffic volume and road capacity. The model uses speed and acceleration distribution tables (available for interstate highways, ramps, arterials, and signalized intersections) to estimate modal activities. With these entire input variables, emissions for each link and sub-zone are estimated.

The VT-Micro vehicle fuel consumption and emission model was developed at Virginia Tech to compute mode specific fuel consumption and emission rates (Ahn et al. 2002; Rakha et al. 2003a; Rakha et al. 2000). The model predicts the instantaneous fuel consumption and emission rates of HC, CO, CO_2, and NO_x of individual vehicles based on their instantaneous speed and acceleration levels. The initial intent behind the development of this model was not to capture all elements that affect vehicle fuel consumption and emission rates, but simply to develop a model in which vehicle dynamics are explicit enough to account for speed and acceleration variations, two elements that have been shown to have significant impacts on vehicle fuel consumption and emission rates.

Equation [1] describes the general format of the VT-Micro model that predicts instantaneous fuel consumption and emission rates of individual vehicles.

$$\mathrm{MOE_e} = \begin{cases} \sum_{i=0}^{3}\sum_{j=0}^{3} \exp(k_{i,j}^{e} \cdot v^{i} \cdot a^{j}) \text{ for } a \geq 0, \\ \sum_{i=0}^{3}\sum_{j=0}^{3} \exp(l_{i,j}^{e} \cdot v^{i} \cdot a^{j}) \text{ for } a < 0 \end{cases} \tag{1}$$

Where MOE_e is the instantaneous fuel consumption or emission rate (L/s or mg/s), a is the instantaneous acceleration (km/h/s), v is the instantaneous speed (km/h), $k^e_{i,j}$ are vehicle-specific acceleration regression coefficients for MOE_e, and $l^e_{i,j}$ are the vehicle-specific deceleration regression coefficients for MOE_e. As can be observed, Equation [1] utilizes two sets of coefficients, one set for the acceleration mode and one set for the deceleration mode. The dual regime was introduced to account for differences in the emission rate sensitivity within the acceleration and deceleration modes of travel. Another important feature is the use of the exponent to ensure that non-negative fuel consumption and emission rates are produced by the models.

The general format of Equation [1] was first developed by testing the ability of various regression models to adequately model the observed steady-state fuel consumption and emission behaviors of eight vehicles that were tested by the Oak Ridge National Laboratory (ORNL) in 1995 (West et al. 1997). Further refinement and validation of the microscopic models were made using data that were collected by the EPA in 1997. This second dataset included dynamometer fuel consumption and emission measurements under ambient conditions for 43 normal-emitting light-duty vehicles and 17 normal-emitting light-duty trucks. A comparison of the fuel consumption and emission rates produced by the microscopic models against in-laboratory measurements made on a dynamometer using the ARTA drive cycle for various groups of vehicles within the EPA database further confirmed the validity of the microscopic model, as described by Rakha *et al.* (2003b). Rakha *et al.* (2004a) used Classification and Regression Tree (CART) algorithm on the 60 normal vehicles to group them into homogenous categories. LDV vehicles were categorized using the CART algorithm considering a number of independent variables that included the vehicle model year, engine size, vehicle mileage, vehicle power-to-weight ratio, and federal emission standard (tier0 and tier1). The statistical analysis indicated that vehicle model year, engine size, and vehicle mileage, were critical in categorizing vehicles, as demonstrated in Table 1. These three variables resulted in 5 LDV and 2 LDT categories considering a minimum of 5 vehicles per category. The vehicle categorization that was developed by Rakha *et al.* is also used in the proposed VT-Meso framework, as will be described later.

2.1 *INTEGRATION modeling framework*

The INTEGRATION software (M. Van Aerde & Assoc., 2002a; M. Van Aerde & Assoc., 2002b; Van Aerde *et al.*, 1996; Van Aerde *et al.*, 1988a; Van Aerde *et al.*, 1988b) was employed for this study for several reasons. First, the software combines car-following, vehicle-dynamics, lane-changing, energy, and emission models. Thus, mobile-source emissions can be estimated from instantaneous speed and acceleration levels. Second, the traffic and emission modeling modules have been tested and validated extensively. For example, the software, which was developed over the past two decades, has not only been validated against standard traffic flow theory (Rakha *et al.*, 2002a; Rakha *et al.*, 1996) but has also been utilized for the evaluation of real-life applications. Furthermore, the INTEGRATION software offers unique capability through the explicit modeling of vehicle dynamics by computing the tractive and resistance forces on the vehicle each deci-second (Rakha and Lucic, 2002b; Rakha et al., 2001c; Rakha et al., 2004c). It should be noted that the procedures described in this paper are general and could be applied to other commercially available software applications if they combine the modeling of various resistance and tractive forces acting on a vehicle with accurate vehicle fuel consumption and emission models.

The INTEGRATION software uses car-following models to capture the longitudinal interaction of a vehicle and its preceding vehicle in the same lane. The process of car following is modeled as an equation of motion for steady-state conditions plus a number of constraints that govern the behavior of vehicles while moving from one steady state to another (decelerating or accelerating). The first constraint governs vehicle acceleration behavior, which is typically a function of vehicle dynamics (Rakha and Lucic, 2002b; Rakha et al., 2004c). The second and final constraint ensures that vehicles maintain a safe position relative to the lead vehicle in order to ensure asymptotic stability within the traffic stream. A more detailed description of the longitudinal modeling of vehicle motion is provided by Rakha et al. (Rakha et al., 2004c) In addition, lane-changing behavior describes the lateral behavior of vehicles along a roadway segment. Lane-changing behavior affects the vehicle car-following behavior, especially at high-intensity lane-changing locations such as merge, diverge, and weaving sections.

The software also models vehicle fuel consumption and emission rates using the VT-Micro framework (Rakha et al., 2004c). The VT-Micro model was developed from experimentation with numerous polynomial combinations of speed and acceleration levels. Specifically, linear, quadratic, cubic, and quadratic terms of speed and acceleration were tested using chassis dynamometer data collected at the Oak Ridge National Laboratory (ORNL). The final regression model included a combination of linear, quadratic, and cubic speed and acceleration terms because it provided the least number of terms with a relatively good fit to the original data (R^2 in excess of 0.92 for all Measures of Effectiveness (MOE)). The ORNL data consisted of nine normal-emission vehicles, including six light-duty automobiles and three light-duty trucks (LDT). These vehicles were selected to produce an average vehicle consistent with average vehicle sales in terms of engine displacement, vehicle curb weight, and vehicle type. The data collected at ORNL contained between 1,300 and 1,600 individual measurements for each vehicle and MOE combination depending on the envelope of operation of the vehicle, which has a significant advantage over emission data collected from few driving cycles since it is impossible to cover the entire vehicle operational regime with only a few driving cycles. Typically, vehicle acceleration values ranged from −1.5 to 3.7 m/s^2 in increments of 0.3 m/s^2 (–5 to 12 ft/s^2 in 1 ft/s^2 increments). Vehicle speeds varied from 0 to 33.5 m/s (0 to 121 km/h or 0 to 110 ft/s) in increments of 0.3 m/s (Ahn *et al.*, 2004; Ahn et al., 2002; Rakha *et al.*, 2004a). In addition, the VT-Micro model was expanded by including data from 60 LDVs and LDTs. Statistical clustering techniques were applied to group vehicles into homogenous categories using Classification and Regression Tree algorithms. The 60 vehicles were classified into five LDV and two LDT categories. In addition, high-emission vehicle (HEV) emission models were constructed using second-by-second emission data. The HEV model was found to estimate vehicle emissions with a margin of error of 10% when compared to in-laboratory bag measurements (Ahn et al., 2004; Rakha et al., 2003b).

The INTEGRATION software computes the effective tractive force as the minimum of two forces: the maximum engine tractive force (F_e) and the maximum frictional force that can be sustained between the vehicle wheels and the roadway surface (F_{max}) (Ahn et al., 2002; Rakha and Ahn, 2004a; Rakha et al., 2004b; Rakha et al., 2001a). The aerodynamic resistance (R_a), rolling resistance (R_{rl}), and grade resistance (R_g) are also computed each deci-second. Subsequently, the maximum vehicle acceleration is then computed as

$$a = \frac{\min(F_e, F_{\max}) - (R_a + R_{rl} + R_g)}{m} \tag{2}$$

Here m is the mass of the vehicle and a is the maximum possible acceleration.

3 VT-MESO MODEL FRAMEWORK

The proposed modal model is primarily intended for use after the traffic demand has been predicted and assigned to the network to estimate link-by-link average speeds, number of vehicle stops, and stopped delay, as illustrated in Figure 1. The model utilizes these link-by-link input parameters to construct a synthetic drive cycle and compute average link fuel consumption and emission rates. Total link fuel consumption and emissions are then computed by multiplying the average fuel consumption and emission rates by its corresponding vehicle-kilometers of travel. Finally, system-wide parameters are estimated by summing across all links within a network. The first step involves the construction of a synthetic drive cycle that produces consistent average speed, number of vehicle stops, and stopped delay estimates. After constructing the drive cycle, the model estimates the proportion of time that a vehicle typically spends cruising, decelerating, idling and accelerating while traveling on a link. A series of fuel consumption and emission models are then used to estimate the amount of fuel consumed and emissions of HC, CO, CO_2, and NO_x emissions for each mode of operation. Subsequently, the total fuel consumed and pollutants emitted by a vehicle while traveling along a segment are estimated by summing across the different modes of operation and dividing by the distance traveled to obtain distance-based average vehicle fuel consumption and emission rates.

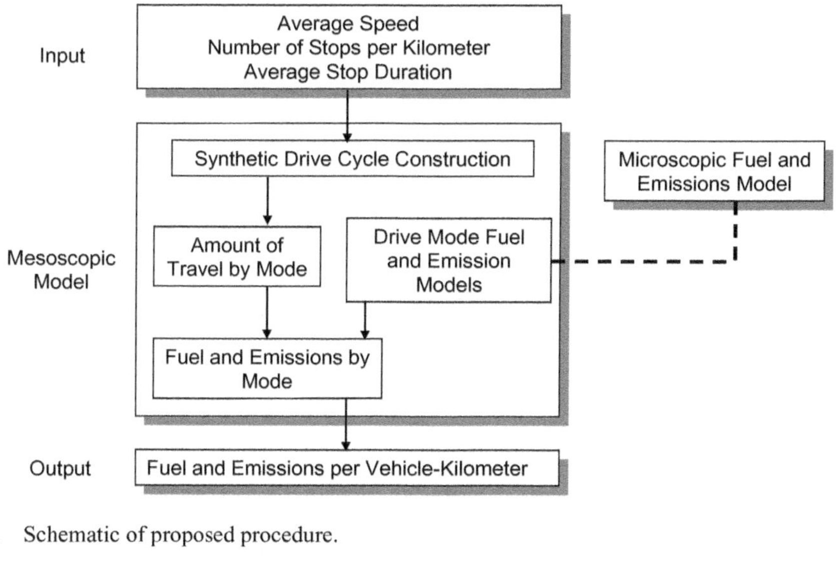

Figure 1. Schematic of proposed procedure.

The proposed modal model relies on three input variables, namely the average trip speed, the number of vehicle stops, and the stopped delay, to construct a synthetic drive cycle that produces equivalent vehicle fuel consumption of emission rates. Regression models were developed to estimate the time and distance traveled during each mode of operation. Using these regression models the synthetic drive cycle can be easily constructed. Currently, the model does not distinguish between freeway and arterial facilities in constructing a drive cycle; however, further research will investigate the merits of such an enhancement.

Figure 2 demonstrates how the synthetic drive cycle is constructed using the three input parameters. It is assumed that the vehicle makes on average three stops of a given duration over the roadway segment. To account for these three stops, the trip speed profile is constructed by fitting three identical stop-and-go cycles over the segment. Within each cycle, the cruise speed is computed to ensure that the input average speed is maintained. As a result, the cruise speed is equal to the average speed when there are no stops. In all other cases, the cruise speed is higher than the average speed, with increasing values for higher number of stops per unit distance and longer stop durations.

3.1 *Scenario development*

In order to validate the VT-Meso model energy and emission estimates, four simulation networks were constrcuted. The INTEGRATION software was utilized to simulate these four networks

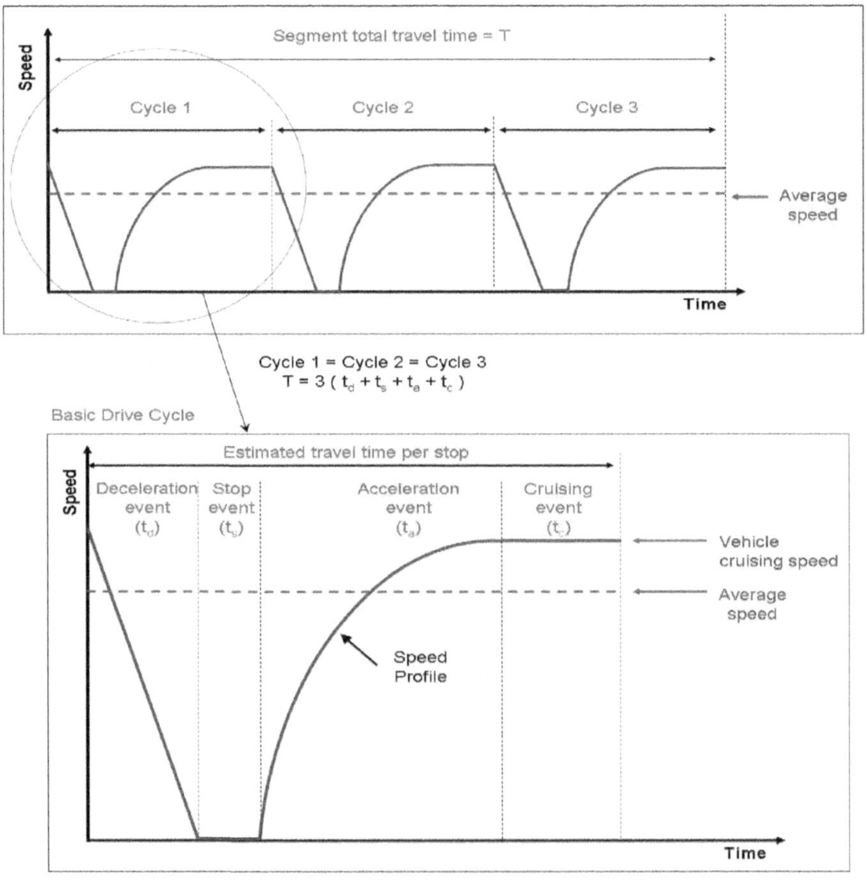

Figure 2. Example synthetic speed profile.

considering 12 different vehicle types. Output from the INTEGRATION software, which include link specific average travel speed, traveled distance, stopped time, and number of stops were used as input to the VT-Meso model. The VT-Meso emission and fuel consumption estimates were compared to those computed by the INTEGRATION software based on second-by-second speed and acceleration measurements in an attempt to validate the procedure.

3.1.1 *Stop sign control scenario*

The objective of this scenario is to compare the VT-Meso and VT-Micro estimates for a stop-and-go condition. The simulation network is a 2-km, single-lane roadway with a single stop sign located after 1 km. This scenario involves vehicle deceleration and acceleration from a complete stop considering different cruising speeds. The cruising free-flow speeds are varied from 32 km/h (20 mi/h) to 64 km/h (40 mi/h) at increments of 8 km/h (5 mi/h) to analyze the impact of stop-and-go driving conditions on fuel consumption and vehicle emission rates for both models. The traffic volumes used for this scenario include a single vehicle vehicle and a demand of 300 veh/h.

3.1.2 *Signal control scenario*

This scenario compares the VT-Meso and VT-Micro estimates for different traffic signal offsets. The network used in this analysis is composed of three signalized intersections along a 2-km roadway segment. Signalized intersections are located after 500 m, 1,000 m, and 1,500 m. The cycle length at each of the three intersections is set at 60 s with offsets varying from 0 to 50 s at 10-s increments. All three signals are controlled by two-phase timings with a 70:30 phase split (east/west versus north/south), and only eastbound traffic exists. The free-flow speed of the network is 64 km/h (40 mi/h) with a lane saturation flow rate of 1,600 veh/h. Traffic demands of this simulation scenario include a single vehicle scenario and a demand of 800 veh/h.

3.1.3 *Partial stop scenario*

Partial stops have a large impact on vehicle fuel consumption and emission rates. The objective of this scenario is to test how the VT-Meso model performs for driving cycles with partial stops. The fuel consumption and emission rates estimated by the model are compared against the rates estimated by the VT-Micro model. The simulation network is a 2.5-km roadway. The first 1-km segment and the last 1-km segment has the same speed limit of 110 km/h, and the 0.5-km segment in the middle has a lower speed limit. The decrease in speed limit varies from 10 km/h to 50 km/h in increments of 10 km/h. A single vehicle and 900 veh/h were simulated for this scenario.

3.1.4 *Ramp scenario*

Weaving volume and weaving section length have significant impacts on the performance of a weaving section. This scenario simulates a Type A weaving section for different weaving volumes and weaving section lengths. The Weaving volume varies from 0 to 2,500 veh/h in increments of 500 vehicles, and the weaving section length varies from 150 m to 450 m in increments of 50 m. The free-flow speed at the ramp section and weaving section is 80 km/h (50 mi/h) and 112 km/h (70 mi/h), respectively.

3.2 *Results*

This section describes the fuel consumption and vehicle emission rate comparison for the four simulation scenarios. This comparison starts with the stop sign control scenario followed by the partial stop scenario, the signal control scenario, and the ramp scenario.

3.2.1 *Stop sign control scenario*

This section describes the results for the stop sign scenario for both normal- and high-emitting vehicles. The scenario tests the performance of the VT-Meso model for conditions involving vehicle stop-and-go maneuvers. As discussed earlier in the methodology section, five different speed limits are considered in this scenario. The speed limits are varied from 32 km/h (20 mi/h) to 64 km/h (40 mi/h) at increments of 8 km/h (5 mi/h).

In the case of normal light-duty vehicles, Figure 3 illustrates the speed profile and emission comparison for the single-vehicle scenario for LDV1s. Figure 4 illustrates the average travel speed and emission comparison for the 300-veh/h scenario for LDV1s. The total fuel consumption, HC, CO, and CO_2 emission rates are higher at lower speeds while the NO_X emission rates shows a slight increase when the free-flow speed increases. Given that a vehicle traveling at lower speeds spends more time traveling the 2-km roadway section, despite the lower time-based fuel consumption and emission rate, the total fuel consumption and emissions are higher at lower speed levels.

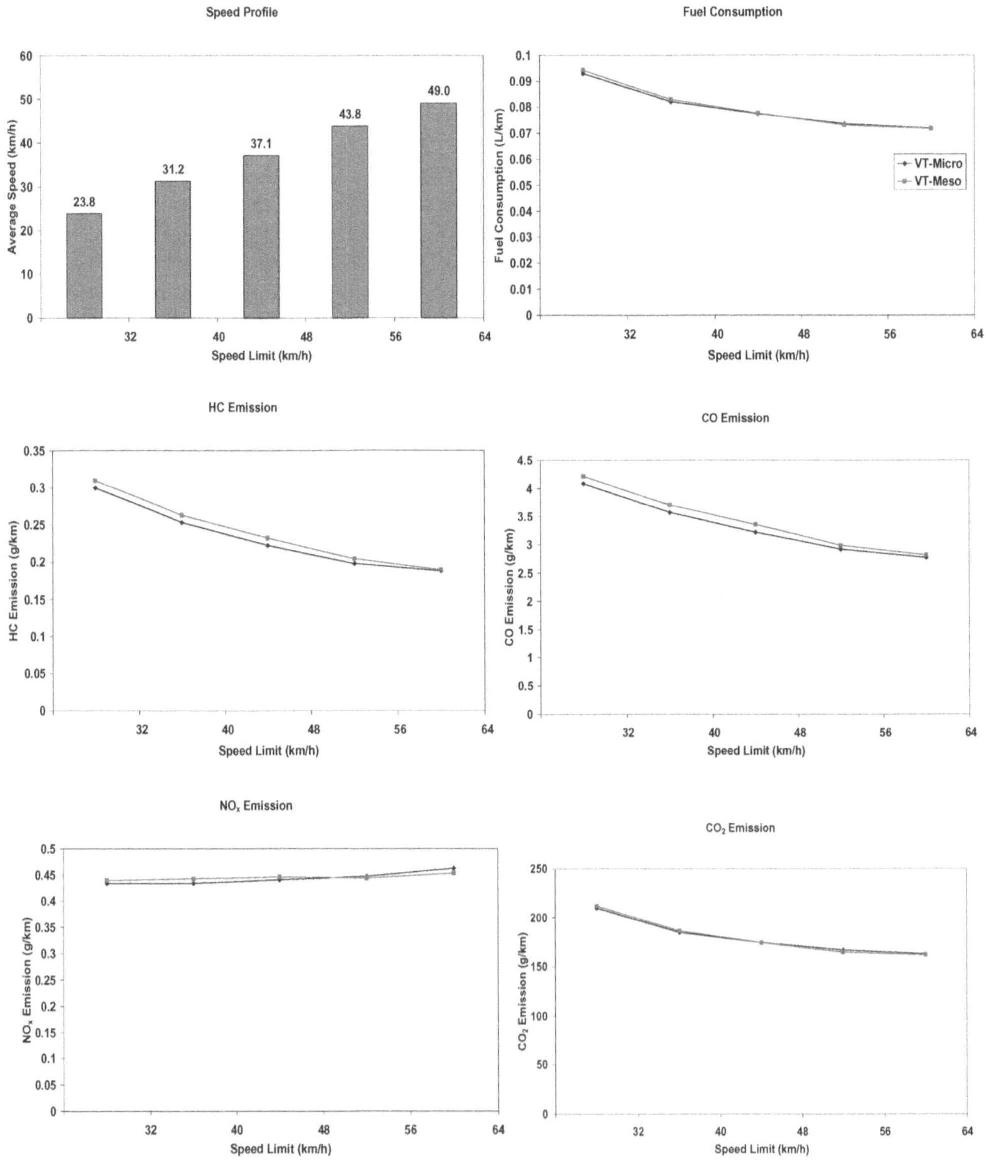

Figure 3. 300 veh/h LDV1 for stop sign scenario.

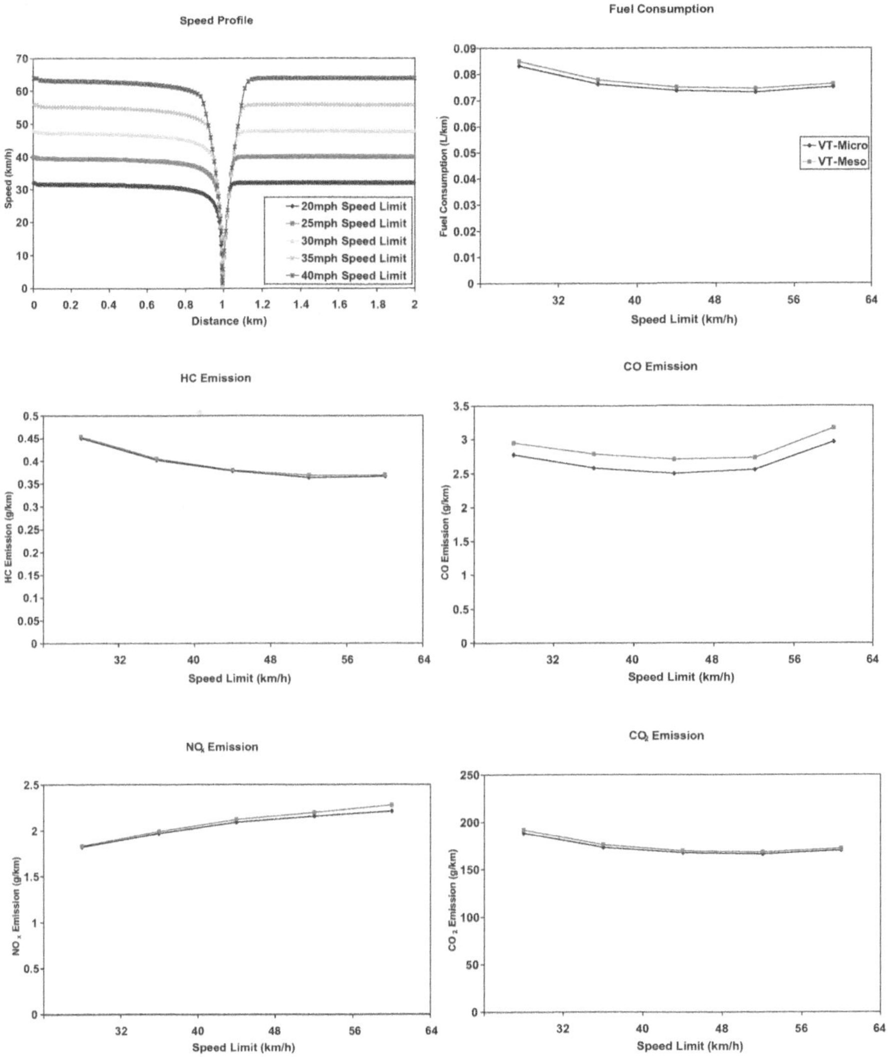

Figure 4. Signal HEV1 vehicle for stop sign scenario.

As a vehicle's speed increases, the time-dependent rate also increases, though more slowly than the travel time rate of increase. Consequently, the distance-based fuel consumption and emission rate decreases until the rate of increase in the time-dependent rate exceeds the rate of decrease in time spent in the system. A more detailed description and explanation of these behaviors can be found elsewhere in the literature (Rakha et al., 2004b).

These figures also illustrate that HC and CO emission are more sensitive to travel speeds. Comparison between the single-vehicle scenario and the 300-veh/h scenario shows that vehicles consume more fuel and generate more emissions for the 300-veh/h scenario, which has a lower travel speed. Comparison between the single-vehicle scenario and 300-veh/h scenario also illustrates that the difference between VT-Micro and VT-Meso estimations is smaller for high traffic volumes because the vehicle interaction reduces the random variability.

The results for HEV1 were similar trend to the LDV1 except for the CO emissions. The result for CO emissions shows a bowl-shaped relationship with the minimum emission rate at 48 km/h

(30 mi/h) while the CO emission for LDV1 has a minimum value at higher speeds. The absolute values of the HEV1 fuel consumption and CO_2 emission rates are close to those of the normal LDV1. In case of the NO_X emissions, the mass emissions for the HEV1 are significantly greater than those for the normal LDV. As is the case for the LDV1 vehicle, the 300-veh/h scenario results in a better VT-Meso estimate in comparison to the single-vehicle scenario.

3.2.2 *Signal control scenario*

This section describes the results for the signal control scenario for both normal- and high-emitting vehicles. The VT-Meso and VT-Micro model fuel consumption and vehicle emission estimates are varied as a function of the traffic signal offsets.

Figure 5 and illustrate the comparison of the VT-Micro and VT-Meso estimations as a function of signal offsets for LDV1s and HEV1s, respectively. As can be seen from these figures, higher average travel speeds are associated with a lower number of stops. The optimal offset for the 300-veh/h scenario is 40s, and the worst offset is approximately 0s regardless of the MOE that is considered in optimizing the signal offsets. The percentage changes in MOEs for optimal and worst offsets are calculated. The fuel consumption, HC, CO, NO_X, and CO_2 emissions for the worst offset are 17%, 18%, 25%, 4%, and 16% higher, respectively, than those of the optimal offsets. This result shows that the fuel consumption and vehicle exhaust emissions are generally low when the network is operating at the optimal signal offset. The figures demonstrate that the VT-Meso

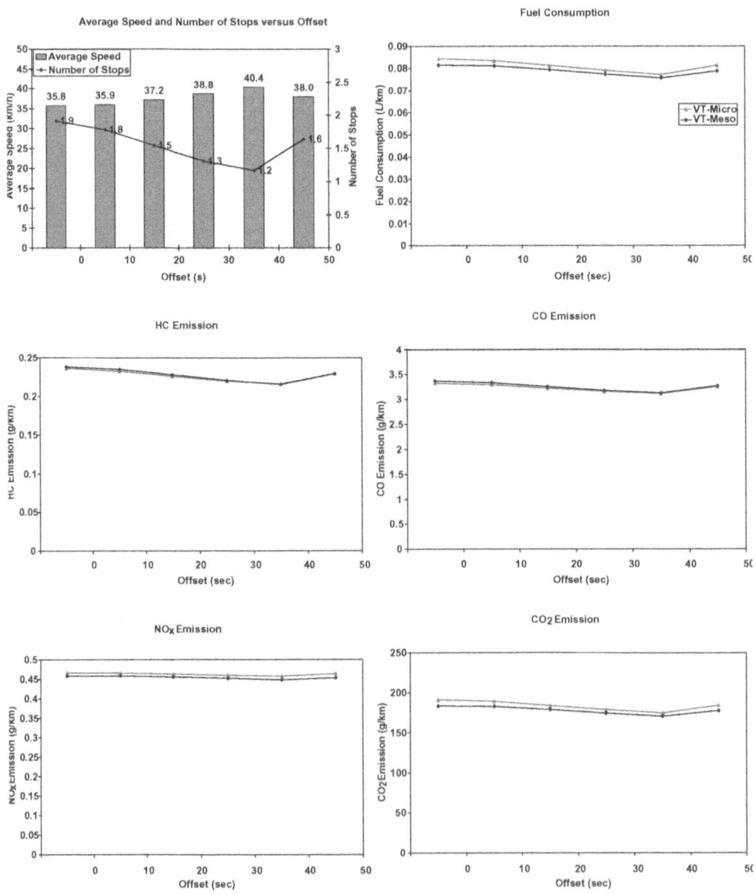

Figure 5. 800 veh/h LDV1 for signal control scenario.

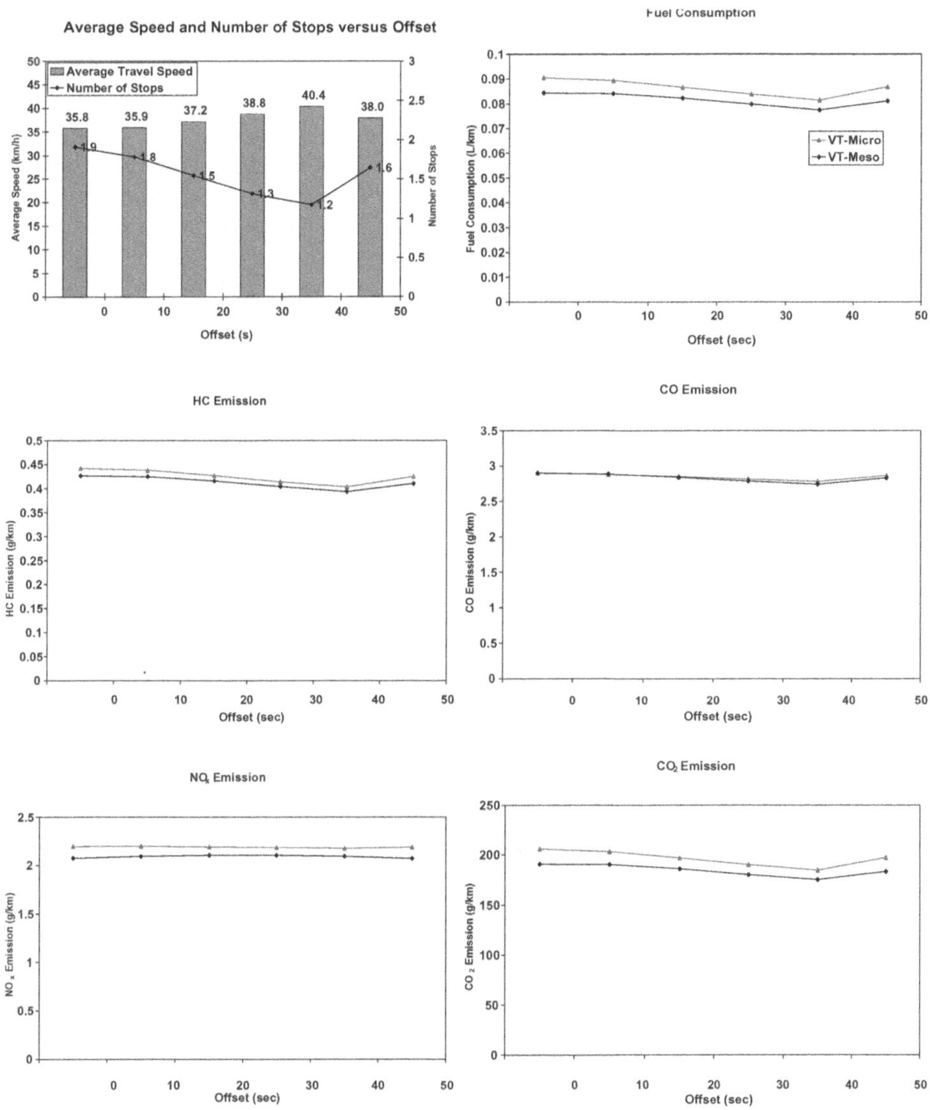

Figure 6. 800 veh/h HEV1 for signal control scenario.

model gives a good estimation for fuel consumption and CO_2 emissions. The estimations for HC, CO, and NO_X emissions are not as close as those for fuel consumption and CO_2 emission, but the VT-Meso model correctly predicts trends of increasing or decreasing HC, CO, and NO_X emissions. Specifically, the model produces the highest and lowest estimates for the same offsets for which the highest and lowest VT-Micro model estimates are obtained.

The variation in fuel consumption and emission rates for the HEV1 as a function of offsets is very similar to the LDV1 results. Similarly, the MOE rates are lowest at the optimal signal offsets. However, the differences in absolute values for the HEV1 are much greater than those of the normal LDV. The percentage changes in MOE for optimal and worst offsets are calculated. The fuel consumption, HC, CO, NO_X, and CO_2 emissions for the worst offset are 9%, 8%, 6%, 0%, and 9% greater, respectively, than those for the optimal offsets. As can be seen, the differences in fuel consumption and vehicle emissions are lower than those of the LDV1. Again the VT-Meso model produces similar trends to those produced using a microscopic model.

3.2.3 *Partial stop scenario*

The main objective of the partial stop scenario is to test the performance of the VT-Meso model for drive cycles with high speed variability. As discussed in the previous section, speed limit difference varied from 10 km/h to 50 km/h in increments of 10 km/h. The comparison between mesoscopic and microscopic models evaluates the effect of fractional stops on fuel consumption and vehicle emissions.

The analysis of LDV1s and HEV1s shows similar results. Figure 7 illustrates the comparison of VT-Micro and VT-Meso estimations as a function of speed limit difference for LDV1 at different traffic volumes. As for the single-vehicle scenario, it is observed that the model currently tends to underestimate NO_X and CO emissions. Alternatively, the fuel consumption, HC emission, and CO emission appear to be fairly symmetric about the line of perfect correlation.

It is hypothesized that these differences are attributable to differences between the actual and constructed drive cycles. Further investigations of the test results indicate that the variability observed in each diagram could also be attributable to the way partial stops are converted into equivalent full stops when aggregating each drive cycle into an average speed, an average number of stops, and an average stop duration. Figure 8 compares a synthetic drive cycle to an actual drive

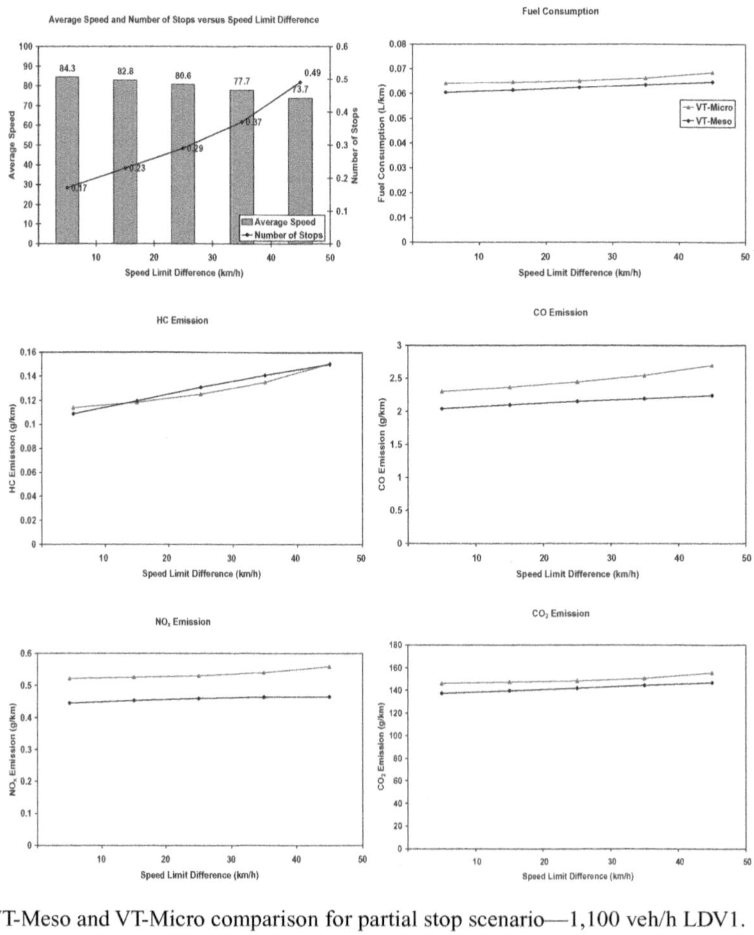

Figure 7. VT-Meso and VT-Micro comparison for partial stop scenario—1,100 veh/h LDV1.

cycle. Both cycles have the same average travel speed, deceleration time, acceleration time, and cruising time. This synthetic drive cycle will create problems for emissions that are more sensitive to high travel speed. Hence, this will cause the VT-Meso model to underestimate fuel consumption and emission rates for drive cycles with frequent partial stops.

As for the 1,100-veh/h scenario, it is also observed that the VT-Meso model provides a better match in terms of both absolute estimations and cyclic trends with the microscopic model estimates. This could be attributed to high traffic volume minimizing the high variability of the single vehicle. Also, high traffic volume is associated with lower travel speed, and the lower travel speed will help better estimate emissions, which are more sensitive at high speeds.

3.2.4 *Ramp scenario*

This section compares the VT-Meso and VT-Micro models for a ramp scenario which entails aggressive accelerations at high speeds (i.e. high engine loads). Two independent variables were included in the ramp scenario: weaving section length and weaving volume. Weaving volumes varied from 0 veh/h to 2,500 veh/h in increments of 500 veh/h, and the weaving section length was varied from 150 m to 450 m in increments of 50 m. Input values of two independent variables ensured that the over-capacity condition and the steady-state condition were simulated.

When the weaving volume is lower than 1,500 veh/h, the simulation network is under capacity. Increasing the weaving volume and/or reducing the weaving section length reduces the weaving section capacity and thus results in over-saturated conditions; this can be seen from the difference in the average travel speed. As for the fuel consumption and vehicle emissions, the plots on the left side show the VT-Micro model estimations, and the plots on the right side show the VT-Meso model estimations. As can be seen from these plots, the VT-Meso model estimations have an excellent match both in terms of absolute values and cyclic trends for fuel consumption, HC, and CO_2 emissions and a good fit for NO_X emissions. The VT-Meso model tends to underestimate CO emission in terms of absolute values, however the trends are consistent.

3.2.5 *Overall analysis*

Table 1 summarizes the differences between the VT-Meso and VT-Micro model estimates. The table generally indicates that the mesoscopic model estimates appear to be consistent with the microscopic estimates. Most of the differences between VT-Meso and VT-Micro estimations are generally less than 10% for the stop sign and signal control scenarios. In these cases, the differ-

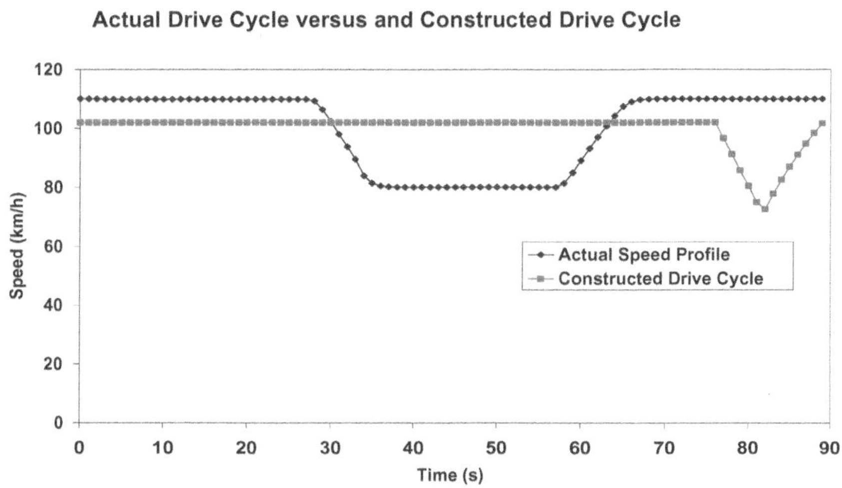

Figure 8. Comparison between synthetic drive cycle and actual drive cycle.

Table 1. VT-Meso and VT-Micro estimations differences.

Stop Sign Scenario

Speed Limit (km/h)	HEV1 Fuel	HEV1 HC	HEV1 CO	HEV1 NO_X	HEV1 CO_2	LDV1 Fuel	LDV1 HC	LDV1 CO	LDV1 NO_X	LDV1 CO_2
32	0.4%	–0.4%	2.3%	–2.3%	0.1%	1.4%	3.3%	3.1%	1.3%	1.2%
40	–0.1%	–0.6%	3.5%	–1.6%	–0.4%	1.1%	3.8%	3.8%	2.0%	0.9%
48	–1.4%	–1.3%	4.2%	–2.3%	–1.7%	0.4%	4.5%	4.5%	1.3%	0.1%
56	–2.5%	–1.4%	2.7%	–2.4%	–2.9%	–0.7%	3.4%	2.4%	–0.8%	–1.0%
64	–2.6%	–2.8%	3.6%	0.2%	–2.8%	–0.2%	0.9%	1.6%	–2.0%	–0.5%

Signal Control Scenario

Signal Offset (s)	HEV1 Fuel	HEV1 HC	HEV1 CO	HEV1 NO_X	HEV1 CO_2	LDV1 Fuel	LDV1 HC	LDV1 CO	LDV1 NO_X	LDV1 CO_2
0	–6.9%	–3.6%	–0.2%	–5.8%	–7.3%	–3.5%	0.7%	1.1%	–1.8%	–4.0%
10	–5.9%	–3.0%	0.2%	–5.0%	–6.3%	–2.9%	1.0%	1.2%	–1.5%	–3.3%
20	–5.1%	–2.6%	–0.3%	–4.2%	–5.5%	–2.3%	0.7%	0.9%	–1.5%	–2.7%
30	–4.8%	–2.4%	–0.9%	–3.8%	–5.2%	–2.2%	0.2%	0.5%	–1.8%	–2.6%
40	–4.8%	–2.5%	–1.5%	–3.7%	–5.1%	–2.0%	–0.2%	0.2%	–2.2%	–2.4%
50	–6.6%	–3.5%	–1.1%	–5.3%	–7.1%	–3.3%	–0.1%	0.5%	–2.2%	–3.7%

Partial Stop Scenario (Single Vehicle)

Speed Limit Difference (km/h)	HEV1					LDV1				
	Fuel	HC	CO	NO_X	CO_2	Fuel	HC	CO	NO_X	CO_2
10	–2.7%	–5.5%	–12.2%	–4.6%	–2.3%	–3.7%	–10.7%	–7.1%	–0.8%	–2.6%
20	–6.0%	–5.9%	–12.1%	–7.9%	–5.7%	–7.0%	–15.2%	–13.8%	–10.2%	–6.1%
30	–5.5%	–0.4%	4.3%	–8.0%	–5.6%	–7.5%	–9.7%	–18.1%	–15.4%	–6.6%
40	–4.9%	2.6%	1.4%	–6.6%	–4.9%	–8.3%	–5.0%	–24.6%	–20.6%	–7.0%
50	–5.3%	1.5%	–13.7%	–4.1%	–4.6%	–8.7%	–6.4%	–30.5%	–25.6%	–7.1%

Partial Stop Scenario (1,100 veh/h)

Speed Limit Difference (km/h)	HEV1					LDV1				
	Fuel	HC	CO	NO_X	CO_2	Fuel	HC	CO	NO_X	CO_2
10	–4.7%	1.5%	17.4%	–3.9%	–5.4%	–5.8%	–4.6%	–11.5%	–14.7%	–6.1%
20	–2.9%	5.0%	28.1%	–2.8%	–3.9%	–5.0%	0.8%	–11.3%	–13.8%	–5.3%
30	–1.2%	6.9%	25.6%	–1.5%	–2.3%	–4.2%	4.5%	–12.1%	–13.4%	–4.3%
40	–0.9%	6.6%	13.2%	–0.5%	–1.8%	–4.1%	4.3%	–13.8%	–14.2%	–4.1%
50	–3.3%	3.2%	–2.9%	–0.7%	–3.7%	–5.7%	–0.5%	–17.0%	–16.9%	–5.6%

ences in results between the microscopic and mesoscopic estimates can be attributed to differences between the actual and constructed drive cycles. Other parameters that could affect the mesoscopic model estimates include the various assumptions made by the mesoscopic model when generating synthetic drive cycles, particularly regarding the assumed deceleration and acceleration rates. In general the mesoscopic model appears to estimate fuel consumption, HC, and CO_2 emissions consistently with the microscopic estimates; however, the model tends to underestimate NO_X emissions.

As for the partial stop scenario, the VT-Meso estimations are lower than those of the VT-Micro model. This underestimation is attributed to the higher fuel consumption usually required to increase the speed of a vehicle from 50 to 60 km/h than from, say, 30 to 40 km/h because accelerations at higher speeds usually result in higher loads being exerted on a vehicle's engine. Because of the nonlinearity of fuel consumption and emission rates as a function of speed, a series of partial stops will often produce higher fuel consumption and emission rates than a corresponding equivalent number of full stops.

3.3 *Conclusions*

This paper compares the VT-Meso and VT-Micro model fuel consumption and emission estimates for four simulation scenarios including stop sign, signal control, partial stop, and ramp scenarios. The results demonstrate that the VT-Meso model provides an excellent match to the microscopic model estimates in terms of both absolute emission rates and cyclic trends. Specifically, this model provides close estimates for the stop sign, signal control, and ramp scenarios. On the other hand, the model predictions were less accurate in the partial stop scenario, especially in the case of CO emissions. The inaccuracies are partly attributed to the non-linear relationship between vehicle emissions and vehicle speed and errors in the construction of synthetic drive cycles.

The estimates from the VT-Meso model cannot be expected to always match estimates from microscopic models because of differences in the underlying drive cycles (the VT-Meso model constructs simplistic drive cycles). The model, however, constitutes an interesting alternative to existing models for cases in which detailed speed and acceleration data are not available. These results indicate that the VT-Meso model framework is sufficient for operational level comparisons. The study also demonstrates the need to perform additional tests to evaluate more precisely the effect of fractional stops on vehicle emissions or to more accurately convert partial stops into equivalent full stops for the purpose of estimating vehicle fuel consumption and emission rates.

REFERENCES

Ahn, K., Rakha, H., Trani, A., and Van Aerde, M. (2002). "Estimating vehicle fuel consumption and emissions based on instantaneous speed and acceleration levels." *Journal of Transportation Engineering*, 128(2), 182–190.

An, F., Barth, M., Norbeck, J., and Ross, M. (1997). "Development of a Comprehensive Modal Emissions Model: Operating Under Hot-Stabilized Conditions." *Transportation Research Record* (1587), 52–62.

An, F., and Ross, M. "Model of fuel economy and driving patterns." *International Congress and Exposition*, Detroit, MI, USA, 63-79 930328.

Bachman, W., Sarasua, W., Hallmark, S., and R., G. (2000). "Modeling regional mobile source emissions in a geographic information system framework." *Transportation Research Part C: Emerging Technologies*, 8(1), 205–229.

Barth, M., An, F., Younglove, T., Scora, G., Levine, C., Ross, M., and Wenzel, T. (2000). *Comprehensive modal emission model (CMEM), version 2.0 user's guide*, Riverside, CA.

CARB. (1991). "Methodology to Calculate Emission Factors for On-Road Vehicle." California Air Resource Board, Sacramento, CA.

EPA. (2002). "User's Guide to Mobile 6, Mobile Source Emission Factor Model." *EPA420-R-02-001*, Ann Arbor, Michigan.

INRO Consultants. (1996). "EMME/2 User's Manual, Release 8." INRO Consultants, Montreal, CA.

Koupal, J., Cumberworth, M., Michaels, H., Beardsley, M., and Brzezinski, D. (2002). "Draft Design and Implementation Plan for EPA's Multi-Scale Motor Vehicle and Equipment Emission System (MOVES)." *#420-P-02-006*, U.S. EPA.

Rakha, H., and Ahn, K. (2004). "Integration modeling framework for estimating mobile source emissions." *Journal of Transportation Engineering*, 130(2), 183–193.

Rakha, H., Ahn, K., and Trani, A. (2003a). "Comparison of MOBILE5a, MOBILE6, VT-MICRO, and CMEM models for estimating hot-stabilized light-duty gasoline vehicle emissions." *Canadian Journal of Civil Engineering*, 30(6), 1010–1021.

Rakha, H., Ahn, K., and Trani, A. (2004a). "Development of VT-Micro model for estimating hot stabilized light duty vehicle and truck emissions." *Transportation Research, Part D: Transport & Environment*, 9(1), 49–74.

Rakha, H., Ahn, K., and Trani, A.A. "The VT-micro framework for modeling of hot stabilized light duty vehicle and truck emissions." *82nd Annual Meeting of the Transportation Research Board*, Washington DC.

Rakha, H., and Ding, Y. (2003). "Impact of stops on vehicle fuel consumption and emissions." *Journal of Transportation Engineering*, 129(1), 23–32.

Rakha, H., Kang, Y.-S., and Dion, F. (2001a). "Estimating vehicle stops at undersaturated and oversaturated fixed-time signalized intersections." *Transportation Research Record*, n 1776, 128–137.

Rakha, H., and Lucic, I. (2002). "Variable power vehicle dynamics model for estimating maximum truck acceleration levels." *Journal of Transportation Engineering*, 128(5), 412–419.

Rakha, H., Lucic, I., Demarchi, S.H., Setti, J.R., and Van Aerde, M. (2001b). "Vehicle dynamics model for predicting maximum truck acceleration levels." *Journal of Transportation Engineering*, 127(5), 418–425.

Rakha, H., Snare, M., and Dion, F. (2004b). "Vehicle dynamics model for estimating maximum light-duty vehicle acceleration levels." *Transportation Research Record*, n 1883, 40–49.

Rakha, H., Van Aerde, M., Ahn, K., and Trani, A.A. (2000). "Requirements for evaluating traffic signal control impacts on energy and emissions based on instantaneous speed and acceleration measurements." *Transportation Research Record*, n 1738, 56–67.

Snare, M. (2002). "Dynamic model for predicting maximum and typical acceleration rates of passenger vehicles," Virginia Tech, Blacksburg, VA.

The Seider Group. (1997). "MINUTP Technical User's Manual." The Seider Group, Palo Alto, CA.

The Urban Analysis Group. (1992). "TRANPLAN, Version 7.1." The Urban Analysis Group, Danville, CA.

West, B.H., McGill, R.N., Hodgson, J.W., Sluder, C.S., and Smith, D.E. "Development of data-based light-duty modal emissions and fuel consumption models." *Proceedings of the 1997 International Fall Fuels & Lubricants Meeting & Exposition*, Tulsa, OK, USA, 233–239.

Efficient Transportation and Pavement Systems – Al-Qadi, Sayed, Alnuaimi & Masad (eds)
© 2009 Taylor & Francis Group, London, ISBN 978-0-415-48979-9

A three-stage procedure for validating microscopic simulation models

Y.E. Hawas
Roadway, Transportation and Traffic Safety Research Center, UAE University, UAE

M. Abdel Hameed & M. Cheqfah
RTTSRC

ABSTRACT: This paper introduces a novel approach to the validation of microscopic simulation models. Most of the validation efforts of microscopic simulation models were to achieve agreements between "aggregate" observed values and model outputs on variables such as volume served, average travel time, average speed, density, and average and maximum vehicle queue length. The validation based on averages of traffic variables however has several shortcomings.

Hawas (2007) presented a microscopic simulation model for incident management in urban networks, i-SIM-S. This paper describes the methodology adopted to validate the model. The methodology combines field data-simulation spatial-temporal analysis, simulation-simulation spatial-temporal analysis, and simulation-simulation aggregate measures validation. A three-stage procedure is devised for model validation. The purpose of the first stage is to validate a benchmark simulator (NETSIM) using field data. The purpose of the second stage is to validate the microscopic traffic patterns extracted from the ⓘ-SIM-S vis-à-vis those extracted by the benchmark simulator NETSIM. This is particularly done by comparing the spatial-temporal distribution of the vehicles along links and accounting for various factors such as traffic volume, link speeds, signal timing, etc. The purpose of the third stage is to validate the aggregate measures of traffic patterns (e.g. travel time, and delay) extracted from the ⓘ-SIM-S vis-à-vis those extracted by the benchmark simulator NETSIM. This is to ensure the adequacy of the procedures and formulae used to estimate such measures.

Keywords: Microscopic Models, Traffic Simulation, Validation

1 INTRODUCTION

Real-time traffic simulators are very effective tool in replicating traffic conditions and identifying optimal control strategies. They have emerged as important evaluation tools for Intelligent Transportation System (ITS) strategies. Macroscopic traffic simulations can play an active role in assisting the planning of ITS strategies. However, due to the macroscopic nature, those models and programs lack the capability of modeling detailed functions and features of traffic control and management techniques and are unable to meet the requirements of evaluating ITS technology benefits. Many existing microscopic simulation models have the modeling details for evaluating ITS technologies. Nonetheless, it is equally important that while improving their logic, that we apply more rigorous validation procedures to a simulation model to ensure its credibility and enhance user confidence.

The purpose of validation is to ensure that the simulation model accurately represents the real-world system so that decisions can be made on the real-world system based on the simulation. Microscopic models have unique characteristics because of the interactions among drivers, vehicles, and the roadway. The lack of data on one hand and the correlations among traffic stream elements further complicates the modeling and validation process. No systematic validation procedure for microscopic traffic simulation has been commonly accepted and performed in the past although more and more people have realized its importance (Hawas 2002).

Microscopic models depend extensively on car following models. The rationale is to utilize observed distributions (e.g. acceleration/ deceleration distributions) to predict the driver's behavior (selection of inter-spacing from the leading car, acceleration, deceleration, etc) in response to a specific stimulus. Car following models are very effective, specially, in modeling traffic in a confounded space (e.g. single lane traffic) where the driver actions become limited. Nonetheless, data available for model calibration and verification are limited. Car following models commonly utilize three functions to describe the perception of the driver to a specific stimulus, the decision making process in response to stimuli, and the final control or reaction by the driver. Car following models differ in their representation of the control-stimuli function. Some models assume a linear relationship where control is expressed as a multiplier of the stimulus, or a nonlinear model, or a more generalized model.

Among the most popular existing microscopic traffic simulation programs are CORSIM (FHWA 2001), INTEGRATION (Van Aerde 1995), WATSIM (Lieberman et al 1996), TRANSIMS (Smith et al 1995), MITSIM (Yang and Koutsopoulos 1996), PARAMICS (Duncan 1994), VISSIM (Fritzsche 1994), and AIMSUN (Casas, Ferrer and García, 1995). They all found their respective applicability in certain types of applications and enjoyed some degree of success, and validation studies have been conducted with many of those simulation programs (Milam and Choa 2001; Wang and Prevedouros 1998; Kim and Rilett 2003; Park and Schneeberger 2003).

The recent evolution of advanced traffic management systems stimulated the traffic research community to revisit the development of the simulation models and assess their applicability vis-à-vis the functionalities that such advanced systems support. Hawas and Mahmassani (1997) discussed the data needs for development, calibration and validation of dynamic simulation models and algorithms for ATIS/ATMS applications. Shin et al. (1999) stressed the immediate need for a tool to evaluate and validate the recent traffic ATIS/ATMS developments.

Validating simulation models in general and microscopic models in particular is a real challenging task due to rareness of validation data (Hawas, 2002), the complexity of the car following and the lane changing process, and the interaction among the car following/ lane changing parameters mentioned earlier. The major source of the difficulty is that the data (in addition to sacristy) is usually available in aggregate measurements of traffic characteristics (e.g. flows, speeds and occupancies at sensor locations, travel times, queue lengths).

Most of the validation efforts were to achieve agreements between "aggregate" observed values and model outputs on variables such as volume served, average travel time, average speed, density, and average and maximum vehicle queue length. The validation based on averages of traffic variables however has several shortcomings. First of all, it lacks a microscopic element. Microscopic traffic simulation should be validated differently from the way a macroscopic simulation is validated, and simply investigating certain macroscopic variables or ignoring the microscopic nature of the logic is certainly a deficiency. Secondly, in many cases, averages alone are not sufficient to validate the model, and there are needs to look at the distributions of the variables over time and space.

Overall, a systematic validation approach is needed to validate a microscopic simulation model. Hawas (2007) presented a microscopic simulation model for incident management in urban networks, ⓘ-SIM-S. The model is built to rectify most of the known deficiencies in the well-known microscopic models. Furthermore, was built with extended capabilities to model incidents accurately, and to capture their existence through loop detectors. These capabilities were utilized to develop logic for incident detection at intersections. More details on the algorithmic procedure and functions could be found in Hawas (2007). This paper describes in details the methodology adopted to validate the model, and applies such methodology to validate the ⓘ-SIM-S model. The methodology combines field data-simulation spatial-temporal analysis, simulation-simulation spatial-temporal analysis, and simulation-simulation aggregate measures validation. It is worthy noting that this paper is rather to be regarded as case study of the validation procedure. More resources and time are needed to generalize the conclusions made in this paper.

2 APPROACH FOR MICROSCOPIC MODEL VALIDATION

The well-known **NETSIM** simulation model was utilized to generate several validation scenarios for comparative analysis vis-à-vis ⓘ-SIM-S. **NETSIM** operates with a car-following model and has the ability to model detectors and offers good flexibility in specifying the detectors locations and the frequency of readings. This feature was utilized together with the counter feature in ⓘ-SIM-S to compare the spatial and temporal distribution of vehicles along the links with different traffic intensities.

NETSIM is widely known by its accuracy in modeling normal traffic operations in urban networks, and by far it is one of the mostly validated microscopic simulation models. It is also capable of simulating incident scenarios in urban street networks. **NETSIM** has the capability of simulating incidents as long term events. Furthermore, it can simulate detectors and as such it can be used to capture the spatial-temporal evolution of traffic streams. Nonetheless, there are several limitations in **NETSIM** incident modeling among which:

1. **NETSIM** cannot model an incident at a specific point in the network. Incidents are modeled as long term events on selected links. The specific locations (on the links) of these events are selected randomly by the model. Furthermore, the incident can cause a blockage to one lane only; simultaneous lane blockage cannot be modeled.
2. **NETSIM** does not account for simultaneous incidents/accidents on the network. One incident has to finish before another one starts.
3. The car following model of **NETSIM** does not account for the changes in the driving behavior as a result of an accident. For example, the slowing down of the vehicles as they pass by the accident location to have a look.

A three-stage approach is suggested to validate ⓘ-SIM-S. The approach first validates **NETSIM** vis-à-vis real field data at normal traffic conditions. The spatial-temporal detector data of ⓘ-SIM-S is then compared versus those developed by **NETSIM** at various recurrent and non-recurrent conditions. Finally, the aggregate traffic measures (travel time) of both models are compared at different scenarios.

2.1 *Stage I: Field Data-**NETSIM** Spatial-Temporal Analysis*

Due to the huge resources needed to collect "microscopic" field data that can be used to validate the simulator and the limited human and resources allotted to this study, it was determined that it is almost impossible to carry on the simulation validation using only field data. Furthermore, field data if available, is only limited to specific traffic conditions (volumes, controls, etc). Also, there is almost no microscopic detailed data to support the validation of "non-recurrent" incident situations, not to mention the complexity of designing field experiments to "emulate" these conditions and the difficulty of capturing the microscopic traffic stream patterns associated with such incidents. For all these reasons, it was thought that an appealing approach would be to set a "benchmark" simulator and validate it using real field data (even roughly to pin point the efforts associated with such task). Upon its validation, the benchmark simulator may then be used to generate the "microscopic" patterns that can be used further to calibrate or validate other microscopic models. It is worthy noting that the idea here is not to validate the benchmark model itself, but to introduce a conceptual framework for the validation of microscopic simulators. If resources are available, the introduced concept can be generalized to gather more field data to validate the benchmark model.

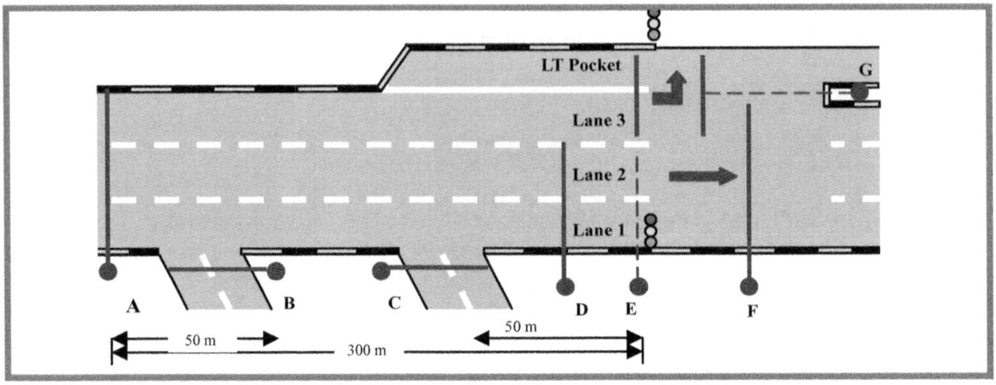

(A) Field Intersection Layout and Observer Locations

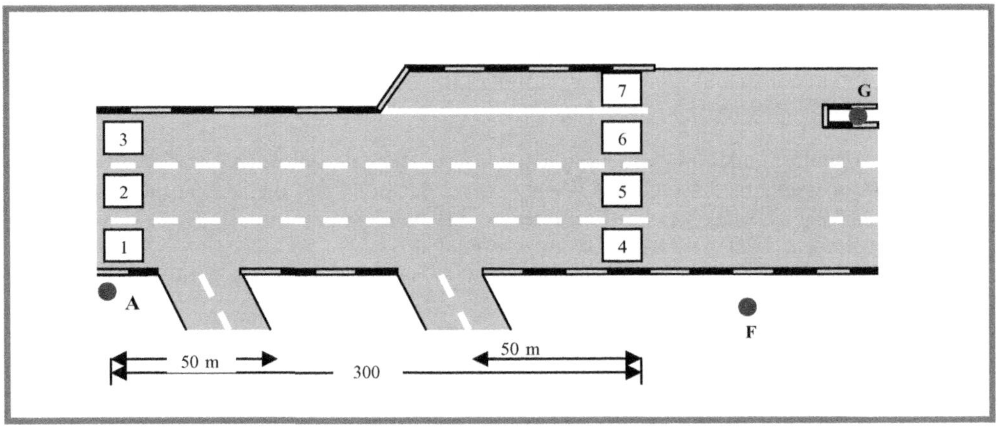

(B) NETSIM Intersection Layout and Detectors

A: Observer records number of vehicles/min at 300 m upstream the intersection stop line;
B: Observer records number of vehicles/min entering and exiting from side street (250 m upstream the intersection stop line);
C: Observer records number of vehicles/min entering and exiting from side street (50 m upstream the intersection stop line);
D: Observer records number of vehicles/min queued on the 1st and 2nd lanes;
E: Observer records number of vehicles/min queued on the 3rd and LT lanes;
F: Observer records number of through vehicles/min from 1st, 2nd and 3rd lanes
G: Observer records number of LT or U-Turn vehicles/min from the 3rd and LT Pocket

Figure 1. Layout of the intersection used for (A) field data (B) **NETSIM** spatial temporal analysis.

The purpose of this stage is to validate the benchmark simulator (**NETSIM**) using field data. Collecting the spatial-temporal field data is quite cumbersome, and entails huge resources. Fig. 1 illustrates the layout of the intersection used for validation. Due to the complexity of data collection methodology, stemming from the fact that it would be used to characterize the microscopic spatial-temporal pattern, a total of seven observers (A through G) participated in the data collection. Data collection of this nature is quite expensive as it would require huge resources. It was essential to include the human element in counting rather than using automatic detectors to estimate the turning percentages and types of vehicles (passenger car, single unit,

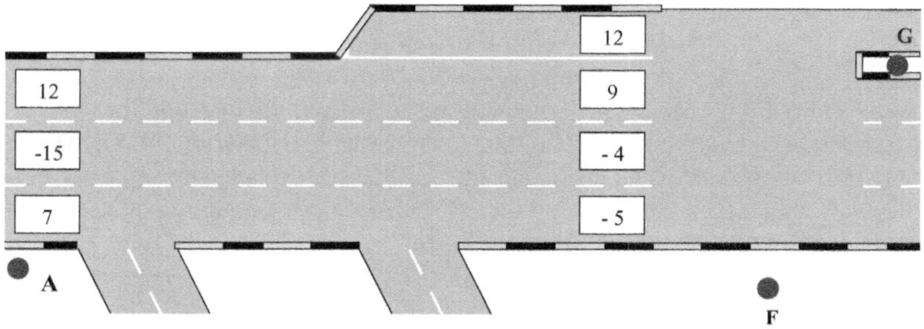

(A) Average Discrepancy of Accumulative Counts* (10 minutes)

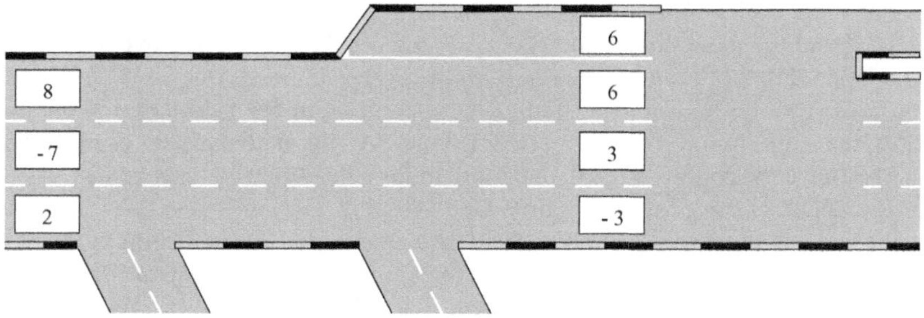

(B) Average Discrepancy of Accumulative Counts (20 minutes)

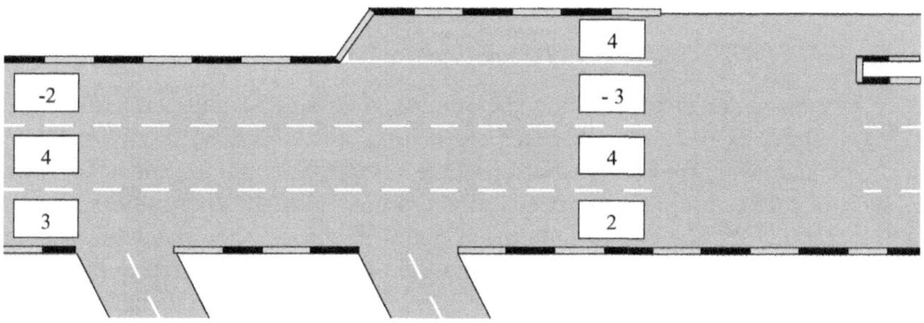

(C) Average Discrepancy of Accumulative Counts (30 minutes)

* [Field Count- NETSIM Count)/ Field Count* 100%]

Figure 2. Field data-**NETSIM** spatial temporal comparative analyses.

truck). Furthermore, to estimate the exact vehicle movement count out of lanes designated to serve more than one movement.

As shown on the figure, seven observers were needed to cover such a simple leg. The role of each observer is indicated on the figure. The approach counting was synchronized to account for the time needed to travel from one observer to the other. The same intersection was modeled using **TSIS** software (**NETSIM** model) to replicate the exact geometry, turning percentages, volumes, and signal timing. The downstream traffic light is operating using a fully actuated controller. The green time (and the varying cycle) of the downstream traffic light was also recorded by a separate observer. The green time did not exhibit significant variations from one cycle to the other (on average about 26 seconds), with a max of 2 seconds difference among cycles. As such, the downstream controller was modeled as a pre-timed signal with green time equal to the average green duration of the phase observed during the counting period (about 26 seconds).

The intersection was modeled with exact field details (geometry, traffic volumes, turning percentages, and control timing). The field observers were "represented" by simulation passage detectors with the frequency of accumulative reading set to one minute (same as field interval). The field and simulation-based data collected were utilized to develop the spatial temporal patterns of both field and simulation. The intersection was batch run by **NETSIM** for ten times and the detector readings were then averaged. Fig. 2 illustrates average discrepancy in percentage between the field accumulative counts and the **NETSIM** detector reading counts. The percentage of discrepancy is estimated using the following form:

Average discrepancy at any location at time (t) = [field accumulative count of vehicles passing this location (point) at time t- **NETSIM** accumulative count for the detector residing this location (point) at time t]/field accumulative count of vehicles passing this location (point) at time t * 100%

The average values of discrepancy are illustrated in the small rectangles in Fig. 2. The negative values indicate higher **NETSIM** counts, while the positive values indicate higher field counts. The values are exhibited for three time instants; 10, 20 and 30. At time 10, the average discrepancy is highest as compared to the 20 and 30 cases. Initially, as the approach is least congested, **NETSIM** tends to allocate more vehicles to the middle lanes, resulting in overestimation to the middle lane counts and underestimation to the right and left lanes. As the network gets more congested, the vehicles are reallocated spatially to replicate more or less the observed field values. The minimum values for discrepancy were observed at time 30.

Despite the fact that the comparison with field data was limited due to the limitation of resources, and the significant efforts that would be needed to capture other scenarios of the spatial/temporal distribution of vehicles in the field, it is evident that **NETSIM** could be used as a benchmark simulator to validate other simulation models. Future work for this stage might consider having more field cases for validation.

2.2 *Stage II: **NETSIM**-ⓘ-SIM-S Spatial-Temporal Analysis*

The purpose of this stage is to validate the microscopic traffic patterns extracted from the **ⓘ-SIM-S** vis-à-vis those extracted by the benchmark simulator **NETSIM**. This is particularly done by comparing the spatial-temporal distribution of the vehicles along links and accounting for various factors such as traffic volume, link speeds, signal timing, etc. The distribution of vehicles along the links as represented by the two simulators is captured via passage detector readings. Virtual detectors are installed on the underlying links and their corresponding accumulative readings are compared. This validation is carried out on two phases; in a normal recurrent traffic conditions, and in incident non-recurrent conditions. This phase of validation is carried out as follows:

1. Comparative analysis of the **ⓘ-SIM-S** and **NETSIM** under recurrent traffic conditions.
2. Comparative analysis of the two models' detectors under various non-recurrent traffic conditions. However, due to the limitations of the **NETSIM** model in modeling incidents, the following procedure is devised:
   a. Introduce long term events (incidents) to **NETSIM** and extract the randomly selected incident location, and associated **NETSIM**'s detector readings for such cases
   b. Use **ⓘ-SIM-S** to replicate the same incidents introduced by **NETSIM** and extract the **ⓘ-SIM-S**'s corresponding detector readings.
   c. Conduct comparative analysis of the detector readings in (a) and (b).

A sample intersection was developed using both **NETSIM** and **ⓘ-SIM-S**. The intersection has four approaches; each with three lanes. Three to four detectors are inserted on each lane. The downstream node is operated by a pre-timed signal of a total cycle time of 140 seconds. The phase timing is changed based on the studied scenario.

Preliminary observations indicated that the detectors' patterns may differ for various approach lengths (m), approach speeds (km/hr), approach green times (sec), approach volumes (veh/hr), detectors locations on approaches, occurrence of incidents, and incident duration, location and starting time (if any). Six validation cases were selected to validate **ⓘ-SIM-S** under various conditions

Table 1. Summary of detector counters validation cases.

Property		Approach EB (D1)	Approach NB (D2)	Approach WB (D0)	Approach SB (D)
Case I—Each lane has three detectors					
Approach Length		300 m	300 m	300 m	300 m
Approach Speed		80 km/hr	80 km/hr	80 km/hr	80 km/hr
Approach Green Time		30 sec	30 sec	30 sec	30 sec
Approach Volume		500 vph	1000 vph	1500 vph	2000 vph
Detectors	First	10 m	10 m	10 m	10 m
Locations on	Second	150 m	150 m	150 m	150 m
Approaches	Third	290 m	290 m	290 m	290 m
Accident on Approach		NO	NO	NO	NO
Accident Duration		N/A	N/A	N/A	N/A
Accident Location		N/A	N/A	N/A	N/A
Accident Starting Time		N/A	N/A	N/A	N/A
Case II (similar to case I with different approach length and detector settings)—Each lane has four detectors					
Approach Length		600 m	600 m	600 m	600 m
Detectors	First	10 m	10 m	10 m	10 m
Locations on	Second	200 m	200 m	200 m	200 m
Approaches	Third	400 m	400 m	400 m	400 m
	Fourth	590 m	590 m	590 m	590 m
Case III (similar to case I with different approach length and signal settings)—Each lane has three detectors					
Approach Length		600 m	600 m	600 m	600 m
Approach Green Time		20 sec	30 sec	40 sec	50 sec
Case IV (similar to case I with different link speed)—Each lane has three detectors					
Approach Speed		60 km/hr	60 km/hr	60 km/hr	60 km/hr
Case V (similar to case II with accident on second lane of E-W approach)—Each lane has four detectors					
Accident on Approach		NO	NO	YES	NO
Accident Duration		N/A	N/A	600 sec	N/A
Accident Location		N/A	N/A	Second Lane 375 m	N/A
Accident Starting Time		N/A	N/A	900 sec	N/A
Case VI (similar to case III with accident on first lane of N-S approach)—Each lane has three detectors					
Accident on Approach		NO	NO	NO	YES
Accident Duration		N/A	N/A	N/A	900 sec
Accident Location		N/A	N/A	N/A	First Lane 191 m
Accident Starting Time		N/A	N/A	N/A	1200 sec

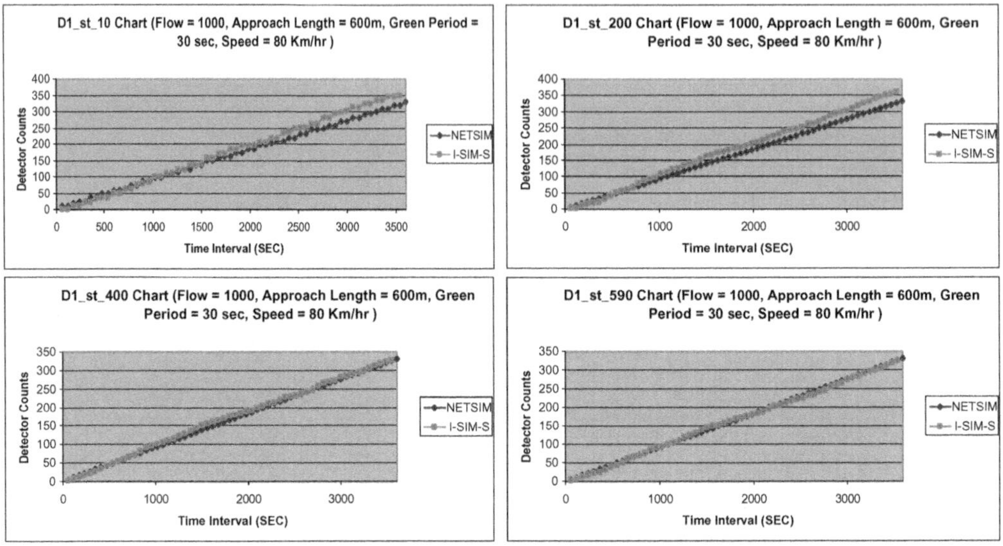

Figure 3. Comparative detector readings (NETSIM and ⓘ-SIM-S) for case II: Second lane.

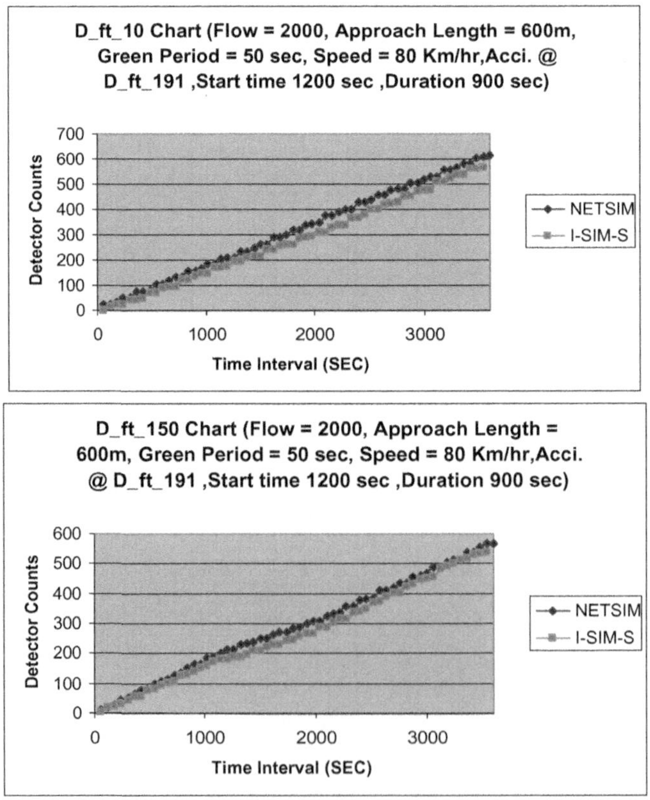

Figure 4. Comparative detector readings (NETSIM and ⓘ-SIM-S) for case VI on the first lane of the 2000 vph approach.

as shown in table 1. The cases were selected to assess the ⓘ-SIM-S modeling accuracy under various situations.

Figs. 3 and 4 illustrate a sample of the detector readings of the studied cases. Fig. 3 illustrates the accumulative detectors readings for the two simulation models (in a non-incident case). It is evident that the two simulators exhibit more or less the same patterns for all the detectors over all the lanes. The case illustrated in Fig. 3 (approach volume = 1000 vehicles/hour) represents a case of medium to high congested situation.

Fig. 4 illustrates the accumulative detector readings for the VI study case (incident case). The incident in such case is introduced on the most congested approach (volume = 2000 vph). The effect of the incident on the accumulative pattern is not noticeable in this case. The following section highlights the effect of the volume on the incident patterns. The figure also indicates the similarity among the detector readings of the two models. All the other results indicated very close similarity. Their results are not indicated due to the paper size limitations.

2.3 *Stage III: NETSIM-ⓘ-SIM-S aggregate measures validation*

The purpose of this stage is to validate the aggregate measures of traffic patterns (e.g. travel time, delay, intersection delay, etc) extracted from the ⓘ-SIM-S vis-à-vis those extracted by the benchmark simulator **NETSIM**. This is to ensure the adequacy of the procedures and formulae used to estimate such measures. Four validation cases were designed as shown in Table 2. The cases differ in the total volume of vehicles to be served ranging from light to heavily congested approaches. The analysis is done by comparing the accumulative total number of vehicles served by each approach at any time instant, total number of vehicles served by the intersection at any time instant, the average travel time on each approach and for the intersection, the average delay time on each approach and for the intersection, and finally the average signal delay on each approach and for the intersection. The four cases were modeled as pre-timed intersections with green times calculated using Webster formula.

Table 3 shows the aggregate measures extracted from both simulators by the end of the simulation period (set to one hour). The table indicates some variations of the aggregate measures, yet with no consistent pattern. The overall intersection aggregate measures are relatively

Table 2. NETSIM -i-SIM-S aggregate measures validation cases.

Case property	Approach EB	Approach NB	Approach WB	Approach SB
Case 1				
Approach Length	300 m	300 m	300 m	300 m
Approach Speed	80 km/hr	80 km/hr	80 km/hr	80 km/hr
Approach Volume	500 vph	1000 vph	1500 vph	2000 vph
Green Period	10	19	29	38
Case 2				
Approach Volume	700 vph	1200 vph	1700 vph	2200 vph
Green Period	12	20	28	36
Case 3				
Approach Volume	1000 vph	1500 vph	2000 vph	2500 vph
Green Period	14	21	27	34
Case 4				
Approach Volume	1200 vph	1700 vph	2200 vph	2700 vph
Green Period	15	21	27	33

Table 3. Comparative aggregate measures of **NETSIM-ⓘ-SIM-S** (estimated at end of simulation)*.

		NETSIM				**ⓘ-SIM-S**			
Approach	Case	Total volume served (vph)	Average travel time (min)	Average delay time (min)	Average signal delay (min)	Total volume served (vph)	Average travel time (min)	Average delay time (min)	Average signal delay (min)
EB	1	497	1.303	1.08	0.977	500	1.625	1.367	1.358
	2	639	3.288	3.072	2.958	658	2.011	1.755	1.744
	3	684	7.078	6.872	6.145	755	5.271	5.017	5.005
	4	771	7.243	7.035	6.188	828	4.803	4.55	4.537
NB	1	945	2.767	2.547	2.45	964	1.543	1.282	1.268
	2	968	5.425	5.21	4.545	1052	2.902	2.646	2.632
	3	1005	5.753	5.538	4.7	1110	3.558	3.303	3.288
	4	1034	5.938	5.725	4.82	1126	3.519	3.266	3.251
WB	1	1406	2.685	2.468	2.232	1487	1.155	0.898	0.882
	2	1365	4.073	3.858	3.185	1441	2.667	2.411	2.395
	3	1254	4.567	4.353	3.56	1457	2.771	2.517	2.502
	4	1288	4.722	4.508	3.637	1446	2.772	2.518	2.502
SB	1	1808	2.80	2.58	2.043	1946	1.2	0.943	0.926
	2	1710	3.45	3.24	2.48	1848	1.941	1.687	1.67
	3	1608	3.64	3.43	2.662	1766	2.186	1.932	1.916
	4	1589	3.862	3.647	2.817	1703	2.298	2.044	2.027
Entire Intersection	1	4656	1.020	0.926	0.822	4897	1.297	1.04	1.024
	2	4682	2.216	2.099	1.797	4999	2.362	2.106	2.091
	3	4551	3.162	3.034	2.565	5088	3.11	2.857	2.841
	4	4682	3.584	3.444	2.875	5103	3.108	2.854	2.839

* Results are estimated by the end of simulation period (one hour).

close. It is worthy noting that although the traffic volumes are set ahead of simulation are exactly the same, the two simulators differ on the actual number of vehicles got served and as such the aggregate measures. This is particularly attributed to the difference in the modeling approach. The ⓘ-SIM-S simulated cases could be optimized for minimum discrepancy in the aggregate measures with **NETSIM** using the so-called Alpha value. For more information on the effect of the Alpha parameter on results, the reader is referred to (Hawas 2006). The results shown here does not reflect optimized discrepancy results for the individual cases, but rather an average Alpha value for all the cases.

Fig. 5 illustrates sample of the aggregate intersection results of the two simulators (for the Case 3) at different instants (2-minute intervals) as they evolve over the entire simulation period. The figure indicates fairly close results with the selected average Alpha value. At the initial stages of simulation (the first ten minutes as the network is least congested), ⓘ-SIM-S exhibited slightly higher travel times and delays as compared to **NETSIM**. This difference gaps out as the networks becomes more congested, resulting in very close performance towards the end of the simulation period.

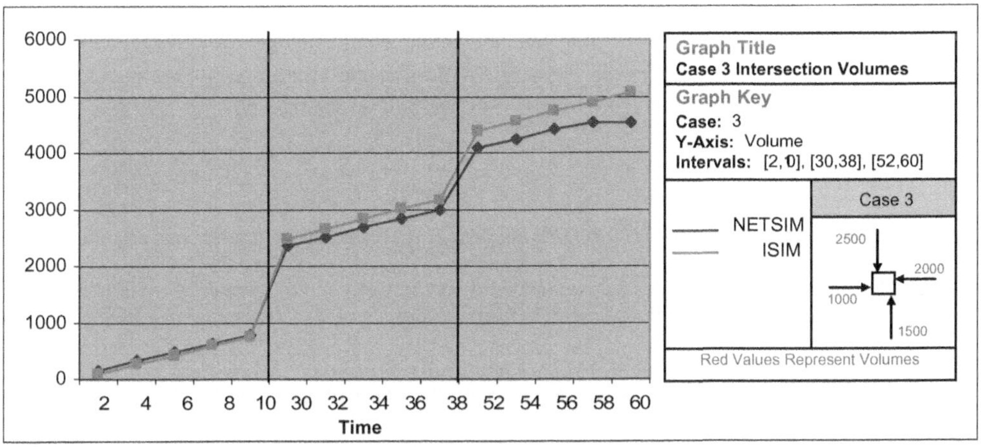

(A) Comparison of Accumulative Vehicles Served by Intersection at Various Time Instants

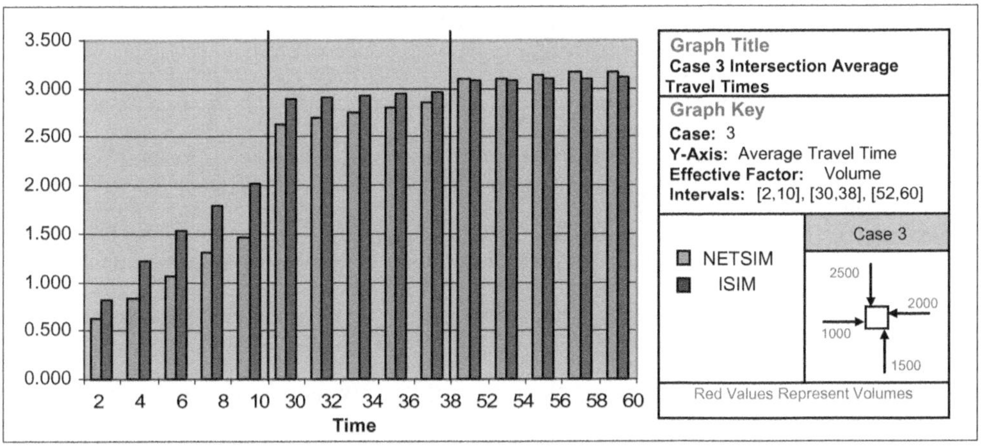

(B) Comparison of Intersection's Average Travel Time at Various Time Instants

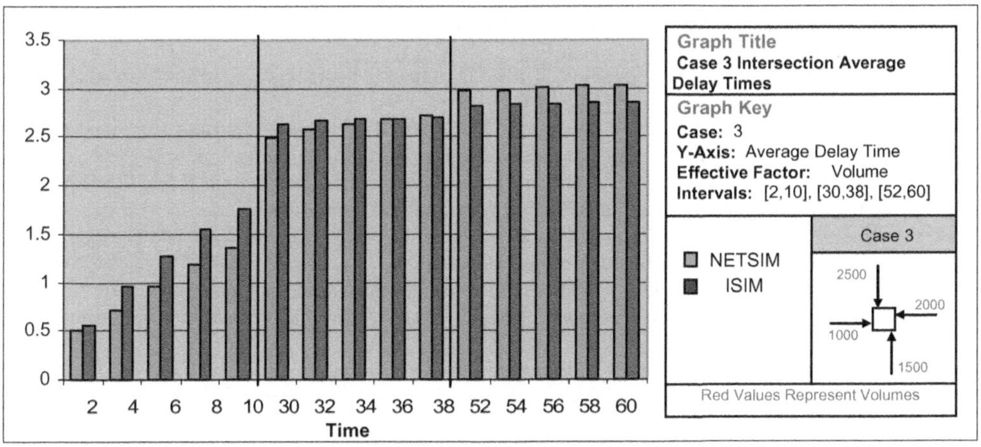

(C) Comparison of Intersection's Average Delay Time at Various Time Instants

Figure 5. Sample results of **NETSIM-Ⓘ-SIM-s** aggregate measures variations at various time instants.

ACKNOWLEDGEMENT

This research has been funded by the Research Affairs at the United Arab Emirates University under funding grants # 01-01-7-11/03, 02-01-7-11/04, and 06-01-07-11/05.

REFERENCES

Brackstone, M. and McDonald, M. (1999) "Car-Following: A Historical Review." Transportation Research Part F, 2, pp. 181–196.

Bruce, L.M., Balraj, N., Zhang Y., Yu S. "Automated Accident Detection in Intersections via Digital Audio Signal Processing," Transportation Research Record, Journal of the Transportation Research Board, January 2003.

Casas, J., Ferrer, J.L. and García, D. (1995) Microscopic Traffic Simulation: A Tool For The Design, Analysis And Evaluation Of Intelligent Transport Systems. Universitat Politècnica de Catalunya.TSS-Transport Simulation Systems.

Chassiakos, A. and Stephanedes, Y. (1993) "Smoothing Algorithms for Incident Detection," Transportation Research Record 1394, pp. 8–16.

Chen, S., Sun, Z.P. and Bridge, B. (1997) Automatic traffic monitoring by intelligent sound detection, Proc. IEEE Conf. on Intelligent Transportation Systems. pp. 171–6.

Corby, M.J. and Saccomanno, F.F. (1997) Analysis of Freeway Accident Detection. Transportation Research Record. Issue 1603, pp. 80–89.

Duncan, G. "PARAMICS-MP Final Report," project report, Edinburg Parallel Computing Center, University of Eidinburg, Edinburg, Scotland, October 1994.

Federal Highway Administration. Traffic Software Integrated System, User's Guide, Version 5.0. 2001.

Forren, J.F. and Jaarsma, D. (1997) Traffic monitoring by tire noise, Proc. IEEE Conference on Intelligent Transportation Systems, pp. 177–82.

Fritzsche, H.-Th. (1994) A model for traffic simulation. Traffic Engineering and Control 35(5), 317–321.

Harlow, C. and Wang, Y. (2001) Automated Accident Detection, Proc. Transportation Research Board 80th Annual Meeting, pp. 90–93.

Hawas, Y.E. and Mahmassani, H. S. (1997) "Data Requirements for Development, Calibration of Dynamic Traffic Models and Algorithms for ATMS/ATIS." Proc., 78th Annual Meeting of the Transportation Research Board, Washington D.C.

Hawas, Y.E. (2002) "Calibrating Simulation Models for ATIS/ATMS Applications" Journal of Transportation Eng, 128(1), pp. 80–88.

Hawas, Y.E. (2007) A Microscopic Simulation Model for Incident Modeling in Urban Networks," Transportation Planning and Technology, Vol. 30, No. 2, pp. 289–309.

Jin, X., Cheu, R.L. and Srinivassan, D. (2002) Development and Adaptation of Constructive Probabilistic Neural Network In Freeway Incident Detection. Transportation Research. Part C: Emerging Technologies. Vol. 10, No. 2, pp. 121–147.

Judicky, D. and Robinson, J. (1992) "Managing Traffic During Nonrecurring Congestion," Institute of Transportation Engineers Journal, Vol. 62, No. 3, pp. 21–26.

Kamijo, S., Matsushita, Y., Ikeuchi, and K., Sakauchi, M. (2000) "Traffic monitoring and accident detection at intersections", Proc. IEEE Transactions on Intelligent Transportation Systems, Vol. 1, No.2 pp. 108–18.

Kim, K. and Rilett, L. Simplex Based Calibration of Traffic Micro-Simulation Models Using ITS Data. Transportation Research Board 82nd Annual Meeting, Washington, D.C. January 2003.

Lieberman, E., et al, "WATSIM: Wide Area Traffic Simulation Model for Freeways and Surface Streets," paper presented at the 75th TB Annual Meeting, Washington, D.C., January 1996.

Milam,R. and Choa, F. Recommended Guidelines for the Calibration and Validation of Traffic Simulation Models. 8th TRB Conference on the Application of Transportation Planning Methods, Corpus Christi, Texas. April 2001.

Park, B. and Schneeberger, J. Microscopic Simulation Model Calibration and Validation: A Case Study of VISSIM for a Coordinated Actuated Signal System. Transportation Research Board Annual Meeting. Washington, D.C. January 2003.

Persaud, B.N., Hall, F.L. and Hall, L.M. (1990) "Congestion Identification Aspects of the McMaster Incident Detection Algorithm," Transportation Research Record, 1287, pp. 67–74.

Shin, M.S., Ran, B., Oh, R.S. and Choi, K. (1999) "Developing An Evaluation Tool for A Real-Time DTA System." Proc., 78th Annual Meeting of the Transportation Research Board, Washington D.C.

Smith, L. et al, "Overview of TRANSIMS: the Transportation Analysis and Simulation System," Los Alamos National Laboratory, May 1995.
Stephanedes, Y. and Chassiakos, A. (1993) "Application of Filtering Techniques for Incident Detection," ASCE Journal of Transportation Engineering, Vol. 119, No. 1, pp. 13–26.
Van Aerde, M. (1995) "INTEGRATION Release II: User's Guide," Transportation Systems Research Group, Queen University, Kingston, Canada.
Wang, H. and Prevedouros, P. Comparison of INTEGRATION, TSIS/CORSIM, and WATSim in Replicating Volumes and Speeds on Three Small Networks. Transportation Research Board Annual Meeting, Washington, D.C. January 1998.
Yang, Q. and Koutsopoulos, H.N. (1996) "A Microscopic Traffic Simulator for Evaluation of Dynamic Traffic Management System," Technical Report.

Efficient Transportation and Pavement Systems – Al-Qadi, Sayed, Alnuaimi & Masad (eds)
© 2009 Taylor & Francis Group, London, ISBN 978-0-415-48979-9

Platoon identification and accommodation at isolated traffic signals

N.A. Chaudhary & H.A. Charara
Texas Transportation Institute, TX, USA

ABSTRACT: Isolated operation of traffic signals is the most prevalent form of active traffic control at roadway intersections. At many such intersections, platoons of vehicles on major intersection approaches are frequently forced to stop in order to serve a few vehicles on a minor approach. Such conditions are undesirable because they cause excessive stops, unsafe conditions, and driver frustration. This paper describes a real-time traffic-adaptive control system to remedy such situations. The paper also presents preliminary results of its field deployment in Texas.

1 INTRODUCTION

Isolated traffic signal operation is the most common form of active traffic control at intersections of major and minor roadways in the U.S. and many other countries. This type of traffic control efficiently serves traffic during light demand conditions, but often becomes sub-optimal when demand increases on one or both major roadway approaches of an intersection. These sub-optimal conditions are further magnified when platoons of approaching vehicles are forced to stop at the intersection to serve one or few vehicles on a minor approach, resulting in excessive stops, unsafe driving conditions, driver frustration, and excessive wear and tear to pavement in the vicinity of the intersection. To provide a remedy for such situations, our research team at the Texas Transportation Institute (TTI) developed a platoon identification and accommodation (PIA) system. The original PIA system (PIA-1), described in detail by Chaudhary et al. (2003 and 2006), provided priority treatment to traffic on a selected main-street direction. It provided this functionality by detecting platoons of vehicles before they arrived at the stopbar and overriding normal controller operation to progress detected platoons. The system used signal preemption as the controller override mechanism. PIA-1 provided user configurable parameters to prevent excessive delays to conflicting traffic movements. Recently, we redesigned the PIA system to provide: (1) priority control in both arterial directions, (2) refinement to system components, and (3) new traffic adaptive features. The enhanced system was deployed in the field and was extensively tested. This paper describes the development and field-testing of the enhanced PIA (PIA-2) system. Readers interested in additional details may review Chaudhary et al. (2008) and Charara et al. (2008).

2 ARCHITECTURE OF ENHANCED PIA SYSTEM

PIA-2 system uses off-the-shelf hardware and custom software to provide platoon detection and platoon progression functions. Figure 1 shows hardware architecture of the PIA-2 system. As shown in this figure, the PIA-2 system consists of a field hardened personal computer (PC) and detection infrastructure for providing advance detection at the two priority approaches. The PC

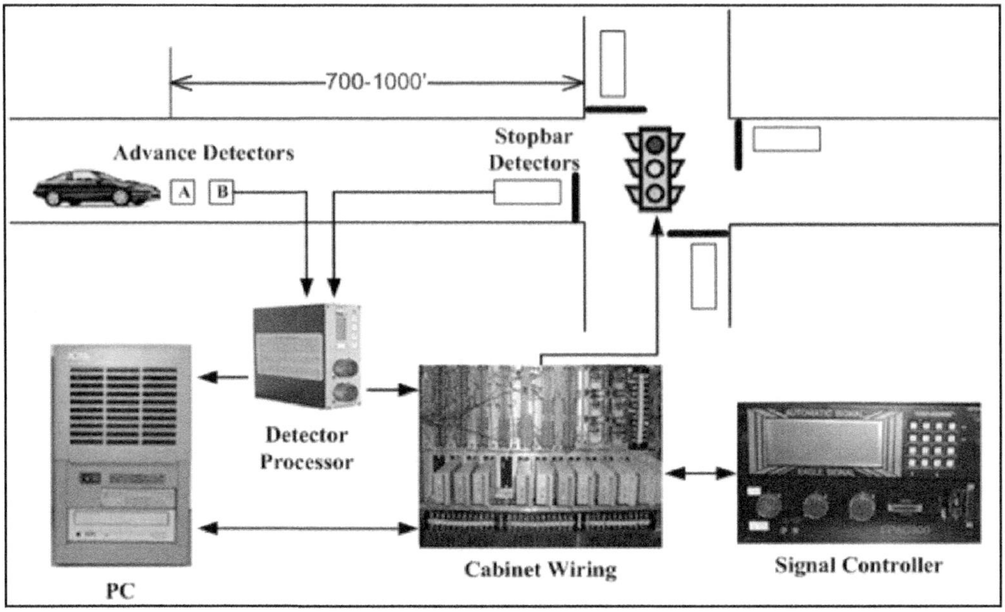

Figure 1. Hardware architecture of PIA-2 system.

resides in the signal controller cabinet or an auxiliary cabinet if the controller cabinet does not have sufficient room. The PIA-2 system uses custom software to provide real-time control. The PC interfaces with advance detectors to obtain information needed by the system software for platoon detection. The PC also interfaces with the controller cabinet to monitor status of signal phases and stopbar detectors in real-time. This interface also provides means through which the software sends commands to override normal controller operation. The PIA-2 system provides two types of interfaces: hardwire and/or serial communication to allow its installation in two types of controller cabinets commonly used in Texas and many other states in the U.S. If needed, digital input-output (DIO) cards are installed in the PC to provide hardwire communications. Up to two DIO cards may be required depending on the number of advance plus stopbar detectors and the number of signal phases at the site. The system provides for controller override via preemption and phase hold features of traffic controllers.

The system requires that each priority approach be equipped with advance detection capable of detecting speeds of individual vehicles on a per lane basis. Any reliable detection technology can be used. The PIA-2 system has been designed to accommodate two broad categories of detection technologies. The first category includes detectors that provide contact closure (on and off) signals from the detector. This category includes traditional inductive loop detectors (ILD), magnetometers, and some video-based detection systems. Reliable calculation of vehicle speed from contact closure signals requires a pair of detectors, called a speed trap. In Figure 1, the pair of boxes labeled as A and B illustrate an ILD-based speed-trap for traffic approaching the intersection from the west. In the following text these advance detectors are referred to as ADA and ADB. The difference in detection times of a vehicle passing over these two detectors combined with known design distance between the two detectors is used to calculate the speed of that vehicle. Furthermore, calculated vehicle speed together with the occupancy of one of the two detectors can be used to calculate the length of that vehicle. The second category includes detectors which directly measure vehicle speeds. These systems include radar-based detectors referred to as smart sensors in this paper. Smart sensors provide detection times and speeds of individual vehicles using a serial message, which can be deciphered using vendor's protocol. Location of advance detection depends on approach speed. Chaudhary et al. (2003) recommended that advance detection be installed at 700 ft upstream of the stopbar for approach speeds of up to 45 mph and 1000 ft upstream of stopbar for approach speeds higher than 55 mph.

3 SYSTEM SOFTWARE AND ALGORITHMS

Figure 2 identifies the four major modules of the PIA software. The following subsections describe algorithms and procedures used by these modules.

3.1 *Software classifier (SC) module*

The SC module in PIA-2 has been added to replace a hardware classifier used by the PIA-1 system. This modification reduces space requirement in the controller cabinet and reduces system installation cost by at least US $3000. The SC module calculates three quantities described below.

3.1.1 *Vehicle counts*

The program computes separate vehicles counts for vehicles passing over advance detectors ADA and ADB. At present, these counts are not used during real-time PIA operation. Daily counts are optionally logged for offline use and can also be displayed in real time.

3.1.2 *Vehicle speeds*

The system uses projected arrivals time of vehicles at the intersection stopbar calculated using detected vehicle speeds. As described below, the program computes two values of speed for each vehicle passing over the speed trap.

$$Speed\,1 = (D \times 1000) / \Delta_{On} \tag{1}$$

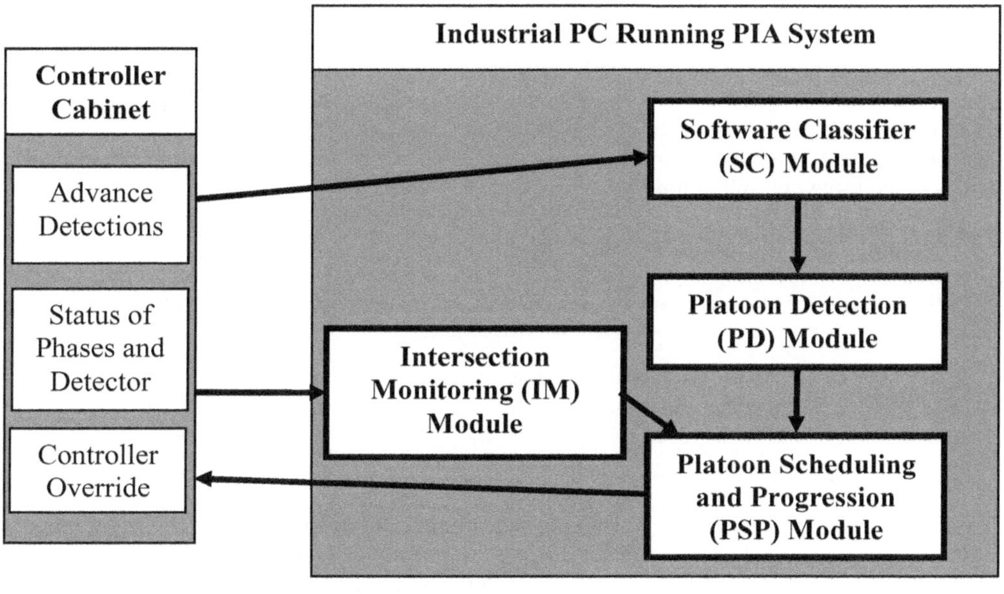

Figure 2. Four major modules in the PIA-2 software.

$$Speed\ 2 = (D \times 1000) / \Delta_{Off} \quad (2)$$

Where:

D = Distance between leading or trailing edges of ADA and ADB (16 ft for standard configuration).

Δ_{On} = Time difference between the event a vehicle hits the leading edge of ADA and the event the same vehicle hits the leading edge of ADB (milliseconds).

Δ_{Off} = Time difference between the event a vehicle hits the trailing edge of ADA and the event the same vehicle hits the trailing edge of ADB (milliseconds).

Note that the multiplier value of 1000 converts time from milliseconds to seconds. The program makes consistency checks on detector "On" and "Off" events before calculating speeds. If data for either one of them fail the test, the speed is flagged to be "bad." If both speed values pass the goodness test, it calculates the average speed. If only one value is good, it uses the good speed. If both values are bad, it ignores the detection.

3.1.3 *Vehicle lengths*

As stated earlier, the occupancy of a detector is a function of detection length (length of vehicle plus detector length) and vehicle speed. Using this information, the program calculates vehicle length (L) using the following relationship:

$$L = (Speed \times Occupancy \times 1.47) - Detector\text{-}Length \quad (3)$$

In this calculation, the program uses the average value of occupancies for ADA and ADB if both values are good. If only one value is good, it uses that value. If both occupancies are bad, the detection is ignored.

3.2 *Platoon detection (PD) module*

For each vehicle detected at the advance detector, the PD module receives speed, detection time, and lane identification (ID) information. Using known distance of the advance detector from the stopbar, it first calculates each vehicle's arrival time at the stopbar. In this calculation, it assumes that the subject vehicle will stay in the lane in which it was detected and maintain a safe headway (say, 2 seconds) between the current and the immediately preceding vehicle in the same lane. Thus, if the projected arrival time of a vehicle is same or less than the immediately preceding vehicle, the module adjusts the projected arrival time of the current vehicle to the projected arrival time of the preceding vehicle plus the minimum headway value. The module uses projected arrival times of vehicles at the intersection for platoon detection and progression. Figure 3 provides a simplified logic of the platoon detection algorithm. This logic consists of two stages described in details in the following subsections.

3.2.1 *Initial platoon identification stage*

This stage is active when there is no detected platoon. During this stage, the program identifies if a smallest acceptable platoon exists. This portion of the program uses the following user-specified parameters:

- number of vehicles in the smallest acceptable platoon (n),
- cumulative headway threshold (T_c), and
- preemption advance (P_a).

In real time, the program carries out the following calculation steps:

1. Upon detection of a new vehicle, find the difference (t) between the projected arrival times of the first vehicle and the last vehicle in the group of last n consecutive vehicles.
2. If t is less than or equal to T_c a platoon meeting the user-specified density and size criteria has been detected, go to Step 4. Otherwise, go to step 3.

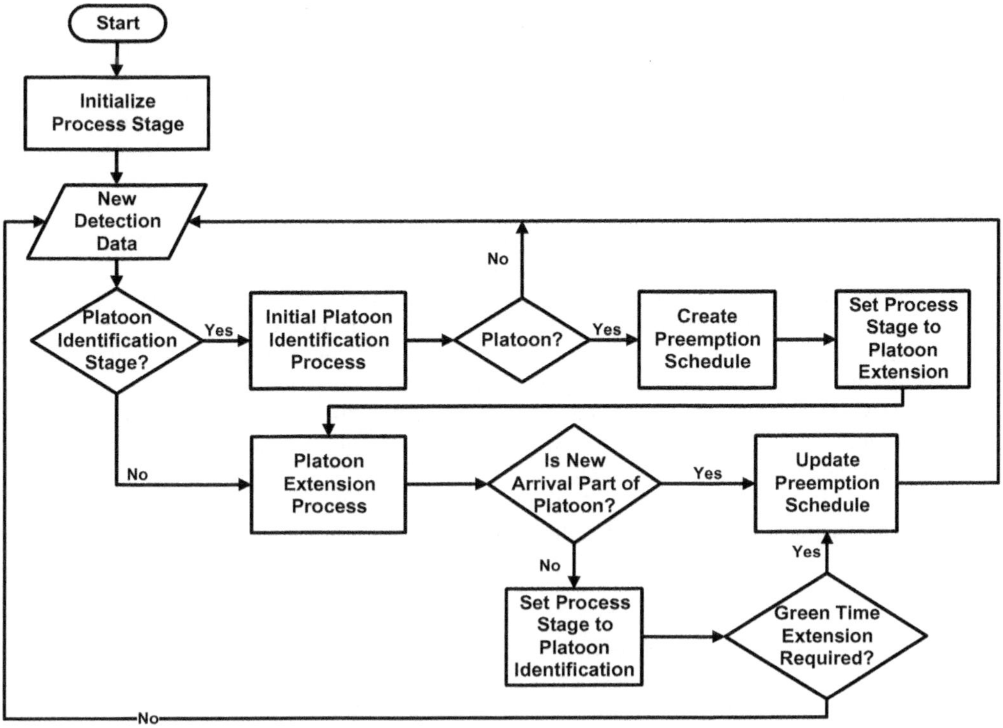

Figure 3. Platoon detection logic.

3. Otherwise, remove the oldest vehicle from the group of n vehicles. If the signal is red at the projected arrival time of this vehicle, add one to the number of vehicles predicted to stop and go to Step 1.
4. Create a preemption schedule, which consists of start and end times. These times are the predicted arrival times of the first and the last vehicles, respectively, in the group of n vehicles meeting the acceptable platoon criterion. Then, make the following two adjustments to the start time of schedule, and activate the platoon extension stage described in the next subsection:
   a. Advance the time to account for the number of vehicles predicted to stop. This adjustment provides queue clearance time before a predicted platoon arrives.
   b. Advance the start time by an additional amount equal to P_a. This factor allows the user to make adjustments to accommodate site-specific factors.

It should be noted that the number of approach lanes, speed limit, and driver behavior play an important role in the selection of values for algorithm parameters n and T_c.

3.2.2 *Platoon extension stage*

This stage activates as soon as a platoon has been detected, and remains active as long as the platoon schedule has not expired. During this stage, the PD routine evaluates each additional vehicle to determine if it is part of the previously detected platoon or if it will be in its dilemma zone at the scheduled preemption end time. If either of these conditions is true, the algorithm extends the preemption termination by an appropriate amount of time. This stage of algorithm uses the following three user-specified parameters:

- average headway threshold (T_h),
- extension to last vehicle in the platoon (T_e), and
- preemption clearance (P_c).

The PD routine uses the first two thresholds to assess if a new vehicle is part of the current platoon. In this mode, the program performs the following calculation steps:

1. Wait for a new detection. If the platoon progression schedule expires during this wait, switch to the initial platoon detection stage described above. Note that a platoon progression schedule expires if the current time is larger than the scheduled end time. If a new vehicle arrives before expiry of the progression schedule, go to Step 2.
2. Perform the following two calculations:
   a. Calculate the average headway for all vehicles in the platoon, including the new vehicle. If this value is less than or equal to T_h, the new vehicle is part of the platoon.
   b. Calculate the headway between the new vehicle and the last vehicle in the current platoon. If this value is less than or equal to T_e, the new vehicle is part of the platoon.
3. If the new vehicle is part of the current platoon, change the platoon progression end time in the schedule to the projected arrival time of the new vehicle, and go to Step 1.
4. If the new vehicle does not meet extension criteria 2a and 2b, compare the vehicle's projected arrival time with the end time in the scheduler with P_c to determine if any additional clearance time is needed. Change the scheduled end time by the calculated adjustment and switch to the initial platoon identification stage. Note that setting P_c equal to 2.5 seconds is equivalent to providing dilemma zone protection to this last vehicle.

3.3 *Intersection monitoring (IM) module*

The IM module in PIA-2 monitors real-time status of phases and stopbar detectors and calculates measures that can be used to provide adaptive functionality to the system. Although this module calculates several real-time performance measures, the current system only uses real-time phase utilization information for non-priority phases. The real-time utilization information for a signal phase is calculated at the termination of green interval for the subject phase. The IM module records the following information for the subject phase:

- phase duration was equal to minimum (*Min*) green setting in the controller;
- phase duration reached maximum (*Max*) green setting in the controller (phase maxed out);
- phase duration was larger than *Min* green and smaller than 0.5 × (*Max* + *Min*); or
- phase duration was larger than or equal 0.5 × (*Max* + *Min*) and smaller than *Max*.

As described later in subsection 3.5, the system uses this real-time termination status information to constrain operation of the platoon scheduling and progression (PSP) module. The IM module also calculates and logs daily frequencies of the above events for off-line analysis.

3.4 *Platoon scheduling and progression module*

As in the PIA-1 system, the PSP module uses real-time phase status, stopbar detector status, and platoon detection information for each main-street priority phase to make platoon progression decisions. However, this module was enhanced during its redesign to make it suitable for PIA-2 system. The new module has the ability to exercise control via the following options:

1. phase holds only,
2. low-priority preempts only, or
3. a combination of low-priority preempts and phase holds.

Option 1 is the least intrusive in that it takes over control to progress platoons only when the desired priority phase is green. As such, it guarantees that the platoon phase does not terminate before serving vehicles in a platoon approaching the intersection. When option 2 is enabled, the program uses preemption to switch platoon phase from red to green or to hold the phase if it is already green. Under this option, the program continues to use preemption until the override termination condition becomes true. Under option 3, the program operates as follows:

- If the platoon phase is red, it issues a preemption signal. As soon as the subject phase becomes green, the program places a hold signal and removes preemption signal, ensuring continuity of controller override.
- If the platoon phase is green, the program places a hold.

It should be noted that the removal of a phase hold signal provides snappier termination of controller override than the removal of a preemption signal. Thus, option 3 is more desirable than option 2. Because a phase hold is used only when a priority phase is green, only two phase-hold inputs need to be used. However, three preempt inputs are needed to handle different platoon progression cases. Two inputs are needed to handle the two cases when a platoon is approaching from only one direction. The third input, which should have higher priority than the other two inputs, is used to handle simultaneous platoon progression in both directions.

3.5 *Constraints on platoon accommodation*

In the presence of heavy demand on priority phases, platoon progression may significantly increase the red time for conflicting minor phases, resulting in longer than normal queues. Such taxation, although built into the PIA-2 system by design, may be severely detrimental to the intersection operation. The PIA-system has two protection mechanisms to minimize such detrimental actions. The first protection mechanism is static in nature and is implemented by using a maximum (*Max*) timer. The system uses this user-specified *Max* timer to restrict the length of controller override (preempt and/or phase hold) in the presence of demand at any conflicting phase. This feature is similar to the *Max* timers for phases in modern traffic controllers. The second protection mechanism is adaptive in nature. It uses real-time phase utilization information for controlling when the *Max* timer activates. This adaptive feature dynamically controls when, and to what extent, a nonpriority phase can be taxed to provide favorable treatment to conflicting priority phases. This mechanism uses three user-defined parameters—minimum delay (D_{min}), maximum delay (D_{max}), and delay increment (I)—for each nonpriority phase. Given these parameters, the algorithm calculates a real-time delay value (D) for each nonpriority phase using the following logic:

1. set D equal to associated minimum delay value,
2. wait for the phase (g) to end, then proceed to Step 3,
3. change D as per the following logic and go to Step 2:

   If $[g < 0.5 \times (Max + Min)] \rightarrow D = D + I < D_{max}$
   Else If $[g \geq 0.5 \times (Max + Min)] \rightarrow D = D - I > D_{min}$
   If [Phase maxed out twice in a row] $\rightarrow D = D_{min}$
   If [Phase terminated by PIA while demand] $\rightarrow D = D - (2 \times I)$

The PIA-2 system ignores any vehicle call on a nonpriority phase for duration equal to the current delay value. Thus, minor phases with light demand are taxed more than minor phases with heavy demand.

4 FIELD IMPLEMENTATION AND TESTING

We installed and tested the PIA-2 system at two intersections in Texas. This section describes the results of field implementation at one of these sites; a T-intersection located about 90 miles from TTI headquarters. Figure 4 shows geometric features and controller timing parameters. For each signal phase (Ø), numbers in parentheses are the Min and Max times programmed in the traffic controller. At this site, the system was operated in a shadow mode for several weeks prior to its full operation. This operation mode, where all system components except controller override worked as normal, provided ample time to test and tweak various system components. As a result of this testing, Max times for phases 4 and 5 were increased to 40 and 30 seconds, respectively. This change was made to ensure that these phases cleared any additional queue buildup due to longer red times resulting from PIA system's override of traffic controller to progress platoons on phases 2 and 6. This testing also allowed us to tweak system parameters, especially those used for platoon detection. Table 1 provides the final values of these parameters. While the system was running in a shadow mode, its IM module was also enabled to write daily data logs.

Full operation of PIA-2 system at this site began on August 8, 2007. We observed traffic operation at the intersection on that day and were satisfied with results to leave the system running without our presence at the site. Because it was not feasible to travel to the site frequently,

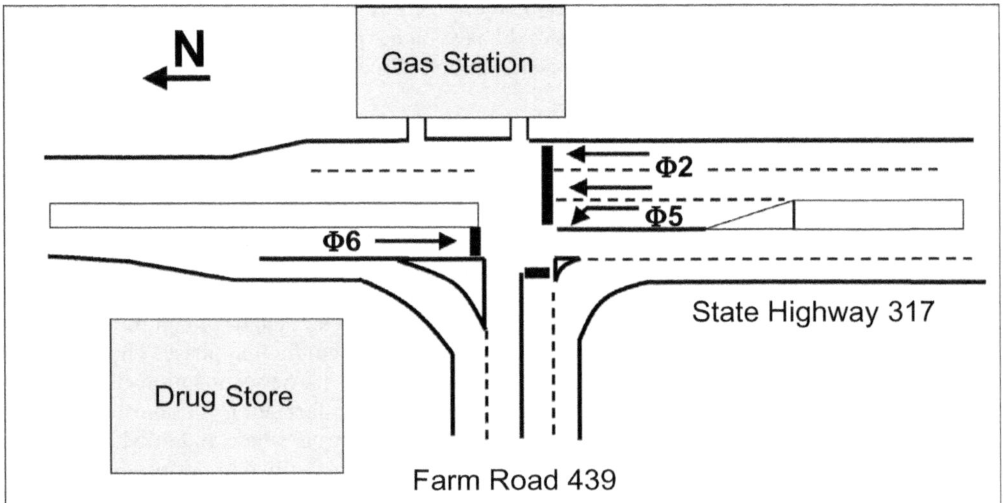

	Barrier 1		Barrier 2
Ring 1	Northbound Through+ Ø2 (10, 30)		Eastbound Left Ø4 (8, 40)
Ring 2	Northbound Left Ø5 (5, 30)	Southbound Through* Ø6 (10, 40)	

Notes:
+ Priority phase serving two through lanes receiving traffic from an upstream traffic signal
* Priority phase serving one through lane without an upstream signal (an isolated approach)
Northbound left turn served by a protected (Ø5) plus a permissive phase

Figure 4. Intersection geometry and controller phase settings at the test site.

Table 1. Platoon detection parameters and constraints.

	Phase 2	Phase 6
Controller Override Max (seconds)	65	65
Smallest Platoon Size, n (vehicles)	6	6
Cumulative Headway Threshold, T_c (seconds)	18	18
Average Headway Threshold, T_h (seconds)	2.5	2.5
Extension Threshold, T_e (seconds)	3	3

Delay (D_{max}) for phases 4 and 5 set to 7 and 0 seconds, respectively.

a wireless remote monitoring system was installed to monitor the operation of PIA-2 software. During August 2007, we traveled to the site two more times. During each of these visits, we observed system operation for several hours and were satisfied with the system operation.

A subset of data saved by the PIA-2 system in daily logs was synthesized for comparing the before and after operation of traffic signal at this site. The before case used data from five consecutive days (Friday through Tuesday) staring on August 3, 2007. The after data used was for the same five days starting on August 17, 2007. Table 2 shows a comparison of before and after signal cycles for these days for a 24-hour period. As identified in this table, Ø2 was used as a reference point for computing the lengths of these cycles. A review of this table shows that the operation of PIA-2 system decreased the number of daily signal cycles for all phases. This is an expected result.

Table 3 shows the utilization of Ø4. From this table, the reader can observe that the frequency of phase Min (8 seconds) reduced. Also, phase terminations by Max-out were almost eliminated. Thus, phase utilization improved. Similar results were observed for Ø5.

Table 4 provides key platoon detection and progression statistics for the two priority phases (Ø2 and Ø6). For each phase, the table provides total number of platoons detected, percent of these platoons actually progressed, and percent of daily demand served by progressing these platoons. For instance, on a Friday (August 17, 2007) the system detected 540 platoons on Ø2, 69% of which were progressed. Furthermore, the number of vehicles progressed by Ø2 amounted to 50% of the total (24-hour) demand for that day. In general, the system progressed 69 to 84 percent of detected platoons, helping 41 to 53 percent of daily traffic demand on Ø2 during these five days.

As shown in Table 4, the number of platoons detected on Ø6 was significantly less than Ø2 during all five days studied. This is an expected result because there is no upstream signal on this intersection approach. The presence of platoons on this approach confirms our pervious observations regarding vehicle bunching on isolated signal approaches. Also note that the system progressed only 47 to 63 percent of these platoons, because Ø6 was constrained by two phases (Ø4 and Ø5) as opposed to only one conflicting phase (Ø4) for the other priority approach (Ø2). Lastly, platoon

Table 2. Comparison of before and after signal cycles.

	Frequency of Cycles	Friday	Saturday	Sunday	Monday	Tuesday
	Before	1027	1019	910	1009	1033
Ø2	After	937	942	848	895	919
	% Change	–9	–8	–7	–11	–11

Table 3. Utilization of phase 4.

	Ø4	Friday	Saturday	Sunday	Monday	Tuesday
Before	Max	2	0	3	2	2
	> Half*	11	4	3	10	9
	< Half	40	35	30	41	43
	Min	48	60	64	47	46
After	Max	0	0	0	0	1
	> Half	4	1	1	4	6
	< Half	52	44	42	53	52
	Min	40	51	54	39	38

* Half = 0.5 × (Max + Min). As shown in Figure 4, Max and min times for this phase were 8 and 40 seconds, respectively.

Table 4. Platoon detection and progression statistics.

		Friday	Saturday	Sunday	Monday	Tuesday
Ø2	Platoons Detected	540	403	314	488	446
	% Platoons Progressed	69	80	84	72	75
	% Daily Demand Served	50	45	41	53	48
Ø6	Platoons Detected	424	263	194	358	365
	% Platoons Progressed	47	59	63	48	49
	% Daily Demand Served	36	32	30	37	35

progression (data in the last line in the table) helped approximately one third of total daily demand for this phase. These numbers are not as large as those for Ø2. Nonetheless, they are significant.

5 CONCLUSIONS AND RECOMMENDATIONS

This paper described the enhanced platoon identification and accommodation system recently developed and deployed by our research team at two isolated intersections in Texas. Preliminary analysis of data from one of these sites shows that the enhanced PIA system is working as intended. A subset of analysis results presented in this paper show that the system detects and progresses a significant number of platoons in both priority directions. Because the PIA system also provides dilemma zone protection for vehicles at the tail end of each progressed platoon, it also improves intersection safety. Such metrics are difficult to measure without long term monitoring. From the analysis of phase utilization data, it appears that the priority treatment of main street through phases did not produce any significant adverse effect for traffic on minor phases. However, traffic studies must be conducted to fully quantify such effects.

So far, the PIA system has been tested at a limited number of real intersections. System installation and testing at additional sites with different geometric and traffic characteristics will enable better understanding of its benefits and limitations. Two additional installations are planned under a follow-up implementation project sponsored by Texas Department of Transportation (TxDOT).

The PIA system collects signal and detector data as events occur. It also processes this information to produce useful real time information. However, not all of this information is currently used by the system. Such information can be used to increase system adaptability to changing traffic conditions. For instance, vehicle classification and changing demand at intersection approaches can be used to make adjustments to platoon detection parameters in real time. It is recommended that the feasibility of implementing such enhancements be further investigated.

ACKNOWLEDGMENTS

The research and development effort described in this paper was funded by TxDOT under Project 0-5507. We appreciate TxDOT staff's assistance in system installation at test sites. We would also like to thank the anonymous referees for reviewing this paper and providing constructive comments.

REFERENCES

Chaudhary, N.A. Abbas, M.M. Charara, H.A. & Parker R. 2003. Platoon Identification and Accommodation System for Isolated Signals on Arterials. Research Report FHWA/TX-05/0-4304-2, Texas Transportation Institute, Texas A&M University System, College Station, Texas.

Chaudhary, N.A., Abbas, M.M. & Charara, H.A. 2006. Development and Field Testing of a Platoon Identification and Accommodation System. Transportation Research Record, 1978, Transportation Research Board, Washington, D.C., pp. 141–148.
Chaudhary, N.A. Charara, H.A., Sunkari, S.R., & Longmire, R.R. 2008. Enhancements to PIA System for Real-Time Control at Isolated Traffic Signals. Report FHWA/TX-08/0-5507-1, Texas Transportation Institute, Texas A&M University System, College Station, Texas.
Charara, H.A., Chaudhary N.A., Sunkari, S.R., & Longmire, R.R. 2008. PIA System Installation and User Guide. Report FHWA/TX-08/0-5507-P2, Texas Transportation Institute, Texas A&M University System, College Station, Texas.

Efficient Transportation and Pavement Systems – Al-Qadi, Sayed, Alnuaimi & Masad (eds)
© 2009 Taylor & Francis Group, London, ISBN 978-0-415-48979-9

Traffic adaptive signal control at roundabouts

N.A. Chaudhary & P. Songchitruksa
Texas Transportation Institute, TX, USA

ABSTRACT: Modern roundabouts are designed to provide safe and efficient movement of competing traffic through roadway junctions. Under light-to-moderate demand and balanced traffic conditions, these roundabouts operate as designed. However, certain combinations of geometric and traffic conditions may compromise both safety and efficiency of a roundabout, requiring signalization to regulate traffic flow. Because signalization increases delay and stops at roundabout approaches, signal control should be activated only when needed and deactivated otherwise. This paper describes the conceptual framework of an adaptive system, which switches between signalized and unsignalized control based on prevalent traffic conditions. The paper also provides the results of computer-based traffic simulation used to verify this idea.

1 INTRODUCTION

Although modern roundabouts have been in use in some countries for decades, they are gaining increased popularity in many other countries. There are two main reasons for this increasing popularity: safety and operational efficiency. A modern roundabout is much safer than a typical intersection because it reduces frequency and severity of accidents. Reduced conflict points, slow approach speeds, and transformation of crossing conflicts into merging conflicts are the main factors contributing to increased safety. Improved operational efficiency of roundabouts stems from the fact that the only control regulation is for the entering vehicles to yield to circulating traffic. Thus, under a majority of traffic conditions, entering vehicles do not even need to stop, reducing delays and stops, which also reduce fuel consumption and vehicular emissions. However, as Krogscheepers & Roebuck (2000) pointed out, operational efficiency of a roundabout is compromised under certain origin-destination and approach volume combinations. Previous studies by researchers such as Stevens (2005) and Akcelik (2005) and have shown that these inefficiencies may be reduced by imposing active traffic control. Examples of such control include metering and signalization. However, active control strategies should be used only when necessary because they may be counterproductive under many traffic conditions.

Time of day control is usually sufficient for locations where traffic conditions are predictable on a day to day basis. For such locations, periods for active control can be determined by conducting field studies. However, fixed-time control may not provide overall benefits at locations with unpredictable traffic conditions. Traffic adaptive control is more suitable for these types of locations.

This paper describes our initial efforts to develop traffic adaptive control strategies for use at roundabouts experiencing unpredictable traffic conditions. We used VISSIM (2007)—a microscopic traffic simulation program—to study the benefits and disadvantages of a simple queue-based strategy for providing adaptive control at a two-lane roundabout with three different traffic patterns. One advantage of VISSIM is its vehicle actuated programming (VAP) feature, which allows the development and testing of non-standard control logic, which may be simple or extremely complex.

2 STUDY DESIGN

In this preliminary study, we simulated traffic flow at a four-approach two-lane roundabout shown in Figure 1. In the figure, approaches are labeled from 1 to 4 for later reference. Each approach has two lanes entering the roundabout. To keep this study simple, we created free right turn lanes on all approaches to allow all right turn vehicles to move unimpeded. As shown in this figure, each simulated approach had two types of detectors, created to implement the desired traffic control. As described later, queue detectors were used to trigger traffic signal control under the adaptive strategy. Demand detectors, located at the stopbar, were used to provide actuated signal control.

To accomplish the objectives of this study, we investigated three scenarios: (1) heavy demand on adjacent approaches 2 and 3, (2) heavy demand on opposing approaches 2 and 4, and (3) heavy demand at all four approaches. The simulation for each case had three distinct periods, which included a warm-up period with light traffic demand, a period with heavy demand, and a flush period with low demand. The origin-destination (OD) percentages were kept unchanged during these periods. Tables 1, 2, and 3 provide demand and OD data used in the three simulation scenarios. In each table, the third row provides percentages of input (demand) for each approach distributed to left (L), through (T), right (R), and U-turn (U) movements.

We simulated three control scenarios with five replications. These control scenarios, implemented through a VAP, were (1) no control, (2) an actuated signal operation during the entire simulation, and (3) adaptive control. Each actuated phase used minimum (Min) and maximum (max) green times of 3 and 90 seconds, respectively, with a three-second yellow interval. The programmed control sequence provided green phase to the four approaches in a clockwise order. Phases 1 through 4 served approaches 1 through 4, respectively. Each phase used a gap time of 2 seconds. Thus, an active phase terminated if the associated demand detector was unoccupied for 2 seconds.

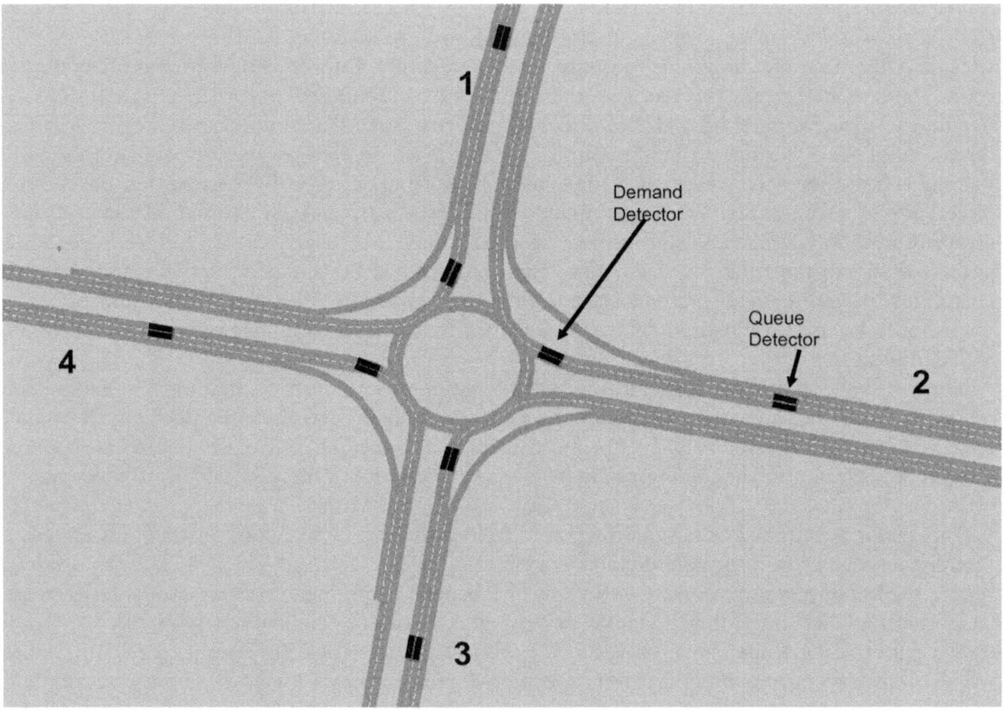

Figure 1. Geometry of simulated roundabout.

Table 1. Heavy demand on northbound and westbound (adjacent) approaches.

Simulation Time (sec)	1: SB input (vph)				2: WB input (vph)				3: NB input (vph)				4: EB input (vph)				Total Input (vph)
	L	T	R	U	L	T	R	U	L	T	R	U	L	T	R	U	
	25	25	45	5	25	45	25	5	25	25	45	5	30	30	35	5	
0–1500	800				1200				1200				800				4000
1500–3300	800				1800				1800				800				5200
3300–4200	400				400				400				400				1600

Table 2. Heavy demand on eastbound and westbound (opposite) approaches.

Simulation Time (sec)	1: SB input (vph)				2: WB input (vph)				3: NB input (vph)				4: EB input (vph)				Total Input (vph)
	L	T	R	U	L	T	R	U	L	T	R	U	L	T	R	U	
	35	35	25	5	45	25	25	5	35	35	25	5	45	25	25	5	
0–1500	800				1200				800				1200				4000
1500–3300	800				1800				800				1800				5200
3300–4200	400				400				400				400				1600

Table 3. Heavy demand on all approaches.

Simulation Time (sec)	1: SB input (vph)				2: WB input (vph)				3: NB input (vph)				4: EB input (vph)				Total Input (vph)
	L	T	R	U	L	T	R	U	L	T	R	U	L	T	R	U	
	35	35	25	5	35	35	25	5	35	35	25	5	35	35	25	5	
0–1500	1000				1000				1000				1000				4000
1500–3300	1300				1300				1300				1300				5200
3300–4200	400				400				400				400				1600

The adaptive strategy used a combination of no control and actuated control depending on occupancy of queue detectors. Under this strategy, actuated control began if any one of the four queue detectors was continuously occupied for eight seconds. The signal started its cycling sequence (1→2→3→4→1) from the approach whose queue detector triggered signal control. Adaptive control switched to no control when all four queue detectors were continuously unoccupied for four seconds. This exit strategy ensured that any queue forming because of signalization cleared before switching back to no control. For comparison purposes, we conducted five replications for each case and averaged performance measures from five replications. The next section provides simulation results.

3 STUDY RESULTS

Figures 2 through 5 show queue profiles for all four approaches under demand Scenario 1. The maximum points in these figures are identified with circles. The three plots in each figure are for the three control strategies simulated. Furthermore, each plot represents the average of five simulation replications. Under the no control scenario (dashed lines in these figures), approach 2 had the largest queue length, extending beyond 3200 feet (approximately 0.6 miles) upstream of the roundabout. This happened because heavy demand from approach 3, getting the opportunity to enter the roundabout first, impeded vehicles on approach 2 from entering the roundabout. Even though approach 3 had higher demand than approach 1, the queues on these two approaches were the same, probably because of a similar, but significantly less effect due to reduced circulating traffic at approach 1 merge point. Approach 4 has the smallest queues, which are almost negligible. If it were not for the heavy circulating traffic at the merge point, the expected queue at approach 1 would have been similar to that of approach 4. In these figures, thin and bold solid lines show queue profiles for actuated signal control (Strategy 2) and adaptive control (Strategy 3), respectively. A review of these plots shows that signal control resulted in a significant reduction in the length of queue on approach 2. In the case of adaptive control the queue length was cut to almost a third. However, as expected, both these strategies produced queues on all other approaches. Thus, signal control was able to produce reasonable equity between approaches. This is the primary reason for signalization at roadway intersections. Furthermore, a comparison of Strategies 2 and 3 shows that adaptive control produced better results (lower queues) than full time signal control. Figure 6 provides delay profile for Approach 2 and Figure 7 provides average roundabout delay profile, with maximum values identified with circles. In Figure 7, pairs of vertical lines identify the earliest and the latest times of adaptive mode activation and deactivation for the five simulation replications conducted for this scenario.

In Figure 7, the vertical bars show 90% confidence intervals of average delays for adaptive control. Table 4 provides a summary of key results for Scenario 1. In essence, this table shows the peak values in the above plots (and in delay plots not included here) highlighted using circles. The reader should note that the second, fourth, and sixth numbers from top in column 4 (1:WB) correspond to circled values in Figure 7. Same numbered entries in the last column correspond to circled entries in Figure 7. As can be seen from the fourth column, adaptive control was better than full time signal control and significantly reduced the worst-case conditions (Maximum queue, and maximum 5-minute delay) for the westbound approach. In this case, the maximum delay reduced from 628 seconds (over 10 minutes) to 214 seconds (about 3.5 minutes). Drivers' tolerance to delay has a maximum limit. Any delay higher than their tolerance may force them exhibit road rage. Although such driver behavior is not possible to measure in a simulated environment, this type of dramatic reduction may be used as a surrogate measure. Queue and delay data for other approaches show that this reduction was achieved at additional expanse (that is increased queues and delays). This is an expected result. However, as indicated by the last column, the overall roundabout operation improved as well. Tables 5 and 6 show similar results for Scenarios 2 and 3. A perusal of these results will show that the overall results for these two scenarios are not as good as those for the first scenario.

Statistical t-tests were conducted to determine if the differences of average delays between adaptive signal strategy and other strategies are statistically significant. Tables 7, 8, and 9 provide these additional results. The numbers in parentheses are the standard deviations of delay. Positive t-statistics denote the lower average delay of the adaptive strategy versus the strategy in comparison and vice versa for the negative t-statistics. The p-values indicate the statistical significance of the test. The p-values of 0.05 or less indicate that the differences are statistically significant at 95% confidence level. For example, in Scenario 1, the p-value of 0.0006 indicates that the "adaptive signal" strategy yields lower average delay than the "no signal strategy" at 95% confidence level.

The results in Tables 7 to 9 show that the adaptive signal strategy yielded lower average delays for all demand scenarios at 95% confidence level except for the following cases:

- In scenario 2, the "adaptive signal" strategy was found to give slightly higher average delay than the "no signal" strategy but the difference is not statistically significant.
- In scenario 3, the "adaptive signal" strategy was found to increase the average delay versus the "no signal" strategy. This increase is marginally significant at 95% confidence level.

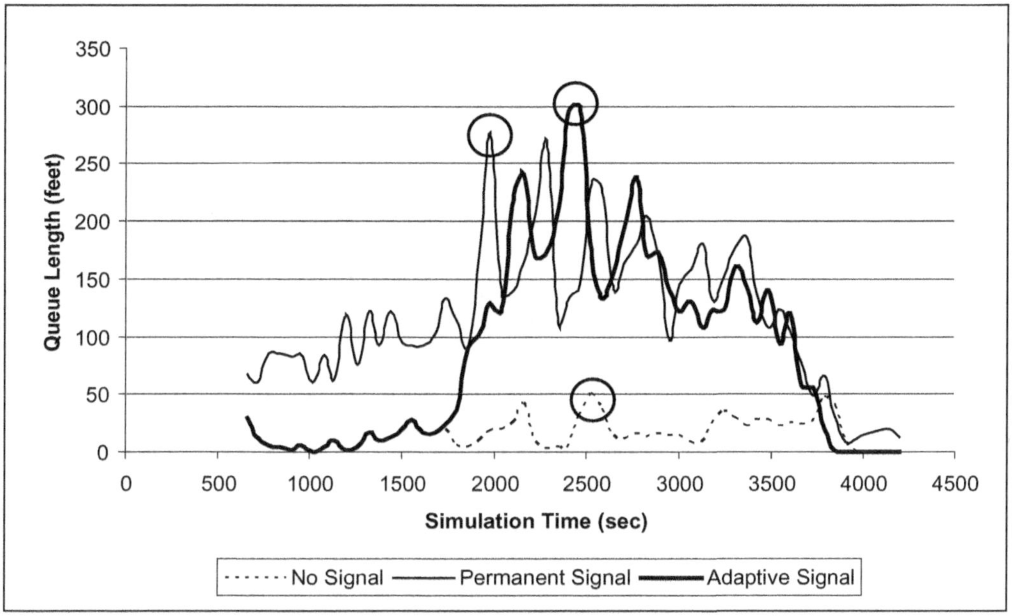

Figure 2. Approach 1 (southbound) average queue length for scenario 1.

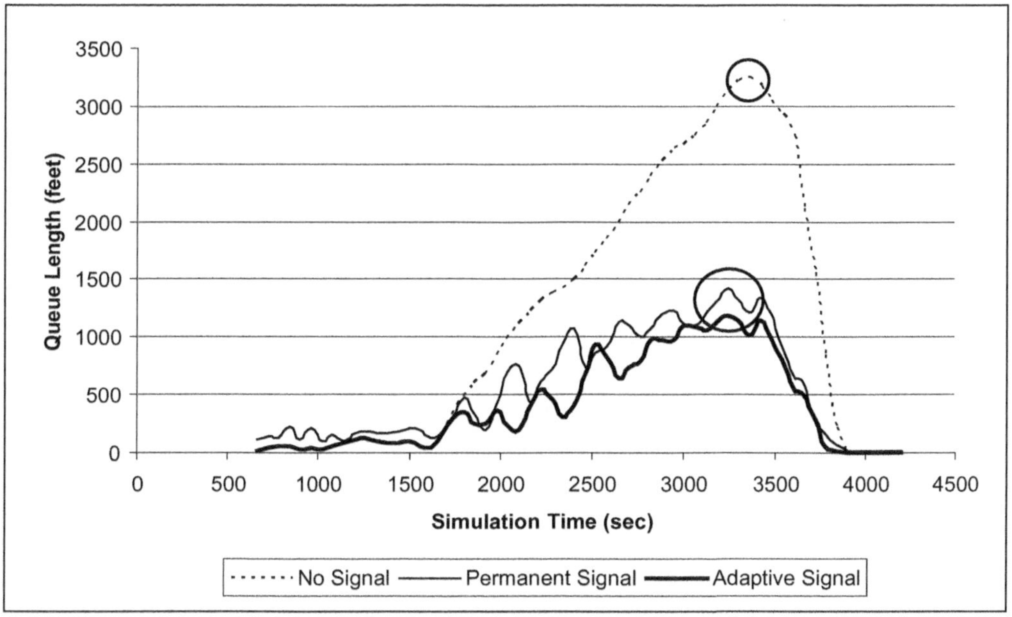

Figure 3. Approach 2 (westbound) average queue length for scenario 1.

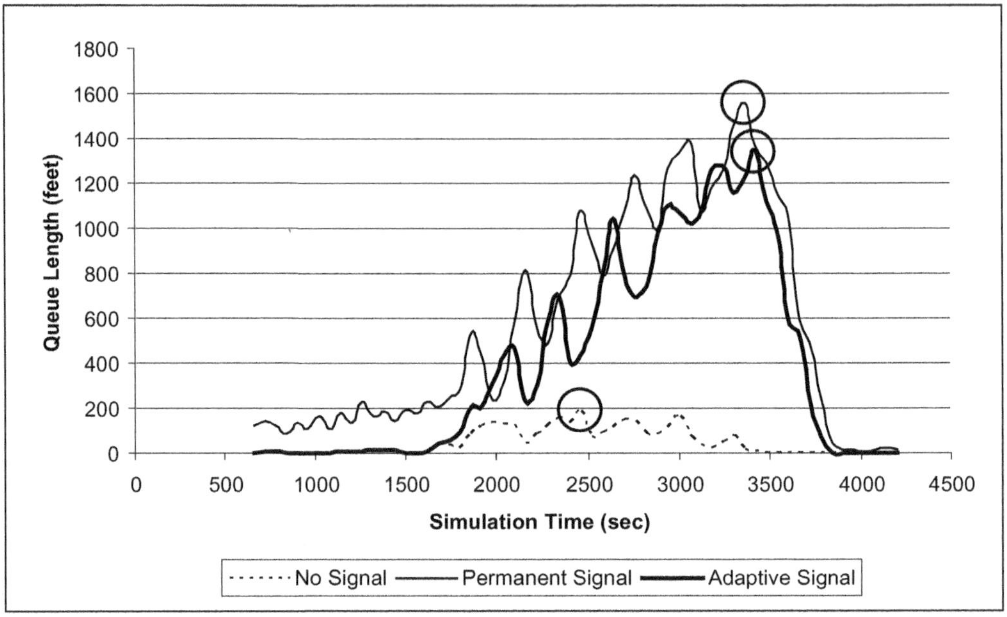

Figure 4. Approach 3 (northbound) average queue length for scenario 1.

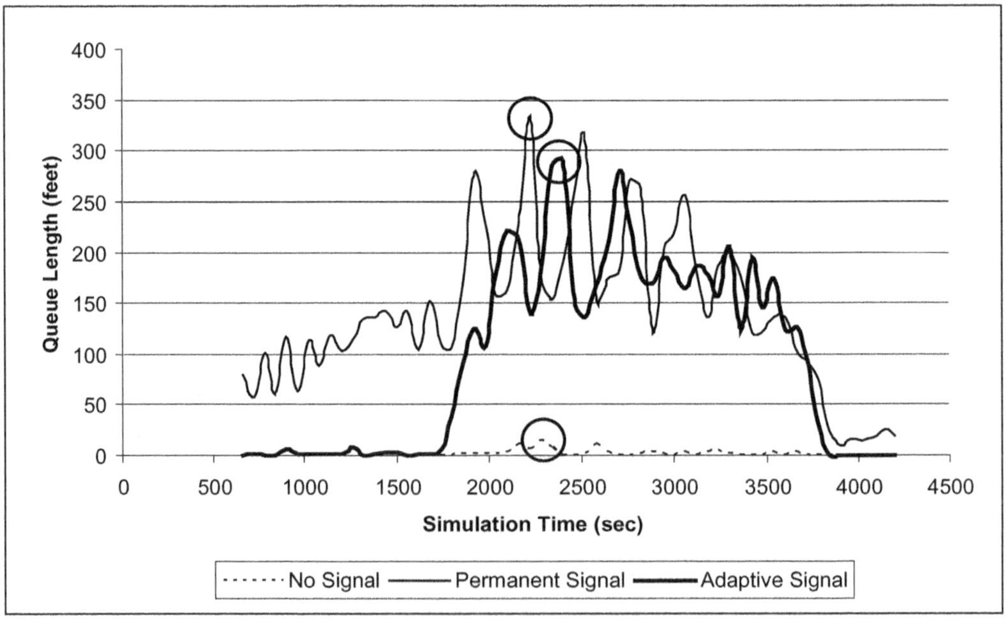

Figure 5. Approach 4 (eastbound) average queue lengths for scenario 1.

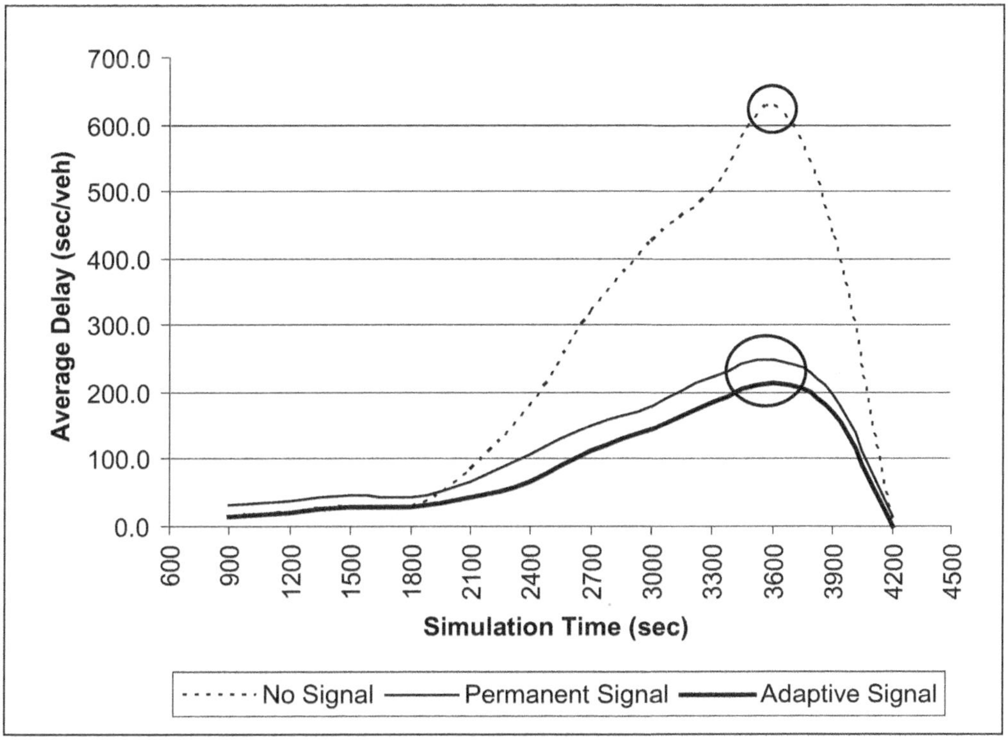

Figure 6. Average delay on approach 3.

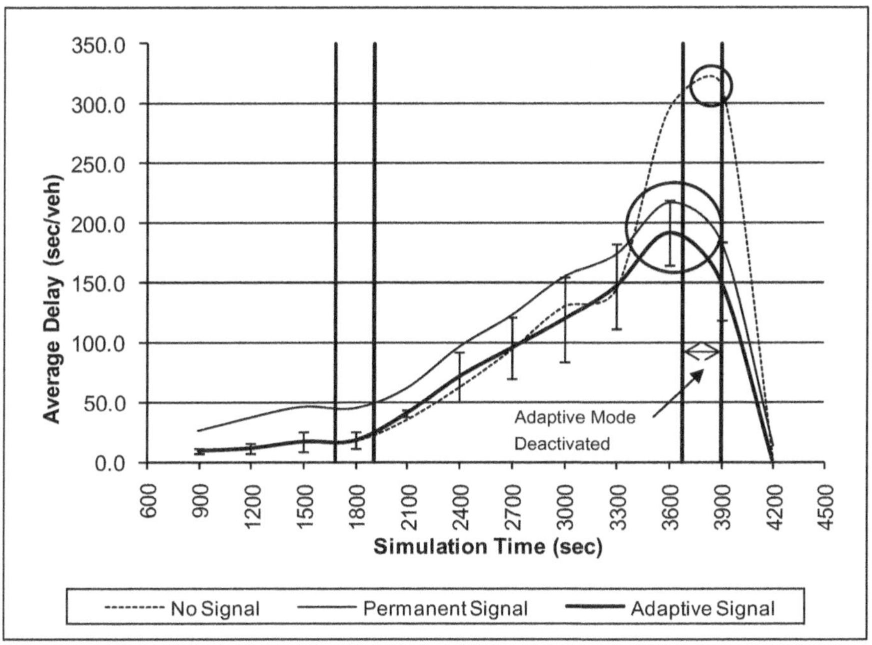

Figure 7. Average roundabout delay.

Table 4. Queue and delay summary for scenario 1.

Strategy	Simulation Outputs	1:SB	2:WB	3:NB	4:EB	Roundabout
No Signal	Max Average Queue Length (ft)	50	3249	186	14	3249
	Max 5-Minute Average Delay (sec/veh)	40	628	28	12	314
Permanent Signal	Max Average Queue Length (ft)	277	1424	1554	333	1554
	Max 5-Minute Average Delay (sec/veh)	78	249	289	90	216
Adaptive Signal	Max Average Queue Length (ft)	299	1182	1353	289	1353
	Max 5-Minute Average Delay (sec/veh)	84	214	257	90	191

Table 5. Queue and delay summary for scenario 2.

Strategy	Simulation Outputs	1:SB	2:WB	3:NB	4:EB	Roundabout
No Signal	Max Average Queue Length (ft)	71	1533	47	3112	3112
	Max 5-Minute Average Delay (sec/veh)	34	270	29	623	408
Permanent Signal	Max Average Queue Length (ft)	330	2311	332	2355	2355
	Max 5-Minute Average Delay (sec/veh)	108	433	110	509	390
Adaptive Signal	Max Average Queue Length (ft)	332	1969	336	2184	2184
	Max 5-Minute Average Delay (sec/veh)	111	429	110	474	376

Table 6. Queue and delay summary for scenario 3.

Strategy	Simulation Outputs	1:SB	2:WB	3:NB	4:EB	Roundabout
No Signal	Max Average Queue Length (ft)	1039	733	988	1160	1160
	Max 5-Minute Average Delay (sec/veh)	288	167	254	326	270
Permanent Signal	Max Average Queue Length (ft)	1264	1171	772	1234	1264
	Max 5-Minute Average Delay (sec/veh)	377	285	175	331	310
Adaptive Signal	Max Average Queue Length (ft)	1128	1088	772	1135	1135
	Max 5-Minute Average Delay (sec/veh)	316	305	177	304	283

Table 7. Average delay and statistical test results for scenario 1.

	Average Delay (sec/veh)*					t-test versus adaptive signal strategy**	
Strategy	1: SB	2: WB	3: NB	4: EB	Roundabout	t-statistics	p-value
No Signal	15.2 (3.1)	277.6 (9.6)	18.9 (4.7)	7.6 (0.7)	100.4 (2.8)	5.41	0.0006
Permanent Signal	52.3 (3.9)	125.3 (16.8)	132.9 (17.3)	62.0 (2.6)	104.0 (9.7)	4.48	0.0021
Adaptive Signal	41.4 (3.1)	94.3 (8.2)	97.0 (28.3)	44.4 (2.0)	77.1 (9.2)	N/A	N/A

* Standard deviations of the corresponding average delays are shown in the parentheses.

** t-tests were conducted to determine if the differences between adaptive signal strategy and other strategies are statistically significant.

Table 8. Average delay and statistical test results for scenario 2.

Strategy	Average Delay (sec/veh)*					t-test versus adaptive signal strategy**	
	1: SB	2: WB	3: NB	4: EB	Roundabout	t-statistics	p-value
No Signal	19.2 (1.5)	136.4 (39.1)	19.7 (2.1)	297.2 (24.1)	148.3 (13.0)	–0.47	0.6500
Permanent Signal	84.5 (6.4)	225.6 (15.0)	81.0 (2.9)	241.5 (6.6)	180.6 (5.7)	7.20	0.0001
Adaptive Signal	66.3 (4.3)	185.5 (26.2)	67.5 (4.7)	207.9 (23.2)	151.4 (7.1)	N/A	N/A

* Standard deviations of the corresponding average delays are shown in the parentheses.
** t-tests were conducted to determine if the differences between adaptive signal strategy and other strategies are statistically significant.

Table 9. Average delay and statistical test results for scenario 3.

Strategy	Average Delay (sec/veh)*					t-test versus adaptive signal strategy**	
	1: SB	2: WB	3: NB	4: EB	Roundabout	t-statistics	p-value
No Signal	106.9 (41.1)	64.1 (21.4)	93.5 (27.1)	127.8 (21.3)	98.1 (17.6)	–2.14	0.0503
Permanent Signal	182.4 (30.0)	140.0 (20.9)	103.3 (11.6)	157.7 (35.4)	145.8 (11.1)	6.26	0.0000
Adaptive Signal	127.0 (22.1)	112.9 (31.3)	86.2 (7.9)	127.0 (25.6)	113.3 (9.7)	N/A	N/A

* Standard deviations of the corresponding average delays are shown in the parentheses
** t-tests were conducted to determine if the differences between adaptive signal strategy and other strategies are statistically significant.

4 SUMMARY, CONCLUSIONS AND RECOMMENDATIONS

This paper described the results of computer simulation comparing the performance of no control, full time actuated traffic control, and adaptive signal control on the performance of a two-lane four approach modern roundabout. This preliminary analysis used three demand scenarios: heavy demand at adjacent approaches, heavy demand at opposing approaches, and heavy demand at all approaches. Combined with fixed OD percentages, these demand scenarios account for a small percentage of demand patterns present at many urban roundabouts during the course of a normal weekday. Furthermore, the simulation testbed was not calibrated to produce realistic merging and branching behaviors. Lastly, a simple signal phasing sequence was used in the signalized control strategies.

Simulation results showed that the adaptive control strategy, based entirely on the presence of a long queue on any one of the roundabout approaches, produced significant improvement in the overall roundabout operation. It reduced the worst case maximum queue, which reached a length of almost 0.6 miles, by a third. In addition, it reduced the worst case delay from 10 minutes to about 3.5 minutes. Such dramatic improvements may have significant safety benefits stemming from reduced driver frustration. Such benefits are not possible to measure in a simulated environment. In the second scenario, the degradation of the adaptive strategy may be due to the fact that the simple signal control sequence, similar to split phasing at regular signalized intersections, caused wastage of roundabout capacity. Furthermore, realistic calibration of merge behavior may prove that the actual benefits of an adaptive strategy are even higher than those demonstrated in this paper.

The preliminary results presented here are significant in the sense that they have encouraged us to expand this work. Such future work is anticipated to include more sophisticated signal control strategies and adaptive algorithms. One example of a different signal control strategy that may have worked better for Scenario 2 (heavy opposing traffic) uses a single phase for both instead of separation, which is similar to inefficient split phasing used at standard intersections for safety reasons. Other approaches may include signalization of some approaches, but not others. Finally, other common demand and OD patterns also warrant further investigation.

ACKNOWLEDGMENTS

We would like to thank the anonymous referees for their constructive criticism and suggestions.

REFERENCES

Krogscheepers, J.C. & Roebuck, C.S. 2000. Unbalanced Traffic Volumes at Roundabouts. In Werner Brilon (ed.), *Proc. Fourth International Symposium on Highway Capacity*. Transportation Research Board, National Research Council, Washington, D.C.: 446–458.

Stevens, C.R. (2005) Signals and Meters at Roundabouts. In Proceedings of the 2005 Mid-Continent Transportation Research Symposium, Iowa State University. http://www.ctre.iastate.edu/pubs/midcon2005/index.htm, accessed 2/15/08.

Akcelik R. (2005). Capacity and Performance Analysis of Roundabout Metering Signals. *Transportation Research Circular Number E-C083*. Transportation Research Board, Washington, D.C.

VISSIM 4.30, User Manual. (2007). PTV, Karlsruhe, Germany.

Efficient Transportation and Pavement Systems – Al-Qadi, Sayed, Alnuaimi & Masad (eds)
© 2009 Taylor & Francis Group, London, ISBN 978-0-415-48979-9

Systematic analysis of capacity of weaving sections

Yihua Zhang & Hesham Rakha
Virginia Tech, Blacksburg, VA, USA

ABSTRACT: The paper first validates the INTEGRATION model for estimating the capacity of weaving sections. Specifically, comparisons are made to field data and the Highway Capacity Manual (HCM) procedures. Subsequently, the paper presents a systematic analysis of the factors that potentially impact the capacity of freeway weaving sections, which includes the length of the weaving section, the weaving ratio, the percentage of heavy vehicles, and the speed differential between freeway and ramp traffic. The study reveals some questionable capacity estimates by the CORSIM software and a gap acceptance procedure proposed in the literature. Specifically, for some sites, the results obtained from these two methods demonstrate counter-intuitive trends. The study also proves that the weaving ratio, which is the ratio of the lowest weaving volume to the total weaving volume, has a significant impact on the capacity of weaving sections, of which can be as high as 326 veh/h/lane. In addition, the study demonstrates that the length of weaving section has a larger impact on the capacity of weaving sections for short lengths and high traffic demands. Furthermore, the study shows that not enough evidence exists to conclude that the speed differential between freeway and ramp traffic has a significant impact on weaving section capacities. Finally, the study demonstrates that the HCM procedures for accounting for heavy duty vehicle impacts weaving section capacities, which is achieved by using a heavy vehicle factor, appear to be reasonable.

1 INTRODUCTION

The freeway weaving analysis procedures in the 2000 Highway Capacity Manual (HCM) are based on research conducted in the early 1970s through the early 1980s *(1)*. Subsequent studies have shown that the methods' ability to predict the operation of a weaving section is limited *(2, 3)*, which is most probably because of the outdated database. As to capacity estimation of freeway weaving sections, some other methods, such as the gap-acceptance based methods and simulation based methods, have been used as alternatives *(2, 3, 4, 5)*.

The research effort that is presented in this paper first validates the INTEGRATION software for estimating the capacity of freeway weaving sections utilizing three testbeds from the literature *(3)*. Subsequently, the paper utilizes the INTEGRATION software to conduct a systematic analysis of critical variables that impact the capacity of weaving sections. The simulation results are compared to the HCM procedures in an attempt to validate and identify the limitations of the current HCM procedures.

2 STATE-OF-THE-ART WEAVING ANALYSIS PROCEDURES

A limited number of publications were found in the literature that was deemed related to this study. For example, Zarean and Nemeth *(6)* utilized the WEAVSIM microscopic simulation model, to investigate the effect of the different arrival speeds on the operation of weaving sections. Subsequently, the researchers developed a regression model for the modeling of weaving sections based on the simulation results. The simulation results demonstrated that the speed differential between the mainline and on-ramp arrivals had a significant effect on the operation of weaving sections, which was not considered in the 1985 HCM procedures *(7)* and is not considered in the current HCM procedures *(8)*.

Skabardonis *et al.* *(9)* applied the INTRAS microscopic simulation model to evaluate the operation of a few major freeway weaving sections. INTRAS was modified to predict the speeds of weaving and non-weaving vehicles and was applied to eight major freeway weaving sections. Vehicle speeds within the weaving sections were compared to a few analytical procedures that included the 1985 HCM procedure, Leisch's procedure, JHK's procedure, Fazio's Procedure, and the Polytechnic Institute of New York (PINY) procedure. The researchers concluded that the INTRAS speed predictions were closer to the field measurements than the analytical procedure speed predictions. Consequently, the researchers concluded that simulation tools could be utilized with field data to enhance existing state-of-the-art analytical procedures for the modeling of weaving section operations.

Stewart *et al.* *(4)* evaluated the capability of INTEGRATION version 1.50 for the modeling of weaving sections. The study showed that both the 1985 HCM procedure and INTEGRATION offered identical conclusions for a given sample problem. However, the study demonstrated differences between the two approaches on critical design parameters of weaving sections. Specifically INTEGRATION identified the number of lanes in the core area as a critical factor affecting weaving section capacity, which was not captured in the HCM procedures. Alternatively, while the HCM procedures demonstrated that the length of the core area was critical in the design of weaving sections, the INTEGRATION results demonstrated that this factor was critical only for short lengths and was less critical as the length of the weaving section increased.

Vermijs *(10)* reported on the efforts in developing the Dutch capacity standards for freeway weaving sections using FOSIM (Freeway Operations SIMulation), a microscopic simulation software developed in the Netherlands. Specifically, a total of 315 Type A weaving sections with different configurations and traffic factors were simulated. All simulation runs were repeated 100 times using different random seeds. The simulation results demonstrated that the weaving section capacity appeared to be normally distributed with a standard deviation in the range of 200 ~ 400 veh/h/lane.

Finally, Lertworawanich and Elefteriadou *(2, 11)* proposed an analytical capacity estimation method for weaving sections based on gap acceptance and linear optimization techniques. The gap acceptance model, however, makes a number of simplifying assumptions and does not include some critical variables thus limiting the potential applicability of the procedures. For example, the procedures do not capture the effect of the weaving section length on the capacity of weaving sections.

3 TEST SITES AND FIELD DATA DESCRIPTION

This section briefly describes the test sites and field data that were utilized for the validation effort. These test sites are described in further detail in the literature *(3)*. The test sites include three weaving sections along the Queen Elizabeth Expressway (QEW) in Toronto, Canada. These weaving sections include a Type B and two Type C weaving sections, denoted B1, C1, and C2, respectively. The posted speed limit on the QEW was 100 km/h at the time of the study with a 10 percent heavy vehicle population. The three selected sites operated under congested conditions because of the intense lane changing behavior within the weaving sections.

The section capacities and total traffic demand classified by on-ramp, off-ramp, upstream mainline, and downstream mainline flows were recorded in the data set. The weaving section capacity was computed as the maximum observed pre-breakdown 15 min flow rate. The data set included 10 days of data for Site B1, 10 days of data for Site C1, and 16 days of data for Site C2. Since the Origin-Destination (O-D) demand varied from one day to another, the capacity of each site also varied accordingly.

Because of a lack of Type A weaving sections within the original data set, Site A1 was added to the dataset. Though no field data were available for this site, it was deemed helpful to include

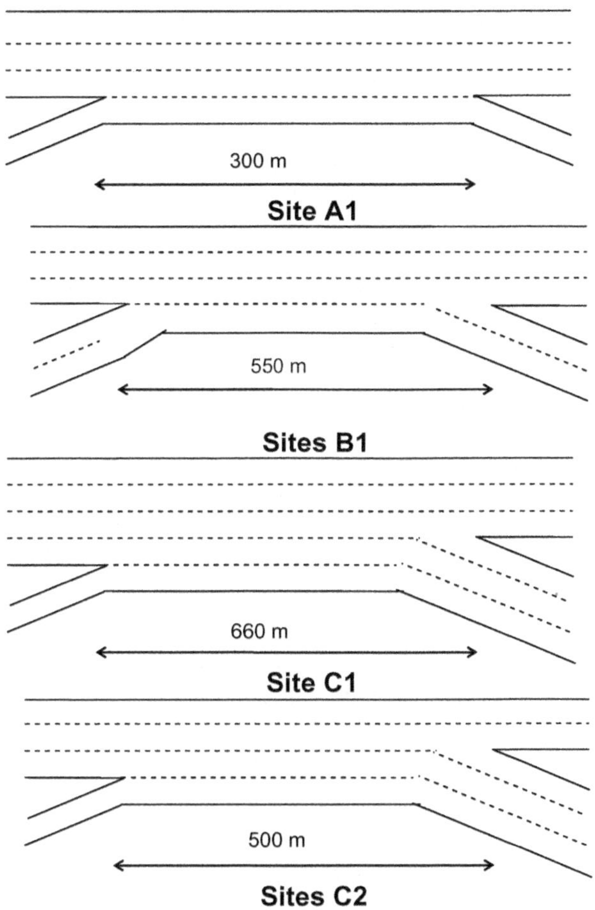

Figure 1. Configurations of test weaving sections.

a Type A site to the sensitivity analysis. The geometric layout and details of these four sites are shown in Figure 1.

In order to capture the pre-segregation that occurs upstream of a weaving section, the upstream freeway and on-ramp links were set at sufficiently long lengths. INTEGRATION's internal car-following and lane changing logic then ensured that vehicles segregated themselves prior to entering the weaving area. The literature demonstrates that the INTEGRATION software is capable of achieving efficient pre-segregation behavior, which is demonstrated in the literature *(21)*.

4 EXPERIMENTAL DESIGN

As was mentioned earlier, the study utilizes the INTEGRATION software to conduct the analysis. The INTEGRATION software is a microscopic traffic simulation and assignment model that represents traffic dynamics in an integrated freeway and traffic signal network. The model has been successfully applied since the early 1990s in North America and Europe *(12, 13, 14, 15, 16, 17, 18, 19, 20)*. Earlier versions of the model (version 1.50) were tested and validated against weaving section field data. However, it was deemed essential to validate the 2.30 version of the model since significant changes have been made to the car-following and lane-changing logic. It should be noted that the

INTEGRATION 2.30 lane-changing logic was described and validated in an earlier paper *(21)*. This paper extends the previous research efforts by validating the estimates of roadway capacity that are derived from the INTEGRATION software as a result of lane-changing behavior within weaving sections. It should be noted that the user inputs an ideal roadway capacity, however, the internal friction and turbulence within the traffic stream that results from lane-changing behavior, produces reductions in the roadway capacity that varies dynamically as the intensity of lane changing behavior increases.

The first step in this study was to calibrate the model to the three test Sites B1, C1, and C2. The calibration of the INTEGRATION software involves the calibration of the traffic demand (O-D tables) and the calibration of the steady-state car-following behavior by estimating four parameters, namely the free-speed, the speed-at-capacity, the ideal capacity, and the jam density. Subsequently, the impact of the random seed on the capacity of weaving sections was investigated because Vermijs *(11)* demonstrated that the random seed resulted in significant differences in weaving section capacities in the range of 200 to 400 veh/h/lane.

The analysis considers a number of factors that are hypothesized to significantly impact the capacity and operations of freeway weaving sections. The geometric and traffic characteristic parameters that are considered in this study are summarized in Table 1. The HCM 2000 defines the weaving ratio as the ratio of the smaller of the weaving volumes to the total weaving volume. The weaving ratio may be viewed as a measure of the distribution of the weaving volume between the mainline and on-ramp flows. However, the HCM 2000 procedures ignore the effect of the weaving ratio on the weaving section capacity. Consequently, the study investigates the impact of this factor on the capacity of weaving sections by maintaining a constant volume ratio (weaving volume/total volume) while varying the weaving ratio, as demonstrated in Table 1.

Another factor that is hypothesized to impact the capacity of weaving sections is the length of the weaving section and has produced differing results across various studies. The HCM 2000 considers the maximum length of a weaving section to be 750 m for all configuration types and beyond these lengths, the HCM recommends the modeling of merge and diverge sections separately. In this study, the impact of weaving length on weaving section capacity for different volume ratios is studied. The study considers weaving section lengths that range from 150 to 750 m, as summarized in Table 1. The considered volume ratios include the full range that is presented in the HCM 2000 capacity tables, which differ according to the type of weaving sections. For example, for Type A weaving sections, the volume ratio ranges from 0.10 to 0.35, while for Type B weaving sections, the volume ratio ranges from 0.10 to 0.80.

Because of the lower geometric design standards for on- and off-ramps, vehicle speeds on these facilities are typically lower than their speeds on freeways. A number of studies have indicated that the lower speeds of vehicles on on- and off-ramps affect the operation of weaving sections significantly. Consequently, as part of this study the impact of the speed differential between freeway and ramp traffic on the capacity of weaving sections is analyzed in a systematic fashion. Specifically, the speed differential between freeway and on-ramp and the speed differential between freeway and off-ramp traffic are considered separately.

The percentage of trucks within a traffic stream is generally considered an important capacity-impacting element because trucks occupy more space than passenger cars and do not share the

Table 1. Geometrical and traffic factors.

Parameter	Values considered
Weaving section type	Type A, Type B and Type C
Weaving ratio	0.0, 0.1, 0.2, 0.3, 0.4, 0.5
Weaving section length	150, 300, 450, 600, and 750 m
Speed differential between freeway and on-ramp traffic	0, 5, 10, 15, 20, 25, and 30 km/h
Speed differential between freeway and off-ramp traffic	0, 5, 10, 15, 20, 25, and 30 km/h
Percentage heavy duty vehicles	0, 5, 10, 15, 20, and 25%

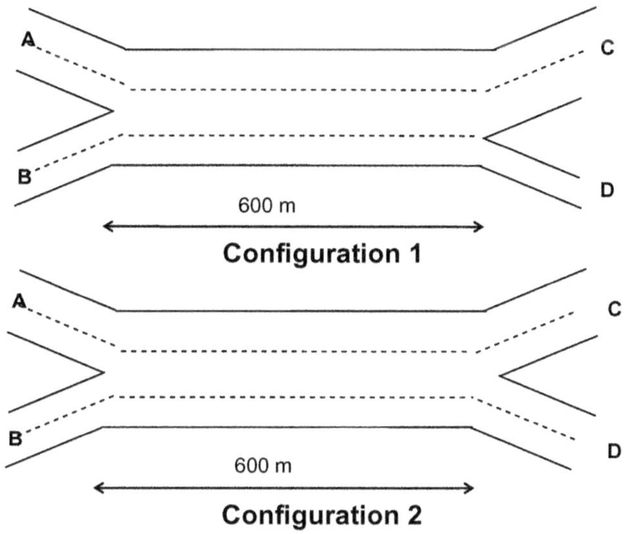

Figure 2. Configurations of alternative type B weaving sections.

same acceleration and deceleration capabilities as other vehicles. The HCM procedures attempt to capture the effect of trucks through the consideration of a heavy truck adjustment factor, as is currently done in many procedures within the HCM. Consequently, the study investigates the impact of different percentages of heavy vehicles on the capacity of weaving sections, as demonstrated in Table 1.

Finally, Lertworawanich and Elefteriadou *(3)* suggested that the two configurations of Figure 2 should be modeled differently because they involved different lane-changing behavior although they are both categorized as Type B weaving sections according to the HCM 2000. Specifically, the second configuration does not require any lane-changing for the weaving vehicles. Unfortunately, no justification for this suggestion was presented. Consequently, the study investigates differences in the capacities of both configurations to warrant differentiating both configurations.

5 SIMULATION RESULTS

This section presents the results of a sensitivity analysis that was conducted as part of this study. Initially, the weaving section capacity estimates derived from the INTEGRATION software are validated against field data capacity measurements. Subsequently, the results of the various hypotheses tests are presented.

Based on the conclusions of the sensitivity analysis, all results are averaged over 30 random repetitions. The use of 30 repetitions ensures that the sample standard error is less than 10 veh/h/lane (55/sqrt(30)) given that the standard deviation of data is 55 veh/h/lane (165 divided by 3 standard deviations).

5.1 *Model Validation*

In order to validate the appropriateness of the INTEGRATION software as a simulation tool for this study, a validation exercise was conducted. The geometric configurations for Sites B1, C1 and C2 were input into the INTEGRATION software. In addition, O-D tables were constructed from the observed volume counts on the mainline downstream and upstream the weaving section, the on-ramp, and off-ramp. The simulation runs were executed by increasing the traffic volumes from

70 to 130 percent of the field observed capacities. Detectors were located within the simulation model as was observed in the field. The maximum 15 min traffic flow rates were then utilized as an estimate of the weaving section capacity. The validation results for Sites B1, C1, and C2 are presented in Figure 3.

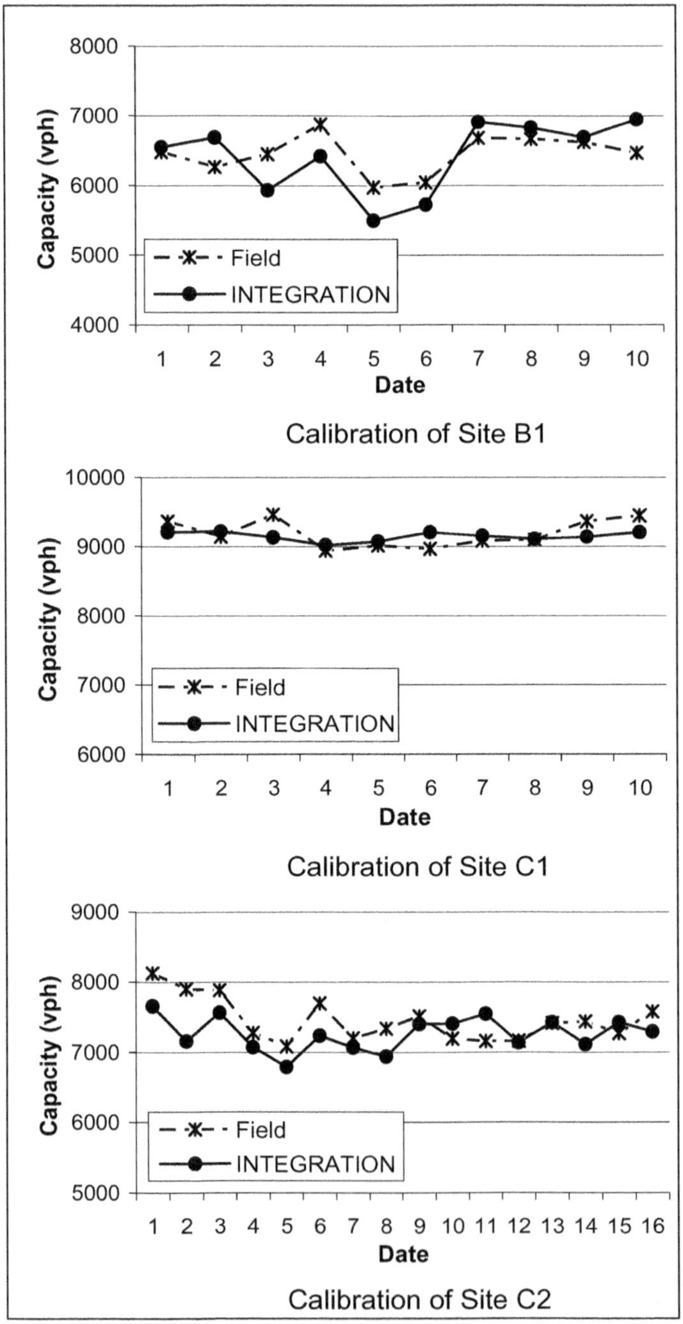

Figure 3. Validation results for sites B1, C1, and C2.

The results clearly demonstrate a close match between the simulation and field capacity estimates both in magnitude and temporal variation over the 10, 10, and 16 analysis days, respectively. Two error measures can be used to estimate the mean magnitude of simulation errors; mean relative error (MeRE) and maximum relative error (MaRE). The MaRE values for Site B1, C1, and C2 are 8.00, 3.51, and 9.42 percent, respectively, while the corresponding MeRE values are 4.98, 1.61, and 3.75 percent, respectively. In summary, the MeRE for all three sites is below 5 percent and MaRE for all the sites is below 10 percent, which demonstrates the validity of the model to estimate the capacity of weaving sections.

$$MeRE = \sum_{i=1}^{n} \left(|y_i - \hat{y}_i| / y_i \right) \Big/ n \tag{1}$$

$$MaRE = \max\left(|y_i - \hat{y}_i| / y_i \right) \tag{2}$$

Where

$\hat{y}_i$: Simulated capacity
y_i: Field observed capacity
n: Number of observation days for each site

5.2 *Sensitivity to Random Seed*

Microscopic simulation software model the behavior of individual vehicles in both space and time. Within the INTEGRATION software, the temporal generation of vehicles may be deterministic, fully stochastic (negative exponential inter-vehicle temporal headways), or partially stochastic (shifted negative exponential inter-vehicle temporal headways). In addition, the level of driver aggressiveness may be varied through a random process. Temporal inter-vehicle headways are generated using a sequence of random numbers. The sequence of random numbers may be varied by altering the random number seed.

The results indicate that the maximum variation in weaving section capacity estimates range in the order of 11 percent with a maximum difference of 660 veh/h, which is equivalent to a difference of 165 veh/h/lane. The results that are presented in this study demonstrate a lower level of variability in the weaving section capacity, which is less than what was observed in an earlier study (Vermijs, 1998). Specifically, the Vermijs study, which was based on 100 random simulations, concluded that the standard deviation of the weaving section capacity had a standard deviation of 200 to 400 veh/h/lane.

5.3 *Model Comparison*

Further validation of the model was conducted by performing a sensitivity analysis on Sites B1, C1, and C2 using a number of software and analytical formulations. The freeway and on-ramp volume ratios were systematically varied for each of the sites in an attempt to compare the models for a wide range of traffic characteristics. Specifically, the capacity estimates derived by the INTEGRATION and CORSIM software, the HCM 2000 procedures, and a gap acceptance procedure developed by Lertworawanich and Elefteriadou (2004) were compared. The results of the four methods for Sites B1, C1, and C2 are illustrated in Figure 4, respectively.

In the case of Site B1, Figure 4 demonstrates that the capacity of a weaving section tends towards the base lane capacity of 2300 veh/h/lane as the weaving volume tends to zero (freeway and on-ramp volume ratio of zero). The results of the HCM procedures and the INTEGRATION simulation results demonstrate that the weaving section capacity decreases consistently as the mainline volume ratio increases. Alternatively, the CORSIM and gap acceptance results

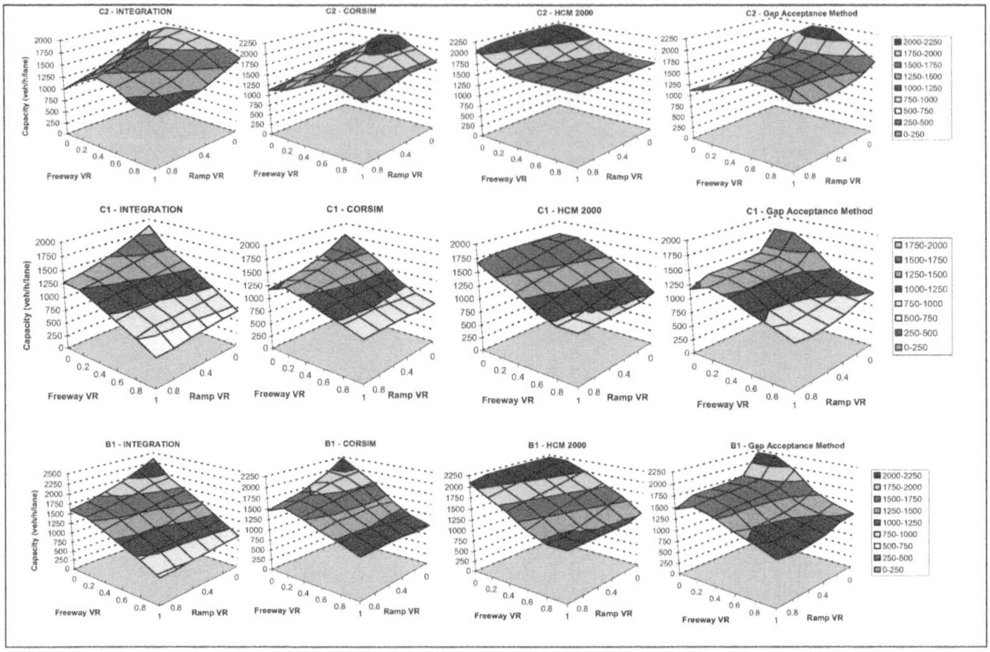

Figure 4. Capacity surfaces for sites B1, C1, and C2.

demonstrate a slight increase in the weaving section capacity as the mainline volume ratio increases from 0 to 20 percent. The increase generated by CORSIM is in the range of 656 veh/h, while in the case of the gap acceptance method is in the range of 124 veh/h. These counter intuitive results raise significant concerns about the adequacy of the CORSIM and gap acceptance procedures for the estimation of weaving section capacity because they indicate that the roadway capacity increases with an increase in the mainline weaving volume. As would be expected, the results of the INTEGRATION, HCM, and gap acceptance procedures demonstrate a decrease in the weaving section capacity as the on-ramp volume ratio increases (percentage of on-ramp weaving vehicles increases). Alternatively, the CORSIM results exhibit counter intuitive behavior with an increase in the weaving section capacity as the percentage of on-ramp weaving vehicles increases.

Figure 4 clearly demonstrates that for either Site C1 or Site C2, the results from the HCM 2000 procedures exhibit a different behavior in comparison to the other three methods. In general, the behavior exhibited by the INTEGRATION and CORSIM software appear to be consistent for Sites C1 and C2. It is interesting to note that the INTEGRATION and CORSIM models demonstrate an increase in the weaving section capacity with an increase in the mainline volume ratio (i.e. by introducing more FR vehicles in addition to the FF vehicles). The reason for this increase in the weaving capacity by introducing the FR O-D demand is caused by the fact that the introduction of the FR demand introduces an additional lane to the freeway vehicles given that the shoulder lane only provides access to the off-ramp. Consequently, the observed increase in weaving section capacity is expected. Unfortunately, the HCM procedures do not capture these intricate effects. It should be noted that Lertworawanich and Elefteriadou used paired-*t* tests to compare the shapes of the CORSIM, HCM, and gap acceptance procedures, and concluded that HCM 2000 procedures yields different results from the other two methods. Here the different behavior of the HCM weaving procedures and simulation methods are verified once again.

5.4 *Impact of Weaving Ratio*

As was mentioned earlier, the weaving ratio is defined as the ratio of the smaller of the weaving volumes to the total weaving volume. In this study, the weaving volume on the on-ramp was kept smaller than the weaving volume on the mainline while maintaining a constant total weaving volume. Consequently, the weaving section capacity estimated by the HCM 2000 procedures remained constant given that the weaving volume was held constant throughout the various scenarios, as illustrated in Figure 5.

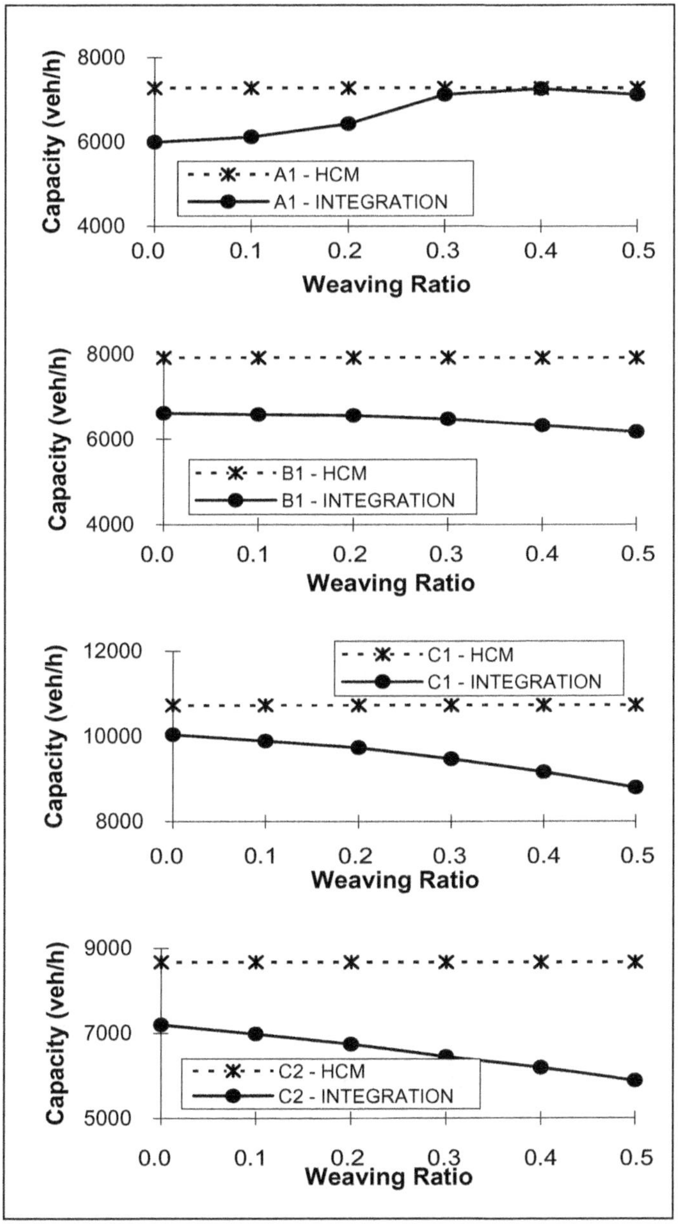

Figure 5. Impact of weaving ratio on capacity.

The simulation results clearly demonstrate that the weaving ratio does have an impact on the capacity of weaving sections even if the total weaving volume remains constant. For example, Site A1 demonstrates an increase in the weaving section capacity as the weaving ratio increases. In the case of Site A1 an increase in the weaving ratio results in a more balanced distribution of the weaving volume between the mainline and on-ramp demands. Since, in the case of Site A1, weaving vehicles are required to make a single lane change to reach their destination, a balanced weaving volume distribution results in a more efficient utilization of the gaps. Alternatively, in the case of Sites B1, C1, and C2, as the on-ramp weaving volume increases, the capacity at the core area decreases. This can be explained by the fact that for a constant weaving volume, more on-ramp weaving vehicles requires more lane change maneuvers within the weaving section and thus increases the turbulence within the weaving section. For the four sites investigated in this study, the variation in the weaving section capacity is in the range of 320 veh/h/lane.

5.5 *Impact of Weaving Section Length*

The effect of weaving section length on weaving section capacity is a controversial issue that has resulted in significant debate over the past years. The study investigates the impact of weaving section length on the capacity of weaving sections using the INTEGRATION software and the HCM procedures, as demonstrated in Figure 6. The results of the two approaches for the four sites demonstrate significantly differing trends. Specifically, the simulation results, unlike the HCM procedures, demonstrate that the impact of the weaving section length on the capacity of a weaving section increases as the traffic demand increases. Clearly, the simulation results appear to be more intuitive.

It is worthy to note that in Figure 6, the simulation results demonstrate that as the weaving volume increases, the weaving section capacity increases initially and then decreases. After a close look at the geometric layout of Sites C1 and C2 in Figure 1 the simulation results appear to be very reasonable. For example, at Site C1, a freeway volume ratio of zero requires that the freeway-to-freeway (FF) vehicles initially travel through a 3-lane segment followed by a 2-lane segment within the weaving section. Alternatively, the ramp-to-ramp (RR) vehicles initially travel on a single lane followed by a 3-lane segment within the weaving section. Consequently, the FF vehicles, unlike the RR vehicles, encounter a bottleneck within the weaving section. Alternatively, if the FF vehicles switch to FR vehicles, the FR demand is then able to utilize a number of off-ramp lanes that are not available for the FF demand. Noteworthy is the fact that the HCM 2000 procedures indicate that the weaving section capacity decreases as the volume ratio increases, which does not appear to be reasonable.

Based on the converging lines of Figure 6 we can conclude that, in general, as the weaving section length increases, its impact on the weaving section capacity decreases. For example, the decrease in the weaving section capacity resulting from an increase of 150 m for a 150 m weaving section is significantly different than its impact on a 600 m weaving section.

5.6 *Impact of Speed Differential between Mainline and Ramp Vehicles*

A number of studies have indicated that the lower speeds of vehicles on on- and off-ramps affect the operation of weaving sections significantly. Consequently, as part of this study the impact of the speed differential between freeway and ramp traffic on the capacity of weaving sections is analyzed in a systematic fashion. Specifically, Sites B1, C1, and C2 are analyzed for three weaving intensities, namely low, medium, and high. In the case of Site B1 the three volume ratios that are considered are 10, 40, and 80 percent while in the case of Sites C1 and C2 volume ratios of 5, 25, and 50 percent are considered. These values were selected based on the maximum recommended values for Type B and C weaving sections for the HCM 2000 procedures.

Statistical analysis of the results (average of 30 simulation runs) using the Kruskal-Wallis test for *K* independent samples revealed that, at a level of significance of 5 percent ($\alpha = 0.05$), there does not exist enough evidence to conclude that the speed differential between the freeway

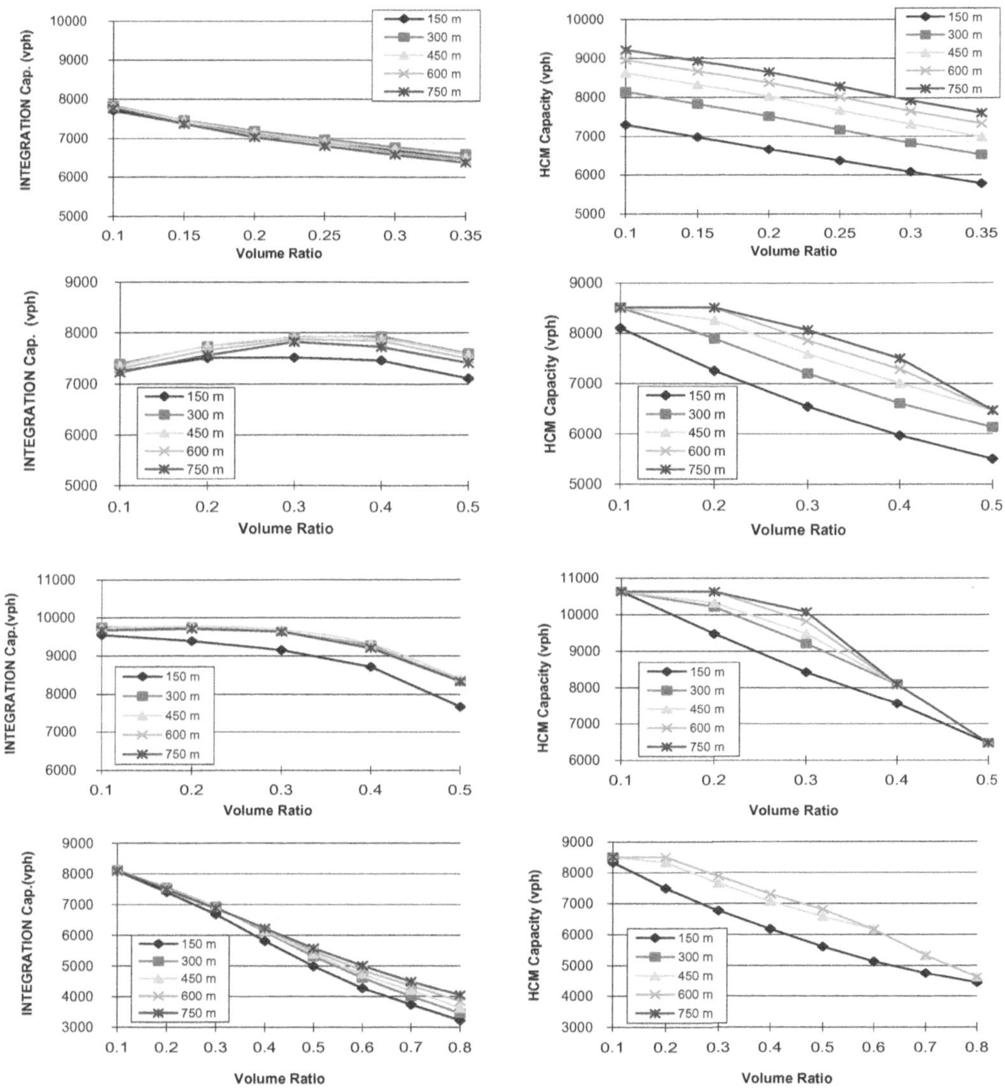

Figure 6. Impact of weaving section length on capacity (presented as A1, B1, C1, and C2).

mainline and the on- and off-ramps affects the capacity of freeway weaving sections. The p-values of the test are demonstrated in Table 2. For each cell in the table, a Kruskal-Wallis test is performed for six samples: differentials of 0, 5, 10, 15, 20, 25, and 30 km/h.

5.7 *Impact of Heavy Vehicles*

In this section the impact of heavy vehicles on the capacity of weaving sections is analyzed using the INTEGRATION software and the HCM procedures, as illustrated in Figure 7. Figure 7 demonstrates a high degree of consistency between the simulation and HCM results for all four sites, although the simulation capacities tended to be lower than the HCM capacity estimates. Consequently, the results of this sensitivity analysis demonstrate the adequacy of the HCM procedures in capturing the impacts of heavy vehicles on the capacity of weaving sections.

Table 2. P-values of kruskal-wallis tests of speed differentials.

	B1		C1		C2	
	On-F	Off-F	On-F	Off-F	On-F	Off-F
H	0.365	0.615	0.944	0.447	0.521	0.120
M	0.066	0.748	0.056	0.305	0.083	0.774
L	0.494	0.956	0.058	0.886	0.097	0.334

On-F: Speed differential between on-ramp and freeway

Off-F: Speed differential between off-ramp and freeway

H: high VR conditions (0.80 for Site B1 and 0.50 for Sites C1 and C2)

M: medium VR conditions (0.40 for Site B1 and 0.25 for Sites C1 and C2)

L: low VR conditions (0.10 for Site B1 and 0.05 for Sites C1 and C2)

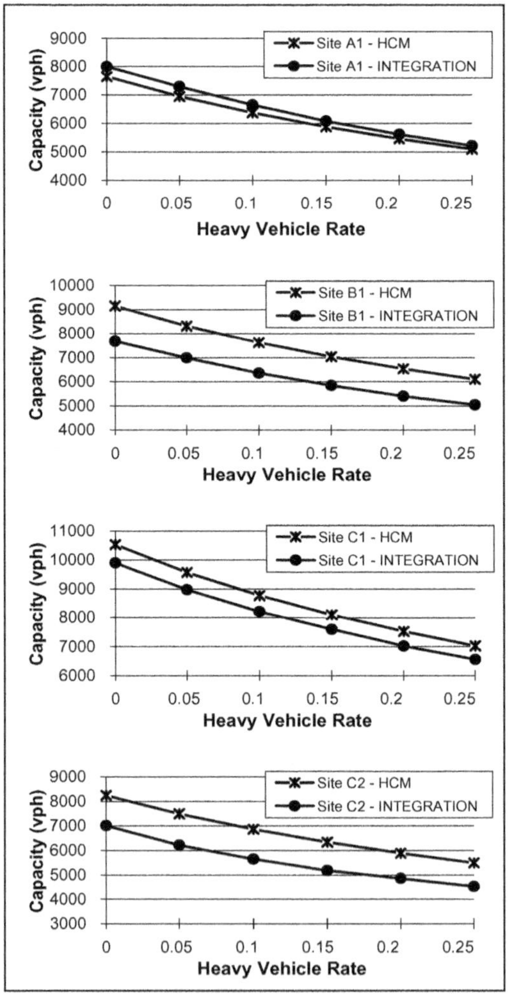

Figure 7. Impact of heavy duty vehicles on capacity.

5.8 *Differentiation between Type B Configurations*

As stated in Section 4, this study investigates whether differences in the capacities of both configurations of Figure 2 warrant considering different configuration types. The two configurations are considered for a weaving length of 600 m using an identical O-D demand, with a weaving ratio of 30 percent and an on-ramp volume of 30 percent the total in-coming total demand. The simulation results that are illustrated in Figure 8 demonstrate that the capacity of configuration 1 is typically less than that of configuration 2, especially for high volume ratios. The lower capacity of configuration 1 can be attributed to the fact that freeway-to-ramp (FR) vehicles are not required to execute any lane changes for the second configuration, which is not the case for the first configuration. Consequently, the simulation results verify the suggestion of separating the two configurations into two different weaving section types.

FINDINGS AND CONCLUSIONS

The research presented in this paper examined one of the most important aspects of analysis of freeway weaving sections, namely the capacity analysis. The findings and conclusions of the study can be summarized as follows:

a. The study demonstrated the validity of the INTEGRATION software for the analysis of weaving section capacities.
b. The study demonstrated some questionable capacity estimates by the CORSIM software and a gap acceptance procedure proposed in the literature. Specifically, the results demonstrated an unrealistic increase in roadway capacity with an increase in the mainline weaving volume for a Type B weaving section.
c. The study demonstrated that the random number seed resulted in a weaving section capacity standard deviation of 65 veh/h/lane.
d. The weaving ratio, which is the ratio of the lowest weaving volume to the total weaving volume, has a significant impact on the capacity of weaving sections. Specifically, differences in the range 326 veh/h/lane were observed in this study. Unfortunately, the weaving ratio is not considered in the HCM 2000 procedures.
e. The length of a weaving section has a larger impact on the capacity of weaving sections as the length of the weaving section decreases and the traffic demand increases.
f. There does not exist enough evidence to conclude that the speed differential between freeway and ramp traffic has a significant impact on weaving section capacities.
g. The HCM procedures for accounting for heavy duty vehicle impacts on weaving section capacities appear to be reasonable.

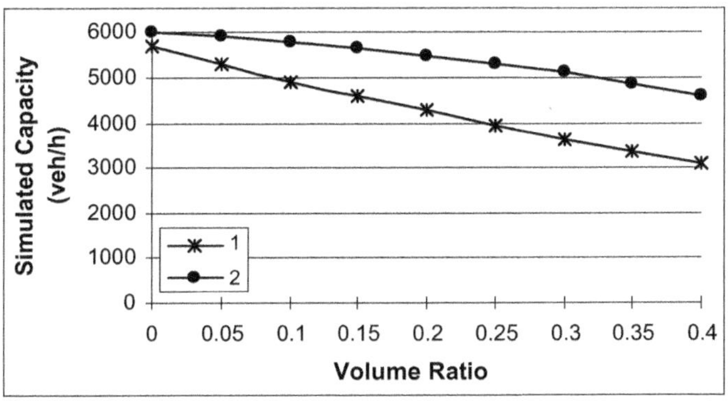

Figure 8. Capacity of both type b configurations.

h. Weaving sections requiring no lane changing by weaving vehicles should not be considered Type B weaving sections.
i. Simulation is a very useful tool for the capacity analysis of freeway weaving sections.

ACKNOWLEDGEMENTS

The authors acknowledge the financial support of the Mid-Atlantic University Transportation Center (MAUTC) in conducting this research effort.

REFERENCES

1. Roess, R. and Ulerio, J., Weaving Area Analysis in Year 2000 Highway Capacity Manual, Transportation Research Record, n 1710, 2000, pp. 145–153.
2. Lertworawanich, P. and Elefteriadou, L., Capacity Estimations for Type-B Weaving Areas Based on Gap Acceptance, Transportation Research Record 1776, TRB, National Research Council, Washington, D.C. 2002.
3. Lertworawanich, P. and Elefteriadou, L., Evaluation of Three Freeway Weaving Capacity Estimation Methods and Comparison to Field Data, Freeway Capacity, TRB 2004 Annual Meeting CD-Rom.
4. Stewart, J., Baker, M. and Van Aerde, M., Evaluating Weaving Section Designs Using INTEGRATION, Transportation Research Record, n 1555, 1996, pp. 33–41.
5. Kwon, E., Lau, R., and Aswegan, J., Maximum Possible Weaving Volume for Effective Operations of Ramp-Weave Areas, Transportation Research Record, n 1727, 2000, pp. 132–141.
6. Zarean, M. and Nemeth, Z.A., WEASIM: A Microscopic Simulation Model of Freeway Weaving Sections, Transportation Research Record, n 1194, 1988, pp. 48–54.
7. Special Report 209: Highway Capacity Manual. TRB, National Research Council, Washington, D.C., 1985.
8. Highway Capacity Manual (HCM) 2000—Weaving Segments, Chapter 24, Transportation Research Board, National Research Council, Washington, D.C. 2000.
9. Skabardonis, A., Cassidy, M., May, A.D., and Cohen, S., Application of Simulation To Evaluate the Operatio*n of Major Freeway Weaving Sections*, Transportation Research Record, n 1225, 1989, pp. 91–98.
10. Vermijs, R., *New Dutch Capacity Standards for Freeway Weaving Sections Based on Micro Simulation*, Third International Symposium on Highway Capacity, 1998, pp. 1065–1080.
11. Lertworawanich, P. and Elefteriadou, L., *Methodology for Estimating Capacity at Ramp Weaves Based on Gap Acceptance and Linear Optimization*, Transportation Research B: Methodological Vol. 37, 2003, pp. 459–483.
12. Bacon, V., et al., *Use of INTEGRATION Model to Study High-Occupancy-Vehicle Facilities*, Transportation Research Record No. 1446, 1994, pp. 8–13.
13. Gardes, Y. and May, A.D., Simulation of IVHS on the Smart Corridor Using the INTEGRATION Model: Initial Investigation. PATH Research Report, UCB-ITS-PRR-93-3, 1993.
14. Hellinga, B., and Van Aerde, M., *An Overview of a Simulation Study of the Highway 401 Freeway Traffic Management System*, Canadian Journal of Civil Engineering, Vol. 21, 1994.
15. Rakha, H., Van Aerde, M., Bloomberg, L., and Huang, X, *Construction and Calibration of a Large-Scale Micro-Simulation Model of the Salt Lake Area,* Paper presented at the 77th Transportation Research Board Annual Meeting, Washington, D.C., January 11–15, 1998.
16. Rakha, H., Medina, A., Sin, H., Dion, F., Van Aerde, M., and Jenq, J., *Coordination of Traffic Signals Across Jurisdictional Boundaries: Field and Simulation Results*, Transportation Research Board 79th Annual Meeting, Washington DC, January, 2000, CD-ROM [Paper # 00-1560].
17. Dion, F., Rakha, H., and Zhang, Y. (In press), *Evaluation of Potential Transit Signal Priority Benefits Along a Fixed-Time Signalized Arterial*. ASCE Journal of Transportation Engineering.
18. Rakha, H. and Ahn, K. (2004), *The INTEGRATION Modeling Framework for Estimating Mobile Source Emissions*. ASCE Journal of Transportation Engineering, March Issue.
19. Rakha, H. and Crowther, B. (2003), *Comparison and Calibration of FRESIM and INTEGRATION Steady-state Car-following Behavior*, Transportation Research, 37 A, pp. 1–27.
20. Rakha, H., Pasumarthy, P., and Adjerid, S. (2004), *Car-Following Models: Formulations, Issues, and Practical Considerations*, Submitted to the Journal of Transportation Research.
21. Rakha, H. and Zhang, Y., *The INTEGRATION 2.30 Framework for Modeling Lane-Changing Behavior in Weaving Sections*, TRB, 2004, paper # 04-3422.

Efficient Transportation and Pavement Systems – Al-Qadi, Sayed, Alnuaimi & Masad (eds)
© 2009 Taylor & Francis Group, London, ISBN 978-0-415-48979-9

Vehicular traffic volume characteristics of urban roadways in Kuwait

Ahmad H. Aljassar
Department of Civil Engineering, Kuwait University, Kuwait

Mohammed A. Ali
D A Watt Consulting Group Ltd., Calgary, Canada

Omar I. Al-Saleh
Kuwait University, Kuwait

Sadon A. Al-Khaldi
General Traffic Department, Ministry of Interior, Kuwait

ABSTRACT: There are many uses for traffic characteristics and traffic volume data. Major uses of data vary from pavement design, roadway geometric design, traffic signal design, and pavement management to advanced traveler information systems. Users of this data vary from planning and road design consultants to commercial roadside advertisers. Temporal and spatial variations in vehicular traffic have been known for several decades. It is only in the past two decades that the availabilty of modern technology allowed the traffic professionals to collect enough data to begin understanding these characteristics. This paper presents results of the continuous count stations of the urban roadways in Kuwait. Peak hour volume, annual average daily traffic volume, hourly variations, daily variations, and monthly variations are presented in the paper. The results of this study are expected to be of immense benefit to many agencies which need traffic information as input to their planning and/or engineering analyses. Such agencies include the Ministry of Pubic Works, Ministry of Interior, Municipality, Public Authority for Housing Care, as well as Planning and Engineering Consultants in the Arabian Gulf countries in general and in Kuwait in particular.

1 INTRODUCTION

Traffic count information is extremely important to traffic planning, design, and operation. This data is regularly requested and used by developers, consultants, realestate agents, homeowner associations, and government agencies. Geometric and structural designs of roads depend among other factors on traffic volumes, their mix characteristics, loads, and distribution with time of the day, day of the week, and month of the year. Traffic volumes are measured using the average annual daily traffic (AADT). This terminology implies that the count is a representative of the average traffic conditions for the whole year. During some parts of the year, traffic may be higher than the AADT, while in others it may be lower than this value. However, the AADT provides a typical daily traffic volume at any location, usable for most situations where traffic counts are needed as an input to a planning and/or engineering analysis.

Many transport and highway authorities worldwide have established permanent traffic count stations. In this section, a review of some selected case studies is presented. Vermont State in the USA has established a Continuous Traffic Counting (CTC) program throughout the state (Vermont DOT 1991). Monthly Average Daily Traffic (MADT), AADT, and Average Annual Weekday

Traffic (AAWDT) were calculated at each CTC station. Nevada's DOT (2001) administers about 5400 miles of roads in the state. During 2001, hourly traffic volumes were monitored continuously at 71 locations statewide. In addition, traffic volumes were collected in short periods (7 days) and factored to Annual Average Daily Traffic (AADT). In the small city of Lloydminster in Canada (Stefanuk 2000), a counting program is in place since 1990 utilizing both the automatic tube counters and manual intersection traffic volumes. In the state of Maine, the DOT has established two distinct programs for traffic count. The continuous traffic count program consists of 57 permanent recorder sites located throughout the state, monitoring traffic volumes 365 days on an hourly basis (Maine DOT 2000). In Hampshire County in the UK, traffic data has been collected by three methods (Hampshire County Council 2001): Manual classified counts, temporary, and permanent count sites. A team of enumerators, who use hand-held capture devices to classify traffic (usually over 12-hour period), undertakes manual classified counts. Automatic temporary traffic counts are undertaken by means of pneumatic tubes. The data is also collected from 100 permanent count sites using inductive loops cut into the carriageway. Four PTCS were established in the state of Andhra Pradesh in India by the Ministry of Surface Transport using loop and dynamic axle sensors (DYNAX) (Chari 1999). The purpose of the study was to get the characteristics of the National and State highways and to develop a procedure for traffic forecasting on these highways.

In order to obtain the traffic characteristics in Kuwait Road Network, a research project was sponsored by the Research Administartion of Kuwait University in 2002. The Ministry of Interior, Kuwait, supported the project by allowing to use some of their permanent count staitons and providing very useful input in the selection of count stations, and logistics and support for the installation of equipment. This paper presents some important charactreitics of the roadways obtained from the results of this study.

2 METHODOLOGY

2.1 *Study area characteristics and site selection*

Kuwait is a rapidly developing country with a population of more than 3 million. It has an excellent road infrastructure comparable to that of any industrialized country. The roads in urban areas are classified into the following functional categories (Aljassar 1998):

- Special Road Network (SRN),
- Primary Road Network (PRN),
- Secondary Roads (SR), and
- Local Roads (LR).

Special Road Networks include motorways and expressways that are major through-routes for traffic with grade-separated junctions and full acceleration and deceleration lanes. Roads in this category have at least one hard shoulder per carriageway and do not have U-turn facilities. The accessibility to SRN is restricted to motor vehicles only. Speed limit on such roads is usually 120 km/hr. Traffic directions are divided by raised concrete barriers. SRN roads are mainly ring roads or radial roads. A total length of about 900 km or 16.3% of total road network consists of SRN.

Primary Road Networks include traffic routes that are usually of a lower design standard than SRN routes. Roads under this category have more frequent at-grade junctions. Speed limit on these roads is usually 80 km/hr. Traffic directions are divided by paved or landscaped islands. A total length of about 1654 km or 30% of total road network consists of PRN.

Secondary Roads are used to distribute local traffic through a district and perhaps to serve a place of importance within a local community. Such roads usually have U-turn facilities. Speed limit on these roads is 60 km/hr and islands generally divide traffic directions. Secondary Roads usually run between blocks in a district to collect traffic from local roads and distribute them on SRN or PRN roads. About 1113 km or 20.2% of the total road network consists of SR.

Local Roads include those, which provide access to individual commercial or local residential units. They run within blocks in a district and distribute traffic on secondary roads. Speed limit on Local Roads is 45 km/hr, and traffic directions are not divided. A Total length of about 1845 km or 33.5% of total road network consists of LR.

Fifteen representative sites from among the four roadway types in Kuwait were selected for continuous monitoring of traffic for this project. The count sites were selected based on the functional classification of the roadways as well as the observed traffic pattern. The count sites were finalised after review meetings with the MOI and the MPW representatives. The selected locations are shown in Figure 1.

2.2 *Instrumentation*

Inductive Loop type detectors were selected for the study, keeping in view the advantages it offers on other technologies. ADR 3000, automatic data recorder (from PEEK TRAFFIC) and TCC 540 traffic counters (from International Road Dynamics) were selected for monitoring of traffic on the selected locations. Inductive loops and piezoelectric sensors were used to collect classified traffic volume data and speeds by lane.

Figure 1. Location of permanent traffic count stations in Kuwait city and its suburbs.

2.3 *Installation of equipment*

Installation of the ADR-3000 counters is preceded by the construction of concrete foundations, installation of protection posts and security housings, and cutting out of 2m x 2m loops and piezos in asphalt in all the lanes of the selected locations. After completing the setup, the programming of counters follows to collect the information on traffic by lane, by class, and by speed.

2.4 *Data uploading and analysis*

Data was uploaded from all the locations on a fortnightly basis, this allowed the research team to minimize the 'data outage' or occurrence of missing data. The AASHTO algorithm (13) was utilized for the data analysis.

2.5 *Quality assurance surveys*

Manual classified traffic count surveys were performed during randomly selected periods to validate the data. Live feeds of traffic were captured using video technology for one hour each at the randomly selected time periods and manual counts were performed in the office.

3 RESULTS AND ANALYSIS

The PTCS data were analysed in four groups as per the functional classification of roads. Figures 2 to 5 show typical traffic variation on a primary road in Kuwait.

3.1 *Special road network (SRN)*

The AADT for SRN was observed to be 129,777, and the Average Weekday Travel (AWDT) was 137,184. The AWDT was about 5.7% more than the AADT. The average peak hour volume was over 8,000 vehicles (6% of AADT). The hourly expansion factor varied from 15.7 in peak hour at 8:00 am to 108.8 at 4:00 am. The daily volume was rather unifrom throughout the week with a DEF varying between 6.5–8.3. The monthly factor varied from 0.9 to 1.1. Even though a number of people leave Kuwait for summer holidays during the months of July-September, the traffic on the SRNs remains consistent as these roads are the main carriers of traffic and function as major corridors in the network.

3.2 *Primary road network (PRN)*

The AADT for PRN was observed to be 126,342, and the AWDT was 133,567. The average peak hour volume was about 8,700 vehicles (6.5% of AADT). A typical hourly variation is shown in Figure 2. Two peak periods typical to Kuwait are very clear from this figure. Morning peak (7–8 am) occurs in westbound direction and afternoon (1–2 pm) peak occurs in the eastbound direction on this road. Due to these two peaks in opposite directions the total traffic is uniform between 8am2pm. Typical urban tidal flow on these roads is evident from the results where peaks shift directions between morning and afternoon peak hours. Friday (weekend) traffic is very different from a weekday as shown in Figure 3. There is only one peak occurring between 6–9 pm. Typical daily variation in a week is shown in Figure 4. The daily volume is rather uniform throughout the week and low during weekends, being the lowest on Friday as expected. It should be noted that in Kuwait, Thursday used to be a rest day (during the time of the study) in Government but is a working day in most of the private sector, hence the lowest traffic is observed on Friday (official weekend), rather than on both the weekend days. The daily volume was rather unifrom with a DEF at about 6.6 during the weekdays (Sat-Wed). It was 7.302 on Thursdays, which was a rest day for public sector and a half working day for most of the private sector in Kuwait. Friday DEF was

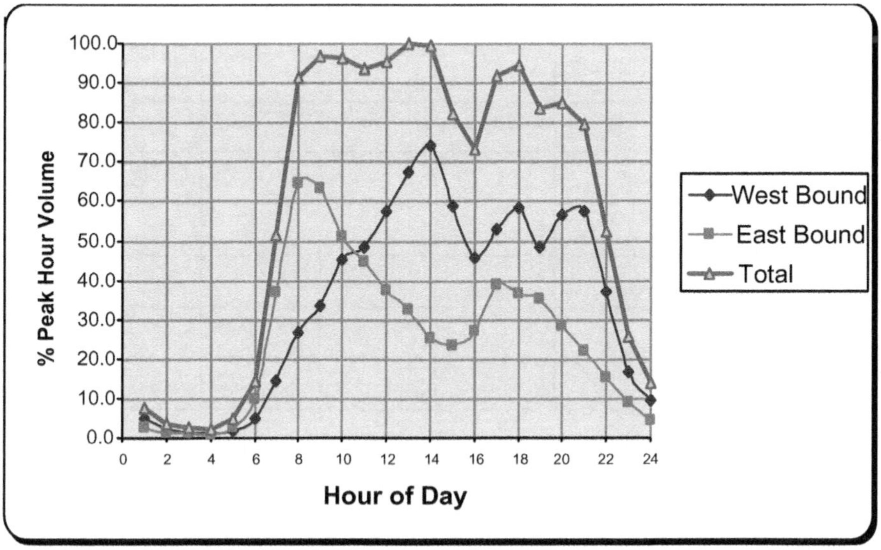

Figure 2. PRN—Weekday hourly variation.

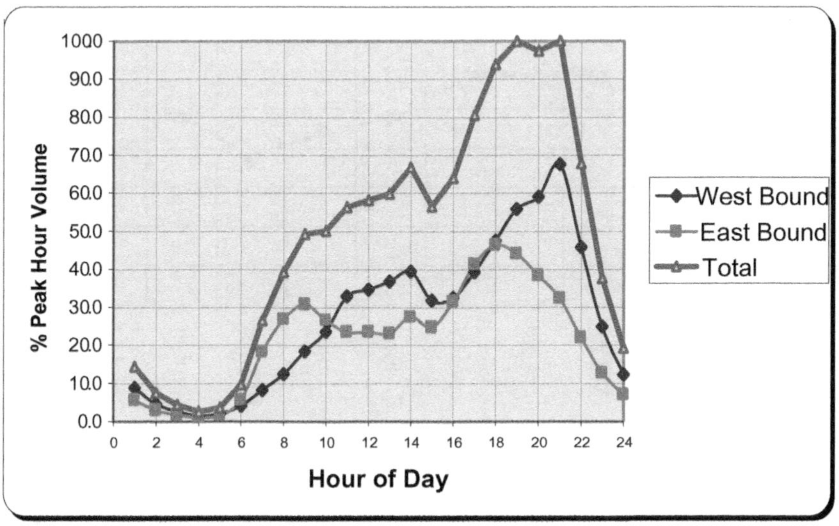

Figure 3. PRN—Friday hourly variation.

observed to be 9.458. The monthly factor varied from 0.915 in April to 1.141 in the summer month of August. The traffic on these roads and other lower level roads show a decrease in traffic volume during summer holiday season. Figure 5 shows a typical monthly variation of traffic on PRN.

3.3 *Secondary roads (SR)*

The AADT for SR was observed to be 49,461, and AWDT was 54,845. The average peak hour volume was about 3,800 vehicles (7% of AADT). The HEF was 13 at 9:00 pm and 557 at 5:00 am. The

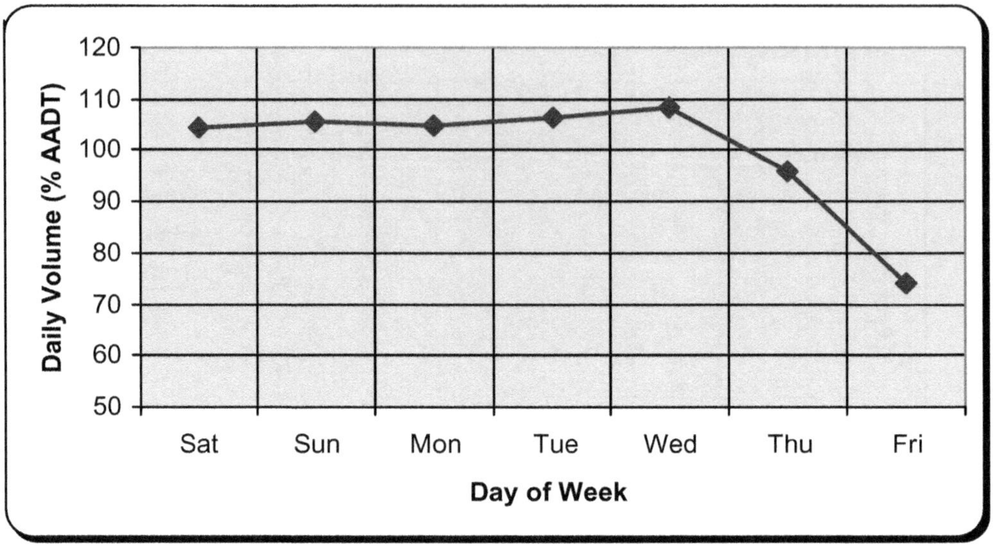

Figure 4. PRN—Daily variation.

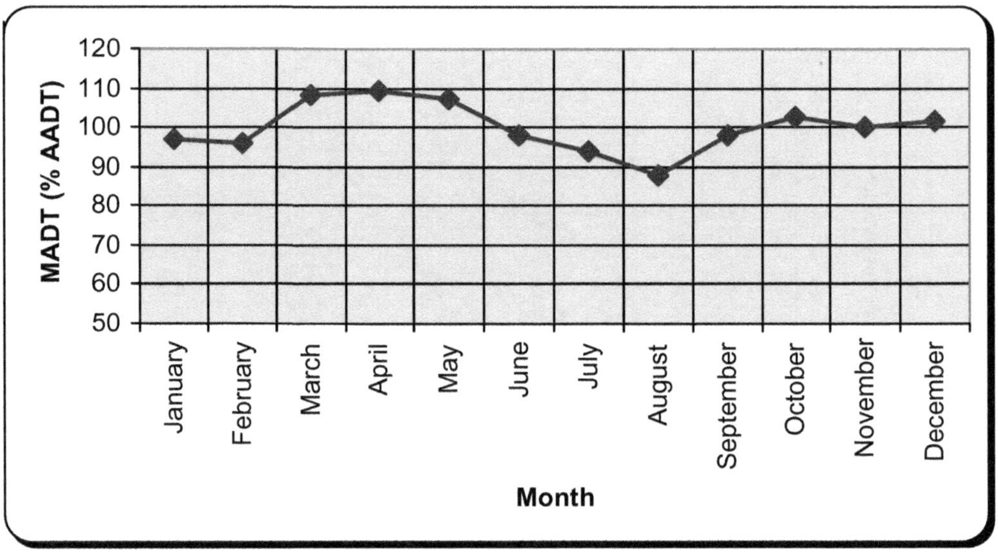

Figure 5. PRN—Monthly variation.

daily volume was rather unifrom throughout the weekdays with a DEF varying between 6.4–6.6, and over 10 on Fridays. The monthly factor varied from 0.914 to 1.086. The monthly variation was observed to be similar to that of the PRN and LR.

3.4 *Local roads (LR)*

The AADT for LR was 2,032. The average peak hour volume was about 166 vehicles (8% of AADT). The HEF varied from 11.8 at 7:00 pm to 537 at 4:00 am. The daily volume was rather unifrom throughout the weekdays with a DEF ranging from 6.7 to 7.6. Interestingly, the daily volume on Thursdays was significntly higher than weekdays due a large number of local (social and shoping) trips made on Thursdays.

4 CONCLUSIONS

This study establishes permanent traffic counting stations for the *first time* in Kuwait. A review of literature has indicated that in several countries around the world, the continuous traffic counting programs are used to collect various traffic characteristics such as traffic volumes, composition, LOADS, traffic variation by day, week, month, and year. One of the main objectives of this study was to develop road-specific expansion factors, which may be used to estimate AADT from short-term counts in Kuwait. Hourly, daily, and monthly expansion factors were developed for the four functional classes of roads in Kuwait. SRN and PRN carried high peak hour volumes of more than 8,000 vph (about 6% of AADT) with an AADT of more than 120,000 vehicles. Daily expansion factors are rather uniform in the weekdays for both of these classes of roadways. Seconday roads carried about 50,000 vehicles per day with about 7% in peak hour. The AADT on local roads was about 2,000 with 8% occuring in peak hour.

ACKNOWLEDGEMENTS

The authors wish to express their sincere thanks to the Research Administration of Kuwait University for having sponsored the research project no. EV-06/00. Thanks are also due to the Ministry of Interior, and Ministry of Public Works, Kuwait, for supporting this project.

REFERENCES

Aljassar, Ahmad H., Abdulaziz A. Al-Kulaib, El-Sayed W. Metwali, and Khaled N. Helali, "Performance of Roads in Kuwait", Proceedings, 1st International Conference on Performance of Roads, Bridges and Airport Pavements in Hot Climates, Dubai, United Arab Emirates, April 28–29, 1998.

Chari, S.R. and B.P. Chandrasekhar, "Establishment of Permanent Traffic Counting Stations in A.P.", Final Report, Regional Engineering College, Warangal, India, 1998.

Hampshire County Council, "Hampshire Local Transport Plan 2001–2006 Section 9.1—Monitoring and performance Indicators", Hampshire, UK, 2001.

Maine DOT, "2000 Traffic Count Book", traffic Engineering Division, Maine, 2000.

Nevada DOT, "The Annual Traffic report", Traffic Information Division, Carson City, Nevada, 2002.

Stefanuk, G. "City of Lloydsminster—Annual Traffic Volume Survey", The City of Lloydsminster, Engineering and Public Works Department, B.C., Canada, Dec. 2000.

Vermont DOT "Continuous Traffic Counter Grouping Study and Regression Analysis Based on 2001 Traffic Data", Vermont, USA, 2001.

Efficient Transportation and Pavement Systems – Al-Qadi, Sayed, Alnuaimi & Masad (eds)

Ras laffan traffic operations, circulation and simulation study

Kalari T. Ajilesh, Abdul-Rahman I. Hamad & Seela Sudarsana
Parsons, Doha, Qatar

ABSTRACT: Ras Laffan Industrial City (RLIC) is one of the fastest growing industrial cities in the world. RLIC is located on Qatar's north gas field and houses some of the elite names in the world of energy such as Shell, Dolphin energy and Qatar's own—Qatargas and RasGas LNG production facilities. Currently RLIC is facing severe traffic congestion on roads due to ongoing and upcoming industrial projects.

The objective of this study was to evaluate current and short-term projected traffic operations within RLIC and recommend any required mitigations to enhance the flow of traffic and reduce travel delay. Micro-simulation models were developed for existing road network within RLIC and calibrated to the field conditions to evaluate the traffic operations. This analysis revealed that most of the major intersections will operate with large delays and unsatisfactory level of service.

Mitigation measures were sought to enhance traffic operation at critical junctions and circulation of labor and professional force within RLIC. The study results show that implementing the proposed mitigation measures will reduce the delay between 1 and 10% on the overall network. Mitigation measures developed in a way that can be implemented within short time and cost effectively. From the statistics of the simulation models, savings in man-hours and reduction in delays, travel times and overall roadway network performances were estimated.

1 INTRODUCTION

Ras Laffan Industrial City (RLIC), situated along the northeast coast of Qatar, has been in recent times deemed as one of the fastest growing industrial cites in the world. The RLIC represents one of the world's most significant industrial export locations and covers an area of 106 square kilometers.

Currently RLIC is facing severe traffic congestion on roads due to ongoing and upcoming industrial projects. In order to complete the construction of planned projects, it is forecasted that a huge manpower (reaching over 105,000) will be required at peak construction periods. All this labor force has to be accommodated within internal labor camps of RLIC.

In its full operation, RLIC will have:

- 105,000 workers in labor camps in different areas of RLIC that need to be bused four times a day
  - from labor camp to work sites in the morning
  - from work site to labor camp for lunch
  - from labor camp back to work
  - from work site back to labor camp at end of shift.
- Sub contractors with labor camps just outside RLIC limits who will have few thousands of workers being bused also four times a day
- About 2500 vehicles from Doha and other areas south of RLIC that bring professional staff to work daily
- 24-hour a day truck circulation from RLIC port to work sites

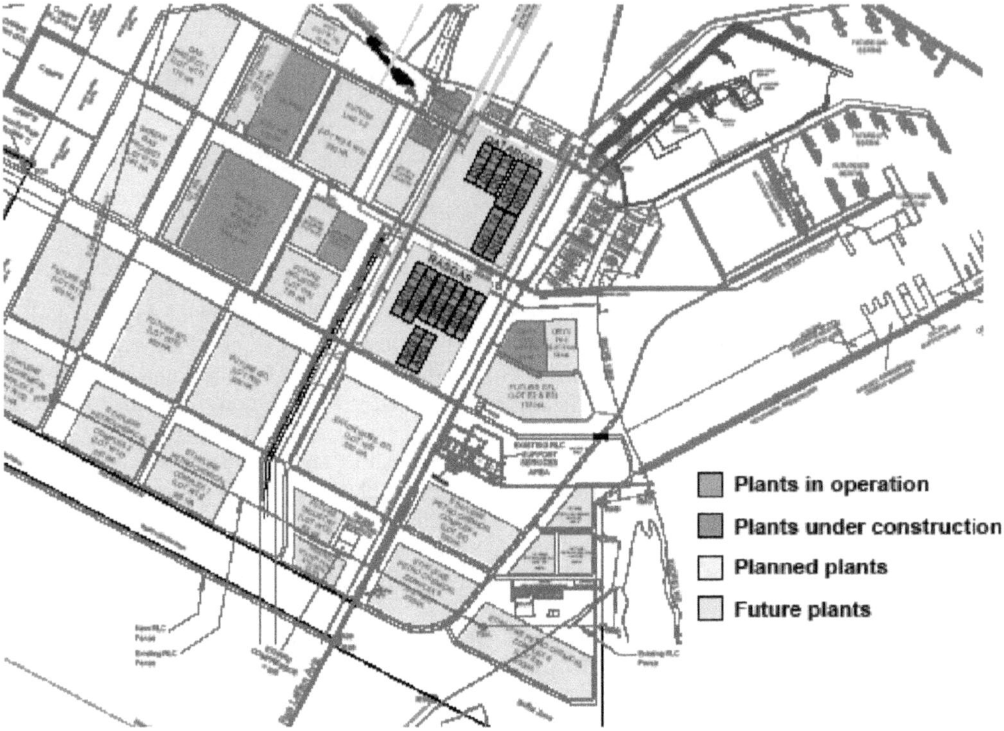

Figure 1. Ras laffan industrial city—site map.

- Special load hauling minimum twice a week from port to work site that require complete closure of the road
- Trucks from Doha and other areas outside RLIC

Due to all these, traffic on RLIC roads have grown considerably in the past few quarters and it demands a revisit the capacity of roads and intersections, and correction measures that should be taken, if necessary, to optimize traffic flow.

2 TRAFFIC DATA COLLECTION

An extensive traffic counting program was carried out in RLIC consisting of automatic traffic counts (ATCs), manual classification counts (MCCs) and manual turning movement counts (TMCs). The other surveys included travel time survey, intersection geometries, signal timing and phasing.

2.1 *Automatic traffic counts*

Automatic traffic counts were carried out for 24 hours, 7 days a week on 8 major roads/streets in RLIC as shown in Figure 2. ATC data was analyzed and summary of results are presented in Figure 3. Analysis revealed that Ras Laffan Avenue has highest average daily traffic (ADT) of 17,902 vehicles at ATC-1 and then Hamad Street and Khalifa Street have 15,067 and 11,822 vehicles at locations ATC-5 and ATC-7, respectively.

Figure 2. Location of traffic counting stations.

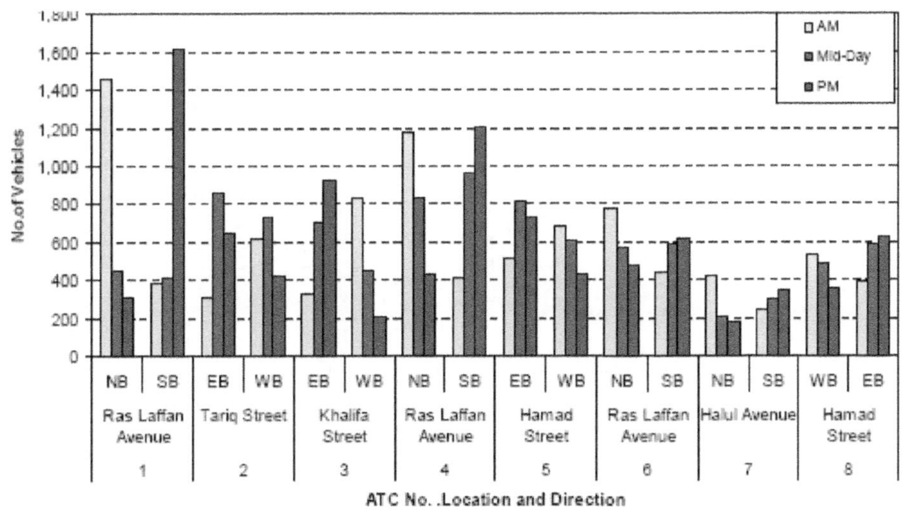

Figure 3. Peak hour volumes.

2.2 *Manual classification counts*

Manual classified counts (MCC's) were conducted at four locations as shown in Figure 2. MCC data was collected for 14 hours at each location and average vehicle composition was determined and presented graphically in Figure 4. The analysis indicates MCC-1 located on Ras Laffan Avenue after main gate shows highest cars percentage of 65% in northbound direction and 61% in southbound direction. Highest percentage of buses was observed at MCC-2 with 33% and 36% in northbound and southbound direction, respectively. MCC-4 indicates highest percentage of trucks with 32% and 26% in eastbound and westbound direction respectively.

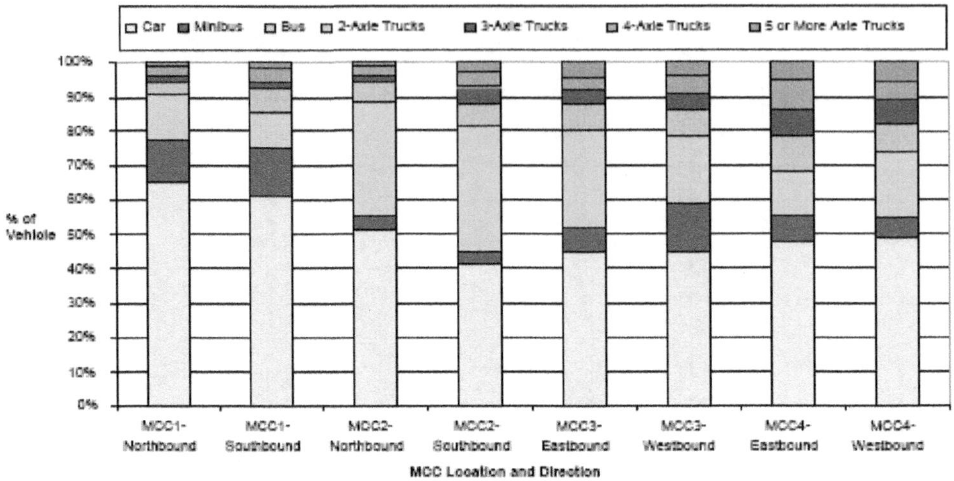

Figure 4. Vehicle composition at MCC locations.

2.3 *Turning movement counts*

The turning movement counts were carried out at 11 major junctions in RLIC. TMC data has been analyzed for peak hour volume at each junction during AM, Mid-day and PM peak periods. These counts revealed that AM, Mid-day and PM peak hours as 6:00–7:00 hrs, 11:00–12:00 hrs and 17:00–18:00 hrs, respectively. Based on the field counts, the total incoming junction volume for each junction during peak hours are presented in Figure 5. Among all, Intersection-5 shows highest volumes during AM, Mid-day and PM peak with 2494 veh/hr, 2387 veh/hr and 2563 veh/hr, respectively. It is evident from Figure 2 that to access any plant or construction site in RLIC, vehicles should pass from Intersection-5.

3 FUTURE VEHICLE FORECAST

In addition to the evaluation of existing roads, future road condition was studied. First quarter of year 2008 was considered to be the future scenario. RLIC projects data included construction manpower within camps of RLIC and also from external camps located outside, operating staff, project management staff and visitors. The review of construction manpower monthly estimates revealed that highest manpower is noticed in July 2007 as 84,072; with 38,500 workers in east camp, 35,045 in west camp and 10,526 in external camps. As agreed upon with RLIC, the construction peak manpower for RLIC is estimated to reach 105,000 in the first quarter of year 2008, which includes:

- 40,000 in east camp
- 50,000 in west camp
- 15,000 in external camps outside RLIC

These numbers are agreed upon for this study and segregation of workforce by camp.

3.1 *Estimation of vehicle generation*

Based on the manual classified counts in RLIC and field investigations, the traffic composition for the trips originating from camps within RLIC are assumed as follows:

- 85% of traffic as bus with passenger average occupancy of 45

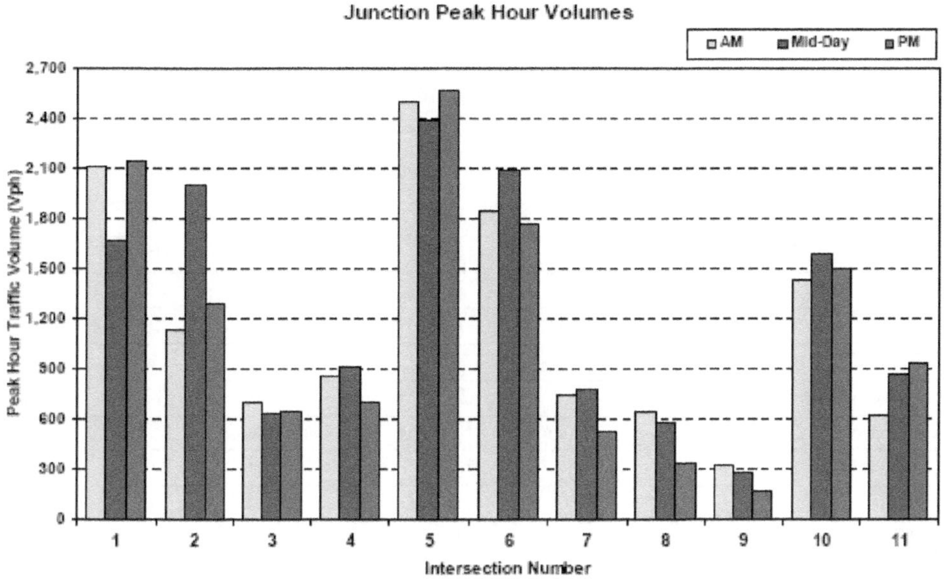

Figure 5. Turning movement counts.

Table 1. Estimation of vehicle generation from east camp.

Name of the camp	Name of the internal camp	Manpower (No. of persons)	Vehicle Generation Bus	Minibus	Car	Total
East Camp	RLC Camp	4,921	93	55	37	185
	QGX (QGII)	10,000	189	113	75	376
	Laffan Refinery	679	13	8	5	26
	CSP	1,500	28	17	11	56
	RLPEP	900	17	10	7	34
	QCS (QG3&4)	18,000	340	203	135	678
	QGTC dry dock	4,000	76	45	30	151
	Total	*40,000*	*756*	*450*	*300*	*1506*
West Camp	RGX (RGIII)	11,366	215	128	85	428
	RGX offplots	4,608	87	52	35	173
	AKG 2	6,700	127	75	50	252
	CCWP ph.2	1,576	30	18	12	59
	Pearl GTL	22,000	416	248	165	828
	IPP 3	3,750	71	42	28	141
	Total	*50,000*	*944*	*563*	*375*	*1882*
Grand total		**90,000**	**1700**	**1013**	**675**	**3388**

- 13.5% of traffic as minibus with passenger average occupancy of 12 and
- 1.5% of traffic as car with passenger average occupancy of 2

Using this traffic composition and passenger occupancy, number of vehicles that are anticipated to generate from each camp has been calculated and presented in Table 1.

3.2 *Operating staff*

The operating staff of all industrial plants in RLIC is estimated to be 6,184 persons in the first quarter of year 2008. It is assumed that trips generated by operating staff will be originated from

Table 2. Estimation of vehicle generation for operating staff.

Name of the Industrial Plant	Operational workforce in yr 2008-1Q	Vehicle Generation		
		Car	Minibus	Total
QP/RLC	633	190	63	253
QP/others	48	14	5	19
QP	**681**	**204**	**68**	**272**
QG I (target 2007)	1,070	321	107	428
QG II	250	75	25	100
QG 3	88	26	9	35
QG 4	13	4	1	5
Laffan Refinery	113	34	11	45
QG opco	**1,533**	**460**	**153**	**613**
RG (2004)	1,186	356	119	474
RG II	120	36	12	48
RG III	225	68	23	90
AKG I	150	45	15	60
AKG II	45	14	5	18
RG opco	**1,726**	**518**	**173**	**690**
Dolphin	560	168	56	224
Barzan	0	0	0	0
Oryx GTL	260	78	26	104
Pearl GTL	250	75	25	100
Exxon Mobil GTL	0	0	0	0
RC2 cracker	125	38	13	50
Others	300	90	30	120
Industrial plants	**5434**	**1,630**	**543**	**2,174**
Support services/industries	**750**	**225**	**75**	**300**
TOTAL	**6184**	**1855**	**618**	**2474**

outside RLIC. In order to estimate number of vehicle that can generate from operating staff, it is assumed that 30% of operating staff will commute by car and 70% by minibus to RLIC. The passenger occupancy of car and bus has been assumed as 1 and 30 respectively. The number of vehicles that can generate from the operating staff has been presented in Table 2.

3.3 *PMT staff and EPC contractor staff*

Project management staff and EPC contractor staff is considered as 3% and 5% of total construction workforce respectively. It is assumed that trips generated by these staff are originated from outside RLIC. It is also assumed that 50% of these staff will commute by car and 50% by minibus. The number of vehicles that can generate has been estimated and presented in Table 3 below.

3.4 *Visitors*

Number of visitors to RLIC is also significant and it is estimated to be 3,155 people per day in the first quarter of year 2008. The trips generated by visitors will be originating from outside RLIC.

Table 3. Estimation of vehicle generation for PMT and EPC contractor staff.

Name of the staff	No. of persons	Vehicle generation		
		Car	Minibus	Total
PMT staff	3150	1575	225	1800
EPC contractor staff	5250	2625	375	3000

It is assumed that all visitors will commute by car to RLIC with passenger occupancy of 1 that will attract 3,155 cars per day.

3.5 *Truck traffic*

Trucks entering RLIC are classified under general cargo (offshore rigs, operation and maintenance equipments), rock and fill for port expansion project, and bulks (fill material, aggregates, sand, cement). These are estimated to contribute a total of 1,226 trucks in the first quarter of year 2008.

4 FUTURE TRAFFIC VOLUMES

Future traffic volumes were forecasted considering:

- Existing traffic flow patterns along existing road network, at all junctions, and on major roads during AM, Mid-day and PM peak hours
- Future construction manpower estimates
- Future operational manpower, PMT and subcontractor staff
- Future road network expansion
- Traffic entering from main gate
- Traffic from east camp to work sites
- Routes followed by vehicles to access the respective work places for existing scenario and future scenario with additional roads and junctions
- Future traffic from west camp to work sites
- Truck traffic to RLIC

Considering all these parameters and utilizing common sense and engineering judgment, peak hour forecasted volumes were developed. The volumes for the first quarter of year 2008 were forecasted for AM, Mid-day and PM peak hours. Figure 6 shows the future traffic volume for AM peak hour.

5 SIMULATION MODEL

Simulation models were developed using VISSIM software for all three peaks for two scenarios, viz;

a. Existing road network (November 2006)
b. Future road network—Base (1st Quarter of year 2008)

5.1 *Existing road network (Nov.2006) model development*

Road network inside Ras Laffan Industrial City can be classified as grid pattern. The main access to RLIC is through Ras Laffan Avenue and this road feeds traffic to all connecting roads and the port. Existing road network includes eight signalized junctions namely intersection numbers 1, 2,

Figure 6. Traffic volumes—first quarter of year 2008—AM peak hour.

5, 6, 8, 9, 10 and 11 and two roundabouts namely Intersection-3 & 7 as shown in Figure 2. Roads connecting Ras Laffan Avenue include Khalifa Street, Hamad Street, Ayesha Street and Tariq Street. These roads serve as main roads distributing the traffic within study area.

Based on the traffic count data collection carried out in November 2006, traffic flow patterns within RLIC have been analyzed and location of camps and routes followed to access each plant were also taken in to consideration. Since traffic counts were carried out on different days, minor adjustments were carried out to the traffic volumes on existing roads wherever necessary to balance the network. These adjusted volumes were used in the model development.

Simulation model development has been carried using VISSIM for AM, Mid-day and PM peak hour operation. The type of control at each location was coded as in the field such as signal phasing, priority rules and access conditions etc. Based on the manual classified counts, the vehicle composition at each input point has been specified separately for all the peak hours in the model to represent existing operation. For each peak hour simulation, four runs were carried out by changing the random seed for the simulation to obtain average results of the statistics. Existing road network in the November 2006 was coded as it is in VISSIM and run with an initial 30 minutes warm-up time. Figure 7 shows the screenshot of the road network developed in VISSIM.

5.1.1 *Calibration and validation*

Calibration and validation form a crucial element of the simulation task through which confidence in the model results can be ascertained. Because of the stochastic nature of the traffic, variations

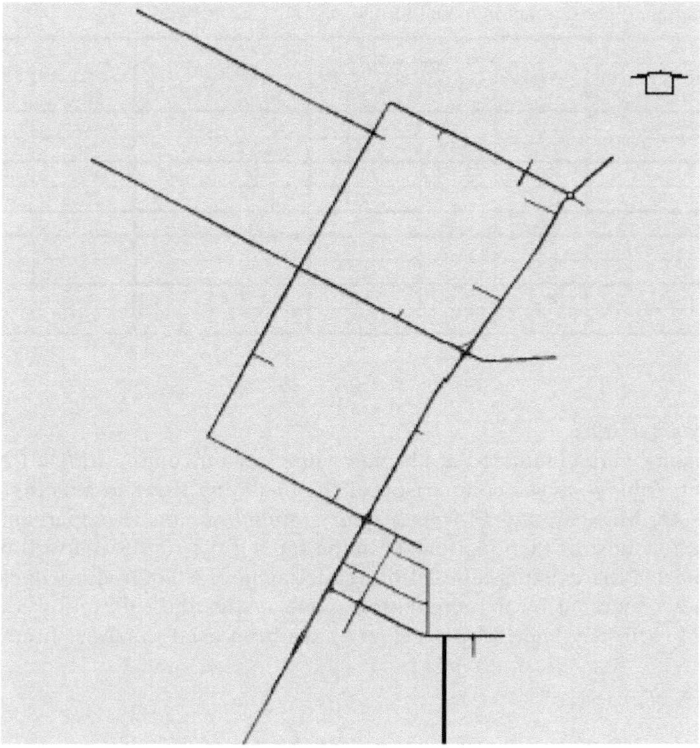

Figure 7. Coded existing road network in VISSIM.

between model and observed data is always expected and the onus is upon the model user to establish the desired reliability level and the validation effort required to achieve it. The calibration process for VISSIM follows similar procedures to conventional traffic models with the implementation of a two phase process covering a thorough check of the input data and comparing modeled results with observed data. Comparison of modeled and observed data is possible for operational analysis where an existing system is being studied. The GEH statistic, a modified chi-squared statistic that incorporates both relative and absolute differences, in comparison of modeled and observed volumes. Generally the GEH static should be used in comparing hourly traffic volumes only. It is represented by the equation below:

$$GEH = \sqrt{\frac{(M-O)^2}{0.5*(M+O)}}$$

Where:
M = Simulated flows
O = Observed flows
Various GEH values give an indication of a goodness of fit as outlined below:
GEH < 5 Flows can be considered as good fit
5 < GEH < 10 Flows may require further investigation
10 < GEH Flows can not be considered to be a good fit
Once model has been calibrated for the existing situation it can then be used to model future scenarios.

Table 4. Calibration of the simulation model for the existing road network.

Intersection No.	AM			Mid-day			PM		
	Observed Flow	Modeled Flow	GEH	Observed Flow	Modeled Flow	GEH	Observed Flow	Modeled Flow	GEH
1	2205	2156	1.05	1916	1931	0.34	2220	2244	0.51
2	1357	1325	0.87	2146	2087	1.28	1496	1399	2.55
3	900	905	0.17	733	744	0.40	683	608	2.95
5	2531	2497	0.68	2407	2416	0.18	2605	2624	0.37
6	1788	1847	1.38	1931	2087	3.48	1853	1763	2.12
7	687	750	2.35	739	779	1.45	550	517	1.43
8	576	657	3.26	605	679	2.92	386	347	2.04
10	1215	1428	5.86	1520	1584	1.62	1525	1502	0.59
11	609	630	0.84	856	870	0.48	943	930	0.42

5.1.2 *Flow comparisons*

Flow comparisons were conducted at all major junctions incoming traffic flow to know the goodness of fit. Table 4 shows comparison of the incoming flows at existing major junction in RLIC for AM, Mid-day and PM peak hour simulation runs. Comparison revealed that, GEH statistics for most of the junctions is in the rage of 0.5 to 3.5 and well below the value of 5. This indicates the existing simulation model achieved required goodness of fit to rely on the statistics generated by the simulation model. After the calibration, these simulation models for AM, Mid-day and PM peak hours have been used to model future road network scenarios.

5.2 *Future road network model development (1st Quarter of Yr 2008)*

Future model development was carried out by using the existing models for AM, Mid-day and PM peak hours separately. Future models include:

i. Future road network expansions
ii. Forecast volumes in AM, Mid-day and PM peak hours for 1st quarter of year 2008
iii. New intersections and planned control measure at the junctions
iv. New camp locations such as west camp
v. Routes followed by vehicles from each camp
vi. Existing junctions and control measure at each junction such as signal phasing, time settings etc.
vii. Access to the Lay down areas
viii. Planned modifications to the existing junctions which includes
    a. Modification of right-in, right-out junction near east camp to a four legged signalized junction
    b. Connection of this new junction to Al Rayyan Avenue

Future road network has 11 signalized junctions, 2 roundabouts and 5 priority junctions.

Future network models developed for AM, Mid-day and PM peak hours were run for simulation and results of junction evaluation and travel times are extracted. Table 5 shows the results of future network junction evaluation with junction incoming volume (vehicles per hour), average delay (in seconds) and maximum queue (in meters) for each intersection during AM, Mid-day and PM peak hours.

The evaluation of simulation results for future road network revealed that in AM peak hour, Intersection-1, 5, 6 and 10 are failing with high delays and long queues. During Mid-day peak hour, intersection no. 2A, 5, 10, 11 are operating with high delays and long queues. In the PM peak hour, Intersection-5, 6 and 10 are showing high delays and queues exceeding 400 m.

Table 5. Future network simulation results—Base case.

Int. No.	AM peak hour			Mid-day peak hour			PM peak hour		
	Junction incoming volume	Average Delay	Maximum Queue	Junction incoming volume	Average Delay	Maximum Queue	Junction incoming volume	Average Delay	Maximum Queue
1	2239	117.5	511.5	1591	26.9	336.3	2098	22.5	178.6
2	674	20.6	74.5	1027	14.1	86	659	8	28.5
2A	1060	55.2	203.3	1065	71.7	462.2	952	37.4	147.1
3	418	0.8	13.5	392	1	13.6	316	1	13.4
5	2673	111.5	511.5	2261	45.2	505	2618	88.5	511.5
6	2326	92.1	511.5	2280	47.3	259.2	2179	91.8	428.4
7	1317	10.1	102.4	1278	6.4	52.1	1022	4.2	58.3
8	1161	31.8	188.3	958	25.7	142.6	733	19.3	67.1
10	1784	80.3	515.6	1487	268.8	515.9	1563	140.7	512.2
11	1473	35.2	242.6	1042	265.9	515.9	1442	27.9	99.9
12	1101	15.8	130.2	1007	20.5	163.6	1158	27.4	137.3
13	532	0.9	14.8	743	2.5	46.9	786	2.5	46.7
14	1212	19.6	182.1	1102	17.7	106.5	1189	17.8	105.6
15	1345	12.5	315.5	1101	4.8	132.6	1234	6.2	98.5
16	1685	17.6	297	1102	4.9	125.1	1616	8.5	79.2
17	685	2.6	27.3	611	2.4	26.5	490	1.3	13

6 FUTURE MITIGATION MEASURES IDENTIFICATION AND EVALUATION

6.1 *General*

As traffic operational analysis results identified the intersections with overall capacity/delay performance "F" for the period of 1st Quarter Year 2008, and based on the agreement of acceptable targeted Level of Service "E", due to the short nature of construction period thus improvements as proposed would cater for optimizations of existing roadway network with minimal infrastructure expenditures.

The method used in the development the mitigation measures was based on the integration of limited upgrades that could also be physically achieved (constructed) before the expected peak traffic period of 1st Quarter Year 2008.

The following are a list of suggested mitigation measure types;

- Signal cycle optimization
- Geometrical upgrade by introduction of right turn slip lane at intersection leg(s)
- Geometrical upgrade by introduction of right turn slip lane with a 50m storage pocket lane at intersection leg(s)
- Geometrical upgrade by enhancing capacity of the inner Left Turn Lane (2nd Left Lane)
- Increase left turn storage lane capacity at intersection leg(s)
- Upgrade by introducing signalization
- Upgrade by introducing civil Infrastructure for signalization

Mitigation measures as recommended were verified on conceptual level only, further detailed design investigation is recommended to evaluate their applicability's in the field. The mitigation measures as listed are expected to include the following sub-improvements;

6.1.1 *Signal cycle optimization*

The existing signals are operating on actuated signal mode, with one time plan set for all times of the day, and days of the week, i.e., full functionalities of actuated signals are not utilized at present. The recommendation of signal optimization is suggested to include; first, verifying the functionality of all existing Inductive loop detectors, second, addition of different operation signal time plans to cater for different peaks/off peak periods of the day.

6.1.2 *Right turn slip lane*
As existing and proposed intersections analyses depict the proximity of capacity at saturation level especially for the peak period of 1st Quarter Year 2008, and due to the high percentage of heavy vehicles requiring safe large turning radius, the recommendation of introducing right-turn slip ramp is advantageous. It is also observed that the existing slip right turns within the current operating intersections are demonstrating a valuable increase in intersections capacity, and promoting the choice of selecting a right turn movement route concepts in determining the different route choices as available.

6.1.3 *Upgrade of inner left turn lane*
It was observed that buses currently avoid utilizing the inner left lane at intersections with dual left lane. This results in under utilization of existing capacity. Since increasing the capacity is critically needed for such movements, it was concluded that setback of stop lines are required as enhancements, in addition to the need for modifications of middle islands termination location and shape. These geometrical changes will also require relocations of signal poles and their relevant infrastructure.

6.1.4 *Upgrade by introducing Signalization*
Level of Service (LOS) capacity analysis was performed for yield or stop controlled intersections using projected 1st Quarter Year 2008 traffic volumes. The LOS analysis results clearly show high delays for certain movements thus requiring traffic control type upgrade. Signalization of these intersections was then performed and resulted in acceptable performances.

6.1.5 *Upgrade by introducing civil infrastructure for signalization*
Recommendation to upgrade intersections control type to signals was based on MUTCD traffic signal warrant criteria analysis. For intersections showing potential signal upgrades but still lack proper warrants, it was recommended as a mitigation measure to introduce the required signal civil infrastructure (underground conduit crossing), specifically for planned future intersections.

6.2 *Development of mitigation measures*

Traffic operational analysis of intersections was conducted for the eighteen intersections within Ras Laffan Industrial City using projected 1st Quarter Year 2008 traffic volumes. The base condition as defined included all geometrical upgrades for the intersections as planned to be constructed before 1st Quarter Year 2008.

The standard criteria for determining improvement interventions as proposed was based on general consensus of parties involved in the scheduled progress meetings whereby and due to the short term peak periods of construction activities, long term high level improvements will not be justified. As a conclusion overall performance of intersections at Level of Service "E" would be acceptable.

Based on international criterion and norm, Level of Service for both signalized and unsignalised priority junction types is determined using the thresholds. The method followed to define intersection mitigation measures was based on first, introduction of improvements of non-intrusive nature, where signal timing optimization was analyzed; as a result of which if satisfactory, then the suggested mitigation measure would be limited to such intervention. Otherwise intrusive type measures were introduced where as each type of measure and its effectiveness in improving Level of Service was analyzed for arriving at minimum mitigation measure(s) as targeted for acceptable Level of Service E.

The list of intersections as shown in the Table 6 below identifies proposed mitigation measures per intersection, and its applicable location.

Table 6. Proposed mitigation measures at intersections.

Intersection Number	Intersection Control Type	Proposed Mitigation Measure Types	Measure Location Leg
1	Signalized Actuated coordinated	a & d	
2	Signalized Actuated coordinated	a	
2A	Signalized Actuated coordinated	a & b	b) On all inbound Intersection Legs
3	Roundabout		
4	Priority (Two-way Stop)		
5	Signalized Actuated coordinated	a & d	d) On Ras Laffan Ave North & South Appr's
6	Signalized Actuated coordinated	a & e	e) On Ras Laffan Ave South Appr
7	Roundabout		
8	Signalized Actuated	a	
9	Priority (Two-way Stop)		
10	Signalized Actuated	a & b & c	b) On Halul Ave North Legs c) On Hamad Str East & West Legs
11	Signalized Actuated	a & c & d	c) On Ras Gas Entrance East Leg d) On Ras Gas Entrance East Leg
12	Signalized Actuated	a & c & d	c) On Hilal Ave North Appr c) On Khalifa Str East Appr d) On Hilal Ave North Appr
13	Priority (Two-way Stop)	g	
14	Signalized Actuated	a	
15	Signalized Actuated	a & c & d & f	c) On Hilal Ave North Appr c) On Hamad Str East Appr d) On Hilal Ave North Appr
16	Signalized Actuated	a & c & d & f	c) On Pearl Access Rd South Appr c) On Hamad Str West Appr d) On Pearl Access Rd South Appr
17	Priority (Two-way Stop)	g	

Type of Mitigation	Description of the Mitigation Measure
a>	Signal Cycle Optimization
b>	Geometrical upgrade by introduction of Right turn slip lane on intersection leg(s)
c>	Geometrical upgrade by introduction of Right turn slip lane with a 50m Storage Pocket lane on intersection leg(s)
d>	Geometrical upgrade by enhancing capacity of the inner Left Turn Lane (2nd Left Lane)
e>	Increase Left Turn Storage Lane capacity on intersection leg(s)
f>	Upgrade by introducing Signalization
g>	Upgrade by introducing Civil Infrastructure for Signalization

The analysis revealed that 8 intersections namely intersection no. 1, 2A, 5, 6, 10, 11, 15, 16 which were failing in future traffic operation for base condition. These intersections are expected to operate at Level of Service E or better with the proposed mitigation measures. Intersections no. 8 and 12 which were operating with Level of Service E in future base condition operation are expected to operate at Level of Service D or better with the mitigation measures. Conceptual drawings were developed to each individual intersection with proposed mitigation measure(s) to validate its applicability and to clarify its impact on right-of-way. The proposed intersection improvements for each intersection are listed in Table 6. Further field investigations and proper design would be required to determine potential conflicts.

7 MEASURES OF EFFECTIVENESS

Based on the traffic operational analysis using SIDRA INTERSECTION software program for the period of 1st quarter year 2008, the intersection operational performance was measured for its overall average reduction in delay per vehicle. Mitigation measures as stated above were limited in nature and resulted in an average vehicle delay reduction per intersection as presented in Table 7.

The delay reductions should not be the only indicator used to determine the required justification for implementing the mitigation measures as suggested. Other operational issues such as split phasing versus Lead/Lag left turn type phasing flexibility needs to be introduced, this is key

Table 7. Reduction in delay for each intersection.

Intersection # 1 - Percentage Delay Reduction (Per Vehicles, In All Peaks, All Moves) =	**-7.7** %
Intersection # 2A - Percentage Delay Reduction (Per Vehicles, In All Peaks, All Moves) =	**-52.1** %
Intersection # 5 - Percentage Delay Reduction (Per Vehicles, In All Peaks, All Moves) =	**-33.4** %
Intersection # 6 - Percentage Delay Reduction (Per Vehicles, In All Peaks, All Moves) =	**-22.3** %
Intersection # 8 - Percentage Delay Reduction (Per Vehicles, In All Peaks, All Moves) =	**-21.7** %
Intersection # 10 - Percentage Delay Reduction (Per Vehicles, In All Peaks, All Moves) =	**-25.9** %
Intersection # 11 - Percentage Delay Reduction (Per Vehicles, In All Peaks, All Moves) =	**-59.5** %
Intersection # 12 - Percentage Delay Reduction (Per Vehicles, In All Peaks, All Moves) =	**-47.8** %
Intersection # 15 - Percentage Delay Reduction (Per Vehicles, In All Peaks, All Moves) =	**-78.3** %
Intersection # 16 - Percentage Delay Reduction (Per Vehicles, In All Peaks, All Moves) =	**-90.9** %

Table 8. Overall delay reduction on all intersections for all peaks.

Average Delay (Per Vehicle, All Peaks, for All Moves, base Condition) =	**74.8** Secs
Average Delay (Per Vehicle, In All Peaks, for All Moves, With mitigation measures) =	**36.8** Secs
Percentage Delay Reduction (Per Vehicles, In All Peaks, All Moves) =	**-50.7** %

tool to use as traffic volumes increase to capacity levels. Furthermore, right turn slip lanes are advantageous and should be recognized not only from a capacity point of view but also from a safe maneuverability one.

In summary the average delay per vehicle at all the addressed signalized junctions for the period of 1st quarter of year 2008 is expected to be 74.8 seconds with no improvements (base condition) and with the proposed improvements is expected to be reduced to 36.8 seconds. This indicates an average percentage delay reduction of approximately 51% as shown in Table 8.

8 FUTURE MITIGATION MEASURES IDENTIFICATION AND EVALUATION

The proposed mitigation measures at each intersection were incorporated in VISSIM simulation model with future road network for AM, Mid-day and PM peak hours separately. These models were simulated for a period of 90 minutes and simulation statistics were collected for period of 60 minutes (after a 30 minute warm-up period) representing the full peak hour.

The overall network operational performance was measured using the following indicators:

- Vehicle-Kilometer Traveled (VKT)
- Vehicle-Hours Traveled (VHT) and
- Mean System Speed (MSS)

Vehicle-Kilometer Traveled (VKT) is computed as the product of the number of vehicles traversing a link and the length of the link, summed over al links. Increase in VKT indicates that more vehicles would be moving through the network.

Vehicle-Hours Traveled (VHT) provides an estimate of the amount of time expended traveling on the road network. The decrease in VHT generally indicates improved system performance.

Mean System Speed (MSS) is an indicator of overall system performance. Higher speeds generally indicate reduced travel times and costs for the road users. The MSS is computed from VKT and VHT as follows:

$$\text{Mean System Speed} = \text{VKT/VHT}$$

These key indicators were determined from the VISSIM network performance for future operation without improvements (Base Condition) and future operation with mitigation measures and presented in Table 9.

Table 9 shows increase in VKT for future network with mitigation measures in all peak hours with the highest being 6.4% in PM peak hour. The VHT for future network with mitigation were reduced significantly in all the peak hours with the highest being 9.8% in Mid-day peak hour. As a result the MSS did increase in comparison with base condition producing a better operational performance with mitigation measures. Thus, even though travel times increased for few sections, the overall network performance is anticipated to improve significantly with proposed mitigation measures.

9 ESTIMATION OF COST OF IMPROVEMENT MEASURES

Preliminary engineering cost estimate was conducted to all proposed mitigation measures. Standard unit price rates are used to determine the overall cost of each individual measure site location. Table 10 shows estimated quantities of proposed works and its applicable unit price along with

Table 9: Network performance comparison for future network base condition and with mitigation measures.

	AM Peak Hour			Mid-Day Peak Hour			PM Peak Hour		
Performance Indicator	Base Condition	With mitigation measures	% increase/ decrease	Base Condition	With mitigation measures	% increase/ decrease	Base Condition	With mitigation measures	% increase/ decrease
VKT	46,034	47,595	3.4%	56,673	59,999	5.9%	45,185	48,093	6.4%
VHT	1,747	1,693	−3.1%	2,359	2,128	−9.8%	1,670	1,663	−0.4%
MSS	26.3	28.1	6.8%	24.0	28.1	17.1%	27	28.9	7.0%

Table 10. Estimated cost for proposed improvements.

Intersection Number / Roadway	Description of Works						Total Estimated Cost (Q.R.)
	Asphalt related		Curb		Signal Upgrade	Delineation	
	Area	E. Cost (Q.R.)	L.M.	E. Cost (Q.R.)	E. Cost (Q.R.)	E. Cost (Q.R.)	
1	957	287,100	750	112,500	200,000	30,000	629,600
2					50,000	40,000	90,000
2A	1,825	547,500	310	46,500	50,000	40,000	684,000
3						15,000	15,000
4						40,000	40,000
5	922	276,600	1,145	171,750	200,000	40,000	688,350
6	693	207,900	380	57,000	200,000	40,000	504,900
7						60,000	60,000
8					50,000	40,000	90,000
9						40,000	40,000
10	1,766	529,800	260	39,000	50,000	40,000	658,800
11	806	241,800	290	43,500	100,000	30,000	415,300
12	2,070	621,000	948	142,200	100,000	30,000	893,200
13		58,000	100	15,000			73,000
14					50,000	30,000	80,000
15	3,232	969,600	780	117,000	700,000	30,000	1,816,600
16	2,130	639,000	1,490	223,500	700,000	30,000	1,592,500
17		58,000	100	15,000			73,000
Rayyan Ave Road Extension	3,500	1,050,000	410	61,500		100,000	1,211,500
		4,436,300		1,044,450	2,450,000	675,000	9,655,750

estimated cost. Justification of expenditure based on the effectiveness of suggested improvement could be analyzed in more than one way. Some of which are tangible items to compare with, example to that, would be the time delay reductions as analyzed and others intangibles issues could include; first, the operation near capacity conditions that could lead to serious unforeseen traffic operational delays with no improvements as suggested, secondly accidents associated with saturated traffic condition again generating major delays and rerouting of traffic. It is then recommended that intangible benefits be recognized by decision makers in evaluating the real effectiveness of such measures.

10 CONCLUSION

With the expected growth of Ras Laffan Industrial City as observed, continuous traffic integration and simulation/analysis would be a fundamental tool to adopt that is to foresee and plan mitigation measures in a timely manner. This is due to network capacity constrains manifesting from more expected build-up projects coming online. The investment in such a tool would be a cost effective measure in comparison with potential cost generating from productivity lost (delays) of operations and construction activities. Thus the list of mitigation measures as proposed in this traffic study are highly recommended and its timely execution is key to its success in addressing the expected traffic peak flow of 1st quarter of year 2008.

ACKNOWLEDGEMENT

The authors would like to acknowledge the RLIC staff, Cheyoda-Technip Joint Venture (CTJV) staff and various other consultants involved in this project.

REFERENCES

Parsons International Limited. 2007. Ras Laffan Industrial City: *Traffic Impact Study.* Cheyoda-Technip JV. Doha. Qatar.

Transportation Research Board. 2000. *Highway Capacity Manual.* National Research Council. Washington DC. USA.

PTV. 2007. *Visum User Manual*—V.9.44. Karlsruhe. Germany.

Akcelik, R., et al. 2002. *aaSidra 2.0 Software.* Akcelik & Associates Pty Ltd, Victoria. Autralia http://en.wikipedia.org/wiki/GEH.

Road and traffic safety

Efficient Transportation and Pavement Systems – Al-Qadi, Sayed, Alnuaimi & Masad (eds)

Evaluation of Edmonton's intersection safety camera program

Tarek Sayed
Civil Engineering, University of British Columbia, Vancouver, BC, Canada

Paul de Leur
Road Safety Engineer, de Leur Road Safety Consulting Ltd, Vancouver, BC, Canada

ABSTRACT: Enforcing traffic signal compliance in urban areas can be a difficult task as the process is often limited by police resources and by the traditional enforcement methods. Therefore, conventional traffic enforcement can be supplemented with advanced technologies, such as intersection safety cameras. Intersection safety cameras, or often referred to as red light cameras, are being used increasingly to help communities enforce against deliberate red light running. Since the 1970's, Europe, Australia and North America have been using photo enforcement technology to improve safety at intersections.

The City of Edmonton has used automated photo enforcement as part of the overall traffic enforcement activity. The Intersection Safety Camera Program started as a pilot project in 1998, which quickly expanded into a comprehensive program for the entire City. Since a significant period of time has elapsed from the inception of the program, the Edmonton Police Commission wanted to complete a study to determine the overall safety effect of the photo enforcement program.

This paper describes the evaluation of Edmonton's Intersection Safety Camera (ISC) Program. The paper describes the methodology that was used in the evaluation, the data collection/compilation effort required for the evaluation and finally, the results of the program, measured in terms of reducing the number of collisions at the intersections that were treated with photo enforcement cameras. Also included in the paper is a summary of the literature associated with the deployment of red light cameras.

Keywords: Red light Cameras, Intersection Safety Cameras, Photo Enforcements Initiatives, Photo Radar Programs

1 INTRODUCTION

The Edmonton Police Service and the City of Edmonton's Transportation and Street Department launched the Intersection Safety Camera Program as a pilot project in 1998. Shortly thereafter and after the passage of some provincial legislation, a comprehensive Intersection Safety Camera Program was implemented for the City. The pilot project initially started with one location in 1998 but expanded considerably to include up to 60 different locations throughout the City, with a total of 24 enforcement cameras available for deployment at these 60 locations.

Like many other jurisdictions, the Edmonton Intersection Camera Safety Program is dynamic, meaning that the location of the cameras moves between the various enforcement sites. This way, the enforcement sites are not at fixed locations such that the motoring public could adapt their driving behavior only at the locations where the ISC program is deployed. The purpose of the rotation of the cameras between enforcement sites is to attempt to change driver behavior over a larger area and not just at fixed locations. The intent is to make drivers think that an intersection safety camera could be deployed at a site and to avoid a violation, they should modify their driving behavior and not attempt a hazardous maneuver of running a red light.

To assist the motoring public in allowing them to modify their driving behavior and as required by provincial legislation, a fixed sign is provided on the approach to an intersection to warn them that the intersection is a photo enforcement location. The warning sign consists of images of vehicles entering the intersection contrary to the signal indication for their travel direction.

Since a significant period of time has elapsed from the inception of the photo enforcement activities, the Edmonton Police Commission wanted to have a study completed to determine the overall safety effect of the photo enforcement initiative. This paper describes the project to evaluate Edmonton's Intersection Safety Camera (ISC) Program. Included in the paper is a brief summary of the literature associated with intersection safety cameras, or often referred to as red light cameras. The paper will also describe the methodology that was used in the evaluation, the data collection/compilation effort required for the evaluation and finally, the results of the program, measured in terms of reducing the number of collisions at the intersections that were treated with photo enforcement cameras.

2 LITERATURE REVIEW

A comprehensive literature review of intersection photo enforcement programs was completed as part of this assignment. The literature review focused on the enforcement and safety effects resulting from the implementation of red light cameras. The literature review also focused on information considered to be the most reliable, including studies that deployed a reliable evaluation methodology, which were supported with the availability of good quality data.

From the literature review, it was determined that automated intersection enforcement systems can be very a very effective enforcement tool. Most studies have shown that the violation frequency is significantly reduced after the implementation of a red-light camera (RLC) system at the intersection safety camera locations and at non-camera locations. A summary of the literature reviewed is provided in Table 1.

The findings of studies that investigated the safety effects resulting from red-light camera programs are somewhat varied but in general, these camera enforcement systems can produce a slight reduction in total collisions and a significant reduction in high severity collisions. Furthermore, the results of these studies generally suggest that the more severe angle-type collisions are significantly reduced, however, a slight increase in the less severe, rear-end type collisions can often occur. The literature also discusses the potential for a "spillover" effect, where the effect of an intersection safety camera location may extend to nearby, untreated intersection.

Table 1. Effectiveness of red-light cameras as enforcement tools.

Author	Year	Country	Effectiveness measure & estimates
Oei et al.	1997	Netherlands	56% reduction in driver violations
Thompson et al.	1989	UK	One site reported a 22% reduction, another reported a 13% increase in driver violations
Arup	1992	Australia	78% reduction in driver violations at camera sites 67% reduction in driver violation at non-camera sites
Chin	1989	Singapore	42% reduction in driver violations at camera sites 27% reduction in driver violation at non-camera sites
Retting et al.	1999a	USA	44% reduction in driver violations at camera sites 34% reduction in driver violation at non-camera sites
Retting et al.	1999b	USA	40% reduction in driver violations at camera sites 50% reduction in driver violation at non-camera sites
Chen et al.	2001	Canada	69% reduction in driver violations after 1 month 38% reduction in driver violations after 38 month Both reductions were at non-camera sites

There are a few studies that provide evidence contrary to the view that red light cameras will reduce collisions and in fact, some studies suggest that red light cameras will show an overall deterioration in safety. Several studies tried to explore the economic consequence of installing RLC but the analysis is based on several assumptions and how the authors perceive the cost/per collision estimates.

The details of the literature review of the safety impacts of red light camera programs are summarized in Table 2 on the following page.

Based on the literature reviewed, it appears that an automated enforcement system at intersections can be a very effective safety initiative. Most studies have shown that the violation frequency is significantly reduced after the implementation of a red-light camera system. It is suggested that this behavioral change could translate into a safety benefit in the form of reduced collisions frequency and/or collision severity.

Table 2. Effectiveness of red-light cameras for as enforcement tools.

Author	Year	Country	Effectiveness measure & estimates		
Persaud et al.	2005	USA		Total crashes	Injuries
			Right angle	–25%	–16%
			Rear end	15%	24%
Council et al.	2005	USA	Using aggregated economic safety costs, a $28,000 to $50,000 economic benefit per site per year.		
Burkey & Obeng	2004	USA	40% increase in total crashes 40 to 50% increase in PDO and injury crashes		
Hillier et al.	1993	Australia	8% reduction in total crashes 26% reduction in injury crashes 29% increase in total right angle crashes 108% increase in total rear end crashes		
South et al.	1988	Australia	7% reduction in injury crashes 32% reduction in injury right angle crashes 31% reduction in injury rear end crashes		
Andreassen	1995	Australia	7% increase in total crashes 13% reduction in total right angle crashes 20% increase in total rear end crashes		
Mann et al.	1994	South Australia	6% increase in total crashes 8% increase in total right angle crashes 12% increase in total rear end crashes 20% reduction in injury crashes 26% reduction in injury right angle crashes 1% reduction in injury rear end crashes		
Ng et al.	1997	Singapore	9% reduction in injury crashes 10% reduction in injury right angle crashes 6% increase in injury rear end crashes		
Retting & Kyrychenko	2002	USA	7% reduction in total crashes 32% reduction in total right angle crashes 3% increase in total rear end crashes 29% reduction in injury crashes 68% reduction in injury right angle crashes		
Office of road Safety	1991	South Australia	8% reduction in total crashes 38% reduction in total right angle crashes 14% increase in total rear end crashes 23% reduction in injury crashes 54% reduction in injury right angle crashes 25% increase in injury rear end crashes		
Queensland Transport	1995	Australia	48% reduction in total crashes 46% reduction in injury crashes		

3 RELIABLE EVALUATION OF ROAD SAFETY INITIATIVES

In reviewing the literature associated with photo enforcement initiatives, it becomes evident that a reliable evaluation is critical to make definitive conclusions concerning the safety effect of these initiatives. The specific effect of safety initiatives such as photo enforcement is not self-evident, even to safety professionals with considerable experience.

There are many factors that must be considered and addressed when attempting to evaluate a safety initiative. These factors, referred to confounding factors, are serious obstacles in conducting a reliable evaluation and if not adequately addressed, then the results from the evaluation cannot be trusted. Three primary confounding factors that needed to be addressed to ensure reliable results for the evaluation of Edmonton's Intersection Safety Camera Program included the following:

1. *History*: The possibility that other factors other than intersection safety camera may have caused all or part of the change in safety at a location.
2. *Maturation*: The long-term collision trend effects that may contribute to changes in safety performance at sites treated with an intersection safety camera.
3. *Regression Artifacts*: The tendency of collisions to oscillate from high counts to low counts, even if no change has occurred within the system.

4 METHODOLOGY TO EVALUATE THE ISC PROGRAM

The methodology to evaluate Edmonton's Intersection Safety Camera (ISC) Program employs the use of collision prediction models. Collision prediction models (CPMs) are mathematical models that relate the collision frequency experienced by a road entity to the various traffic and geometric characteristics of this entity. The CPMs are developed using certain statistical techniques and have several applications such as evaluating the safety of various road facilities, identifying collision-prone locations, and evaluating the effectiveness of safety improvement measures.

Historically, two statistical modeling methods have been used to develop collision prediction models: 1) conventional linear regression, and 2) generalized linear regression. Conventional linear regression assumes a normal distribution error structure, whereas a generalized linear modeling approach (GLM) assumes a non-normal distribution error structure (usually Poisson or negative binomial). Recently generalized linear regression modeling (GLM) has been used almost exclusively to develop CPMs since conventional linear regression models lack the distributional property to adequately describe crashes. The inadequacy is due to the random, discrete, non-negative, and typically sporadic nature, that characterize collision occurrence.

The prediction model structure used in this study relates the frequency of collisions to the product of traffic flows entering the intersection. The model form is shown below in equation 1.

$$E(\Lambda) = a_0 V_1^{a_1} V_2^{a_2} \tag{1}$$

Where: $E(\Lambda)$ = Expected collision frequency (collisions/3 years)
V_1, V_2 = Major/minor road traffic volume (AADT)
a_0, a_1, a_2 = Model parameters.

The variance of the expected collision frequency is given by equation 2.

$$Var(\Lambda) = \frac{E(\Lambda)^2}{\kappa} \tag{2}$$

Where: κ = The negative binomial parameter of the CPM.

The reduction in the number of collisions at the treatment sites can be calculated using the Odds Ratio (*O.R.*) according to equation 3. The effect of the treatment is determined by subtracting 1 from the Odds Ratio, as shown below in equation 4.

$$O.R. = \frac{A/C}{B/D} \tag{3}$$

$$Treatment\ Effect = O.R. - 1 \tag{4}$$

Where: A = Safety at the comparison site in the before period;
B = The EB safety estimate at the treated sites if no treatment occurred;
C = Safety at the comparison sites in the after period; and
D = Safety at the treatment sites in the after period.

It should be noted that all quantities in the Odds Ratio are observed quantities (with assumed Poisson distribution), with the exception of quantity *B*, which is calculated. Therefore, the major work involved in evaluating the benefits of a certain treatment is basically determining the quantity *B*. This quantity is calculated by utilizing CPMs and the Empirical Bayes (EB) refinement procedure. The Empirical Bayes safety estimate and its variance for a location *i* are calculated using equations 5 and 6 as follows:

$$(EB_i)_b = \gamma_i \cdot E(\Lambda_i) + (1-\gamma_i)\cdot(y_i), \quad VAR(EB_i)_b = \gamma_i \cdot (1-\gamma_i)\cdot E(\Lambda_i) + (1-\gamma_i)^2 \cdot (\gamma_i) \tag{5}$$

$$\gamma_i = \frac{E(\Lambda_i)}{E(\Lambda_i) + VAR(\Lambda_i)} = \frac{1}{1 + \frac{VAR(\Lambda_i)}{E(\Lambda_i)}} \tag{6}$$

Where: κ = The negative binomial parameter of the CPM; and
y_i = The observed collisions in the before period for location *i*.

The value *B* in the Odds Ratio, from equation 3, is calculated by using equation 7 and following Sayed et al. (Sayed, 2001):

$$B = (EB_i)_a = (EB_i)_b \times \frac{E(\Lambda_i)_a}{E(\Lambda_i)_b} \tag{7}$$

Where: $(EB_i)_a$ = The EB safety estimate of treated site *i* in the "after" period had no treatment taken place.
$(EB_i)_b$ = The EB safety estimate of treated site *i* that occurred in the "before" period.
$E(\Lambda_i)_a$ = The collision frequency given by the CPM for treated site *i* using its traffic flows in the "after" period.
$E(\Lambda_i)_b$ = The collision frequency given by the CPM for treated site *i* using its traffic flows in the "before" period.

To get the expected value and variance of the Odds Ratio, the method of statistical differentials is used as follows:

$$\mathrm{E}\{Y\} = Y + \left[\sum_1^n (\partial^2 Y/\partial X_i^2)\ \mathrm{VAR}\{X_i\}\right] \Big/ 2 \tag{8}$$

$$\mathrm{VAR}\{Y\} = \left[\sum_1^n (\partial Y/\partial X_i)^2\ \mathrm{VAR}\{X_i\}\right] \tag{9}$$

By applying equations 8 and 9 to the Odds Ratio as defined in equation 3, the following two equations (10 and 11) for the Odds Ratio can be obtained:

$$E(O.R.) = \left(\frac{A/C}{B/D}\right) \times \left(1 + \frac{VarB}{B^2} + \frac{VarC}{C^2}\right) \qquad (10)$$

$$Var(O.R) = \left(\frac{A/C}{B/D}\right)^2 \times \left(\frac{VarA}{A^2} + \frac{VarB}{B^2} + \frac{VarC}{C^2} + \frac{VarD}{D^2}\right) \qquad (11)$$

5 ISC PROGRAM: DATA COLLECTION AND COMPILATION

With any time-series evaluation of a safety treatment, the success of the evaluation is governed by the quality and quantity of reliable data. Based on the discussion with staff from the City of Edmonton, it was believed that the data available for the ISC Program would be sufficient to allow for an accurate evaluation of the program.

Data was required for three distinct groups of sites when completing the evaluation of Edmonton's Intersection Safety Camera Program. The three distinct groups of sites include the following:

1. Treatment Group Sites:
   - These are the group of sites (intersections) that have an intersection safety camera installed and deployed.
2. Comparison Group Sites:
   - These are a separate group of sites that are subjected to the same traffic and environmental conditions as the treated sites but do not have an ISC installed at the intersection.
3. Reference Group Sites:
   - This is a large group of sites that are considered of similar character to the treatment sites and used to develop the collision prediction model used in the evaluation.

5.1 *Treatment group sites*

A total of 25 sites were selected to serve as the treatment group of sites (i.e., the site where the intersection safety cameras were installed). This sample of locations would allow for a successful evaluation of the ISC Program and would satisfy the data needs for the before and after time periods. All locations in the treatment group of sites were selected to be typical, four-leg, intersections with either 3 of 4 lanes on each approach and with a posted speed of 50 kph.

Accurate traffic volume and collision data was available for each treatment group location in both the before and after time periods (i.e., before and after the intersection safety camera was installed). The before collision data corresponded to a 3 year time period before the implementation of the ISC and the after collision data dates ranged from 2 to 3 years after the ISC was implemented. Considerable effort was undertaken by City staff to collect reliable traffic volume data for the before and after time periods. Table 3 provides a list of the treatment group sites, including the implementation date when the intersection safety camera was installed.

5.2 *Comparison group sites*

The control group of sites is used to correct for the time trend effects, including the confounding factors of history and maturation. City staff selected the control group sites, ensuring that the control group sites had similar traffic and environmental conditions as the treated sites. Care must be exercised in selecting the control group sites to ensure that the time periods for the treatment and control sites are similar and that the factors influencing safety are similar between the treatment and control sites.

Table 3. Treatment group sites.

ID No.	Major road	Minor road	ISC Implementation date
1	Calgary Trail	23 Ave	September 1, 1999
2	23 Avenue	50 Street	August 16, 2002
3	34 Avenue	111 Street	June 27, 2003
4	66 Street	34 Avenue	November 28, 2003
5	91 Street	34 Avenue	February 1, 2002
6	99 Street	63 Avenue	February 1, 2002
7	75 Street	82 Avenue	October 18, 2002
8	178 Street	87 Avenue	January 13, 2004
9	170 Street	87 Avenue	March 21, 2001
10	75 Street	98 Avenue	September 11, 2002
11	50 Street	101 Avenue	September 1, 1999
12	Groat Road	111 Avenue	March 21, 2001
13	97 Street	127 Avenue	September 1, 1999
14	97 Street	137 Avenue	March 21, 2001
15	137 Avenue	127 Street	January 31, 2002
16	50 Street	137 Avenue	November 28, 2003
17	153 Avenue	82 Street	November 28, 2003
18	75 Street	Argyll Road	September 1, 1999
19	97 Street	167 Avenue	July 2, 2003
20	Jasper Avenue	116 Street	September 1, 1999
21	Albert Trail	137 Avenue	September 1, 1999
22	Stony Plain Road	156 Street	June 27, 2003
23	114 Street	University Ave	January 31, 2002
24	Yellowhead Trail	149 Street	March 21, 2001
25	Yellowhead Trail	156 Street	September 1, 1999

Staff from the City's Traffic Operations Department identified a total of 47 suitable control group locations for the treatment sites. The control group sites were matched to the treatment sites based on the implementation date of the ISC at each treatment site. This was to ensure that the time periods for the before and after time periods for the traffic and collision data matched between treatment and control sites. A total of four control groups were established as follows:

- Treatment Sites Implemented in 1999: 12 Control Sites
- Treatment Sites Implemented in 2001: 7 Control Sites
- Treatment Sites Implemented in 2002: 14 Control Sites
- Treatment Sites Implemented in 2003: 13 Control Sites

It is also noted that there were no treatments sites selected that were implemented in 2000, and thus, there were no control group for 2000. An attempt was made to differentiate between comparison group sites that are in close proximity to treatment sites and those that are not, in order to investigate the potential "spillover" effect, which extends the effects of RLCs to nearby, untreated intersections. Unfortunately, it was difficult to select comparison sites that are not in close proximity to treated sites since RLC intersections are scattered throughout the City of Edmonton. Therefore, it is difficult to determine whether a "spillover" effect exists. However, it should be noted that the presence of "spillover" would likely mean that the results presented here underestimate the safety impact of RLC.

For each of the control sites, staff from the City of Edmonton's Traffic Operations Department provided the requisite traffic volume and collision data. The before traffic volume and collision data included a 3 year time period and the after traffic volume and collision data dates ranged from 2 to 3 years.

5.3 *Reference group sites*

The third data set that is required for the Intersection Safety Camera evaluation was the data for the reference group sites. City staff provided the required traffic volume and collision frequency data for 100 reference group locations. The traffic and collision data was provided for a three year time period before the implementation of the intersection safety cameras.

The reference group is a large group of sites of similar character to the treatment sites, used to develop a predictive model that can estimate the collision frequency at an intersection based on the amount of traffic entering the intersection. The collision prediction model (CPM) developed from the reference group is used to correct for the problems created by the regression to the mean. It is also used to determine the value of "B" in the Odds Ratio, as discussed in the methodology section of this paper. CPMs for total collisions, severe collisions, and property damage only (PDO) collisions, and for collision types were developed using the reference group of sites.

5.4 *Collision data in Edmonton*

Since collision databases between jurisdictions can be quite different, it is important to understand how Edmonton's collision data should be interpreted. The City of Edmonton's Transportation Department maintains an annual computerized file, namely the Motor Vehicle Collision Inventory System (MVCIS), which contains the details of reportable motor vehicle collisions within the City. The Transportation Department uses the data to:

- Recognize trouble spots;
- Determine probable causes and recommend engineering solutions;
- Assess effect of remedial engineering measures;
- Provide input to selective enforcement programs; and
- Provide market information for safety training program and campaigns.

Collisions, as used by the Transportation Department, refer to reportable on-street collisions that result in $1000 property damage and/or injury, do not occur on private property, and must include at least 1 motor vehicle. The MVCIS database does not include non-vehicular collisions (i.e., cyclist hitting a pedestrian). The MVCIS database includes collisions that are forwarded to the Transportation Department by the Police Service within a specified time.

Intersection collisions include collisions that occur inside and 10 meters past the legally defined limits of the outer crosswalk lines of intersecting roads and includes right-turn cut-offs. Mid-block collisions include collisions that occur on a section of roadway between two intersections. Mid-block include the balance of collisions that do not occur in an intersection as defined above.

The Province of Alberta owns the Alberta Collision Report Form (ACRF) and has designed the collision form to include general information that can be shared by enforcement and transportation departments federally, provincially and by the municipalities. Since many parts of the ACRF form provide limited information to analyze collisions for transportation purposes, the Transportation Department of the city of Edmonton supplements several of the choices in the ACRF, in particular the Primary Event box or Collision Type. Important to the evaluation of the intersection safety camera program is the following assignments that are made by City staff:

1. An incident coded with a cause that is specified as "Follow to Close" is considered to be equivalent to *rear end* type of collision.
2. An incident coded with a cause that is specified as "Left turn across path" is considered to be equivalent to *right angle, head on* type of collisions.
3. An incident coded with a cause specified as "Failed to Observe Traffic Signal" is considered to be equivalent to *right angle, head on* collision type.

It is duly noted that the evaluators of Edmonton's Intersection Safety Camera program has accepted the collision type assignment assumptions listed above and have not made any attempts to verify the link between the collision cause and collision type. Furthermore, comparisons with other studies that cite changes in collision type may not be relevant due to the definitional issues noted.

To maintain accuracy of the collision details, logic checks are built into the MVCIS program. In some cases the police may fill out the ACRF improperly, with conflicting logic. In such cases, the transportation staff will choose information that makes the most logical sense and therefore, record a different cause than the police staff entered in the primary box.

6 RESULTS: ISC PROGRAM

The results of the Intersection Safety Camera Program and is divided into two parts. The first section presents the result of the development of the collision prediction models, necessary for the evaluation. The second section presents the changes in safety caused by the implementation of the ISC Program.

6.1 *Development of collision prediction models (CPMs)*

Using the volume and collision data from the 100 reference group sites, a total of four CPMs were developed. The CPMs can predict the number of collisions at signalized intersections in Edmonton, with a CPM for each of the following types of collisions:

- CPM 1: Predicts the frequency of severe collisions (fatal and injury);
- CPM 2: Predicts the frequency of property damage only collisions (PDO);
- CPM 3: Predicts the frequency of angle collisions (right-angle, head-on)[1];
- CPM 4: Predicts the frequency of rear-end collisions[2]; and,

The functional form used to development the CPMs are based on the product of the two volumes entering the intersection.

$$E(\Lambda) = a_o V_1^{a_1} V_2^{a_2} \tag{12}$$

Where: $E(\Lambda)$ = Expected collision frequency,
V_1, V_2 = Major/minor road traffic volume (AADT),
a_o, a_1, a_2 = Model parameters.

The various collision prediction models and the corresponding model parameters that have been developed for the evaluation of Edmonton's Intersection Safety Camera Program are shown in Table 4.

Two statistical measures were used to assess the significance and goodness of fit of the prediction models, including the *Pearson* χ^2 statistic (equation 13) and the scaled deviance (SD) (Equation 14). Both the *Pearson* χ^2 and the scaled deviance statistics are asymptotically χ^2 distributed with *n-p-1* degrees of freedom and therefore for a well-fitted model, the expected value of the *Pearson* χ^2 and the SD will be approximately equal to the number of degrees of freedom (Maycock and Hall, 1984).

$$Pearson\ \chi^2 = \sum_{i=1}^{n} \frac{[y_i - E(\Lambda_i)]^2}{Var(y_i)} \tag{13}$$

$$SD = 2\sum_{i=1}^{n}\left[y_i \ln\left(\frac{y_i}{E(\Lambda_i)}\right) - (y_i + \kappa)\ln\left(\frac{y_i + \kappa}{E(\Lambda_i) + \kappa}\right)\right] \tag{14}$$

[1] Right-angle, head-on collisions ("angle" collisions) is equivalent to "Failed to observe traffic signal" and "Left-turn across path" based on assumptions by Edmonton's Transportation Department.

[2] A "rear-end collision" is assumed to be equivalent to the collision cause of "Following too closely", an assumption made by the Transportation Department at the City of Edmonton.

Where: y_i = Observed number of collisions at an intersection;
$E(\Lambda_i)$ = Predicted number of collisions obtained from CPM; and,
$Var(y_i)$ = The variance of the observed claims.

The details of the goodness of fit measures for the developed CPMs are shown in Table 5 below. All models showed a very good fit to data.

6.2 *Safety effect of the ISC program*

Using the control groups established for the various treatments sites, which were based on the implementation year for the treatment site and using the various collision prediction models developed and presented above, the safety effect caused by Edmonton's Intersection Safety Camera Program can be determined.

The safety effect results from the ISC Program are summarized according to each treatment site and combined for all sites evaluated. The change in the safety performance at each treatment site is summarized in Table 6 on the following page.

As can been seen from the results, 20 of the 25 locations experienced a reduction in the total number of collisions after the ISC was implemented. The changes in the total collisions are based on the sum of the changes in PDO and F+I collisions. The magnitude of the reductions per treatment site ranged from a low of 3.7% to a high of 41.4%. A total of 17 of the 25 sites experienced a reduction in severe collisions, with the decreases ranging from 3.5% to 51.4%. Similarly, the PDO collisions were reduced at 19 of the 25 sites, with percent reductions ranging from 0.7% to 45.6%.

An investigation of the collision types reveals similar results. The "angle" collisions were reduced at 18 of the 25 sites investigated, with decreases ranging from 1.7% to 85.2%. In terms of the "rear-end" collisions, 18 sites from the total of 25 treated sites experienced a reduction of "rear-end" collisions, with the magnitude of the reduction ranging from a low of 8.2% to a high of 55.2%. The safety performance results indicate than the majority of treatment sites show a very positive overall effect due to the implementation of ISCs.

The results for the total reduction in collisions at each treatment site are illustrated graphically in Figure 1. It is noted that one treatment site has experienced a large increase in total collisions (+45%), specifically, Site 15: 137th Avenue at 127th Street. It is unknown whether there are other conditions or circumstances that may be contributing to the increase of collisions at this

Table 4. Collision prediction models for different collision severity and types.

CPM reference	Collision type	Model	κ
CPM-1	Severe (F+I)	$Col/3yrs = Exp(-8.9957)\ V_1^{0.779}\ V_2^{0.4464}$	6.266
CPM-2	PDO	$Col/3yrs = Exp(-10.172)\ V_1^{0.9368}\ V_2^{0.4482}$	8.163
CPM-3	Angle	$Col/3yrs = Exp(-7.3445)\ V_1^{0.6137}\ V_2^{0.4488}$	5.171
CPM-4	Rear End 1	$Col/3yrs = Exp(-13.9331)\ V_1^{1.223}\ V_2^{0.5100}$	3.997

Table 5. Goodness of fit measures for collision prediction models.

	a_0		a_1		a_2				
CPM	STD. Error	T-Stat.	STD. Error	T-Stat.	STD. Error	T-Stat.	SD	χ^2	$\chi^2_{(0.05)}$
Severe (F+I)	1.35	−6.66	0.13	5.60	0.108	4.10	114.74	86.50	119.87
PDO	1.17	−8.62	0.12	7.57	0.09	4.95	111.56	101.58	119.87
Angle	1.52	−4.80	0.15	3.99	0.11	3.88	116.84	91.99	119.87
Rear End 1	1.64	−8.46	0.17	7.06	0.12	4.09	108.50	95.42	119.87

Table 6. Summary of results: Individual intersections for the ISC program.

	Treatment site		Change in collisions[3]				
ID No.	Major road	Minor road	Total	Severe	PDO	Angle	RE
1	Calgary Trail	23 Ave	–3.7%	17.9%	–14.7%	–5.2%	–13.3%
2	23 Avenue	50 Street	16.0%	–13.5%	50.5%	26.8%	5.1%
3	34 Avenue	111 Street	6.3%	–11.5%	20.9%	–37.8%	13.8%
4	66 Street	34 Avenue	–5.9%	–17.7%	4.9%	–12.5%	–19.1%
5	91 Street	34 Avenue	26.2%	31.3%	21.6%	68.5%	6.6%
6	99 Street	63 Avenue	–20.8%	43.8%	–39.5%	–11.7%	–31.4%
7	75 Street	82 Avenue	–23.6%	–17.3%	–27.8%	–1.7%	–27.2%
8	178 Street	87 Avenue	–22.9%	–41.4%	–6.5%	20.4%	–50.5%
9	170 Street	87 Avenue	–9.0%	–6.1%	–11.3%	–53.7%	4.8%
10	75 Street	98 Avenue	–12.6%	–25.0%	–3.4%	–8.5%	–18.1%
11	50 Street	101 Avenue	–16.0%	16.4%	–34.1%	–10.1%	–30.6%
12	Groat Road	111 Avenue	–14.2%	–11.7%	–15.7%	–10.3%	–23.2%
13	97 Street	127 Avenue	–37.8%	–32.5%	–41.5%	–51.7%	–24.9%
14	97 Street	137 Avenue	–8.2%	–8.9%	–7.8%	–32.4%	3.9%
15	137 Avenue	127 Street	44.8%	86.0%	19.1%	61.3%	38.2%
16	50 Street	137 Avenue	–8.1%	11.9%	–18.9%	–13.9%	–13.4%
17	153 Avenue	82 Street	–4.6%	–10.2%	–0.7%	7.2%	–20.6%
18	75 Street	Argyll Road	–9.4%	2.3%	–14.0%	3.7%	–15.3%
19	97 Street	167 Avenue	4.6%	29.8%	–9.8%	9.4%	–8.7%
20	Jasper Ave.	116 Street	–41.4%	–51.4%	–35.9%	–39.1%	–55.2%
21	Albert Trail	137 Avenue	–3.8%	–32.4%	15.5%	–28.8%	2.1%
22	Stony Plain Rd.	156 Street	–30.7%	–8.0%	–45.6%	–26.2%	–38.3%
23	114 Street	University Ave	–9.4%	–3.5%	–13.4%	–24.1%	–8.2%
24	Yellowhead Trail	149 Street	–37.0%	–50.7%	–28.3%	–85.2%	–21.1%
25	Yellowhead Trail	156 Street	–25.4%	–16.0%	–30.7%	–55.2%	–23.6%

[3] A negative value represents a reduction in the collision frequency and conversely, a positive value represents an increase in collisions.

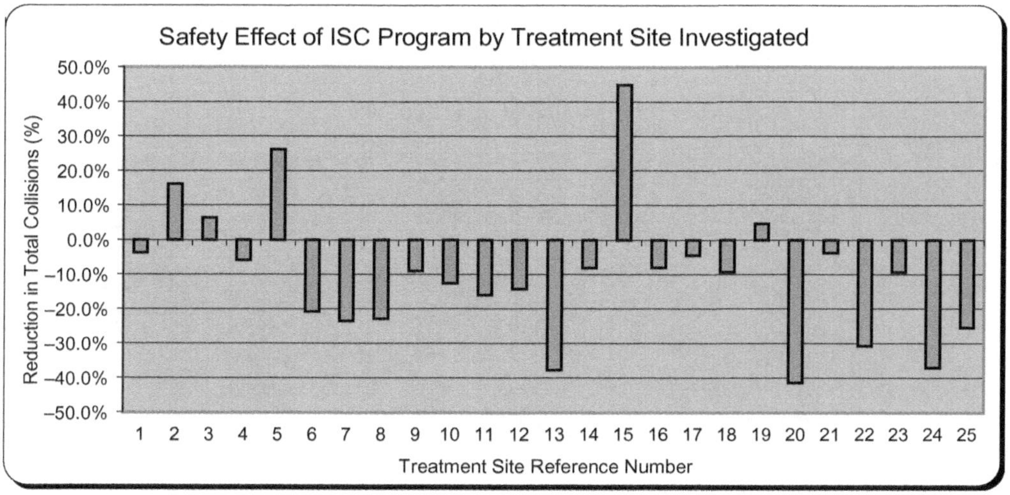

Figure 1. Safety effect [4] of ISC program at each treatment site reviewed.
[4] A negative value represents a reduction in the collision frequency and conversely, a positive value represents an increase in collisions.

intersection. The result at Site 15 does have an impact on the overall safety effect of the 25 sites examined, as will be discussed later in this section.

The safety performance results from the Intersection Safety Camera evaluation is also presented in terms of the overall program effectiveness, considering all of the 25 treatment sites that were evaluated. The results are presented in Table 7, which includes the percent reduction by collision severity level or type. Also in the table is a significance value used to determine if the collision reduction values are statistically significant.

From Table 7, the reduction in total collisions is estimated to be 11.1%. Overall, severe collisions were reduced by 6.1%, and PDO collisions were reduced by 14.3%. "Angle" type collisions were reduced by 17.2% and "rear-end" collisions were reduced by 12.4%.

The results also indicate that the collision reductions are statistically significant at the 95% level, except for severe collision. This is largely attributed to Site 15 (37th Avenue at 127th Street), which showed a +86.0% increase in severe incidents after the implemen-

Table 7. Summary of results considering 25 sites evaluated.

Collision category	Change in collisions[4]	p-value	Significant at 95%
Severe (F+I)	–6.1%	16.8	NO
PDO	–14.3%	0.00	YES
"Angle"	–17.2%	0.00	YES
"Rear End"	–12.4%	0.01	YES
Total	–11.1%	0.00	YES

Table 8. Summary of results: considering 24 sites evaluated.

Collision category	Change in collisions[5]	p-value	Significant at 95%
Severe (F+I)	–9.3%	3.25	YES
PDO	–15.5%	0.00	YES
"Angle"	–20.2%	0.00	YES
"Rear End 1"[6]	–14.0%	0.00	YES
Total	–13.0%	0.00	YES

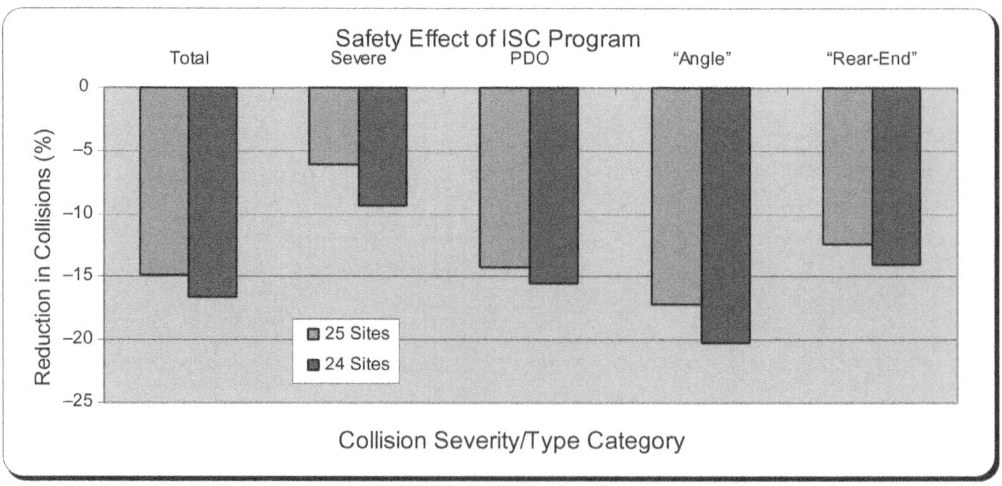

Figure 2. Summary of safety effect of ISC program.

tation of the intersection safety camera. If Site 15 (137th Avenue at 127th Street) were removed from the assessment and only 24 treatment sites were reviewed, the results would improve slightly and the estimated collision reductions for all collision categories would be statistically significant at the 95% confidence level. These results are shown below in Table 8.

The results for the overall program effectiveness is illustrated Figure 2. As can be seen from the figure, there is a reduction in collisions after the implementation of the Intersection Safety Camera at a site. Included in the figure below are the results from all 25 sites evaluated, as well as the results when Site 15 (137th Avenue at 127th Street) were removed from the treatment group (i.e., 24 treatment sites).

7 CONCLUSION OF EDMONTON'S ISC PROGRAM

The methodology used to evaluate Edmonton's Intersection Safety Camera Program is considered to yield accurate and reliable results. Furthermore, the data required to support an effective evaluation was available, including data for the treatments sites, control group sites and reference group of sites. This data allowed for the evaluation to address the threats to a successful evaluation, namely the confounding factors of history, maturation and regression artifacts.

The results show that the introduction of the ISC Program was effective in reducing the collisions at locations where the cameras were used. The majority of the 25 sites investigated showed an overall collision reduction, with only one site, showing a significant increase in collisions (it is noted that a few other sites showed an relatively small increase in collisions). Overall, all collision severity and collision type categories (Total, Severe, PDO, "Angle" and "Rear-end") showed significant collision reductions, with total collision reductions exceeding 10%, noting that in the results, that the collision type is based on collision cause.

ACKNOWLEDGMENTS

We would like to acknowledge and thank the efforts by staff from the Edmonton Police Commission/Police Department and staff from the City of Edmonton's Transportation Department, who were very helpful in preparing the significant amount of data required for the evaluation and responding to the needs of the evaluation.

REFERENCES

1. Andreassen, D., 1995. A Long Term Study of Red Light Camera and Accidents, ARRB Report ARR 261, Australian Road Research Board, Victoria, Australia.
2. Arup, 1992. Red Light Camera Evaluation Study—Implementation in Brisbane, Report No. 6221, Arup Transportation Planning for Queensland Transport, Melbourne, Australia.
3. Burkey, M., and Obeng, K. A Detailed Investigation of Crash Risk Reduction Resulting from Red Light Cameras in Small Urban Areas. Report for USDOT, July 2004, 60 pp.
4. Chen, G., Wilson, J., Meckle, W., and Casey, R., 2000. General Deterrence Effects of Red Light Camera and Warning Signs in Traffic Signal Compliance in British Columbia, J. Traffic Med., Vol. 29, pp. 46–53.
5. Chin, H.C., 1989. Effect of Automatic Red-Light Cameras on Red-Running. Traffic Engineering and Control (30), pg. 175–179.
6. Council, F.M., Persaud, B., Lyon, C., Eccles, K., Griffith, M., Zaloshnja, E., and Miller, T. 2005. Economic Analysis of the Safety Effects of Red Light Camera Programs and the Identification of Factors Associated with the Greatest Benefits. Paper presented at the 84th annual meeting of the Transportation Research Board, Washington DC.
7. Hillier, W., Ronczka, J., and Schnerring, F. 1993. An Evaluation of Red Light Cameras in Sydney, Research Note RN 1/93, Road Safety Bureau, Roads and Traffic Authority, New South Wales, Australia.

8. Mann, T., Brown, S., and Coxon, C., 1994. Evaluation of the Effects of Installing Red Light Cameras at Selected Adelaide Intersections, Report Series 7/94, Office of Road Safety, South Australian Department of Transport, Walkerville, South Australia.
9. Maycock, G. and Hall, R.D., 1984. Accidents at 4-arm Roundabouts, Transport and Road Research Laboratory. TRRL Laboratory Report 1120.
10. Ng, C.H.,Wong,Y.D., Lum, and K.M., 1997. The Impact of Red-Light Surveillance Cameras on Road Safety in Singapore, J. Road Transport Res., Vol. 6, p. 2.
11. Oei, H.L., Catshoek, J.W.D., Bos, J.M.J., and Varkevisser, G.A., 1997. Project Red-Light and Speed (PRO-ROS), SWOV Report R-97-35, Institute for Road Safety Research, Leidschendam, the Netherlands.
12. Persaud, B., Council, F.M., Lyon, C., Eccles, K., and Griffith, M., 2005. A multi-Jurisdictional Safety Evaluation of Red Light Cameras. Paper presented at the 84th annual meeting of the Transportation Research Board, Washington DC.
13. Queensland Transport, 1995. Queensland's Road Toll—Queensland Transport Submission to the Parliamentary Travelsafe Committee, Queensland Transport, Queensland, Australia.
14. Retting, R.A., and Kyrychenko, S.Y., 2002. Crash Reductions Associated with Red Light Camera Enforcement in Oxnard, California, Am.J.Public Health, Vol. 92, pp. 1822–1825.
15. Retting, R.A., Williams, A.F., Farmer, C.M., and Feldman, A.F., 1999a. Evaluation of Red Light Camera Enforcement in Fairfax, Virginia, ITE J., Vol. 69, pp. 30–34.
16. Retting, R.A.,Williams, A.F., Farmer, C.M., and Feldman, A.F., 1999b. Evaluation of Red Light Camera Enforcement in Oxnard, California, Accident Analysis and Prevention, Vol. 31, pp. 169–174.
17. Retting, R.A., Ferguson, S.A., and Hakkert, A.S., 2003. Effects of Red Light Cameras on Violations and Crashes: A Review of the International Literature. Traffic Injury Prevention, 4:17–23.
18. Sawalha, Z., and Sayed, T., 2001. Evaluating the Safety of Urban Arterial Roadways. Journal of Transportation Engineering, ASCE, Vol. 127, No. 2, pp. 151–158.

Efficient Transportation and Pavement Systems – Al-Qadi, Sayed, Alnuaimi & Masad (eds)
© 2009 Taylor & Francis Group, London, ISBN 978-0-415-48979-9

Safety evaluation of alternative guardrail system for exclusive motorcycle lanes

A.B. Ibitoye
University of Ilorin, Ilorin, Nigeria

A.M.S. Hamouda
Qatar University, Doha, Qatar

R.U. Sohadi & S.V. Wong
University Putra Malaysia, Malaysia

ABSTRACT: The tremendous increase in number of motorcycles and fatalities has become a major concern to road safety experts as affect the safety of motorcyclists in relation to their collision with guardrail. In this current study, three-dimensional computer software application was used to model a motorcycle and guardrail as Multi Body system models and Finite elements model respectively. Some typical qualitative results on injury criteria and acceleration due to head, thorax, and femur are presented. The predicted rider's injury risk criteria were used to assess safety of guardrail response on motorcyclists. The obtained results confirmed that the existing w-beam guardrail is not safe to motorcyclist, especially for the head injury at impact speed 48 km/h and impact angle of 45 degree. The study then used design optimization to test twenty-four alternative models in order to propose best alternative model for consideration as the alternative guardrail design.

1 INTRODUCTION

Motorcycle usage in Malaysia and most Asian countries is a traditional norm and has become dominant mode of transportation. The uncontrolled number of motorcycle usage in these countries has resulted in high casualties among the motorcyclists. This has constituted into a major portion of the total fatalities in these countries; making motorcycle to be associated with a higher risk of death or injury than any other form of transportation. This may be due to relative less stability of motorcycle and little protection to motorcyclists, compared to a four-wheeled vehicle. In Malaysia, motorcycle accounts for 49% of registered vehicles (Royal Police of Malaysia, PDRM, 2003). About 68% of all road accident injuries involved motorcyclists with overall relative risks of about 20 times higher than that of passenger cars (Radin et al., 1995). As a result of continuous increase in the total number of registered motorcycle, which results in increasing number of motorcycle crashes, the government of Malaysia introduced exclusive motorcycle lanes to reduce safety risk to motorcyclists.

Globally, frequency of motorcycle crashes with guardrail has been reported to cause severe injuries to motorcyclists due to inadequate safety measures. In California, Ouellet (1982) reported that out of 900 motorcycle accidents, collision with barrier accounted for 63 somatic region injuries and 37 head-neck region injuries, totalling 98 injuries overall. Similarly in Los Angeles, it was found that bodily impacts with W-beam crash barrier resulted in AIS3+ injuries that occurred in 66% of head impacts with barrier (Ouellet, 1982). In West Germany, 66% of motorcyclists suffer very severe trauma after impacting with guardrails (Koch and Schuler, 1987). A real world data regarding motorcyclist impacts with crash barriers in Germany put severity of injuries as three deaths out of 50 motorcyclists impacting crash barrier, 31 seriously injured and less than one third escaped with minor injuries (Domhan, 1987).

FEMA (2000) reported a study conducted in France between 1993 and 1995 on motorcycle accident with guardrail accounting for 8% of all motorcycle fatalities and 13% of fatalities on rural roads. In a similar study in France, Quincey et al. (1988) compares motorcycle-barrier crashes with other types of motorcycle crashes and found that motorcyclists are over 5 times more likely to be killed or seriously injured in impacts with crash barrier than in other types of crashes. In Denmark, as reported in FEMA (2000), 10% of motorcyclists that run-off the road hit a crash barrier and 20% of these died because of the crash barrier while 60% of them were seriously injured.

However, the existing w-beam guardrail was designed to be strong enough to withstand high axial tensile and bending stresses that occur during heavy vehicle impacts (up to 4000 kg). Its posts provide rigidity to the entire system, while the attached blocks prevent snagging of the posts. The block-outs provide restraining forces above the centre of gravity of vehicle to prevent vehicle roll-over. In addition, ends of w-beam are provided with anchorages to enable the system to develop its full tensile strength. The stiffness and energy absorption capacity of w-shape guardrail limit displacements of the rail to around 1 m (Bank and Gentry, 2001). The energy is absorbed through plastic flexural deformation of the rail, through deformation of posts and block-outs and through plowing of posts through the soil.

These guardrail features that are vital to car and truck occupant may be fatal to motorcyclists due to the shape of the impacting area (Sala and Astori, 1998). The guardrails present certain hazards to motorcyclists as they are tested using only four wheel vehicles (EEVC, 1993). It leaves a space between the road surface and bottom of the rail that may allow fallen rider to slide under it into contact with roadside hazards. Exposed parts of guardrail support posts concentrate impact forces on the motorcyclist's body, and the edges of horizontal beams may cause laceration injuries (Viner 1995). Non-continuous surface and rigidity of this w-beam guardrail present edges that concentrate impact forces to result in more severe injuries to motorcyclists (Duncan et al., 2000). During front-on collision, the impacting rider may virtually absorb all the kinetic energy as his body is not protected and the weight of motorcycle is about 10 times lower than that of a light vehicle.

In view of danger likely to be caused to motorcyclists by the existing guardrail, this paper first investigates the safety implications of the existing guardrail system and then uses the findings to develop alternative guardrail designs that are safer for protecting motorcyclists along exclusive motorcycle lane. Thus, potential injury risks to rider are assessed based on the assumed typical crash scenario that motorcyclist remains on an upright motorcycle during collision with guardrail.

2 MATERIAL AND METHOD

This study was based mainly on a computer simulation of motorcycle crash event with guardrail as a tool for evaluating safety implication of guardrail collision to motorcyclist and for developing alternative guardrail design for safer protection of motorcyclist during such collision. This process involved overall system modeling, crash simulation and development of alternative guardrail models.

2.1 *Overall system modelling*

Simulation process involved modelling of four systems including road as inertia reference space as well as crash interaction of these systems. The other three systems are the motorcycle, dummy and guardrail models, details of which are described in Ibitoye et al (2006) and Ibitoye et al (2007). The complete simulation model is as shown in Figure 1.

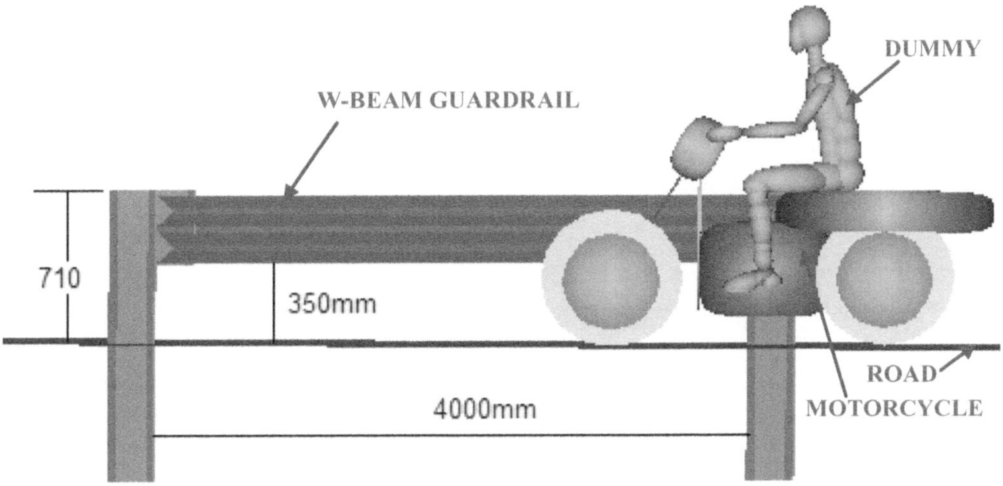

Figure 1. Complete simulation model.

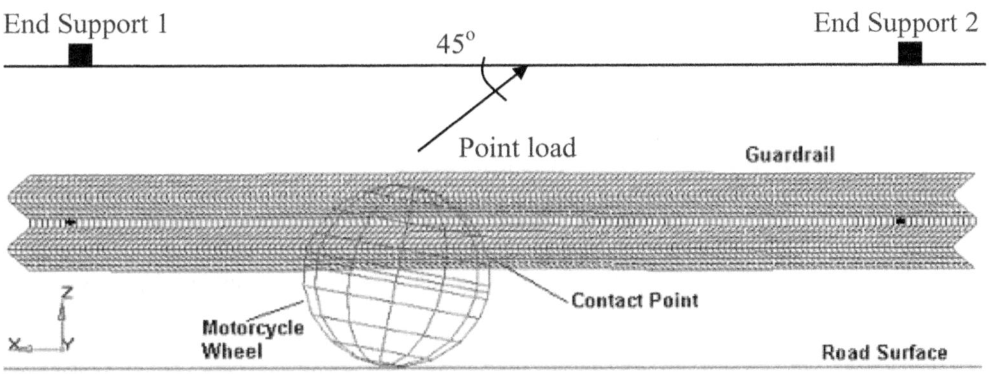

Figure 2. Motorcycle wheel impact force on guardrail nodes.

2.2 *Crash simulation*

Motorcycle crash is a multifarious event involving the interaction of rider, guardrail and the motorcycle itself. In this study, modelling and simulation of motorcycle crash was carried out using MADYMO software. MADYMO software was used because it combines in one simulation program the capability offered by the multi-body and finite element techniques. The eighteen impact conditions considered in this simulation were based on the outcome of literature review on the previous crash simulations and the reported real life accidents. For instance, EEVC (1993) reported that majority of motorcycle collisions take place at speeds between 30 and 60 km/hr. The impact angles of 15, 30 and 45 degree considered was based on the suggestion in literature, especially in Gibson and Benetatos (2001). The post spacing of 2 m and 4 m used are in accordance to the existing standard for installing w-beam guardrail. The impact point was taken as the mid span of the rail for maximum deflection and in regard to the conducted crash test for validation purpose (Figure 2).

2.3 *Developing and optimization of alternative guardrail models*

The main concept behind the development of alternative designs is to reduce sustainable injuries to motorcyclist to at least a tolerance level equivalent to the required biomechanical limits. Since

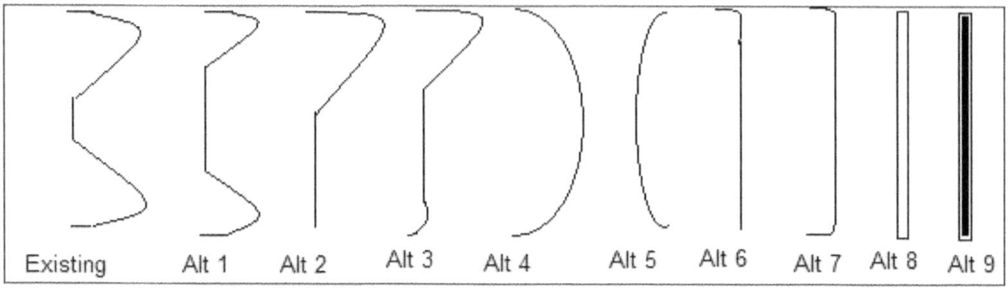

Figure 3. Schematic drawing of modified guardrail shapes considered.

the existing guardrail has been identified as being too stiff and dangerous to motorcycle impacts, this study concentrates on the modification of the rail section in order to minimize injury risks. Ten (10) rail cross sections (Figure 3) were considered in this study before arriving at the three rail shapes (Existing, Alt3 and Alt4) that were selected for design optimization.

Researchers believed that guardrail safety performance can be improved by giving priority to improvement of impact area (shape) through the modification of distance between the lower and upper peaks of the rail surface. These changes increase the effective depth and allow the rail elements to distribute the impact load and redirect the motorcycle to reduce severity of impact response on the rider. The study also considered variation of rail thickness to enable verification of thickness effect on rider's safety. The thickness of the rail was reduced from 2.67 mm (12 gauge) to 2.28 mm (13 gauge) as the only available thickness for steel material. The use of composite material was also considered to test the advantages of better energy absorption of composite material compared to steel. E-glass lamina composite properties as reported in Bilingardi et al (1998) to test the material behaviour under impact loads as well as the Pultruded fibre glass composite properties as reported in Bank and Gentry (2001) and Sala and Astori (1998) were used. The rail thickness of 3.0 mm as well as 2.67 mm and 2.28 mm were

considered for the design optimization involving combination of the design factors to formulate 24 design alternatives.

Optimization of these alternative models involved using of the standard biomechanical limits for head injury risks; HIC = 1000 and head acceleration = 80 g. The acceptance or rejection of each alternative was based on these limits with 2 m post spacing and only the accepted alternatives were further optimised with 4 m post spacing using the same impact speed and angle. Effects of these alternative models to rider's safety were also evaluated using the standard injury criteria for HIC, and head acceleration while the outcome of these analyses were ranked according to their degree of satisfaction.

3 RESULT AND DISCUSSION

Simulation result to investigate safety implication of existing guardrail found that the kinematics of rider for all impact conditions are similar during the initial stage with the dummy having leg contact with the guardrail surface and projecting with head forward. But, the dynamics of rider towards the landing depend on the impact speeds and angles.

As illustrated in Figure 4, for all impact conditions, the rider fall to ground with head except at impact speed of 60 km/h for impact angle 30 at 4 m post spacing as well as impact angle 45 for both 2 m and 4 m spacing. This indicates that higher impact speed can cause high vaulting of rider and thus resulting in no head contact with the ground. This result was validated using a physical crash test with dummy mounted a motorcycle and impacting w-beam guardrail which was orientated at 45° angle to the travel direction.

Therefore, the study has been able to establish the fact that the severity of impact increases with speed as well as angle of impact and with reduced post spacing. The study also established that the

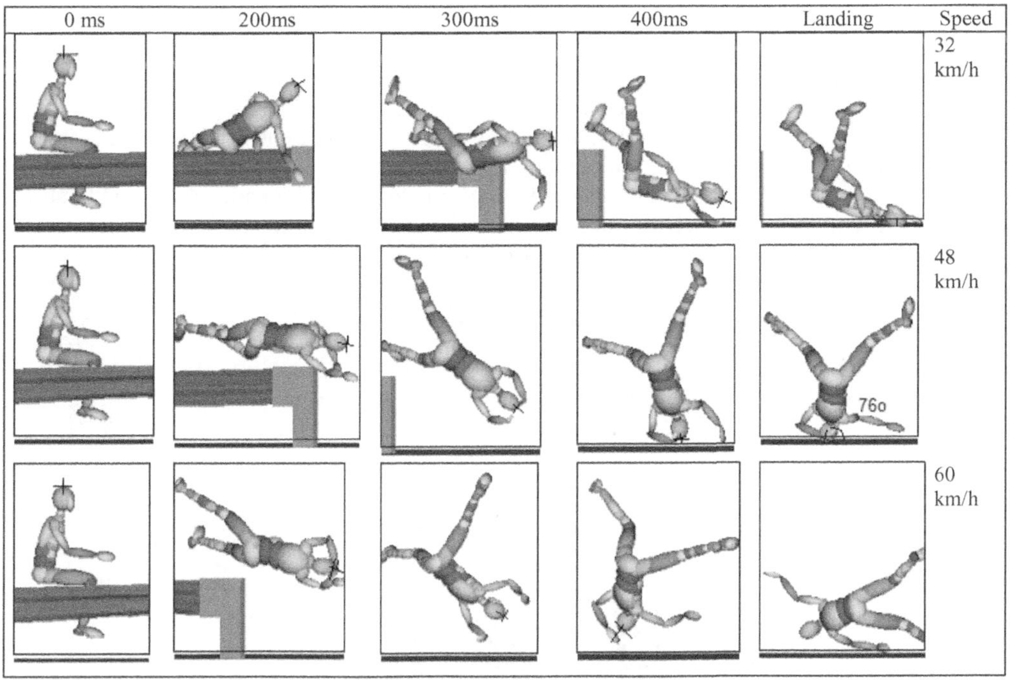

Figure 4. Comparing rider's kinematics at 45 degree for 4 m spacing.

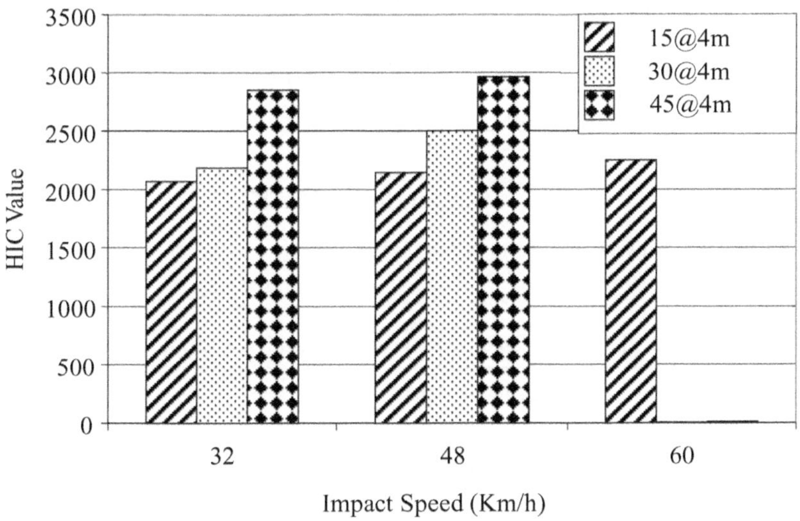

Figure 5. Comparison between HIC values for 4 m post spacing.

likely impact speed along the exclusive motorcycle lanes in Malaysia is 48 km/hr which fall within the reported mean speed 45 km/h to 61 km/h for low volume traffic on these roads. The obtained higher head injury value at impact angle 45 degree conforms to the requirement by ISO (1996) for motorcycle impact study. However, a relationship between the head injury risks values with effect to impact angle and speed are established as presented in the Figures 5 and 6. It can be observed from these figures that these injury patterns vary with speed and angle of impact. Highest values are obtained at impact angle 45 degree and impact speed 48 km/h while impact at 60 km/h exhibits lower values due to higher vaulting of rider that caused landing with hand or foot rather than head to ground. This effect resulted from very high decelerating rate of rider.

Evaluation of alternative guardrail models involved evaluation of guardrail dynamic responses and effect of these responses on the reduction of potential injuries to motorcyclist. Thus, the ability of the proposed alternative to distribute impact loads, to deform adequately and to absorb impact energy was examined. This process allowed for the investigation of the suitability of each alternative model as a replacement to the existing guardrail.

3.1 *Effect of alternative models on guardrail response*

Effects of alternative models on the dynamic response of guardrail to impact compared to those of the existing guardrail are summarized in Tables 1 and 2.

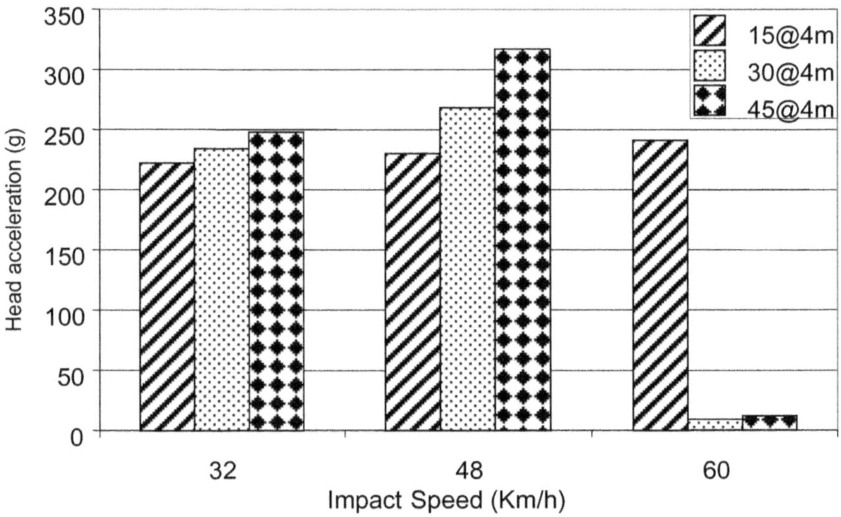

Figure 6. Comparison between head acceleration for 4 m spacing.

Table 1. Predicted guardrail dynamic response for proposed guardrail (4 m).

Response parameters	Existing design	Alt. 2	Alt. 4	Alt. 13
Initial Peak force (N)	14606	6374	10576	8285
Max. deflection (m)	0.16	0.12	0.17	0.15
Energy absorbed (N-m)	685	914	576	660

Table 2. Predicted guardrail dynamic response for proposed guardrail (2 m).

Response parameters	Existing design	Alt. 2	Alt. 4	Alt. 13
Initial Peak force (N)	7264	7340	7139	3113
Max. deflection (m)	0.10	0.07	0.11	0.14
Energy absorbed (N-m)	341	504	281	407

Effect of alternative models on guardrail dynamic response indicates that guardrail at 2 m span exhibit very low values of maximum deflection and energy absorption capacity compared to guardrail with 4 m span. This implies that the guardrail at 2 m spacing is too stiff to respond to impact force and that more energy may be dissipated to the motorcyclist to cause severe injury. Comparing the three alternatives, alternative 2 at 4 m and 2 m respectively exhibits lowest maximum deflection (0.12 m and 0.07 m) but highest energy absorption capacity (914 N-m and 504 N-m). This indicates better capacity of this alternative to distribute impact loads. Therefore, based on the guardrail response alternative 2 of 4 m span as shown in Table 1 is preferred. This result confirms that the geometry of W-beam guardrail has great impact on the safety of motorcyclist.

3.2 *Effect of alternative models on rider's safety*

The three best ranked alternatives were further modified to include covering of exposed guardrail edges and evaluated based on their potential injury risks as shown in Table 3. Outputs at this stage were ranked based on significant injury risk reduction of HIC and Head acceleration with consideration to local availability of material and expertise for fabrication, installation and maintenance of the selected alternative.

This result shows that modification of design configuration could lead to some advantages that would benefit the motorcyclists (Figure 7). These advantages include; deeper effective depths that can enable distribution of impact load, flatter surface that can distribute loads as impact progresses, thereby absorbing more energy leading to gradual ejection of rider from motorcycle, use of reduced steel thickness to aid better flexibility of the rail to deform, contain and redirect the rider from having head contact with ground and ease of fabrication and installation comparable to the existing guardrail system.

Table 3. Effect of alternative models on life-threatening injuries.

Injury risks parameters	Existing design	Alt. 2	% age Red.	Alt. 4	%age Red.	Alt. 13	%age Red.
HIC	2963	119	96.0	169	94.3	141	95.2
Head_acc	317	30	90.5	41	87.1	32	89.9
Ranking		**1**		**3**		**2**	

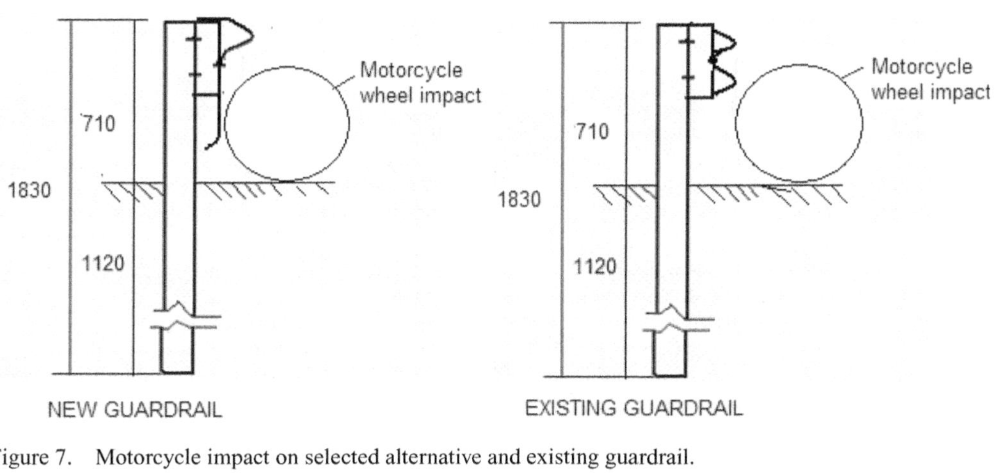

Figure 7. Motorcycle impact on selected alternative and existing guardrail.

These advantages are indications that potential injury risks to motorcyclists can be reduced by modifying existing guardrail configuration and reducing the steel thickness. The flat and wide surface of the new design with deeper effective depth provides similar function as the installation of additional beam to lower section of the existing guardrail as reported in Sala and Astori (1998) as well as in Koch and Scheuler (1987). This solution agrees with Morgan and Ogden (1999) which found that it is only upon impact with large surface areas that impact are not localized and severe. A buffer zone provided at top and bottom of the new design would aid in energy absorption as expected from impact attenuator or damper which was reported as not withstanding the rodent attack from the environment (Sala and Astori, 1998).

Thus, the problem of motorcyclist absorbing virtually all of the kinetic energy on impact (Duncan et al., 2000) has been addressed. Also, the problem of fallen rider sliding under the guardrail to impact roadside hazard has been addressed by the extension of rail depth to a distance of 200 mm from ground level. The obtained significant reduction in head injury risks is an indication that brain damage can be prevented with the new design. The non head contact to ground by rider due to its soft redirection agrees with the proposal of Ruan and Prasad, (1995) that reduction of severity of head contact is associated with reduction of brain shear stress.

4 CONCLUSION

In conclusion, this paper found that variation of design factors (thickness, shape and material) of the existing guardrail would have potentiality to reduce injury risk to motorcyclist. Also, optimization of developed alternative design indicates that the existing steel guardrail is not safe to motorcyclists. However, modification of guardrail shape was found to have greater potentiality of reducing injury risks to motorcyclists. Thus, the study outcome reflects the importance of variation in design factors to the safety enhancement of motorcyclists. It also implies that use of computer simulation and modelling techniques combined with design optimization can serve as important tools for evaluating safety implication of existing guardrail and for proposing a safer guardrail for better protection of motorcyclists.

ACKNOWLEDGMENTS

The authors are indebted to the Ministry of Science, Technology and Environment for funding this work through an IRPA Grant. We would like to express our deep appreciation to the following companies: Modenas, United Akrab, IGC- Industrial Galvanizers and Transplus Sdn Bhd for their various supports.

REFERENCES

Bank, L.C. and Gentry, T.R. (2001) Development of Pultruded Composite material Highway Guardrail, Composites Part A: Apply Science and Manufacturing, Vol. 32 pp. 1325–1338.

Belingardi, G., Gugliotta, A. and Vadori, R. (1998) Numerical Simulation of Fragmentation Composite Material Plates due to Impact. International Journal of Impact Engineering, 21(5) 335–347.

Domhan, M. (1987) Crash barriers and Passive Safety for Motorcyclists Proceedings of the STAPP Car Crash Conference, SAE Paper No. 870242.

Duncan, C. et al. Motorcycle and Safety Barrier Crash-Testing: Feasibility Study, ISBN 0 642 25556 3. Accident Research Centre Monash University, Victoria 3800. 2000.

European Experimental Vehicles Committee. 1993. *Report on Motorcycle Safety*, pp. 45–59.

European Experimental Vehicles Committee. EEVC (1993). Report on Motorcycle Safety. 45–59.

Federation of European Motorcyclists Association; FEMA (2005). The Road to Success: Improving Motorcyclists' Safety by Improving Crash Barrier http://www.fema.ridersright.org on 15/8/2005).

Gibson, T. and Benetatos, E. Motorcycles and Crash Barrier. A report prepared for NSW Motorcycle Council, 2000. Human Impact Engineering Australia.

Ibitoye, A.B., Wong, S.V., Radin, R.S. and Hamouda, A.M.S. (2007): "Roadside Barrier And Passive Safety Of Motorcyclists Along Exclusive Motorcycle Lanes" Journal of Engineering Science and Technology. Vol. 2 No 1: 1–20.

Ibitoye, A.B. (2007): "Design and Develpment of Roadside Safety Barrier (Guardrail) for Exclusive Motorcycle Lanes" PhD Thesis University Putra Malaysia, Malaysia.

Ibitoye, A.B., Wong, S.V., Radin, R.S. and Hamouda, A.M.S. (2006): "Simulation of Motorcyclist's Kinematics during Impact with W-Beam Guardrail" International Journal of Advance In Engineering Software. 1(37): 56–61.

International Standard, ISO (1996). Motorcycles—Test and Analysis Procedures for Research Evaluation of Rider Crash Protective Devices Fitted to Motorcycles (1st ed.) ISO 13232-6 & 7.

Koch, H. and Schueler, F. (1987). Reduction Of Injury Severity Involving Guardrails by the Use of Additional W-Beams, Impact Attenuators and Sigma Posts as a Contribution to The Passive Safety of Motorcyclists. Eleventh international Technical Conference on Experimental safety Vehicles. pp. 878–883. Washington DC.

MADYMO Theory Manual 2004 TNO Automotive, Crash-Safety Research Centre, Delft, The Netherlands.

Ouellet, J.V. "Environmental Hazards in Motorcycle Accidents". American Association for Automotive Medicine annual conference, 26th, October 4–6, Ottawa. Proceedings, 1982, pp. 117–129.

Quincy, R., Vulin, D. and Mounier, B. Motorcycle Impacts with Guardrails. In *Transportation Research Circular: International Roadside Safety Hardware Research*, No. 341, pp. 23–28, Dec. 1988.

Radin Umar, R.S., Mackay, G.M. and Hill, B.L. (1995) Preliminary Analysis of Motorcycle Accidents: Short Term Impacts of the Running Headlights Campaign and Regulation. Journal of Traffic Medicine. United Kingdom, 23 (1): pp. 17–28.

Royal Malaysia Police (PDRM), 2003 Statistical Report on Road Accidents, Malaysia. Traffic Branch, Bukit Aman, Kuala Lumpur, Malaysia.

Ruan, J. and Prasad, P. (1995) Skull Isostresses and Brain Isostresses Responses of a Finite Element Human Head Model. Proceedings of the 1995 Bioengineering Conference, ASME BED, Vol. 29.

Sala and Astori 1998. "New Concepts And Materials For Passive Safety Of Motorcyclists" Proceedings of the IRCOBI Conference, Goteborg.

Viner, J.G. (1995). The Roadside Safety Problem Transportation Research Circular No. 435, Transportation Research Board 17–29.

Efficient Transportation and Pavement Systems – Al-Qadi, Sayed, Alnuaimi & Masad (eds)

The viability of real-time prediction and prevention of traffic accidents

Mohamed Abdel-Aty & Anurag Pande
University of Central Florida, Department of Civil & Environmental Engineering, FL, Orlando, USA

ABSTRACT: This paper addresses a novel idea of real-time traffic safety improvement on freeways. First by predicting traffic crashes on the freeway mainline using on-line loop detector data, then by proposing ITS remediation strategies to reduce the crash risk in real-time. The results of the predictive models were found to be consistent with the probable mechanisms of crashes under different flow regimes. Using micro-simulation software and the crash potential determined through the models developed in this research to associate traffic characteristics with the risk of a crash occurring, several mitigation strategies were attempted to improve safety in real-time. The before (pre-ITS) and after case were then compared to examine the effects of the crash mitigation methods. The results show that Variable Speed Limits and Ramp metering have positive effects on lowering the crash potential.

Keywords: Accident prediction, real-time safety risk, variable speed limit, ramp metering, micro simulation

1 INTRODUCTION

In recent years substantial growth has been observed in traffic management and information systems. The capability of storing and analyzing data has grown significantly and considerable amounts of data are being collected and stored for ITS applications. These data include speed, vehicle counts and occupancy collected from loop detectors installed beneath freeway pavement. While this growth has been around for some time, research efforts directed toward the application of freeway loop detector data for traffic safety have gained momentum only recently. Since traffic flow would be measured in terms of loop data collected in real-time, the approach diverges from traditional safety studies aimed at estimation of crash frequency or rate on freeway sections through aggregate measures of traffic flow (e.g., AADT or hourly volumes). Application of loop detector data for traffic management has been limited to incident detection algorithms. These algorithms are developed by analyzing historical post-incident loop data and attempt to detect incidents as quickly as possible. It is essentially a reactive strategy which is being rendered irrelevant with wide spread use of mobile phones and surveillance cameras. This research is an effort in the direction of proactive traffic management that would involve anticipating impending incidents such as crashes.

The problem of discriminating crash prone conditions from normal freeway traffic is set up as a classification problem in this research. To this end we also collect "non-crash" loop detector data representing 'normal' traffic conditions. Traffic parameters should then be used as inputs to the binary (crash *vs.* non-crash) on-line classification models. This research effort is aimed at the development of models that involve estimating the crash likelihood on the mainline freeway, then devising ITS strategies to reduce this likelihood.

Our group at University of Central Florida (UCF) has also been actively involved in research linking crash patterns with loop detector data. Various modeling methodologies have been explored e.g., Probabilistic neural networks (PNN) (Abdel-Aty and Pande, 2005), matched case-control Logistic Regression (Abdel-Aty et al., 2004), multi-layer perceptron (MLP)/radial basis function (RBF) neural network architectures (Pande, 2003) and Generalized Estimation Equation (Abdel-Aty and Abdalla, 2004). The data for these studies were collected from 13.2-mile central corridor of Interstate-4 in Orlando.

2 MATCHED CASE-CONTROL LOGISTIC REGRESSION MODELS

Loop detector data on Freeways usually include speed, vehicle counts and lane occupancy provided every 30 seconds. There have been few studies dealing with attempting to predict crashes using loop detector data. Traffic conditions measured as coefficient of variation in speed and lane occupancy have been found to be significant freeway crash precursors (e.g., Lee et al., 2002). These studies have developed crash prediction models using real-time values of these precursors obtained from underground freeway loop detectors located upstream and/or downstream of crash sites. However, these models do not take into consideration geometric and environmental factors such as horizontal curve and season. Furthermore, crash precursors are measured from loop detectors in the neighborhood of crash location at time duration prior to *crashes* only.

The accuracy of a real-time crash prediction model may be increased if the model utilizes information on traffic flow characteristics for both crash and non-crash cases while controlling for other external factors. This can be achieved using a within stratum analysis of a binary outcome variable Y (crash or non-crash) as a function of traffic flow variables $X_1, X_2, \ldots, X_p$ from matched crash-non-crash cases where a matched set (henceforth referred to as stratum) can be formed using crash site, time, season, day of the week, etc., so that the variability due to these factors is controlled. This is known as matched case-control study while each case refers here to a crash and control to a non-crash.

In one of our earlier studies, performed using this methodology, a logistic regression model to predict crashes on a 13-mile segment of Interstate-4 in Orlando metropolitan area was developed (Abdel-Aty et al., 2004). The model achieved satisfactory crash identification and demonstrated the feasibility of predicting crashes in real-time. The model was developed using data from a small, dense urban segment of the freeway with the crashes spanning a short period of time (seven months). The traffic, geometric and even crash characteristics on the segment remain largely uniform (i.e., same *AADT*/peak hour, little or no variation in the geometry along the segment, and mostly rear-end crashes caused by frequent formation and dissipation of ephemeral queues). In order to develop better predictive models the crash database was expanded to include all 1528 multi-vehicle crashes that occurred during 4-year period (from 1999 through 2002) on the 36-mile instrumented corridor of Interstate-4. The corridor under consideration along with the surrounding roadways is shown in the map depicted in Figure 1. Since crashes under moderate-to-high and low speed regimes are

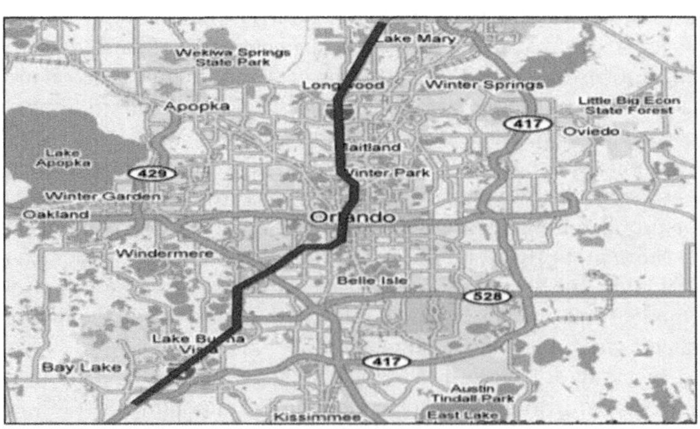

Figure 1. Map of the interstate-4 study section and adjacent roadways.

expected to follow different occurrence mechanism they were analyzed separately using the within stratum case-control logistic regression (Abdel-Aty et al., 2005).

3 STUDY AREA AND AVAILABLE DATA

The following data on Interstate-4 are collected every 30 seconds; average vehicle counts, average speed, and lane detector occupancy. Sixty nine stations in each direction spaced at approximately 0.5 mile, report these parameters from the three through lanes. The crash data were collected from the FDOT crash database for the years 1999 to 2002. First, the location for all the crashes that occurred in the study area during this period was identified. For every crash, the loop detector station nearest to its location was determined and referred to as the station of crash.

3.1 *Data preparation*

Traffic data were extracted for the day of crash and on all corresponding (non-crash) days to the day of every crash to form matched sampling strata. The correspondence here means that, for example, if a crash occurred on April 12, 1999 (Monday) 6:00 PM, I-4 Eastbound and the nearest loop detector was at station 30, data were extracted from station 30, four loops upstream and two loops downstream of station 30 for half an hour period prior to the estimated time of the crash for all Mondays of the same season in that year at the same time. This matched sample design controls for most of the critical factors affecting crash occurrence such as driver population composition, season, day of week, location on the freeway, etc. (thus implicitly accounting for these factors). Hence, this case will have loop data table consisting of the speed, volume and lane occupancy (percent of time the loop is occupied by vehicles) values for all three lanes from the loop stations 26–32 (on eastbound direction) from 5:30 PM to 6:00 PM for all the Mondays of that season of year 1999, with one of them being the day of crash (crash case). More details of this sampling technique, application of this methodology, time of crash determination and data cleaning could be found in earlier studies by the authors (Abdel-Aty et al., 2004; 2005).

The 30-second data have random noise and is difficult to work with in a modeling framework. Therefore, the 30-second raw data were combined into 5-minute level in order to get averages and standard deviations. Thus for 5-minute aggregation half an hour period was divided into 6 time slices. The stations were named as "*B*" to "*H*", with "*B*" being farthest station upstream and so on. It should be noted that "*F*" is the station closest to the location of the crash with "*G*" and "*H*" being the stations downstream of the crash location. Similarly the 5-minute intervals were also given "IDs" from 1 to 6. The interval between time of the crash and 5 minutes prior to the crash was named as slice 1, interval between 5 to 10 minutes prior to the crash as slice 2, interval between 10 to 15 minutes prior to the crash as slice 3 and so on. The final dataset was created by aggregating data over all lanes and averages and standard deviations were obtained by using parameter (speed, volume and occupancy) values over the three lanes. Hence, averages (and standard deviations) at 5-minute level were based on 30 ($10 * 3$ lanes) observations. For each of the seven loop detectors (*B* to *H*) and six time slices (1–6) mentioned above, the values of means (*AS, AV, AO*), standard deviations (*SS, SV, SO*), and coefficients of variation (CVS, CVV, CVO) of speed, volume and lane occupancy, respectively, were available for all crashes and the corresponding non-crash cases. These parameters from various stations/time slices were identified with four/five letter names with each letter representing, (i) A/S/CV (average/standard deviations/coefficient of variation), (ii) S/V/O (speed/volume/occupancy), (iii) stations B/C/D/E/F/G/H, and (iv) time slices 1/2/3/4/5/6, respectively. Due to data availability issues, there were different numbers of controls (non-crash cases) for each case (crash). To carry out matched case-control analysis, a symmetric data set was created (i.e., each crash case in the dataset has the same number of non-crash cases as controls) by randomly selecting five non-crash cases for each crash in the dataset. The type of crash information available in the *FDOT* crash database was utilized to retain only multi-vehicle crashes since the ambient traffic characteristics are more likely to affect crashes involving interaction among vehicles rather than the single vehicle crashes (which were removed from the dataset). The resulting dataset had 1528 matched strata available for analysis. It was ensured that the non-crash controls in the database did not experience any crash during the time data from surrounding loops was collected.

4 PRELIMINARY ANALYSIS

As part of preliminary data analysis, distributions of traffic parameters were explored. Figure 2 shows the histogram distribution of the variable *ASF*1. "*ASF*1" is the average of speeds measured from the three lanes at the station closest to the crash location (*Station F*) during the 5-minute period leading to the crash (*Slice* 1).

The histogram distribution appears to have the shape of two adjacent approximately mound-shaped distributions. The overlapping frequencies are observed over *ASF*1 values ranging between 35 to 40 mph (Figure 2). The two relative peaks of this histogram (approximately at 25 and 55 miles per hour) suggest the need for two separate models: One model for predicting crashes that formed the histogram with peak near *ASF*1 = 25 and the other to predict crashes that formed the histogram with peak near *ASF*1 = 55. The two models may be different in the sense of containing different sets of crash predictors. Hence, the value of *ASF*1 = 37.5 MPH was chosen as the cut-off and two subsets of the main dataset were created. The first subset consists of all crashes with *ASF1* ≤37.5 and their matched non-crash controls. The remaining crashes and their matched non-crash controls formed the second subset of data. In the two following sections the logistic regression models for the two groups of crashes are described.

5 MODEL FOR LOW-SPEED REGIME

The results from our previous work (Abdel-Aty et al., 2004) and some preliminary analysis (Pande and Abdel-Aty, 2005) led us to evaluate parameters from only five stations (*Station D-H*, two upstream stations, two downstream stations and station of the crash itself) as potential explanatory variables. It was also decided not to go beyond 10–15 minute slice (*Slice* 3) based on the results from the same studies (Abdel-Aty et al., 2004, Pande and Abdel-Aty, 2005). In these studies we found no evidence that the conditions 15–20 minutes prior to crash occurrences were significantly

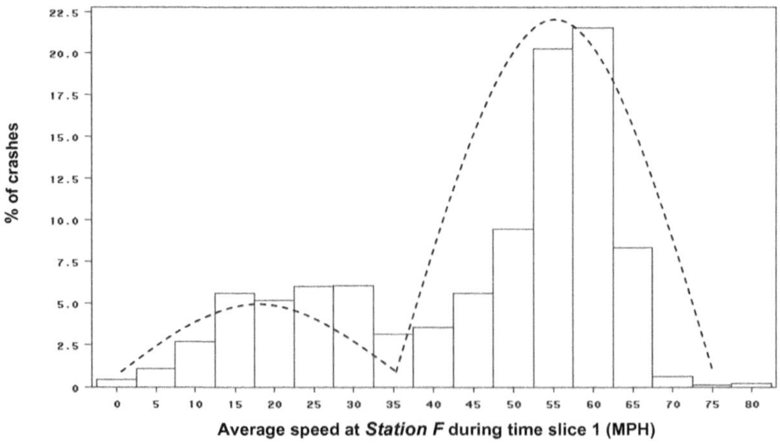

Figure 2. Distribution histogram of *ASF1* for all crashes.

Table 1. Final model developed for the low-speed regime.

Variable	Paramete estimate	Standard error	χ^2	$Pr > \chi^2$
Log*CVSF*2	2.64827	0.49216	28.9539	<.0001
Log*CVSF*3	0.88842	0.47859	3.4460	0.0634
Log*AOE*2	1.33966	0.46218	8.4018	0.0037
Log*AOH*3	0.97766	0.55210	3.1357	0.0766
*SVF*2	–0.43603	0.12289	12.5883	0.0004

different than 'normal' freeway traffic. It left two time slices, *Slice* 2 and *Slice* 3 to analyze. Note that parameters from *Slice* 1 were not considered as an input to the model because *Slice* 1 is too close to the actual time of crash occurrence (and might be late anyway to try to reduce the crash risk).

The parameter estimates and related statistical summary of coefficients for the low speed regime crash prediction model is provided in Table 1. This model may be used for classification of real-time patterns under low-speed regime. Both coefficients of variation in speed (measured during time *Slice* 2 as well as 3; Log*CVSF*2 and Log*CVSF*3) from the station closest to the crash location (*Station F*) remain in the model, with one during time *Slice* 2 (5–10 minutes) being much more significant. Both Log*CVSF*2 and Log*CVSF*3 have positive model coefficients implying highly varying speeds (possibly due to frequent formation and dissipation of queues) prior to the crash occurrence. This high Log*CVSF*2 is also coupled with low standard deviation in volume (indicated by negative coefficient of *SVF*2) implying that the number of cars on three lanes remains fairly equal over time. The other factors in the model are *AOH*3 and *AOE*2, both with positive coefficients. High occupancy at one mile downstream (during 10–15 minutes *Slice*) and half a mile upstream (during 5–10 minutes *Slice*) indicates the backward propagation of congested flow regime. The interpretations (indicating congested conditions with frequent formation and dissipation of queues) largely fit into the mechanism of rear-end crashes, which are the most common type of collisions on freeways, at least under low-speed regime.

6 MODEL FOR MODERATE-TO-HIGH SPEED REGIME

For this model the procedures used to identify significant variables were similar to the low speed regime model but, as expected, resulted in almost entirely different set of predictors. Parameter estimates and related statistical summary of the coefficients for the high speed regime model with final set of predictors is provided in Table 2. The most interesting aspect of the model shown in Table 2 is that it does not include the coefficients of variation in speed as one of the factors. One possible reason could be that since the model is for the moderate-to-high speed regime, the coefficient of variation in speed may not be able to capture the variation in speed due to the large denominator. This led us to estimate the effect of standard deviation of speed, which also turned out to be insignificant. Another important feature of this model is that the coefficient of Log*AOF*2 is negative indicating 'smooth' operating conditions at the station of the crash during 5–10 minutes before the crash (this is confirmed with the high average volume at *Slice* 3–*AVE*3). The only other occupancy variable entering in the model is Log*AOH*3. With a positive coefficient it indicates some cause of congestion about one-mile downstream of the crash site (again confirmed with low *AVG*2 and *SVH*2).

It was found that the classification accuracy of the model for these conditions was inferior to the model for the low-speed crashes. The possible reason may be that in 'smoother' traffic conditions (high-speed regime) the errors on the drivers' part would contribute more significantly towards crash occurrence. Of course, there was no way to account for the pre-crash behavior of the drivers involved in historical crashes and the performance of this model suffers more due to this

Table 2. Final model developed for the moderate-to-high-speed regime.

Variable	Parameter estimate	Standard error	χ^2	$Pr > \chi^2$
Log*AOF*2	−0.93423	0.39970	5.4632	0.0194
Log*AOH*3	1.14584	0.34156	11.2541	0.0008
*SVH*2	−0.22878	0.07181	10.1494	0.0014
*AVG*2	−0.10055	0.03376	8.8723	0.0029
*AVE*3	0.05932	0.03419	3.0100	0.0828

limitation than the low-speed model where the contributions of traffic conditions towards crash occurrence is more significant. A more detailed discussion on model building and parameter coefficient interpretations may be found in Abdel-Aty et al. (2005).

7 MODEL BUILDING: CONCLUSIVE REMARKS

In this section of the paper we have addressed the development of models for the identification of crash prone conditions on the freeway. The analysis of pre-crash traffic surveillance data (obtained from loop detectors) along with the matched non-crash cases is used towards that aim. The section describes the separate logistic regression models developed for crashes that occur under low-speed and moderate-to-high speed traffic conditions. These models are otherwise generic in nature as they include crashes initiated by all harmful events (rear-end, sideswipe, etc.,). Note that in some of our more recent studies (Pande and Abdel-Aty, 2006a, b) we have developed models separated by crash type as well. However, since the logistic regression models do provide us with a measure of relative risk under moderate-to-high-speed and low-speed traffic conditions, in the following sections these models are used to evaluate effectiveness of ITS strategies such as ramp metering and variable speed limits towards the reduction of real-time risk.

8 ITS STRATEGIES TO REDUCE THE CRASH RISK

In this study the micro-simulation software PARAMICS has been chosen to test different ITS strategies to reduce the real-time crash risk on the I-4 freeway. The use of PARAMICS to approximate variable speed limits (Lee et al., 2004) and ramp metering (Chu et al., 2004) has been successfully used in previous research. VSL is usually modeled with Variable Message Signs (VMS), which warn drivers to reduce their speed as well as changing the posted speed limits on various links as the simulation progresses. Ramp metering is also relatively simple to produce in PARAMICS using traffic signals on the freeway ramps along with complex signal phasing tools that are available.

To calibrate the freeway network, literature on similar freeway simulation models were consulted to find out the best method of determining the appropriate calibration parameters (mean headway, mean reaction time, queuing speed, and queuing distance). In most sources, the flow and travel time were used as the main factors of choice for calibration (Bertini et al., 2002; and Trapp, 2002). However, these works did not refer to the values used for the calibration parameters listed previously. Gardes et al. (2002) and Lee et al. (2002) stated their values for mean headway and driver reaction time as 1s and 0.6s and 0.615s and 0.415s, respectively. This was used as the base point for our calibration procedure. Calibration was performed in order to match the locations of congestion along the freeway as well as speeds at the detector stations since these factors affect the crash likelihood equations based on the models presented above in Tables 1 and 2. The values that produced the minimum errors in validation were a mean headway of 1.0 sec, a mean reaction time of 0.42 sec, a queuing speed of 8 mph, and a queuing distance of 8 ft. These parameters were close to the values that were used in other work, therefore, were considered appropriate.

8.1 *Crash likelihood*

Because of the complex nature of traffic collisions, micro-simulation software usually cannot be used to evaluate safety unless a surrogate measure is used. Abdel-Aty et al. (2005) and Pande et al. (2005) all created statistical models that can approximate the level of risk on a freeway based on the real-time traffic parameters. The models presented above to assess risk are two separate models—one to determine risk during the low-speed flow regime (speed < 37.5 mph) and another to determine risk during the moderate-to-high-speed flow regime (speed ≥ 37.5 mph). The low-speed model is shown in (1) and involves the speed, volume, and occupancy. The moderate-to-high-speed model, shown in (2) uses just the occupancy and volume as the input variables.

$$\text{Crash\_Likelihood} = 2.64827\text{Log}CVSF2 + 0.88842\text{Log}CVSF3 + 1.33966\text{Log}AOE2 + 0.97766\text{Log}AOH3 - 0.43603SVF2 \quad (1)$$

Where Log*CVSF*2 = Log of the standard deviation of the speed divided by the average speed 5–10 minutes before the time of interested at the location of interest; Log*CVSF*3 = Log of the standard deviation of the speed divided by the average speed 10–15 minutes before the time of interested at the location of interest; Log*AOE*2 = Log of the average occupancy 5–10 minutes before the time of interest 0.5 mi upstream of the detector of interest; Log*AOH*3 = Log of the average occupancy 10–15 minutes before the time of interest 1 mi downstream of the detector of interest; *SVF*2 = Standard deviation of the speed divided by the average speed 5–10 minutes before the time of interest at the station of interest.

$$\text{Crash\_Likelihood} = -0.93423\text{Log}AOF2 + 1.14584\text{Log}AOH3 - 0.22878SVH2 - 0.10055AVG2 + 0.5932AVE3 \quad (2)$$

Where Log*AOF*2 = Log of average occupancy 5–10 minutes prior time of interest at the detector of interest; Log*AOH*3 = Log of average occupancy 10–15 minutes prior to time of interest 1 mi downstream of detector of interest; *SVH*2 = Standard deviation of volume 5–10 minutes prior to the time of interest 1 mi downstream of the detector of interest; *AVG*2 = Average volume 5–10 minutes prior to the time of interest 0.5 mi downstream of the detector of interest; *AVE*3 = Average volume 10–15 minutes prior to the time of interest 0.5 mi upstream of the detector of interest. These models provide a measure of the likelihood of a crash occurring and how that likelihood (risk) changes over time at a particular location as the traffic flow changes. A decrease in the value given by the model over time shows a decrease in risk of a crash occurring.

9 TESTING VARIABLE SPEED LIMITS

Variable speed limits are typically used to reduce variances in speed on freeways (Borrough, 1997; Pilli-Sivola, 2004). By reducing the speed variability, the number of extremely short headways is reduced and the mean speed is typically lowered (Ha et al., 2003). Variable speed limits are usually implemented in order to reduce congestion and lower the average travel time. Additionally, the study by Borrough (1997) has found that VSL application also reduces the number of crashes that are observed. Borrough (1997) found that VSL implementation along with strong enforcement of the speed limits led to a reduction in crashes by almost 28% over 18 months. Although this safety benefit is not a real-time prevention measure, this reduction in crashes shows that variable speed limits have the potential to increase safety along freeways. Lee et al. (2004) performed a study in which VSL was applied to a micro-simulated network in an effort to reduce the crash potential. However, this study only focused on a 1 mile network that included only one ramp. This study, instead, simulates nearly a 20 mile stretch of roadway which will allow for increased flexibility of the implementation strategies that are tested. This flexibility includes the location of speed limit changes, the amount of the change and method of changing the speed limit (lowering, increasing, or simultaneous combination of the two).

9.1 *Effects of VSL on safety*

Preliminary runs involving VSL were performed by lowering the upstream speed limit near a detector for 30 minutes during the simulation. The change in the crash risk, from the before and after condition, was noted to evaluate the potential safety benefits. The preliminary runs showed that variable speed limits had little to no effect on the crash risk index during the low-speed condition. This is most likely caused by the fact that during the low-speed scenario vehicles are traveling at congestion well below the posted speed limit and, therefore, the change in speed limit on the roadway will not effectively change the speed the vehicles are traveling at. Therefore, the analysis of variable speed limits solely focused on the moderate- to high-speed scenario where its implementation is expected to have a considerable effect on traffic flow.

The moderate- to high-speed implementation of VSL considered the upstream lowering of the speed limit, the downstream raising of the speed limit and the combination of the two. It is logical that the downstream raising of the speed limit and the upstream lowering of the speed limit together would reduce the crash risk on the freeway. Previous research (Abdel-Aty et al., 2004; 2005) has shown that crash potential increased because of the formation of queues that create a backward forming shockwave of lower speeds while speeds were still high upstream. Therefore, in order to help dissipate these queues, the downstream speed limit was increased with the goal of moving the cars out of the queue quicker. The upstream speed limit was lowered to slow the vehicles just before the queue in order to reduce the speed variation at the shockwave interface.

For this study, a 3.25-hr simulation period was used which included a 15 minute warm-up period. To simulate the VSL strategies, speed limits were changed 30 minutes after the warm up period and were maintained for 30 minutes before reverting back to the original speed limit. The experiment included the testing of several variables related to the VSL implementation including the pattern of speed limit change, the amount of change, the location of the change, the length of the speed limit change, the progression of the change, and the gap distance between the speed limit changes. Several cases were analyzed and it was determined that a case with abrupt 15 mph change both up and down stream showed the most significant improvement in the crash potential. This proves the effectives of the combination of upstream lowering of the speed limit and down-stream raising of the speed limit over using just one method.

This scenario was then chosen to test the length of the speed limit changes. After extensive testing, the length of the changes varied from 2 to 6 miles upstream of the detector of interest and 2 to 5.5 downstream of the detector of interest. The results are about the same in each case, therefore a case where minimum change affected the freeway was chosen. Changing the speed limits 2 miles upstream and 2 miles downstream of the detector of interest affected the minimum length of roadway. Using this scenario, the progression of the change was then tested to determine the best method. This included varying the time step for the change (either an abrupt change in the speed limit or changing the speed limit in small increments every 5 or 10 minutes until the desired speed limit is achieved), the speed step (the amount of the change for each time step), and the number of time steps required to achieve the desired speed limit. All tested cases show a reduction in the crash potential but a 5 minutes time step and 5 mph speed step proved to be the most efficient as the crash potential is lowered more in this case.

The last variable to determine the optimal VSL implementation was the gap distance between the upstream lowering of the speed limit and the downstream raising of the speed limit. The previous best method was tried using gaps of 0 to 3 miles. The results of these runs are shown in Figure 3 which is a plot of the crash potential vs. time. As shown in Figure 3, the best case found was a gap distance of 0 miles (case 1). In fact, using a gap distance of anything other than 0 miles

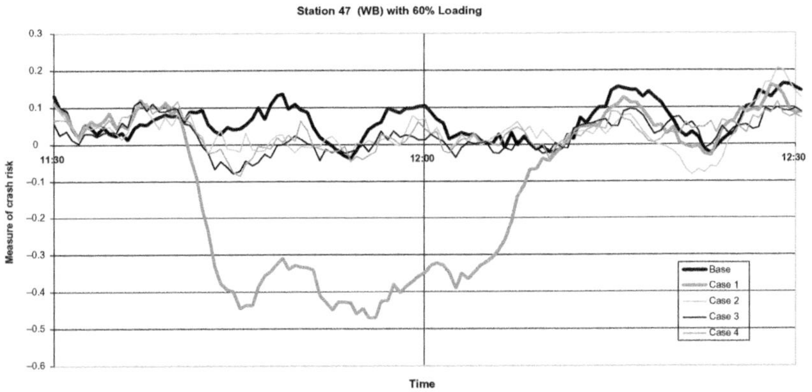

Figure 3. Crash potential vs. time at station 47 for gap distance cases.

resulted in minimal reductions in the crash risk and, therefore, would not be effective at increasing the safety on the freeway. Please note that plots similar to Figure 3 were created for each test case in order to determine which case provided the best results.

10 TESTING RAMP METERING

Ramp metering is typically used to limit the amount of disruption caused to the mainline traffic stream by vehicles entering from on-ramps. By allowing vehicles to enter based on the traffic conditions, vehicles are delayed at the meters but the freeway speeds remain higher (and with less variability) and the overall traffic volume increases. In fact, a study in Minnesota has shown that ramp metering increases speeds by 13 to 26% and decreases travel time by 6 to 16% (Cambridge Systematics, 2001). Currently, ramp metering is used throughout the United States in California, Minnesota, and New York, as well as in many countries throughout Europe. The signals that are used to control the ramps during ramp metering can be pre-timed (fixed) or use complex algorithms that take into account the mainline traffic flow (reactive). Fixed ramp metering limits the flow (vph) of vehicles entering the ramp, regardless of the mainline traffic situation. Reactive ramp metering takes into account the mainline flow and allows a higher rate of vehicles onto the freeway in free flow situations and reduces the flow during congestion. Though there have not been many studies done relating ramp metering to freeway safety specifically, there is plenty of evidence to show that ramp metering should have a positive safety affect. The primary purpose of ramp metering is to increase speed and, therefore, flow by reducing the speed variation on the mainline. Since the variance in speed has been found to be a significant factor in the determination of crash risk (e.g., Pande et al., 2005), any reduction in this variance should reduce the overall crash potential. Also, by implementing ramp metering, fewer extremely short headways would occur which could decrease the potential for rear-end crashes.

10.1 *Effects of ramp metering on safety*

Because variable speed limits did not show any effect at lowering the crash potential in the low-speed situation, ramp metering was considered as a viable option. To simulate the effect of meters on ramps in PARAMICS traffic signals were placed at on-ramp entrances. Preliminary analysis involved the manual metering of ramps by visual inspection. During this process the simulation was viewed and the metered ramp was turned on and off manually based on the level of congestion that was witnessed on the mainline. After performing multiple runs of this nature, the safety benefits of ramp metering became evident.

For the test cases, fixed ramp metering was considered. Fixed ramp metering involves using pre-timed traffic signals, at the ramp entrance to the freeway in order to control the number of vehicles that enter the mainline. By adjusting the cycle length and the green time per cycle, the rate at which vehicles are allowed to enter the freeway can be controlled. The experiment considered tested the cycle length, green time per cycle and the number of ramps that were to be metered to determine the best case scenario. Eight experimental design cases were created and are shown in Table 3.

Overall, the improvement in the crash risk was found to be minimal when metering only one ramp. Only a 3.6% decrease in the crash potential at the station of interest (station 43) was noted. However, at the station upstream of the ramp (station 42) an added 6.3% decrease in crash potential was discovered. When metering is also employed on 7 ramps, the reduction in the crash potential becomes more evident. The results from one of the best cases are shown in Figure 4. These results show the summation of the crash potential throughout the length of the simulation for the stations that were affected by the ramp metering. However, both the single ramp and seven ramp best-case scenarios indicate that there is an increase in the crash potential about 3 miles upstream and downstream of station 43. Additionally, when the crash risk for a single station is examined for the duration of the ramp metering implementation at 30 sec intervals, the best case shows that there are periods where the crash risk is actually higher than the base case. Therefore, while fixed ramp metering shows potential to mitigate the probability of crashes occurring overall, at times this type of metering increases the crash potential as well.

Table 3. Tests cases for ramp metering.

Case	Cycle length	Green time/phase	Number of ramps metered
1	50	25	1
2	50	15	1
3	25	10	1
4	25	15	1
5	50	25	7
6	50	15	7
7	25	10	7
8	25	15	7

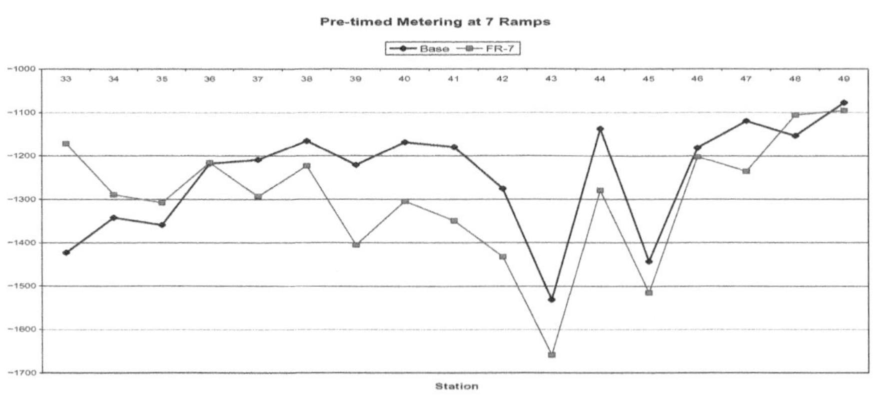

Figure 4. Crash potential vs. station for single ramp metering cases.

However, this is most likely due to the fact that fixed ramp metering does not consider the mainline traffic when allowing vehicles to enter the freeway. Vehicles enter regardless of the congestion that exists, just at a more controlled rate than when ramp metering is not used. Therefore, in order to fully realize the potential of ramp metering as a safety measure, a feedback ramp metering system should be examined (Dhindsa, 2006). The ALINEA (Papageorgiou et al. (1991)) feedback ramp metering algorithm had been used and closely replicated the effects of manual metering by using detectors in the roadway to change the metering rate as the traffic flow changes. Dhindsa (2006) has shown increased safety benefit seen on the mainline freeway due to feedback ramp metering.

11 CONCLUSIONS

The results show that using the stratified case-control analysis, the log odds of crash occurrence may be obtained for a given value of certain traffic flow parameters. Hence, the potential "crash location" created due to ambient traffic conditions may be identified to warn the motorists about the impending hazard, to attempt to influence the speed to reduce its variation. The distribution of the 5-minute traffic speed measured at the station closest to the location of the crash before it occurred suggested the need for multiple models depending on the freeway operation characteristics. Two separate models were developed by splitting the whole crash data into two datasets depending on the 5-minute average speed observed just before the time of the crash. Although the procedure used in model building for low and moderate-to-high-speed models were similar, different parameters entered in the two models. This is not surprising because the crashes that occur under moderate-to-high speed traffic regime differ distinctly from their low speed counterparts not only in terms of severity but in terms of the conditions as well. The low speed crashes mostly occur in persisting congested conditions where queues form and dissipate quite frequently. In contrast, for moderate-to-high speed crashes freeway operation is usually smooth at the crash location before the crash and some disruptive conditions originating downstream and propagating backwards arguably cause drivers to make errors, thereby increasing crash potential. The time duration analyzed here is 5–15 minutes prior to the time of the crash and that's why in the moderate-to-high speed crash model more parameters from the downstream stations appear to be significant. The models developed here may be combined with the real-time application strategy to provide an effective tool to identify "real-time black-spots" on the freeway. While it is evident that the modeling technique proposed here is promising in terms of application in driver warning systems on freeways, a detailed and careful effort is required in order to study the field deployment of such a model.

In addition, this study demonstrates that the on-line mainline traffic flow data can be used as real-time surrogate measures of traffic volume on ramps and traffic conditions near ramps, and their consequent effect on the probability of ramp crashes. The findings in this study suggest the important traffic flow parameters on the mainline that should be considered in the design and management of freeway ramps to improve their safety. Although the suggestion of specific countermeasures is beyond the scope of this study, the speed management on off-ramps (e.g., in-road lights) and the warning of an impending queue using VMS on on-ramps are potential countermeasures that can be suggested based on the findings.

Finally, this study has shown significant safety benefits of using Intelligent Transportation Systems to reduce the crash likelihood on an urban freeway. The ITS measures implemented in this work, Variable Speed Limits and Ramp Metering, both help to reduce the safety risk along the mainline freeway. By simulating their effects using the PARAMICS micro-simulation, various strategies were tested to determine the best method of reducing the crash risk. For the variable speed limits, abrupt lowering of the speed limit by 15 mph 2 miles directly upstream and raising of the speed limit of 15 mph 2 miles directly downstream of the station of interest reduces the crash potential the most efficiently for moderate-to-high-speed situations (although more modest changes in speed also produced positive results). For low-speed situations, allowing vehicles to enter the mainline using smaller amounts of green time during shorter cycle lengths for 7 consecutive on-ramps in the network provides the best results. However, using a fixed ramp metering system does not react to the changing flow of traffic. Therefore, in order to fully realize the potential of ramp metering, a feedback algorithm (such as ALINEA) should be implemented to see if the reduction in crash potential can be more consistent.

REFERENCES

Abdel-Aty M. and Pande A. (2005) Identifying crash propensity using specific traffic speed conditions, *Journal of Safety Research*, Vol. 36 No.1, pp. 97–108.

Abdel-Aty, M., Uddin, N., and Pande, A., (2005) Split models for predicting multi-vehicle crashes under high speed and low speed operation conditions on freeways. *Transportation Research Record 1908*, pp. 51–58.

Abdel-Aty, M., Uddin, N., Abdalla, F., Pande, A., and Hsia, L., (2004) Predicting freeway crashes based on loop detector data using matched case-control logistic regression. *Transportation Research Record* 1897, pp. 88–95.

Abdel-Aty M. and Abdalla F. (2004) Linking Roadway Geometrics and Real-Time Traffic Characteristics to Model Daytime Freeway Crashes Using Generalized Extreme Equations for Correlated Data, *Journal of the Transportation Research Board,* No. 1897, pp. 106–115.

Bertini, R., Lindgren, R. and Tantiyanugulchai, S. (2002) Application of PARAMICS Simulation at a Diamond Interchange," Research Report PSU-CE-TRG-02-02. Portland State University, Transportation Research Group.

Borrough, P. (1997) Variable Speed Limits Reduce Crashes Significantly in the U.K, The Urban Transportation Monitor, March 14, 1997.

Cambridge Systematics (2001) Twin Cities Ramp Meter Evaluation Phase II—Interim Ramp Meter Strategy, Report Prepared for Minnesota DOT, November 2001.

Chu, L., Liu, X. and Recker, W. (2004) Using Microscopic Simulation to Evaluate Potential Intelligent Transportation System Strategies under Non-recurrent Congestion, Transportation Research Record 1886.

Dhindsa, A. (2006) Evaluating Ramp Metering and Variable Speed Limits to Reduce Crash Potential on Congested Freeways using Micro Simulation. Master's Thesis, University of Central Florida, Orlando.

Dilmore J. (2005) Implementation Strategies for Real-Time Traffic Safety Improvements on Urban Freeways, Master's Thesis, University of Central Florida, Orlando.

Gardes, Y., May, A., Dahlgren, J. and Skabardonis, A. (2002) Freeway calibration and application of the PARAMICS model, Presented at the Transportation Research Board 81st Annual Meeting, Washington, DC.

Lee C., Hellinga B. and Saccomanno F. (2004) Assessing Safety Benefits of Variable Speed Limits, *Transportation Research Record 1897*, pp. 183–190.

Lee, C., Saccomanno, F., and Hellinga, B., (2002) Analysis of crash precursors on instrumented freeways. *Transportation Research Record 1784*, pp. 1–8.

Pande A. and Abdel-Aty M. (2006a) Assessment of freeway traffic parameters leading to lane-change related collisions, *Accident Analysis and Prevention* (forthcoming).

Pande A. and Abdel-Aty M. (2006b) A Comprehensive Analysis of the Relationship between Real-time Traffic Surveillance data and Rear-end Crashes on Freeways, *Transportation Research Record* No. 1953.

Pande, A., Abdel-Aty, M., and Hsia, L., (2005) Spatio-temporal variation of risk preceding crash occurrence on freeways. *Transportation Research Record 1908*, pp. 26–36.

Pande A. and Abdel-Aty M. (2005) A Freeway Safety Strategy for Advanced Proactive Traffic Management, *Journal of Intelligent Transportation Systems*, Vol. 9 Issue 3, pp. 145–158.

Pande, A., Classification of real-time traffic speed patterns to predict crashes on freeways. MS Thesis, *University of Central Florida*, 2003.

Papageorgiou, M., Hadj-Salem, H., and Blosseville, J-M. (1991) ALINEA: A Local Feedback Control Law for On-Ramp Metering, *Transportation Research Record 1320*, pp. 58–64.

Pilli-Sivola, Y. (2000) State of the Art in Finland Concerning RWIS and Variable Message Signs," Finnish National Road Administration, February 2000.

Trapp, R. (2002) Microscopic traffic flow modeling of large urban networks—approach and techniques at the example of the City of Cologne, Transportation Research Board 81st Annual Meeting, Washington, DC.

Efficient Transportation and Pavement Systems – Al-Qadi, Sayed, Alnuaimi & Masad (eds)

A probabilistic framework for vision-based road safety analysis

N. Saunier & T. Sayed
Department of Civil Engineering, University of British Columbia, Vancouver, BC, Canada

ABSTRACT: The advent of powerful sensing technologies, especially video sensors and computer vision techniques, has allowed for the collection of large quantities of detailed traffic data. They allow us to further advance towards completely automated systems for road safety analysis. This paper presents a comprehensive probabilistic framework for automated road safety analysis. Building upon traffic conflict techniques and the concept of the safety hierarchy, it provides computational definitions of the probability of collision for road users involved in an interaction. It proposes new definitions for individual road users and aggregated measures over time. This allows the interpretation of traffic from a safety perspective, studying all interactions and their relationship to safety. New and more relevant exposure measures can be derived from this work, and traffic conflicts can be detected. A complete vision-based system is implemented to demonstrate the approach, providing experimental results on real world video data.

1 INTRODUCTION

Road safety is characterized by the absence of accidents, i.e. collisions between road users. The safety is traditionally measured by the number of collisions, or rather its expected number at a given time. Traffic safety diagnosis has been traditionally undertaken using historical collision data. However, there are well-recognized problems of availability and quality associated with collision data. Additionally, the use of collision records for safety analysis is a reactive approach: a significant number of collisions has to be recorded before action is taken.

Therefore there has been considerable interest in research dealing with surrogate safety measures (Gettman and Head, 2003). The observation of traffic conflicts has been advocated as an alternative or complementary approach to analyze traffic safety from a broader perspective than collision statistics alone (Brown, 1994, Hydén, 1987, Sayed et al., 1994, Sayed and Zein, 1999, Svensson and Hydén, 2006). Traffic conflicts are interactions with very similar processes to collisions, but without a collision. A conflict is defined as "an observational situation in which two or more road users approach each other in space and time to such an extent that a collision is imminent if their movements remain unchanged" (Amundsen and Hydén, 1977). The concept of collision course is derived from this widely accepted definition of traffic conflicts. In (Svensson, 1998), users are defined to be on a collision course when, "unless the speed and/or the direction of the road users changes, they will collide". Deciding if two road users are on a collision course thus depends on extrapolation hypotheses. The definition of (Svensson, 1998) uses the common hypothesis of extrapolation with constant velocity, i.e. speed and direction. Some definitions of traffic conflicts also include that at least one of the road users involved takes an evasive action, often in emergency.

The relationship between traffic conflicts and collisions must be established to use traffic conflicts as surrogates to collisions for safety analysis. Most road users move freely in the traffic, without having to take into account other road users. Otherwise, road users interact with each other. An interaction is defined as a situation in which two or more road users are close enough in space and time, and their distance is decreasing. Many researchers, especially in

Scandinavian countries (Hydén, 1987, Svensson and Hydén, 2006), assume that all interactions can be ranked in a safety hierarchy, with collisions at the top. The interactions located next to the collisions in the safety hierarchy, very similar to collisions, but without an actual collision, can be called quasi-collisions. The interactions can thus be recursively ranked in the safety hierarchy. One can imagine "quasi-quasi-collisions", and so on...The pyramid shape of the hierarchy stands for the number of events. The further from collisions, the less severe and more frequent the events.

For this concept to be operational, the safety hierarchy is transferred into measurable parameters based on certain assumptions. For each interaction in the hierarchy, a severity can be estimated, matching its position in the hierarchy, i.e. measuring the proximity to the potential occurrence of a collision, which can be interpreted as the probability of collision. Many severity indicators, such as the Time-To-Collision (TTC) and the Post-Encroachment Time (PET), have been developed to evaluate the distance in space and time between the vehicles involved and their evasive action(s) (Archer, 2004, van der Horst, 1990). Traffic Conflict Techniques (TCTs) involve observing and evaluating the frequency and severity of traffic conflicts at an intersection by a team of trained observers.

Several automated systems, using mostly video sensors, have been and are being developed for traffic monitoring (Beymer et al., 1997, Hu et al., 2004a, Kastrinaki et al., 2003, Laureshyn and Ardö, 2007, Maurin et al., 2005). In a previous paper (Saunier and Sayed, 2007), the authors have shown that traffic conflicts can be detected in video sequences using HMM-based semi-supervised machine learning techniques. This paper presents an extension of the earlier work (Saunier and Sayed, 2006, 2007): a comprehensive probabilistic framework relying on the concept of the safety hierarchy. It provides a computational definition of severity as the probability of collision that is suited for an automated system. Being able to compute the probability for any road user to collide at a given time also allows to detect traffic conflicts and estimate detailed exposure measures.

The rest of the paper proceeds as follows. The next section presents related work in traffic safety. The third section will present the probabilistic framework and the computation of the collision probability. In the fourth section, experimental results will illustrate the approach on real traffic data provided by the vision-based system developed in our group.

2 RELATED WORK

The basis for this work draws heavily from the traffic conflict literature, and the concept of safety hierarchy, as shown in the introduction. Considerable work has been accomplished to validate the TCTs, involving especially the development of severity indicators, in order to obtain more objective judgments from observers. The relationship between traffic conflicts and collisions is not simple (Brown, 1994). TCTs try to identify the subset of the most serious traffic conflicts that are the closest to collisions. Calibration conferences (Malmö, 1983 and Trautenfels 1985) compared most of the different TCTs developed so far, and found them to mostly agree (although the definitions of the considered traffic conflicts differ). Researchers and practitioners use various combinations of severity indicators to detect traffic conflicts and estimate their severity.

Among TCTs, The Swedish TCT is one of the best-known, and is still being actively used for everyday safety assessments. It relies on the Time to Accident (TA), defined as "the time that is remaining from when the evasive action is taken until the collision would have occurred if the road users had continued with unchanged speeds and directions. Its value can be calculated based on the estimates of the distance to the potential point of collision and the speed when the evasive action is taken" (Anonymous, 2005). This speed and the TA are used to determine the border between "serious conflicts" and "non-serious conflicts". This TA indicator is the value at a special instant of the interaction of the general TTC severity indicator, which is defined as long as a collision course exists. Another severity indicator is the PET, defined as the time measured from the moment the first road user leaves the potential collision point to the moment the other road user enters this point. Other indicators include speed and its derivatives, and subjective elements of the observers' judgment of the chance of a collision, incorporating "speed, proximity of vehicles, apparent control of the driving task and environmental conditions such as visibility or a wet road surface" (Brown, 1994).

The various severity indicators provide cues to the estimation of the severity of interactions. Yet the quest for the right "border" between traffic conflicts and non-conflict interactions doesn't appear so relevant when considered into the general framework of the safety hierarchy which places all interactions on a continuum along the severity dimension. The potential for the use of all interactions was investigated in (Svensson, 1998, Svensson and Hydén, 2006). They studied the shape of the distribution of interactions according to their severity, and concluded that different shapes stand for different safety situations and that there is a severity threshold, under which interactions indicate normal road users' interactions, and above which interactions are hints of safety issues.

If one considers the whole continuum of traffic events, the concept of exposure comes into play. Exposure is typically defined as a "measure of spatial or temporal duration in the traffic system in relation to the number of dynamic system objects, road-users, vehicles (axles), etc" (Archer, 2004, Hakkert and Braimaister, 2002). Common measures are a number of inhabitants or some amount of travel, either in distance traveled (road user-hours), or in time traveled (road user-kilometers) that takes into account the speed of road users. Exposure was introduced to make comparisons more fair between different situations, for example annual numbers of collisions for countries with different number of inhabitants, or different car ownership levels. Dividing the number of collisions by the corresponding exposure yields a collision rate, relative to that measure of exposure, that is routinely used to compare varied situations (countries, period of observation, group of users...). However, this should be done with care since there is no reason to believe a priori that the relationship between the safety and the exposure is linear (***Error! Reference source not found.***). More generally, any situation that is necessary for a collision to happen can be considered as exposure to collision. All interactions in the safety hierarchy are exposure to collision, more or less close according to their severity. Estimating the severity of interactions provides exposure measures that are more detailed and relevant than mere volume counts, giving more insight into the processes that lead to collisions.

There have been a few attempts at building a system for automated road safety analysis (Atev et al., 2005, Hu et al., 2004b, Kamijo et al., 2000, Laureshyn and Ardö, 2006, Messelodi and Modena, 2005). To our knowledge, the system presented in (Saunier and Sayed, 2007) is the first automated system for traffic conflict detection. There is limited research in automated systems that can provide severity indicators, except for specific situations and type of traffic conflicts. Only (Hu et al., 2004b) describes a system that can measure the probability of collision for any two interacting road users, but it is validated only on a few experiments with toy cars. The formulas to compute the collision probability for two interacting road users are taken from this work. This paper presents extensions for the collision probability of one road user and aggregated measures over time.

3 THE PROBABILISTIC FRAMEWORK

3.1 *Re-thinking the collision course*

Any situation that is necessary for a collision to happen can be considered as exposure to collision. If there is no exposure, e.g. one stays at home, or is the only road user, no collision with another road user is possible. There are obviously physical limits to possible road user movements (bounded acceleration, maximum angle of turning movement, road users' reaction times...), so that it is not possible for all road users to collide in the future. If a collision between two road users is possible, i.e. there is a possible chain of events that can lead to a collision, the probability of collision can be considered. The collision probability must take into account all the possible movements of the road users, which will return the severity, or proximity to a potential collision.

The definition of a collision course (Svensson, 1998) takes into account only "unchanged road user movements", i.e. movements without road users' intervention. This raises many questions and is difficult to properly define. It is often used experimentally with simplified hypotheses for road users' movements. However a collision course can be better defined as an interaction in which the collision probability is non-zero at a given time. The severity of the collision course at a given instant is the collision probability, summing the probability of all chain of possible events that can lead to a collision. This implies the existence of a probability distribution over all traffic events, more precisely over road users' movements, and requires a practical way to estimate this distribution for real-world use.

3.2 *The collision probability for two road users*

The formulas presented in this part are based on (Hu et al., 2004b), and to a lesser extent on (Messelodi and Modena, 2005). The collision probability for a given interaction between two road users can be computed at a given instant by summing the collision probability over all possible motions that lead to a collision, given the road users' states. This requires the ability to generate for each road user at any instant a distribution over its possible future positions given its previous positions. A possible future motion, i.e. a temporal series of predicted positions, defines an extrapolation hypothesis. The collision probability is approximated by a discrete sum when taking into account a finite number of the most probable extrapolation hypotheses.

First the collision probability at time t_0 for two road users A_1 and A_2 with respective observed trajectories $Q_{1,t \le t0}$ and $Q_{2,t \le t0}$ (before t_0) is defined when considering only one extrapolation hypothesis for each, respectively H_i and H_j. The predicted positions according to the hypotheses H_i and H_j are computed for a number of time steps: the predicted time of the collision $t_{i,j}$ is the first instant at which the road users would be in contact. The larger $\Delta_{i,j} = t_{i,j} - t_0$, the more likely the road users can react and avoid the collision. This time takes into account speed and distance and is directly measurable against the road users' reaction times. The formula of the probability of collision given hypotheses H_i and H_j is taken from (Hu et al., 2004b)

$$P(Collision(A_1, A_2) \mid H_i, H_j) = e^{-\frac{\Delta_{i,j}^2}{2\sigma^2}} \tag{1}$$

where σ is a normalizing constant. It is estimated in (***Error! Reference source not found.***) that this probability should change when the elapsed time $\Delta_{i,j}$ is close to the road user reaction time. Therefore σ is chosen to be equal to an average user reaction time (a value of 1.5 seconds is chosen for the experiments described in this paper). As a consequence, the number of predicted positions computed can be limited to 2σ, as the resulting probability is very close to zero when $\Delta_{i,j}$ reaches that value. Based on (***Error! Reference source not found.***), the collision probability for two road users A_1 and A_2 at t_0 is

$$P(Collision(A_1, A_2) \mid Q_{1,t \le t_0}, Q_{2,t \le t_0}) = \sum_{i,j} P(H_i \mid Q_{1,t \le t_0}) P(H_j \mid Q_{2,t \le t_0}) e^{-\frac{\Delta_{i,j}^2}{2\sigma^2}} \tag{2}$$

where $P(H_i \mid Q_{1,t \le t_0})$ is the probability of road user A_1 to move according to extrapolation hypothesis H_i (same for A_2 and H_j). The sum is done over a variety of extrapolation hypotheses, although this number must be limited to maintain reasonable computation times. This formula is illustrated in a simplified example in Figure 1. In a traditional TCT, one could choose a threshold on collision probability and other indicators to define traffic conflicts. In the new approach described in this paper, road safety can be automatically analyzed in detail by computing continuously the collision probability of all interactions.

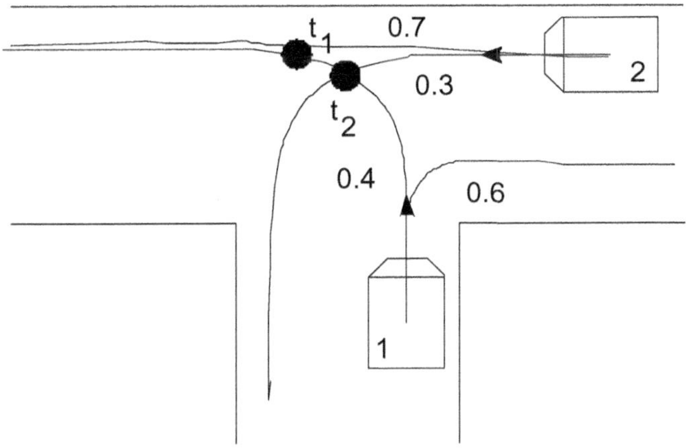

Figure 1. In this simplified situation, two vehicles approach a T intersection at time t_0. Only two extrapolation hypotheses are considered for each vehicle. Vehicle 1 is expected to turn left or right, with respective probabilities 0.4 and 0.6. Vehicle 2 is expected to go straight or turn left, with respective probabilities 0.7 and 0.3. There are two potential collision points, that can happen at times t_1 and t_2. The collision probability at time t_0 is computed as

$$P(Collision) = 0.4 \times 0.7 \times e^{-\frac{(t_1-t_0)^2}{2\sigma^2}} + 0.4 \times 0.3 \times e^{-\frac{(t_2-t_0)^2}{2\sigma^2}}.$$

3.3 *Aggregating over time*

The previous definitions deal only with one road user or one interaction between two road users at a given instant. The collision probability for two road users in interaction can be used for the detection of traffic events relevant to safety. However, to characterize a given period of time at a location, one needs a method to accumulate the indicators over all interactions that occurred in the monitored area during this period of time, or over all road users that went through the monitored area during this period of time.

The first aggregation level is the interaction or the road user. This indicator should reflect the highest collision probability over time, but also the amount of time during which this collision probability was high. This should therefore be similar to an integral of the instantaneous collision probability over time. However, the tracking errors and the noise produce unstable measures of collision probability over time which are truncated and noisy. This would make difficult to compare fairly the interactions. Consequently it is preferred to use the average of the n largest values. Let *SeverityIndex*(A_1, A_2) and *SeverityIndex*(A_1) be the averages of the n largest values taken respectively by the collision probability $P(Collision(A_1, A_2) \mid Q_{1,t\leq t_0}, Q_{2,t\leq t_0})$ over the time that the two road users A_1 and A_2 interacted in the monitored area, and by the collision probability $P(Collision(A_1) \mid Q_{1,t\leq t_0}, Q_{2,t\leq t_0}, \ldots, Q_{n,t\leq t_0})$ over the time that the road user A_1 has spent in the monitored area. The values can subsequently be summed over time for all interactions or road users. The severity indices for the time interval $[t_1\ t_2]$ are

$$InteractionSeverityIndex([t_1 t_2]) = \sum_{\substack{(i,j)\text{ such that } A_i \text{ and } A_j \text{ are observed} \\ \text{in Interaction during } [t_1\ t_2]}} SeverityIndex(A_i, A_j) \quad (3)$$

$$UserSeverityIndex([t_1 t_2]) = \sum_{\substack{i \text{ such that } A_i \text{ is observed} \\ \text{during } [t_1\ t_2]}} SeverityIndex(A_i) \quad (4)$$

4 OVERVIEW OF A VISION-BASED AUTOMATED SYSTEM

This framework is used in a complete automated system for road safety analysis. Such a system requires a high level understanding of the scene and is traditionally composed of two levels of modules (see Figure 2):

1. A video processing module for road user detection and tracking,
2. Interpretation modules for interaction analysis and traffic conflict detection.

For road safety applications, the approach relies on the building of two databases: a trajectory database, where the results of the video processing module are stored, and an interaction database, where all interactions between road users within a given distance are considered, and for which various indicators, including collision probability and other severity indicators, are automatically computed. Identifying traffic conflicts and measuring other traffic parameters becomes the problem of mining these databases.

The road user detection and tracking module used in the system described in this paper relies on a feature-based tracking method that extends to intersections the method described in (Beymer et al., 1997). In this approach, distinguishable points or lines in the image are tracked: a moving object may have multiple features, which must be grouped for each object. A detailed description of the tracking algorithm is presented in (Saunier and Sayed, 2006). The algorithm relies on world coordinates through the estimation of the homography matrix. The tracking accuracy for motor vehicles has been measured between 84.7% and 94.4% on three different sets of sequences (pedestrians and two-wheels may also be tracked, but less reliably). This means that most trajectories are detected by the system, although overgrouping and oversegmentation still happens and creates some problems. The most important limitation for traffic conflict detection is the inaccuracy in the estimation of vehicle sizes.

Trajectories provided by the first module are used in subsequent modules to extract relevant information. A first system was developed to detect directly traffic conflicts using HMM-based semi-supervised machine learning techniques (Saunier and Sayed, 2007). To apply the probabilistic

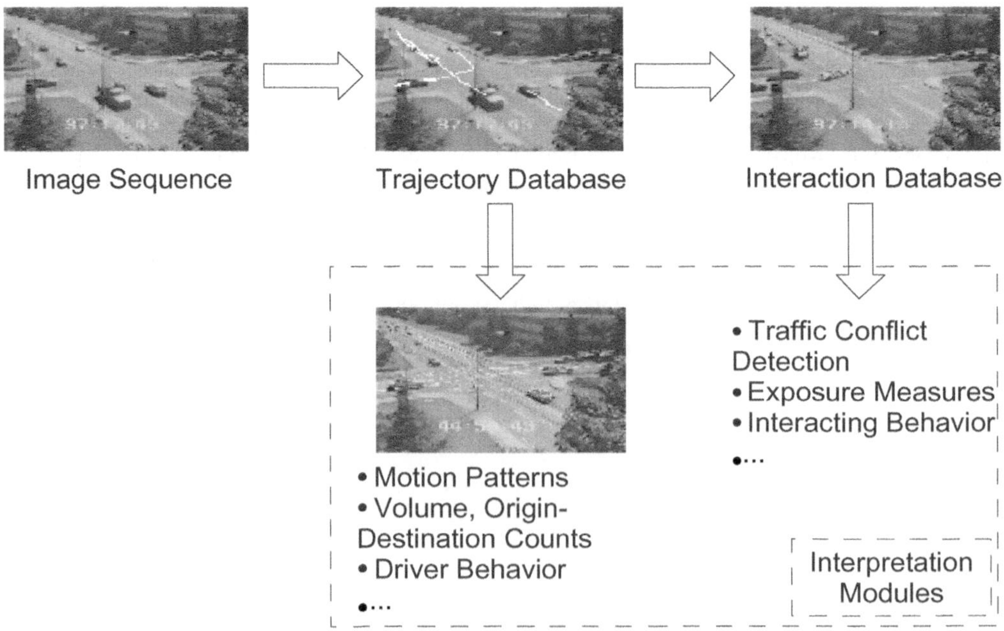

Figure 2. Overview of a modular system for vision-based automated road safety analysis.

safety framework presented in this paper, it is necessary to be able to predict road users' future positions. Motion patterns, represented by actual prototype trajectories without any special pre-processing, are learnt incrementally using the Longest Common Sub-sequence Similarity (LCSS) (Vlachos et al., 2005). The description of the motion pattern learning algorithm is beyond the scope of this paper and is described in detail in (Saunier et al., 2007). The motion pattern probabilities are computed by matching all trajectories over a given period of time using LCSS, and can be updated continuously in a real-time application, as traffic patterns change in time. When computing the collision probability, the partial trajectories of each considered road user at each time are matched against the set of learnt prototypes using the LCSS.

5 EXPERIMENTAL RESULTS

The core architecture of the system has been implemented, using the Intel OpenCV library. On the contrary to (**Error! Reference source not found.**) which uses toy cars, the present work is tested on real traffic video data, and a few traffic conflict instances identified by trained traffic conflict observers.

Three sets of data are used. The first is a set of traffic sequences on the same location initially used for the training of traffic conflict observers in the 1980s. Their length ranges from 10 seconds to 60 seconds. This "Conflict" set contains 2941 feature trajectories of a minimum length of 40 frames, and 327 vehicle trajectories. The second dataset is composed of two long sequences, each close to one hour long, recorded at an intersection in the Twin Cities (United States), in Minnesota. This "Minnesota" set contains 88255 feature tracks of a minimum length of 40 frames, and 11734 vehicle trajectories. The third dataset is composed of 6 sequences, each about 20 minutes long, recorded in Reggio di Calabria (south Italy). This "Italy" dataset contains 138009 feature tracks of a minimum length of 40 frames, and 9849 vehicle trajectories.

First the motion patterns are learnt from the feature trajectories, which are smoothed using a Kalman filter beforehand. It is difficult to evaluate such an unsupervised task. The learnt prototypes for the datasets are presented in Figure 3. The visual examination of the motion patterns suggests a plausible division of the trajectory space. Traffic patterns are well identified, and the traffic volumes are consistent with observation.

5.1 *Traffic conflict study*

Since only a few traffic conflict instances are available in the Conflict dataset, only preliminary results obtained for the three detectable traffic conflict instances are reported in this paper. It appears that the prototype trajectories are well suited for the computation of the collision probability. An example of movement prediction is presented for one conflict in Figure 4.

The curves of the computed collision probability as a function of time, for the three traffic conflicts, are displayed in Figure 5. For each of these instances, one vehicle is over-segmented, resulting in two trajectories, and thus two traffic events (and two curves). It appears that the collision probability shows an expected evolution over time, starting with low values, increasing until the probability of collision reaches a maximum, to decrease afterward, often truncated due to tracking errors and disrupted trajectories.

Over-segmentation of tracked vehicles can cause major problems. The same vehicle detected twice can entail the detection of an interaction between two very close "imaginary" vehicles, often with very high computed collision probability. Fortunately, these interactions are mostly filtered out by testing for the similarity between the trajectories of interacting vehicles using the LCSS distance. In two of the three sequences containing traffic conflicts, querying interactions for which the severity index is superior to 0.1 returns only the traffic conflicts. For the third sequence, it returns the traffic conflict and some interactions between vehicles in traffic moving in opposite directions. Querying the other sequences that contain no detectable traffic conflicts also return these "normal" interactions that can be easily identified. This shows that traffic conflict detection

Figure 3. Motion patterns learnt respectively on sequences of the Conflict dataset (top), the Minnesota dataset (bottom left) and the Italy dataset (bottom right), resulting respectively in 58, 128 and 58 prototype trajectories. The tracks are displayed in color, from white to red indicating the number of matched trajectories in the sequence for each pattern, i.e. the traffic volume along these patterns.

Figure 4. An example of movement prediction in a real traffic conflict situation (Sequence 1, See top plot in Figure 5). The vehicle trajectories are red and blue, with a dot marking their position, and the future positions are respectively cyan and yellow.

can be achieved by computing the collision probability. Adding other severity indicators will further improve the detection results.

5.2 *Severity indices*

The severity indices of all interactions are computed for the Minnesota and Italy datasets, which are both more than one hour long. The distributions of the interaction according to their severity indices are represented individually for each sequence of the two datasets in the Figure 6. As expected, the distributions exhibit the shape of the safety hierarchy, with the frequency of events decreasing as the severity increases. The different sequence in each dataset have different distributions. For example, more interactions for all level of severity are observed in the sequence 2 in the Minnesota dataset. This type of analysis could be performed to compare different situations, for example in before and after studies. It is also possible to study interactions by their locations, by building severity maps, and therefore analyze particular problems in the intersection.

6 CONCLUSIONS AND FUTURE WORK

This paper presents a comprehensive probabilistic framework for automated road safety analysis. It provides computation definitions of the probability of collisions for road users involved in an interaction, extending the work of (Hu et al., 2004b) to an individual collision probability and aggregated measures over time. By integrating this framework into a complete system for vision-based road safety analysis, it is shown that these definitions are suitable for an automated system. This provides detailed severity measures, exposure estimates and a method to detect and study traffic conflicts. The system is demonstrated using real traffic data, including some traffic conflict instances, illustrating the approach and its usefulness.

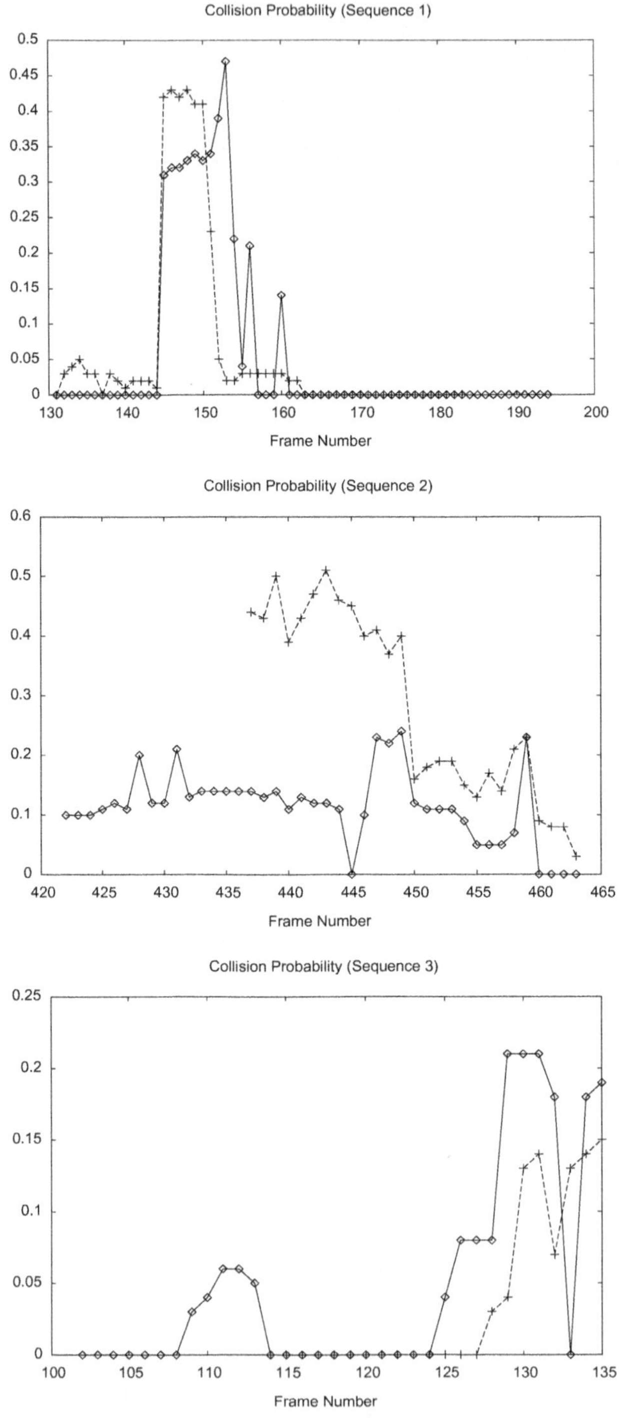

Figure 5. The collision probability for the three traffic conflicts, as a function of time (counted in frame numbers). In all sequences, vehicle 1 travels south-bound through the intersection and vehicle 2 comes from an opposing approach. Vehicle 2 turns left in sequence 1 (top) (See Figure 4), right in sequence 2 (middle) and stops in sequence 3 (bottom).

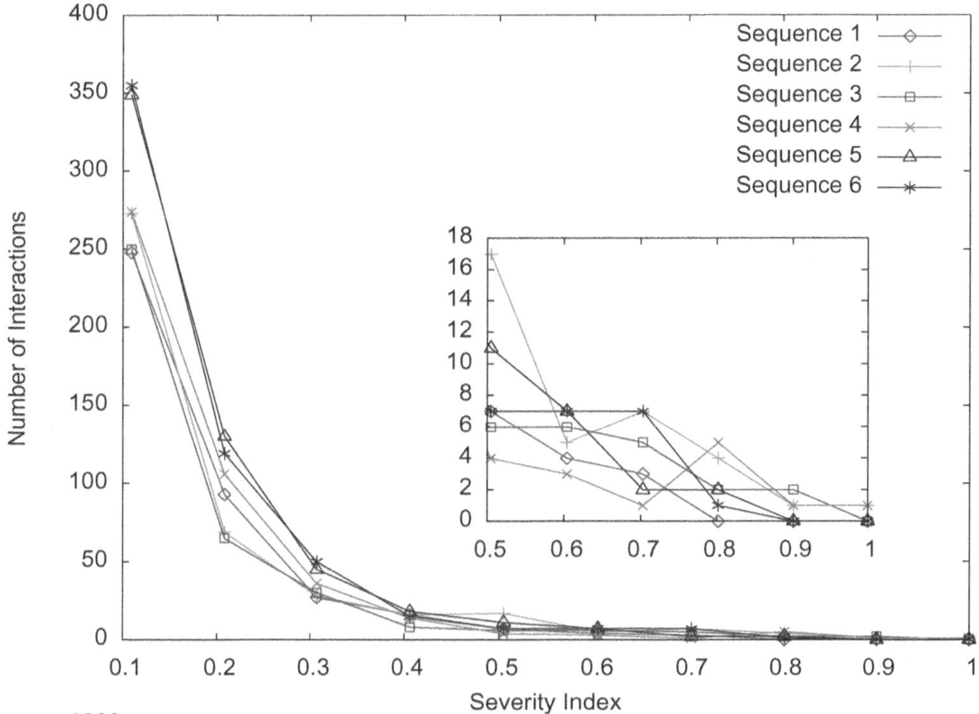

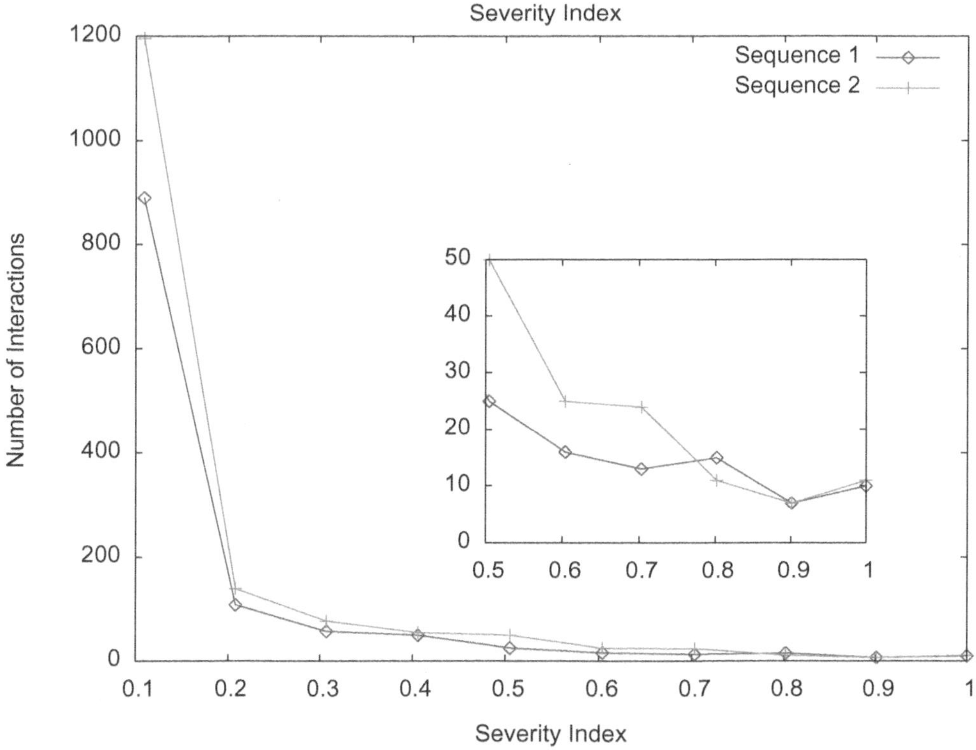

Figure 6. Distribution of the interactions according to their severity indices (with a zoom on the higher severities), quantified by 0.1 (the point at severity index x stands for the number of interactions with severity index between x-0.1 and x), for the Minnesota dataset (top) and the Italy dataset (bottom).

New data is currently being collected to expand the results and validate the computed measures. Further research is needed to investigate and validate the relationship of collision probability to safety.

REFERENCES

Amundsen, F. and Hydén, C., editors (1977). *Proceedings of the first workshop on traffic conflicts*, Oslo, Norway. Institute of Transport Economics.

Anonymous (2005). The swedish traffic conflict technique. http://www.tft.lth.se/rapporter/Conflict1.pdf.

Archer, J. (2004). *Methods for the Assessment and Prediction of Traffic Safety at Urban Intersections and their Application in Micro-simulation Modelling*. Academic thesis, Royal Institute of Technology, Stockholm, Sweden.

Atev, S., Arumugam, H., Masoud, O., Janardan, R., and Papanikolopoulos, N.P. (2005). A vision-based approach to collision prediction at traffic intersections. *IEEE Transactions on Intelligent Transportation Systems*, 6(4): 416–423.

Beymer, D., McLauchlan, P., Coifman, B., and Malik, J. (1997). A real-time computer vision system for measuring traffic parameters. In *Proceedings of the 1997 Conference on Computer Vision and Pattern Recognition (CVPR '97)*, pages 495–501, Washington, DC, USA. IEEE Computer Society.

Brown, G.R. (1994). Traffic conflict for road user safety studies. *Canadian Journal of Civil Engineering*, 21: 1–15.

Gettman, D. and Head, L. (2003). Surrogate safety measures from traffic simulation models, final report. Technical Report FHWA-RD-03-050, Federal Highway Administration.

Hakkert, A. and Braimaister, L. (2002). The uses of exposure and risk in road safety studies. Technical Report R-2002-12, SWOV, Leidschendam.

Hauer, E. (1995). On exposure and accident rate. *Traffic Engineering + Control*, 36(3) :134–138.

Hu, W., Tan, T., Wang, L., and Maybank, S. (2004a). A survey on visual surveillance of object motion and behaviors. *Systems, Man and Cybernetics, Part C, IEEE Transactions on*, 34(3): 334–352.

Hu, W., Xiao, X., Xie, D., Tan, T., and Maybank, S. (2004b). Traffic Accident Prediction using 3D Model Based Vehicle Tracking. *IEEE Transactions on Vehicular Technology*, 53(3): 677–694.

Hydén, C. (1987). *The development of a method for traffic safety evaluation: The Swedish Traffic Conflicts Technique*. PhD thesis, Lund University of Technology, Lund, Sweden. Bulletin 70.

Kamijo, S., Matsushita, Y., Ikeuchi, K., and Sakauchi, M. (2000). Traffic monitoring and accident detection at intersections. *IEEE Transactions on Intelligent Transportation Systems*, 1(2): 108–118.

Kastrinaki, V., Zervakis, M., and Kalaitzakis, K. (2003). A survey of video processing techniques for traffic applications. *Image and Vision Computing*, 21(4): 359–381.

Laureshyn, A. and Ardö, H. (2006). Automated video analysis as a tool for analysing road safety behaviour. In *ITS World Congress*, London.

Laureshyn, A. and Ardö, H. (2007). Straight to video? *Traffic Technology International*, pages 107–109.

Maurin, B., Masoud, O., and Papanikolopoulos, N.P. (2005). Tracking all traffic: computer vision algorithms for monitoring vehicles, individuals, and crowds. *Robotics & Automation Magazine, IEEE*, 12(1): 29–36.

Messelodi, S. and Modena, C.M. (2005). A computer vision system for traffic accident risk measurement: A case study. Technical Report ITC-irst T05-06-07, ITC.

Saunier, N. and Sayed, T. (2006). A feature-based tracking algorithm for vehicles in intersections. In *Third Canadian Conference on Computer and Robot Vision*, Québec. IEEE.

Saunier, N. and Sayed, T. (2007). Automated Road Safety Analysis Using Video Data. In *Transportation Research Board Annual Meeting Compendium of Papers*, Washington, D.C. 07-0211.

Saunier, N., Sayed, T., and Lim, C. (2007). Probabilistic Collision Prediction for Vision-Based Automated Road Safety Analysis. In *The 10th International IEEE Conference on Intelligent Transportation Systems*, Seattle.

Sayed, T., Brown, G.R., and Navin, F. (1994). Simulation of Traffic Conflicts at Unsignalised Intersections with TSC-Sim. *Accident Analysis & Prevention*, 26(5): 593–607.

Sayed, T. and Zein, S. (1999). Traffic conflict standards for intersections. *Transportation Planning and Technology*, 22: 309–323.

Svensson, A. (1998). *A Method for Analyzing the Traffic Process in a Safety Perspective*. PhD thesis, University of Lund. Bulletin 166.

Svensson, A. and Hydén, C. (2006). Estimating the severity of safety related behaviour. *Accident Analysis & Prevention*, 38(2): 379–385.

van der Horst, R. (1990). *A time-based analysis of road user behavior in normal and critical encounter.* PhD thesis, Delft University of Technology.
Vlachos, M., Kollios, G., and Gunopulos, D. (2005). Elastic translation invariant matching of trajectories. *Machine Learning*, 58(2–3): 301–334.
Gettman, D. and L. Head. Surrogate safety measures from traffic simulation models, final report. Technical Report FHWA-RD-03-050, Federal Highway Administration, 2003.
Brown, G.R. Traffic conflict for road user safety studies. Canadian Journal of Civil Engineering, 21: 1–15, 1994.
Hydén, C. The development of a method for traffic safety evaluation: The Swedish Traffic Conflicts Technique. PhD thesis, Lund University of Technology, Lund, Sweden, 1987. Bulletin 70.
Sayed, T., G.R. Brown, and F. Navin. Simulation of Traffic Conflicts at Unsignalised Intersections with TSC-Sim. Accident Analysis & Prevention, 26(5): 593–607, 1994.
Sayed, T. and S. Zein. Traffic conflict standards for intersections. Transportation Planning and Technology, 22: 309–323, 1999.
Svensson, Å. and C. Hydén. Estimating the severity of safety related behaviour. Accident Analysis & Prevention, 38(2): 379–385, March 2006.
Amundsen, F. and C. Hydén, editors. Proceedings of the first workshop on traffic conflicts, Oslo, Norway, 1977. Institute of Transport Economics.
Svensson, Å. A Method for Analyzing the Traffic Process in a Safety Perspective. PhD thesis, University of Lund, 1998. Bulletin 166.
Archer, J. Methods for the Assessment and Prediction of Traffic Safety at Urban Intersections and their Application in Micro-simulation Modelling. Academic thesis, Royal Institute of Technology, Stockholm, Sweden, December 2004.
van der Horst, R. A time-based analysis of road user behavior in normal and critical encounter. PhD thesis, Delft University of Technology, 1990.
Beymer, D., P. McLauchlan, B. Coifman, and J. Malik. A real-time computer vision system for measuring traffic parameters. In Proceedings of the 1997 Conference on Computer Vision and Pattern Recognition (CVPR '97), pages 495–501, Washington, DC, USA, 1997. IEEE Computer Society. ISBN 0-8186-7822-4.
Hu, W., T. Tan, L. Wang, and S. Maybank. A survey on visual surveillance of object motion and behaviors. Systems, Man and Cybernetics, Part C, IEEE Transactions on, 34(3): 334– 352, August 2004a.
Kastrinaki, V., M. Zervakis, and K. Kalaitzakis. A survey of video processing techniques for traffic applications. Image and Vision Computing, 21(4): 359–381, April 2003.
Laureshyn, A. and H. Ardö. Straight to video? Traffic Technology International, pages 107–109, 2007.
Maurin, B., O. Masoud, and N.P. Papanikolopoulos. Tracking all traffic: computer vision algorithms for monitoring vehicles, individuals, and crowds. Robotics & Automation Magazine, IEEE, 12 (1): 29–36, March 2005.
Saunier, N. and T. Sayed. Automated Road Safety Analysis Using Video Data. In Transportation Research Board Annual Meeting Compendium of Papers, Washington, D.C., January 2007.
Saunier, N. and T. Sayed. A feature-based tracking algorithm for vehicles in intersections. In Third Canadian Conference on Computer and Robot Vision, Québec, June 2006. IEEE.
Anonymous. The swedish traffic conflict technique. http://www.tft.lth.se/rapporter/Conflict1.pdf, 2005.
Hakkert, A.S. and L. Braimaister. The uses of exposure and risk in road safety studies. Technical Report R-2002-12, SWOV, Leidschendam, 2002.
Hauer, E. On exposure and accident rate. Traffic Engineering + Control, 36(3): 134–138, 1995.
Atev, S., H. Arumugam, O. Masoud, R. Janardan, and N.P. Papanikolopoulos. A vision-based approach to collision prediction at traffic intersections. IEEE Transactions on Intelligent Transportation Systems, 6(4): 416– 423, December 2005.
Hu, W., X. Xiao, D. Xie, T. Tan, and S.J. Maybank. Traffic accident prediction using 3d model based vehicle tracking. IEEE Transactions on Vehicular Technology, 53(3): 677–694, May 2004b.
Kamijo, S., Y. Matsushita, K. Ikeuchi, and M. Sakauchi. Traffic monitoring and accident detection at intersections. IEEE Transactions on Intelligent Transportation Systems, 1(2): 108–118, June 2000.
Laureshyn, A. and H. Ardö. Automated video analysis as a tool for analysing road safety behaviour. In ITS World Congress, London, 2006.
Messelodi, S. and C.M. Modena. A computer vision system for traffic accident risk measurement: A case study. Technical Report ITC-irst T05-06-07, ITC, 2005.
Vlachos, M., G. Kollios, and D. Gunopulos. Elastic translation invariant matching of trajectories. Machine Learning, 58(2–3): 301–334, February 2005.
Saunier, N., T. Sayed, and C. Lim. Probabilistic collision prediction for vision-based automated road safety analysis. In The 10th International IEEE Conference on Intelligent Transportation Systems, Seattle, October 2007.

Efficient Transportation and Pavement Systems – Al-Qadi, Sayed, Alnuaimi & Masad (eds)
 ISBN 978-0-415-48979-9

Linear regression crash prediction models: Issues and proposed solutions

H. Rakha
Charles Via, Jr. Department of Civil and Environmental Engineering, Blacksburg, Virginia, USA

M. Arafeh, A.G. Abdel-Salam, F. Guo & A.M. Flintsch
Virginia Tech Transportation Institute, Blacksburg, Virginia, USA

ABSTRACT: The paper develops a linear regression model approach that can be applied to crash data to predict vehicle crashes. The proposed approach involves novice data aggregation to satisfy linear regression assumptions; namely error structure normality and homoscedasticity. The proposed approach is tested and validated using data from 186 access road sections in the state of Virginia. The approach is demonstrated to produce crash predictions consistent with traditional negative binomial and zero inflated negative binomial general linear models. It should be noted however that further testing of the approach on other crash datasets is required to further validate the approach.

1 INTRODUCTION

The current state-of-the-art for developing Crash Prediction Models (CPMs) is to adopt General Linear Models (GLMs) considering either a Poisson or a negative binomial error structure (Lord et al. 2004; Lord et al. 2005; Sawalha et al. 2006). Recently, researchers have also proposed the use of zero inflated negative binomial regression models in order to address the high propensity of zero crashes within typical crash data (Shankar et al. 1997; Shankar et al. 2003). The use of Linear Regression Models (LRMs) is not utilized because crash data typically do not satisfy the assumptions of such models, namely: normal error structure and constant error variance.

The objectives of the research presented in this paper are two-fold. First, the paper demonstrates how through the use of data manipulation it is possible to satisfy the assumptions of LRMs and thus develop robust LRMs. Second, the paper compares the LRM approach to the traditional GLM approach considering a negative binomial error structure to demonstrate the adequacy of the proposed approach. The objectives of the paper are achieved by applying the models to crash, traffic, and roadway geometric data obtained from 186 freeway access roads in the state of Virginia in the U.S.

In terms of the paper layout, initially a brief background of CPMs is presented. Subsequently, the unique characteristics of the crash, traffic, and roadway geometry data that are utilized to validate the proposed approach are described. Next, the two modeling approaches are described and applied to the access road data. Finally, the study conclusions are presented.

2 BACKGROUND

An earlier publication (Lord et al. 2004) indicated that "there has been considerable research conducted over the last 20 years focused on predicting motor vehicle crashes on transportation facilities. The range of statistical models commonly applied includes binomial, Poisson, Poisson-gamma (or Negative Binomial), Zero-Inflated Poisson and Negative Binomial Models (ZIP and ZINB), and Multinomial probability models. Given the range of possible modeling approaches and the host of assumptions with each modeling approach, making an intelligent choice for modeling motor vehicle crash data is difficult at best." The authors further indicate that "in recent years, some researchers have applied "zero-inflated" or "zero altered" probability models, which assume that a dual-state process is responsible for generating the crash data." The authors indicated that "these models have been applied to capture the 'excess' zeroes that commonly arise in crash data—and generally have provided improved fit to data compared to Poisson and Negative Binomial (NB) regression models."

Lord et al. (Lord et al. 2004) conducted a simulation experiment to demonstrate how crash data may give rise to "excess" zeroes. They demonstrated that under certain (fairly common) circumstances excess zeroes are observed—and that these circumstances arise from low exposure and/or inappropriate selection of time/space scales and not an underlying dual state process. They concluded that a careful selection of the time/space scales for analysis, including an improved set of explanatory variables and/or unobserved heterogeneity effects in count regression models, or applying small area statistical methods (observations with low exposure) represent the most defensible modeling approaches for datasets with a preponderance of zeros. We partially agree with these conclusions, however modelers my not have much choice in their time/space scale selection given the limitation of traffic and crash data.

In this paper we present an alternative approach that combines novice data aggregation with LRMs to address the challenges of crash data that were described earlier.

3 DATA DESCRIPTION

As was mentioned earlier, the two approaches for developing CPMs are demonstrated using a database of crash, traffic, and geometric data obtained from 186 randomly selected arterial access roads connected to freeway ramps. The sections that were considered included the following:

1. Areas designated as rural (79 observations) and others designated as urban (107 observations),
2. Sections with (76 observations) and without acceleration lanes (110 observations),
3. Arterial facilities with a median (121 observations) versus without (65 observations),
4. With 2, 4, or 6 lanes (57, 95, and 34 observations, respectively),
5. First intersection either signalized, stop-sign controlled, or no control on the arterial, and
6. Sections with a left turn bay (78 observations) versus without (108 observations).

The length of the access roads varied from 3 to 1110 m with an average length of 169 m while the distance to the first intersection varied from 6 to 2285 m with an average value of 298 m. The Average Annual Daily Traffic (AADT) varied from 112 to 117,314 with an average AADT of 19,456 veh/day. The number of crashes over 5 years varied from 0 to 169 crashes in a section with an average number of crashes of 12 over the 186 study sections.

The crash data were extracted from the Highway Traffic Records Information System Crash Database (HTRIS) for the years 2001 through 2005. The traffic data were obtained from the Road inventory VDOT database. The data represents the AADT of the section with a few exceptions were the traffic data represent a 24 hour count.

After fusing the crash, traffic, and geometric data it was possible to plot the data, as illustrated in **Error! Reference source not found.** Specifically, the figure demonstrates a general increase in the number of crashes as the facility Average Annual Daily Traffic (AADT) increases. The figure also illustrates a high cluster of data at the short access road distances with minimum observations for access roads in excess of 400 m. Similarly, AADTs in excess of 8,000 veh/day are a rare occurrence. The figure does illustrate a number of sections with high AADTs and short access roads with a small number of crashes. Conversely, observations with high crashes are also observed for low AADTs and long access roads.

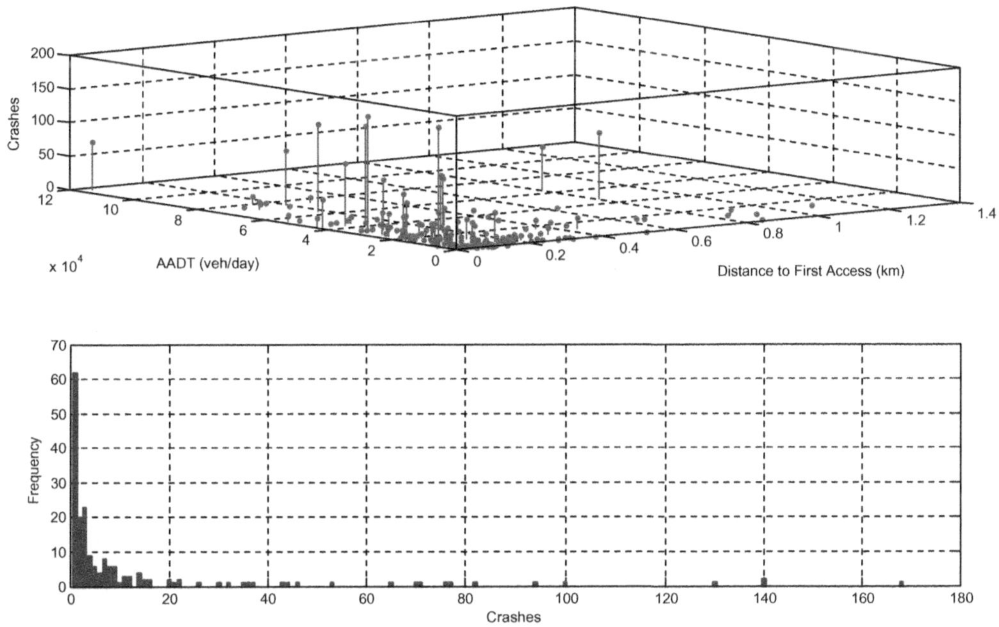

Figure 1. Data distribution.

The crash frequency distribution demonstrates that the number of crashes ranges from a minimum of 0 to a maximum of 169 crashes over 5 years, as illustrated in **Error! Reference source not found.** The higher frequency of zero crashes is typical of crash data when the exposure rate is low as was described in the literature (Lord et al. 2004). The higher propensity of zero observations has lead some researchers to apply zero-inflated negative binomial models to the modeling of crashes considering two underlying processes (Shankar et al. 1997; Shankar et al. 2003). The frequency distribution is consistent with a negative binomial distribution with a consistent decrease in the frequency as the number of crashes increases.

A statistical analysis of the data demonstrated that the type of road (rural versus urban), number of lanes, the availability of a median, the type of signal control at the nearest intersection, and availability of an acceleration lane were not statistically significant. The details of these statistical tests are beyond the scope of this paper, but are provided elsewhere (Medina Flintsch et al. 2008).

4 MODEL DEVELOPMENT

This section describes the two approaches that were tested in the paper for developing crash prediction models. The first approach is the common approach that is reported in the literature, which is based on the use of Poisson, Negative Binomial (NB), Zero Inflated Poisson (ZIP), and Zero Inflated Negative Binomial (ZINB) regression models. An alternative approach that is developed in this paper is the use of LRMs. Initially, a discussion of the NB and ZINB regression approach is presented followed by a discussion of the proposed LRM approach. Both approaches are then compared using crash data from 186 access roads in the state of Virginia in the U.S.

Prior to describing the various models, the model structure is discussed. Specifically, the study considers a crash rate that is formulated as

$$CR = \frac{C}{L_2 V^p} \cdot \frac{10^6}{(365 \times 5)^p} = \exp(\beta_0 + \beta_1 L_1), \tag{1}$$

where CR is the crash rate (million vehicle crashes per vehicle kilometer of exposure over a 5 year period), C is the total number of crashes over the study section of length L_2 in the 5-year analysis period (crashes), L_2 is the length of the section which is the distance between the freeway off-ramp and the first intersection (km), L_1 is the distance between the freeway off-ramp and the first access road (may equal L_2 if the first access road is an intersection) (km), V is the section AADT (veh/day), B_0 and B_1 are the model constants.

The model of Equation [1] can then be manipulated to produce a linear model of the form

$$C = \frac{(365 \times 5)^p}{10^6} \cdot \exp(\beta_0 + \beta_1 L_1 + \ln(L_2) + p\ln(V)) + E \tag{2}$$

Here E is a random error term that accounts for the error that is not captured in the model. The advantage of this model is that (a) it is linear in structure after applying a logarithmic transformation; (b) it ensures that the crashes are positive (greater than or equal to zero); and (c) it produces zero crashes when the exposure is set to zero (i.e. when L_2 or V is zero).

4.1 *Poisson or Negative Binomial model approach*

The Poisson distribution is more frequently applied to models with count data. The probability mass function of a Poisson crash random variable (λ is the mean crashes per unit time) is given by

$$P(C) = \frac{\lambda^C}{C!} e^{-C}; \quad C = 0,1, \tag{3}$$

One feature of the Poisson random variable (C) is the identity of the mean and variance. The NB model overcomes this limitation by considering the λ parameter to be a random variable that is distributed following a Poisson distribution. Consequently, the negative binomial distribution has the same support ($C = 0, 1, 2, \ldots$) as the Poisson distribution but allows for greater variability within the data (variance can be greater than the mean). Consequently, a good alternative model for count data that exhibit excess variation compared to a Poisson model is the negative binomial model.

The structure of the GLM model that computes the expected number of crashes (E(C) can be written as

$$\mathrm{E}(C) = \left(\frac{(365 \times 5)^p}{10^6} \right) L_2 \cdot \exp(\beta_0 + \beta_1 L_1 + p\ln(V)). \tag{4}$$

Two sets of Poisson and two NB models were developed, one a standard model (NB and Poisson) and one a zero inflated model (ZINB and ZIP), as summarized in Table 1. Unfortunately, the Poisson regression model suffered from over-dispersion as indicated by the value of the deviance divided by the degrees of freedom which was much greater than a value of 1.0 (was 184.7). Consequently, a Modified Poisson Regression was also applied using an over-dispersion parameter. The model produces the same parameter values; however, the deviance is reduced to 1.0 and thus is valid from a statistical standpoint. In the case of the NB model the value of the deviance divided by the degrees of freedom was close to a value of 1.0 (1.2063) and thus demonstrating the adequacy of the negative binomial error structure. It should be noted that the model predictions for the zero-inflated models are computed my multiplying the results of Equation [4] by $1 - \exp(\theta)/(1+\exp(\theta))$ as

$$\mathrm{E}(C) = \left(\frac{(365 \times 5)^p}{10^6} \right) L_2 \cdot \exp(\beta_0 + \beta_1 L_1 + p\ln(V)) \times \left(1 - \frac{\exp(\theta)}{1 + \exp(\theta)} \right) \tag{5}$$

Table 1. Summary results of regression models.

Parameter	LRM	NB	Poisson	ZINB	ZIP
B_0	4.27	4.76	3.42	4.76	3.83
B_1	–6.88	–3.64	–6.90	–3.64	–2.00
p	0.86	0.81	0.92	0.81	0.84
θ	0.00	0.00	0.00	–16.00	–1.69
SSE	365608	389589	437926	389589	254307
SSE (%)	0%	7%	20%	7%	–30%
Slope	0.472	0.512	0.549	0.512	0.339
Error (%)	2.12	1.95	1.82	1.95	2.95

The results of Table 1 demonstrate that the model parameters a practically identical for both the NB and ZINB models except for the θ parameter. Given that the θ parameter is much less than zero the predictions of the NB and ZINB are very similar. In the case of the Poisson models (Poisson and ZIP) the model parameter values are significantly different, as demonstrated in Table 1.

4.2 *Linear Regression Modeling approach*

In this section we consider a linear regression approach for the development of a model. If we consider the number of crashes per unit distance as our dependent variable the model of Equation [2] can be cast as

$$\frac{\mathrm{E}(C)}{L_2} = \mathrm{E}(C') = \frac{(365 \times 5)^p}{10^6} \cdot \exp(\beta_0 + \beta_1 L_1 + p \ln(V)) \tag{6}$$

Equation [6] is a linear model with two independent variables: V and L_1. It should be noted that an analysis of crashes per unit distance ensures that the data are normalized across the different section lengths. The development of a LRM using the least squares approach requires that the data follow a normal distribution. A statistical analysis of the data revealed that there was insufficient evidence to conclude that the data were normal. Furthermore, the dispersion parameter, which measures the amount of variation in the data, was significantly greater than 1.0 indicating that a negative binomial model would be appropriate for the data.

Here we present an approach for normalizing the data in order to apply a least squared LRM to the data. The approach involves sorting the data based on one of the independent variables and then aggregating the data using a variable bin size to ensure that the second independent variable remains constant across the various bins. Data transformations can then be applied to the data to ensure normality and homoscedasticity (equal variance). Once the parameters of the first independent variable are computed, the data are sorted on the second independent variable. The data are then aggregated in order to ensure normality and homoscedasticity and then linear models are fit to the data to compute the variable coefficient. The approach is demonstrated using the access road crash data in the following sub-sections.

4.2.1 *Selecting exposure measures*

The typical exposure measure for crashes is million vehicle-miles or million vehicle-kilometers of travel. However, researchers have argued that the exponent of the volume variable (V) in the exposure measure is not necessarily equal to 1.0 [7, 8]. Consequently, the first step in the analysis was to compute the exponent of V (denoted as p).

Given that the data vary as a function of two variables L_1 and V, it was important to normalize one of the variables while analyzing the second variable. In order to estimate the volume exponent, the data were sorted based on their AADT values and aggregated using variable bin sizes to ensure that the L_1 variable remained constant across the various bins, as illustrated in Figure 1. The figure demonstrates that by performing a linear regression of L_1 against V that the slope of the line is insignificant ($p > 0.05$) and thus there is insufficient evidence to conclude that the L_1 variable varies across the aggregated data.

In estimating crash rates it is important that the measure of exposure ensures that the data are normalized. In doing so a multiplicative crash adjustment factor (F_i) for each bin i was computed as

$$F_i = \frac{\min_i \left[\max_j \left(\frac{C_{ij}}{L_{ij}} \right) \right]}{\max_j \left(\frac{C_{ij}}{L_{ij}} \right)}, \tag{7}$$

where C_{ij} is the number of crashes for section j in bin i and L_{ij} is the length of section j in bin i. The F_i correction factor ensures that the maximum number of crashes remains constant (equal to the minimum number of crashes) across the various bins, which is by definition what an exposure measure is. The correction factor is also equal to

$$F_i = \alpha V_i^{\beta} \tag{8}$$

where V_i is the mean AADT volume across all observations j in bin i and α and β are model coefficients. By solving Equation [7] and [8] simultaneously we derive

$$CR = F_i \cdot \max_j \left(\frac{C_{ij}}{L_{ij}} \right) = \alpha V_i^{\beta} \cdot \max_j \left(\frac{C_{ij}}{L_{ij}} \right) = \alpha \cdot \frac{C_i}{L_i V_i^{-\beta}} = \frac{\text{Crashes}}{\text{Length} \cdot \text{AADT}^p}. \tag{9}$$

Consequently, β is equivalent to $-p$ and can be solved for by fitting a regression line to the logarithmic transformation of Equation [8] as

$$\ln(F_i) = \ln(\alpha) + \beta \ln(V_i). \tag{10}$$

After applying a least squared fit to the data, the model residual errors were tested for normality. As illustrated in Figure 2, although in the case of the original non-transformed data the residual error did not pass the normality test, the log-transformed residual errors did pass the test ($p = 0.355$).

A least squares LRM was then fit to the log-transformed data producing an R^2 of 0.89, as illustrated in Figure 3. The model was statistically significant ($p << 0.005$) and both the intercept and slope coefficients were significant ($p = 0.02$ and 0.00, respectively). Consequently, the exponent

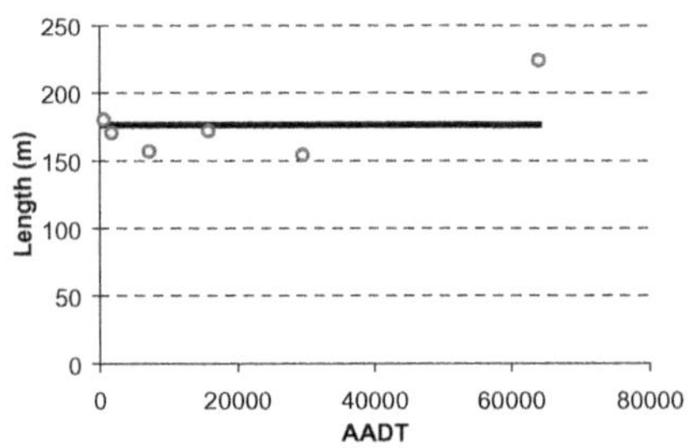

Regression Statistics	
R^2	0.509
R^2 (Adj)	0.386
Obs.	6

ANOVA

	df	SS	MS	F	Significance F
Reg.	1	17608.38	17608.38	4.14	0.11
Residual	4	17003.46	4250.86		
Total	5	34611.84			

	Coeff.	SE	t Stat	P-value	Low 95%	Up 95%
Intercept	530.039	35.820	14.80	0.00	430.59	629.49
AADT	0.002	0.001	2.04	0.11	0.00	0.01

Figure 1. Effect of access section length within AADT binning.

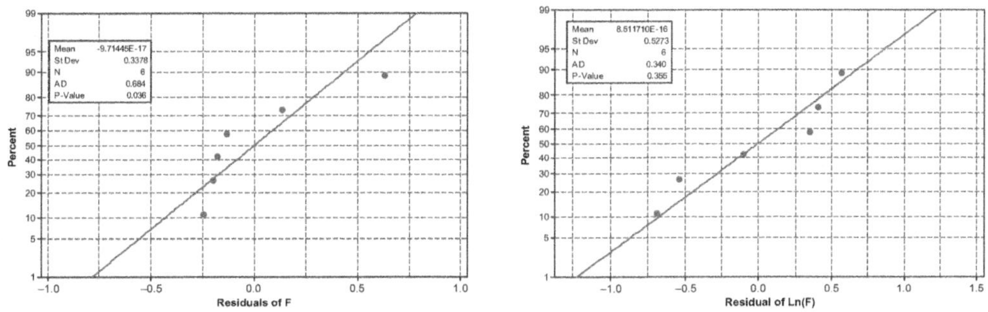

(a) Normality Test on F_i (b) Normality Test on Log-transformed F_i

Figure 2. Test of normality of AADT adjustment factors.

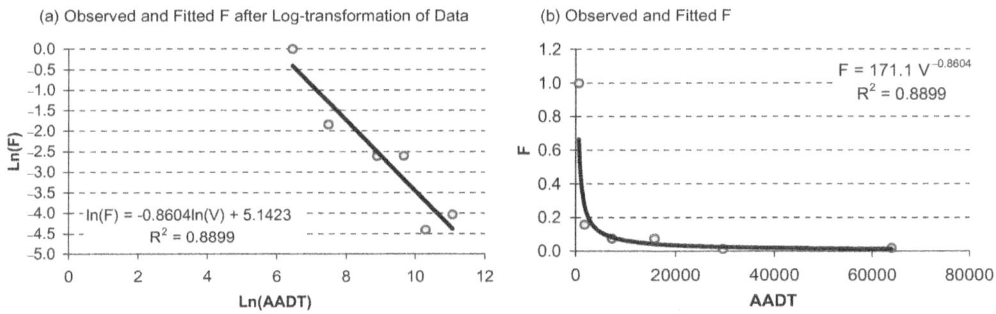

Regression Statistics	
R^2	0.890
R^2 (Adj)	0.862
Obs.	6

ANOVA

	df	*SS*	*MS*	*F*	*Sig. F*
Reg.	1	11.24	11.24	32.33	0.00
Residual	4	1.39	0.35		
Total	5	12.63			

	Coeff.	*SE*	*t Stat*	*P-value*	*Low 95%*	*Up 95%*
Intercept	5.1423	1.38	3.73	0.02	1.31	8.97
ln(AADT)	−0.8604	0.15	−5.69	0.00	−1.28	−0.44

Figure 3. Computation of exposure measure.

of the AADT for utilization in the exposure measure is 0.86, which is very similar to what was derived from the negative binomial fit to the data (p = 0.81).

Once the exponent of the AADT was estimated, the crash rate was computed for each of the 186 study sections. A linear regression model in a single independent variable of the form

$$\ln(CR) = \beta_0 + \beta_1 L_1 \tag{11}$$

was fit to the data. In order to satisfy data normality, the data were sorted based on L_1 and aggregated into equally sized bins of 8 observations. It should be noted that the typical approach to binning is to use equal intervals for binning as opposed to equally sized bins. This unique data aggregation approach is equivalent to considering a longer analysis period (in this case considering an analysis period of 8 × 5 = 40 years). The data aggregation increases the level of exposure and thus reduces the number of zero crash observations (in this case zero observations are removed), given that it is highly unlikely to have no crashes over a 40-year period. For each bin the average section length (L_1) and crash rate (CR) was computed. As demonstrated in Figure 4 there was insufficient evidence to reject the data error normality and homoscedasticity assumption for the log-transformed data ($p = 0.479$) and thus a least squares GLM could be applied to the data.

A robust linear regression was applied to the data to derive the model parameters and remove outlier data. This procedure dampens the effect of observations that would be highly influential if least squares were used [9]. The robust linear regression fit uses an iteratively re-weighted least squares algorithm, with the weights at each iteration calculated by applying the bi-square function to the residuals from the previous iteration. This Matlab algorithm gives lower weight to points that appear to be outliers. Data that should be disregarded are given a weight of zero. Consequently, the regression model is less sensitive to outliers in the data as compared with ordinary least squares regression. Data observations with zero weights were removed from the analysis (in this case a single observation was removed).

The results of the analysis demonstrate a statistically significant model ($F = 51.56$ and $p << 0.0005$) with an R^2 of 0.72. The intercept and L_1 coefficients are statistically significant ($p << 0.0005$ and $p << 0.0005$, respectively) with values of 4.269 and –6.879, respectively.

Similarly, a regression model was fit to the data considering the independent variable as the distance to the first intersection. A similar robust regression was applied to the data to derive the model intercept and slope. Given that the intercept confidence limits included the value of intercept of the first model, the intercept was kept constant in both models. A regression was then performed to estimate the optimum slope. The model is significant ($F = 111.44$ and $p << 0.0005$) with an R^2 of 0.85. The slope of the line is significant ($p << 0.0005$) with a value of –4.135.

In summary, the final models that were developed are of the form

$$E(C) = \frac{(365 \times 5)^p}{10^6} \cdot \exp(\beta_0 + \beta_1 L_1 + \ln(L_2) + p \ln(V)), \text{ or} \tag{12}$$

$$E(C) = \frac{(365 \times 5)^p}{10^6} \exp(\beta_0) \cdot \exp(\beta_1 L_1) \cdot L_2 V^p = \gamma \cdot \exp(\beta_1 L_1) \cdot L_2 V^p. \tag{13}$$

The expected number of crashes in a single year (C'') can be computed by adjusting the model intercept by the $(p-1) \times \ln(5)$ as

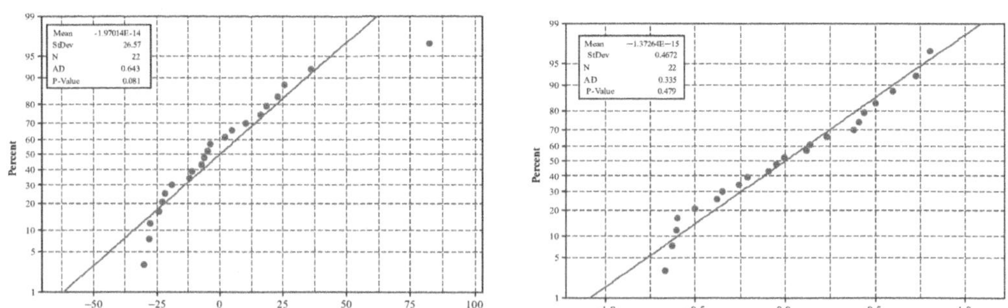

(a) Normality Test on Crash Rate

(b) Normality Test on Log-transformed Crash Rate

Figure 4. Test of normality for crash rate data.

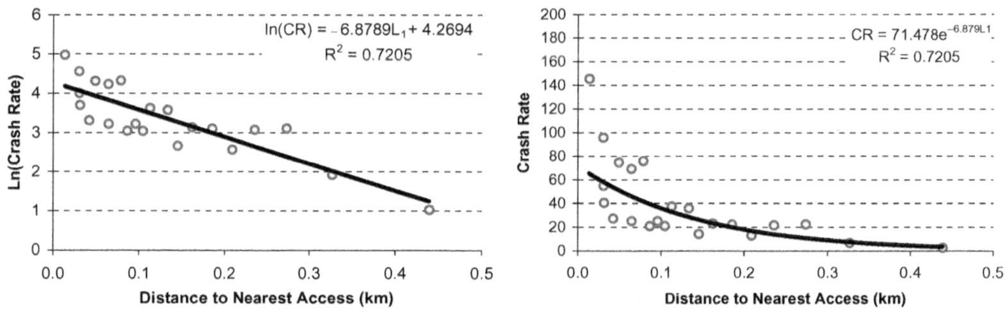

Regression Statistics	
Multiple R	0.849
R^2	0.721
R^2 (Adj)	0.707
SE	0.479
Obs.	22

ANOVA

	df	*SS*	*MS*	*F*	*Sig. F*
Regression	1	11.819	11.819	51.563	0.000
Residual	20	4.584	0.229		
Total	21	16.403			

	Coeff.	*SE*	*t Stat*	*P-value*	*Low 95%*	*Up 95%*
Intercept	4.269	0.163	26.222	0.000	3.930	4.609
L_1	-6.879	0.958	-7.181	0.000	-8.877	-4.881

Figure 5. Crash prediction model considering nearest access point.

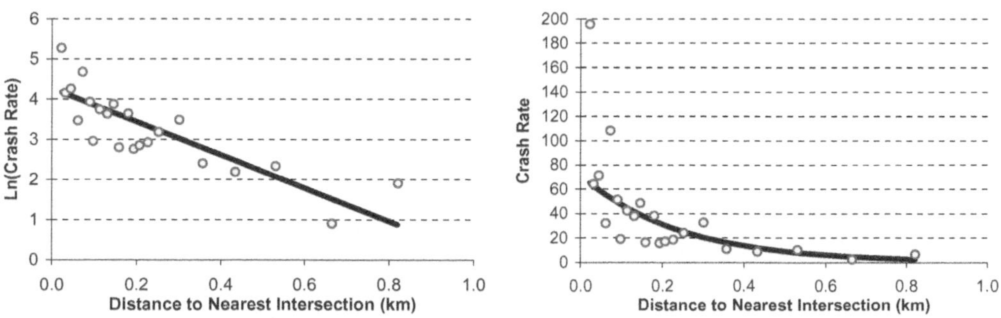

Regression Statistics	
Multiple R	0.921
R^2	0.848
R^2 (Adj)	0.801
SE	0.558
Obs.	22

ANOVA

	df	*SS*	*MS*	*F*	*Sig. F*
Regression	1	36.552	36.552	117.436	0.000
Residual	21	6.536	0.311		
Total	22	43.088			

	Coeff.	*SE*	*t Stat*	*P-value*	*Low 95%*	*Up 95%*
Intercept	4.269	#N/A	#N/A	#N/A	#N/A	#N/A
L_2	-4.135	0.382	-10.837	0.000	-4.928	-3.341

Figure 6. Crash prediction model considering nearest intersection.

$$E(C'') = \frac{365^p}{10^6} \cdot \exp(\beta_0 + (p-1)\ln 5 + \beta_1 L_1 + \ln(L_2) + p\ln(V)). \quad (14)$$

The expected crash rate for a single year in million vehicle kilometers where the traffic volume is raised to the exponent p (CR') can be computed as

$$E(CR') = \exp(\beta_0 + (p-1)\ln 5 + \beta_1 L_1) \quad (15)$$

The expected crash rate in vehicle-miles traveled (VMT) considering an exponent of 1.0 (CR'') is computed as

$$E(CR'') = \exp(\beta_0 + (p-1)\ln 5 + \beta_1 L_1) \times 1.6(365V)^{p-1}, \quad \text{or}$$

$$E(CR'') = \exp(\beta_0 + \ln 1.6 + (p-1)\ln 5 + (p-1)\ln 365 + \beta_1 L_1 + (p-1)\ln V) \quad (16)$$

A number of researchers have argued for the need to calibrate the exposure AADT exponent. Consequently, a sensitivity analysis was conducted to study the impact of alternative exponents on the CPM predictions. Exponent values ranging from 0.6 to 1.2 were evaluated based on values reported in the literature [7], as illustrated in Figure 7. The results clearly indicate that the crash predictions increase as the exponent increases, however the variation in crash predictions as a function of V and L_1 remains fairly consistent.

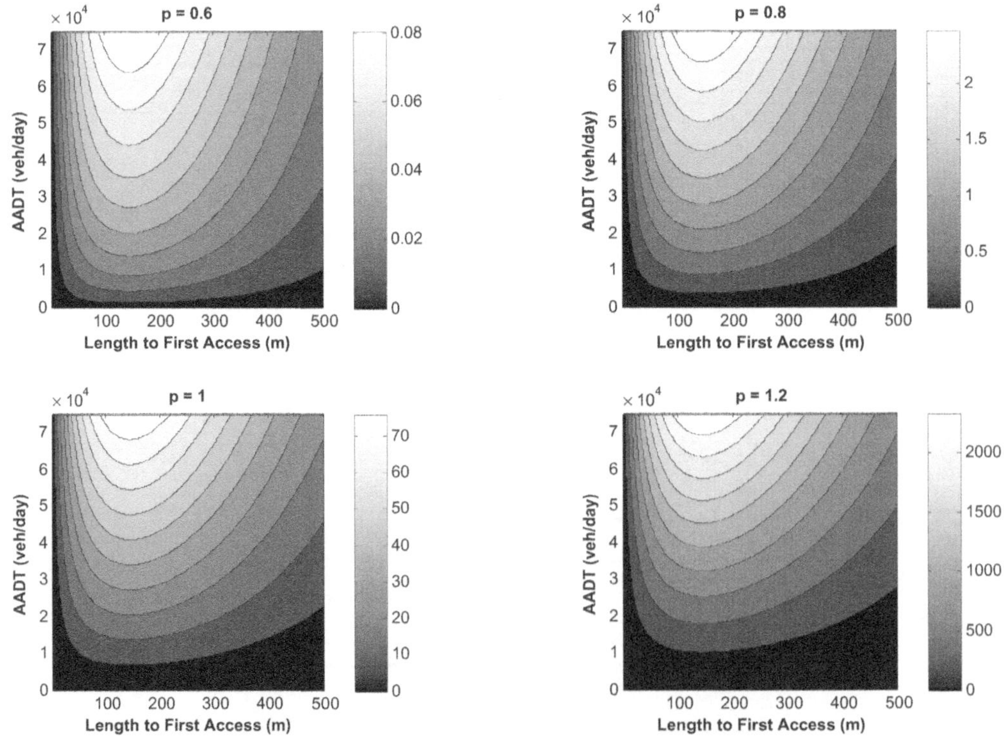

Figure 7. Variation in expected crashes as a function of aadt exponent.

It should be noted that for a constant AADT the expected crashes initially increases as the length of the spacing between the freeway ramp and the first access road section increases before decreasing again, as demonstrated in Figure 8. The maximum expected crashes occurs at an access road spacing of approximately 150 m (500 ft). The observed behavior might appear to be counter intuitive at first glance, however can be explained by the fact that as the study section increases the expected number of crashes per unit distance decreases, as illustrated in Figure 9 and Figure 10, while the level of exposure increases. Initially, the rate of increase in the level of exposure exceeds the rate of decrease in the crash rate producing an increase in the number of crashes. Consequently, decisions should be made using either a crash rate or the expected number of crashes per unit distance, as illustrated in Figure 10. Noteworthy is the fact that the expected number of crashes for an access road spacing of 30 and 150 m is highlighted to demonstrate the current criteria for access road spacing.

The average number of crashes across all 186 study sections was 2.45 crashes/year with an average AADT of 19,456 and an average access road spacing of 169 m (550 ft). The expected number of crashes derived from the model for the same AADT and access road spacing is estimated by the linear regression model at 2.43 crashes/year (highlighted in Figure 8) and thus demonstrating the validity of the model results. Alternatively, the both the NB and Poisson models over-estimate the expected number of crashes to be 2.96 and 2.66 crashes/year, respectively.

L (m)	L (ft)	AADT (veh/day)										
		5000	10000	15000	20000	25000	30000	35000	40000	45000	50000	75000
0.0	0	0.00	0.00	0.00	0.00	0.00	0.00	0.00	0.00	0.00	0.00	0.00
15.3	50	0.19	0.35	0.49	0.63	0.76	0.89	1.02	1.14	1.26	1.38	1.96
30.6	100	0.34	0.62	0.88	1.13	1.37	1.61	1.83	2.06	2.28	2.49	3.53
45.9	150	0.46	0.84	1.20	1.53	1.86	2.17	2.48	2.78	3.08	3.37	4.77
61.2	200	0.56	1.01	1.44	1.84	2.23	2.61	2.98	3.34	3.69	4.04	5.73
76.5	250	0.63	1.14	1.62	2.07	2.51	2.93	3.35	3.76	4.16	4.55	6.45
91.7	300	0.68	1.23	1.75	2.24	2.71	3.17	3.62	4.06	4.49	4.92	6.97
107.0	350	0.71	1.29	1.83	2.35	2.85	3.33	3.80	4.26	4.72	5.17	7.32
122.3	400	0.73	1.33	1.89	2.42	2.93	3.43	3.91	4.39	4.86	5.32	7.54
137.6	450	0.74	1.35	1.91	2.45	2.97	3.47	3.96	4.45	4.92	5.39	7.63
152.9	500	0.74	1.35	1.91	2.45	2.97	3.47	3.97	4.45	4.92	5.39	7.64
168.2	550	0.74	1.34	1.89	2.43	2.94	3.44	3.93	4.41	4.88	5.34	7.57
183.5	600	0.72	1.31	1.86	2.38	2.89	3.38	3.86	4.33	4.79	5.24	7.43
198.8	650	0.71	1.28	1.82	2.33	2.82	3.30	3.76	4.22	4.67	5.12	7.25
214.1	700	0.68	1.24	1.76	2.26	2.73	3.20	3.65	4.09	4.53	4.96	7.03
229.4	750	0.66	1.20	1.70	2.18	2.64	3.08	3.52	3.95	4.37	4.79	6.78
244.6	800	0.63	1.15	1.63	2.09	2.53	2.96	3.38	3.79	4.20	4.60	6.52
259.9	850	0.61	1.10	1.56	2.00	2.42	2.83	3.24	3.63	4.02	4.40	6.23
275.2	900	0.58	1.05	1.49	1.91	2.31	2.70	3.09	3.46	3.83	4.19	5.94
290.5	950	0.55	1.00	1.41	1.81	2.20	2.57	2.93	3.29	3.64	3.99	5.65
305.8	1000	0.52	0.95	1.34	1.72	2.08	2.43	2.78	3.12	3.45	3.78	5.36
321.1	1050	0.49	0.89	1.27	1.62	1.97	2.30	2.63	2.95	3.26	3.57	5.06
336.4	1100	0.46	0.84	1.20	1.53	1.86	2.17	2.48	2.78	3.08	3.37	4.78
351.7	1150	0.44	0.79	1.13	1.44	1.75	2.04	2.33	2.62	2.90	3.17	4.50
367.0	1200	0.41	0.75	1.06	1.36	1.64	1.92	2.19	2.46	2.72	2.98	4.23
382.3	1250	0.39	0.70	0.99	1.27	1.54	1.80	2.06	2.31	2.55	2.80	3.96
397.5	1300	0.36	0.66	0.93	1.19	1.44	1.69	1.93	2.16	2.39	2.62	3.71
412.8	1350	0.34	0.61	0.87	1.11	1.35	1.58	1.80	2.02	2.24	2.45	3.47
428.1	1400	0.32	0.57	0.81	1.04	1.26	1.47	1.68	1.89	2.09	2.29	3.24
443.4	1450	0.29	0.53	0.76	0.97	1.17	1.37	1.57	1.76	1.95	2.13	3.02
458.7	1500	0.27	0.50	0.71	0.90	1.09	1.28	1.46	1.64	1.81	1.99	2.82

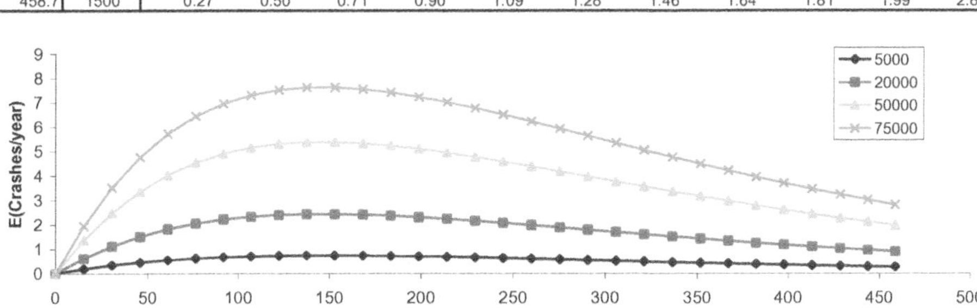

Figure 8. Variation in the expected number of yearly crashes as a function of the access section length and AADT.

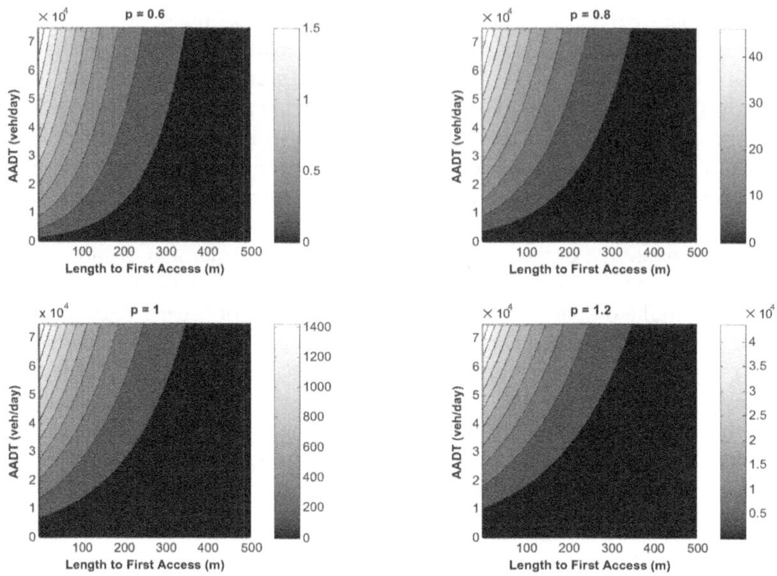

Figure 9. Variation in expected crashes/km as a function of AADT exponent.

L (m)	L (ft)	AADT (veh/day)										
		5000	10000	15000	20000	25000	30000	35000	40000	45000	50000	75000
0.0	0	#N/A	#N/A	#N/A	#N/A	#N/A	#N/A	#N/A	#N/A	#N/A	#N/A	#N/A
15.3	50	12.49	22.67	32.14	41.16	49.88	58.35	66.62	74.73	82.70	90.55	128.35
30.6	100	11.25	20.42	28.94	37.07	44.91	52.54	59.99	67.29	74.47	81.54	115.57
45.9	150	10.13	18.38	26.06	33.38	40.44	47.31	54.02	60.60	67.06	73.42	104.07
61.2	200	9.12	16.56	23.47	30.06	36.42	42.60	48.64	54.57	60.38	66.11	93.71
76.5	250	8.21	14.91	21.13	27.06	32.79	38.36	43.80	49.13	54.37	59.53	84.39
91.7	300	7.39	13.42	19.03	24.37	29.53	34.54	39.44	44.24	48.96	53.61	75.99
107.0	350	6.66	12.09	17.13	21.95	26.59	31.11	35.52	39.84	44.09	48.27	68.42
122.3	400	6.00	10.88	15.43	19.76	23.94	28.01	31.98	35.88	39.70	43.47	61.61
137.6	450	5.40	9.80	13.89	17.79	21.56	25.22	28.80	32.31	35.75	39.14	55.48
152.9	500	4.86	8.83	12.51	16.02	19.41	22.71	25.93	29.09	32.19	35.25	49.96
168.2	550	4.38	7.95	11.26	14.43	17.48	20.45	23.35	26.19	28.99	31.74	44.99
183.5	600	3.94	7.16	10.14	12.99	15.74	18.42	21.03	23.59	26.10	28.58	40.51
198.8	650	3.55	6.44	9.13	11.70	14.18	16.58	18.93	21.24	23.51	25.74	36.48
214.1	700	3.20	5.80	8.23	10.53	12.76	14.93	17.05	19.13	21.17	23.17	32.85
229.4	750	2.88	5.23	7.41	9.49	11.49	13.45	15.35	17.22	19.06	20.87	29.58
244.6	800	2.59	4.71	6.67	8.54	10.35	12.11	13.83	15.51	17.16	18.79	26.63
259.9	850	2.33	4.24	6.01	7.69	9.32	10.90	12.45	13.96	15.45	16.92	23.98
275.2	900	2.10	3.82	5.41	6.93	8.39	9.82	11.21	12.58	13.92	15.24	21.60
290.5	950	1.89	3.44	4.87	6.24	7.56	8.84	10.09	11.32	12.53	13.72	19.45
305.8	1000	1.70	3.09	4.38	5.62	6.81	7.96	9.09	10.20	11.28	12.35	17.51
321.1	1050	1.53	2.79	3.95	5.06	6.13	7.17	8.19	9.18	10.16	11.12	15.77
336.4	1100	1.38	2.51	3.56	4.55	5.52	6.46	7.37	8.27	9.15	10.02	14.20
351.7	1150	1.24	2.26	3.20	4.10	4.97	5.81	6.64	7.44	8.24	9.02	12.79
367.0	1200	1.12	2.03	2.88	3.69	4.47	5.23	5.98	6.70	7.42	8.12	11.51
382.3	1250	1.01	1.83	2.60	3.33	4.03	4.71	5.38	6.04	6.68	7.31	10.37
397.5	1300	0.91	1.65	2.34	2.99	3.63	4.24	4.85	5.44	6.02	6.59	9.34
412.8	1350	0.82	1.49	2.11	2.70	3.27	3.82	4.36	4.89	5.42	5.93	8.41
428.1	1400	0.74	1.34	1.90	2.43	2.94	3.44	3.93	4.41	4.88	5.34	7.57
443.4	1450	0.66	1.20	1.71	2.19	2.65	3.10	3.54	3.97	4.39	4.81	6.82
458.7	1500	0.60	1.08	1.54	1.97	2.39	2.79	3.19	3.57	3.96	4.33	6.14

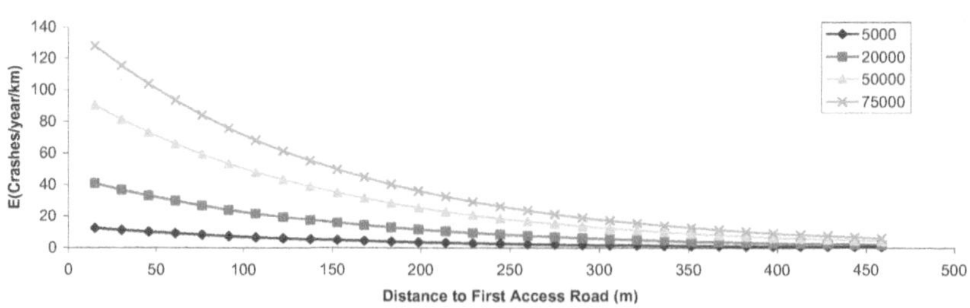

Figure 10. Variation in the expected number of yearly crashes per kilometer as a function of the access section length and AADT.

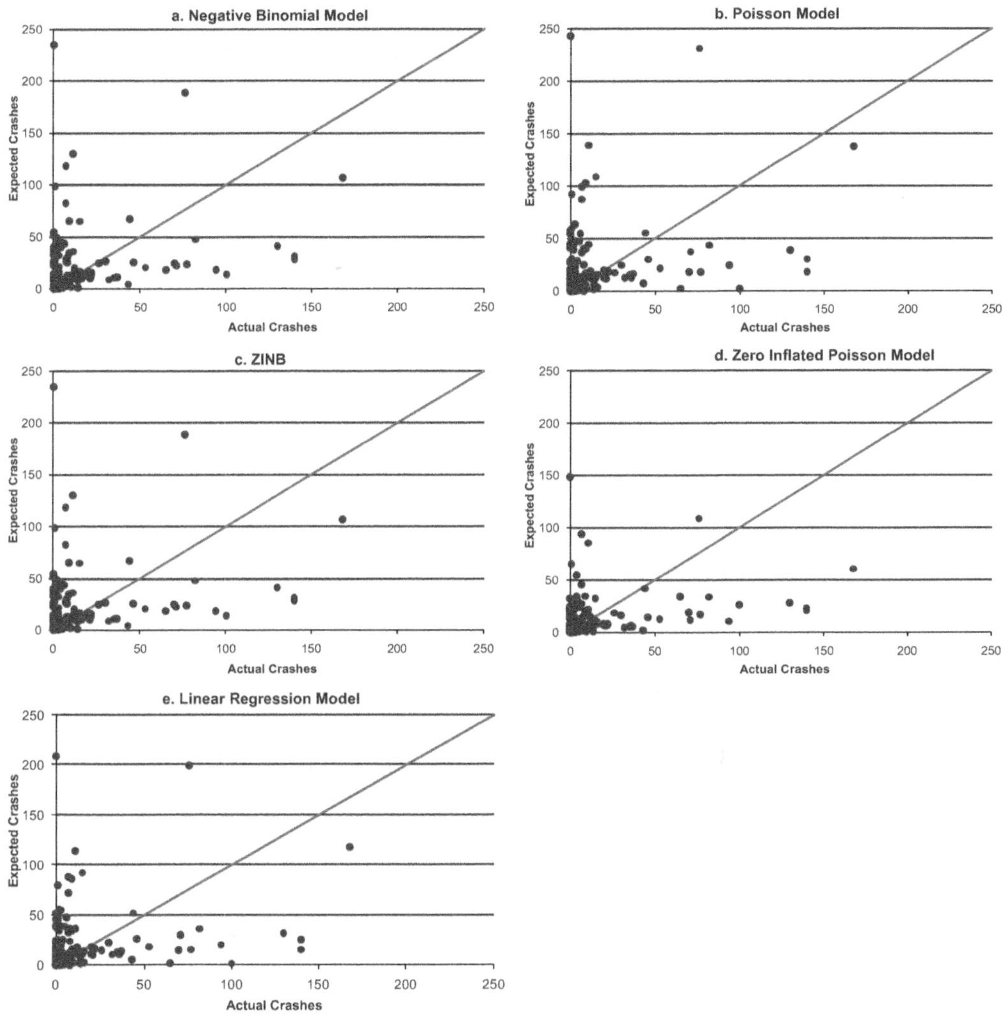

Figure 11. Comparison of actual and expected crashes over a 5-year period (all 186 sites).

In an attempt to validate the developed model, the AADT and access road spacing parameters for each of the 186 sites were input to the various models and the expected number of crashes was estimated. A comparison between the observed and estimated crashes revealed a reasonable level of correlation (Pearson correlation coefficient of 0.24) between the observed and linear regression model estimated crash rates, as illustrated in Figure 11. However, a high level of variability is observed in the data. The figure also clearly demonstrates that all models tend to under-estimate the expected number of crashes is slopes ranging from 0.339 to 0.549, as summarized in Table 1. The results of Table 1 demonstrate that the ZIP model produces the least Sum of Squared Error (SSE) between the estimated and observed number of crashes followed by the proposed LRM model. The ZIP model, however, in reducing the SSE results in a slope that is only 0.339 and thus greatest under-estimation error compared to the other models. Alternatively, the Poisson model produces a slope that is closest to 1.0 (0.549), however, the SSE is 20% higher than the proposed LRM model SSE. Consequently, the proposed LRM model appears to offer the best compromise in terms of SSE and model prediction.

Table 2. Impact of access road spacing on annual crash rate (AADT = 20,000 veh/day).

Distance to First Access Road				
L (ft)	L (m)	Crashes per 10^6 VMT	Relative	Relative
0	0.0	10.07	1.00	8.14
50	15.2	9.07	0.90	7.33
100	30.5	8.17	0.81	6.60
150	45.7	7.35	0.73	5.94
200	61.0	6.62	0.66	5.35
250	76.2	5.96	0.59	4.82
300	91.4	5.37	0.53	4.34
350	106.7	4.83	0.48	3.91
400	121.9	4.35	0.43	3.52
450	137.2	3.92	0.39	3.17
500	152.4	3.53	0.35	2.85
550	167.6	3.18	0.32	2.57
600	182.9	2.86	0.28	2.31
650	198.1	2.58	0.26	2.08
700	213.4	2.32	0.23	1.88
750	228.6	2.09	0.21	1.69
800	243.8	1.88	0.19	1.52
850	259.1	1.69	0.17	1.37
900	274.3	1.53	0.15	1.23
950	289.6	1.37	0.14	1.11
1000	304.8	1.24	0.12	1.00

Distance to First Intersection				
L (ft)	L (m)	Crashes per 10^6 VMT	Relative	Relative
0	0.0	10.07	1.00	3.53
50	15.2	9.46	0.94	3.31
100	30.5	8.88	0.88	3.11
150	45.7	8.34	0.83	2.92
200	61.0	7.83	0.78	2.74
250	76.2	7.35	0.73	2.57
300	91.4	6.90	0.69	2.42
350	106.7	6.48	0.64	2.27
400	121.9	6.08	0.60	2.13
450	137.2	5.71	0.57	2.00
500	152.4	5.36	0.53	1.88
550	167.6	5.04	0.50	1.76
600	182.9	4.73	0.47	1.66
650	198.1	4.44	0.44	1.55
700	213.4	4.17	0.41	1.46
750	228.6	3.91	0.39	1.37
800	243.8	3.67	0.36	1.29
850	259.1	3.45	0.34	1.21
900	274.3	3.24	0.32	1.13
950	289.6	3.04	0.30	1.07
1000	304.8	2.86	0.28	1.00

5 SUMMARY FINDINGS

A key desire of departments of transportation is to identify the minimum distance from a freeway ramp to provide access to local businesses. The model developed as part of this research effort was utilized to compute the crash rate associated with alternative section spacing, as summarized in Table 2. The results demonstrate an eight-fold decrease in the crash rate over an access road spacing ranging from 0 to 300 m. An increase in the minimum spacing from 90 m (300 ft) to 180 m (600 ft) results in a 50% reduction in the crash rate.

6 CONCLUSIONS

The paper demonstrates that a least square LRM approach can be applied to crash data to develop crash prediction models. The proposed approach involves creative manipulation of the data to satisfy the least square LRM assumptions; namely normality and homoscedasticity. The approach can be summarized as follows:

a. Consider the use of an exponential function. This function ensures that the number of crashes equals zero when the exposure is zero; that the number of crashes are always positive; and that the model reverts to a linear function after performing a logarithmic transformation.
b. Sort and aggregate the data based on the AADT using a variable bin size while ensuring that the second independent variable remains constant across the various bins.
c. Compute crash adjustment factors to normalize the maximum number of crashes across the various bins.
d. Perform a logarithmic transformation on the crash adjustment factors to compute the AADT exponent using a LRM while ensuring that the data satisfy the LRM assumptions of normality and homoscedasticity.
e. Compute crash rates using the AADT exponent that was computed earlier and then sort and aggregate the crash rate data based on the second independent variable using an equally sized bin structure (equal number of observations in each bin).
f. Compute the average dependent and independent variable for each bin.
g. Perform a logarithmic transformation of the data and ensure normality and homoscedasticity to develop the final crash prediction model.

The proposed approach was tested and validated using data from 186 access road sections in the state of Virginia. The approach was demonstrated to be superior to traditional negative binomial models because it is not influenced (through data aggregation) by the prevalence of the large number of zero observations that are typical of crash data.

Further testing of the proposed approach on other datasets is needed to validate the proposed approach. Furthermore, a sensitivity analysis of the sensitivity of results on different binning approaches for task (b) on the model outcomes is required.

ACKNOWLEDGEMENTS

The authors acknowledge the financial support from the Virginia Department of Transportation. The authors also acknowledge the work of Ivy Gorman, Pengfei Li, and Dhruv Dua in extracting the data from the HTRIS and GIS Integrator. Furthermore the authors are grateful to the input and feedback received from Eugene Arnold, Stephen Read, Travis Bridewell, Hari Sripathi, and Stephen Bates.

REFERENCES

Ivan, J. (2004). "New approach for including traffic volumes in crash rate analysis and forecasting." *Transportation Research Record* **1897**: 134–141.

Lord, D., S. Washington, et al. (2004). Statistical Challenges with Modeling Motor Vehicle Crashes: Understanding the Implications of Alternative Approaches, Center for Transportation Safety, Texas Transportation Institute.

Lord, D., S.P. Washington, et al. (2005). "Poisson, Poisson-gamma and zero-inflated regression models of motor vehicle crashes: balancing statistical fit and theory." *Accident Analysis & Prevention* **37**(1): 35–46.

Medina Flintsch, A., H. Rakha, et al. (2008). *Safety Impacts of Access Control Standards on Crossroads in the Vicinity of Highway Interchanges*. Submitted to 87th Transportation Research Board Annual Meeting, Washington D.C., TRB.

Montgomery, D.C., E.A. Peck, et al. (2001). *Introduction to Linear Regression Analysis*.

Qin, X., J. Ivan, et al. (2004). "Selecting exposure measures in crash rate prediction for two-lane highway segments." *Accident Analysis & Prevention* **36**(2): 183–191.

Sawalha, Z. and T. Sayed (2006). "Traffic accident modeling: some statistical issues." *Canadian Journal of Civil Engineering* **33**: 1115–1124.

Shankar, V., J. Milton, et al. (1997). "Modeling accident frequency as zero-altered probability processes: an empirical inquiry." *Accident Analysis & Prevention* **29**(6): 829–837.

Shankar, V.N., G.F. Ulfarsson, et al. (2003). "Modeling crashes involving pedestrians and motorized traffic." *Safety Science* **41**(7): 627–640.

Efficient Transportation and Pavement Systems – Al-Qadi, Sayed, Alnuaimi & Masad (eds)
© 2009 Taylor & Francis Group, London, ISBN 978-0-415-48979-9

Turning the world's roads into forgiving highways preventing needless deaths

Michael G. Dreznes
International Road Federation, Washington Road Safety Task Force

1 INTRODUCTION

Imagine the worldwide alarm if a fully loaded Boeing 747 airplane was crashing everyday somewhere around the world. This would be the number one topic in every legislative body in every country, at every dinner table and the lead story on every news channel. Countless amounts of money, energy and time would be spent to come up with some solution to this tragic situation that would threaten to shut down economies from the United States to Germany to Japan to South Africa to Australia to Chile. Finding an answer to stop these airplanes from crashing would be the single most important issue for all of mankind. This would be the correct response because this issue would be so important.

Statistically speaking, six to six and a half fully loaded Boeing 747 airplanes crash everyday... on the roads around the world. The World Health Organization estimates that as many as 1,200,000 people are killed every year on the roads. Too often unless the person in the accident is someone close to us or someone famous, no one notic es the death and the carnage continues with seemingly little concern by anyone other than the immediate family and friends. This is wrong and something must be done to stop this carnage on the roads.

Unfortunately, it is impossible to completely eliminate all accidents around the world. As long as humans are driving the vehicles, accidents will happen on the roads. All humans make mistakes. When a person makes a mistake with a steering wheel in his or her hand, the result can be a very serious traffic accident. While these accidents will never go away, it is possible to design highways to use today's technology to make these impacts less severe. In effect, this technology is forgiving motorists when they make a mistake, and not making the motorist pay for his or her mistake with capital punishment. All of the roads in the world should be "forgiving highways."

Highways are often called a country's arteries. It is a deserving description. Just as a body uses veins and arteries to circulate blood, highways are used to circulate people throughout a country. The challenge highway engineers in the twenty-first century and beyond face is to utilize state of the art technology to provide kilometers of roads in very small areas near, or in cities around the world. This is where people want to live and this is where the roads are needed. One of the inevitable results of these new highway designs, through no fault of the designs themselves, just the lack of ideal geometries, will be black spots, or dangerous potential accident areas. These typically are areas where drivers need to make decisions. When making a decision, the driver can be either right or wrong.

Approximately thirty percent of the fatal accidents on the road will be single vehicle, non-pedestrian (SVNP) accidents where a car will run off the road and impact a rigid roadside object. These rigid roadside hazards include bridge abutments, bridge piers in the median, median barrier terminals, bridge rail ends, sign supports, railroad crossing signal arms, or the barrier ends located in the aptly named "gore areas" at exits, to name just a few.

Locating a black spot is not difficult. Ask any traffic policeman where additional roadside hazard protection is needed, and he or she will quickly start to tell you when he or she last used the "Jaws of Life" to free a mangled body from a crashed vehicle that impacted a rigid object near the road. Ask an experienced highway design engineer to unfold new highway drawings, and he or she will undoubtedly be able to identify a location with poor geometries that could be a problem. Ask a safety auditor in England or Australia to discuss an existing highway that they reviewed, and he or she will be aware of many roadside locations that could be made safer with improved crash protection.

Most qualified experts in the highway safety industry could travel any road in any country around the world and identify multiple dangerous roadside hazards that are not properly shielded. The experts may also identify stopped or slow moving trucks in work zones that can be extremely dangerous to motorists, even when these trucks are fitted with arrow boards, lights and variable message signs. They may also point out inadequate protection for workers and motorists due to the use of cones or barricades in these work zones instead of properly installed barriers that would provide positive protection.

Not correcting a dangerous condition on the highway or using a cheap, unproven meathod to correct the hazard can prove to be a much more costly option than treating the site with a properly designed and tested crash protection. Utilizing proper crash management is proving on a daily basis around the world that it is a highly economic tool that must be used to improve roadway safety.

2 BACKGROUND

In the 1950's and 1960's, the number of vehicles and the kilometers of roads grew around the world. As they grew, so did the number of fatalities on the roads. About one third of these fatalities were due to single vehicle, non-pedestrian (SVNP) collisions. Most of these accidents were motorists impacting rigid objects near the side of the road. Unfortunately, little was done in the 1950's and 1960's to prevent this carnage on the highways anywhere in the world. The prevailing theory was that if you errantly leave the road, you must face the consequences because it was your own fault.

Many road administrators at that time argued that the SVNP accidents were often the result of excess speed or alcohol consumption. They contended that drivers involved in these SVNP accidents were "authors of their own misfortunes". In some countries, imprecise reporting made it difficult to identify the hazard that was the cause of the accident. Therefore, the dangers posed by a particular site were not clear, so neither were the possible solutions.

Most road administrators in the 1950's and 1960's believed that alert and competent drivers could achieve satisfactory levels of roadway safety if they stayed on the highways that were designed using conventional engineering. By the mid 1960's, it was obvious that this logic was no longer valid. Between 1956 and 1966 the number of motor vehicles on the American roads alone had grown 47% to over 96 million. In 1966, the number of deaths on the United States roads alone had reached 53,000 and more than 2,000,000 people were subjected to disabling injuries. Many countries around the world had similar fatality levels and vehicle growth figures. Something needed to be done to make the roads safer.

In the late 1960's, concentrated efforts to create a "Forgiving Highway,"or a safe road that would not make driver error a capital offense, were initiated in many parts of the world.

Single vehicle, non-pedestrian accident

The United States Federal Highway Administration (FHWA) was at the forefront of the efforts to develop the "Forgiving Highway" concept in the United States. During the 1970's and into the 1980's, many countries evaluated the success of the Federal Highway Administration programs and incorporated some of the American ideas, along with remedial actions that were more appropriate to the needs of the motorists in their countries. In every case, the goal was to create a "Forgiving Highway."

Today road authorities around the world recognize the benefits of a "Forgiving Highway" for the motoring public. They understand that ideally the road design for a "Forgiving Highway" should incorporate the clear zone concept by removing hazards that are near the road. This is not always practical, especially in high volume urban roads that are surrounded by a variety of complex highway structures, traffic signs or lighting columns.

If removing the hazard is not practical, road authorities should move the hazard further away from the road to reduce the possibility of impacts. Long masted luminaire supports would be good examples of this solution.

Removing or moving the hazard is definitely the most effective and desirable action to correct a dangerous roadside situation. However, if designing out or removing the hazard is not possible, then the next option is to design these hazards that are near the road so they will "breakaway" when impacted. Making the hazard less rigid will lessen the severity of the impact for the occupants of the vehicles.

If a breakaway feature is not feasible, the hazard should be shielded using a crash barrier (e.g., steel guardrail, concrete barrier or cable barrier) or a crash cushion. The remainder of this paper will concentrate on the last option; shielding the hazard using a crash barrier or a crash cushion.

3 TYPICAL ROADSIDE HAZARDS

Most high-risk sites on a highway occur at a point when the driver must make a decision. If an unprotected rigid object is in the driver's decision area, and if the driver makes the wrong decision, the results can be disastrous. The hazards that cannot be removed, moved, or made breakaway, and therefore must be shielded from errant vehicles, are similar in all countries.

Table 1 illustrates the similarities between the most frequently impacted hazards involved in SVNP fatal accidents in the United States and Europe. Successful concepts used to create a "Forgiving Highway" can be translated between countries.

3.1 *Trees*

Trees are the most frequently impacted objects in Europe and the United States during the SVNP accidents. Roads were initially built through forests and trees were removed to make the road. Obviously that meant that the roads would be lined with the trees that were not cut down. The trees that were close to the roads initially were a good thing because, they provided shade for the horses or the oxen that were pulling the carriages and wagons. Trees also provided a guide to allow motorists to know where a road was located in areas that experienced significant levels of snow. None of these reasons for planting trees near a road make sense today. Unfortunately, it continues to be common practice to "beautify" roads by planting trees too close to the traveled way. The general public and government officials often remain reluctant to remove existing trees despite the fact that motor vehicle impacts into trees resulted in the greatest number of fatalities in the United States as well as in Europe, and this trend is likely to continue as long as trees are located close to the road.

Indiscriminate removal of all trees from the adjacent roadside is not the answer because it is not cost-effective. Rather, a firm policy should be implemented and followed to remove trees that were either frequently impacted or considered, by experience, to be potentially dangerous. Should it be decided not to remove a hazardous tree, then shielding may be the last

Table 1. Single vehicle non-pedestrian fatalities based on first harmful event.

Object	United states (2006) Percentage	Europe (2006) Percentage
Unknown/none	Not included	42.0
Trees	50.0	17.0
Pole/post	12.0	9.0
Guardrail/barrier/fence	11.0	6.0
Culvert/curb/ditch	7.0	8.0
Other	20.0	18.0

Source: *RISER,2006 IIHS, 2006.*

Trees are the number one roadside killer

resort to provide safety to occupants of errant vehicles that collide with the tree. This can be accomplished through the use of guardrail (crash barrier) or crash cushions or a combination of both.

3.2 *Poles*

It has been estimated that in the United States alone about 130 million utility poles, sometimes referred to as telegraph or telephone poles, exist. Millions more luminaire supports also are installed in the United States. More than 100 million of these posts are installed in close proximity to the traveled way of highways, roads, and streets. Of these, more than 50 million can be considered hazardous roadside objects. Literally thousands of lives are lost each year by vehicle impacts with poles and posts around the world.

One product that could make these poles and sign posts safer is a breakaway device. A variety of breakaway devices are available. The Transportation Research Laboratory (TRL) in Great Britain invented one of the first devices in the early 1960's.

Breakaway poles sever when impacted and allow the vehicle to pass through with a minimum change in velocity. Although the breakaway pole and signpost concept was invented in Great Britain, they are not extensively being used in Great Britain, but are being used in other countries.

One breakaway concept, the breakaway timber utility pole is being used in the United States. This pole, designed to break away upon vehicle impact, is intended for use in situations where a rigid pole is known to be in a potentially hazardous location and its removal or relocation is not possible or practical.

These poles were field tested in nineteen roadside installations in Massachusetts over a two-year period. It is noteworthy that in five hits, no serious injuries were reported to vehicle occupants, at no time has utility service been interrupted, and no litigation was filed against any of the participants in the evaluation.

Slip base design breakaway post

Breakaway timber utility pole

Guy wires are required for certain utility pole applications. These guy wires can be very dangerous to motorists. New concepts are being developed to allow these guy wires to break away when impacted to prevent vehicles from being vaulted or rolling over.

In the late 1980's the collapsible pole was introduced in Sweden. The pole is designed to capture an impacting vehicle and slowly decelerate it. The pole has been successful in the Scandinavian countries and highway agencies in other countries are currently evaluating this concept. The lattice posts also are being used very successfully throughout Europe.

Another method used to provide motorist protection from a utility pole installed near a multi-road intersection is to install a wrap-around crash cushion of small diameter collapsible high-density polyethylene tubes. When the tubes are impacted, the tubes crush and the vehicle is slowly decelerated to a stop. These systems are recommended when anticipated vehicle impact speeds of 70 kilometers per hour (43 moles per hour) or less are envisaged. A variety of other crash cushions are also available to provide head-on and angle protection for motorists from roadside posts.

4 CRASH BARRIERS

Many different types of roadside barrier systems, also known as guardrails, guide rails, longitudinal barriers or crash barriers, are being used around the world. Most steel barriers are considered semi-rigid or semi-yielding barriers because they are designed to have limited deflection when hit on an angle by an errant vehicle. This deflection will typically provide for a less severe impact for the motorists in the vehicle. Care must be taken to ensure that these steel barriers have a proper, flat clear zone directly behind them when they are installed to allow for the anticipated deflection. The post spacing can have a significant effect on the amount of deflection of the steel barriers. Others use a steel box beam or a double corrugated steel railing or triple corrugated steel railing as its longitudinal member fastened to steel or wood posts.

Cable barriers have become very popular for a variety of applications in areas where clear areas are present for the anticipated deflection when they are impacted. Cable barriers are considered flexible barriers. Typically motorists are subjected to less severe consequences when a cable barrier is impacted due to the deflection. They provide good aesthetics and can be easily maintained after an impact.

Concrete barriers that utilize various designs of safety shapes also are being used where no deflection of the barrier is acceptable. Concrete barriers are very popular in medians due to their limited, and often non-existent deflection characteristics.

Errant vehicles colliding with barriers kill thousands of people each year around the world. Motorcyclists are overrepresented in the number of fatalities when barriers are impacted. The overall number of annual fatalities caused by impacts into barriers have been reduced in recent years by improvements such as the use of improved performance concrete shapes, crash cushions and terminals to protect the rigid ends, block-outs, shorter spacing of guardrail posts, proper installation and the use of the triple corrugated railings (thrie beam).

Many of these crash barriers have extremely dangerous ends. A variety of efforts have been made to treat the ends over the years, including turned down ends, breakaway cable terminals (BCT), and modified eccentric loaded terminals (MELT) in the 1970's, 1980's and 1990's. None of these solutions are acceptable today. They do not perform with today's vehicle fleet.

The turned down or ramped end showed itself to be extremely dangerous by causing the vehicle to ramp and the driver to lose control upon impact. In June of 1990, the United States recognized this danger and outlawed the use of ramped ends on high speed, high volume roads. In 1998 these turned down ends were prohibited for use on the upstream end of any barriers on the National Highway System in the United States. Disastrous experiences with turned down ends in multiple countries have many road authorities looking for options to improve their barrier terminals.

The small car that became prevalent on European and American roads in the 1970's created a real problem for the breakaway cable end terminal. Guardrail ends using the breakaway cable terminal have been known to penetrate or spear smaller cars during impacts causing motorist injuries and deaths. In 1993, the FHWA in the United States outlawed the use of the BCT for new roads. The performance is fatally unacceptable. The MELT met a similar fate when NCHRP 350 was implemented in 1998 because it did not perform consistently with the larger test vehicle, which was a pick up truck.

A variety of crashworthy terminals that do not spear, roll or vault an impacting vehicle are commercially available. Their use has been increasing in the United States in recent years and the trend is being followed in Europe and other countries around the world.

Approximately 750,000 terminals, some crashworthy, some not crashworthy are currently installed on the approach ends of crash barriers in the United States. About 15,000 of these terminals are impacted every year. Not using crashworthy end treatments to shield the ends of the crash barriers cannot be professionally justified.

One reason often given for not undertaking a program to replace dangerous turned down ends or breakaway cable terminals with crashworthy terminals and crash cushions relates to liability.

Typical turned down end installation

The ABC Terminal

The ET 2000 Terminal

The X-Tension Terminal

Road authorities are concerned that if they admit that one turned down end is dangerous, and they replace it with a crashworthy terminal, and then an errant vehicle hits another similar turned down end, the highway agency will be subjected to serious liability issues for allowing this admittedly dangerous turned down end condition to exist. Lawyers will argue that the road authority knew the turned down end was dangerous because it was replaced at other sites.

Legal authorities agree that road authorities do not need to correct every similar hazard once they correct one hazard. This is financially unrealistic. Legal precedent in certain countries has shown that if a highway agency has a written plan that is realistic based on financial restraints and time parameters, the courts will rule that the agency is doing everything in its power to correct the problem. The important issue is that the agency recognizes the problem and implements a realistic action plan.

5 INTRODUCTION TO CRASH CUSHIONS

Crash cushions, also called impact attenuators, are passive restraint systems that are designed to reduce the severity of an impact. A crash cushion reduces the consequences of an accident by slowly decelerating an errant vehicle before it impacts the rigid hazard. It is a means for "extending the time of the crash event", or simply decreasing the severity of the impact by reducing the rate of deceleration.

Crash cushions function in many ways like a parachute for your car. A person jumping from an airplane at 1,000 meters (3,300 feet) with no parachute will impact the ground at a very high speed. All of his or her bones will be broken and his or her internal organs will be destroyed due to the extremely high deceleration level he will experience. However, if this same person employs a parachute during the jump, he or she will slowly descend and softly land on the earth. His or her body will experience very low deceleration levels and he or she will walk away after hitting the ground.

A car traveling at 100 km/hour (62.5 mph) that impacts a rigid roadside object will come to a sudden violent stop. The passengers in the vehicle, who for a brief millisecond will still be traveling at 100 km/hour (62.5 mph), will then be thrown forward into some part of the car, the windshield, the steering wheel or hopefully a seat belt. Finally, all of the passenger's internal organs will impact his or her chest wall, causing internal bleeding and internal injuries. The deceleration levels will be incredible and the person will probably die.

However, if this same car traveling at the same speed impacts a properly designed crash cushion instead of the rigid object, the results will be much different. The vehicle will be brought to a controlled, safe stop as the impact is extended over time. Just as the parachute extended the time of the fall, the crash cushion extends the time of a crash. The vehicle still goes from 100 km/h (62.5 mph) to zero. What is different is the time used to go from 100 km/h (62.5 mph) to zero. By extending the time of the event, the deceleration forces on the people in the vehicle are reduced. The passengers will experience far lower deceleration rates, possibly allowing them to walk away from an accident that otherwise might have killed them.

6 HISTORY OF CRASH CUSHIONS

The first crash cushions were developed in the 1960's in the United States. They consisted of empty oil drums that were systematically arranged in front of a roadside object. When impacted, these drums crushed, transferring the energy from the vehicle into the drums and extending the time of the event.

Modifications to this basic concept were made over the years. Instead of empty steel barrels, plastic drums filled with varying weights of sand were employed. The sand in these inertial barriers was elevated to maintain the same center of gravity with the impacting vehicles. This constant center of gravity ensured that the car stayed on the ground during the event.

Crash cushions that used water as the energy-absorbing element followed the inertial barriers. These water systems provided redirection by guiding errant vehicles impacting at an angle back safely into the original traffic flow. While these water systems worked well for larger vehicles, they did not always provide safe deceleration levels for occupants of a small vehicle during head-on impacts.

In the late 1970's, small cars became more prevalent on the world's roads, and a new energy-absorbing element was needed. In 1981, a new concept, which used foam to decelerate the vehicle, was introduced. Boxes filled with this foam material were placed between sliding fender panels to safely decelerate vehicles weighing between 820 kg and 2040 kg (1800 to 4500 pounds) traveling at speeds up to 113 km/h (70 mph). This concept could be used to shield hazards as narrow as 610 mm (24 inches) and as wide as 5 meters (16.5 feet) providing great versatility to road authorities. Approximately 60% of the components in these systems could typically be reused after a design impact.

In recent years a variety of innovative crash cushions from countries around the world have been developed. In the United States, crash cushions are tested to meet the performance requirements for crash cushions as prescribed in NCHRP 350. This document updated previous testing criteria to reflect the current United States vehicle mix and to utilize increased knowledge learned about crash cushions and road safety. The crash cushions tested to NCHRP 350 criteria are designed to safely decelerate 820-kg (1,800 pounds) cars up to 2,000-kg (4,500 pounds) pickup trucks at speeds up to 100 km/h (62.2 mph).

In Europe, crash cushions are tested to the European EN-1317-3 criteria. Crash cushions have been tested for hazards as narrow as 610 mm (24 inches) to 2.3 meters (90 inches) and at speeds up to 110 km/h. The test vehicles used in Europe weigh between 900 kgs. (1,985 pounds) to 1500 kgs. (3,300 pounds).

These new generation products such as the QuadGuard System, the TAU, the TAU II, the TRACC, the VecuStop and the Quest can protect hazards as narrow as 450 mm (18 inches) to

Inertial barrier arrays

Hi-Dro crash cushions

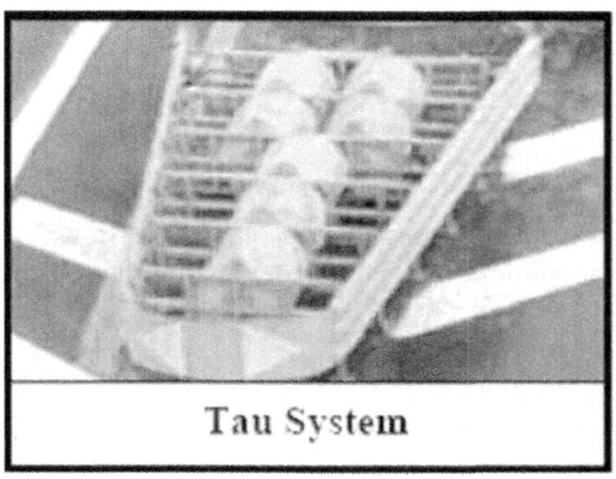

Tau System

Vecu-Stop System

an unlimited width. They have been designed to work with vehicles traveling between 50 km/s (30 mph) to 110 km.h (68.9 mph) givings road authorities the luxury of many excellent choices for their crash cushion needs.

Sixty to sixty-five percent of many of these crash cushions can typically be reused after a design impact. Repair is simple and typically can be completed on site in less than two hours by a two-man crew.

Some crash cushion sites are very dangerous for maintenance crews or the crash cushions are hit very often. In these cases, road authorities often prefer to use crash cushions with greater than 60% + reusability. They are willing to spend more money initially because the crash cushion life cycle cost will justify the additional expense. In the long run, this could prove to be the best use of the taxpayers' money.

A variety of these reusable systems are available including the HEART System, the QuadGuard LMC System, the SMART System, the QuadGuard ELITE System and the REACT 350 System. Up to ninety percent of these systems can be reused after a design impact. After most, but not all design impacts, some of these systems partially restore themselves to provide protection during the next impact. Maintenance departments appreciate the long-term value of these reusable crash cushions. All of these crash cushions do require limited maintenance after a design impact.

Unfortunately, design engineers and maintenance engineers have separate budgets. The design engineers typically want the system with the cheapest initial price that meets the standards. The maintenance engineers want systems that will be easy and inexpensive to repair with low life cycle costs. Both engineering groups should take into consideration site specific information to be sure the crash cushion with the lowest possible life cost is used.

It is very important that road authorities first try to identify why these crash cushions are being impacted so often. Typically improved signage, reduced speed limits, better delineation or

QuadGuard System

TRACC System

CrashGuard System

TAU II System

REACT 350 reusable crash cushion

Crash Cushion with Chevron of Flexible Delineators

modified pavements can significantly reduce the number of impacts, and all of these activities should be considered before a high initial cost reusable crash cushion is installed.

Another cost effective option is be to install a partially reusable crash cushion and form a chevron of flexible delineator posts in front of the crash cushion. This advanced warning will help to reduce the number of impacts into the crash cushion. This often is an interim step to ensure a higher initial cost reusable system is justified at a site.

7 CRASH CUSHIONS TODAY

Crash cushions have become a common site on roads in the United States and Europe, with over 50,000 systems currently installed. Road authorities require these appurtenances to be tested to meet NCHRP or CEN criteria before they are allowed on the roads. Their insistence on the use of tested and proven crash cushions has resulted in tens of thousands lives being saved and hundreds of thousands of serious injuries prevented on the world's roads since 1969. One report from England noted that during a twenty-six-month evaluation period, one minor injury was experienced at sites where crash cushions had been installed. At these same sites during the seven years before crash cushions were installed, five people had been killed, four others had been subjected to serious injury and seven others had experienced minor injuries due to impacts with these rigid roadside obstacles.

Crash cushions are also saving lives on roads in Asia, South America, the Middle East, Australia and Africa. These crash cushions are performing very well in these countries. In January of 2008 at the Transportation Research Board (TRB) Annual Meeting in Washington, D.C., the AFB20(2) Roadside Safety Design Subcommittee on International Research Activities unanimously passed a proclamation that stated:

The AFB20(2) Roadside Safety Design Subcommittee on International Research Activities recommends that road authorities in all countries should only specify roadside safety hardware, i.e. longitudinal safety barriers, crash cushions, terminals and transitions that has met either NCHRP 350 or EN 1317 criteria (or their updates).

Using properly tested crash cushions, terminals and barriers will make roads safer. The Annual Report on Highway Safety Improvement Programs published by the FHWA indicates that crash cushions have an 8.0 Benefit-Cost Ratio. These life saving devices not only are effective, they are cost effective and need to be used on all roads in every country,

8 CONCLUSION

Highways will always be dangerous. As long as humans are driving cars, accidents will happen. Since accidents cannot be eliminated, efforts must be made to design the highway so the ultimate results of the accidents are less severe. This would create Forgiving Highways that forgive the driver when he or she makes a mistake while driving.

Crash cushions, crashworthy terminals, crash barriers and breakaway devices are cost effective passive restraint devices that will not stop accidents, but they will reduce the severity of these accidents. Not installing these products could result in motorists suffering deaths that could have been avoided. It is impossible and unfathomable to try to justify these "needless deaths."

Today most countries with sophisticated highway systems and a desire for safer roads are incorporating these passive restraint devices into their highway design. Their goals continue to be a cost effective means to create "Forgiving Highways." These products are tools that can be, and must be used to help them reach their goals.

REFERENCES

1. American Association of State Highway and Transportation Officials (AASHTO). *Roadside Design Guide,* AASHTO, Washington, D.C., 1989.
2. Dreznes, M. *Makes Roads Safer Using Crash Cushions: The German Example,* IRF Regional Conference for Europe, Belgrade, 1991.
3. Dreznes, M. *The Importance of Using Properly Tested, Proven Crash Cushions,* Institut National De Recherche Routiere Symposium on Road Development and Safety, Luxembourg, 1989.
4. Federal Highway Administration (FHWA). *The 1992Annual Report on Highway Safety Improvement Program,* FHWA, Washington, D.C., 1992.
5. Federal Highway Administration (FHWA). *Supplemental Information for Use with the Roadside Computer Program,* FHWA, Washington, D.C., 1992.
6. Gilleran, B.F. *"The Breakaway Timber Utility Pole.",* Unpublished, Office of Highway Safety, Room 3407 Federal Highway Administration, Washington, D.C., 1992.
7. Griffin III, L. 1. "How Effective are Crash Cushions in Reducing Deaths and Injuries?" *Public Works,* March, 1984.
8. International Road Federation (IRF). *World Road Statistics 1985–1989,* IRE, Washington, D.C., 1990.
9. Kircher, J.R. "Impact Attenuators: A National Survey—Part H." *Public Works,* January, 1986.
10. Lawson, S.D. "Single-Vehicle Collisions with Roadside Objects: The Problem and Its Need of Treatment." *Traffic Engineering Control* 26 (October, 1985).
11. Lawson, S.D., *Crash Protection and Behavioral Aspects of Single-Vehicle Accidents.* Symposium Proceedings, University of Birmingham, England: University of Birmingham, January 1987.
12. National Highway Traffic Safety Administration (NHTSA). *Fatal Accident Reporting System* 1990. NHTSA, Washington, D.C., 1991.

13. Proctor, S., M. Belcher. The *Development of Roadside Crash Cushion in the UK* unpublished, 1990.
14. Proctor, S., D. Grieves, T. Berensford and J. Bowling. "Cushioning the Blow Brummie Style." *Surveyor.* 21 January, 1988: 20–22.
15. Tamanini, F. J. *Impact Attenuators: An Overview of their Characteristics and Effectiveness,* unpublished, Chicago, 1988.
16. Tamanini F.J.M. Dreznes. *Roadside Hazards: The American Experience.* IRF Regional Conference Proceedings, Riyadh, 1988.
17. Ross, E., D. Sicking, J. Rollins. *Estimating Frequency Severity and Costs of Roadside Accidents,* Texas Transportation Institute, College Station, 1986.
18. Sweroad. *Justification for the Use of Crash Cushions in Saudi Arabia,* unpublished, Riyadh, 1988.
19. Wisconsin. U.S. Department of Transportation, Federal Highway Administration, Wisconsin Division. *Crash Cushions—Accident Experience and Maintenance Practices.* Wisconsin: Department of Transportation, July 1986.
20. Federal Highway Administration (FHWA). *The 1996Annual Report on Highway Safety Improvement Program,* FHWA, Washington, D.C., 1996.
21. Ray, M. Ivey, D. Alberson, D. *Expanding the Safety Performance of Guardrail End Treatments.* Texas Transportation Institute, College Station, 1998.
22. Wald, Matthew. "Motor-Vehicle related Deaths Will Increase, Study Predicts." *The New York Times,* April 7, 2004.

Efficient Transportation and Pavement Systems – Al-Qadi, Sayed, Alnuaimi & Masad (eds)
© 2009 Taylor & Francis Group, London, ISBN 978-0-415-48979-9

Road accidents in Kuwait: Causes and consequences

Eisa Al-Enezi
Jahra Maintenance Department, Ministry of Public Works, Jahra, Kuwait

Ahmad H. Aljassar
Department of Civil Engineering, Kuwait University, Kuwait

Mohammed A. Ali
D A Watt Consulting Group Ltd, Canada

Fahad Al-Rukaibi
Department of Civil Engineering, Kuwait University, Kuwait

ABSTRACT: Traffic accidents on roads are a major cause of deaths, injuries, and property damages in the State of Kuwait. In an attempt to identify causes and consequences, reported traffic accidents of one year in Kuwait were analyzed using a sample of 2271 reports. Main types of information from such reports were extracted, coded, and statistically analyzed. Important results were obtained from frequency analyses and cross-classification analyses of many factors related to traffic accidents including accident severity, time and location of occurrence, geometry of the road, and cause of accident. The paper presents the results of such analyses which include tabular and graphical presentations of frequency and cross-classification analyses, in addition to an accident spot map for traffic accidents in Kuwait. The paper is concluded with recommendations to improve traffic safety and further studies to analyze traffic accidents.

Keywords: Traffic Safety, Traffic Accidents, Accident Analysis, Accident Types, Road Safety, Traffic Law, Kuwait

1 INTRODUCTION

Worldwide deaths, disabilities, and injuries from road accidents have reached epidemic proportions. More than 1.17 million people die in road crashes and about fifteen million are injured on the world's roads every year. Next to circulatory diseases and cancer, road accidents are probably the third major cause of death for men in the developed world (O'Flaherty 1997). Each year, around 3,500 people are killed in Britain and around 33,000 people are injured in road accidents (Anderson 2000). Reviews of accident statistics in the United States reveal that in 1996, over 43,000 people were killed on highways in a staggering 11 million accidents (NSC 1996).

Kuwait is at the top of list of the countries with alarmingly high traffic accident fatality rates. A summary of accident type statistics for the year 2004 for each Governorate is presented in Table 1. About 55,000 accidents occurred in that year. Over 97% of accidents were vehicle collisions/crashes. About 33% of the total accidents occurred in the Capital Governorate followed by 27% in Hawalli Governorate.

Around 398 fatalities occ urred in the year 2004. The overwhelmingly high proportion of younger drivers involved in traffic accidents is a serious concern. Over 50% of the fatalities for 2004 were in the age group of 18–40 (MOI 2004). Kuwait has implemented new, more stringent penalties under the new traffic laws in November 2001 in order to minimize traffic violations and thereby reduce traffic accidents. Ministry of Interior (MOI) statistics show that the main cause of accidents

Table 1. Number of accidents and injuries in governorates for 2004 (MOI 2004).

	Accident type					Injuries			
Governorate	Collision	Running over	Overturning	Drugs	Total	Minor	Major	Total	Deaths
Capital	17571	110	80	75	**17836**	100	51	**151**	**59**
Hawally	14473	64	111	111	**14759**	77	21	**98**	**57**
Farwaniya	8473	129	99	84	**8785**	113	61	**174**	**71**
Jahra	3254	77	211	74	**3616**	87	43	**130**	**90**
Ahmadi	6054	95	200	87	**6436**	99	37	**136**	**95**
Mubarak	3298	37	86	25	**3446**	108	27	**135**	**26**
Total	**53123**	**512**	**787**	**456**	**54878**	**584**	**240**	**824**	**398**
Percentage	**96.8%**	**0.9%**	**1.4%**	**0.8%**	**100%**	**1.1%**	**0.4%**	**1.5%**	**0.7%**

Table 2. Breakdown of traffic accidents (1999–2004).

Year	1999	2000	2001	2002	2003	2004
Total number of accidents	26635	27696	31028	37650	45376	54878
Collision accidents	24887	26069	29277	35989	43612	53123
Accident of overrun/ overturning/drugs/others	1748	1627	1751	1661	1764	1755
Total injuries	1743	1125	1566	2249	1332	824
Minor injuries	1178	736	1043	1641	927	584
Major injuries	565	389	523	608	405	240
Fatal accidents	333	331	300	315	372	398
Average no. of accidents per day	72.97	75.88	85.01	103.15	124.32	150.35
Average no. of injuries per day	4.78	3.08	4.29	6.16	3.65	2.26
Average no. of fatalities per day	0.91	0.91	0.82	0.86	1.02	1.09

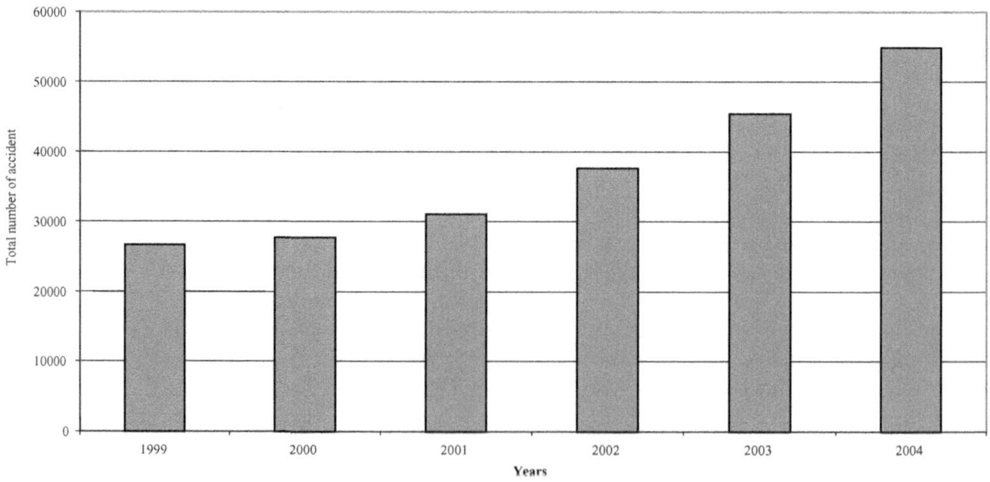

Figure 1. Total number of traffic accidents in Kuwait (1999–2004).

is careless driving which represents about 93% of the total accidents. The second main cause of accidents according to the statistics is collision into an object which represents about 4%. The rest of the reasons together represent about 3% of total accidents. The statistics show that all but few accidents are blamed on careless driving. The results of this study shed more light on accidents of 2004.

Statistics show that accidents in Kuwait are increasing exponentially as shown by the numbers of reported accidents through the years 1999–2004 which are presented in Table 2, and shown graphically in Figure 1. The number of accidents is increasing steadily over the years. Collisions represent the larger proportion of the total number of accidents. It is also increasing at a high rate. The average number of accidents per day increased to more than double between the year 1999 and 2004 (from 73 to 150). Accidents involving personal injuries fluctuate between a daily average of 2 to 6 instances. Fatal accidents which include at least one death occurre at a daily average of about 1; i.e., at least one life is lost daily because of road accidents.

The study presented in this paper is a product of a graduate research work done in the Civil Engineering Department of Kuwait University (Al-Enezi 2007).

2 METHODOLOGY AND DATA COLLECTION

The study relies mainly on accident reports archived in MOI departments of different governorates. Considerable effort was spent to collect the sample size used in this study. Accident reports were collected from the General Administration of Investigations departments of General Traffic Claim and its different branches located in different governorates. All 2271 accident forms were collected randomly without any specifications or criteria.

3 ANALYSIS AND RESULTS

Analyses were performed on the data collected from the General Administration of Investigation of the MOI. The raw data is in the form of traffic accident reports for the year 2004. The analyses performed include frequency analysis, cross-classification tabulation analysis, and construction of accident spot maps.

3.1 *Frequency analysis*

Frequency analysis was performed for the following variables: time, day, weather, road condition, accident type, location description A, location description B, number of lanes, road type, car movement, accident cause A, accident cause B (rear collision), accident cause C (side collision), and accident severity.

Road type is included only in new accident forms of MOI. Accident distributions according to road type are shown in Table 3. It is noticed that 45.6% of accidents (from new forms) are in internal (local) roads. Cumulative percentage shows that 43% of accidents occur on ring roads, expressways, and main roads. This is likely due to high speeds and high traffic volumes on these types of roads. The second highest is the main-roads category, accounting for 23.6% of accidents. This is probably because of more frequent entrances, exits, and intersections on such type of roads. Figure 2 is a graphical representation of the frequency distribution of Table 2.

3.2 *Cross-classification analysis*

The interrelationships between some of the variables were analyzed. The variables are selected on the basis of the expected relationship between them. These interrelationships are studied by utilizing the cross-classification analysis capability of the SPSS program used. It should be noted that any cross-classification including "governorate" as a variable did not include Jahra governorate because there were almost no accident reports for Jahra in the sample.

Table 3. Accident distribution according to road type.

Road type	Percent	Cumulative percent
Ring road	8.6	8.6
Expressway	10.8	19.4
Main	23.6	43.0
Internal	45.6	88.6
External	0	89
Tow way	0.6	89.5
Roundabout	5.2	94.6
Bridge	0.3	94.9
Intersection	1.5	96.4
Signalized intersection	4	100
Total	100%	

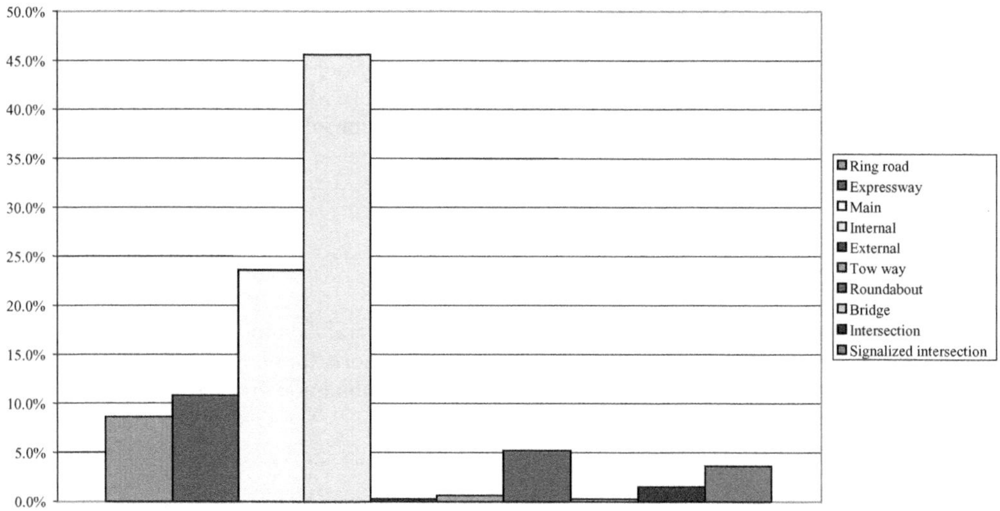

Figure 2. Frequency distribution of accidents based on road type.

Table 4. Accident type versus governorate cross tabulation.

	Governorate					
Accident type	Capital	Hawally	Farwaniya	Ahmadi	Mubarak Alkabir	Total
Car collision	6.8%	38.8%	26.0%	18.8%	9.5%	100.0%
Collision into light pole	0.0%	4.8%	19.0%	66.7%	9.5%	100.0%
Collision into trees	9.1%	18.2%	9.1%	45.5%	18.2%	100.0%
Collision into wall	6.3%	37.5%	12.5%	37.5%	6.3%	100.0%
Collision into street furniture	9.1%	20.5%	17.0%	46.6%	6.8%	100.0%
Overturning	4.5%	11.4%	2.3%	50.0%	31.8%	100.0%
Falling down	0.0%	33.3%	0.0%	66.7%	0.0%	100.0%
Running over	3.4%	6.9%	24.1%	37.9%	27.6%	100.0%
Others	0.0%	100.0%	0.0%	0.0%	0.0%	100.0%

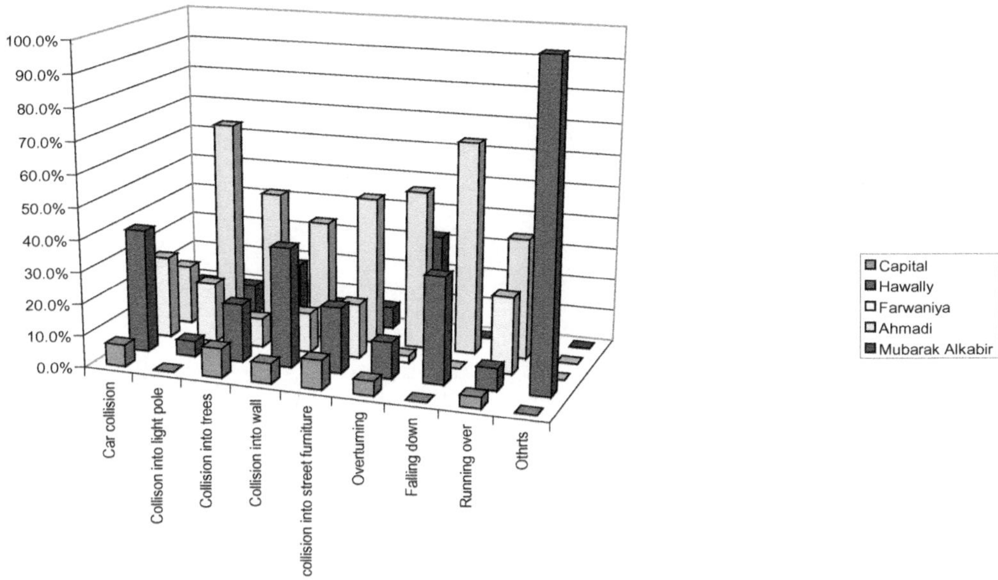

Figure 3. Cross classification of accident type versus governorate.

Table 4 shows the cross tabulation of Accident Type and Governorate. It is noticed from the table that car collision which has the highest percentage of accident occurrence has it highest percentage of 38.8% in Hawally Governorate. This may be attributed to the high traffic volume and low speed in Hawally Governorate districts. The other types of accidents are noticed to have their highest percentage of occurrence in Ahmadi Governorate.

Figure 3 is a graphical representation of the Governorate-versus-Accident Type cross classification analysis results.

4 TYPICAL ACCIDENTS

As a result of reviewing the accident reports, typical traffic accidents were observed. They were identified statistically as the most frequent types of accidents. The most frequent accident type is the one caused by sudden stop which is shown in the diagram of Figure 4. It is responsible for 25.6% of traffic accidents. This usually occurs in traffic congestion where vehicles are moving slowly and traffic is mainly in a stop-and-go situation, since drivers usually do not keep enough space headway with other vehicles. It also happens when a driver at a relatively high speed is surprised with traffic congestion ahead.

5 ACCIDENT SPOT MAPS

Accident spot maps represent a very important output from accident analysis studies. They show locations of frequent accidents on the road network. Such maps present distribution and severity of reported accidents. This is important information for decision-makers to identify locations that require improved safety measures or at least monitoring and further investigation to pinpoint exact causes of accidents as a first step to reduce risk and thus improve safety to the driving public. Figure 5 shows an aggregated accident spot map for the State of Kuwait.

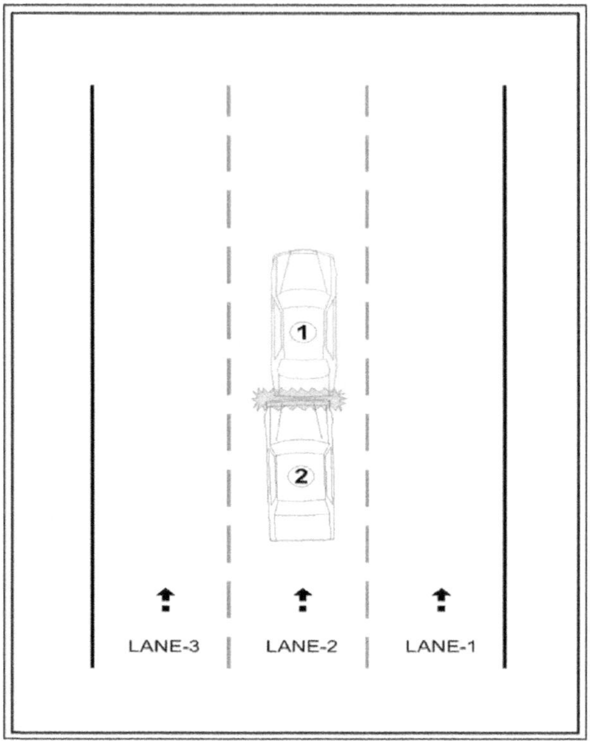

Figure 4. Accidents caused by sudden stop.

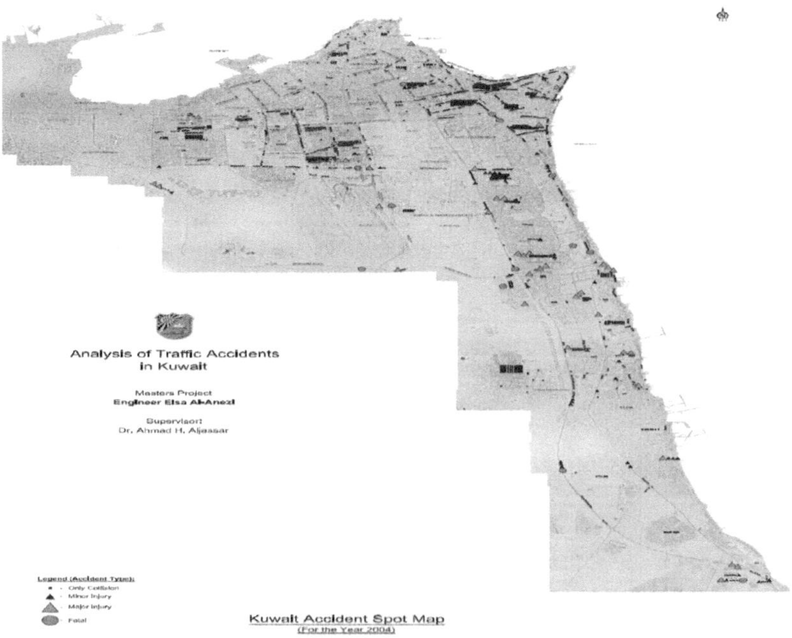

Figure 5. Aggregated accident spot map for the state of Kuwait.

6 CONCLUSION AND RECOMMENDATIONS

The presented study is considered a first step towards analysing traffic accidents in Kuwait. Cross-classification analyses and spot maps of accidents provide many important types of information that are required by engineers and decision-makers to improve safety for the driving public. Such analyses must be performed regularly (annually). If accident reports were stored electronically, all reported accidents may be analysed and then more accurate conclusions may be reported.

During the review process of accident reports, some recommendations were suggested. The following are some of these recommendations:

- Improve sight distances which are usually reduced by rising vegetations at intersections and roundabouts.
- Increase awareness of traffic signals ahead by installing warning signs before signalised intersections.
- More stringent control should be exercised to reduce violations to traffic regulations, especially at intersections.
- Study (and redesign if necessary) entrances and exits to and from main roads. Some of these locations may need provision of acceleration and deceleration lanes to reach appropriate speeds for entering a main road or exiting from it.
- Provide more pedestrian crossings and overpasses to reduce illegal road crossings that may be responsible for many run-over accidents.
- Accident locations need to be more accurately identified in order to construct more accurate database and accident spot maps. This can be easily achieved nowadays by using available GPS technology to reference an accident location on a GIS map.
- Collection of site accident data on electronic forms will make storage, retrieval, and analysis of accident data easier, more accurate, and less time and labour consuming.
- Increase awareness of traffic lanes and regulations through media.
- Teach traffic principles and ethics in schools. Traffic lessons may be introduced at secondary school levels.
- Improve public transit and encourage its use to reduce traffic congestions and accidents.
- More stringent control should be exercised to prevent distraction of drivers' attentions such as considering "driving while holding a mobile phone" a traffic violation.
- More enforcement should be exercised on the use of seatbelts for front-seat passengers. Seat-belts do not prevent accidents, but it is well proven that they reduce injury levels significantly.
- Raise awareness of drivers to observe maximum as well as minimum posted speed limits. Higher differentials in driving speeds are expected to be responsible for many accidents.
- Encourage researchers to perform similar and more extensive studies to draw a better and clearer picture of traffic accidents in Kuwait, and pinpoint exact causes of accidents to provide effective corrective measures.
- Road geometry in Kuwait should be reviewed on the national level and corrections have to be applied at many locations that are not up to standards.

REFERENCES

1. O'Flaherty, C.A., "Transport Planning Traffic Engineering", Arnold Publications, London, England, 1997.
2. Ministry of Interior (MOI), "*Annual Statistics Report for 2004*", General Traffic Administration, Kuwait, 2004.
3. Anderson, Robin, "*Introducing Speed Cameras ACT- How to Win Friends and Influence People*", ACT Urban Service, Accident Facts, Canberra, Australia, 2000.
4. National Safety Council (NSC), "*Accident Facts*", USA, 1996.
5. Al-Anezi, E., "Analysis of Traffic Accidents in Kuwait", Master's Project Report, College of Graduate Studies, Kuwait University, February 2007.

Efficient Transportation and Pavement Systems – Al-Qadi, Sayed, Alnuaimi & Masad (eds)
© 2009 Taylor & Francis Group, London, ISBN 978-0-415-48979-9

Research priorities for the transportation and traffic safety center in Kuwait

Abdel-Aziz Al-Kulaib
Under Secretary of Ministry of Public Works, Kuwait
Member of the Board of Trustees, Transpoortation and Traffic Safety Center, Kuwait

Sheikh Thamer A.A. Al-Sabah
The Government Performance Follow up Agency—Council of Ministers, Kuwait
Member of the Board of Trustees, Transpoortation and Traffic Safety Center, Kuwait

Fahad S. Al-Rukaibi
Civil Engineering Department, Kuwait University, Kuwait
Director Member of the Board of Trustees, Transpoortation and Traffic Safety Center, Kuwait

Ahmad H. Aljassar
Civil Engineering Department, Kuwait University, Kuwait
Member of the Board of Trustees, Transpoortation and Traffic Safety Center, Kuwait

Mohamed Eleiche
Transpoortation and Traffic Safety Center, Kuwait

ABSTRACT: The transportation sector has witnessed rapid developments in the State of Kuwait and in the Arabian Gulf region in general. With continuously increasing growth of the economy and population, the demand on the transportation sector is increasing as well. This results in evolving problems in the transportation field, including traffic jams, traffic safety, and environmental problems, to name a few. The relevant scientific research has a two-fold deficiency; one is in the form of insufficient research tackling the newly developed problems; the other is in the form of non-prioritized topics of research. Therefore, one of the main priorities in the Transportation and Traffic Safety Center in Kuwait is to organize a research database for the available performed studies serving the transportation sector, and to prioritize its research activities depending on the level of its need and importance. This paper reviews the transportation system in the State of Kuwait, gives a background on the Transportation and Traffic Safety Center, and identifies the research priorities of the Center to maximize benefits from such research studies.

1 INTRODUCTION

The urban transportation is a major driving factor in any economy and provides the infrastructure in any society for movement of goods and people, and ties the economic components together.

The transportation network in any state indicates the level of its economy and its level of development. Countries with advanced transportation networks and varieties of transportation means between their cities are considered developed countries, while countries with difficulties in transportation between their cities and towns are under- developed.

The Transportation and Traffic Safety Center in Kuwait University was established to address all the transportation challenges in the State of Kuwait from a scientific point of view, in order to achieve high service levels for transportation, and guarantee the positive role of transportation towards the planned objectives of Kuwaiti economy.

This paper describes the structure of the center, its role and strategy, and its policy and targets. It also addresses the main transportation challenges in the State of Kuwait, and the research priorities and plans of the center.

2 THE TRANSPORTATION AND TRAFFIC SAFETY CENTER

Established in June 2007, the Transportation and Traffic Safety Center has a clear and concise vision regarding its role in the State of Kuwait. The Center is a scientific entity inside the Faculty of Engineering and Petroleum in Kuwait University, and is composed of a board of trustees, center manager, and center units. The board of trustees represents ministries and agencies which have direct relation with transportation and traffic. The board includes representatives from the Ministry of Interior (MOI), the Ministry of Public Works (MPW), the Citizen-To-Government Organization (c2g), Kuwait Foundation for the Advancement of Sciences (KFAS), in addition to members from the academic staff from the Faculty of Engineering and Petroleum in Kuwait University. The main role of the board of trustees is to establish the strategy of the center, monitor its progress, and support the center activities within the represented ministries and organizations. The manager of the center is a member of the board of trustees and is the liaison person between the board of trustees and the center.

2.1 *Strategy of the center*

The center has long-term and short-term goals. The long-term goals are based on strategic directives such as the establishment of a national traffic strategy for the State of Kuwait in order to achieve high levels of service and traffic safety for transportation; evaluation of the current status of the road network; provision of measures for development and achievement of its goals; estimation of the demand on transportation for the next 15 years; and planning the transportation development to accommodate this expected increasing demand.

The short-term objectives of the center are conforming to the strategic goals and planning to achieve them. Among the short-term objectives is introducing the center to ministries and organizations in the transportation and traffic field, in addition to the establishment of a research activities agenda and focusing on the increased traffic safety concerns.

2.2 *Structure of the center*

The center has 3 main units; the first is the unit of Transportation and Traffic Studies, which focuses on the study of traffic and transportation in the State of Kuwait. The second unit is Unit of Material Studies which is concerned with the studies in the road materials and asphalt mixes. The third unit is for Traffic Safety Studies.

The Transportation and Traffic Safety Center is based on a foundation of academic staff with proven research record in transportation and traffic safety, and geospatial databases and transportation and traffic databases of the State of Kuwait. The center also has specialized laboratories for traffic counting and analysis, and material laboratories for testing pavement and road materials.

3 TRANSPORTATION IN KUWAIT

3.1 *Kuwait geography and demography*

The State of Kuwait is situated in Southwest Asia with a strategic location at the head of Arabian Gulf between Saudi Arabia and Iraq. It is located at the far northwestern corner of the Arabian Gulf. It has an area of about 18,000 square kilometers, and a population of about 3 million. Kuwait has around 200 kilometers of coast borders in the Arabian Gulf to the east in addition to nine Arabian Gulf islands. Kuwait Bay, which indents the shoreline for about forty kilometers, provides natural protection for the port of Kuwait and accounting for nearly one half of the shoreline. To the south and west, Kuwait shares a long border of 250 kilometers with Saudi Arabia. In the north and northwest, there is a borderline of about 240 kilometers with Iraq. The developed areas of The State of Kuwait extend mainly along the shoreline. They represent a minor percentage of the total state land.

3.2 *Transportation infrastructure in Kuwait*

Kuwait has a modern network of roads, with all-weather highways running north to Iraq and south to Saudi Arabia. Roadways extend 6,000 km, including 5,000 km of paved roads. Land transport accounts for a significant share of Kuwait's imports and exports.

Kuwait has five ports, including a cargo port at Al-Shuwaikh on Kuwait Bay, and an oil port at Mina Al-Ahmadi that is equipped with a huge pier at which eight large tankers can be loaded simultaneously. Kuwait has regular calls from ocean shipping, and local sailing craft carry goods between Kuwait and the neighboring states, Iraq, and Saudi Arabia. Marine transport accounts for most of Kuwait's foreign trade.

There are seven airports, three of which have paved runways. The principal airport, Kuwait International Airport, is located south of the City of Kuwait.

Kuwait has a modern, well-maintained transportation system. However, the entire system suffered extensive damage in the 1990 Gulf War. But by 1993 repairs had brought most facilities back to their pre-war condition.

3.3 *Transportation challenges in Kuwait*

The State of Kuwait has different transportation facilities. In land transportation, there are the paved and unpaved road networks. In marine transportation, there are 12 marine ports. In the aviation side there is the main civil Kuwait International Airport in addition to several other military small airports.

The main challenge for transportation in Kuwait is the traffic congestion and the traffic safety. The levels of service on roads at peak hours are low. And around 500 deaths annually are reported due to traffic accidents. In 2006, about 1500 fatalities and severely-injured cases were reported due to traffic accidents.

The State of Kuwait suffers from the nearly full dependence on private transportation (private auto and taxi) which overloads the urban transportation network and increase congestion, especially in peak periods.

4 RESEARCH ACTIVITIES

In order to face the transportation challenges in the State of Kuwait, and to conform with the political strategy that plans to make Kuwait an international financial center by the year 2015, the Transportation and Traffic Safety Center established a national strategic five years plan to help improve the transportation and traffic performance, with emphasis on traffic safety.

4.1 *Research infrastructure*

Many researches and studies were conducted for the transportation in the State of Kuwait in the last few years by ministries and agencies in diversity of applications, and these valuable studies were not gathered in a unique pool for archiving and storing for future use and to be the base for the research. Also, the geo-database of transportation of the State of Kuwait is essential for the studies of the Center.

Due to this situation, the Center started a short-term plan to establish an information system with geospatial dimension to collect and store all the related data, maps, information, statistics and studies for the transportation research in the State of Kuwait. This information system will form the main pool where all the previous researches and studies are stored and archived to be easily accessible for the planning of future research activities.

The information system will be based on three main components: the first is the database of the previous researches and studies related to the transportation and traffic in the past 15 years. The second is the geo-database of the Geographic Information System (GIS) which contains the transportation networks and demographic data of the State of Kuwait. The third is the traffic statistics and traffic-count studies which form a valuable source of information for the future research.

4.2 *Research areas*

Currently, there are many challenges facing the transportation and traffic system in the State of Kuwait, and these challenges are the core of research interests for the Center. The first of all is the widespread traffic congestion problem. This problem is increasing not only in the State of Kuwait, but also in all the Gulf countries, and known to have negative impact on the economy, environment, and health. There are also the transit transport alternatives, which are considered to be at the main focus of all GCC country states as rationale solution for traffic congestion.

Another aspect is in the safety side as the Kuwaiti society is suffering from high number of casualties and severe accidents causing injuries. In 2006, there were 500 casualties and 1000 severe-injury traffic accidents. This problem is of high priority in the research field.

The strategic national transportation projects such as railway and metro networks or the establishment of new highways need to be precisely evaluated by scientific procedures to ensure their success and achievement of goals. The Center also provides scientific consultancy for transportation projects.

5 RESEARCH PRIORITIES

The Center organized an ambitious plan to address the transportation challenges in the State of Kuwait. It will span over 5 years such that a one-year short plan starts in the current year, then a mid-term plan to be completed in the next four years.

5.1 *Short-term research priorities*

The short-term plan includes the following items:

- Establish the strategy of the Transportation and Traffic Safety Center:
  This strategy will be the base for the work of the Center for the next five years, and includes the aims, goals and objectives, and the policies and plans to achieve them.
- Establish a national traffic strategy for the State of Kuwait:
  The State of Kuwait needs a long-term strategy for the planning of the traffic in order to conform to the economical and developmental growth. The Center will play a major role alongside other organizations in creating this strategy.
- Establish an information system for the Kuwaiti researches in the transportation field:
  Create a unique pool for archiving and storing the researches and studies conducted for the transportation in the State of Kuwait in the last few years.
- Analysis of traffic accidents in the State of Kuwait for years 2005 and 2006:
  The traffic accidents are increasing in the State of Kuwait, and casualties and severe injuries are causing pressure on the society. The Center started a study to analyze the reasons of these accidents and how to prevent them.
- Design a state-of-the-art information system to collect the accident data:

The collection of the accident data is vital for the analysis of their causes. The Center initiated a study to establish a state-of-the-art information system based on GIS/GPS to collect the accurate data of the accident in digital format, which will enable the creation of accident spot map and precise analyses.

- Evaluation of large transportation projects in the State of Kuwait:
  The State of Kuwait initiated strategic transportation projects such as the metro network. These projects require cost/benefit studies, and measurements of their impact on the traffic. The Center has a defined role in this process.

5.2 *Long-term research priorities*

The long-term plan includes the following items:

- Integrated transportation information system for the State of Kuwait:
  As previously mentioned, the proposed information system for the Center is the foundation for all research activities, and it is a continuous task, that needs to be created, maintained and update regularly. Following are the main components of the proposed information system:
  - Geo-database for road network and traffic infrastructure:
    The detailed data and maps of the road network are essential request for the researches of the Center, and a complete geo-database for the road network and traffic facilities and description such as cross data, and camera locations will provide a valuable resource for the research activities.
  - Collection of data related to transportation:
    Collecting the data related to transportation in the State of Kuwait, such as number of vehicles, bus routes, accident locations, traffic statistics, etc.
  - Traffic survey:
    The demand of the traffic is dynamic and in continuous change. In order to forecast the demand accurately, updated traffic survey needs to be conducted to update the traffic data and specify the reasons of congestions. The Center has plans to conduct updated traffic survey in the State of Kuwait to be the input for the equilibrium models.
- Application of mathematical models for traffic network equilibrium:
  The mathematical models for the traffic network equilibrium for congested urban transportation networks provide the scientific measurements for the levels of service of roads and the efficiency of the proposed solutions. The Center has modern GIS and transportation software which will be used with the updated traffic data and geo-database for the measurement of the levels of service.
- Scientific evaluation of transportation alternatives:
  The transit transportation systems are considered the favored alternative for congested urban traffic. The evaluation of the proposed transit should be examined accurately to ensure the success of its role in reducing the congestion. The Center with its academic experiences is capable of playing a main role in this evaluation.
- Forecast of future demands on transportation:
  The diagnostic of traffic problems is based on the present and future demand on the transportation system, which needs to be determined via traffic surveys and traffic counts for current transportation demand levels. In cooperation with the official authorities in the State of Kuwait, the Center will plan traffic surveys and counts in order to prepare the input data to the transportation models.
- Planning of future transportation needs:
  The future demands on transportation request early planning and development that are proportional to its growth in order to avoid future congestion or increase current traffic congestion. The Center will prepare plans in coordination with interested organizations, which will estimate the impact and requirement of the evolution of transportation demand.

6 CONCLUSION

The challenges in transportation and traffic in Gulf Cooperation Council (GCC) states are evolving, and the State of Kuwait is in the core of these challenges. The Transportation and Traffic Safety Center research is focusing on facing these challenges in Kuwait in particular, and in the GCC countries in general. The main interest of the Center in the current stage is to achieve solutions for better levels of service for roads at peak hours, reduce traffic congestion, thorough analysis of severe road accidents and ways to reduce them, increase of roads safety by road safety auditing systems, and increasing awareness to consider road safety as a main parameter while designing roads. The Center has ambitious research plan to enhance the transportation and traffic safety in Kuwait and in the GCC region.

REFERENCES

Alrukaibi, Fahad, "General Strategy for Transportation and Traffic Safety Center", Transportation and Traffic Safety Center, Kuwait University, September 2007.

Boyce, David, "Forecasting Travel on Congested Urban Transportation Networks: Review and Prospects for Network Equilibrium Models", TRISTAN V: The Fifth Triennial Symposium on Transportation Analysis, Le Gosier, Guadeloupe, June 13–18, 2004.

Ministry of Interior, "Report on Traffic Accidents in Year 2006", Kuwait, 2007.

Ministry of Interior, "Report on Traffic Accidents in Year 2005", Kuwait, 2006.

Aljassar, Ahmad H., Omar I. Al-Saleh, Mohammed A. Ali, "Traffic Analysis on Kuwait Road Network", Final Report of Project No. EV-06/00, Research Administration, Kuwait University, June 2005.

Efficient Transportation and Pavement Systems – Al-Qadi, Sayed, Alnuaimi & Masad (eds)
© 2009 Taylor & Francis Group, London, ISBN 978-0-415-48979-9

Optimal coordination of traffic signals to cope with traffic congestion

N.A. Chaudhary & C.-L. Chu
Texas Transportation Institute, Texas, USA

ABSTRACT: In a significant number of metropolitan areas worldwide, traffic congestion continues to grow at a rapid pace. Minimizing adverse effects of such congestion requires transportation agencies to proactively manage their transportation systems by utilizing all available means at their disposal. Optimal coordination of traffic signals is one such method. However, achieving the desired benefits requires proper identification of the root cause(s) of congestion, selection of appropriate subset of optimization objectives, and utilization of the most appropriate optimization tool to derive optimal signal timings. This paper contains traffic signal coordination guidelines for use by practitioners. As such, it characterizes different types of traffic conditions, identifies various control objectives and when to use them, and describes the strengths and weaknesses of various computer-based optimization and analysis tools available to practitioners.

1 INTRODUCTION

In recent years, there has been a sharp increase in the number of metropolitan areas facing severe traffic congestion in their signalized systems. Furthermore, the severity and duration of traffic congestion on signalized roadways has continued to increase at a steady pace. A recent study conducted by National Transportation Operations Coalition (2007) to assess the health of signal systems in the United States (U.S.) estimated that poor signal timing accounts for 5–10% of all traffic delays. Citing previous findings that the benefits-to-cost ratio of signal retiming is 40:1 or better, this study recommended routine timing updates to improve the current low grade.

Increasing urban traffic congestion, however, is a result of flourishing cities. A few years ago, Taylor (2002) made the following proposition: "Traffic congestion is evidence of social and economic vitality; empty streets and road are signs of failure." He further stated that efforts to manage congestion should accept the fact that automobiles are central to metropolitan life, and that short-lived relief is not proof that adding capacity is a bad idea. In the context of traffic signals, the last statement could be extended to include optimal utilization of existing signal capacity. These statements point out the fact that congestion will remain an issue for modern cities, and continued efforts will be required to minimize its detrimental effects, even if such improvement efforts turn out to be short-lived. Tackling a specific congestion scenario requires that its nature and source(s) are properly characterized. Many times however, this is not the case.

This paper provides general guidelines for coordination traffic signals to manage congestion. As such, it characterizes different types of traffic conditions, distinguishes the sources of congestion, identifies various control objectives and when to use them, and describes various computer-based optimization and analysis tools available to users. The paper is organized as follows. Section 2 provides a review of literature. Section 3 discusses congestion, oversaturation, and differences between them. Section 4 presents guidelines for providing relief from congestion in signal systems. Finally, Section 5 presents concluding remarks.

2 LITERATURE REVIEW

Signal optimization for congested conditions has been studied since the 1960s. Gazis & Potts (1963) proposed a "store and forward" strategy for dealing with oversaturated traffic signals. This strategy, refined and presented later (D'Ans & Gazis 1976), used time varying traffic demands and a mathematical programming approach to optimally store and dissipate queues at signals where demand exceeds capacity. This store and forward approach did not account for the effects of queue variation within a cycle and offsets. Rahmann (1973) argued that queuing would be a norm during peak periods and presented the idea of designing signals as storage/output devices, even during undersaturated conditions. Pignataro et al. (1978) attempted to define congestion and oversaturation in terms of their causes and scope, and proposed guidelines for dealing with such conditions. They proposed three solutions: (1) minimal-response signal policies (i.e. cycle length selection matched to block length), (2) queue-actuated responsive control (Lee et al. 1975), and (3) other non-signal treatments (i.e. enforcement and prohibition of right-turn-on-red) in a signalized environment. The primary objective of the queue-control policy was to delay or eliminate intersection blockage. Lieberman & Messer (1990) developed mathematical models for optimizing signal timings based on an internal metering policy. Internal metering ensures that no link receives more traffic than it can hold without spilling back traffic, requiring the determination of best green splits and offsets. The objective of such models is to maximize throughput. Other researchers have continued to refine these ideas. For instance, Gal-Tzur et al. (1993) developed an iterative signal-timing optimization method, which linked a custom mathematical program with TRANSYT signal-timing optimization program. In this scheme, the mathematical model was used to determine the green splits and the level of metering, whereas TRANSYT was used for simulating the dynamic processes within the cycle and for offset optimization. Other instances include the works of Choi (1997) and Lieberman & Chang (2005) who refined the internal metering models and demonstrated their real-time application. Kim (1990) and Kim & Messer (1992) developed mathematical models for controlling congested interchanges and arterials with single critical intersections. These models managed queues at external approaches to a system, while preventing spillback and reducing delay on the interior links.

Research conducted by Chaudhary & Balke (1997) found that driver expectancy plays an important role in determining headways and variations in headways increase for vehicles further back in the queue. The implication of this finding is that capacity may be lost if very long cycles are used. They studied five coordination strategies using computer simulation. The results of this study showed that coordination of actuated signals for progressing traffic flow in the congested direction produces lower delays, fewer stops, and shorter queues. In his work with congested interchanges, Messer (1998a) found that undersaturated systems might become congested because of poor signal timing and deficient spacing between the signalized intersections. By providing an upper bound on signal control delay for oversaturated arterial operations, he showed that congestion can be characterized and modeled. Messer (1998b) extended an existing model for analyzing the operational impacts of queue spillback on the capacity and delay of closely spaced signalized intersections. Chaudhary & Chu (2000) and Chaudhary et al. (2000) developed two models for coordinating diamond interchanges with adjacent signals. Their first model is a simple capacity analysis procedure that uses Webster's method (Webster & Cobbe 1966) for calculating green splits for the interchange and adjacent signals. However, Webster's green split calculation method, which treats each signalized intersection separately, may not be valid when queued traffic from one intersection starts to interfere with an adjacent signalized intersection. To accommodate such situations, the second model uses an iterative method for simultaneously calculating green splits and offsets for all signals in the system. The procedure automatically determines if the system is undersaturated or oversaturated. If the system is oversaturated, it pushes all excess demand to the user-defined boundary of the system, while keeping the interior links clear of detrimental queues (Chu & Chaudhary 2005). Although this method needs refinements, it has shown to produce excellent results for small systems.

Other researchers have continued to explore the development of methods and models to deal with congested traffic signal systems. Abu-Lebdeh & Benekohal (2003) developed a model for maximizing system throughput by managing queues. Liu & Masao (2001) developed a model based on the method of cumulative flows. The objective of this model is to minimize delay by allocating green times to various phases. Ahn & Machemehl (1997) developed a traffic simulation model to provide a methodology for traffic signal timing in oversaturated urban arterial networks. They considered two control objectives: maximizing throughput, and preventing or minimizing queue spillback. They found that offset was a dominating factor. Other factors found to be important were link length and cross street green time. Abu-Lebdeh & Benekohal (2000) developed models for estimating the capacities of oversaturated arterials. Khatib et al. (2001) proposed that the goal of timing in congested conditions should be to allow higher-volume approaches to discharge more traffic than the lower-volume approaches, so that the intersection can return to a normal condition as quickly as possible. They developed and simulation-tested mathematical models that optimize maximum green intervals for achieving this goal by preventing spillback and overflow on signalized arterials with actuated signals. Girianna & Benekohal (2002) developed an algorithm that determines green times and offsets to manage/distribute queues in time and space.

While researchers continue their quest for improved congestion management strategies and models, practitioners must continue attempt to solve their problems using whatever means available to them. The literature provide a glimpse of such attempts, including: (1) a trial and error methods, which kept shifting congestion from one location in the system to another Click (2007), (2) after failing mitigate severe congestion on a 2.72-mile arterial stretch via optimization of signal timing parameters, resorting to a flush strategy that ignored cross streets (Sabra 2007), (3) successful development of a strategy to improve congestion and safety in a small corridor, but failure in implementation due to inter-jurisdictional barriers (Chamberlin 2007), (4) success by fixing downstream problems, using appropriate cycle length, double servicing heavy phases, running closely-spaced signals using a single controller, and doing whatever it takes to coordinate all closely-spaced intersections with the problematic signal (De Camp 2007), and (5) success by using flexible timing, efficient placement of detectors, queue detection, efficient gap settings to minimize duration of oversaturation, variable walk time, and use of other advanced controller features (Nelson 2007).

3 CAUSES OF TRAFFIC CONGESTION AND SOLUTIONS

A critical review of literature reveals that often the terms "congestion" and "oversaturation" are misclassified and used interchangeably without a clear distinction between them. Some examples of how researchers have defined oversaturation include:

- A condition where traffic queue persists from cycle to cycle either due to insufficient green splits or because of blockage.
- A condition where demand during a cycle exceeds the capacity of green.
- A case when splits are correct, maximum cycle length is used, and at least one signal approach does not clear during a cycle.

The first definition is a characterization of congestion, but not necessarily oversaturation. The last two definitions are incorrect, because these conditions are often observed in undersaturated conditions. Nonetheless, there is general agreement in the traffic engineering community regarding the importance of queue management as a key congestion mitigation strategy. To gain a better understanding of all relevant issues, it is worthwhile to examine traffic flow and behavior at traffic signals.

Traffic signals are installed at roadway intersections to provide safe and equitable right of way to competing traffic movements. To accomplish this objective, a traffic signal cycles through a sequence of green indications for each group of compatible movements, while displaying red to all competing movements. Thus, queuing is a design feature of traffic signals. For a given demand scenario, longer red time means longer queue. Figure 1 illustrates this point by comparing two time-space diagrams (Cases 1 and 2), where vehicles arrive at the same uniform arrival rate, stop at the red signal, and move on when the signal turns green. Case 2 has the same relative

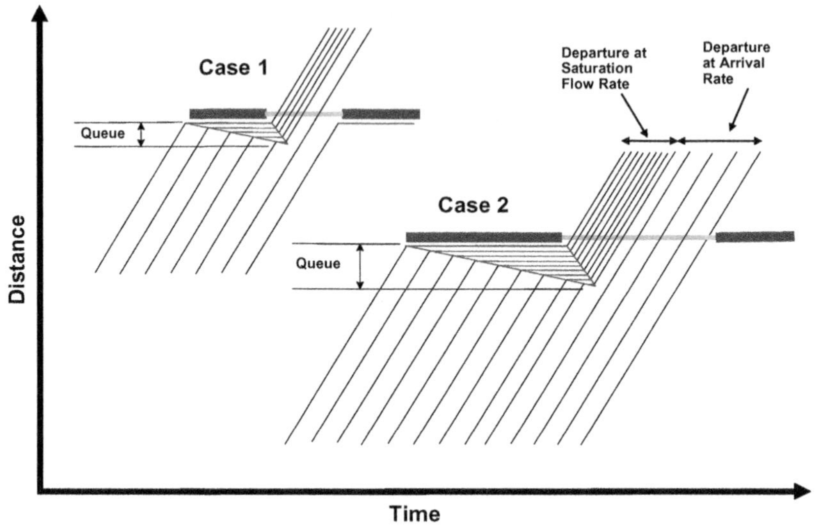

Figure 1. Effect of cycle length on cycle by cycle queue length.

proportions of red and green intervals as in Case 1, but the cycle length (the sum of red and green intervals) is longer. This causes longer queue and more delay in Case 2.

Figure 1 illustrates a simplified arrival-departure scenario at an isolated approach. This figure provides the following additional insights:

- During the first part of green phase, queued vehicles depart at the saturation flow rate. Once the queue has cleared, vehicles depart at the arrival rate. This is the characteristic of an undersaturated approach, whose capacity is higher than demand.
- In Case 1, a larger proportion of green is used at the saturation flow rate than in Case 2.
- Case 2 results in larger delay per cycle, identified by the area of the triangle created by the queue formation process, than Case 1.
- In Case 2, more vehicles pass the signal without stopping than in Case 1. However, if arrival rate (demand) starts to increase, this advantage of Case 2 will start to diminish.

As illustrated, signal timings (that is, the lengths of green and red periods) control the values of two conflicting performance measures: delay and number of stops. Thus, engineering judgment must be exercised to select the preferred measure. In general, timings that minimize stops (requiring a larger cycle length) are preferred over delay minimization for intersections with high speed approaches. It should be noted that the length of a green signal is determined based on demand and control objective for the traffic stream (also called movement) to be served by that green, while the length of red is controlled by similar factors for all conflicting movements. Because increase in green for one traffic movement means more red (and consequently more delay and longer queues) for the conflicting movements, there is a practical limit on the length of any green indication. A signal green reaches a saturated state when it is entirely consumed by the departure of a standing queue, and an oversaturated state when the queue fails to clear. The condition when a queue fails to clear by the end of green is called a cycle failure. Cycle failures may occur due to saturation, oversaturation, suboptimal timings, or other factors. Frequent cycle failures, especially for multiple traffic streams at an intersection are signs of congestion, which is characterized by excessive delays. If the spatial impact of congestion is limited to the vicinity of the intersection, it is considered local. Congestion is considered severe when its influence spreads to such an extent that the delays reach unacceptable levels. At isolated intersections,

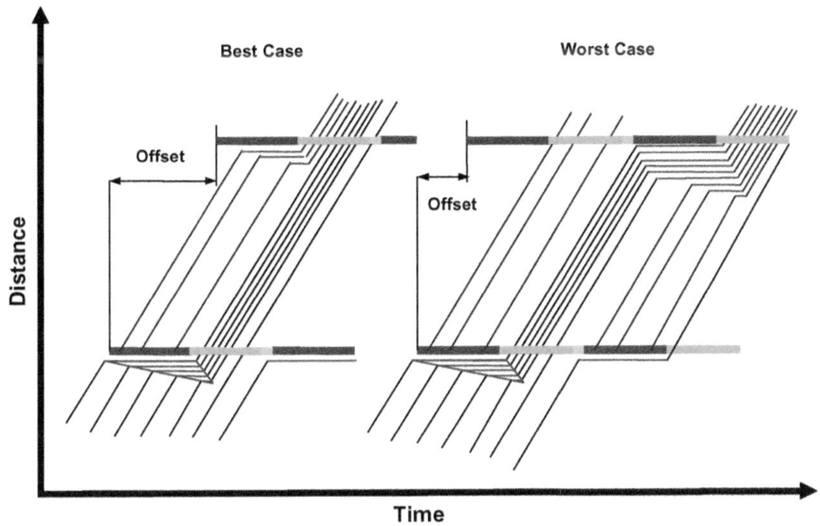

Figure 2. Examples of best and worst signal coordination.

this situation generally arises when demand exceeds intersection capacity for sufficiently long periods of time.

If multiple signalized intersections exist in close proximity, traffic flow characteristics become more complicated. Figure 2 illustrates two extreme cases of these traffic flow characteristics by extending the previous example to include a downstream signal. In the first (Best) case, the platoon of vehicles from the upstream signal arrives at the downstream signal when the queue has just cleared, and passes through this signal without stopping. This case produces minimum delay, no stops to through traffic, and maximum progression. In the second (Worst) case, all vehicles arriving from the upstream signal are forced to stop. This case produces the longest queue and no progression. Note that the only difference between the two scenarios is the offset between the two signals. Offset is a signal timing parameter that establishes time relationship between the beginnings of greens at adjacent traffic signals. Together with signal cycle lengths with a common base, offsets are used to synchronize a system of traffic signals.

Undersaturated traffic signal systems, such as the one illustrated in Figure 2, can be synchronized to minimize systemwide delay or stops, maximize arterial progression, or optimize a compromised combination of multiple conflicting objectives. Depending on several factors, optimal synchronization may not be possible to achieve best flow for all conflicting traffic streams. During peak demand conditions, unmanaged or long queues may make traditional coordination objectives of delay minimization or progression maximization irrelevant or infeasible to attain. Under such congested conditions, the objective of coordination changes to throughput maximization, which requires optimal management of queues.

As illustrated in Figure 3, geometric deficiencies or constraints combined with cyclic queues can cause capacity loss. This illustration shows that a short left turn bay can be the source of two types of problems depending on demand pattern. Left side of this figure shows a case where the left-turn phase is red, through phase is green, and queue spillback from the left-turn bay is preventing through vehicles in the left lane from moving. In this case, a portion of the through capacity is lost. In the second case (on the right) left turn phase is green, but the queue of through vehicles stopped due to red signal is preventing left turn vehicles from reaching the turn bay and using their green phase. Both of these cases are causing starvation, where a phase is green, but vehicles are prevented from using it. Depending on demand levels for left turn and through streams,

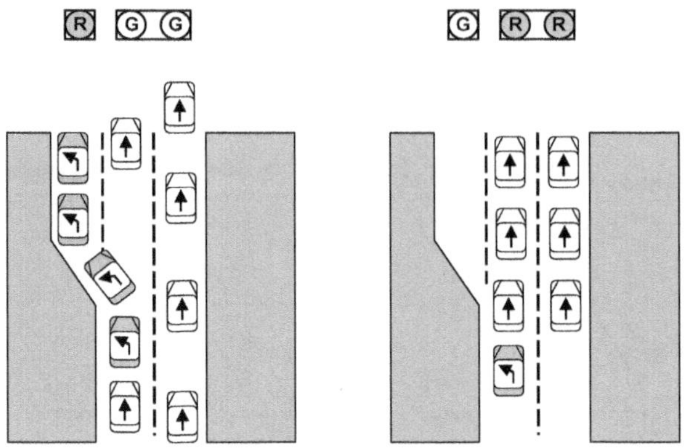

Figure 3. Spillback and starvation caused by a short left-turn bay.

Figure 4. Congestion caused by suboptimal signal timings.

double cycling of the left-turn green, lagging the left-turn green, or simultaneous greens to left and through vehicles may be used to solve these types of problems.

If allowed to continue, however, local congestion shown in Figure 3 can quickly spread to adjacent intersections. Figure 4 shows such a situation where the through signal is green, but vehicles are prevented from moving because the downstream link is full. The source of congestion in this case is not oversaturation, but lack of optimal coordination between adjacent traffic signals, which is also hurting side street traffic in this case. As illustrated here, there is a clear distinction between congestion and oversaturation, an assertion also made by Urbanik (2007).

Figure 5 shows a screen capture from the computer simulation of a real system in Texas. As shown in this figure, this three-signal system experienced severe congestion in one direction during the morning rush hour. Existing signal timing optimization methods, which calculate green times for all signal assuming them as isolated entities, could not solve this problem. However, as demonstrated by Chaudhary et al. (2000), a system-based external metering (gating) approach eliminated this congestion. Figurc 6 shows simulation results of optimized timings calculated for the same system by Chaudhary et al. (2000). In fact, the retimed system even had excess capacity to handle additional demand.

Although not verified for a larger system, the system-based approach used to solve the above problem is simple from a conceptual point of view. Figure 7, which shows a three-signal system, is

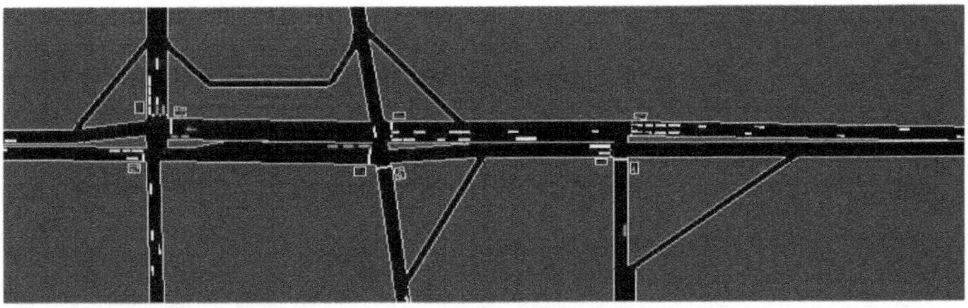

Figure 5. Existing timing plan for Ayers street system in Corpus Christi, Texas.

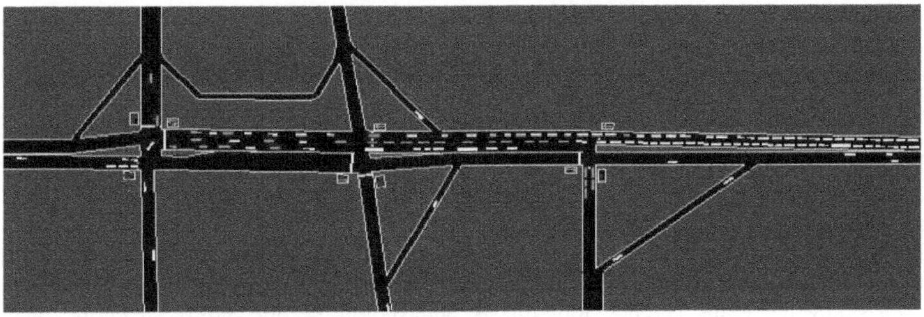

Figure 6. Optimized timing plan for Ayers street system in Corpus Christi, Texas.

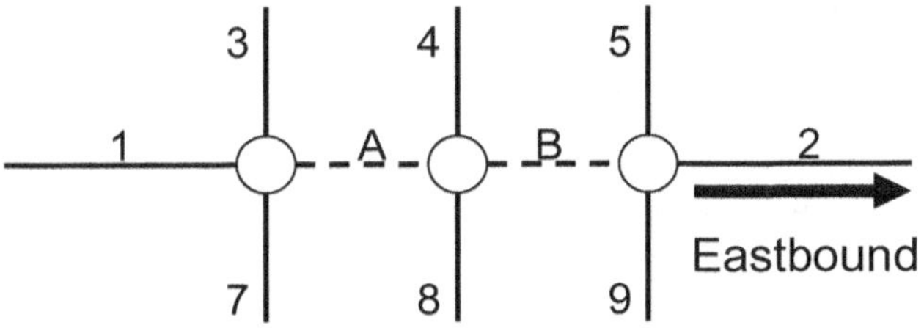

Figure 7. Three-signal system.

used to illustrate this approach. In this approach, links are classified as internal (links A and B) and external (links 1 through 9). The timing calculation methodology ensures that internal links remain uncongested by giving sufficient green times to these links. In addition, the methodology coordinates signals to efficiently move traffic without compromising available internal capacity. Note that the movement of eastbound traffic is only possible if eastbound through green for link A has enough capacity to accommodate demand arriving from links 1, 3, and 7 and eastbound through green for link B has enough capacity to accommodate demand arriving from links 4, 8, and A. Thus, calculations of all these greens are interdependent. In addition, the formation of queues on links A and B must be managed to prevent any loss of capacity due to blocking. Such queue management must also ensure that external to external traffic (i.e. from link 3 to link 7, from link 1 to link 3, etc.) is not blocked by queues. The calculations become more complicated when the opposite direction is added.

Finally, it should be noted that methodologies such as the one described above can only harness capacity if the system has excess capacity available. Such methodologies cannot eliminate congestion caused by oversaturation, where demand is more than the capacity of one or more critical subsystems in the roadway network. However, optimal timing and coordination strategies to manage queues can eliminate queue spillbacks and blocking, which are causes of secondary congestion in oversaturated systems. If successfully implemented, such strategies can significantly reduce the severity and duration of congestion in oversaturated systems.

4 CONGESTGION MANAGEMENT GUIDELINES

As discussed in the previous section, the primary cause of congestion in a signalized roadway network may be oversaturation or inefficient utilization of available capacity. Both situations can be improved by optimal timings and coordination of traffic signals. This section provides guidelines on how to eliminate or mitigate congestion in traffic signal systems. Achieving these objectives requires the availability of resource, which include: staff trained to perform all required tasks and appropriate analysis/optimization tools. Assuming that all necessary resources are available, the following steps should be taken:

Step 1: Identify the highest priority location needing congestion mitigation. The initial identification may be the result of driver complaints. However, for better assessment of the situation, a professional must inspect the situation in person. The inspection task could be done from a tall building or by driving through the facility. The objective of this inspection is to assess the cause and scope of the problem as illustrated in Figure 5. Based on this initial assessment, schedule a more detailed data collection/observation.

Step 2: Conduct detailed data collection called for by the data collection plan. The data collection may need two or more people per intersection in the study area. To understand where congestion starts, how it spreads, and how long it lasts, data should be collected for the entire congestion period plus at least 30-minute periods before and after. Using this data, identify the causes and durations of primary and secondary congestion locations, and growth/dissipation rates and durations of queues. Also collect demand data with as much accuracy as possible. To do that, count the number of vehicles serviced by a green interval during each cycle as well as the growth in the size of any queue. For a given time period, the demand for the subject green will be the total number of serviced vehicles plus the growth in queue during that period.

Step 3: Analyze the data, determine what needs to be done to fix the problem, and implement change in signal timings. The fix may be as simple as re-allocation of time between conflicting phases or changing the offset on the spot. In more complicated cases, data will have to be further processed in the office. Such processing usually involves the use of computer-based optimization and analysis tools.

Step 4: Repeat all previous steps to determine the improvements achieved. The above process would be considered successful if the severity and duration of congestion has reduced. This process also enables the implementation of additional improvements by accounting for causes masked due to the severity of congestion studied earlier.

As mentioned above, there may be a need to use computer-based tools to identify improvement alternatives. Three general types of tools are available for this purpose. The first category includes signal timing optimization models, such as: TRANSYT 7F, Synchro and PASSER V. These models are simple to use. They have the ability to generate and evaluate alternate signal control scenarios, and recommend the best timings based on a specified objective. Which model is selected depends on whether the objective of timing is to minimize delay and stops or maximize progression bandwidth. TRANSYT 7F (T7F) and Synchro are primarily designed to minimize delay and stops, but T7F allows its objective to be changed to the maximization of progression opportunities. PASSER V (P5) can maximize progression bands or minimize delay. In addition, mesoscopic traffic models of T7F and P5 have the ability to account for the effects of queue spillback and blocking. It should be noted that most optimization programs use the isolated intersection model for calculating green times, which may be inappropriate for congested signal systems.

The second category includes microscopic simulation models such as: CORSIM, SimTraffic, VISSIM, and AIMSUN. Models in this category provide features to allow detailed analyses of complex situations. However, the use of these tools is time consuming even with significant training and experience. Also, simulation models do not have optimization capabilities. As such, they can only evaluate exact scenarios specified by the user. Thus, all alternate scenarios must be generated by the analyst using either a signal timing optimization model or expert judgment, evaluated one at a time, and manually compared to select the best alternate. Microscopic simulation models are recommended for complex situations where optimization models have failed.

The third type includes mesoscopic traffic assignment models, such as: VISSUM and DYNASMART. These models have simpler traffic models with the added capability to account for changing route decisions. These models are similar to previously identified simulation models in that they can analyze given scenarios only, requiring users to identify and evaluate alternates scenarios one by one. In addition, these models require detailed origin-destination data. Collection of such data accurate enough for use in operational analysis is costly and time-consuming.

5 CONCLUSIONS

The focus of this paper is congestion mitigation in signalized roadway systems. The paper provided a detailed review of relevant literature, characterized different sources of congestion using examples, and provided guidelines for dealing with such congestion. It also provided an overview of tools available for the analysis and optimization of traffic signal timings. It should be noted that this paper only addressed optimal management of existing supply by retiming traffic signals. This is the shortest term and the least costly approach to dealing with urban traffic congestion, but may not cure severe congestion problems. When that happens, one or both of the following two more expansive and long term approaches are warranted.

The first approach is to add capacity. Techniques in this category include: changing lane assignments at signal approaches, converting two-way roads to one-way, access management, extending turn bays, adding lanes, and grade separation. The second approach is to manage demand. These approaches include encouraging employers to implement staggered hours, encouraging car pooling by providing high occupancy lanes for them, and tolling.

ACKNOWLEDGMENTS

This paper presents preliminary findings of an ongoing research project funded by the Texas Department of Transportation (TxDOT) under contract number 0-5998. We would like to thank TxDOT staff for their support and feedback. We would also like to thank the anonymous referees for providing constructive criticism of our paper.

REFERENCES

Abu-Lebdeh, G. & Benekohal, R. 2000. Signal coordination and arterial capacity in oversaturated conditions. *Transportation Research Record*. 1727:68–76.

Abu-Lebdeh, G. & Benekohal, R. 2003. Design and evaluation of dynamic traffic management strategies for congested conditions. *Transportation Research Part A*. 37(2):109–127.

Ahn, G.-H. Machemehl, R. 1997. *Methodology for traffic signal timing in oversaturated arterial networks*. Report SWUTC/98/465510-1. University of Texas. Austin. Texas.

Chamberlin, R. 2007. Ongoing efforts to coordinate an oversaturated corridor in Lebanon, New Hampshire. Presented at January 2007 Transportation Research Board Workshop on Operating Traffic Signal Systems in Oversaturated conditions. Washington, D.C.

Chaudhary, N.A. & Balke, K.N. 1997. *Real-time coordinated-actuated traffic control during congested conditions*. Research Report 1288-S. Texas Transportation Institute. College Station, Texas.

Chaudhary N.A. & Chu, C.-L. 2000. *Guidelines for timing and coordinating diamond interchanges with adjacent traffic signals*. Research Report 4913-2. Texas Transportation Institute. College Station, Texas.

Chaudhary, N.A., Chu, C.-L., Balke, K.N. & Kovvali, V.G. 2000. *Coordination of diamond interchanges with adjacent traffic signals*. Research Report 4913-1. Texas Transportation Institute. College Station, Texas.

Choi, B.-K. 1997. *Adaptive signal control for oversaturated arterials*. Ph.D. Thesis. Polytechnic University. New York.

Chu, C.-L. & Chaudhary, N.A. 2005. Coordination of diamond interchange with adjacent traffic signals. 2005 Annual Transportation Research Board Meeting CD ROM.

Click, S. 2007. Retiming of NC 54 at I-40 and Farrington Road in Durham, NC: a case study in oversaturated control. Presented at January 2007 Transportation Research Board Workshop on Operating Traffic Signal Systems in Oversaturated conditions. Washington, D.C.

D'Ans, G. & Gazis, D. 1976. Optimal control of oversaturated store-and-forward transportation networks. *Transportation Science*. 10(1):1–19.

De Camp, G. 2007. A Michigan's secret for operating traffic signals during oversaturation. Presented at January 2007 Transportation Research Board Workshop on Operating Traffic Signal Systems in Oversaturated conditions. Washington, D.C.

Gal-Tzur, A., Mahalel, D. & Prashker, J. 1993. Signal design for congested networks based on metering. *Transportation Research Record*. 1398:11–118.

Gazis, D. & Potts, R. 1963. The oversaturated intersections. *Proceedings of the 2nd international symposium on traffic theory*. 221–237.

Girianna, M. & Benekohal, R. 2002. Dynamic signal coordination for networks with oversaturated intersections. *Transportation Research Record*. 1811:122–130.

Khatib, Z., Abdel-Rahim, A., Judd, G. & Jagarapu, K. 2001. *Actuated coordinated signal system, Phase I-oversaturated conditions, Phase II-cycle-by-cycle analysis*. Final Report No. 1–20. National Institute for Advanced Transportation Technology. University of Idaho.

Kim, Y. 1990. *Development of optimization models for signalized intersections during oversaturated conditions*. Ph.D. Thesis. Texas A&M University. College Station, Texas.

Kim, Y. & Messer, C. 1992. *Traffic signal timing models for oversaturated signalized interchanges*. Interim Report FHWA/TX-92/1148-2. Texas Transportation Institute. College Station. Texas.

Lee, B., Crowley, K. & Pignataro, L. 1975. Better use of signals under oversaturated flows. *TRB Special Report*. 153:60–72.

Lieberman, E. & Chang, J. 2005. Optimizing traffic signal timing through network decomposition. *Transportation Research Record*. 1925:167–175.

Lieberman, E. & Messer, C.J. 1990. *Internal metering policy for oversaturated networks*. Report TR-238. KLD Associates. Huntington Station, New York.

Liu, H. & Masao, K. 2001. An approach on network traffic signal control under the real-time and oversaturated flow condition. In 80th Annual Transportation Research Board Meeting CD ROM. Washington, D.C.

Pignataro, L., McShane, W., Crowley, K. & Casey, T. 1978. *Traffic control in oversaturated street networks*. NCHRP Report 194. Transportation Research Board. Washington, D.C.

Messer, C.J. 1998a. Simulation studies of traffic operations at oversaturated, closely spaced signalized intersections. *Transportation Research Record*. 1646:115–123.

Messer, C.J. 1998b. Extension and application of Prosser-Dunne model to traffic operation analysis of oversaturated, closely-spaced signalized intersections. *Transportation Research Record*. 1646:106–114.

National Transportation Operations Coalition 2007, *National traffic signal report card: executive* summary 2007. http://www.ite.org/reportcard/NTSRC%20Exec%20Summary%20final.pdf, accessed 02/18/2008.

Nelson, E. 2007. Timing strategies for managing oversaturation in Harris County. Presented at January 2007 Transportation Research Board Workshop on Operating Traffic Signal Systems in Oversaturated conditions. Washington, D.C.

Rahmann, W. 1973. Storage/Output design of traffic signals. *Australian Road Research*. 5(1):38–43.

Sabra, Z. 2007. Strategies to mitigate gridlock on high volume arterials in Sacramento County, CA. Presented at January 2007 Transportation Research Board Workshop on Operating Traffic Signal Systems in Oversaturated conditions. Washington, D.C.

Taylor, B.D. 2002. Rethinking Congestion. *Access*. 21:8–16.

Urbanik, T. 2007. Oversaturation: What do we really know? Presented at January 2007 Transportation Research Board Workshop on Operating traffic signal systems in oversaturated conditions. Washington, D.C.

Webster, F.V. & Cobbe, B.M. 1966. Traffic signals. *Road Research*. Technical Paper No. 56. Her Majesty's Stationary Office. London. England.

Efficient Transportation and Pavement Systems – Al-Qadi, Sayed, Alnuaimi & Masad (eds)
© 2009 Taylor & Francis Group, London, ISBN 978-0-415-48979-9

Dynamic early merge and dynamic late merge at work zone lane closures

A.E. Radwan
College of Engineering and Computer Science, University of Central Florida, Orlando, FL, USA

R.C. Harb
CATSS University of Central Florida, Orlando, FL, USA

ABSTRACT: To improve traffic safety and efficiency in work zones, several states in the U.S. introduced and tested the Dynamic Lane Merge System (DLM), an intelligent work zone traffic control system. The dynamic lane merging, can take two forms; dynamic early merge and dynamic late merge. Previous DLM applications demonstrated the strengths and weaknesses of each merging scheme. However, up to date the literature lacks of a comparison between the two aforementioned dynamic merging schemes. This manuscript reviews previous DLM applications and suggests two simplified dynamic merging systems; Simplified Dynamic Early Merging (SDEM) and Simplified Dynamic Late Merging (SDLM) for testing on short term work zones.

1 INTRODUCTION

1.1 *Early and late merging concepts*

The idea behind the early merge strategy is to create a no-passing zone to promote merging into the open lane before reaching the end of a queue caused by the congestion. Moreover, drivers will be prohibited from using the closed lane to pass vehicles in the queue and merge into the open lane ahead of them. The concept behind late merge is to make more efficient use of roadway storage space by encouraging drivers to use all available traffic lanes to the merge point. Once the merge point is reached, the drivers in each lane take turns proceeding through the work zone.

1.2 *Work zone lane merging strategies*

Early lane management schemes took the static forms of early and late lane merging. The static form does not change in real time in response to traffic conditions. Typically, the static early merge includes additional "LANES CLOSED" static sings placed upstream of lane closure on average at 1-mile intervals (McCoy and Pesti, 2001). The static late merge comprises static signs calling for "USE BOTH LANES TO MERGE POINT" 1.5 miles upstream of the work zone and "MERGE HERE TAKE YOUR TURN" near the beginning of the taper (McCoy and Pesti, 2001).

One of the listed advantages of the static early merge is that it reduces the frequency of forced merges especially at higher traffic volume (Nemeth and Rouphail, 1982). However, other studies stated that it may confuse drivers under uncongested condition (Datta et al., 2004) and increase travel time through the work zone (Mousa et al., 1990). As for the static late merge, previous studies stated that the implementation of the static late merge strategy resulted in 75% fewer forced merges (McCoy et al., 1999), an increase in capacity from 1,460 pcph to 1730 pcph (McCoy et al., 1999), a delayed onset of congestion by 14 minutes (Byrd, 1999) and a reduced queue length from 7,800 ft to 6,000 ft (Byrd, 1999). Another study stated that static late merge may confuse drivers at the merge point during low congestions (McCoy and Pesti, 2001).

1.3 *Dynamic early merging applications*

The dynamic early merge system creates a NO-PASSING zone upstream of a work zone taper based on real-time measurements of traffic conditions (Tarko and Venugopal, 2001). The system consists of queue detectors and "DO NOT PASS WHEN FLASHING" signs that would be triggered by the queue detectors. When a queue is detected next to a sign, the next closest sign's flashing strobes, upstream, are activated creating the NO-PASSING zone. This system makes queues jumping an illegal task. Figures 1 and 2 illustrate this system.

The Indiana Lane Merge System (ILMS) was tested in the field in the 1997 construction season by the Indiana Department of Transportation. It was found that the system smoothes the merging operations in advance of the lane closures. Drivers merged when they were supposed to merge, the flow in the open lane was uniform, and rear-end accident rates decreased. However, this system did not increase the throughput and the results of a simulation study conducted by Purdue University indicated that travel times through work zones with ILMS are larger (Tarko, 1998).

In 1999, the University of Nebraska conducted a study of the ILMS on I-65 in the vicinity of Remington, Indiana. This study was limited to a four day data collection exclusively under

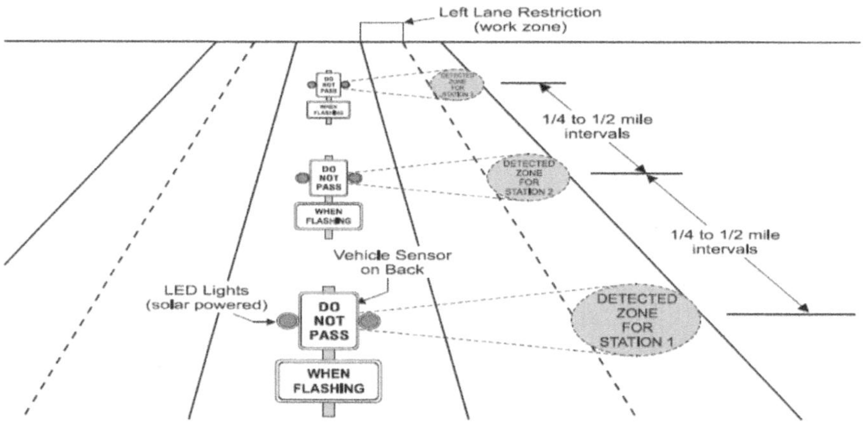

Figure 1. Indiana lane merge system (*Source: Beacher et al., 2004*).

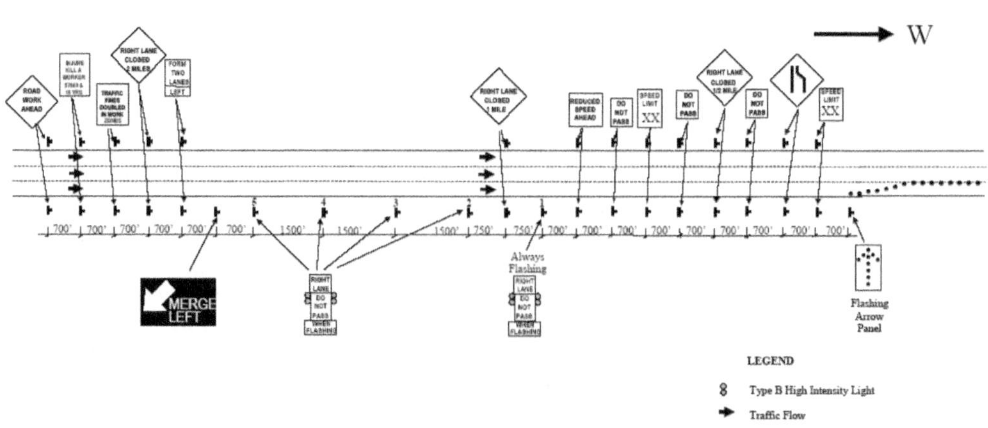

Figure 2. Dynamic early lane merge traffic control system used in michigan (*Source: Datta et al., 2004*).

uncongested conditions. In this project, the right lane was closed and the data collected (by video cameras and laser speed gun) and extracted included traffic volumes, speeds, conflicts, lane distributions, flows, and time headways. Comparing the ILMS with the standard MUTCD merge control, the results showed that the ILMS increased the capacity to some extent (from 1,460 to 1,540 vphpl). As for the safety aspect of the ILMS, since the data collected was limited to uncongested conditions and to 16 hours of video data, it was not clear whether the ILMS improve safety in terms of number of forced merges (McCoy et al., 1999).

The ILMS was also studied by Purdue University and the results were detailed in a report published in 2001. This system was studied on I-65 near West Lafayette, Indiana. This project entailed extensive data collection under both congestion and uncongested conditions for a duration of four months in 1999. Multiple loop detectors and two cameras were used for data collection. Purdue University studied both the safety effects of the ILMS by developing conflict frequency models as well as capacity effects of the ILMS. The results of the analyses showed that the ILMS decreases the capacity by 5%. The Authors mentioned that the decline in the capacity may be due to the unfamiliarity of the drivers with the system (Tarko and Venugopal, 2001).

The Wayne State University conducted a study to assess a similar system to the ILMS commonly referred to as Michigan Lane Merge Traffic Control System (LMTCS). This study compared four sites where the system was installed to four control sites where traditional MUTCD merge was implemented. The "DO NOT PASS WHEN FLASHING" signs were activated manually by personnel on the four sites. The lane closure configuration and geometry of freeway sections were homogeneous in the test and control sites for consistency. The data collected included aggressive driver behavior, location of merging, presence of law enforcement. In addition to that, the floating car method was utilized to record travel times and delays. According to their results the LMTCS increased the average operating speed, decreased the delays (49 vehicle hours of delay per hour), decreased the number of aggressive driving maneuvers during peak hours (from 73 to 33) (Wayne State University, 2001).

1.4 *Previous dynamic late merging applications*

McCoy and Pesti (2001) proposed a dynamic late merge in which the late merge would be employed only at times of high congestion. McCoy and Pesti (2001) stated that the late merge can reduce congestions and delays, whereas the early merge increased congestions and delays. Beacher et al., (2004) applied the dynamic late merge system in Tappahannock, Virginia and conducted a before and after study to explore the benefits of the system. Figure 3 shows the site diagram with the dynamic late merge system. According to their results, the percentage of vehicles in the closed lane increased significantly from 33.7 to 38.8 percent when comparing the late merge to the MUTCD treatment. The throughput volumes showed no statistical difference between the MUTCD treatment and the late merge. Time in queue was not significantly different between the two types of traffic control. According to Beacher et al., (2004) the lack of improvement in throughput and time in queue may be attributable to the relatively low percentage of heavy vehicles.

In June 2003, the University of Kansas, in cooperation with the Kansas Department of Transportation and the Scientex Corporation deployed the Construction Area Late Merge (CALM) system in Kansas (Scientex; Meyer 2002). This system is the dynamic version of the Late Merge Concept introduced by PennDOT. This system employs traffic detectors to sense congestion upstream of a construction lane closure. The traffic data is communicated in real-time to a central controller where proprietary software algorithms determine the critical thresholds of traffic density and speed to activate real-time messages directing motorists to remain in their lanes until they approach the lane closure, where they merge alternately by taking turns. The results showed that the average volume through the work zone was enhanced after the drivers were accustomed with the system. However, the net change in volume did not show a significant improvement over baseline values.

Maryland's Dynamic Late Merge (DLM) System comprises a set of 4 portable CMS and 3 RTMS detectors that are added to the standard static traffic control devices utilized at construction lane

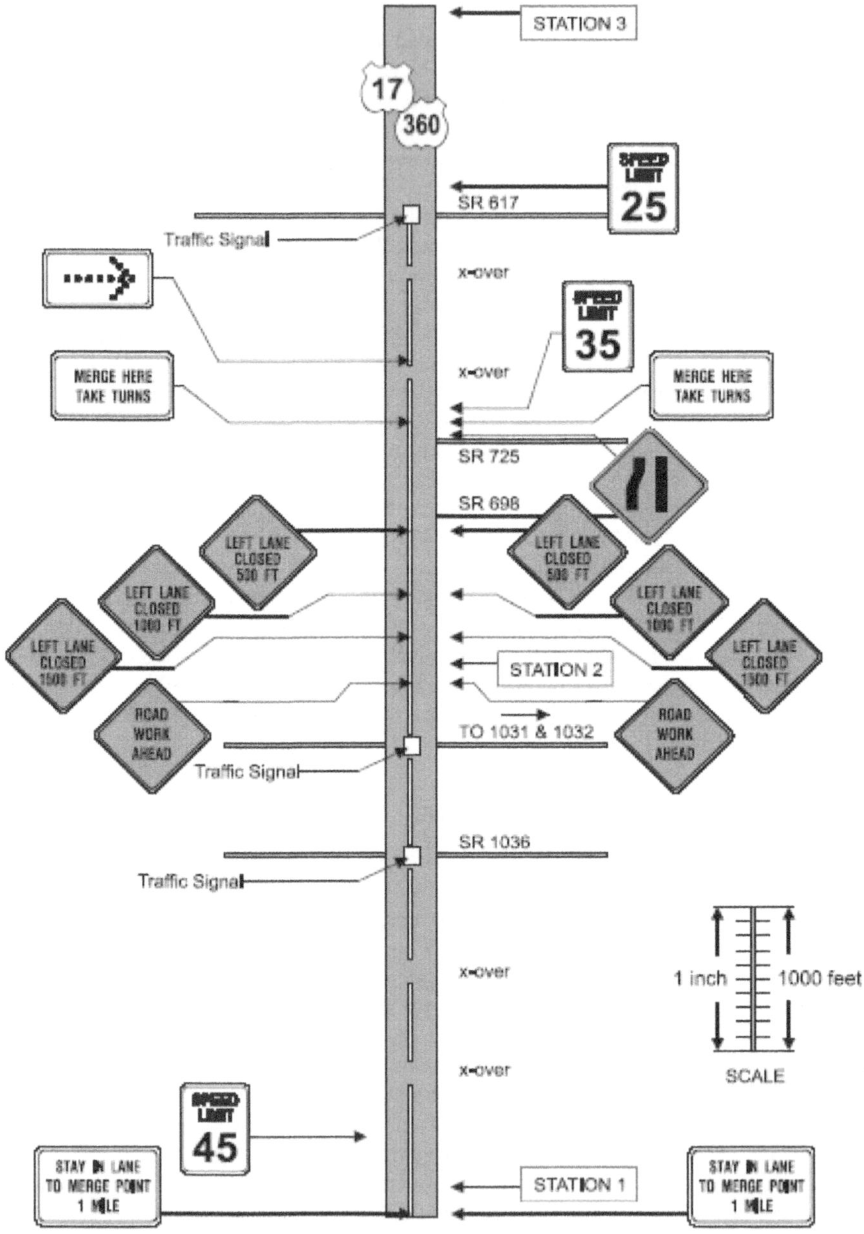

Figure 3. Tappahannock, virginia site diagram (Beacher et al., 2004).

closures. The CMS furthest upstream (~1.5 miles) from the taper alternated between the messages "USE BOTH LANES" and "TRAFFIC BACKUP". The next two CMS located at approximately ½ mile and ¼ mile from the taper itself, the final CMS alternated between messages "TAKE YOUR TURN" and "MERGE HERE". The University of Maryland, College Park conducted the evaluation of the system by utilizing one day of baseline (or control) data where the road closure utilized only the standard static traffic control signs. This was followed by 4 days with the DLM system activated. The results showed that more vehicles were in the discontinuous lane. Many drivers were

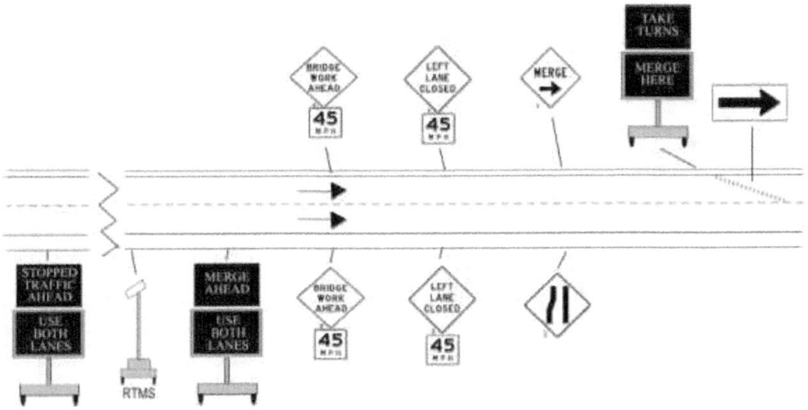

Figure 4. Minnesota's DLM.

observed merging before the designated merge location during the evaluation period. These early merges resulted in multiple merging points and appeared to result in some confusion on the proper place to merge. The queue lengths were observed to be reduced between 8% and 33% during the 4 days evaluation with the activation of the DLM System. Unfortunately, numerous traffic conflicts were observed between the two-lane traffic. These conflicts resulted in conditions of stop and go traffic. The authors finally stated that the advantages of the DLM system are increased throughput, shorter queue lengths, and more uniform distribution of lane use before the taper.

The Minnesota Department of Transportation (MnDOT) evaluated the Dynamic Late Merge System (DLMS) which consists, in addition to the standard orange and black warning signs placed in advance of the lane closure, of three Changeable Message Signs (CMS) and a Remote Traffic Microwave Sensor (RTMS) detector (See Figure 4). When congestion begins to form, the signs are activated to provide lane use instructions to drivers. The CMS farthest from the work zone displays the message "STOPPED TRAFFIC AHEAD-USE BOTH LANES". The next CMS sign reads "USE BOTH LANES-MERGES AHEAD". The sign closest to the work zone will show alternating messages of "TAKE TURNS-MERGE HERE". When traffic speeds increase as congestion dissipates, the signs will turn off and the system will return to the typical static work zone traffic control that encourages early merging (Dynamic Late Merge System Evaluation).

2 PROPOSED SDEM AND SDLM STRATEGIES

2.1 *Florida's Motorist Awareness System (MAS)*

Currently the Motorist Awareness System (MAS) is employed by FDOT on short term work zones. Figure 5 shows the MAS system in use according to Index- 670 from the FDOT standards. The MAS consists of an advanced warning arrow panel at the lane taper, one portable changeable message sign (PCMS), two portable regulatory signs (PRS) with flashing beacons, two radar speed display units (RSDU), and other regulatory signs.

2.2 *Motivation behind the simplified systems*

In order to simplify the dynamic lane merging equipment relocation, installation and calibration, this research proposes the addition of only one portable changeable message sign and one sensor trailer to the current MAS system. Therefore, instead of moving, installing and

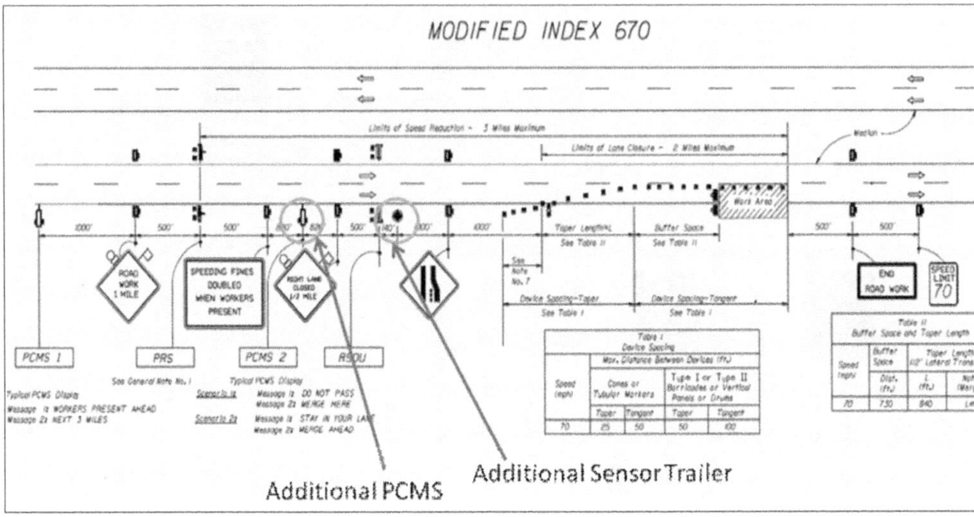

Figure 5. MAS from Index 670 (FDOT) and additional PCMS and sensor trailer.

calibrating multiple traffic sensors and PCMS, which may be a lengthy process for a short term work zone, one may relocate one PCMS and one sensor trailer every time the work zone moves. As shown in Figure 5, this research proposes the addition of one Portable Changeable Message Sign (PCMS) and one traffic sensor trailer. It should be noted that the location of these two new pieces of equipment will remain the same for both scenarios (DLM late and early).

2.3 *Proposed operation of the SDEM and SDLM Systems*

The system operation is based on real-time speed data acquired from the traffic detection zone (perpendicular to the RTMS) aggregated over 3-minute time interval. To test the simplified dynamic early merging (SDEM), the PCMS displays "DO NOT PASS" followed by "MERGE HERE" alternately. To test the simplified dynamic late merge (SDLM) scheme, the PCMS displays "STAY IN YOUR LANE" followed by "MERGE AHEAD" consequently. It should be noted that the additional PCMS is activated when the average speed of all lanes over 3-minutes time stamp drops below 50 mph. The deactivation threshold is selected to be 55 mph. For instance, if the average speed over the next three minutes time-stamp exceeds 55 mph, the PCMS is deactivated; otherwise the PCMS remains activated. It should also be noted that the minimum activation time of the PCMS may not be less than 5 minutes.

2.4 *Preliminary system testing in the field*

The preliminary analysis shows the temporal variations of the volumes and average speeds for 3 minutes intervals in the vicinity of the RTMS for the SDEM and SDLM setups. Examining Figure 6 and 7, one can see that the volume in the open lane is higher than the volume in the closed lane in the vicinity of the RTMS for both setups. Comparing Figures 6 and 7, one can observe that the volume in the closed lane for the early merge setup (SDEM) is higher than the volume in the closed for the late merge setup (SDLM) in the vicinity of the RTMS. This suggests that drivers are abiding by the PCMS messages. Figures 6 and 7 also show the time at which the PCMS is activated (average speed drops below 50 mph).

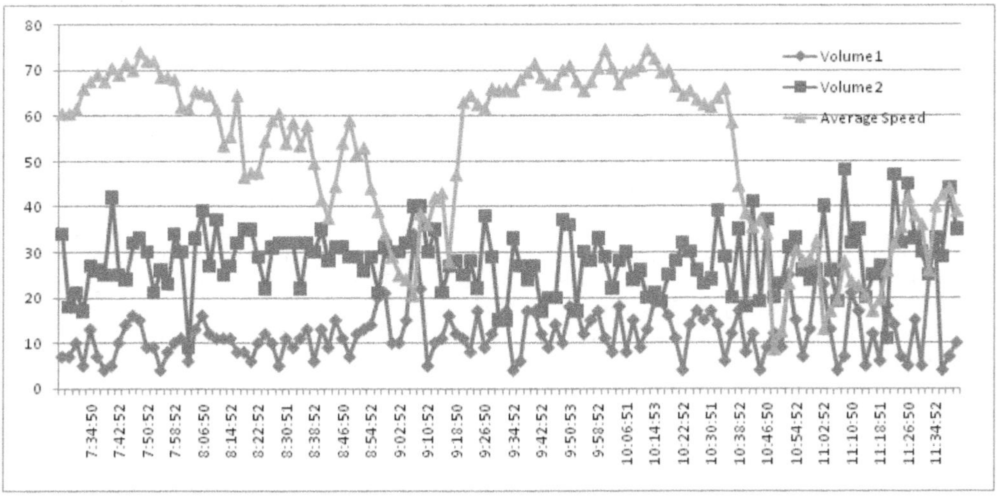

Figure 6. Vol.1 (closed lane), Vol.2 (open lane), and average speed (**early merge**).

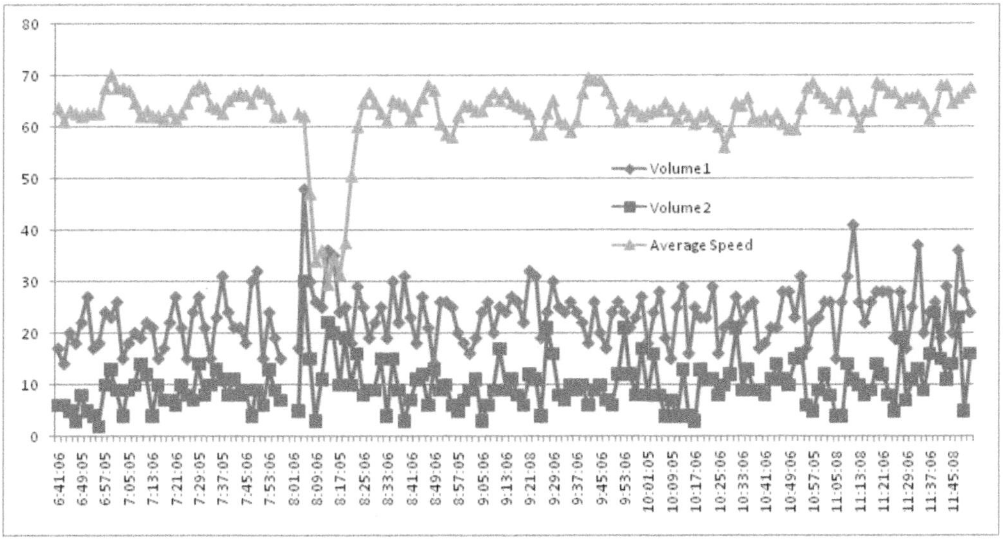

Figure 7. Vol.2 (closed lane), Vol.1 (open lane), and average speed (**late Merge**).

2.5 *Field observations and future research*

For the early merge setup (SDEM), it was noticed that drivers often adhere to the message displayed on the PCMS therefore drivers merged earlier. However, when one driver disobeys the message and decides to drive further down the closed lane, a platoon of vehicles follows the latter which defeats the purpose of the messages and early merging. As for the late merge setup (SDLM), since the traffic volumes were not high enough to trigger the messages, the authors cannot make any recommendation based on the field observations. This part of the analysis is still preliminary and data are currently being collected. Future conclusions and recommendations will be made available once data collection and analyses are completed.

REFERENCES

Beacher, A.G., Fontaine, M.D. and Garber, N.J. Evaluation of the Late Merge Work Zone Traffic Control Strategy. Report VTRC-05-R6 Virginia Transportation Council, August 2004.

Byrd, P.S., Pesti, G., Jessen, D.S. and McCoy, P.T. Driver Survey of the Late-Merge Work Zone Traffic Control Strategy. Paper presented at the 78th Annual Meeting of the Transportation Research Board, Washington, D.C., 1999.

Datta, T., Schattler, K., Kar, P. and Guha, A. Development and Evaluation of an Advanced lane Merge Traffic Control System of an Advanced Dynamic Lane Merge Traffic Control System for 3 to 2 Lane Transition Areas in Work Zones. Report Number RC-1451, Michigan Department of transportation, 2004.

McCoy, P.T., Pesti, G. and Byrd, P.S. Alternative Information to Alleviate Work Zone Related Delays. SPR-PL-1(35) P513. University of Nebraska, Lincoln, 1999.

McCoy, P.T. and Pesti, G. Dynamic Late Merge Control Concept for Work Zones on Rural Interstate Highway. Paper presented at the 80th Annual Meeting of the Transportation Research Board, Washington, D.C., 2001.

Meyer, E. Construction Area Late Merge (CALM) System (2004). Midwest Smart Work Zone Deployment Initiative http://www/matc.unl.edu/research/MwSWZDI Dynamic Late Merge System Evaluation, Initial Deployment on I-10. Summer 2003.

Minnesota Department of Transportation http://www.dot.state.mn.us/trafficeng/research/data/DynLateMerge.pdf

Mousa, R.M., Rouphail, N.M. and Azadivar, F. "Integrating Microscopic Simulation and Optimization: Application to Freeway Work Zone Traffic Control". Transportation Research Record 1254. 1990.

Nemeth, Z.A. and Rouphail, N.M. "Lane Closures at Freeway Work Zones: Simulation Study". Transportation Research Record 869.1982.

Tarko, A. and Venugopal, S. Safety and Capacity Evaluation of the Indiana Lane Merge System. FHWA/IN/JTRP/-2000/19. Purdue University West Lafayette, Ind., 2001.

Tarko, A. Kanipakapatman, S. and Wasson, J. Modeling and Optimization of the Indiana Lane Merge Control System on Approaches to Freeway Work Zones. Final Report, FHWA/IN/JTRP-97/12. Purdue University, West Lafayette, Indiana. 1998.

Wayne State University, Department of Civil and Environmental Engineering. Dynamic Lane Merge Traffic Control System: A Strategy for Alleviating Aggressive Driving Behavior at Work Zones in Michigan, 2001.

Highway design and transportation planning

Efficient Transportation and Pavement Systems – Al-Qadi, Sayed, Alnuaimi & Masad (eds)
© 2009 Taylor & Francis Group, London, ISBN 978-0-415-48979-9

Evaluation of warning signs and markings to reduce speeds on curves

N. Stamatiadis
University of Kentucky, Lexington, KY, USA

A. Vest
Kittelson & Assocaites Ft. Luaderdale, FL, USA

ABSTRACT: Sharp horizontal curves can pose dangers to the driver when dealing with speed adjustment, vehicle placement, and judgment of the appropriate operating speed. Roadway designers use various warning methods to aid drivers in such situations. Two primary methods of achieving this are warning signs and pavement markings. There is a suspicion however that these devices are often misinterpreted or disregarded by drivers. The objective of this study is to evaluate the use of several warning signs and pavement markings at problematic rural horizontal curves, and to evaluate their effectiveness in relation to speed reduction. Several types of warning signs and pavement markings were used to determine methods and combinations that could reduce operating speeds more effectively. The results indicate that the most promising treatments in reducing operating speeds are flashing lights and transverse lines. Another treatment that also showed some potential for reducing speeds is the new combination sign.

1 INTRODUCTION

Roadway designs should provide for a safe and efficient facility. Frequently other constraints, such as financial and geographic, may require solutions that will prohibit the development of the ideal roadway design. Such deviations from the desirable design can lead to the use of geometric conditions that may require sharper curves, limited sight distances, steeper grades, and other issues that could affect the driver's ability to follow the intended design. The critical problem with such designs is that they often do not provide any information or clues to the driver as to what is the appropriate operating speed. Sharp horizontal curves can pose dangers to the driver when dealing with speed adjustment, vehicle placement, and judgment of the appropriate operating speed. Roadway designers have introduced many warning methods to aid drivers in realizing and using the appropriate operating speed at dangerous roadway locations. Therefore, a prime location to test some of these warning methods is at horizontal curves that have some of these undesirable characteristics.

The primary methods of conveying roadway information to the driver are warning signs and pavement markings. According to the Manual on Uniform Traffic Control Devices (MUTCD), "warning signs call attention to unexpected conditions … to situations that might not be readily apparent to road users" and "alert road users to conditions that might call for a reduction of speed or an action in interest of safety and efficient traffic operations" (MUTCD, 2000). Also according to the MUTCD, "markings on highways have important functions in providing guidance and information for the road user" and can be "used to supplement other traffic control devices". The MUTCD notes that an important characteristic of the pavement markings as opposed to the warning sign is that they allow the driver to focus on the roadway but still acknowledge the warning.

There is a suspicion however that warning signs and pavement markings are often misinterpreted or disregarded by drivers. In these cases the effectiveness of signs and markings is reduced and often the intended reduction in operating speeds is not achieved. Moreover, the absence of adjusting the operating speeds may some times lead to a crash. Thus, safety concerns regarding the effectiveness of these devices arise. A recent review of safety and speeds noted that there is a higher likelihood for a crash when traveling at higher or lower speeds than the average speed (Stutser et al 1998).

The objective of this work is to evaluate the use of several warning signs and pavement markings at problematic rural horizontal curves, and to evaluate their effectiveness in relation to speed reduction. The specific tasks undertaken to complete this study was an evaluation of standard warning signs and pavement markings, installation of a variety of treatments and speed measurement, analysis of the collected data, and development of recommendations regarding the effectiveness of the various treatments.

2 LITERATURE REVIEW

A literature review was completed to gain a better understanding of the effectiveness of warning signs and pavement markings at reducing operating speeds. There have been many innovative approaches in the implementation of warning signs and pavement markings and these advances assisted in determining what the best measures to apply in this study were.

2.1 *Pavement markings*

The two most common types of pavement markings are longitudinal (i.e., center and edge line markings) and transverse markings (i.e., crosswalk lines, intersection stop lines, etc.) (MUTCD, 2000). Pavement markings come in many shapes, sizes, and functionalities. Regardless of their immediate purpose, pavement markings are used to inform and warn drivers, pedestrians, and bicyclists of local and federal regulations and potentially hazardous locations. The MUTCD states that the most inherent function of pavement markings is that they allow motorists to focus on the roadway where the danger is located, as opposed to signs or lights located off the roadway (MUTCD, 2000). Typical pavement markings are placed in advance of the impending roadway hazard to allow motorists to react accordingly and provide them with a sufficient amount of time to determine their proper reaction. Normally, the redesign and reconstruction of the roadway is the most efficient means of addressing potential hazards, but when redesign and reconstruction are not feasible, pavement markings can be used to alleviate or moderate these situations (Storm, 2000).

Transverse pavement markings (also known as optical speed bars) are stripes located at horizontal curve tangents, roundabout approaches, intersection approaches, construction areas, and freeway off ramps (Meyers, 1999). The goal of transverse markings is to reduce speed and improve safety at potentially hazardous locations. The markings are placed in advance of the location in question and perpendicularly to the path of traffic, to decrease vehicle speed before the location is reached. The spacing between stripes is reduced and they decrease in thickness, as they get closer to the location (Griffin and Reinhardt, 1996). The purpose of these markings is to create an optical illusion, which would force drivers to slow down. The line spacing and size is intended to give the driver a sense of acceleration, regardless whether the vehicle is actually accelerating. This impression of acceleration will give drivers the indication they are traveling faster than intended, which in turn will force them to decrease their operating speed.

A set of three applications of transverse pavement markings conducted by Enuston (1972) examined their effectiveness on operating speeds. Each application was at a different type of facility and included approach to a construction zone at an Interstate facility, a curve approach at a two-lane rural highway, and an approach to an overpass. A different roadway length and number of lines was used in each application to address the specifics of each site. Speed measurements were taken at the approach and along the treatment and comparisons were made before and after the installation. Mixed results were obtained for each site regarding the effectiveness of the transverse lines in reducing operating speeds. For the work zone approach, the results indicate a minimal speed reduction which decreased with time and attributed it to a "novelty" effect. The second site also included rumble strips and this combination had a larger initial speed reduction, but eventually the average speed began to return to the initial average speed. Moreover, the rumble strips reduced speeds dramatically, and the average speed increased considerably when the rumble strips were removed. In the third study, the average speeds were reduced following the treatment installation without any change in speed variation.

Backus (1976) implemented transverse pavement markings across two-lanes of traffic on a four-lane highway, approaching a horizontal curve and the speed was measured 100 feet from the point of curvature. It was determined that before the insertion of the pavement markings, the 35 mph speed limit was exceeded 60 percent of the time, and 18 percent of the traffic exceeded 40 mph. After the installation of the markings, the percentage of traffic exceeding 35 mph decreased by 35 percent, and the percentage of traffic exceeding 40 mph decreased by 10 percent. The experiment also yielded a decrease in average mean speed of 2.5 mph, which Backus concluded was statistically significant.

A study that experimented with transverse markings at the approaches of 5 separate intersections has also been completed (Jarvis, 1989). The markings were a yellow thermoplastic material, 2 feet wide and a different distance between lines was used where the distance was reduced closer to the intersection. The results indicate that speeds decreased significantly as drivers entered the marked area, but after reaching their maximum speed reduction, drivers gradually returned to normal speeds, showing no signs of improvement. An innovative conclusion that Jarvis hypothesized is that the markings acted merely as a hazard warning—this can be attributed to drivers reducing speeds at the beginning of the pattern and then returning to normal speeds—and not a tool of affecting driver operating speeds. Another conclusion of the research was that the gradual declination in the distance between the transverse markings was not necessary and that equal spacing of the markings would allow a larger number of markings to be used at the area first observed by the driver.

2.2 *Warning signs*

A main objective of warning signs is that they give a sufficient amount of time for drivers to react to forthcoming roadway hazards (MUTCD, 2000). The application of warning signs can be based on an engineering study or engineering judgment. If the warning sign placement is performed from an engineering study, then the required time for a proper reaction needs to be considered. This time is the total time needed to react to a warning sign based on Perception, Identification (understanding), Emotion (decision making), and Volition (execution of decision) (PIEV). The PIEV times can vary accordingly, based on the dimensions of the roadway, posted or 85th-percentile speed, and the hazards associated with the roadway.

The most common type of warning sign in advance of a curve is the horizontal alignment sign. This sign is often accompanied by an advisory speed plaque, which is located below the horizontal alignment sign. The common function of this warning sign is to alert drivers of the impending change in the horizontal curvature of the roadway. The advisory speed plaque suggests a safe speed, other than the posted speed limit, that should be used to safely negotiate the curve. The excessive use and commonality of the horizontal alignment sign is probably the reason that the sign is often ignored. It is so often used that it "tends to breed disrespect for all signs" (MUTCD, 2000). Therefore, drivers will pay less attention to warning signs if they are used too frequently, thus creating an unsafe environment.

Hawkins (1994) conducted a study to examine the effects of supplemental warning plaques in addition to standard signing. The study used pedestrian crossings and railroad crossings to conduct the research. Drivers were surveyed after traversing sites with the sign alone and with supplemental plaques in addition to the sign. The purpose of this study was to determine whether the principles on which the current system of warning signs is based should be changed to improve driver comprehension. According to Hawkins, supplemental plaques can provide drivers additional information such as distance to the potential hazard, length of potential hazard, direction or location of hazard, recommended speed, and other miscellaneous identification or response information.

The combination horizontal alignment/advisory speed sign is a relatively new sign that combines the horizontal alignment sign with the advisory speed plaque onto a single sign. This sign is used to supplement the horizontal alignment sign with advisory speed plaque and is installed at the point of curvature, after the horizontal alignment sign. This signing reiterates the warning conveyed from the horizontal alignment sign as the driver approaches the curve. The sign duplication (2 warning signs) is envisioned to work as a stronger indication of the potential hazard. The one-direction large arrow sign is most commonly used to demarcate an upcoming change in the horizontal alignment of the roadway. The large arrow sign should be placed at the beginning of the curve, perpendicular to traffic. The sign should be placed at a location that allows the sign to be seen for a sufficient distance from the tangent of the curve. The ample distance will provide drivers an adequate amount of time to make a decision based on the change in alignment.

Common warnings located on warning signs are flags, flashing lights, and spotlights. The goal of these types of warnings is to give the driver a different warning perspective. For instance, bright, orange flags on a horizontal alignment sign are definitely not a usual occurrence. The typical warning sign is frequently ignored, but if something atypical was attached to the sign, it could possibly alert drivers in an uncharacteristic manner, forcing them to slow down or alter their driving behavior. Several studies have been performed to determine the effectiveness of warning signs accompanied by flashing lights. Lyles (1980) examined the use of flashing lights that supplemented warning signs at a rural intersection with poor sight distance. The flashing lights caused a 1.6 to 3.2 mph speed reduction compared with only a 0.8 mph speed reduction without the flashing lights. In another study, Lyles (1981) used flashing lights with the existing warning signs that warned drivers of construction zones on rural highways. The flashing lights resulted in a 3 to 4 mph speed reduction for short work zones and a 7.5 mph speed reduction for long work zones. Zegeer (1975) studied a situation where flashing lights were used with school zone speed restriction signs. The flashing lights reduced average speeds by 3.6 mph, and on roads with speed limits of 55 mph the average speed was reduced by 10 mph. Hanscome (1976) studied a situation where flashing lights were used to warn of the possibility of skidding due to wet weather. The flashing lights reduced average speeds by 9 percent for wet conditions.

According to the MUTCD, "delineators are particularly beneficial at locations where the alignment might be confusing or unexpected, such as... curves" (MUTCD, 2000). Delineators are good methods of guidance[1] especially at night, because they are reflective and are at a comparable height to the headlights of vehicles. It is essential that delineators be spaced at a constant distance with several delineators visible at all times, when used at locations of changing horizontal alignment. The study by Zador et al (1986) found that speeds increased by approximately 1.5 mph at horizontal curves after the installation of post delineators. The study also found that vehicles tend to move towards the centerline of the roadway after the installation of post delineators on right horizontal curves, and have no placement effect for left horizontal curves. The authors concluded that an argument could be made that the speed increases found in the chevron and post delineator cases, reflect the adaptation of the drivers to an increased level of information about the upcoming roadway conditions, giving them an advantage in maneuvering through the curves.

Recent work on establishing the links between safety and speed indicated that there is such a relationship and the likelihood to be involved in a crash increases with departure from the average travel speed (Stuster, 1998). The use of means that could reduce such variance among speeds was a recommendation of the study. Therefore, the proper use and installation of warning signs could assist on achieving this goal.

[1] The MUTCD claims that delineators are guidance devices and not warning devices.

2.3 *Literature review summary*

Operating speeds can effectively be reduced if warning signs and pavement markings are installed at hazardous roadway locations. The literature review showed the following.

1. Pavement markings can reduce operating speeds effectively. These markings act as a visual warning, they alter human perception, and they enable drivers to pay attention to the roadway without having to look off to the side of the roadway to see a warning sign.
2. Warning signs have also been found to reduce operating speeds at dangerous roadway sections and thus affect safety, and they seem to be even more beneficial if coupled with other warning signs or devices. Typical warning signs (i.e., curve warning signs, speed plaques, chevrons, etc.) are often overlooked due to their frequent use, but if additional warning signs or devices (i.e., combination horizontal alignment/advisory speed sign, flashing lights, flags, etc.) are used with the commonly used warning sign, drivers will often times acknowledge the warning sign when normally they would not, or they may react quicker to the warning.
3. Speed reductions due to warning signs and pavement markings vary from site to site, so it is very difficult to accurately predict what kind of results will occur.
4. The literature dealing with warning signs and flashing lights explained that where flashing lights are used and the hazard is not obvious, that regardless of the type of accompanying sign, a speed reduction of 2 to 3 mph can be expected; where the hazard is more clearly explained by the sign, the speed reduction is likely to be greater and the driver will probably pay closer attention.

3 METHODOLOGY

The research plan focused on identifying potential sites where different treatments were to be introduced and speed measurements would be taken to estimate the effectiveness of each treatment. A request for candidate sites was made to each Kentucky Transportation Cabinet District office and a list of potential sites was developed. Each site proposed was evaluated through a site visit where the alignment was examined and documented. The existing warning signs and pavement markings were noted and any particular elements of the sites were recorded. A list of sites was developed and sites were selected based on their traffic volume, geometry, and crash history. Three sites were selected and the results are presented here.

The speeds were measured for existing and newly treated conditions at three locations throughout the curve approach. The devices were placed throughout the tangent and curve section on the curve approach, and a time and speed for each vehicle that passed over them was measured. This allowed for following individual vehicles throughout the system and then determining their speed reduction as they progressed through the study area. The location for the speed measurement devices differed for each site because of the existing geometry and traffic control. A contributing factor was the distance from the existing warning sign to the point of curvature. This distance dictated the position of the first speed measuring device in advance of the existing warning sign. Regardless of this distance, the three devices were placed in advance of the exiting warning sign, at the warning sign, and at the point of curvature at equal distances. Automated speed measuring devices (HI-STAR) were used and data reduction software was developed to correctly identify and track individual vehicles through the curve. Speed data was collected for free-flowing vehicles, i.e. trailing vehicles were eliminated from the data base and only the speed of the lead vehicle was used.

All treatments were given a five-day waiting period before the speed was measured. The treatment was removed after the data collection and the next treatment was installed with a new five-day waiting period. The waiting periods between treatments was implemented so that local traffic could be more familiar with the treatment and in turn, not give false speed-readings due to potential novelty effects. For instance, if a local driver navigates the same road every day, and then sees something different, then this driver is likely to slow down more than usual. If the drivers are given a few days to become familiar with the new situation, the recorded speeds will be more accurate and allow for a better evaluation of the effectiveness of the treatment.

Several types of warning signs and pavement markings were considered for use in this study to determine which methods and combinations are those that could reduce operating speeds more effectively (Figure 1). All of the sites that were studied had an existing horizontal alignment sign with speed plaque in advance of the curve. The advisory speed was set following the procedures outlined in the MUTCD using a ball bank indicator and it was assumed that they were properly set. The common treatments for all sites include: 1) addition of flags to the exiting curve warning sign; 2) addition of flashers to the existing curve warning sign; 3) installation of the combinational

Typical existing conditions

New combination sign

New combination sign with flashers

Warning sign with flags

Rumble strips

Transverse lines

Figure 1. Sample of treatments.

horizontal alignment/advisory speed sign placed at the point of curvature in addition to the exiting curve warning sign; 4) addition of flashing lights to both signs (existing and new); 5) installation of post delineators on the right hand side; and 6) installation of transverse pavement lines. For one of the sites, large arrows were installed because they were not part of the existing treatment.

To test for differences among various treatments and determine which treatment has the potential for a greater speed reduction, a series of statistical tests were used. The general null hypothesis is that no treatment has any effect on the speed reduction. To test this, two different tests were employed. The first tests the difference in average speeds, and the second examines the variances of the speed distributions. The test for the average speeds allows for simple comparisons between averages and is achieved with a z-test. The second test examines whether the treatments have impacted the distribution of the speeds by forcing more drivers to drive at similar speeds, i.e. reducing the variance among speeds.

4 RESULTS

The goal of this project was to evaluate the effectiveness of the treatments used on these sites. The data was analyzed using the average speeds, average speeds for day and night, overall changes in average speeds from one measurement location to another, percentage changes in average speeds from one measurement location to another, 85th percentile speeds, over the 85th percentile speeds, variances, and frequency distributions. The results for the average speeds were not very consistent among the three sites (Table 1).

The data for Site 1 indicate that there is almost no speed difference among the speeds at each measurement location, and for some treatments speeds increased. The percentage change in the average speed shows that the biggest change can be seen at the second measurement location

Table 1. Average speeds by site and measurement location.

		Speeds (mph)				Percent change		
Site	Treatment	1	2	3	Total reduction	1	2	3
1	Existing	51.1	46.6	43.4	–7.7	–	–	–
	Flags	50.9	46.9	43.3	–7.6	0.0	0.0	0.0
	Arrow	51.4	47.4	43.9	–7.5	1.0	1.1	1.4
	New sign	52.0	47.9	44.3	–7.7	2.2	2.1	2.3
	Delineators	52.1	48.2	44.1	–8.0	2.4	2.8	1.8
	Lines	51.8	47.8	44.1	–7.7	1.8	1.9	1.8
2	Existing	52.2	48.8	45.8	–6.3	–	–	–
	Flags	51.3	47.6	44.5	–6.9	0.9	1.5	2.7
	Flasher	52.7	48.8	45.6	–7.1	3.6	4.0	5.3
	New sign	52.5	48.7	45.0	–7.5	3.2	3.7	3.9
	Both flashers	53.0	48.4	44.3	–8.7	4.1	3.2	2.3
	Delineators	52.8	49.4	45.3	–7.5	3.7	5.2	4.6
	Lines	52.1	47.4	43.1	–9.0	2.4	1.1	–0.4
3	Existing	53.4	53.4	48.5	–5.0	–	–	–
	Flags	53.9	53.4	48.5	–5.4	5.9	13.8	12.0
	Flasher	55.4	51.2	49.6	–5.9	8.9	9.1	14.5
	New sign	54.6	50.6	48.7	–5.9	7.2	7.9	12.5
	Both flashers	54.6	50.3	48.8	–5.7	7.2	7.2	12.7
	Chevrons	53.6	52.7	48.3	–5.3	5.3	12.5	11.6
	Delineators	52.8	48.9	48.6	–4.2	3.7	4.2	12.2
	Lines	52.9	50.2	49.6	–3.2	3.9	7.0	14.6

(at the existing warning sign). The statistical analysis revealed that the large arrow sign, the combination warning sign, the post delineators, and the transverse lines significantly affected the average speed for this site. A potential reason for this difference could be that the new treatments increased the driver comfort level, thus allowing the drivers to navigate the curve faster. Another possibility could be that the drivers' disregard for new warning signs and markings had no effect on attracting the drivers' attention.

The second site showed noteworthy speed reductions at the second measurement location, i.e. at the location of the existing curve warning sign. The flag treatment and transverse lines treatment experienced an average speed reduction of 2.6% (1.3 mph) and 2.9% (1.5 mph), respectively. The statistical test indicated that the warning sign with flags, both signs with flashing lights, and the transverse lines significantly affected the average speed. Therefore, it can be concluded that these three treatments had a direct impact on the reduction of speeds when compared to the existing conditions.

At the third site, the second measurement site showed some relative speed reductions for three of the treatments. The warning sign with flashing lights had a decreased average speed of 4.2% (2.2 mph), the combination horizontal alignment/advisory speed sign had a decreased average speed of 5.3% (2.8 mph), and both warning signs with flashing lights had a decreased average speed of 5.8% (3.1 mph). These results may indicate that the drivers did not recognize or see the warnings, until after they passed the first measurement site. A possible reason for the sudden reduction in average speed for the second measurement could be the high level of speed (the average speed for all treatments combined was approximately 55 mph) associated with the first measurement. The statistical analysis showed that the warning sign with flashing lights, the new combination warning sign, and both signs with flashing lights significantly affected the average speed for this site as compared to the existing conditions.

Table 2. 85th and over 85th percentile speeds by site and measurement location.

		85th percentile Speeds (mph)				Over 85th percentile speeds (mph)			
Site	Treatment	1	2	3	Difference	1	2	3	Difference
1	Existing	55.7	51.2	46.5	–9.2	60.4	55.3	50.0	–10.4
	Flags	56.4	52.0	47.3	–9.1	60.8	55.4	49.9	–10.9
	Arrow	56.2	51.4	47.2	–9.0	60.2	53.4	47.7	–12.5
	New sign	56.9	52.6	47.7	–9.2	60.1	53.3	48.2	–11.9
	Delineators	57.1	52.8	47.4	–9.7	61.0	53.8	48.5	–12.5
	Lines	56.8	52.5	47.6	–9.2	60.6	53.7	48.1	–12.5
2	Existing	57.1	53.1	49.5	–7.6	61.0	55.9	50.7	–10.4
	Flags	56.1	51.5	47.7	–8.4	60.3	53.9	48.7	–11.5
	Flasher	57.4	53.4	49.7	–7.7	61.3	55.5	50.5	–10.8
	New sign	57.5	53.1	48.7	–8.7	61.8	55.6	50.4	–11.3
	Both flashers	57.3	53.1	47.9	–9.4	61.0	54.4	48.8	–12.2
	Delineators	57.7	54.3	49.7	–8.0	61.3	55.9	49.6	–11.7
	Lines	56.8	51.8	47.7	–9.1	60.4	54.0	47.9	–12.5
3	Existing	57.5	57.0	52.2	–5.3	61.1	58.8	52.4	–8.7
	Flags	58.1	57.6	52.3	–5.8	62.2	59.4	52.7	–9.5
	Flasher	59.6	55.2	53.4	–6.2	63.1	56.3	53.8	–9.3
	New sign	59.2	54.7	52.4	–6.8	63.2	56.8	53.5	–9.7
	Both flashers	59.2	54.3	52.4	–6.8	63.2	56.6	53.6	–9.5
	Chevrons	58.4	56.3	52.4	–6.0	62.5	56.4	52.8	–9.7
	Delineators	57.2	53.4	52.7	–4.5	61.2	55.5	54.3	–6.9
	Lines	56.8	53.9	53.4	–3.5	60.2	55.6	53.7	–6.5

An approach that can be used to examine the effectiveness of treatments is to observe the 85th percentile speeds. It could be hypothesized that even if a treatment did not affect the average and 85th percentile speeds, it may influence the speeds of vehicles exceeding the 85th percentile speed. This could be considered as a positive indication, since these drivers could be considered as the ones that may have a larger crash potential. It should be noted here that the analysis was performed using the vehicles that were exceeding the 85th percentile speed at the first measurement location by examining their speed change as they proceeded to the other two measurement locations. The 85th percentile speeds are larger than the average speeds, as would be expected (Table 2).

The total reduction in speed however is significantly higher for all three sites. The measurement location that showed the largest change in 85th percentile speeds should be a good indication of where drivers decide to change their speed the most dramatically in order to navigate the curve safely. For the first two sites this is the second measurement location while for the third site the last measurement location shows the biggest change in the 85th percentile speeds. The data indicate that for some treatments the 85th percentile speeds increased: a phenomenon that could not be explained in any other way other than the presence of the treatments may have created a feeling of a "safer" condition and the treatments had the exactly opposite effect.

The data for the speeds of motorists exceeding the 85th percentile speed showed a more dramatic speed reduction by the third measurement location for the first two sites and the second measurement for the third site (Table 2). Among the common treatments, the new combination sign, the transverse lines, and the delineators produced the highest speed reduction at the respective measurement locations for each site. The large arrow that was used only at the first site, since it was already in place at the other locations, also showed a significant impact on these speeds.

5 CONCLUSIONS

The problem with inconsistently designed roadways is that they do not lend any clues to the driver as to the appropriate action to take at hazardous or unexpected curves. Two methods of conveying necessary roadway information to the driver are warning signs and pavement markings. For this study, several warning signs and pavement markings were implemented at rural curves to evaluate their effectiveness to reduce operating speeds. A literature review was performed to evaluate past experiences with similar situations and to potentially determine which warning signs and pavement markings are the most effective. Several curves were chosen as potential study sites and these sites were narrowed down to sites that had a curve related crash history. Speeds were measured at three locations at each site involving several treatments. The speed data was then analyzed to determine what treatments were the most effective.

The data from the three sites gave mixed results as to the effectiveness of treatments in reducing operating speeds. The data for Site 1 indicates no major speed reductions for most of the treatments. In fact, most treatments experienced a slight increase in all speed measures. However, all treatments at this site reduced the speed variance considerably compared to the existing conditions. The data for Site 2 showed considerable speed reductions for all treatments. Moreover, the two treatments that experienced the most significant speed reductions were the flashing lights on both warning signs and the transverse lines. The data for Site 3 also showed speed reductions from some of the treatments. The treatments that experienced the most significant speed reductions were the existing warning sign with flashing lights, the new combination warning sign, and both warning signs with flashing lights.

A noteworthy finding of this work was that for all three sites, the average of the speeds over the 85th percentile speed showed a reduction indicating that most treatments have the potential to affect high speeds more than the average or 85th percentile speed. It can be concluded that some of the warning signs and pavement markings do moderately reduce the operating speeds of vehicles. The warning signs with flashing lights can reduce speeds and the new combination warning sign can also be quite effective with the addition of flashing lights. The transverse lines showed considerable speed reduction for Site 2 and probably would have seen similar results at Site 1 if the pavement pattern was longer (it was adjusted after the Site 1 data collection to provide for longer warning period).

The objective of this work was to determine if anything can be done to reduce operating speeds by providing additional warning information to the driver. Based on these findings, some of the treatments have shown promising results but there are possible options that could even enhance these treatments. For example, the flashing lights were only working at nighttime, which might explain their increased effectiveness as compared to the existing conditions. Therefore, the use of lights that could be visible also at daytime could have the same impact. The use of rumble strips in addition to the transverse lines or a longer pattern of transverse lines might be two additional treatments that may have a significant effect. Rumble strips are an option that could accompany transverse lines or be used strictly by themselves. Based on past research, transverse lines over a greater length are more effective at reducing speeds. Transverse lines could also be carried out through the curve, instead of stopping at the point of curvature. Finally the use of larger signs, especially for the new combination sign, may also improve the effectiveness of the treatment because drivers could see it from a further distance away.

The results of the study indicate that there are some promising treatments that have the potential to impact operating speeds and particularly high speeds. The new combination curve warning and suggested speed seems to have a positive effect on reducing speeds and its use is encouraged. However, it is recommended that it should be used with caution to avoid overuse and thus be disregarded by the drivers. The use of flashing lights is recommended for most sites, since they at least have the potential to impact operating speeds at night. Even though the transverse lines showed mixed results, it is expected that they are promising and further research is warranted.

REFERENCES

Backus, R. (1976) *Optical Illusion Improves Safety, Western ITE*. 30 (3).

Enustun, N. (1972) *Three Experiments with Transverse Pavement Stripes and Rumble Bars, Report No. STD-RD-216-72*. Traffic and Safety Division, Michigan Department of State Highways and Transportation, Lansing, Michigan.

Federal Highway Administration (2000) *Manual on Uniform Traffic Control Devices for Streets and Highways*. U.S. Department of Transportation.

Griffin, L. and Reinhardt, R. (1996) *A Review of Two Innovative Pavement Patterns That Have Been Developed to Reduce Traffic Speeds and Crashes, Texas Transportation Institute*, 59 pp.

Hanscome, F. (1976) *Evaluation of Signing to Warn of Wet Weather Skidding Hazard, Transportation Research Record 600*. Transportation Research Board, Washington, D.C., pp. 20–27.

Hawkins, H.G. (1994) *Use of Supplemental Plaques To Improve Effectiveness of Warning Signs, Transportation Research Record 1456*. Transportation Research Board, Washington, D.C., pp. 20–26.

Jarvis, J. (1989) *The Effect of Yellow Bar Markings on Driver Behaviour, Report No. 173*. Australian Road Research Board, Victoria, Australia.

Lyles, R. (1980) *An Evaluation of Signs for Sight Restricted Rural Intersections, FHWA/RD-80-002*. Federla Highway Administration, U.S. Department of Transportation.

Lyles, R. (1981) Alternative *Sign Sequences for Work Zones on Rural Highways, FHWA/RD-80-163*. Federal Highway Administration, U.S. Department of Transportation.

Meyers, E. (1999) *Application of Optical Speed Bars to Highway Work Zones, Transportation Research Board 78th Annual Meeting*, Washington, D.C.

Storm, Richard. (2000) *Pavement Markings and Incident Reduction, 2000 Transportation Scholars Conference*. Center for Transportation Research and Education, Iowa State University, Ames, Iowa.

Stuster, J., Coffman, Z. and Warren, D. (1998) *Synthesis of Safety Related Research to Speed and Speed Limits* FHWA-RD-98-154, federal Highway Administration, Washington, DC.

Zador, P., Stein, H., Wright, P. and Hall, J. (1986) *Effects of Chevrons, Post-Mounted Delineators, and Raised Pavement Markers on Driver Behavior at Roadway Curves*, Insurance Institute for Highway Safety.

Zegeer, C. (1975) *The Effectiveness of School Signs with Flashing Beacons in Reducing Vehicle Speeds, Report 429*, Division of Research, Kentucky Bureau of Highways, Frankfort, KY.

Efficient Transportation and Pavement Systems – Al-Qadi, Sayed, Alnuaimi & Masad (eds)
© 2009 Taylor & Francis Group, London, ISBN 978-0-415-48979-9

Implementing new vehicle stability-based minimum radius guidelines for simple and reverse horizontal curves: Is it cost effective?

Amir Abd El Halim
Stantec Consulting, Waterloo, Ontario, Canada

Said M. Easa
Ryerson University, Toronto, Ontario, Canada

Abd El Halim O. Abd El Halim
Carleton University, Ottawa, Ontario, Canada

ABSTRACT: A new and improved design approach for determining the minimum radius requirements for simple and reverse horizontal curves has been developed using advanced vehicle simulation software. The approach, which considers vehicle stability and three-dimensional (3D) alignments, produces larger radii than those presented in current geometric design guides. The larger radii (flatter curves) of the new approach provide a smoother transition between the tangent and curves of the highway and results in a reduction of the overall length of the alignment. In turn, the reduction in highway alignment results in savings in construction, maintenance, and collision costs. This paper evaluates the costs and benefits associated with implementing the new design guidelines for the minimum radius of both simple and reverse horizontal curves with intermediate tangents. Two different cases were evaluated: (1) a new alignment involving construction of the flatter horizontal curves based on the new guidelines relative to the curves of the current design guides and (2) an existing alignment for which the flatter horizontal curve is to be constructed to improve the existing alignment. The results illustrate that implementing the new minimum radius guidelines generates a more cost-effective highway alignment, in addition to satisfying vehicle stability requirements.

1 BACKGROUND

Current geometric design guides of the Transportation Association of Canada (TAC) and the American Association of State Highway and Transportation Officials (AASHTO) establish minimum radius requirements for horizontal curves based on a simple formula known as the point-mass model (TAC 1999, AASHTO 2001). The model simplifies a vehicle to a single point and does not consider vehicle stability requirements or the effect of three-dimensional (3D) alignment. This highly simplified approach to modeling a vehicle as it traverses a horizontal curve is based on the laws of mechanics and kinematics. This model was adopted by North American design guides due to its simplicity (Hassan et al. 1998).

To overcome the limitations of the point-mass model, a new and improved simple horizontal curve design approach that considers vehicle stability and the effect of 3D alignment on the required minimum radius was developed (Easa and Dabbour 2003). The approach was also used to develop new minimum radius requirements for reverse horizontal curves (Easa and Abd El Halim 2006). The new radius requirements were developed using the Vehicle

Dynamics Models Roadway Analysis and Design (VDM RoAD). VDM RoAD is advanced vehicle simulation software developed at the University of Michigan Transportation Research Institute (Sayers 1999). It is a user-friendly integrated set of computer tools for simulating and analyzing the braking and handling behavior of vehicles on selected road and highway designs. The software simulates a vehicle running down a specified highway alignment and can analyze lateral acceleration, directional control, roll stability, and stopping sight distance. The VDM software has been validated using field data. The results of the study revealed that an increase in the minimum radius is required to satisfy vehicle stability requirements on 3D alignments.

Several studies have revealed that a large proportion of traffic-related fatalities occur on curved sections, especially on horizontal curves of two-lane rural highways. A study by Talarico and Morrall (1993) reported that curved highway sections increase the demands placed on a driver, and these sections of highway have significantly higher collision rates than tangent sections. Wooldridge et al. (2000) evaluated the effects of horizontal curvature on driver visual demand and found that visual demand was significantly higher on curves with sharper radii. Collision rates on curves have been found to be 1.5 to 4 times greater than those on tangent sections of similar highways. Statistics have also shown that sharper curves are associated with higher collision rates than milder curves (Lam et al. 1999).

As the radius of a horizontal curve is increased, the curve becomes flatter, and a smoother transition between the tangent and the curve is achieved. Increasing the radius of a horizontal curve is one of the most effective means of decreasing collision rates on horizontal curves, and thus improves highway safety. Several benefits and costs savings arise from an increase in curve radius. When the radius is increased, the overall alignment length is reduced resulting in construction saving, collision reduction, and vehicle operating saving. The purpose of this paper is to evaluate the benefits and costs associated with new minimum radius requirements for simple and reverse curves compared with those of current design guides. A traditional reverse curve consists of a simple curve in one direction (to the right or the left), immediately followed by another simple curve in the reverse direction. Reverse horizontal curves, or simply reverse curves, are advantageous in hilly and mountainous areas, and are common on two-lane highways and interchange ramps.

The following sections describe the geometry of the existing and new horizontal curves and the proposed evaluation methodology. The cost-effectiveness of the new minimum radius guidelines and sensitivity analysis are then presented, followed by the concluding remarks.

2 HIGHLIGHTS OF NEW GUIDELINES

The new guidelines were developed using VDM RoAD that, as previously mentioned, takes into consideration the dynamics of a vehicle and the geometric characteristics of the roadway. To establish the new guidelines, design vehicles (passenger car, WB-15, WB-20, and 3a/2a) were run through the test alignments (horizontal curves) for different design speeds and the corresponding minimum radii. The peak lateral acceleration that the vehicle experienced as it traversed the curve was recorded and compared with the AASHTO lateral acceleration value used in the point-mass model. Due to the effects of vehicle stability and the 3D alignment, the peak lateral acceleration values obtained from VDM RoAD were consistently higher. To compensate for these effects, the minimum radius was incrementally increased and the vehicles were re-run through the alignment. This was an iterative and time-consuming procedure that was continued until the peak lateral acceleration obtained from VDM RoAD was equal to the AASHTO lateral acceleration. The new guidelines for simple and reverse horizontal curves require, respectively, up to 20% and 27% larger radii than those recommended in the current guidelines (Easa and Dabbour 2003, Easa and Abd El Halim 2006). Table 1 shows the recommended minimum radius guidelines for simple horizontal curves for a maximum superelevation $e_{max} = 0.04$.

Table 1. Recommended 3D minimum radius guidelines for simple horizontal curves for $e_{max} = 0.04$ (5).

Design speed (km/hr)	TAC Design minimum radius (m)	3D Required minimum radius (m)[a]	Required increase in radius (%)	3D Design radius[c] (m)
(a) WB-15				
60	150	161	7.3	170
70	200	219	9.5	220
80	280	311	11.1	320
90	380	428	12.6	430
100	490	568	15.9	570
110	680	791	16.3	800
(b) WB-20				
60	150	166	10.7	170
70	200	224	12.0	230
80	280	323	15.4	330
90	380	444	16.8	450
100	490	562 (574)[b]	17.1	580
110	680	805 (812)[b]	19.4	820
(c) 3a/2a				
60	150	162	8.0	170
70	200	221	10.5	230
80	280	319	13.9	320
90	380	434	14.2	440
100	490	572	16.7	580
110	680	808	18.8	810

[a] To maintain the same level of lateral acceleration.
[b] To avoid rollover.
[c] Rounded values.

3 GEOMETRY OF CURRENT AND NEW HORIZONTAL CURVES

The proposed new minimum radius guidelines are flatter than those of the TAC and AASHTO design guides for given conditions. In effect, implementing the new guidelines would be similar to a technique known in geometric design improvements as "*Curve Flattening*". This technique involves reconstructing a horizontal curve to make it less sharp, and as a result the curve becomes longer. Figure 1 shows the geometry of flattening a simple horizontal curve for typical situations, along with the current TAC or AASHTO curve (sharper) and the new curve (flatter). The current and new curves are assumed to have the same tangents, and therefore have different points of curvature (PC) and points of tangent (PT) and the same central angle. Although the length of the new curve (L_n) is longer than that of the current curve (L_0), the overall length of the new alignment is reduced. The reduced alignment length produces several benefits and cost savings. From Figure 1, it is observed that by increasing the radius (R_1) of the sharper curve to the radius of the new curve (R_2), the resulting length (L_2) becomes longer than that of the TAC curve. However, the degree of curvature of the new curve DC_2 is less than that of the TAC curve DC_1. The degree of curvature DC is given by:

$$DC = \frac{5729.6}{R} \tag{1}$$

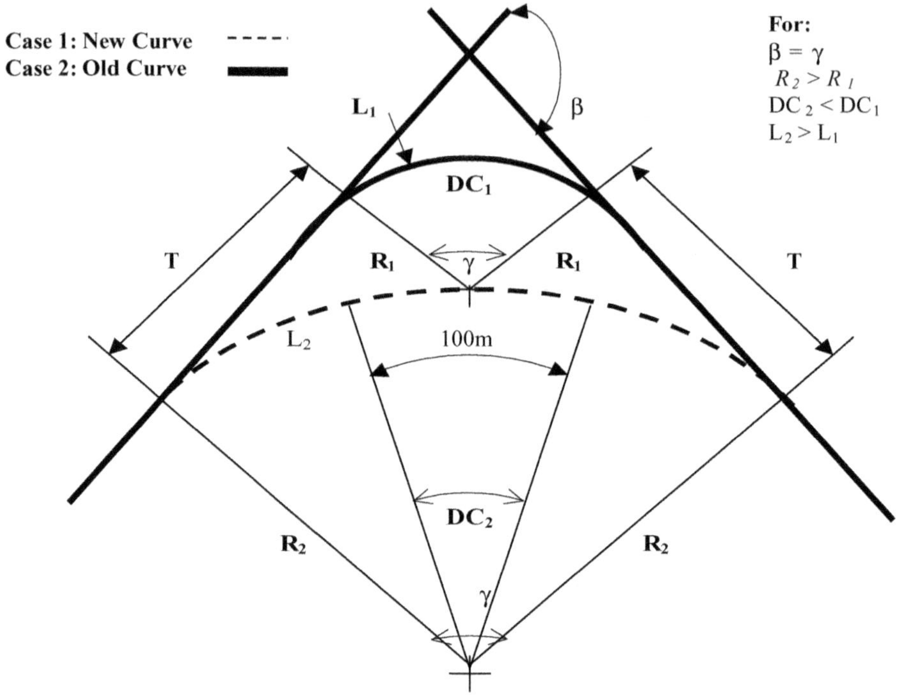

Figure 1. Geometry of curve flattening for typical situations (PC and PT not same).

When the radius of a sharp horizontal curve is increased, the curve length increases as follows:

$$L_2 = \frac{DC_1}{DC_2} L_1 \tag{2}$$

where L_1 = length of the TAC sharper curve (m), L_2 = length of the new curve (m), DC_1 = degree of curve of the TAC curve, °/100 m, and DC_2 = degree of curve of the new curve, °/100 m.

Despite the increase in the length of the new curve, the total length of the alignment for the TAC curve is longer since a portion of the tangent section on either end of the TAC curve becomes part of the TAC curve, if it were to be constructed, relative to the flatter curve. Thus, the reduced tangent length is the difference between the length of the new flatter curve and the length of the TAC curve plus the two tangent portions eliminated due to curve flattening. Therefore, the length of tangent eliminated can be calculated from:

$$T = \frac{5729.6}{1000}\left(\frac{1}{DC_2} - \frac{1}{DC_1}\right)\tan\left(\frac{\gamma}{2}\right) \tag{3}$$

where T = tangent length eliminated due to curve flattening (km) and γ = central angle (∘). The lengths of the TAC alignment L_0 (km), and the new alignment, L_n (km) are given by:

$$L_0 = L_1 + 2T \tag{4}$$

$$L_n = L_2 \tag{5}$$

Thus, the reduction in the total length of the alignment due to curve flattening ΔL, is given by:

$$\Delta L = \left[11.4592\tan\left(\frac{\gamma}{2}\right) - 0.10\gamma\right] * \left(\frac{1}{DC_2} - \frac{1}{DC_1}\right) \tag{6}$$

where ΔL = change in length (km).

4 EVALUATION METHODOLOGY

The reduced alignment length of Equation [6] results in savings in construction cost, collision reduction, maintenance and rehabilitation, travel-time, winter maintenance, and vehicle operation. To examine the cost-effectiveness of the new simple and reverse horizontal curves with intermediate tangents, it is necessary to contrast the costs and benefits of constructing the TAC and new curves. The methodology used in this analysis for evaluating the cost-effectiveness of the new guidelines is similar to that used by Talarico and Morrall (1993) who conducted a comprehensive study to evaluate the cost-effectiveness of curve flattening in Alberta. Some of the parameters used in their methodology have been revised to reflect Ontario conditions.

4.1 *Construction cost saving*

When a horizontal curve is made flatter, the total length of an alignment is decreased by an amount ΔL given by Equation [6]. This decreased alignment length produces a shorter roadway alignment and can be considered a construction saving. Assuming that the average costs of a 9.0-m and 11.0-m wide highway in Ontario is $720,000 and $880,000 per km, respectively (MTO, 2003 dollars), any reduction in alignment length can be considered a benefit due to reduced construction, material, and labor costs. The construction cost of a horizontal curve (simple or reverse) can be estimated from:

$$C_s = \frac{R\Delta\Pi}{180} * U_c \tag{7}$$

where C_s = construction cost ($), R = curve radius (m), Δ is the central angle (∘), and U_c = unit cost ($/km). The cost saving attributed to constructing the flatter horizontal curve is then given by:

$$C = \Delta L * U_c \tag{8}$$

where C = construction cost saving (*$*).

4.2 *Collision cost saving*

Research has shown that a relationship exists between the degree of curvature and the collision rate of a horizontal curve. The general trend outlined by Lamm et al. (1999) states that as the degree of curvature increases (radius of curve decreases), the collision rate increases. Implementing the new flatter curve would obviously decrease the degree of curvature. With the decrease in *DC,* the collision rate on the alignment decreases and the level of safety is improved. To quantify the benefits attributed to a reduction in collisions, consider the following collision model used by Talarico and Morrall (1993):

$$\Delta A_t = AR_s(\Delta L)AADT + 0.0102(DC_1 - DC_2)AADT \tag{9}$$

where ΔA_t = annual reduction in collisions, AR_s = collision rate on a tangent section (collisions per million vehicle-kilometers), ΔL = change in length (km), $AADT$ = traffic volume in year t (in millions), and DC_1 and DC_2 = degree of curvature of the TAC and new horizontal curves, respectively. Typically, AR_s equals 0.560 collisions per million vehicle-kilometers. The annual saving due to collision reduction, K_t is given by:

$$K_t = \Delta A_t * U \tag{10}$$

where ΔA_t = annual reduction in collisions calculated from Equation [9] and U = average cost per collision ($\$$). The cost of a collision varies with the severity of the collision, and collisions generally consist of 2% fatalities, 12.5% moderate injuries, and 73% property damage (Talarico and Morrall 1993). In Alberta, the costs (adjusted to 2003 dollars) are as follows (10): $875,513 (fatality), $586,036 (serious injury), $5,729 (moderate injury), and $2,284 (property damage). Using these collision costs and data, a weighted average cost per collision was calculated as $93,148.

4.3 *Vehicle running cost saving*

The vehicle operating costs are a function of vehicle type and speed, and depend on the vertical profile of the highway alignment (Wooldridge et al. 2000). The vehicle running cost saving is given by:

$$W_t = [P_{ht}(H_{ht}) + P_{\text{sub}}(H_{\text{sub}}) + P_{\text{arv}}(H_{\text{arv}})](\Delta L) \tag{11}$$

where W_t = vehicle running cost saving in dollars per 1,000 vehicles, P_{ht} = percentage of heavy trucks, P_{sub} = percentage of single unit trucks and busses, P_{arv} = percentage of autos and recreational vehicles, H_{ht} = running cost of heavy trucks, H_{sub} = running cost of single-unit trucks and buses, and H_{arv} = running cost of autos and recreational vehicles. Note that all costs are in dollars per 1,000 vehicle-km (based on 2003 dollars). The vehicle running costs in Alberta per 1,000 vehicle-km are estimated as (2003 dollars), $147 (autos), $440 (single unit trucks and busses), and $375 (heavy trucks) (Howery 1991). The annual vehicle running cost saving, W, is then given by:

$$W = W_t \; AADT \tag{12}$$

4.4 *Speed-change cost saving*

The cost saving associated with the speed change from the TAC sharper curve to the new flatter curve varies with the type of vehicle. The speed change saving can be estimated using the following formulas (Howery 1991):

$$Z_{ht} = (0.327V_i + 2.8)^{1.74} - (0.327V_f + 2.8)^{1.74} \text{ (Heavy trucks)} \tag{13}$$

$$Z_{\text{sub}} = (0.1936V_i + 1.8)^{1.9} - (00.1936V_f + 1.8)^{1.9} \text{ (Single-unit trucks and busses)} \tag{14}$$

$$Z_{\text{arv}} = (0.0588V_i + 1.4)^{1.9} - (0.0588V_f + 1.4)^{1.9} \text{ (Automobiles and recreational vehicles)} \tag{15}$$

where Z_{ht} = speed-change cost in dollars per 1,000 heavy trucks, Z_{sub} = speed-change cost in dollars per 1,000 single-unit trucks and busses, Z_{arv} = speed-change cost in dollars per 1,000 autos and recreational vehicles, V_i = speed of the TAC sharper horizontal curve, and V_f = speed of the new flatter horizontal curve. Note that all the costs are adjusted for 2003. The speeds V_i and V_f are calculated from:

$$V_{85} = 104.82 - \frac{3574.51}{R} \tag{16}$$

where V_{85} = 85th percentile operating speed (km) and R = curve radius (m). The annual speed-change cost saving, Z (per 1,000 veh), is then given by:

$$Z = [P_{ht}(Z_{ht}) + P_{\text{sub}}(Z_{\text{sub}}) + P_{\text{arv}}\ (Z_{\text{arv}})]\text{AADT} \tag{17}$$

4.5 *Net present value of annual savings*

The savings outlined previously are annual savings and must be calculated for the entire life (duration) of the project. The present value of a series of uniform end-of-period savings is a function of the interest rate, i, the life of the project, N, and the annual saving A (Newnan et al. 2008). For this analysis, a 20-year project life and an interest rate of 3% were assumed.

5 ECONOMIC BENEFITS OF NEW GUIDELINES

The economic benefits of the new flatter horizontal curve guidelines (Easa and Dabbour 2003, Easa and Abd El Halim 2006) are evaluated. The analysis first focuses on the case where the TAC and new curves have the same tangents, and thus the same deflection angle (Figure 1). For purposes of this analysis, a deflection angle of 45° was used. The evaluation was made for two possible situations of implementing the proposed flatter curves: new alignment and an existing alignment. For a new alignment, the following three cases were evaluated:

1. Case S-N: Using flatter simple horizontal curve for a new alignment
2. Case R-N: Using flatter reverse horizontal curve for a new alignment
3. Case RT-N: Using flatter reverse horizontal curve with intermediate tangent for a new alignment

Three similar cases were also evaluated for flattening the curve of an existing alignment: Case S-E, Case R-E, and Case RT-E. The special case where the TAC and new curves pass through the same PC and PT of the horizontal curve (with different tangents) is then evaluated. The geometric design parameters for various cases are shown in Table 2.

5.1 *Curves have the same tangents: New alignment*

5.1.1 *Case S-N: Using flatter simple horizontal curve for a new alignment*

Considering the design parameters of this case (Table 2), the TAC guidelines recommend a minimum radius of 680 m, while the new design approach suggests a minimum radius of 810 m (Easa and Abd El Halim 2006), an increase of 130 m. There are several benefits associated with the larger radius. For horizontal curves of radii 680 m and 810 m, the degree of curvature according to Equation [1] is *8.43° and 7.07°, respectively. As a result of the increase in radius, the total length*

Table 2. Geometric design parameters for each case.

Parameter	Case S-N Case S-E	Case R-N Case R-E	Case RT-N Case RT-E
V_D	110 km/h	70 km/h	70 km/h
$R_{\min}$ *(1)*	680 m	190 m	190 m
R_{req} *(5,6)*	810 m	230 m	221 m
$e_{\max}$	0.04	0.06	0.06
Central angle	45°	45°	45°
AADT	7,000	7,000	7,000
Ratio	–	1.0	1.0
Tangent length	–	0 m	100 m

of the alignment is decreased by ΔL. Substituting for $DC_1 = 8.43^o$, $DC_2 = 7.07^o$, and $\gamma = 45^o$ into Equation [6], then $\Delta L = 0.0429$ km, or 43 m. Thus, by using the flatter radius, the total length of the alignment decreases by 43 m. This reduction in alignment length produces the following benefits:

- The cost of constructing the two horizontal curves can be estimated using the alignment length and unit construction cost ($/km). Considering that the average cost of a 9.0-m wide highway in Ontario is $720,000/km (MTO 2003), the construction of the flatter horizontal curve results in a construction saving of $30,917.
- For the collision cost saving, substituting for $\Delta L = 0.0429$ km, $DC_1 = 8.43$, $DC_2 = 7.07$, $AR_s = 0.560$, and $AADT = 7{,}000$ into Equation [9], the annual reduction in the total number of collisions ΔA_t is 0.09681. The annual saving due to collision reduction, K_t, is calculated from Equation [10]. Assuming that the average cost of a collision U is *$93,149* (in 2003 Dollars), K_t, in the first year is $9,006.
- The vehicle running cost saving is calculated by assuming that autos make up 75% of the total vehicle traffic, single unit trucks and busses make up 7%, and heavy trucks make up 18% of the total share. Then, the running cost saving per thousand vehicles in each year is calculated from Equation [11] as $8.95 per 1,000 vehicles. Then, the annual vehicle running cost saving in the first year is calculated using Equation [12] as $22,859.
- The estimated speeds for the TAC and new curves calculated using Equation [16] are 99.6 km/h and 100.4 km/h, respectively. The speed-change cost saving for heavy trucks, single-unit trucks and buses, and autos and recreational vehicles is calculated using Equations [13–15] as $12.33, $8.85, and $1.03 per 1,000 vehicles, respectively. The annual speed-change cost saving is $9,225.

The net present value of the annual operating cost saving can be calculated by multiplying the annual saving and benefits by the present-worth factor (Newman et al. 2008). Assuming a project life of 20-years, and an interest rate of 3%, the present-worth factor is 14.88. Therefore, over the life of the project, the total savings in collision cost, speed-change cost, vehicle running cost, and construction cost will be $134,005, $137,274, $340,141, and $30,917, respectively, resulting in a total saving of $642,337. The costs and benefits for other design speeds (60 km/h to 110 km/h) are presented in Table 3 (A) for $e_{max} = 0.04$.

When implementing a flatter simple horizontal curve for a new alignment, the net present worth of the benefits decreases for a design speed of 60 km/h to 80 km/h, and increases from 80 km/h to 110 km/h. A summary of all savings over the 20-year project life for all design speeds is presented in Table 3 (A) for $e_{max} = 0.04$.

5.1.2 *Case R-N: Using flatter reverse horizontal curve for a new alignment*

Considering the design parameters of this case (Table 2), the TAC guidelines recommend a minimum radius of 190 m, while the new design approach suggests a minimum radius of 234 m. Over the 20-year life of the project, $467,289 would be saved in collision reduction saving, $1,049,458 would be saved due to speed change saving, $230,249 would be saved due to reduced running

Table 3. Net present value of costs and benefits of implementing new 3D minimum radii for simple and reverse horizontal curves with intermediate tangents.

V_D (km/h)	TAC DC_1	Imp. design DC_2	Total annual collision saving, K_t ($)	Total vehicle operating saving, W ($)	Total speed change saving, Z_t ($)	Const. saving, new curve ($)	Const. cost, exisiting curve ($)	New design NPW of benefits ($)	Existing design NPW of benefits ($)
60	38.20	33.70	175,425	52,329	397,773	4,756	96,132	630,284	529,395
70	28.65	24.91	154,627	78,494	346,731	7,134	130,062	586,988	449,791

(*Continued*)

Table 3. (Continued).

V_D (km/h)	TAC DC_1	Imp. design DC_2	Total annual collision saving, K_t ($)	Total vehicle operating saving, W ($)	Total speed change saving, Z_t ($)	Const. saving, new curve ($)	Const. cost, exisiting curve ($)	New design NPW of benefits ($)	Existing design NPW of benefits ($)
80	20.46	17.90	118,596	104,658	246,308	9,513	180,955	479,077	288,608
90	15.08	13.02	113,572	156,988	202,822	14,269	248,814	487,653	224,570
100	11.69	9.88	124,495	235,482	181,685	21,404	327,982	563,067	213,681
110	8.43	7.07	134,004	340,141	137,274	30,917	458,044	642,337	153,376
40	104.17	85.52	1,363,650	62,795	2,223,643	5,708	75,775	3,655,796	3,574,312
50	63.66	52.56	826,585	99,426	1,730,242	9,037	123,276	2,665,290	2,532,977
60	44.07	36.04	618,732	151,755	1,392,051	13,794	179,825	2,176,333	1,982,714
70	30.16	24.49	467,289	230,249	1,049,458	20,928	264,648	1,767,924	1,482,349
80	22.92	18.13	432,280	345,374	915,976	31,393	357,387	1,725,022	1,336,242
90	16.85	13.17	390,337	497,130	722,850	45,186	491,973	1,655,501	1,111,835
100	13.02	10.29	350,890	612,254	545,621	55,651	629,952	1,564,415	878,813
110	9.55	7.53	356,882	842,503	408,719	76,579	860,670	1,684,682	747,433
40	104.17	97.11	515,473	20,932	799,214	1,903	66,727	1,337,522	1,268,892
50	63.66	58.47	385,924	41,864	795,454	3,805	110,835	1,227,047	1,112,406
60	44.07	39.79	327,902	73,261	734,445	6,659	162,860	1,142,268	972,748
70	30.16	25.93	346,203	162,221	780,061	14,745	249,945	1,303,230	1,038,540
80	22.92	19.69	286,993	214,551	615,587	19,501	329,113	1,136,632	788,018
90	16.85	14.32	261,221	313,976	495,153	28,539	452,389	1,098,889	617,961
100	13.02	11.04	246,695	413,402	394,706	37,576	586,975	1,092,379	467,828
110	9.55	8.02	259,502	596,555	308,111	54,224	807,514	1,218,391	356,653

costs, and $20,928 would be saved in construction costs due to a reduced alignment length. In total, assuming a 20-year project life, and an interest rate of 3%, $1,767,925 in saving would be incurred by constructing the new reverse horizontal compared with the TAC design curve.

When implementing a flatter reverse horizontal curve for a new alignment, the net present worth of the benefits decreases for a design speed of 40 km/h to 100 km/h, and increases from 100 km/h to 110 km/h. A summary of all savings over the 20-year project life for all design speeds is presented in Table 3 (B) for $e_{max} = 0.06$.

5.1.3 *Case RT-N: Using flatter reverse horizontal curve with intermediate tangent for a new alignment*

Considering the design parameters of this case (Table 2), the existing reverse curve has radii of 190 m, and will be increased to a radius of 221 m. Over the 20-year life of the project, $346,203 would be saved in collision reduction saving, $780,060 would be saved due to speed change saving, $162,221 would be saved due to reduced running costs, and $14,745 would be saved in construction costs due to a reduced alignment length. In total, assuming a 20-year project life, and an interest rate of 3%, constructing the improved reverse horizontal curve compared with the TAC curve would produce $1,303,230 in savings. The costs and benefits for other design speeds (60 km/h to 110 km/h) are presented in Table 3 (C) for $e_{max} = 0.06$.

When implementing a flatter reverse horizontal curve with an intermediate tangent for a new alignment, the net present worth of the benefits decreases for a design speed of 40 km/h to 60 km/h, increases from 60 km/h to 70 km/h, and then decreases from 70 km/h to 110 km/h. A summary of all savings over the 20-year project life for all design speeds is presented in Table 3 (C) for $e_{max} = 0.06$.

5.1.4 *Case S-E: Flattening of existing simple horizontal curve*

In this case, the existing simple curve has a radius of 680 m that will be increased to 810 m. Over the 20-year life of the project, the savings in collision reductions, speed change cost, and reduced running cost would be the same as in case S1 ($134,006, $137,274, and $340,141, respectively). The cost of increasing the radius of the existing curve calculated from Equation [7] is $458,044. Therefore, the net present worth of the project can be calculated as:

$$NPV_{total} = -458{,}044 + 134{,}006 + 137{,}274 + 340{,}141 = \$153{,}377$$

Since the net present value is positive, increasing the radius of the existing horizontal curve is cost-effective.

5.1.5 *Case R-E: Flattening of existing reverse horizontal curve*

In this case, the existing reverse curve has a radius of 190 m, and will be increased to a radius of 234 m. Over the 20-year life of the project, $467,289 would be saved in collision reduction saving, $1,049,458 would be saved due to speed change saving, and $230,249 would be saved due to reduced running costs. The cost of increasing the radius of the existing reverse curve is $264,648. The net present worth of the project is $1,482,348, and thus increasing the radius of the reverse horizontal curve is cost effective.

5.1.6 *Case RT-E: Flattening of existing reverse horizontal curve with intermediate tangent*

In this case, the existing reverse has curve radii of 190 m, and will be increased to a radius of 221 m. Over the 20-year life of the project, $346,203 would be saved in collision reduction saving, $780,061 would be saved due to speed change cost, and $162,221 would be saved due to reduced running cost. The cost of increasing the radius of the existing reverse curve is $249,945. The net present worth of the project is $1,038,540, and thus increasing the radius of the reverse horizontal curve with intermediate tangent is cost effective.

5.2 *Special scenario*

The previous sections evaluated the benefits of the new design approach given the assumption that the TAC and new curves have the same tangents. Furthermore, the new larger and TAC curves do not pass through the same PC and PT. In situations where alignment conditions constrain the geometry of the PC and PT, a similar procedure is undertaken, however a few modifications to the analysis are required. First, the deflection angle of the new and TAC curve will no longer be the same, and therefore, must be calculated. Assuming the deflection angle (β) of the TAC curve is 45°, the deflection angle (α) of the new curve can be derived from:

$$\alpha = 2\sin^{-1}\left(R_1 \sin(\beta/2)/R_2\right) \tag{18}$$

where R_1 is the radius of the TAC curve, R_2 is the radius of the new curve, and β is the deflection angle of the TAC curve. The geometry of this case is presented in Figure 2.

Consider a simple horizontal curve with the design speed V_D = 110 km/h, superelevation rate e_{max} = 0.04, and a central angle of 45°. The TAC guidelines recommend a minimum radius of 680 m, while the new design approach suggests a minimum radius of 810 m. The NPW of the benefits for constructing the new simple horizontal curve compared with the TAC curve is $231,013 assuming a 20-year project life. This amount is significantly lower than the saving amount found for Case S-N (NPV = $642,337), which assumes that the two curves do not pass through the same PC and PT.

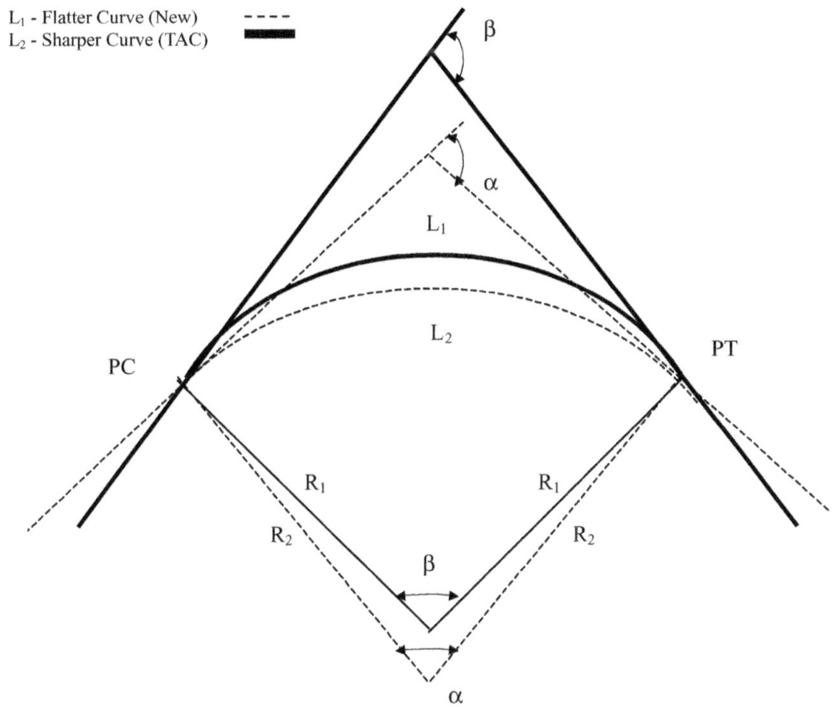

Figure 2. Geometry of curve flattening for special scenario (PC and PT are the same).

6 APPLICATION EXAMPLES

To illustrate the economic benefits of implementing the new larger radius guidelines for simple horizontal curves, consider a simple curve with a design speed of 100 km/h, and a TAC design radius of 680 m. According to the new guidelines (Easa and Abd El Halim 2006), the new larger radius is 810 m. Assume that the life of the project is 20 years, interest rate is 3%, $e_{max} = 4\%$, and assuming the typical values for the various cost saving parameters. For an alignment with a low traffic volume (AADT = 1,000 veh/day), the NPW of the savings equals-$380,000 (a negative value). This indicates that increasing the radius of the TAC curve to that of the new guidelines would involve incurring some cost. Now consider the same simple horizontal curve ($V_D = 100$ km/h, $R = 680$ m), but with traffic volume that is significantly higher (AADT = 13,000 veh/day). In this case, the NPW of the savings equals $700,000 and implementing the new guidelines would result in a safer and more cost effective highway alignment. In both cases, the flatter curve of the new guidelines satisfies vehicle stability requirements.

The same trends are observed for reverse horizontal curves. However, the savings are larger due to the larger required minimum radius of the reverse curve and because the reverse curve consists of two simple curves (i.e. twice the benefits). Regardless of the amount of saving, the new larger minimum radius guidelines for both simple and reverse horizontal curves should be implemented since they satisfy vehicle stability requirements, and produce a safer highway alignment.

7 SENSITIVITY ANALYSIS

It is important to note that the net present worth of any project highly depends on the AADT of the highway. It was found that the net benefits increased as the AADT increases. For lower

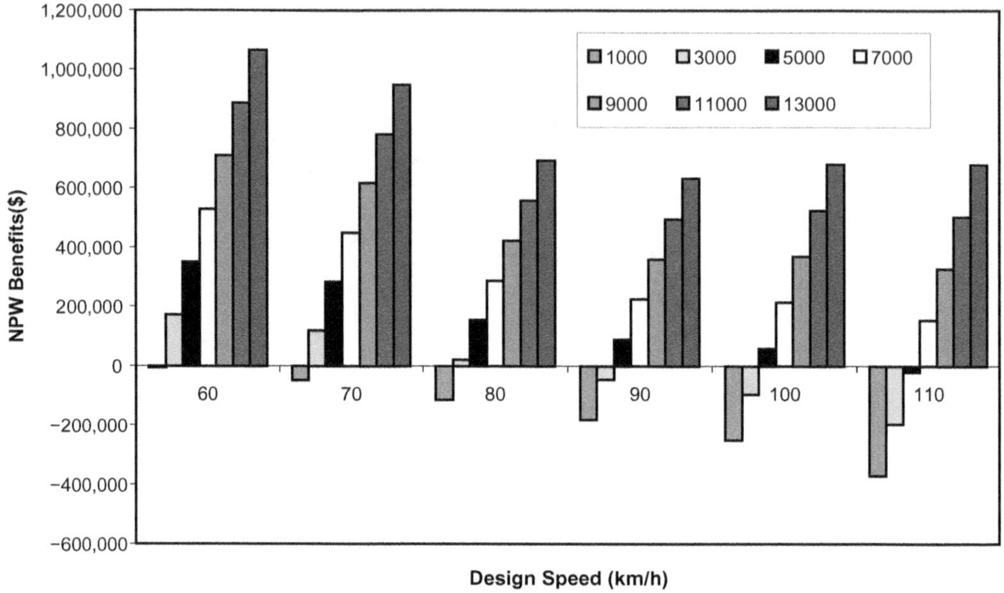

Figure 3. NPW-Benefits of increasing radius of an existing TAC horizontal curve for various design speeds and AADT.

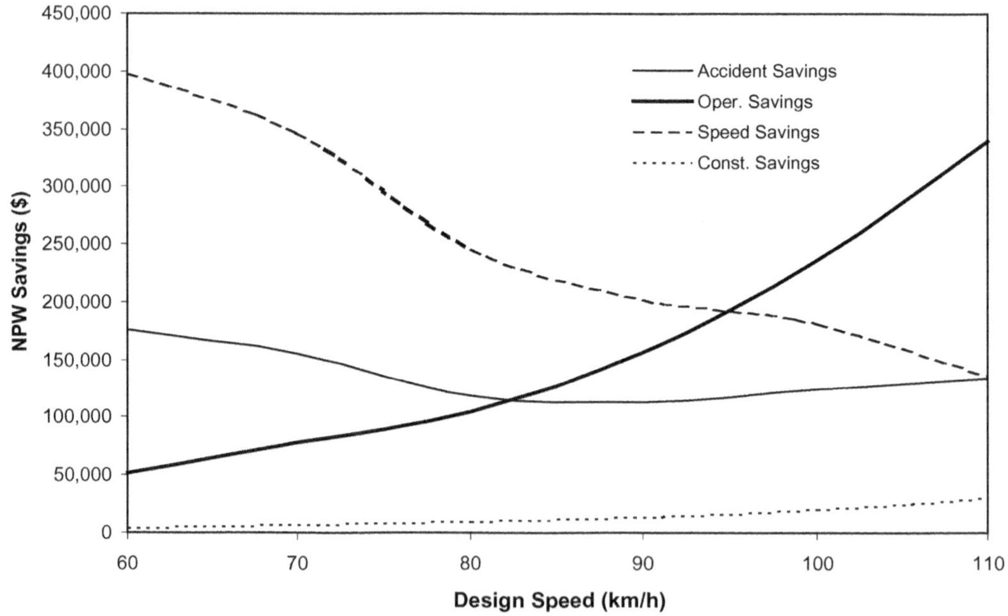

Figure 4. Variation of different savings types with design speed (e_{max} = 0.04, AADT = 7,000).

AADT values (e.g., less than 1,000), the net present worth of the project is negative; indicating that increasing the radius of an existing horizontal curve is not cost-effective (even though warranted based on vehicle stability considerations). The sensitivity of the total net present worth of the new guidelines to AADT and the unit construction cost was conducted. Since most of the cost-saving models are functions of AADT, it was observed that AADT has a significant impact on the net present worth of a project. Figure 3 presents the net present worth (for various levels of AADT) of increasing the radius of an existing TAC simple horizontal curve to the radius of the new guidelines (e_{max} = 0.04, V_D = 60 km/h to 110 km/h). It is observed that as AADT increases, the net present worth of a project increases. For AADT = 5000 or greater the NPW is positive for all design speeds. For AADT = 1000, the NPW of the project is negative for a design speeds 60 km/h or greater. For AADT = 3000, the NPW of the project is negative for a design speed of 90 km/h or greater.

The variation of all savings with design speed is presented in Figure 4 (e_{max} = 0.04, AADT = 7,000). It is observed that the operating saving increases as design speed increases, the speed change saving decreases as design speed increases, and the construction saving increases design speed increases. The accident saving exhibited a U-shaped trend in which the saving first decreases for increasing design speed (60 to 90 km/h) and then increases (90 km/h to 110 km/h). It is clear that the construction cost saving represent a relatively small portion compared with other types of saving. In addition, the operating saving is more dominant at higher design speeds, while the speed change saving is more dominant at lower design speeds.

8 CONCLUDING REMARKS

If a highway designer is looking for a solution to a sharp horizontal curve with a high collision rate, the use of a larger radius horizontal curve, or *curve flattening*, may provide a cost-effective and safe design alternative. Several studies have shown that the collision rate on sharp horizontal curves is up to 4 times higher than that on straight tangent sections. The collision rate on horizontal curves was found to increase as the degree of curvature increases. Implementing the new flatter horizontal curve guidelines instead of the sharper curves of current design guides produces several cost savings, such as construction cost saving (due to reduced alignment length), collision cost saving, and vehicle operating cost saving. The NPW of the project would be generally positive, except at very low levels of AADT. Based on this research, the following comments are offered:

1. The new horizontal curve design approach that considers the effects of vehicle stability and 3D highway alignment requires larger radii than those of the TAC design guide. Regardless of the unit construction cost, the new horizontal curves are more economical to construct for cases where the central angle of the new and TAC curves are the same. In addition to cost-effectiveness, the new simple and reverse horizontal curves address vehicle stability requirements and improve safety compared with existing guidelines.
2. Implementing the new guidelines for simple horizontal curves results in savings approximately in the range \$0.5M–\$0.6 M for new designs and \$0.2M–\$0.5 M for existing designs. For reverse horizontal curves, implementing the new guidelines results in savings in the range \$1.5 M–\$3.5 M for new designs and \$0.7 M–\$3.5 M for existing designs. For reverse horizontal curves with an intermediate tangents (T = 100 m), implementing the new guidelines results in savings in the range \$1.1M–\$1.3 M for new designs and \$ 0.2M–\$0.5 M for existing designs (AADT = 7,000, 20-year project life).
3. The new horizontal curve guidelines were developed using VDM RoAD simulation software. Three types of trucks (WB15, WB20, NHTSA 3a/2a) were used as the design vehicle. Using these design vehicles results in a more conservative design compared with using passenger cars. The savings associated with the guidelines based on trucks as a design vehicle are greater

since the required minimum radius of the horizontal curve is larger for trucks due to vehicle stability requirements.

4. For newly constructed highway alignments, implementing the new larger radius guidelines is always more cost effective due to the shorter alignment length. However this is not always the case for existing alignments. When an existing curve is made flatter, the cost of constructing the new curve must be considered, and a detailed cost-benefit analysis must be undertaken to examine the cost effectiveness of the project.
5. In circumstances where the TAC and new curves are forced to pass through the same PC and PT, implementing the new guidelines produces reduced savings compared with the case where the tangents and deflection angle are the same. A similar analysis procedure can be undertaken, but some slight modifications to the presented methodology would be required.
6. In addition to the benefits considered in this analysis, several other benefits such as time saving, fuel saving, and winter maintenance saving (de-icing agents) exist. However, these benefits were not included in the analysis as the savings are difficult to quantify. Also, land acquisition, the right-of-way, and possibly earthwork costs were not considered in this analysis as they are highly variable and site specific. If data for these benefits and costs are available for specific cases, they can be easily incorporated in the presented evaluation methodology.

ACKNOWLEDGEMENTS

The study is supported by a Discovery Grant from the Natural Sciences and Engineering Research Council of Canada.

REFERENCES

Abd El Halim, Amir, A. and Easa, S.M. 2004. Economic Benefits of Implementing New Minimum Radius Guidelines for Simple Horizontal Curves. CD-Rom *Proc, 5th CSCE Transportation Specialty Conference*, Saskatoon, Saskatchewan.

American Association of State Highway and Transportation Officials. 2001. *A Policy on Geometric Design of Highways and Streets. AASHTO*, Washington, D.C.

Easa, S.M. and Abd El Halim, Amir 2006. Radius Requirements for Trucks on Three-dimensional Reverse Horizontal Curves with Intermediate Tangents. *J. Transportation Research Record* 1961, TRB, National Research Council, Washington D.C., 83–93.

Easa, S.M. and Dabbour, E. 2003. Design Radius Requirements for Simple Horizontal Curve on 3D Alignments. *Canadian Journal of Civil Engineering*, 30(6): 1022–1033.

Hassan, Y., Easa, S.M. and Abd El Halim, A.O. 1998. Highway Alignment: Three Dimensional Problem and Three Dimensional Solution. *Transportation Research Record 1616*, TRB, National Research Council, Washington D.C., 17–25.

Howery, K.E. 1991. *Benefit Cost Analysis*. Report prepared for Alberta Transportation and Utilities, Edmonton, Alberta.

Lamm, R., Psarianos, B. and Mailaender, T. 1999. *Highway Design and Traffic Safety Engineering Handbook*. McGraw-Hill Companies, Inc., New York, N.Y.

Newnan, D.G., Eschenbach, T.G. and Lavelle, J.P. 2008. Engineering economic analysis. Oxford University Press, New York.

Sayers, M.W. 1999. Vehicle *Dynamics Programs For Roadway and Roadside Studies*. University of Michigan Transportation Research Institute, UMTRI-98-20-1, Michigan.

Talarico, R.J. and Morrall, J.F. 1993. The Cost-Effectiveness of Curve Flattening in Alberta. *Canadian Journal of Civil Engineering*, 21: 285–296.

Transportation Association of Canada. 1999. *Geometric Design Guide for Canadian Roads*, TAC, Ottawa, Ontario.

Wooldridge, M.D., Fitzpatrick, K., Koppa, R. and Bauer, K. 2000. Effects of Horizontal Curvature on Driver Visual Demand. *Transportation Research Record 1737*, TRB, Washington D.C., 71–77.

Efficient Transportation and Pavement Systems – Al-Qadi, Sayed, Alnuaimi & Masad (eds)
© 2009 Taylor & Francis Group, London, ISBN 978-0-415-48979-9

The effect of geometric layout on the operational performance of multilane roundabout

Z. Khatib
Department of Civil and Environmental Engineering, University of Sharjah, UAE
Currently Engineering Consultants, Dubai, UAE

G. Abu-Lebdeh
Department if Civil Engineering, American University of Sharjah, UAE

ABSTRACT: Roundabouts have distinctive operational and safety features that make them the choice for traffic control when multiple roadways meet and form a junction. Especially when volumes are moderate and distribution of traffic among approaches is about even, roundabouts can function efficiently and safely. When demand is high and/or not evenly distributed among entry approaches, excessive delay and aggressive driving behavior may emerge thus compromising both operational characteristics of roundabouts. This study focuses on the impact of the "shape" of the roundabout (i.e., circular versus rectangular with half circle ends, henceforth called semi-circular). Empirical data was used in this research to evaluate the impact of the geometry of the roundabout on traffic operations and to draw conclusions that may help to design different shapes such that the impact of the negatives are lessened. The data collected revealed that the shape of the roundabout can have an impact on traffic operations and, by implication; the shape of a roundabout might be used to bring about certain desirable behavior and/or mitigate negative aspects of their operations.

1 INTRODUCTION AND STUDY MOTIVATION

Roundabouts are a common form of traffic control in many urban areas; other areas use them as a mean to calm traffic. There are known to bring about certain desirable operational and safety returns. However, roundabouts can also be less than desirable if certain conditions such as heavy and uneven flow on the entry approaches. Others argue that roundabout are not particularly friendly to pedestrians and can be confusing to elderly and unfamiliar drivers.

The influence of the geometry of roundabouts, specifically radius, on traffic operations and safety has been studied before. The impact of the shape of a roundabout (circular versus other) is less understood. Intuition suggests that the shape of a roundabout, just any other geometric feature of a roadway, may have influence on the behavior of motorists using the roundabout. If such is the case, one can then argue that agencies should adopt designs such that certain desirable aspects of the roundabout operations are reinforced, while other undesirable aspects are dampened or eliminated. The work described in this paper follows in this spirit: it is aims at identifying the impact of the shape of a roundabout on traffic operations for the purpose of then using different shape roundabouts to bring about certain desired operational and safety benefits.

2 FILED DATA AND SETUP

Data on volumes and speeds was collected at two roundabouts in the city of Sharjah. One is circular and the other one is rectangle with half circles at both ends (called semi-circle in this paper).

Data was aggregated for every 15-minute interval; therefore each of the data points represents an average of the respective values recorded in the 15-minute interval. Land use and the intensity of land development surrounding the sites were practically similar. Geometric dimension such as lane widths and posted speed limits were similar for both sites.

Data on speed and volumes was collected using specialize traffic counters. The counters were placed on all circulating and entry lanes as shown in Figures 1 and 2.

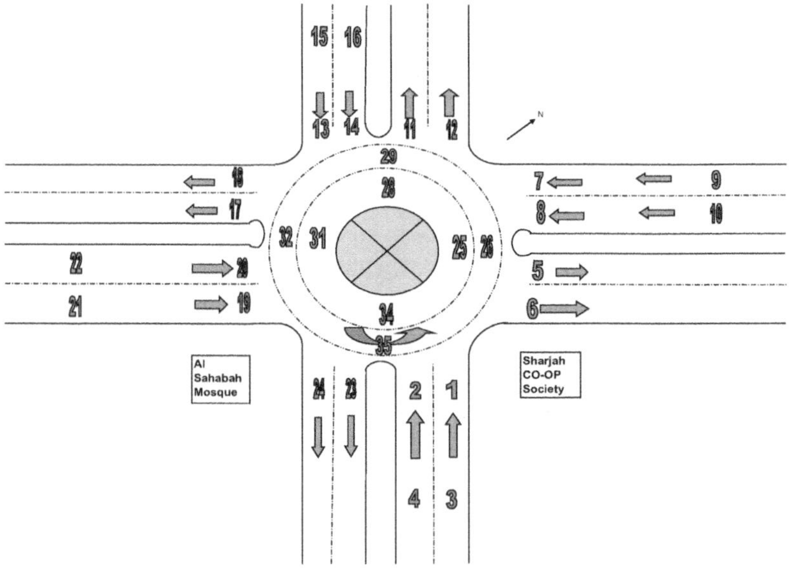

Figure 1. Circular roundabout.

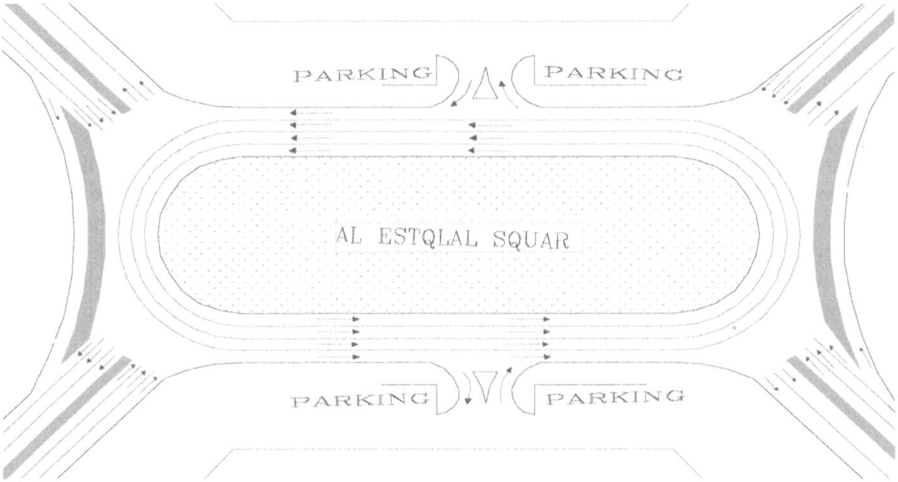

Figure 2. Semi-circular roundabout.

The circular roundabout has regular shape with diameter of 58.5 m (192 ft). This roundabout serves four approaches; each has two entry lanes with standard lane width of 3.65 m (12 ft). There are four free right turn lanes one on each corner.

The irregular (non circular) shape, it has an oval shape, almost a rectangular shape. The middle rectangular island has a length of 180 m (590 ft) and a width of 50 m (164 ft). This roundabout serves four approaches; each approach has three entry lanes, but only two approaching lanes. There are free right turn lanes at the short edge. However, inside the roundabout, almost at the middle of the longitudinal edge, there is an entry/exit access point. See Figure 2. Traffic data were collected for consecutive 15 hours starting at 2:15 pm. Data were recorded every 15-minute increment.

3 ANALYSIS AND RESULTS

Speed of both traffic entering the roundabout and those circulating were both assessed vis-à-vis traffic with which both would conflict with.

3.1 *Speed of entry traffic vs. circulating volume*

In this case speeds of entering traffic, both in the curb and median lane, were compared for different volumes of circulating traffic. The same comparison was done for the circular and semi-circular roundabout sites.

Figure 3 shows the trend in speeds of entering traffic versus circulating traffic for the circular roundabout. The speed of entry lane volumes is initially higher when the volume of circulating traffic is lower, which is expected. But the speed of the same entry traffic got higher again for the higher range of circulating traffic. This results may initially appears odd but it is explainable: as the volume of circulating traffic increases and the number of acceptable gaps decrease, drivers become more aggressive and accept shorter gaps but then make up for the shortness of gaps by increasing their exit speed. The trend of initial high-lower-then high again observed for the curb entry lane traffic is less clear for the median lane (Figure 4). For both curb and median lane, the response of the entering drivers are different to the very high circulating volume: the curb-lane drivers compensated for the lack of sufficient safe gaps by departing at increase speed; those in the median lane could not and hence showed an overall decline in average speed as the circulation volumes became high. These trends and observations apply equally to all sites, which is a clear demonstration of their consistency and significance.

When the trends for curb and median-lane traffic of the semi-circular are compared to those of the circular roundabout, there is much less variation, actually remarkable uniformity in the speeds of entering traffic of the semi-circular traffic (Figures 5 and 6); the speeds were around 20 mph

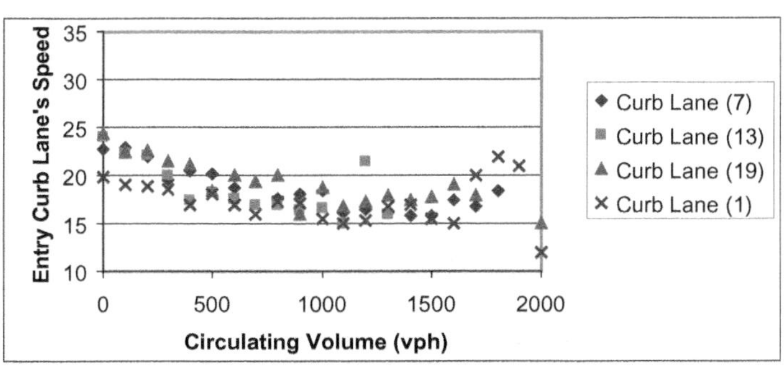

Figure 3. Total circulating volume *vs.* curb entry lane's speed.

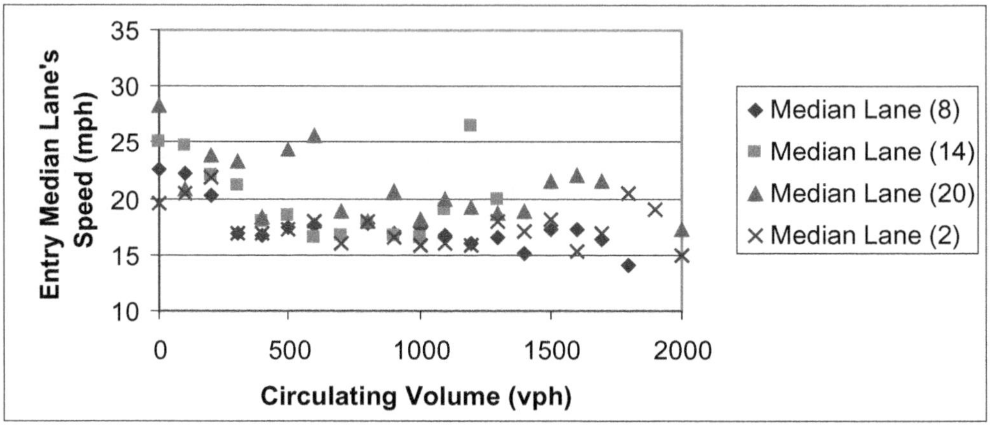

Figure 4. Total circulating volume *vs.* entry median lane's speed.

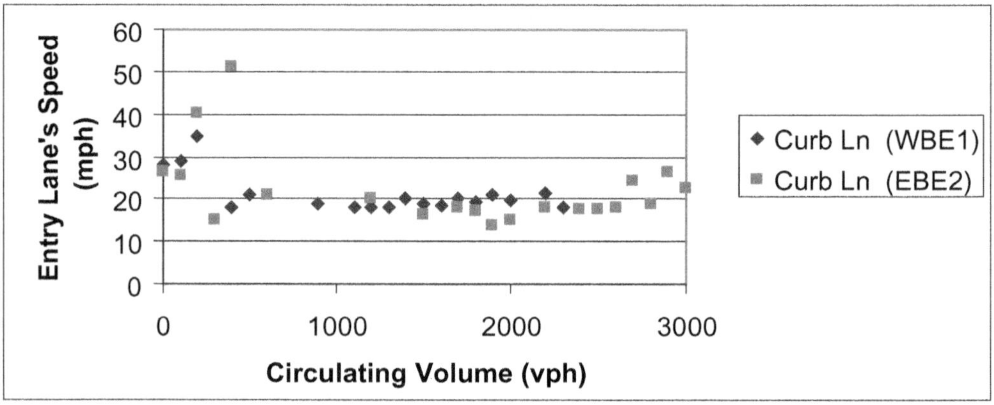

Figure 5. Total circulating volume *vs.* entry curb lane's speed.

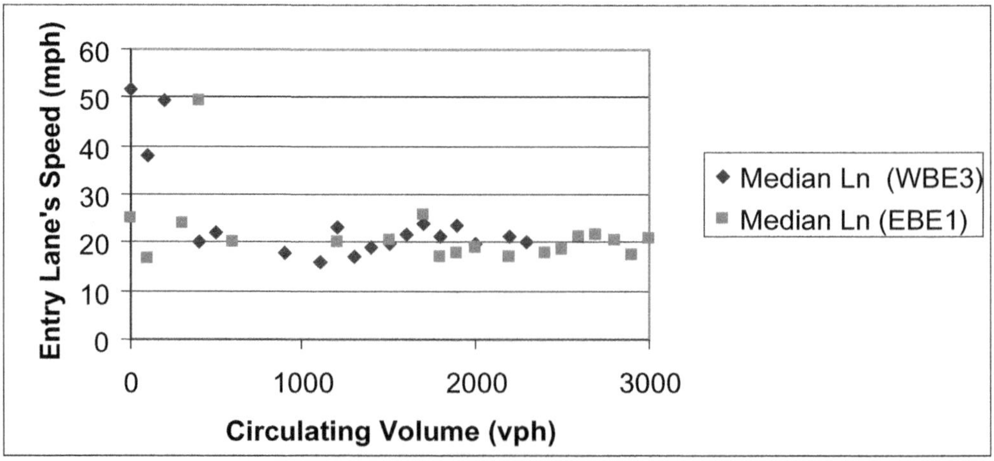

Figure 6. Total circulating volume *vs.* entry median lane's speed.

for the semi-circular traffic but ranged from 25 mph to 15 mph and then back to over 20 mph, respectively, for lower, med, and high range volumes of circulating traffic. These results suggest that the shape of the roundabout can potentially influence the quality of traffic operations and, by implication, its safety. This observation has significant practical implication.

It is known that if certain approaches of a roundabout are dominant (because of heavier flow rate) then it is likely to unduly restrict entry of traffic from other less dominant approaches. Based on the above observations one can see that the shape of the roundabout can be designed to "temper" the influence of dominant approach, or approaches, to achieve more equitable opportunities for other traffic to enter the roundabout and complete its movement without undue extra delay.

3.2 *Speed of circulation traffic vs. entry volume*

In this round of analyses the influence of entry volume or flow rate on the operating speed of circulating traffic is examined. This was done for both shapes of roundabouts, first, the circular roundabout.

Figure 7 shows that the speed of circulating inside-lane traffic did not change much for the range of volumes entering the roundabout, where it remained around 23 mph. The same was true for the speeds of traffic in the outside lane (Figure 8).

For the semi-circular roundabout (Figures 9 and 10) and except for the cases of very low entering volume, speeds of circulating traffic of both the inside and outside lanes did not change much with the volume of entering traffic, which is consistent with expectations. In that sense, this observation is similar to the circular traffic. But unlike the circular roundabout case, the speeds of the circulating traffic were significantly higher for low entry volumes. This can be explained by the fact that in this case the absence of heavy entering volume, circulating traffic has much milder curvature (higher radius curve) to negotiate and hence it is more conducive to higher speed; drivers simply reacted to favorable geometric conditions and operated at a higher speed that they felt safe enough for the geometry of the road. This in not necessarily negative except if one is to think of pedestrians wanting to cross. This highlights an important practical point.

The geometry and shape of the roundabout can be designed to effect the desired combination of conditions such that both circulating and entering traffic and other users of the system (bicyclists and pedestrians) can use the roundabout safety. Perhaps an extension of this thought is that the geometry and shape of the roundabout may need to become a function of the surrounding land

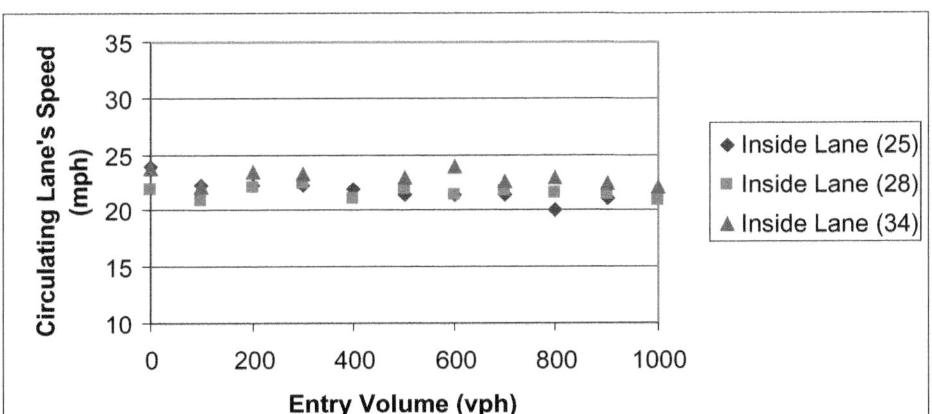

Figure 7. Total entry volume *vs.* circulating inside lane's speed.

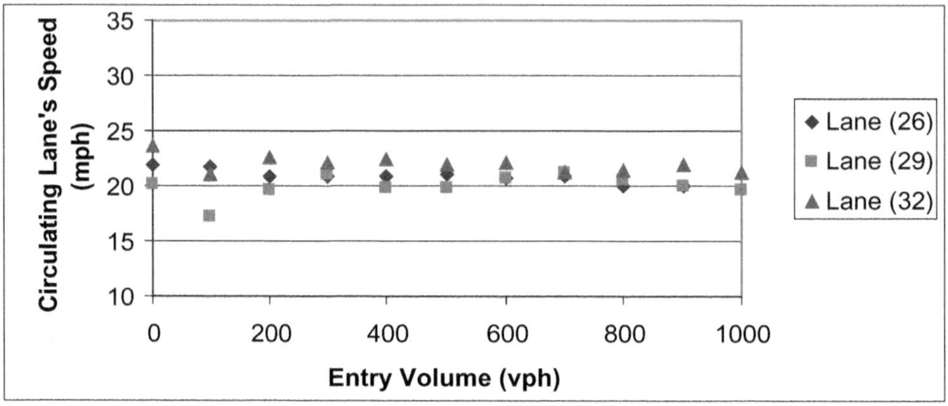

Figure 8. Total entry volume *vs.* circulating outside lane's speed.

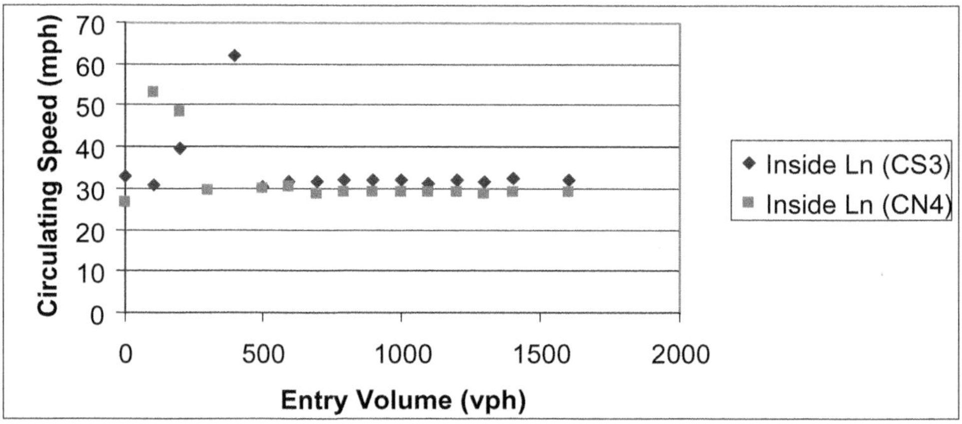

Figure 9. Total entering volume *vs.* circulating lane's speed.

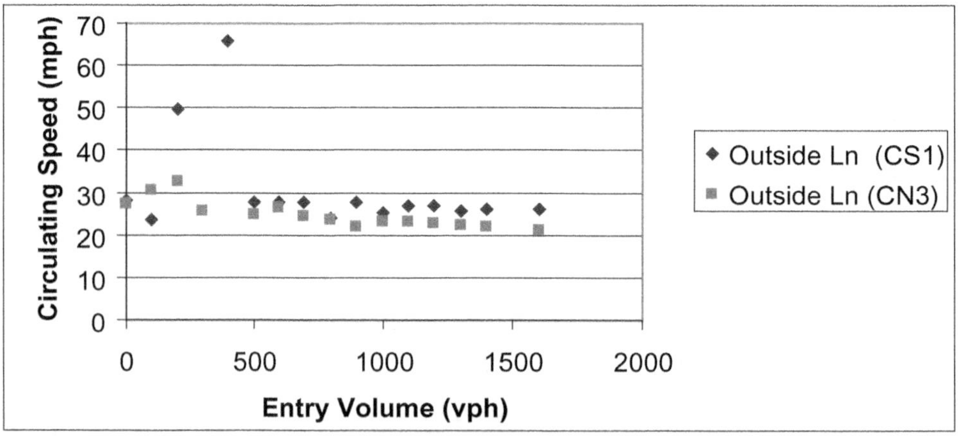

Figure 10. Total entering volume *vs.* circulating lane's speed.

use: in rural areas with low and uniform volumes circular-shape roundabouts might be a good choice whereas in urban areas, were non-vehicular traffic may, and where volumes are higher and more variable (less uniform) among the approaches, non-circular roundabout designs might be use to mitigate some of the undesirable attributes such as the less-friendliness to non motorized modes, and the domination of uneven entry volumes that may cause excessive delay and long queues at approaches to the same roundabout.

The results presented above reveal some interesting findings that can have practical uses. Other sites and more data will have to be evaluated. Surround land use and factors that could influence traffic operations were similar for the two sites used in this study. Other factors were controlled for by categorizing the analysis appropriately. In future studies there are a need to continue to account for factors that could be at play but not quite related to the shape of the roundabout.

4 CONCLUSIONS

This paper has shown that the geometric shape of roundabout has an impact on its operational performance. For instance, as the volume of circulating traffic increases and the number of acceptable gaps decrease, drivers become more aggressive and accept shorter gaps but then make up for the shortness of gaps by increasing their exit speed. Drivers behave differently based on which entry lane they occupy, which leads to the importance of analyzing lane-by-lane entry flows not aggregated as one approach. These results suggest that the shape of the roundabout can potentially influence the quality of traffic operations and, by implication, its safety.

REFERENCES

1. "Roundabouts: An Informational Guide" US Department of Transportation, Federal Highway Administration. Publication No. FHWA-RD-00-067. June 2000.
2. "Guide to Traffic Engineering Practice Part 6—Roundabouts" AUSTROADS 1993.
3. "Estimating negotiation radius, distance and speed for vehicles using roundabouts" Rahmi Akcelik. 24th Conference of Australian Institute of Transport Research. 2002.
4. "NCHRP 572 Roundabouts in the United States" Transportation Research Board, Washington DC 2007.

Efficient Transportation and Pavement Systems – Al-Qadi, Sayed, Alnuaimi & Masad (eds)
© 2009 Taylor & Francis Group, London, ISBN 978-0-415-48979-9

Development of acceptance measures for long term performance of BOT highway projects

Hozayen A. Hozayen
Civil Engineering, Public Works Department, Cairo University, Cairo, Egypt

Fahad Alrukaibi
Civil Engineering Department, Kuwait University, Kuwait
Director of the Transportation and Traffic Safety Center—Kuwait University, Kuwait

ABSTRACT: This paper presents an approach for the development of acceptance measures for satisfactory long term performance of Built-Operate-Transfer (BOT) highway projects. In this development, pavement evaluation measures are used to accurately assess the construction quality and to identify areas with high potential for future problems. Both highway agencies and the private sector of paving industry will benefit from implementing these criteria. To achieve this goal, performance measurements were taken on a rural highway to provide information concerning surface roughness and pavement distresses (raveling, rutting, and cracking) overtime of the surfacing. This highway (total, 572 km) was divided during construction into three different sections (contractors). The length of the road sections 1, 2, 3 used in the analysis are 390, 139, and 43 km respectively. Each section was achieved by a different contractor [contractor (1) for section (1), contractor (2) for section (2), and contractor (3) for section (3)]. The three sections were treated separately due to the different expected pavement performance. Results indicated that regression relationships between pavement raveling and surface roughness have been established for the three sections of the highway with a coefficient of determination (R^2) ranging from 0.946 to 0.962. As well, the relationships between pavement condition rating (PCR) and roughness or rutting or cracking may be modeled by a second-degree polynomial regression function. New performance models developed during this investigation support the impact of construction variability on the roughness index and surface distress of asphalt pavements.

Keywords: Build-Operate-Transfer (BOT), Pavement Condition Rating (PCR), Road Surface Tester (RST), Pavement Serviceability Rating (PSR), International Roughness Index (IRI)

1 INTRODUCTION

1.1 *Background*

A major component of existing infrastructure is road which in fact provides merits for developing and enhancing other infrastructure components such as sewer, telephone and electricity cables, etc. It is essential, therefore, to maintain and expand the existing road network. Required capabilities to meet this challenge are not available to the public sector alone. The private sector of the paving industry should share such responsibility.

Pavement industry includes contractors, producers, suppliers, public agencies, consultants, equipment manufactures and researchers. In order to be competitive, the industry must address a number of environmental, health, social, public policy, technical and economic issues. The private paving sector should also develop and implement new technologies, employ skilled and trained personnel in order to achieve better quality.

1.2 *Scope of paper*

Infrastructure in many parts of the world is deteriorating due to excessive demands, lack of maintenance and rehabilitation, as well as expenditure cuts. Because of increased pressure for sustainable and cost-effective infrastructure services, governments are searching for alternative methods for managing and financing these essential services by adopting various Public-Private-Partnership privatization schemes [5].

In order for the private paving industry to take a role in construction and operating transport facilities, the relationship between public sector and government agencies needs to be formulated and defined. (1) Clear definitions and reliability for the responsibilities expected from the private sector should be made; (2) What kind of performance measures should the owner (probably, the public sector) implement to ensure acceptable riding quality that meets the public justification and its well-being? On the other hand, contractors will be faced by long term responsibilities and guarantees that they should provide for the pavements they have constructed. In other words; and (3) The contractor should convince himself by the proposed design for the facility whose future maintenance will be the contractor responsibility.

"Understanding "why" some pavements perform better than others is a key to building and maintaining a cost-effective highway system. Pavement distress is an important indicator of pavement performance. Evaluation of all the pavement data items is of key concern, particularly that related to the road users.

The major purpose of the performance related pavement evaluation is to determine the current condition of the pavement structure. Four key measures can be used to characterize or define the condition of the pavement over a life cycle period [3]:

1. Roughness (as related to serviceability or ride comfort)
2. Surface distress
3. Deflection (as related to structural adequacy)
4. Surface friction (as related to safety).

Theses measures, of course, have to be carefully and objectively defined, and capable of being quantified with a reasonable degree of accuracy, repeatability and consistency [11].

1.3 *Objectives of the paper*

A reliable design alternative requires accurate and reliable input data which affect to a large extent both initial and maintenance costs of the road facility. This paper will focus on developing means to incorporate input data variability on road design and predicted performance on which criteria for in-service related specifications can be established. More specifically, the objectives of the paper include the following:

1.3.1. Identification of future privatization prospects available to the paving industry in relation to public and other governmental agencies,

1.3.2. Outline trends and new developments in determining data and variables required for the pavement design process,

1.3.3. Assess sources of variability that result in uncertainty in the decision making process for construction and maintenance of asphalt pavements, and

1.3.4. Develop regression relationships between aggregate reveling, pavement roughness, rutting, cracking, and pavement condition rating based on logical and statistical criteria.

2 LONG TERM PERFORMANCE GURANTEED PAVEMENTS

The control of construction is a vital part of construction management and of the management system as a whole. Construction quality can be controlled in many ways, all of which involve plans, specifications and certain measurement techniques for ensuring that the final pavement is constructed according to the design that previously been selected and described in the plans and specifications. There are three types of specifications [9]:

2.1 *Traditional specifications*

In this concept, the highway agency specifies the exact materials and procedures for the contractor to follow. These specifications typically include material proportioning and mixing limits, and the proper procedures to follow for a job to be acceptable. Variability in material properties and construction techniques is generally ignored. As long as the contractor adheres to the prescribed methods, full pay can be expected. However, quality of compliance to specifications and quality of performance are two different things and are not always directly related to each other [13]. This approach to construction specification development is predicted on the assumptions that the owner, or the owner's agent, fully understands the relationships between the construction process and quality of the product, and the primary repository of the technical knowledge needed to link the two.

2.2 *Statistical end-result specifications*

End-result specifications are ones in which the contractor and the contractor's suppliers are responsible for quality control; the purchasing agency is responsible for describing the level of quality desired in the end product and the procedures that will be used to judge quality and acceptance, and for determining acceptability through a program of sampling, testing, and decisions based on statistical principles. The application of statistical methods provide a basis for dealing with the problem of materials and construction variability as well as a defensible technique for assessing specification compliance in a manner that optimized the risk to both the agency and the contractor. Thus, while end-result specifications may guarantee improved compliance and improved evidence of compliance, in themselves they do not guarantee improved performance, which depends on a better understanding of the relationship between the factors controlled during construction and the performance and worth of the finished product.

2.3 *Performance-related specifications*

Pavement performance is a function of its relative ability to serve traffic over a period of time. Performance-related specifications in the highway industry are specifications for material and elements of construction that have been demonstrated to correlate significantly with performance of finished work. They are based on quantified relationships between attributes measured at the time of construction (through either process control or acceptance testing) and performance. They include sampling acceptance (or rejection) criteria. They may also include payment schedules, with both positive and negative price adjustments, that are directly related through quantitative models to changes in worth of the finished work, anticipated as a result of departure from the quality level defined as acceptable.

The development of performance-related specifications for hot-mix asphalt concrete is a realistic and implemental goal. But that before such a specification can be used a replacement for current end-results specifications, additional or refined pavement performance prediction models must be developed [1].

3 RESEARCH PROGRAM

Performance measurements were taken on a rural highway to provide information concerning surface roughness and pavement distresses (raveling, rutting, and cracking) overtime of the surfacing. The highway (total, 572 km) was divided into three different sections. The length of the road sections 1, 2, 3 used in the analysis are 390, 139, and 43 km respectively. Each section was achieved by a different contractor. The three sections were treated separately due to the different expected pavement performance.

3.1 *Data collection*

The Laser Road Surface Tester (RST), travels in an annually basis on the entire rural network of the Kingdom of Saudi Arabia. Each highway section is defined and recognized by the RST as segments (1 Km long) which is further broken into 50 m portions (1 km involves 20 portions) to be surveyed by the RST. The rating given for the segment represents the average pavement condition of the 20 readings in one kilometer. The RST vehicle travels in such a manner that the wheels of the vehicle follow the same wheel path as the normal traffic.

Surveyed distresses include pavement raveling, roughness, pavement condition rating (PCR), rutting, and cracking. Theses distresses are transferred from the mounted devices (see Table 1) on the RST to an ORACLE database in the vex environment of the Ministry of Communications [8]. Pavement condition rating can then be calculated based on deduct values set for each distress type.

Data collected for each 1 km of road segment is presented in a friendly-interface ORACLE files. Each output includes the road number and class, surveying direction (direction of travel), surveyed lane (usually the right lane), pavement condition rating (PCR), roughness (mm/m), rutting (mm), cracking index (0–35), skid resistance (friction coefficient), and raveling as percent of coarse aggregate loss. Pavement condition ratings are based on a Pavement Serviceability Rating (PSR) system. The PSR scale runs from 0.0 to 5.0, with 5.0 being the best. Factors considered in determining the PSR for a given section of roadway are ride quality in terms of the International Roughness Index (IRI), average rut depth and age of the surface course. IRI values and average rut depth are taken directly from profiler data.

3.2 *Data analysis*

Pavement roughness has long been recognized as a primary indicator of pavement performance. Also, surface distress which can be broadly classified into three groups; fracture, distortion, and disintegration is "Any indication of poor or unfavorable pavement performance or signs of impending failure; any unsatisfactory performance of a pavement short of failure". Thus, surface distress will be somewhat related to roughness (the more cracks, distortion and disintegration—the rougher the pavement will be) as well as structural integrity (surface distress can be a sign of impending or current structural problems).

Table 1. Summary of devices mounted in the Road Surface Tester (RST) vehicle [10].

No.	Description	Function
1	Eleven laser sensors mounted on the front beam (2.5 m long)	Measure the distance from the support beam to the surface of road
2	Two accelerometers	Measure disturbances in the longitudinal profile caused by the suspension of the vehicle
3	Two inclinometers	Measure the slope of the vehicle in relation to the horizon
4	Pulse transducer mounted on the right front wheel	Measure the distance traveled by the vehicle
5	Key pad consisting of (5) five keys, four of which operate each time they are pressed. The 5th key remains in the pushed down mode as being pressed once	Manually register the data, such as: existence of edge cracking, bleeding, etc.

3.3 *Observation domain*

Observations on pavement raveling, roughness, pavement condition rating (PCR), rutting, and cracking have been carried out for the three sections [section (1) for contractor (1), section (2) for contractor (2), and section (3) for contactor (3)] of the highway using the Road Surface Tester (RST) vehicle. A summary of observed data is presented in Table 2.

Collected data for section (3) have small sample size as the road length is less. It could be noted that the rut depth for section (3) is smaller than those of sections (1) and (3) as shown in Table 2(c). Also, section (3) experienced relatively low raveling percent compared to other sections as shown in Table 2(a).

Table 2. Summary of observed data statistics.

(a) Raveling-roughness relationships

Highway section	Variable	No. of observations	Average	Max.	Min.	Standard deviation
1	Raveling	461	6.883557	25	0.00	6.060213
	Roughness		2.180477	7.49	1.01	0.870675
2	Raveling	386	6.214275	23.44	0.00	5.788414
	Roughness		2.153212	10.05	1.02	0.907912
3	Raveling	118	6.484831	22.45	0.00	5.95163
	Roughness		2.178305	5.33	1.14	0.847309

(b) PCR-roughness relationships

Highway section	Variable	No. of observations	Average	Max.	Min.	Standard deviation
1	PCR	440	57.93325	89.31	18.51	20.4846
	Roughness		2.091318	4.09	1.01	0.699066
2	PCR	377	63.62698	89.93	23.08	19.88282
	Roughness		2.120186	5.28	1.02	0.746201
3	PCR	118	70.5072	89.85	41.21	13.55791
	Roughness		2.17830	5.33	1.14	0.847309

(c) Rutting– PCR relationships

Highway section	Variable	No. of observations	Average	Max.	Min.	Standard deviation
1	Rutting	447	7.060246	16.98	3.8	2.691932
	PCR		76.17678	88.84	56.13	7.921805
2	Rutting	368	6.841005	17.03	2.82	3.072679
	PCR		66.8453	87.5	23.08	17.21826
3	Rutting	219	2.326575	6.25	1.14	0.894004
	PCR		63.30132	87.5	12.13	14.96558

PCR: Pavement Condition Rating (%).

(*Continued*)

Table 2. (Continued).

(d) Cracking—PCR relationships

Highway section	Variable	No. of observations	Average	Max.	Min.	Standard deviation
1	Cracking	345	12.40293	24.98	0.05	7.6403
	PCR		63.24574	89.57	24.21	15.76043
2	Cracking	295	7.361559	25	0.03	6.377252
	PCR		70.80675	87.5	21.62	14.28229
3	Cracking	103	8.426796	25	0.03	8.6955
	PCR		66.80825	87.5	41.21	10.97754

PCR: Pavement Condition Rating (%).

Table 3. Regression models considered for the analysis.

Dependent variable	Independent variable	Relationship	Model type
Raveling	Roughness	Raveling = a + b (roughness)	Linear
		Raveling = a e $^{b\,(\text{roughness})}$	Exponential
		Raveling = a + b (roughness) + c (roughness)^2	Polynomial
PCR	Roughness	PCR = a + b (roughness)	Linear
		PCR = a e $^{b\,(\text{roughness})}$	Exponential
		PCR = a + b (roughness) + c (roughness)^2	Polynomial
PCR	Rutting	PCR = a + b (rutting)	Linear
		PCR = a e $^{b\,(\text{rutting})}$	Exponential
		PCR = a + b (rutting) + c (rutting)^2	Polynomial
PCR	Cracking	PCR = a + b (cracking)	Linear
		PCR = a e $^{b\,(\text{cracking})}$	Exponential
		PCR = a + b (cracking) + c (cracking)^2	Polynomial

PCR: Pavement Condition Rating (%).

3.4 *Types of models*

Three types of models were calibrated for the twelfth data sets of highway sections (1, 2, and 3). Each set includes observations for two directions of traffic, that is, in fact each direction has been treated as a separate pavement section. It is recognized that the pavement condition in one direction may differ from the condition on the other direction [4]. Therefore, a highway may be fitted to more than one model type depending on the direction. For illustration purpose, the model having higher R^2 (coefficient of determination) value was selected regardless of the direction considered.

Table 3 presents the dependent and independent variables considered for each model type where a, b, c, and d are constants derived from regression analysis. The research focuses on the relationship between pavement roughness and asphalt concrete raveling as well as PCR vs. roughness, PCR vs. rutting and PCR vs. cracking relationships.

4 REGRESSION ANALYSIS

The performance of pavement system can be estimated through the deterioration models developed based on the actual predictions of response parameters considered and how far the model represents the behavior of the pavement [12]. These models are utilized in the system to predict

Table 4. Summary of regression analysis results.

Raveling –IRI

Highway Section (S)	Linear Model	R^2	Exponential Model	R^2	Polynomial Model	R^2
1	$y = 6.906x - 8.0605$	0.96	–	–	$y = 8E\text{-}05x^2 + 0.0063x - 0.303$	0.96
2	$y = 7.0305x - 8.6786$	0.94	–	–	$y = 0.0002x^2 - 0.0127x + 0.854$	0.94
3	$y = 0.1618x - 3.1422$	0.86	–	–	$y = -0.2203x^2 + 7.7612x - 9.1418$	0.96

PCR-IRI

S	Linear Model	R^2	Exponential Model	R^2	Polynomial Model	R^2
1	$y = -26.215x + 112.76$	0.80	$y = 161.6e^{-0.5259x}$	0.83	$y = 4.223x^2 - 36.784x + 127.668$	0.91
2	$y = -0.1636x + 94.555$	0.80	$y = 103.03e^{-0.0029x}$	0.75	$y = 9.57x^2 - 70.772x + 165.341$	0.90
3	$y = -0.3621x + 92.051$	0.83	$y = 95.249e^{-0.0054x}$	0.81	$y = 6.235x^2 - 46.566x + 142.123$	0.85

PCR-Rutting

S	Linear Model	R^2	Exponential Model	R^2	Polynomial Model	R^2
1	$y = -0.1119x + 82.964$	0.64	$y = 87.163e^{-0.0021x}$	0.62	$y = 0.91769x^2 - 21.046x + 153.982$	0.90
2	$y = -0.1x + 85.288$	0.38	$y = 86.455e^{-0.0016x}$	0.31	$y = 0.4298x2 - 10.0516x + 115.413$	0.85
3	$y = -0.138x + 78.478$	0.34	$y = 80.94e^{-0.0026x}$	0.30	$y = 0.343x^2 - 12.771x + 95.117$	0.83

PCR-Cracking

S	Linear Model	R^2	Exponential Model	R^2	Polynomial Model	R^2
1	$y = -0.1156x + 83.246$	0.53	$y = 86.628e^{-0.002x}$	0.49	$y = -0.04804x^2 - 0.28823x + 76.7846$	0.71
2	$y = -1.5496x + 82.214$	0.47	$y = 84.106e^{-0.0269x}$	0.48	$y = -0.06563x^2 - 0.0974x + 77.478$	0.67
3	$y = -0.8385x + 73.874$	0.44	$y = 73.906e^{-0.0136x}$	0.48	$y = -0.0656x^2 - 0.8228x + 68.7723$	0.59

PCR: Pavement Condition Rating (%).
IRI : International Roughness Index.

future performance of a pavement section, to identify the rehabilitation needs, and to estimate the network conditions after the implementation of different rehabilitation activities [2].

A regression analysis was applied on the data for each section to develop section-specific performance model. Three model types mentioned above were used in order to arrive at the best fitted relationship for each highway section. Table 4 shows that each highway section (3 sections) involves one location and one direction and that the models (3 types) were applied to the four relationships (Raveling vs. IRI, PCR vs. IRI, PCR vs. rutting and PCR vs. cracking). The number of examined mathematical relationships is, therefore, 36 models formula (see Table 4). The international roughness index (IRI) is a measurement of pavement roughness that is widely accepted for evaluating pavement serviceability, especially its riding quality. Besides, pavement surface distress is a key performance measure used in asset management systems for monitoring the current condition of a pavement and predicting future criteria [6].

Fig. 1 illustrates raveling–roughness relationships for highway sections (contractors) 1, 2, and 3 respectively. It is noted that raveling is directly proportional to roughness. This means that, an increase in roughness leads to higher dynamic axle loads, which in turn can lead to a tangible acceleration in pavement distress [7]. Because raveling is a separation of the aggregate (stone) from the pavement surface, this will give it a rough texture. As shown in Fig. 1, a high degree of correlation was observed between raveling and roughness with R^2 (coefficient of determination) values ranging from 0.95 to 0.96 using a second-degree polynomial relationship. The three models for raveling-roughness relationship involve a constant regression intercept (a) ranging from −9.14 to 0.85. This means that even there is no raveling observed in the pavement surface, a roughness international index value still exists. The same conclusion was reported by Hozayen [4].

The graphs of pavement condition rating (PCR) versus pavement roughness for each section are displayed in Fig. 2. As shown, PCR is inversely proportional to roughness. Also, a high degree of correlation was observed between PCR and roughness with R2 values ranging from 0.85 to 0.91 depending on highway contractors. Also, the three models for PCR-roughness relationship involve a constant regression intercept (a) ranging from 127.67 to 165.34. This means that even at no PCR, a given roughness index should be considered.

Additional analyses were completed to investigate the relationship between PCR and rutting performance. As shown in Fig. 3, a second-degree polynomial model correlates PCR-rutting performance with a coefficient of determination (R^2) ranging from 0.84 to 0.90 depending on highway construction. Again, the three models for PCR-rutting performance relationship involve a constant regression intercept (a) ranging from 95.11 to 153.98. This means that even there is no PCR, a rutting value still exists.

The relationships between PCR and surface cracking for the different sections are illustrated in Fig. 4. The coefficients of determination (R^2) for these relationships are high (0.69–0.78) due to the fact that PCR depends on cracking distress and rutting performance. As well, the relationship between PCR and cracking distress shows a high R2 values as PCR versus rutting performance.

5 ESTABLISHING ACCEPTANCE MEASURES

The quality control for a BOT project cannot be left to the contractor where poor quality work will inevitably result in poor returns for investors as well as poorer levels of service for road users. The International Roughness Index (IRI) can be used as a criterion for accepting/rejecting pavement quality.

One of the main problems with this type of contract is to ensure that the work conducted during the concession period is "fit for purpose" This is particularly important in that the interventions are all specified based on the successful expected life or performance of the work conducted during the concession term. A possible solution is to tie pavement surface quality into an extended maintenance period with a minimum acceptable IRI trigger values based on models such as those developed in this paper.

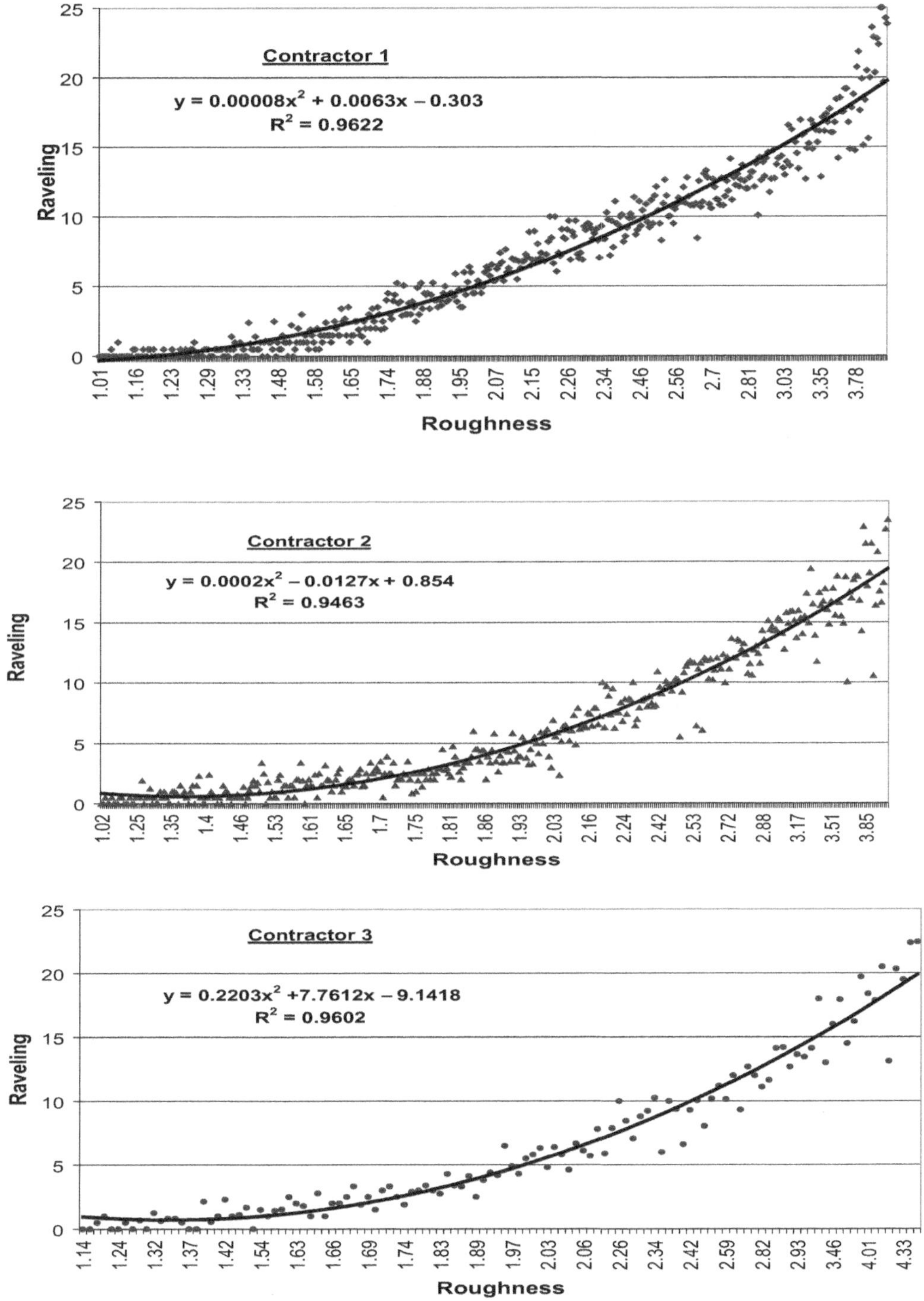

Figure 1. Raveling–Roughness polynomial relationship for contractors (1), (2) and (3).

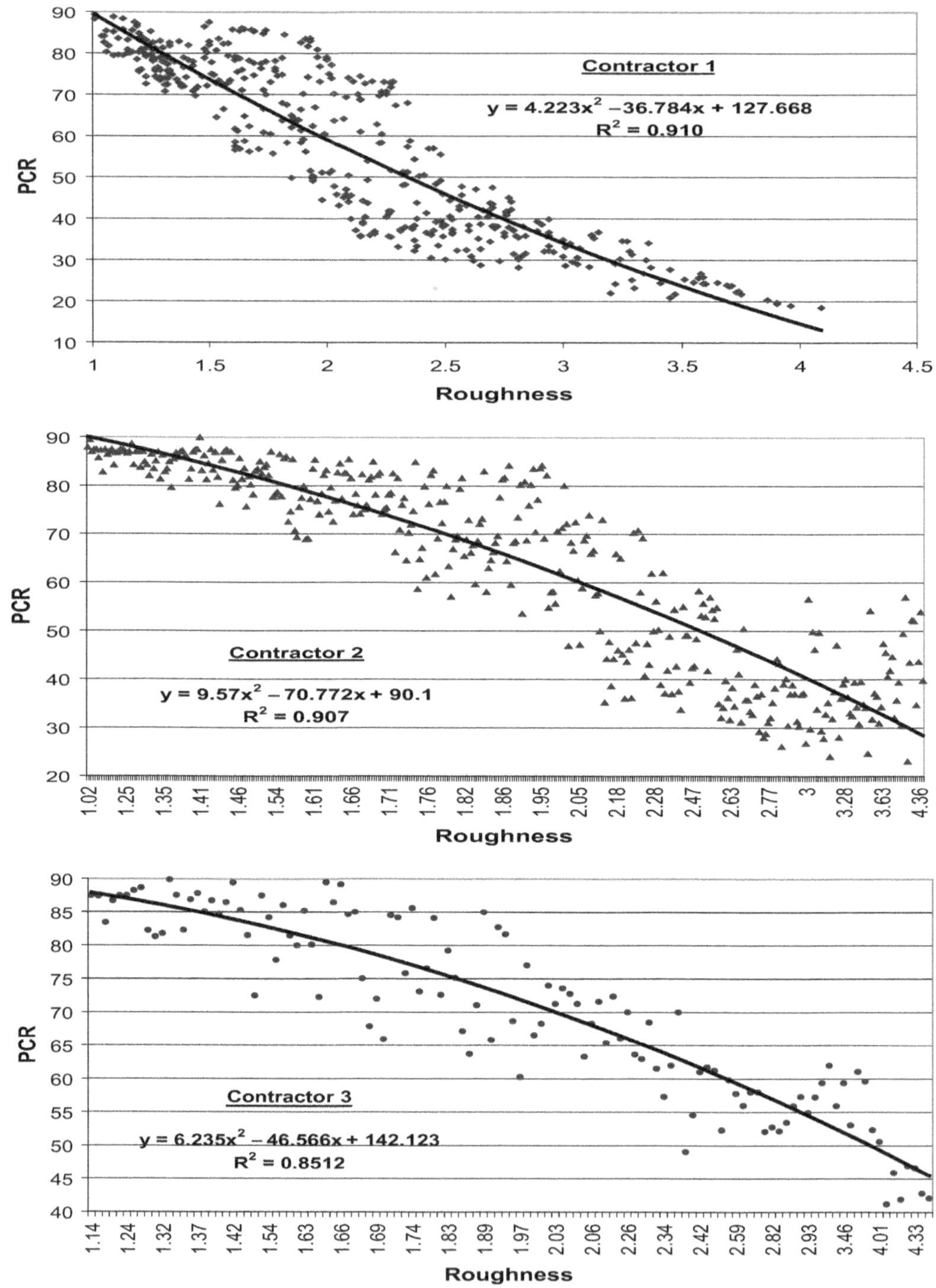

Figure 2. PCR–Roughness polynomial relationship for contractors (1), (2) and (3).

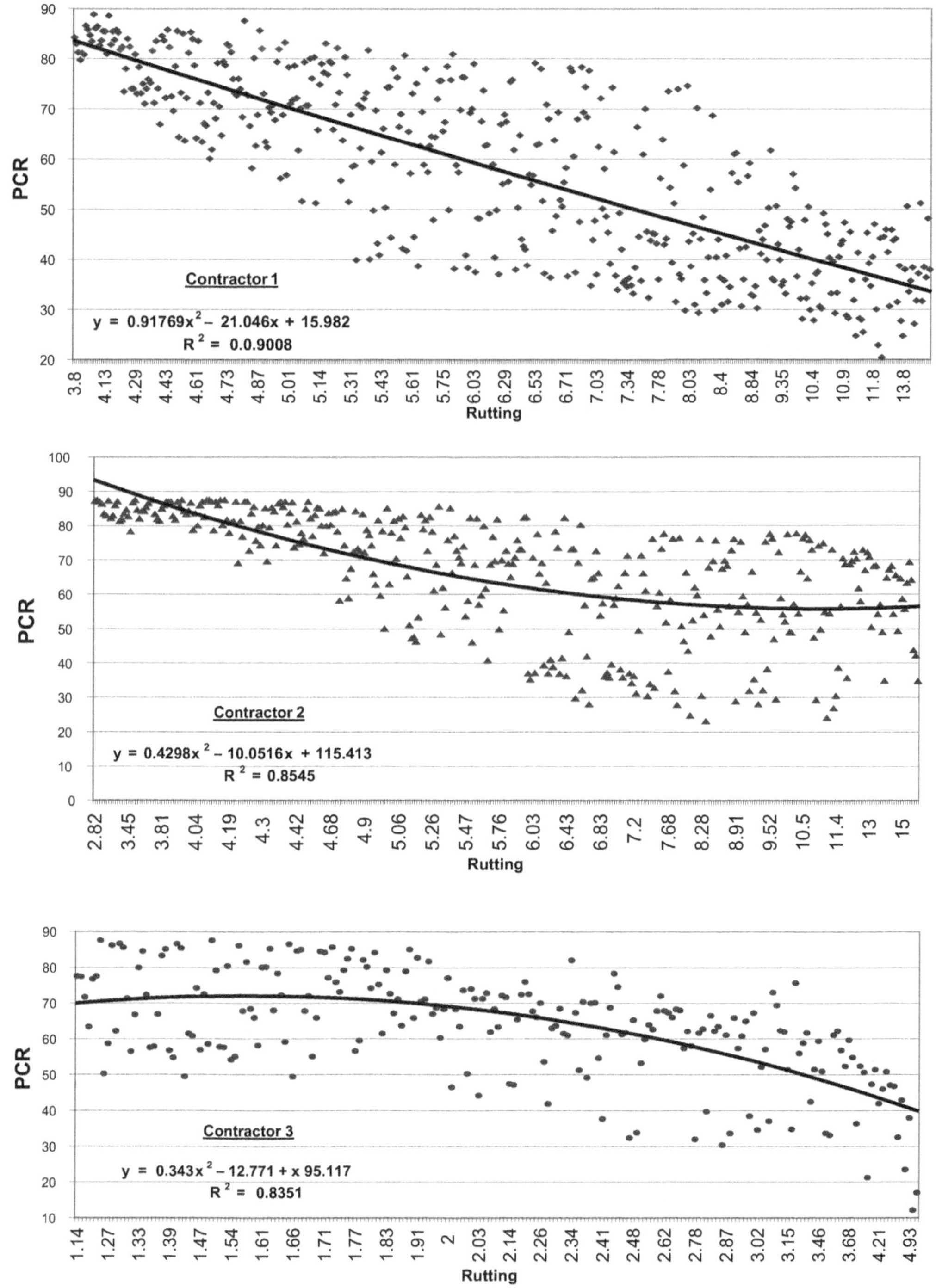

Figure 3. PCR–Rutting polynomial relationship for contractors (1), (2) and (3).

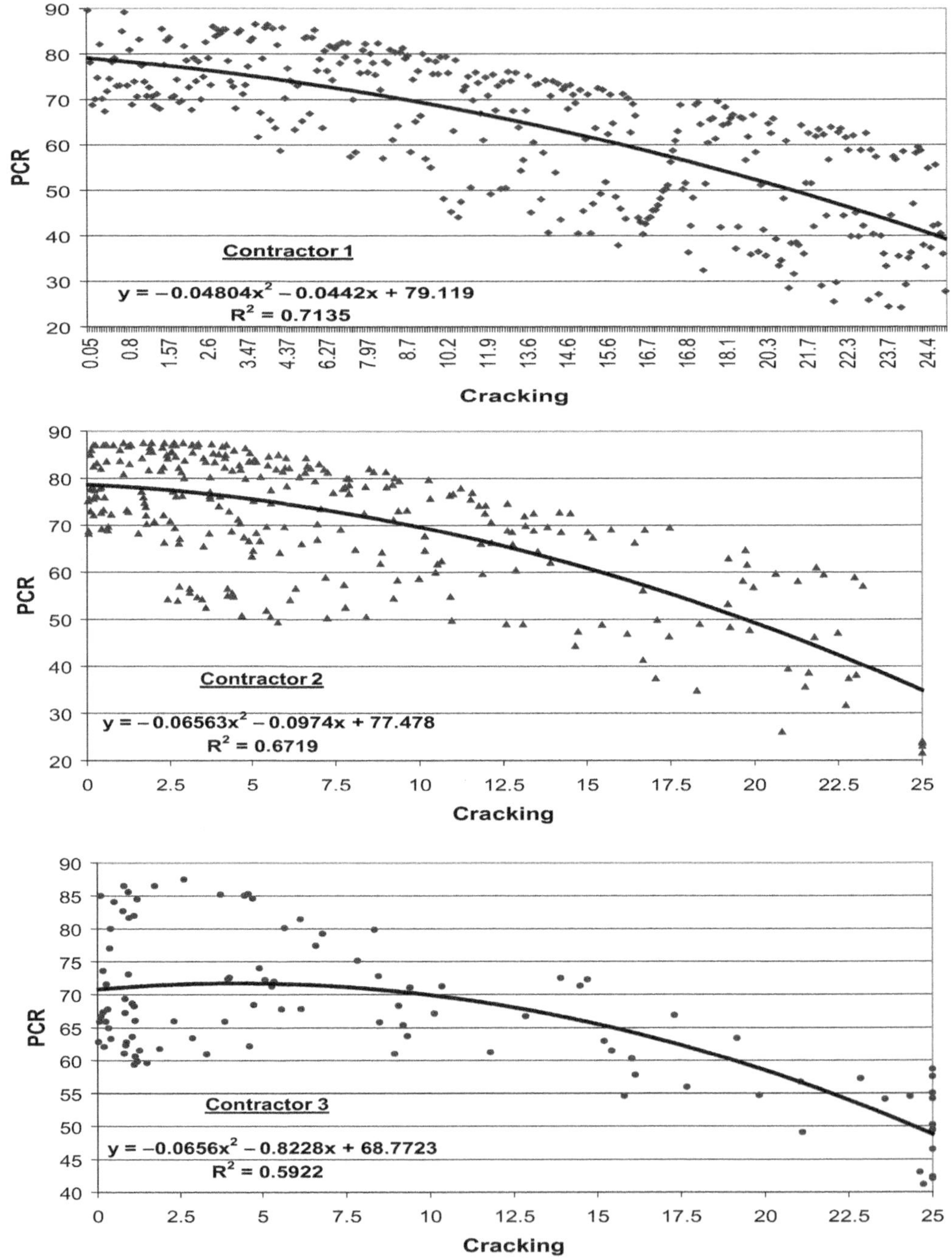

Figure 4. PCR–Cracking polynomial relationship for contractors (1), (2) and (3).

Table 5. Relationships for highway sections.

Highway section	Relationship	Model formula	R^2
1	Raveling-Roughness	Raveling = – 0.303 + 0.0063 (roughness) + 0.00008 (roughness)^2	0.9622
	PCR-Roughness	PCR = 127.668 – 36.784 (roughness) + 4.223 (roughness)^2	0.910
	PCR-Rutting	PCR = 153.982 – 21.046 (rutting) + 0.91769 (rutting)^2	0.9008
	PCR-Cracking	PCR = 76.7846 – 0.28823 (cracking) – .04804 (cracking)^2	0.7135
2	Raveling-Roughness	Raveling = 0.854 – 0.0127 (roughness) + 0.0002 (roughness)^2	0.9463
	PCR-Roughness	PCR = 165.341 – 70.772 (roughness) + 9.57 (roughness)^2	0.907
	PCR-Rutting	PCR = 115.413 – 10.0516 (rutting) + 0.4298 (rutting)^2	0.8545
	PCR-Cracking	PCR = 77.478 – 0.974 (cracking) – 0.06563 (cracking)^2	0.6719
3	Raveling-Roughness	Raveling = – 9.1418 + 7.7612 (roughness) – 0.2203(roughness)^2	0.9602
	PCR-Roughness	PCR = 142.123 – 46.566 (roughness) + 6.235 (roughness)^2	0.8512
	PCR-Rutting	PCR = 95.117 – 12.771 (rutting) + 0.343 (rutting)^2	0.8351
	PCR-Cracking	PCR = 68.7723 – 0.8228 (cracking) – 0.0656 (cracking)^2	0.5922

PCR: Pavement Condition Rating (%).

However, a trigger (IRI) roughness value often results in many arguments on the exact scope of work of the contractor. Another option is to establish a strong independent concessionaire from the start that will not be influenced by a direct relationship with the contracting bodies.

6 CONCLUSIONS

This paper presented the long-term performance framework for developing some measures of BOT highway projects decision. The analysis of a field-generated data base for roughness index and surface distress of asphalt pavements was also conducted. Based on the materials presented, the following conclusions may be drawn:

1. The international roughness index (IRI) may be used as a criterion for accepting performance of BOT highway projects.
2. Regression relationships between pavement raveling and surface roughness have been established for the three sections of the highway with a coefficient of determination (R2) ranging from 0.946 to 0.962.
3. A regression intercept of surface roughness exists suggesting that if no raveling distress observed a given roughness index value should be considered in establishing future accepting criteria.
4. New performance models developed during this investigation indicated the impact of construction variability on the roughness index and surface distress of asphalt pavements.
5. The relationships between pavement condition rating (PCR) and roughness or rutting or cracking provide an approach towards the development of acceptance criteria for long term performance measures for BOT highway projects.
6. The currently quality control specifications of constructed pavements lack the measures needed to assess the pavement performance. The performance of pavement can be easily and accurately monitored by conducting roughness index and surface distress measurements regularly and comparing the results with the as-built measurements.

ACKNOWLEDGEMENTS

The Authors would like to acknowledge the cooperation of highly officials of Ministry of Transport of Saudi Arabia for enabling data collections.

REFERENCES

1. Anderson, D.A, Luhr, D.R., and Antle, C.E., NCHP Report 332 (1990). "Framework for Development of Performance-Related Specifications for Hot-Mix Asphaltic Concrete", Transportation Research Board, National Research Council, Washington, D.C.
2. Bekheet, W., Helali, K., Abd El Halim, A., and Springer, J. (2005). "A Comprehensive Approach for the development of Performance Models for Network-Level PMS using LTPP Data", TRB 84th Annual Meeting, Transportation Research Board, Washington, D.C.
3. Haas, R., Hudson, W.R. and Zaniewski, J. (1994). "Modern Pavement Management", Krieger Publishing Company, Malabor, Florida.
4. Hozayen, A.H. (2002). "Effect of Asphalt Concrete Raveling on Pavement Roughness", Proc. 30th Annual Conference, Canadian Society for Civil Engineering (CSCE), Montreal.
5. Kassab, M., Hegazy, T., and Hipel, K.W. (2005). "Towards a Comprehensive Decisions Support System for Privatization Decisions", TRB 84th Annual Meeting, Transportation Research Board, Washington, D.C.
6. Landers, S., Falls, L.C., and Cheetham, A. (2001). "Development of a Pavement Performance Measure for the British Columbia Pavement Management System", Proc. 5th Int. Conf. on Managing Pavements, Vol. 1, Paper No. 49, Seattle, Washington.
7. Lee, D., and Chatti, A. (2001). "Development of Roughness Thresholds for Preventive Maintenance using PMS Data from In-Service Pavements", Proc. 5th Int. Conf. on Managing Pavements, Vol. 1, Paper No. 148, Seattle, Washington.
8. Ministry of Communications (2001). "Pavement Data System—System Description", Riyadh, Kingdom of Saudi Arabia.
9. NCHRP Synthesis of Highway Practice 212 (1995). "Performance-Related Specifications for Highway Construction and Rehabilitation", Transportation Research Board, National Research Council, Washington, D.C.
10. Novak, R.L. (1985). "Swedish Laser Road Surface Tester", Road Evaluation Workshop, Proceedings, Report No. FHWA-85–2100.
10. Phang, W.A. and Haas, R. (1992). "Future Prospects for a Competitive Paving Industry", Proc. 1992 TAC Annual Conference, Quebec.
12. Veeraragavan, A., and Rao, M.R. (2001). "Application of Reliability Concepts in Pavement Maintenance Management", Proc. 5th Int. Conf. on Managing Pavements, Vol.1, Paper No. 78, Seattle, Washington.
13. Zaghloul, S.M. (1997). "Development of Performance Based Quality Control Specifications for Asphalt Pavements", Proc. 5th Int. RILM Symposium MTBM Lyon 97, pp. 591–598.

Efficient Transportation and Pavement Systems – Al-Qadi, Sayed, Alnuaimi & Masad (eds)
© 2009 Taylor & Francis Group, London, ISBN 978-0-415-48979-9

Probabilistic framework for calibrating of highway geometric design models

Karim Ismail & Tarek Sayed
Department of Civil Engineering, University of British Columbia, Vancouver, BC, Canada

ABSTRACT: An often overlooked aspect of highway geometric design is to the proper account for uncertainty in design models, input variables, and design parameters. This paper is a motivation to transform the current design models from deterministic to probabilistic. Due to specific aspects of highway geometric design, this transformation requires new theoretical understanding, interpretation, and development of underlying safety concepts.

Reliability theory offers a widely acceptable framework for tracing the propagation of uncertainty through various design models. After defining a rejection domain for design outputs, the probability that a certain dimensioning decision of a road element leads to unacceptable driving conditions can be used as a design safety metric. The design safety metric tests the design of a road segment under a hypothetical group of driving situations that are generated based on the elementary knowledge of variability in design inputs. It is evident that the safety quantified in this research is different from the conventional definition of safety as being associated with collisions. In this research we strive to draw the line between the two safety definitions. Hopefully, with further development of various geometric design models, a link can be realized between.

The literature includes several probabilistic analysis of dimensioning decision such as median width, horizontal curve radius, and roadway and railway intersection design. There is however an absence of research that extends these isolated studies to cover the entire spectrum of geometric design cases. In this research we attempted to formalize and integrate probabilistic analysis into standard design models. Several of the ideas and concepts demonstrated in this research were inspired by reliability-based structural design.

1 INTRODUCTION

A fundamental objective of highway geometric design is to build safe roads through proper dimensioning of various road elements. Various geometric elements such as vertical alignment, horizontal alignment, and cross section elements are designed within an acceptable domain for which critical extremes are set by standard design guides. The development of standard codes for geometric design involves the adoption and formalization of analytical models that adequately describe real road environment along with suggested values for various design parameters. Although current design codes are the outcome of an evolving process of improvement, there some theoretical issues that have been generally overlooked. This paper is an attempt to advance the process of code development so that it account for uncertainty in design parameters and variables in a structured approach.

Most of the design decisions in the field of civil engineering are not based on the presence or absence of knowledge *in toto*, but rather the degree of knowledge. The design of civil infrastructure, with no exception of highway geometric design, is in the essence taking engineering decisions under various degrees of uncertainty. It is unavoidable that the knowledge of design inputs that are associated with human factors or vehicle operation is imperfect and carry a discrete source of uncertainty, and a risk of design inadequacy, into the design process. It is conceivable that for a safe functioning of designed road elements that design parameters be selected to cover an acceptable range of a representative population. In addition, the more the sources of uncertainty are identified, quantified, and factored into the standard design models, the higher the safety level. A straightforward application to this principle is deterministic design in which the mean values of design inputs are scaled using established safety factors in order to ensure that the designed road elements will provide safe conditions for an acceptable percentage of operational instances.

The two fundamental steps toward the design of a civil engineering system are: system idealization and performance prediction. The first step involves mathematical simplification of a rather complex road environment into a workable analytical model. The second step is to limit the system performance within an acceptable operation frame during the feasible life of the engineering system. The design process does not traditionally involve the selection of ideal design alternatives that entirely mitigate potential failures—otherwise the performance prediction would be a matter of common observation. A potential system failure is not entirely avoidable, but the consequences of system failure are minimized through the design process. The standardization of the design process, in order to account for the risk of failure, requires in part a rational assessment of the underlying uncertainties. Since design parameters follow a wide range of statistical distributions and design models can be highly non-linear, a rigorous treatment of the variability in design parameters can be carried out using reliability theory.

Reliability theory concerns the analytical tools necessary for tracing the propagation of uncertainty throughout the design process, beginning with random distributions of design inputs until final design output. The major target of reliability analysis is to estimate the probability that an engineering system fails to perform its stated purpose under anticipated operational conditions. The probability of failure is widely adopted as an indicator of the overall safety level of design in several disciplines of civil engineering, e.g. structural safety (Melchers, 1999), water supply (Dolezal and Haimes, 1993), and road safety (Navin, 1990). In highway geometric design, the term probability of non-compliance P_{nc} is used instead of probability of failure due to the absence of a physical failure in real operation. Based on the foregoing discussion, P_{nc} represents a distinct indicator of risk that involves only the uncertainty in design components. The safety level that corresponds to P_{nc} is termed in this paper "design safety", as a distinction from operational safety, where the latter is based on observed or predicted collisions.

It is noteworthy design safety is not suggested to supplant the conventional use of collision history as an indicator of road safety. In reality, despite the presence of other approaches for assessing road safety such as traffic conflicts and subjective risk assessment (De Leur and Sayed, 1990), road collisions, either observed or predicted, stand out as an objective and quantifiable measure of road safety—precisely operation safety. Rather, design safety is obtained as a product of a more meaningful representation of uncertainties and the use of a rigorous framework for tracing the flow thereof. Design safety reflects the aggregate level of confidence in various dimensioning scenarios within the scope of pre-defined standard design model and random distribution of input parameters. The fact whether a relationship between observed crash record for a functioning road element and its P_{nc} exists is largely speculative and depends on the amount of reduction in uncertainty achieved by employing a robust design process. That is, the degree at which the statistical distributions of design parameters represent their real populations or how the idealization of road environment is a realistic depiction of the real system. It is plausible that if, hypothetically, uncertainty is completely reduced, by means of exhaustive identification and quantification of variability in design parameters, and all road environments are identical across designed sites, design safety should be positively related to, and a valid predictor of, collision risk. In fact, the degree of agreement between design safety and operation safety is, presumably, suggestive of how comprehensive and representative the design model is.

From a different outlook, design safety can be viewed as a parallel approach of comparing the safety level of different road elements. Such a means of assessing safety level, particularly through further refinement, can prove valuable in the practical cases in which it is difficult to base the safety evaluation on the number of road collisions. These practical limitations include insufficiency of collision data, discrepancies or anomalies in collision records due to unforeseen confounding factors, or lack of relevant collision prediction models. Moreover, inadequacies in highway design may not necessarily result in observed collisions, but rather induce uncomfortable driving conditions or increased driver workload.

The calculation of P_{nc} requires that variability in design parameters be quantified. The selection of the probability distributions that characterize variability in design parameters is a potential subject of inconsistency if left to the designer. Accordingly, the calculation of P_{nc} should be performed at by the code-developer in order to maintain design consistency. The main proposition that underlies the analysis in this paper is that the design safety level, represented by P_{nc} and generated by standard design models, needs to be consistent and close to a pre-specified level. This paper presents a general analytical method for satisfying the previous proposition. The paper also suggests methods of selecting this pre-specified or acceptable design safety level. For demonstration, the method is applied to the design model of crest vertical curves provided in AASHTO *Green Book* 2001.

1.1 *Stochastic components in highway design*

Highway geometric design is a multi-phased process, with each phase requiring a specific body of knowledge, expertise, and analysis in order to create a solid foundation of engineering decisions. Similar to other civil engineering disciplines, each design phase entails some assumptions, estimates, and predictions that contribute to an aggregate magnitude of uncertainty in the design process. Highway operation consists of three primary components: the driver, the vehicle, and the road. The designable component is the roadway and/or the roadside environment with the objective of delivering safe and comfortable operation. Unlike the majority of civil engineering disciplines, the human component is central to the safety and efficiency of the system. Human factors, such as driver behavior, expectations, aesthetic perception, mental workload, and potential errors, are essential factors that lend themselves to the design of all features of a highway system. Accordingly, risk-based highway geometric design can be viewed as the geometric delineation, alignment, and proportioning of roadway and roadside features in order to accommodate forgivable navigation or control errors that involve driver and/or vehicle. Evidently, the major stochastic parameters used in the design process are related to human factors and vehicle performance.

The driving mechanism can be divided into three primary tasks: control, guidance, and navigation, Lunenfeld, et al. (1990). The significance of human factors in the design process is underscored in several driving situations: high cognitive or perceptive demand on drivers, erroneous information of road cues, or too little demand on drivers, Alexander, et al. (1985). The complexity of introducing comprehensive human factor models to the design process bears mainly on two facts. First, the driving mechanism puts specific demand on drivers that entails complex cognitive, psychological, and perceptive processes. A thorough understanding and formulation of the gathering, processing, and interpretation of driving information is an area of ongoing research, and such comprehensive models are not present in the literature, Kantowitz (1992). During recent decades, road safety practitioners and researchers have become increasingly aware of the need for better understanding and consideration for human factors in the current design practice, Kanellaidis, et al. (1997). Second, Highways are intended for public service, thus highway operation affects a relatively broad segment of society. The individual difference across the driving population is an additional source of variability, Taoka (1989). Some aspects of individual variability are explained by age differences, Marmor (1982) and Fox (1989). Other aspects are attributed to the natural diversity in driving performance.

It can be inferred that a considerable source of uncertainty in the design process bears on three driver-related issues: the need for a comprehensive understanding of human factors in the driving process, individual differences among road users, and the anticipated number of road-users throughout the feasible life of the designed highway. In addition, traffic flow prediction brings to focus issues related to transportation planning, an area of research that involves different spectra of variability and uncertainty. The vehicle component in highway operation poses a different source of variability in terms of vehicle dimension, mechanical characteristics, and operational requirements. In sum, uncertainty in highway geometric design is a multifaceted input that requires a rational and scientifically defensible modeling framework.

1.2 *Probability interpretation in the context of highway geometric design*

The broad objective of probabilistic design is to identify, quantify, and mathematically characterize uncertainties throughout the various design phases in order to manage and regulate the risk associated with design outputs. The analytical means of risk-based design bears significantly on reliability theory. There are various types of mathematical models for characterizing uncertainty—probabilistic, fuzzy, and convex, Ben-Haim (1994). Probabilistic uncertainty is the type used in the course of this study due to the availability of the necessary analytical tools. Before delving into the mechanics of reliability analysis, it is necessary to provide a brief discussion of the interpretation of probability within the scope of highway geometric design.

In highway geometric design, the majority of input parameters regarding human factors and vehicle characteristics can be classified as stochastic processes. A stochastic process yields outcomes that are too variable to be described by ordinary deterministic rules, Harr (1987). Since its mathematics first appeared, the interpretation of probability within the realm of civil engineering has continuously been a controversial issue, Fishburn (1964), and Lind (1996). Salmon (1966) reports the following possible interpretations of probability:

Classical (frequentist): probability is the ratio between favorable outcomes of an event that is equality likely to occur and all different outcomes. Prediction according to this interpretation implies that a future event depends on past observations of similar events, i.e. past observation can reliably predict the future despite the fact that the physical variability is not explained.

Subjective: probability is a mathematical scaling of the degree of belief associated with the outcome of an event. The application of subjective, judgmental, or Bayesian probability is not restricted to events amenable to repeated sampling or observations. This interpretation can be broadly defined to include antique or one-time occurrence, the validity of an uncertain proposition or hypothesis, and an uncertain state of nature. In addition, subjective probability can more comprehensibly employ Bayesian updating for amending reliability assessments in order to account for engineering judgment, additional information, or qualitative assessment of risk level, Lindley (1972).

The main output of the reliability analysis presented in this study is P_{nc}, which reflects the probability that a traveling vehicle will operate beyond the acceptable domain set out by design models. The latter is widely recognized as probability of failure in the realm of civil engineering, which refers to any performance of an engineering system beyond a prescribed domain, Petroski (1996). However, the term failure is commonly perceived within the context of civil engineering as related to the physical collapse of a facility. Navin (1990) used the term non-compliance to supplant the term failure, in light of the absence of physical collapse within the context of highway and traffic engineering. For consistency, the term non-compliance is used in the course of this paper, although the term failure can be correctly used.

Zeeger, et al. (1989) were the first to introduce a frequentist probabilistic definition of road safety. According to that definition, probability of non-compliance (P_{nc}) can be interpreted as the relative frequency of collision occurrence in reference to the total number of vehicles traversing a highway element. However, there are several concerns in regard to the applicability of the previous interpretation within the frame of risk-based design:

1. The occurrence of an unfavorable driving situation, represented by a high P_{nc}, may not necessarily lead to a collision, but rather evasive maneuvering or an insecure driving experience, Zheng (1997). In addition, collisions are complex phenomena that involve the occurrence of numerous possible circumstances, some of which are unforeseeable at the design stage. The currently available body of knowledge does not permit an accurate and exhaustive association between specific driving circumstances and the occurrence of a collision. Moreover, P_{nc} is

calculated for design models that do not necessarily factor in all the variables that influence driving, but the design models rather rely on mathematical idealization of actual conditions. It follows that P_{nc} is essentially representative of design safety, and as long as the link between the latter and operation safety is not investigated, an interpretation of P_{nc} that involves collisions is largely speculative.
2. The currently available statistical distributions that can be used to describe the variability of input parameters are based on a limited sample of the population of drivers and vehicles. Hence, there is an unavoidable degree of approximation related to the used statistical distributions as a representative of the variability in driving conditions over all future designs.

Road collisions are generally random and rare events that, in more than 90% of the cases, involve human error, Wong and Nicholson (1992), and Sayed and Navin (1995). These human errors can come in the form of slips, lapses, violation, error in anticipation or judgment, failure to read road information, and irresponsibility, Zheng (1997). Human errors that result in collisions involve gross mistakes in guidance or control that cannot be calculated a priori, but rather statistically associated with specific road features based on collision observations over a period of time. Accordingly, the association between P_{nc} and collisions is unattainable without collision history. The relationship between design safety and the observed system performance is a recent area of discussion in other fields in civil engineering, e.g. Lind (2005). Based on the previous discussion, it is commendable that P_{nc} be interpreted as a subjective probability—rather than a frequentist probability. Therefore, it can be viewed as an indication of the overall factor of safety built into the design output that compensates for lack of knowledge or reflect a perceived risk.

1.3 *Reliability theory*

Theory of reliability studies the probabilistic and mathematical models required to calculate the ability of a system to perform its stated purpose according to the anticipated operation conditions, Gertsbakh (1989). The concept of integrating probability theory into engineering design first appeared as probabilistic mechanics, Mayer (1926). The introduction of reliability analysis to the realm of civil engineering, and structural engineering in particular, was intended to replace the concept of factor of safety with a more rational framework for addressing uncertainty in design, Freudenthal (1954). Reliability theory became a widely accepted engineering discipline that addresses the probabilistic nature of engineering design within a sound analytical framework, Cornell (1981). The conventional procedure of applying reliability methods is:

1. Defining the analytical model that predicts system performance,
2. Defining uncertainty in analysis inputs by means of probability distributions, and
3. Calculating the probability that the system performance is within an acceptable limit. In addition, the relative influence of inputs on the system performance is a secondary output of some reliability methods. It is expected that the soundness of reliability analysis depends on the accuracy of the input probabilistic information, Gerstbach (1989).

The basic elements of reliability analysis are an N-dimensional vector of input variables $X = x_1, x_2, \ldots, x_n$ and limit state function or performance function $G(X)$. The performance function G is constructed in that it yields positive outputs when the system performance is acceptable or safe, and yields negative values for unfavorable system performance. The performance function is conventionally written in terms of the difference between supply R and demand S, as follows:

$$G = R - S \tag{1}$$

In highway geometric design, supply represents the group of input parameters that are related to design requirements that concern safe and comfortable driving conditions. Demand represents driver or vehicle requirements that need to be accommodated. Demand and supply are uncertain, and their intersection represents conditions of probable non-compliance, as shown in Figure 1.

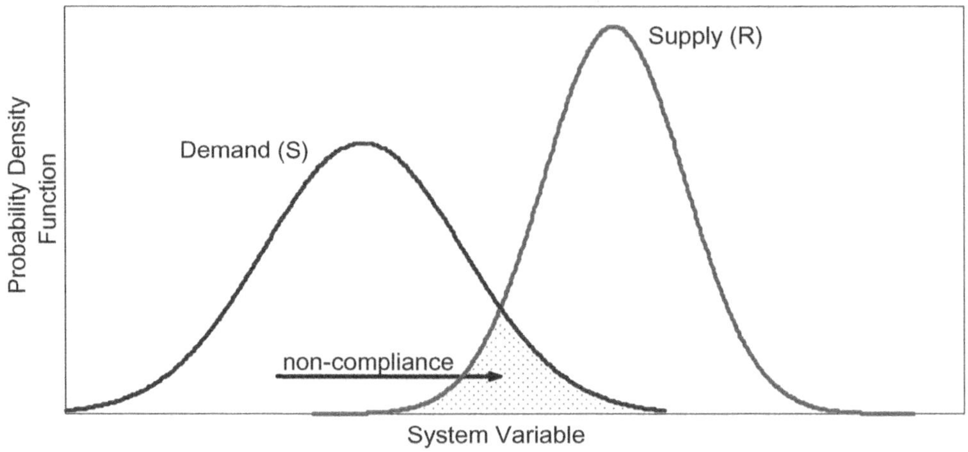

Figure 1. Basic components of performance function.

The uncertainty regarding input variables can be characterized by assigning each input variable x a probability density function $f(x)$. The reliability of a system—the probability of its being in the acceptable performance domain is obtained as:

$$P_c = 1 - P_{nc} = \Pr ob(G > 0) = 1 - \int \ldots \int_{G(x) \leq 0} f(x)dx \tag{2}$$

where

P_c = probability of compliance or reliability

P_{nc} = probability of compliance or reliability

The previous integration can be performed by, Melchers (1999):

1. Direct integration (applicable in limited number of special cases),
2. Numerical integration, such as Monte Carlo simulation, and
3. Avoiding integration through transforming the integrand to a multi-normal joint probability density function, e.g. First Order Second Moment (FOSM) and First Order Reliability Method (FORM).

The performance function in the reliability analysis presented in this paper is not a closed-form function, but rather an iterative algorithm. Monte Carlo simulation was not selected due to the anticipated costly calculations and the need for investigating the relative influence of input variables—which is not offered by the latter method. FORM analysis was selected over FOSM due to the non-normality of some input parameters, the more accurate results offered by FORM, and the availability of an analytical platform, FERUM, for the implementation of FORM analysis, Hawkaas and Kiureghian (2000).

FORM analysis attempts to facilitate the calculation of probabilities by means of transforming the random variables X to a standardized normal space Y according to the following expression:

$$p = F_x(x) = \Phi(y) \tag{3}$$

where

p = some probability content associated with $X = x$

$F_x(x)$ = marginal cumulative distribution function of X

$\Phi(y)$ = cumulative distribution function for the standardized normal random variable Y

The reliability of a system is often expressed in terms of a safety (or reliability) index β, a unitless indicator that is directly proportional to the safety level. The reliability index β can be calculated as follows:

$$P_{nc} = \Phi(-\beta) \tag{4}$$

The prevalence of this expression in the literature of reliability analysis may be explained by two factors: 1) reliability index offers a more meaningful and interpretable numerical value in comparison to probability values which are normally of a low order of magnitude, Augusti, et al. (1984), 2) most reliability methods include probability transformation to a standardized normal space, hence it is convenient to express variability in the analysis output within the same space, and 3) Reliability index β has a particular geometric significance that is central to FORM analysis. Figure 2 shows the relationship between P_{nc} and reliability index.

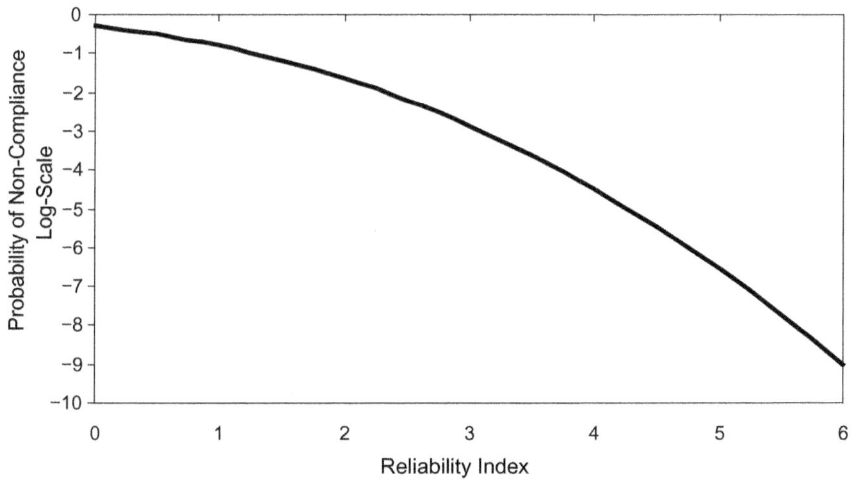

Figure 2. The relationship between reliability index and logarithm the probability of non-compliance.

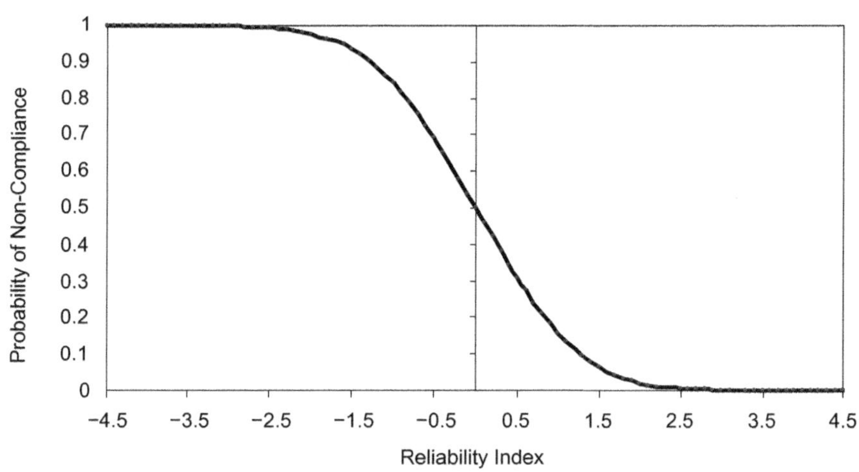

Figure 3. The relationship between reliability index and probability of non-compliance.

1.4 *Development of risk-based design*

In order to compensate for imperfectness in knowledge about system behavior, designers opted to maintain a specific magnitude of overdesign. Prior to the emergence of reliability methods, the magnitude of overdesign was characterized in terms of factor of safety. Factor of safety is analytically the ratio between the expected supply to the expended demand. Factor of safety concentrates design safety level into a single quantity, which is usually selected based on past experience and/or subjective assessment, Rao (1992). The concept of factor of safety that emerged essentially in applied mechanics, found feasible application in other areas: medicine, Butterfield (1963), Jones (1982) and Taylor et al. (1982), and business, Solomon et al. (1983). In the realm of structural engineering, factor of safety was subject to numerous revisions and criticism regarding its analytical and practical validity, Freudenthal (1956), Carter (1997), Elishakoff (1983, 1999). Despite some early skepticism, e.g. Bolotin (1964), reliability methods are expected to be integrated into standardized design in the foreseeable future, Oden et al. (2003). One of the key steps in developing code-writing is the integration of reliability methods into the formulation of design models as a rational framework for addressing uncertainty, Ellingwood (1994). The ultimate objective of calibrating the design code is to eliminate the lack of equal probability of failure, in such it becomes close to a premeditated and acceptable level.

1.5 *Acceptable risk level*

The selection of an acceptable risk level, or probability of non-compliance, is a paramount step in standardized probabilistic design. It appears that the majority of research in the field of reliability methods was devoted to the analytical details of calculating probability rather than the selection of an acceptable value, Grandori (1998). In road safety, risk is expressed as the product of the probability of unfavorable outcome and the cost of consequences, Haight (1986). Determination of acceptable risk is an area of continual research that involves a risk-benefit trade-off for individuals and society, Keeney (1980a, 1980b), Vrijling et al. (1998) and Rackwitz (2002). Ditlevsen (1997) proposed a rational framework for selecting a target design safety level β_t based on minimizing expected cost of failure, which includes probability of failure and societal cost of failure, and different types of construction cost. At the core of the last methods lies the relationship between the cost of system failure and β_t. Without investigating the link between design safety and operational safety and due to the aforementioned assumption regarding the subjective nature of design safety in the highway geometric design, a target design safety level ought to be selected based on a method that does not involve predicting number of collisions. Lind (1978) postulated that existing codes are optimal if a need for revision is not present. The novelty of the previous principle lies in the assumption that standardized design is in the long run self-optimizing. Roësset (2002) referred to a similar fashion in calibrating some design models in which β_t was selected based on past experience rather than optimization process. Similarly in highway geometric design, past designs that are deemed optimal in respect to construction cost and safety performance can be used to estimate β_t-itself is considered a random variable. A relevant example of the self-optimization postulate is evident in the last release of AASHTO design guide. Several concerns were raised regarding the cost-effectiveness of vertical curves design according to AASHTO. Either this is attributed to an improved vehicle performance or rising cost of materials. The design model was revised in order to reduce the lengths of designed vertical curves with insignificant safety implications, Fambro et al. (1997). The latter approach toward code-calibration is adopted in the course of this paper.

2 PREVIOUS RELIABILITY-BASED RESEARCH

Moyer and Berry (1940) undertook the pioneering work of incorporating probabilistic methods into highway design by developing a method to determine the safe speed on highway curves.

The method included calculating the margin of safety as the ratio between ball-bank reading recorded on a curve and a generally accepted safe value. Further, the operating speed was recognized as a random variable and hence selected based on specific percentile values—85th percentile for design speed of 30 mph or less and 90th percentile for 35 mph. Navin (1990) cited several concerns about the implicit and qualitative safety level in geometric design standards. In response, the concept of margin of safety based on standard design equations was introduced as a meaningful and quantitative measure of the safety level built into isolated highway components. The margin of safety was calculated based on assuming that variables are normally distributed and independent. The reliably method used was FOSM, in which obtained values of reliability index β represented the margin of safety.

Easa (1993) developed a probabilistic model based on FOSM reliability method for the design of the intergreen interval in signalized intersections. The main objective is to eliminate the zone where a driver faced with yellow signal fails to either stop or clear the intersection. In achieving so, the stopping sight distance is equated with intersection clearing distance. The variability in the input parameters was calculated based on some assumed values of the corresponding coefficient of variation. Two probabilities of non-compliance values were assumed and new design charts were reconstructed in order to obtain intergreen times. It is noteworthy that the developed probabilistic model was closed-form; hence the reconstruction of the calibrated design chart did not involve any numerical minimization—unlike the calibration process introduced in this study.

Faghri (1988), Easa (1994 a), and Easa (1999) studied the problem of sight distance at uncontrolled road intersections and road-railway grade crossings under different operational conditions. The main objective of intersection design is to permit an adequate sight distance for an approaching driver to safely stop prior to entering an intersection in order to preclude potential collision. The moments of the input variables were obtained from relevant studies or by assuming some values for the coefficient of variation and assigning specific percentiles for design values. The probability of non-compliance was calculated at difference values of available sight distance using FOSM reliability method.

Richl and Sayed (2006) studied the effect of median width along curved highway segments in order to understand the risk of sight distance restriction. If a median is potentially narrow, a driver may fail to stop within the available sight distance due to sight restriction caused by the median barrier. The probability distributions of input variables were all obtained from relevant studies. The reliability method used was Monte Carlo simulation. For two highway alignments, the probabilities of non-compliance were calculated for several scenarios of sight distance restriction. The obtained probabilities were compared to the maximum obtainable probability of non-compliance for a horizontal curve design based on the same design speed of the two highway alignments.

Sarhan and Hassan (2008) calculated used a Monte Carlo simulation to calculate P_{nc}, renamed in that paper as Probability of Hazard, for a three-dimensional sight distance model. Monte Carlo simulation is a well-established branch of reliability analysis that numerically simulates the distribution of the limit state function. The method of reliability analysis, as presented later, used in this is paper is more efficient and yields sensitivity measures that cannot be obtained by Monte Carlo simulation.

By in large, all previous work did not progress much beyond calculating P_{nc}. The work presented in this paper, part of a Master thesis work (Ismail, 2006), proposes a new framework for integrating reliability analysis into standard design codes. The work was performed using two-dimensional and three-dimensional sight distance models.

3 CALIBRATION METHOD AND APPLICATIONS

Let β_t be the target reliability index that represents an acceptable risk level, D_f be the feasible scope of input parameters that will be considered in the calibration process, and U_f is the scope of non-standard input parameters. In keeping with the proposition of risk consistency, a penalty function is used to quantify the difference between each β_i associated with design output i and β_t.

It is noteworthy that penalty function is a key element in the calibration process that can integrate socioeconomic aspects of design. Penalty function can be formulated in order to differentiate between cases of overdesign and underdesign, Lind (1977). For example, the following formulation takes into account construction and running costs of a highway element:

$$p = P_c(\beta - \beta^t) + \int_0^T U_s(\beta - \beta^t) \cdot e^{-r \cdot t} d_t \tag{5}$$

where

T = expected number of years after which the geometric design may be reviewed

P_c = construction cost as a function of the overall design risk level

U_s = estimated annualized collision costs,

r = estimated continuous discount rate during the next T years.

For demonstration purpose the penalty function used in the next analysis is assumed to be symmetrical as follows:

$$p = \sum_{i \in D_f, U_f} (\beta_i - \beta_t)^2 \tag{6}$$

In order to minimize the penalty value, a calibration factor C_t is added to the standard design model in order to make risk level close to β_t. In order to avoid zero values, an exponential form is assumed as follows:

$$C_t = e^{A^T D + B^T U} \tag{7}$$

Where A and B are vectors of constants multiplied by corresponding input parameters. For simplicity, a single calibration factor is multiplied by the input parameter $x^* \in X$ that possesses the highest correlation with β. The calibration model is simply the following minimization problem:

$$C_t = \underset{A,B}{\text{Min}}\{p\}; \quad \forall\, X = \{x_1, \ldots, x^* \cdot C_t, \ldots, x_n\} \tag{8}$$

The main steps of the calibration process are summerized as follows, Melchers (1999):

1. Define the scope of the code, which is the range of input parameters that will be considered in the calibration process. Input parameters beyond this range are not necessarily expected to yield a consistent risk level.
2. Select design cases from the scope of the input parameters.
3. Select a target reliability index β^t.
4. Conduct reliability analysis for the design cases to find the corresponding risk levels.
5. Select calibration factors that will minimize the scatter of the reliability indices around the target value.

4 CALIBRATION OF 2D CREST VERTICAL CURVE DESIGN MODEL

This section presents an application of the foregoing calibration method to the design model of crest vertical curve located on a tangent. A 2D projection of the vertical curve is sufficient space of calculation due to the absence of horizontal curvature. It is noteworthy the accuracy of the calibrated design charts depends to great extent on the statistical distributions and the various assumptions included in the calibration process. Therefore, the reconstruction of the design charts presented in this section is for demonstrative purpose.

4.1 *Defining input parameters*

The crest vertical curves are designed to provide adequate sight distance to allow the driver enough time to take specific decisions. The length of this sight distance depends on the complexity of the driving decision. This model will consider only stopping sight distance and the analysis can similarly be applied to other sight distance requirements. The deterministic inputs are vertical curve length L and algebraic difference A. The probabilistic input parameters are provided in Table 1.

Several studies in the literature, Navin (1990), Easa (1993, 1999), used design speed instead of operating speed in the reliability model. It is more meaningful to use operating speed in formulating a reliability models that involve highway design. Drivers build their selection of a comfortable and safe driving speed based on their own interpretation of roadway and roadside environment. Accordingly, operating speed includes some human factors and it should be taken into account along with other driver-dependent parameters, viz.: PRT and deceleration rate.

The model uncertainty consider in the current application is caused by the fact that roadway longitudinal grade varies along a stopping vehicle traversing the vertical curve, Taignidis (1998, 2001). The value of the entering grade g is taken as an input parameter as a compensation for not calculating the exact required stopping sight distance. The required *SSD* is calculated by means of summing over finite constant grades. The speed reduction during these increments is kept constant at 0.1 k/h. SSD is evaluated as follows:

$$SSD = 0.278 \cdot V \cdot PRT + \sum_{i=0}^{n} \frac{2V_i \cdot \Delta V}{25.92 \cdot \left(a + \frac{g_i \cdot 9.81}{100}\right)} \tag{9}$$

where, $V_i = V - i \cdot \Delta V$, ΔV is the speed increment which is taken 0.1 k/h, g_i is the longitudinal grade at the point on the vertical curve that corresponds to iteration i, and n is the last iteration at which $V_i \leq 0$.

The minimum available sight distance (ASD) is calculated as follows:

$$ASD = \begin{cases} ASD^* & \text{if } L > ASD^* \\ \left(L + 200 \cdot \dfrac{k}{2L} \cdot \left(\sqrt{h_1} + \sqrt{h_2}\right)^2\right) & \text{if } L \leq ASD^* \end{cases} \tag{10}$$

where, h_1 is the driver eye height (m), h_2 is the object height (m), k is the rate of vertical curvature which equals L/A.

The initial grade g_0 at which a vehicle starts braking is calculated according to Taignidis (2001).

Table 1. Distribution parameters.

Parameter	Mean	Standard deviation	Distribution	Design values	Percentile value
PRT	1.5 s	0.4 s	Log Normal[a]	2.5 s[d]	98.1
Driver eye height	1.14 m	0.055 m	Normal[b]	1.08 m	10.4
Driver deceleration	4.2 m/s^2	0.6 m/s^2	Normal[b]	3.4 m/s^2	9.1
Object height	–	–	Deterministic	0.6 m	–
Operating speed[e]	105.1–149.69 A/L	5.57 k/h	Normal[c]	–	85
Operating speed[f]	103.24–3676/R	4.47 k/h	Normal[c]	–	

[a] Obtained from Lerner (1995).
[b] Obtained from Fambro et al. (1997).
[c] Obtaind from Fitzpatrick et al (2000).
[d] AASHTO *Green Book* (2001).
[e] Crest vertical curves with k ≤ 43.
[f] Crest vertical curves with k ≤ 43 combined with horizontal curves.

4.2 *Calibration model*

The performance function of the stopping sight distance is as follows:

$$g = ASD - SSD \tag{11}$$

The scope of the calibration process in this analysis is chosen as: $V \in \{70, 75, \ldots, 100\}$, $A \in \{4, 5, \ldots, 9\}$, $g \in \{A - 6, A - 5, \ldots, 6\}$. For simplicity, the input parameters are assumed to be statistically independent. For 341 design cases, β values are calculated and their distribution is presented in Figure 4.

4.3 *Target reliability index*

The selection of β_t is analogous to the determination of an acceptable quality of design for specific highway elements. The latter is an important decision that relies on the general policy of highway design. In other areas of civil engineering, several methods that are developed for the determination of β_t involve minimizing the expected societal cost as well as construction cost, Ditlevsen (1997) and Vrijling et al. (1998). As an alternative to quantifying a relationship between β and collision frequency, β_t can be inferred from existing vertical curves. For a relatively large group of sites, a multi-criteria assessment of design quality can be undertaken. The criteria of evaluation may include: collision record, the ratio between actual construction cost and construction cost of the vertical curve had it been dimensioned at design lengths, an expert evaluation of the in situ driving conditions, and an expert assessment of the cost-effectiveness of design. The sites can be ranked according to the previous evaluation. A percentile ranking has to be selected in order to define the pool of sites that possess acceptable cost for society. The target reliability index β_t can be calculated based on the average β values that are inferred for the acceptable pool of sites. The main shortcoming in the previous method is that because it relies on existing vertical curves, it is difficult to study the cost-effectiveness of the expectedly small percentage of vertical curves that are dimensioned lower than standard requirements. The main advantage of the previous method is that it offers a rational approach for integrating subjective risk assessment and some aspects of cost-effectiveness into standard highway design.

As a first step toward a code-calibration, β_t can be selected based on current design standards. In this respect, β_t can readily be taken as the average of current design standards. One of the important issues related to the application code-calibration in highway design is whether β_t is specific for each

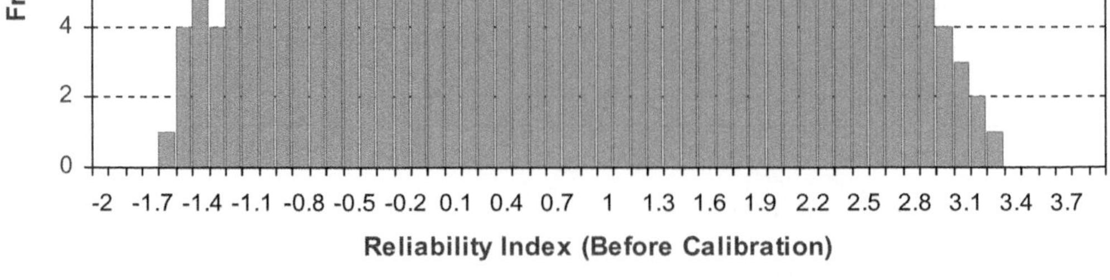

Figure 4. Reliability indices scatter before calibration of 2D design model.

design speed. The exact definition of design speed and the methods used for its selection was the focus of many studies, Barnett (1936), Leisch et al. (1977), McLean et al. (1976, 1978, 1979). It is evident that the difference between operating speed and design speed is recognized the literature, McLean (1978), Messer et al. (1981), Fitzpatrick (2000), and Jessen et al. (2001). The definition of design speed evolved from being considered as a maximum safe speed to a selected design parameter used principally in dimensioning various highway features and reflects the general quality of highway design, Fitzpatrick et al. (2003). Based on the subjective interpretation of P_{nc} it is inferred that β_t is proportional to design speed and can be viewed in part as a probabilistic surrogate for design speed. Based on the following discussion, β_t is calculated for each design speed as follows:

$$\beta_t = -\Phi^{-1}\left(\sum_{i \in D_{fV}, U_{fV}}^{n} \frac{P_{nci}}{n}\right) \tag{12}$$

where D_{fV} and U_{fV} are the feasible scope of input parameters at design speed V, and P_{nci} is the probability of non-compliance at design point i. The calibrated input parameter x^* is selected to be SSD.

4.4 *Discussion of results*

Based on the method described in Equations 5–8, the calibration factor C_t is calculated as follows:

$$C_t = e^{a \cdot g + b \cdot A + c \cdot V} \tag{13}$$

where, a, b, and c are calibration coefficients. Table 2 shows the results of applying the minimization process in Equation 8. Negative calibration coefficients are associated with input parameters that are directly proportional with pre-calibration β values. The signs of the calibration factor are consistent with the previous argument. The absolute values of the calibration coefficients are proportional with the pre-calibration scatter in β values. The results in Table 2 are consistent with the previous argument given that the pre-calibration variance of β values is 4E-3 at design speed 70 km/h while it is 8.4E-4 at design speed 100 km/h.

The sensitivity of the penalty function is presented in Table 2. The sensitivity is calculated by independently perturbing the input parameters by 1% and observing the percentage of change in the minimum value of the penalty function. As shown in Table 2, the penalty function is comparatively sensitive to all calibration coefficients. It can be concluded that the selection of the parameters in the calibration process is feasible.

Table 2. Perception and reaction time estimates from different studies.

		Calibration coefficients		
Design speed (km/h)	Target reliability	a (10^2)	b (10^2)	c (10^4)
70	−1.14	−3.23	3.39	−18.3
75	−0.68	−2.80	2.81	11.60
80	−0.28	−2.23	2.00	−9.10
85	0.17	−2.08	1.90	−6.45
90	0.67	−2.00	1.75	−5.52
95	1.19	−1.71	1.42	−4.14
100	1.73	−1.58	1.26	−3.49
Average Sensitivity Coefficients*		1	1.67	0.78

*The average sensitivity factors for the given range of design speeds is benchmarked to the sensitivity of coefficient a for comparison.

As a sample, the before-and-after distribution of β values at design speed 70 km/h is presented in Figure 5 and Figure 6. The overall distribution of β values after calibration is presented in Figure 7. Using the calibration factors shown in Table 2, the AASHTO design chart

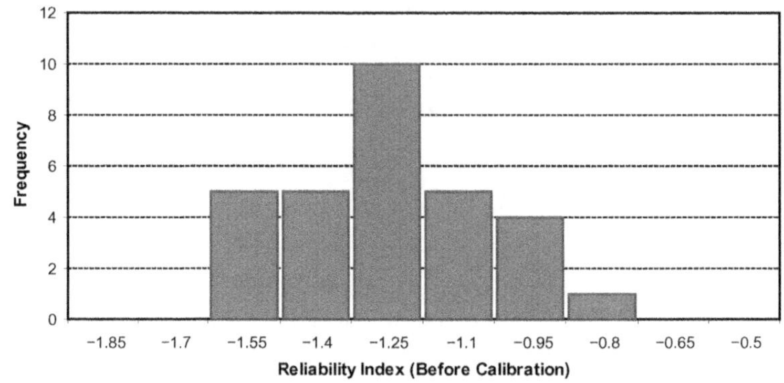

Figure 5. Sample distribution of reliability indices before calibration of 2D design model. Design Speed is 70 km/h.

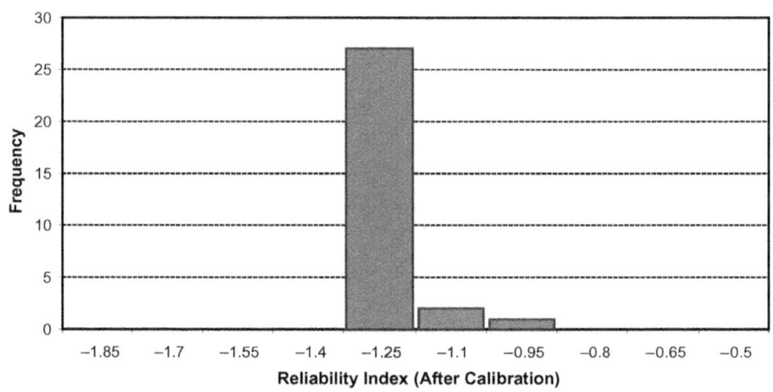

Figure 6. Sample distribution of reliability indices after calibration of 2D design model. Design speed is 70 km/h.

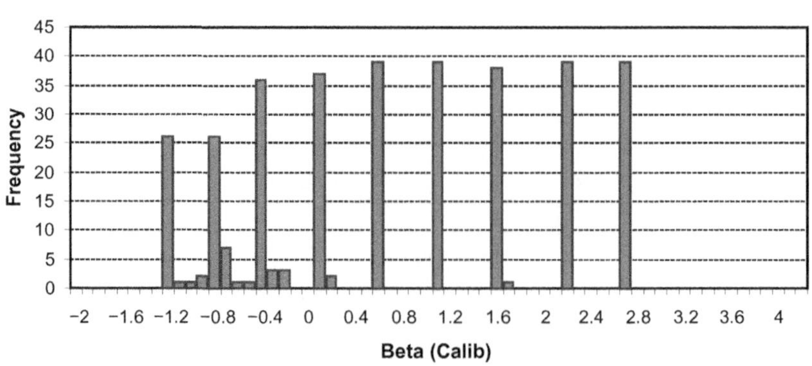

Figure 7. Reliability indices scatter after calibration of 2D design model.

is reconstructed as shown in Figure 8 at entering grade of 0%. Similar design charts can be produced for other values of entering grade. In order to compare design outputs before and after calibration, the vertical curve lengths obtained from both cases over the aforementioned scope of input parameters are presented in Figure 9. It is noteworthy that the percentage of change in curve length due to calibration ranges from 36.5% to –24.4%.with an average 1.4%. The insignificant in the change in average curve length can be explained by selecting β_t based on the average β values before calibration. The vertical grouping of the points shown in Figure 9 is due to neglecting slope variation in calculating the pre-calibration curve lengths.

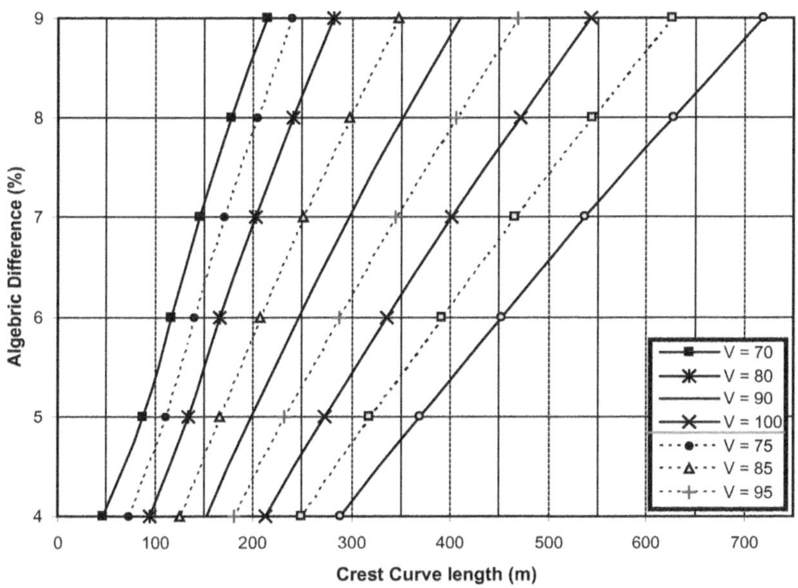

Figure 8. Reconstructed 2D design chart after calibration. The value of entering grade is 0% for all design speeds.

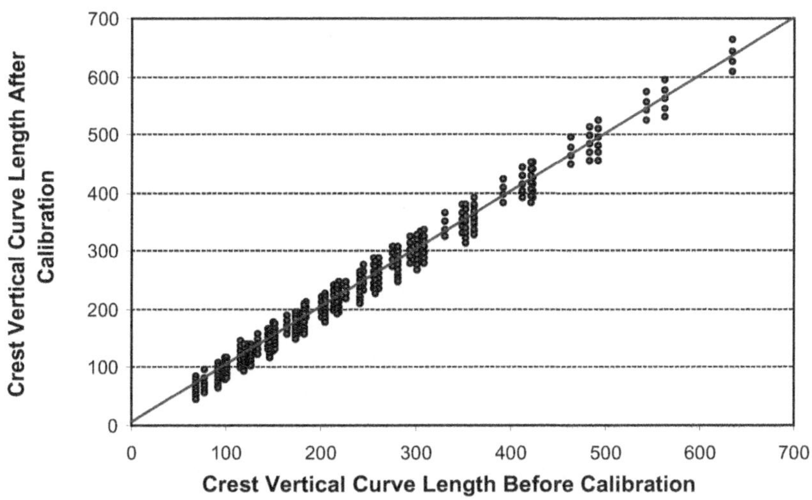

Figure 9. Comparison between standard design outputs before-and-after calibration of 2D design model.

4.5 *Calibration of 3D crest vertical curve design model*

The main objective of this section is to reconstruct the design model of crest vertical curve such that it accounts for the effect of horizontal curvature. Based on the hypothesis presented in Ismail and Sayed (2008), the most critical combination of horizontal and vertical curves can be obtained for. The critical combination of horizontal and vertical curves provides the least available sight distance. The components of horizontal curvature needed for the composition of the most critical combination are: the horizontal curve radius, the dimensions of cross section elements, and the side slope value. Several design charts can be constructed for a feasible range of horizontal curvature components. For demonstration, the most critical combination will be calculated based on the base value presented in Ismail and Sayed (2008). The operating speed model is presented in Table 1, where the most significant input parameter is horizontal curve radius. The elimination of vertical curvature from the operating speed model is in part a result of the fact that horizontal curvature significantly controls driving behavior than vertical curvature. As a result of superimposing horizontal curvature on the highway alignment, operating speeds are closer to design speeds—compared to the presence of only vertical curvature. Based on the foregoing discussion regarding the selection of β_t, the same target values presented in Table 2. The main difference between 2D and 3D calibration is:

1. ASD is calculated based on the algorithm presented in chapter 3 and Appendix I. The selection of horizontal curvature components is based on the hypothesis presented in chapter three.
2. Operating speed is calculated according to Table 1 for a combined crest vertical curve and a horizontal curve.
3. Without compromising accuracy, the convergence coefficients of FORM analysis are relaxed in order to reach to a reliability index value in a time efficient way.
4. The calculated 3D ASD is calculated to the closest 10 cm. Accordingly, the step length in calculating numerical differentiations is increased in order to capture any variation in the limit state function.

Figure 10 and Figure 11 show the distribution of β values before and after calibration respectively. Figure 12 shows the reconstructed design charts for combined vertical and horizontal curves.

For 3D design model, the distribution of β values for a complete list of design speeds is included in Appendix V. Figure 13 shows a comparison between the calibrated vertical curve

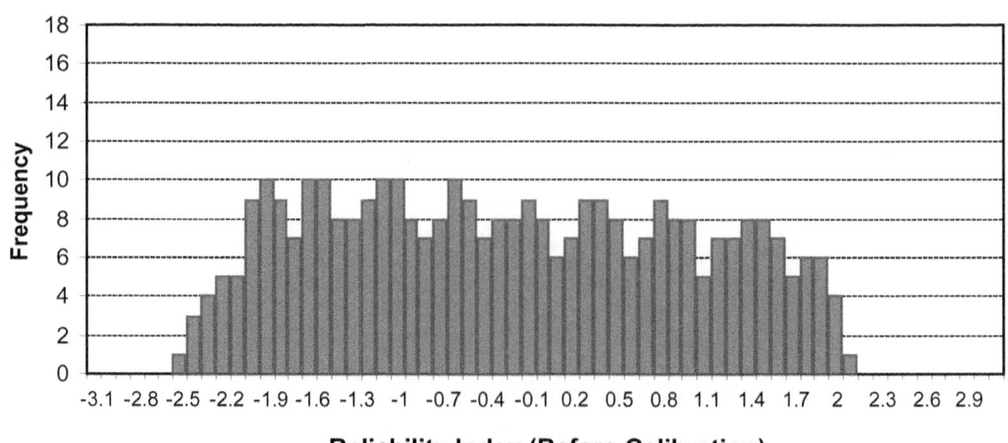

Figure 10. Reliability indices scatter before calibration of 3D design model.

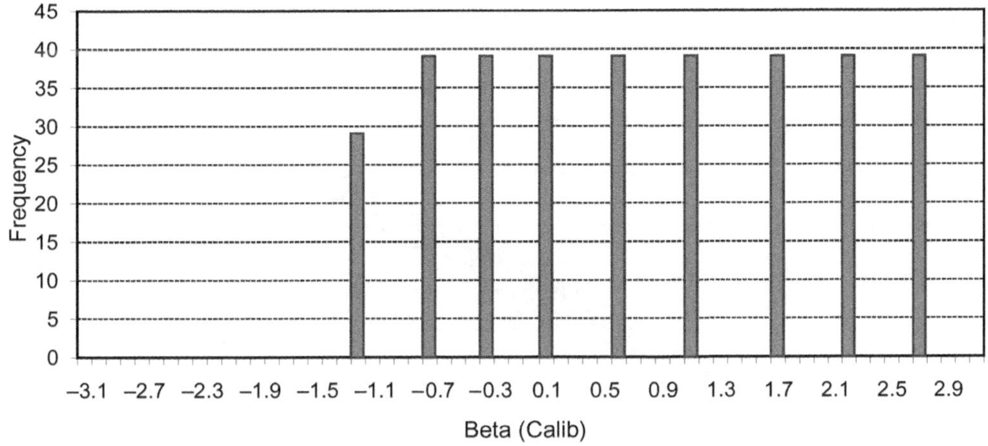

Figure 11. Reliability indices scatter after calibration of 3D design model.

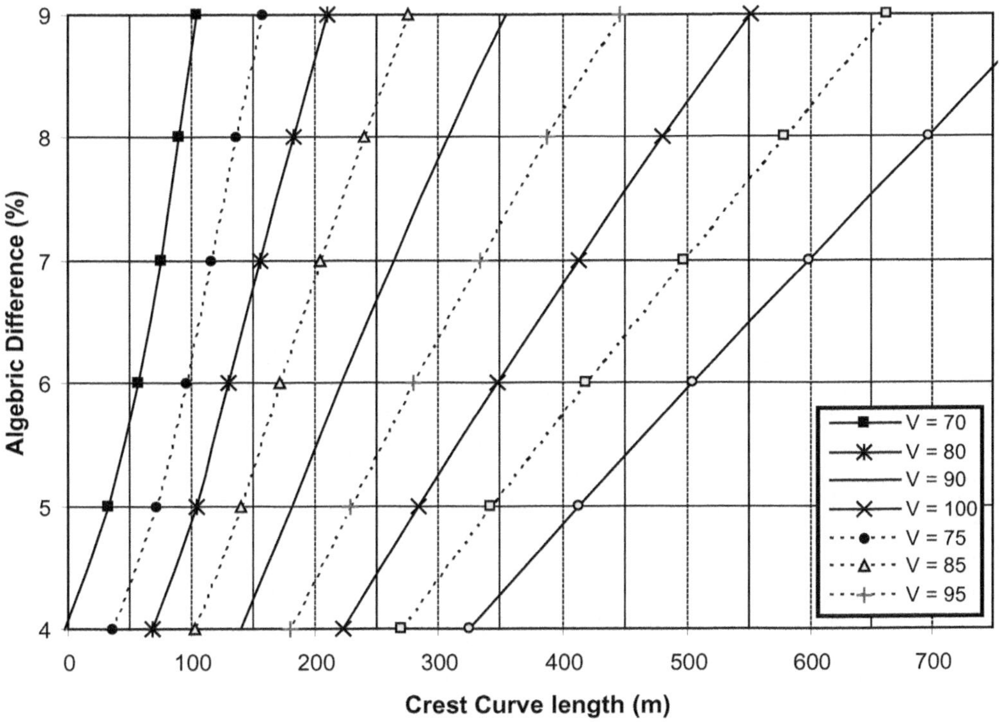

Figure 12. Reconstructed 3D design chart after calibration. The value of entering grade is 0% for all design speeds.

lengths obtained from 2D and 3D design models. The decrease in operating speed due to the superimposition of horizontal curvature predominates over the corresponding reduction in ASD. It is manifested in Figure 13 that vertical curve lengths obtained from 2D calibrated design model are longer than those obtained from 3D design model.

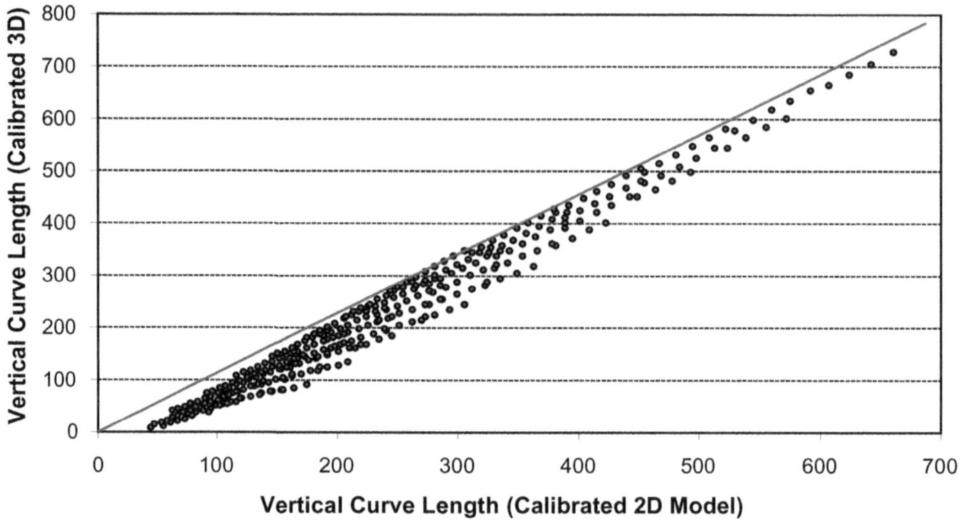

Figure 13. Relationship between calibrated vertical curve lengths obtained from 2D and 3D design models.

5 CONCLUSIONS

The analysis presented in this paper is based on a distinct level of safety, or conversely risk level. The presence of this risk level stems from the uncertainty in the design model and the input parameters. Design risk is quantified by the probability of non-compliance to design requirements. The main premise that underlies the analysis presented in this paper is that the risk level associated with standard design outputs has to be consistent and close to a premeditated level. This premise is addressed by means of calibrating the standard design models. The calculation of design risk is based on reliability analysis. Due to the reliance of reliability analysis on several initial assumptions, the code calibration process is a responsibility of the code-developer. A theoretical discussion was presented in order to reach a pertinent interpretation of the probability of non-compliance within the context of highway geometric design.

A general method for code calibration was presented. The method is based on multiplying some input parameters by calibration factors. The mathematical form of the calibration factors was constructed so that it compensates for the pre-calibration distribution of risk level. The degree distribution of risk level can be quantified by a penalty function. The method requires a target or acceptable risk level be determined. A general method for selecting target risk level is proposed. The method is based on evaluating the quality of design of a representative group of existing sites. The target risk level can be calculated as the average of a specific percentage of the representative group that are deemed as exhibiting an acceptable and cost-effective safety level. The method offers an approach for integrating subjective evaluation of risk into standard design. A preliminary method for selecting target risk levels was proposed and was adopted in the ensuing analysis. It was proposed that there is a unique target risk level for a selected design speed. Accordingly, target risk level can supplant design speed as a probabilistic means of measuring the overall design class.

In principle, the proposed calibration method can be applied to all geometric design models. As a case study, the proposed calibration method was applied on the standard model of designing crest vertical curves. The statistical distribution of the input parameters was taken from studies in the literature. Input parameters were assumed statistically independent, although there is a possibility of correlation, e.g. perception reaction time and deceleration rate. The distribution of the pre-calibration probability of non-compliance is relatively wide, ranging from 7.7% to 98.6%.

In addition, there is an evident that the higher group of design speeds exhibited disproportionately low risk level without any corresponding difference in collision rate. It is suggested that the curves at higher design speeds are over-designed—rather than curves at lower design speeds are under-designed. The previous suggestion cannot be emphasized solely based on this paper without empirical validation. However, it may be in part supported by the absence of a conclusive relationship between available sight distance on crest vertical curves and both collisions and operating speed in the literature.

Based on this paper, two main research areas are potentially important for code calibration: acceptable risk and statistical distributions. The literature shows several studies about the determination of socially and individually acceptable public risk in hazardous engineering projects. However, further research is required in the area of road safety. The importance of quantifying randomness in design parameters by means of statistical distributions should receive more focus. It is evident that the majority of studies that involve field measurements of design parameters focus on single percentile value, e.g. mean or 85th percentile, rather than the entire statistical distribution.

REFERENCES

1. Alexander, G., and Lunenfeld, H. (1985). *The Role of Driver Expectancy in Highway Design and Traffic Control.* Federal highway Administration, Washington, D.C.
2. Augusti, G., Baratta, A., and Casciati, F. (1984). *Probabilistic Methods in Structural Engineering.* Chapman and Hall, London.
3. Barnett, J. (1936). Safe Friction Factors and Superelevation Design. *Sixteenth Annual Meeting Highway Research Board*, National Research Council, Washington, D.C., Proceedings, pp. 69–80.
4. Ben-Haim, Y. (1994). *A Non-Probabilistic Concept of Reliability.* Structural Safety, Vol. 14, pp. 227–245.
5. Bolotin, V.V. (1964) *Mechanics of Solid Bodies and the Theory of Reliability.* In Mechanics of Solids, Proceedings of the Second All-Union Congress on Theoretical and Applied Mechanics, Vol. 3, pp. 68–82.
6. Butterfield, B. (1963). *From Donor to Patient-Another Safety Factor.* Transfusion, Vol. 3 (2), p. 125.
7. Carter, A.D.S. (1997). *Mechanical Reliability and Design.* Wiley, Macmillan, London.
8. Cornell, C.A. (1981). *Structural Safety: Some Historical Evidence that it is a Healthy Adolescent.* Structural Safety and Reliability, (Moan, T. and Shinozuka, M., eds.), Elsevier Scientific Publishing Company, Amsterdam, pp. 19–29.
9. De Leur, P. and Sayed, T. Development of a Road Safety Risk Index. *In Transportation Research Record: Journal of the Transportation Research Board, No. 1784*, TRB, National Research Council, Wahington D.C., 2003, pp. 33–42.
10. Ditlevsen, O. (1997). *Structural Reliability Codes for Probabilistic Design—a Debate Paper Based on Elementary Reliability and Decision Analysis Concepts.* Structural Safety, Vol. 19 (3), pp. 253–270.
11. Easa, S.M. (1993). *Reliability-Based Design of Intergreen Interval at Traffic Signals.* Journal of Transportation Engineering, ASCE, Vol. 19 (2), pp. 255–271.
12. Easa, S.M. (1994 a). *Reliability-Based Design of Sight Distance at Railroad Grade Crossings*, Transportation Research Part A, Vol. 28 (1), Pergamon Press plc, pp. 1–15.
13. Easa, S.M. (1994 b). *Design Considerations for Highway Reverse Curves.* Transportation Research Record 1445, Transportation Research Board, National Research Council, Washington, D.C., pp. 1–11.
14. Easa, S.M. (1999). *Reliability Approach to Intersection Sight Distance Design.* Transportation Research Record No. 1701, Transportation Research Board, National Research Council, Washington D.C., pp. 42–52.
15. Elishakoff, I. (1983). *How to Introduce Initial-Imperfection Sensitivity Concept into Design.* In Collapse: The Buckling of Structures (Thompson J.M.T., and Hunt G.W., eds.), Cambridge University Press, pp. 345–357.
16. Elishakoff, I. (1999). *Probabilistic Theory of Structures.* Dover, New York (first edition: Wiley, 1983), pp. 243–246.
17. Ellingwood, B. (1994). *Probability-Based Codified Design: Past Accomplishments and Future Challenges.* Structural Safety Vol. 13 (3), pp. 159–176.
18. Fambro, D.B., Fitzpatrick, K., and Koppa, R.J. (1997). *Determination of Stopping Sight Distances.* NCHRP Report 400, Transportation Research Board, Washington, D.C.
19. Fishburn, P.C. (1964). *Decision and Value Theory*, John Wiley & Sons.

20. Fitzpatrick, K., Carlson, P., Brewer, M.A., Wooldridge, M.D., and Miaou, S. (2003). *Design Speed, Operating Speed, and Posted Speed Practices*. NCHRP Report No. 504. Transportation Research Board, National Research Council, Washington, D.C.
21. Fitzpatrick, K., Elefteriadou L., Harwood, D., Collins, J., McFadden, J., Anderson, I., Krammes, R., Irizarry, N., Parma, K., Bauer, K., and Passetti, K. (2000). *Speed Prediction for Two-Lane Rural Highways*. Publication No. 99–171, Federal Highway Administration, US Department of Transportation, Washington, D.C.
22. Fox, M. (1989). *Elderly Driver's Perceptions of their Driving Abilities Compared to their Functional Visual Perception Skills and their Actual Driving Performance*. Taira, E. (Ed.). Assessing the Driving Ability of the Elderly New York: Heyworth Press.
23. Freudenthal, A.M. (1956). *Safety and Probability of Structural Failure*. Transactions ASCE, Vol. 121, pp. 1337–1375.
24. Freudenthal, A.M. (1956). *Safety and Probability of Structural Failure*. Transactions ASCE, Vol. 121, pp. 1337–1375.
25. Gertsbakh, I.B. (1989). *Statistical Reliability Theory*. Marcel Dekker, New York.
26. Grandori, G., Guagenti E., and Tagaliani A. (1998). *How to Measure Uncertainty?* Computers and Structures, Vol. 67, pp. 47–51.
27. Haight, F.A. (1986). *Risk, Especially Risk of Traffic Accidents*. Accid Anal & Prev., Vol. 18 (5), pp. 359–366.
28. Harr, M. (1987). *Reliability-Based Design in Civil Engineering*. McGraw-Hill, New York.
29. Hawkaas, T., and Der Kiureghian, A. (2000). *A Computer Program for Nonlinear Finite Element Reliability* Analysis. ICOSSAR *'01, 8th International Conference on Structural Safety and Reliability*, June 17–21, Newport Beach, California.
30. Ismail, K. and Sayed, T. New Algorithm for Calculating 3D Available Sight Distance J. Transp. Engrg., Volume 133, Issue 10, pp. 572–581, 2007.
31. Jessen, D.R., Schurr, K.S., McCoy, P.T., Pesti, G., and Huff, R.R. (2001). *Operating Speed Prediction on Crest Vertical Curves of Rural Two-Lane Highways in Nebraska*. Transportation Research Record No. 1751, Transportation Research Board, National Research Council, Washington, D.C., pp. 67–75.
32. Jones, J.L., and Jones R.E. (1982) *Determination of Safety Factors for Defibrillator Waveforms in Cultured Hearts*. American Journal of Psychology, Vol. 242 (4), pp. 662–670.
33. Kanellaidis, G., and Sakki, M. (1997). *Human Factors Issues in Highway Design*. Paper presented at the Human Factors Session, XIIIth I.R.F. World Meeting, Toronto, Ontario, Canada.
34. Kantowitz, B.H. (1992). *Selecting Measures for Human Factors Research*. Human Factors, Vol. 34 (4), pp. 387–398.
35. Keeney, R.L. (1980a). *Evaluating Alternatives Involving Potential Fatalities*. Operational Research, Vol. 28, pp. 188–205.
36. Keeney, R.L. (1980b). *Utility Functions for Equity and Public Risk*. Management Science, Vol. 26, pp. 345–353.
37. Leisch, J.E., and Leisch, J.P. (1977). *New Concepts in Design-Speed Application*. Transportation Research Record No. 631, Transportation Research Board, National Research Council, Washington, D.C., pp. 4–14.
38. Li, D., Dolezal, T., and Haimes, Y.Y. Capacity reliability of water distribution networks. *In Reliability Engineering & System Safety*, Vol. 42, No. 1, 1993, pp. 29–38.
39. Lind N.C. (1977). *Reliability Based Structural Codes, Practical Calibration*. Safety of Structures under Dynamic Loading, Trondheim, Norway, pp. 149–60.
40. Lind, N.C. (1978). *(i) Reliability-Based Structural Codes. Optimization Theory; (ii) Reliability-Based Structural Codes. Practical Calibration; (iii) Safety Level Decisions and Socio-Economic Optimization.* Safety of Structures under Dynamic Loading, Holland, Kavlie, and Moe (eds.), Vol. 1, Tapir, Trondheim, pp. 135–175.
41. Lind, N.C. (1996). *Validation of Probabilistic Models*. *Civil Engg. Systems*, Vol. 13 (3), pp. 175–183.
42. Lind, N.C. (2005). *Book Review*, Structural Safety, Vol. 27 (3), pp. 282–283.
43. Lindley, D.V. (1972). *Bayesian Statistics, a Review*, SIAM, Philadelphia, PA.
44. Lunenfeld, H.J., and Alexander, G. (1990). *A User's Guide to Positive Guidance* (3rd Edition) FHWA SA-90-017, Federal Highway Administration, Washington, D.C.
45. Marmor, M. (1982). *Aging and the Retina*. In R. Sekuler, D. Kline, and K. Dismukes (eds.). Aging and Human Visual Function, New York: Alan R. Liss.
46. Mayer, M. (1926). *Die Sicherheit der Bauwerke und ihre Berechnung nach Grenzkraeften anstatt nach zulaessigen Spanungen*. Springer, Berlin (In German).
47. McLean, J.R. (1976). *Vehicle Speeds on High Standard Curves*. Traffic Engineering: Sessions 20–27 Proceedings of the Eighth ARRB Conference, University of Western Australia, Perth. Australian Road Research Board, Vol. 8, Part 5, Session 21, (August 1976), pp. 1–8.

48. McLean, J.R. (1978). *Review of the Design Speed Concept.* Australian Road Research, Vol. 8 (1), pp. 3–16.
49. McLean, J.R. (1979). *An Alternative to the Design Speed Concept for Low Speed Alignment Design.* Transportation Research Record No. 702, Transportation Research Board, National Research Council, Washington, D.C., pp. 55–63.
50. Melchers, R. *Structural Reliability Analysis and Probability*, Wiley, Chichester, New York, 1999.
51. Messer, C.J., Mounce, J.M., and Barckett, R.Q. (1981). *Highway Geometric Design Consistency Related to Driver Expectancy.* Volume II, Research Report No. FHWA-RD-81-036, Federal Highway Administration, U.S. Department of Transportation.
52. Moyer, R.A., and Berry, D.S. (1940). *Marking Highway Curves with Safe Speed Indicator.* Presented at 20th Annual Meeting of Highway Research Board, Washington, D.C., pp. 339–428.
53. Navin, F.P.D. (1990). Safety Factors for Road Design: Can they be Estimated? Transportation Research Record No. 1280, Transportation Research Board, National Research Council, Washington, D.C., pp. 181–189.
54. Navin, F.P.D. Safety Factors For Road Design: Can They Be Estimated? *In Transportation Research Record: Journal of the Transportation Research Board, No. 1280*, TRB, National Research Council, Washington, D.C., 1990, pp. 181–189.
55. Oden, J.T., Belytschko T., Babuška I., and Hughes T.J.R. (2003). *Research Directions in Computational Mechanics.* Computer Methods in Applied Mechanics and Engineering, Vol. 192, pp. 913–922.
56. Petroski, H. (1996). *Invention by Design.* Harvard University Press, Cambridge, MA.
57. Rackwitz, R. (2002). Structural *Optimization and Risk Acceptability based on the Life Quality Index.* Structural Safety, Vol. 24 (2–4), pp. 297–331.
58. Rao, R.S. (1992). *Reliability-Based Design.* McGraw-Hill, New York.
59. Richl, L., and Sayed, T. (2005). *Evaluating the Safety Risk of Narrow Medians Using Reliability Analysis*, In print, Journal of Transportation Engineering, ASCE, Vol. 132 (5), pp. 366–375.
60. Roësset, J., and Yao, J.T.P. (2002). *State of the Art of Structural Engineering.* Journal of Structural Engineering, Vol. 128 (8), pp. 965–975.
61. Salmon, W. (1966). *The Foundation of Scientific Inference.* Univ. of Pittsburgh, Pittsburgh.
62. Sarhan, Mohamed; Hassan, Yasse. Three-Dimensional, Probabilistic Highway Design: Sight Distance Application. TRB 87th Annual Meeting Compendium of Papers DVD. Washington DC. 2008.
63. Sayed, T., and Navin, F.P.D. (1995). *Identifying Accident-Prone Locations using Fuzzy Pattern Recognition.* Journal of Transportation Engineering, ASCE, Vol. 121 (4), pp. 352–358.
64. Solomon, K.I., Chazen C., and Miller R.L. (1983). *Compilation and Review—the Safety Factors.* Journal of Accountancy, Vol. 156 (1), p. 50.
65. Taignidis, I. (1998). *Aspects of Stopping-Sight Distance on Crest Vertical Curve.* Journal of Transportation Engineering, ASCE, Vol. 124 (4), pp. 335–342.
66. Taignidis, I., and Kanellaidis, G. (2001). *Required Stopping Sight Distance on Crest Vertical Curves.* Journal of Transportation Engineering, ASCE, Vol.127 (4), pp. 275–282.
67. Taoka, G.T. (1989). *Brake Reaction Times of Unalerted Drivers.* ITE Journal, Washington, D.C., Vol. 59 (3), pp. 19–21.
68. Taylor, A.E., Martin D., and Parker J.C. (1982). *Tissue Resistance Safety Factors.* International Journal of Microcirculation, Vol. 1 (3), pp. 223–233.
69. Vrijling, J.K., Van Hengel, W., and Houben, R.J. (1998). *Acceptable Risk as a Basis for Design.* Reliability Engineering and System Safety, Vol. 59, pp. 141–150.
70. Wong, Y.D., and Nicholson, A. (1992). *Driver Behaviour at Horizontal Curves: Risk Compensation and the Margin of Safety.* Accid. Anal. Prev., Vol. 24 (4), pp. 425–436.
71. Zegeer, C., Twomey, J., Heckman, M., and Hayward, J. (1989). *Safety Effectiveness of Highway Design Features Volume II, Alignment.* Report No. FHWA-RD-91-045, Federal Highway Administration, Washington, D.C.
72. Zheng, Z.R. (1997). *Application of Reliability Theory to Highway Geometric Design.* PhD Thesis, University of British Columbia.

Efficient Transportation and Pavement Systems – Al-Qadi, Sayed, Alnuaimi & Masad (eds)
© 2009 Taylor & Francis Group, London, ISBN 978-0-415-48979-9

Analysis of mode split in the greater Toronto area: Long-range temporal trends and underlying travel behaviour

Amer Shalaby
Department of Civil Engineering, University of Toronto, Toronto, Ontario, Canada

Austin Shih
Urban Systems Ltd., NE Calgary, AB Canada

ABSTRACT: The objective of this study is to conduct an empirical analysis of long-range modal split trends (1986–2001) in the Greater Toronto Area (GTA) for the morning peak period (6–9 am) with examination of underlying factors. Furthermore, network service characteristics (travel times) were examined and results from both analyses were compared. The analysis revealed a widespread trend of declining transit modal shares in the GTA. Upon analysis of travel-related factors, certain factors such as population aging and the rapid growth of auto-dependent suburbs account in part for this trend. However, further analysis of mode split changes revealed persistent reductions in transit modal shares despite accounting for those 2 factors. The study of travel time data generated in EMME/2 showed that auto travel times have deteriorated at a faster pace than transit travel times, with severe traffic congestion a prevalent phenomenon in recent years. The effects of congestion are substantiated with evidence of peak spreading.

1 BACKGROUND AND OBJECTIVES

From 1986 to 2001, the GTA has seen substantial growth in population, employment and person travel. Over the same time period, the overall GTA road network has undergone limited capacity expansion. Considering that vital sections of the road network operated at or near capacity during the daily peak periods in 1986, the combined effect of travel growth and limited road capacity expansion has since led to a steady deterioration of traffic flow conditions particularly in the peak periods. Similarly, the transit network has undergone limited capacity expansion. Many transit routes have been operating at capacity, particularly during peak periods, and as a result travel times by transit have also deteriorated. However, it is believed that this rate of deterioration has occurred at a slower pace than auto travel times, particularly along major transit routes. These routes continued to have fully or partially exclusive right of ways, protecting the overall attractiveness of transit relative to the car.

The above trends suggest the overall attractiveness of auto travel (measured by travel times and reliability) has declined at a faster rate than transit travel. However, the Transportation Tomorrow Survey (TTS) data of 1986 through 2001 has revealed that auto travel continued to grow at a faster rate than transit, resulting in a modal split increasingly skewed in favour of auto. This trend has been observed in the morning peak (6:00 am to 8:59 am) modal split, where auto-driver shares increased from 57% in 1986 to 59% in 2001 and the auto-passenger share increased from 10% to 12%. Over the same time period, transit shares dropped from 22% to 16%. The seemingly conflicting trends of modal split and service characteristics need to be further investigated and confirmed, hence is the nature of this study.

This paper attempts to develop a better understanding of the underlying factors and the travel behavioural implications with respect to transportation policies. An investigation into the temporal trends of modal split by a number of travel related factors was performed first using the TTS data, followed by an analysis of temporal trends with regard to network service characteristics in EMME/2.

2 ANALYSIS OF TEMPORAL CHANGES IN TRAVEL-RELATED FACTORS

Before analyzing long-range mode split trends, it is important to understand the changes that occurred in various factors that could have played a role in effecting the observed modal split trends.

2.1 *Demographic and socio-economic characteristics*

This section is concerned with changes in the number and distribution of GTA households, housing type, household size, population, gender and age, as well as employment characteristics.

2.1.1 *Households*

The number of households in the GTA increased to just below 2 million in 2001. The majority of this growth occurred in the 905-Belt (the "suburban" regions surrounding the City of Toronto), particularly in York and Peel regions which are just north and south of the City of Toronto, respectively. In spite of this growth, Toronto continues to hold a significant portion of households in the GTA (47.8% in 2001).

Smaller-sized households (either one- or two-person households) have remained dominant since 1986 due to the increasing dominance of such households, more specifically, the growing supremacy of apartment dwellings in Toronto. This substantial imbalance towards apartment dwellings offset the increased growth of the housing market in the 905-Belt, which accounted for nearly 78% of the growth in the GTA households between 1986 and 2001.

2.1.2 *Population*

The GTA population increased by 12.1% from 1986 to 1991, followed by 8.6% from 1991 to 1996, and by 9.5% from 1996 to 2001. York had the highest growth rate among the six regional municipalities. Between 1986 and 2001, the 905-Belt has emerged as a major area for development and growth. The share of the GTA's population is now 56% in favour of the 905-Belt in 2001 over 52.5% in favour of Toronto in 1986.

Toronto's stagnating population growth rate does not bode well for the transit mode. Since the 905-Belt regions have inferior transit systems compared to Toronto, with the increased growth in suburban regions will only contribute to further reduction of overall transit mode share.

2.1.3 *Age*

As shown in Exhibit 1. a), in 2001 the "Baby Boom" generation (BBG: those born between the years 1946 and 1964) reached their peak driving age and whose travel behaviour will likely have a more significant effect on travel patterns. With an aging BBG, well into their forties and fifties, the population is becoming predominately older. This fact is further supported in Exhibit 1. b) whereby the median age in every region has increased substantially since 1986, with the GTA median age rising to 35.7 years old. This "Baby Boom Effect" is a factor to keep in mind as it is likely to cause an increase in auto-driver trips.

2.1.4 *Gender*

The older population (i.e. 25+) has a higher percentage of females. However, from ages 0 to 25 the male population size exceeds that of the female population, indicating that in coming years the gender balance will tip towards the male population. The expected result is increased male auto-driver mode split, as males are more prone to drive than females.

2.1.5 *Household workers*

After a steady decline and reaching a low of 1.34 workers per households in 1996, by 2001 the average number of household workers in the GTA rebounded to 1.39, despite the fact that household size has declined in general. It stands to reason that the GTA began to the recover from the recession in the early 1990s.

2.1.6 *Employed labour force participation rate (ELFPR)*

The Employed Labour Force Participation Rate (ELFPR), measured as the percent employed of the population between 11 and 98 (the age range considered in the TTS), declined from 1986 to 2001. Most of the decline was in full-time jobs, whereas participation in part-time and work-at-home jobs increased overall during the same time period. The decline in ELFPR was primarily due to the economic recession that hit most of Canada in the early 1990's. Similar to the trend in full time worker households, post recession recovery of the local economy spurred the ELFPR to 59.7%.

2.2 *Urban activity system characteristics*

Travel demand is a function of the urban activity system and the distribution of activities within that system. This section examines changes in spatial interaction between residential and employment locations in the GTA.

2.2.1 *Spatial interaction*

In this section, the following five spatial markets are employed among workers:

i. employed in PD 1, which includes Toronto's CBD (PD 1 employment market);
ii. residing and working in Toronto (Toronto—Internal);
iii. residing and working in 905 Belt (905—Internal);
iv. residing in 905-Belt and working in Toronto (905-Belt to Toronto); and
v. residing in Toronto and working in the 905-Belt (Toronto to 905-Belt).

The employment markets in PD1 and Toronto have declined steadily between 1991 and 2001. The stark decline can be attributed to the remarkable growth in the 905-Belt—Internal market. In the two five-year periods, the 905– Internal market reached a level of over a million workers (41.8% of GTA workers). The percentage of workers residing outside Toronto but working in it declined between 1996 and 2001, and the number and percent of workers residing in Toronto but working outside increased. This result indicates the potential increase of reverse-commuting between Toronto and the 905-Belt regions.

Based on these results one would expect trips originating from the 905-Belt to increase, therefore, intra-regional trips as well as inter-regional trips originating from the 905-Belt destined for Toronto to increase. Combining an aging population in a region of increasing employment participation, it is expected these travellers to have greater access to a vehicle thereby promoting auto-driver trips.

2.2.2 *Mobility characteristics*

Mobility characteristics include factors that affect one's access to a vehicle. In this section the characteristics of licenced drivers and vehicle ownership are examined.

2.2.3 *Licence ownership*

The driver's licence possession rate in Toronto remains the lowest in the GTA at 74% while the GTA has gradually peaked in 2001 at 81%. In spite of continual growth in the number of licenced drivers in Toronto, the proportion of the GTA licenced drivers who reside in Toronto declined steadily while the proportion of licenced drivers residing in York and Peel increased.

2.2.4 *Household vehicle ownership*
As shown in Exhibit 2. a), between 1986 and 1991, the majority of growth in GTA households was by households having 2 or more vehicles. In Toronto, the number of households having one vehicle increased more than any other household type. In contrast, the number of households having two vehicles increased in each 905-Belt region more than any other household type.

From Exhibit 2. b), between 1991 and 1996, the increase in the number of households having 0 or 1 vehicle accounted for most of the growth in the GTA households. In Toronto, households having no vehicles accounted for most of the local growth while households having two or more vehicles significantly dropped in number. From Exhibit 2. c), households with 2 or more vehicles constituted the largest growth in GTA as well as individual regional households from 1996 to 2001.

The above changes led to a reduction in the number of vehicles per household in all regions between 1986 and 1996. By 2001, the average number of vehicles per household increased in all regions, again likely due to post-recessionary recovery. This result obviously has great implications for travel behaviour. With increasing vehicular availability, the greater the likelihood of trips being made via the auto-mode.

2.2.5 *Trip characteristics*
Only trips with both ends located in the GTA were analysed. The study did not include trips by persons 10 years of age or younger since information on such trips was not available (except in 1986 when the TTS collected information on trips by persons aged between 6 and 10). Unlike analyses reported in previous sections, data collected in 1991 were not included.

2.2.6 *Number of trips*
The post-recessionary trends combined with accelerated growth in housing and population in the 905-Belt has resulted in an increased trip rate. The total number of person trips made in the GTA on a typical weekday during the morning peak period travel increased by 16.2% from 1996 to 2001 compared to only 13% between 1986 and 1996.

2.2.7 *Trip start time*
Trip start times are analysed between the hours of 5:00am and 9:59am. From 1986 to 2001, trip shares from the hour 8:00am to 8:59am remained stagnant near 40% of all GTA trips during the hours of 5:00am to 9:59am. The non morning peak period hours (i.e. 5:00am–5:59am and 9:00am–9:59am) experienced the highest growth rate in trips between 1986 and 2001; in particular, the growth rate in the hour 5:00am to 5:59am was over 90%, followed by 56% for hour 9:00am to 9:59am. This result hints at the existence of peak spreading, a phenomenon relating to congestion that is discussed in further detail in a following section. For now, it is worth noting this behaviour.

2.2.8 *Trip purpose*
Trip purpose is identified here by the purposes of both trip ends. This analysis considers four major trip purposes, which are further decomposed where appropriate. These trip purposes are: (i) home-based work, (ii) home-based school, (iii) home-based discretionary, and (iv) non home-based. The proportion of home-based work trips declined from 1986 to 2001, during which the proportion of home-based discretionary and non home-based trips increased. Between 1986 and 1996, the number of home-based work trips decreased sharply, though by 2001 the number of home-based work trips rebounded registering over 140,000 new trips and had the highest growth rate during this period (40%). This sharp increase in home-based work trips in the GTA translates into a possible increase in auto-driver trips generated in the morning peak.

The distribution of trips heavily favours home-based work trips by those residing in Toronto. However, this number has decreased from 1986 to 2001 (over 11%) and is now lower than the proportion of home-based work trips in the 905-Belt. York is the sole region to increase its home-based work trip share during this time.

Home-based work trip shares have decreased during the morning peak period, most severely between 7:00 am and 7:59 am. Home-based school trips are most concentrated within the hour 8:00 am and 9:00 am, which stands to reason since residence-to-school distances are typically short on average. Home-based discretionary and non home-based trips are spread between 7:00 am and 9:59 am. Only between 9:00 am and 9:59 am are there more trips of home-based discretionary type than home-based work.

2.2.9 *Peak spreading*

Exhibit 3. a) shows the change in the distribution of trip types by trip start time; as such an analysis of potential peak spreading can be performed. Peak spreading is a phenomenon which results largely from trip start times of trip makers altered in order to avoid periods of high congestion during the morning peak period.

Exhibit 3. a) reveals the possible existence of peak spreading between 1986 and 1996 centred on hour 7:00 am to 7:59 am, during which the share of home-based work trips declined. Subsequently, between hours 6:00 am–6:59 am and 8:00 am–8:59 am, only minor reductions in home-based work trip shares were observed. During non morning peak period hours, 5:00 am–5:59 am and 9:00 am–9:59 am, home-based work trip shares increased suggesting that congestion may very well be affecting mode split. Between 1996 and 2001, the peak spreading phenomenon is not as apparent, though still exists on a smaller scale.

In order to fully establish this theory, a travel time analysis by hour is performed later in the report.

2.2.10 *Trips by spatial market*

Based on previous sections, it is expected that more trips during the morning peak will be generated from within the 905-Belt, specifically in York and Peel. From Exhibit 3. b), by 1996, the percentage of trips made internal to Toronto fell to just below 42%. The majority of this loss was absorbed by the 905-Belt where internal trips and trips destined to Toronto from the 905-Belt increased. This trend continued into the final five-year period where Toronto-Internal person trip share was only 39% whereas the 905-Internal person trip share increased to 42%, surpassing Toronto as the largest trip generating market.

The influence of intra-regional trips in the 905 region has increased significantly between 1986 and 2001. The thrust of this growth is centred within the York and Peel regions. The two regions comprised nearly 73% of growth in trips within the 905-Belt between 1986 and 2001. For trips from the 905-Belt to Toronto, York and Peel represent the two highest GTA trip shares since 1986. However, only York region experienced consistent growth between 1986 and 2001. In contrast, Peel region trip shares have declined during the same period. Switching directions (i.e. trips from Toronto to the 905-Belt), York and Peel continue to generate the largest proportion of trips among the 905-Belt. In Peel and York, trip shares have dropped slightly by 2001.

The significance of these results is in the fact that growth of the 905-Belt is breeding self contained trips. Self contained trips within the 905-Belt is not beneficial for the transit mode since, as stated previously, the transit network in the 905-Belt is underdeveloped. Thus, this "905 Effect" is an important factor to consider when analysing transit mode split.

2.2.11 *Trip distance*

The weighted average straight-line distance of GTA trips increased slightly to 10.8 km between 1986 and 2001. However, the median straight line distance decreased from 6.1 km to 5.7 km during the same period. These results suggest that on average trips made during the morning peak period are longer in 2001 than they were in 1986, but the number of shorter distance trips being made has increased during the same period. This result can directly be related to the increase in intra-regional trips being made in the 905-Belt as explained in the previous section.

3 ANALYSIS OF CHANGES IN MODE SPLIT

Upon reviewing changes in travel related factors in the GTA, the next step is an analysis of mode split with respect to these factors. The mode split analysis is performed for the morning peak period alone (6:00 am to 8:59 am), for travellers between ages 11 and 98 years, living in either an apartment or house, and residing within one of the following regions, Toronto, Durham, York, Peel, Halton or Hamilton-Wentworth.

This section examines in further detail the changes in mode split by demographic, socio-economic, urban activity, mobility and traffic characteristics in order to deduce whether this trend of reduced transit mode split between 1986 and 2001 is a widespread or a localised phenomenon.

3.1 *Mode split by demographic and socio-economic characteristics*

3.1.1 *Gender*

Transit mode split has declined for both genders since 1986. Among males, mode split fell over 3% while among females the change was more substantial, nearly 10%. The majority of these lost shares were absorbed by the auto-mode, either auto-driver or auto-passenger. For males, the auto-driver mode split fell by little over a percent during the 15-year period, while auto-passenger, walk and other modes increased by 2.6%, 1.1% and 1% respectively. Female auto-driver mode split increased dramatically since 1986, nearly 8%. The auto-passenger mode experienced an increase of nearly 2% during the same period while remaining modes changed little.

3.1.2 *Age*

To develop a broad understanding of the trends in mode split with respect to age, critical age groups were combined to represent a separate stage in a typical traveller's life where travel behaviour is likely to change. These groups were: 11–15, 16–25, 26–40 and 41+. The majority of trips made in the morning peak period for travellers 26 years and older are made using the auto-mode (nearly 80%). Between 11 and 15 years of age, where the auto-driver mode is not available due to licence ownership restrictions, the largest mode split belongs to the walk mode (over 37% in 2001).

Between 1986 and 2001, auto-passenger shares increased among ages 11–15 and 16–25 to stand at 30% and 21.6% respectively. The transit mode split decreased in all cohorts, notably among 11–15 (10%). The latter two age cohorts (26–40 and 41+) experienced little change in auto-passenger and transit mode shares. The auto-driver mode experienced gains in age cohorts 26–40 and 41+ of 2% and 6% respectively, however the corresponding shares for age cohort 16–25 decreased by 4% during the same period. Walk and other mode shares, only prevalent among young travellers below 25 years, changed little save for a decline in the walk mode by 6% for the 11–15 age cohort and a 2% increase among 16–25 year old age cohort.

As shown in Exhibits 4. a) and b), from 1986 to 1996, the transit mode share declined in every age cohort between 11 and 80 years, significantly among the younger generation, between 11 and 24, as well as in older groups, between 55 and 79 years of age. In contrast, the auto-passenger mode split growth was most considerable between 11 and 19 and 66 and 79 years of age. For the auto-driver mode, beyond 35 years of age the magnitude of growth steadily increased.

From 1996 to 2001, the transit mode share was still declining in certain age cohorts but not to the extent found in the previous period. Younger travellers saw their transit shares decrease near 3%. Older age cohorts, those 65 and over, saw transit shares drop as much as 6%. However, transit shares for the interior age cohorts did not fare as poorly. Similarly, the auto-driver mode increased its mode share, with the age cohort 30–34 the only group to experience a decline. The most significant increase in auto-driver mode share occurred again in the older population, between 65 and 79. The auto-passenger mode suffered a steady decay in its mode share.

3.1.3 *Baby boom generation*

As mentioned previously, one important factor to consider when examining the trends in mode split over a 15-year period is the effect of the "baby boom" generation (BBG). Exhibit 4. c) shows the mode split by age, separating the baby boomers from non baby boomers. The mode split of the BBG is highly in favour of the auto-mode with the percentage of trips increasing from 75% in 1986 to 85% in 2001. Of this change, the auto-driver mode gained shares from both auto-passenger and transit. Auto-passenger shares fell from 8% to 6.5% while transit shares fell sharply by 9% from 22% to 13%. The non BBG has a far more distributed mode split. The auto-driver mode split has fluctuated between 1986 and 2001 while the auto-passenger mode has seen its share increase from 11% in 1986 to 16% by 2001. The transit mode share experienced a 4% decline to rest at 18% in 2001. The walk and other modes combine for roughly 20% of GTA trips, a significantly higher figure than its counterpart within the BBG (which combine for only 3% of GTA trips).

Overall, the BBG has a higher propensity for auto use than the non BBG. Nevertheless, in both age categories the same trend is seen, transit shares decreasing while auto shares increasing. Exhibit 4. d) illustrates the changes in mode split isolating the BBG within the regions of Toronto and 905-Belt to account for the effect of urban sprawl as well. The mode split of trips made by the BBG living within Toronto steadily leans towards the auto-mode. In 1986, the auto-driver mode split was 56%, and by 2001, this value increased to 66%. Transit mode shares decreased proportionally from 32% in 1986 to 22% in 2001. The remaining modes stayed constant during the analysis period. Turning towards the BBG living in the 905-Belt, the auto-driver mode share was already extremely high in 1986 (79%), but continued to rise above 90% in 2001, while the non BBG located within Toronto utilize the auto-mode the least, less than 52%.

The expectation is for the transit mode to be more competitive with the auto-mode within the City of Toronto. Furthermore, having discovered that non BBG has a lower propensity for auto use, it is more likely that this age group, located within Toronto alone, will have a large proportion of trips dedicated to transit and walk modes. In contrast, the BBG living within the 905-Belt are likely to travel using the auto-mode. Overall, the auto-mode usage has not decreased, even in competitive transit markets (i.e. Toronto) with low propensity auto travellers (non BBG). Moreover, transit usage continues to decrease, albeit at a slower rate, in these same regions. Hence, besides the factors accounted for previously, there are other unexplained factors that are adversely affecting transit mode shares.

3.1.4 *Employment status*

Employment status is stratified among full-time, part-time and home-based workers. Among full time workers, the auto-driver mode share increased by over 6% between 1986 and 2001 at the expense of both the auto-passenger and transit modes (whose mode shares fell 1% and 5% respectively). Walk and other modes possess an insignificant share of total trips. The auto-driver and auto-passenger modes for part-time workers increased from 41.2% and 12.3% in 1986 to 53.2% and 14.4% respectively in 1996. However, between 1996 and 2001, the auto-driver share fell by nearly 2%, and the auto-passenger share increased by less than 1%. The transit mode share experienced a steady decline since 1986, falling from 27.5% to 18.6% by 2001. The walk mode share fell from 13.4% to 8.6% by 1996, but rebounded to reach 10% in 2001.

3.1.5 *Mode split by urban activity characteristics*

The effect of spatial interaction on mode split is the focus of this section.

3.1.6 *Intra-regional trips*

Exhibit 5. a) examines the changes in mode split with respect to intra-regional trips by region of residence. Toronto possesses the most intricate transit system in the GTA, if not all of Canada, thus it comes as no surprise that the transit mode share is far superior in Toronto than any region within the 905-Belt. In spite of superior infrastructure, transit mode split for Toronto intra-regional trips between 1986 and 2001 fell from 35.6% to 30% during this time. In contrast, the auto-driver mode split remained relatively steady near 45% and the auto-passenger and walk modes saw their shares increase by 2% and 3% respectively.

Unlike Toronto, where the auto-mode share combined for only 55%, in the 905-Belt the auto-mode share ranged between 70% and 80% in 2001. In particular, by 2001 the auto-driver mode in York and Peel increased by 2% each, while the auto-passenger mode rose by 4% each. These increases came at the cost of all competing modes of travel: transit, walk and other.

3.1.7 *Inter-regional trips*

Exhibits 5. b) and c) illustrate the trends in mode split by inter regional trips from the 905-Belt to Toronto and Toronto to the 905-Belt respectively. Exhibit 5. b) shows that transit mode share has increased in all regions save York between 1986 and 2001. The auto-driver and auto-passenger mode shares in the 905-Belt, aside from York, fell. Exhibit 5. c) reveals the dominance of the auto-mode with respect to trips originating in Toronto destined for the 905-Belt. The transit shares fell steadily between 1986 and 2001 for most 905-Belt destinations, save Hamilton-Wentworth. The most severe decline occurred in York where transit shares fell from 11.6% in 1986 to 8.6% in 2001. In contrast, Hamilton-Wentworth experienced a moderate increase in transit mode shares from 1996 to 2001. In every region besides Hamilton-Wentworth, auto-driver and auto-passenger mode splits increased between 1986 and 2001.

3.1.8 *Straight line distance*

Between 1986 and 2001, the transit mode had not been competitive with the auto-mode for short-distance trips. The Transit shares for trips between 0–3 km and 4–15 km in distance decreased 7% each. For trips 16–30 km in distance, transit shares decreased to finish at 18.8% in 2001, while trips 31+ km increased in transit shares peaking at 17% in 2001. For trips less than 4 km in distance, mode split for auto-drivers and passengers increased by nearly 6% and 2% repectively. For trips between 4–15 km in distance, the auto-driver and passenger mode shares increased by 4% and 1.7% respectively. For trip distances 16–30 km and 31+ km, auto-driver shares increased by 2.6% and 2% respectively.

Between 1986 to 2001, the general trend observed is a shift from transit use to auto use. In the first period, the majority of this shift went to the auto-passenger mode. In the second period, this shift was towards the auto-driver mode. Thus transit has consistently been unable to compete with the automobile in short distance trips.

3.2 *Mode split by mobility characteristics*

Two markets ideal for transit use is examined in the following section: unlicensed drivers and households with no vehicles. Both expect to possess a large portion of transit mode split.

3.2.1 *Licence ownership*

From Exhibit 6. a), between 1986 and 1996 among licensed drivers, the auto-driver mode split increased from 72% to 75%. Auto-passenger shares increased mildly by less than 1%. Transit shares fell from 16.2% to 12.7%. Walk and other mode split remained stable. From 1996 to 2001, mode shares changed less than 1% for all modes. For unlicensed drivers, the most notable change is the dramatic decline in transit mode share, from 43% in 1986 to 26% in 2001. By 2001, the auto-passenger, walk and other mode shares all benefited from the declining transit shares.

For licensed drivers, the majority of trips made during the morning peak utilize the automobile (over 80% in 2001). However, for unlicensed drivers the majority of transit share loss was absorbed by the auto-mode. The most reasonable explanation for this result is the effect of the recession, whereby cheaper forms of travel benefited from cost-cutting travelers.

3.2.2 *Household vehicles*

As shown in Exhibit 6. b), between 1986 and 1996, the transit mode share declined sharply in households with no vehicles, from 74.1% to 65.3%, during which time auto-passenger, walk and other mode shares increased by 1%, 6% and 1.5% respectively. For one-vehicle households, transit shares fell 27.2% to 21.5% while auto-driver, auto-passenger, walk and other mode shares increased by 1%, 2.7%, 1% and 1% respectively. In multi-vehicle households, transit shares fell from 12.4% to 8.4%. Auto-driver, walk and other mode shares increased less than 1% each, while auto-passenger shares increased by 2.6%.

Between 1996 and 2001, mode split changed little in households with no vehicles. Among households with one vehicle, the auto-driver mode split increased from 50.4% to 51.5%. The remaining modes remained steady. Of multi-vehicle households, no changes of mode split over 1% was observed.

Another target market for transit service providers are those travellers living in households with no vehicles. From 1986 to 2001, a significant drop in transit mode share occurred which mirrors the trend among unlicensed drivers. During this time, the walk mode share improved, suggesting similar causal events, the recession. In one-vehicle households, a shift towards a cheaper form of travel occurred where auto-passenger mode shares increased at the expense of transit. This coincides with the straight line distance analysis made previously where the auto-passenger mode made large strides in mode split at the expense of transit when in direct competition.

3.3 *Mode split by trip characteristics*

This section presents mode split by trip factors, i.e. trip purpose and trip start time. A deeper look at the changes in mode split by trip start time may uncover more evidence of peak spreading.

3.3.1 *Trip purpose*

Exhibit 7. a) shows mode split changes with respect to trip purpose. Among home-based work trips, the auto-driver mode share increased from 63.2% in 1986 to 67.4% in 2001 at the expense of the transit mode share which fell from 23.5% in 1986 to 20.3% in 2001 and the auto-passenger mode share which fell from 9.2% in 1986 to 8.4% in 2001. The walk and other mode shares constitute a small amount of mode split and changed little.

For home-based school trips, both transit and auto-driver mode shares fell sharply from 31% and 9.4% respectively in 1986 to 19.6% and 7.2% in 2001. The remaining modes, auto-passenger, walk and other saw their shares increase as a result. This reflects a growing trend of students being driven to school as opposed to taking transit or walking. Home-based discretionary and non-home based mode shares of transit decreased slightly from 1986 to 2001. In general, the auto-mode dominates these two markets, in both cases accounting for more than 90% of all trips in 2001.

3.3.2 *Trip start time*

As shown in Exhibit 7. b), the distribution of mode split by trip start time remains relatively constant in all three periods. The transit mode share peaks during the hour 7:00 am-7:59 am. Between 8:00 am and 8:59 am, the auto-mode share is at its lowest while the walk mode share is at its highest.

As shown in Exhibit 7. c), from 1986 to 1996 the transit mode share for every hour between 5:00 am and 9:59 am declined, most severely between 5:00 am and 5:59 am which fell over 6%. Peak spreading is a possible explanation for the loss in auto-passenger and walk mode shares between 5:00 am and 6:59 am as the majority of the new home-based work trips beginning during this time are likely by auto-drivers. The auto-driver mode share increased in every hour, peaking during hours 5:00 am to 5:59 am and 6:00 am to 6:59 am.

Exhibit 7. d) shows the changes in mode split between 1996 and 2001 by trip start time. In every one-hour period, the auto-driver mode share increased and the transit mode share fell. Auto-passenger mode share remained relatively constant throughout while walk and other mode shares increased between 7:00 am and 7:59 am and decreased between hours 8:00 am to 8:59 am and

9:00 am to 9:59 am. The peak spreading phenomenon is clearly represented in Exhibit 7. d). The peak period discussed in Part 1 centres on 7:00 am and 7:59 am. Auto-driver mode share did not change during this time, though in the hours preceding and following, this mode share increased significantly.

4 ANALYSIS OF CHANGES IN TRANSPORTATION SYSTEM CHARACTERISTICS

4.1 *Changes in the transportation network*

Miller and Shalaby (2003) state that the addition of Hwy407, an electronic toll highway, was the only major capacity increase in the GTA highway network in recent time. Although used by many, the highway has not attracted much traffic from alternate roads and freeways (most notably Highway 401), mainly because of the relatively high toll during the morning peak. The transit network has also experienced modest growth, with the addition of the Sheppard subway line the most notable capacity addition. Thus, it is safe to state that transit and auto networks have not experienced high degrees of capacity expansion in the last fifteen years.

4.2 *Changes in network service characteristics*

Studies have shown that the levels of congestion on the road network, particularly on freeways, have been rising due to the increasing levels of travel demand and limited freeway network expansion. According to Shalaby (1998), a study reported that in 1986, 40% of the total GTA freeway network operated under severely congested conditions. This percentage increased to 60% in 1991. Although these conditions eased between 1991 and 1996 reflecting reduced numbers of work trips under recessionary conditions, but it is suggested these levels are noticeably increasing again as the economy recovers (Irwin et al.,1998).

Assuming insignificant changes in the transit level of service, the increased levels of congestion on the road network would likely mitigate any increase in auto-driver mode split and encourage public transit use and car-pooling.

4.2.1 *Estimated travel times*

In this study, the traffic assignment modules of EMME/2 were used to estimate auto and transit travel times in the GTA in 1986 and 2001. These estimated times were used to: 1) evaluate the changes in auto and transit level-of-service, 2) identify any links that may exist to corroborate the changes in mode split discussed in the previous section, and 3) substantiate the existence of peak spreading suggested previously in the paper.

It is important to note that transit travel times developed in this study are not *total* travel times but *in-vehicle* travel times. The total travel time includes all supplementary travel times between the origin and destination such as wait time, boarding time etc. Therefore, the results presented are not representative of the total time taken for a trip, but only time spent in the vehicle. The main reason for focusing on in-vehicle travel times is the difficulty associated with accounting for the weights of the different trip time components. Furthermore, auto travel times do not account for wait times of car-poolers or time spent travelling from parking locations to destinations.

4.2.2 *Travel time results*

Based on Exhibit 8. a), auto travel times have deteriorated at a faster rate than transit. The weighted average travel time increased by 65% for the auto-mode while only 12% for transit. In 1986, the weighted average travel time for auto trips in the GTA was nearly three times faster than transit; by 2001 it was less than double. This result is not surprising since it was shown previously that overall trips generated between 1996 and 2001 have increased significantly. However, in spite of increasing travel times, the auto-mode is increasingly the dominant means of travel in the morning peak period.

Exhibit 8. b) shows the changes in travel times by spatial market for both auto and transit modes. The auto-mode outperforms transit in nearly every spatial market. However, the relative gap in performance visibly narrowed for trips between the 905-Belt and PD1; the travel time from each region more than doubled between 1986 and 2001 except York. The changes in transit travel time were minor in comparison, mainly because such market is well served by high-speed rail services (subway and commuter rail), with their exclusive rights of way. The auto-mode is far more competitive intra-regionally than transit. For trips from the 905-Belt to Toronto, the rate of deterioration in the auto travel times is again higher than transit. Although transit remains less efficient, with continued growth and development of the 905-Belt, it is likely this gap will continue to shrink. Another result worth noting is the changes in travel times for the reverse commute. Trips from Toronto to the 905-Belt have increased (though their percent share of GTA trips remains relatively insignificant), but the effects of the reverse commute may be a factor to consider in future studies.

These results provide slight insight into the trends uncovered in the mode split analysis. The auto-mode, although still the most dominant means of travel, has seen its level of service (mainly travel times) deteriorated at a much faster pace than transit. Furthermore, this is a pattern witnessed in all spatial markets and not an isolated phenomenon. With the continued growth of the 905-Belt combined with the effects of the post-recessionary recovery period, it is obvious that congestion along major routes within and between these regions will increase, with little to no increase in road capacity.

4.2.3 *Peak spreading analysis*

Peak spreading is a phenomenon that is likely to occur with long distance trips, in this case, trips between the 905-Belt and Toronto. It is evident from the exhibit that on average, trips made during between 8:00 am and 8:59 am are faster than trips made in the previous two time periods. With home-based discretionary or non-home based trips peaking at this time combined with the recent increase in auto-driver mode share for trips less than 10 km, it is likely a large number of short distance trips occur during this hour. As shown in Exhibit 9. b), between 7:00 am and 7:59 am travel times are consistently the worst in every spatial market. As expected, for inter-regional trips the most rapid increase in travel times occurred between 8:00 am and 8:59 am. Furthermore, for trips from the 905-Belt to Toronto, the time periods preceding and following 7:00 am and 7:59 am experienced faster growth in every region save York. These quantitative results confirm the existence of peak spreading among inter-regional auto trips between 1986 and 2001.

5 SUMMARY AND CONCLUSIONS

This study has initially examined changes in travel-related factors in the GTA between 1986 and 2001. In general, the discussed changes imply increased total travel, increased proportion of discretionary travel, increased trip rate, increased dispersion of travel, longer travel times, increased shares by auto and GO Rail and reduced local transit mode split. Critical factors discussed include the Baby Boom and 905 effects, post-recessionary recovery, as well as evidence of peak spreading.

The study then examined changes in mode split with respect to the travel-related factors. In general, the trend of reduced transit mode share was a widespread phenomenon that extended throughout all travel-related factors in the GTA. Among individual spatial markets, transit shares suffered the most reduction in Toronto. However, between 1996 and 2001, the rate of decline in transit mode shares, has slowed.

The effect of the Baby Boom Generation on mode split was shown to be quite significant. In 2001, the baby boomers were in their prime driving years, thereby creating an influx of auto-driver trips. The mode split for the BBG revealed a sharp rise in auto-driver modal share, mirrored by an equally sharp decline in transit shares. The next step saw a mode split analysis on the non-BBG alone. In spite of removing the BBG, there still existed a decline in transit splits, albeit not as severe. Between 1986 and 2001, the 905-Belt experienced startling growth in population, households and employment. The result of this was an increase in the proportion of trips made by 905 residents, where now the 905-Belt generates more trips than the Toronto region. However, with inferior transit systems, there is a growing propensity among 905-Belt residents to use the auto-driver mode. This 905 effect is a growing trend and does not show any signs of slowing. As such, both the BBG effect and 905 effects were factored out and a mode split analysis was performed on the remaining population. It was shown that the trend of reduced transit modal shares persisted. This result is likely due to the onset of the recession era. Since the economic downturn in the early 1990s, fewer trips were being generated, causing less congestion in the road network, in-turn the auto-mode split increased. Already, Irwin, 1998 remarks this trend is reversing. Finally, evidence from the mode split analysis suggested the existence of peak spreading which was further investigated through a travel time analysis.

The study then examined the network service characteristics by identifying changes in the transportation system followed by an analysis of estimated travel times for both auto and transit modes. Firstly, it was identified that since 1986, besides highway 407, there have been no other major additions to the road network. Similarly the transit infrastructure has experienced modest increase in the 15-year period. The estimated travel times show that between 1986 and 2001, auto travel times have deteriorated far more rapidly than transit; a 65% increase in travel time versus 12%. However, in spite of increasing travel times, the auto-mode is still the most dominant means of travel in the morning peak period, outperforming transit in nearly every spatial market. The results lend credence to the post-recessionary theory; with the economic recovery will come an increase in congestion. This is most vividly shown for trips between the 905-Belt and PD1; save York, the travel time from each region more than doubled. Changes in transit travel times were minor in comparison. Among intra-regional trips, the auto-mode is far more competitive than transit, which is unsurprising given the transit systems in the 905-Belt are far less sophisticated than the one in Toronto. Another noteworthy result is the reverse commute and their increasing average travel times. As shown, trips from Toronto to the 905-Belt have increased (though their percent share of GTA trips remains small in comparison) but the effects of the reverse commute may be a factor to consider in future studies.

Based on the distribution of travel times within the morning peak period, there was sufficient evidence to suggest the existence of peak spreading among inter-regional auto trips between 1986 and 2001. This spreading centred upon at hour 7:00 am to 7:59 am, and travel times increased rapidly in the following hour, 8:00 am-8:59 am, for the entire GTA. Therefore, it stands to reason that congestion has played a significant role in the current trends in mode split.

Considering the following factors: the rate of decline of transit mode split is slowing, the 905-Belt will continue to grow and develop, the effects of the latest recession are waning, the baby boom generation is continuing to age while the baby boom echo (children of the baby boom generation) are on the rise, and the continued lack of capacity expansion of the road network; if one were to assume *ceteris paribus*, then there is a likelihood of an equilibrium effect occurring in mode split in the near future, followed by a period of transit share growth. However, these speculations are only qualitative in nature supported by some quantitative observations, and are by no means conclusive.

The study has shown that the phenomenon of declining transit mode share with increasing auto mode shares over the years has occurred in most stratifications (by demographic, socio-economic, urban activity and mobility characteristics) despite the fact that auto level of service has declined at a faster rate than transit level of service, narrowing the gap between the two modes with respect to level of service. This may suggest that other factors are responsible for such decline. One possibility is an increasing tolerance of auto travelers to congestion and declining tendencies to switch modes compared to transit riders.

REFERENCES

EMME/2 User's Guide Manual Release 9, INRO Consultants INC., 1998.

Irwin N., and Rubinstein M. GTA Transportation Plan Part 1: "Choices for the Future"—A Strategic Assessment. MTO. pp. 1.10–1.11. 1998.

Miller E.J., and Shalaby P.E. Evolution of Personal Travel in Toronto Area and Policy Implications. J. Urban Planning & Development, p. 11. 2003.

Shalaby P.E. Exploring Personal Travel Trends in the Greater Toronto Area. NSERC, IBI Group. 1998.

Shih, A. Analysis of Mode Split in the Greater Toronto Area: Long-Range Temporal Trends and Underlying Travel Behaviour. Thesis of Masters of Applied Science, Department of Civil Engineering, University of Toronto, 2004.

Welding, P. (1957) "The Instability of Close Interval Service." Operational Research Quarterly, Vol. 8, No. 3, pp. 133–148.

LIST OF EXHIBITS

EXHIBIT 7

a. Mode Split by Trip Purpose
b. Mode Split by Trip Start Time
c. Changes in Mode Split by Trip Start Time 1986–1996
d. Changes in Mode Split by Trip Start Time 1996–2001

EXHIBIT 8

a. Estimated GTA Travel Times
b. Estimated Travel Times by Spatial Market

EXHIBIT 9

a. Estimated GTA Travel Times by Hour
b. Estimated Hourly Travel Times by Region

EXHIBIT 1

a) Age profile for the GTA

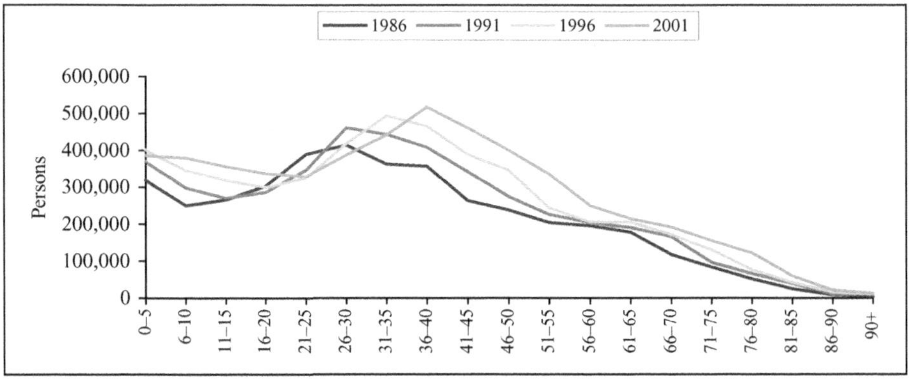

b) Median age of regional populations

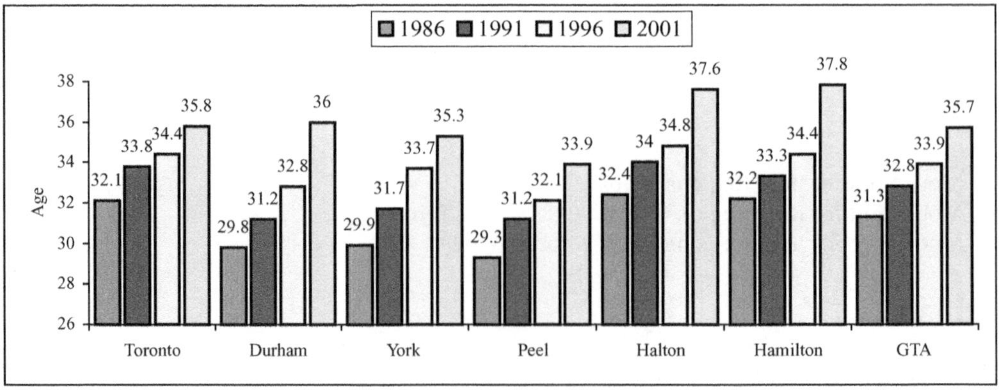

EXHIBIT 2

a) Change in households by number of household vehicles 1986–1991

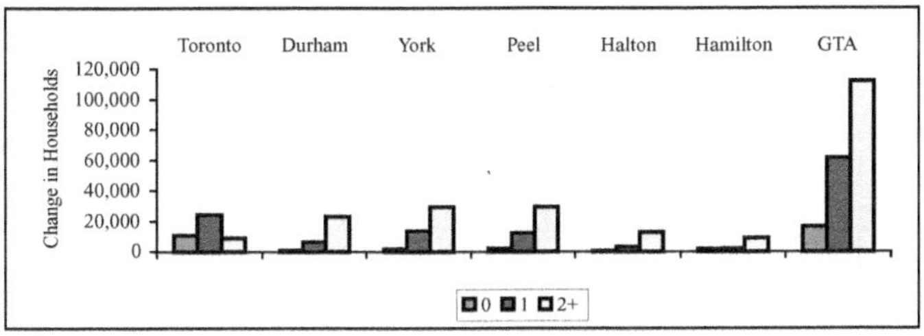

b) Change in households by number of household vehicles 1991–1996

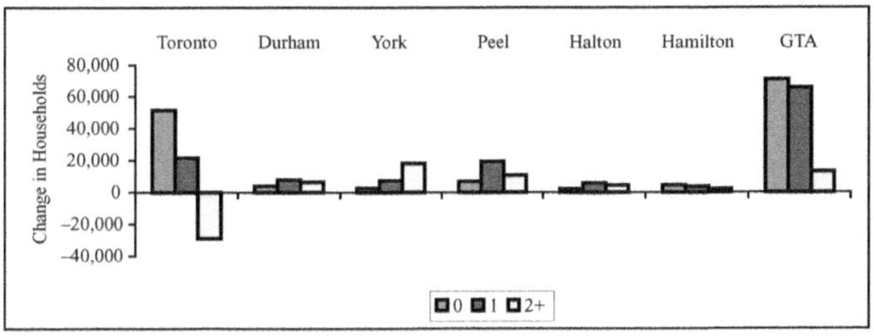

c) Change in households by number of household vehicles 1996–2001

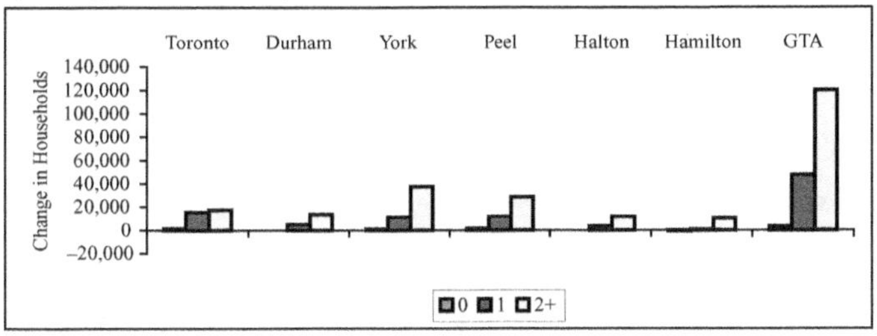

EXHIBIT 3

a) Change in distribution of trip purpose by trip start taime

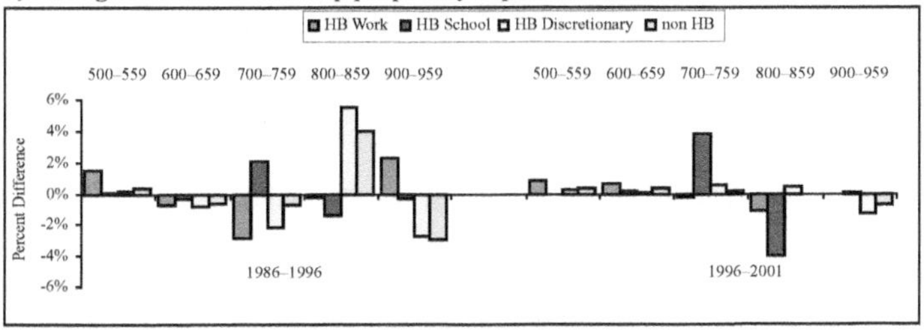

b) Distribution of GTA trips based on spatial market

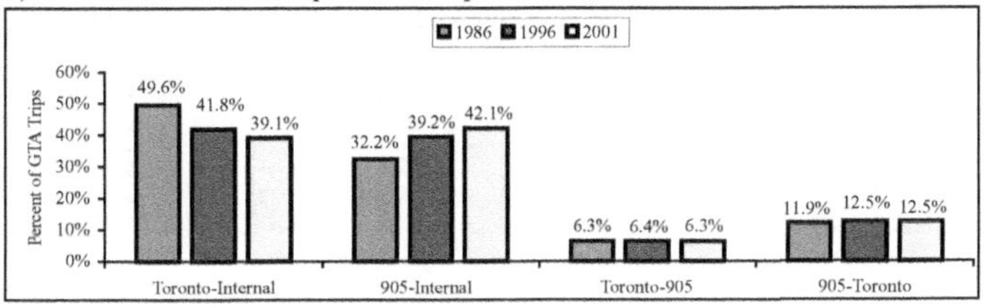

EXHIBIT 4

a) Changes in mode split by age 1986–1996

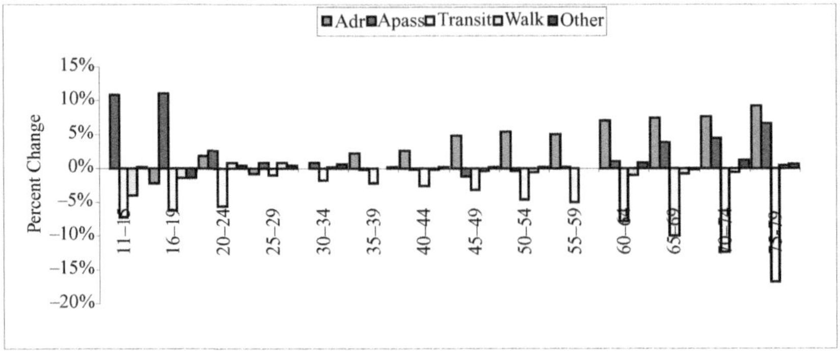

b) Changes in mode split by age 1996–2001

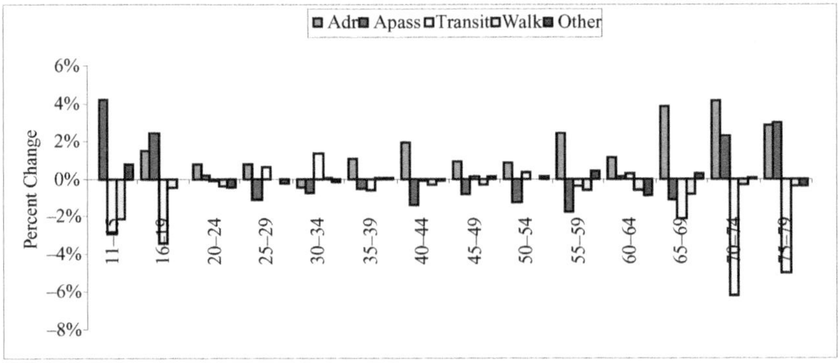

c) Changes in mode split by baby boom generation

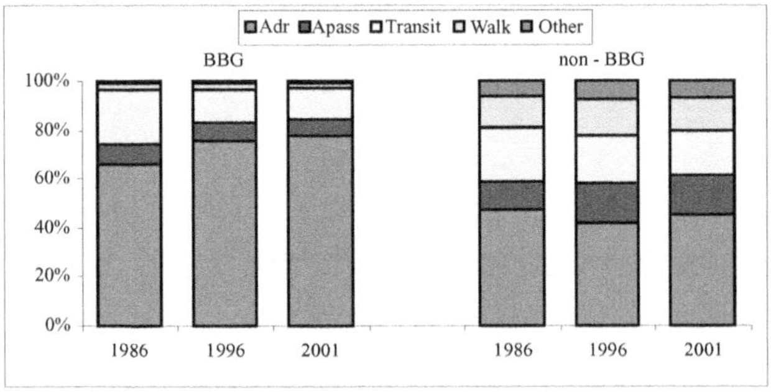

d) Changes in mode split by baby boom generation by region of residence

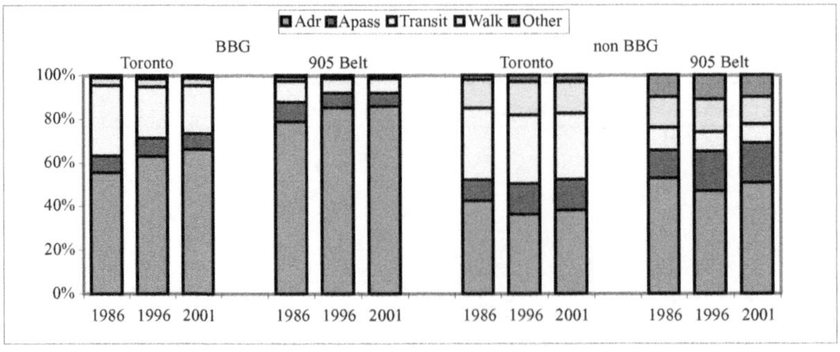

EXHIBIT 5

a) **Mode split for intra regional trips**

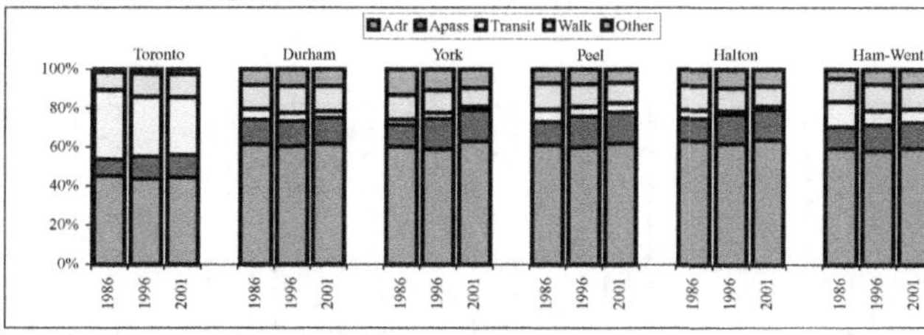

b) **Mode split for inter regional trips– 905 belt to toronto**

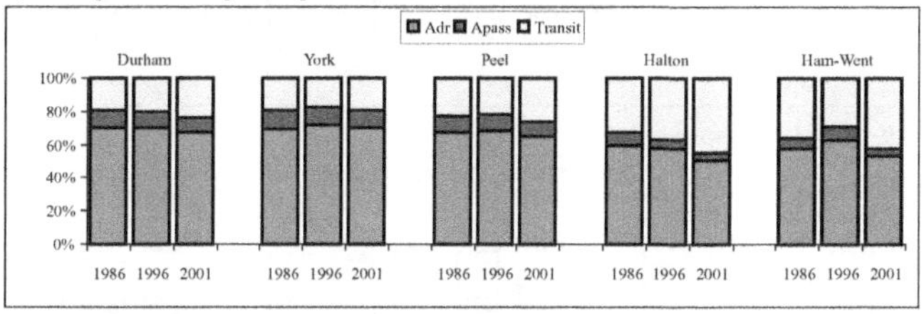

c) **Mode split for inter regional trips– toronto to 905 belt**

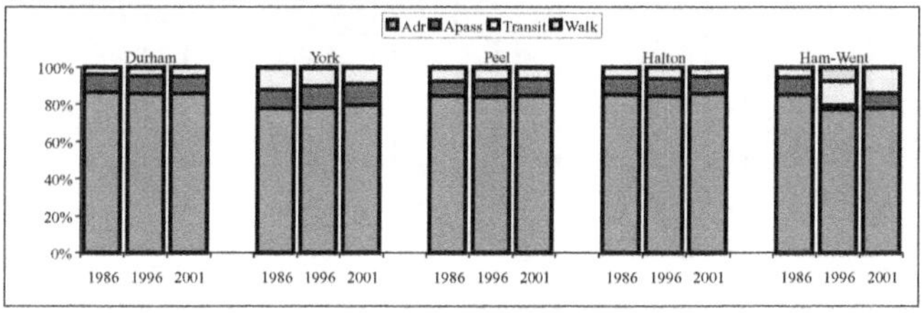

EXHIBIT 6

a) **Mode split by licence ownership**

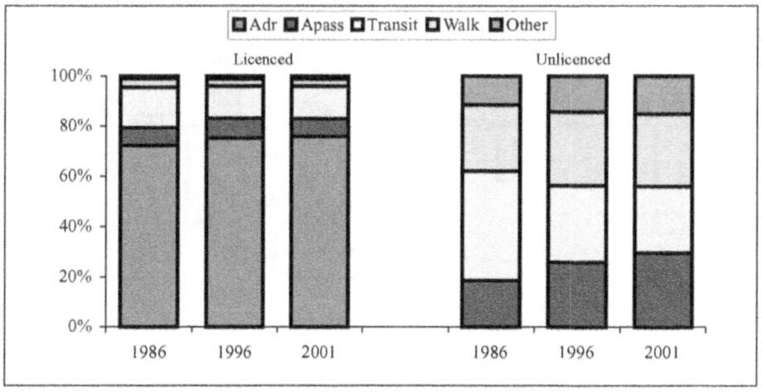

b) **Mode split by vehicles per household**

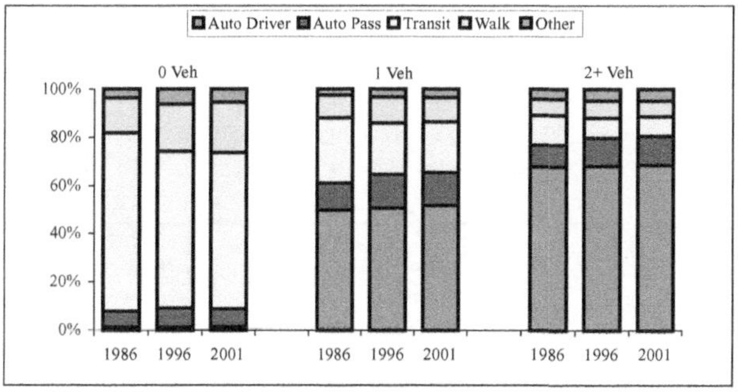

EXHIBIT 7

a) Mode split by trip purpose

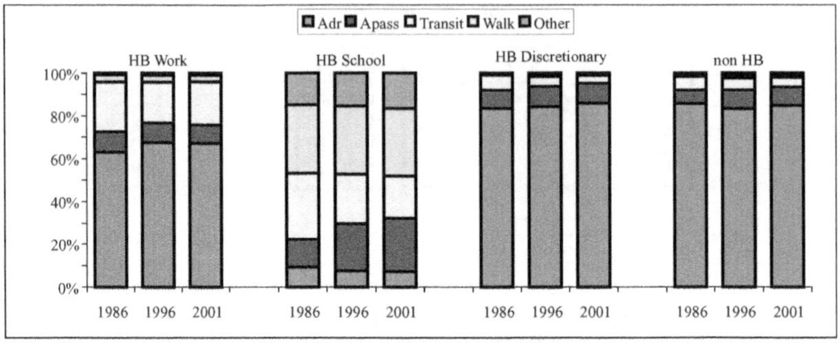

b) Mode split by trip start time

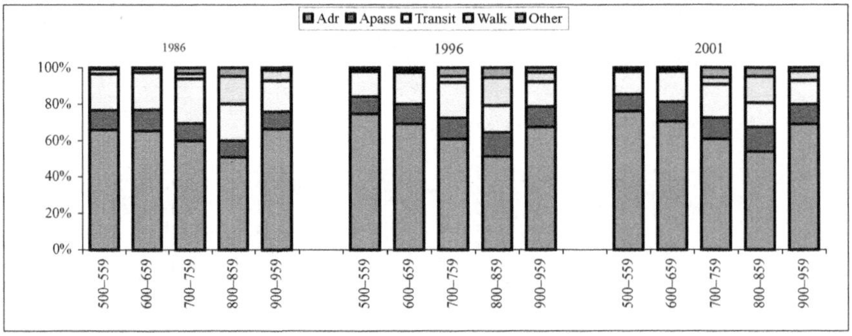

c) Changes in mode split by trip start time 1986–1996

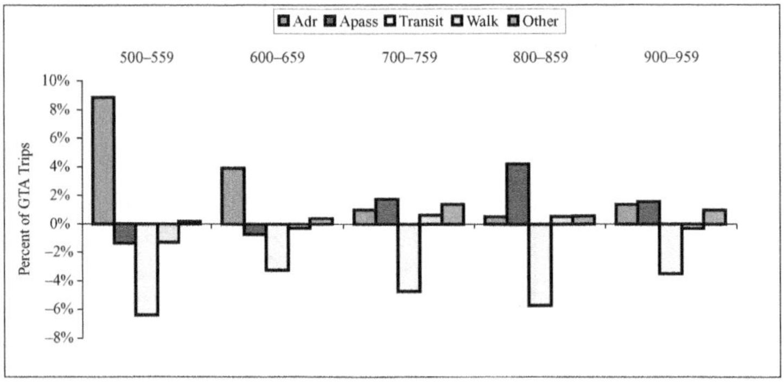

d) Changes in mode split by trip start time 1996–2001

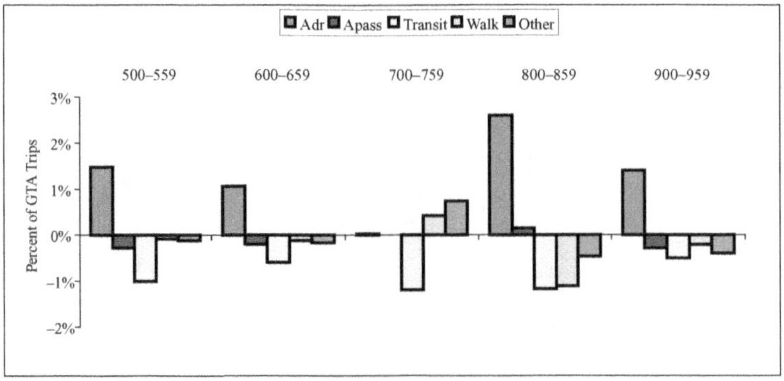

EXHIBIT 8

a) Estimated GTA Travel Times

	Auto			Transit		
	1986	**2001**	**% Change**	**1986**	**2001**	**% Change**
Weighted Average Travel Time	13.2	21.8	64.9%	36.8	41.1	11.8%
Average Travel Time	14.4	24.4	70.2%	36.8	46.6	26.7%
Median Travel Time	12.4	18.9	52.5%	31.4	37.7	20.1%

b) Estimated Travel Times by Spatial Market

			Self Containment			Toronto to 905			905 to Toronto			GTA to PD1		
		Region	**1986**	**2001**	**%Δ**	**1986**	**2001**	**%Δ**	**1986**	**2001**	**%Δ**	**1986**	**2001**	**%Δ**
Auto	Weighted Average Travel Times	Toronto	10.2	15.5	51%	-	-	-	-	-	-	12.8	29.4	130%
		Durham	9.4	12.7	36%	24.8	28.6	15%	29.1	69.7	140%	40.3	100.3	149%
		York	12.3	14.9	21%	15.1	21.9	45%	20.4	34.9	72%	29.0	42.6	47%
		Peel	9.3	14.5	56%	17.7	27.7	56%	19.3	42.6	121%	25.3	63.5	151%
		Halton	8.0	10.9	37%	28.9	41.3	43%	32.1	71.3	122%	31.3	84.0	168%
		Ham-Went	9.1	10.0	9%	43.7	54.0	24%	48.6	97.6	101%	40.6	109.3	169%
	Median Travel Times	Toronto	10.0	14.3	44%	-	-	-	-	-	-	12.9	31.3	143%
		Durham	7.7	11.9	53%	24.1	26.8	11%	28.1	69.7	148%	37.8	103.2	173%
		York	11.7	13.9	19%	15.1	21.5	42%	19.9	33.5	68%	27.0	70.5	161%
		Peel	9.2	14.4	57%	17.6	25.9	48%	19.7	42.4	115%	25.2	67.6	168%
		Halton	7.4	10.1	36%	27.4	39.5	44%	32.3	74.1	130%	36.1	83.5	131%
		Ham-Went	8.9	9.8	10%	43.0	52.6	22%	49.3	101.2	105%	49.8	109.5	120%
Transit	Weighted Average Travel Times	Toronto	33.1	35.8	8%	-	-	-	-	-	-	35.9	40.0	11%
		Durham	15.8	24.7	56%	75.6	91.0	20%	80.4	119.4	49%	79.5	125.3	57%
		York	16.4	18.9	15%	42.4	47.6	12%	62.5	101.3	62%	72.6	75.6	4%
		Peel	17.5	25.6	47%	57.9	59.4	3%	67.8	76.8	13%	69.5	71.0	2%
		Halton	14.3	25.4	78%	96.8	108.4	12%	91.5	114.9	26%	86.0	113.6	32%
		Ham-Went	14.3	20.1	41%	87.5	101.3	16%	78.5	91.5	17%	68.8	88.5	29%
	Median Travel Times	Toronto	29.7	31.7	7%	-	-	-	-	-	-	34.7	38.9	12%
		Durham	13.2	21.9	66%	81.0	88.5	9%	79.5	118.5	49%	79.1	122.6	55%
		York	11.4	13.8	22%	38.2	44.9	18%	64.8	70.2	8%	71.4	75.7	6%
		Peel	14.1	22.9	62%	56.3	57.7	3%	66.9	76.6	14%	68.4	79.2	16%
		Halton	12.6	19.0	51%	100.8	106.7	6%	89.4	116.1	30%	87.7	116.4	33%
		Ham-Went	12.8	18.8	47%	87.9	96.7	10%	70.3	89.8	28%	67.3	88.7	32%

Note - The median travel time was taken from OD pairs that generated demand; any OD pairs without demand were not included.

EXHIBIT 9

a) Estimated GTA travel times by hour

	Time Period	1986	2001
Weighted Average Travel Time	600–659	17.3	19.7
	700–759	18.7	21.5
	800–859	12.0	14.6
Median Travel Time [1]	600–659	36.3	39.5
	700–759	46.7	46.7
	800–859	36.2	40.9
Median Travel Time [2]	600–659	15.2	16.3
	700–759	16.1	17.0
	800–859	10.8	11.9

[1] Median from all estimated travel times,

[2] Median from travel times with corresponding demand.

b) Estimated hourly travel times by region

		Self Containment			Toronto to 905			905 to Toronto		
		Weighted Average Travel Time			Weighted Average Travel Time			Weighted Average Travel Time		
Region	**Time Period**	**1986**	**2001**	**%Δ**	**1986**	**2001**	**%Δ**	**1986**	**2001**	**%Δ**
Toronto	600–659	12.5	12.2	–2%	-	-	-	-	-	-
	700–759	15.2	13.9	–9%	-	-	-	-	-	-
	800–859	9.7	9.4	–3%	-	-	-	-	-	-
Durham	600–659	11.3	13.9	24%	27.7	26.5	–4%	36.8	47.3	29%
	700–759	10.4	12.0	16%	26.8	28.0	4%	43.1	51.7	20%
	800–859	7.3	8.2	13%	23.0	24.5	6%	29.7	39.9	34%
York	600–659	15.2	14.9	–2%	17.6	17.7	0%	26.0	26.0	0%
	700–759	13.6	13.5	–1%	18.2	19.7	8%	28.3	29.5	4%
	800–859	9.0	9.3	3%	14.9	18.4	23%	19.5	21.9	12%
Peel	600–659	10.5	11.4	9%	19.9	20.1	1%	21.7	26.9	24%
	700–759	11.5	12.8	11%	22.3	26.3	18%	29.8	34.4	15%
	800–859	7.9	8.3	6%	18.0	20.5	14%	19.3	24.5	27%
Halton	600–659	9.3	10.6	14%	28.5	30.9	8%	38.5	43.6	13%
	700–759	3.8	10.2	170%	35.2	37.5	6%	49.8	54.9	10%
	800–859	7.0	7.1	2%	29.7	33.8	14%	34.5	40.7	18%
Ham–Went	600–659	10.4	10.3	–1%	44.0	49.1	11%	56.2	61.9	10%
	700–759	10.4	10.0	–4%	49.9	49.9	0%	75.4	80.2	6%
	800–859	8.1	7.5	–7%	46.7	49.5	6%	54.7	61.4	12%

Efficient Transportation and Pavement Systems – Al-Qadi, Sayed, Alnuaimi & Masad (eds)
 ISBN 978-0-415-48979-9

Intermodal connectivity and the movement of goods for Dubai

Carl Berkowitz & Bassem Younes
Academy for Safety and Security, Dubai, UAE

ABSTRACT: Efficient intermodal transportation connectivity is critical to Dubai of the 21st century. The very foundation of this Emirate's infrastructure requires an intermodal transportation system that is fully capable of high efficiency and reliability for the movement of goods (food, raw materials and consumer products).

Intermodal connectivity points (also known as terminals) are provided and maintained in Dubai by both the public and the private sector. The public sector provides the basic infrastructure (roads, bridges, tunnels, transfer facilities, traffic signals, regulation enforcement, and etc.) and the framework that permits intermodal connectivity; and the private sector provides most of the vehicles, terminals and related infrastructure necessary to provide transportation services.

For Dubai to create the most efficient level of intermodal connectivity requires the development of policies and programs with comprehensive operating guidelines. Intermodal transportation is enormously complex, and affected by the social, political and economic aspects of society. It is difficult to identify common features between the different modes of transportation and the different operators and there are numerous players with many different perspectives of the best methods of doing business.

This paper suggests guidelines and best practices in the operation of intermodal connectivity points within and between modes, so that each can work properly and freight can easily connect; and offers an opportunity to think "out of the box" and encourages a fresh look at the connectivity problem. It must be remembered that freight is becoming increasingly mode invisible with performance (time, cost, and reliability) determining the choice of mode; and where the ability to interchange between modes and vehicles should be in a timely cost-effective manner where performance is measured on the ability to interconnect in order to optimize the end-to-end movement with the most efficient system.

1 BACKGROUND

Dubai's current population of 1.4 million is projected to grow to 5.4 million by 2020; and recent transportation studies indicate that the road system is already congested and lacks the capacity to move traffic. As a congestion countermeasure, Dubai plans to invest heavily in upgrading the existing transportation infrastructure by focusing on: optimizing existing resources, improving public transportation (light rail and bus), airport expansion and a new international airport at Jebel Ali, expanding the port facilities, establishing new intermodal transportation and logistic centers, expanded warehouse and distribution center capacity, and expanding the existing road and bridge network.

Currently Dubai's International Airport handles more than 1.4 million tons of air cargo. Over the past two years (2005 to 2006), the airport has experienced a 6 per cent growth in cargo, and this is projected to further increase in 2007. Expansion plans for the airport include a new logistics center that will separate freight and cargo operations and increase the cargo capacity to 4.5 million tons by 2020.

The new Jebel Ali International Airport will have an integrated logistics and intermodal transportation center (located next to the world's third largest free zone) serving the new airport and the nearby port facilities. The airport will be capable of handling 12 million tons of freight and accommodate the new A380–800 super freighter.

Dubai's Ports at Port Rashid (35 berths) and Jebel Ali (71 berths) handled 110 million tons of cargo in 2006 (inbound, outbound and transshipment) an increase of 19 per cent over the previous year. The port experienced a 23 per cent growth in container traffic during the first five months of 2007 and presently ranks among the top ten in the world, having handled 4.2 million TEU's.

2 OVERVIEW

Each transportation mode or vehicle change involves system-waiting time and experience has shown that perceived penalties from delay are often greater than the actual delay or the cost. Since everything ultimately is translated in time and money, it is important to lower the time/cost penalty and reduce the risk of uncertainty through better knowledge and good modal and intermodal interchange.

One of the key concerns of intermodal connectivity is in the area of facility responsibility: who owns and operates the interchange and if improvements are to be made who is responsible, if a new interchange is needed who will build it and how will improvements of the existing and the development of new interchanges be financed.

Although highly visible to the industry, the general public and government officials are not fully aware of the intermodal connectivity needs of the system. To move forward, there must be a union of thought . . . that it is important to have an intermodal transportation system where there is:

- A wider availability of information that is shared on a collaborative basis.
- Constant consultation with public agencies, industry groups interested or affected and the general public.
- Partnerships between the public and private sector.

Intermodal connectivity is not an end unto itself, it is only one aspect of an integrated transportation policy, and other aspects include land-use, environmental concerns, public safety, national security and coordination with other systems. The intermodal connectivity goal is to establish a set of principles and develop guidelines for the planning, design, management and operation of facilities with a focus on schedule coordination, layout and design of connectivity points, establish information systems to promote best operating and management practices in order to improve what already exists and direct the creation of new facilities that maximize available resources.

An intermodal connectivity point is best described as a facility where transportation services transfer between one or more modes is practical and cost-effective, or can be made so. The physical consideration is the infrastructure which has a wide range of features, including quality of access from the user's point of view, transportation network linkages, availability of surrounding land use for complementary development and improvements and the quality of the surrounding area.

Some general guidelines when considering intermodal connectivity points include:

1. Serving the heaviest point to point demand without mode transfer (direct and as fast as possible).
2. Existing routes and services should be adjusted to offer cost and time effective increased interchange opportunities.
3. Attention should be given to improving or creating connectivity points where high demand routes meet.
4. Traffic and congestion management measures should be adopted to provide improved access.
5. Major connectivity points should be located at or near major destinations.
6. Interlinking of schedules should be emphasized in relation to overall travel time and improving predictability of the total trip.

7. Connectivity arrangements should be tightly specified and all operators should cooperate and respect the schedules of each service.
8. Planning of schedule and schedule changes should be a matter of policy consistent with commercial considerations.
9. At each connectivity point one organization should be responsible to ensure that connections are made.
10. Information to assist the system users is needed at every stage of the goods movement whether a transfer is involved or not.
11. Connectivity points should be located to allow for convenient interchange between the different modes involved where there is sufficient land to accommodate the facility

3 SECURITY AND CONNECTIVITY

The impact of terrorism is unthinkable but it is now part of the intermodal connectivity lexicon and terminals are now faced with planning and operating under the potential of a future terrorist disruption to operations. The intermodal system is resilient but not sufficiently elastic to sustain new constraints or periodic disruptions and, as a result of this new reality, new choke points have been added to the connectivity equation. Intermodal connectivity must now address the issue that it will no longer be business as usual and transportation organizations home and abroad must add "what if" scenarios to their planning and operating process.

The intermodal transportation system and businesses have in recent years moved to lean manufacturing and just-in-time inventory management where any disruption can result in serious problems to the logistics and supply chain system. Logistics planners in response to terrorism are beginning to rethink just-in-time delivery systems that were predicated on parts, materials and goods arriving, as they are needed and now reconsidering keeping higher levels of inventory to guard against future disruptions (just-in-case inventory).

New issues have emerged with missed deliveries, goods stuck in the pipeline affecting well-established global buying practices. Ports are becoming a bottleneck and will continue to be a problem until security procedures are streamlined.

3.1 *Information technology the key to connectivity*

The most important business for an intermodal connectivity point is their communication system delivering information that is accurate, timely and available wherever and whenever needed. This type of system allows on-line, real-time communication with all the management elements throughout the system and provides all the primary and supporting information required by anybody and everybody.

Most intermodal connectivity point information systems are not integrated and are stand-alone or business specific. Unfortunately very few facilities have information networks that can give flexibility, reliability and consistency of service to all the different customers using the facility; intermodal facilities have to be effective and efficient.

Intermodal connectivity points are often referred to as a knowledge-based operation that generates the information to make smart, cost-effective decisions and plan day-to-day operations for long-term profitability. Every facility can be best described as an information organization, as such, information is the glue that holds the operation together and therefore information accounts for the facilities competitive advantage and operating efficiency over traditional non-intermodal operations.

To have successful intermodal connectivity it is important to understand information and the process that generates it, as it is the essential ingredient to good decision-making. Successful connectivity relies on organizational agility and the ability to increase the speed, accuracy and efficiency of gathering important information and get the collected information to the right person at the right time and at the right place.

Information is more than data, computers or communication systems, and is dependent upon skillfully acquired technology and services that is managed, understood, and used to build organizational proficiency. The key idea is to work smarter not harder, work at remote locations, keep informed and get closer to the customer.

Access to data is the key word and the key objective is to use information to improve the infrastructure through integration, flexibility, agility and standardization in order to reduce costs and increase throughput. A byproduct of quality information management is increased control, better information, better integration and an improved quality and level of service.

In developing information systems, legacy data applications have to be taken into consideration. These applications can be bridged into web-based solutions enabling access by a wide variety of users at scattered locations and any person or organization is able to view in real-time all the information they need.

A large number of networks can be reduced to one network, which can deliver reliability, consistency, serviceability and access to a wide variety of applications. With a single infrastructure supporting all different applications sets (data, voice and video) while meeting the security requirements of the system users, connectivity can be delivered which gives multiple customers access to a common set of data, is adoptable to technology and business needs changes, and results in major improvements in service levels and efficiencies in a single high-performance network.

3.2 *Connectivity*

Experience has demonstrated that the success of an intermodal connectivity point is dependent on functionally integrating all the system components, no matter what level of technology is utilized. To achieve high levels of connectivity, it is necessary to analyze the functions and operational relationship of the system's individual parts. The connectivity point must be viewed as a total system and the various functions as subsystems.

3.2.1 *System performance*

The first step is the process of transforming operational needs into a description of a system performance configuration through a process of definition, synthesis, analysis, design, testing and evaluation. System performance essential components are operating characteristics and growth assumptions, processing and flow performance, level of service performance and space criteria, operating capacity and environmental, schedule and cost constraints.

- Operating characteristics and growth assumptions—the basic composition of the facility population, visitors and the various types of employees that move through the facility must all be defined. These characteristics are not just percentages of originations or terminations at the connectivity point but also must include how many vehicles use the facility and arrival time distributions.
- Processing and flow performance—design and operational optimization studies to address processing rates for the major elements and security checkpoints.
- Level of service, performance and space criteria—the performance measures for the entire facility are often somewhat subjective since they are influenced by policies of the governing authorities and generally include maximum waiting time, maximum transfer time, space per container, space per chassis, etc.
- Operating capacity—demand assessment is the fundamental element of the system and forecasting demand is the common element. Typically simulation models can be used to analyze the operational capacity to determine the projected growth rate that can be sustained and to establish the constraints that would affect the facility capacity. (Capacity analysis is critical throughout the design, construction and operational phases.)
- Environmental, schedule and cost constraints—these constraints are broad in definition and not suited to eliminating options at the beginning of the systems analysis process; they form the basic framework for the key evaluation criteria.

3.2.2 *Functional subsystems*

The subsystem's function is to integrate the related technical parameters and program interfaces in a manner that optimizes the intermodal connectivity point. Subsystems are defined as the component parts of the facility and the functional elements that comprise the overall transportation access and circulation facilities that include: access roads, vehicle processing, and storage facilities. Also, it is critical to understand the interaction between functional subsystems, since performance constraints on one subsystem have a ripple effect throughout the system.

3.2.3 *Operational results*

Operational results are to specify the reliability, maintainability, safety and survivability, human and other subsystem factors and integrate them into the total facility operation effort to meet cost, schedule and performance objectives.

It is essential to evaluate all the major functional areas within an intermodal connectivity point and to assess the facilities performance. A careful evaluation would include the following areas: demand estimates, performance analysis, operational capacity analysis, interface/infrastructure assessment, lifecycle costing and coordination of facilities and subsystems.

- Demand estimates—analysis of the flow of activities on the subsystem elements over the entire period of doing business covering the facility during peak and off-peak. (All subsystems may not peak at the same time and may have peak demand at completely different times.)
- Performance analysis—addresses subsystem variables, such as level of service, frequency of service, equipment sizing, throughput and operational speeds.
- Operational capacity analysis—determined from assessing the system and subsystem demands and performance.
- Interface/infrastructure assessments—physical interfaces and infrastructure need to be evaluated, including the space that is available for equipment, maintenance, utilities, and etc.
- Lifecycle costing—an estimate of capital cost, operation and maintenance cost, as well as a detailed analysis of cash flow and return on investment.
- Coordination of facilities and subsystems—increasing the size and capacity of the subsystem, ability to reconfigure, the subsystem safety, hazard assessment and flow rates.

With this information in-hand, the following intermodal connectivity point issues can be addressed:

1. How does the connectivity point relate to the overall transportation system?
2. What is the ability of the connectivity point to accommodate short-term and long-term expansion?
3. What is the facilities adaptability to current and future technology advances?
4. Is there flexibility for alternative equipment handling and storage modes?
5. What is the accessibility to various modes?
6. Is the facility user friendly for operators and customers?
7. What is the cost-effectiveness and space efficiency?

4 THE IMPORTANCE OF INTERMODAL FREIGHT

Intermodal transportation is an essential economic activity and should be targeted for special treatment by policy planners in order to develop and improve connectivity. Today, trucking dominates Dubai's intermodal freight transportation.

In order to improve connectivity points, it is important to fully understand the intermodal system, its role, its complexities and how cargo travels from the shipper to the receiver. The shipper has several options:

1. Use their own trucks to deliver the goods, usually directly to the receiver and insure reliability, control, customer relations, speed of delivery, flexibility and cost.

2. Retain a commercial carrier to pick-up the freight and deliver it to the receiver. This trucker will collect the freight from the shipper and if it is a full truckload, deliver it directly to the receiver (door-to-door operation). If the freight is less than a full truckload, the trucker may first deliver it to their terminal where it is consolidated with other shipments going to the same receiver or to the same general geographic area so that several shipments will be delivered on one truck. Within one Emirate, there is generally one terminal involved, but if the freight is traveling from one Emirate or Gulf State to another (referred to as line haul), there is generally a terminal operation involved at the destination. At the terminal end of the journey, the freight is deconsolidated and the shipments are individually delivered to the end receiver. The trip between terminals can be by truck, or transferred to another mode to move by sea or air. (A proposed rail freight system for the United Arab Emirates and Gulf States offers even greater potential for economies of scale and a way to avoid heavy road congestion.)
3. Secure the services of a third party (freight forwarder, logistics company, and etc.) and outsourcing the transportation function between the shipper, carrier and receiver. The third party may own transportation or may arrange service on behalf of a client. The primary role of the third party is to consolidate freight from various shippers in order to take full advantage of the economy of scale.

To optimize intermodal connectivity points the system needs to address several key aspects, such as:

1. Large number of participants (shippers, third parties, transportation firms, drivers, terminal operators, transportation authorities, government agencies, customers) in the process some concerned with: supply and others with demand, the way in which goods move within and to and from one location to another and that one single freight movement may be associated with several truck trips. Many of the connectivity issues are on the supply side and the key to understanding the process is on the demand side. The demand for freight arises from the demand for goods and the sole function of the intermodal system is to move goods from origin to destination to be consumed, processed, repaired, modified, stored, disposed of, reshipped, or used.
2. Intermodal transportation provides a means of serving a given level of demand and the supply of transportation is in response to the demand for goods. The transportation equipment, connectivity points and labor pool are the resources needed to provide the transportation supply in response to the demand by business, government and individuals. The level of supply is determined by decisions affecting the size and capacity of the fleet, the transportation network, connectivity points, level of services, scheduling, routing, road congestion, etc.

5 VARIABLES THAT AFFECT CONNECTIVITY

The intermodal system is essentially concerned with the efficient and reliable movement of goods and its impact on connectivity will depend on the commodity type, state (solid, liquid, bulk, frozen, etc.), mass, volume, value, number of items, if perishable, if fragile, degree of hazard, other special characteristics, such as: just-in-time, time sensitivity, special attention and frequency of delivery.

Connectivity can be adversely impacted by the types of buildings permitted, the availability and the type of loading and unloading facilities, special storage area requirements, access and egress conditions and the size of the facilities. All these controls will determine the throughput capacity of the intermodal connectivity point.

The key component on the supply side is the fixed infrastructure dedicated to the movement of freight (connectivity points, roads, bridges, tunnels, etc.). Some of the factors that impact connectivity are the infrastructure quality, distances between facilities, travel time, traffic control, equipment available, bridge clearances, load limits, restrictions, physical capacities, environmental restrictions, adjoining land uses, special needs, on-line movements, inter-line movements, intermodal connections, container loading or unloading, material handling equipment, vehicle servicing, throughput capacity and hours of operation.

All freight is transported locally by truck, and its modal characteristics affect intermodal connectivity points. The most important characteristics for trucks include: configuration, body type, mass, dimensions, number of wheels, axels and axel spacing, load and volume capacity, type of ownership, operational control, any special features associated with the use of the truck and the need for special permits. There are similar requirements for the non-road vehicles; consideration is also given to trip purpose, time of day, origin and destination characteristics, movement scheduled or on-demand, unit-load or multi-shipments and whether it is a single or a multi-origin-destination movement.

6 PRIVATE SECTOR AND GOVERNMENT

The role of planning for intermodal connectivity is essentially in the hands of the private sector. This sector is primarily composed of the trucking companies, third parties, facility operators, shipping companies, airlines, and the support operators and is composed of a large number of different size individual firms with an ability to serve its market. Marine ports and airports are operated by public authorities, which may have more in common with the private sector then with the government sector. The private sector when it takes into account its cost of doing business is only concerned with its cost and not the external costs.

The intermodal industry and the people directly and indirectly involved (shippers, transporters, customers) generally accept the inefficiencies of the system infrastructure and realize they have minor influence over government to foster major efficiencies in the system. The private sector has the view that it is a level playing field and that any major improvements in the transportation infrastructure system efficiency goods movement has no major impact on profit margins. Because of the competitive nature of the industry, any cost reductions are passed along to the customer.

To establish optimal intermodal connectivity points, the Dubai government must play a more active role in the area of infrastructure development and continue to play a minor role in the day-to-day operations of the private sector; thus ensuring that the market place will continue to operate competitively and efficiently.

A government's role is inevitable in intermodal connectivity because of its influence on economic activities and transportation policies. Government policy interests are most significant at intermodal connectivity points that are publicly owned, such as marine ports and airports; since the responsible agency has operational and investment interests concerned with various aspects of location, investment, design, operating efficiency, connecting to the highway system and other related areas.

Government is also responsible for road traffic management and right-of-way design to accommodate the physical dimensions of the vehicles using the transportation infrastructure; and attention is given to infrastructure design features, such as lane width, pavement depth, overhead clearances, road gradients, and to the elements of infrastructure related to the operation of truck traffic.

The movement of goods on the road network is important and from the intermodal connectivity point of view has to receive greater government involvement in order to facilitate truck movements, or controlling truck routes, introduce intelligent transportation system technology and other traffic management efficiency techniques.

7 INTERNATIONAL TRADE AND CONNECTIVITY

The growing importance of global trade to Dubai's economic position depends upon the efficient movement of goods through the region's intermodal network. Disruptions and delays to goods movement will place regional economies at a competitive disadvantage to other regions of the Gulf and to the Dubai economy. Intermodal connectivity is so critical and must be at the top of the list for efficient access into marine ports and airports which includes enhanced access and local traffic circulation and intermodal transfer facilities by eliminating unnecessary congestion delays and inefficient routings.

Every effort should be made to minimize the intermodal system's external operating costs. To improve connectivity and operating costs, the quality of management in the industry needs to be upgraded and a realization that government and private sector should not accept current congestion levels even if there is no competitive advantage as a result of a lower level of system congestion.

8 CONGESTION AND CONNECTIVITY

Congestion has an enormous impact upon the intermodal connectivity system, through its impact on productivity and operations. Road congestion is a major problem in Dubai and, wherever possible, companies have tried to ensure that others carry the cost of congestion. Roadway congestion has forced intermodal connectivity point operators and customers to operate over longer hours, limiting the movement of intermodal equipment.

Unfortunately when highway construction projects are being considered, the impact of transportation congestion on goods movement operations are not considered in the analysis. Government planning, analysis and design are focused on reducing the work-trip travel time and in most instances overlook the costs incurred by the intermodal industry in trying to avoid congestion. In terms of intermodal connectivity, the key factors to consider are time costs and wages, vehicle operating costs, accident costs, costs related to stress, lack of schedule reliability and cost of damaged goods.

Delays and inefficiencies can also be increased by deficiencies in the infrastructure that are grouped into traffic and congestion management issues and to the configuration as well as the capacity of the system. Areas that are not receiving sufficient government attention include: substandard roadway lane widths, inadequate traffic control systems and pavement markings, poor signage, inadequate maintenance, substandard roadway and intersection geometrics, lack of intersection visibility, tight turning radius, incident management, work zone management, intersection operations, and restricted streets (no truck traffic). Attention to these design deficiencies is one opportunity for the government to improve intermodal connectivity points, which is relatively easy and inexpensive to deal with, often only requiring recognition that goods movement vehicles have different requirements than passenger vehicles.

All cargo has to be loaded and unloaded at some location in the intermodal system and is transferred from one mode to another in an efficient manner, these operations within the intermodal connectivity point and deliveries are generally not an issue. However there are other operations that are not efficiently performed, such as deliveries to the city center where there is a lack of off-street parking or to delivery sites where access is difficult or loading and unloading connectivity points are inadequate. There is potential for productivity gains by improving operations at the shipment end point and some of the issues include: loading and unloading delays, truck parking, delays at facilities, hours available for delivery and site access and egress.

9 CONCLUSION AND RECOMMENDATIONS

Roadway goods movements have always been recognized as part of the general traffic stream; unfortunately, this recognition has been limited and until recently, the public sector has all but ignored their special needs. The public sector can influence intermodal connectivity point efficiency by introducing connectivity strategies at the system/facility, parking/loading and unloading level. An ideal connectivity strategy would involve the selection of specific routes for truck use and to make these routes attractive to the vehicle operators.

Government has the monopoly on providing roadway space and nearly all goods movements take place on this network. Many opportunities are available to improve connectivity on the public network, including rationalizing the space and patterns of roadway use, improving access to freight generating areas, upgrading the road system in congested corridors, redesigning intersections to ease the flow of trucks, providing truck friendly turning lanes (continuous flow intersections), increasing truck capacity and road speed and keeping the network surfaces well maintained.

BIBLIOGRAPHY

1. Berkowitz, C., "Advance Virtual Simulation Model for Intermodal Terminals," Proceedings, 6th Asia Pacific Transportation Development Conference and 19th International Chinese Transportation Professional Association, Hong Kong and Macau, May–June, 2006.
2. Berkowitz, C. and C. Bragdon, "Advanced Virtual Simulation Technology: Case Study," Proceedings, Government Security Expo and Conference, Washington, DC, April, 2006, Joint Author.
3. Berkowitz, C. and C. Bragdon, "Advance Simulation Technology Applied to Port Safety and Security," Proceedings, American Society of Civil Engineer, Ninth International Conference on Applications of Advance Technology in Transportation, Chicago, August, 2006, Joint Author.
4. Berkowitz, C. and C. Bragdon, "Advance Simulation Technology Applied to Terminal Safety and Security," Passenger Terminal World Magazine, November, 2006.
5. Berkowitz, C. and M. Hadi, "Congestion Reduction Strategy for Dubai, Road and Transport Authority," Dubai, UAE, January, 2006.
6. Berkowitz, C. and B. Younes, "Intermodal Connectivity and the Movement of Goods" Unpublished Research, January 2006 to March 2008.

Efficient Transportation and Pavement Systems – Al-Qadi, Sayed, Alnuaimi & Masad (eds)
© 2009 Taylor & Francis Group, London, ISBN 978-0-415-48979-9

Factors affecting the service performance of public buses in Dubai

Amal Zuhair AbuKuwaik
Emirates Bank, UAE

Akmal Abdelfatah
Civil Engineering, American University of Sharjah, Sharjah, UAE

ABSTRACT: The public transport system must be aligned with the growth rate of the city. An effective management system of Dubai's public transport network will lower expenses, simplify procedures, improve safety and increase efficiency.

In this study a methodology is developed to identify the significant factors affecting the public bus system in Dubai. The procedure is conducted through the collection of experts' opinions, which are expressed by relative weights and obtained through specially designed questionnaire.

The most significant factors list produced can be used as a tool to manage the bus transit system in Dubai. The research revealed that the journey time in comparison with other modes of transit, punctuality to schedule and real time information at bus stops are among the major factors affecting the public bus system in Dubai. The top most significant factors were studied and recommendations were proposed.

1 INTRODUCTION

With rapid urbanisation, the need for transportation systems has increased in most cities of the UAE. The inadequate public transport system of the cities, accompanied by lack of public transit management strategies is promoting the use of private vehicles. On one hand, Dubai has experienced a significant increase in its individual private vehicle population over the last few decades. On the other hand, the public transport system did not grow at a comparable rate. This phenomenon resulted in a very high congestion levels all over Dubai.

Another factor that complicated the problem is the inefficient management of transit system, which causes Dubai to face the drain on the exiting resources such as cost of not fully occupied buses, more employees and bus drivers, etc. This situation calls for innovative, cost effective, safe and reliable transport strategies. This paper aimed to produce a list of the most significant factors impacting public bus system performance in Dubai. The list can be used as a component in a systematic evaluation procedure for the bus routes to facilitate the decision making in order to improve the quality of the operation. A bus questionnaire survey was carried out to assess transportation experts' views into factors impacting bus service performance.

2 OBJECTIVES

The main task of this project is to identify the most significant factors affecting the operation of the public bus system in the city of Dubai. This will assist in formulating an efficient strategy to improve the existing management system (such as controlling bus capacity, frequency, etc) of the public buses in Dubai.

3 LITERATURE REVIEW

Many researchers internationally proposed management solutions and models to manage bus transit systems.

A study explored managing bus fleet by performance targets in 2005. It explained how changes in management style, consumer rights legislation, contractual obligations and other factors have combined to make management-by-targets increasingly common in the public sector. Facing increasing demand for transport, government would increase provision of infrastructure to meet expected demand. The paper concluded that traditional way of facing demand became untenable as rising demand began to outstrip any conceivable increase in capacity. The study also stated that targets must relate to the key outcomes that need to be achieved rather than to those that are easiest to measure.[4]

This project methodology aligns with this research and proposes solutions to Dubai city based on management tools rather than new infrastructure.

Post implementation of management system, validation and improvements to the system is a must to ensure effectiveness and efficiency. In 2002 in Kuwait, a study was conducted to determine levels both of management awareness of passenger priorities, as well as passenger satisfaction with the current performance of the bus service. The research designed two structured questionnaires to study bus transit system serving metropolitan Kuwait. The results ranked levels of noise inside the bus, low travel speed, and lack of air conditioning as the top three deficiencies in need of improvement.[5] Another research in 1995 in Bangkok, were the public bus transit system is one of the largest in the world, studied trip frequency scheduling taking into account limited bus fleet size, lack of parking spaces, the fluctuating traffic conditions throughout the day, and the special feature that buses and crews are not pre-assigned to specific trips by the beginning of the daily operation. It concluded that trip frequency scheduling is an important issue for the Bangkok bus transit system and that savings can be made without significantly reducing the service level. "It appeared that over crowdedness of buses is not only caused by a limited bus fleet and by traffic congestions, but also by insufficient planning."[6].

4 BACKGROUND: PUBLIC TRANSIT SYSTEM IN DUBAI

Dubai has an established network of buses, which is managed by the Roads & Transport Authority (RTA). "Around 240,000 passengers travel on the system per day. The bus routes span the whole length and breadth of 3,885 square kilometres area of the Emirate of Dubai."[1]

The fleet consists of top quality custom-built buses prepared with individual seats, air-conditioning, electronically operated destination display system and computerized fare equipment. The bus capacity has an average of 51 seating and 10 standing passengers and recently, smaller capacity buses were acquired to service inner routes and a few low-density corridors. The bus network consists of 516 quality buses, 62 bus routes, 5500 bus trips daily, avg. 240,000 passengers per day, 153 wayside passenger shelters (plan for 500), 1500 bus stops, 3 bus maintenance depots, Electronic Destination display, Computerised Fare collection system and Contact less Smart Card (card for fare payment with 3 options monthly pass, pre-paid and student pass).[1]

5 DATA COLLECTION AND ANALYSIS

Various factors impacting bus systems based on the detailed literature review were identified in the first stage of research. Personal interviews were also conducted with key stakeholders associated with Dubai transportation planning and public transport services to get their feedback on the research topic. From the previous research, 17 major factors impacting bus systems were found to have significant impact on transit system.

A total of 30 transportation experts working closely in Dubai transportation field participated in this study and provided their expert opinion in the form of their response by completing the questionnaire surveys. Those 30 responses were taken for analysis and the result of the analysis was summarized in the second questionnaire and distributed again to the 30 experts. The second questionnaire results were further analyzed and conveyed consistent results which terminated the data collection cycle.

6 ANALYSIS OF ROUND I SURVEY

The collected data was statistically analyzed to shortlist most significant factors among the 17 factor based on experts' opinions. The 17 factors included are punctuality to schedule, frequency, access/egress time, distance between stops, transfer points with other bus routes, availability of mode change points, journey time in comparison with other modes, boarding times, information provision, real time information at bus stops, route maps on buses, facilities available at bus stop, crowdedness of the bus, low floor access bus, handicapped facilities, safety and security and fare. They were the result of a detailed literature review and several meetings with stakeholders and transport authorities of Dubai.

The results of the analysis showed that punctuality to schedule, real time information at bus stops and journey time compared with other modes are ranked the highest by respondents for which the delay caused by traffic congestion in Dubai as well as major cities in the UAE is responsible. Some other factors were hardly considered effective such as handicapped facilities and low floor bus access.

In order to shortlist the most significant factors, a test of hypothesis was performed to all 17 factors. For the weight to be accepted, it should be higher than the equally weighted result, i.e., should be greater than 1/17. Data were tested for 95%, 90%, 85% and 80% degree confidence, and the results of 85% were selected. Accepted factors were: punctuality to schedule, frequency, journey time in comparison with other modes, information provision, real-time information at bus stops, route maps on buses and crowdedness of the bus.

7 ANALYSIS OF ROUND II SURVEY

The collected data from Survey II were analyzed to get the average weight of the 7 factors which were short listed from Round I. A test of hypothesis was conducted to decide on acceptance or rejection as well as ranking of factors based on z value.

Based on the results of the statistical analysis performed and since all factors were accepted, the final weighted factors was obtained and presented below.

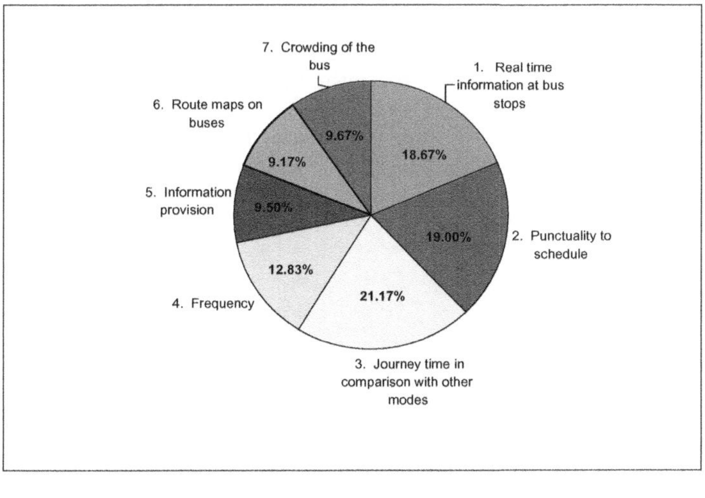

The weighted factors list of the most significant factors impacting public bus system in Dubai.

8 CONCLUSION AND RECOMMENDATIONS

This research has studied best practices in the public bus management systems and gathered local experts' views on the most important factors impacting Dubai's public buses. The methodology used was Delphi method through direct questionnaires and several personal interviews with the stakeholders.

The result of this research was producing a weighted factors list that listed the 7 factors affecting the system and their impact out of 17 factors. The 17 factors included in this study were obtained through extensive literature review and personal interviews with stakeholders. It is concluded that real time information at bus stops (18.67%), punctuality to schedule (19.00%), journey time in comparison with other modes of transit (21.17%), frequency (12.83%), information provision (9.50%), route maps on buses (9.17%) and crowdedness of the bus (9.67%) are the 7 most significant factors impacting the Dubai public bus system.

It is also concluded that access/egress time, distance between stops, transfer points with other bus routes, availability of mode change points, boarding times, facilities available at bus stops, low floor access bus, handicapped facilities, safety and security and fare are not significant to the city. These results provide stakeholders with clear bounds for their management solution, for example, it is very recommended to increase number of express routes since journey time is a significant factor while distance between bus stops is insignificant.

The most significant factors were studied to suitable management solution to enhance the efficiency of the system and encourage the use of public bus over private vehicles. Summary of recommendations is:

Recommendation 1: Introduce real time announcements in all major bus stops and interchanges.
Recommendation 2: Introduce real time announcements on buses.
Recommendation 3: Bus priority lanes are highly recommended in the city.
Recommendation 4: Transit signal priority is recommended in the congested corridors.
Recommendation 5: Introduce more express bus routes. This would be more efficient if combined with recommendations 3, 4 and 6 rather than being individually implemented.
Recommendation 6: In the short term, RTA should consider the potential of replacing low frequency bus services with high frequency shared taxi services.
Recommendation 7: RTA needs to develop and consult on a customer directed information strategy for buses. This can include the following elements:

- A multi-modal computerized journey planner for Dubai available on the internet with terminals available at all major interchanges, shopping centres, key public buildings and eventually at all bus stops.
- Extending the number of stops that list key information and destinations such as petrol stations, hospitals, shopping centres, parks, leisure centres and how to get there.

Recommendation 8: Route maps showing every stop, interchange possibilities and key destination points (i.e. public buildings, tourist attractions, etc) should be provided on buses.

Recommendation 9: Dedicate first or at least few seats of the Bus for females exclusively.

REFERENCES

1. Dubai Road and Transport Authority.
2. National academy press, "Bus Route Evaluation Standards A synthesis of transit practice," *TRB-NRC report TCRP Synthesis 10*, Washington, D.C. 1995.
3. Ioannis Minis, Eric Keys, Theodore Athanasopoulos, "Contribution to the design of the Athletes Bus Network during the Athens 2004 Olympic Games," *Transportation Research*, Part A 40 (2006) 776–791.
4. Greg Marsden, Peter Bonsall, "Performance targets in transport policy," *Transport Policy*, 13 (2006) 191–203.

5. P.A. Koushki, O.I. Al-Saleh, M. Al-Lumaia, "On management's awareness of transit passenger needs," *Transport Policy*, 10 (2003) 17–26.
6. Dirk L. van Oudheusden, William Zhu, "Trip frequency scheduling for bus route management in Bangkok," *European Journal of Operational Research*, 83 (1995) 439–451.
7. Maria P. Boile′, "Estimating technical and scale inefficiencies of public transit systems," *Journal of Transportation Engineering*, May/June, 2001.
8. Wall, G., McDonald, M., "Improving bus service quality and information in Winchester," *Transport Policy*, (2007), doi:10.1016/j.tranpol.2006.12.001.
9. Martin Lowson, "Idealised models for public transport systems," *International Journal of Transport Management*, 2 (2004) 135–147.
10. Jonathan C., Levin C., Gwo-Wei Torng, "Dwell-time effects of low-floor bus design," *Journal of Transportation Engineering*, Vol. 120, No. 6, November/December, 1994.
11. Sanja Durmisevic, Sevil Sariyildiz, "A systematic quality assessment of underground spaces—public transport stations," *Cities*, Vol. 18, No. 1, pp. 13–23, 2001.
12. Fabiana Scapolo, Ian Miles, "Eliciting experts' knowledge: A comparison of two methods," *Technological Forecasting & Social Change*, 73 (2006) 679–704.
13. Caroline J. Rodier, Robert A. Johnston, David R. Shabazian, "Evaluation of advanced transit alternatives using consumer welfare," *Transportation research part C*, 6 (1998) 141–156.
14. Katrin Dziekan, Karl Kottenhoff, "Dynamic at-stop real-time information displays for public transport: effects on customers," *Transport*, Res. Part A (2007), doi:10.1016/j.tra.2006.11.006.
15. Mendenhall, Beaver, Beaver, *Probability and Statistics*, Thomson, 12th edition, 2006.
16. Reebu Zachariah Koshy, V. Thamizh Arasan, "Influence of Bus Stops on Flow Characteristics of Mixed Traffic," *Journal of Transportation Engineering*, August, 2005.

Efficient Transportation and Pavement Systems – Al-Qadi, Sayed, Alnuaimi & Masad (eds)
© 2009 Taylor & Francis Group, London, ISBN 978-0-415-48979-9

Selecting Syria new highway corridors utilizing GIS

Abdul-Rahman I. Hamad
Parsons, Doha, Qatar

Rabi Dawahra
Hi-Tech House, Damascus, Syria

ABSTRACT: A study for utilizing build, operate and transfer (B.O.T) concept in financing road projects in Syria was prepared.

Main objectives included evaluating economic & financial feasibility of finance through getting tolls from road users to cover costs of road construction, maintenance & operating, and ensuring the necessary information for making decisions about adoption of BOT system for the two main axes of this study, and for future axes to be financed.

GIS was utilized to conduct alternatives alignment analysis for the two new roadway corridors. This involved building a GIS information bank with various layers from available digital mapping and demographic sources. Each layer was given a virtual cost (weight) based on importance which was used to build a cost surface. From that various solutions were generated then narrowed to three for each corridor. The three alternatives were furthered analyzed and one preferred alternative for each corridor was selected.

1 INTRODUCTION

With the Syrian Arab Republic's significant progress in the last 35 years towards road transport infrastructure, high traffic density has emerged on many roads. This high density was felt especially on two main axis roads: the axis that connects the North and South of Syria (Turkish border to Jordanian border), the axis that connects the West and East of Syria (Seaports of Syria to Iraqi borders and vice versa), which both became congested during last few years. There are few internal roads also between Syrian cities, where traffic congestion is being felt due to increased number of vehicles and bad road conditions. At present there is an imperative need and lot of pressure on government authorities to improve the traffic operating conditions on these roads. At the same time it was determined that the self financing of road for these high density corridors of traffic will not be feasible for the Syrian government, at the present juncture. As an alternative to financing the roads through public money it was considered to develop these roads on Build-Operate-Transfer (BOT) basis where the costs of financing the construction, maintenance & operating of roads will be covered by the investor, who will recover these costs through tariffs (tolls) from road users according to the framework agreed upon with the organizing authority PERC (i.e. Public Establishment for Road Communications).

The Main Objectives of the overall Study include:

- Evaluating the economic & financial feasibility of finance through getting tolls (tariffs) from road users, in covering the costs of road construction, maintenance & operating.
- Ensuring the necessary information for making decisions about the adoption of this system for the two axes on the one hand, and for the future axes to be financed according to the said system on the other hand.

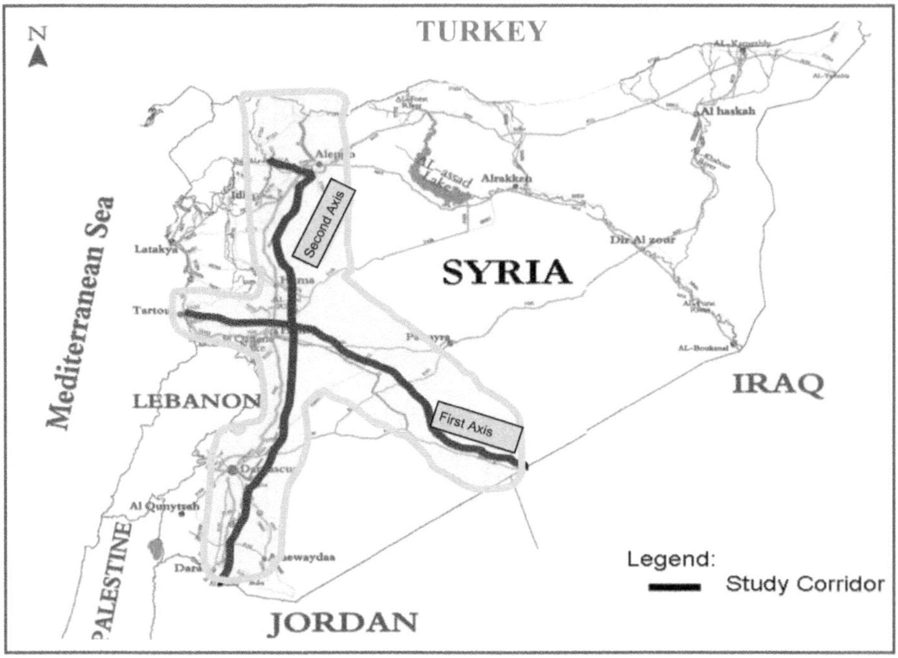

Figure 1. Study corridors.

The objective of this paper is to describe the process utilized to conduct the selection of the highway alignment utilizing GIS, which is one of the work elements conducted towards achieving these goals, and does not cover other aspects of the study.

The study includes the two corridors containing the following two existing axis roads:

- The First Axis: Tartous—Al Tanaf—Iraqi Borders
  Starts from the sea-side City of Tartous and goes into Homs governorate, to end at the Iraqi Borders (after Al-Tanaf) for an approximate length of 378 km.
- The Second Axis: Turkish-Syrian Borders to Jordanian-Syrian Borders
  Starts form the northern Syrian-Turkish borders and goes to the south across Idleb-Hama-Homs-Damascus-Dar'a governorates, to reach the Syrian- Jordanian borders for an approximate length of 500 km.

2 COLLECTING TOPOGRAPHICAL, GEOLOGICAL AND SATELLITE MAPS

All required background mapping was obtained from the appropriate official sources in cooperation with PERC. These maps include the following:

2.1 *Aerial and Satellite photos*

Two Landsat Satellite photos of 15 m resolution had been collected from Center of Remote Sensing; the coordinate projection system for these two photos is UTM on WGS84 in TIFF format and with Red, Green and Blue (RGP) spectrum. These two photos cover:

- First photo covers Nasseeb—Bab El Hawa axis, and includes the areas through which the first axe will pass (North—South), this photo is a mosaic made up of group of photos show parts of Daraá- Damascus countryside—Damascus—Homs, Hamma, Idleb and Aleppo governorates.

- Second photo covers Tartous—Altanaf axis, and includes areas through which the second axe will pass (Coast—Desert), this photo is a mosaic made up of group of photos show parts of Damascus countryside –Homs, Hamma, and Tartous governorates.

2.2 *Digital topographical maps*

These maps are produced by Public Survey Establishment, and their scale is 1/100,000. The used projection system in these maps is the Syrian Stereographic System saved in shapefile format. Twenty-two digital maps had been obtained to cover all the possible areas through which road axes might pass. The maps identify the following layers:

- Contour lines layer
- Vegetative cover layer
- Residential areas layer
- Roads and railways layer
- Water Resources, lakes, water streams layer
- Landmark features layer (such as countries and cities borders, telecom & electrical lines, oil lines, gas lines, etc . . .).

2.3 *Geological maps*

These maps are paper maps highlighting soil categories and they had been obtained from the General Establishment of Geology and Mineral Resources. They consist of 10 maps of 1/200,000 scale to cover all possible areas which road axes might pass through.

3 PLANNING PRIMARY ROUTES FOR AXES USING ARCGIS

Conducting this analysis included the following work steps:

3.1 *Preparation works*

Certain steps were taken to check the accuracy and comprehensiveness of the collected data, which was gathered in single Data Bank according to the following procedures:

3.1.1 *Gathering and evaluating received data*

Compiled data for the project (aerial photos and digital topographical maps) was evaluated to check the age, variety and coverage extent for the project area. Based on this evaluation insufficient coverage points or data accuracy or modernity had been defined. All these mentioned points had been discussed with PERC, and agreed on the necessity to compile additional data which would enhance the data bank. Therefore suitable available alternates had been used.

Figure 2 shows a sample of the original data obtained with highlighted dated limits of urban area and dated limits of an artificial lake.

3.1.2 *Alternative data resources*

The following resources had been used as an aid to gain additional data:

- Web loaded Landsat Satellite photographs of 15 m resolution had been used; these pictures covered expanded areas, noting that they were relatively old (between 1999 and 2002).
- Land Use Maps produced by FAO (Food and Agriculture Organization) had been used; the coordinate projection system for these maps is UTM on WGS84 saved as .shapefile format. These maps had been produced before 2000.

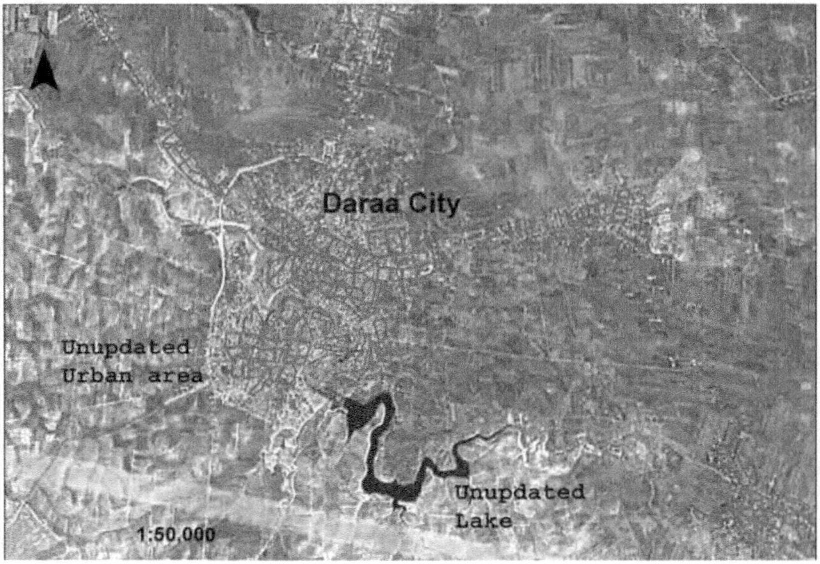

Figure 2. sample of un-updated data.

- Developed areas boundary data have been gained from paper maps related to Ministry of Local Administration and Environment, the coordinate projection system in these maps is the Syrian Stereographic System. The boundaries have been digitized and added to the data bank.

3.1.3 *Unifying data coordinate system*

The Syrian Stereographic System had been adopted as the coordinate system for all data.

3.1.4 *Project's geographical information bank built up*

All the abovementioned data have been merged in single unified Geographical database for the purpose of applying all future processing works and analysis on it; as represented in Figure 3.

3.2 *Building and preparing of geographical layers*

The ArcGIS software was used to build and prepare the geographical layers in accordance with the following steps:

3.2.1 *Topographical data*

The digital topographical map contours have been adopted to produce the Digital Elevation Model (DEM) for the study area. The DEM was then used to generate 3D Model for the natural Land starting from the contour layers. 3D analysis has been carried out on land relief in accordance with the following conditions:

- Gathering contour layers within the geographical database and unifying them in single geographical layer.
- Verifying the elevation values in the contour layers and exclude the odd ones.
- Changing the measurement unit of the digital layers from kilometer scale unit to meter scale unit.
- Eliminate discrepancies between the survey adjacent sheets to ensure the continuity of the contours by using Spatial Adjustment tools which are available within ArcGIS.

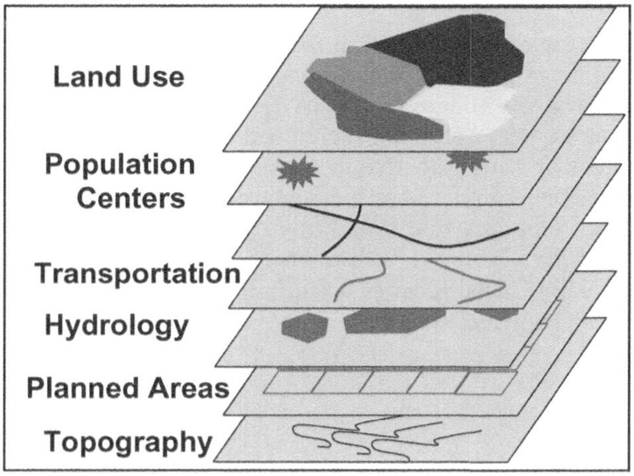

Figure 3. Geographical information bank.

- Produce Triangulated Irregular Network (TIN) layer for the studied area depending on the contour sheets that have been handled as described above and layers which represent break lines
- Produce Digital Elevation Model (DEM) for the studied area starting from the TIN layer.
- Produce slope layer based on Digital Elevation Model (DEM) that has been generated.
- Overlay the Satellite pictures on Digital Elevation Model (DEM).

3.2.2 *Active communities areas*

Human communities play a very significant role in choosing the optimal required path. The paths are identified for all proposed alternative roadway options on the condition that neither the current existing communities nor their future expansion limits should be penetrated by any proposed alignment. On the other hand, this is done while making sure that these roads are close enough to the communities in a way that allow it to be one of economical and human development factors for these communities.

To accomplish this, two new layers were developed:

- First layer represents the main cities along the two axes of the study.
- Second layer represents the industrial cities along the two axes of the study.

3.2.3 *Land use data*

The compiled data from FAO (Food and Agriculture Organization) had been adopted in sorting out land classification layers based on land utilization. The layers were sorted out and arranged according to main classifications and in-line of their importance in choosing the optimal route. Then the data was prepared to be used in routes scenarios according to the following categories:

- Bare Land
- Forests
- Grass Land
- Hydro Areas
- Rocky Land
- Seasonal Crop
- Developed Areas
- Trees

3.2.4 *Hydro data*

Hydro and stream areas are considered critical factors in choosing the optimal route and making decisions; either in avoiding the passage of the proposed road over the hydro areas, or in additional cost estimation for constructing bridges in case the design requirements necessitate the passage of the road over some hydro areas. In addition, there is a need to define the places of drainage facilities for rain water and other water resources to cross the proposed route. The available data from the digital topographical maps produced by the Public Survey Establishment had been adopted to sort out the hydro layers, according to the following steps:

- Arranging the hydro layers in the geographical information bank in a way that allows separating the necessary layers to implement this stage. The available layers consist of:
  - Rivers
  - Lakes
  - streams
  - Dams
- Sorting out the previous layers according to their importance and then, performing the merging operations required between the layers that have similar geometric characteristics. Under this operation, the layers are classified according to the main rows required to import them in routes scenario within layers in Geographical Information Bank of the project according to the following:
  - Hydro areas layer (Lakes, Reservoir Dams, Artificial Lakes, Swamps . . .)
  - Hydro streams layer (Rivers—Seasonal streams, Valleys, Channels . . .)
- Using Arc Hydro software, based on the previously derived DEM aiming to produce the following layers:
  - hydro streams—flow basins,
  - collective flow basins, and
  - the longest stream within the flow basin

3.2.5 *Building of objective images*

In this stage, the final layers derived from the set and built up geographical layers phase, were transferred from Vector graphic form to Raster graphic pixel form in a way that makes the produced layers as following:

- Land Slope layer
- Developed areas layer
- Land Use layer, with the following classifications:
  - Bare Land
  - Forests
  - Grass Land
  - Hydro Areas
  - Rocky Land
  - Seasonal Crop
  - Developed Area
  - Trees
- Hydrology Data layer that consists of the following data
  - Hydro Streams
  - Flow Basins
- Main Cities layer
- Industrial Cities layer
- Hydro areas layer that consist of the following classification:
  - Lakes
  - Rivers

4 BUILDING OF ALTERNATIVE ROUTE SCENARIOS

4.1 *Layers weighing*

Building route scenarios requires reclassification of all previous mentioned layers through giving weight to each class within each layer. The weight must coincide with the layer priority in the route scenarios where the following principles followed up in weighing process:

- Each layer has been given 0–10 weight or classified as no data, whereas:
  - weight 0 represents the most favored areas for passing (or the least cost for passing), and
  - weight 10 represents the least favored areas for passing (or highest cost for passing).
- No Data represents the areas that the routes are not able to pass through (developed areas and lakes).

Determining the weight to be applied for each layer was developed as an iterative process. First, the analysis team, based on their international experience with similar project and local experience with the conditions in Syria developed weights for each of the factors. These weights were then presented and discussed with PERC to reach consensus. PERC suggested some changes based on their in-depth experience of developing roads in Syria and another round was done to update the weights, and the final agreed upon weights were used in the analysis.

4.2 *Building of cost layers*

All the weighted layers were gathered in one layer, which represents the final weighted cost that will be used in finding out the optimal route. This layer will be developed as Raster graphic type, where the value of every single cell in it will represent the total cost needed to move through this cell.

5 DERIVATION OF ROUTE OPTIONS

5.1 *Derivation and analysis of various alternatives*

After reaching the total cumulative cost of a layer in the previous stage, it is now possible to find out the least cost to move from origin to destination, and drawing the route generated by ArcGIS software according to the weighting system adopted, thus many routes were produced by changing weight of each layer (according to its importance for the route) and giving each layer weight of 1 to 10 or no data.

Figure 4 represents 2 examples of primary weighting that have been used. First Example focused on the importance of the contour and Land Use layers, so 50% of the weight or importance was given for the contour layer, 20% for the land use layer, 30% for hydro areas and streams layer and 0% for other layers. The second example focused more on the importance of hydro areas and streams layers, so 45% of the weight is considered for it, 35% for the contour layer, 10% for land use layer and 10% for big cities layer.

Figure 5 illustrates group of scenarios that had been generated using GIS for the First Axis (Tartous- Altanaf—Iraqi Borders), whereas Figure 6 illustrates group of scenarios that had been generated for the Second Axis (Turkish Border—Jordanian Border).

5.2 *Deriving the three most suitable scenarios for each axis*

A manual engineering review of the generated solution scenarios was carried out for these routes in order to identify the weighting and weights that produce better options in line with adopted design criteria for the new routes related to horizontal and vertical curves, design speed, balancing cut and fill areas, minimizing structural work, etc . . . This process resulted in defining three (3) alternative scenarios 3 for each of the two study axes. Table 1 represents the parameters (weighting—weights) that had been used to generate the 3 options for the first axis (Tartous—Altanaf—Iraq borders). A similar table has been produced to derive the 3 options for the second axis (Turkish Border—Jordanian Border).

	Run1	Run2
	Land Use(0.2)	Land Use(0.1)
Land Use (FAO)	Parameters	Parameters
Bare Land	1	1
Forest	10	10
Grass Land	1	1
Hydro	NoData	NoData
Rocky Land	8	8
Seasonal crops	3	3
Setellement	NoData	NoData
Trees	6	6

	Run1	Run2
	Slope(0.5)	Slope(0.35)
Slope	Parameters	Parameters
0–2	1	1
2–5	2	2
5–10	4	4
10–12	6	6
12–15	8	8
15–25	9	9
25–40	10	10
40–<	No Data	No Data

Organization Plan(0)		
Organization	No Data	No Data
Other	0	0

Water Data(0.05)		
Water body	10	10
Other	0	0

	Hydro Stream(0.1)	Hydro Stream(0.2)
Distance To Stream (m)	Parameter	Parameter
0–500	10	10
500–2000	8	8
2000–4000	6	6
4000–6000	4	4
6000–10000	3	3
10000–13000	2	2
13000–<	1	1

S_Lake (0)		
Lake body	No Data	No Data
Other	0	0

HYdro Catch(0.15)	HYdro Catch(0.2)

	M_Cities(0.0)	M_Cities(0.1)
0–5000	0	10
5000–10000	0	0

Figure 4. Examples for generated alternatives.

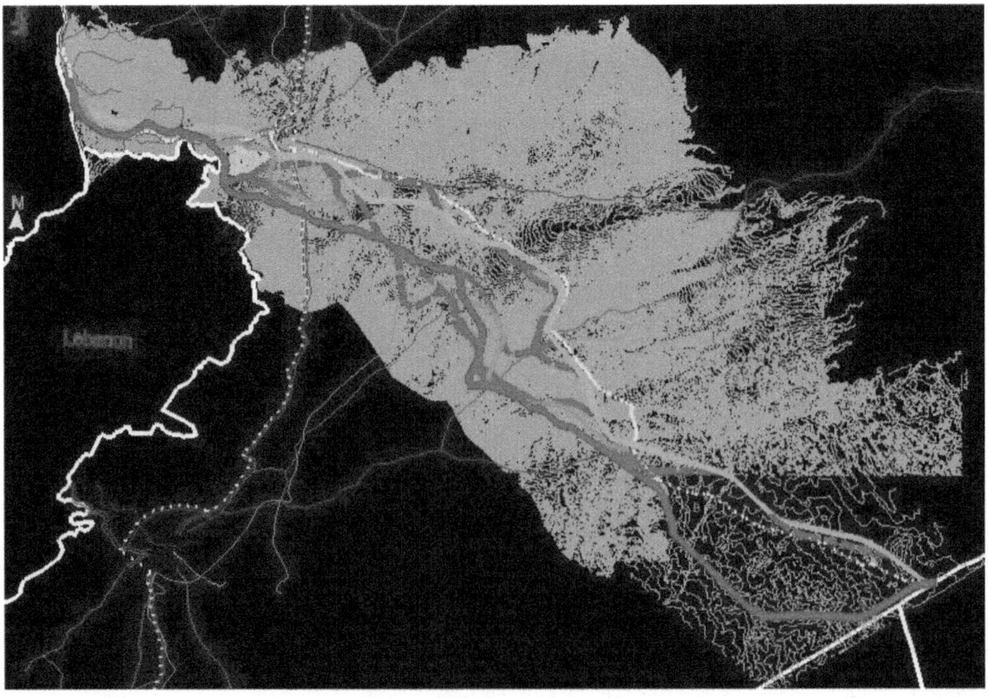

Figure 5. Alternative scenarios for axis 1.

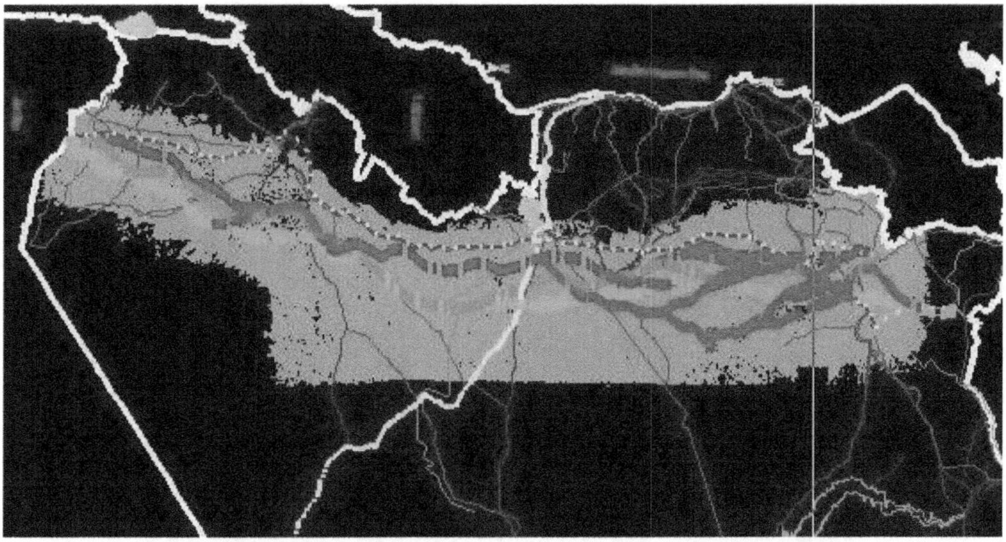

Figure 6. Alternative scenarios for axis 2.

Table 1. Solution parameters for axis 1.

Layer Name	Layer Importance		
	Alt 1	Alt 2	Alt 3
Natural ground slopes	(0.50)	(0.35)	(1)
	Cost	Cost	Cost
0–2	1	1	1
2–5	2	2	4
5–10	4	4	9
10–12	6	6	No Data
12–15	8	8	No Data
15–25	9	9	No Data
25–40	10	10	No Data
40–<	No Data	No Data	No Data
Catchments areas	(0.15)	(0.20)	(0)
Surface water Streams	(0.10)	(0.20)	(0)
Dist.Betw. Route & stream	Cost	Cost	Cost
0–500	10	10	10
500–2000	8	8	8
2000–4000	6	6	6
4000–6000	4	4	4
6000–10000	3	3	3
10000–13000	2	2	2
13000–<	1	1	1
Lakes	(0)	(0)	(0)
Lake body	No Data	No Data	No Data
Other	0	0	0
Rivers and streams	(0.05)	(0.05)	(0)
River body	10	10	10
Other	0	0	0

Layer Name	Layer Importance		
	Alt 1	Alt 2	Alt 3
Land Use (FAO)	(0.20)	(0.10)	(0)
	Cost	Cost	Cost
Bare Land	1	1	1
Forest	10	10	10
Grass Land	1	1	1
Hydro	NoData	NoData	NoData
Rocky Land	8	8	8
Seasonal crops	3	3	3
Settlement	NoData	NoData	NoData
Trees	6	6	6
Main Cities	(0)	(0.10)	(0)
0–5000	10	10	10
5000–10000	0	0	0
10000–<	6	6	6
Other Populated areas	(0)	(0)	(0)
0–5000	0	0	0
5000–10000	3	3	3
10000–<	8	8	8
Development Plans	(0)	(0)	(0)
Developed Areas	No Data	No Data	No Data
Other	0	0	0

6 FURTHER REFINMENT OF SCENARIOS

After defining the three alternative scenarios, a 3D Model was designed for the proposed scenarios and then a virtual flight over the routes was developed to evaluate the results of the GIS analysis and identify any required adjustment. Also geometric refinement of the scenarios was carried out to account for the fact that the GIS solutions have been built up by using the ArcGIS software based on current situation for all geometric items to choose the route, such as slopes and natural land levels, without taking into consideration the possibility of developing these slopes by excavation or embankment works. The same was done for when analyzing the route to avoid hydro

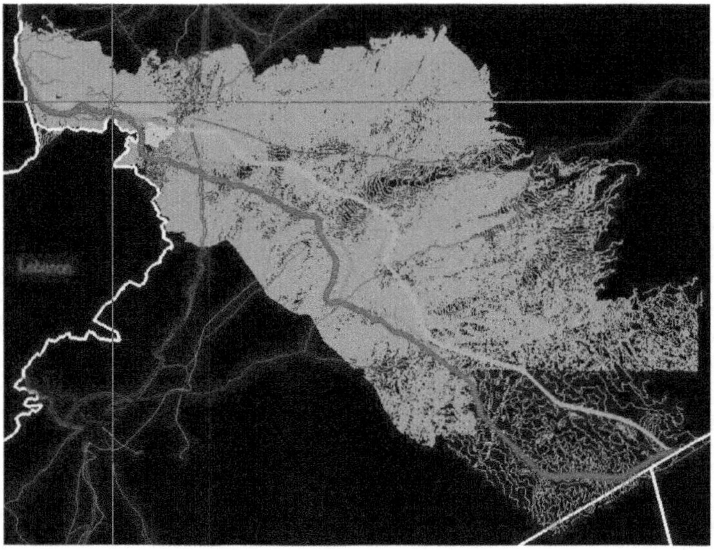

Figure 7. Preferred alternatives for axis 1.

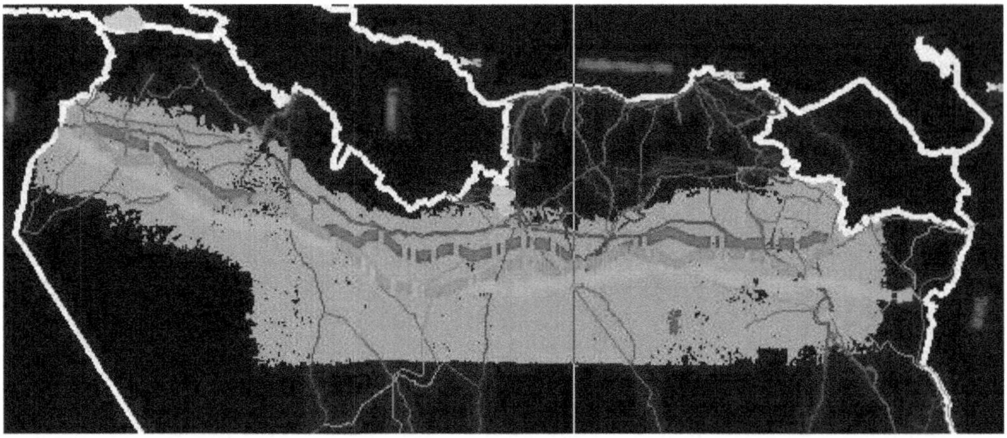

Figure 8. Preferred alternatives for axis 2.

streams without taking into consideration the possibility to change their direction. The refinement was done so an engineering decision could be introduced where it makes sense, and to introduce artificial earth works or structural works to enhance the route selection.

Figures 7 and 8 illustrate the three preferred scenarios, after refinements, for axis one and axis two, respectively.

7 CONCLUSIONS

Follow-up field investigations of the selected routes resulted in minor refinements by updating features that were introduced after the date which the topographical and aerial maps were developed (e.g. new dams, town expansion, etc.). However for the majority of the length of the selected route alignments, the filed investigations confirmed the conclusions of the GIS solutions.

Therefore, the utilization of GIS proved very valuable. It saved a lot of time in handling, developing, and evaluating numerous alternatives for very long routes while producing credible solutions that fits the requirements of the study.

ACKNOWLEDGEMENT

The authors would like to acknowledge the Public Establishment for Roads Communications in Syria and various other public departments involved in this project.

REFERENCES

Parsons International Limited. 2007. Syria BOT Feasibility Study for Roads in Syria, Task 3 Report, Doha. Qatar.

ArcView/ArcGIS, ESRI Corporation, U.S.A.

Pavements

Binder characteristics and their effect on HMA and pavements

Efficient Transportation and Pavement Systems – Al-Qadi, Sayed, Alnuaimi & Masad (eds)
© 2009 Taylor & Francis Group, London, ISBN 978-0-415-48979-9

Performance of modified asphalt binder and mixtures under increased truck loading limits

H.U. Bahia, R. Delgadillo & A. Motamed
University of Wisconsin—Madison, Madison, Wisconsin, USA

D. Christensen
Advanced Asphalt Technologists, USA

ABSTRACT: In many developing countries the truck loading limits are not well enforced. It is also an economic challenge to enforce truck loading limits when demand for rapid construction and limited transportation facilities are available. Design of asphalt mixtures to resist the higher loads carried by trucks is a real challenge. The information regarding sensitivity of asphalt binders and mixtures to high stresses is scares. This study is a laboratory investigation for the response of selected asphalt binders and mixtures to increased stresses under repeated loading at pavement temperatures in the range of warm climates, such as the climate in the Gulf area. Advanced testing procedures including the Dynamic Shear Rheomter for binders and the Simple Performance Testing Device for mixtures was used to compare basic rheological behavior under repeated loading. The loading levels were varied within a wide range and the deformation was analyzed in terms of recoverable and permanent. The results clearly show that the asphalt binders plays a major role in the non-linear stress dependency of mixtures. The effect of modification type is also found to affect performance. A discursion of the need for considering and modeling of stress dependency of permanent deformation is presented. The paper points out the need to simulate actual loading ranges in testing of paving materials and the possible errors that could result from underestimating loads or using Linear Visco-elastic properties to characterize asphalt binders and mixtures.

1 INTRODUCTION AND BACKGROUND

1.1 *Background*

Since the emergence of Superpave equipment and specification in the pavement industry in 1993, a tool was provided for researchers and practitioners to understand the pavement characteristics and behavior in better way. By using SHRP equipments, it became possible to capture binder and mixture properties attribute to their performance in the field.

Permanent strain or rutting is one of the most important pavement deteriorations. Due to its complexity and importance, many studies were conducted to understand and alleviate this problem. Rutting study is divided to two parts; Binder and Mixture.

Regarding the binder part, the SHRP study (Anderson et al. 1994) introduced the Dynamic Shear Rheometer DSR, which was used to capture the properties of binder that contribute to mixture rutting resistance. At that time, it was assumed that total dissipated energy causes rutting. Consequently, by integration of the stress-strain curve, total dissipated energy in each cycle was calculated. As a result, $G^*/\text{Sin}\delta$ was introduced as rutting parameter. Although, this approach was a major advancement in asphalt technology, after some years and by using modified asphalt, it was seen that this parameter can not capture the permanent strain performance characteristic of modified binders as well as conventional binders (Delgadillo et al. 2006, D'Angelo et al. 2007).

To fill the gap in SHRP specification, the NCHRP 9–10 research program introduced a new procedure. In this new procedure, loading pattern is changed from cyclic to repeated creep and recovery (irreversible) and a new parameter was selected and introduced for characterizing the ability to resist rutting of bitumen (Bahia et al. 2001). This parameter which was called viscous component of the creep stiffness G_v, derived from using the Burger's four element spring-dashpot model. Using repeated creep and recovery, instead of oscillation, made it possible to differentiate permanent and recoverable strains (Bahia et al. 2001).

On the mixture side, many studies were also conducted to capture the effect of different component of the mixture on the pavement rutting and mixture stress sensitivity to the load level. These studies on the mixture rutting showed there is a nonlinear relationship between pavement permanent deformation and load magnitude, especially at high stress levels and temperatures. The rutting model developed by Kaloush & Witczak early during NCHRP Project 1–37 (2000) and the model developed in the early 1990's by Leahy and Witczak (1991) are two models which show this nonlinearity. But, since, these two models are only based on the laboratory's data and have not been calibrated by field data, they cannot take into account effect of tire pressure. As a result of NCHRP Project 1–37, the Mechanistic Empirical Pavement Design Guide (MEDG) model is developed (Design Guide 2002, Christensen & Bonaquist 2005). This model was not only based on the lab data, but also calibrated with field data. Therefore, it is reasonably accurate and suitable for analyzing the relationship between tire pressure and rutting in HMA pavements.

1.2 *Nonlinearity in mixture rutting (mixture stress sensitivity)*

As it is mentioned in the pervious section, MEPDG model is one of the most important models that relates the permanent pavement deformation to stress level. Here, this model is discussed to illustrate mixture rutting's nonlinearity.

Stress level close to the pavement surface has a great effect on the pavement permanent deformation; obviously, by increasing the stress level, it will increase. As a first estimate, the stress near the surface and contact area considered equal to the tire inflation pressure. In the latest version of the MEDG, the following equation is used to model permanent deformation in flexible pavements (Design Guide 2002, Christensen & Bonaquist 2005):

$$\varepsilon_p = \varepsilon_r k_1 \times 10^{-3.4488} T^{1.5606} N^{0.479244} \tag{1}$$

Where:

ε_p = accumulated plastic strain at N repetitions, in/in

ε_r = resilient strain in the HMA mixture under the given conditions, in/in

T = temperature, °F

k_1 = a factor that depends upon total thickness of HMA layers and depth of point of interest. The value of k_1 is calculated using the following equation:

$$k_1 = (C_1 + C_2 Z) 0.328196^Z \tag{2}$$

$$C_1 = -0.1039h^2 + 2.4868h - 17.342 \tag{3}$$

$$C_2 = 0.0172h^2 - 1.7331h - 27.428 \tag{4}$$

Where:

Z = depth of point of interest, in.

h = total thickness of HMA layers, in.

Since linear elastic analysis is used to determine ε_r, it can be assumed to be proportional to tire pressure p. The constant of proportionality is the resilient modulus of the surface material.

Therefore, for any given pavement structure at any given temperature, plastic strain can be rewritten:

$$\varepsilon_p = pk_2 N^{0.479244} \tag{5}$$

Where k_2 is a factor encompassing k_1, and the elastic modulus ($Er = p/\varepsilon_r$), the relation between ε_r and p, and $10^{-3.4488} * T^{1.5606}$. Comparing two tire pressures that results in same value of ε_p, the effect of tire pressure on number of passes (*N*) is as follows:

$$p_1 N_1^{0.479244} = p_2 N_2^{0.479244} \tag{6}$$

By dividing that by p_2 and raising both sides to the power of 2.09 (inverse of 0.479244):

$$N_2 = \left(\frac{p_1}{p_2}\right)^{2.09} N_1 \tag{7}$$

This equation shows that if the pressure, p_2 increases to two times of its initial value p_1, the number of passes (*N*) which will cause the same damage will decrease to a quarter or on the other hand it will cause four times more than what caused by its initial value. As it can be seen, this nonlinearity is making the load limitation on the roads more critical.

If the average truck tire pressure is assumed to be 100 lb/in^2 (as was assumed during the development of the Superpave system), Equation 8 then leads directly to the equation for tire pressure factor:

$$TP = \left(\frac{p}{100}\right)^{2.09} \tag{8}$$

Besides, it has been seen that using different equipment such as the Hamburg Wheel Tacking, or Asphalt Pavement Analyzer (APT), to measure mixture rutting will result into different rutting measurement and ranking of the pavement (D'Angelo et al. 2006). This difference is related to different stress level applied on the sample. Therefore, it is proven that rutting in asphalt mixture is stress sensitive.

The cause of the nonlinearity in mixture rutting can be assumed to be aggregates as well as binders. This assumption is based on the fact that many studies on granular materials have shown that granular materials are stress sensitive. In fact one of the main factors varied in measuring resilient modulus of aggregates for base courses according to the AASHTO T307 standard protocol is stress level. In addition, recent studies on binder response under repeated creep have shown that binders, particularly modified binders show significant stress sensitivity above a certain stress level. The question has been raised, however, about the level of stress and strain in binder domain of mixture under normal traffic conditions. A recent study at the University of Wisconsin at Madison was conducted to determine the range of binder stresses. In this study, by using image analysis and finite elements, strain and stress distribution in the mixture were determined. Then, by knowing the strain and stress distribution in the mixture, stress level for binder tests were changed accordingly, to capture the real behavior of binder in the field.

1.3 *Strain distribution within the mix and idealized mastic*

In this research asphalt concrete cross-sectional images were converted into finite element mesh, Figure 1. Then, by assigning appropriate mechanical properties to different constituent of mixture (aggregate, binder, air void) and analyzing the model under constant height shear loading, strain distribution in mastic domain was found (Kose 2001).

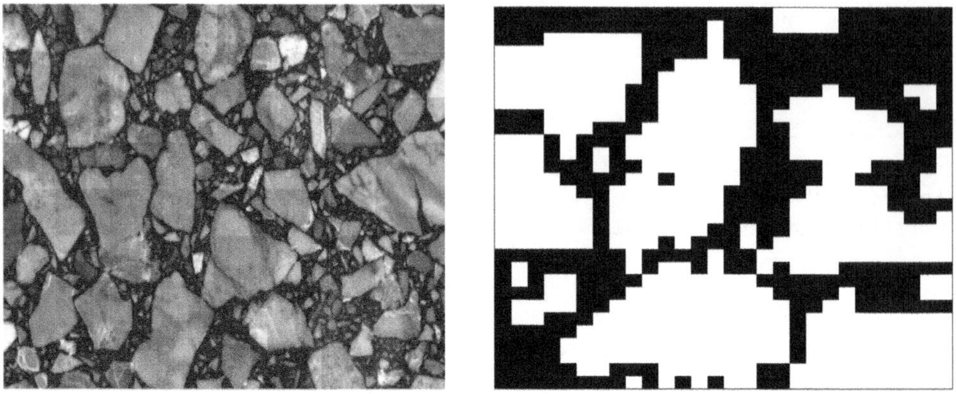

Figure 1. Finite element mesh of coarse mix (Kose 2001).

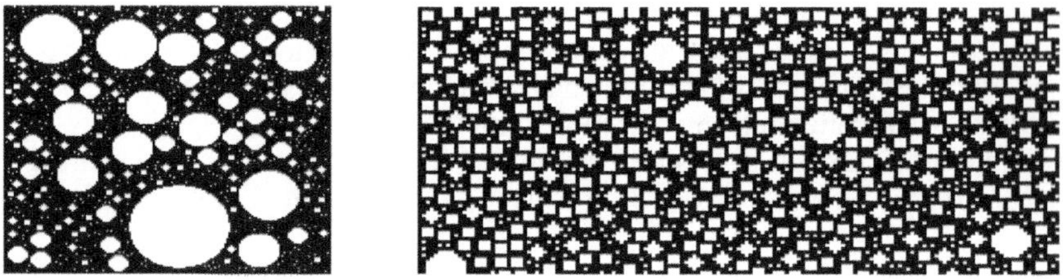

Figure 2. Idealized mastic: Coarse gradation (left) and fine gradation (right) (Kose 2001).

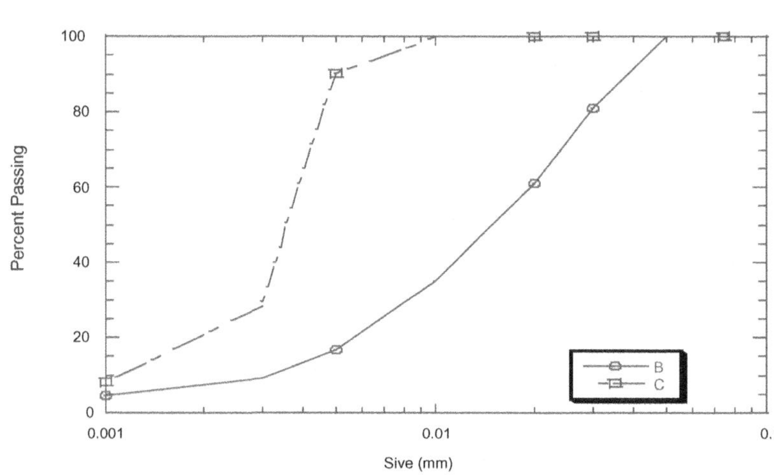

Figure 3. Gradation curve for "coarse" and "fine" idealized mastic (Kose 2001).

This analysis was conducted by assuming the plane strain distribution, ensures that the image and the finite element mesh are identical, and linear elastic properties for mastic and aggregate. Assigned Young's Modulus and Poisson's ratio to aggregate was 25000 MPa and 0.25. These properties for mastic were 5 Mpa and 0.45.

After assigning material properties, boundary condition were defined to simulate the testing condition of constant height shear loading. Finally, by applying horizontal displacement on the top of the sample to obtain predetermined bulk shear strain, strain density function in the mastic was found. Analysis shows that the average shear strain in the mastic is approximately 4.3 times more than bulk strain and also it could be as high as 85 times of the applied bulk strain.

In the second step, idealized mastics with two different gradation, coarse and fine, were prepared, Figure 2. Coarse gradation is made up of particles ranging from 1 to 50 microns while fine gradation is composed of particles range from 1 to 10 microns. Figure 3 depicts the gradation for both fine and coarse fillers.

Based on the actual gradation and image gradation analysis, it was calculated that mastic in images must include approximately 40% dust and 60% binder by area. Based on this proportion, finite element simulation was prepared for the idealized mastics. Since, the geometry of the simulated mastic can have an effect on the strain distribution with in the mastic, two aspect ratios were simulated and analyzed under the same bulk strain, 0.1%. Analysis shows that strain distribution changes with gradation, but the average strain remain the same. These analyses have also showed that binder strain within the mastic ranges between one to six times the mastic strain, with the average value of 1.8. Therefore, it can be concluded that strain in binder is 7.8 (4.3*1.8) times more than bulk strain of mixture, in average. Also, strain in binder can reach as high as 510 (85*6) times the mixture bulk strain (Kose 2001).

The limitations of that research were maximum resolution of scanner and maximum number of elements that FE software could solve. This research resulted into changing in binder stress level.

2 BINDER STRESS SENSITIVITY (NONLINEARITY IN BINDER BEHAVIOR)

2.1 *Selected binders*

For evaluating the non linearity of the RCR test, 8 binders were selected. The high temperature PG grades of the chosen binders were 64 or 70. Six of the binders were polymer modified (SBS, SB and Elvaloy) and two were oxidized. The binders were RTFO aged before testing. The selected binders are shown in Table 1.

2.2 *Experimental procedures*

DSR samples were prepared according to the AASHTO TP5 standard procedure. The stress-controlled rheometer is programmed to run a repeated creep test of a total of 100 cycles of 1 second loading and 9 seconds unloading. Each binder was tested at six different stress levels: 25 Pa, 100 Pa, 400 Pa, 1600 Pa, 6400 Pa and 10000 Pa. Two testing temperatures were used. For representing the summer pavement temperatures in Wisconsin, a temperature of 46°C was used. A second temperature equal to 64°C was also used for simulating more extreme summer conditions.

Table 1. Selected binders for repeated creep and recovery testing.

PG Grade	Modifier
PG 64 – 28	SBS
PG 64 – 40	SB
PG 64 – 28	Elvaloy
PG 64 – 28	None (Oxidized)
PG 70 – 28	SBS
PG 70 – 34	SB
PG 70 – 34	Elvaloy
PG 70 – 28	None (Oxidized)

Two replicates were tested for each binder. The total strain after 100 cycles was recorded for each combination of temperature and stress. In case of good repeatability, the average of the two replicates was used for the analysis of results. Good repeatability was assumed when the differences between the two replicates were lower than 15%. In case of larger differences between replicates, a third replicate was run. Good repeatability was achieved for all binders except for the PG 70-34 Elvaloy, where the replicates varied by around 18%. Figure 4 shows the testing sequence for each sample.

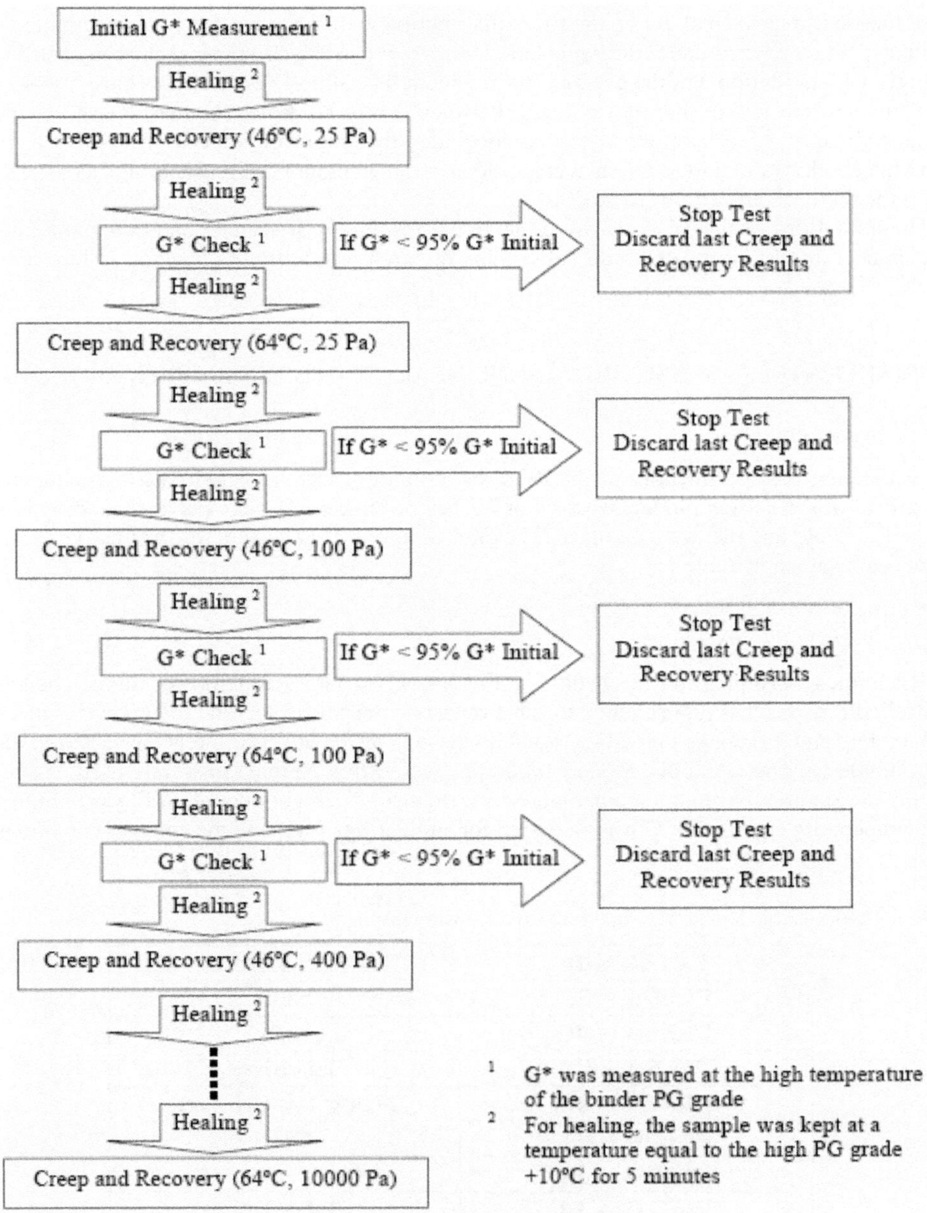

Figure 4. Creep and recovery testing sequence for each binder sample.

As shown in Figure 4, a healing (resting) period was included between tests during which the sample was kept at a temperature equal to the high temperature of the PG grade plus 10°C (HT+10°C) for 5 minutes. To make sure that no damage in the sample has occurred because of the high stress applied, G* test was carried out several times in between the RCR tests. When a drop of 5% or more in G* at 10 Hz was observed, the sample was assumed to be damaged. At this point, the testing was stopped and the results from the last RCR test were discarded.

From the testing results, the cumulated strain at 100 cycles (AS100) was recorded at each testing condition. Table 2 shows an example of the recorded data for the PG 70-28 SBS. Table 2 also shows the normalized strain at 100 cycles (NAS100) for each test condition. The NAS100 is calculated by dividing the AS100 by the corresponding stress level used in the testing. If the response is linear and there is no stress dependency the NAS100 should be the same. Any changes in this parameter are considered non-linearity or stress dependency.

The NAS100 allows estimating the stress level where the binders start showing non-linearly. Figure 5 shows the results of NAS100 for eight tested binders, all stress levels at 46°C and 64°C.

Table 2. RCR testing results for the PG 70-28 SBS binder.

Binder	Temp.	Stress	Replicate 1		Replicate 2	
	°C	Pa	AS_{100}	NAS_{100}	AS_{100}	NAS_{100}
PG 70-28 SBS	46	25	11.3	0.45	10.7	0.43
		100	46.0	0.46	43.6	0.44
		400	178	0.44	172	0.43
		1600	700	0.44	679	0.42
		6400	3256	0.51	3290	0.51
		10000	6920	0.69	6920	0.69
	64	25	147	5.90	137	5.48
		100	606	6.06	567	5.67
		400	2715	6.79	2630	6.58
		1600	13304	8.32	13600	8.50
		6400	89300	13.95	88300	13.80
		10000	179000	17.90	181000	18.10

Table 3. Percent change in NAS100 (RCR test) for different stress levels.

Binder	Temp.	% Change in Normalize Strain		
	°C	1600 Pa	6400 Pa	10000 Pa
PG 70-34	46	-10%	-20%	-20%
SB	64	30%	120%	210%
PG 70-34	46	-10%	-20%	-20%
Elvaloy	64	0%	0%	50%
PG 70-28	46	-10%	30%	80%
Oxidized	64	30%	220%	fail
PG 70-28	46	0%	20%	60%
SBS	64	50%	140%	220%
PG 64-28	46	0%	0%	0%
SBS	64	-10%	30%	110%
PG 64-40	46	20%	50%	80%
SB	64	120%	360%	560%
PG 64-28	46	0%	0%	10%
Elvaloy	64	0%	40%	90%
PG 64-28	46	0%	30%	80%
Oxidized	64	30%	60%	fail

From Figure 5 we can observe that the different binders showed different stress sensitivities. At 46°C four of the binders (PG 64-28 Oxidized, PG 70-28 oxidized, 64-40 SB and 70-28 SBS) showed some important stress sensitivity while the rest of the binders did not show important changes in NAS100 at 46°C. At 64°C, however, all binders showed some stress sensitivity. By comparing Figure 5 (a) and (b), it can be seen that the linear range for each binder changes with temperature. In terms of stress, the linear limit for 64°C is smaller than for 46°C for all binders. This can be seen more clearly in Table 3 which shows the percentage increase in NAS100 for three stress levels: 1600 Pa, 6400 Pa and 10000 Pa. The numbers in Table 3 are rounded to the nearest 10% for simplicity.

Some binders showed small increases in the NAS100 with stress level at 46°C, as shown in Table 3. It is very likely that the NAS100 of these binders will decrease if the stress is increased beyond 10000 Pa. Also it is possible that these binders show strain hardening, which is a phenomenon known for cross linked polymer during which the resistance to deformation (stiffness) increases as strain is increased.

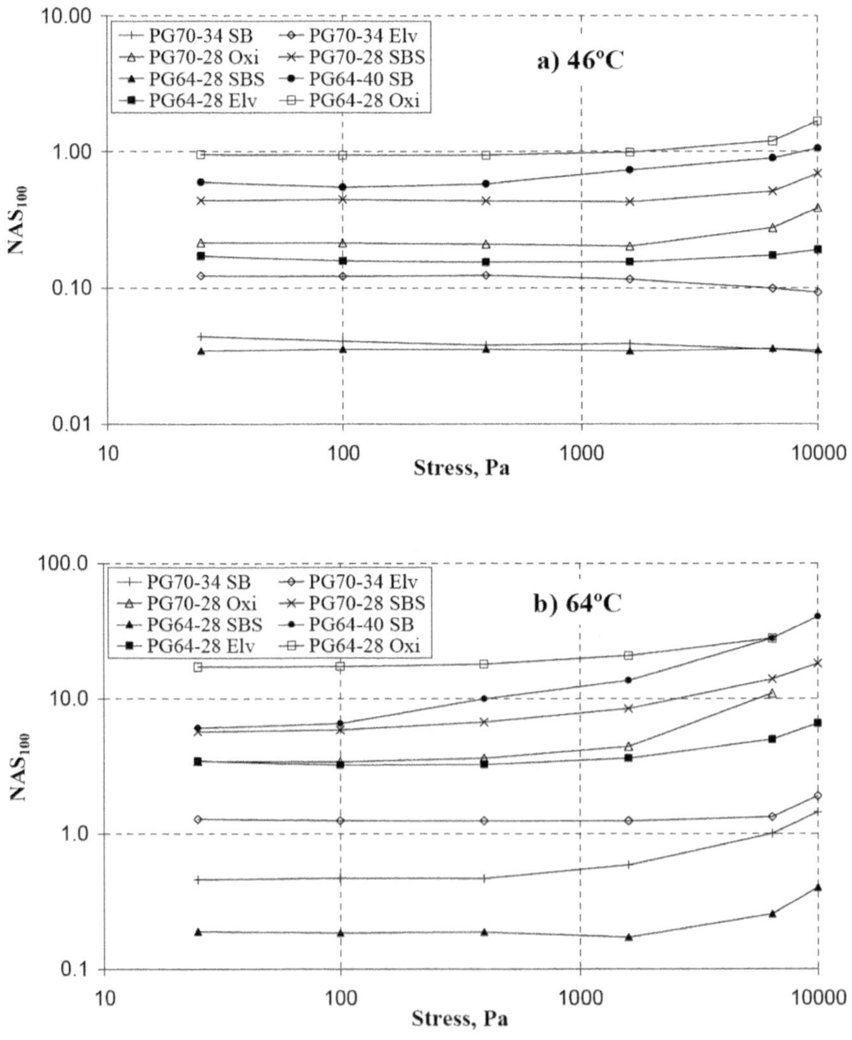

Figure 5. Normalized accumulated strain at 100 cycles (NAS100) at 46°C (a) and 64°C (b).

2.3 *Using stress sweep to determine binder's stress dependency*

Stress sweep testing in cyclic loading can be carried out to determine the stress sensitivity of binders. This is considered a simpler test since most asphalt rheometers used in practice allow changing stress level but many are not designed to conduct creep and recovery. The motivation of this part of the study was to evaluate if a simple stress sweep would allow the same distinction between binders' stress sensitivity.

Stress sweep testing at 10 Hz between 1 and 50000 Pa was carried out for the same binders at 46 C and 64 C. Figure 6 shows the variations in G* for the stress sweep at 46°C and 64°C. The plots confirm that the stress sensitivity of binders is different at different temperatures. The linear range (no change in G*), in term of stress, is wider at 46°C than at 64°C. For example, binders PG 70-28 SBS and PG 64-40 SB did not show important G* variations for the stress range used at 46°C. However at 64°C the same binders collapsed before 10000 Pa.

To allow comparing the stress sweeps results with the RCR testing results, Table 4 was developed which shows the percentage decrease in G*sinδ for three different stress levels: 1600 Pa, 6400 Pa and 10000 Pa. The loss modulus G*sinδ was chosen instead of G* to provide a better comparison. The loss modulus is related to the permanent deformation in the binder, so it can be compared with NAS100. The numbers in Table 4 are rounded to the nearest 5%. By comparing Tables 3 and 4 it can

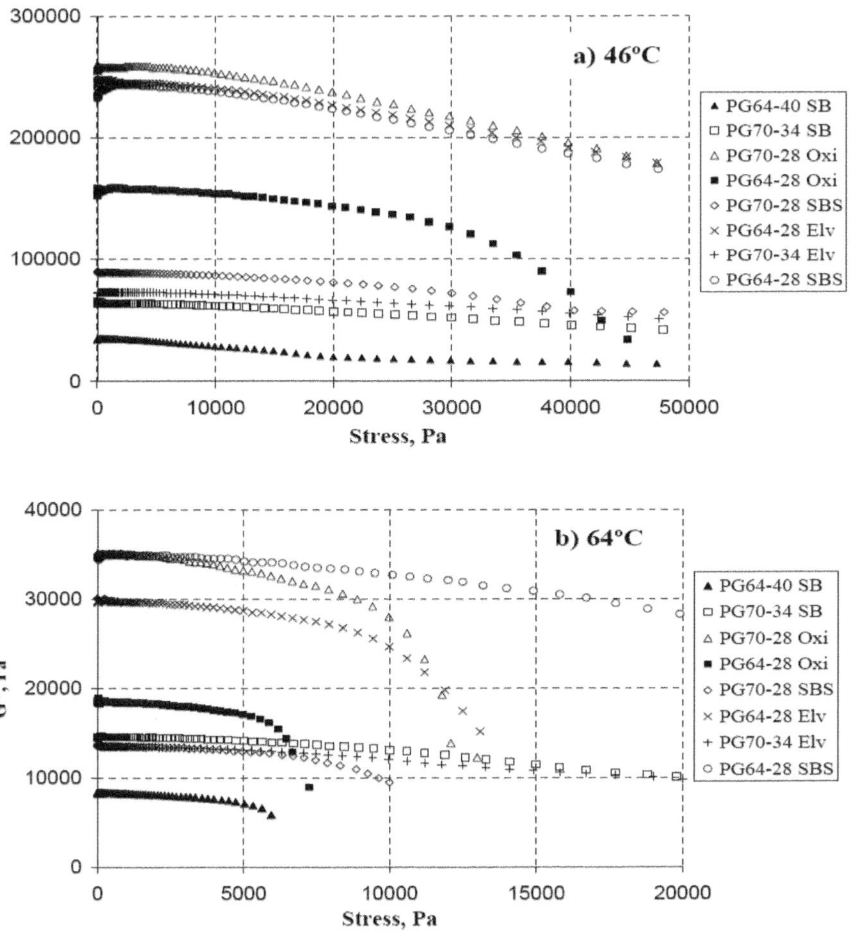

Figure 6. Stress sweep test at 46°C (a) and 64°C (b).

Table 4. Percent decrease in G*sinδ (stress sweep test) for different stress levels.

Binder	Temp. °C	% Change in Loss Modulus		
		1600 Pa	6400 Pa	10000 Pa
PG 70-34	46	0%	0%	5%
SB	64	0%	5%	5%
PG 70-34	46	0%	0%	0%
Elvaloy	64	0%	5%	10%
PG 70-28	46	0%	0%	0%
Oxidized	64	0%	5%	20%
PG 70-28	46	0%	0%	5%
SBS	64	0%	5%	25%
PG 64-28	46	-5%	-5%	0%
SBS	64	0%	0%	5%
PG 64-40	46	0%	10%	15%
SB	64	5%	30%	40%
PG 64-28	46	0%	0%	0%
Elvaloy	64	0%	5%	20%
PG 64-28	46	0%	0%	0%
Oxidized	64	0%	20%	50%

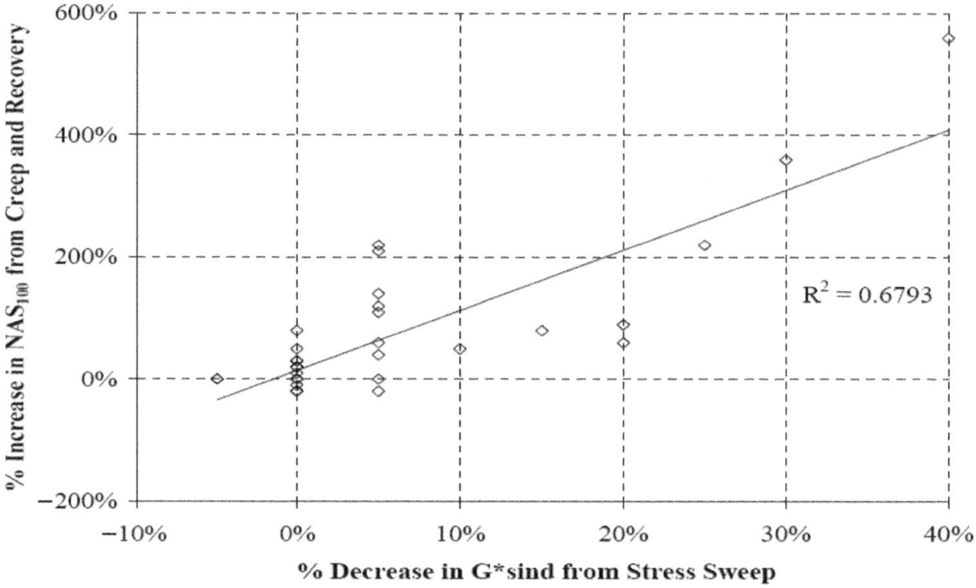

Figure 7. Comparison between stress sensitivity predicted with stress sweep and RCR binder testing.

be seen that both show a similar trend in ranking binders with respect to stress sensitivity. Figure 7 shows the percentage increase for NAS100 versus the percentage decrease in G*sinδ for the same stress level. A relatively good relationship is shown between the stress sensitivity measured by the two methods. The relationship is however not perfect and a wide scatter can be observed at changes of less than 20% in G*sinδ. The scale along the two axes is very different. It seems that the NAS100 is a more sensitive parameter to evaluate non linearity compared to the decrease in G*sinδ.

Another conclusion that can be made here is by increasing load, the ability of binder to resist deformation, G*, will decrease. This loss of resistance is not linear and after a certain point G^* falls tremendously. Decreasing G^* on one hand and increasing the load level on the other hand will exacerbate the problem of deformation of pavement under heavy loads.

3 MIXTURE RUTTING TESTING

3.1 *Test description*

Mixture rutting testing using unconfined compression with repeated creep and recovery loading was conducted. The reasons for the selection of the testing type were: creep and recovery loading represents well the loading pattern that the asphalt pavement experiments during service; it is a well known test, commonly used for the evaluation of mixture rutting susceptibility; and, it has been demonstrated that the test correlates well with asphalt rutting in the field (Witczak et al. 2002). A cylindrical sample of 100 mm diameter and 150 mm height is placed in an environmental chamber at the desired temperature and then loaded. LVDT's are fixed to the samples for measuring the axial and transverse deformation. The sample setup is shown in Figure 8. Ten thousand cycles of loading and unloading are applied to the each sample, with the loading phase lasting

Figure 8. Mixture sample setup for rutting test.

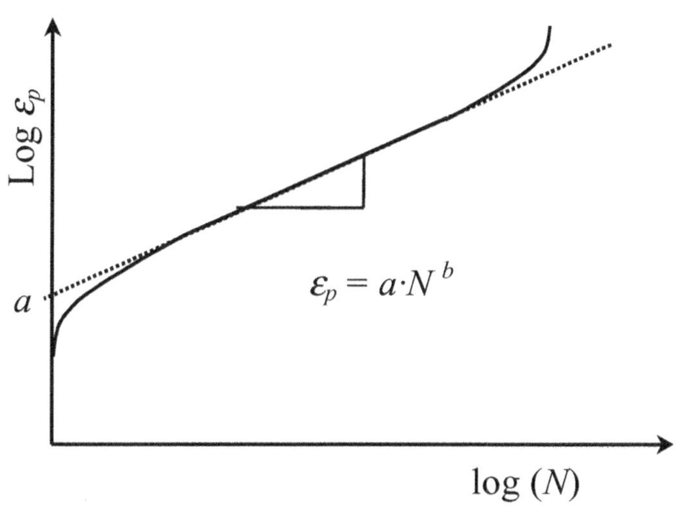

Figure 9. Schematic of mixture permanent deformation v/s loading cycle.

0.1 seconds and the unloading phase lasting 0.9 seconds. The vertical permanent deformation (ε_p) after the recovery phase of each loading cycle (N) is recorded. The shape of the plot of ε_p versus N normally resembles a creep curve. Primary, secondary, and tertiary creep usually can be identified from the plot. Figure 9 shows the schematic for the ε_p versus N plot in a log–log scale.

Because of the shape of the ε_p versus N curve, the analysis of the testing results usually involves the fitting of a power law curve and the determination of the intercept (*a* in Fig. 9) and the slope (*b* in Fig. 9). There is also another parameter usually considered in the analysis of the rutting test results, which is called flow number (*Fn*). *Fn* corresponds to the loading cycle number where tertiary deformation starts. There are several methods for determining this point; the one used in this research was the one suggested by Biligiri et al. (2007).

3.2 *Mixture types and testing conditions*

The asphalt mixtures commonly used in the United States for pavements can be generally divided in coarse mixtures and fine mixtures, depending on the predominant aggregate size of the gradation. For this reason, two different aggregate gradations were used in the mixture sample preparation: one coarse and one fine. Figure 10 shows the aggregate gradations that were used. The gradations are expressed in percentage passing, which means the percentage of the total aggregate particles (in weight) that is smaller than the corresponding sieve size.

The type of aggregate used was crushed limestone, which is one of the most common kinds of aggregates for highway construction because of its natural availability in the United States. Two sources of aggregates were considered: a limestone from central Kentucky and a limestone from Wisconsin. The Kentucky source was selected because this is the aggregate commonly used by the Asphalt Institute for standard testing. Its properties are known nationwide so, by using this aggregate, the results of the present work have more meaning nationwide. The aggregate from Wisconsin was considered to make the results of the research locally relevant. The temperature of testing for the mixtures was 46°C, which is the average pavement temperature during the summer months in Wisconsin.

In order to analyze the effect of different vehicle loadings on the pavement, three different stress levels were considered in the mixture testing: 0.689 MPa (100 psi), 1.03 MPa (150 psi), and 1.38 MPa (200 psi). The first stress level is similar to the tire pressure of large trucks on highways. The highest stress is included to represent high stresses on airfield pavements due to the high airplane tire pressures. The original testing plan for mixtures also considered 2.06 MPa (300 psi) of testing stress. However, preliminary testing determined that the failure of the samples using this

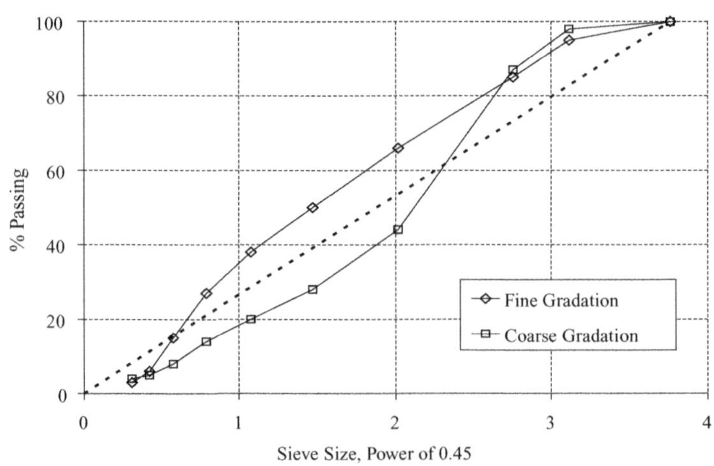

Figure 10. Aggregate gradations used for mixture samples.

stress level was too quick and gave no valuable information, so the maximum stress was kept at 1.38 MPa.

3.3 *Experimental design*

The repeatability of mixture rutting testing is known to be not as good as the repeatability of binder testing. For this reason, three replicates were considered for each testing condition. The different testing conditions (stress, source, gradation) were distributed between the different batches in order to eliminate testing error. Table 5 and Table 6 show the experimental design for all the mixture samples, which make a total of 36 samples. The first letter in the sample ID refers to the gradation: F for fine

Table 5. Fine mixture samples and experimental design.

Sample ID	% Air voids	Batch	Source	Testing stress (MPa)
F-N3-100-A	7.6%	IV	Limest KY	0.689
F-N3-100-B	6.9%	III	Limest WI	0.689
F-N3-100-C	7.1%	V	Limest WI	0.689
F-N3-150-A	6.9%	IV	Limest WI	1.03
F-N3-150-B	5.9%	I	Limest WI	1.03
F-N3-150-C	6.1%	V	Limest KY	1.03
F-N3-200-A	6.1%	III	Limest WI	1.38
F-N3-200-B	6.1%	I	Limest WI	1.38
F-N3-200-C	6.8%	II	Limest WI	1.38
F-B5-100-A	7.9%	IV	Limest WI	0.689
F-B5-100-B	8.1%	III	Limest WI	0.689
F-B5-100-C	7.3%	V	Limest KY	0.689
F-B5-150-A	6.7%	IV	Limest KY	1.03
F-B5-150-B	7.1%	I	Limest WI	1.03
F-B5-150-C	7.6%	V	Limest WI	1.03
F-B5-200-A	6.3%	III	Limest WI	1.38
F-B5-200-B	6.9%	I	Limest WI	1.38
F-B5-200-C	7.2%	II	Limest WI	1.38

Table 6. Coarse mixture samples and experimental design.

Sample ID	% Air Voids	Batch	Source	Testing stress (MPa)
C-N3-100-A	7.3%	IV	Limest WI	0.689
C-N3-100-B	6.6%	III	Limest WI	0.689
C-N3-100-C	9.6%	V	Limest KY	0.689
C-N3-150-A	10.1%	IV	Limest KY	1.03
C-N3-150-B	6.3%	I	Limest WI	1.03
C-N3-150-C	8.0%	V	Limest WI	1.03
C-N3-200-A	5.0%	III	Limest WI	1.38
C-N3-200-B	6.1%	I	Limest WI	1.38
C-N3-200-C	7.3%	II	Limest WI	1.38
C-B5-100-A	6.6%	IV	Limest KY	0.689
C-B5-100-B	7.1%	III	Limest WI	0.689
C-B5-100-C	7.4%	V	Limest WI	0.689
C-B5-150-A	7.3%	IV	Limest WI	1.03
C-B5-150-B	6.5%	I	Limest WI	1.03
C-B5-150-C	8.1%	V	Limest KY	1.03
C-B5-200-A	6.3%	III	Limest WI	1.38
C-B5-200-B	5.6%	I	Limest WI	1.38
C-B5-200-C	7.1%	II	Limest WI	1.38

or C for coarse. The 2nd code stands for the binder: N3 for the unmodified or B5 for the polymer-modified. The 3rd code stands for the testing stress: 100 psi, 150 psi, or 200 psi. Finally, the three samples tested at each condition were named A, B, and C.

3.4 *Test results: Flow number and simple power law model*

Preliminary analyses were performed on the mixture testing results using the conventional procedures used in the asphalt industry. First the flow number Fn was compared and analyzed for all testing conditions. Then, using a power law model for fitting the rutting data, the two parameters of the model (initial strain and rutting slope) were analyzed. Figure 11 presents the results of the

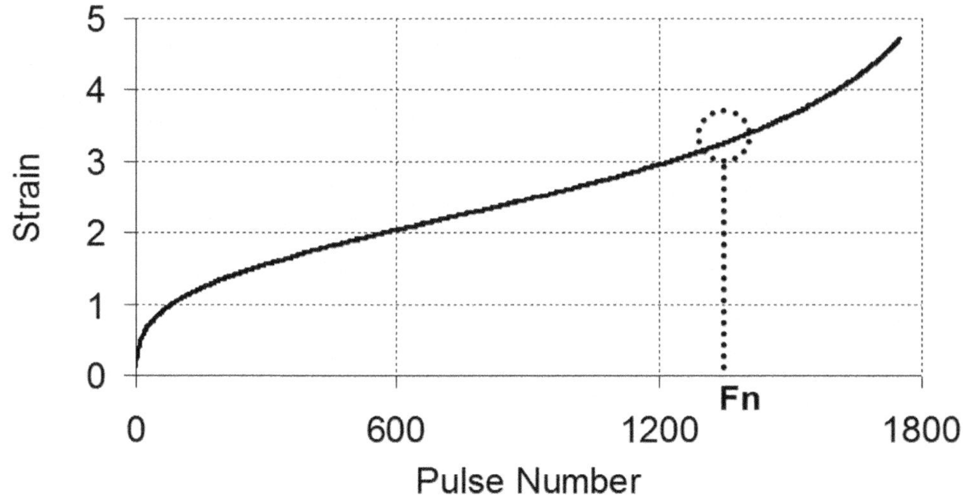

Figure 11. Axial strain versus pulse number for sample C-B5-100-C.

Table 7. Flow number test results.

Sample ID	Stress (MPa)	*Fn*	Sample ID	Stress (MPa)	*Fn*
F-N3-100-A	0.689	2076	C-N3-100-A	0.689	994
F-N3-100-B	0.689	10000	C-N3-100-B	0.689	3330
F-N3-100-C	0.689	1799	C-N3-100-C	0.689	627
F-N3-300-A	1.03	746	C-N3-300-A	1.03	260
F-N3-300-B	1.03	1026	C-N3-300-B	1.03	485
F-N3-300-C	1.03	1065	C-N3-300-C	1.03	416
F-N3-200-A	1.38	656	C-N3-200-A	1.38	829
F-N3-200-B	1.38	331	C-N3-200-B	1.38	264
F-N3-200-C	1.38	441	C-N3-200-C	1.38	149
F-B5-100-A	0.689	1596	C-B5-100-A	0.689	1719
F-B5-100-B	0.689	10000	C-B5-100-B	0.689	2016
F-B5-100-C	0.689	3595	C-B5-100-C	0.689	735
F-B5-300-A	1.03	486	C-B5-300-A	1.03	85
F-B5-300-B	1.03	281	C-B5-300-B	1.03	142
F-B5-300-C	1.03	271	C-B5-300-C	1.03	62
F-B5-200-A	1.38	181	C-B5-200-A	1.38	116
F-B5-200-B	1.38	70	C-B5-200-B	1.38	26
F-B5-200-C	1.38	129	C-B5-200-C	1.38	21

permanent axial deformation on sample C-B5-100-C versus loading cycle. The Fn is shown schematically in the Figure 11.

3.4.1 *Flow number*

The results for the *Fn* of the tested samples are shown in Table 7. All samples were tested at 46°C. A preliminary analysis of Table 7 indicated the presence of some outliers. Using analysis of variation with 95% confidence showed that one of the batch number III was responsible for most of the outliers. It is believed that the reason for this was that the samples in batch III were compacted twice. First, the samples were used in a trial to determine the number of gyrations required to reach the target air voids (6% to 8%). After that, the samples were re-melted and compacted to the required number of gyrations. Because of the double heating cycle, the binder could have been over aged and excessively hardened. For this reason, the samples of batch III (F-N3-100-B, F-B5-100-B, C-N3-100-B, C-B5-100-B, C-N3-200-A, C-B5-200-A, F-N3-200-A, F-B5-200-A) were removed from the analysis that follows. It was also determined that the influence of the aggregate source on the *Fn* results was insignificant.

The average *Fn* results (without outliers) are plotted for each condition in Figure 12. The standard deviation is also included in the graph.

The effect of stress can be clearly seen in the results shown in Figure 12. For example, for binder N3 mixed with the fine aggregate gradation (Fine N3), reducing the stress to one half (from 1.38 to 0.689 Mpa) resulted in increasing the Fn number from an average of 414 to 1650 cycles. This is an increase of 4 folds which agrees very well with the model shown in equation (8) and (9) of this paper. For other combinations of binders and aggregates the changes are more or less than what is predicted by equation (8) but they are not very different.

The performance of both binders can be compared by examining Figure 12 which shows that for 0.689 MPa, the samples prepared with the polymer-modified binder B5 showed higher *Fn* than the samples prepared with the unmodified binder N3, for the same aggregate gradation. For the higher stress levels, however, N3 was the binder that showed better performance. It is also evident

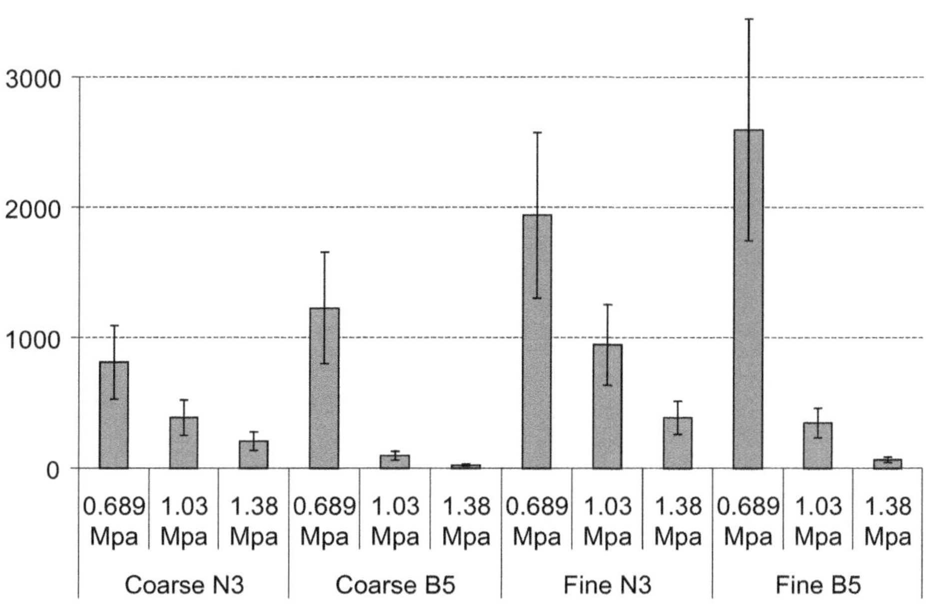

Figure 12. Average flow number results for each condition.

that when the stress was increased, the decrease in *Fn* for the samples prepared with B5 was more pronounced than the decrease observed in the samples prepared with N3. This suggests that the higher nonlinearity of B5 directly affects the mixture performance in terms of *Fn*. Therefore, it can be concluded that by increasing the load level the flow number will decrease. This decrease is not linear and it may follow an exponential function. The rate of decrease is mostly related to the binder type; neat binder has a lower rate.

3.4.2 *Simple power law model*

As explained earlier in this report, *Fn* is defined as the cycle number when the deformation slope starts increasing, which means that tertiary flow is starting. For some mixtures, however, this increase in slope is very gradual, and they can resist many cycles of loading after *Fn* without considerable increases in deformation. To further analyze the influence of the binder characteristics in the rutting performance of mixtures, the deformation results were studied using a power low model.

$$\varepsilon_p = \varepsilon_{p1} \cdot N^S \qquad (9)$$

Where: ε_p = accumulated strain
ε_{p1} = initial strain
N = cycle number
S = slope of deformation

Figure 13 shows the fitting of the vertical strain with the power law model for the sample F-N3-200-B. The two parameters of the model were analyzed for all the testing conditions. In order to study the linearity of the response, the ε_{p1} parameter was normalized by the stress level. The ε_{p1} parameter is usually mostly influenced by the binder properties, so the normalization allows to study the effect of the nonlinearity of the asphalt binders on the mixture behavior. The samples from batch III were excluded for the analysis because of the reasons explained before.

Figure 14 shows the average normalized initial strain $\varepsilon_{p1\,Norm}$ for all the testing conditions, with standard deviations. The Figure 14 shows a clear influence of the binder on $\varepsilon_{p1\,Norm}$. For both gradations, the $\varepsilon_{p1\,Norm}$ values were higher for the B5 binder. For both binders, the $\varepsilon_{p1\,Norm}$ value

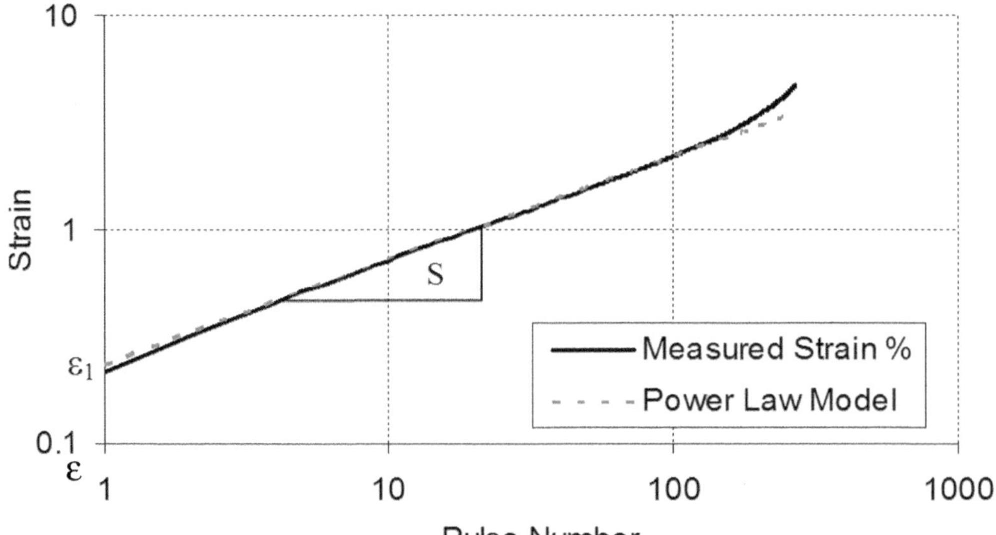

Figure 13. Measured and fitted mixture strain versus cycle number N.

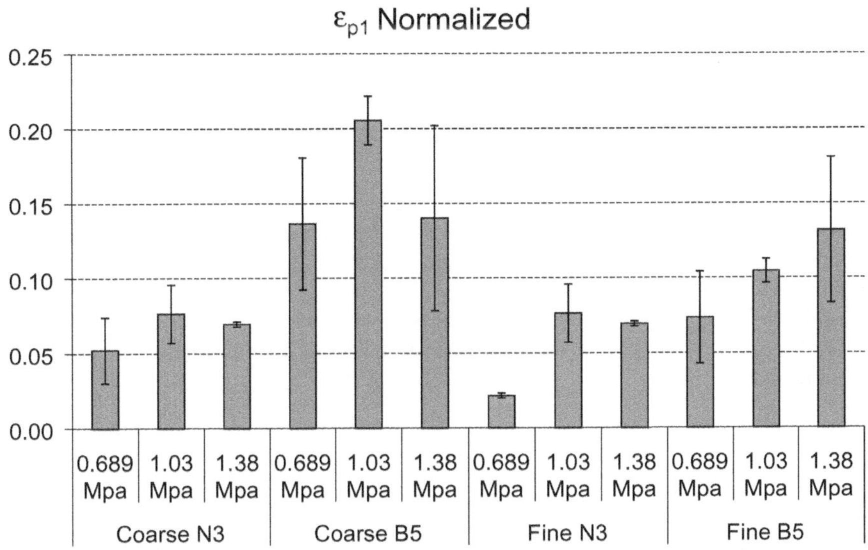

Figure 14. Normalized initial strain.

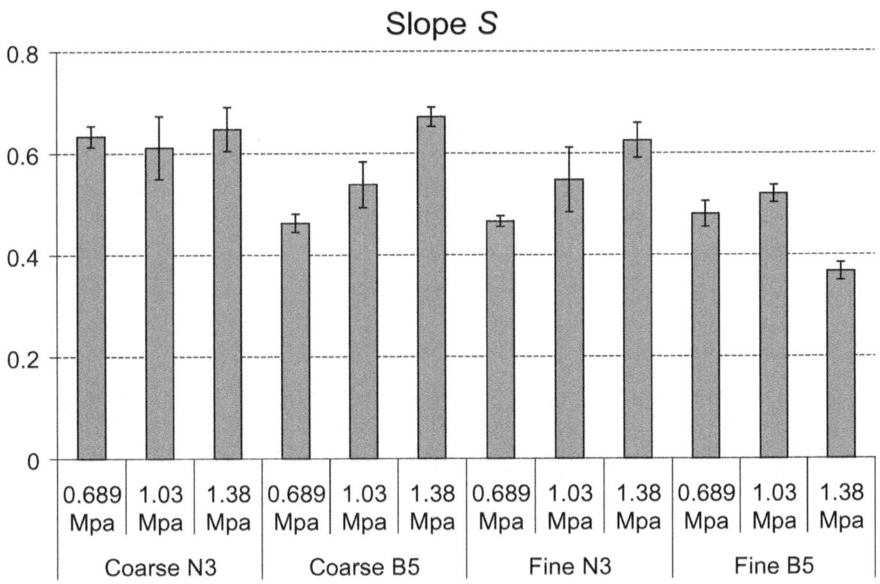

Figure 15. Permanent strain slope S.

increased significantly when the stress was changed from 0.689 MPa to 1.03 MPa. The changes in the parameter were less clear when the stress was increased from 1.03 MPa to 1.38 MPa. It appears that increasing stress results in increase strain ($\varepsilon_{p1\ Norm}$) but not in a clear and consistent trend.

The trends for the slope S are shown in Figure 15 which indicates some degree of aggregate gradation dependence for S. It is seen that the S values for the fine mixes are slightly smaller than the S values for the coarse mixes. It also appears that the S values for the modified binder B5 were slightly smaller than the ones for the unmodified binder N3, but the differences were not very significant. The stress influences on *S* do not have a clear trend.

4 SUMMARY OF FINDINGS

This study has shown that permanent strains in asphalt mixtures and binders increase with increasing stresses. These increases are not proportional to the stress and follow a nonlinear trend. The binder type and modification as well as aggregate gradation, have important effects on the rate of increase in permanent strains.

These results clearly point out that higher stresses can cause disproportionately higher rutting. Limited laboratory data show that the approximate effect of increasing loads can be estimated using the following equation:

$$N_2 = \left(\frac{p_1}{p_2}\right)^{2.09} N_1 \tag{7}$$

Where N is the number of load repetition or truck passes and p is the tire pressure. The power value of 2.09 needs further validation using different polymers and aggregate sources. To alleviate the problem of flexible pavement rutting due to higher truck loads, two remedies are proposed:

- First, load limitation on truck loads, especially for strategic highways are needed. The limits should be based on actual analysis of how asphalt mixtures sensitive to stress increase. It is clear that mixtures can be designed to be less sensitive than others to increase in loading. Thus a mixture design procedure that includes evaluation of "mixture stress sensitivity" is required.
- Second, modified binders that are more resistant to higher stresses (less sensitive to stress increase) should be specified. In order to be able to do so, a binder test procedure that can mimic field conditions in terms of stresses, temperatures and traffic volume is needed together with a reasonable model to correlate the binder properties to mixture rutting, is required which needs further investigations.

REFERENCES

Anderson, D.A, D.W. Christensen, H.U. Bahia, R. Dongre, M.G. Sharma, C.E. Antle and J. Button. 1994. Binder Characterization and Evaluation, Volume 3: Physical Characterization. Strategic Highway Research Program, Report SHRP-A-369, National Research Council, ISBN 0-309-05767-1, Washington DC.

Bahia, H.U., D.I. Hanson, M. Zeng, H. Zhai, M.A. Khatri and R.M. Anderson. 2001. Characterization of Modified Asphalt Binders in Superpave Mix Design. Publication NCHRP 459, National Academy Press, Washington, D.C.

Biligiri, K., K. Kaloush, M. Mamlouk and M. Witczak. 2007. Rational Tertiary Flow for Asphalt Mixtures. Proceedings of the *2007 Transportation Research Board Annual Meeting*. Washington, DC.

Christensen, D.W. and R.F. Bonaquist. 2005. *NCHRP Projects 9–25 and 9–31 Revised Final Report*, Sterling, VA: Advanced Asphalt Technologies, LLC, Submitted to the NCHRP for publication.

D'Angelo J., R. Dongré and G. Reinke. 2006. Evaluation of Repeated Creep and Recovery Test Method as an Alternative to SHRP+ Requirements for Polymer Modified Asphalt Binders. Proceedings of the Fifty First Annual Conference of the Canadian Technical Asphalt Association (CTAA). pp.143–162.

D'Angelo, J., R. Kluttz, R. Dongre, K. Stephens and L. Zanzotto. 2007. Revision of the Superpave High Temperature Binder Specification: The Multiple Stress Creep Recovery Test. *Journal of The Association of Asphalt Paving Technologists*, Vol 76.

Delgadillo R., K. Nam and H.U. Bahia. 2006. Why do we need to change G*/sinδ and How? *Road Materials and Pavement Design*, Vol.7, Issue 1, Hermes Science Publications.

Design Guide 2002. July 2004. developed for NCHRP Project 1–37a by ERES Division, Applied Research Associates and Arizona State University.

Kaloush, K.E. and M.W. Witczak. September 2000. "Development of a Permanent to Elastic Strain ratio Model for Asphalt Mixtures," *NCHRP Project 1–37 A Inter-Team Technical Report*, University of Maryland.
Kose, S. 2001. Development of a Virtual Test Procedure for Asphalt Concrete. PhD Thesis, UW Madison.
Leahy, R.B., and M.W. Witczak. 1991. "The Influence of Test Conditions and Asphalt Concrete Mix Parameters on Permanent Deformation Coefficients Alpha and Mu," *Journal of the Association of Asphalt Paving Technologists*, Vol. 60, p. 333.
Witczak, M., K.E. Kaloush, H. von Quintus. 2002. Pursuit of the Simple Performance Test for Asphalt Mixture Rutting. *Journal of the Association of Asphalt Paving Technologists.* Vol. 71, pp. 671–691.

Efficient Transportation and Pavement Systems – Al-Qadi, Sayed, Alnuaimi & Masad (eds)
© 2009 Taylor & Francis Group, London, ISBN 978-0-415-48979-9

Steady shear properties of a class of aged bitumens

S. Anjan Kumar & J. Murali Krishnan
Department of Civil Engineering, Indian Institute of Technology Madras, Chennai, India

A. Veeraragavan
Indian Institute of Technology Madras, Chennai, India

ABSTRACT: Aging is the gradual changes in the properties of bituminous binders over time, mainly due to heat, oxidation, ultra-violet radiation and loss of volatile components. The most significant aging of bitumen takes place during mixing where the thin film is exposed to high temperatures together with increased airflow. Also, the bitumens continue to age throughout their in-service life.

In this investigation, six modified bitumens with control unmodified bitumen were tested for their steady shear properties during their various stages of aging. The rolling thin film oven test (RTFOT) and pressure aging vessel (PAV) were used to simulate aging of bitumens during mixing and in-service respectively. It was found that long term aging has more severe effect on the steady shear properties of the bitumens than short term aging.

Keywords: Aging; Modified Bitumens; Dynamic Shear Rheometer; Steady Shear; Apparent Viscosity

1 INTRODUCTION

The viscoelastic response of bitumen is dependent on both time and temperature. Bitumen which imparts viscoelastic nature to the aggregate mixes when used as a binder for pavement construction undergoes both reversible and irreversible change in its internal structure composition. This irreversible change is termed as aging, which is due to oxidation of bitumen constituents and results in significant change in structure and composition of bitumen in service (Siddiqui and Ali 1999). Formation of highly polar and strongly interacting functional groups of oxygen due to aging results in this change in internal structure of bitumen (Petersen 1984). The development of internal structure and transitory nature in multi-constituent material like bitumen is the prime concern in understanding its mechanical behaviour (Krishnan and Rajagopal 2005). Increased traffic level and climatic variation pose variety of thermo-mechanical demands on bitumen used for paving applications. In this regard many types of modifiers are used as additives to improve the mechanical and thermodynamic properties of bitumen (Lu and Isacsson 1995), which makes the characterization more complex.

This work evaluates the viscous properties with due consideration to internal structural change of six different types of modified bitumens and unmodified bitumen by conducting steady shear experiments using a Dynamic Shear Rheometer (DSR). The irreversible change in the internal structure due to various levels of aging (RTFOT & PAV) is assessed by observing the stress overshoot as well as in the variation of apparent viscosity during a specific shear rate. The dependency of aging index on testing conditions was also investigated.

2 LITERATURE REVIEW

Modification of asphalts by polymer addition develops stronger secondary structure which has the ability to show resistance to applied stresses and temperature variations (Chen et al. 2002). Wekumbura et al. (2007) studied the destruction and reformation of the internal structure of polymer-modified asphalts by an interrupted shear test. Results showed that in polymer modified asphalts, the stress overshoots during steady shear and reaches a steady value, and the same was not observed in case of unmodified asphalts. They also concluded that the disturbed structure of polymer modified asphalts during shear, can reform with time and impart the ability of self-healing. Lu and Isacsson (1997) studied the effect of polymer content/structure and bitumen type on viscosity characteristics of SBS polymer modified bitumens using a calibrated glass capillary viscometer with an accurately reproducible head. The study indicated that modification with sufficiently high polymer content increases the non-Newtonian behaviour of the bitumens and also influenced the correlations between the kinematic and dynamic viscosities. Chowdary et al. (2007) investigated the evolution of internal structure of both modified and unmodified asphalts. Results clearly showed the non-linear behaviour of asphalts and also the significance of polymer modification in exhibiting the transition from shear thickening behavior to shear thinning behavior. Lu and Isacsson (2000) studied the physical hardening of bitumen and they indicated that the creep response and hardening index of the bitumens is changed considerably. Their results also indicated that degree and kinetics of physical hardening were dependent on the base bitumens, and in most cases, the effect of polymer modification was insignificant.

The physical and rheological properties of asphalt were improved by SBS copolymer modification (Cortizo et al. 2004). They investigated the effect of aging on SBS copolymer modified bitumen using rotational viscometers and thermo-oxidative degradation by using size exclusion chromatography and infrared spectroscopy. Increase of higher molecular size due to polar compounds explains the structural modification on behavior of polymer modified asphalt during their aging. Information on the degradation likely to occur in polymer modified bitumens during field service was studied by using modified RTFOT and DSR (Oliver and Tredrea 1997). It was found that long term laboratory exposure at a temperature below the maximum encountered in pavement service resulted in large changes in the rheological behavior of the modified bitumens and these changes were likely to reduce the advantage which fresh modified bitumens have over unmodified bitumens in service. Temperature susceptibility of asphalt depends on crude source; processing method and blend proportion (Rajan et al. 2008). They investigated effect of crude source, processing method and blend proportion on the viscous properties of the asphalt blends using rotational viscometer and DSR, they showed that proper blending process, selection of crude source and proportioning reduces the temperature susceptibility of asphalts.

Aging index can be applied to evaluate the performance of modified and unmodified bitumens (Mahrez and Karim 2003). They studied the effect of aging on the rheological performance of crumb rubber modified bitumen in comparison with 80/100 unmodified bitumen using DSR. Mahrez and Karim. (2003) pointed out that the linear viscoelastic parameters such as complex modulus and phase angle changed drastically. Lu and Isacsson (1998) investigated the aging properties of SBS polymer modified bitumens using Dynamic Mechanical Analysis, Gel Permeation Chromatography and Fourier Transform Infrared Spectroscopy. In all the cases, the aged modified bitumens showed better rheological properties than aged base bitumens. The study also indicated that the aging index was largely influenced by temperature and frequency. Huang et al. (2006) discussed the increase of 'stiffness' by addition of crumb rubber to various types of bitumen and showed that the base properties of bitumen is a main factor. Huang et al. (2006) also investigated the rheological properties of crumb rubber modified asphalts during aging and found that the addition of crumb rubber to bitumens reduced viscosity buildup with aging. In addition, rubber modification dramatically increased the elasticity of bitumens.

Most of the above mentioned investigations except those of Wekumbura et al. (2007) and Chowdary et al. (2007) have primarily looked at the linear viscoelastic properties of asphalts and modified asphalt during various stages of aging. Typically, during steady shear, three viscometeric functions are normally defined for any viscoelastic fluids: these are the shear rate dependent viscosity and the functions associated with first and second normal stress coefficients. Due to the transient nature of the steady shear experiments, benchmarking and the suitability of any constitutive model is normally evaluated with respect to these experiments, especially to predict the stress growth during steady shear (Attane et al. 1988). However, in asphalt literature one rarely comes across experimental data related to steady shear and the stress change. This investigation reports some very detailed investigations for a wide variety of modified and unmodified bitumen during several stages of aging.

3 EXPERIMENTAL INVESTIGATIONS

3.1 *Materials*

Six modified bitumens viz., two grades of SBS polymer modified bitumen i.e., PMB-40 and PMB-70, two grades of crumb rubber modified bitumen viz., CRMB-55 and CRMB-60, natural rubber modified bitumen (NRMB-70) and recycled plastic modified bitumen (RPMB-60) were used and compared with 60/70 grade unmodified bitumen as control bitumen. The numbers next to the type of modifier used indicates the penetration grade of the bitumens.

3.2 *Aging*

The RTFOT equipment which met ASTM (D 2872-88) requirements was used to stimulate short term aging conditions. A pressurized aging vessel system was used for the accelerated aging of bitumens. As soon as the time of RTFOT of bitumen samples was over, 50 ± 0.5 g of each bitumen sample was immediately poured into pans and placed together in the same PAV for a single test run. The test conditions adopted were temperature of 100°C with an air pressure of 2.1 ± 0.1 MPa and 20 hours of aging for a single run. After 20 h, the pans were removed and samples were de vacuumed to remove the air inside within 30 min.

3.3 *Dynamic Shear Rheometer (DSR)*

The dynamic shear rheometer used for the experimentation is the Physica MCR 301 manufactured by Anton Paar. It measures torque with an accuracy of 10^{-6} Nm and normal force with an accuracy of 0.002 N. Moreover, it can maintain temperature with an accuracy of 10^{-2} °C and can maintain the gap at 1 mm with 10^{-4} mm accuracy. The instrument comes with software—*Rheoplus*. For experiments a steady shear of 0.1 RPM was selected and maintained constant for all levels of aging to obtain meaningful comparisons using parallel plate geometry with 8 mm diameter. The apparent viscosity values were picked at 360, 420 and 480 sec's.

The specific experiment conducted here is essentially what is called as "stress growth upon the inception of steady shear flow" (Bird et al. 1987). A schematic of this test is shown below in figure 1. For all time previous to time $t = 0$, the sample is kept at rest, all components of stress are hence zero and the shearing is begun at time $t = 0$. For times $t \geq 0$, the sample is subjected to constant velocity gradient $\dot{\gamma}_0$. The main objective of this experiment is to see how the material functions $\eta^+(t, \dot{\gamma}_0)$ reach the steady state value. Essentially the idea is to see how much time the material

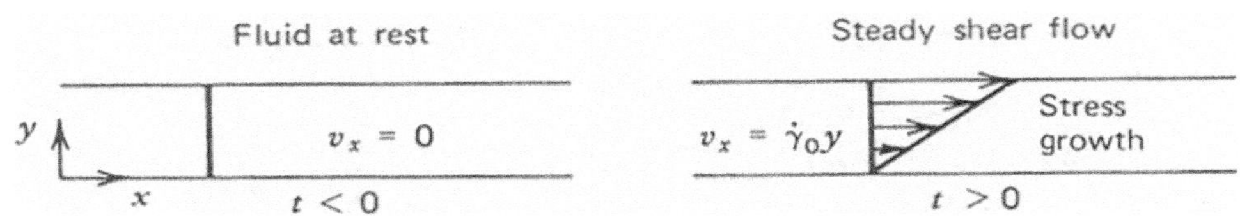

Figure 1. Stress growth upon the inception of steady shear flow.

takes to reach the steady state value and this in a sense describes the nature and evolution of the internal structure of the material (Bird et al. 1987).

4 RESULTS AND DISCUSSION

4.1 *Viscous properties of unaged bitumens*

In DSR the viscous properties of both modified and unmodified bitumens are evaluated at three temperature viz. 60°C, 65°C and 70°C. Steady shear experiments were conducted at a shear rate of 0.1 rpm at all temperature levels for 10 min using 8 mm parallel plate geometry. A condition time of 5 min was left to attain temperature equilibrium.

Figure 2 a, b, and c shows the effect of temperature on viscous properties of unaged bitumens. As can be seen, the increase in temperature is related to the decrease in apparent viscosity. The viscous properties varied significantly with temperature in unmodified bitumen (60/70), NRMB-70 and RPMB-60 when compared with the polymer modified bitumens. Higher viscosity values of 41833 and 21700 Poise were observed in PMB-40 and CRMB-55 at 60°C.

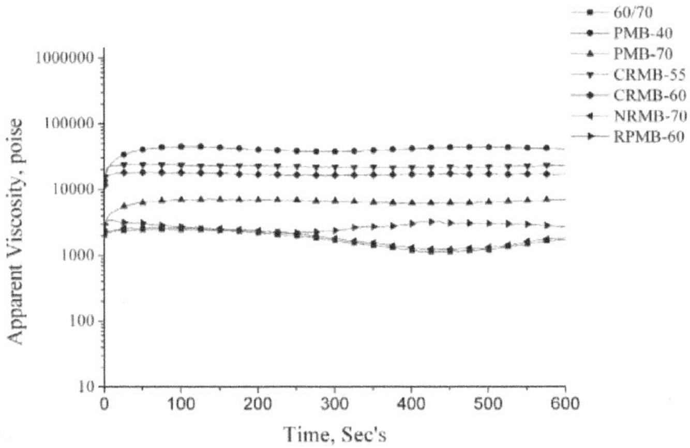

(a)

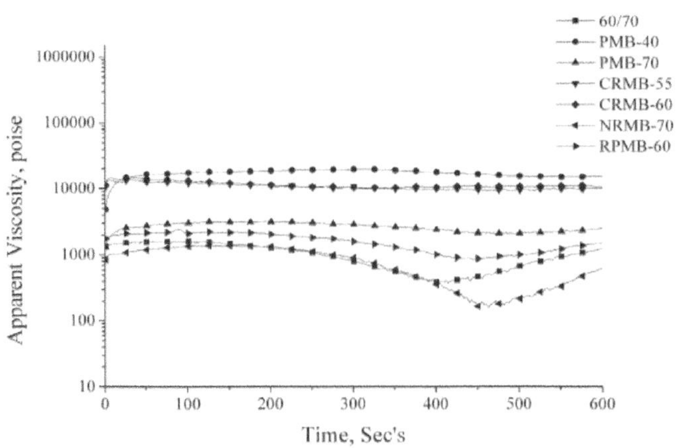

(b)

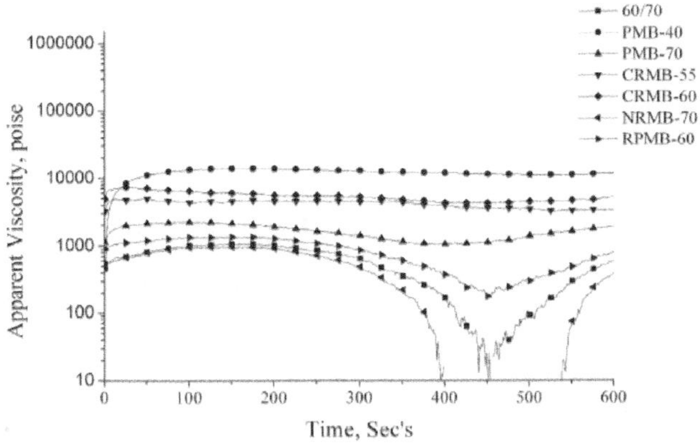

(c)

Figure 2 a, b, c. Viscous properties under unaged condition at 60°C, 65°C & 70°C.

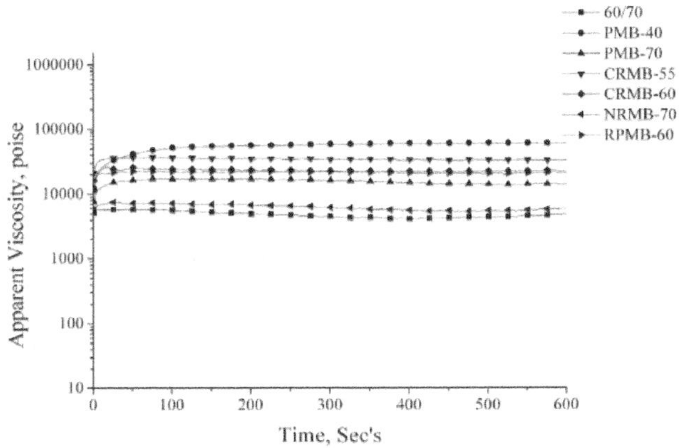

(a)

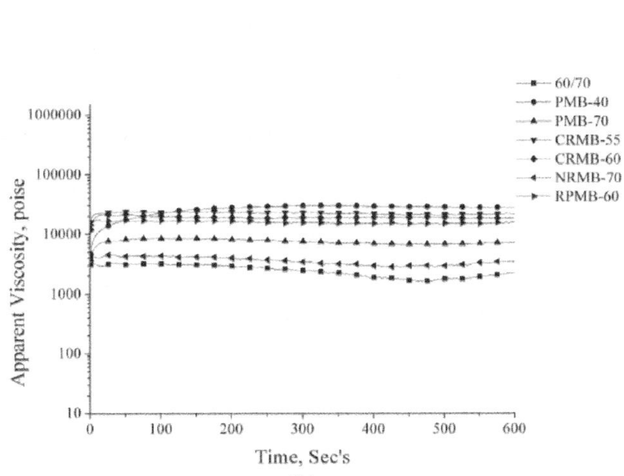

(b)

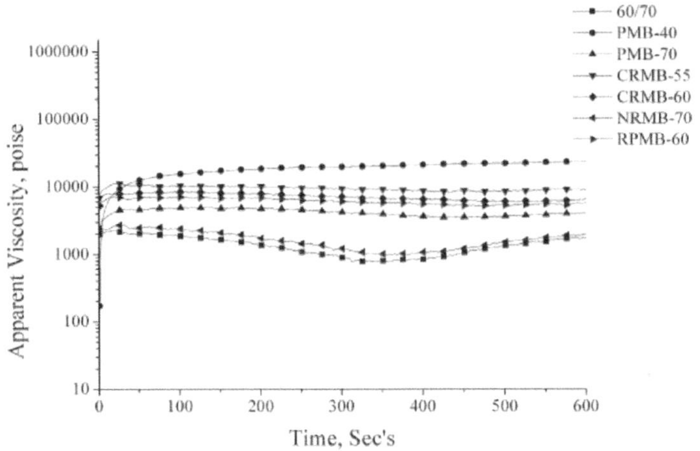

(c)

Figure 3 a, b, c. Viscous properties under RTFOT aged condition at 60°C, 65°C & 70°C.

4.2 *Viscous properties of RTFOT aged bitumens*

The RTFOT was aimed at simulating short-term aging of bitumen during mixing, transport, laying and compaction. The variation of viscous properties of RTFOT aged bitumens can be seen in figure 3 a, b, and c. It can be observed that the apparent viscosity of RPMB is more than PMB-70, which was lesser in unaged condition at all temperature levels. This shows that RPMB is more susceptible to aging than other modified and unmodified bitumens. Also one can observe that the time taken by PMB-40 to reach a steady value is 200 sec and 10 sec for other bitumens at 60 and 65°C. In PMB-40 the apparent viscosity values keep on increasing at 70°C up to 10 min, which clearly questions the measurement of dynamic viscosity at 6, 7 and 8th min as per ASTM standard. The same PMB-40 has reached a monotonically steady value at 70°C under unaged conditions; this clearly shows the change in internal structure of bitumen subjected to aging.

4.3 *Viscous properties of PAV aged bitumens*

Steady shear experiments were carried out at different temperatures on PAV aged bitumens as explained earlier. Figure 4 a, b and c evidently shows the change in internal structure, as the time taken to reach a steady value is 50 sec, which was 10 sec under unaged and RTFOT aged conditions. One can also observe that the viscosity overshoots up to 1.4×10^6 Poise in crumb rubber modified bitumens which was 4×10^4 Poise and 2.4×10^4 Poise under RTFOT and unaged conditions respectively. The viscous properties of RPMB have increased more than PMB-70 when subjected to PAV aging, which was lower at unaged conditions. This clearly shows that polymer modification results in very little evolution in internal structure when compared with other types of modification used in this study.

In an effort to understand the behavior of different types of modifiers during aging, the steady shear experimental data collected is shown in a detailed manner in the following figures. From figure 5, one can evidently observe the change in apparent viscosity of unmodified and polymer modified bitumens. The variation in viscosity of PMB subjected to various levels of aging is very less compared to unmodified bitumen as can be seen. It can be observed that RTFOT and PAV aged PMB has not reached a steady state even after 200 sec but, the same has reached a steady state under unaged conditions. Also the variation in viscosity of RTFOT and PAV aged unmodified bitumen is very higher compare to PMB.

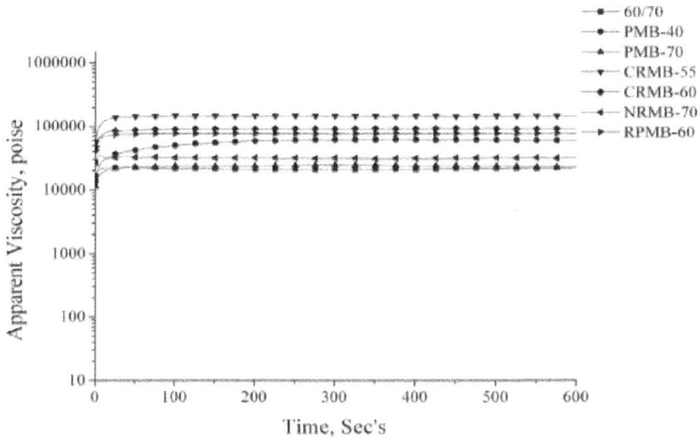

(a)

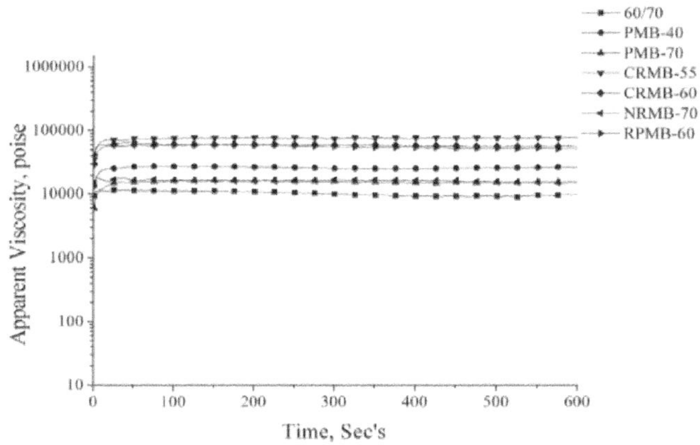

(b)

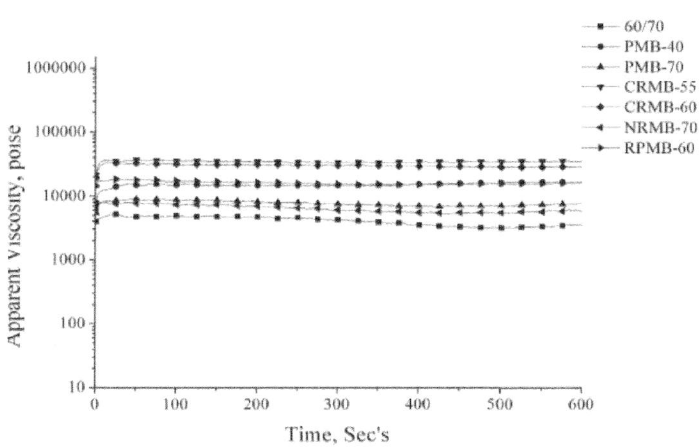

(c)

Figure 4 a, b, c. Viscous properties under PAV aged condition at 60°C, 65°C & 70°C.

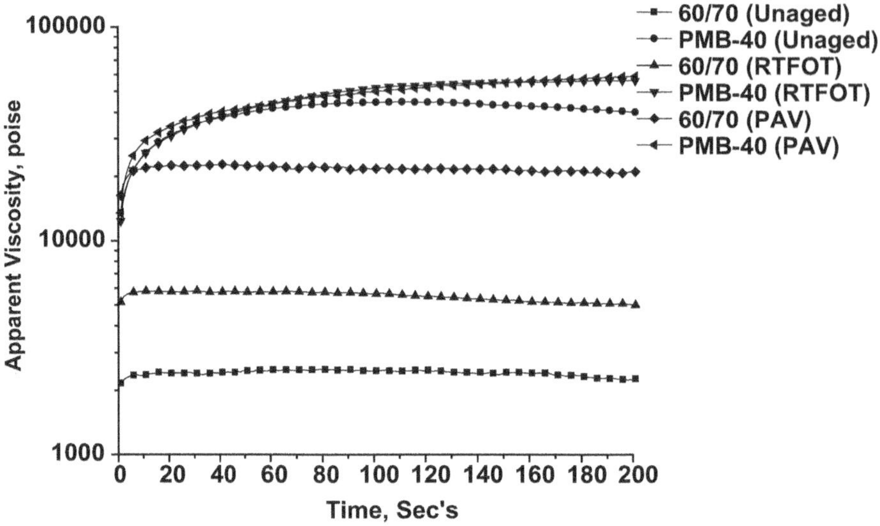

Figure 5. Variation in Apparent Viscosity of unmodified and polymer modified bitumens @ 60°C subjected to various levels of aging.

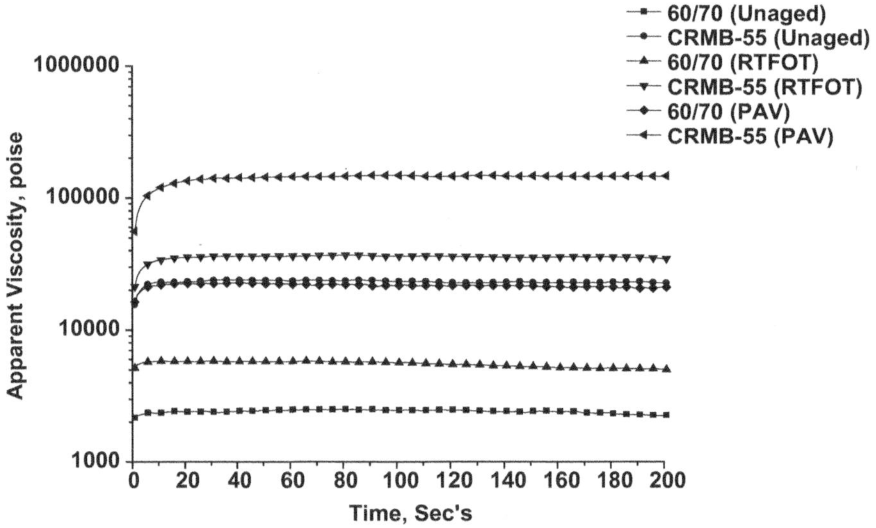

Figure 6. Variation in apparent viscosity of unmodified and crumb rubber modified bitumens @ 60°C subjected to various levels of aging.

The viscosity of PAV aged unmodified bitumen is almost same as that of unaged CRMB as seen from figure 6. The variation in viscosity under RTFOT and unaged conditions is less, whereas when the same material is subjected to PAV aging the viscosity spikes up to 1400000 Poise.

From figure 7, it can be observed that under unaged conditions the viscosity of unmodified and NRMB is almost the same. When the materials are subjected to RTFOT aging the difference is less, whereas under PAV aged conditions there is a distinct difference in viscosity values. Also

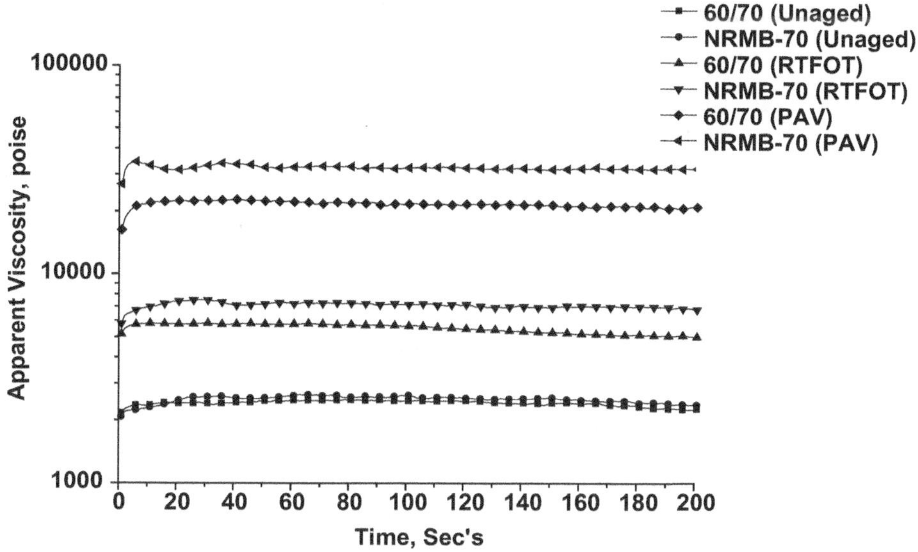

Figure 7. Variation in apparent viscosity of unmodified and natural rubber modified bitumens @ 60°C subjected to various levels of aging.

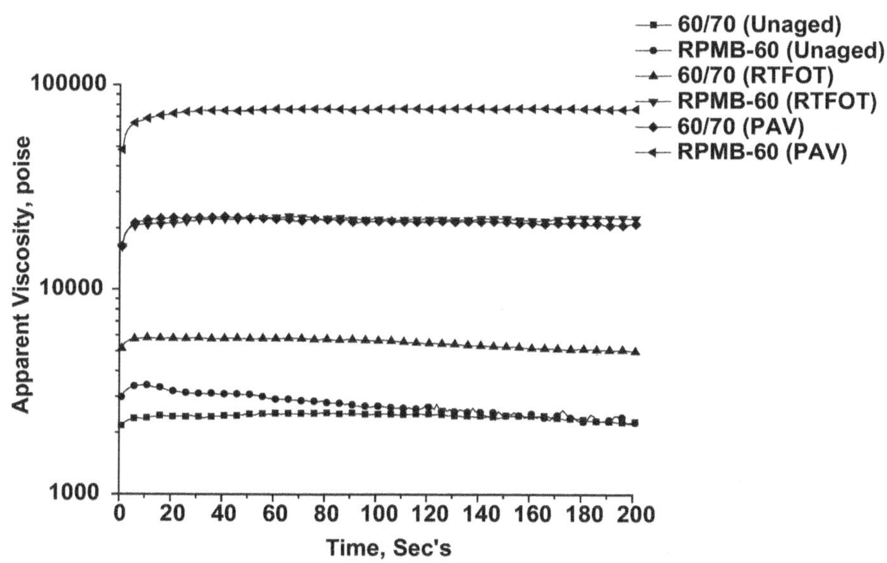

Figure 8. Variation in apparent viscosity of unmodified and recycled plastic modified bitumens @ 60°C subjected to various levels of aging.

one can observe from figure 8 that the viscosity values of RTFOT and PAV aged RPMB bitumen is higher than unmodified bitumen, which questions the role of recycled plastic as a modifier in improving the mechanical behaviour of base bitumen. One can clearly observe that the viscosity of PAV aged unmodified bitumen is lesser than that of RTFOT aged RPMB. This clearly shows the degradation of the modifier has more effect on hardening than the oxidation of base bitumen.

Table 1. Variation in aging index of different bitumens with levels of aging.

	Viscosity aging index, @ 60°C		Viscosity aging index, @ 65°C		Viscosity aging index, @ 70°C	
Bitumens	RTFOT	PAV	RTFOT	PAV	RTFOT	PAV
60/70 (A)	3.49	17.22	3.74	18.98	6.48	22.98
PMB-40 (B)	1.44	1.46	1.67	1.48	1.24	1.32
PMB-70 (C)	2.36	3.83	2.98	6.50	3.25	6.23
CRMB-55 (D)	1.56	6.73	2.17	7.85	2.28	9.00
CRMB-60 (E)	1.34	5.49	1.18	5.46	1.43	6.60
NRMB-70 (F)	4.13	23.23	8.71	47.20	6.55	32.20
RPMB-60 (G)	6.98	25.41	14.61	52.58	15.40	43.23

4.4 *Viscosity aging index*

Bitumen aging is traditionally evaluated by means of the aging index, which is defined as the ratio viscosity of the aged bitumen to that of the unaged bitumen. It can be seen from table 1, that the aging index of unmodified bitumen is higher than polymer and crumb rubber modified bitumens. One can also observe from table 1 that for NRMB and RPMB the aging index is greater than unmodified bitumen in all testing conditions (60°C, 65°C and 70°C) and also at all levels of aging.

5 CONCLUSIONS

In this study, steady shear experiments have indicated that the bitumens undergo considerable degradation during aging resulting in irreversible change in internal structure. Increase in viscous properties of bitumens subjected to various levels of aging has confirmed embrittlement of bitumen in service. Apparent viscosity measurement has indicated those crumb rubber, recycled plastic used as modifiers, are not inert additives and increases the viscous properties of modified bitumen considerably after all levels of aging. A somewhat marked increase in viscous properties is observed in PAV aged rubber and recycled plastic modified bitumens even when compared to unmodified bitumen. Also steady shear experiments shows that modification with a sufficiently high SBS content increases the degree of non-Newtonian behaviour of bitumens. The information obtained from these studies is useful in understanding the compatibility of modification additives used to improve the temperature susceptibility of base bitumens.

REFERENCES

1. Attane, P., Turrel, G., and Pierrard, J.M. (1988), On the Use of Transient Data for the Evaluation of Integral Constitutive Equations, Journal of Rheology, 32(l), 23–46.
2. Bird, R.B., Armstrong, R.C., and Hassager, O. (1987). "Dynamic of polymeric fluids." Fluid Mechanics, Wiley-Interscience Publication, New York.
3. Chen, J.S., Liao, M.C., and Shiah, M.S. (2002) Asphalt Modified by styrene butadiene-styrene triblock Copolymer: Morphology and Model, Journal of Materials in Civil Engineering, 14(3): 224–229.
4. Cortizo, M.S., Larsen, D.O., Bianchetto, H. and Alessandrini, J.L. (2004) Effect of the Thermal Degradation of SBS copolymers during the Aging of Modified Asphalts, Fuel, 86: 275–282.
5. Huang, S-C., Pauli, T.R., and Robertson, R.E.. (2006) Rheological Changes in Crumb Rubber-Modified Asphalts with Long-Term Aging, TRB, Annual meeting CD-ROM.
6. Krishnan, J.M., and Rajagopal, K.R. (2005) On the Mechanical Behaviour of Asphalt, Mechanics of Materials, 37(11):1085–1100.
7. Lu, Xaiohu and Isacsson, Ulf. (1995) Testing and Appraisal of Polymer Modified Road Bitumens—State of the Art, Materials and Structures, 28: 139–159.

8. Lu, Xaiohu and Isacsson, Ulf. (1997) Influence of Styrene-Butadiene-Styrene Polymer Modification on Bitumen Viscosity, Fuel, 76: 1353–1359.
9. Lu, Xaiohu and Isacsson, Ulf. (1998) Chemical and rheological evaluation of aging properties of SBS polymer modified bitumens, Fuel, 77: 961–972.
10. Lu, Xaiohu and Isacsson, Ulf. (2000) Laboratory study on the low temperature physical hardening of conventional and polymer modified bitumens, Construction and Building Materials, 14: 79–88.
11. Mahrez, A. and Karim, M.R. (2003) Rheological Evaluation Of Aging Properties of Rubber Crumb Modified Bitumen, Journal of the Eastern Asia Society for Transportation Studies, 5: 820–833.
12. Oliver, J.W.H. and Tredrea, P.F. (1997) The Changes in Properties of Polymer Modified Binders with Simulated Field Exposure, AAPT, 570–592.
13. Petersen, J, C. (1984) Chemical Composition of Asphalt as Related to Asphalt Durability-State of Art, Transport Research Record, TRB, National Research Council, Washington, D.C., 999: 13–30.
14. Rajan N. K., Selvavathi V. , Sairam B. and Krishnan J.M. (2008) Influence of Crude Source on the Viscous Properties of Blended Asphalt, International Journal of Road Materials and Pavement Design, In Press.
15. Siddiqui, M.N. and Ali, M.F. (1999) Studies on the aging behavior of the Arabian asphalts, Fuel, 78: 1005–1015.
16. Venkaiah Chowdary., Thushara, V.T., Ahmed Asif, K., and Krishnan J.M. (2007) Experimental Verification of Evolution of Internal Structure of Asphalt and Modified Asphalt, 4th International SIIV Congress, Palermo (Italy).
17. Wekumbura, C., Stastna, J., and Zanzotto, L. (2007) Destruction and recovery of internal structure in polymer-modified asphalts, Journal of Materials in Civil Engineering, 19(3): 227–232.

Efficient Transportation and Pavement Systems – Al-Qadi, Sayed, Alnuaimi & Masad (eds)
© 2009 Taylor & Francis Group, London, ISBN 978-0-415-48979-9

Influence of surface interactions between bitumen and mineral fillers on the rheology of bitumen-filler mastics

R. Taylor
Shell Bitumen, UK

G.D. Airey
University of Nottingham, Nottingham, UK

ABSTRACT: This paper describes a study of the rheological behaviour of ten fillers in three different bitumen types. The fillers were laboratory manufactured from clean single-sized aggregates and were grouped according to their broad petrologic type. Five of the fillers were produced from aggregates, which had calcium carbonate as the principal mineral, and five fillers were manufactured from aggregates that contained significant quantities of silica. The binders consisted of 40/60 and 10/20 penetration grade bitumens manufactured from a Venezuelan crude source and a synthetic polymer-modified resin. An extensive characterisation of the fillers was undertaken to account for particle size distribution, particle shape, bulk volume, surface area and surface energy. The bitumens where rheologically characterised using standard penetration and softening point tests and more complex creep compliance testing using a Dynamic Shear Rheometer. Additionally, the bitumens in the study were characterised for surface energy using a Sessile Drop technique allowing calculations of adhesion, W_{12}, to be determined for the bitumen-filler combinations. The creep compliance of the bitumen-filler mastics over a range of solid volume fractions (φ) was measured with the relative creep compliance being derived for each of the mastics using the ratio of the creep compliance of the pure bitumen to the creep compliance of the mastic. The maximum packing fraction (φ_{max}) was determined for each bitumen-filler combination using a simple two-point projection technique based on the reciprocal relative creep compliance. The intrinsic viscosity (from the Krieger & Dougherty rheological model) of each filler type was derived in each bitumen type and this in turn compared to the index properties obtained during the classification of the materials. In addition the product of intrinsic viscosity and maximum packing fraction in the Krieger-Dougherty equation, which describes the stiffening effects in the mastic outside of the bulk volume filling relation, φ/φ_{max}, was found to be closely related to the surface free energy of the mastics provided.

1 INTRODUCTION

Fillers modify the properties, increase the performance of, and provide improved durability to composites, polymers, rubbers, adhesives, coatings and construction materials (such as concrete and asphalt mixtures). Fillers are used to lower the cost of materials, increase rigidity and give special properties to a material (such as colour or fire retardancy). The filler properties have considerable effects on the processing characteristics of materials such as mixing, pumping and compacting. The effects of fillers are therefore of vital importance.

Fillers are typically fine powders with a particle size distribution in the range of 0–100μm. They can be naturally occurring materials such as calcium carbonate, manufactured fillers (e.g. carbon black), or derived from industrial wastes such as fly ash from power stations. Other common fillers include silica, kaolin ("China Clay"), mica, feldspar and diatomite. Filler can be defined as "*solid material capable of changing the physical and chemical properties of materials by surface interaction or its lack thereof and by its own physical characteristics*". A comprehensive list of definitions can be found elsewhere (Wypych 1999). The above definition implies the existence of two ways in which fillers modify a system. Firstly, the way in which the filler's shape, particle size and particle size distribution affect the system through filling of the liquid with solid particles. Secondly, the way in which interactions between the solid and liquid phases of the mixture affects the material. The second interactions can vary from strong chemical bonds or physical interactions leading to strongly reinforced materials, to almost no interaction at all.

The most frequently used filler in asphalt mixtures is limestone (calcium carbonate), which is derived from the consolidation of minute micro-organisms during the formation of the earth's crust. Limestone is the general term for rocks where calcite, a form of calcium carbonate, is the predominant mineral. Limestone may also contain a proportion of magnesium carbonate, silica, clays, iron oxides and organic material. Other materials commonly used as fillers in asphalt include Portland Cement and hydrated lime, which possesses well documented properties with regard to mixture durability and increased resistance to moisture damage in asphalt (Little & Epps 2001). Additionally, recycled fillers in the form of so-called "baghouse" fines have also been frequently used in asphalt mixtures. The performance of baghouse fines was the subject of several key studies on the behaviour of fillers in asphalt mixtures following changes in the Clean Air Act 1970 in the United States (Kandhal 1980, Anderson et al. 1982).

When bitumen is combined with mineral filler, a mastic is formed. This mastic can be viewed as the component of the asphalt mixture that binds the aggregates together and also the component of the asphalt that undergoes deformation when the pavement is stressed under traffic loading. The characteristics of the filler can significantly influence the properties of the mastic, and thus the filler properties can have significant effects on asphalt mixture performance. The importance of fillers, and the resulting mastics, has been studied for a century (Richardson, 1907). Arguably, the key advance in the understanding of asphalt fillers was made shortly after the Second World War when Rigden (1947) proposed that the volume of fixed bitumen that a filler could accommodate in its compacted state influenced the stiffening behaviour of fillers in mastics. Rigden also proposed that the volume of "fixed bitumen" that the system could accommodate scaled the effects of a given volume of filler. Since that time, several other researchers have found Rigden's approach reasonably successful in predicting the stiffening effect of various filler types in a given bitumen type (Heukelom & Wijga 1971, Kandhal 1980, Anderson et al. 1992a, 1992b, Kavussi & Hicks 1997, Kandhal et al. 1998, Cooley et al. 1998).

However, the traditional approach of examining a set of fillers with varying properties in a single bitumen and then relating the properties of the mastic to index properties of the fillers invariably leads to the conclusion that the effects of stiffening are primarily a result of the ratio between the solid volume fraction and the maximum packing fraction, with the voids in compacted filler being used as a proxy for the maximum packing fraction (Rigden 1947, 1954, Craus et al. 1978, 1981, Ishai et al. 1980). This approach does not account for the interaction between the filler and the bitumen, which can be seen when fillers stiffen different bitumens to different extents due to the level of interaction between the two phases. This paper introduces an alternative approach to quantifying the effects of fillers and bitumens on the properties of bitumen-filler mastics using a combination of rheological models and surface energy measurements of the components compared to the simple volume filling relationship described above.

2 RHEOLOGICAL MODELS

Mastics can be considered as suspensions of solid filler particles in a liquid (i.e. bitumen). Rheological models have been used to describe the behaviour of solid particles in a liquid since the early part of the twentieth century starting with Einstein's model (1906) in Equation 1:

$$\eta = \eta_0 \left(1 + [\eta]\varphi\right) \tag{1}$$

where η = the viscosity of the suspension; η_0 = the viscosity of the liquid phase; $[\eta]$ = the intrinsic viscosity of the solid phase; and φ = the solid phase volume.

Relative viscosity (η_{rel}) describes the increase in viscosity of the suspension as a ratio to the viscosity of the liquid phase. The Einstein model predicts a linear change in relative viscosity with regard to increases in solid concentration (φ). As solid volume fraction increases, the magnitude of the surface area and particle-particle contacts increase and effects outside of simple volume filling become more prevalent.

Thus, in reality, increasing additions of solid particles leads to exponentially increasing relative viscosity up to the point where flow is not possible and relative viscosity becomes infinite. The solid volume fraction at this point is termed the *maximum packing fraction* (φ_{max}) and alongside relative viscosity, is a key term used in modelling and understanding the behaviour of suspensions. In line with Rigden's observations relating to voids within compacted filler, the effects of solid volume on viscosity are scaled by φ_{max} and this ratio is an important variable controlling the rheology of suspensions. Later models developed to describe the behaviour of suspensions (Mooney 1957, Krieger & Dougherty 1959, Chong 1971) include the term φ/φ_{max} and predict exponentially increasing stiffening as the solid volume fraction, φ, approaches φ_{max}. At low filler concentrations, such models tend to reduce to the Einstein equation (Barnes 2000). The Krieger and Dougherty equation has the form:

$$\eta = \eta_0\left(1 - \frac{\varphi}{\varphi_{\max}}\right)^{-[\eta]\varphi_{\max}} \tag{2}$$

A further variable featured in rheological models used to describe the behaviour of suspensions is *intrinsic viscosity*, $[\eta]$, of the filler. For spheres in dilute suspensions, Einstein proposed that the value of intrinsic viscosity is 2.5 and this is referred to as the Einstein Coefficient. Intrinsic viscosity is typically attributed to shape, but encompasses all effects experienced by the suspension outside of the bulk filling relationship of φ/φ_{max}. Interaction at the solid liquid interface would be included in the term intrinsic viscosity according to the models used to describe the rheology of suspensions. The product of intrinsic viscosity and maximum packing fraction found in the Krieger-Dougherty equation is reported as being close to the value of 2 for a variety of filled systems (Kitano et al. 1981, Barnes 2000) and this coefficient is referred to as the Marron-Pierce-Kitano Coefficient (MPK) (Kitano et al. 1981, Rides 2005).

For asphalt-filler mastics the typical approach to research has been to consider the change in penetration, softening point or viscosity of the bitumen (Heukelom & Wijga 1971, Anderson et al. 1982, Kandhal 1980, 1998). Later researchers have examined change in complex modulus of bitumen when adding different filler types to a single bitumen type (Cooley et al. 1998, Harris & Stuart 1995). These types of measurements, for example relative complex modulus, are akin to the relative viscosity component of rheological models used to describe the behaviour of suspensions.

For asphalt fillers, the relative viscosity, or related term, has been compared to several index properties of fillers, typically particle size distribution, particle shape, surface activity and the void content of filler in its compacted state. Most researchers have found that the voids in the compacted filler is the property of the filler most suited to predicting the relative viscosity (or other relative rheological measures) of bitumen-filler mastics (Heukelom & Wijga 1971, Kandhal 1980, Anderson et al. 1992a, 1992b, Kavussi & Hicks 1997, Kandhal et al. 1998, Cooley et al. 1998). As the maximum packing fraction is related to the voids in the compacted filler and the effects of solid volume fraction, as proposed in the rheological models outlined previously are scaled by the maximum packing fraction, this conclusion appears reasonable.

However, several researchers have noted that different fillers stiffen different bitumen types to a different extent (Dukatz & Anderson 1980, Anderson et al. 1982, Shashidar & Romero 1998) and this has been a limitation to the approach of measuring the filler properties in air. Extreme cases have been noted such as hydrated lime where stiffening massively exceeds the potential estimated by voids in the mineral filler (Shashidar & Romero 1998). Crucially, voids in the compacted filler are measured in air (Rigden Voids) and do not account for interactions between the bitumen and the filler in the mastic. Interactions of fillers in different bitumen types may lead to changes in different ways. For example φ_{max} may differ in different bitumens as a result of different levels of dispersion of the filler, or alternatively the filler may cause changes in the viscosity of the liquid phase as a result of restructuring, chemical changes or other such effects.

3 MATERIALS

3.1 *Fillers*

During the experimental programme, ten fillers were examined in three different binder types. The fillers were divided into two sub-sets based on their petrologic type. Five fillers were manufactured from aggregates where calcium carbonate was the principal mineral and five were derived from igneous aggregates containing a significant proportion of silica. According to Wypych (1999), calcium carbonate can be considered "inert" as it does not possess functional groups at the surface. On the other hand, silica based fillers do possess functional groups of their surface and can therefore form bonds and be considered "active". It follows that there should be differences between the behaviour of fillers derived from calcium carbonate aggregates and fillers derived from siliceous aggregates. A summary of the petrologic type of the fillers is given in Table 1.

The intention was to provide a set of fillers with a relatively small range of physical properties, such as particle size distribution and shape, but with variations in surface characteristics. To this end, fillers were ground in the laboratory to produce fillers of essentially the same particle size distribution and particle size. The ten fillers were ground from natural aggregates and all fillers had particles substantially smaller than 100μm.

The particle size distributions of the fillers by volume, by mass and specific surface area were determined using a laser diffraction technique. Additionally, several descriptors of the particle size distribution were derived, Fineness Modulus, $D_{10}V$ (diameter at which ten percent by volume of particles are smaller than), and the Uniformity Coefficient D60/10 (the diameter at which 60% by mass of particles are smaller than divided by the diameter at which 10% by mass of particles are smaller than). A summary of the particle size distribution results is given in Table 2.

All fillers were substantially finer than 100 microns in particle size. As a general trend, Fillers S1–5 are coarser than Fillers C1–5. Filler C5, which has a significant proportion of clay within its matrix, produced the finest particle size distribution of the fillers.

Following filler manufacture, a series of detailed classification tests were undertaken on the fillers. Maximum packing fraction, φ_{max}, was estimated by using compacted dry density of the filler as related to the specific gravity, while shape was quantified by optical microscopy and the calculation of aspect ratios. In addition to physical characterization of the fillers, tests were carried out to calculate the surface energy components of each filler using dynamic vapour sorption tests (Good & Van Oss 1992, Hefer 2004, Bhasin 2006). γ^{TOTAL}, γ^{LW}, γ^{AB}, γ^{+} and γ^{-} were derived for each filler. Specific surface area was obtained by the BET equation (Brunauer et al. 1938) using octane

Table 1. Petrologic type of the fillers manufactured for the study.

Limestone fillers		Silica fillers	
C1	Calcium Carbonate	S1	Granodiorite
C2	Calcium Carbonate	S2	Granite-Gneiss
C3	Dolomitic Limestone	S3	Basalt
C4	Dolomitic Calcite	S4	Amphibolite
C5	Mudstone	S5	Gneiss

Table 2. Particle size distribution data for the fillers.

Filler	Passing 100μm	Passing 60μm	Passing 20μm	D(v,0.90) μm*	D(v,0.50) μm*	D(v,0.10) μm*
C1	94.3	84.1	64.2	79.2	7.8	1.1
C2	97.5	89.0	64.4	63.0	9.9	1.3
C3	97.0	87.6	64.2	66.8	9.5	1.1
C4	96.5	87.9	65.5	66.8	9.5	1.2
C5	96.9	90.1	74.5	59.8	4.5	0.9
S1	91.3	75.3	45.1	95.2	25.0	2.3
S2	88.1	68.9	35.9	106.4	35.0	3.6
S3	95.3	84.2	57.8	76.3	14.1	1.8
S4	92.8	77.1	47.6	89.7	22.5	1.8
S5	93.4	81.3	53.9	85.0	17.0	2.0

*Diameter where x% by volume is smaller.

Table 3. Summary of the key index parameters of the fillers

	Physical			Surface Area		Surface Energy (mN/m^2)		
Fillers	φ_{max}	Aspect Ratio	Fineness Modulus	BET (m^2/g)	MBV	γ^{LW}	γ^+	γ^-
C1	0.398	1.31	4.1	1.44	0.83	84.0	10.7	0.4
C2	0.449	1.34	4.1	1.23	0.61	54.6	0.4	3.0
C3	0.467	1.35	4.2	0.97	0.15	81.5	7.3	0.1
C4	0.603	1.73	4.1	1.93	1.78	73.8	7.1	1.8
C5	0.482	1.75	4.9	8.27	9.24	70.6	23.9	31.5
S1	0.583	2.14	3.0	0.71	1.1	77.2	32.5	6.5
S2	0.507	1.74	2.5	0.67	0.87	81.2	4.9	0.1
S3	0.513	1.54	3.6	2.21	3.44	61.4	35.0	4.8
S4	0.431	2.34	3.2	0.62	1.4	71.5	2.6	0.2
S5	0.562	2.14	3.4	1.85	1.35	71.6	5.4	7.4

gas and finally, a Methylene Blue Value (BS EN 933-9 1999) was measured on each filler as an indicator of surface area. A summary of the key filler properties is given in Table 3.

3.2 *Binders*

Three different binder types were chosen for the experimental work:

- Binder B1: 40/60 Penetration Grade Venezuelan bitumen
- Binder B2: 10/20 Penetration Grade Venezuelan bitumen
- Binder B3: A polymer-modified synthetic resin

The three binders were chosen to represent a range of both chemical and rheological types. Binders B1 and B2 are derived from the same crude oil source and should be chemically similar. Binders B1 and B3 are rheologically similar (similar penetration and softening point values) but chemically very different. Whereas Binders B2 and B3 are dissimilar both chemically and rheologically. An extensive characterization was carried out on the three binders prior to the mastic stage of the experiments. Penetration (BS EN 1426 2000) and Softening Point (BS EN 1427 2000) values were determined for the three binders and presented in Table 4.

Table 4. Penetration and softening point values of binders.

Binder	Penetration (dmm)	Softening Point (°C)
B1	48	53.0
B2	16	68.0
B3	50	60.0

Table 5. Creep compliance as a function of temperature for the binders.

	Creep Compliance ($1/J \times 10^6$)		
Temperature (°C)	B1	B2	B3
45	255	29	515
49	515	37	1084
53	975	49	2219
57	1770	66	3917
61	3056	90	5691
65	4902	120	7213
69	6035	169	8360

Table 6. Surface energy of binders determined using sessile drop technique.

Binder	γ^{TOTAL}	γ^{LW}	γ^{AB}	γ^{+}	γ^{-}
B1	31.66	29.55	2.11	3.52	0.32
B2	46.43	43.20	3.29	3.59	0.76
B3	40.46	37.86	2.60	1.85	0.91

In addition, the creep compliance behaviour of the binders over a temperature range of 45°C to 69°C was measured using a dynamic shear rheometer (DSR) in creep testing mode. The creep response of the binders was measured over 1 second loading with a stress level of 2 kPa. Measurements were first taken at a temperature of 45°C and then at 4°C intervals up to a final temperature of 69°C. A thermal equilibrium of 15 minutes was applied between temperature measurements. A summary of the test results are given in Table 5.

In order to determine the interfacial free energy of the two phases (binder and filler), γ_{12}, and the thermodynamic work of adhesion, W_{12}, (Equation 3), it is necessary to obtain surface energy components of both phases.

$$W_{12} = \gamma_1 + \gamma_2 - \gamma_{12} \tag{3}$$

As outlined above the surface energy components of the fillers in the study were measured using dynamic vapour sorption, while the surface energy components of the binders were determined using the sessile drop method (Thelen 1958, Adamson & Gast 1997, Hefer 2004). The surface energy components of the three binders are given in Table 6.

3.3 *Mastics*

The ten fillers were tested in each of the three different binder types in terms of their creep compliance measured using a DSR. The mastics were tested over the same range of temperatures from

45°C to 69°C used for the binders at three levels of solid volume addition, 0.3φ, 0.4φ and 0.5φ. Relative creep compliance was derived for each bitumen-filler mastic according to Equation 4.

$$J(t)_{rel} = \frac{J(t)_{bitumen}}{J(t)_{mastic}} \tag{4}$$

The maximum packing fraction of each bitumen-filler combination, φ_{max}, was calculated at each temperature using a two point projection technique based on the relative creep compliance recorded at 0.4φ and 0.5φ. Finally, master curves of the stiffening effect (relative creep compliance) of the fillers were produced for all the fillers in each binder type as a function of the solid volume fraction normalized by the maximum packing fraction, φ/φ_{max}.

4 MAXIMUM PACKING FRACTION

The combined effects of particle density, particle shape and particle size distribution can be captured in a single property known as the maximum packing fraction. This property can be estimated in several ways, including examining the packing characteristics of the filler in air. In asphalt technology, this approach is expressed as "Rigden Voids" which is the air void content of the filler obtained under standard test conditions (BS EN 1097-4 1999) and which relates approximately to 1—the maximum packing fraction.

There is a general consensus amongst researchers that the effects of fillers in bitumen are a result of two factors—the physical volume concentration of solid particles and interactions between the filler surface and the bitumen (Rigden 1947, 1954, Heukelom & Wijga 1971, Anderson et al. 1973, 1982, Shashidar & Romero 1999). Despite this, the vast majority of filler research has focussed on linking the stiffening effect of different fillers to a simple index property of the filler in air (Crauss et al. 1978, 1981, Dukatz & Anderson 1980, Harris & Stuart 1998, Kandhal et al. 1998, Cooley et al. 1998). However, the determination of φ_{max} in air fails to account for important interactions between the two phases. Additionally, the dispersal of the filler will differ in different bitumen types leading to a range of values for φ_{max}, whilst tap density, Rigden voids, etc are typically measured in a single medium, air.

An alternative approach would be to combine the volume concentration effects of the filler with the interaction between filler and bitumen by determining φ_{max} not in air but in the actual mastic. Maximum packing fraction, φ_{max}, is defined as the solid volume fraction at which viscosity of a suspension becomes infinite, i.e. the point at which there is no flow. In terms of creep compliance, φ_{max} would relate to the solid volume fraction at which relative creep compliance becomes infinite—i.e. $1/J(t)_{rel}$ approaches zero.

Researchers have proposed a relative simple two-point projection method to determine φ_{max} for ceramic pastes (Hurysz & Cochrane 2004). This procedure has been adapted to determine φ_{max} of bitumen-filler mastics. The adapted method of determining φ_{max} is to plot the reciprocal of relative creep compliance against volume concentration and extrapolate to zero. This point represents the reciprocal infinite relative creep compliance, i.e. creep compliance of the mastic $J(t)$ approaching zero. At this point the volume of filler would be sufficient to reduce the creep compliance level to zero. Taking the data from the creep recovery tests at 0.4φ and 0.5φ, this procedure was carried out for each filler combination and the results presented in Table 7 alongside φ_{max} determined in air.

There are no correlations between the values of φ_{max} determined in one type of binder and the values of φmax obtained in other binder types. Additionally, the values of φ_{max} are not well correlated to the single value of φ_{max} determined from the compacted bulk density of the filler in air (Rigden Voids). The same filler has different maximum packing fraction values in different bitumen types and also differs sometimes significantly from the Rigden Voids determination. This indicates that estimating maximum packing fraction from the volumetric properties of the filler in

Table 7. Maximum packing fraction of different fillers.

Filler	Air	Binder B1	Binder B2	Binder B3
C1	0.398	0.534	0.523	0.547
C2	0.449	0.534	0.558	0.534
C3	0.467	0.579	0.558	0.604
C4	0.603	0.578	0.590	0.517
C5	0.482	0.555	0.572	0.624
S1	0.583	0.547	0.505	0.537
S2	0.507	0.576	0.604	0.583
S3	0.513	0.554	0.510	0.577
S4	0.431	0.517	0.553	0.565
S5	0.562	0.568	0.550	0.553

air is unlikely to be useful in all situations and highlights the importance of including the interactions between filler and binder in mastic characterisation.

5 INTRINSIC VISCOSITY

Researchers have long proposed that the effects of solid volume fraction φ are scaled by φ_{max} (Rigden 1947, 1954, Heukelom & Wijga 1971) and Rigden referred to this relation as the "true volume fraction". The term φ/φ_{max} is a feature of several models used to describe the rheology of suspensions (Mooney 1957, Krieger & Dougherty 1959, Chong 1971). By normalizing relative creep compliance to φ_{max}, calculated using the two-point projection method from the creep tests, a master curve of stiffening can be formed for each binder.

It is important to note that each binder has its own distinct master curve indicating the importance of the interaction between filler and binder and the important role of the binder. If we consider the three master curves derived for the binder-filler mastics, the order of stiffening observed is B1>B3>B2 (See Fig. 1). The synthetic Binder B3 stiffens to an intermediate extent compared to the two "traditional" bitumens, with the B3 curve being closer to the curve of the hard bitumen (B2) rather than the curve of Bitumen B1 with which B3 is rheologically closer.

Two primary factors can be used to describe the creep properties of binder-filler mastics, the solid volume fraction expressed as a percentage of the maximum packing fraction, φ/φ_{max}, and the intrinsic viscosity of the filler $[\eta]$ with intrinsic viscosity relating to any effects outside of the bulk volume filling relation φ/φ_{max}.

Using the modified Krieger-Dougherty model (Equation 5) and known values of $J(t)_{rel}$, φ and φ_{max}, values for $[\eta]$, the intrinsic viscosity of the filler, can be calculated. Intrinsic viscosity has classically been related to particle shape. Einstein proposed that for spheres with perfect adhesion $[\eta] = 2.5$. Spheres are proposed to have the lowest intrinsic viscosity, increasing as the shape turns to grains, cubes, plates and rods (fibres).

$$J(t)_{rel} = \left(1 - \frac{\varphi}{\varphi_{\max}}\right)^{-[\eta]\varphi_{\max}} \tag{5}$$

The differences of the master curves can be directly related to differences in the average values for $[\eta]$ of the fillers in binders B1 (4.43), B2 (3.26) and B3 (3.74). In general the intrinsic viscosities of the fillers are lower in Bitumen B2 than in Bitumen B1, which mirrors the reduction in relative creep compliance for the mastics.

The relationship between traditional filler index properties and intrinsic viscosity (calculated from the Krieger Dougherty equation) is generally very poor with no significant correlations

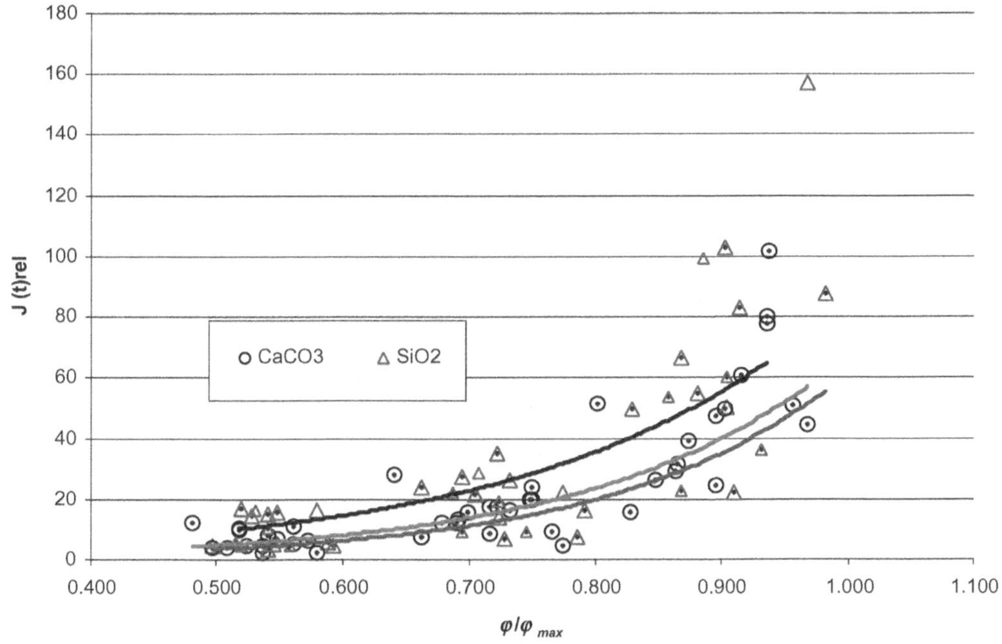

Figure 1. Stiffening of mastics in different bitumen types, normalised for maximum packing fraction.

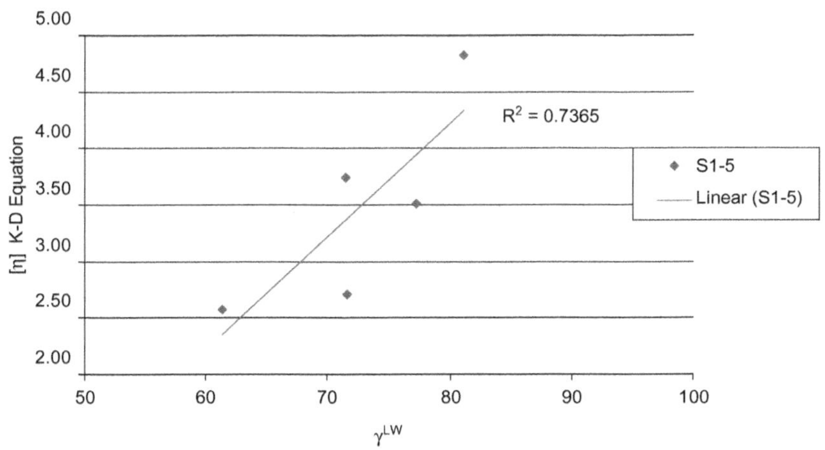

Figure 2. Intrinsic viscosity versus lifshitz van der waals component of surface energy of the fillers S1-5, Bitumen B2.

being found. However, correlations do exist between the values of $[\eta]$ and the surface energy components of the fillers as shown in Figure 2.

In general for the "active" siliceous fillers, at low concentrations, the aspect ratio has the strongest correlation with intrinsic viscosity; this confirms classical thinking as proposed by Barnes (2000). At higher concentrations, there is a significant increase in inter-particle interactions and the role of the surface appears more important than the shape of the particles.

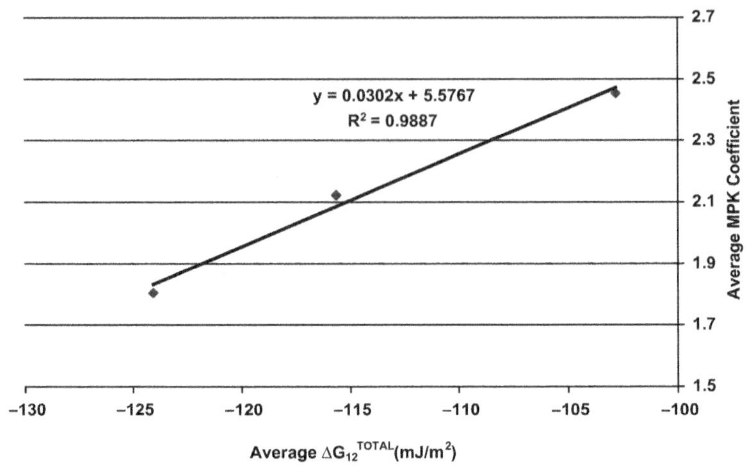

Figure 3. Average Marron-Pierce-Kitano coefficient as a function of average work of adhesion for the three binders.

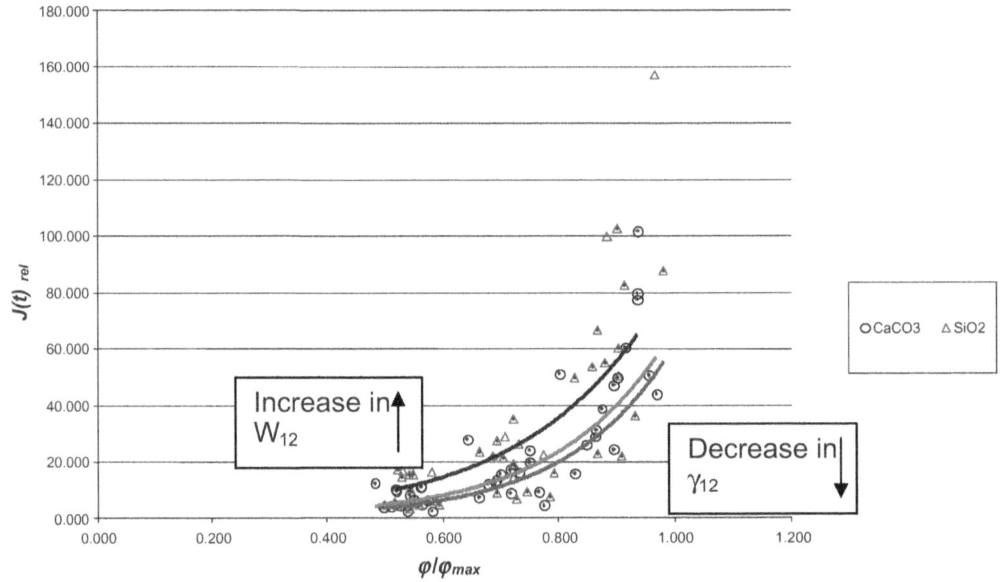

Figure 4. Shifts in master curves of filler stiffening by binder type—effects of surface free energy.

6 BOND ENERGY

Intrinsic viscosity or the MPK coefficient can also be related to the work of adhesion between filler and binder in the different mastic combinations. The average values of the MPK coefficient for the three binders and the average work of adhesion based on the surface energy components of the binders and fillers are plotted in Figure 3.

The above infers that surface free energy could explain the shift in the curves of stiffening with different binder types. As the interactive energy, γ_{12}, between the two phases decreases the overall,

W_{12}, work of adhesion (bond strength) increases and leads to higher levels of stiffening for the same ratio of φ/φ_{max}. This can be seen in Figure 4 where the stiffness master curves are presented as a function of increasing bond strength between the filler and binder combinations.

7 CONCLUSIONS

This paper presents an alternative approach to the characterisation of bitumen-filler mastics and the means of quantifying the effects of fillers and binders on the properties of mastics. Using a combination of rheological models, surface energy measurement techniques and detailed rheological properties, a more robust procedure compared to the traditional Rigden Voids (volume concentration) procedure is presented.

Mastics using ten different fillers and three different binder types have been manufactured and tested using creep compliance tests at different levels of solid volume addition. The term relative creep compliance was introduced to describe the stiffening effects of the fillers at different solid volume fractions in different binders. A two-point projection technique was used to estimate the maximum packing fraction, φ_{max}, and master curves of stiffening were derived by normalising the solid volume fraction to the maximum packing fraction, φ/φ_{max}.

Intrinsic viscosity and the MPK coefficient based on the Krieger-Dougherty equation has been used to describes the level of stiffening outside of simple volume filling and solid volume filling as a ratio of maximum packing fraction. Using a thermodynamic approach, the work of adhesion was used to describe the interfacial conditions in the mastic for each binder-filler combination. On average, the work of adhesion was strongly correlated to the MPK coefficient and shifts in stiffness master curves could be linked to changes in the interfacial conditions of the mastics, with higher levels of interaction leading to higher levels of stiffening for the same ratio of φ/φ_{max}.

REFERENCES

Adamson, A.W. & Gast, A.P. 1997. *Physical Chemistry of Surfaces* (6th edition). New York: John Wiley & Sons.

Anderson, D.A., Bahia, H. U. & Dongre, R, 1992a. Rheological properties of mineral filler—asphalt mastics and their relationship to pavement performance, *Effects of aggregates and mineral fillers on asphalt mixture performance*: ASTM STP 1147, Richard C. Meininger, editor, American Society for Testing and Materials, Philadelphia.

Anderson, D.A., Dongre, R., Christensen, D.W. & Dukatz, E.L. 1992b. Effects of minus 200-sized aggregate on fracture behaviour of dense-graded hot-mix asphalt, *Effects of aggregates and mineral fillers on asphalt mixture performance*: ASTM STP 1147, Richard C. Meininger, editor, American Society for Testing and Materials, Philadelphia.

Anderson, D.A. & Goetz, W.H. 1973. Mechanical Behaviour and reinforcement of mineral filler—asphalt mixtures. *Proceedings of the Association of Asphalt Paving Technologists* 42: 37–66.

Anderson, D.A., Tarris, J.P. & Brock, J.D. 1982. *Dust collector fines and their influence on mixture design.* Report No NAPA QIP 102.

Barnes, H.A. 2000. *A handbook of elementary rheology.* Institute of Non-Newtonian Fluid Mechanics, University of Wales, ISBN 0-9538032-0-1.

Bhasin, A. 2006. *Development of methods to quantify bitumen-aggregate adhesion and loss of adhesion due to water.* PhD Thesis, Texas A&M University.

Brunauer, S., Emmett, P.H. & Teller, E. 1938. Adsorption of gases in multimolecular layers. Journal of the American Chemical Society 60: 309–319.

Chong, J.S., Christiansen E.B. & Baer A.D. 1971. Rheology of concentrated suspensions. Journal of Applied Polymer Science, 22: 2007–2021.

Cooley, L.A., Stroup-Gardiner, M., Brown, E.R., Hanson, D.I. & Fletcher, M.O. 1998. Characterisation of asphalt-filler mortars with Superpave Bitumen tests. *Association of Asphalt Paving Technologists* 67: 42–65.

Craus, J., Ishai, I. & Sides, A. 1978. Some Physico-Chemical Aspects of the Effect and the Role of the Filler in Bituminous Paving Mixtures. *Proceedings of the Association of Asphalt Paving Technologists* 47: 558–590.

Craus, J., Ishai, I. & Sides, A. 1981. Durability of bituminous paving mixtures as related to filler properties. *Proceedings of the Association of Asphalt Paving Technologists* 50: 291–318.

Dukatz, E.L. & Anderson, D.A. 1980. The effect of various fillers on the mechanical behavior of asphalt and asphaltic concrete. *Proceedings of the Association of Asphalt Paving Technologists* 49: 530–549.

Einstein, A. 1906. On the movement of small particles suspended in stationary liquids required by the molecular-kinetic theory of heat. *Annalen der Physik* 19: 289–386.

Good, R.J. 1992. Contact angle, wetting and adhesion: A critical review. Journal of Adhesion Science and Technology 6: 1269–1302.

Good, R.J & van Oss, C.J. 1992. *The modern theory of contact angles and the hydrogen bond components of surface energies.* Modern Approach to Wettability: Theory and Application, Plenum Press, New York.

Harris, B.M. & Stuart, K.D. 1995. Analysis of mineral fillers and mastics used in stone matrix asphalt. *Association of Asphalt Paving Technologists* 64: 54–80.

Hefer, W.H. 2004. *Adhesion in bitumen-aggregate systems and quantification of the effects of water on the adhesive bond.* PhD Dissertation, Texas A & M University.

Heukelom, W. & Wijga, P.W.O. 1971. Viscosity of dispersions as governed by concentration and rate of shear. *Proceedings of the Association of Asphalt Paving Technologists* 40: 157–187.

Hurysz, K.M. & Cochrane, J.K. 2004. Modelling paste properties with minimum experimentation. *Journal of Ceramic Processing Research* 5(3): 191–195.

Ishai, I., Craus, J. & Sides, A. 1980. A model for relating filler properties to optimal behaviour of bituminous mixtures. *Proceedings of the Association of Asphalt Paving Technologists* 49: 416.

Kandhal, P.S. 1980. *Evaluation of baghouse fines in bituminous paving mixtures.* Pennsylvania Department of Transport, Project No 79–23.

Kandhal, P.S., Lynn, C.Y. & Parker, F. 1998. *Characterization tests for mineral fillers relating to performance of asphalt paving mixtures.* NCAT Report No. 98–2.

Kavussi, A. & Hicks R.G. 1997. Properties of bituminous mixtures containing different fillers. *Proceedings of the Association of Asphalt Paving Technologists* 66: 153–176.

Kitano, T., Kataoka, T. & Shirota, T. 1981. An empirical equation of the relative viscosity of polymer melts filled with various inorganic fillers. *Rheol. Acta* 20: 207–209.

Krieger, I.M. & Dougherty, T.J. 1959. A mechanism for non-newtonian flow in suspensions of rigid spheres. *Trans. Soc. Rheol.* 20: 137–152.

Little, D.N. & Epps, J.A. 2001. *The benefits of hydrated lime in hot mix asphalt.* National Lime Association, US.

Mooney, M. 1957. The viscosity of a concentrated suspension of spherical particles. *Journal of Colloidal Science* 6: 162–170.

Richardson, C. 1907. *The Modern Asphalt Pavement.* John Wiley and Sons, New York.

Rides, M. 2005. *Rheological characterisation of filled materials: a review.* National Physical Laboratory Report DPEC-MPR 013, UK.

Rigden, D.J. 1947. The use of fillers in bituminous road surfacings. *Journal of the Society of Chemical Industry* 66: 23–32.

Rigden, D.J. 1954. "*Road Research Technical Paper No. 28.*", Road Research Laboratory, Hammondsworth, Middlesex, HMSO, London 1954.

Shashidhar, N. & Romero, P. 1998. Factors affecting the stiffening potential of mineral fillers. *Transportation Research Record Volume 1638*, Transportation Research Board of the National Academies, ISSN 0361-1991.

Thelen, E. 1958. Surface energy and adhesion properties in asphalt-aggregate Systems. HRB Bulletin 192 63–74. Washington D.C.: Highway Research Board.

Wypych, G. 1999. *Handbook of Fillers.* ISBN 1–895198 19–4, ChemTec Publishing, Toronto, Canada.

Efficient Transportation and Pavement Systems – Al-Qadi, Sayed, Alnuaimi & Masad (eds)
© 2009 Taylor & Francis Group, London, ISBN 978-0-415-48979-9

Barrier-Based Fuel Resistant Binder (B^3)

G. Mohammed Memon
PhalTech Corporation, Ashburn, VA, USA

ABSTRACT: The Barrier-Based Fuel Resistant Binder was designed in response to the toxic substances associated with coal tar while providing superior resistance to non-aromatic petroleum fuels and aliphatic solvents, such as are found at airfields, truck depots, loading facilities, parking lots, service stations, fast food restaurants and auto service centers.

Barrier-Based Fuel Resistant Binders generate an aggressive level of stiffness at high temperatures which retards deformation or rutting, while maintaining resistance to cracking at low temperatures. They can withstand enormous "point loads" exerted by aircraft or heavy vehicles, as well as withstand premature aging or oxidation.

The Barrier-Based Fuel Resistant Binder generates its strength and structure by utilizing our specialized cross-linking system, combining the adhesion properties of concentrated polymer modified asphalt with the carbon black network structure of crumb rubber. When properly applied, a resistant barrier forms protecting the sub-soils underneath from petroleum fuel contamination.

This ability of this kind of binder to resist fuel lies in a special blend of stone as an aggregate or synthetic rubber (i.e. polymer or crumb rubber, or other related recycled products like roofing shingles, etc…) as a modified binder ingredient, which also help give the modified binder an added strength when it is put under pressure by salvation or thermal or mechanical stress. Binders produced this way can effectively be used in variety of applications like potholes, colour pavings, bridge pavings, bridge decks, ferry ramps, side walks, intersections, play grounds, bicycle path etc.

This type of binder has shown special elastic behaviour, with a strong structural network. The strong covalent adhesion between the binder and the aggregate shows improved strength in the pavement.

Keywords: Modified Binder, Fuel Resistant Binder, Color Binder, Polymer Modified Binder, Cross-linked Binder, Storage Stable Binder

1 INTRODUCTION

Signs of gasoline, diesel fuels, and jet fuels are observed on pavements and runways, showing possible leaks and spills which cause not only soil and environmental damage but it damage to pavements as well. Oil and fuel spills can also contribute to the current global warming issue.

In the United States, the EPA is talking tougher measures for the spills from asphalt and fuels storage tanks, and are currently seeking further measures to prevent spills from these tanks from damaging the air and especially the soil content under and around them.

A common practice for providing a fuel-resistant surface has been the use of a coal tar slurry seal. While providing a good fuel-resistant surface, coal tar is detrimental as a known carcinogen. The US Government and States have been eliminating its usage.

Polymer Modified Asphalt (PMA) was developed by Netherlands (Khan, 1999), a coal tar-free jet fuel resistant known as Sealoflex was applied at Kuala Lumpur Airport and was observed in excellent condition till 2004, i.e. showed excellent resistant towards jet fuel spill on the runway.

The polymer modified asphalt developed by Citgo, known as Citgoflex (Rooijen et al. 2004) was developed and used to replace the coal tar sealant and tested at La Guardia Airport (USA) in August

2002 and a visual inspection was carried out in October, 2003 with no measurable rutting observed. However, results from laboratory test showed dramatic collapse after immersion into jet fuel.

This new development of a barrier-based fuel resistant binder can be used with little modification as color binder, and it can also be successfully applied for potholes. This brings dramatic and sensational news to a market where fuels resistance and colored binders are needed. The introduction of a new product of Polymer Modified Asphalt (PMA) in combination with colored or non-colored binder as Hot or Cold Mix can results in a fuel resistant binder with both high and low temperature rheological properties and very strong tensile strength.

This product was designed for Hot or Cold Mix asphalt that could be used to make pavement under Asphalt, Fuel Tanks, Colored Playgrounds, Bus Lanes, Bicycle Lanes, Parking Lots, Cross Walks, Cross-Roads for Intersection. In this process at the initial stage colored roads can be paved for Sightseeing, Potholes, Roofing, Bridge Decks and Deceleration Belts. The process begins by generating structure or strength into the paving system. Use of PMA and Color Binders in combination with our specialized Cross-Linking system forms a unique structural network in the pavement. The PMA section is capped with a colored or non-colored Binder, and then the mix is made from both the aggregate and/or recycled tire rubber. The recycled tire rubber generates a very strong elastic pavement.

2 EXPERIMENTAL

Tha laboratory studies were conducted to conclude the jet fuel resistance on Base Binder, Emulsion, PMA, PMA with Carrier and X-Linking agent.

This new kind of binder can be made out of emulsion as well as hot asphalt, with an addition of a specialized carrier and cross-linking agent [Pt-743] (Patent in pending Memon, 2007) in a continuous stirring mode. The samples were made from base asphalt of PG 64-22 to a PMA of PG 94-22 (Memon 2004) and the resulting PMA showed that the polymer molecule was linked and form a polymer network with asphalt molecule without any gel or lumps formation and with excellent storage stability.

However, the major problem with the polymer modified asphalt is that it did not pass the ASTM D3320 and showed more than 1% resistance towards jet fuel. This shows that the cross-linked polymer modified asphalt prepared this way is not holding the jet fuel spill on the pavement, which results in reducing the life of the pavement.

The intensity of life reduction in pavements depends upon the damage caused by the jet fuel. The emulsion as well as the hot asphalts were applied on two types of aggregates; ¼" stone and 6 mesh crumb rubber as an aggregate. The trials with ¼" stone aggregate were made at an asphalt terminal (May 2007) with a heavy vehicular traffic, whereas with 6 mesh rubber, the trials were made at a sport ground (August, 2007). (Memon, 2007).

The homogeneity or separation characteristics (AASHTO PP5-93) were assessed by the fluorescent microscopy method.

Separation by Fluorescent Microscopy:

A microscope slide was prepared by placing a small amount of asphalt onto the slide and then covering it with a cover slip. It was then smeared by heating the slide at 90–110°C for 15 seconds. A fluorescent light was turned on in the microscope and the image was adjusted under different levels of magnification (normally 400 magnification power). After adjustment of the image capture it was viewable within the real time digital camera. The captured image was then transferred to the computer screen and then saved for further analysis.

3 RESUTLS AND DISCUSSION

The polymer modified asphalt was prepared to achieve the PG 94-22 and their rheological properties are shown in Table 1. This asphalt was used for this invention to build the network strength

Table 1.

Asphalt	PG-grade	DSR value	Creep stiffness or S-value	m-value	Elastic recovery
Base	64-22	1.17	180	0.339	14%
PMA	94-22	2.878	145	0.324	81%

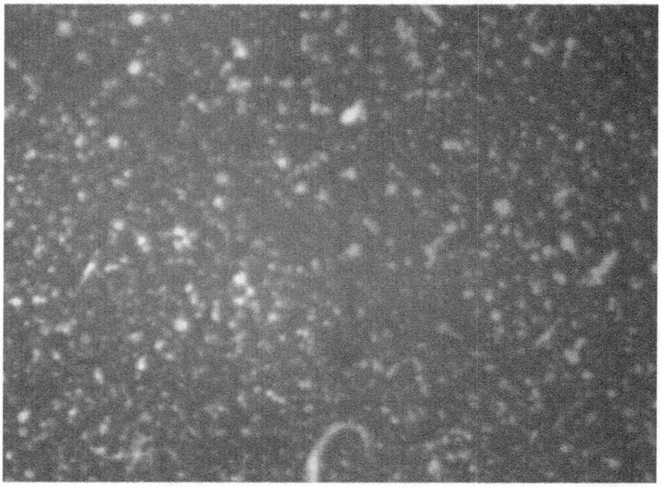

Figure 1. Fluorescent microscopic slide of PMA before cross-linking.

with sufficient stiffness to react with carrier curing agent in the presence of cross-linking agent (PT-743) prepared as per US patent # 6,818,687 B2.

Figure 1 is a fluorescent microscopic slide showing the improper distribution of SBS without any networking in the asphaltic system. This type of polymer distribution results in a problem of raveling in the pavement, which was a big problem of asphalt industry, because this kind of product will be unstable in storage, but after the development of linking agents that problem is taken care of. PhalTech's new generation technology (PT-743) was used to link SBS to solve these problems.

The figure 2 (a fluorescent microscopic after x-linking)) is conforming the formation of proper polymer networking and its complete linking with the asphalt functional groups and shows no more fluorescence in the PMA. This also confirms the proper action of PT-743 linking agent/activator.

The Figure 3 shows the performance of different asphalt in resisting jet fuel by using ASTM D3320, and it was observed that the base asphalt (Pg 64–22) was almost 11%, whereas, the emulsion and the PMA prepared as per (Memon 2004) process shows resistance towards jet fuel by 1.5 and 1% respectively.

The resistance to Jet fuel as per ASTM method can be determined by keeping asphalt specimen dipped or immersed in Jet Fuel by 24 hours and then by measuring the weight loss through weighing before and after the immersion of the specimen into the jet fuel.

This figure (3) shows that the base binder can not sustain the jet fuel oil on the Airport Pavements, whereas, the emulsion and PMA made by Memon's Method (Memon 2004), shows significant improvement towards the Jet Fuel resistant, but still it is not a complete cure. Whereas, the

Figure 2. Fluorescent microscopic slide of PMA after cross-linking.

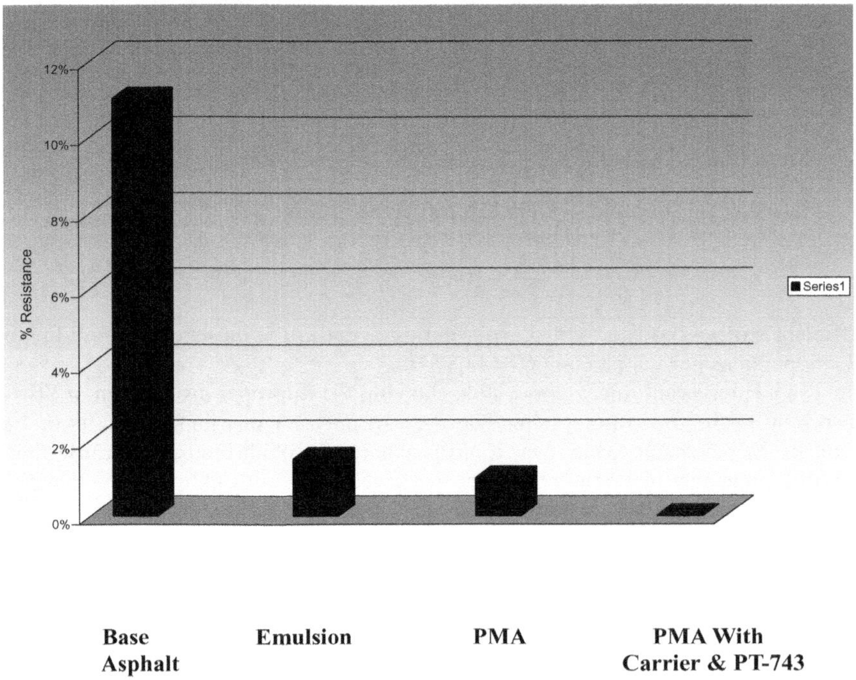

Figure 3. Percent resistance of binder immersed in jet fuel.

PMA with carrier (with curing agent) and PT-743 shows 0% resistance towards the immersion of the specimen into the Jet Fuel oil for 24 hours.

Based on these findings field trial was made in May, 2007 at Hudson Asphalt Terminal, having heavy vehicular traffic of heavy trucks fill with asphalt with some spill of diesel oil here and there and presently there remains no any crack or damage to the lay down. This field testing was a kind of stress testing for this kind of barrier based binder.

Figure 4 is the product made with above produced material mixed with recycled roofing shingle to produce the material for filling the pot-wholes, produced through PMA with Carrier and PT-743 (X-Linking Agent). The figure one already showed that this kind of binder has shown some significant improvement to resist the penetration of Jet Fuel in the specimen.

Figure 5 is showing a lay-down of the material (PMA with Carrier and X-Linking agent PT-743) produced by the above formulation to resist the penetration of Jet Fuel into the specimen.

This product was checked out as per ASTM D3320 for four weeks and showed an average loss of almost 0%, as well as being immersed into Jet Fuel for 24 hours and showing 0% resistance towards the penetration of Jet Fuel.

Therefore, the modified binder produced by Memon, 2007 showed with jet fuel an average loss of zero percent as is showed in picture one and two. This kind of binder has an ability to resist fuel be in a special blend of stone as an aggregate or synthetic rubber i.e. polymer or crumb rubber or other related recycled products like roofing shingles etc as a modified binder ingredients. This also help give the modified binder an added strength to sustain the thermal and mechanical cracking when it is put under pressure by salvation or thermal or mechanical stress.

The binder so produced can effectively be used in variety of applications including potholes, colour pavings, bridge pavings, bridge decks, ferry ramps, side walks, intersections, playgrounds, bicycle path, gasoline storage tanks, air fields etc.

Barrier Based Fuel Resistant Binder generates its strength or structure by utilizing our specialized cross-linking system, combining the adhesive as well as cohesive bond properties of concentrated polymer modified asphalt and carbon black network structure of crumb rubber. This will also enhance not only the ease of pavement application, but will help to increase the jet fuel resistance in the pavement. In this system when recycled rubber is being used as an aggregate, it will generate carbon black from the rubber as a good source of reinforcement. When properly applied, it can be successfully implemented as a resistant barrier form, protecting sub-soil underneath from petroleum fuel contamination.

Figure 4. Sample prepared from emulsion using recycled shingles.

Figure 5. Trial lay down For PMA with carrier and PT-743.

Figure 6. Colored playground with PMA with carrier and PT-743.

The figure 6 is a colored playground made out of the newly developed Barrier Based Binder (B^3) with carrier and cross-linking agent (PT-743) and iron oxide as a coloring agent.

The figure 7 is a colored patio made out of the newly developed Barrier Based Binder (B^3) with carrier and cross-linking agent (PT-743) and iron oxide as a coloring agent. The binder so

Figure 7. Colored patio from PMA with carrier and PT-743.

produced has also shown special elastic and ductile behaviour, with a strong structural network between the binder and the aggregate (stone or crumb rubber). The strong cohesive or adhesive bond between the binder and the aggregate (stone or crumb rubber) shows improved strength in the pavement and allows the pavement to last longer and increase the overall pavement lifespan. This kind of binder with its special properties can retard reflective and low temperature cracking better than any other product available in the market.

4 CONCLUSION

1. This kind of binder has a very special elastic property, which can build a strong structural network.
2. The barrier based binder due to its better rheological properties for both high and low temperature and better visco-elastic behavior shows excellent resistant towards jet fuel.
3. This kind of binder has improved components, which impede exposure to ultraviolet radiation generated by sunlight, resulting in an increase in the pavement life.
4. The construction of pavement is simple and adaptable.
5. The barrier based binder has been proven to lower the noise levels due to high amounts of crumb rubber.
6. After adding color to the barrier based binder creates light colored surfaces, useful for bridge paving and ferry ramps.
7. This kind of binder can be easily used to refurbish old roads and improve its life.
8. Due to its durability this binder can successfully applied pot holes, cracks etc.
9. It is easy and economical to maintain.

REFERENCES

1. Khan S.K. (1999), "Sealoflex for Airport Pavement Construction and Rehabilitation", Sealoflex World Conference, Tokyo, Japan.
2. Rooijen R.C., de Bondt A.H. & Corun R.L., "Performance Evaluation of Jet Fuel Resistance Polymer Modified Asphalt for Airport Pavements" 2004 FAA World Wide Airport Technology Transfer Conference Atlantic City, New Jersey, USA.

3. Memon G.M., "Modified Asphalt With carrier and Activator" US Patent # 6, 818, 687 issued on Nov. 16, 2007.
4. ASTM, Annual Book of ASTM Standard, part 15: Road, Paving, Bituminous Materials, Traveled Surface Characteristics, 1982.
5. Memon G.M., "Barrier Based Fuel Resistant Binder" Patent in Pending, 2007, USA. The AASHTO Standard Test Method, no. TP5-93, "Method for Determining the Rheological properties of Asphalt Binder, using a Dynamic Shear Rheometer (DSR)".

Efficient Transportation and Pavement Systems – Al-Qadi, Sayed, Alnuaimi & Masad (eds)
 ISBN 978-0-415-48979-9

Performance-based selection guidelines for roadway bituminous-based crack sealants

Imad L. Al-Qadi, Shih-Hsien Yang & Elham Fini
Department of Civil and Environmental Engineering, University of Illinois at Urbana-Champaign, Urbana, IL, USA

J.-F. Masson
Institute for Research in Construction, National Research Council of Canada, Ottawa, ON, Canada

Kevin K. McGhee
Virginia Transportation Research Council, USA

ABSTRACT: The long-term performance of pavements depends in good part on the quality and frequency of maintenance. Appropriate maintenance protects the pavement from deterioration, corrects deficiencies, and ensures safe and smooth riding. Crack sealing is practiced on a routine basis as preventive maintenance and as part of corrective maintenance prior to an overlay or a greater rehabilitation project. A timely and properly installed sealant adds several years of service life to the pavement at a relatively low cost. As a consequence, the selection of an appropriate sealant in a maintenance project becomes an important issue. Current sealant selection is based on ASTM standards that consist of quality control tests, not of performance indicators. These standards do not consider the changes in mechanical properties due to aging or the differences in local service temperatures. Given the breadth of temperatures in North America and its yearly variation, there is an urgent need for performance-based indicators of sealant performance. In this paper, a series of tests that provides a systematic approach to help highway agencies selecting proper sealants is proposed. These include an accelerated aging test, an apparent viscosity test performed at sealant application temperatures, a dynamic shear rheometer (DSR) test to assess flow in summer temperature, a crack sealant bending beam rheometer (CSBBR) and a crack sealant direct tension test (CSDTT) for cohesive properties at sub-zero temperature, and a blister test for adhesive properties. Using identified thresholds for these tests, performance-based guidelines for the selection of hot-poured crack sealants were developed.

1 INTRODUCTION

ASTM standard D5535 defines a sealant as a material that possesses both adhesive and cohesive properties to form a seal, which prevents liquid and solid from penetrating into the pavement system. Crack sealing has been widely accepted as a routine preventative maintenance practice. Given a proper installation is achieved, crack sealant can extend pavement service life by a period ranging from three to five years (Chong and Phang, 1987). Numerous studies also demonstrated the cost effectiveness of crack sealants (Joseph, 1990; Cuelho et al., 2002, 2003; Fang et al., 2003; Ward, 2001; Chong and Phang, 1987; Chong, 1990).

Crack sealant is produced so that it keeps its shape as applied and hardens through chemical and/or physical processes to form a viscoelastic rubber-like material that withstands extension or compression (crack movement) and weathering (Al-Qadi et al., 2007). However, in many cases, premature failure of crack sealants may be observed in one of the following scenarios. During the sealant installation, if the viscosity of the sealant is too high, sealant might not be able to fill the crack properly; hence, it will affect the interface bonding between sealant and pavement substrate.

If the viscosity is too low, sealant might flow out from the cracks. In the field, a sealant extends at low temperature and compresses at high temperature because pavement crack opening increases with decreasing temperature and decreases with rising temperature. At high service temperature, sealant may fail due to pull out from the crack by tire passing. At low service temperature, the crack opening may increase from 10% to more than 90% depending on the environmental location; hence, one of the two mechanisms might be observed: cohesive or adhesive failure. The former occurs in the sealant, while the latter occurs at the sealant-pavement crack wall interface. At low temperature, sealant becomes more brittle due to physical hardening; and is subjected to short-duration loading due to crack movements associated with stick-slip motions and truck trafficking as well as long periods of environmental loading.

Standards and specifications for selecting crack sealant have been established by several organizations, including American Society for Testing and Materials (ASTM), American Association of State Highway and Transportation Officials (AASHTO), and U.S. and Canadian federal, state, provincial, and municipal agencies. The objective of the specifications is to select materials that have the necessary properties to perform adequately in the field. However, these specifications are generally empirical and do not measure sealant fundamental properties. Hot-poured bituminous crack sealants are typically selected based on standard, but empirical tests such as cone penetration, resilience, flow, and bond to cement concrete briquettes (ASTM D6690). ASTM Standard D5329-04 summarizes most of these tests.

These tests are used by most state highway agencies in selecting their crack sealing materials; but the tests and specification limits may vary from one state to another. These differences create difficulties for crack sealant suppliers because many states with the same environmental conditions specify different limits for the measured properties. These tests were also reported to poorly characterize the rheological properties of bituminous-based crack sealants and to predict sealant performance in the field. This may lead to premature sealant failure when used in unfavorable conditions.

2 PERFORMANCE-BASED SPECIFICATION

To develop performance-based guidelines for the selection of hot-poured crack sealants that meet the aforementioned requirements and minimize the cost of possessing new testing equipment, the research group made use of the SuperPave™ binder performance grading (PG) equipment. Modifications to the existing viscosity test, bending beam rheometer, and direct tension test devices, specimen size and preparation, and testing procedures were made to accommodate the testing of crack sealants. In addition, new tests for sealant aging and sealant evaluation at high service temperatures were introduced. Upon the completion of test validation, test measured performance parameters were recommended for implementation as part of the newly developed "Sealant Grade" (SG) system.

2.1 *Apparent viscosity*

Sealant viscosity is among the parameters that affect initial bonding. Therefore, applying a sealant at the appropriate viscosity provides a better crack filling and enhances interface bonding. Several factors affect the measured viscosity of hot-poured crack sealant. Therefore, it is essential to identify the material characteristics that influence the rheological behavior of hot-poured crack sealant at installation. These characteristics need to be set at reasonable limits, to simulate field installation as closely as possible. While standard tests to examine sealant consistency exist, these standard tests have not been proven to predict field performance. As part of an effort to bridge the gap between sealant fundamental properties and field performance, a test procedure was developed to measure apparent sealant viscosity using the same rotational viscometer equipment used in the SuperPave™ PG system. The Brookfield Thermosel system (as adopted from SuperPave™) was found to be sufficient for the sealant testing. After extensive testing, a Brookfield rotational viscometer was adopted; modification of the test procedure and equipment was implemented (Figure 1). SC4-27

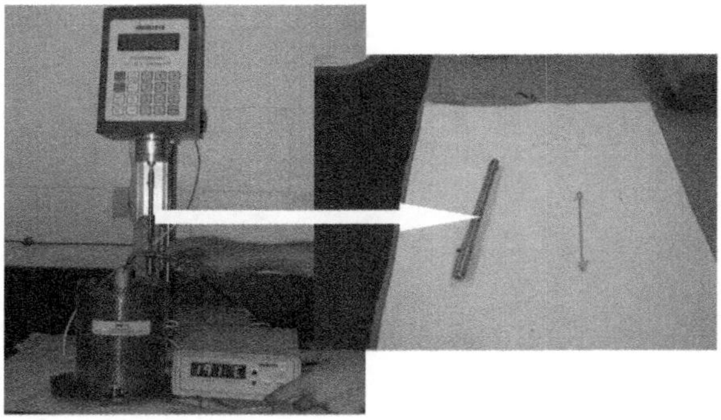

Figure 1. Brookfield thermosel system and rigid rod used for crack sealant testing compared to the rod used for asphalt binder.

spindle at a speed of 60 rpm (shear rate of 20.4 s^{-1}) at the recommended installation temperature is used. The spindle is attached to a newly developed rigid rod. The rod is a replacement for the current hook; it prevents a rubber particle from disturbing the spindle rotation which results in better test repeatability. A conditioning time of 20 min and a waiting time of 30s before collecting data are also recommended to ensure that the measured apparent viscosity has stabilized. The development of this procedure is described in detail elsewhere (Al-Qadi et al., 2008b).

Because viscosity plays an essential role in predicting field performance of hot-poured crack sealant, upper and lower apparent viscosity limits are recommended. An upper limit of 3.5 Pa.s ensures that sealant is liquid enough to pour, whereas a lower limit of 1 Pa.s controls the potential of using excessively fluid sealant. Hence, the sealant apparent viscosity should be between 1.0 and 3.5 Pa.s when measured at recommended installation temperature.

2.2 *Sealant aging*

To mimic the effect of ageing on sealants, several accelerated aging methods were compared (alone or in combination) after various aging periods and temperatures, including small-kettle aging, microwave aging, pressure aging, oven aging, and vacuum oven aging. The results of several physico-chemical analyses, which include gel permeation chromatography (GPC), fourier transform spectroscopy (FTIR), thermogravimetry analysis (TGA), and dynamic shear rheometry (DSR), of sealants weathered in the field were compared to those of sealants aged quickly in the laboratory (Figure 2). As expected, sealants with good performance contain components resistant to weathering, whereas sealants with poor performance oxidize quickly.

Among the various accelerated aging method, vacuum oven aging proved to be the most appropriate method to simulate sealant weathering. In this method, sealants are cut into slices and placed on a stainless steel pan; each pan contains 35 g of sealant. The pan is transferred into a conventional temperature controlled oven which is preheated to 180°C for approximately 5 min to allow sealant to melt and form an approximately 2 mm film. The sealant is then removed from the oven and cooled to room temperature. Once it cools, sealant is placed in a vacuum oven preheated to 115°C for 16 hr. After 16 hr, the vacuum is released and sealant is placed in an oven at 180°C for 5 min or until the sealant is fluid enough to pour.

2.3 *Sealant flow and deformation*

Bituminous sealants applied to cracked pavements sometimes fail due to deformation under the combined action of shear stresses and high service temperatures (Masson et al., 2007). A dynamic shear rheometer (DSR) was used to define the performance parameter for the tracking resistance.

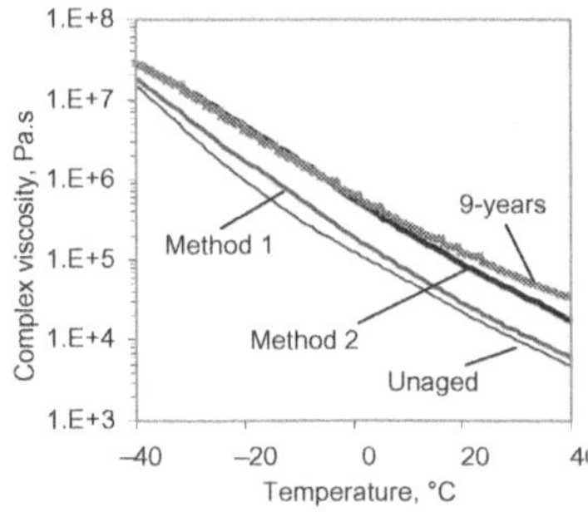

Figure 2. Complex viscosity of a field-aged sealant compared to that after accelerated aging.

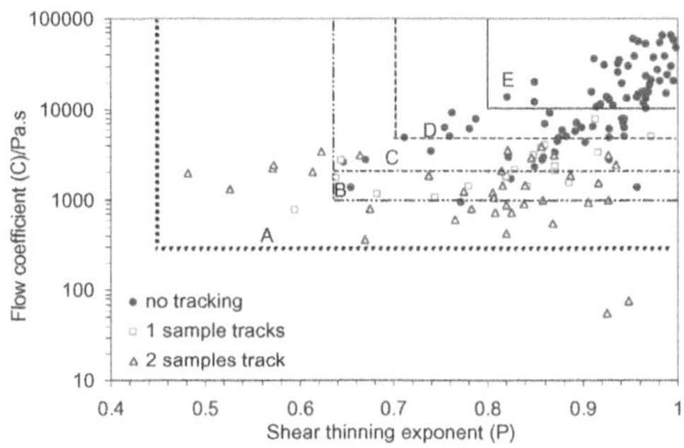

Figure 3. Semi-log plot of the Ostwald parameters and possible performance limits.

The sealant was subjected to increasing stresses at temperatures between 46°C and 82°C. These conditions were meant to mimic the effects of various traffic levels and maximum temperature in the field.

For each creep and recovery cycle, the shear rate ($\dot{\gamma}$) is determined then plot log(σ) against log($\dot{\gamma}$) to obtain the isotherm curve. The isothermal curves were interpreted based on the Ostwald power law model. This model provides two parameters: a flow coefficient (C) and a shear-thinning coefficient (P). These coefficients correlated well with sealant pseudo-field performance as measured by tracking (Collins et al., 2007). Figure 3 shows the relationship between these factors and performance during the pseudo-field test. The solid markers indicate the sealants that did not fail during the pseudo-field test, and the open markers show those that failed. The semi-log scale in Figure 3 serves to highlight the high-failure regions (open markers). Limiting values for P and C can be established to limit the risk of sealant failure. The selection criteria was recommended to be $C \geq 4000$ Pa.s and $P \geq 0.70$ which provide for a failure risk of only 3% only among the all tested sealants.

2.4 *Flexural creep*

The bending beam rheometer (BBR) is used in most pavement laboratories nowadays to measure binder stiffness at low temperature. A modified BBR test, a crack sealant bending beam rheometer (CSBBR), was introduced to measure the flexural creep of crack sealant at temperatures as low as –40°C. The development of this procedure is described in detail elsewhere (Al-Qadi et al., 2008d). The test modification was mainly categories into three folds: First, the specimen thickness was doubled from 6.35 to 12.7 mm to overcome the excessive deflection due to the softness of crack

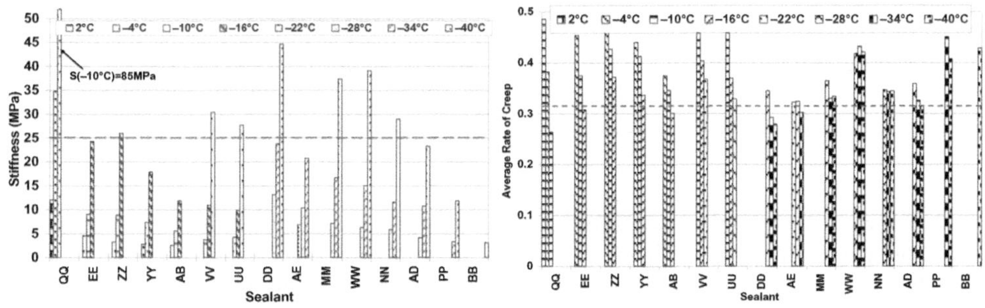

(a) (b)

Figure 4. (a) Stiffness at 240s and (b) Average creep rate at various testing temperatures for 15 sealants.

sealant. Second, the device was modified to accommodate the new specimen geometry; therefore, the specimen support and calibration kits of the BBR were modified. Third, the silicon-based release agent was used to replace the Mylar strip because the Mylar strip melted at sealant pouring temperature. The test is conducted in the following manner that 35 g of sealant is first heated at its recommended pouring temperature and then poured into an aluminum assembled mold. A rectangular sealant beam is cast with dimension of 12.7 mm in height, 12.7 mm in width, and 102 mm in length. The beam is then placed in a fluid environmental chamber and the specimen is placed on a two-point support and subjected to a point creep loading. The specimen is exposed to a creep loading for 240s then followed by 480s unloading. The load and deflection of the sealant beam is recorded during the period of loading and unloading. The results were used to calculate performance parameters which are stiffness at 240s (**Error! Reference source not found.**a) and average creep rate (**Error! Reference source not found.**b).

In addition, it was found that the critical loading time for crack sealant material at low temperature after 5 hr of loading. If the temperature superposition principle is applied, the stiffness at 240s for a given temperature can be used to predict the stiffness after 5 hrs of loading at a temperature of approximately 12°C greater. Given the variation in sealant response to temperature change, 6°C shift is deemed appropriate. The CSBBR test results recommend two performance criteria for application: stiffness at 240s and average creep rate (Figure 4). The recommended thresholds, which are temperature independent, for the two criteria are maximum 25MPa and minimum 0.31, respectively.

2.5 *Low temperature tensile properties*

To investigate whether a sealant can survive in a particular service conditions, the SuperPave™ Direct Tension Test (DTT) was considered and modified for crack sealants. The development of crack sealant direct tension test (CSDTT) is described in detail in elsewhere (Al-Qadi et al., 2008c). The principle of the CSDTT is to slowly pull a crack sealant specimen in tension until it breaks. The dog-bone shaped specimen used in the DTT has a rectangular cross section. Its ends are enlarged so that when crack sealant is poured into the mold, it has a large adhesive area between the crack sealant and end tabs. The end tabs are made from Phenolic G-10 material, to provide good bonding. The SuperPave™ DTT specimen geometry can only extend the sealant specimen up to 32% strain. This may be significantly smaller than the crack sealant extension in the field. Therefore, for crack sealant testing, the specimen geometry was modified. The new specimen dimensions are the following: 24-mm-long, 6-mm-wide and 3-mm-thick; the effective gauge length is 20.3 mm. The maximum extension that can be achieved using this specimen is 19 mm which is equivalent to approximately 94% strain. This meets the extreme service conditions that sealants may experience in the field. Table 3 presents the geometry comparison of SuperPave™ binder DTT specimen and CSDTT specimen. Regardless of specimen geometry,

Table 1. Thresholds for crack sealant extendibility at various temperatures.

Temperature (°C)	–4	–10	–16	–22	–28	–34	–40
Extendibility (%)	10	25	40	55	70	85	85

high-polymer-content sealants have shown equivalent peak stress at its maximum elongation state. Hence, the length effect is negligible in the sealant tensile strength when high-polymer-content products were used (Al-Qadi et al., 2007). The specimen preparation procedure was also modified to accommodate various sealant compositions to improve the workability while pouring the sealant into a mold. The mold was heated to 50°C lower than sealant pouring temperature prior to pouring the sealant. Right after the sealant was poured into the mold, a spatula was used to slightly tap the sealant to ensure that sealant filled the mold.

The study recommends using the DTT as a standard test to evaluate the bituminous-based hot-poured crack sealant at low temperature. The performance parameter, extendibility, was recommended for use in the specification. The threshold for the extendibility depends on the sealants' lowest application temperature and is presented in Table 1. In addition, because the test is conducted under a relatively higher deformation rate compared to real crack movement, the research team recommends a +6°C shift in the crack sealant grading system. For instance, if the lowest application temperature is determined as –16°C, the test would then be conducted at –10°C. If the extendibility of such sealant is over 25%, the sealant passes the criteria and is approved for use.

2.6 *Adhesive properties at low temperature*

The adhesion capability of hot-poured bituminous sealants is usually evaluated using a standard test of an empirical nature (ASTM D5329). There is, however, no indication that the results of this test pertain to field performance. In addition, this test examines adhesion of sealant to Portland cement concrete, and the test result does not account for aggregate composition, which is the main component of HMA. Therefore, a reliable test method, which is based on sealant rheology, accounts for aggregate composition, and correlates with field performance, is urgently needed. This study proposes three laboratory tests to predict interface bonding of crack sealant to aggregate at service temperatures ranging from –4°C to –40°C. The three tests are designed to address the needs of manufacturers, transportation organizations, including contractors, transportation agencies and consultants, and researchers, respectively.

The first laboratory approach addresses the compatibility of sealant with a specific substrate, by measuring the free energy of the bond, *work of adhesion*. The second test makes use of the *direct tension* test (DTT) device. The third test is a fracture type test that utilizes a fracture mechanics approach to derive a fundamental property of the bond, *interfacial fracture energy* (IFE). The first approach, *work of adhesion*, that sealant is heated and mixed at the manufacturer's recommended installation temperature and poured onto an aluminum sheet to form a thin, smooth surface. The sealant is cooled at room temperature to solidify and make thin plates. A five-micrometer pipette is used to manually apply liquid drops from three probe liquids (water, formamide, and glycerol) onto the sealant plate. The image of each drop is captured by microscope within 15s after it is applied. The resulting contact angle is used to determine the work of adhesion between sealant and substrate.

The second approach, direct bond test, that sealant is pulled to detach from its aggregate counterpart by applying a tensile force. A new test fixture that simulates sealant pouring condition and loading mechanism in the field was developed. The briquette assembly consists of two aluminum half-cylinders of 25 mm diameter and 12 mm thickness; aluminum is conservatively selected as a substrate reference material. Each aluminum briquette is confined within an aluminum grip designed to work with the DTT sitting posts. The assembly has a half cylinder mold, open at the upper part. The mold is placed between the two aluminum half cylinders on an even surface.

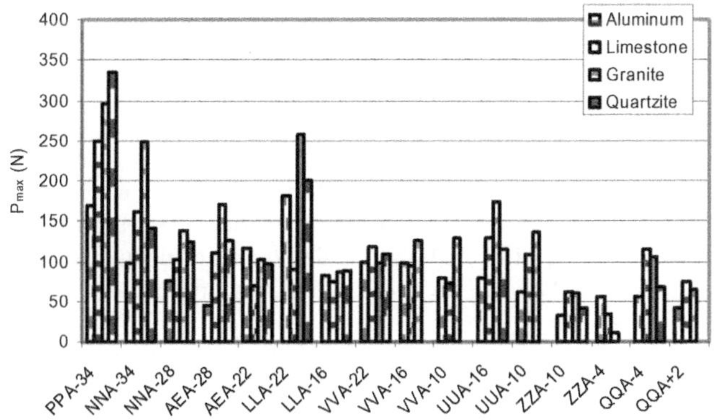

Figure 5. Maximum load measured for bonding between sealants and substrates.

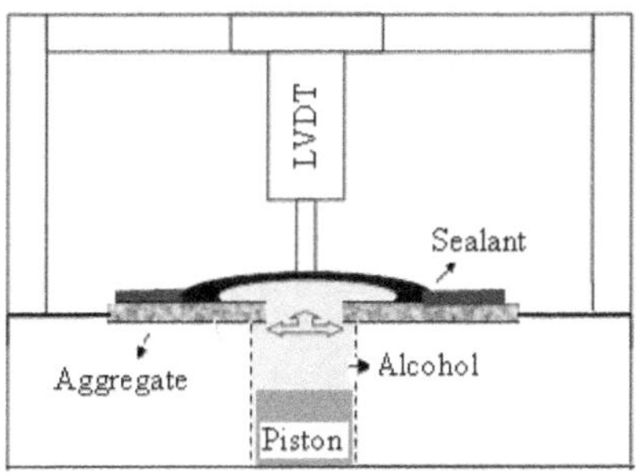

Figure 6. A schematic of blister apparatus.

In order to ensure that adhesive failure occurs, and to define failure's location, a notch is made at one side of the sealant-aggregate interface. A 12.5 × 2 mm shim is placed at one aggregate-sealant interface. The assembly then is placed in the DT machine; the notched side is placed at the non-moving side of the DT machine.

To conduct the test, sealant is heated at its recommended installation temperature and poured into the half cylinder mold. After one hour of annealing at room temperature, the specimen is trimmed and placed in the cooling bath for 15 min. The specimen is then removed from the bath, demolded, and placed back in the bath for another 45 min before testing. Using the DT device, the end pieces are pulled apart by moving one of the end pieces at a speed of 0.05 mm/s. The maximum load, an indication of bond strength, was selected as the performance parameter (Al-Qadi et al., 2008a). Figure 5 shows the variation of this parameter for several pairs of sealant-aggregate. Using comparison between test results of laboratory-aged specimens and field data, a 50 N minimum at graded temperature was selected as the performance threshold.

The third approach, blister test, uses a servo-hydraulic system to displace a piston at a constant rate and the upward movement of the piston injects a liquid medium (alcohol) at a constant rate of 0.1 L/hr through a channel that is connected to the specimen (Figure 6). The specimen is composed of an annular (donut-shaped) substrate plate (aggregate or a standard material) covered with binder or sealant on one side. Alcohol pushes the adhesive (binder or sealant) away from the substrate creating a blister which continues to grow until the adhesive separates from the substrate. The blister height and the pressure are recorded during the test and they are used to calculate the IFE. In simplified form, IFE can be calculated as half of the product of the maximum pressure and the corresponding blister height. In addition to IFE, adhesive modulus can be determined from this test using the test data before debonding occurs.

3 PRELIMANARY FIELD VALIDATION

3.1 *Montreal field site*

Five sealants (A, B, E, G and J) with previously documented field performance were installed in Montreal, Quebec, Canada in fall 1990 and were used in this study to validate the crack sealant performance criteria. Field samples were collected during the visual survey at years 1, 3, 5, and 9. A long-term sealant performance survey was conducted four years after installation (Table 2). The long-term performance is mainly affected by the sealant weathering and stiffening. A performance index (PI) was suggested based on the level of de-bonding and pull-out. The PI is calculated as follows (Masson et al., 1999):

$$PI = 100 - (D + nP) \tag{1}$$

where,

PI = sealant performance index;
D = percent debonded length of the sealant;
P = percent pull-out length; and
n = an integral that accounts the effect of pull-out over de-bonding on performance.

The n value was assigned 4 in Masson et al. (1999) study; the suggested value is based on the reasoning that loss of 1 m of sealant may allow the intrusion of sand and stone into the pavement which can damage the pavement during its expansion and contraction (Peterson 1982). Loss of sealant also allows more water to penetrate into the pavement. This damage is more critical than de-bonding over the same length. The higher PI value, the better the sealant performance is.

3.2 *ASTM specification vs. performance-based specification*

The test results of five Montréal field sealants were tested in the laboratory according to ASTM D6690 Type II test specification and the results of the five sealants are reported in Table 3

Table 2. Four-year sealant performance (after Masson et al. 1999).

Sealant	De-bonding (%)	Pull-out (%)	PI (%)
A	11	14	33
B	22	1	74
E	20	2	72
G	36	14	8
J	13	12	39

Table 3. Montreal sealant standard test results (after Masson 1999).

Sealant	Penetration (25 °C) (<90 dmm)*, †	Flow 60 °C (<3 mm)*	Resilience (25 °C) (>60%)*	Bond (–29 °C) (3 cycles)*	Result
A	86	0.5	57	F	Fail
B	68	0.5	64	P	Pass
E‡	104	1	73	P	Fail
G	50	0.5	51	F	Fail
J	66	6	48	P	Fail

*ASTM D3405 requirements. Levels beyond acceptable limits are underlined.
†1 dmm = 0.1 mm.
‡Meets ASTM D1190 specification.

Table 4. Montreal sealant performance based test results.

Test	DSR		CSBBR			CSAT	
Performance parameter	C	P	Stiffness (MPa)	Avg. creep rate	CSDTT λ (%)	Max load (N)	
Criteria	>4.0 kPa	>0.70	<25MPa	>0.31	>85%	>50 N	
Temp (C)	58	58	–34	–34	–34	–34	Result
A	2.97	0.67	20.87	0.298	11.32	58.99	Fail
B	38.29	0.98	21.63	0.308	21.16	156.99	Fail
E	19.26	0.94	3.1	0.438	93.04	No debond	Pass
G	7.01	0.94	126.3	0.235	0.4	50.11	Fail
J	7.64	0.94	602.1	0.164	0.69	N/A	Fail

(Masson 1999). The results reveal that only sealant B passes the specification. Sealant A, E, G, and J cannot pass the specification. Compare to the field performance, sealant B and E show superior performance than the rest of sealants.

The test results of same five Montréal field sealants were also tested in the laboratory according to the newly developed crack sealant performance grade test specification and the results of the five sealants are reported in Table 4. The sealants were test in accordance with the service condition of Montreal, which has high summer temperature of 58°C and low winter temperature close to –40°C. Therefore, the DSR tested was conducted at 58°C and low temperature cohesive and adhesive properties tests were conducted at –34°C. According to the specifications, the low temperature tests are conducted at 6°C higher than service temperature. The superior sealant was identified as sealant E which shows good tracking resistance at summer temperature and low stiffness, high average creep rate and no debonding was observed. Sealant B also passes the DSR, CSBBR and CSAT test; however, it does not pass the CSDTT test at –34°C. However, if the test was shifted to the higher grade (–34°C) that the sealant would be tested at –28°C, the sealant could pass the specification. Sealant A did not pass the tracking for summer temperature, average creep rate, extendibility and adhesion tests at low temperature. Sealants G and J show poor laboratory performance at low temperature. They show high stiffness, low average creep rate and little extendibility and weak bonding.

4 CONCLUSION

New performance-based sealant tests were developed. The newly developed testing procedures are, in most cases, measure fundamental sealant properties that include: apparent viscosity at the recommended installation temperature, vacuum oven aging to simulate sealant weathering in the field, a DSR test to assess sealant's tracking resistance at high service temperatures, the CSBBR test to evaluate sealant's creep properties at low temperatures, the CSDTT to characterize sealant's low temperature extendibility, and low temperature adhesive (surface energy, direct adhesion, and blister) tests to evaluate the bonding between sealant and its substrate. Thresholds for the aforementioned tests were identified. These tests are used in cascade to assess the suitability of sealants for use in the diverse climatic regions of North America. This study presents needed *Performance-based Guidelines for the Selection of Hot-poured Crack Sealants*, based on the rheology of sealants, to replace the existing empirical approach for sealant selection. Although the new test procedures were developed using more than 15 widely used sealants in North America, more field performance data are needed in order to fine-tune the crack sealant performance-based specification criteria.

REFERENCES

1. Al-Qadi, I.L., Fini, E., Masson, J.-F, Loulizi A., Elseifi, M., and McGhee, K. *Development of Apparent Viscosity Test for Hot-Poured Crack Sealants*. Final Report, Submitted to Virginia Transportation Research Council, VDOT, Charlottesville, VA, 2008(b).
2. Al-Qadi, I.L., Fini, E., Figueroa, H.D. Masson, J.-F, and McGhee, K. *Development of Adhesion Tests for Hot-Poured Crack Sealants*. Final Report Submitted to Virginia Transportation Research Council, VDOT, Charlottesville, VA, 2008(a).
3. Al-Qadi, I.L., Yang, S.-H., Dessouky, S., and Masson, J.-F. Low Temperature Characterization of Hot-Poured Crack Sealant Using Modified SHRP Direct Tensile Tester. *Transportation Research Record: Journal of the Transportation Research Board*, No. 1991, Oct 2007, pp. 109–118.
4. Al-Qadi, I.L., Yang, S.-H., Dessouky, S., and Masson, J.-F. Low Temperature Characterization of Hot-Poured Crack Sealant Using Modified SHRP Direct Tensile Tester. *Transportation Research Record: Journal of the Transportation Research Board*, No. 1991, Oct 2007, pp. 109–118.
5. Al-Qadi, I.L., Yang, S.-H., Masson, J.-F., Elseifi, M., Dessouky, S., Loulizi, A., and McGhee, K. *Characterization of Low Temperature Creep Properties of Crack Sealants Using Bending Beam Rheometry*. Final Report, Submitted to Virginia Transportation Research Council, VDOT, Charlottesville, VA, 2008(d).
6. Al-Qadi, I.L., Yang, S.-H., Masson, J.-F., Dessouky, S., Loulizi, A., Elseifi, M., and McGhee, K. *Characterization of Low Temperature Mechanical Properties of Crack Sealants Utilizing Direct Tension Test*. Final Report, Submitted to Virginia Transportation Research Council, VDOT, Charlottesville, VA, 2008(c).
7. Cuelho, E., Ganeshan, S.K., Johnson, D.R., Freeman, R.B., and Schillings, P.L. *Relative Performance of Crack Sealing Materials and Techniques for Asphalt Pavements*. Third International Symposium on Maintenance and Rehabilitation of Pavements and Technological Control, Guimarães, Portugal, 2003, pp. 327–337.
8. Cuelho, E., Johnson, D.R., and Freeman, R.B. *Cost-Effectiveness of Crack Sealing Materials and Techniques for Asphalt Pavements*. Final Report, FHWA/MT-02-002/8127, Montana Department of Transportation, Helena, MT, 2002, 247 pp.
9. Chong, G.J. and Phang, W.A. Consequences of Deferred Maintenance Treatment on Transverse Cracks in Cold Regions, Proceedings of Paving in Cold Areas Mini-Workshop, Ottawa, Ontario, Canada, 1987, pp. 638–686.
10. Chong, G.J. and Phang, W.A. Improved Preventive Maintenance: Sealing Cracks in Flexible Pavements in Cold Regions. Final Report, PAV-87-01, Ontario Ministry of Transportation, Downsview, ON, 1987.
11. Chong, G.J. Rout and Seal Cracks in Flexible Pavements—A Cost-Effective Preventive Maintenance Procedure. *Transportation Research Record: Journal of the Transportation Research Board*, TRR, No. 1268, National Research Council, Washington D.C., 1990, pp. 8–16.
12. Fang, C., Galal, K.A., Ward, D.R., Haddock, J.E., and Kuczek, T. Cost-Effectiveness of Joint and Crack Sealing. Presented at the 82nd Annual Meeting of the Transportation Research Board, Washington, D.C., 2003.
13. Masson, J.-F., Collins, P., Perraton, D., Al-Qadi, I., Rapid assessment of the tracking resistance of bituminous crack sealants. Canadian Journal of Civil Engineering 34, 2007, pp. 126–131.
14. Masson, J.-F., and Lacasse, M.A. Effect of Hot-Air Lance on Crack Sealant Adhesion. Journal of Transportation Engineering, ASCE, Vol. 125, No. 4, 1999, pp. 357–363.
15. Peterson, D.E. Resealing joints and cracks in rigid and flexible pavements. NCHRP Synthesis of Highway Practice 98, Transportation Research Board, Washington, D.C., 1982.
16. Ward, D.R. Evaluation of the Implementation of Hot-Pour Sealants and Equipment for Crack Sealing in Indiana. Final Report, FHWA/IN/JTRP-2000/27, FHWA, Washington, DC, 2001, 189 pp.

Efficient Transportation and Pavement Systems – Al-Qadi, Sayed, Alnuaimi & Masad (eds)
© 2009 Taylor & Francis Group, London, ISBN 978-0-415-48979-9

Stress relaxation behavior of asphalt emulsion residue

S. Jeya & J. Murali Krishnan
Department of Civil Engineering, Indian Institute of Technology, Chennai, India

ABSTRACT: It is well known that the mechanical behavior of the binder available in a pavement completely controls the mechanical response characteristics of the asphalt layers. Within the context of cold mix asphalt construction, it assumes considerable significance due to the complex emulsion breaking process and the resulting asphalt emulsion residue left in the pavement. This investigation involves using two different asphalt emulsion residue recovery process in the laboratory and testing the stress relaxation behavior of the residue. Two standard emulsions, namely slow setting and rapid setting type were used for asphalt residue recovery. Two recovery processes were used to recover asphalt from the asphalt emulsion, the first one involved air drying at room temperature, and the second one involved the use of rolling thin film oven at a temperature of 60 degree C for 120, 240 and 360 minutes. Additionally PAV experiments were conducted on the slow setting and rapid setting residues obtained from RTFOT recovery process and compared with the PAV aged parent asphalt under identical conditions. The recovered residue was tested for its stress relaxation behavior in a dynamic shear rheometer for different strains and temperatures. It was seen that the emulsion residue recovery process plays a major role on the rheological properties. It was also seen that emulsified asphalt residue exhibit considerably 'stiffer' behavior when compared with parent asphalt.

Keywords: Asphalt Emulsions; Slow Setting; Rapid Setting; Residue Recovery Process; Stress Relaxation

1 INTRODUCTION

More than 10% of paving grade asphalt is used in emulsified form throughout the world. While the use of emulsions as a prime coat, tack coat and as cold mix in location with adverse weather conditions is well known, due to the renewed interest in the reduction of energy consumption and emissions worldwide recently, cold mix technologies are being looked as a potential alternative to hot mix asphalt.

In a typical cold asphalt mix application, the brown colored emulsion is initially mixed with the aggregates and spread. Once all the water has evaporated, the emulsion is said to be broken and this is indicated by the change in color of asphalt emulsion to black (Redelius and Walter 2005). The asphalt binder that remains after the emulsion has broken is called the asphalt emulsion residue. The study of asphalt emulsion and asphalt emulsion residue is essential to know the difference between the rheological properties of different emulsions and also the residue that is available in field. While considerable investigations have been carried out related to storage and transport stability of emulsions, emulsion break time and emulsion breaking process, the quality of the binder after the emulsion has broken has received very little attention.

Also, very few investigations have been conducted till now comparing the rheological properties of emulsion residue available in the cold asphalt mix with asphalt available in the hot mix asphalt construction. One major impediment is related to the emulsion residue recovery process that can be conducted in the laboratory mimicking the real life cold asphalt pavement's emulsion breaking process. Even today the standards have not been formulated towards the same (Hazlett 2007). Also very few investigations have actually compared the rheological behavior of emulsion residue recovered through different process.

2 LITERATURE REVIEW

The currently available (ASTM D244 (2004), AASHTO T59 (1990)) use only the efflux time viscometers for characterizing asphalt emulsions. Clearly the focus of these standards is more on the pumping and spraying application requirement and not related to the rheology of emulsions as such.

Heukelom and Wijga (1966) conducted investigations on various types of emulsions and deduced relationship between relative viscosity of dispersion with the rate of shear and volume concentration. King et al. (1996) were among the first researchers to indicate the performance of emulsions made using polymer modified asphalts using performance grading system. Takamura (2000) proposed a new residue-recovery procedure that uses airflow under ambient temperatures to reasonably simulate field conditions; he also proposed an accelerated mechanism for use in quality control activities. Salomon and Zhai (2002) used a dynamic shear rheometer and conducted various tests on emulsified asphalts.

Lesueur (2003) and Lesueur et al. (2003) investigated the rheological properties of asphalt emulsions using two experimental techniques. The first experimental technique involved the use of conventional efflux time viscometer and the second experimental technique involved the use of a steady state rheometer. Here the asphalt emulsions were modeled as Caisson-type fluids. The interesting issue in this study was the need to consider the complete spectrum of the rheological behavior of asphalt emulsions. Marasteanu and Clyne (2005) investigated the characterization of asphalt emulsion typically used in cold in-place recycling applications using the AASHTO specifications for asphalt binders. The emulsions were cured using air curing and rolling thin film oven test curing. The air-cured samples were also aged in the pressure aging vessel. The residues were tested using the bending beam rheometer and the direct tension test at low temperatures and the dynamic shear rheometer at high and intermediate temperatures. It was found that the mechanical properties of the emulsion residue were always inferior to that of the parent asphalt. It is to be pointed out here that Marasteanu and Clyne(2005) used the yardstick of performance grading system to base this observation.

The transportation research board of United States released recently transportation circular E-C122 titled "Asphalt Emulsion Technology-Review of Emulsion Residue Procedures" (Salomon 2007) in which all the issues related to emulsion residue recovery process are discussed in detail. In fact the major concern is related to the rheological properties of emulsion residue available in the field with the main stumbling block being the lack of an acceptable recovery method suitable to all emulsions (Salomon 2007).

Most of the emulsion residue recovery process available can be broadly classified into thin film oven methods and methods based on chemical treatments (Gueit et al. 2007). The thin film oven methods consists of using the existing rolling thin film oven equipments and simulating the short term aging of the binder. One difficulty associated with the residue recovered from these tests is the necessity to manipulate and homogenize them before conducting any rheological tests. The issues are considerably difficult when working with polymer modified emulsions.

Considering the increasing percentage of cold mix asphalt throughout the world, it is surprising to see very little investigation conducted on emulsions and emulsion residue. Also within the Asian context no such investigation has been conducted. Due to the significant influence of the parent asphalt on the rheological properties of emulsion produced this aspect assumes considerable importance.

3 TEST PROGRAM

3.1 *Materials used for testing*

Two asphalt emulsions were investigated in this study manufactured by Hindustan Colas Ltd, India. They are slow setting emulsion and rapid setting emulsion. The parent asphalt used for manufacturing these emulsions was also tested.

3.2 *Emulsion residue recovery*

The asphalt residue was recovered from emulsions using two curing procedures. The first curing procedure involves air-drying the material. A 50 g emulsion sample with a thickness less than 1 mm was placed in a pan and allowed to stand for 24 h at room temperature. The remaining residue was used for testing. The second curing procedure involved the use of rolling thin film oven test (RTFOT). In this procedure, 60 g of emulsion sample was poured in each RTFOT bottle for retrieving a reasonable amount of residue due to the consideration of loss in water content. The sample was dried at a temperature of 60 °C for 120, 240 and 360 min. with a forced airflow of 4 L/min. The remaining residue was poured and used for further testing. Additionally, the RTFOT dried samples of RS and SS emulsion residues were subjected to aging in pressure aging vessel using the standards ASTM protocol.

3.3 *Equipment used*

The stress relaxation tests were performed using dynamic shear rheometer using 8 mm parallel plate geometry at two intermediate temperature and two different strains. The dynamic shear rheometer used for the experimentation is the Physica MCR 301 manufactured by Anton Paar. The Rheoplus software of the dynamic shear rheometer was used for data acquisition.

3.4 *Testing protocol*

When a material is subjected to constant strain and then when the strain is maintained constant, the material shows an instantaneous increase in stress and then the stresses gradually relax with time. The materials are classified as fluid-like and solid-like based on whether the stresses relax to zero or a non-zero value respectively for a specific period of observation. In the current investigation, the stress relaxation experiment was conducted by first subjecting the specimen to a constant strain rate for 0.5 second. Once the allowable strain is engendered in the specimen, the control system was switched to the constant strain mode and the stress relaxation was measured. This experiment is normally called "*stress relaxation after cessation of steady shear flow*" (Bird et al.1987). The schematic sketch is shown in figure 1. For each condition, two trials were conducted to check the repeatability.

In this investigation, the stress relaxation experiments were conducted to characterize the rheological behavior of emulsion residue and parent asphalt. It is possible to test for steady shear, creep and recovery and oscillatory shear properties of asphalts and asphalt emulsion residue. Our main interest was to bench mark for the first time the difference in the response of emulsion

Figure 1. Stress relaxation after cessation of steady shear flow (from Bird et al. 1987).

residue vis-a-vis parent asphalt in terms of stress relaxation behavior when subjected to a constant strain. Typically we are interested to know whether the manufacture of emulsion with its attendant mechanical and chemical interaction resulted in a much stiffer material in field after construction. The stress relaxation test is also much easier to conduct due to the ease of control and the good repeatability with the equipment in hand. Also, the stress relaxation comes handy here in terms of characterizing the relaxation time for a specific initial stress.

The residue recovery process consists of air drying as well as drying in the RTFOT oven for a specific period of time and the loss of moisture content was measured after each drying process. In case of air drying process, it was found that after 24 hours, there was no significant reduction in moisture content. The RTFOT air drying process was conducted in such a manner that as soon as the moisture content reduced to the levels of the air drying process, the experiments were stopped. From table 1, it can be seen that 360 minutes of RTFOT drying more or less simulates the 24 hours of air drying process. Since at this point of time, we were not sure of the possible loss of volatiles during the RTFOT drying process, we did not proceed beyond the limit we got from the air drying process. The stress relaxation experiments were conducted on the recovered residues for both the emulsions and the parent asphalt at two temperatures 20°C and 30°C for each at 25% and 50% strain.

4 RESULTS AND DISCUSSION

Figures 2 through 11 show the stress relaxation results obtained in this investigation. Specifically figure 2 and 3 shows the stress relaxation at 20°C for slow setting and rapid setting emulsion of air dried and RTFOT dried emulsions with different duration. It is seen that, among the two residue recovery process, the air dried material shows stiffer response than RTFOT dried material for both the emulsions. The analysis of the experimental results here is based on two parameters. The first parameter is associated with the stress level of the material for the same constant strain and the second parameter is associated with the stress remaining in the sample at a specific instant of time when compared with other samples subjected to a different recovery process. For instance it

Table 1. Loss of moisture content in the residues from two recovery process.

Material	SS RTFOT			RS RTFOT			SS air dried	RS air dried
Time	120 min.	240 min.	360 min.	120 min.	240 min.	360 min.	24 hours	
Moisture content loss (%)	23	30.52	33.12	22	28.46	31.79	35	30

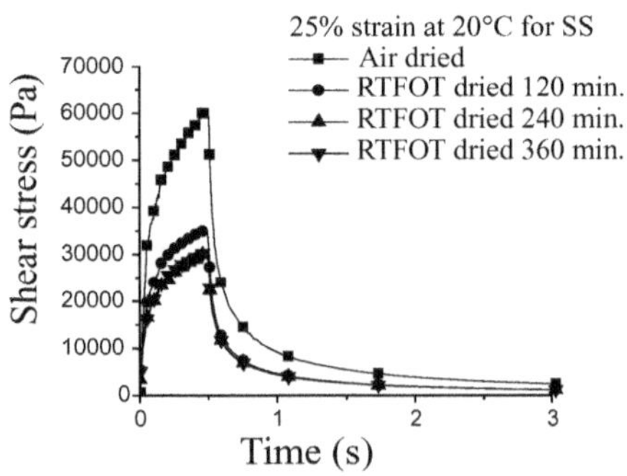

Figure 2. Stress relaxation pattern for SS at 20°C for 25% strain.

is seen that in figure 2, the air dried sample requires considerably more stress for the same strain and at the end of 2 second the air dried sample still has some stress remaining in the material when compared with RTFOT dried material. The relaxation of residue recovered by RTFOT process for different duration shows similar responses indicating that the difference in moisture content does not influence the rheological behavior during stress relaxation. The effective removal of moisture content takes place during 120 min. and further removal of moisture content by increasing the RTFOT duration will not significantly affect the stress relaxation behavior based on the experimental observations.

Figures 4 and 5 show the influence of temperature for slow setting and rapid setting emulsions. Figures 6, 7, 8 and 9 show the influence of different strains for slow setting emulsion and rapid setting emulsion obtained from RTFOT for different time durations. In figures 6 and 8, it is noted that for the same strain, the SS and RS exhibits different peak stress. The slow setting residue obtained from RTFOT shows higher peak stress than the rapid setting residue. It shows the removal of moisture content in SS from RTFOT is more effective than compared to rapid setting residue. However, a reverse trend is seen when the strain is increased to 50% (figures 7 and 9).

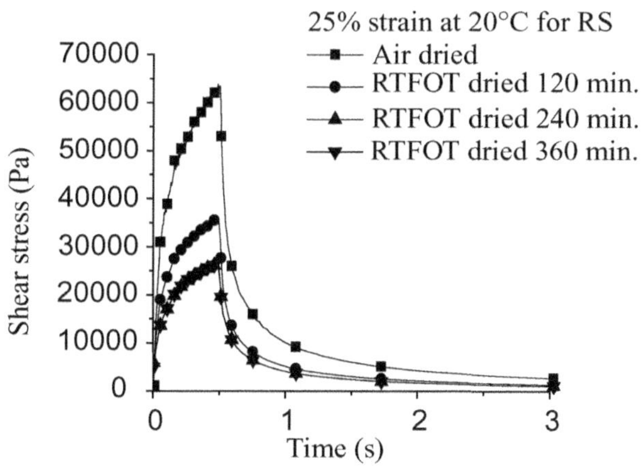

Figure 3. Stress relaxation pattern for RS at 20°C for 25% strain.

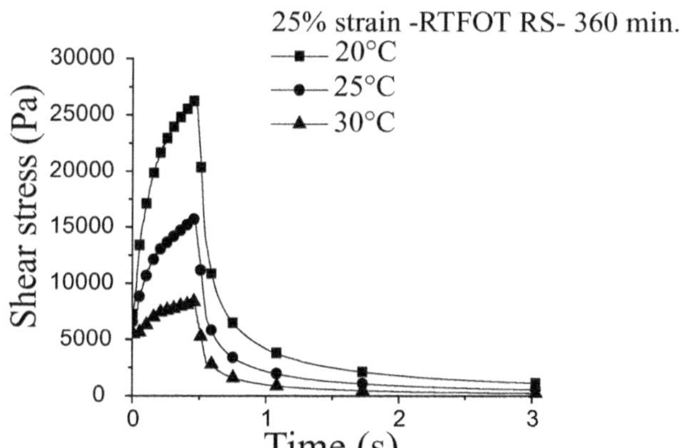

Figure 4. Stress relaxation pattern for RS for 25% strain at three temperatures.

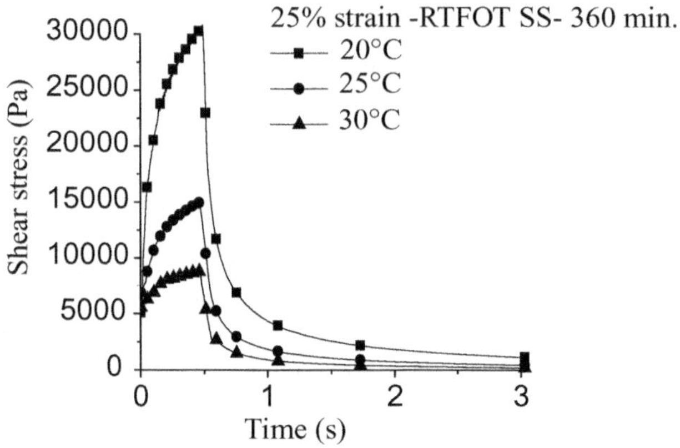

Figure 5. Stress relaxation pattern for SS for 25% STRAIN at three temperatures.

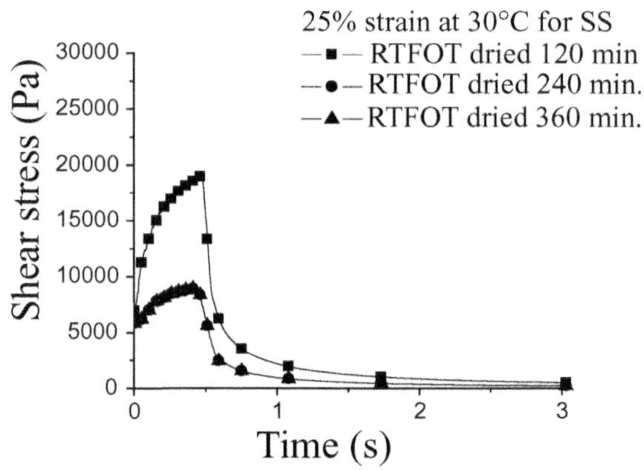

Figure 6. Stress relaxation pattern for SS at 30°C for 25% strain.

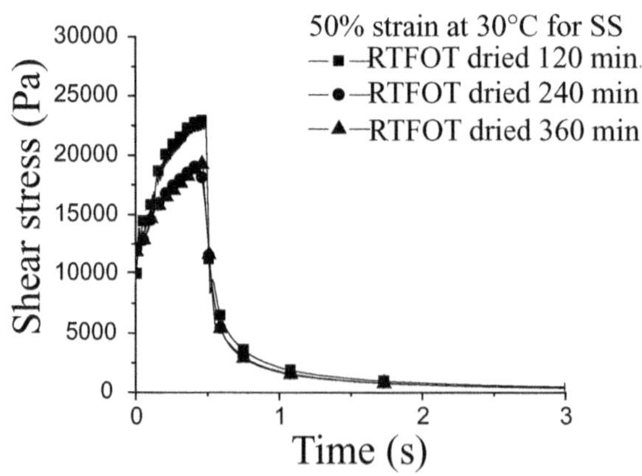

Figure 7. Stress relaxation pattern for SS at 30°C for 50% strain.

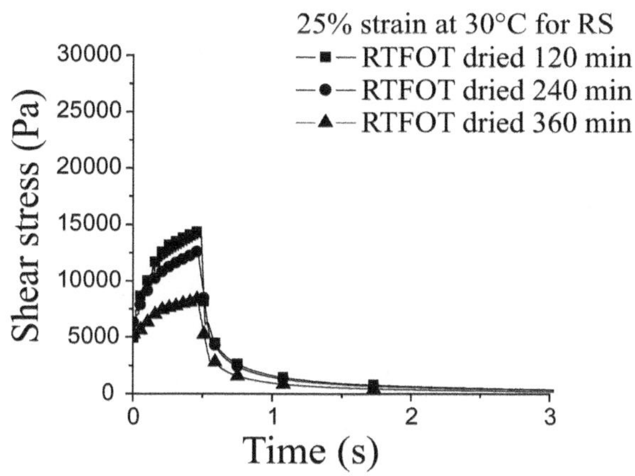

Figure 8. Stress relaxation pattern for RS at 30°C for 25% strain.

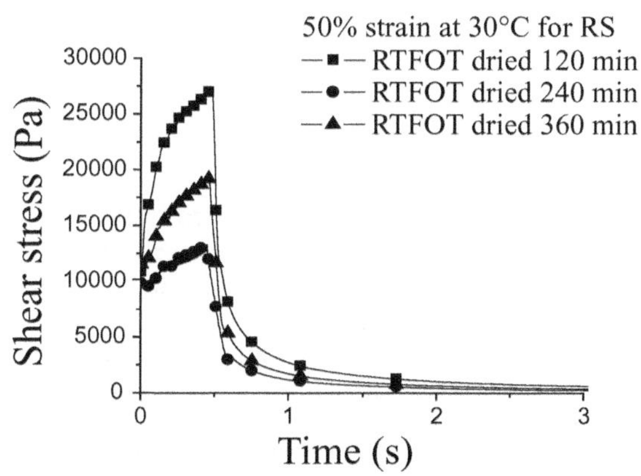

Figure 9. Stress relaxation pattern for RS at 30°C for 50% strain.

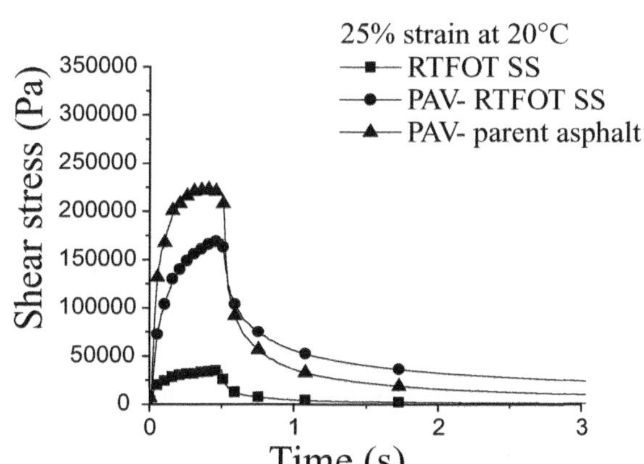

Figure 10. Stress relaxation pattern at 20°C for 25% strain for SS obtained from PAV.

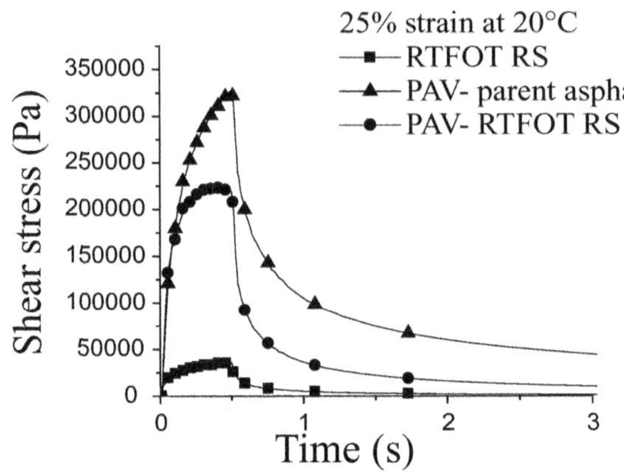

Figure 11. Stress relaxation pattern at 20°C for 25% strain for RS obtained from PAV.

Also, PAV aging of RS and SS residue was conducted for the 120 min. dried samples only. It is seen that the parent asphalt exhibits stiffer behavior than the slow setting and rapid setting residue. The results are shown in figures 10 and 11.

5 CONCLUSIONS

In this investigation, two types of emulsion were investigated along with the parent asphalt used in their manufacture. An air drying and RTFOT drying process was formulated to recover the residue from the emulsions. Stress relaxation was conducted at three temperatures and two strain level on the emulsion residues as well as on parent asphalt.

It was found that the air dried material shows stiffer response than RTFOT dried material for both the emulsions. The relaxation of residue recovered by RTFOT process for different duration shows the similar responses indicating the difference in moisture content does not influence the rheological behavior beyond a specific period of RTFOT drying. Also the influence of different temperature and strains for slow setting and rapid setting emulsions are shown.

This is a preliminary investigation on the rheological properties of asphalt emulsion residue and it is seen that considerable challenges exists both in terms of development of residue recovery process as well as in terms of developing appropriate test protocols to quantify the rheological response of the material. Investigations are currently being conducted to understand the steady shear properties of emulsions and emulsion residues.

REFERENCES

1. ASTM D244 (2004). "Standard Test Methods and Practices for Emulsified Asphalts." Annual Book of Standards, 4, West Conshohocken, PA.
2. AASHTO T59 (1990). "Standard method of test for Testing emulsified asphalts." Washington, D.C.
3. Bird, R.B., Armstrong, R.C., and Hassager, O. (1987). "Dynamic of polymeric fluids." Fluid Mechanics, Wiley-Interscience Publication, New York.
4. Gueit, C., Robert, M., and Durand, G. (2007). "Characterization of the Different Phases in the Life Cycle of the Binder in a Bitumen Emulsion: Recovery Methods", Transportation Research Circular E-C122, 1–10.
5. Hazlett, D. (2007). "Emulsion Residue Recovery Techniques." Transportation Research Circular E-C122, 15–23.

6. Heukelom, W., and Wijga, P.W.O. (1966). "Viscosity of dispersions as governed by concentration and rate of shear." Proc., the Association of Asphalt Paving technologists, 40, 418–437.
7. James, A. (2007). "Asphalt Emulsion Technology." Transportation Research Circular E-C102, 7–20.
8. King, G., Lesueur, D., King, H., and Planche, J.P. (1996). "SHRP test evaluation of high float and polymer modified bitumen emulsion residues." Proc., Eurobitume/Euroasphalt congress, Strasbourg.
9. Lesueur, D. (2003). "The Rheological Properties of Bitumen Emulsions. Part I Theoretical Relationships between Efflux Time and Rheological Behavior." Road Materials and Pavement Design, 4(2), 151–168.
10. Lesueur, D., Ezzarougui, M., Herve, G., and Odie, L. (2003). "The Rheological Properties of Bitumen Emulsions Part II Experimental Characterization." Road Materials and Pavement Design, 4(2), 169–184.
11. Marasteanu, M.O., and Clyne, T.R. (2006). "Rheological characterization of asphalt emulsions residues". Journal of Materials in Civil Engineering, 18(3), 398–407.
12. Redelius, and Walter, (2005). "Emulsions and Emulsion Stability." Surfactant Science, Sjoblam, eds., CRC Press, New York, 132, 383–418.
13. Salomon, D. (2007). "Asphalt Emulsion Technology- Review of Emulsion Residue Procedures." Foreword to Transportation Research Circular E-C122.
14. Salomon, D., and Zhai, H. (2002). "Rheological measurements of asphalt road emulsions." 3rd World Congress on Emulsions, France, 24–27.
15. Takamura, K. (2000). "Improved Fatigue Resistance of Asphalt Emulsion Residue Modified with SBR Latex." Proc., ISSA/AEMA, 1–17.

Efficient Transportation and Pavement Systems – Al-Qadi, Sayed, Alnuaimi & Masad (eds)
© 2009 Taylor & Francis Group, London, ISBN 978-0-415-48979-9

Characterisation and modification of a hard binder for pavements in hot climates

A.A.A. Molenaar, R.N. Khedoe, F. Arce & M.F.C. van de Ven
Delft University of Technology, Delft, The Netherlands

ABSTRACT: Pavements in hot climates suffer from permanent deformation. When these deformations become severe, they result in hazardous driving conditions. Using a hard binder usually increases the resistance to permanent deformation significantly but application of these binders might have a negative effect on crack resistance and workability. This paper discusses the potential use of a special hard binder called C-Fix. The binder is characterized by means of rheological tests as well as direct tension tests. Also test results obtained on mastics made with C-Fix binder and two types of filler are discussed. Furthermore attention is paid to the effect of an SBS modification on the characteristics of C-Fix and mastics made with C-Fix. The results show that application of a C-Fix binder gives excellent characteristics to asphalt mixtures in terms of stiffness, resistance to permanent deformation and strength at high temperatures. At low temperatures however, which occur at night in desserts and during wintertime in areas with a typical land climate, C-Fix binders behave rather brittle and are sensitive to cracking. The Altraflex modification certainly adds crack resistance to the C-Fix binder but the effect of the filler seems to be even more important.

Keywords: C-Fix, Altraflex, rheology, relaxation test, tensile test, filler

1 INTRODUCTION

The most important damage types observed on asphalt pavements in hot climates are permanent deformation and cracking. Permanent deformation, also called rutting, is caused by plastic and/or viscous deformation of the asphalt mixture. Next to mixture composition, high wheel loads and high contact pressures together with high temperatures are the factors influencing the formation of rutting. Using well compacted stone skeleton mixtures and harder binders usually solves much of the rutting problem. Hard binders however are less flexible and mixtures made of them usually have a lower fatigue resistance than mixtures made with softer binders. Furthermore they usually show a brittle type of behaviour. In itself this is not so much of a problem in case of bottom up fatigue cracking since this type of cracking can effectively be controlled by increasing the thickness of the asphalt layer. Also the fact that hard binders allow higher bitumen contents to be used, helps in reducing bottom up fatigue cracking. Many pavements however show a significant amount of top down cracking. Although the mechanism of top down cracking is not yet fully understood, it is clear that wearing course mixtures with hard binders are much more prone to this type of cracking than mixtures with a softer binder (Gerritsen et al. 1987, Groenendijk 1998).

Cracking in hot climates is quite often not only due to traffic loads. In hot desert climates, daily pavement temperatures can easily range from 60°C in the afternoon to 5°C in the early morning. Such temperature variations induce thermal stresses in the pavement especially when a hard binder is used. Temperature cracking becomes an even bigger problem when the temperature variation during the year is taken into account. Pavement temperatures varying between 60°C in the summer and –20°C in winter time are not unusual in areas with a typical land climate. In such cases, pavement damage as shown in figure 1 is not uncommon.

Figure 1. Typical crack pattern observed in an area with hot dry summers and cold winters.

All this implies that the selection of the most appropriate binder for asphalt mixtures to be used in hot climates is not a simple task especially not when significant temperature variations occur.

This paper describes some of the research done at the Delft University of Technology on C-Fix, a special hard binder that exhibits excellent resistance to permanent deformation. First of all some characteristics of C-Fix will be presented. Next, test results are presented on the rheological behaviour of C-Fix. Then some attention will be paid to modification of C-Fix in order to improve the low temperature characteristics. The paper will conclude with some recommendations.

2 C-FIX

As a by-product of refined crude oil, a very dense and carbon rich residue is produced. This is used to create a low quality fuel normally used in intercontinental shipping vessels or thermo-electric power stations. Burning this fuel consequently results in a high CO_2 emission.

At Shell this by-product was used as feedstock to produce a binder for mineral aggregates. This implies that carbon no longer is burned and brought to the atmosphere, but fixed in the binder, protecting and keeping the environment clean. Asphalt mixture specimens produced with C-Fix appear to have a flexural strength at room temperature of 4–9 MPa, which is similar to the flexural strength of cement concrete, 5–9 MPa. These values are not observed for asphalt mixtures with conventional bitumen at room temperature (Khedoe 2006).

Some characteristics of C-Fix are shown in table 1. It should be noted that in the remainder of this paper a C-Fix binder with a penetration of 5 resp. 9 is called C5 resp. C9.

The mixing temperature of C-Fix binder is 200°C.

Asphalt mixtures with C-Fix are very stiff at service temperature and only small deformations occur when the pavement is subjected to a load with long loading time. This is the case in

Table 1. C-Fix properties compared with penetration grade bitumen (http://www.c-fix.com/default.aspx).

	Bitumen pen 100	Bitumen pen 10–20	C-Fix
Penetration 25°C [dmm]	100	10–20	5–6
Penetration 40°C [dmm]			15–28
Ring & ball [°C]	42	60–78	84–88
Viscosities [Pa.s]			
150°C			1,5
160°C	0,1		0,9
170°C			0,4

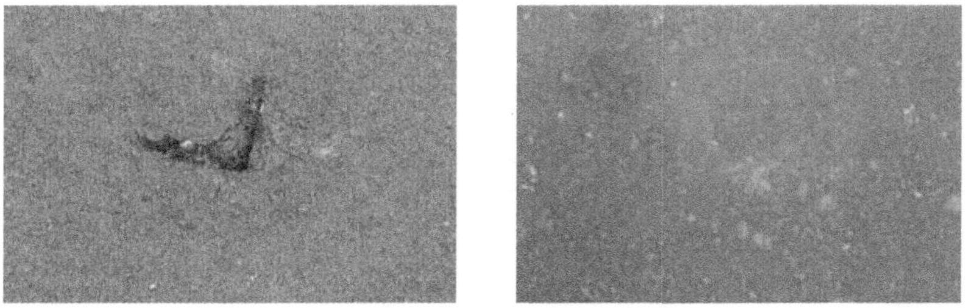

Figure 2. Deformation due to a point load in a mixture with a polymer modified bitumen (left) and a C-Fix binder (right) (http://www.c-fix.com/default.aspx).

container storage areas where the pavement is subjected both to heavy loads of long duration and to scrapping of containers. Figure 2 shows the excellent resistance to permanent deformation of a C-Fix mixture when compared to a polymer modified mixture.

3 TEST PROGRAM

Several tests were performed to characterize the C-Fix binder. Amongst these tests were Dynamic Shear Rheometer (DSR) tests, Direct Tension (DT) tests and relaxation tests. The tests were performed on pure C-Fix as well as C-Fix/filler mastics using different types of filler. For comparison also a B70/100 penetration bitumen was tested as well as B 70/100/filler mastics (see table 2 for details on the composition of the mastics). The most important test results will be discussed here-after.

3.1 *DSR test results*

Figure 3 shows the master curves for the complex shear modulus G^* and the phase angle δ as determined for the B70/100 and C9 binder. C9 is a C-Fix binder with a penetration of 9.

Figure 3 clearly shows that the C9 binder has a much higher stiffness and a much lower phase angle than the B70/100 binder. This figure points out the much more elastic behaviour of the C-Fix binder when compared to the reference B70/100 binder.

3.2 *Relaxation test results*

The elastic behaviour of the C-Fix binder was also evaluated during the relaxation test program. In these tests not only C9 (a C-Fix binder with penetration 9) and B70/100 were tested but also mastics made with these binders. To obtain the mastics, the binders were mixed with an artificial

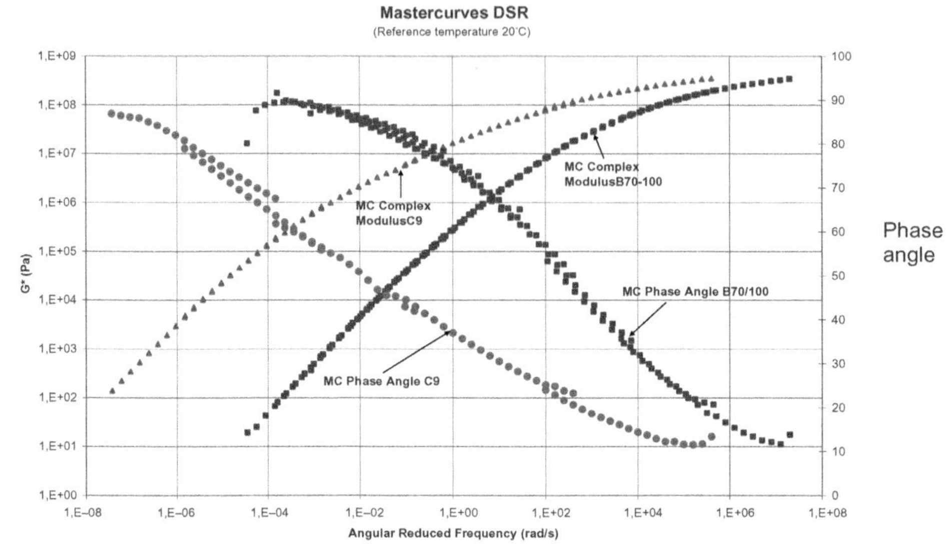

Figure 3. Complex modulus and phase angle master curves of C9 binder and 70/100 bitumen at reference temperature 20°C.

Table 2. Properties of the fillers used.

	Wigro 60 K	Duras 15
Type	Medium type filler with hydroxide	Very weak
Producer	Ankerpoort Winterswijk	Ankerpoort Maastricht
Components	Limestone 65–75% Calcium hydroxide 25–35%	Limestone 100%
Gradation		
• 2 mm	100% (m/m)	100% (m/m)
• 0,125 mm	87–97% (m/m)	90–100% (m/m)
• 0,063 mm	80–90% (m/m)	76–86% (m/m)
Density	2,48–2,68 [Ton/m³]	2,60–2,80 [Ton/m³]
Water sensitivity	≤10% (m/m)	≤10% (m/m)
Specific surface	4.000–6000 [cm²/gr]	5.330 [cm²/gr]
Rigden voids	44–50	28–34

filler called Wigro 60 K. This filler contains hydrated lime. Different filler/binder ratios (f/b ratios) were used. Details about the Wigro 60 K filler are shown in table 2.

Table 2 also shows properties of the Duras 15 filler. This type of filler was used later on in the research next to the Wigro 60 K filler. The gradations of both fillers are shown in figure 4.

Some results of the relaxation tests are shown in table 3 and figure 5.

The relaxation tests were performed using the DT device. The specimens were subjected to a specific displacement rate until a certain displacement was obtained. Then the specimens were kept at that deformed state and the relaxation of the stress in the specimen was measured. Table 3

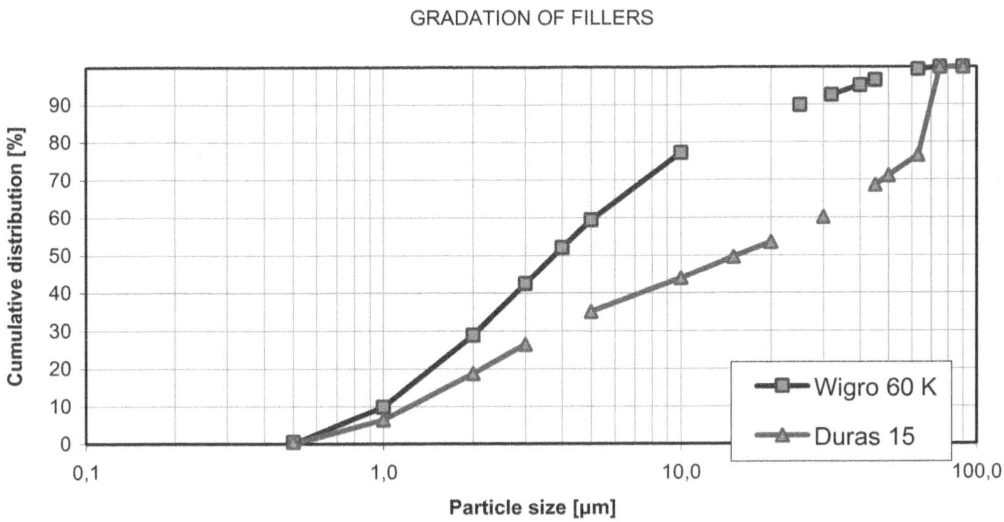

Figure 4. Gradation of the wigro 60 K and duras 18 filler.

Table 3. Relaxation of different binders and mastics at –10°C.

Material	f/b	Displacement rate* [mm/min]	Displacement [mm]	Max. Stress [MPa]	Relaxation (t = 600s) [%]
B70/100	0	3	0,127	0,76	100,0
B70/100	1	3	0,127	2,04	100,0
C9	0	0,3	0,073	1,05	55,7
C9	0,5	0,3	0,054	1,05	34,8
C9	1	0,3	0,063	1,29	25,4
C9	1,2	0,3	0,059	1,46	22,4

shows that higher displacement rates were used for the B70/100 samples than for the C9 samples. This was done because tests at lower displacement rates yielded unreliable results for the B70/100 samples. Because of limitations in the capacity of the DT device in terms of load levels that could be applied, the high displacement rate could not be applied on the C9 samples.

In figure 5, relaxation is expressed in terms of the ratio of the stress that occurs at a given moment in time over maximum occurring stress.

The results clearly show that the C9 C-Fix binder shows only limited relaxation when compared to the reference B70/100 binder. This implies that temperature stresses that are induced in the pavement will relax much slower in asphalt layers with a C-Fix binder than in layers with a traditional pen grade binder. This can imply that at low temperature conditions, the combined effect of temperature and traffic induced stresses can result in premature failure.

One should however be aware of the fact that the B70/100 and B70/100 mastic samples were not subjected to some kind of aging protocol prior to testing. In reality aging will harden the B70/100 binder significantly. This implies that in practice the relaxation behaviour of the B70/100 and B70/100 mastic will be not as good as shown in figure 5. It will be clear that only very limited hardening will take place in the C-Fix binder.

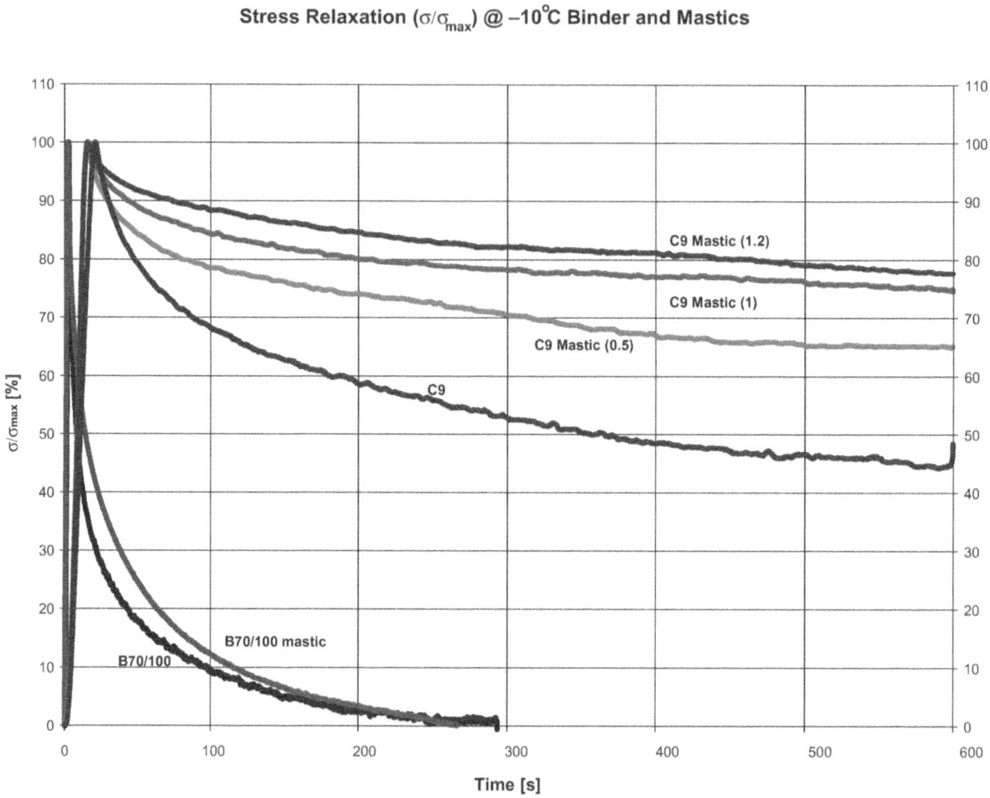

Figure 5. Results of relaxation tests on B70/100 and C9 as well as B70/100 and C9 mastics at −10° C.

4 DISCUSSION OF THE TEST RESULTS OBTAINED ON C-FIX AND B70/100 BINDERS AND MASTICS

Table 4 (Khedoe 2006) shows that at temperatures of 10°C and higher, a porous asphalt concrete mixture made with 4.5% C5 binder (note that in this case a C5 binder was used) had a superior tensile strength when compared to mixtures with a B70/100 binder (same binder content). Both mixtures were made using a Wigro 60K filler and had approximately 20% voids. The results obtained at 10°C however show that because of its lower fracture toughness, the mixture made with C-Fix is brittle compared to the mixture made with B70/100 bitumen because of its lower fracture toughness. The expectation is that at temperatures lower than 10°C, the differences in tensile strength between the C5 and B70/100 mixture will become even smaller while the fracture energy of the B70/100 mixture will be significantly higher than that of the C5 mixture.

These results together with the results of the DSR and relaxation tests indicate that although mixtures made with C-Fix binders will have a superior behaviour at high temperatures, the combination of temperature and traffic induced stresses together with the limited relaxation capacity of C-Fix binders might result in cracking during cold nights and/or cold winters.

5 MODIFICATION OF C-FIX WITH ALTRAFLEX

Because of the brittle behaviour of C-Fix at low temperatures, it was decided to investigate to what extent a polymer modification could improve this. The modifier selected was Altraflex. This is a

Table 4. Indirect tension test results obtained on porous asphalt concrete mixtures made with a C5 and B70/100 binder.

	5.1% Binder		4.5% Binder				
	C5		C5		B70/100		Temperature
Dry	ITS [MPa]	Energy [KJ/mm]	ITS [MPa]	Energy [KJ/mm]	ITS [MPa]	Energy [KJ/mm]	[°C]
	2.23	0.408	2.20	0.387	1.32	0.806	10°C
	–	–	1.67	0.508	0.405	0.320	25°C
	–	–	1.15	0.481	0.134	0.100	40°C

Table 5. Characteristics of altraflex modified C-Fix.

	Bitumen 70/100 [3]	C-Fix-C5	C-Fix +6% Altraflex	C-Fix +12% Altraflex
Penetration 25°C [0,1 mm]	64	5	10	10
Penetration 40°C [0,1 mm]	301	15	20	22
Penetration 60°C [0,1 mm]	–	38	47	49
Softening point [°C]	45,6	84	97	106
Penetration Index	–1,8	0,5	2,9	3,8
Density [Ton/m^3]	1030	1090	–	–

special type of modifier that can be added to the mixture in the pugmill in a batch plant during the mixing process. Altraflex consists for about 50% of SBS, some active filler containing $Ca(OH)_2$, and an additive that helps to absorb free radicals and so reduces the sensitivity for oxidation and aging. Furthermore some adhesion improvers are added. At room temperature, Altraflex has the shape of small grains having a white to beige colour (see also figure 8).

The big advantage of this modifier over other modifiers is that it can be added directly and mixed in the pugmill. This allows relatively small amounts of polymer modified mixtures to be produced and avoids the need for continuous stirring the binder which is needed to prevent segregation in case use is made of ready made polymer modified bitumens.

Mixing of the Altraflex with C-Fix was carried out with a heavy duty laboratory high shear mixer Silverson model L4R. The rotation speed was around 6.000 rpm at 200°C for 10 minutes for the two amounts of Altraflex studied. For sample preparation, 300 grams of C-Fix was heated to 200°C and poured into a 900 ml cylindrical flask. Under controlled temperature and with the high shear mixer, a pre-weighted amount of Altraflex was added gradually to the binder. Mixing was performed then for 10 minutes. Table 5 shows some characteristics of Altraflex modified C-Fix.

Table 5 shows that adding Altraflex indeed makes the C-Fix binder softer at lower temperatures since the penetration values at 25, 40 and 60°C of the modified binders are higher than those of pure C-Fix. Figures 6 and 7 show the master curves for G^* and δ for C-Fix and the Altraflex modified C-Fix.

Figure 6 shows that modifying C-Fix with 6% Altraflex results in a G^* which is 50–60% of the G^* of pure C-Fix. The phase angles of the 6% Altraflex modified material however are about the same as those for the pure C-Fix. From these results it can be concluded that the temperature stresses in the 6% Altraflex modified binder will be 50–60% of those in a binder of pure C-Fix. This is because:

$$\sigma_t = E * \alpha * \Delta T$$

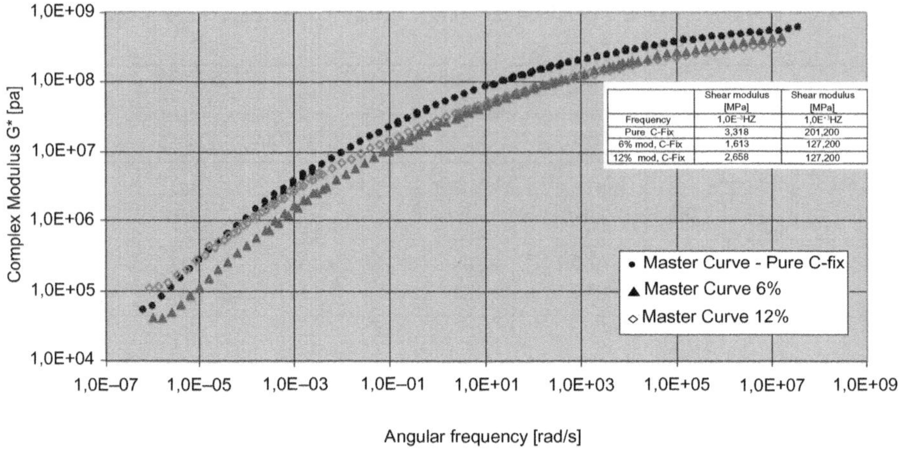

Figure 6. Master curve for G* for C-Fix and altraflex modified C-Fix (reference temperature is 20°C).

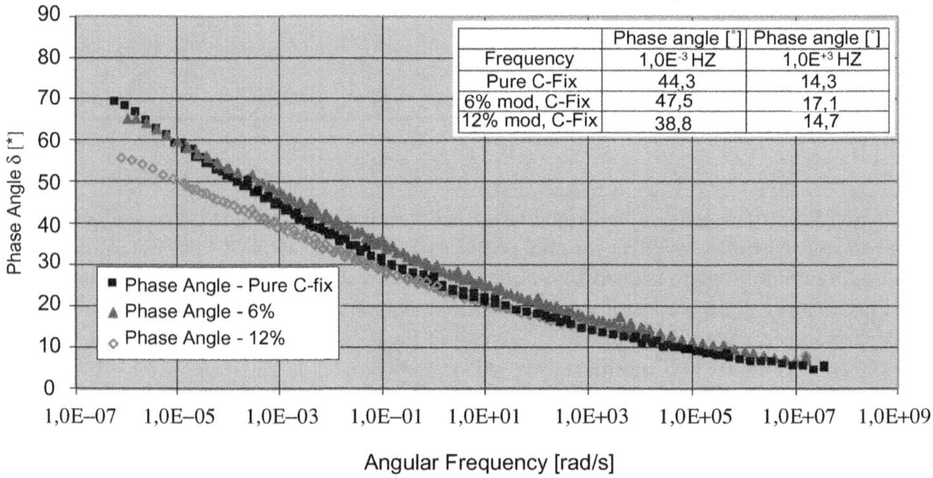

Figure 7. Master curve for δ for C-Fix and altraflex modified C-Fix (reference temperature is 20°C).

where σ_t is the temperature stress, E is the stiffness modulus, α is the coefficient of linear thermal expansion and ΔT is the drop in temperature.

From the fact that the master curves for δ are almost the same for pure C-Fix and 6% Altraflex modified C-Fix, one can conclude that the relaxation behaviour of both binders will be the same. By means of relaxation tests it was proven that this was indeed the case.

DT tests were performed according to the ASTM D6723-02 (ASTM 2001) aiming to determine the stress/strain curve and the stress/strain at failure. This test was done for the pure C-Fix binder, 6% Altraflex modified binder and mastics made with 6% Altraflex modified binder, two types of filler and two filler/binder ratios. In addition, tests with mastic using filler Duras 15 (f/b = 1,0) and no Altraflex were performed. The tests were not done on the 12% Altraflex modified binders since it appeared that it was difficult to get a homogeneous dispersion of the Altraflex when using that amount.

Table 6. Direct tension test results obtained on C-Fix and altraflex modified C-Fix binders as well as obtained on mastics made of altraflex modified C-Fix and two types of filler.

	DTT						
	Temperature [°C]	σ_f mean [Mpa]	σ_f Std. Dev.	ε_f mean [%]	ε_f Std.Dev.	Toughness [GJ/m^3]	# samples tested
Pure C-Fix (C5)	–10°C	1,060	0,13	0,560	0,12	4,318	11
6% Altraflex modif. C-Fix		1,082	0,16	0,798	0,10	5,535	14
Mastic Wigro 60 K-f/b = 1,0		2,475	0,43	0,656	0,21	15,548	9
Mastic Wigro 60 K-f/b = 0,5		2,208	0,24	0,775	0,07	9,960	14
Mastic Duras 15-f/b = 1,0		2,467	0,30	0,826	0,16	13,820	6
Mastic Duras 15-f/b = 0,5		1,659	0,11	0,829	0,05	7,700	10
Mastic Duras 15-f/b = 1,0—NO POLYMER		2,551	0,23	0,883	0,06	13,157	5
Pure C-Fix (C5)	0°C	1,112	0,15	1,167	0,13	8,440	6
6% Altraflex modif. C-Fix		1,250	0,07	1,807	0,02	11,557	8
Mastic Wigro 60 K-f/b = 1,0		2,782	0,78	1,048	0,20	23,986	9
Mastic Wigro 60 K-f/b = 0,5		2,076	0,31	1,295	0,20	18,000	11
Mastic Duras 15-f/b = 1,0		2,415	0,60	1,187	0,37	26,160	5
Mastic Duras 15-f/b = 0,5		1,651	0,09	1,377	0,17	14,330	6
Mastic Duras 15-f/b = 1,0—NO POLYMER		2,709	0,53	1,361	0,17	24,025	6
Pure C-Fix (C5)	+10°C	1,001	0,01	2,916	0,10	15,300	6
6% Altraflex modif. C-Fix		0,846	0,06	3,362	0,56	21,736	10
Mastic Wigro 60 K-f/b = 1,0		2,998	0,32	1,578	0,10	26,964	10
Mastic Wigro 60 K-f/b = 0,5		1,717	0,00	2,065	0,00	31,050	14
Mastic Duras 15-f/b = 1,0		2,465	0,05	2,598	0,15	37,240	6
Mastic Duras 15-f/b = 0,5		1,478	0,24	3,689	0,44	37,310	8
Mastic Duras 15-f/b = 1,0—NO POLYMER		2,480	0,21	2,804	0,45	47,540	8

The test results are shown in table 6. Table 6 clearly shows that adding 6% Altraflex increases the toughness of the C-Fix binder. It is recalled that toughness is defined as the energy needed to fail the specimen. Toughness values are determined by calculating the area under the stress strain curve.

Remarkable however is the large effect of the filler. It seems that the effect of the filler is overwhelming the effect of the Altraflex. Especially the effect of the inert Duras filler on the tensile strength and toughness of the C-Fix/Duras mastic (no Altraflex was used in this mortar) is surprising.

5 DISCUSSION OF THE TEST RESULTS OBTAINED ON ALTRAFLEX MODIFIED C-FIX

From the very beginning of the test program it was realized that mixing a very hard binder like C-Fix, having a much lower maltene content than pen grade bitumens, with SBS might be

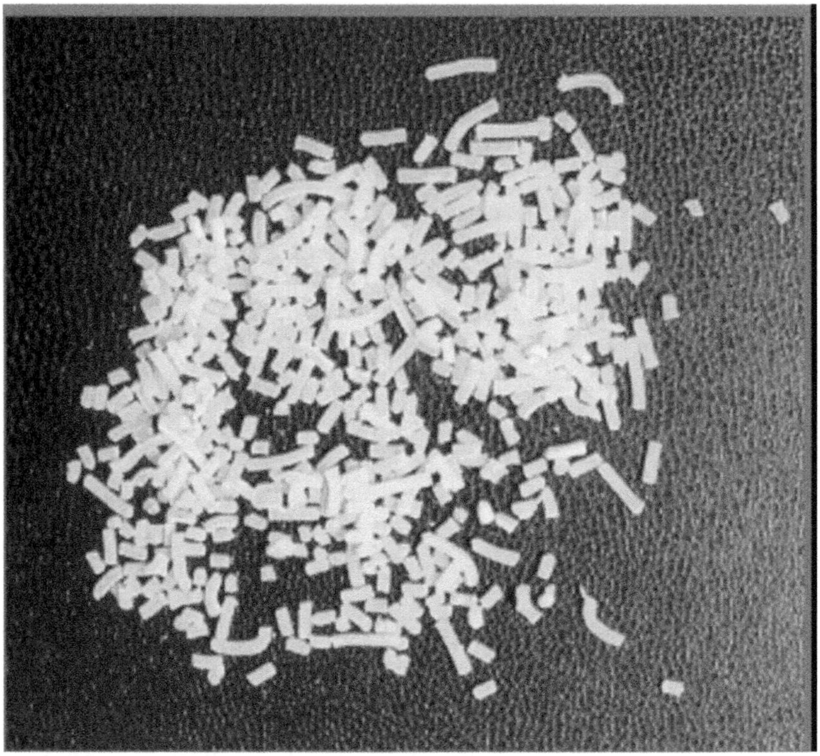

Figure 8. Altraflex grains.

complicated. One of the reasons is that the polymer and the asphaltenes 'compete' for the solvency power of the maltene phase and if insufficient maltenes are available, phase separation may occur. The other problem is related to the grain size of the Altraflex particles. Figure 8 shows the Altraflex particles as they were added to the C-Fix binder. The particle length varies between 0.5 to 10 mm while the diameter varies between 1 and 1.5 mm. During mixing it was sometimes observed that the Altraflex grains didn't fully dissolve in the binder. It is believed that using Altraflex grains of filler size will significantly improve dissolving of the Altraflex grains.

A very important result of the research was the pronounced effect of the filler. It was expected that the active filler Wigro 60K could also react with the maltene phase making the C-Fix even more brittle.

This however was not the case. Application of the Wigro 60K filler significantly improved the strength and toughness of the mastic, the higher filler/binder ratios gave the best results. Remarkably though was the effect of the inert filler Duras 15 especially when *not* used in combination with Altraflex. At all temperatures the toughness of the C-Fix/Duras mastic was 3 times higher than that of pure C-Fix and at all temperatures, the tensile strength was about 2.5 times higher. These findings clearly indicate that the effect of fines should be taken into account when evaluating the effect of binder modifications. One should also keep in mind that a modification of the C-Fix binder with 6% Altraflex in fact means a modification with only 3% SBS. In a mastic with a filler/binder ratio of 1, this means that there is 33.3 times more filler in the mastic than SBS. Depending on the type of filler this could imply that the effect of the SBS is masked by the effect of the filler. The results shown in table 6 give reasons to believe that this was indeed the case.

6 CONCLUSIONS

All in all it is concluded that application of C-Fix binders will result in asphalt mixtures with a high resistance to permanent deformation and excellent crack resistance characteristics at elevated temperatures when compared with mixtures made with pen grade bitumens. It is also concluded that modifying C-Fix binders with SBS type modifications will help in improving the crack resistance of asphalt mixtures made with C-Fix binders. Higher SBS contents than used in this study might be needed. Furthermore it is concluded that Altraflex should become available in smaller grain sizes in order to improve dissolving it in C-Fix.

REFERENCES

ASTM International. 2001. *Standard test method for determining the fracture properties of asphalt binder in Direct Tension.* Standard ASTM D6723-02. West Conshohocken.

Arce, F. 2007. *The potential of Altraflex to improve low temperature behaviour of C-Fix mixtures.* MSc thesis. Delft University of Technology. Delft.

Gerritsen, A.H., C.A.P.M. van Gurp, J.P.J. van der Heide, A.A.A. Molenaar & A.C. Pronk. 1987. *Prediction and prevention of surface cracking in asphaltic pavements.* Proc. 6th International Conference on Asphalt Pavements. Ann Arbor.

Groenendijk, J. 1998. *Accelerated testing and surface cracking of asphaltic concrete pavements.* PhD thesis. Delft University of Technology. Delft.

Khedoe, R.N. 2006. *Possible use of C-Fix in porous asphalt.* MSc thesis. Delft University of Technology. Delft.

http://www.c-fix.com/default.aspx

Efficient Transportation and Pavement Systems – Al-Qadi, Sayed, Alnuaimi & Masad (eds)
© 2009 Taylor & Francis Group, London, ISBN 978-0-415-48979-9

Determination of percent working binder in recycled asphalt pavement materials

I.L. Al-Qadi, S.H. Carpenter, G.L. Roberts & H. Ozer
University of Illinois at Urbana-Champaign, Urbana, IL, USA

ABSTRACT: The amount of working binder available in RAP materials is unknown. A research plan was developed to determine the amount of working binder using complex modulus measurements of mixes containing stockpile RAP and the aggregate and binder recovered from the original RAP. The mixtures made from the recovered RAP materials were used as a basis of comparison to determine the amount of working binder. The testing was not able to determine an amount of working binder. However, when investigating the mix design method, it was found that the RAP mixes were behaving as if 100% of the binder was working. It was also determined that double bumping the asphalt grade softened the mix. Future use of RAP in mixes should not require special considerations in mix design, but the new binder should be modified to offset an increased stiffness produced with the use of RAP.

1 INTRODUCTION

In recent years the desirable use of reclaimed asphalt pavement (RAP) in construction of new hot mix asphalt (HMA) pavements has become more widespread. The addition of RAP to HMA mixes has long been known to change the properties of the mix. The magnitude of this change depends on the amount of working RAP binder in the mix. The amount of working RAP binder is important since it has been found to be significantly stiffer than virgin binder due to aging (Kemp and Predoehl 1981). The processing and stockpiling of RAP may also increase the aging due to increased oxidation of the pavement during stockpiling (Smiljanic et al. 1993).

Because RAP binder has an increased stiffness, the determination whether all of the RAP binder combines with the new binder is important to know. If no RAP binder is working then no modifications to the mixture is needed because the mixture stiffness will be controlled entirely by the new binder being used. However, if all the stiff RAP binder is working then the amount or type of new binder being used in the mixture will require some modification to offset the increased stiffness produced by the combination of the two binders in the mixture. The amount of working binder observed in RAP has never been precisely determined, but several previous studies have noted approximate amounts of working RAP binder. It has previously been determined that the stiff binder effect is not noticeable at RAP percentages up to 20% (Kennedy et al. 1998). Investigations have also examined high RAP percentage mixes containing 40% RAP to determine the working binder in these mixes. It was found that the working binder was significantly different from both the 0% and 100% working binder situations (McDaniel et al. 2000). These findings would lead one to expect mixture differences related to the RAP binder in high RAP mixes.

2 OBJECTIVE

The objective of this research was to demonstrate the ability to characterize the amount of working RAP binder during the mixing process. The desired outcome of the research was to determine the amount of blending in a recycled mix in a manner that could be readily implemented into the mix design procedure.

3 TESTING METHODOLOGY

3.1 *Testing plan*

In order to investigate the amount of working binder in RAP, mixes containing 0, 20, and 40% fractionated stockpile RAP were prepared and tested. The complex modulus of the mixtures was tested for each of the RAP percentages. Specimens were also prepared using recovered aggregates and binders from the original RAP at the same RAP percentages to simulate specific working binder levels. The simulated working binder levels prepared with the recovered RAP materials were designed to simulate 0, 50, and 100% working binder. Two aggregate and RAP sources were used in mixing. The first aggregate and RAP source was obtained from the Illinois Department of Transportation (IDOT) District 1, and will be referred to "D1" in this paper. The second aggregate and RAP source was obtained from IDOT District 4 and will be referred to "D4" in this paper.

The complex modulus specimens were prepared with PG 64-22 binder. In order to investigate the effect of double bumping on the performance of the mix, samples containing 40% RAP were prepared with a PG 58-28 binder. The double bumping is expected to offset the effect of the stiff RAP binder and give similar performance to a HMA mix prepared with PG 64-22 binder.

Shear complex modulus, G*, of the binder used in mixing was also determined to illustrate what amount of the stiffness increase observed in the mix complex modulus was due to purely binder effects and what may be caused from another aspect of the mix. The dynamic Shear Rheometer (DSR) testing was done on virgin binder, 20%/80% and 40%/60% blends of recovered to virgin binder.

3.1.1 *Mix complex modulus testing plan*

The complex modulus testing followed a modified AASHTO TP 62-03 test suite. An additional frequency, 0.01 Hz, was added for extra data at the low loading frequencies. Master curves were prepared using the time temperature superposition method using 20°C as a reference temperature. The master curves were then compared to determine if the performance of the mix changed with the addition of RAP.

3.1.2 *Binder complex modulus testing plan*

In addition to the mix complex modulus, the binder shear complex modulus was also determined. The binder complex modulus was determined to illustrate any effect of binder on the increasing mix complex modulus results compared other stiffening effects observed from the addition of RAP materials. The binder tested was prepared to simulate both the virgin and short term aging conditions expected to be encountered in the HMA mixes prepared.

The following blend/test conditions were tested:

- Original Blend, and RTFO processing
  - tank PG 64-22 binder
  - blend of 80% virgin binder and 20% recovered RAP binder from D1 RAP
  - blend of 80% virgin binder and 20% recovered RAP binder from D4 RAP
  - blend of 60% virgin binder and 40% recovered RAP binder from D1 RAP
  - blend of 60% virgin binder and 40% recovered RAP binder from D4 RAP
- Original Blend only, no processing
  - recovered binder from D1 RAP
  - recovered binder from D4 RAP.

The binder samples were tested at the following temperatures and frequencies. The testing temperatures ranged from 16°C to 64°C at the SuperPave binder grade temperatures. The testing frequencies were 0.1 Hz, 0.2 Hz, 0.4 Hz, 0.8 Hz, 1.6 Hz, 3.2 Hz, 6.4 Hz, 13 Hz, and 27.3 Hz. After testing master curves were constructed at a reference temperature of 20°C.

3.2 *Mix design*

The mix designs were prepared using current IDOT practices. IDOT currently assumes that 100% of the RAP binder is working. Using this assumption the mix design was prepared using stockpile RAP materials to determine optimum asphalt content. Optimum asphalt content was defined as the asphalt content which produced 4.0% air voids at 50 gyrations. The mix designs were then checked with the specimens containing recovered aggregates substituted for RAP, but no recovered RAP binder.

The mix designs were constructed to hold the gradations as constant as possible. Gradation charts for the mixes containing the District 1 aggregate source can be found in Figure 1. The inlaid table in Figure 1 shows the percentage of each aggregate type in the mixture. The material designations are as follows: CM16–coarse aggregate, RAP–stockpile RAP, FM20–manufactured sand, FM02–natural sand, and Filler–mineral filler.

The mix design used in this research was later investigated in order to determine the effect of RAP on the mix design strategy that should be employed when using RAP. The SuperPave mix design strategy was employed to determine if any effects were manifested from the addition of RAP. If the optimum asphalt required additional binder then it was assumed that the RAP binder was not fully working. However, if no modification to the optimum asphalt content was required it can be assumed that the RAP binder is 100% working or at least acting as if the binder were 100% working.

The mix gradation for the District 4 aggregate source is given in Figure 2. As before, a table containing the percentages of each aggregate in the mix was inlaid into Figure 2. The material designations are slightly different that those above. The material designations for this aggregate source are as follows: CM 13–coarse aggregate, RAP–stockpile RAP, FM21–manufactured sand, FM02–natural sand, and Filler–mineral filler.

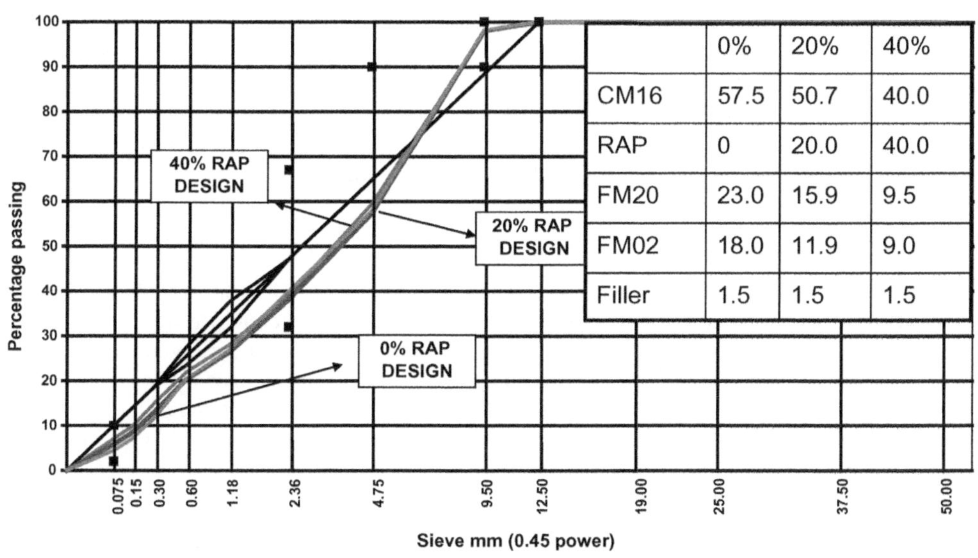

	0%	20%	40%
CM16	57.5	50.7	40.0
RAP	0	20.0	40.0
FM20	23.0	15.9	9.5
FM02	18.0	11.9	9.0
Filler	1.5	1.5	1.5

Figure 1. Mix gradation of District 1 mixtures.

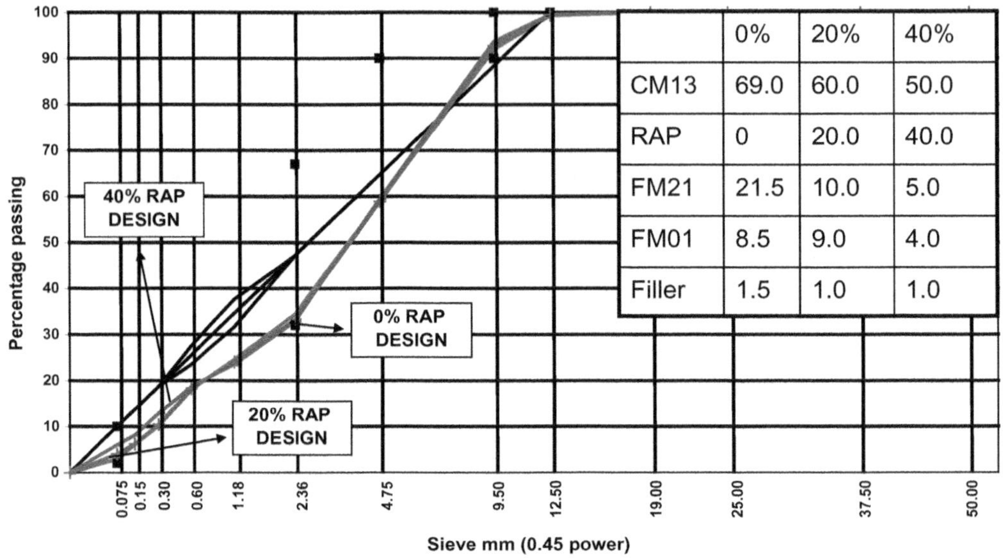

	0%	20%	40%
CM13	69.0	60.0	50.0
RAP	0	20.0	40.0
FM21	21.5	10.0	5.0
FM01	8.5	9.0	4.0
Filler	1.5	1.0	1.0

Figure 2. Mix gradation District 4 mixtures.

As can be seen from Figures 1 and 2, the mix gradations are very similar across the District 1 materials and District 4 materials. The similar gradation of the mixes minimizes any affect that may be caused by aggregate structure changes on the complex modulus results.

4 MATERIALS AND SPECIMEN PREPARATION

4.1 *Materials used*

Specimens were prepared using virgin aggregates and RAP. Two sources of RAP and virgin materials were sampled from two different sources within Illinois. The first aggregate source was sampled from IDOT District 1. The RAP materials obtained from District 1 were fractionated to decrease variability. The RAP material obtained from District 4 contained a large amount of agglomerated particles. These particles were heated and broken down and then fractionated to reduce mixture variability. The base binder used in the mixtures was a PG 64-22. A PG 58-28 was also used to investigate the effect of double bumping the binder grade.

4.2 *Extraction process*

The RAP binder and aggregates were recovered following the Rotovapor extraction method AASHTO TP2. The solvent used was an n-propyl bromide. At the conclusion of the extraction process as outlined in the standard, the filter was removed from the extraction apparatus and cut open to recover as many fines from the filter as possible. This preserved as much of the original gradation of the RAP materials remained after the extraction. The recovered aggregates were covered in water during cleaning of the extraction vessel. These aggregates were placed in the oven overnight at 80°C. The aggregates were then removed and covered with alcohol and water and placed back in the oven overnight at 80°C. The alcohol allowed for the solvent absorbed into the aggregates to come to the surface and evaporate off leaving clean aggregates to be used later in mixing. After the second oven drying cycle the recovered aggregates were fractionated and used in mixing.

4.3 *Specimen sets*

Four unique specimen sets were prepared in order to investigate the working binder in RAP. Set 1 contained the mixes prepared with stockpile RAP materials. Sets 2, 3, and 4 contain recovered binder and aggregates with no stockpile RAP being used. The recovered RAP aggregate and binder were varied to simulate exact working binder percentages.

The Set 2 specimens simulate the 0% working binder case. The binder used in this specimen set was only virgin binder. Any binder assumed to be contributed from RAP was replaced with virgin binder. This mixing strategy replicated the case where the RAP materials act as a black rock and contribute no binder to the mix.

The Set 3 specimens simulate the 50% working binder case. The specimens contained both virgin binder and recovered binder. In this case half the binder assumed to be contributed from the RAP was replaced with virgin binder, while the other half was recovered binder added back into the mix, and virgin binder completed the total asphalt content.

The Set 4 specimens simulate the 100% working binder case. The specimens contained both virgin and recovered binder. In this case all of the binder assumed to be contributed from the RAP was added as recovered binder with the remainder being virgin binder. This mixing strategy replicated the case in which all of the binder on the RAP particle accepts and mixes with the virgin binder being added.

5 EXPIREMENTAL RESULTS AND DISCUSSION

5.1 *Mix complex modulus*

The mixes were tested and complex modulus values were measured. The curves shown in the figures below are average curves of two specimens. In some cases more than two specimens were measured in order ensure repeatability of the test. The results are presented and discussed in this section. Figure 3 contains the complex modulus master curves for the District 1 mixes containing 20% RAP. The results of the complex modulus testing from the District 1 20% RAP mixes show

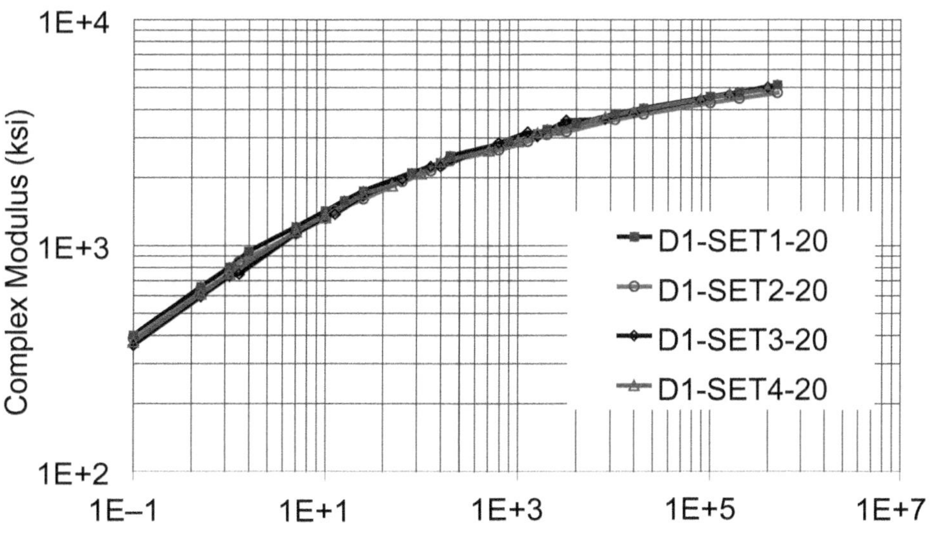

Figure 3. District 1 20% RAP mix complex modulus master curves.

very little differentiation. These results agree with previous findings that the addition of 20% RAP or less do not adversely affect the HMA performance to the extent that any binder modification is necessary.

The District 4 20% RAP complex modulus results show similar results to District 1. Figure 4 contains the District 4 20% RAP complex modulus master curves. From both 20% RAP results one can confirm the previous findings that the inclusion of 20% RAP does not significantly alter the mix performance to the extent that binder modification is required. Considering only the complex modulus results one can propose that no special considerations need to be paid to mixes containing 20% RAP. However, fracture and/or fatigue tests should be conducted on these mixes in order to determine if the addition of RAP adversely affects the thermal cracking resistance or fatigue life of the mix.

The District 1 40% RAP complex modulus results contain more interesting results than the 20% RAP mixtures. Figure 5 contains the District 1 40% RAP complex modulus master curves. The master curve of the Set 1 samples which contain stockpile RAP are significantly higher than the Set 4 samples which simulate the 100% working binder situation. The significantly greater stiffness can not be explained solely by the stiff binder effect. If the stiff binder were the only mechanism affecting stiffness and 100% blending was occurring, then the master curves for the Set 1 and Set 4 data would lie on top of each other. It is also important to note that the three sets containing recovered aggregates converge at high frequencies whereas the set containing stockpile RAP does not. This is also indicative of other mechanisms affecting the stiffness.

The District 4 40% RAP complex modulus results again show similar results to those of the District 1 specimens. Figure 6 contains the master curves for the District 4 40% RAP complex modulus results. The increased stiffness observed with the stockpile RAP in the District 1 specimens were also observed with the District 4 specimens. As seen before the specimens with stockpile RAP did not converge to the same point as those with recovered aggregates. However, in the District 4 Set 1 specimens do appear to come closer to converging at high frequencies. This indicates that the binders in the two materials (D1 and D4) are different from one another and will exhibit different results in a recycling operation.

The stiff binder effect is not enough to account for the complex modulus results observed. Some other mechanisms are responsible for the increased stiffness. One possible mechanism is the

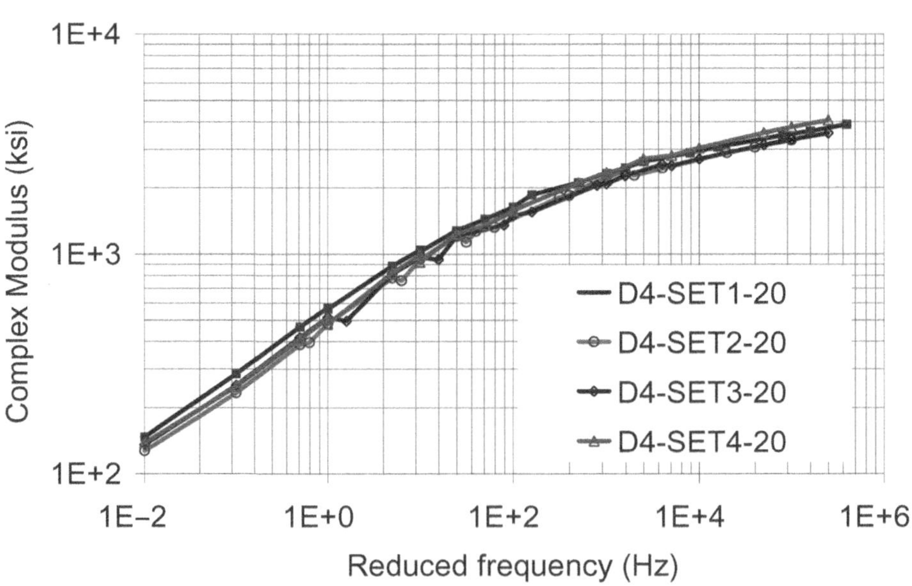

Figure 4. District 4 20% RAP mix complex modulus master curves.

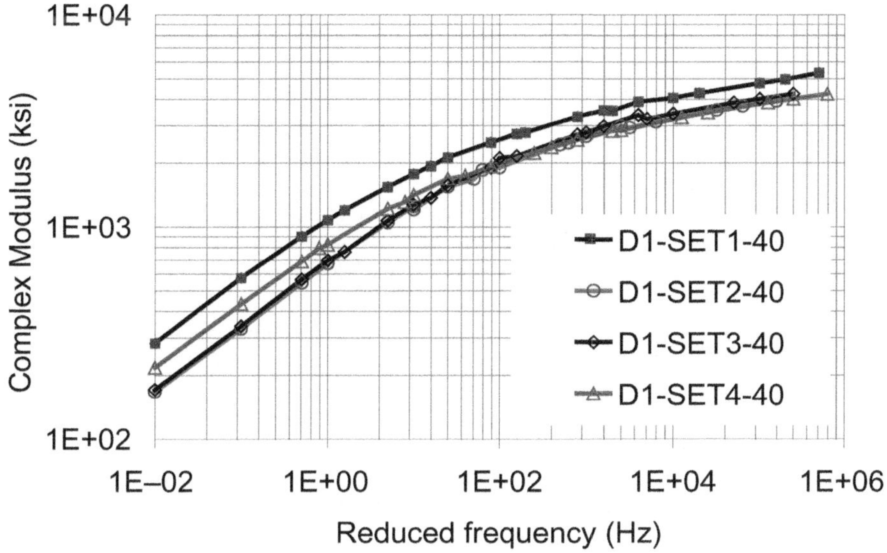

Figure 5. District 1 40% RAP mix complex modulus master curves.

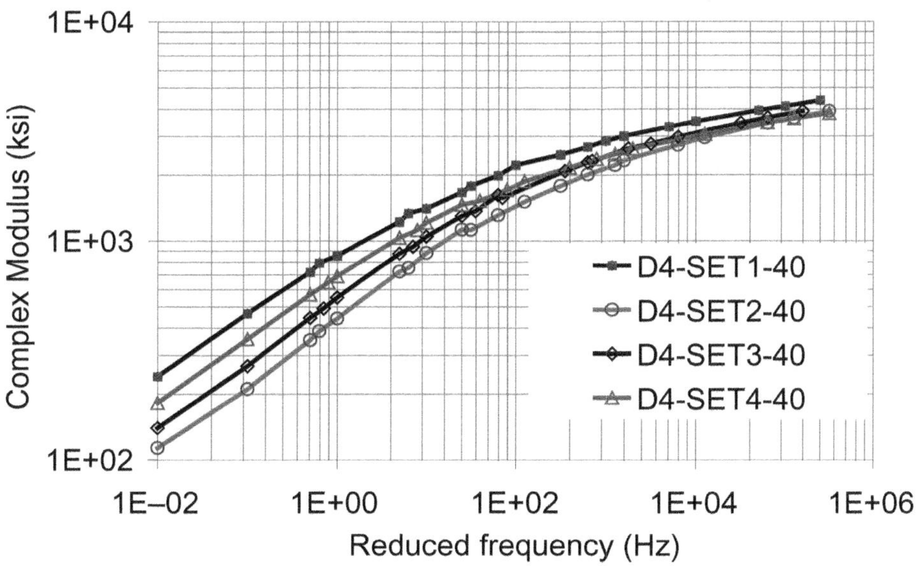

Figure 6. District 4 40% RAP mix complex modulus master curves.

long term diffusion of asphalt into the surface of an aggregate. After a virgin HMA is prepared a certain amount of or component of the binder is absorbed into the surface of the aggregates. This absorption of asphalt into the aggregate will continue over time. This diffusion affects the mixture by stiffening the interface of the aggregate and mastic. Binder has been found to be stiffer the closer it is to the aggregate particle (Huang et al. 2005). The diffusion effect could be manifested in high RAP complex modulus results since it is unlikely that during normal mixing all of the binder at the RAP aggregate interface can be used to combine with the added virgin binder. This stiffer interface affects the bulk properties of the mix.

In the specimens containing recovered aggregates, the diffusion effect was eliminated by washing all of the binder off of the RAP particles prior to mixing. The increased stiffness localized in the aggregate-mastic interface was erased when the binder and aggregates were separated from each other. Therefore the maximum stiffness observed in the Set 4 mixtures will not reach that of the Set 1 stockpile RAP mixtures since the stiff interface cannot be reproduced with the method used.

Another possible mechanism causing the high stiffness is the aggregate skeleton of the mixtures. If 100% of the binder is not working then the mix gradation assumptions are compromised. When the binder is working it is assumed that the fine particles contained in the mastic are also released and mix with the virgin materials. Thus if less than 100% of the binder is working, the gradations will not be constant between the stockpile RAP and recovered aggregate mixes. The recovered aggregate mixes will contain more effective fine particles which in turn modifies the mastic properties. These modified mastic properties will then result in a modified complex modulus value causing an overall stiffness increase.

It is likely that both the diffusion effect and modified aggregate structure have some affect on the increased stiffness. However, it was not possible to determine how much of the overall increased stiffness was contributed by the stiff binder, diffusion, and aggregate structure. Further investigation is needed to determine how much each of the three mechanisms affects the observed complex modulus values.

5.2 *PG 58-28 binder complex modulus*

The technique of double bumping the binder for RAP mixtures containing more than 20% RAP was also investigated during the project. The 40% RAP mixes used in this study are generally considered to be high RAP mixes. Therefore, one would anticipate seeing benefits of using a softer binder in offsetting the effect of the stiff RAP binder.

The results for both District 1 specimens containing PG 58-28 binder can be found in Figure 7. The master curves show that double bumping does not exactly return the 40% RAP mixture to the

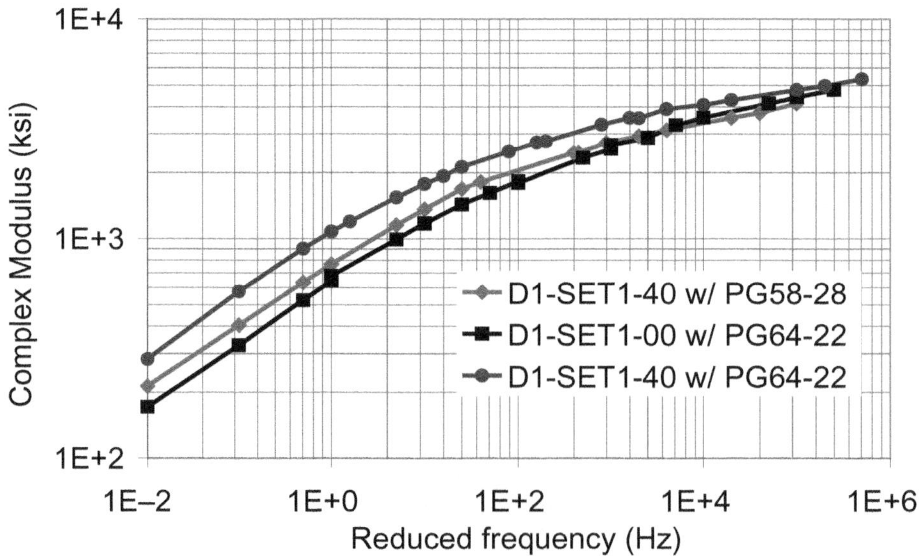

Figure 7. Effect of double bumping on district 1 high RAP mixes.

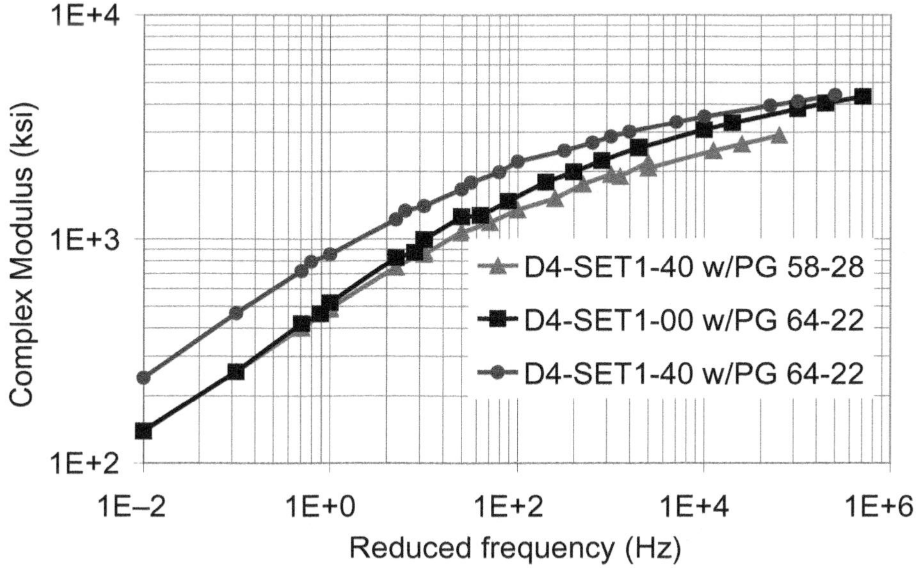

Figure 8. Effect of double bumping on district 4 high RAP mixes.

performance of a 0% mixture. However, it does show that at high frequencies the PG 58-28 mixture becomes softer than a 0% mix. From these results one can see the softening that occurs from bumping the low-temperature grade exceeds the soft modulus in the original binder, but the high temperature bump does not quite bring the stiffness back to what would be achieved with a PG 64-22 for this particular RAP sample.

The same procedure was repeated for the District 4 40% RAP mixtures. The complex modulus master curves can be found in Figure 8. The District 4 mix containing the PG 58-28 binder shows that double bumping creates a mix that is softer than the original 0% RAP mix with a PG 64-22 binder. The difference between the District 1 and 4 results are likely due to the RAP sources containing binders of different aging or original binder grade.

The effects of double bumping can clearly be seen to significantly soften the mix. However, depending on the RAP source double bumping may actually create a weaker mix than would result from not bumping the virgin binder grade. Therefore, before double bumping, the RAP source should be studied to determine the proper binder modifications for the specific RAP used. It may only be necessary to decrease one of the binder grades to offset the RAP effect. While, blindly double bumping may not be a good practice for all high RAP mixtures, modifications to the binder are generally beneficial and justified for all high RAP mixtures.

5.3 *Binder shear complex modulus*

As discussed in the previous sections, the binders in the D1 and D4 RAP materials are expected to be different. G* testing on the binders was conducted as presented earlier. Figure 9 shows the G* master curves for original PG 64-22 (subjected to RTFO) and RAP binder blends. In addition to measured 20 and 40% RAP binder blends a calculated 30% blend was also evaluated. The 30% blend curve was created from the existing measured values (by averaging the 20 and 40% blends). The 30% RAP binder blend is closer to the percentage of RAP binder in a 40% RAP mixture. As shown with the master curves, RAP binder blends are indeed much stiffer than PG 64-22 for both Districts' binders. It is interesting to note that the effect of increasing District 1 RAP binder in the

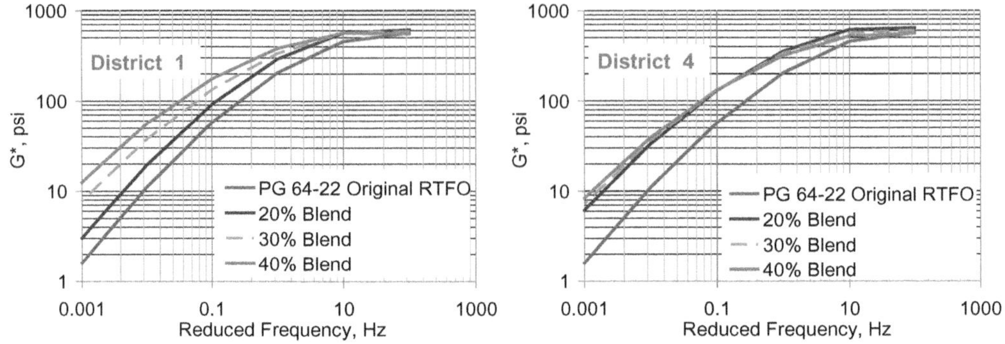

Figure 9. Binder blend complex modulus master curves.

Table 1. Optimum asphalt content of mixes.

Mix type	Optimum asphalt content (%)
D1 0% Stockpile RAP	5.9
D1 20% Stockpile RAP	5.7
D1 40% Stockpile RAP	5.65
D1 20% Recovered aggregates	5.7
D1 40% Recovered aggregates	5.55
D4 0% Stockpile RAP	5.9
D4 20% Stockpile RAP	6.0
D4 40% Stockpile RAP	6.0
D4 20% Recovered aggregates	5.9
D4 40% Recovered aggregates	5.9

blend is clearly manifested whereas 20 and 40% District 4 RAP binder blends do not show significant difference. This may raise the question of compatibility issues between District 4 recovered binder and PG 64-22, which may have prevented homogeneous blending. Finally, a comparison between District 1 and 4 binder blends can also be made. It is apparent from these two figures that District 1 recovered binder is stiffer than District 4 recovered binder.

5.4 *Mix design*

It has long been understood that the addition of RAP affects the complex modulus of HMA mixes. However, the topic of how RAP affects the mix design strategy has not been as well researched. In this project the effect of RAP on the mix design process was investigated. Mix design was investigated to determine if additional binder would be required in order to compact a specimen to 4.0% air voids with 50 gyrations. The need for additional binder indicates that the 100% working binder assumption is not valid. If equal or less binder was required then the 100% working binder assumption could not be invalidated. The optimum asphalt contents required for each mix can be found in Table 1.

The optimum asphalt contents shown in Table 1 either hold constant or decrease as increasing amounts of RAP are added to the HMA mix. This District 1 samples showed a decrease in optimum asphalt content. Thus one can conclude that the binder in the District 1 RAP source is acting as if it were nearly 100% working. It is important to note that one cannot directly make assumptions about the amount of working binder from considering only the mix design data, rather one can make a judgment of how the binder is behaving.

The District 4 optimum asphalt contents indicate a similar behavior to the District 1 mixes. The optimum asphalt content is approximately the same for all District 4 mixtures. The slight increase in the 20% and 40% stockpile RAP mixes are likely due to experimental variation. Again considering the District 4 results, one can assume as if the binder were acting as if 100% working binder was occurring. Therefore, when considering all the results, it can be assumed that since the binder is acting as if 100% of the binder is working the mix design strategy currently employed by IDOT is valid and does not require any modification. However, as was shown in the complex modulus discussion, while the amount of binder does not need to be modified, the binder grade used does need to be examined in order to ensure proper mix performance.

6 CONCLUSIONS

During this investigation into the amount of working binder in RAP mixtures several important findings were determined. The first and perhaps most important conclusion was that the addition of RAP materials does not require any additional binder to reach optimum binder content if a 100% working binder assumption is used. The optimum asphalt contents remained relatively constant with the inclusion of RAP. The RAP mixes were acting as if the 100% working binder assumption was valid. However, this assumption of 100% blending was not able to be tested directly.

The presence of RAP did not affect the stiffness of mixtures made with 20% RAP, verifying the current procedure of not modifying binders for these low percentage recycled mixtures. The use of high RAP percentages did affect the stiffness of mixes containing 40% RAP. The increased stiffness can be offset by bumping the high temperature grade to achieve a mixture with stiffness closer to that in an HMA made with an original PG binder. Double bumping the binder by also increasing the low-temperature grade may soften the mix, and produce a mix that is softer at low temperatures then an original HMA mixture. The general practice of double bumping should be examined carefully before being performed.

Binder differences can have an impact on asphalt blending, and the resulting modulus changes, and a general extension of findings to all RAP materials should not be made. Testing recovered and re-blended binders may provide insight into how effective RAP binders will be in the recycling process.

The complex modulus results indicate that mechanisms other than the sole effect of stiff binder are acting in RAP mixes. The stiffness of the mixes containing RAP was stiffer than those containing recovered aggregates and binder. Thus, other mechanisms such as the binder diffusion effect and aggregate structure of the mix may be present. Future high RAP usage should not require any special considerations during the mix design process. However, binder considerations should be made to give comparable performance to HMA mixes.

ACKNOWLEDGMENT

This publication is based on the results of ICT-R27-11, **Determination of Usable Residual Asphalt Binder in RAP.** ICT-R27-11 was conducted in cooperation with the Illinois Center for Transportation; the Illinois Department of Transportation, Division of Highways; and the U.S. Department of Transportation, Federal Highway Administration.

DISCLAIMER

The contents of this paper reflect the view of the authors, who are responsible for the facts and the accuracy of the data presented herein. The contents do not necessarily reflect the official views or policies of the Illinois Center for Transportation, the Illinois Department of Transportation, or the Federal Highway Administration. This report does not constitute a standard, specification, or regulation.

REFERENCES

Huang, B., G. Li, D., Vukosavljevic, X., Shu, X., and B.K. Egan, "Laboratory Investigation of Mixing Hot-Mix Asphalt with Reclaimed Asphalt Pavement," *Transportation Research Record: Journal of the Transportation Research Board*, No. 1929, Washington, D.C., 2005, pp. 37–45.

Kemp, G.R., and N.H. Predoehl, "A Comparison of Field and Laboratory Environments on Asphalt Durability," *Proceedings of the Association of Asphalt Paving Technologists*, Vol. 50, 1981, pp. 492–537.

Kennedy, T.W., W.O. Tam, and M. Solaimanian, "Optimizing Use of Reclaimed Asphalt Pavement with the SuperPave System," *Journal of the Association of Asphalt PavingTechnologists*, Vol. 67, 1998, pp. 311–333.

McDaniel, R.S., H. Soleymani, R.M. Anderson, P. Turner, and R. Peterson, *Recommended Use of Reclaimed Asphalt Pavement in the SuperPave Mixture Design Method*, NCHRP Final Report (9–12), TRB, Washington, D.C., 2000.

Smiljanic, M., J. Stefanovic, H.-J. Neumann, I. Rahimaian, and J. Jovanovic, "Aging of Asphalt on Paved Roads—Characterization of Asphalt Extracted from the Wearing Courses of the Belgrade-Nis Highway," *Journal of Erdol and Kohl*, Vol. 46, No. 6, Hamburg, Germany, 1993.

Efficient Transportation and Pavement Systems – Al-Qadi, Sayed, Alnuaimi & Masad (eds)
© 2009 Taylor & Francis Group, London, ISBN 978-0-415-48979-9

Influence of tack coats on the adhesion properties of asphalt pavements

C. Raab & M.N. Partl
Department of Road engineering and sealing components, Empa, Duebendorf, Switzerland

ABSTRACT: The paper summarizes the results of a laboratory study in which a common cationic tack coat was used between two layered slabs. The slabs were produced with a roller compactor and the adhesion of cores taken from the slabs was evaluated by means of direct shear testing according to Leutner. Furthermore, a common cationic, but not polymer modified, tack coat was compared to a polymer modified one. To evaluate the limits of the application of tack coats dirt and moisture was applied between the two layers before and after the application of a tack coat. For comparison the tack coat was applied on one side of the slab only, when the other stayed untreated. Besides, the influence of tack coats on fine milled surfaces and the influence of curing time on the adhesion were also investigated.

1 INTRODUCTION

In road construction tack coats have been used for many years to improve the adhesion between the different layers of a pavement. Although the benefit of using a tack coats is not debated (Uzan et al, 1978), (Hachiya et al, 1997), (Recasens et al, 2003), (Mohammad, 2002), (Canestrari et al, 2005) it is not clear how much tack coats effectively contribute to the adhesion properties. In most cases tack coats are applied empirically based on the practical experience of contractors and consultants. Parameters, such as surface characteristics, cleanness of the underlying base course surface and the age of the underlying layer play an important role, although they are often neglected in reality.

The choice of the right tack coat and its appropriate application (amount of tack coat, manner of application) is an important condition for the quality of the bond between pavement layers. Depending on tack coat type and manner of tack coat application nowadays ecological and working hygienic aspects are gaining more and more importance. Today good quality tack coats based on cationic emulsions are available for a variety of applications. Nevertheless, criteria for the use (necessity of application, type of product to use, amount of application) which would be stated in a standard or in an official testing specification are missing. A fact that is opposed to a correct handling of the products.

In order to provide for these deficiencies a laboratory study (Raab & Partl, 2007) was conducted in which a common cationic tack coat was used between two layered slabs. The slabs were produced with a roller compactor and the adhesion of cores taken from the slabs was evaluated by means of direct shear testing according to Leutner (Leutner, 1979). Different parameters such as surface characteristics and the influence of curing time were investigated and their influence on the adhesion between the layers was determined. To evaluate the limits of the application of tack coats dirt and moisture was applied between the two layers before and after the application of a tack coat.

2 SLAB CONSTRUCTION

Although the investigation was a laboratory study, the slab construction should be as close to reality as possible. After trials with other laboratory compactors the MLS roller compactor constructed at the University of Stellenbosch, South Africa was used.

The compactor (Figure 1) consists of a steel roller with a width of about 91 cm and a diameter of about 35 cm and a metal frame with a rail on which the steel roller can be displaced horizontally.

In the vertical direction a winder enables the steel roller to be moved; furthermore an automatic vibration system can be applied. During compaction the steel roller is spayed with water and prior to compaction it is heated with a heater reflector. The steel roller is pushed manually back and forth in the longitudinal direction of the slab. A piece of concrete within the metal frame builds the base for compaction.

For the construction of the asphalt slabs a wooden frame with the dimension of 0.95 on 1.08 m was applied and the hot mixture was filled into it and compacted.

The slabs were constructed in two layers. The base layer which consisted of asphalt concrete 22 (AC T 22) and had a thickness of 6 cm, while the surface layer consisting of asphalt concrete 11 (AC11) was 4 cm thick. The mixtures followed the European standard 13108-1 (EN 13108-1 Bituminous mixtures, 2008). Both mixtures consisted of an unmodified penetration grade binder 50/70. The slabs had a dimension of 0.99 m × 0.86 m. After the construction of the base layer 200 g/m^2 of tack coat were sprayed on the surface on one side of the slab using a spray gun and the surface layer was applied thereafter. The tack coat was an unmodified bituminous cationic emulsion C 50 with a nominal bitumen content of 50% according to the European standard, EN 13808 (European Standard EN 13808, 2005). For the investigation of the influence of polymer modified tack coats in one case the tack coat C 60 P with a nominal bitumen content of 60% and a polymer modification was used.

Two days after construction 12 cores were taken on either side of the slab (with and without tack coat). Figure 2 shows a schematic of coring for one slab:

Although the construction with the MLS roller compactor proved to be more close to the compaction on a real construction site, there were still some difficulties regarding the air void content of the surface layer. While the air void content of the base layer with 5 to 7 Vol-% was quite evenly distributed over the slab, considerable differences from 2 to 10 Vol-% could be found for the air

Figure 1. Laboratory steel roller compactor.

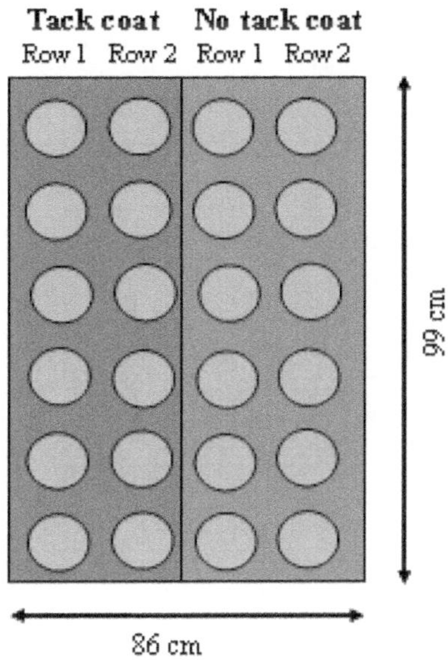

Figure 2. Slab with coring pattern.

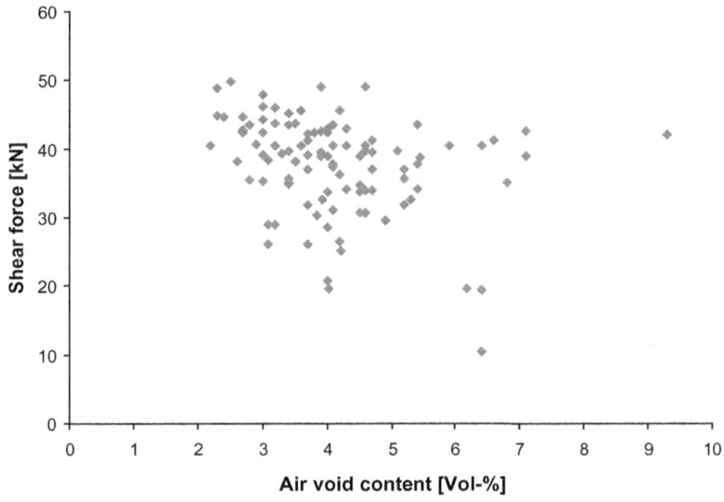

Figure 3. Air void content of the surface layer AC 11and shear force.

void content of the surface layer. That there is no correlation between the shear force and the air void content can be illustrated by Figure 3, although a slight tendency of low air void contents towards high shear forces and vice versa cannot be neglected. In this Figure shear forces and air void content of cores from different slabs (with and without tack coat, but only for untreated slabs, e.g. slabs without water, dirt or milling) are depicted.

To minimize the influence of the air void content on the bond between the layers, slabs with an air void content of more than 8 Vol-% were rejected and replaced. The majority of tested cores had an air void content of the surface layer between 3 and 5 Vol-%.

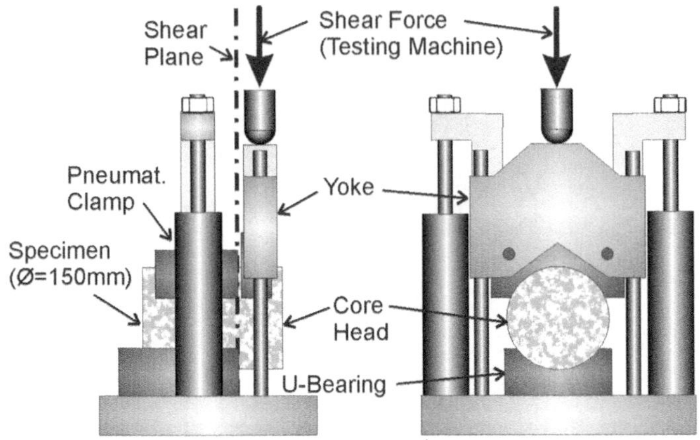

Figure 4. LPDS test device.

3 ADHESION TESTING

The Layer-Parallel Direct Shear (LPDS) test device (Figure 4) is an EMPA modified version of equipment developed in Germany by Leutner (Leutner, 1979). The modified LPDS test device fits into an ordinary servo-hydraulic Mar-shall testing machine and allows testing of cores with a diameter of about 150 mm; (Raab & Partl 1999), (Partl & Raab, 1999). One part of the core (up to the shear plane to be tested) is laid on a circular u-bearing and held with a well defined pressure by a semicircular pneumatic clamp. The other part, the core head, in Figure 4, remains unsuspended. Shear load is induced to the core head by a semicircular shear yoke with a deformation rate of 50 mm/min, thus producing fracture within the pre-defined shear plane of 2 mm width.

The cores were conditioned in a climate chamber for 8 hours and all tests were conducted at a temperature of 20°C. Since 2000 the LPDS is incorporated into the Swiss Standard (Swiss Standard, 2000).

4 TEST PROGRAM

The test program is depicted in Table 1.

Slab 1, Original state: In order to have a slab, which could be used for comparative purposes, the first slab was untreated. On one half of the slab the cationic tack coat C 50 was applied at room temperature (ca. 21 to 23°C), while the other had no tack coat.

Slab 2, Polymer modified tack coat: The second slab had also no special treatment, but this time one side of the slab was sprayed with a polymer modified tack coat C 60 P at 60°C according to the product specification.

Slab 3 and slab 4, Surface characteristic: Slabs 3 and 4 were constructed to determine the influence of the surface characteristic. Both slabs were fine milled. This was done using a small milling machine. After the milling, which removed up to 5 mm of the existing surface, a tack coat was applied on slab 4, while slab 3 had no further treatment.

Slab 5, Cleanness: The purpose of the treatment was the simulation of a dirty road surface *before* the application of tack coat. Fine filler, representing dirt on the road during construction, was put on the surface of the lower layer, before the application of tack coat on one half of slab 5. This procedure was done with the help of a sieve, the application rate was 360 g/m^2.

Table 1. Test program.

Slab No.	Parameters	Tack coat	Treatment of base layer surface and other remarks
1	Original state	C 50	No treatment
2	Polymer modified tack coat	C 60 P	No treatment
3	Surface characteristic	None	Fine milling of base layer surface
4	Surface characteristic	C 50	Fine milling of base layer surface
5	Cleanness	C 50	Dirt (filler) was applied *before* tack coat application
6	Cleanness	C 50	Dirt (filler) was applied *after* tack coat application
7	Cleanness and moisture	C 50	Wet dirt (filler) was applied *before* tack coat application
8	Cleanness and moisture	C 50	Wet dirt (filler) was applied *after* tack coat application
9	Curing time and age	C 50	Slab was stored outside for 2 months (summer conditions) before testing
10	Curing time and age (comparison)	C 50	Slab was cored and directly tested after construction

Table 2. Test results.

	Shear force with tack coat				Shear force without tack coat			
	kN				kN			
Slab No.	min.	max.	mean	stdev.	min.	max.	mean	stdev.
1	31.0	49.0	40.3	4.7	20.9	46.3	34.0	8.3
2	35.0	45.3	40.0	3.8	39.3	49.9	44.0	3.3
3	–	–	–	–	1.9	36.3	26.0	4.1
4	3.0	24.7	12.6	6.9	–	–	–	–
5	10.7	32.7	24.6	7.8	0.7	19.4	8.4	5.2
6	3.4	20.4	10.2	5.2	0 (broken)	8.1	2.8	3.1
7	1.9	11.1	7.5	3.9	0 (broken)	11.1	2.9	3.9
8	0 (broken)	6.9	5.1	2.7	0 (broken)	5.7	1.0	1.9
9	37.9	49.1	42.1	3.8	26.2	43.5	34.9	5.2
10	23.9	40.3	33.5	4.3	27.7	43.9	35.4	5.4

Slab 6, Cleanness: The purpose of the treatment was the simulation of a dirty road surface *after* the application of tack coat. For slab 6 the same procedure as for slab 5 was used, the only difference was, that one half of the slab had been sprayed with tack coat before the filler (dirt) was applied.

Slab 7, Cleanness and moisture: The purpose of the treatment was the simulation of a dirty and wet road surface *before* the application of tack coat. Fine filler was mixed with water with a relationship of 2.5:1.

5 RESULTS OF ADHESION TESTING

Table 2 depicts the test results for all slabs, given the minimum and the maximum shear force, the mean value of the shear force and the standard deviation for each slab. From all slabs 12 cores with and 12 cores without tack coat were tested. In case of slabs 3 and 4 the investigation was based on 24 cores each.

5.1 *Influence of unmodified tack coat*

The comparison of slab 1 and slab 10, which were prepared and tested the same way, showed that the application of tack coat in the laboratory can have a positive result on the bonding of the layers; although in the lab a tack coat seems not to be absolutely necessary. For the first slab a mean value of 34 kN without tack coat and a mean value of 40 kN with tack coat was determined, while slab 10 received a mean value of 35 kN with no tack coat and 33 kN when a tack coat was used.

5.2 *Influence of polymer modified tack coat*

The application of a polymer modified tack coat as opposed to a non modified one did not seem to have a positive influence on the adhesion between the two layers: In case of the polymer modified tack coat a mean value of 40 kN was measured, while the mean value for the cores without tack coat came up to 44 kN.

5.3 *Influence of surface characteristics*

Figure 5 depicts the surface of the base layer after fine milling: When comparing the fine milled slabs with and without cationic tack coat, in case of tack coat, the mean value was dramatically lower (12.6 kN) than for no tack coat, where the mean value achieved 26 kN.

5.4 *Influence of cleanness before tack coat application*

In order to simulate the cleanness of the base layer surface, before the application of a tack coat about 360 g/m^2 of fine corundum was put on this surface as shown in Figure 6. Then the cationic unmodified tack coat was sprayed and the surface layer was applied.

Here, the results clearly revealed the potential of a tack coat to ensure the bonding properties even in a critical situation: When the mean value for the non treated cores came up to only 8.4 kN, when using a tack coat the mean value reached nearly 25 kN.

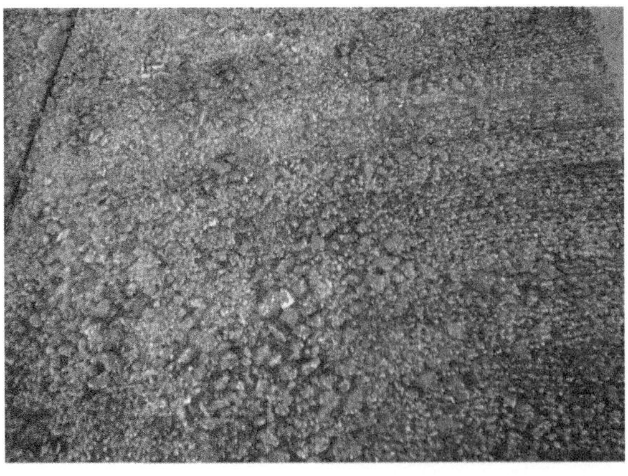

Figure 5. Surface of the base layer after fine milling.

Figure 6. Surface of the base layer after the application of dirt (fine corundum).

5.5 *Influence of cleanness after tack coat application*

The importance of cleanness after the application of a tack coat was demonstrated by the results of the LPDS testing. Even in this situation the use of tack coats could prevent the worst: When no tack coat was applied most cores were already destroyed by coring and the mean value only reached 2.8 kN, whereas in case of tack coat all cores were still intact and the mean value was measured with 10 kN.

5.6 *Influence of cleanness and moisture before tack coat application*

In order to simulate an unclean and wet base layer surface, fine corundum was mixed with water and brushed on the base layer before a tack coat was sprayed (see Figure 7).

This experiment showed that the presence of water on an unclean base layer surface was even more critical, than when no water was present. In this case the two layers were very weakly bonded when no tack coat was applied. LPDS tests determined a mean value of 2.9 kN, while many cores were already broken before testing. When using a tack coat the mean value of LPDS testing achieved 7.5 kN.

5.7 *Influence of cleanness and moisture after tack coat application*

The combination of dirt and water when the tack coat was already applied proved to be the most critical case. For the situation of an untreated surface similar results as shown above (mean value of 2 kN, with many of the cores already destroyed by coring) were achieved. But even when a tack coat was applied on the dirt and wet surface the adhesion properties were extremely weak. The mean value only came up to 5 kN, while many of the cores already lost their bond during coring.

5.8 *Influence of curing time and age*

In order to simulate the influence of curing time and age, slab 9 was constructed, but left for 2 months outside before the cores were taken and the testing effectuated. Since the whole experiment took place during summer time (June to August), the temperatures were between 11°C and 35°C and occasionally heavy rain fall occurred.

When looking at the comparison of the mean values with and without the application of a tack coat it seemed clear that curing and aging had a positive effect on the bonding properties. When the mean value without tack coat was determined to be 35 kN, the mean value in the case of cationic tack coat received a value of 42 kN.

Figure 7. Surface of the base layer after the application of wet dirt (wet fine corundum).

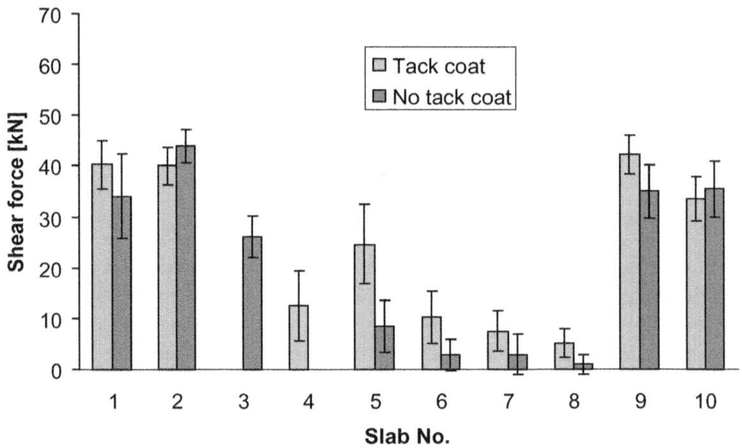

Figure 8. Mean force values from LPDS testing and standard deviation for all slabs.

5.9 *Comparison of results*

Figure 8 shows a summary and a comparison for all tested slabs. The Figure depicts the mean force values for the LPDS testing of 12 cores and also depicts the standard deviation. In case of slabs 3 and 4 (influence of surface characteristics) the mean value was calculated from 24 cores.

6 SUMMARY AND DISCUSSION

6.1 *General remarks*

The research summarized in this paper aimed at the investigation of tack coats and their influence on the bond between asphalt pavements. Although it was a laboratory study the construction of the slabs (asphalt laying and compacting) tried to simulate the construction on the road as good as possible.

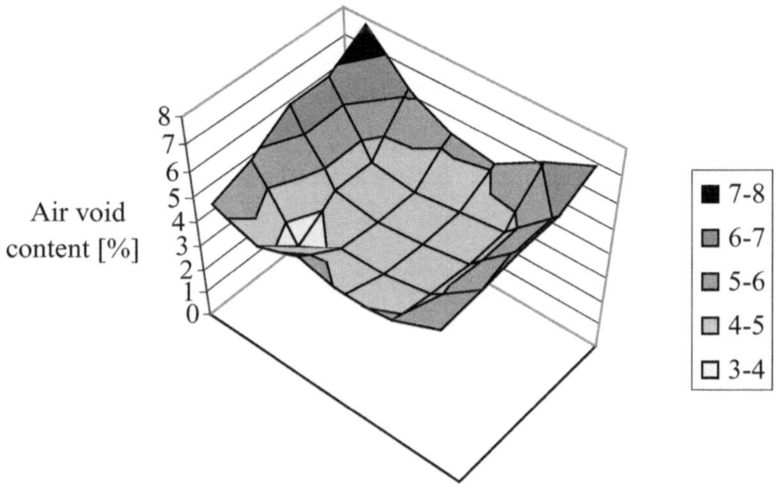

Figure 9. Air void distribution of the surface layer over the surface of the specimen.

That the situations in the lab and on the road all the same show severe differences cannot be neglected. Not only did the laboratory steel roller compactor not lead to a uniform compaction and revealed considerable differences in the compaction over the slab surface, as shown by the contribution air void content of the surface layer over the slab surface in Figure 9.

6.2 *Influence of polymer modified tack coat*

As shown by the results of this research using a polymer modified tack coat does not seem to have an advantage compared to the use of an unmodified one. Although one has to keep in mind that the asphalt mixture which was used in these experiments consisted of unmodified bitumen. The reason for the stated fact could also be found in a different bonding mechanism of polymer modified tack coats, which according to their chemical structure need a longer curing time, before the bond is totally developed.

6.3 *Influence of surface characteristics*

That fine milling does not necessarily have a positive effect on the adhesion properties determined by LPDS testing had already been shown in other publications (Raab C. and Partl M.N., 2006). Furthermore, the application of tack coat on fine milled asphalt concrete surfaces seems even to have a negative influence on the determined adhesion values, resulting in considerable lower shear forces when compared to slabs where no tack coat was applied.

Here again, the mechanism of bonding could be different. Since fine milling removes the binder on the asphalt concrete surface, the time needed for bonding is considered to be longer than on a non milled surface.

6.4 *Influence of cleanness*

In case of a dirty base layer surface the advantage of the application of a tack coat is clearly revealed. Especially, when the tack coat is applied on a dirty surface its potential to restore the bond between the layers and to lead to acceptable adhesion force values becomes clear. When applying the tack coat before the pollution takes place, its potential is limited, but still the advantaged when compared to the situation without tack coat is obvious.

6.5 *Influence of cleanness and water*

Also in case of the presence of wet dirt the use of tack coats proves to be favorable regarding the adhesion properties of asphalt pavements. Although, the potential of tack coats is smaller than in case of dry dirt, still a rest bond can be achieved, when without tack coat in extreme cases wet dirt might cause total separation between the layers. As shown above, the potential of tack coats to ensure the bond between layers is greater if tack coat was applied before the pavement was polluted.

6.6 *Influence of curing time and age*

The curing time and the age have as shown by the high adhesion values a positive effect on the bonding properties of asphalt pavements when tack coats were applied. When the curing and aging is done according to the procedure described earlier (in moderate to hot climate) the adhesion values for the coated slabs increase by about 10 kN.

7 CONCLUSIONS

It could be shown that especially in the case of dirt and moisture—common phenomena on construction sites—tack coats have a great potential to secure and improve adhesion properties. When—in case of "no tack coat"—the cores already broke during coring, the cores with tack coat achieved shear forces up to 10 kN. Although, there is a difference if the tack coat is applied before or after the dirt/moisture treatment, the use of tack coats is still beneficial.

Curing time and age are also important factors for the improvement of bonding on coated surfaces.

Opposed to the benefit of tack coats on dirty and wet surfaces as well as in case of longer curing times, the advantage of tack coats on fine milled surfaces is doubtable. Also, the use of polymer modified tack coats seems not to be beneficial when the pavement is constructed using ordinary e.g. non polymer modified binders. In both latter cases the different bonding mechanism might be responsible for the stated behaviour and lead to a different testing regime in these cases. So, adhesion testing not directly after construction but after some time might lead to different (higher) adhesion values.

Again, it is important to mention that a laboratory study might give first hints, but that the situation in the lab is not totally comparable to the situation on a construction site. Even if, parameter such as cleanness or moisture influence might be simulated in laboratory tests, the compaction itself and its differences from in situ compaction seem to be important factors.

Therefore, it is recommended that some of the parameters and their influences should be looked at in 1:1 field tests. Especially since in the framework of the research project many of the investigated aspects (fine milling, aging, polymer modified tack coats) were only touch on very briefly and therefore require a broader inspection.

REFERENCES

Canestrari, F., Ferrotti, G., Partl, M.N. & Santagata, F., (2005). Advanced Testing and Characterization of Interlayer Shear Resistance. CD Proceedings of the 84th TRB Annual Meeting, Washington DC, USA.

European Standard EN 13808 (2005). Bitumen and bituminous binders—Framework for specifying bituminous cationic emulsions European Committee for Standardisation, Brussels or Schweizer Norm SN 670205 NA, (2007): Bitumen und bitumenhaltige Bindemittel Rahmenwerk für die Spezifizierung kationischer bitumenhaltiger Emulsionen, Schweizerischer Verband der Straßen- und Verkehrsfachleute VSS, Zürich, (in German and French).

European Standard EN 13108-1 (2008). Bituminous mixtures, European Committee for Standardisation, Brussels or Schweizer Norm EN SN 640431-1b NA Asphaltmischgut. Schweizerischer Verband der Straßen- und Verkehrsfachleute VSS, Zürich, (in German and French).

Hachiya, Y. & Sato, K. (1997). Effect of tack coat on bonding characteristics at interface between asphalt concrete layers. 8th International Conference on Asphalt Pavements, Seattle Washington, USA.
Leutner, R. (1979). Untersuchungen des Schichtenverbunds beim bituminösen Oberbau. Bitumen. Journal No 3, 84–91, (in German).
Mohammad, L.N., Raqib, M.A., & Huang, B. (2002). Influence of Asphalt Tack Coat Materials on the Interface Shear Strength. Louisiana Transportation Research Center, Baton Rouge, USA.
Partl, M.N. & Raab, C. (1999). Shear Adhesion between Top Layers of Fresh Asphalt Pavements in Switzerland. Proceedings of 7th CAPSA Conference on Asphalt Pavements for Southern Africa. Victory Falls, Zimbabwe, 29 August–2 September, Victory Falls, Zimbabwe, 5.130–5.137.
Raab, C. & Partl, M.N. (1999). Methoden zur Beurteilung des Schichtenverbunds von Asphaltbelägen. ASTRA—Project VSS 12/94, Report No. 442, (in German).
Raab, C. & Partl, M.N. (2006). Adhesion Testing of Rehabilitated Concrete Pavements. 10th International Conference on Asphalt Pavements, Paper Nr. 79, Quebec, Canada.
Raab, C. & Partl, M.N. (2007). Prüfung von Haftklebern (Testing of Tack Coats), ASTRA—Project VSS 1999/277 Report No 1196 (in German).
Recasens, M., Jiménez, P., Gonzalez, B. & Manuel J. (2003). Evaluation of the effect of tack coat s. LCB shear test, 6th RILEM Symposium PTEBM'03, Zurich, Switzerland.
Swiss Standard SN 671961 (2000). Bituminöses Mischgut, Bestimmung des Schichtenverbunds (nach Leutner), Schweizerischer Verband der Straßen- und Verkehrsfachleute VSS, Zürich, (in German and French).
Uzan, J., Livneh, M. & Eshed, Y. (1978). Investigation of adhesion properties between asphaltic-concrete layers, Asphalt Paving Technology, Vol. 47.

HMA characteristics and evaluation

Efficient Transportation and Pavement Systems – Al-Qadi, Sayed, Alnuaimi & Masad (eds)
© 2009 Taylor & Francis Group, London, ISBN 978-0-415-48979-9

Applying fundamental materials engineering to solve pavement infrastructure problems

D.N. Little
Texas A&M University, USA

ABSTRACT: Fundamental material and engineering properties of asphalt mixture components can be used to evaluate the performance of asphalt mixtures and to form the baseline by which to understand the fundamental mechanisms of damage. This paper illustrates how two material properties, surface free energy (*SFE*) and molecular properties of the asphalt binder (e.g., shape and size), are used to understand and assist in the performance modeling of the fatigue damage process including the balance between crack propagation during loading and crack healing during rest periods. The paper also explains how *SFE* can be used to select asphalt mixture components (aggregate and asphalt binder) that are the most compatible with each other among the available candidates so that the resulting mixture will have the highest probability to resist moisture damage.

1 IMPACT OF MATERIAL PROPERTIES ON PERFORMANCE OF ASPHALT MIXTURES

1.1 *Role of surface free energy in the fracture process*

The thermodynamic criterion for unstable crack growth states that the energy stored in a specimen should be sufficient to supply the energy needed to create the new surfaces that define the crack. For the case of elastic materials, Griffith (1920) showed that this criterion is satisfied when the critical strain energy release rate (G_c) is equal to the ideal bond energy or work of fracture (2γ) as shown in Equation 1:

$$G_c = 2\gamma \tag{1}$$

Where γ is the surface free energy of the material.

For the case of adhesive fracture for purely elastic materials, the criterion for unstable crack becomes:

$$G_c = \gamma_1 + \gamma_2 - \gamma_{12} \tag{2}$$

Where γ_1 is the SFE for material 1, γ_2 is the SFE for material 2, and γ_{12} is the interfacial SFE of the interface between materials 1 and 2.

It is well recognized that inelastic materials experience irreversible energy dissipation during the fracture process, which makes the practical fracture energy much higher than the ideal work of fracture calculated using the right hand sides of Equations 1 and 2. The practical work of fracture (W_P) can be expressed as follows:

$$W_P = W_A + W_{PL} + W_{VE} \tag{3}$$

Where W_A is the ideal work of fracture calculated from the right hand sides of Equations 1 and 2, W_{PL} is the dissipated plastic energy and W_{VE} is the dissipated viscoelastic energy.

Orowan (1945) modified Griffith's criterion for cohesive failure in order to account for plastic energy dissipation at the crack tip as follows:

$$G_c = 2\gamma + W_{PL} \quad (4)$$

Many studies have evaluated the differences between the practical and the ideal work of fracture under this principle. The main finding is that W_P can be much higher than the W_A depending on material properties, geometry of the specimen, thickness of the adhesive, temperature and loading rate for time-dependent materials. However, many eminent researchers have concluded that the plastic (W_{PL}) and viscoelastic (W_{VE}) components of energy dissipation are related to the ideal work of fracture (W_A) and that any change in W_A is reflected in a significant change in the W_P (Okamatsu et al. 2000). In other words, the practical work of fracture is directly related to and controlled by the ideal work of fracture. Some of the key findings relating W_P and W_A found in the literature are summarized in Table 1.

Table 1. Examples of a few empirical relationships between ideal and practical work of fracture based on the work of recognized experts.

Reference	Model	Remarks
Xu et al. (1992)	$W_P = W_A[1 + f(a_T, \dot{a}, c)]$ Where W_P practical work of fracture, W_A is ideal work of fracture, a_T is time-temperature shift factor, *a* is crack growth velocity, and c is the crack length	Practical work of fracture at the crack interface between two viscoelastic materials is related to the ideal fracture energy
Seshadri et al. (2007)	$W_P = W_A[1 + f(v)]$ Where *f(v)* a problem dependent variable with size and geometry	Seshardri et al. found that as practical and ideal work become equal when crack velocities are very low, practical and ideal work become equal. At very high crack velocities—in the glassy region—practical work of fracture is equal to ideal work of fracture multiplied by the ratio of the initial and long-term modulus of the viscoelastic material
Jokl et al. (1980)	$G_c = 2\gamma + W_{PL}(\gamma)$ Where G_c critical strain energy release rate and γ is surface free energy	Jokl et al. found that for crystalline materials, such as metals, plastic work (W_{PL}) is a multiplicative function of the ideal work of fracture
Kaelble (1974)	$W_{PL} = k\gamma^{1/2}$ Where *k* is a constant	Kaelble found that in the case where plastic energy is much higher than ideal work of fracture, a relationship exists between the two quantities
Sharon et al. (1996) and Miller et al. (1999)	$W_P = 91.3\, e^{0.033\, WA}$ Where W_P is practical work and W_A is ideal work	Sharon et al., found that, although under ideal conditions that prevent viscous or plastic deformation $W_P >> W_A$, an exponential relationship exists between ideal and practical work

1.2 *Calculation of bond energy from surface free energy (SFE)*

A widely used theory that explains the origin of surface free energy in solids and liquids is the acid-base theory (van Oss, 1994; van Oss et al., 1988). According to this theory, the total surface free energy of any material is divided into three components based on the type of molecular forces on the surface. These components are the Lifshitz-van der Waals (LW) or non polar component, the Lewis acid component, and the Lewis base component. The total surface free energy of a material is obtained by combining these three components as follows:

$$\gamma = \gamma^{LW} + \gamma^{+-} = \gamma^{LW} + 2\sqrt{\gamma^{+}\gamma^{-}} \tag{5}$$

where, γ is the total surface free energy of the material, γ^{LW} is the LW component, γ^{+-} is the acid-base component, γ^{+} is the Lewis acid component, and γ^{-} is the Lewis base component. If the surface free energy components of the asphalt binders and aggregates are known, then the work of adhesion between these two materials is determined as follows:

$$W_{AS} = 2\sqrt{\gamma_A^{LW}\gamma_S^{LW}} + 2\sqrt{\gamma_A^{+}\gamma_S^{-}} + 2\sqrt{\gamma_A^{-}\gamma_S^{+}} \tag{6}$$

where, the subscripts 'A' and 'S' represent the asphalt binder and the aggregate (or stone), respectively. Similarly, when water displaces asphalt binder from the aggregate surface, the work of debonding is determined as follows:

$$W_{ASW}^{wet} = \gamma_{AW} + \gamma_{SW} - \gamma_{AS} \tag{7}$$

where, the subscripts 'AW', 'SW', and 'AS' refer to the interfacial energy between asphalt binder and water, aggregate and water, and asphalt binder and aggregate, respectively.

For a three phase system in which asphalt binder is displaced from the surface of the aggregate by water, the value of W_{ASW}^{wet} is negative. This indicates that debonding is a thermodynamically favorable phenomenon. Also, the magnitude of this term is a measure of the thermodynamic potential that drives water to displace asphalt binder from its interface with the aggregate surface.

In summary, using equations (6) and (7) it is possible to compute the work of adhesion between the asphalt binder and the aggregate and the work of debonding, or for water to displace asphalt binder from the surface of the aggregate, provided that the surface free energy components of these three materials are known. The aforementioned thermodynamic quantities are used to construct energy parameters that quantify the moisture sensitivity of asphalt mixes. These energy parameters are explained in subsequent sections.

1.3 *Microcrack healing process*

Crack healing occurs immediately after removal of the external load. Figure 1 illustrates the crack tip with the zone of interest where the healing process is concentrated. The terminology used to describe the geometry and stresses in the healing process zone is borrowed from Schapery (1989). The length of the crack or healing process zone over which the intermolecular forces across the crack surfaces are effective in promoting healing is denoted by β. The tensile stress between the crack surfaces in the healing process zone is denoted by σ_b. The rate at which the tip advances in the healing process zone to cause wetting between the crack surfaces is denoted by $\dot{\alpha}_b$.

Based on the extensive work done in polymer healing by Wool and O'Connor (1981) three primary stages of the healing process can be identified:

- wetting of the two faces of a nano crack,
- instantaneous strength gain due to interfacial cohesion between the crack faces, and
- long term strength gain due to diffusion and randomization of molecules from one face to the other.

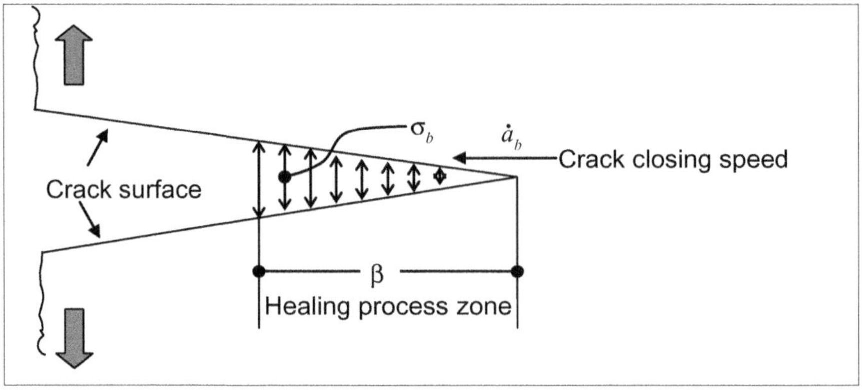

Figure 1. Crack propagation and fracture process/healing zone in mode I loading.

Based on this finding they ingeniously described net macroscopic recovery or healing in a material by combining an intrinsic healing function of the material with a wetting distribution function using a convolution integral as follows (Bhasin et al, 2008):

$$R = \int_{\tau=-\infty}^{\tau=t} R_h(t-\tau)\frac{d\phi(\tau)}{d\tau}d\tau \tag{8}$$

In equation 10, R is the net macroscopic healing function, $R_h(t)$ is the intrinsic healing function of the material, $\phi\,(t, X)$ is the wetting function, and τ is the time variable. Succinctly stated, the convolution integral implies that the rate at which a crack regains its ability to carry load or heal is the net effect of: i) the rate at which the two cracked surfaces wet and ii) the rate at which a wetted crack surface regains strength due to cohesion and inter diffusion. Bhasin et al. (2008) developed an extended expression to define the wetting function, based on the work of Schapery, and the intrinsic healing function.

When the Asphalt Institute (1982) determined the shift between laboratory and field fatigue data, they calculated a shift factor of about 18. Lytton et al. (1993) determined that healing accounts for about one-third of the lab-to-field shift factor. Consequently, healing has a huge impact on the rate of damage. It is important to note that, unlike fracture, where work of adhesion or work of cohesion competes with work dissipated by plastic flow for energy dissipation; in the healing process the work of cohesion (surface energy) does not compete with plastic deformation and therefore its significance in this process is obvious.

Kim et al. (1990) found an empirical relationship between healing and molecular structure. They found that a greater density of longer hydrocarbon chains promotes healing. A possible explanation is that these longer chains promote interdiffusion between crack faces, and in fact promote the process adapted by Bhasin et al. (2008) to explain healing. This process begins with wetting of the crack faces as the cracks are pushed back together by residual stresses in the pavement. After this, the process of healing is an interaction of the wetting process and diffusion across the crack face. Kim et al (1990) used the $-CH_2/-CH_3-$ ratio to quantify the average chain length where a higher $-CH_2/CH_3$ ratio signifies a longer average chain length. They also used the methylene-to-methyl hydrocarbon (MMHC) ratio to identify the level of branching of molecules. They found that more branching is related to poorer healing. Kim et al. (1990) hypothesized that a higher degree of branching restricts freedom of movement and thus restricts diffusion across the micro and/or nano crack faces.

2 EVIDENCE THAT MATERIAL PROPERTIES IMPACT PERFORMANCE

2.1 *Evidence that fracture fatigue damage is impacted by work of cohesion/work of adhesion and SFE*

Table 2 summarizes results from torsional fatigue testing of fine asphalt mixtures using dynamic mechanical analysis (DMA). In this figure, the value of average change in crack radius is a calculated term (Little et al. 2007) representing the increase in crack growth from the initial state of the mix until after the application of the 50,000 load cycles. This value is determined by taking into account the effect of surface energy of the binder and aggregate and material properties determined by conducting a test on the mastic sample using DMA. Although a limited number of mixtures is considered, the data demonstrate the impact and importance of including the work of adhesion or cohesion based on the surface energy of the constituent asphalt binder and aggregate to estimate the crack growth characteristics of asphalt mixes.

2.2. *Evidence that SFE impacts the healing process: Wetting and intrinsic healing*

Figure 2a illustrates the correlation between work of cohesion, W_c, and initial intrinsic healing, R_o, as defined by Bhasin et al. (2008). The value of work of cohesion was determined for three asphalt binders of very different compositions. The W_c values are recorded at the top of the of the three columns on the left side of Figure 2a. The values of R_o, recorded over the three columns on the right of Figure 2a, represent percent healing after a very short rest period of only one-minute. The calculation of percent recovery is based on the shear modulus determined in a dynamic shear rheometer (DSR)—torsional-experiment. In this case two asphalt patties are pushed together with a very low contact pressure of 0.4 N and allowed to interact for one-minute. The interface between the two patties is the "weak point" and therefore defines the shear modulus and is where the effects of healing are felt. The shear modulus of the two patty system is compared to the shear modulus of an intact sample (with interface) of the same size as the two discrete pieces and under the same normal pressure in order to define percent healing. The importance of Figure 2a is the clear relationship between W_c (and thus *SFE*) and the short term intrinsic healing, R_o. Figure 2b compares the percent increase in fatigue life when rest periods are included in the fatigue experiment (compared to when there are no rest periods) with the longer term component of intrinsic healing as defined by Bhasin et al., 2008, R_h. These values (R_h) are printed over the columns on the right side of Figure 2b and represent the shear modulus recovery (percent) after a 4-minute rest period. In Figure 2b the numbers printed over the three columns to the left are the percent increase in fatigue life due to rest periods in fatigue experiments compared to experiments without rest periods. The important fact here is the correlation between healing and the components of healing, R_o and R_h that are directly related to *SFE*.

Table 2. Fatigue cracking of field fixes and crack growth modeled using surface energy measurements.

		Average change in crack radius @ 50,000 cycles			
		Case: 1 Adhesive bond energy		Case: 2 Cohesive bond energy	
Mix	Fatigue cracking reported from qualitative field observations	Avg.	St. Dev.	Avg.	St. Dev.
A	Good	6.5	0.54	5.9	0.57
B	Intermediate	10.4	2.24	7.6	1.17
C	Fair	15.1	1.34	11.1	1.13

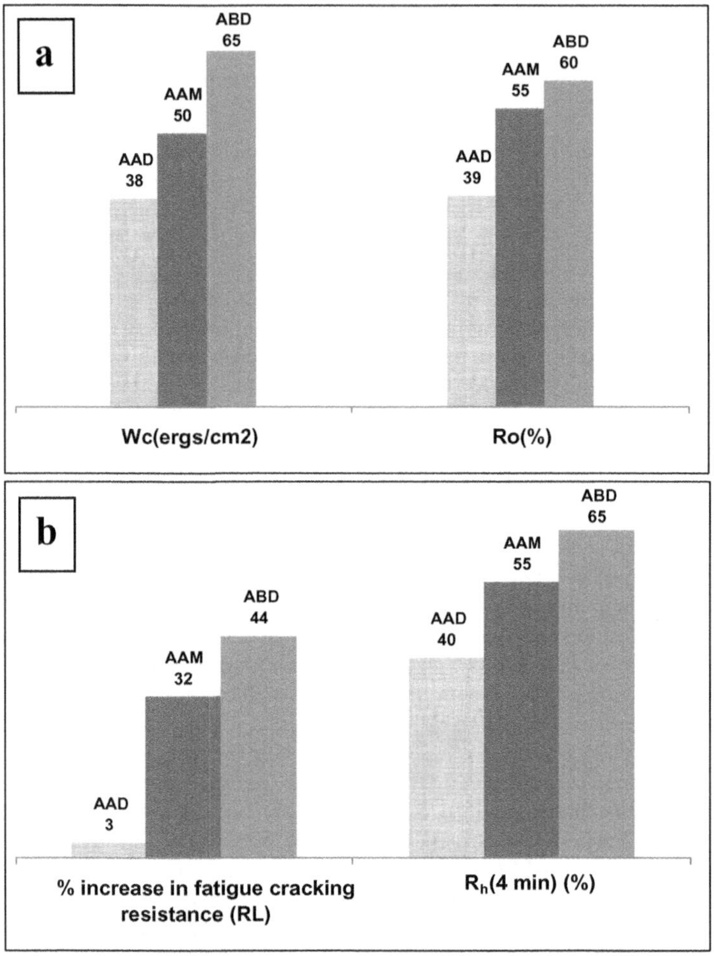

Figure 2. Part "a" compares the rank order of the work of cohesion, W_c, in ergs/cm² to the initial component of intrinsic healing, R_o, determined based on recovery of shear modulus after a one-minute rest period for three very different asphalt binders (compositionally). Part "b" compares the increase in fatigue cracking resistance due to the cumulative application of rest periods for mixtures fabricated with the same three binders compared in part "a" with aggregate "RL" and the shear modulus recovery after a 4-minute rest period.

2.3 *Evidence of the impact of molecular structure on fracture healing*

Recent work at Texas A&M has used molecular modeling to test the Kim et al. (1990) hypothesis. Model compounds were selected to produce a representative ensemble of constituents that represent a repeated amorphous unit of bitumen. Energy minimization methods were used to ensure a stable configuration of the constituents. Self diffusion simulations were performed, Figure 3, for the representative ensembles. Of particular interest is the determination of the rate of diffusion within the bulk of a system comprised of the model representative ensembles and at the interface of such an ensemble. Figure 3 summarizes representative results. In Figure 3 it is clear that, within the bulk, molecular structure has little impact of diffusivity of molecules. However, at the surface or crack interface, diffusivity is strongly impacted by the $-CH_2/CH_3$ ratio. At the interface, asp 226, with the highest saturates content among the three has the highest diffusion coefficient.

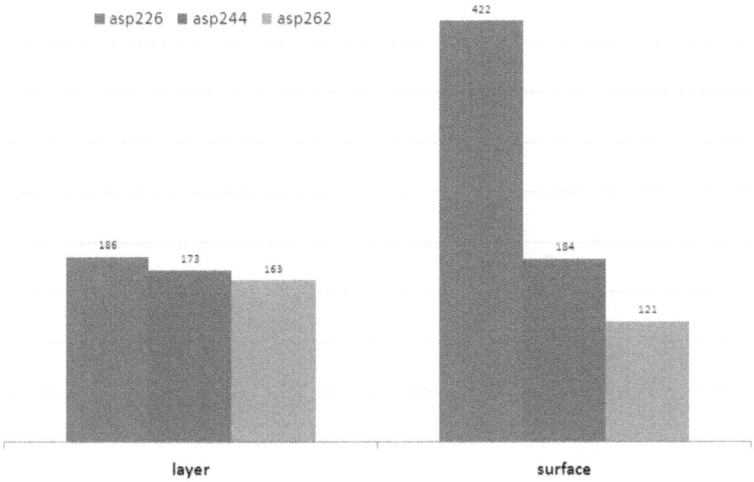

Figure 3. Diffusion coefficient v. saturates percentage. (The notation for each column represents the composition of the asphalt (asp). The three digit number following the designation asp defines the composition of the binder. The first number multiplied by 10 is the asphaltene composition in percent, the second multiplied by 10 is the napthene aromatics composition in percent, and the third number multiplied by 10 is the saturates composition in percent . The term "layer" denotes the diffusion coefficient of the molecules in the bulk layer. The term "surface" denotes the diffusion coefficient of molecules at the surface of a layer. The numbers printed over the columns are the diffusion coefficients in units of mm^2/sec.

2.4 *Evidence of the impact of SFE on moisture damage*

For an asphalt mix to be durable and resist moisture damage, it is desirable that the work of adhesion, W_{AB}, between the asphalt binder and the aggregate be as high as possible. A higher magnitude of work of adhesion implies that more work is required to separate the asphalt binder from its interface with the aggregate. Furthermore, a high magnitude of, W_{ASW}^{wet} indicates a high thermodynamic potential for water to cause debonding. Therefore, it is desirable that this quantity be as small as possible in order to reduce moisture sensitivity. Little et al. (2006) refined the relationship to account for the aforementioned energy relationship as well as specific surface area (SSA) of the aggregate:

$$ER_2 * SSA = \left| \frac{W_{AS} - W_{BB}}{W_{ASW}^{wet}} \right| * SSA \tag{9}$$

where, W_{BB}, is the cohesive bond energy of the bitumen and other terms are as previously described, and SSA is specific surface area.

Figure 4 illustrates the efficacy of the energy term ER_2 * SSA to define damage in terms of the ratio $Fatigue_{wet}/Fatigue_{dry}$, where $Fatigue_{wet}$ is the direct tensile fatigue life of a hot mix asphalt sample after moisture conditioning, and $Fatigue_{dry}$ is the direct tensile fatigue life of a sample of the same composition and fabricated by the same protocol in the dry state (without moisture conditioning). In Figure 4, it is clear that a reasonable correlation exists between damage due to moisture and the energy ratio for the nine mixtures (nine data points) represented. Each data point is the average of three replicate samples of a specific mixture, three different asphalt binders (AAD, ABD, and AAM) were combined with the three aggregates (RA, RK and RL).

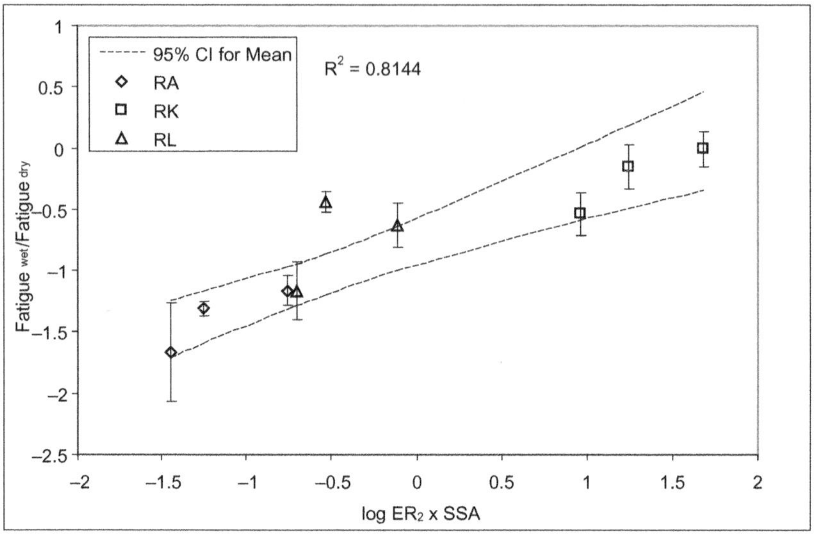

Figure 4. Correlation between fatigue damage in wet and dry conditions.

3 METHODS OF MEASUREING SURFACE ENERGY

Surface free energy components of the asphalt binders were measured using the contact angle approach with the Wilhelmy plate device. Glass slides coated with the asphalt binder were used in these tests. Three components (nonpolar, acid and base) of the low surface energy asphalt binder are determined by calculating contact angles of various liquids on the surface of the asphalt as the asphalt-coated plates are dipped into and extracted from the probe liquids. Surface free energy components of the high energy aggregates are determined by measuring adsorption isotherms of the aggregates with vapors of various liquids using the Universal Sorption Device (USD). As with the Wilhelmy Plate, the USD allows one to determine all three components of SFE required to calculate work of adhesion and work of cohesion. The specific surface area of the aggregates is determined as an inherent part of the test procedure to measure the surface free energy components of the aggregates. The test methodology used in this study to obtain the surface free energy components of the asphalt binder and the aggregate are discussed in detail in other publications (Hefer et al, 2005 and Bhasin et al, 2005).

4 CONCLUSIONS

Fundamental material properties such as *SFE* and molecular shape and size have a profound impact on the fatigue damage process. This paper illustrates two major points. First, *SFE* is an example of a fundamental engineering material property that helps to establish a baseline for fundamental mechanisms. It also helps to explain the fatigue damage process in terms of fracture and crack growth and in terms of the process that balances fracture and crack growth-micro and nano crack healing during rest periods. Second, *SFE* and molecular diffusion, as defined by molecular structure, can serve as screening tools in the selection of asphalt mixture components (asphalt binder and aggregate—coarse and fine) that are compatible and can meet the demands of proposed environment within which they are to function.

REFERENCES

Asphalt Institute. (1982). Research and Development of the Asphalt Institute's Thickness Design Manual Series No. 1.

Bhasin, A., Masad, E., Little, D., and Lytton, R. (2006). Limits on Adhesive Bond Energy for Improved Resistance of Hot Mix Asphalt to Moisture Damage, *In Transportation Research Record 1970, Journal of the Transportation Research Board,* pp. 3–13.

Bhasin, A. and Little, D.N. (2007). Characterization of Aggregate Surface Dnergy Using the Universal Sorption Device. *Journal of Materials in Civil Engineering (ASCE)*, Vol. 133, Number 10, pp. 582–589.

Bhasin, A., Little, D.N., Bommavaram, R., and Vasconcelos, K. (2008). A Framework to Quantify the Effect of Healing in Bituminous Materials Using Material Properties, *Road Materials and Pavement Design an International Journal*, Lavoisier.

Griffith, A.A. (1920). *Transaction of the Royal Society of London*, A221, 163.

Hefer, A.W., Bhasin, A., and Little, D.N. (2006). Bitumen Surface Energy Characterization Using A Contact Angle Approach. *Journal of Materials in Civil Engineering (ASCE)*, Vol. 18, Issue 6, pp. 759–767.

Jokl, M, L., Vitek, V., McMahon, C.J. (1980). A Microscopic Theory of Brittle Fracture in Deformable Solids: A Relation between ideal Work to Fracture and Plastic Work, *Acta Metallurgica*, 28, 1479–1488.

Kaelble, D., H. (1974). A Relationship Between the Fracture Mechanics and Surface Energetics Failure Criteria, *Journal of Applied Polymer Science*, 18, 1869–1889.

Kim, Y.R., Little, D.N., and Benson, F.C. (1990). Chemical and Mechanical Evaluation on Healing Mechanism of Asphalt Concrete. *Proc. Association of Asphalt Paving Technologists*, 59, 240275.

Little, D.N. and Bhasin, A. (2006). Using Surface Energy Measurements to Select Materials for Asphalt Pavement. Texas Transportation Institute, Final Report of Project 9–37, National Cooperative Highway Research Program, Washington, D.C.

Lytton, R.L., J., Uzan, E.G., Fernando, R., Roque, D., Hiltmen, and S., Stoffels (1993) Development and Validation of Performance Prediction Models and Specifications for Asphalt Binders and Paving Mixtures. SHRP Report N° A-357, Strategic Highway Research Program, National Research Council, Washington, D.C.

Miller, O., Freund, L.B., and Needleman, A. (1999). Energy Dissipation in Dynamic Fracture of Brittle Materials, *Modeling Simul. Mater. Sci.*, 7, 573–586.

Okamatsu, T., Yasuda, Y. and ochi, M. (2001). Thermodynamic Work of Adhesion and Peel Adhesion Energy of Dimethoxysilyl-Terminated Polypropylene Oxide/Epoxy Resin System Jointed with Polymeric Substrates. *Journal of Applied Polymer Science*, 80, 1920–1930.

Orowan, E. (1945). *Trans Isntn Engrs Shipbuilders Scot*, 89:89.

Schapery, R.A. (1989). "On the Mechanics of Crack Closing and Bonding in Linear Viscoelastic Media." *International Journal of Fracture*, 39, 163–189.

Seshadri, M., Saigal, S., Jagota, A., Bennison, S. (2007). Scaling of Fracture energy in tensile Debonding of Viscoelastic Film, *Journal of Applied Physics*, 101, 093504.

Sharon, E., Gross, S., and Fineberg, J. (1996). *Physical Review Letters*, 76, 12, 2117–2120.

van Oss, C.J. (1994) *Interfacial Forces in Aqueous Media.* Marcel Dekker Inc., New York.

van Oss, C.J., Chaudhury, M.K., and Good, R.J. Interfacial Lifshitz-van der Waals and Polar Interactions in Macroscopic Systems. *Chemical Reviews*, Vol. 88, 1988. pp. 927–941.

Wool, R. P., and O'Connor, K.M. (1981). A Theory of Crack Healing in Polymers. *Journal of Applied Physics*, 52(10), 5953–5963.

Xu, D, Chung-Yuen, H., and Kramer, E. (1992). Interface Fracture and Viscoelastic Deformation in Finite Size Specimens, *J. Appl. Phys*, 72, 8, 3305–3316.

Efficient Transportation and Pavement Systems – Al-Qadi, Sayed, Alnuaimi & Masad (eds)
© 2009 Taylor & Francis Group, London, ISBN 978-0-415-48979-9

Relationship of asphalt pavement skid resistance to aggregate properties

Eyad Masad
Texas A&M Univsersity at Qatar, Doha, Qatar

Arash Rezaei
Texas A&M Univsersity, TX, USA

ABSTRACT: A method was developed in this study for estimating the skid resistance of an asphalt mixture based on aggregate characteristics and gradation. Asphalt mixture slabs with different combinations of aggregate sources and mixture designs were fabricated in the laboratory and their skid resistance was measured after different polishing intervals. The wheel polishing device developed by the National Center for Asphalt Technology (NCAT) was used for polishing the slabs. Frictional characteristics of each slab were measured by the sand patch method, British Pendulum, Dynamic Friction Tester (DFT), and Circular Texture Meter (CTMeter). Aggregates were characterized using the Micro-Deval abrasion test and texture analysis using the Aggregate Imaging System (AIMS). The aggregate gradation was described using the two-parameter cumulative Weibull distribution function. The data statistical analysis has led to the development of a model for the International Friction Index (IFI). The IFI model's parameters were found to be functions of aggregate texture measured using AIMS and Weibull distribution parameters describing aggregate gradation. This model allows estimating the frictional characteristics of an asphalt mixture during the mixture design stage.

Keywords: Asphalt Mixture, Skid Resistance, Aggregate Gradation, Aggregate Texture, Asphalt Concrete Polishing

1 INTRODUCTION

There has been interest in understanding the polishing process of different aggregates in asphalt pavements. The difference between polishing susceptibility is attributed to differences in the content of wear-resistant minerals, mainly silica (Bloem, 1971). Besides, rocks containing igneous and metamorphic constituents are in general less susceptible to polishing than sedimentary rocks and could improve the overall frictional properties of pavement surface (West et al. 2001). However, aggregates within the same category (i.e., sedimentary) differ considerably in their resistance to polishing depending on their mineralogical composition. Liang and Chyi (2000) found that as the calcite and dolomite contents increase, the polish susceptibility of aggregates decreases.

Several studies attempted to develop laboratory test methods to pre-evaluate the aggregates and develop an aggregate classification system based on this resistance to polishing (Crouch et al. 1996). The British Polish Value (BPV) has been commonly used for this purpose. This test, however, is believed to measure only the microtexture of the pavement or the terminal polished value once the pavement reaches its equilibrium skid resistance (Henry & Dahir, 1979). Crouch et al. (1996) reported that some methods of pre-evaluating the aggregates for asphalt surface courses such as the British Pendulum and chemical or mineralogical methods are only able to classify well-performing aggregates. Recent efforts have focused on developing new methods to pre-evaluate aggregate resistance to polishing. For example, Prowell et al. (2005) suggested using the Micro-Deval as a surrogate to determine an aggregate resistance to weathering and abrasion instead of the sulfate soundness test. Mahmoud and Masad (2007) recommended the use of AIMS to measure loss of aggregate texture after different polishing intervals in the Micro-Deval. Luce et al. (2007) using Mahmoud and Masad's proposed method, demonstrated that aggregate texture had good correlation with skid measurements of a limited number of field sections.

The objective of this study was to develop a method to evaluate aggregate resistance to polishing and estimate pavement skid resistance based on aggregate characteristics and gradation.

2 MATERIALS AND TESTING METHODOLOGY

Five different aggregate sources from Texas were used in this study. Thin sections taken from each aggregate source were impregnated with blue-dyed epoxy (for easy pore delineation) and one-half of the section was stained with Alizarin Red-S to distinguish calcite and aragonite from dolomite. The thin sections were analyzed with a Zeiss Axioskop 40 petrographic microscope equipped with a rotating stage and a Pixelink digital camera. The mineralogy and relative hardness (Mohs hardness scale) of each aggregate are listed in Table 1.

Several methods were used to evaluate aggregate resistance to polishing. The Micro-Deval test evaluates aggregate capability to resist abrasion in a wet environment. In this test, a steel container is loaded with 5000 grams of steel balls and 1500 grams of an aggregate sample in the range of 4.75 mm to 16 mm and 2000 ml of tap water. This material is subjected to 9600 to 12000 revolutions and the sample weight loss is calculated and reported (TxDOT, 2007, AASHTO, 2005). The test was conducted in this study according to Tex-461-A which differs from AASHTO in T 327-05 in the weights of each of aggregate sizes used. The British Polish value (BPV) or Polished Stone Value (PSV) was used to measure aggregate surface friction. This test was conducted according to Tex-438-A procedure (TxDOT, 2007). This test uses a low-speed testing equipment measuring the frictional properties of the aggregate coupons polished in an accelerated polishing machine by swinging a pendulum with a specific normal load and standard rubber pad (TxDOT, 2007). The Aggregate Imaging System was used to measure aggregate texture on images captured at the surface of aggregate particles (Masad et al., 2007). The AIMS was conducted before polishing in the Micro-Deval and after different polishing intervals. The AIMS measurements provide information on rate of aggregate loss of texture.

Thirteen different slabs were compacted using a small steel-wheeled vibratory roller and their frictional characteristics were measured by Dynamic Friction Tester (DFT) and Circular Texture

Table 1. Aggregates analyzed in petrographic study.

Aggregate source	Mineralogy	Mohs hardness	Texas department of transportation (TxDOT) classification
A	Calcite	3	Limestone
B	Quartz, Feldspar, Dolomite, Calcite, Glauconite	7, 6, 3.5–4, 3, 2	Sandstone
C	Zircon, Quartz, Rutile, Feldspar, Dolomite, Calcite	7.5, 7, 6–6.5, 6, 3.5–4, 3	Limestone
D	Quartz, Feldspar, Dolomite Siderite, Calcite	7, 6, 3.5–4, 3.5–4, 3	Gravel
E	Quartz, Feldspar, Dolomite Siderite	7, 6, 3.5–4, 3.5–4	Mostly dolomite with a small percentage of granite

Aggregate F used in preparing asphalt mixtures is a 50/50 blend of aggregates A and B.

Meter (CTMeter). Each slab was 26 inch wide, 60 inch long and 2.5 inch thick. These slabs were polished at three locations by the polishing machine originally developed by NCAT. This machine is capable of polishing a donut shape area with a mean diameter of 284 mm to accommodate DFT and CTMeter measurements (Vollor & Hanson, 2006). Friction and texture measurements were performed before any polishing and after predefined polishing cycles. The maximum number of polishing cycles was 100000 cycles for Type C and Type D and 200000 cycles for PFC mixtures. All Type D mixtures showed signs of severe raveling after 5000 polishing cycles. Consequently, it was decided to discontinue polishing these mixtures and only include measurements before polishing in the data analysis.

According to ASTM E1911-98 the DFT consists of three rubber sliders and a motor that reaches to 100 km/h tangential speed (ASTM, 2006). The rubber sliders are attached to a 350 mm circular disk. By measuring the traction force in each rubber slider the coefficient of friction of the surface is determined. The DFT measurement at 20 km/h is an indication of microtexture (Hall et al., 2006).

The CTMeter test has been described in ASTM E2157 (ASTM 2006). The CTMeter has a Charged Couple Device (CCD) laser displacement sensor mounted on an arm above the surface. A motor at a tangential velocity of 6 mm/min drives the arm. The CCD laser takes 1024 samples of the pavement surface at one round with 0.87 mm spacing. To calculate the MPD (mean pavement depth) the data are divided into eight equal 111.5 mm arcs. The calculated MPD for each segment is averaged and presented as MPD for the test surface (Hall et al. 2006).

3 RESULTS

3.1 *Aggregates and mixtures test results*

The aggregate testing results are shown in
Table 2. The AIMS texture measurements are shown in Figure 1.
In order to describe aggregate resistance to polishing, aggregate texture, as measured in AIMS, is described using the equation proposed by Mahmoud and Masad (2007) shown in Equation (1):

$$Texture(t) = a_{agg} + b_{agg} \bullet e^{(-c_{agg} \cdot t)} \quad (1)$$

where a_{agg}, b_{agg}, and c_{agg} are regression constants while t is the polishing time in the Micro-Deval. In this equation a_{agg}, $a_{agg} + b_{agg}$, and c_{agg} are interpreted as the terminal, initial, and rate of texture change, respectively. Their values, shown in Table 3, are determined using a non-linear regression analysis of the experimental measurements. The DFT was conducted at three different locations on each slab, and two measurements were taken at each location. Figure 2 shows the variation of friction coefficient with the number of polishing cycles.

Figure 2 shows that DF_{20} decreased as polishing cycles increased except for aggregate B which remained almost constant in both type C and PFC mixes. The DF_{20} curves leveled off and reached a terminal value after a certain number of polishing cycles. The CTMeter test was performed at three different locations on each slab, and six measurements were conducted at each location.

Table 2. Measured aggregate characteristics.

Agg. source						
Test procedure	A	B	C	D	E	F
Micro-Deval %Wt. Loss	24	16.2	11.2	2.1	14	20.1
Polish Stone Value (PSV)	25	38	21	28	24	31.5
Texture before Micro-Deval (BMD)	80	265	193	142	269.3	172.5
Texture after Micro-Deval (AMD)	36	222	95	108	192.6	129

Table 3. Regression constants based on three time intervals.

Agg. source Parameter	A	B	C	D	E	F
a_{agg}	39.13	166.7	93.6	105.67	189.1	110.87
b_{agg}	37.46	99.43	99.15	36.33	72.704	47.24
c_{agg}	0.025	0.006	0.041	0.026	0.023	0.018

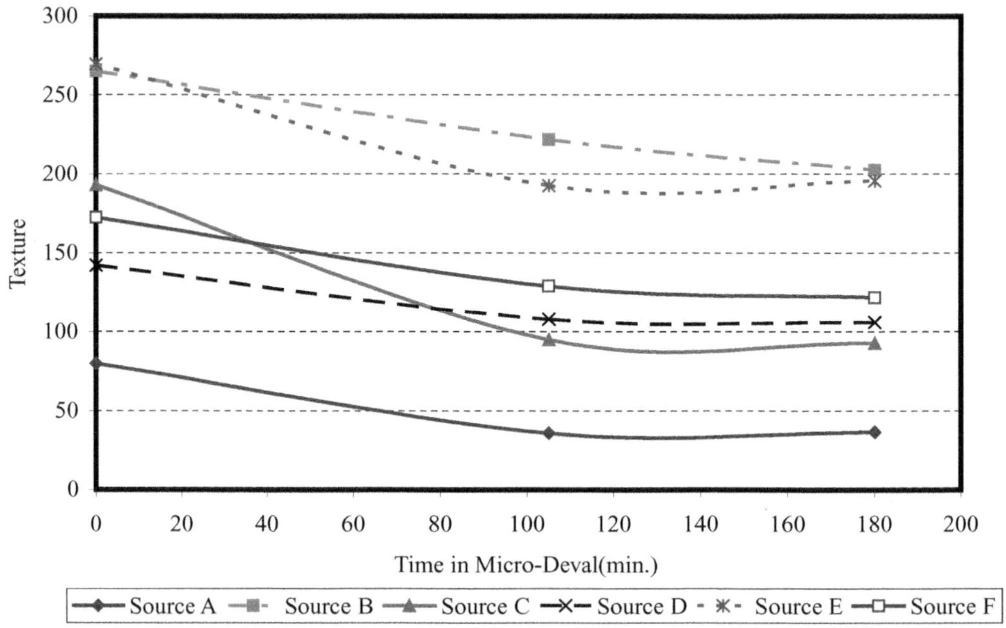

Figure 1. Aggregate texture vs. time in micro-deval.

No particular trend was observed for MPD values from CTMeter of different mixes in terms of polishing cycles.

3.2 *Prediction model for mixture skid resistance*

A model is presented in this section to describe the change of asphalt mixture surface friction as a function of aggregate characteristics and gradation. The mixture friction is quantified by the International Friction Index (IFI) which has recently been proposed in order to develop a universal method for the characterization of pavement surface friction (Henry, 2000). The IFI can be calculated using friction and texture measurements obtained by means of different test methods.

The IFI equation includes calibration factors that depend on the test methods used to measure friction and texture. For DFT and CTMeter measurements, the IFI could be calculated as in Equations (2) and (3):

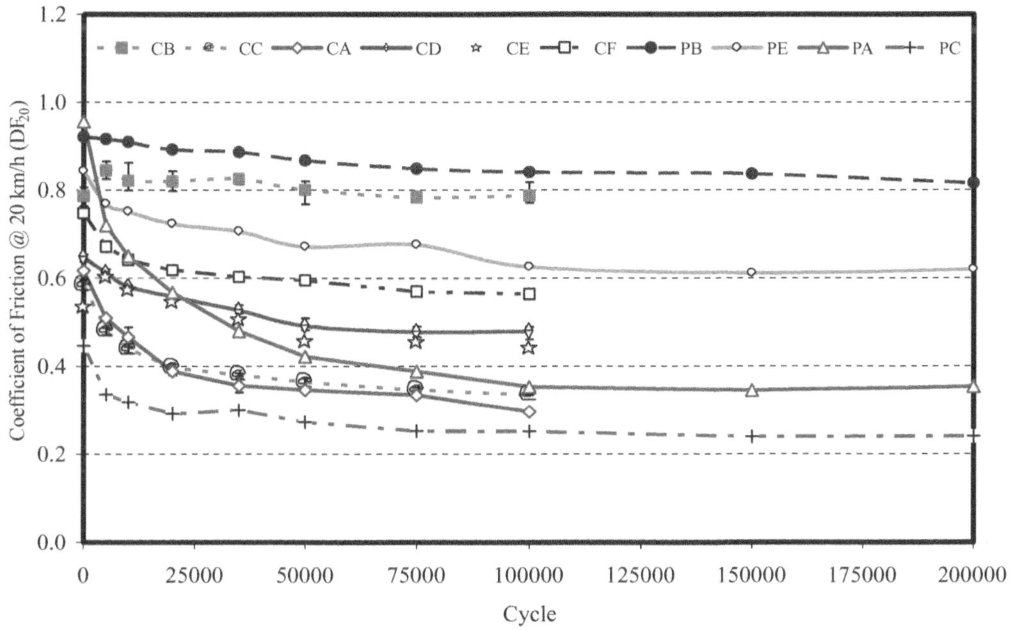

Figure 2. Coefficient of friction at 20 km/h.

$$IFI = 0.081 + 0.732\, DF_{20} \cdot e^{\left(\frac{-40}{S_P}\right)} \tag{2}$$

$$S_P = 14.2 + 89.7\, MPD \tag{3}$$

where, the parameter Sp is related to macrotexture that is measured using the CTMeter, and DF20 is the friction at 20 km/h measured using the DFT. IFI represents a standardized wet friction at 60 km/h (F60) (Hall et. al., 2006).

The calculated IFI values are shown in Figure 3. The equation proposed by Mahmoud and Masad (2007) for describing the loss of aggregate texture (Equation 1) was used to describe the change of IFI. However, the time in Equation (1) is replaced with increments of 1000 polishing cycles as shown in Equation (4):

$$IFI(N) = a_{mix} + b_{mix} \cdot e^{(-c_{mix} \cdot N)} \tag{4}$$

where a_{mix}, $a_{mix} + b_{mix}$, and c_{mix} are the terminal, initial and rate of change in IFI. N is the number of increments of 1000 polishing cycles (e.g., N = 60 for 60,000 polishing cycles). The regression coefficients of Equation (4) were determined and tabulated in Table 4. The R^2 values in Table 4 clearly show that Equation (4) was able to accurately describe the IFI functions for all mixtures.

The effect of different aggregate characteristics on IFI rate of change (c_{mix}) and final value (a_{mix}) were studied in order to explore the possibility of predicting the parameters of the IFI model (Equation 4) based on these aggregate characteristics. A simple linear regression analysis was performed and the coefficient of determination (R^2) and the statistical significance of the regression coefficient p-value were evaluated for each variable (Table 5). An aggregate characteristic with high R^2 and a p-value less than 0.05 is considered significant.

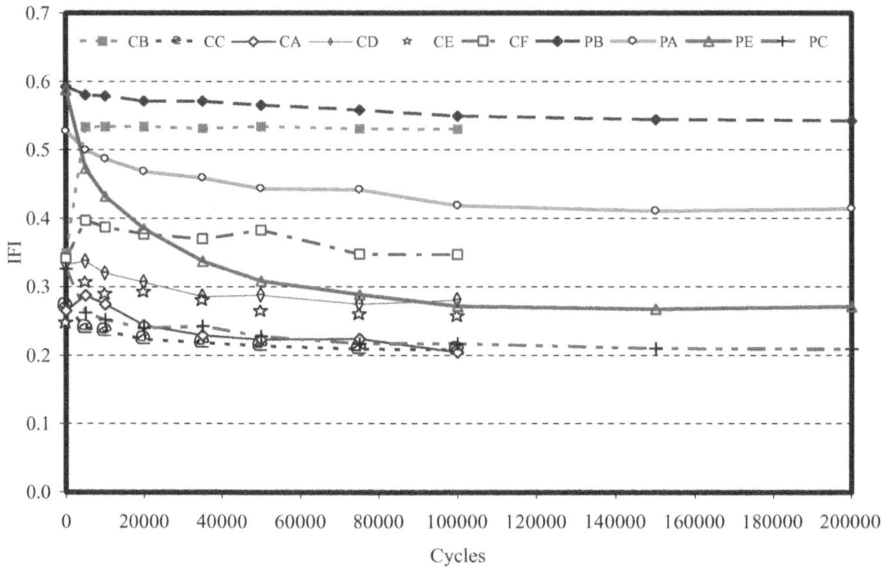

Figure 3. Calculated IFI for different mixes.

Table 4. Values of the regression model parameters (equation 4) for different mixes.

Mix type	a_{mix}	b_{mix}	c_{mix}	R^2 of fitting Eq. (4) to data
CA	0.213	0.097	0.050	0.96
CB	0.474	0.061	0.001	0.98
CC	0.212	0.058	0.102	0.96
CD	0.275	0.065	0.038	0.95
CE	0.250	0.060	0.023	0.95
CF	0.225	0.170	0.003	0.87
PA	0.279	0.288	0.055	0.97
PB	0.539	0.048	0.013	0.97
PC	0.221	0.117	0.155	0.90
PE	0.416	0.101	0.025	0.95

A nonlinear regression analysis was performed to determine the IFI coefficients (coefficients of Eq. 4) as function of aggregate. However, it became apparent that aggregate gradation should be included in the analysis in order to account for the influence of mixture design on skid resistance. For this purpose, a cumulative two-parameter Weibull distribution that has the form of Equation (5) was used to fit the standard aggregate size distribution data:

$$F(x;k,\lambda)=1-e^{-(x/\lambda)^k} \tag{5}$$

where x is the variable (aggregate size in millimeter), and k and λ are model parameters known as shape and scale parameters, respectively. These two parameters change depending on the mixture gradation and they can be included in the regression analysis in order to account for the effect of aggregate gradation on mixture design.

The statistical model that emerged to best represent the IFI's coefficients in Eq. (4) as functions of aggregate characteristics and aggregate gradation are shown in Equations (6–8):

Table 5. Correlation between IFI and aggregate characteristics for different mixtures.

	Type C				PFC			
	IFI rate (c_{mix})		IFI terminal (a_{mix})		IFI rate (c_{mix})		IFI terminal (a_{mix})	
Aggregate property	R^2	p-value	R^2	p-value	R^2	p-value	R^2	p-value
L.A. %Wt Loss	0.00	0.45	0.16	0.22	0.09	0.35	0.09	0.35
Mg Soundness	0.09	0.28	0.01	0.43	0.53	0.13	0.08	0.36
Polish Stone Value (PSV)	0.68	0.02	0.70	0.02	0.51	0.14	0.83	0.04
MD %Wt Loss	0.05	0.33	0.01	0.44	0.13	0.32	0.00	0.48
Acid Insolubility	0.27	0.14	0.35	0.11	0.54	0.13	0.97	0.01
Texture Change BMD and AMD	0.64	0.03	0.44	0.08	0.50	0.15	0.91	0.02
Texture AMD	0.37	0.10	0.53	0.05	0.32	0.22	0.77	0.06

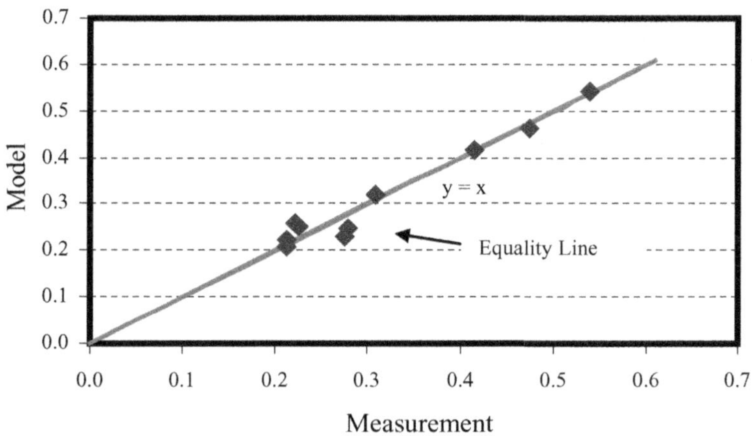

Figure 4. Comparison of terminal IFI values.

$$a_{mix} = \frac{18.422 + k}{118.936 - 0.0013 \times (AMD)^2} \qquad R^2 = 0.96 \tag{6}$$

$$a_{mix} + b_{mix} = 0.4984 \ln\left(5.656 \times 10^{-4}\left(a_{agg} + b_{agg}\right) + 5.846 \times 10^{-2} k - 4.985 \times 10^{-2} \lambda\right) + 0.8619 \quad R^2 = 0.82 \tag{7}$$

$$c_{mix} = 0.765 \cdot e^{\left(-7.297 \cdot 10^{-2} / c_{agg}\right)} \qquad R^2 = 0.90 \tag{8}$$

where, AMD is aggregate texture after Micro-Deval, $a_{agg}+b_{agg}$ and c_{agg} are initial and rate of texture change for corresponding aggregate, and k, λ are Weibull distribution shape and scale factors, respectively. Figure 4 shows that the model does a very good job in representing IFI measurements.

The regression model in Equations (6–8) shows that the terminal mixture friction (a_{mix}) was a function of aggregate texture after polishing (AMD), the initial mixture friction ($a_{mix}+b_{mix}$) was a function of aggregate initial texture ($a_{agg}+b_{agg}$), and a higher rate of aggregate texture reduction (c_{agg}) was associated with a higher IFI change (c_{mix}). In addition, mixes with higher k-values—primarily PFC mixes—had higher terminal IFI values.

4 CONCLUSIONS

This paper presented a method to evaluate the influence of aggregates on the skid resistance of asphalt pavements. A statistical model was developed for predicting the IFI coefficients as functions of aggregate texture characteristics and gradation. The aggregate texture was measured using the Aggregate Imaging System (AIMS) after different polishing times in the Micro-Deval. The IFI model can be used to select mixture design and aggregates in order to enhance the friction of asphalt mixtures. The results showed that the same aggregate can have different contributions to mixture frictional characteristics depending on mixture design.

REFERENCES

AASHTO. 2005. *AASHTO Standard Specifications for Transportation Materials and Methods of Sampling and Testing, Part 2B: Testing*, American Association of State Highways and Transportation Officials, Washington, D.C.

ASTM. 2006. *Annual Book of ASTM Standards*, Vol. 04.03, American Society for Testing and Materials, West Conshohocken, Pennsylvania.

Bloem, D.L. 1971. Skid-resistance ;the role of aggregates and other factors, *National sand and gravel association circulars.*

Crouch, L.K., G. Shirley, G.Head, & W.A. 1996. Goodwin. Aggregate polishing resistance pre-evaluation, In *Transportation Research Record, Journal of Transportation Research Board, No. 1530*, Transportation Research Board of the National Academies, Washington D.C., pp. 103–110.

Hall, J.W., L.T. Glover, K.L. Smith, L.D. Evans, J.C. Wambold, T.J. Yager, and Z. Rado. 2006. Guide for pavement friction. P*roject No.1–43*, Final Guide, National Cooperative Highway Research Program, Transportation Research Board, National Research Council, Washington, D.C.

Henry, J., & Dahir, S. 1979. Effects of textures and the aggregates that produce them on the performance of bituminous surfaces. In *Transportation Research Record, Journal of Transportation Research Board, No. 712*, Transportation Research Board of the National Academies, Washington D.C.

Henry, J.J. 2000. Evaluation of pavement friction characteristics: *NCHRP Synthesis: Topic 30–11*, National Cooperative Highway Research Program, Washington, D.C.

Liang R.Y. & L.L. Chyi. 2000. Polishing and friction characteristics of aggregates produced in Ohio. *FHWA report, FHWA/OH-2000/001*, Federal Highway Administration, Columbus, Ohio, www.dot.sates.oh.us/researc/2000/Materials/14678-FR.pdf Accessed: June 20, 2007.

Luce, A., E. Mahmoud, E. Masad, & A. Chowdhury. 2007. Relationship of aggregate texture to asphalt pavement skid resistance. *Journal of Testing and Evaluation, Vol.36, No.5,* American Society of Testing and Materials (ASTM).

Mahmoud, E., & E. Masad. 2007. Experimental methods for the evaluation of aggregate resistance to polishing, abrasion and breakage. *Journal of Materials in Civil Engineering, ASCE, Vol.19, No.11*, pp. 977–985.

Prowell, B.D., Zhang, J. & Brown, E.R. 2005. Aggregate properties and the performance of Superpave-designed hot mix asphalt. *National Cooperative Highway Research Program Report 539*, Transportation Research Board, National Research Council, Washington, D.C.

Texas DOT.2007. *TxDOT Online Manuals*, Bituminous Test Procedures, 400-A, Texas Department of Transportation. www.dot.state.tx.us/services/general_services/manuals.htm. Accessed April 2, 2007.

Vollor, T.W., & Hanson, D.I. 2006. Development of Laboratory Procedure for Measuring Friction of HMA Mixtures-Phase 1. N*CAT Report 06-06*, Auburn University.

West T.R., J.C. Choi, D.W. Bruner, H.J. Park, & K.H. Cho. 2001. Evaluation of dolomite and related aggregates used in bituminous overlays for Indiana pavements. In *Transportation Research Record, Journal of Transportation Research Board, No. 1757*, Transportation Research Board of the National Academies, Washington D.C., pp. 137–147.

Efficient Transportation and Pavement Systems – Al-Qadi, Sayed, Alnuaimi & Masad (eds)
© 2009 Taylor & Francis Group, London, ISBN 978-0-415-48979-9

Investigation into two procedures of applying gilsonite into HMA mixtures

B. Huang & X. Shu
Department of Civil and Environmental Engineering, The University of Tennessee, Knoxville, TN, USA

ABSTRACT: This paper presents a laboratory study on the effect of two different application procedures of gilsonite (modifying and coating) on the performance of hot-mix asphalt (HMA). First the properties of asphalt binder modified with different percentages of gilsonite were evaluated. Then laboratory experiments were conducted on the HMA mixtures prepared with the two methods. The properties tested included indirect tensile (IDT) strength, tensile strength ratio (TSR), dissipated creep strain energy limit ($DCSE_f$), dynamic modulus and flow time. The results indicated that the addition of gilsonite resulted in a modified asphalt with both increased stiffness and elastic recovery. However, different application procedures would form different microstructures in HMA mixtures, which would consequently lead to the difference in HMA performance. The coating procedure could establish a three-layer microstructure, which was beneficial for improving the moisture susceptibility characteristics of HMA, whereas the mixtures with gilsonite-modified asphalt binder had much improved rut resistance.

1 INTRODUCTION

Gilsonite is a type of natural asphalt and it is similar to hard petroleum asphalt. Due to its higher softening point, the stiffness of gilsonite is about 50 times higher than conventional asphalt under room temperature (Huang et al. 2006). Gilsonite has been used as an asphalt binder modifier to modify and stiffen HMA mixtures (Gaughan 1990, Simpson & Mahboub 1994). Due to the fact that gilsonite is also a kind of asphalt binder in nature, it can be quickly dissolved into asphalt binder and coat aggregate particles during the mixing process. Thus, adding gilsonite into HMA mixtures does not cause any problems to blending, mixing and compacting of HMA mixtures that other asphalt binder modifiers usually cause.

Two procedures have been reported on applying gilsonite into HMA mixtures (Gaughan 1990). The first one is the most often used one and it is to directly apply and dissolve gilsonite into asphalt to modify and stiffen the conventional asphalt binder for the mixtures (called modifying method in this study). The microstructure of the resulting HMA mixture with the gilsonite-modified asphalt binder is the same as that of conventional HMA mixture, which is a two-layered mixture (Figure 1a).

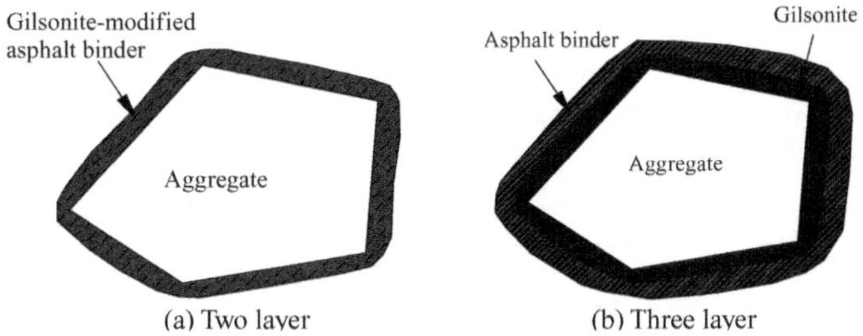

Figure 1. Different microstructures of HMA mixture.

The second procedure is to coat aggregate particles with gilsonite prior to mixing them with conventional asphalt binder (called coating method in this study). Due to the relatively short mixing time (usually less than 60 seconds) and much high softening point of gilsonite (129–204°C), it is hard for gilsonite coating to be blended and dissolved into virgin asphalt binder during the normal mixing process. Hence, the coating method will result in a three-layered HMA mixture (Figure 1b). The gilsonite interlayer serves as cushion between aggregate particle and asphalt binder, which will result in different properties of the two- and three-layered HMA mixtures (Huang et al. 2006).

2 LABORATORY EXPERIMENTS

Laboratory experiments were conducted to evaluate the properties of asphalt binder modified with different percentages of gilsonite and HMA mixtures prepared using two different procedures of applying gilsonite. Conventional asphalt binder PG 64-22 and limestone aggregate were used for the binder and mixture tests. The binder tests included dynamic shear rheometer and elastic recovery tests. HMA mixture tests included Superpave indirect tensile (IDT) tests, tensile strength ratio (TSR) test, and dynamic modulus and flow time tests.

2.1 *Materials*

The granulated gilsonite (passing No.10 sieve) was provided by Ziegler Chemical and Mineral Corporation. The coarse aggregates selected in this study were crushed limestone with nominal maximum size of 12.5 mm. The fine aggregates consisted of No. 10 screenings, natural sand, and manufactured sand. The virgin asphalt binder used in this study was conventional PG 64-22 asphalt. The properties of the materials and the gradations of aggregates can be found in Huang et al. (2006).

2.2 *Mixture design*

The Marshall mix design procedure was employed to design the control mixture. For the HMA mixtures, 50% limestone D-Rock, 15% No. 10 screenings, 25% natural sand, and 10% manufactured sand were selected. The optimum asphalt content was 5.0 percent. Mixture volumetric properties are listed in Table 1. No separate mix design was performed for the two different mixtures modified with gilsonite.

2.3 *Procedures of applying gilsonite*

Two different procedures of applying gilsonite into HMA mixtures were used in this study. For the first procedure, asphalt binder was preheated and gilonite was then added into it. Mechanical mixer was used to speed up the dissolution of gilsonite and to produce a uniform gilsonite-modified asphalt binder. The gilsonite percentage was selected at 10% by weight of the total asphalt binder for HMA mixture because at low gilsonite contents (<5%), there is no signicant change in the properties of gilsonte-modified asphalt binder. And from the published literature (Gaughan 1990, Simpson & Mahboub 1994), gilsonite percentage higher than 10% will cause negative property changes to asphalt mixtures.

Table 1. Volumetric properties of HMA mixture.

AC %	G_{mm}	G_{mb}	Air Voids %	VMA %	Stability kN	Flow mm
5.0	2.456	2.356	4.0	16	11.6	2.77

Note: AC = asphalt content; G_{mm} = maximum specific gravity of loose mixture; G_{mb} = bulk specific gravity of compated specimen; VMA = voids in mineral aggregate.

For the second procedure, coarse limestone 'D-rock' was coated with gilsonite prior to mixing with fine aggregates and asphalt binder. During this laboratory experiment, trichloroethylene was used to dilute gilsonite so that a thin layer could be coated on the surface of aggregate particles. For the practical operation in hot-mix asphalt plant, other coating methods should be employed in a safe and economical way. The calculated film thickness of the gilsonite coating was about 2.5 microns in this study. Figure 2 presents the coating of aggregates with gilsonite.

2.4 *Sample preparation*

HMA mixture specimens were prepared for performance testing. 150-mm diameter cylindrical samples were compacted by the Superpave Gyratory Compactor (SGC) for the Superpave IDT and moisture susceptibility tests. Specimens for the axial dynamic modulus and flow time test were 100 × 150-mm cored in the center and trimmed on both ends from 150 × 170-mm SGC compacted cylindrical specimens. The air voids for the dynamic modulus, flow time and IDT tests were 4 ± 0.5% and those for moisture susceptibility test were 7 ± 1%. The samples were made in triplicate.

2.5 *Binder tests*

Asphalt binders modified with 5 different percentages of gilsonite (0, 2%, 5%, 10%, 15%) were tested to evaluate the effect of gilsonite on the properties of asphalt binder. Master curve and the fatigue parameter of asphalt binder, $|G^*|*\sin\delta$, were obtained using an Anton Paar Physica MCR 501 Rheometer. In addition, elastic recovery test was conducted in accordance with AASHTO T 301 to explore the effect of gilsonite on the ductility and elastic recovery property of asphalt binder.

2.6 *Mixture performance tests*

2.6.1 *Super IDT tests*

The Superpave IDT tests include the resilient modulus, creep, and indirect tensile strength tests and they were performed following the procedures developed by Roque and Buttlar (1992) and Buttlar and Roque (1994). The testing system and associated analysis procedures are described in detail in the above mentioned literature.

2.6.2 *Moisture susceptibility test*

The moisture susceptibility test was conducted in accordance with AASHTO T 283. Six cylindrical specimens are divided into two sets of three specimens. One set is used as the control set, whereas the other set is used for moisture-conditioning. Moisture-conditioning starts with induction of 55 ~ 80% saturation in the specimens followed by keeping the specimens in a water bath at 60 ± 0.5°C for 24 ± 0.5 hours.

Figure 2. Coarse aggregates coated with gilsonite (After Huang et al. 2006).

The indirect tensile strength of both the control and conditioned mixtures is determined at 25°C. The moisture susceptibility is indicated by the tensile strength ratio (TSR) expressed as:

$$TSR = \frac{S_{tm}}{S_{tc}} \tag{1}$$

where S_{tm} = average indirect tensile strength of the moisture-conditioned set, and S_{tc} = average indirect tensile strength of the control set.

2.6.3 *Dynamic modulus test*

The dynamic modulus test was conducted on cylindrical specimens 100 mm in diameter and 150 mm tall at 25°C, under no confining pressure, and at the loading frequencies from 0.01 to 25 Hz by subjecting specimens to a sinusoidal vertical pressure. The strain level was controlled between 75 to 125 με. The dynamic modulus, $|E^*|$, was calculated as the ratio of the amplitude of the sinusoidal stress, σ_0, applied to the specimen and the amplitude of the induced sinusoidal stain, ε_0:

$$|E^*| = \frac{\sigma_0}{\varepsilon_0} \tag{2}$$

2.6.4 *Flow time test*

In this test, the specimen was subjected to a constant axial compressive stress (600 kPa) at a specified temperature (54°C). The resulting permanent axial strain is measured as a function time and numerically differentiated to calculate the flow time. Flow time is defined as time corresponding to the minimum rate of change of permanent axial strain.

3 RESULTS AND DISCUSSION

3.1 *Test results of gilsonite-modified asphalt binder*

Figure 3 presents the master curve of asphalt binder modified with different percentages of gilsonite. It can be seen that the dynamic shear modulus of gilsonite-modified asphalt increased with the increase in the content of gilsonite. However, at low gilsonite content (<5%), the difference in the dynamic shear modulus was not so significant, especially at high frequencies.

When gilsonite is used to modify and stiffen asphalt binder, one big concern is that the fatigue resistance of binder will be compromised. In this study, the fatigue parameter ($|G^*|$*sin δ) of binder was selected to evaluate its fatigue property. The higher the fatigue parameter, the lower the fatigue resistance of asphalt binder (Roberts et al. 1996). Figure 4 presents the fatigue parameter of gilsonite-modified asphalt binder. It appeared that the fatigue parameter of asphalt binder was increased with more gilsonite added into binder, which would result in lower fatigue resistance. Similar to the master curve, only asphalt binder modified with high percentage of gilsonite showed significant difference. It can also be seen from Figure 4 that the modifying effect of gilsonite was more significant at high temperature than at low temperature.

Figure 5 presents the phase angle value of gilsonite-modified asphalt binder. From Figure 5, the phase angle decreased with the increase in the gilsonite content. This means that the binder became more elastic with the incorporation of more gilsonite. Despite the decrease in phase angle, the fatigue parameter, $|G^*|$*sinδ, still increased due to large increase in dynamic shear modulus.

Another concern with the incorporation of gilsonite into asphalt binder is that the ductility and elastic recovery property of asphalt binder probably will be compromised. Thus, the elastic recovery test was conducted to explore the effect of gilsonite on the elastic recovery of binder.

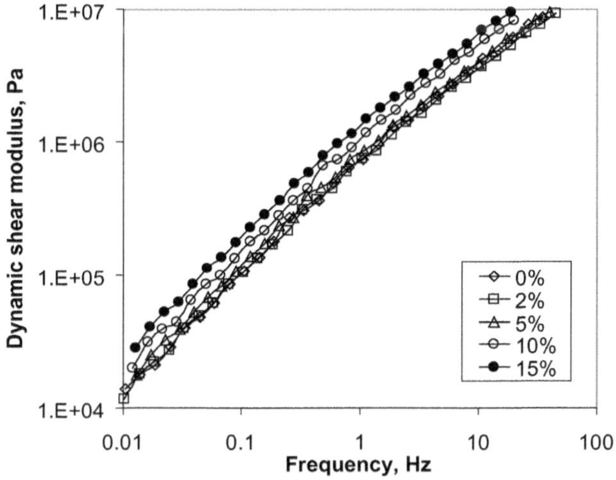

Figure 3. Master curve of gilsonite-modified asphalt binder (25°C).

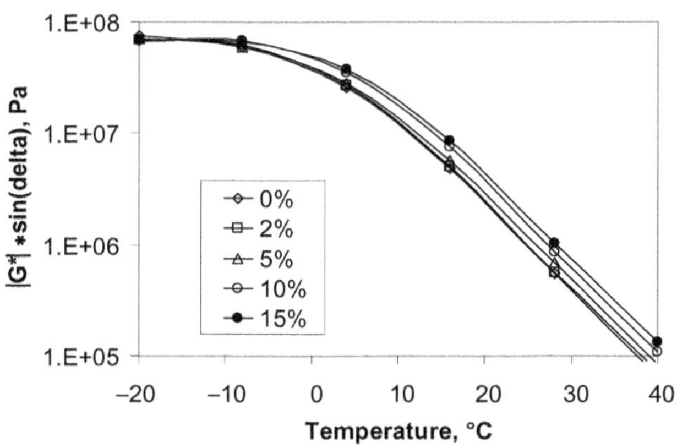

Figure 4. Fatigue parameter of gilsonite-modified asphalt binder.

Figure 6 presents the elastic recovery rate of gilsonite-modified asphalt binder. It can be seen that the elastic recovery rate actually increased with more gilsonite added into asphalt binder. It should be noted that even with increased values, the elastic recovery of gilsonite-modified asphalt binder was still much less than the typical SBS modified asphalt binder. The combined effect of highly increased stiffness and slightly increased elastic recovery will probably lead to decreased resistance of asphalt binder to fatigue and thermal cracking.

3.2 *Test results of HMA mixtures*

Figure 7 shows IDT strength results for HMA mixtures prepared with the two different methods. The gilsonite-modified HMA mixture exhibited higher indirect tensile strength (averaged about 1.58 MPa) than the gilsonite-coated mixture (averaged about 2.23 MPa).

Figure 8 presents the results of TSR from the moisture susceptibility test. It is obvious that the gilsonite-coated HMA mixture exhibited much higher TSR value than the gilsonite-modified mixture. The reason can be attributed to the fact that the coated mixture possesses two different

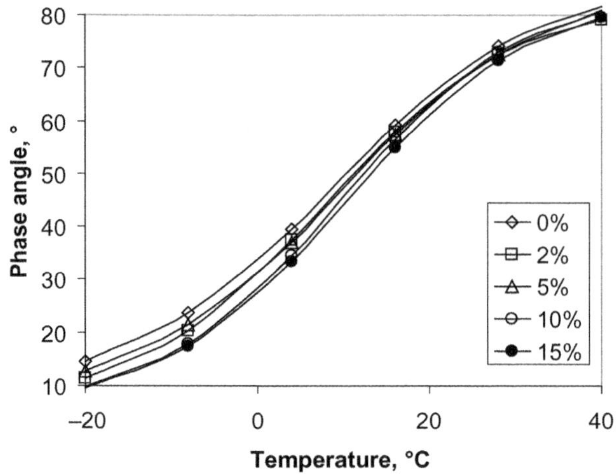

Figure 5. Phase angle of gilsonite-modified asphalt binder.

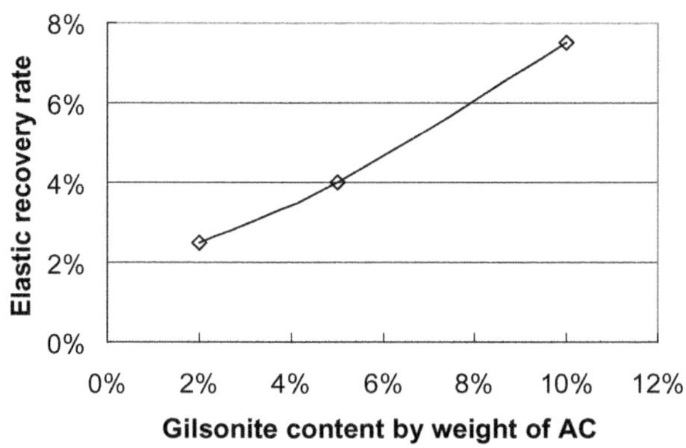

Figure 6. Elastic recovery rate of gilsonite-modified asphalt binder.

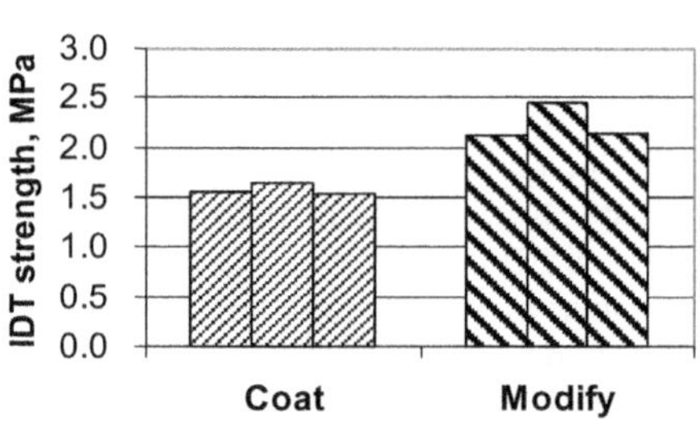

Figure 7. IDT strength results.

interfaces (the aggregate-gilsonite interface and the gilsonite-asphalt interface). These two interfaces can perform better because their good affinity between each other than the single aggregate-asphalt interface of the modified mixture.

Figure 9 compares the results of the dissipate creep strain energy threshold ($DCSE_f$) for the two HMA mixtures. $DCSE_f$ represents the ability of HMA mixtures to absorb fracture energy (Roque

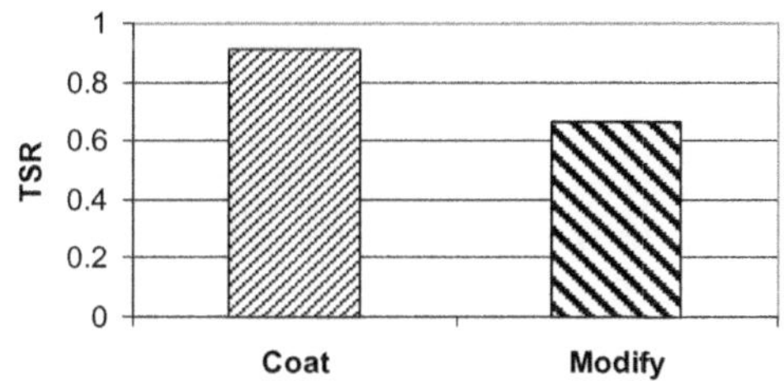

Figure 8. TSR results.

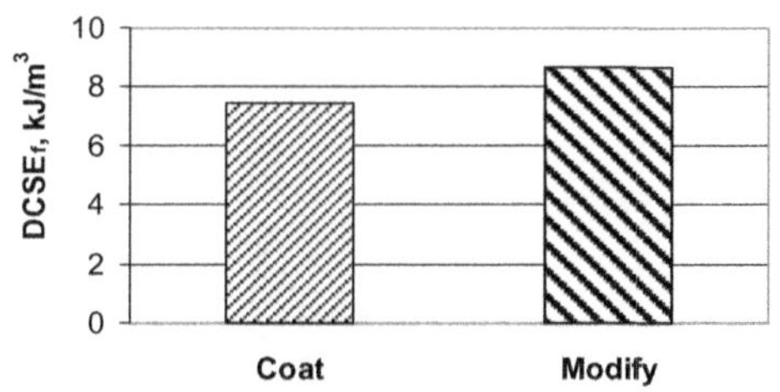

Figure 9. DCSE$_f$ results from superpave IDT tests.

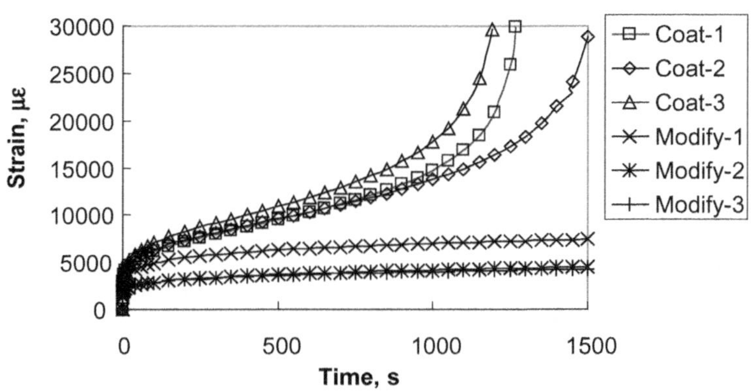

Figure 10. Accumulated axial strain in flow time test.

et al. 2002). The larger the DCSE$_f$ value, the higher the resistance of HMA mixtures to fatigue and fracture. From Figure 9, it can be seen that the modified mixture had higher DCSE$_f$ value than the coated mixture, which indicates that the modified mixture was potentially more resistant to fatigue and fracture than the coated mixture.

Figure 10 illustrates the accumulation of axial strain with time under constant compressive stress in the flow time test. Figure 11 presents the results of flow time for both mixtures. The gilsonite-coated HMA mixture had a much higher strain rate than the modified mixture. The strain development rate was also demonstrated in Figure 11. The coated mixture had a far lower flow time than the modified mixture.

Figure 12 and 13 present the dynamic modulus and phase angle for both mixtures. The gilsonite-modified mixture had higher dynamic modulus and lower phase angle values than the gilsonite-coated mixture. This means that the modified mixture was stiffer and more elastic than the coated mixture, which was consistent with the findings from IDT strength, DCSE$_f$, flow number results.

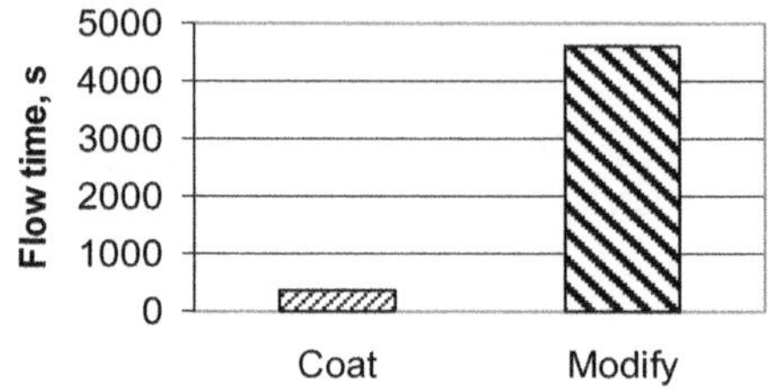

Figure 11. Flow time results.

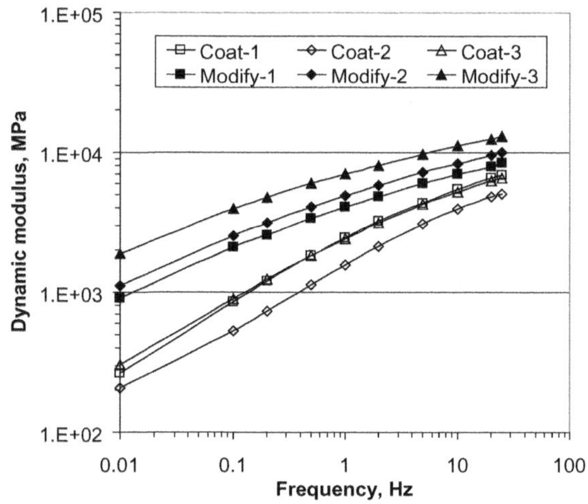

Figure 12. Dyanmic modulus results at 25°C.

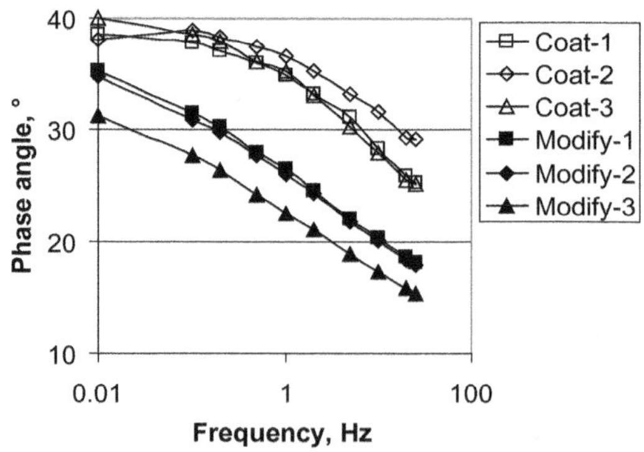

Figure 13. Phase angle results at 25°C.

4 SUMMARY AND CONCLUSIONS

A laboratory study was conducted to explore the effect of different methods of adding gilsonite into HMA on the properties of HMA mixture. Based on the results from this study, the following can be summarized:

- Gilsonite modification resulted in a stiffened asphalt binder. However, the improved elastic recovery was less significant than typical SBS polymer modified asphalt binder.
- Different application procedures of gilsonite would change the microstructure (especially the interface) of HMA mixture, which would consequently lead to the change in the performance of HMA mixture.
- The coating procedure could establish a three-layer microstructure and subsequently two interfaces. This peculiar microstructure was beneficial for improving the moisture susceptibility characteristics of HMA mixtures and flexible pavements.
- The mixtures with gilsonite-modified asphalt binder exhibited much improved rut resistance, which strongly indicated the suitability of gilsonite-modified asphalt binder for high-temperature pavement application.
- The gilsonite-modified HMA mixture exhibited a higher $DCSE_f$ than the coated mixture, which indicates that the modified mixture was potentially more resistant to fatigue and fracture than the coated mixture.
- The results presented in this study only represented the materials under the test conditions in this study. More complete tests were recommended in future study

REFERENCES

Buttlar, W.G. & Roque, R. 1994. Experimental development and evaluation of the new SHRP measurement and analysis system for indirect tensile testing at low temperature. *Transportation Research Record* 1454: 163–171.

Gaughan, R.L. 1990. Development of high strength asphalt mixes. In *Pavements and Materials; Proc. Conf. Australian Road Research Board* Part 2: 41–52.

Huang, B., Li, G. & Shu, X. 2006. Investigation into three–layered HMA mixtures. *Composite Part B: Engineering* 37(7–8): 679–690.

Roberts, F.L., Kandhal, P.S., Brown, E.R., Lee, D.Y. & Kennedy, T.W. 1996. *Hot Mix Asphalt Materials, Mixture Design, and Construction (2nd Ed.)*. Lanham, Maryland: NAPA Education Foundation.

Roque, R., Birgisson, B., Sangpetngam, B. & Zhang, Z. 2002. Hot mix asphalt fracture mechanics: a fundamental crack growth law for asphalt mixtures. *Journal of the Association of Asphalt Paving Technologists* 71: 816–27.

Roque, R. & Buttlar, W.G. 1992. The development of a measurement and analysis system to accurately determine asphalt concrete properties using the indirect tensile mode. *Journal of the Association of Asphalt Paving Technologists* 61: 304–332.

Simpson, A.L. & Mahboub, K.C. 1994. Case study of modified bituminous mixtures: Somerset, Kentucky. In *Infrastructute: New Materials and Methods of Repair; Proc. Matls. Engr. Conf.* 804: 88–96.

Efficient Transportation and Pavement Systems – Al-Qadi, Sayed, Alnuaimi & Masad (eds)
© 2009 Taylor & Francis Group, London, ISBN 978-0-415-48979-9

Asphalt rubber asphalt concrete friction course overlay as a pavement preservation strategy

K. Kaloush, K. Biligiri & M. Rodezno
Department of Civil and Environmental Engineering, Arizona State University, Tempe, AZ, USA

M. Belshe
FNF Construction, Tempe, AZ, USA

G. Way & D. Carlson
Rubber Pavement Association, Tempe, AZ, USA

J. Sousa
Consulpav International, Inc. USA-Portugal

ABSTRACT: The use of Asphalt Rubber (AR) pavements in the USA has been successful by several States. AR binder used in the hot mix asphalt is a mixture of 80 percent hot asphalt and 20 percent ground waste tire crumb rubber. Typically, AR-Asphalt Concrete Friction Course (AR-ACFC) mixes contain 9 to 10 percent asphalt rubber binder and their use has been primarily focused on reducing thermal and reflective cracking, and highway noise. This paper discusses the AR-ACFC benefits as a pavement preservation strategy. It highlights some results of the laboratory material characterization tests, and presents several field performance evaluation outcomes including: highway noise reduction, mitigation of daily thermal variances in Portland Cement Concrete pavements, improved skid resistance, reduced roughness, and reduction of emission rates of tire wear per kilometer driven.

1 ASPHALT RUBBER OVERLAYS IN ARIZONA

The Arizona Department of Transportation (ADOT) began to use Open Graded Friction Courses (OGFC) with conventional asphalt as early as 1954. The primary reason for using this material was to provide a surface with good skid resistance and ride quality. During twenty years of Asphalt Rubber (AR) use, ADOT work evolved from using AR chip seals to utilizing reacted AR as a binder in open and gap graded asphalt concrete. In 1988, ADOT built its first AR mix project, which consisted of a 25 mm layer of an open-graded Asphalt Rubber Asphalt Concrete Friction Course (AR-ACFC) placed on Interstate 19 south of Tucson. This AR-ACFC mix was placed on top of a plain jointed concrete pavement and performed very well to date with very little maintenance. Hundreds of projects have been successfully built to date. A 50 mm structural gap graded Asphalt Rubber Asphalt Concrete (ARAC) overlay was designed and built in 1990 on Interstate 40 near Flagstaff, and it was overlaid with 12.5 mm AR-ACFC [Way 2000]. The purpose of the project was to compare the performance of the overlay on a severely cracked concrete pavement. The asphalt rubber sections built had the least percentage of reflective cracks, one third less than a 100 mm conventional overlay and less than one half a 200 mm overlay.

Construction of an AR pavement involves first mixing and fully reacting the crumb rubber with the hot asphalt cement. Typically 20 percent ground tire rubber that meets specific gradation is added to hot asphalt heated to a temperature of about 190°C and mixed for at least one hour.

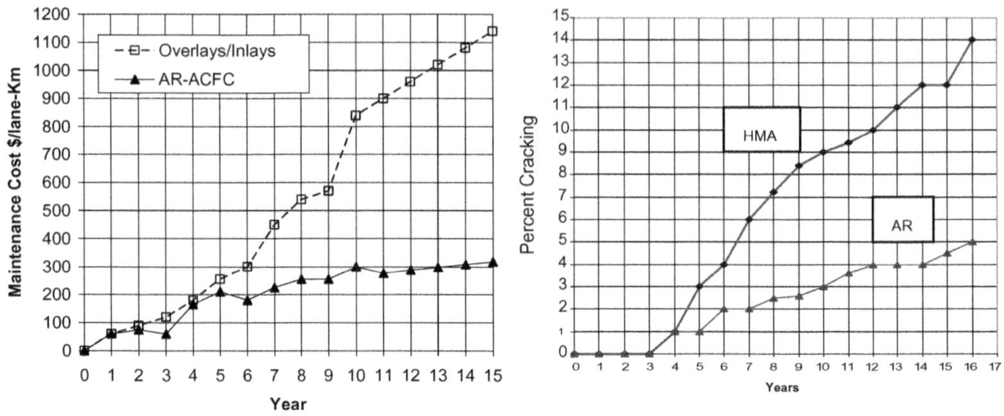

Figure 1. Conventional overlays versus asphalt rubber: a) Maintenance costs; b) Percent cracking.

After reaction the AR is kept at a temperature of about 175°C until it is introduced into the mixing plant.

On cracked pavements, a gap-graded ARAC, generally 37.5 to 50 mm thick, is placed to address cracking. An AR-ACFC may be placed depending upon the traffic volume and type of highway. ARAC mixes are normally one half or less than one half the thickness of conventional Hot Mix Asphalt (HMA) pavements without asphalt rubber. The finished AR product is generally 25 to 50 percent more expensive, but is actually more cost effective when considered in a life cycle cost analysis. This is because AR requires less maintenance costs (as shown in Figure 1a), has low percent cracking (Figure 1b), and therefore has longer service life.

This paper discusses the AR-ACFC benefits as a pavement preservation strategy. While there are several categories that distinguish the selection of a good pavement preservation alternative, the authors selected the following focus areas as the criteria for a good pavement preservation strategy: good mixture and binder characteristics through laboratory performance testing; reduced tire/pavement noise, improved thermal gradient characteristics and interaction with the urban climate; good field frictional characteristics for safety; improved ride quality and comfort for users; less tire wear emissions and therefore better impact on air quality; and finally, cost and energy demand effectiveness through life cycle cost analysis.

2 LABORATORY PERFORMANCE

2.1 Background

The Department of Civil and Environmental Engineering at Arizona State University (ASU) has been involved with several asphalt rubber mixtures characterization studies. The studies are being conducted in cooperation with ADOT and other local and international transportation agencies. The ultimate goal from these studies is implementing a methodology for performance related specifications for asphalt rubber pavements, and developing typical design input parameters for specific conditions.

In these characterization studies, hot AR mixes are obtained from the field during construction. The mixes are reheated and compacted at the air voids level specified for each project. The specific tests used for these studies are: consistency binder tests, repeated load (dynamic creep) for permanent deformation evaluation, Dynamic (complex) modulus for stiffness evaluation, flexural beam test for fatigue cracking evaluation and indirect tensile tests for thermal cracking evaluation (Kaloush et al, 2002, 2003a, 2003b, 2003c, 2004). For brevity, only the consistency binder and dynamic complex modulus tests are presented below for an indication of laboratory performance evaluation.

2.2 Binder tests

ADOT studies have shown that by using asphalt rubber as a binder, the film thickness is increased to a value of 19–36 micrometer compared to the typical dense-graded HMA film thickness of about 9 micrometer [Way 2000]. In Arizona, the grade of asphalt binder used as a base to make AR is a PG 58-22 (AC-10, Pen 85-100), in contrast to the typically stiffer grade of PG 64-16 (AC-20, Pen 60-70) used in the mountains. In the desert the AR base asphalt grade is PG 64-16 (AC-20, Pen 60-70) compared to the PG 70-10 (AC-40, Pen 40-50) typically used for dense graded mixes. The 20 percent ground tire crumb rubber particles change the AR temperature susceptibility as shown in Figure 2 [Kaloush 2002, Sotil 2003]. As it can be observed, the viscosity-temperature susceptibility of the rubber modified binders is better (flatter, lower slope) than the conventional (virgin) binder, both at high and low temperature conditions. At lower temperature conditions, the AR binders are softer than the virgin binder. Higher binder viscosities at high temperatures and lower viscosities at lower temperature are indicative of good overall mix performance characteristics. These characteristics also agree with observed field performance, where AR mixes are known to have better response against permanent deformation, and low-temperature cracking.

2.3 Dynamic complex modulus E* tests

By current practice, dynamic complex modulus testing of asphalt materials is conducted per AASHTO TP 62-03 (Witczak et al, 2002). Before presenting test results of the AR-ACFC mixes, a discussion on the state of stress applied in the laboratory should be presented. The E* tests can be performed either unconfined or with varying confinement levels. When comparing dense, gap and open graded mixes, confined dynamic modulus E* tests should be performed to appropriately rank the mixes. The confined Dynamic Modulus E* test is especially important for the open graded mixes because it represents the true state of stress of in the field (surface layer with high confinement stress under loading). The effect of confinement is clearly shown in Figure 3, where typical master curves for an AR-ACFC mixture test results are presented for unconfined and three levels of confinements: 69, 138, and 207 kPa. The confined test results yield much higher moduli and the difference among the level of confinements continue at high temperatures. Table 1 shows the complex modulus test results for several AR mixtures compared to conventional mixtures (by means of a modular ratio comparison). The results show that the AR-ACFC mixes have better moduli values compared to the conventional mixes. A comprehensive documentation of the test

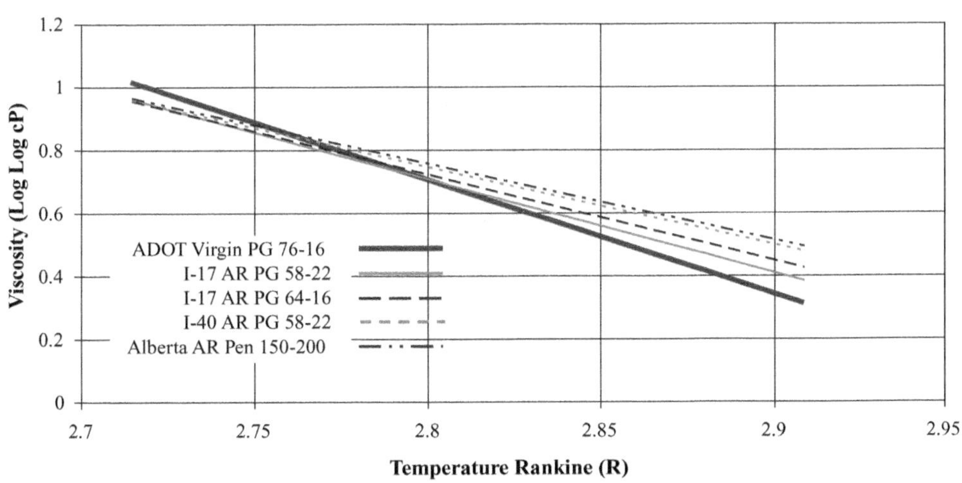

Figure 2. Comparison of the viscosity—temperature relationships for the different binders.

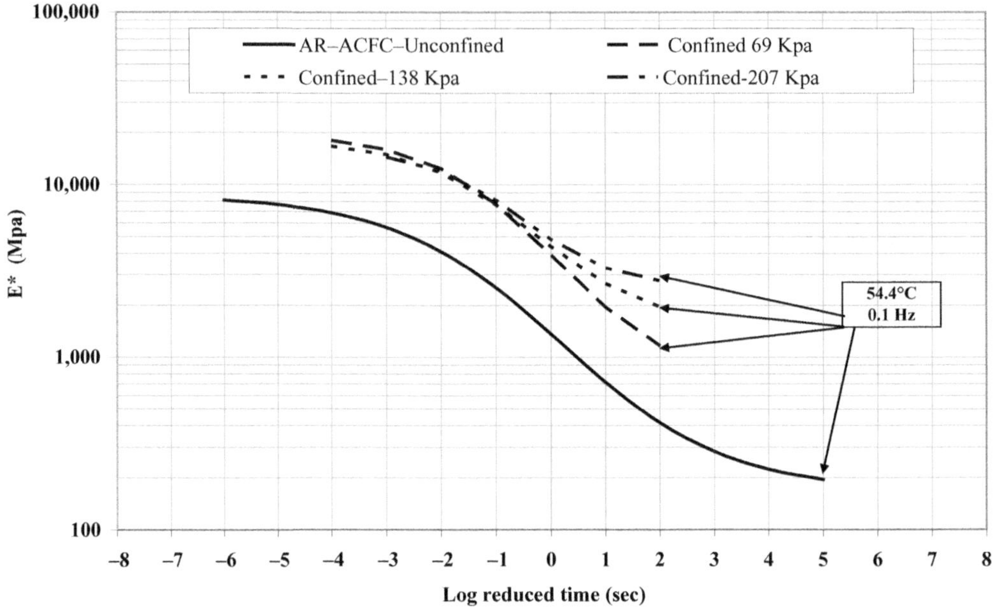

Figure 3. Comparison of E* master curves and effect of confinement for typical AR mixture.

Table 1. Typical modular ratio @ 37.8°C/10 Hz for AR and Conventional mixes, confined testing.

Mix ID	Binder type	AC %	Va %	Nom. aggregate	Modular ratio
ARAC (stiff)	64–16 (AR)	8.9	5.5	19.0-mm GG	1.08
AR-ACFC	58–22 (AR)	8.8	17.6	9.0-mm OG	1.02
ARAC[1]	58–22 (AR)	6.8	10.9	19.0-mm GG	1.00
Conventional	64–22	5.2	6.6	19.0-mm DGM	0.94
Conventional	AC-20	4.1	7.4	37.5-mm DGM	0.77
ARAC (Soft)	Pen 150-200 (AR)	8.9	9.7	19.0-mm GG	0.67

OG = Open graded mixture; GG = Gap graded mixture; DGM = Dense graded mixture; [1]Reference mix.

results on the various asphalt rubber mixtures tested can be found in several publications by the authors [Kaloush 2002, 2003a,b,c and Sotil 2003].

3 TIRE/PAVEMENT NOISE CHARACTERISTICS

Highway traffic noise is generated by engine, exhaust, and tire/pavement interaction. It has been well documented in the literature that the tire/pavement interaction is the dominant source of highway noise. Dominant factors that contribute to the tire/pavement noise include: air pumping, compression of tread block, friction, porosity, absorption, aggregate texture, thickness, age of the pavement and temperature (Biligiri et al, 2008). There are different views among experts on which are the dominant factors or mechanism causing tire/pavement noise. The authors believe that AR-ACFC mixes reduce tire/pavement noise because they act as an acoustic absorber due to the viscoelastic nature of the asphalt mix, and because air pushed through the layer voids (>18%) avoiding the air compression under the tire. Furthermore, the smooth riding surface characteristics and small top size aggregate contribute to less tire deformation with travel and less squeezing of air between the tire and pavement. The viscoelastic characteristics of an AR-ACFC come from much higher asphalt binder content (9–10%) and inclusions of crumb rubber (20% by weight of the binder). When comparing field noise measurements of the rubber versus non-rubber ACFC mixes (as shown in Figure 5), there is reason to believe that rubber particles in the AR-ACFC contribute to less noise due to the sound absorptive characteristic of rubber materials.

Asphalt rubber friction course mixes were shown to be more noise dampening than conventional and other modified mixes in terms of lower field noise measurements. ADOT conducted a pavement preservation experiment on the Interstate—10 (I-10) in 1999 (Scofield 2000). As part of this experiment, 32 replicate test sections were constructed constituting five asphalt concrete pavement wearing courses. The Annual Daily Traffic (ADT) for this highway is about 60,000 with 25% trucks. The five different pavement types included: Permeable European Mixture (PEM), Stone Matrix Asphalt (SMA), Asphalt Rubber Open Graded Friction Course (AR-ACFC), Polymer Modified Open Graded Friction Course (P-ACFC), and ADOT's Standard Open Graded Friction Course (ACFC). The AR-ACFC mix experienced the least cracking and wear after eight years of service with the other test sections showing considerable cracking and wear as illustrated in Figure 4.

On-Board Sound Intensity (OBSI) noise measurements were taken during the Fall 2002 by ADOT as part of the Arizona's Quiet Pavement Program. In addition, Dynatest Inc. obtained new noise measurements in March 2008 as part of a larger California—Arizona highway noise study [Scofield 2000, 2003, CALTRANS 2006]. Figure 5 shows a comparison of average noise readings

Figure 4. Example of pavement surface condition in 2007 for two pavement types after eight years of service.

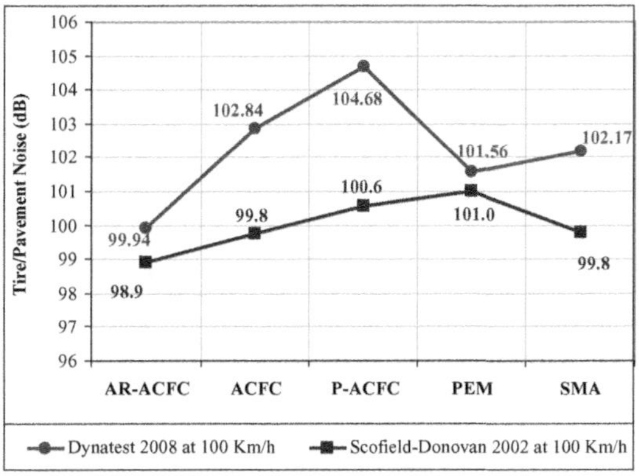

Figure 5. Comparison of average tire/pavement noise (dB) for arizona I-10 test sections.

between the two measurement periods for the five pavement types. The least noise observed for both periods is for the AR-ACFC mixture. This difference agrees with visual distress observations made in the Fall of 2007. Several sections that exhibited higher noise have greater amount of raveling and cracking.

4 TEMPERATURE GRADIENT EFFECT ON PORTLAND CEMENT CONCRETE

Since 2003, ADOT has been placing AR-ACFC mixes over existing Portland Cement Concrete Pavements (PCCP) as part of the Quiet Pavement Program. As of January 2008, approximately 200 kilometers of urban freeway kilometers have been overlaid in the greater Phoenix area. Recently, joint ASU-ADOT studies have been conducted to evaluate the consequences of this paving strategy on the Urban Heat Island (UHI) effect, and the insulating effects of the AR-ACFC mix on PCCP.

Temperature data collected over the past 4 years showed that the darker AR-ACFC surface color increases the surface temperatures of the pavement during daytime. However, the nighttime UHI effect showed a benefit of using the AR-ACFC overlay in reducing the pavement surface temperatures due to the porosity and lower thermal mass of the layer [Golden and Kaloush 2006]. An important consideration is subjecting these surfaces to traffic, which provides the necessary aeration effect. These findings are discussed below.

Several studies have shown that temperature-induced pavement responses are more significant than traffic induced responses (Mahboub et al, 2004). Rigid pavements are affected by a range of environmental factors including temperature. The daily temperature variations generate curling stresses which can ultimately lead to a partial loss of subgrade contact. Temperature gradients in particular in a rigid pavement slab have been shown by some researchers to directly relate to fatigue damage (Masad et al, 1996). A gradient of 0.56°C/25 mm in the slab increases fatigue damage due to truck traffic by a factor of 10 as opposed to a zero-temperature-gradient condition.

Interstate 10 at Ray Road in the Phoenix area offered an opportunity to create excellent modeling conditions for quantifying the effects of AR-ACFC overlays of PCCP. The I-10 pavement north of Ray had been overlaid with AR-ACFC as part of the Quiet Pavements Phase 3 project in May 2005. With two test sites within the vicinity of the start of the overlay, one within the AR-ACFC overlay area and one with only PCCP, pavement temperature data were generated for both conditions of PCCP with and without AR-ACFC (see Figure 6). Also, each site included

Figure 6. Ray road test site at I-10.

Table 2. Comparison of temperature differentials.

Overlay/Traffic case	Max ΔT °C	Min ΔT °C	Range of ΔT °C
With AR-ACFC: Traffic vs. No Traffic	3.5	–2.5	6.0
Without AR-ACFC: Traffic vs. No Traffic	5.0	–1.0	6.0
With Traffic: AR-ACFC vs. No AR-ACFC	4.0	–3.5	7.5
Without Traffic: AR-ACFC vs. No AR-ACFC	4.0	0.5	4.5

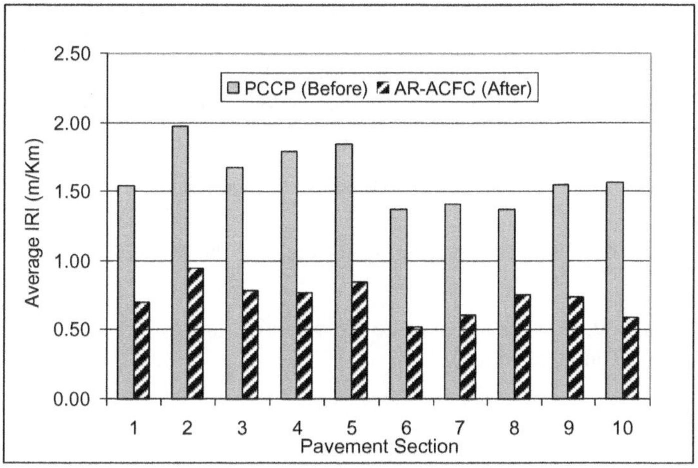

Figure 7. Comparison of roughness measurements before and after the AR-ACFC overlay.

temperature sensor placement located in the shoulder areas of each condition allowing a matrix that included data from areas with and without traffic. In May 2006 the Ray Road I-10 test site was constructed. The existing PCCP was cored and sensors were placed using dowels to insure top to bottom spacing. Sawcut lines were then made from the cores to the shoulder so that future data collection could be made without impacting traffic. Data was recovered for the month of June 2006, and the data were examined in depth.

Table 2 shows comparisons of the average temperature differentials (ΔT) measured for June 2006. Curling stresses for each respective section were calculated. The section with traffic and without the AR-ACFC overlay experiences daytime induced stresses on the magnitude of 25% greater than the section with traffic and with the AR-ACFC overlay. Night time values for the section without AR-ACFC were about 8% higher. When considering that a major portion of the damage to a PCCP structure is thought to result from thermal gradient induced stresses as opposed to traffic loadings, the service life of the PCCP can be significantly extended with the use of AR-ACFC overlays as a pavement preservation strategy.

5 ROUGHNESS

Pavement roughness is measured by ADOT Pavement Management Division using a profilometer. The results are reported in International Roughness Index (IRI). The IRI provides a numeric scale of measuring roughness with a break point between what is considered rough and smooth pavement is often considered to be about 2 m/Km. IRI values less than 1 m/Km are considered very good by Federal Highway Administration guidelines. Typical IRI measurements before and after the AR-ACFC overlay on PCCP pavement sections in the Phoenix area are shown in Figure 7. It is

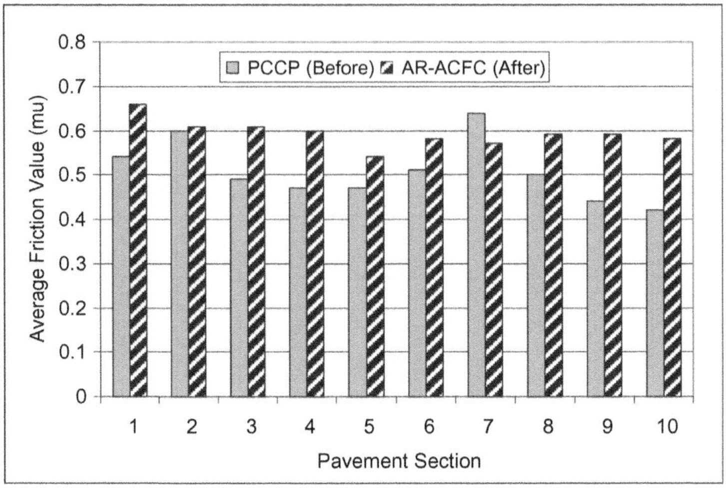

Figure 8. Comparison of friction measurements before and after the AR-ACFC overlay.

observed that the AR-ACFC overlays provide a substantial improvement in ride quality by reducing roughness in half (on the average).

6 FRICTION

Surface friction is also measured by ADOT Pavement Management Division utilizing a MU Meter. Measurements are reported as a skid number (measured value of friction times 100). In Arizona, the intervention level for friction reported for interstate, primary and secondary roadways is 0.34. Similar to the roughness measurements, typical average friction values before and after the AR-ACFC overlay on PCCP pavement sections in the Phoenix area are shown in Figure 8. As expected, and because this is a wet friction test, the data shows an improvement in the skid numbers after the AR-ACFC overlay (except for pavement section 7). In addition, the friction data after the overlay are more uniform as the overlay seems to correct polished surface problem that existed on the original PCCP pavement.

7 TIRE WEAR EMISSIONS

Tire wear contributes to atmospheric Particulate Matter (PM) and is regulated by the United States Environmental Protection Agency because PM has been shown to affect human health. Vehicle emissions are a significant source of both PM2.5 and PM10. Vehicle fleet emissions per mile traveled have been reduced significantly in the last 30 years as a result of improved engine operation and tailpipe controls. However, "zero emission" vehicles will continue to generate PM from tire wear, road wear, brake wear, and re-suspended road dust.

In a 2005 ASU study, aerosol measurement techniques were applied to evaluate tire wear emissions from the vehicle fleet using the Deck Park highway tunnel in Phoenix, Arizona. The Deck Park Tunnel highway surface was PCCP, and was resurfaced with an AR-ACFC layer as part of the ADOT Quiet Pavements Program. This study took advantage of a rare opportunity to sample tire wear emissions at the tunnel before and after the AR-ACFC overlay. The study reported on the measured PM emissions from the on-road vehicle traffic during typical highway driving conditions for the two different roadway surfaces. It presented the analysis of representative tire tread samples for tires wear marker compounds.

The tire wear analysis of representative tire tread samples included tire wear marker compounds, which have been identified and quantified in representative tire wear composite samples from used tires in Arizona, ambient aerosol samples and aerosol samples collected from the two roadway tunnels. Test extraction and separation protocols to determine the amounts of tire wear tracers in tire treads have been also developed as part of this study [Alexandrova et al, 2007].

The emission rates of tire wear tracer compounds (identified as #3 and #4) have been calculated as shown in Table 3. Emission rates of tire wear tracers were found higher at the PCCP road surface than at AR-ACFC road surface. The emission rates of tire wear per kilometer driven at PCCP road surface were 1.4–2 times higher than emission rates of tire wear at AR-ACFC road surface. These findings provided ADOT with revised tire wear emission data for use in their federally-mandated air quality modeling for the Phoenix airshed.

8 COST AND ENERGY CONSIDERATIONS

As it was mentioned earlier, the finished AR product is generally 25 to 50 percent more expensive. However, life cycle cost analysis has shown that a substantial dollar savings can be obtained over the expected life of a project when AR paving strategies are employed [Hicks and Epps, 2000]. This is because AR requires less maintenance costs, has low percent cracking, and therefore has longer service life.

The energy savings by using AR (both gap graded and open graded designs) is quite impressive [Sousa et al, 2006]. Table 4 represents the heat of combustion values for crumb rubber modifier used in AR. The 310267 kJ/Kg represents energy savings in terms of less asphalt concrete overlays is needed over the life of the pavement. It is also the result of using less than one half the thickness of conventional paving material with equal or better field performance. The 310267 kJ/kg energy savings refers to a two inch AR gap graded overlay used as an alternative to a conventional four inch asphalt pavement overlay. This has been a common practice of ADOT for over

Table 3. Tire wear emission rates measured in the deck park tunnel, μg/km.

Tire wear emission rate based on	Experiment 1 (PCCP road surface)	Experiment 2 (AR-ACFC road surface)
Compound #3	354 ± 71	177 ± 35
Compound #4	172 ± 34	120 ± 24

Compounds 3 and 4 are tire wear tracers.

Table 4. Energy utilization (kJ/kg) for asphalt rubber [Sousa et al, 2006].

Process	Energy gain/loss (kJ/kg)
Tire shedding	–1744
Shred transportation	–1744
Granulation	–3586
CRM transportation	–1744
Steel recovery	+1900
Asphalt saved	+209,325 to 465,168
Aggregate saved	+107,860
Total gain/loss	+310,267 to 566,109

135 pavement preservation projects built to date. The 566109 kJ/kg energy savings refers to one inch open graded AR-ACFC mix used on top of a Portland cement concrete pavement in comparison to a five inch conventional asphalt overlay. The 107860 kJ/kg energy savings refers to the mining energy and transport energy associated with using thicker pavements compared to the thinner AR pavements.

9 CONCLUSIONS

This paper discussed specific aspects on why AR-ACFC overlays are considered as good pavement preservation strategy. The focus areas summarized the outcome of several research studies. Many benefits have been identified for the utilization of AR-ACFC overlays. The reduction of cracking has been confirmed through field observations and laboratory testing conducted on the binder and the mixture. The restoration of a smoother ride and the increase in skid resistance were additional benefits identified through several pavement test sections. There is no doubt that AR-ACFC overlays provide the least tire/pavement noise compared to any other road surface type evaluated in these studies. In addition, AR-ACFC overlays have shown to have a significant impact on reducing the induced stresses in PCCP due to thermal gradients. A composite PCCP/AR-ACFC pavement design will be very long lasting. Emission rates of tire wear per kilometer driven on AR-ACFC overlays were reduced by half, having great impact on air quality and human health in urban areas. Life cycle cost analysis and energy considerations were also favorable for this pavement preservation strategy.

REFERENCES

Alexandrova, O., Kaloush, K. and Allen, J. 2007. "Impact of Asphalt Rubber Friction Course Overlays on Tire Wear Emissions and Air Quality Models for Phoenix, Arizona Airshed". *Journal of the Transportation Research Board*, No. 2011, pp. 98–106. Washington, D.C.

Biligiri, K.P., Kaloush, K.E. and Golden, J.S. 2008. Paving Material Properties and Tire/Pavement Noise", accepted for Publication in *Eurasphalt & Eurobitume (E&E) Congress 2008*, Copenhagen, Denmark.

Golden, J.S. and Kaloush, K.E. 2006, "Mesoscale and Microscale Evaluation of Surface Pavement Impacts on the Urban Heat Island Effects". I*nternational Journal of Pavement Engineering*, Volume 7, No. 1, pp. 37–52, Taylor & Francis.

Hicks, R.G. and Epps, J.A. 2000. Life Cycle Costs For Asphalt Rubber Paving Materials, *Proceedings of the Asphalt Rubber 2000 Conference*, Vilamoura, Portugal.

Kaloush, K.E., Witczak, M.W., Way, G.B., Zborowski, A., Abojaradeh, M. and Sotil, A. 2002. Performance Evaluation Of Arizona Asphalt Rubber Mixtures Using Advanced Dynamic Material Characterization Tests. *Final Report*, Arizona State University, Tempe, Arizona.

Kaloush, K.E., Zborowski, A. and Sotil, A. 2003a. Performance Evaluation of Asphalt Rubber Mixtures in Alberta. *Final Report*, Arizona State University, Tempe, Arizona.

Kaloush, K., et al., 2003b. Material Characteristics of Asphalt Rubber Mixtures. *Proceedings Asphalt Rubber 2003 Conference*, ISBN 85-903997-1-0, pp. 129–145, Brasilia, Brazil.

Kaloush, K.E., Witczak, M.W., Sotil, A. and Way, G.B. 2003c. Laboratory Evaluation of Asphalt Rubber Mixtures Using the Dynamic Modulus (E*) Test. CD-ROM Publication. *2003 Annual Transportation Research Board Meeting*. Washington, D.C.

Kaloush, K., et al., 2004. Cracking Characteristic of Asphalt Rubber Mixtures, *Proceedings of the Fifth International Rilem Conference*, PRO 37, pp. 485–492, Limoges, France.

Mahboub, K.C., Liu, Y. and Allen, D. 2004. Evaluation of Temperature Responses in Concrete Pavement, *Journal of Transportation Engineering*.

Masad, E., Taha, R. and Muhunthan, B. 1996. Finite element analysis of temperature effects on plain-jointed concrete pavements, *Journal of Transportation Engineering*.

Scofield, Larry, 2000. Arizona's Pavement Preservation Experiment, Arizona Department of Transportation, Phoenix, Arizona.

Scofield, L. and Donovan, P. 2003. Development of Arizona's Quiet Pavement Research Program, Proceedings of Asphalt Rubber 2003, Brasilia, Brazil.

Sotil, A. 2003 Material Characterization of Asphalt Rubber Mixtures Using the Dynamic Modulus Test. *Master of Science Thesis*. Arizona State University, Tempe, Arizona.
Sousa, J., Way, G.B. and Carlson, D.D. 2006. Energy Consumption of Alternative Scrap Tire Uses, *Proceedings of the Asphalt Rubber 2006 Conference*, Palm Springs, California.
The California Department of Transportation, 2006. Division of Environmental Analysis, Further Development of the Sound Intensity Method of Measuring Tire Noise Performance of In-Situ Pavements, page 9, January 2006.
Way, G.B. 2000. Flagstaff I-40 Asphalt Rubber Overlay Project nine years of success. *TRB 79th annual meeting*, Washington D.C.
Way, G.B. 2000a Asphalt Rubber-Research and Development. 35 years of progress and controversy, *ETRA annual meeting*, Brussels. Belgium.
Witczak, M.W., Kaloush, K.E., Pellinen, T., El-Basyouny, M. and Von Quintus, H. 2002. Simple Performance Test for Superpave Mix Design. NCHRP Report 465. *Transportation Research Board. National Research Council*. Washington D.C. 2002.

Efficient Transportation and Pavement Systems – Al-Qadi, Sayed, Alnuaimi & Masad (eds)

Evaluation of foam asphalt technology for road base construction in the Gulf

Hamad I. Al-Abdul Wahhab, Mirza G. Baig & Isam A. Mahmoud
Department of Civil Engineering, King Fahd University of Petroleum and Minerals, Dhahran, Saudi Arabia

Hisham M. Kattan
Roads and Heavy Equipment Department, Saudi Aramco, Dhahran, Saudi Arabia

Mohammed Al-Mehthel, Saleh F. Al-Idi & Jonathan J. Grosch
Consulting Services Department, Saudi Aramco, Dhahran, Saudi Arabia

ABSTRACT: Foamed Asphalt Technology is a process for expanding the surface area of the asphalt. The process involves adding a small amount of water to hot asphalt create a foaming action that improves the mixing ability of the asphalt (Iowa LTAP 2001) Recycling and treating existing in-place road materials using the foamed asphalt technique is a cost-effective method for producing a base course equal or superior to the original structure at substantial savings compared to conventional rehabilitation procedures or new construction. This technology helps in preserving the supplies of aggregates as it relies on recycling of recycling the existing natural aggregates (http://www.rocktoroad.com/n02foamedasphalt.html) The technology has been successfully employed in Europe, Africa, and the Middle East (Shaybah project) since the late 1980s and is being increasingly adopted in the U.S., Canada, and Australia as its benefits become more widely known.

This paper summarizes the research carried out in the area of Foamed Asphalt Technology that was planned to compare the performance of foamed asphalt pavement mixes with conventional aggregate of road bases. The research work focused on the investigation and evaluation of the feasible use of foamed asphalt technology for Saudi roads using marginal quality construction materials, marl, and Reclaimed Asphalt Pavement (RAP) materials for local applications. Materials included the Ministry of Transport (MOT) granular base class A and B, subbase material class B, and reclaimed asphalt pavement (RAP) material. Foamed asphalt mixes were designed for subbase class B (SB) and RAP material utilizing low percentage of Portland cement. Foamed asphalt mixes were optimized to meet dry and wet indirect tensile strength (ITS) requirements. Designed mixes in addition to granular base class A and B were evaluated for shear strength, angle of internal friction, dynamic resilient modulus at 25°C and wheel tracking test dry at 50°C and soaked at 22°C.

Results indicate that base class A has the lowest rutting followed by base class B then foamed SB class B and finally foamed RAP. Portland cement was effective in reducing ITS loss of foamed asphalt mixes. Resilient modulus testing indicated that SB mix has behavior comparable to base class A. RAP mix has shown the best behavior. Saturation has reduced resilient modulus of all mixes significantly. Foamed asphalt technology can be used successfully to construct road bases from locally available marginal or recycled materials.

Keywords: Foamed Asphalt, Recycling, Stabilization, Granular Base, Subbase

1 BACKGROUND

In eastern Saudi Arabia, there is a scarcity of good quality construction materials. Portland cement or other stabilizers are required to enhance the marginal materials available thereby making the improvement uneconomical and sometimes impractical. Other means of stabilization techniques are needed to upgrade their performance in order to use them for the construction of base or subbase layers in the harsh arid desert climate.

Foamed asphalt mix refers to asphalt concrete mixtures of road-building aggregates and foamed asphalt binder (Figure 1). Although this technology was developed more than 30 years ago and lauded by researchers the world over, it did not gain much acceptance or implementation after its development, mainly because there was no equipment available at that time to produce or apply the product on a commercial scale.

Recently, foamed asphalt technology has gained acceptance as an effective and economical construction material improvement and stabilization technique. This is due primarily to improving aggregate penetration, aggregate coating, handling, and compaction characteristics. Evidence of renewed interest in this technology was observed worldwide recently, largely because new paving equipment has been designed specifically for its application and is easily available. The technology has been successfully employed in Europe, Africa, and the Middle East since the late 1980s and is being increasingly adopted in the U.S., Canada, and Australia as its benefits become widely known (A.A. Loudon & Partners 1996a, 1996b; Bowering and Martin 1976; Csanyi 1957, 1962; Kendall et al. 2001; Lee 1981; Mobil Oil 1973; Muthen 1998a, 1998b; Soter International 1994).

Foamed asphalt does not behave like regular asphalt during the application process. Foamed asphalt mixture acts like soil treatment rather than asphalt concrete mix. It is very moisture sensitive. This moisture sensitivity is the primary reason why foamed asphalt is not appropriate for every road. The technique did not work on trial roads because the water table is high or springs flow underneath these roads. The existence of the water table softens and weakens the subgrade, which has negative effect on the top layer. However, as long as the subgrade is good, foamed asphalt technology will work well (Focus 2003). Recently, the use of foamed asphalt in cold recycling has gained more and more acceptance in Europe, South Africa, and Asia (Chiu-Te et al. 2002).

Asi et al. (1998, 1999, 2002) carried out several laboratory research programs to investigate the feasible use of foamed asphalt technology in Saudi Arabia to improve the prevalent dune sands for possible use as a base or subbase material. Several variables were investigated to evaluate the relative improvement of dune sand as well as to permit the development of design procedures for the future use of foamed asphalt technology in the harsh climatic conditions of eastern Saudi Arabia. Statistical analysis of the results was employed to verify the effects of emulsified asphalt and foamed asphalt treatment, with and without the addition of Portland cement, on the strength characteristics of the treated mixes. The results displayed a significant improvement in the performance of dune sand-foamed asphalt mixes, as compared to that of the emulsified asphalt mixes.

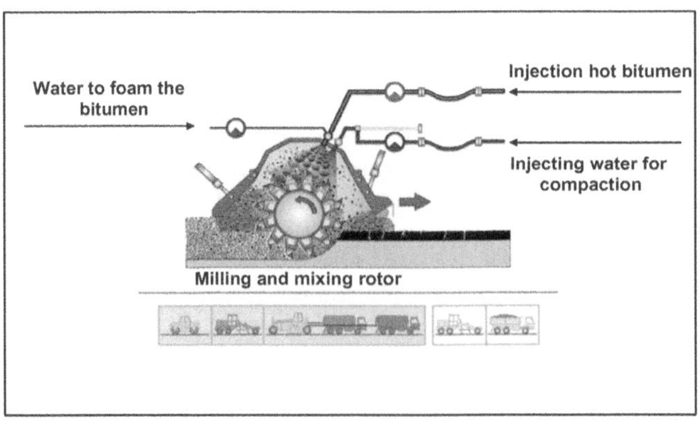

Figure 1. Schematic drawing showing the process in WR 2500, as it accurately meters and sprays asphalt, water and air.

Figure 2. Shaybah access road.

The first use of the foamed asphalt application in Saudi Arabia was in 1997 in Shaybah road. The road was constructed, planed, and the top 200 mm of the existing marl Road was recycled using foamed bitumen and cement before being laid onto the remainder of the marl road (subbase). The recycled road base was rolled and graded to the required profile to produce the foamed asphalt pavement. The foamed asphalt was subsequently surfaced with a slurry seal. The overall assessment of the road's condition and its performance from the data recorded is that the road and the foamed asphalt, in particular, performed remarkably well under the heavy traffic loading and harsh conditions for the intended design life. It has exceeded its original design life as a construction access road for the Shaybah development. The level of deterioration, from recent visual surveys, shows incompetent slurry seal surfacing and it is increasing rapidly and severe in some places (Figure 2), which indicated the need to protect the stabilized layer with a layer of hot asphalt concrete mix.

Nearly all local roads were built according to the old Ministry of Transport (MOT) specifications utilizing aggregate base type A or B and subbase type B (SB). Most of these roads will now need major maintenance or reconstruction. The utilization of base and subbase materials in the reconstruction will result in a major cost saving. In this study, laboratory tests were conducted to compare the performance of foam-based layers with the conventional aggregate-based layers mixes. Old MOT base class A, B and subbase class B were compared to foamed asphalt mixes made with MOT class B subbase and with reclaimed asphalt pavement (RAP) and aggregate.

2 EXPERIMENTAL WORK

2.1 *Material selection*

The selected materials used for this study are MOT class B subbase and MOT class A and B base as per old MOT specifications. The materials were tested for gradation and physical characteristics to assure their conformity to MOT standards.

2.2 *Mix design procedure*

Laboratory mix design procedure for foamed asphalt treated mix was carried out based on the cited literature (Wirtgen 2004) and summarized in Figure 3.

2.3 *Performance testing*

In this study, designed foamed asphalt mixes that include class B subbase and reclaimed asphalt pavement (RAP) and aggregate in addition to the virgin materials that include class A aggregate,

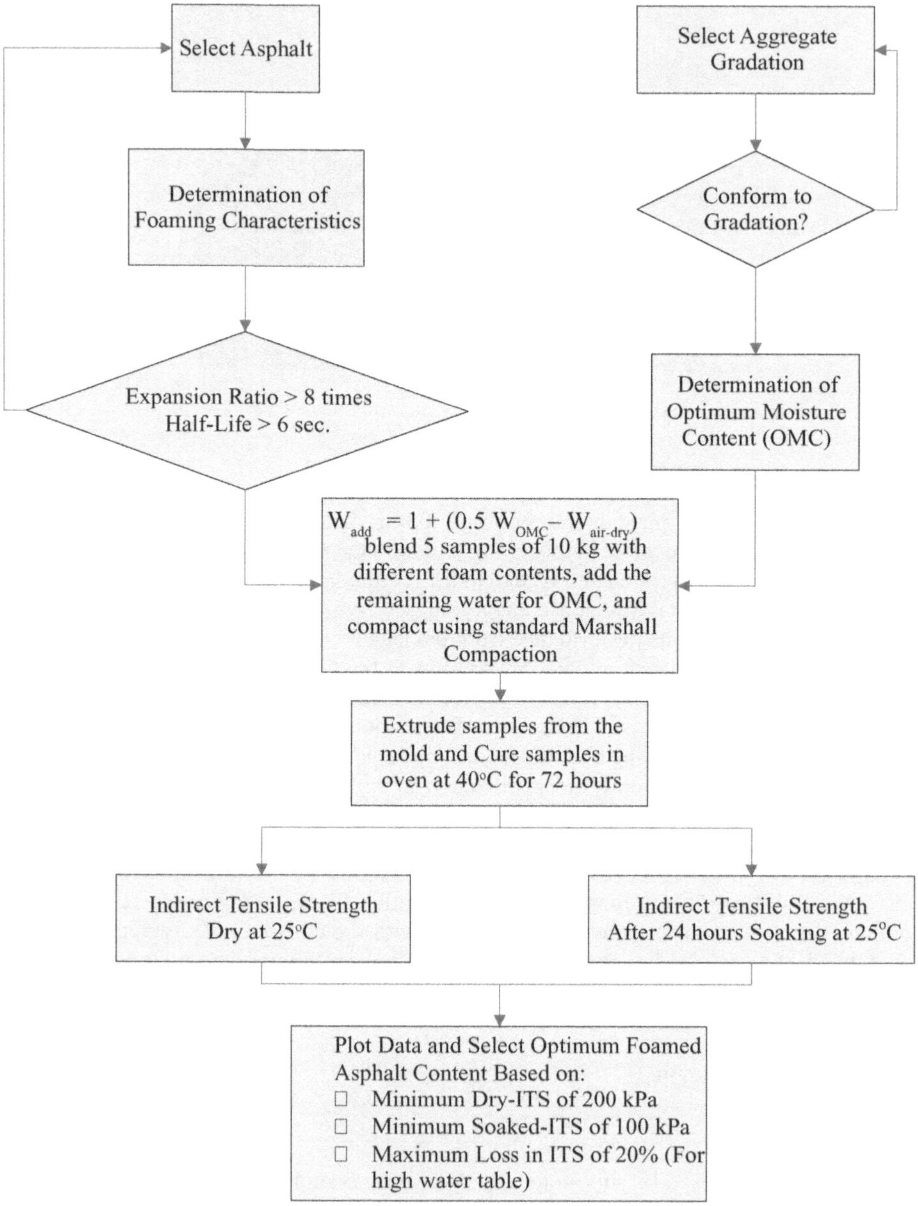

Figure 3. Summary of mix design procedure for foamed asphalt treated mix (Adopted from wirtgen 2004).

class B aggregate and class B subbase (SB) were subjected to indirect tensile strength (ASTM D 4867), California bearing ratio (CBR), dynamic resilient modulus (M_R), and static triaxial tests to evaluate their engineering properties. The study was conducted in three interdependent phases. These test phases included:

1. CBR test for
   - Class A Aggregate
   - Class B Aggregate

- MOT Class B Subbase (SB)
- Reclaimed Asphalt Pavement (RAP) and Aggregate

2. Split tensile strength test for
   - Foamed SB
   - Foamed RAP

3. Resilient modulus and static triaxial test for
   - Class A Aggregate
   - Class B Aggregate
   - Foamed SB
   - Foamed RAP

3 RESULTS

3.1 *Design of foaming characteristic*

Laboratory scale WLB 10 foamed asphalt plant built by Wirtgen Gmbh Company was used to carry out laboratory mix designs and produce foamed asphalt that closely simulates full-scale production, as shown in Figure 4. The unit essentially consists of a kettle to heat the asphalt and calibrated systems for asphalt, water, and air. It enables predetermined volumes of asphalt, water, and air to be injected into the expansion chamber where the foam is formed and is then discharged through a nozzle. The expansion ratio and half-life of the foam can be manipulated by altering the proportion of water that is added to the asphalt and the optimum addition of water determined. Once the design of the foam has been completed, the required volume of foamed asphalt is discharged directly into a sample of aggregate, while it is being agitated in a laboratory mixer. Normally five samples are produced in this way, with varying asphalt contents. Prior to mixing in the foamed asphalt, water is added to bring the material to W_{added} of its optimum moisture content for compaction, where water added (W_{added}) is determined as follows (Wirtgen 2004):

$$W_{added} = 1 + (0.5W_{OMC} - W_{air\text{-}dry}) \tag{1}$$

where

W_{added} = pre-mixing water to be added to the sample
W_{OMC} = optimum moisture content
$W_{air\text{-}dry}$ = water in air dried sample

If required, cement is also added to the aggregate before it is mixed with the foamed asphalt. After mixing with foamed asphalt, water is added to bring the total water content to the optimum

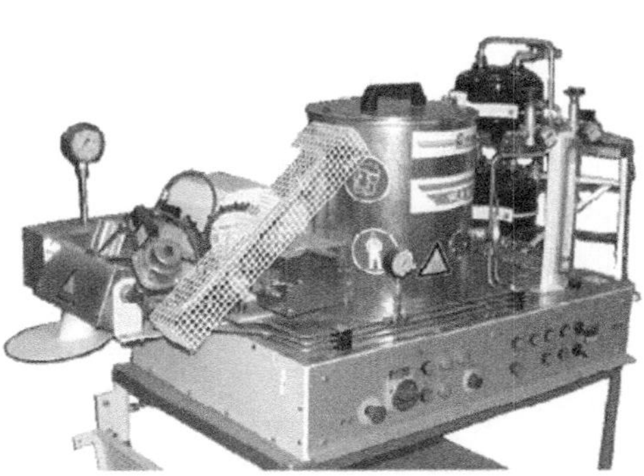

Figure 4. Laboratory scale foamed asphalt plant WLB 10.

moisture content of the aggregate. Both new aggregates and reclaimed asphalt pavement (RAP) were obtained from the field. New aggregates include MOT class A and B base and class B sub-base (SB). The RAP materials were blended together with SB at a ratio of 50/50 by volume, to produce a combined gradation.

Two different foamed asphalt mixes were used in this study, SB and RAP + MOT class B base. Portland cement was evaluated at the ratios of 2 and 4%. Laboratory work carried out on the treated mixes consisted of:

- Plant calibration and optimum foam production.
- Establishing the optimum moisture/fluid content of the two different aggregate mixes.
- Mixing varying percentages of foamed asphalt and water with samples of each mix type.
- Compacting briquette specimens and curing the briquettes prior to testing.
- Carrying out the required engineering tests.

Foamed asphalt was produced using the laboratory foaming machine WLB 10. Calibration of the machine was needed to determine the flow rate of the asphalt at different temperatures and water ratios at a specific pressure (Wirtgen 2004). The asphalt flow rate was measured at different temperatures ranging from 150°C to 180°C since the asphalt flow rate increases with the temperature. The amount of foaming water was varied at each temperature and the expansion ratio and half-life were measured for each water content. Figure 5 shows the variation of the expansion ratio and half-life at temperature of 180°C. It was found that the expansion ratio and half-life increase with increase in temperature. By comparing the foam characteristics at the three temperatures, it was found that the asphalt at 180°C produced the best foaming characteristic. The optimum water content was selected to provide the minimum expansion ratio of eight times and minimum half-life of 6 sec as explained in Figure 6 (Wirtgen 2004). The optimum water content was found to be 3.3% at 180°C.

3.2 *Aggregate gradation and testing*

The aggregate gradation for the supplied materials was selected to meet MOT specifications shown in Table 1 for base material class A and B and SB. Aggregate envelope was selected at the mid-point of the upper and lower limits of the specifications. Aggregate size greater than 19 mm was eliminated since the recommended maximum aggregate size for the foamed asphalt mix design is 19 mm and therefore the aggregate gradation was adjusted accordingly.

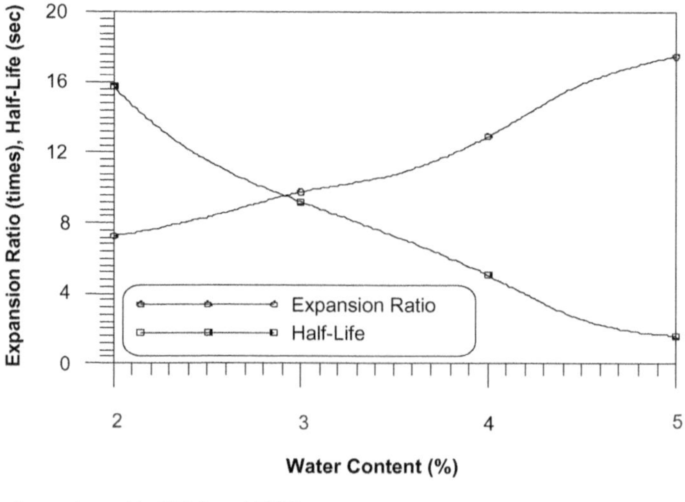

Figure 5. Expansion ratio and half-life at 180°C.

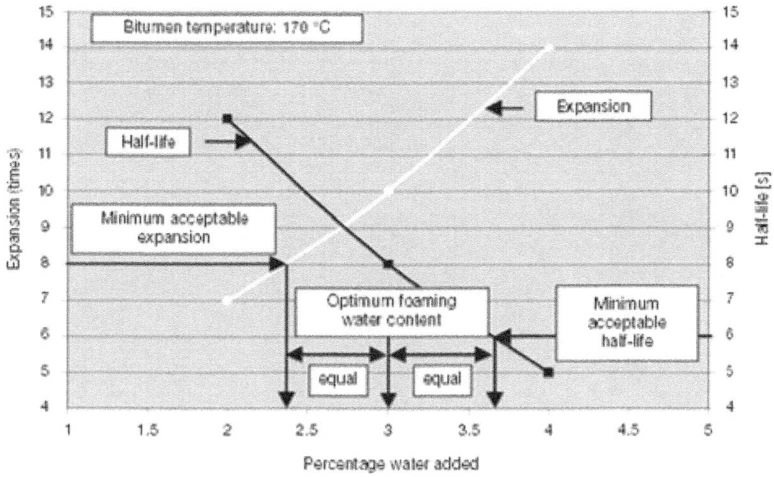

Figure 6. Determination of optimum foaming water content (wirtgen 2004).

Table 1. Selected aggregate gradation.

Sieve No.	Base material class A	Base material class B	Subbase material class B
1	100	100	100
¾	87	96	93
½	69	86	60
⅜	56	75	78
# 4	44	64	63
# 10	29	50	48
# 40	16	30	27
# 200	8	15	13

Aggregate materials were tested to determine the California bearing ratio (CBR) and optimum moisture content (OMC). The OMCs for all aggregates and RAP + MOT class B base blend were determined by using AASHTO T-180. Test results are shown in Figures 7 and 8 and summarized in Table 2. Determination of OMC is a prerequisite for foamed asphalt mix design. Experience has indicated that foamed asphalt and cold asphalt mixes will yield the highest density when compacted at OMC.

3.3 *Optimum foamed asphalt content*

The selected aggregate gradation for SB was mixed with different foamed asphalt percentages to find the optimum foamed asphalt content. Aggregate gradation of base class B was mixed with reclaimed material (RAP) by a ratio of 50:50 by volume. Portland cement was added at the ratios of 2% and 4%. The blended mixtures were compacted by using 75-blow Marshall hammer into standard four-inch samples. The compacted specimens were then tested using indirect tensile strength (ITS @ 25°C) for dry and soaked conditions (24 hrs @ 25°C). The results are presented in Figure 9 for the dry specimens.

Recommendations from experts in the Australian Road Research Board (ARRB) are that the dry indirect tensile strength should be at least 29 psi (200 kPa) and the soaked tensile strength be at least 14.5 psi (100 kPa) For base course layer in the pavement structure (SABITA 1998). The design

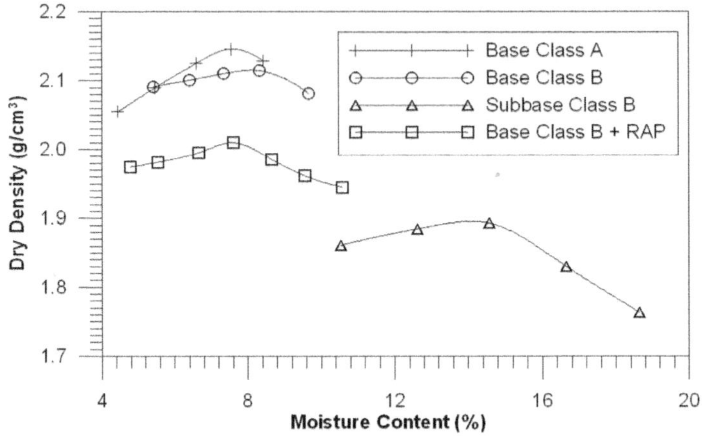

Figure 7. Relation between dry density and moisture content.

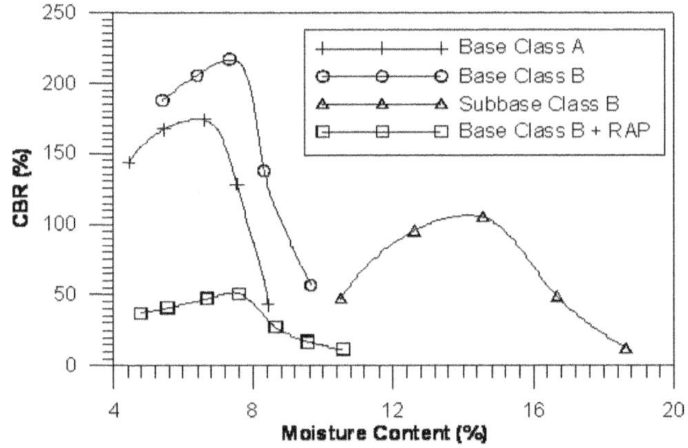

Figure 8. Relation between CBR and moisture content.

Table 2. OMC, Maximum dry density and CBR for the received materials.

Specimen type	Optimum moisture content (%)	Maximum dry density (g/cm³)	CBR (%)
Base class A	7.54	2.15	128
Base class B	7.53	2.11	217
Subbase class B	14.57	1.89	105
Base class B + RAP	7.59	2.01	51

binder content should be selected as the binder content at which the soaked indirect tensile strength is at the maximum. For use as a base course layer in the pavement structure, where the water table is close to the surface, it is believed that an ITS of more than 29 psi (200 kPa) together with more than 80% retained strength of ITS will perform adequately (SABITA 1998). The results show that all mix combinations passed the requirements at the dry condition as shown in Figure 9. Mixes

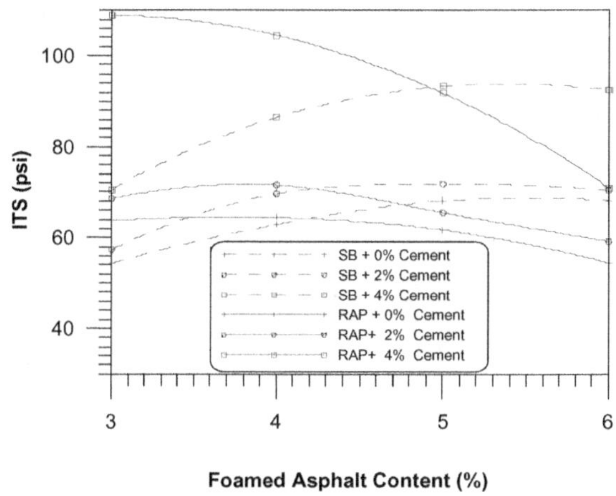

Figure 9. Dry ITS at different cement contents.

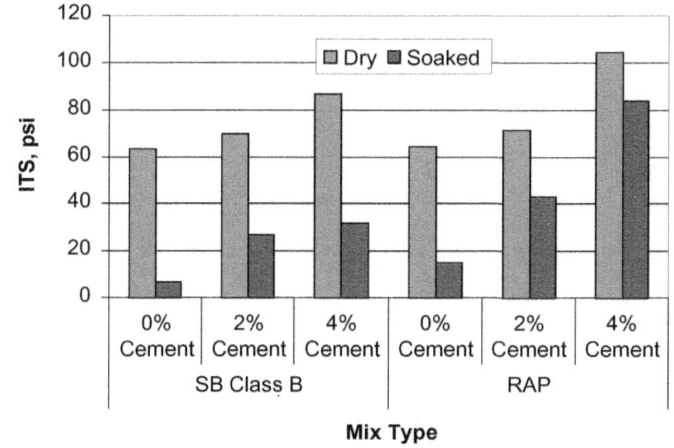

Figure 10. ITS at optimum foamed asphalt content and different cement contents.

without cement appear to be very sensitive to the effect of water as the ITS after soaking dropped below the target limit. Soaked SB at low-foamed asphalt content and 0% cement content showed complete collapse when soaked for 24 hrs. The addition of Portland cement significantly improved the soaked ITS; using lower cement content would be more economical. Therefore, 2% cement with 4% foamed asphalt resulted in an ITS of 26 psi, while the addition of 2% cement with 4% foamed asphalt to the RAP resulted in an ITS of 40 psi. It is noted that the retained ITS of 80% was not possible for SB nor economical. A summary of the ITS results at the optimum foamed asphalt content and different percentages is shown in Figure 10.

3.4 *Dynamic resilient modulus*

Dynamic resilient modulus (AASHTO T-307) was measured using the dynamic triaxial test at 25°C for four mixes including: base material class A, base material class B, SB + 4% foamed asphalt + 2% cement, and RAP + 4% foamed asphalt + 2% cement. The samples were fabricated to have 4 in. diameter and 8 in. height, prepared at the optimum water and foamed asphalt content and compacted to the optimum density using Proctor compactor. Samples were cured for 72 hrs at 40°C. The specimens were tested under different levels of confined pressure (5–20 psi) and deviator stress (10–35 psi) to simulate the traffic loading that the granular base and subbase materials are subjected to in the road. Results fit a logarithmic relation between the bulk stress (deviator + 3x confining) and the resilient modulus with good correlation. The results show that the addition of foamed asphalt to the base class B with RAP improves the M_R significantly as shown in Figure 11. Moreover, granular base class B shows resilient modulus behavior similar

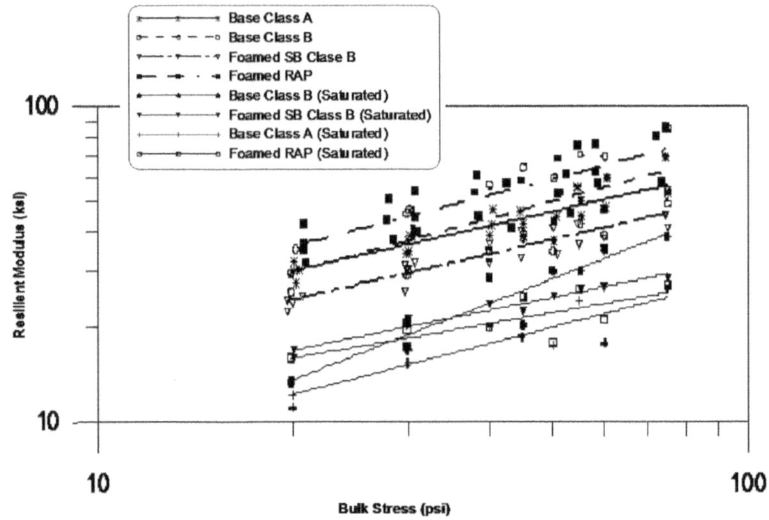

Figure 11. Variation of M_R with bulk stress for all mixes.

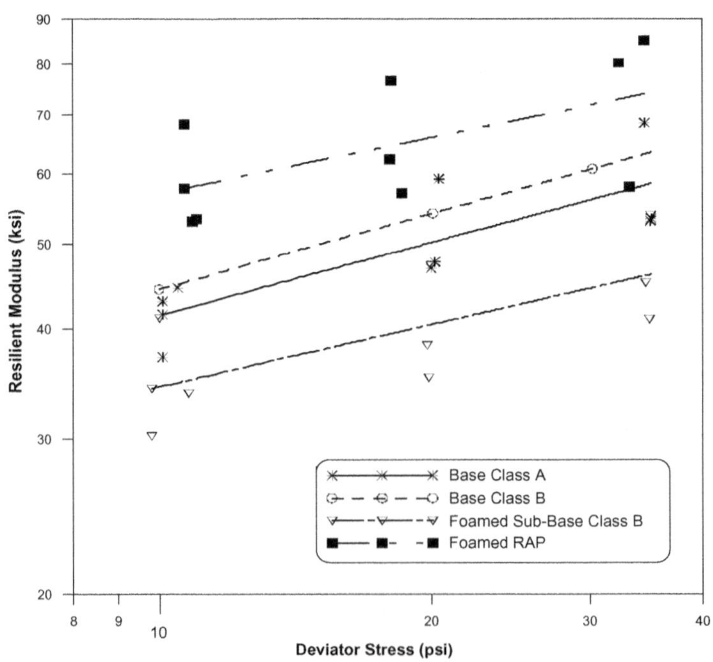

Figure 12. M_R at confined pressure of 20 psi.

to that of granular base class A. Subbase class B provided an M_R of 22 to 45 ksi, depending on the applied bulk stress. All the mixes showed a comparable rate of increase in the M_R. Figure 12 shows a comparison of the material behavior under a confining stress of 20 psi (heavy loading). It can be seen that the RAP material treated with foamed asphalt has the best performance, followed by base course type B and A, and finally by subbase class B treated with foamed asphalt. Subbase class B with foamed asphalt shows an M_R in the range of 35 to 45 psi, depending on the deviator stress (dynamic load stress). Upon soaking, the resilient modulus values dropped by a magnitude of about 50% of the resilient modulus of the dry samples. Foamed asphalt mixes gave values slightly greater than those for base class A.

4 ACCELERATED LOADED WHEEL TEST

Rutting behavior of compacted base mixes was simulated using Wessex engineering wheel track tester. Two slabs for each mix were compacted to the maximum dry density at optimum water content. Eight slabs, 45 cm × 22 cm × 10 cm thick, were prepared from the base material class A, base material class B, subbase material class B + 4% foamed asphalt + 3% cement, and base material class B + 55% RAP + 4% foamed asphalt + 2% cement, and were compacted to the optimum density using dynamic compaction. The slabs were then cured at 40°C for 72 hours and tested dry under a wheel load of 100 psi at 50°C. 50°C has been selected to represent the highest base temperature in Saudi Arabia (Al-Abdul Wahhab and Balghunaim 1994). The results shown in Figure 13 indicate that granular base class A has the lowest rutting followed by granular base

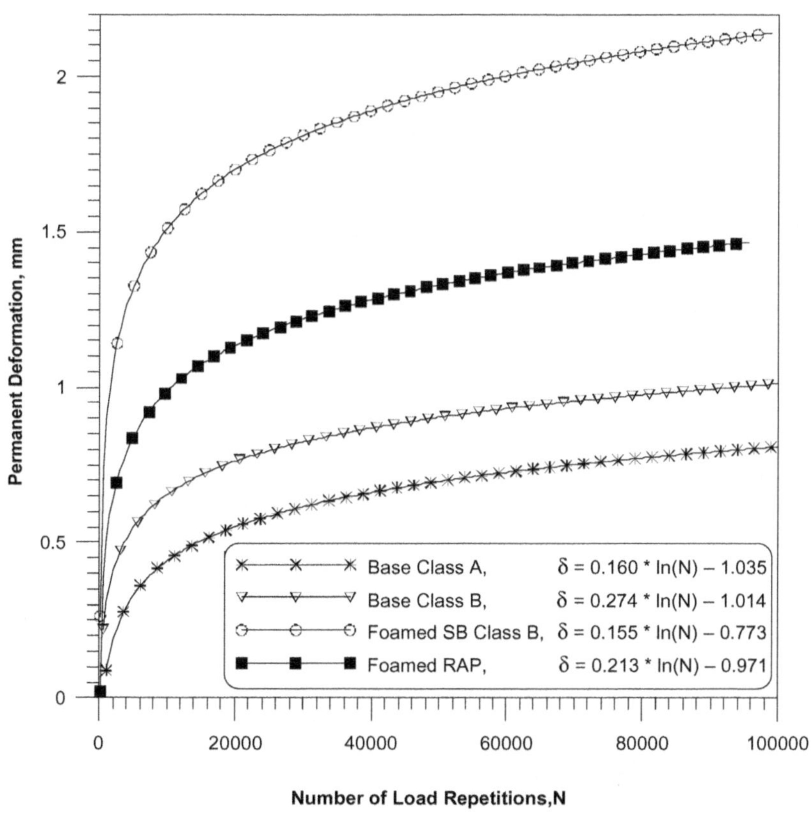

Figure 13. Results of the permanent deformation using wheel track machine, dry at 50°C.

class B. Foamed SB class B material exhibited higher permanent deformation compared to the other mixes while foamed RAP gave less permanent deformation than foamed SB class B but higher than base class B. It must be noted that foamed RAP mix is more sensitive to temperature than the other mixes while foamed SB class B exhibited abrasion rutting as compared to base class A and B which did not show such behavior.

At 100,000 load repetitions, recorded permanent deformations were 0.8, 1.0, 1.5 and 2.1 mm for base class A, base class B, foamed RAP, and foamed SB class B, respectively. It should be noted that an applied stress of 100 psi is two to three times more than that expected in the field. The measured permanent deformation, therefore, may be considered negligible.

5 CONCLUSIONS

The intent of the study was to evaluate RAP and SB materials treated with foam asphalt and compare their behavior to that of granular base class A and B. Observation on the results can be summarized in the following points:

1. Portland cement was effective in reducing ITS loss for foamed asphalt mixes. 2% Portland cement was required to reduce ITS loss to an acceptable value of ITS for roads that are not in a high water table area.
2. Resilient modulus testing indicated that SB mix has behavior comparable to base class A. RAP mix has shown the best behavior. Saturation has reduced resilient modulus of all mixes significantly.
3. Permanent deformation in wheel tracking tests for all foamed asphalt and aggregate base materials was negligible. Rutting was more a function of asphalt thickness than base material.
4. Foamed asphalt mixes should be sealed with layer(s) of hot asphalt concrete mix for best performance.
5. Foamed asphalt can be used successfully to improve the quality of subbase materials and recycled road materials for road base construction.

ACKNOWLEDGMENTS

The authors would like to acknowledge the support provided by King Fahd University of Petroleum & Minerals (KFUPM) and Saudi Aramco for the execution of this research.

REFERENCES

A.A. Loudon & Partners (1996a) In-Place Recycling with Foamed Bitumen, *Series of Technical Bulletins*, South Africa.

A.A. Loudon & Partners (1996b) Foamed Bitumen Mix Design Procedures, *Series of Technical Bulletins*, South Africa.

Al-Abdul Wahhab, H.I., and Balghunaim, F.A. (1994) Asphalt Pavement Temperature Related to the Arid Saudi Environment, *ASCE Journal of Materials in Civil Engineering*, 6(1): 1–14.

Asi, I.M., Al-Abdul Wahhab, H.I., Khan, M.I., and Siddiqui, Z.U. (1998) Stabilization of Dune Sand for Base and Sub-base Applications Utilizing Foamed Asphalt Technology, *1st International Conference on Performance of Roads, Bridges and Airport Pavements in Hot Climates*, Dubai, April 28–29.

Asi, I.M., Siddiqui, Z.U., Al-Abdul Wahhab, H.I., and Khan, M.I. (1999) Stabilization of Sabkha Soil for Road Bases Using Foamed Asphalt Technology, *Fifth Saudi Engineering Conference*, March.

Asi, I.M., Al-Abdul Wahhab, H.I., Al-Amoudi, O.S.B., and Siddiqui, Z.U. (2002) Stabilization of Dune Sand Using Foamed Asphalt, *Geotechnical Testing Journal*, ASTM, Vol. 5, Issue 2.

Bowering, R.H., and Martin, C.L. (1976) Performance of Newly Constructed Full Depth Foamed Bitumen Pavements, *Proceedings of the 8th Australian Road Research Board Conference*, Perth, Australia.

Chui-Te Chiu, Huang, M.U., and Lu, R.C. (2002) A Study on Application of Foamed Bitumen Treated Base in Taiwan, *Final Report*, Department of Civil Engineering, Chung Hua University, March.

Csanyi, L.H. (1957) Foamed Asphalt in Bituminous Paving Mixtures, *Highway Research Board Bulletin*, No. 160, p. 108.

Csanyi, L.H. (1962) Foamed Asphalt for Economical Road Construction, *Civil Engineering*, 32(6), p. 62.

Focus (2003) Foamed Asphalt—A Success on Federal Lands Highway, U.S. Department of Transportation, FHWA, August.

Iowa LTAP (2001). Why experts are talking about foamed asphalt 40 years after its debut, Newsletter of Iowa Local technical Assistance Program, January–February 2001 issue.

Kendall, M., Baker, B., Evans, P., and Ramanujan, J. (2001) Foamed Bitumen Stabilization: The Queensland Experience, *20th Australian Road Research Conference*, March.

Lee, D.Y. (1981) Treating Marginal Aggregates and Soils with Foamed Asphalt, *Proceedings, Association of Asphalt Paving Technologists*, Vol. 50, p. 211.

Mobil Oil Australia Ltd. (1973) Foamed Bitumen Cold Dustless Mixtures, *Technical Bulletin Bitumen*, No. 5.

Muthen, K.M. (1998a) Foamed Asphalt Mixes: Mix Design Procedure, *Contract Report CR-98/077*, Dec. 1998 © 1998 SABITA Ltd. & CSIR Transportek http://foamasph.csir.co.za:81/

Muthen, K.M. (1998b) Foamed Asphalt Mix Design Procedure, *Technical Report*, CSIR Transportek & SABITA Ltd., South Africa.

SABITA Ltd. and CSIR Transportek (1998) Foamed Asphalt Mixes: Mix Design Procedure, *Contract Report CR-98/Draft*, February.

Soter International (1994) Foamed Asphalt Technology Transfer to the Kingdom of Saudi Arabia, *Preliminary Proposal*, Canada.

Wirtgen, GmbH (2004) *Cold Recycling Manual*, 2nd edition, ISBN 3-00-936215-05-7, Windhaven, Germany.

Efficient Transportation and Pavement Systems – Al-Qadi, Sayed, Alnuaimi & Masad (eds)
© 2009 Taylor & Francis Group, London, ISBN 978-0-415-48979-9

Cold Concrete Asphalt: What do we understand about CCA?

J. Hutschenreuther
Ingenieurgesellschaft für bautechnische Prüfungen mbH, Weimar, Germany

ABSTRACT: *ColdConcreteAsphalt* (CCA) shall be used as a road building material and combines the properties and benefits of a concrete and asphalt material—Composite material. Our objective is to produce a construction material that is consistent, "Jointless", and with the ability to be trafficked 24 hours after application. The material shall have load bearing capacities equal or greater than concrete but with the full benefits of a flexible material.

The application of the material can be laid by conventional means i.e. asphalt paver.

1 MATERIAL COMPOSITION

CCA consists of the following

- Bitumen coated aggregates
- Hydraulic binder
- Additives
- Filler
- Sand
- Water

CCA will be produced with a coated aggregate skeleton design, sand and hydraulic binder, modifed by addition of additives. The process of coating the aggregates allows the material to be fully flexible within the finished composition of the design.

- *Bituminum coated aggregates*

The coating process of the aggregates will be carried out with a hot mix which will be replaced by a cold emulsion method. Filler dust is added to the process to avoid the congealing of the aggregates.

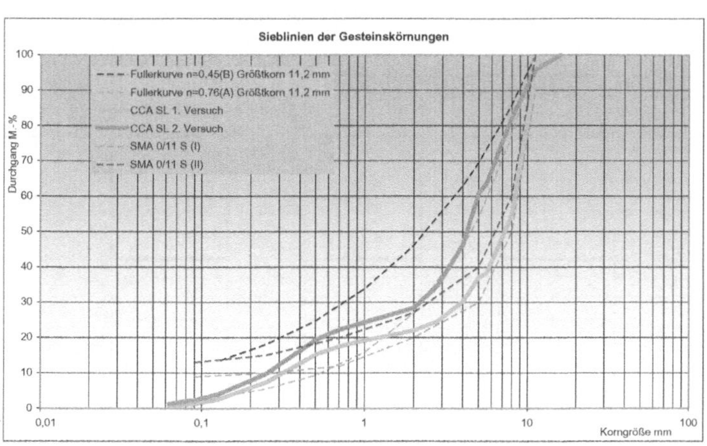

- *Hydraulic Binder*

The hydraulic binder consists of graded cement, micro-silica, fibres, and other additives (super plasticizer, filler powder) and additional additives to prevent shrinkage and enhance workability.

- *Grading detailing the upper and lower parameters*

CCA will be produce with the sieve lines similar to that of a FULLER or SMA grading.
The mix design varies with the binder content and the water binder ratio.

2 MATERIAL BENEFITS

- High stability and resistance to rutting and deformation
- Subject to high point loading
- Short curing process
-

Vergleich verschiedener Kenngrößen		CCA Entwurf 1.Versuchsfeld in N/mm²	CCA Entwurf 2.Versuchsfeld in N/mm²	CCA Entwurf Labtest in N/mm²	Beispiel C30/37 in N/mm²	Beispiel C 35/45 in N/mm²
		CCA			Concrete	
Zylinderdruckfestigkeit	f_{ck}	34	31,5	33,5	30	35
Würfeldruckfestigkeit	$f_{ck, cube}$	40,5	35,0	37,5	37	45
rechnerische Würfeldruckfestigkeit	$f_{ck, cube, calc}$	42,1	39,0	41,5	37,2	43,4
Biegezugfestigkeit	f_{ctm}	4,91	4,61	4,75	2,9	3,2
rechnerisch zu erwartende Biegezugfestigkeit	$f_{ctm, calc}$	3,15	2,99	3,12	2,90	3,21
E-Modul	E_{cm}	20400	16100	17200	31900	33300
rechnerisch zu erwartender E-Modul	$E_{cm, calc}$	33022	32354	32891	31939	33282

-

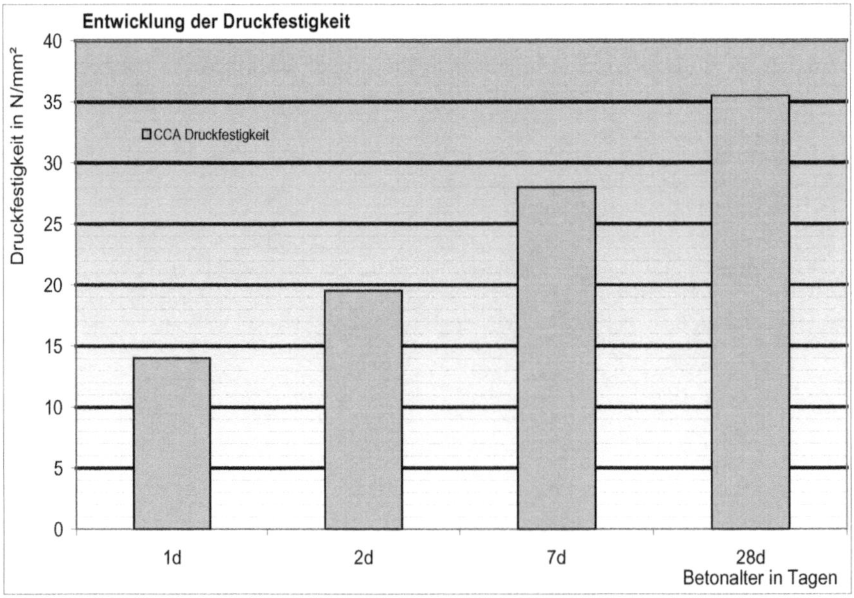

- 
laying

roling/compacting
- 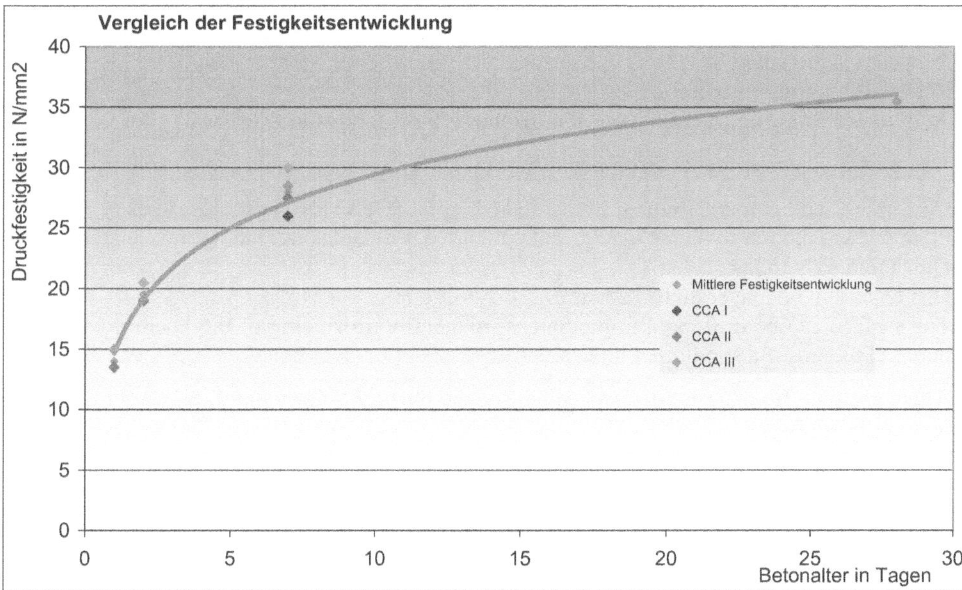

- Jointless system with a choice of texture depth and finishes
- Cold application with ease of paving
- Impermeable or permeable depending on specific use/trafficked after 24 hour period/construction thicknesses of 3 to 5 cm
- Compressive strengths according to the mix designs: 17–35 N/mm^2
- Tensile strength: 2–7 N/mm^2
- E- Modulus: 8.000–12.000 N/mm^2

3 COMMERCIAL APPLICATIONS (NOT LIMITED)

- Heavy duty wearing course for major roads and highways.
- Industrial compounds i.e. Petrol filling stations, waste transfer stations, distributions centres.
- Airport runways, taxiways, and parking aprons.

- Container and Loading yards—Logistic uses.
- Maintenance and repair to existing road and highways.
- Trams and bus lanes.

4 CFA—BITUFOAM

Cold Recycled Foamed Bitumen Stabilised Material (**CFA**) produced to form the foundation or main structural layer of the road pavement.

To realise and develop a **hybrid** Foamix (Bitufoam) material to allow the use of reclaimed construction material within the highway construction process; not so susceptible to weather but with the performance of a tradition bituminous macadam.

Over 90% of FOAMIX (**CFA**) constituents can be recycled materials, thus conserving natural resources.

Components may include:

Road planings
Crushed concrete
Recycled aggregates
Recycled bituminous material
Addition of prime aggregates in various grades (fines, sand, and filler dust)

Equal in engineering but environmentally preferable to traditional hot mix asphalts.

- Cold Recycled Foamed Bitumen Stabilised Material (**CFA**) also has a low density resulting in up to 15% reduction in material required compared with conventional hot mix materials.
- Quick and easy to lay.
- Can be laid in one layer up to 120 mm.
- The surface can be trafficked immediately—generally trafficking of the Foamix will help the material to cure and should not be avoided.

There are specific material designs suited for the purpose of use and these need to be defined before applications.

Examples:

Demolition waste—typical demolition construction waste consisting of concrete, asphalt, kerbs, ceramic tiles & stone. Crushed & screened with no more than 2% organic matter.

Steel Slag—different types of slag can be crushed and screened to a 0/30 mm size to produce good aggregate for a CFA.

VW Terminal, Port of Tyne authority—this is one instance where 27,500 tonnes of blast furnace was shipped into Teesport and with the Wirtgen KMA 200 plant set in close proximity; 28,400 tonnes of Foamix material was produced for 300,000 m^2.

Recycled Sub base—crushed & screened. A Foamix material consists of 90% of recycled material.

Some of the main advantages for using ColdFoamedAsphalt product are:

No expensive plant to apply or insulated vehicles required to maintain temperatures.
Quick and easy to lay, less dense as hot mix which can give material savings.
Can be trafficked immediately and overlaid immediately without the need for tack coat.
Uses recycled and/or marginal aggregates, which saves on waste and saves valuable natural resources.
Energy efficient manufacture which saves on CO^2 emissions.
ISO9002 quality controlled process.

CFA will be used in a combination with **CCA** as the base course.

Efficient Transportation and Pavement Systems – Al-Qadi, Sayed, Alnuaimi & Masad (eds)
© 2009 Taylor & Francis Group, London, ISBN 978-0-415-48979-9

Mix design and rutting resistance of bituminous mixtures containing incinerator bottom ash aggregates

M.M. Hassan & H. Khalid
University of Liverpool, UK

ABSTRACT: Four bituminous mixtures were developed containing 0, 30, 60 and 80% Incinerator Bottom Ash Aggregate (IBAA) by weight. The effect of IBAA on the moisture susceptibility and ageing characteristics of these mixtures was evaluated using a water immersion regime and a simple oven-ageing protocol respectively. In addition, the effect of IBAA on mixtures' rutting resistance was studied using constant strain rate and constant stress compression tests. Results showed that IBAA has a significant effect on the mixtures' retained stiffness after water immersion and oven ageing. The influence on the mixtures' resistance to permanent deformation was found to be dependent on temperature and applied stress level.

1 INTRODUCTION

Incinerator Bottom Ash Aggregate (IBAA) is the most common waste material produced by burning municipal solid wastes in energy from waste plants. It was not uncommon to use IBAA for landfill. However, due to European Union restrictions imposed in 2004 on landfill sites, alternative usage has become necessary. One such alternative is using IBAA in asphalt for roads. Interest in IBAA started in the 1970s and declined in the 1980s then it picked up, again, from 1990 to be a common research field. Properties of IBAA have been studied mechanically, physically and environmentally. These properties likened IBAA to lightweight aggregates. Numerous successful trials were conducted to study IBAA as a road construction material. Nevertheless, higher usage is still desirable.

In this paper, usage of IBAA at high levels in bituminous mixtures was investigated. Bituminous mixtures containing 30, 60 and 80% IBAA by weight, were developed. The effect of IBAA on the moisture susceptibility and ageing characteristics of bituminous mixtures was evaluated. In addition, the effect of IBAA on mixtures' rutting resistance was studied using constant strain rate and constant stress compression tests.

2 BACKGROUND

Visual classification of IBAA fractions shows the presence of metals, slag, stone, ceramic, glass and organic materials. IBAA is a highly porous aggregate, which enhances its water absorption and, consequently, bitumen absorption. IBAA was reported to have high modulus values when it is in a compacted state (Hartlen & Elander 1986). The unit weight, mean Proctor density and optimum moisture content values of IBAA have confirmed its classification as lightweight aggregates (Eighmy et al. 1992). Moreover, It has been shown that its use as unbound aggregate in road construction is safe environmentally (Bruder-Hubscher et al. 2001). As a result, the use of IBAA in numerous road construction applications has been found promising.

Regarding its use in bituminous mixtures, a number of trials have been conducted. It was found that substitution of up to 20% IBAA for virgin aggregates yielded mixtures having aggregate structures that are well developed to resist compaction and rutting (Ogunro et al. 2004). A study by Vassiliadou and Amirkhanian (1999) showed that up to 30% substitution of natural aggregates with IBAA resulted in lower resilient modulus values. Moreover, using 32% IBAA has been suggested as satisfactory for a base course or low quality surface applications (Garrick & Chan 1993). High IBAA replacement has been attempted with success in the US, where 50% was used in a binder course asphalt mixture on a major road trial (Zhang et al. 1999). The moisture susceptibility was investigated for 15% IBAA bituminous mixtures and it was found that the addition of IBAA did not reduce their resistance to moisture damage (Zeng & Kasibati 2003). Despite this positive finding, the effect of water ingress on the mechanical properties of bituminous mixtures containing high levels of IBAA requires further investigations.

This paper presents the results of a laboratory study in which the effect of moisture ingress on the stiffness modulus of bituminous mixtures with various IBAA contents has been evaluated using a simple water immersion regime. Changes in the stiffness due to a common accelerated laboratory ageing procedure have also been measured. The investigated mixtures were then subjected to monotonic constant strain rate and constant stress compression tests at different temperatures to study their resistance to permanent deformation.

3 MATERIALS AND MIX DESIGN

3.1 *Bituminous mixtures' materials*

Three materials were used in this study: limestone, IBAA and bitumen. Limestone from North Wales, UK was chosen as the control aggregate in the bituminous mixtures. It was supplied in six sizes: 20, 14, 10, 6, 3 mm—dust and filler. IBAA from Teesside, UK was supplied in two sizes: 20–10 mm and 10 mm—down. The binder used was 100/150 Pen bitumen sourced from Venezuela.

A limestone control blend was first developed, which meets the BS EN 4987-1 (BSI 2003) grading envelop specifications for a 20 mm binder course. Based on this blend, limestone was replaced by IBAA at different levels: 30, 60 and 80% by weight. The developed blends were then utilised to produce hot bituminous mixtures. As a result, four mixtures were manufactured: mix OA contains 0% IBAA and is the control mix; mix AA, BA and CA contain 30, 60 and 80% IBAA respectively.

3.2 *Mix design*

Prior to mixing, the aggregates and bitumen were heated to a mixing temperature of 160°C. Then the mixtures were laid in the moulds and compacted to slabs using a laboratory roller compactor. Cylindrical specimens of 100 mm diameter and 65 mm height were cored from the slabs. The specimens' volumetric properties were calculated, encompassing the compacted density of the mix (CDM), voids in mineral aggregate (VMA) and theoretical voids in mix (VIM). As can be seen from Figure 1, the IBAA mixtures are lighter than the control limestone material due to the difference in their respective specific gravities, which were 2.42 and 2.72. Subsequently, the indirect tensile stiffness modulus (ITSM) at 20°C was measured in accordance with BS EN 12697-26 (BSI 2004).

The design parameters in Figure 1 indicated that the VIM and ITSM values predominantly decreased with increase in binder content. The CDM reached its maximum values at 4.5, 5.5, 6.5, and 7.5% binder content for mixtures 0A, AA, BA, and CA respectively, while the VMA were at their lowest at the same binder contents. The volumetric approach, i.e. max. CDM and min. VMA values, was adopted to determine the optimum binder content (OBC). As a result, binder contents of 4.5, 5.5, 6.5, and 7.5% were chosen as OBCs for mixtures 0A, AA, BA, and CA respectively. At these OBCs the voids filled with bitumen (VFB) were found to be 55, 64, 56, and 53% which indicates the high absorption properties of the IBAA.

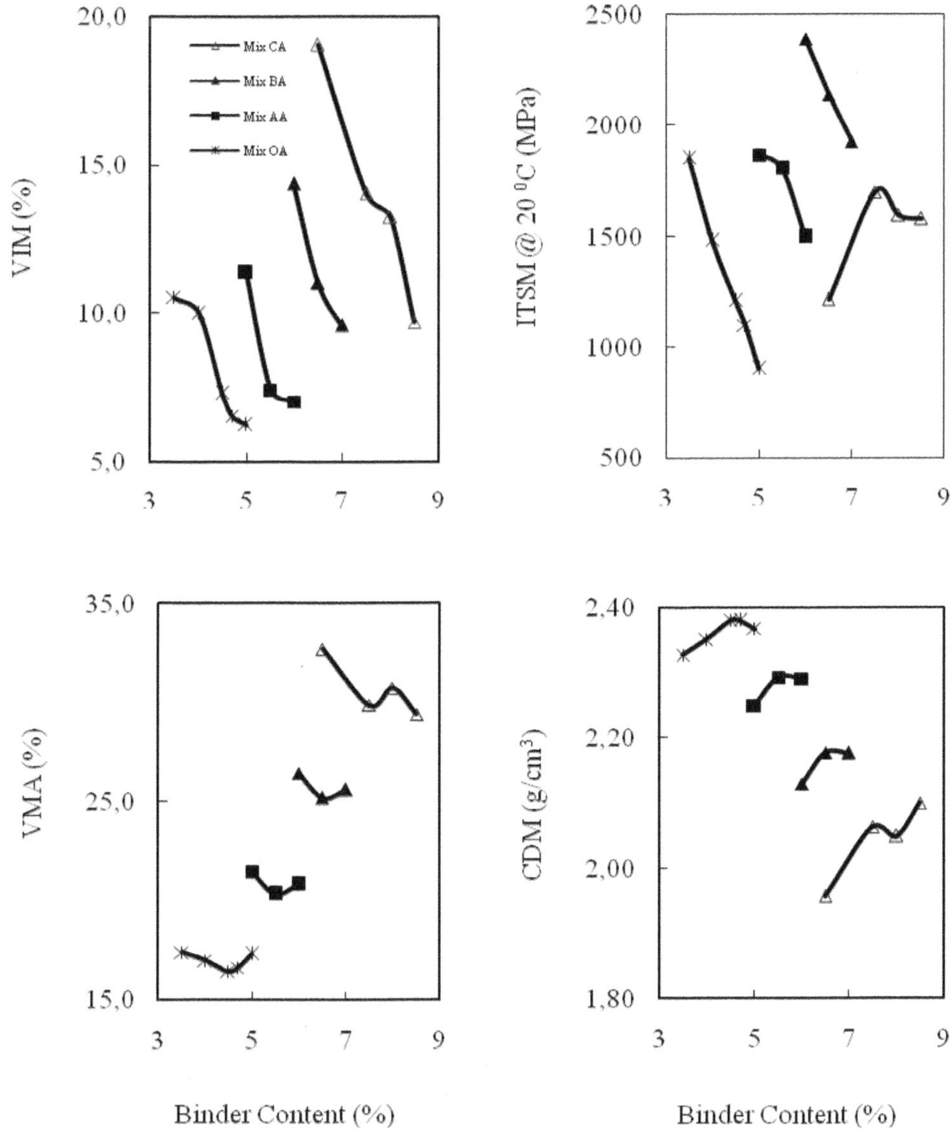

Figure 1. Mix design parameters.

3.3 *Effect of ageing on stiffness*

A laboratory accelerated ageing test was conducted with the aim of relating the in-service stiffness modulus of bituminous mixtures to their initial stiffness modulus measured soon after manufacturing. The protocol used was presented by the UK Highways Agency (Chaddock & Pledge 1994) in which the specimens are mounted in the percentage refusal density (PRD) moulds and then heated in a temperature controlled forced air draft oven at 60°C for 48 hrs. Ordinary 100 mm diameter compaction moulds, which are similar to PRD moulds, were used in this study. The stiffness of the specimens was measured before and after ageing to predict the stiffness equivalent to one year in service, in accordance with Chaddock and Pledge's assumptions.

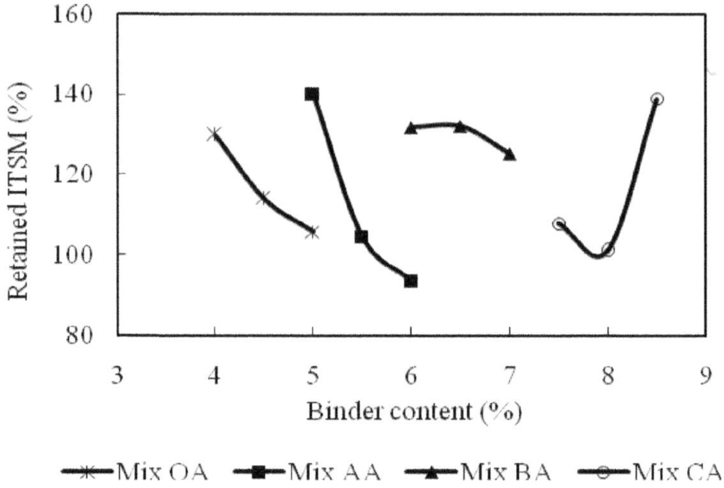

Figure 2. Effect of ageing on stiffness of IBAA bituminous mixtures.

Figure 2 shows that mix AA, at its OBC, underwent least change in stiffness due to ageing, followed by mixes CA, OA and BA at their respective OBCs. This can be attributed to the fact that mix AA has the lowest voids, which aided in resisting ageing effects. Mix CA, although had highest voids, the binder film was thinnest as most of the binder was absorbed by the large proportion of IBAA in the blend. Thus, the effect of the binder was minimal. Mixes OA and BA showed commensurate ageing to their respective voids contents.

The increase in ITSM was in the range of 5 to 15% for mixes OA, AA and CA, at their OBCs, which is good from the durability viewpoint. On the other hand, for mix BA, the increase was about 30%, at its OBC, which is not good at low temperature conditions as the material may become more brittle and fracture can happen earlier. However, the rate of increase of the ITSM is expected to diminish after the first year of mixture production (Said 2005).

3.4 *Effect of IBAA on moisture susceptibility of mixtures*

Moisture-induced damage of bituminous mixture can drastically reduce a pavement's expected design life. This phenomenon is referred to as stripping and results when moisture causes a loss of bond between the aggregate and the binder. To combat this striping, proper mix design is essential. Nonetheless, if a mix is properly designed not compacted correctly, it may still be susceptible to moisture damage because of high air void content that permits water to enter the mixture. Therefore, bituminous mixtures should be tested in a situation where moisture does infiltrate the air voids of the mixture. In most moisture susceptibility tests, coated aggregate is soaked in water under controlled conditions of time and temperature. Stiffness measurements are taken before and after immersion in order to assess the mixture susceptibility to water ingress. One of these regimes is that stipulated by the British Board of Agrément (1998).

The procedure entails the determination of stiffness of the unconditioned specimens. The specimens are then placed in a vacuum dessicator at 20°C and covered with distilled water, and a vacuum of 510 mm Hg applied for 30 minutes. They are then moved from the dessicator and placed in a water bath at 60°C for 6 hours. After that, the specimens are immediately moved to a cold water bath at 5°C for 16 hrs, and then placed in a moderate temperature bath at 20°C for 2 hours. This thermal conditioning cycle, 6 hrs at 60°C, 16 hrs at 5°C and 2 hrs at 20°C, is repeated three times one after the other. At the end of the last cycle, the specimens are conditioned overnight at 20°C and their stiffness determined.

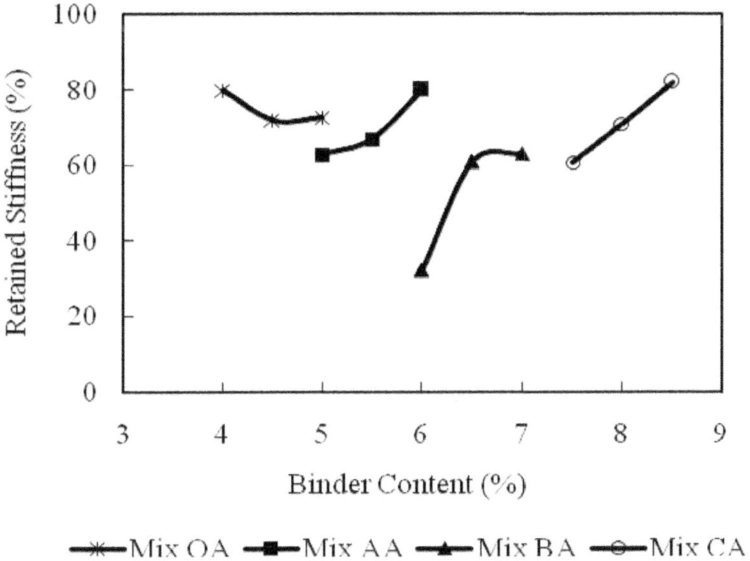

Figure 3. Effect of IBAA on moisture susceptibility of bituminous mixtures.

The retained stiffness results, shown in Figure 3, indicate that all mixtures suffered a loss in their stiffness values due to moisture effects. However, this retained stiffness ranged from 60 to 70% over a narrow range of binder contents. As a result, the IBAA could be considered as having only a minor effect on moisture susceptibility of bituminous mixtures by contrast to the control mixture.

4 EFFECT OF IBAA ON MIXTURE RUTTING RESISTANCE

Rutting is one of the most common distress failure modes in UK flexible pavements. It mainly occurs due to densification and shear deformation. Rutting resistance in bituminous mixtures is influenced by the properties of bitumen and aggregate, volumetric composition of the mix, compaction level, and environmental conditions.

In this paper, the compressive deformation behaviour of the four mixtures was investigated over a wide range of strain rates, stresses and temperatures under monotonic uniaxial compression test conditions. Two types of uniaxial tests were conducted: constant strain rate and constant stress (creep) tests. These two tests were used to determine the steady state axial stresses and strain rates which identify the monotonic permanent deformation behaviour of the tested mixtures. This steady state permanent deformation behaviour was used as a tool to express mixtures' rutting resistance.

4.1 *Constant strain rate test*

In the constant strain rate tests, 67 mm diameter and 134 mm high specimens were tested at temperatures of 20 and 40°C. To ensure uniformity of temperature, samples were conditioned in a temperature control cabinet, at the testing temperature, for at least 12 hrs preceding the test. Specimens were then placed between two polished steel platens, smeared with silicon grease, to reduce friction, and a small pre-load was applied to take out any relaxation in the system. Specimens were then allowed to deform under a uniaxial compression stress. A 10 tonne compression machine was used to apply constant strain rate loads up to failure. Four different strain rates of 0.00005, 0.0001, 0.005, and 0.001 1/s were adopted in this test. For each strain rate, the stress-strain relationship

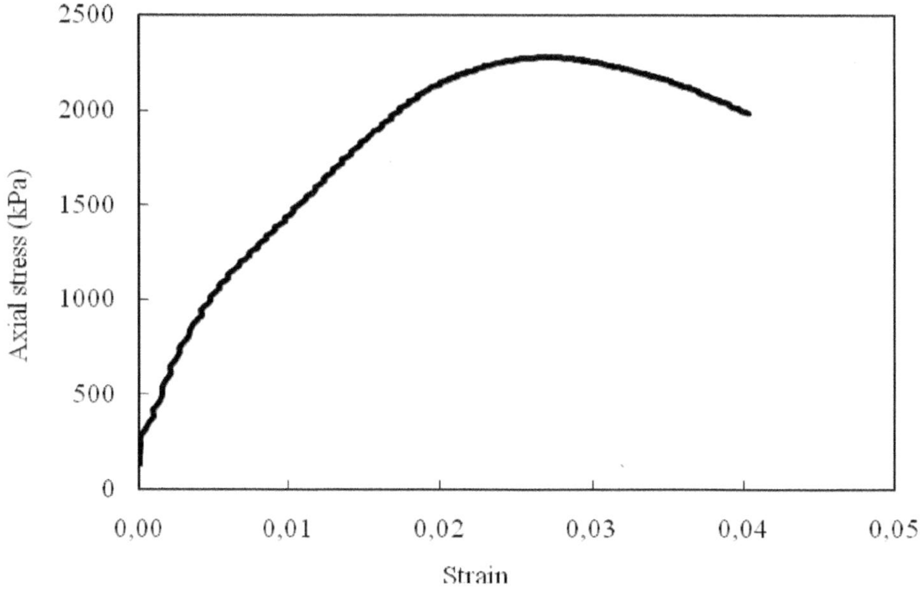

Figure 4. A typical constant strain rate test result; mix AA under 0.001 1/s strain rate and 20°C.

was captured and recorded by a computer. The steady state axial stress corresponding to each strain rate was determined using a method proposed by Ward (1971). In this method, the maximum observed stress, at each strain rate, was defined as the steady state stress at that particular strain rate as shown in Figure 4.

4.2 *Creep test*

In constant stress (creep) tests, the same sample geometry and instrumentation as in the constant strain rate tests were used. Specimens were tested at 5 and 20°C and the x-head was allowed to apply a constant load, ranging between 500 and 3000 kPa, on samples for 1800 sec. This test duration was found to be adequate for mixtures to reach steady state conditions. In each test the deformation strain over time was recorded. Figure 5 shows typical creep test results, from which it can be seen that a creep curve can be divided into two regions. In the first region, primary creep, the material behaved elastically. In the second region, the strain rate was approximately constant. This region was called the secondary creep region and its strain rate was adopted as the steady state strain rate corresponding to a particular value of applied compressive stress. A third region, not visible in Figure 5 but appeared only in some tests, exhibited tertiary creep in which the strain rate increased as the specimen became progressively damaged.

4.3 *Steady state deformation behaviour*

Steady state stresses were plotted against their corresponding steady state strain rates to produce steady state permanent deformation curves for mixtures at each test temperature. The experimental results were found to be adequately captured by a Modified Cross Model (MCM), which was firstly developed by Cross (1965) then modified by Cheung (1995) and Deshpand (1997). The MCM was used successfully, in numerous studies, to predict bituminous mixtures permanent deformation characteristics (Deshpand & Cebon 2000, Collop & Khanzada

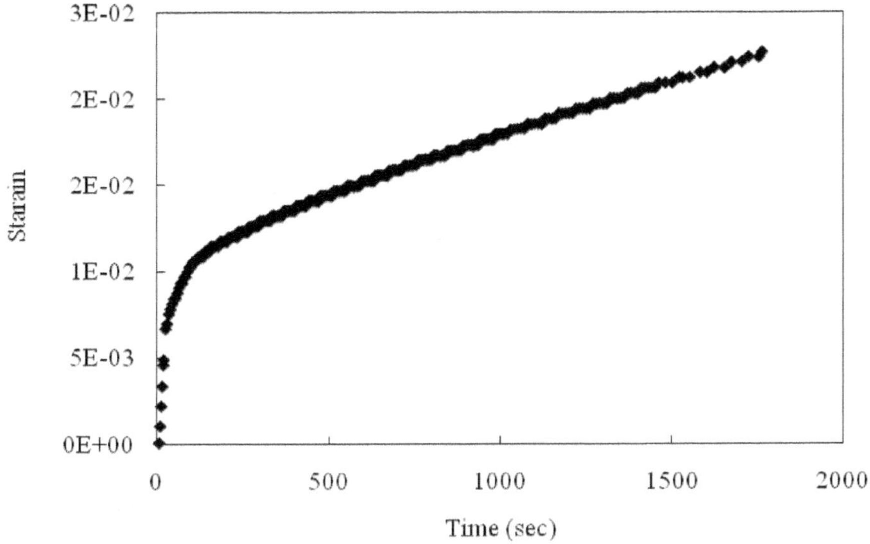

Figure 5. A typical creep test result; mix BA under 2000 kPa stress and 5°C.

2001, Ossa et al. 2004, Ossa et al. 2006). The MCM, expressed in Eqn 1, showed that, for low strain rates the equation reduces to a linear viscous law and, for high values of strain rates, the equation reverts to a non-linear viscous law.

$$\sigma = \frac{S\sigma_0\varepsilon'}{\varepsilon'_p}\left\{\frac{1}{1+\left(\frac{S\varepsilon'}{\varepsilon'_p}\right)^m}\right\} \tag{1}$$

where: σ is the uniaxial stress, ε' is the uniaxial strain rate, S is the stiffening factor, and σ_0, m, ε'_p are material constants for mixtures.

Steady state permanent deformation results, shown in Figure 6, indicated that, at low temperatures, such as 5°C, the higher the IBAA content the higher was the rutting resistance of mixtures, especially at low and medium stress load conditions. At high stress levels, i.e. more than 5000 kPa, IBAAs' effect on rutting resistance was minimal. The reason beyond this behaviour may be attributed to the increase in mixture stiffness and the high angularity of IBAA particles which leads to high particle to particle interlock. These factors were found to have a lesser effect at high stress levels due to the expected crushing that occurs at these levels. At moderate and high temperatures, i.e. 20 and 40°C, the same behaviour was noticed, as in Figures 7 & 8, excluding mix CA which comprised 80% IBAA. This mix showed a slightly improved rutting resistance compared with mix OA but a much lower one than mix BA which contained less IBAA. The reasons behind this behaviour are not entirely known.

Generally, results showed that IBAA had a noticeably positive effect on the mixtures' permanent deformation behaviour and, consequently, their rutting resistance.

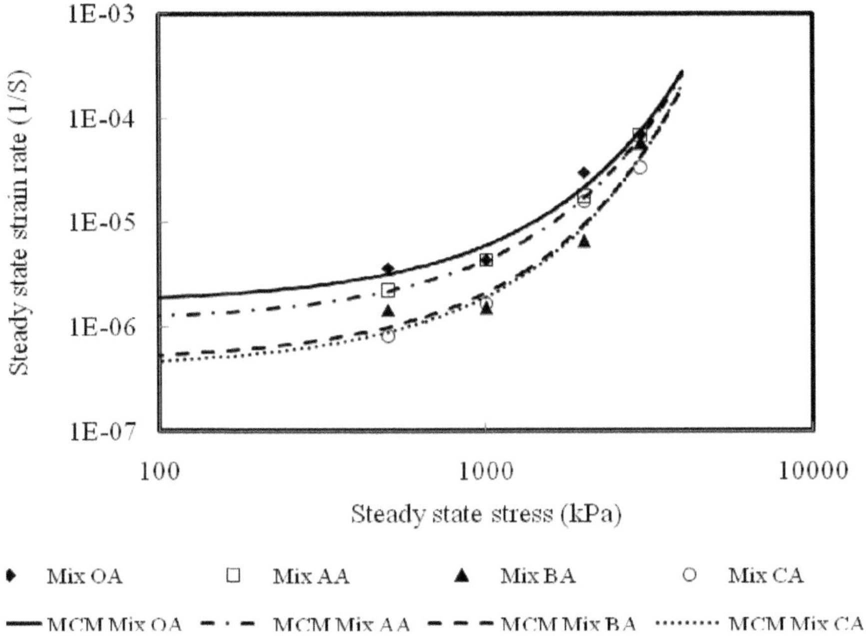

Figure 6. Effect of IBAA on rutting resistance at 5°C.

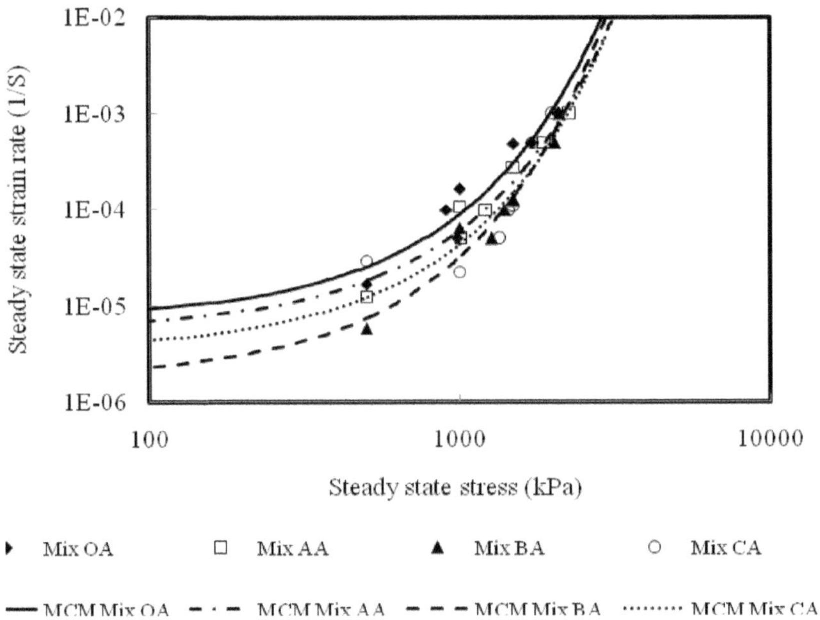

Figure 7. Effect of IBAA on rutting resistance at 20°C.

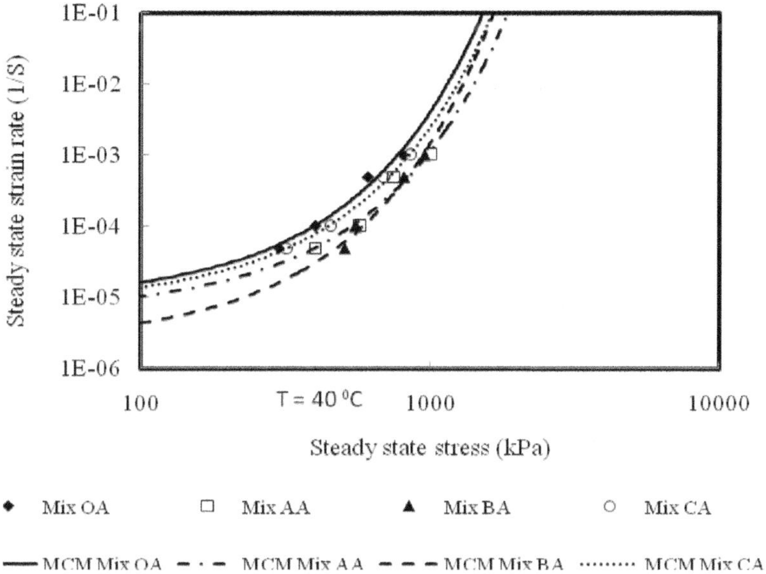

Figure 8. Effect of IBAA on rutting resistance at 40°C.

5 CONCLUSIONS

From the preceding sections, the following remarks can be made:

1. It is clear that the more IBAA was added, the higher the optimum binder content was. Higher IBAA contents led to more voids owing to the high absorption properties of IBAA.
2. Addition of IBAA significantly improves the mixtures' ITSM values, excluding mix CA, at 80% IBAA.
3. The control mix retained stiffness, after a water immersion regime, was found to be 70%. Using IBAA led to a reduction in the retained stiffness of the mixtures. However, this reduction was less than 10% compared with the control mix.
4. IBAA mixtures' stiffness increased when subjected to an ageing protocol. This increase was significantly high at 60% IBAA, whereas the 0, 30 and 80% mixtures underwent insignificant changes to their stiffness values.
5. At low temperatures, the higher the IBAA content the higher was the rutting resistance of the bituminous mixture, especially at low and medium stress conditions. At high temperatures, the same trend of behaviour was noticed except for mix CA which contained 80% IBAA content.

ACKNOWLEDGMENT

The authors are indebted to Ballast Phoenix, Tarmac and Nynas Bitumen for their technical and financial support of the project. The authors are also grateful to the Egyptian government for the award of a study scholarship to pursue this research.

REFERENCES

British Board of Agrément. 1998. Guidelines document for the assessment and certification of thin surfacing systems for highway. SG3/98/160. SG3. *British board of agreement. Highway authorities product approval scheme (BBA/HAPAS).*

Bruder-Hubscher, V. Leroy, F. Coughanowr, C. & Enguehard, F. 2001. Utilisation of bottom ash in road construction: evaluation of the environmental impact. *Waste Management and Research* 19(6): 545–556.
BSI. 2003. Coated macadam (asphalt concrete) for roads and other paved areas: specification for constituent materials and mixtures. *British institution, BS 4987-1:2003.*
BSI. 2004. Bituminous mixtures—test methods for hot mix asphalt—stiffness, *British institution, BS EN 12697-26:2004.*
Chaddock , B. & Pledge, A. 1994. Accelerated and field curing of bituminous road-base. *Project report PR87. Transportation Research Laboratory.*
Cheung, C. 1995. *Mechanical behaviour of bitumens and bituminous mixes.* PhD thesis, Engineering department, Cambridge University.
Collop, A. & Khanzada, S. 2001. Permanent deformation in idealized sand asphalt bituminous mixtures. *International Journal of Road Materials and Pavement Design* 2(1): 7–28.
Cross, M. 1965. Rheology of non-newtonian fluids: a new flow equation for pseudoplastic systems. *Journal of Colloid Science* 20: 417–437.
Deshpand, V. & Cebon, D. 2000. Uniaxial experiments on idealized asphalt mixes. *Journal of Materials in Civil Engineering* 12(3): 262–271.
Deshpand, V. 1997. *Steady state deformation behaviour of bituminous mixes.* PhD thesis, Engineering department, Cambridge University.
Eighmy, T. Grees, D. Zhang, X. Tarr, S. &Whitehead, I. 1992. Bottom ash utilization evaluation for the concord, New Hampshire waste-to-energy facility. *Environmental research interim report. University of New Hampshire.*
Garrick, W. & Chan, K. 1993. Evaluation of domestic incinerator ash for use as aggregate in asphalt concrete. *Transportation Research Record* 1418: 30–34.
Hartlen, J. & Elander, P. 1986. Residues from waste incineration—chemical and physical properties. Report no. SGI VARIA 172. *Swedish geotechnical institute, Linkoping, Sweden.*
Ogunro, V. Inyang, H. Young, D. & Oturkar, A. 2004. Gradation control of bottom ash aggregate in Superpave bituminous mixes. *Journal of Materials in Civil Engineering* 604–613.
Ossa, E.A. Taherkhani, H. & Collop, A. 2006, Compressive behaviour of asphaltic mixtures. *Association of Asphalt Paving Technologists* 620–655.
Ossa, E.A. Deshpand V. & Cebon, D. 2004. Uniaxial monotonic and cyclic behaviour of bituminous mixes. CUED/C-MICROMECH/TR.95. *Cambridge University. Engineering department report.*
Said, S. 2005. Aging effect on mechanical characteristics of bituminous mixtures. *Transportation Research Record* 1901: 1–9.
Ward, M. 1971. Review: the yield behaviour of polymers. *Journal of Material Science* 1071(6) 1397–1417.
Zeng, M. & Kasibati, K. 2003. Evaluation of moisture susceptibility of asphalt mixtures containing bottom ash. *Transportation Research Record* 1832: 25–33.
Zhang, X. Grees, D. Kaprinski, S. & Eighmy, T. 1999. Utilization of municipal solid waste combustion bottom ash as a paving material. *Transportation Research Record* 1652: 257–263.

Efficient Transportation and Pavement Systems – Al-Qadi, Sayed, Alnuaimi & Masad (eds)
© 2009 Taylor & Francis Group, London, ISBN 978-0-415-48979-9

Permanent deformation analysis of HMA mixtures using simple performance tests

Louay N. Mohammad
Department of Civil and Environmental Engineering and Louisiana Transportation Research Center, Louisiana State University, Louisiana, USA

Sandeep Obulareddy
Louisiana Transportation Research Center, Louisiana State University, Baton Rouge, LA, USA

Sam Cooper
Technology Transfer, Louisiana Transportation Research Center, Baton Rouge, LA, USA

Abe Bae
Materials Research Associate, Louisiana Transportation Research Center, Baton Rouge, LA, USA

ABSTRACT: The objective of this study was to evaluate the permanent deformation of Hot Mix Asphalt (HMA) mixtures based on two simple performance tests (dynamic modulus |E*|, flow number F_N) and a Loaded Wheel Tracking test (LWT, Hamburg-type). Four plant-produced HMA mixtures were selected in this study. Those mixtures included two levels of nominal maximum aggregate size (12.5- and 25.0-mm) and two levels of mixture type (Superpave and SMA). Test results indicated that the |E*| values were sensitive to the nominal maximum aggregate size. Larger aggregates tended to have higher |E*| values at high temperatures. Furthermore, the ranking of the mixtures evaluated from the flow number tests was consistent with their use in the field. In addition, good correlation was observed between the results from the flow number and loaded wheel tracking tests.

Keywords: Simple Performance Tests, Permanent Deformation, Hot Mix Asphalt, Dynamic Modulus, Flow Number, Loaded Wheel Tracking

1 INTRODUCTION

Permanent deformation or rutting is a common problem in asphalt pavements, particularly in hot climate regions such as the State of Louisiana. Rutting is the result of a complex combination of densification and shear flow (SHRP 1994). The primary mechanism of rutting is shear deformation, which is caused by large stresses in the upper layers of hot mix asphalt concrete and is affected primarily by temperature. Studies have shown that rutting in asphalt pavements is proportional to the number of load cycles and is limited to the upper 100 mm of the asphalt concrete layer (OECD 1998). While significant rutting may be interpreted as a major structural failure, it is also a serious safety issue for road users because there is a potential for hydroplaning due to water accumulation during inclement weather.

The Strategic Highway Research Program (SHRP) in the United States concluded with the introduction of the Superpave (<u>Su</u>perior <u>Per</u>forming Asphalt <u>Pave</u>ments) mix design and analysis system. The Superpave mix analysis system did recommend mix verification for intermediate and high volume traffic through advanced materials characterization tests utilizing the Superpave Shear Tester test protocols. It was quickly recognized that those test protocols were too complex for routine mix design application and that a simple performance test was needed to complement the Superpave volumetric mix design procedure. In response to this need, National Cooperative Highway Research Program (NCHRP) Project 9–19, Superpave Support and Performance Models Management, recommended three candidate Simple Performance Tests (SPTs) to complement the Superpave volumetric mixture design method (Witczak et al. 2002, Bonaquist et al. 2003). These are flow time (FT), flow number (F_N), and dynamic modulus |E*| tests. In addition, the dynamic modulus test was selected for the HMA materials characterization input utilized in the Empirical and Mechanistic Guide for Design of New and Rehabilitated Pavement Structures, developed under NCHRP Project 1–37 A. This paper presents the findings of a study conducted to evaluate the permanent deformation of HMA mixtures based on two SPTs (|E*|, F_N) and a LWT test.

2 OBJECTIVES AND SCOPE

The objective of this study was to evaluate the permanent deformation and laboratory characteristics of HMA mixtures based on two SPTs (dynamic modulus |E*|, flow number F_N) and a loaded wheel tracking test (Hamburg-type). Four plant-produced HMA mixtures were selected in this study. Those mixtures included two levels of nominal maximum aggregate size (12.5- and 25.0-mm) and two levels of mixture type (Superpave and SMA).

3 PROJECT OVERVIEW

Three asphalt pavement rehabilitation projects in the State of Louisiana were selected in this study. The first project was located on Interstate Highway I-10 near the community of Egan, hereafter designated as I-10 Egan, where both binder course and wearing course HMA mixtures were evaluated. The second project was also located on Interstate Highway I-10 near the city of Vinton, hereafter designated as I-10 Vinton, where the wearing course mixture was selected for evaluation. The third project was on US Highway 190 near the city of Port Allen (US190), where the binder course mixture was evaluated. Figure 1 presents the location of the selected projects as well as mixture designations and their general information.

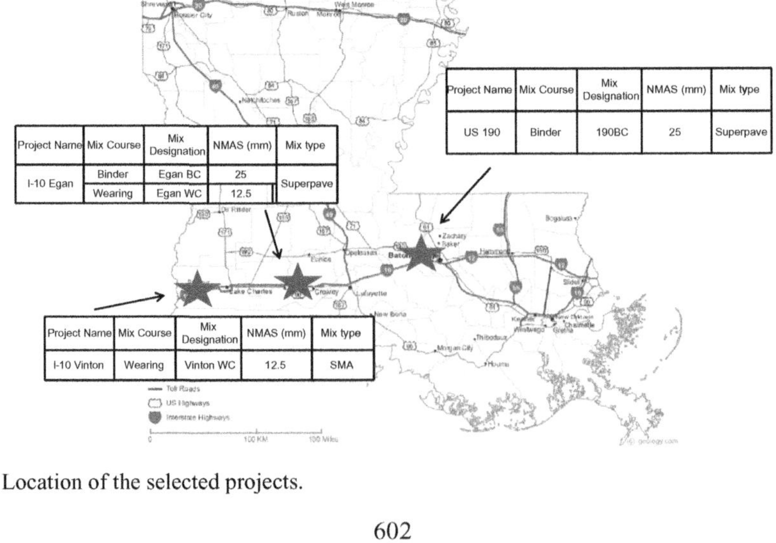

Figure 1. Location of the selected projects.

3.1 *Asphalt binder*

A SBS elastomer polymer-modified binder meeting Louisiana Standard Specifications for Roads and Bridges (2000 Edition) for PG 76-22M (M stands for polymer modified binder) was used for all mixtures in this study. More details about the specification requirements and sample test results for this binder were presented elsewhere (Wu et al. 2004).

3.2 *Mixture design*

Table 1a presents the job mix formula for the selected asphalt mixtures. Both I-10 Egan wearing and binder Superpave mixtures, Mix 1 and Mix 2, respectively, were designed for high volume roads as per the Louisiana Standard Specifications for Roads and Bridges (2000 Edition) at a compaction gyration number of N_{ini}, N_{des}, and N_{max} of 9-, 125-, 205- gyrations, respectively. The I-10 Vinton wearing course, Mix 3, was a typical high-volume Stone Matrix Asphalt (SMA) mixture used in Louisiana. The US190 binder mixture, Mix 4, was also a Superpave mixture designed for intermediate traffic volume at a compaction gyration number of N_{ini}, N_{des}, and N_{max} of 8, 100, and 160 gyrations, respectively.

3.3 *Test factorial and sample preparation*

Three laboratory engineering tests, namely, the dynamic modulus |E*|, flow number (F_N), and Hamburg type load wheel tracking tests were conducted in this study. Table 1b presents the test

Table 1a. Job mix formula.

	Mixture name			
Mix formula content	Mix 1 (I-10 Egan)	Mix 2 (I-10 Egan)	Mix 3 (I-10 Vinton)	Mix 4 (US190)
Mix type	25 mm Superpave binder course	12.5 mm Superpave wearing course	12.5 mm SMA wearing course	25 mm Superpave binder course
Binder type	PG 76–22M	PG 76–22M	PG 76–22M	PG 76–22M
% G_{mm} at $N_{Initial}$	85.4	84.1	N/A	87.9
% G_{mm} at N_{Max}	97.1	97	N/A	97.1
Design binder content, %	4.0	5.0	6.0	3.6
Design air void, %	4.0	4.0	4.0	4.0
VMA, %	12.8	14.5	16.6	11.8
VFA, %	69.5	72	76	67
%* at 37.5 mm (1½ in)**	100	100	100	100
% at 25 mm (1 in)	96	100	100	97
% at 19 mm (¾ in)	87	100	100	84
% at 12.5 mm (½ in)	68	98	93	65
% at 9.5 mm (⅜ in)	59	89	71	52
% at 4.75 mm (No.4)	35	50	30	32
% at 2.36 mm (No.8)	23	29	20	24
% at 1.18 mm (No.16)	17	19	–	20
% at 0.6 mm (No.30)	13	13	15	15
% at 0.3 mm (No.50)	7	10	12	8
% at 0.15 mm (No.100)	4	–	–	4.9
% at 0.075 mm (No.200)	3.6	6.5	8	3.6

*% Passing,
** Metric (U.S.) sieve.

Table 1b. Test factorial.

			Engineering property tests and testing temperatures		
Mixture	NMAS (mm)	Mix type	|E*| −10, 4, 25, 37.8, 54.4°C	Flow Number (F_N) 54.4°C	LWT 50°C
EganBC	25	Superpave	3	3	2
EganWC	12.5	Superpave	3	3	2
Vinton WC	12.5	SMA	3	3	2
190BC	25	Superpave	3	3	2

factorial and the corresponding test temperatures. The numbers in table 1b represent the number of samples used for each test.

Sufficient loose mixtures were secured from the plant production facility and transported to the laboratory for compaction and testing. Samples were compacted to the designed gyration level and density. The |E*| and flow number test samples were cylindrical specimens with a dimension of 100 mm by 150 mm. To fabricate these specimens, a 100 mm core was taken from Superpave gyratory compactor (SGC) compacted samples that were 150 mm in diameter by 170 mm in height. The core was then trimmed into 150 mm in height. The wheel tracking test slabs, 320 mm long by 260 mm wide by 80 mm thick, were fabricated using a linear kneading compactor. The target air voids for all test samples was 7.5 ± 0.5 percent. Test procedures for each test considered are briefly described below.

3.3.1 *Dynamic modulus |E*| test*

This test was conducted in accordance with AASHTO Standard TP 62-03 "Standard Method of Test for Determining Dynamic Modulus of Hot-Mix Asphalt Concrete Mixtures." The test consists of applying a uniaxial sinusoidal compressive stress to an unconfined cylindrical test specimen (100 mm diameter by 150 mm height). The stress-to-strain relationship under a continuous sinusoidal loading for linear viscoelastic materials is defined by a complex number called the "complex modulus" (E*). The absolute value of the complex modulus, |E*|, is defined as the dynamic modulus. The dynamic modulus is mathematically defined as the maximum (i.e., peak) dynamic stress (σ_0) divided by the peak recoverable axial strain (ε_0).

A sinusoidal compressive stress was applied to test specimens at −10, 4, 25, 37.8 and 54.4°C with loading frequencies of 0.1, 0.5, 1.0, 5, 10, and 25 Hz at each temperature. Therefore, each specimen was tested for all of the 30 combinations of temperature and frequency of loading, and corresponding deformations and phase angles at each temperature-frequency combination were recorded. An increasing order of temperature (starting with the lowest temperature and proceeding to the highest one) was maintained throughout the whole test. Testing at a particular temperature began with the highest frequency of loading and proceeded to the lowest one. This temperature-frequency sequence was designed such to cause minimum damage to the specimen before the next sequential test. In addition, the haversine compressive stress was applied on each sample to achieve a target vertical strain level of 100 microns in the unconfined test mode. Triplicate specimens were tested for each mixture type considered in this study.

3.3.2 *Repeated load permanent deformation test (Flow number test)*

This test is performed to determine the permanent deformation characteristic of HMA mixtures by applying a repeated load for several thousand repetitions on a cylindrical asphalt sample. The "Flow Number" is defined as the starting point, or cycle number, at which tertiary flow occurs on a cumulative permanent strain curve obtained during the test.

The F_N test was conducted according to the test method described in Annex B of the NCHRP Report 513 (Bonaquist et al. 2003). This method includes a repeated load for 10000 repetitions with a loading cycle of 1.0 second in duration, which applies a 0.1-second haversine load followed by 0.9-second rest period. Permanent axial strains are recorded throughout the test. The test is conducted at a single effective temperature T_{eff} and design stress levels. In this study, the effective temperature and design stress level selected were 54.4°C and 207 kPa, respectively. An unconfined test mode was used.

3.3.3 *Hamburg-type loaded wheel tracking test*

This test was conducted according to AASHTO T 324-04 "Standard Method of Test for Hamburg Wheel-Track Testing of Compacted Hot-Mix Asphalt (HMA)" to determine the rutting characteristics of HMA mixtures evaluated in this study. This test is considered a torture test that produces damage by rolling a steel wheel across the surface of a slab submerged in water at 50°C. The slabs have a length of 320 mm, a width of 260 mm, and a thickness of 80 mm. They are secured in a reusable steel containers using plaster of paris, and are then placed into the wheel-tracking device.

The device tests two slabs at a time using two reciprocating solid-steel wheels. The wheels have a diameter of 203.5 mm and a width of 47 mm. The load is 710 N. The device operates at 53 ± 2 passes/min. The test continued for 20,000 cycles or 20 mm deformation, whichever was reached first. The final rut depth data were used in the subsequent section of analysis. A maximum allowable rut depth of 6.0 mm at 20,000 passes is recommended by the Louisiana Department of Transportation and Development for high volume traffic mixtures.

4 DISCUSSION OF RESULTS

4.1 *Loaded wheel tracking test results*

Table 2 presents the LWT test results for the four mixtures evaluated. In general, all mixtures, except Mix 4, performed well. Mix 3 (SMA) had the smallest average rut depth, followed by Mix 2 and then Mix 1. The highest average rut depth was found for Mix 4. In summary, LWT test results indicate that Mix 3 is the best performer in resisting permanent deformation, whereas Mix 4 is the poorest.

4.2 *Dynamic modulus |E*| test results*

Figure 2 presents the average dynamic modulus test results at five different temperatures (-10, 4.4, 25, 37.8, and 54.4°C) and six frequencies. It is noted that the coefficients of variation of the |E*| tests in this study were generally less than 20 percent for the three test samples of each mixture. As the temperature increased and the loading frequency decreased, the dynamic modulus |E*| value decreased as shown in figures 2a-e. Figure 2f shows the |E*| master curves for each mixture considered. The complex modulus master curve possessed a general "S" shape. It is observed that all master curves cluster to each other at the two ends (at low and high frequencies) with the largest separation existing in the middle portion between each curves, figure 2f.

It is noted that the highest |E*| test results at −10°C were observed for Mix 1, followed by mixtures 3, 4, and 2 as shown in figure 2a. This numerical order, however, kept changing as the

Table 2. LWT rut depth results (mm).

	Mix 1	Mix 2	Mix 3	Mix 4
Slab1	4.3	3.1	2.3	13.9
Slab2	3.3	3.3	2.3	15.6
Ave.	3.8	3.2	2.3	14.8
Std.	0.7	0.1	0.0	1.2
% C.V.	18.9	4.4	0.6	8.1

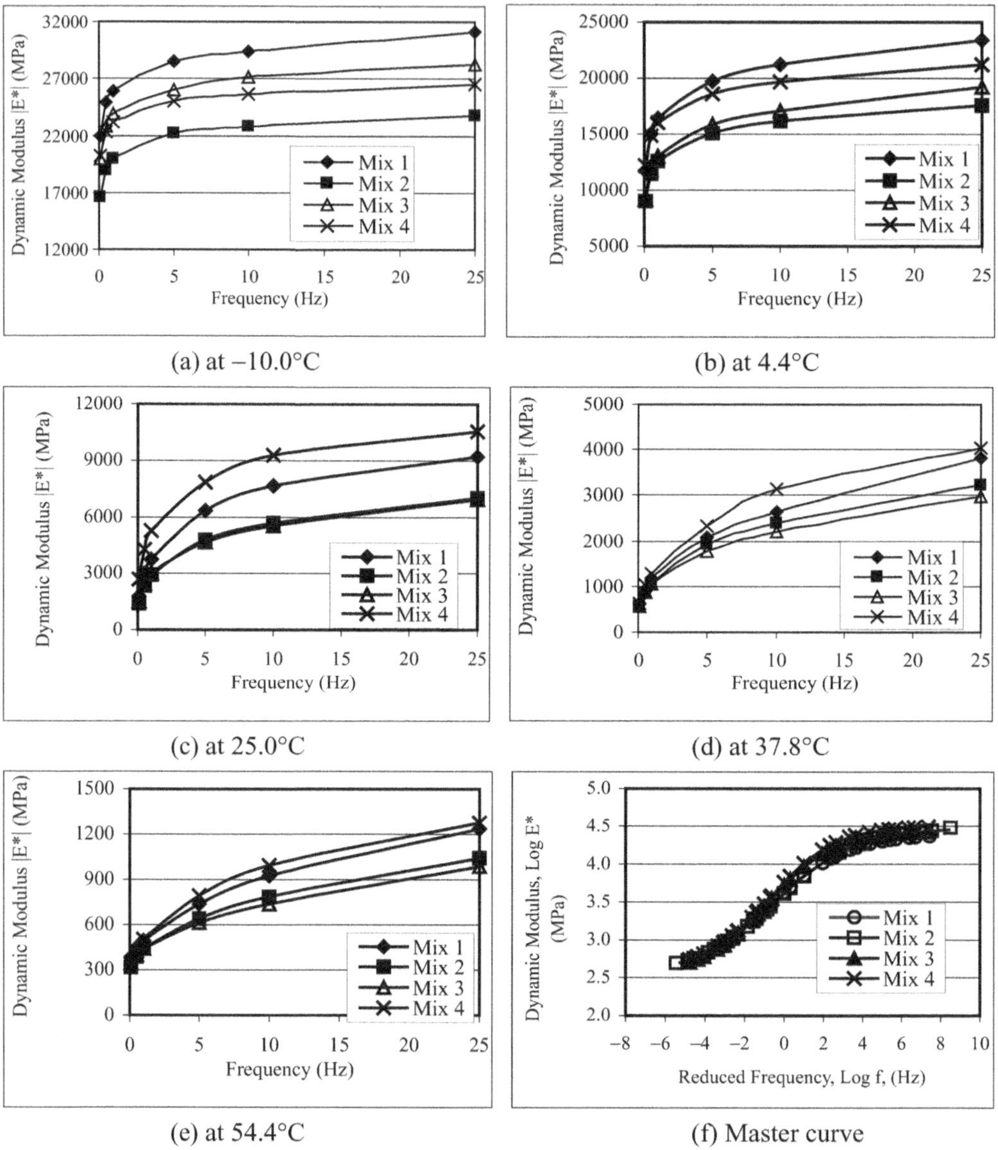

Figure 2. Dynamic modulus |E*| test results.

temperature increased (Figs 2b-e). For example, Mix 4 moved up from the third place at −10°C to the second place at 4.4°C, and then reached the first place for the other three temperatures (25.0, 37.8 and 54.4°C). On the other hand, the |E*| values for Mix 3 showed a reversed decreasing trend, from its initial second place at −10°C to the last place at 54.4°C. Such observations indicate that the elastic property of |E*| for an asphalt mixture is strongly temperature dependent and the rate of change in |E*| values due to the change of temperatures are mixture dependent. It is generally accepted that low |E*| values at low and intermediate temperatures are beneficial for an asphalt mixture to resist the low temperature and fatigue cracking, whereas high |E*| values at high temperatures are desired for rut-resistant mixtures. Interestingly the two binder course mixtures in this study, Mix 1 and Mix 4, tended to have higher dynamic modulus |E*| values at temperatures above

25°C than the two wearing course mixtures, Mix 2 and Mix 3. This observation seems to indicate that the binder course mixtures have better rut-resistance than the wearing course mixtures evaluated in this study.

The high |E*| values for the binder course mixtures at high temperatures may be attributed to the larger aggregate size (NMAS = 25 mm). Large aggregates could form a stronger stone-to-stone contact in an |E*| test and result in high stiffness. High |E*| values alone are not indicative of better rut-resistance for all asphalt mixtures of various designs. As stated earlier, the I-10 Vinton wearing course mixture, Mix 3, is an SMA mixture, which was designed to be rut-resistant at high temperatures and used for high volume roads. The LWT test results showed that this mixture was one of the best performers. However, it showed low |E*| values at high temperatures in this study. Gap-graded mixture with high binder content, such as the Vinton wearing course mixture, may be more suitable to be tested in a confined test mode, similar to the indirect tensile resilient modulus tests as reported by Mohammad et al. (1999).

In summary, the |E*| test was observed to be sensitive to the aggregate nominal maximum size in a mixture. Larger aggregates could result in high |E*| values at high temperatures.

4.3 *Flow number (F_N) test results*

Table 3 presents the flow number (F_N) test results of individual samples along with the average flow number value for each mixture, standard deviation, and percent of coefficient of variation. High flow number (F_N) is desired for a rut-resistant mixture. The highest and lowest percent of coefficient of variations in F_N values were 30.7 percent and 10.5 percent for Mix 1 and Mix 4, respectively. Both Mix 2 and 3 showed very high flow number values (F_N>10,000) under the test condition of compressive stress at 207 kPa used in this study, indicating that both mixtures are the best performers in resisting permanent deformation among the mixtures tested. The binder mixture, Mix 4, on the other hand, had the lowest average F_N value of 3,099. Mix 1, however, showed better performance that Mix 4 (F_N = 5,643). It is worth noting that the order of mixture ranking from this test was similar to the one obtained from the LWT test results. However, the flow number test results seem to disagree with the results observed from dynamic modulus |E*| test in terms of permanent deformation resistance for the four mixtures considered.

It is noted that the flow number test simulates field loading conditions, and provides test parameters with multiple applications. These parameters cover a wide range of material behavior from elastic properties (resilient strain and modulus) to plastic properties (flow number and permanent strain). The authors are concerned that the high test variations in a flow number test may cause difficulties in developing the design criteria for its use in a Superpave mix design.

Table 3. Flow number test results.

Mixtures	NMAS (mm)	Flow number (F_N)	Average	STD.	% C.V.
Mix 1	25.0	6,944 6,304 3,680	5,643	1,730	30.7
Mix 2	12.5	>10,000 >10,000 >10,000	>10,000	N/A	N/A
Mix 3	12.5	>10,000 >10,000 >10,000	>10,000	N/A	N/A
Mix 4	25.0	3,024 2,816 3,456	3,099	326	10.5

4.4 *Permanent deformation analysis*

Statistical analyses of the test results were performed using the Statistical Analysis System (SAS) software. A multiple comparison procedure, Fisher's Least Significant Difference (LSD), was carried out with a 95 percent confidence interval. The multiple comparison procedure ranked the mean test results and placed them in groups designated by "A", "B", and "C". The letter "A" is used to rank the group with the most desired values, i.e., high $|E^*|$ values or low LWT rut depths, followed by the other letter grades in the appropriate order.

A rut factor defined as $|E^*|/\sin\delta|_{5hz\ \&\ 54.4°C}$ was computed and used in statistical analysis. This factor was similar to the one used by Witczak et al. (2002), where the $|E^*|/\sin\delta|_{5hz\ \&\ 54.4°C}$ was considered as the permanent deformation factor for an asphalt mixture. Table 4 presents the statistical grouping for $|E^*|/\sin\delta\,|_{5hz\ \&\ 54.4°C}$, flow number and LWT rut depth. Table 4 also shows the numerical ranking of mixtures based on the average test values of each parameter analyzed.

Test results of $|E^*|/\sin\delta|_{5\,Hz\ \&\ 54.4°C}$ were clustered in only one group, "A", which means the rut factor values were not statistically different within the four mixture types. The statistical analyses on the flow number tests results, however, indicated that the "A" group mixtures included Mix 2 and Mix 3; the "B" group contained only Mix 1; and Mix 4 was ranked as the "C" group. The "A" group mixtures are considered to have better rut resistance than the "B" or "C" group mixtures. The order of statistical ranking from the LWT rut test results was similar to the one of the flow number ranking.

The statistical results for the LWT tests indicate that, in term of rutting resistance, Mix 1, Mix 2, and Mix 3 were ranked in group A, while Mix 4 is in group B, which represents a relatively lower rut performance. The LWT wheel tracking test is a torture test that simulates pavement deformation under wheel loading. Although the rut depth from this test is not a fundamental property for

Table 4. The statistical grouping of test results.

$\|E^*\|/\sin\delta\|_{5\,Hz\ \&\ 54.4C}$		Flow number		LWT rut depth	
Mix 1	A	Mix 2	A	Mix 1	A
Mix 2	A	Mix 3	A	Mix 2	A
Mix 3	A	Mix 1	B	Mix 3	A
Mix 4	A	Mix 4	C	Mix 4	B

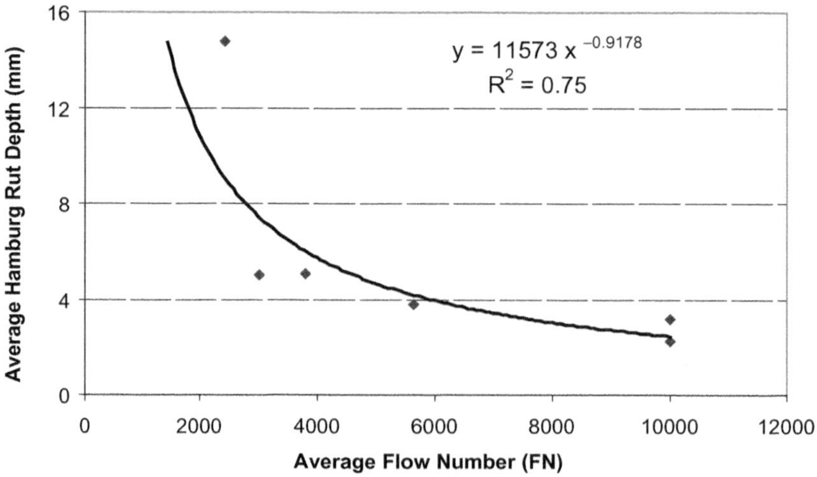

Figure 3. Relationship between LWT rut depth and flow number.

a mixture, it was believed by many researchers that the LWT wheel tracking test does differentiate the best from the poorest mixtures in the resistance of permanent deformation. In addition, a good correlation was observed between the average LWT rut depth and the flow number for the four mixtures tested. As shown in figure 3, the LWT rut depth decreases as the flow number increases.

In summary, by combining the statistic ranking on both the flow number and the LWT test results, the four asphalt mixtures in this study have the following ranking orders in terms of permanent deformation resistance (from high to low):

Mix 2 and Mix 3 → Mix 1 → Mix 4

The ranking list presented above is consistent with the expected field performance of these mixtures. It is noted that Mix 1, Mix 2 and 3 were designed for high volume interstate pavements, whereas Mix 4 was designed for intermediate volume roads.

5 SUMMARY AND CONCLUSIONS

The permanent deformation characteristics of four plant-produced HMA mixtures were evaluated through three laboratory tests: the dynamic modulus |E*|, flow number (F_N), and Hamburg-type loaded wheel tracking tests. The following observations were drawn from the analysis of the test results:

- The |E*| test results were sensitive to the nominal maximum aggregate size (NMAS) in the mixture. Larger aggregates tended to have higher |E*| values at high temperatures.
- The rut factor computed from the |E*| test results could not differentiate the laboratory performance for the four asphalt mixtures evaluated in this study.
- The mixture ranking order obtained from the flow number test results was consistent with the expected field use of those mixtures considered in this study.
- Good correlation was observed between the flow number values and the LWT rut depths, indicating both parameters are sensitive to the permanent deformation characteristics for the mixtures evaluated. However, high test variations in a flow number test may cause difficulties in developing design criteria for its use in a Superpave mix design.

ACKNOWLEDGMENT

This study was supported by the Louisiana Transportation Research Center (LTRC) and the Louisiana Department of Transportation and Development (LADOTD). The authors would like to express thanks to all those who provided valuable help in this study.

REFERENCES

AASHTO TP 62-03. 2003. Standard method of test for determining dynamic modulus of hot-mix asphalt concrete mixtures.

Anderson, R.M. & McGennis, R. 2003. Ruggedness evaluation of the shear frequency sweep test for determining the shear modulus of asphalt mixtures, *Journal of the Association of Asphalt Paving Technologists*, 72: 154–172.

Bonaquist, R.F., Christensen, D.W. & Stump, W. III. 2003. Simple performance tester for Superpave mix design: First-article development and evaluation, National Cooperative Highway Research Program (NCHRP) Report 513, Transportation Research Board, National Research Council, Washington, D.C.

Kaloush, K.E. & Witczak, M.W. 2002. Tertiary flow characteristics of asphalt mixtures. *Journal of the Association of Asphalt Paving Technologists*, 71: 248–280.

Louisiana Standard Specifications for Roads and Bridges. 2000. State of Louisiana, Department of Transportation and Development, Baton Rouge, 2000 Edition.
Mohammad, L.N., Zhang, X., Huang, B. & Tan, Z. 1999. Laboratory performance evaluation SMA, CMHB and dense graded asphalt mixtures. *Journal of the Association of Asphalt Paving Technologists*, 68: 252–283.
OECD. 1998. Heavy trucks, climate and pavement damage, prepared by an OECD scientific experts group, Organization for Economic Co-operation and Development, Paris, France: OECD Publications and Information Center [distributor], Washington, D.C.
SHRP. 1994. Permanent deformation response of asphalt aggregate mixes (SHRP-A-415), Final report, Strategic Highway Research Program, National Research Council.
Witczak, M.W., Kaloush, K., Pellinen, T. & El-Basyouny, M. 2002. Simple performance test for superpave mix design, National Cooperative Highway Research Program (NCHRP) Report 465, Transportation Research Board, National Research Council, Washington, D.C.
Wu, Z., Mohammad, L.N., Wang, L., Raghavendra, A. & Abadie, C. 2004. Sensitivity analysis of permanent deformation on Superpave and SMA mixtures as measured by the asphalt pavement analyzer, *International Journal of Pavements*, 2(3): 50–61.

Efficient Transportation and Pavement Systems – Al-Qadi, Sayed, Alnuaimi & Masad (eds)
© 2009 Taylor & Francis Group, London, ISBN 978-0-415-48979-9

Improving the field compaction of asphalt pavements using X-ray CT and imaging techniques

E. Masad
Texas A&M University, College Station, Texas, USA
Texas A&M University at Qatar, Doha, Qatar

E. Kassem
Texas A&M University, College Station, Texas, USA

ABSTRACT: This paper reports on the findings from a study that focused on evaluating the influence of different compaction patterns on asphalt pavement uniformity in terms of air void distribution. A number of projects with different asphalt mixture types were compacted, and cores were extracted at different locations from these projects. The X-ray Computed Tomography system was used to capture the air void distributions in these cores. The images were used to develop maps of air void distributions across the pavement surface and depth that were useful to study the uniformity of air void distributions. The analysis results have revealed that the uniformity is highly related to the sequence of using different compaction equipment. More importantly, the efficiency of compaction at a point is a function of the location of this point with respect to the roller compactor width. The results could be used to develop the compaction patterns in order to construct pavements with uniform air void distribution.

1 BACKGROUND AND RESEARCH OBJECTIVES

Compaction is the process by which the volume of asphalt mixtures is reduced as a result of applying external forces. The decrease in volume leads to increase in unit weight and aggregate interlock (U.S. Army Corps of Engineers 2000 & Roberts et al. 1996). The performance of asphalt mixtures is highly influenced by the quality of compaction. Poor compaction will lead to permanent deformation, aging and moisture damage.

A number of studies were carried out to investigate the relationship between laboratory compaction methods, field compaction methods and mechanical properties of asphalt mixtures (Consuegra et al. 1989 & Harvey and Monismith 1993). Other studies have attempted to examine the relationship between the internal structure and the mechanical properties of laboratory specimens and field cores (Peterson et al. 2004, Masad et al. 1999 & Tashman et al. 2001). Little effort was devoted to examine the influence of changes in field compaction patterns on both degree of compaction and air void uniformity in asphalt pavements. Air void uniformity within the asphalt pavements results in more uniform properties and improves the overall performance. In addition, there is a need to assess field compactability of asphalt mixtures based on laboratory experimental measurements during the mixture design process.

There are three main objectives in this study. The first objective is to understand the compaction factors that influence air void uniformity in asphalt pavements. The second is to establish a method to quantify the influence of different compaction patterns on asphalt pavements air void uniformity. The third objective is to predict field compactability based on laboratory compaction information.

2 RESEARCH METHODOLOGY

The research objectives were achieved by executing the following tasks:

- Conduct field compaction for a number of asphalt pavement sections with different mixtures.
- Extract field cores from different locations in test sections.
- Use the X-ray Computed Tomography (CT) and image analysis techniques to capture the air void distribution for the extracted field cores.
- Develop horizontal and vertical maps of air void distribution in test sections.
- Quantify the uniformity of air void distribution in test sections.
- Evaluate the influence of compaction patterns on air void distribution.
- Prepare laboratory samples and compare the laboratory compaction information to the field data.

3 FIELD COMPACTION

The researchers conducted field compaction for nine test sections of asphalt pavements in five projects. The field projects included different asphalt mixtures and Table 1 presents a description of these mixtures. The research team recorded the field compaction data such as the type of compaction equipment, sequence of rollers or compaction pattern, number of passes, relative location of each pass, and mat temperature. Different compaction patterns were evaluated and a summary of these patterns is given in Table 2.

Table 1. Summary of mixture designs used in compaction study (Kassem et al. 2008).

Highway ID	Mixture type	Date of field testing	Aggregate (major)	Binder type, PG	Optimum AC %	Max Sp Gr.	Design VMA %	Design air void, %
IH 35, WAC	SMA	May 2006	Traprock & Limestone	76-22	6.0	2.563	18.3	4.0
SH 36, YKM	Type D	July 2006	Limestone	64-22	4.9	2.447	15.1	3.5
US 87, YKM	Type C	Oct 2006	Siliceous RiverGravel	76-22S	4.3	2.460	13.8	4.0
US 259, TYL	Type C	March 2007	Sandstone & Limestone	70-22S	4.3	2.478	13.1	3.0
SH 21, AUS	Type C	June 2007	Limestone	70-22	4.7	2.467	14.3	3.0

Table 2. Description of compaction patterns (Kassem et al. 2008).

	Compaction pattern 1			Compaction pattern 2		
Highway ID	BD	IM	FS	BD	IM	FS
IH 35, WAC	V	N/A	S	N/A	N/A	N/A
SH 36, YKM	V and S	P	S	V	P	S
US 87, YKM	V and S	P	S	P	V	S
US 259, TYL	V and S*	P	P	V and S	P	P
SH 21, AUS	V	P	S	P	V	S

Note. BD = Breakdown, IM = Intermediate, FS = Finish, S = Steel wheel static, P = Pneumatic wheel, V = Vibratory steel wheel
* progressive motion.

Field cores were extracted from different locations within the test sections. These locations include the wheel path, between wheel path, center of lane, and longitudinal joints. In some cases there were some restrictions on the number of extracted cores. An example of coring layout is given in Figure 1.

4 RELATIONSHIP BETWEEN NUMBER OF PASSES AND PERCENT AIR VOIDS

The percent of air voids of extracted field cores was measured using AAHTO T 166 procedure. The number of passes of different rollers was plotted across the width of a test section along with percent air voids of extracted field cores at different locations. Two examples are presented in Figures 2 and 3. Each point represents an average value of percent air voids for at least two cores taken longitudinally.

The correlation between the percent of air voids and number of passes of different rollers was evaluated using the r-squared value (R^2) as shown in Table 3. The results showed poor correlation, if any, between the number of passes and percent air voids at different locations across the mat.

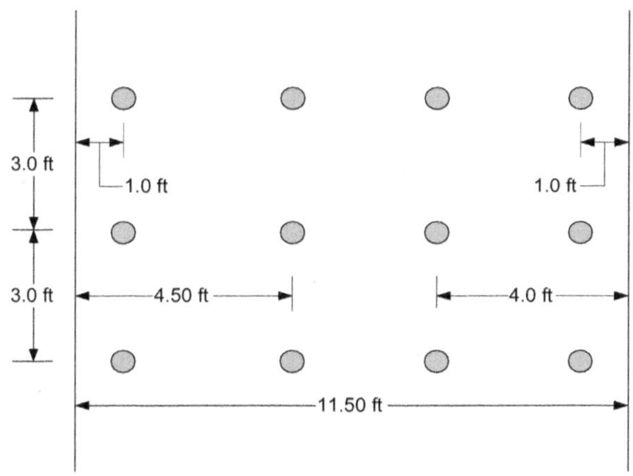

Figure 1. Field coring layout on SH 21 (Kassem et al. 2008).

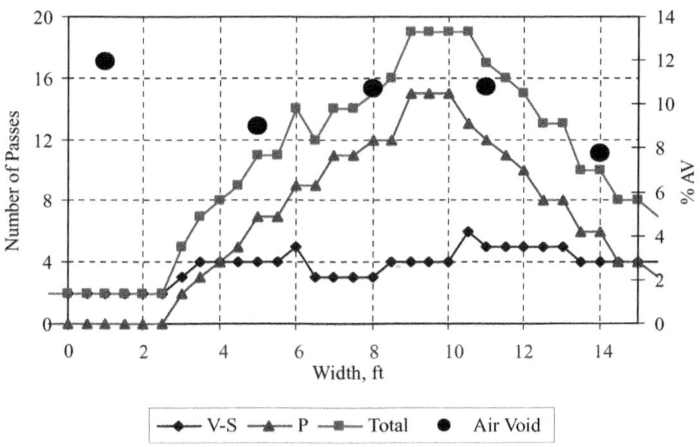

Figure 2. Number of passes and percent of air voids across the mat in the US 259 test section (Kassem 2008).

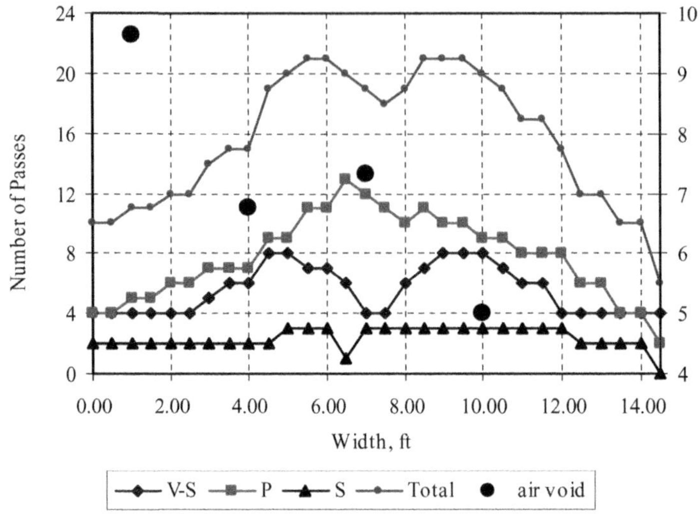

Figure 3. Number of passes and percent of air voids across the mat in the US 87 test section (Kassem 2008).

Table 3. R-squared value (Kassem et al. 2008).

Highway ID	Compaction pattern #	Total number of passes All rollers	Pneumatic roller	Vibratory and static rollers
IH 35	1	0.47		0.47
SH 36	1	(-) 0.19	(-) 0.15	(-) 0.24
	2	(-) 0.02	(-) 0.02	(-) 0.02
US 87	1	0.73	0.25	0.87
	2	0.80	0.47	0.93
US 259	1	0.04	0.02	0.25
	2	0.43	0.32	0.62
SH 21	1	0.44	0.23	0.04
	2	0.24	0.49	0.15
	Average R^2	0.33	0.20	0.34

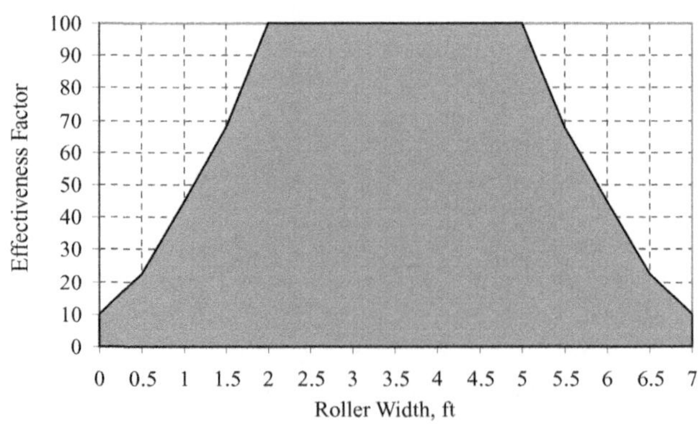

Figure 4. Effectiveness factor of the roller across its width (Kassem et al. 2008).

It was observed that cores compacted near to center of the roller width (static and vibratory) were found to have higher density than cores compacted at the edge of the roller even for field cores were taken from the middle of the mat and away from the joint. The researchers did not observe a similar relationship with respect to pneumatic-tired rollers. Therefore, each roller pass was multiplied with an effectiveness factor which is a function of the location of the field cores with respect to the width of roller. The change in air voids was plotted against the summation of number of passes multiplied by the effectiveness factor of each single pass which is termed as Compaction Index (CI). Researchers evaluated different effectiveness factors across the roller width. Figure 4 shows the effectiveness factors across the roller width which achieved the best average correlation ($R^2 = 0.80$) between the change in percent air voids and CI. The results show that the effectiveness of compaction is higher at the middle of the roller and decreases towards the edge of the roller.

5 HORIZONTAL AND VERTICAL AIR VOIDS DISTRIBUTION

X-ray CT was used to capture the air void distribution for the extracted field cores. The X-ray CT is nondestructive technique which means that the field cores can be used in other tests after X-ray scanning. The field cores were scanned at every 1 mm of depth. Image analysis technique (Image-Pro Plus 1999) was used to analyze the X-ray Images in order to compute the air void distribution across the depth. Consequently, three-dimensional maps of air void distribution in pavement sections were generated. Matlab software (2004) was utilized in this purpose by entering the location of the extracted field cores across the pavement test section and the percent air voids at different heights for the field cores. This application is unique as it provides an estimate of air void distribution horizontally and vertically every 1 mm of the depth. Figures 5 and 6 show the vertical air void distribution in one of the test section. It can be seen from Figure 5 that the middle part of pavement section is more compacted than the top and bottom parts, and this finding was confirmed by the results of other test sections. However it is interesting to note that using pneumatic-tired compactor in the breakdown stage produced more uniform and less percent air voids in the top two thirds of the depth as shown in Figure 6.

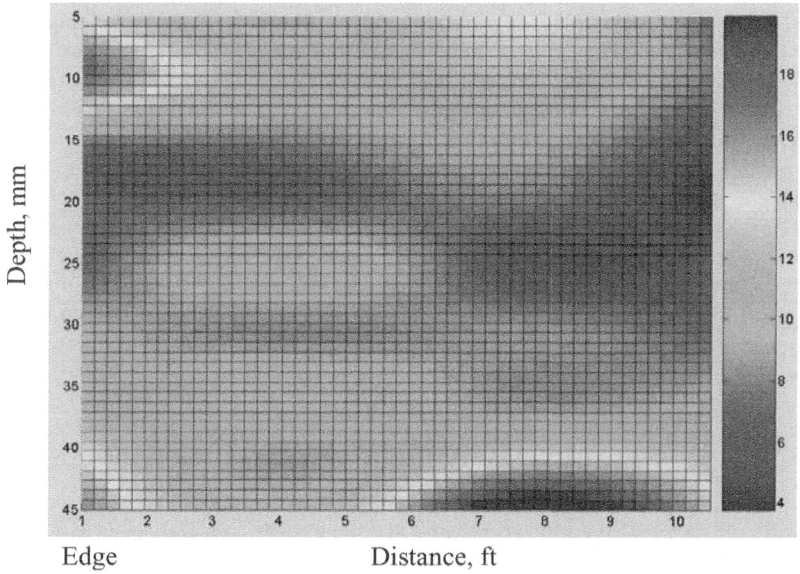

Figure 5. Air void distribution (%) along the depth of the mat for the SH 21 test section (Pattern 1), note: total width of the mat is 11.5 ft (Kassem et al. 2008).

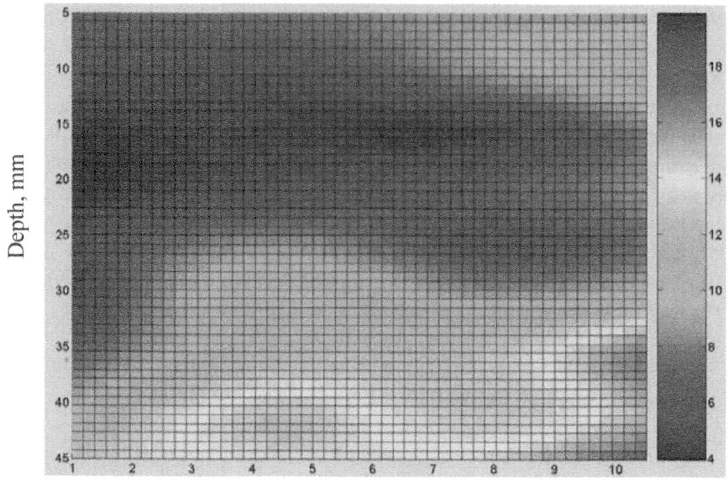

Figure 6. Air void distribution (%) along the depth of the mat for the SH 21 test section (Pattern 2), note: total width of the mat is 11.5 ft (Kassem et al. 2008).

6 APPLICATIONS OF THE COMPACTION INDEX

The relationship between the percent air voids and CI could be used to compact pavements to have uniform CI which produce more uniform air void distribution. This point is illustrated in Figure 7. It can be seen that the CI correlates very well with the percent air void distribution across the test section.

The CI can also be used to determine the compactability of the asphalt mixtures in the field. Figure 8 shows that change in percent air voids at different rates as more compaction effort is applied. The compactability of each asphalt mixture was assessed in the laboratory by compacting SGC specimens (150 mm diameter and approximately 63.5 mm in height) at 1.25° and 2.0° gyration angles. The slope of the number of gyrations in logarithmic scale against change in percent in air voids was calculated. The average slope of laboratory compaction curves was plotted against the field CI at 8 percent air voids for each mixture as shown in Figure 9. It can be seen that the asphalt mixture required more compaction effort (higher CI value) in the field had small slope in the laboratory which indicates that more compaction energy was needed in the laboratory. This relationship can be used to estimate the compaction effort or CI need in the field based on the slope of the laboratory compaction curve.

Another application of the CI is improving the compaction of longitudinal joints. It is accepted that field cores extracted near to longitudinal joints are typically less compacted compared to cores taken towards the center of mat. It was found in this study, as shown in Figure 4, that the effectiveness of the roller pass decreases towards the edge. Consequently, overhanging of roller edge at a distance about 1.5 ft to 2 ft will increase the effectiveness factor and hence improve the compaction at the longitudinal joint.

7 QUANTIFYING UNIFORMITY OF AIR VOID DISTRIBUTION

The uniformity of air void distribution is quantified using mathematical indices termed Uniformity Index (UI). The UI is calculated as follows (Kassem et al. 2008):

a) Plot percent air voids *f(x)* against the core depth x.
b) Fit a fourth order polynomial for *f(x)*.

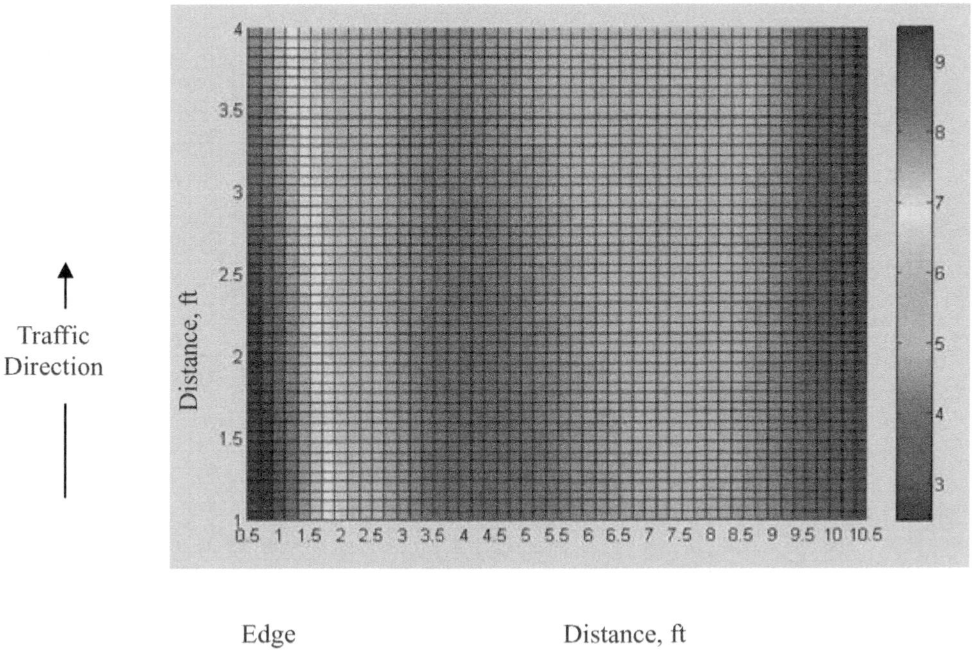

(a)

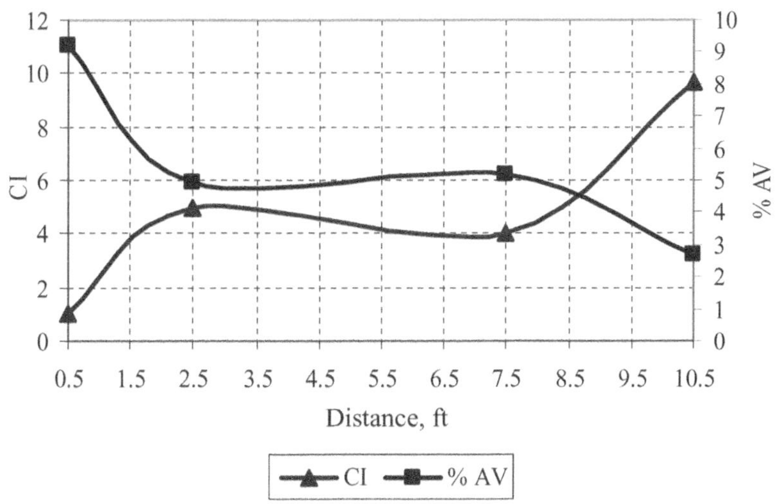

(b)

Figure 7. (a) Air void distribution (%) across the mat for the IH 35 job (b) The CI and average percent of air voids across the mat for the IH 35 test section, note: the total width of the mat is 15 ft (Kassem et al. 2008).

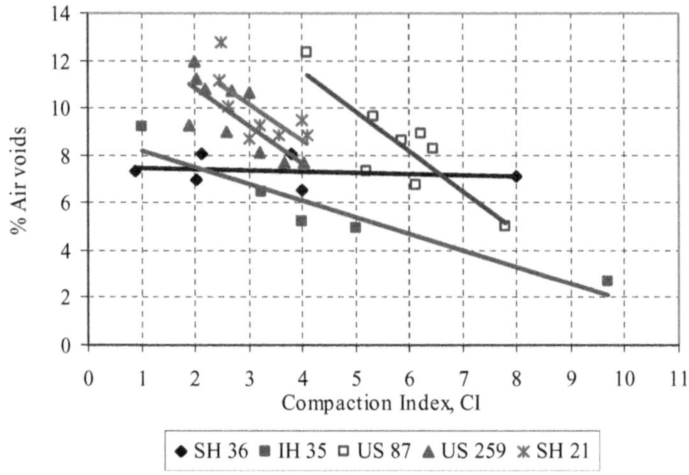

Figure 8. The CI versus the percent of air voids (Kassem et al. 2008).

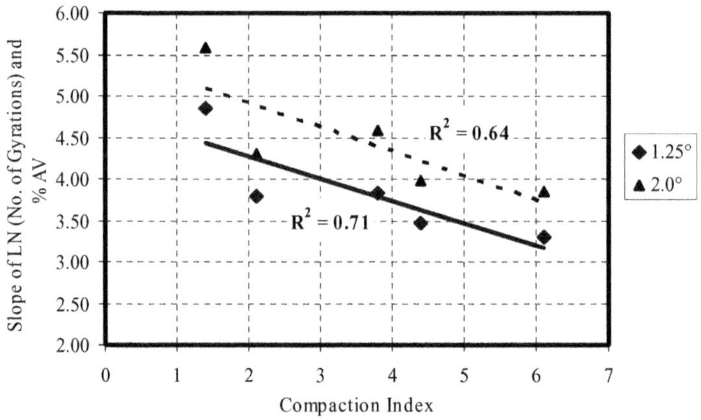

Figure 9. Compaction index versus the slope of LN (No. of gyrations) and percent air voids curve at 8 percent air voids for different mixes (Kassem et al. 2008).

c) Calculate the derivate *f(x)* of the function *f(x)*.
d) Calculate the UI using Equation (1).

$$UI = \frac{1}{b-a}\int_a^b [f'(x)]^2 dx \tag{1}$$

The UI is equal to zero for a straight line function and it increases with the increase in nonuniformity in air void distribution across the field cores. The integration limits depend on the thickness over which the uniformity is quantified. The analysis is conducted for the full depth and top and bottom parts. Figure 10 shows the UI for the two test sections of US 87 project. It is noted that using pneumatic-tired roller in breakdown in the second compaction patterns resulted in more uniform air void distribution or less UI (Figure 10(b)) compared with first compaction pattern (Figure 10(a)).

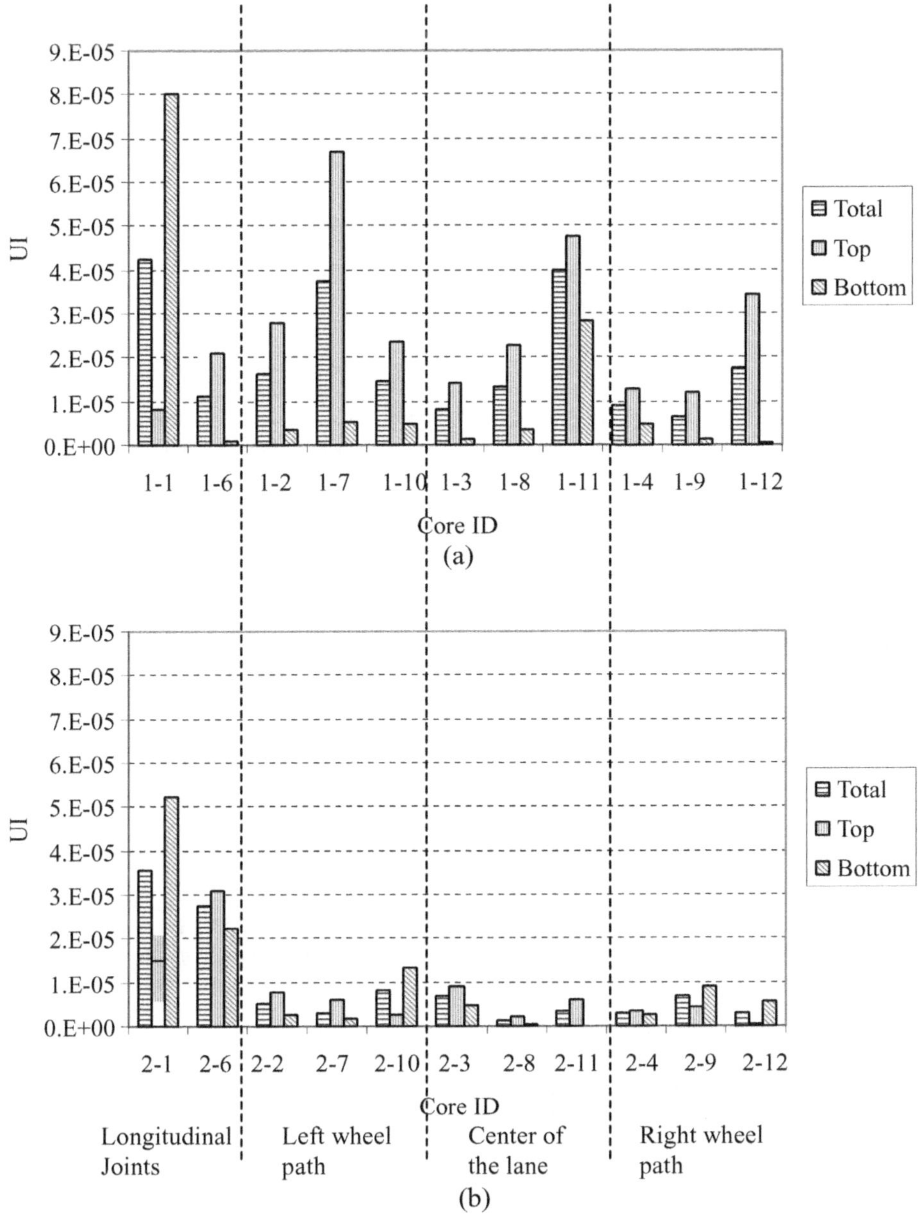

Figure 10. UI for the US 87 test section (a) Pattern 1 (b) Pattern 2 (Kassem et al. 2008).

8 CONCLUSIONS

This study examined some factors that influence air void distribution in the pavements and methods by which the uniformity of air void can be improved. It was found that the efficiency of the compaction at a given point within the mat is highly influenced by the location of this point with respect to the roller width. The efficiency of the compaction was found to be more at the center than at the edge of the roller. Consequently, a new index termed as the Compaction Index (CI) was developed. The CI takes into account number of passes and the effectiveness factor of every single pass. The CI was found to be better correlated than the number of passes with the air void distribution across the mat. The CI was used to assess the compactability of different asphalt mixtures and their sensitivity for the compaction efforts. The CI correlated very well with slope of the laboratory compaction curves which gives the potential to estimate the required CI to achieve a certain percent air void in the field using laboratory measurements.

The CI can be used to improve the compaction at longitudinal joint by increasing the effectiveness of the compaction. The efficiency of the compaction can be improved by overhanging the roller edge at a distance 1.5 ft to 2 ft which increases the effectiveness factor. Using Pneumatic-tired roller in the breakdown stage found to improve air void uniformity across the whole depth in one project and improve the uniformity for the top part in other project.

REFERENCES

Consuegra, A., Little, D.N., Quintus, H.V. & Burati, J. 1989. Comparative evaluation of laboratory compaction devices based on their ability to produce mixtures with engineering properties similar to those produced in the field. In *Transportation Research Record* 1228, TRB, National Research Council. Washington, D.C., 80–87.

Harvey, J., & Monismith, C.L. 1993. Effects of laboratory asphalt concrete specimen preparation variables on fatigue and permanent deformation test results using strategic highway research program A-003A proposed testing equipment. In *Transportation Research Record* 1417, TRB, National Research Council. Washington, D.C. 8–57.

Image-Pro Plus. 1999. *Media Cybernetics*, L.P. Version 4.1. Georgia, MD.

Kassem, E., Masad, E., Chowdhury, A. & Claros G. 2008. Influence of field compaction pattern on asphalt pavement uniformity, *Journal of the Association of Asphalt Paving Technologists*, AAPT, (Accepted for publication).

Masad, E., Muhunthan, B., Shashidhar, N. & Harman, T. 1999. Internal structure characterization of asphalt concrete using image analysis. *Journal of Computing in Civil Engineering (Special Issue on Image Processing),* ASCE, 13(2): 88–95.

Masad, E. 2004. X-ray Computed Tomography of aggregates and asphalt mixes. *Materials Evaluation*/July: 775–783.

Matlab software. 2004. *The MathWorks*, Inc. Version 7. Natick, MA.

Peterson, B., Mahboub, K., Anderson, M., Masad, E. & Tashman, L. 2004. Comparing Superpave Gyratory Compactor data to field cores. *Journal of Materials in Civil Engineering, ASCE,* 16(1): 78–83.

Roberts, F.L., Kandhal, P.S., Brown, E.R., Lee, D. & Kennedy T.W. 1996. *Hot mix asphalt materials, mixture design and construction.* National Asphalt Pavement Association (ed), Research and Education Foundation. Lanham, Maryland.

Tashman, L., Masad, E., Peterson, B. & Saleh, H. 2001. Internal structure analysis of asphalt mixes to improve the simulation of Superpave Gyratory compaction to field conditions. *Journal of the Association of Asphalt Paving Technologists,* 70: 605–645.

U.S. Army Corps of Engineers. 2000. Hot-Mix asphalt paving handbook 2000. AC 150/5370-14 A, *U.S. Army Corps of Engineers* (ed). Washington, DC.

Efficient Transportation and Pavement Systems – Al-Qadi, Sayed, Alnuaimi & Masad (eds)
© 2009 Taylor & Francis Group, London, ISBN 978-0-415-48979-9

Evaluation of coarse-graded HMA pavements in North Dakota using the Asphalt Pavement Analyzer

N. Suleiman
Civil Engineering Department, University of North Dakota, USA

ABSTRACT: A research study was performed on a coarse-graded Superpave overlay pavement in North Dakota. The project aggregate properties, gradation, and mix design information were within the NDDOT Superpave mix specifications. Field cores were used to test the pavement's rutting resistance, air voids, and asphalt permeability. The Asphalt Pavement Analyzer (APA) was utilized to evaluate the core's rutting resistance under dry and wet conditions. The dry tested specimens were satisfactory, but the wet-tested specimens failed the APA's 7.0 mm rut depth specification. Air voids were on the low side of the in-place air voids target while asphalt permeability was considered "practically impermeable".

Keywords: Coarse-graded pavements, Asphalt Pavement Analyzer, Rut-resistance, Asphalt permeability, In-place air voids, Superpave mix design

1 INTRODUCTION

1.1 *Background*

The aggregate gradation used in Superpave hot mix asphalt (HMA) mix design is required to be within control points 0.075 mm (No. 200) and 2.36 mm (No. 8), and the nominal maximum aggregate size (Roberts et al. 1996, Asphalt Institute, 1996). Both coarse- and fine-graded mixtures can be designed within these control points (Roberts et al. 1996, Kandhal et al. 2002). A majority of states accept both coarse- and fine-graded Superpave mixtures if the Superpave volumetric properties such as voids in the mineral aggregate (VMA) and voids filled with asphalt (VFA) are met. However, some states have begun to specify only fine-graded Superpave mixtures whereas others specify only coarse-graded Superpave mixtures. The states that specify coarse-graded mixtures define them as those mixtures with gradations below the maximum density line (Buchanan et al. 2002) and believe that coarse gradation provides a "strong aggregate structure." This belief is not essentially based on any significant mix strength test data. After some coarse-graded Superpave mixtures exhibited premature and excessive rutting (more than the fine-graded mixtures) on WesTrack and exhibited excessive *in-situ* permeability in many other states, some states have started to specify only fine-graded mixtures which were defined generally as those mixtures with gradation above the maximum density line (NCHRP 2002). Aggregate gradation is an important factor that influences the permanent deformation potential of hot-mix asphalt (HMA). One common way of characterizing aggregate gradation is by making a gradation plot on a 0.45 power chart, which also contains the maximum density line.

It is important in the construction of hot mix asphalt (HMA) that the mix is adequately compacted in-place so that the initial permeability is low and there will not be significant additional densification under traffic (Choubane et al. 1998, Cooley et al. 2001). For dense-graded mixes, numerous studies have shown that initial in-place air void content should not be below approximately 3 percent or above approximately 8 percent. Low in-place air voids have been shown to result in rutting, bleeding, and shoving, while high air voids allow water and air to penetrate into the pavement leading to an increased potential for water damage, oxidation, raveling, and cracking. Earlier studies have shown that most conventional dense-graded mixtures become excessively permeable at in-place air voids above 8 percent. However, recent studies indicate that coarse-graded Superpave mixes can be excessively permeable to water at in-place air voids less than 8 percent (Choubane et al. 1998, Cooley et al. 2001). Numerous factors can potentially affect the permeability of HMA pavements. Such factors include aggregate particle size distribution, aggregate particle shape, pavement density (air voids or percent compaction), nominal maximum aggregate size (NMAS), and lift thickness for a given NMAS (Kandhal et al. 2002).

Unlike stone matrix asphalt mixtures (SMA), coarse-graded asphalt mixtures are affected mainly by the percent retained on the 4.75 mm (No. 4) sieve and not much by the nominal maximum aggregate size (NMAS). Furthermore, the use of the 4.75 mm sieve size as a NMAS can be advantageous since thin lifts can be utilized. This has been a hot issue in cost saving especially in maintenance operations.

1.2 *Objectives and scope*

The main objectives of the study were: (1) to evaluate the rutting resistance performance of coarse-graded Superpave HMA pavement cores using the asphalt pavement analyzer; (2) to carry out permeability tests on coarse-graded field specimen cores obtained from different locations on US Highway 81 north of Grafton, ND; and (3) to identify the in-place air void percentage of coarse-graded Superpave HMA pavement from Highway 81 north of Grafton, ND.

The scope of the work consists of obtaining coarse-graded Superpave field cores from three different locations on US Highway 81 north of Grafton, ND and testing their rut resistance performance and permeability. Rut resistance performances were tested under dry and wet conditions using the APA. Permeability and in-place air void content tests were done on the field cores to explore potential durability problems, if any. A total of fifteen 150 mm (6-inch) diameter field cores were extracted from each location to carry out the aforementioned tests.

2 PROJECT SELECTION AND SAMPLING

2.1 *Project selection*

NDDOT project "NH-6-081(053)192" on US Highway 81 north of Grafton, ND was selected for this study. The project entailed a Superpave coarse-graded mix design. The pavement was an overlay job and was placed in August 2005. Field core samples were obtained from centerline locations within three sections of US Highway 81. The first section (S1) started from reference point (RP) 192.4 to RP 196.0. The second section (S2) was located between RP 196.0 and RP 203.0, while the third section (S3) was between RP 203.0 and RP 204.5. Fifteen samples from each section were cored for a total of 45 core samples. Twelve samples out of the 15 from each section were used for APA testing (6 dry and 6 wet) and the remaining 3 cores were used for permeability determination.

The first section entailed an 89 mm (3½ inch) overlay that was placed in two lifts. The top lift was 50 mm (2 inches) thick. In the second section, a 127 mm (5 inches) of overlay was used and placed in 3 lifts where the top lift was 50 mm (2 inch) thick. A 178 mm (7 inch) mine and blend was used in the third section and placed in 4 lifts. The top lift was also paved at 50 mm (2 inches).

2.2 *Project materials*

Two binder grades were used for the project, PG 58-34 for the top 50 mm (2 inches) of the mine and blend section and PG 58-28 for the rest of the pavement sections. Aggregates for the project were taken from Pit # NW ¼ of 10-155-56 owned by Pioneer-Fordville. The consensus and blend properties were within Superpave specification limits except for the percent fine aggregate angularity (FAA). The individual and blend aggregate gradations indicate that the Superpave specifications were met. Most notable, the percent passing the 4.75 mm (#4 sieve) was 40.7 percent which is consistent with typical coarse-graded mixes gradations. Table 1 displays the aggregate gradations for the Superpave blend.

2.3 *Project mix design*

Superpave mix design in accordance with NDDOT section 410 was performed for the project. The voids analysis and mix properties are shown in Tables 2 and 3 below.

Table 1. Aggregate gradations for the Superpave blend.

Aggregate description	Aggregate #	Blend %	Sieve size	Blend gradation	Control points (Superpave) Lower	Upper
Pioneer	1	17	5/8"	100.0	100	100
Pioneer	2	20	1/2"	94.6	90	100
Pioneer	3	25	3/8"	72.8		
Pioneer	4	23	#4	40.7		
Pioneer	5	15	#8	30.1	28	58
Sum of % = 100			#16	20.1		
			#30	11.9		
			#50	7.2		
			#100	5.0		
Nominal Maximum Size = 1/2 inch			#200	4.2	2	7

Table 2. Voids analysis of the mix @ Ndes for various binder contents.

Superpave mix designs	Properties @ different AC contents				
AC content (%)	4.8	5.3	5.8	6.3	6.8
Bulk specific gravity of the mix (Gmb)	2.300	2.314	2.321	2.331	2.349
Percent aggregate	95.2	94.7	94.2	93.7	93.2
Theor. maximum SpG of mix (Gmm)	2.447	2.437	2.418	2.410	2.401
Air voids, Va (%)	6.0	5.1	4.0	3.3	2.2
Voids in mineral agg. (VMA)	14.0	14.0	14.1	14.2	14.0
Voids in mineral agg. filled (VFA)	57.0	63.7	71.5	77.1	84.6

Table 3. Mixture properties at recommended asphalt content.

Mix properties	Coarse-graded Superpave mix	Specification
Optimum AC (%)	5.8	
Density (pcf)	144.8	
Air voids (%)	4.0	3.0–5.0
VMA (%)	14.1	14.0 Min
VFA (%)	71.5	65.0–78.0
%Gmm @ Ninitial	86.5	89.0 Max
%Gmm @ Nmaximum	97.1	98.0 Max
AC film thickness (m)	11.6	7.5–13.0
Dust/Effective AC Ratio	1.0	0.6–1.3
Asphalt absorption (%)	1.49	
Maximum SpG @ Ndes	2.417	
Effective (Gme)	2.644	

The displayed results in the above two tables indicate that the volumetric properties have fallen within specifications. A 5.8 percent asphalt binder was recommended. An average 14.1 percent VMA was determined for the project. VMA seems to be at the low side risking a rutting potential.

2.4 *Sample preparation*

The dimensions of the cored samples were 150 mm (6 inch) in diameter and variable heights. The height depended on the core location since coring was extended to the base layer to ease the core recovery process. The samples were then sawed down to the proper height by cutting the bottom side of the specimen. The original top surface was always maintained intact (without cutting) for all of the core samples (both for APA and permeability testing).

Specimens were trimmed to a 75 mm (3 inch) height for APA testing and to about 50 mm (2 inches) for permeability testing. The main concern for the permeability specimens was to insure the removal of the tack coat which lies at about 50 mm (2 inches) from the top surface. The presence of a tack coat would affect the permeability results. For APA testing, maintaining a level surface at the bottom as well as vertical sides is warranted for the accuracy of the APA results.

3 PERMEABILITY TESTING AND ANALYSIS

3.1 *Introduction*

Adequate in-place compaction during the construction of hot mix asphalt (HMA) is essential in producing low permeability asphalt mixes. For dense-graded mixes, numerous studies have shown that initial in-place air void content should not be below approximately 3 percent or above approximately 8 percent (Choubane et al. 1998). Low in-place air voids have been shown to result in rutting, bleeding, and shoving, while high air voids allow water and air to penetrate into the pavement leading to an increased potential for water damage, oxidation, raveling, and cracking. However, due to problems associated with coarse-graded Superpave mixes, the size and interconnectivity of air voids have been shown to greatly influence permeability (Choubane et al. 1998).

Numerous factors such as aggregate particle size distribution, nominal maximum aggregate size (NMAS), aggregate particle shape, pavement density (or air voids), and lift thickness can affect pavement permeability (Cooley et al. 2001). In this study, the only available variable is the in-place density (or air voids) of the different core samples. Based on literature review of several asphalt permeability studies (Choubane et al. 1998, Cooley et al. 2001, Westerman 1998), the ranges of permeability coefficients (K) that correspond to high, low, and impervious designations are displayed in Table 4 below.

3.2 *Air voids results and analysis*

Air voids for the collected core samples were measured to explore any link between air voids (or field densities) of samples and permeability values. Air void measurements were developed from

Table 4. Permeability designations based on permeability coefficient ranges.

Permeability designation	Range of permeability coefficient, k
High permeability	1×10^{-1} cm/s to 1×10^{-4} cm/s
Low permeability	1×10^{-4} cm/s to 1×10^{-6} cm/s
Practically impervious	1×10^{-6} cm/s to 1×10^{-9} cm/s

samples that later were used for measuring permeability. The average air voids were 4.5, 4.1, and 4.6 for the S1, S2, and S3 core specimens, respectively. The average bulk specific gravity of the mix (G_{mb}) values for the same sections were 2.299, 2.314, and 2.294. Also the average VMA values were 14.9, 14.4, and 15.1 percent for the corresponding sections. Comparing the air voids with G_{mb} and VMA values for S1, S2, and S3 core specimens, one can note that the higher air void values agrees with lower G_{mb} and higher VMA values and vice versa.

3.3 *Asphalt permeability results and analysis*

For this project, the ASTM Standard PS 129-01 was adopted to perform the permeability tests. Generally, low permeability coefficient values were observed for all the test sections including several specimens that yielded zero permeability (K) values. The zero permeability values were observed in specimens with low air voids ranging between 2.4 and 3.9 percent. The average permeability coefficient values for S2, S3, and S1 sections were 1.75×10^{-5}, 3.16×10^{-5}, and 7.54×10^{-5}, respectively. According to Table 4 for permeability designation, the average sections K values correspond to the low permeability category.

4 RUT RESISTANCE TESTING AND ANALYSIS

4.1 *Introduction*

The Asphalt Pavement Analyzer which is a new generation of the Georgia Load Wheel Tester (GLWT) has been used to evaluate rutting, fatigue, and moisture resistance of HMA mixtures (Kandhal et al. 1999). In this study, testing with the APA was conducted according to 63-03 "Standard Method of Test for Determining Rutting Susceptibility of Asphalt Paving Mixtures," a provisional AASHTO designation with modifications to accommodate NDDOT project requirements (NDDOT 2002).

In this study, Superpave specimens were tested at 58°C, corresponding with the high end temperature of the binder's performance grade. Testing was carried out to 8,000 cycles for the dry condition and 25,000 cycles for the submerged (wet) condition. Specimens were conditioned in a 58°C water bath for 24 hours before moisture susceptibility testing. The expression wet here refers to the 24-hour submersion in a 58°C water bath followed by the APA moisture sensitivity testing (also submerged at 58°C). The 58°C was chosen to represent the high end temperature of the top lift binder grade (PG 58-34). All specimens were sawed to a 75 mm height before they were placed in the APA molds.

4.2 *APA results and analysis*

The APA performance (rut resistance) results of dry and wet tests for the coarse-graded core specimens are shown in Table 5 and Figure 1 below. Each of the dry or wet case is broken down by three sections (S1, S2, and S3). The values shown in Table 5 and plotted in Figure 1 represent the rut depth (in mm) under the corresponding wheel of the APA.

The APA rut failure criterion adopted for this study is 7 mm (0.276 inch) for traffic levels between 0.3 and 3 million design equivalent single axle loads (ESALs). The design ESAL for US Highway 81 has been estimated at 0.6 million ESALs, so the 7 mm failure criterion would apply. The terms S1-Dry, S2-Dry, S3-Dry, S1-Wet, S2-Wet, and S3-Wet indicate the section number and APA testing condition. The numbers that appear above the rut depth in Table 5 indicate the core specimen number.

The statistical Analysis of Variance (ANOVA) was performed on the APA results within wet versus dry subsets and for the three sections. The ANOVA statistical analysis of the APA results for the dry and wet conditions is presented in Table 6 below. The null hypothesis is given as, H_0: the means of the results are equal. On the other hand, the alternate hypothesis, H_1: the means of

Table 5. APA rut resistance results.

APA test cases	Left side depth (mm)		Center depth (mm)		Right side depth (mm)		AVE (mm)
S1-Dry	2*	5	7	10	11	15	7.4302
	7.1671	6.82000	7.52457	7.70017	7.33709	8.03214	
S2-Dry	18	21	23	25	28	29	7.1139
	6.58240	8.29390	6.89241	6.406771	6.72889	7.77883	
S3-Dry	32	35	39	41	42	45	6.8190
	7.64977	7.74629	6.12099	6.875604	6.44086	6.08035	
S1-Wet	1	3	6	8	12	14	8.1999
	7.14610	8.44067	6.94199	8.150669	7.82118	10.6989	
S2-Wet	17	19	22	24	26	30	9.0609
	8.89692	9.25488	8.11331	9.355614	9.44857	9.29624	
S3-Wet	31	33	36	38	40	44	8.7193
	8.51671	8.36834	9.39485	10.93232	7.60196	7.50164	

* Core sample numbers.

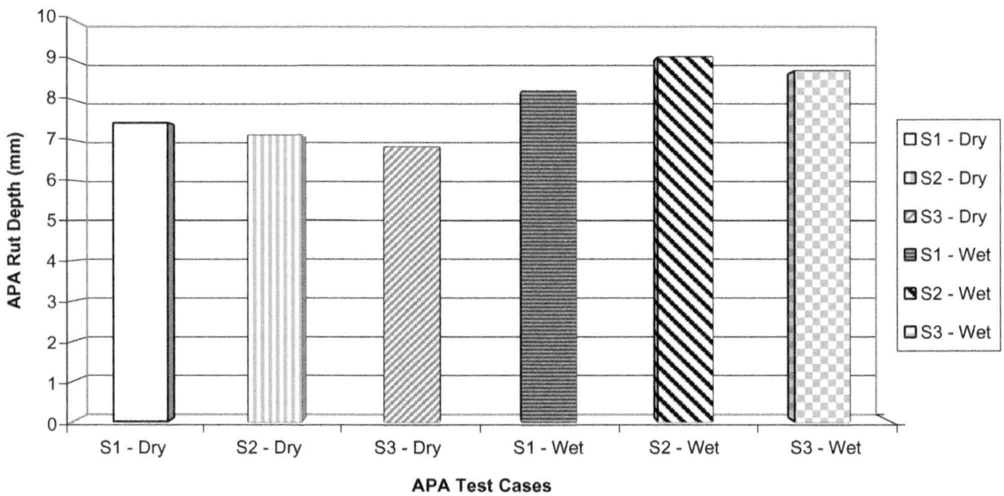

Figure 1. Average APA rut resistance results.

the results are not equal. Rejection of the null hypothesis indicates that the results are significantly different and can be compared.

Observation of the statistical results indicates that the P-value is less than the significance value ($P = 0.000648 < 0.05$); therefore, the null hypothesis is rejected. This means that the results within the subsets of wet versus dry and across the three sections are significantly different and can be compared.

Examination of the results indicates that 9 out of the 18 dry specimens failed the APA rut test, while all but one of the submerged specimens failed the APA rut test. In other words, the average APA rut depths for the dry and submerged specimens were 7.1 mm (0.28 inch) and 8.7 mm (0.34 inch), respectively. So, under dry conditions, the coarse-graded Superpave mixes exceeded the high end of the failure specification by 1.4 percent. Meanwhile, the average rutting under the APA wheels for the submerged specimens was higher than the failure specification value by

Table 6. Analysis of variance (ANOVA) statistics on the APA results.

SUMMARY

Groups	*Count*	*Sum*	*Average*	*Variance*
S1-Dry	6	44.5811	7.430183	0.178867
S2-Dry	6	42.68321	7.113868	0.562868
S3-Dry	6	40.91386	6.818976	0.545813
S1-Wet	6	49.19945	8.199908	1.828279
S2-Wet	6	54.36553	9.060921	0.250996
S3-Wet	6	52.31582	8.719303	1.649836

ANOVA

Source of Variation	*SS*	*df*	*MS*	*F*	*P-value*	*F crit*
Between Groups	24.69447	5	4.938894	5.906991	0.000648	2.533554
Within Groups	25.0833	30	0.83611			
Total	49.77777	35				

24.3 percent which is absolutely unacceptable. In other words, the presence of moisture reduced the coarse-graded mixes rutting resistance by 22 percent when compared to the dry case average.

Further inspection of the APA results shows interesting trends between the APA rutting values and the pavement sections. For the dry cases, APA rut values decreased with an increase in the section number (S1, S2, and S3). This means that, on average, S2 is more rut resistant than S1 and S3 is more rut resistance than S1 and S2. For the wet cases, the APA rut values did not follow the same pattern of the dry cases. So S2 exhibited the highest rut depth followed by S3 and then S1. This means that under submerged conditions, S3 specimens are more rut resistance than S2 specimens and that S1 specimens are more rut resistance than S2 or S3 specimens.

5 SUMMARY AND CONCLUSIONS

A 12 mile long coarse-graded asphalt concrete overlay project [NH-6-081(053)192] was designed and placed on US Highway 81 north of Grafton, North Dakota. The project aggregate properties, gradation, and mix design information indicate that NDDOT section 410 Superpave mix design specifications were achieved.

The project was divided into three sections: S1, S2, and S3. Regarding the structures of the different sections, S1 received a total of 89 mm (3½ inch) overlay, S2 had a 5 inch overlay, and S3 attained a 178 mm (7 inch) overlay in addition to a mine and blend job. S1 and S2 sections received a PG 58-28 binder for the entire overlay, but the S3 section contained a PG 58-34 binder in the top 50 mm (2 inches) of the overlay.

For the study presented in this paper, the rutting resistance of the coarse-graded mix was performed on field core samples and tested using the asphalt pavement analyzer. Asphalt permeability and the air voids of the field cores were also determined. Statistical analysis indicated that the variations within the wet and dry APA results were significant but the variations for APA results within sections, air voids results, or permeability results were insignificant.

The APA results show that 50 percent of the dry tested core specimens across the project sections of the coarse graded mix have exhibited satisfactory rutting resistance. And since the average APA rut depth was 7.1 mm, which is slightly above the 7.0 mm failure specification, the overall assessment of the coarse graded mix under dry conditions can be put near satisfactory with room for improvement. But the effect of moisture takes its toll on the coarse-graded mix's ability to resist rutting. All but one of the 18 wet tested specimens had failed the APA rut depth specification. An average of 8.7 mm APA rut depth is deemed unacceptable. Examination of the APA results across sections showed mixed trends.

The average in-place air voids and permeability coefficient values were determined at 4.7 percent and 4.15 x 10^{-5} cm/sec, respectively. The core specimen air void results were generally on the low side of the 6 to 8 percent in-place air voids target. The permeability results were ranging between very low to totally impermeable. A designation of "practically impermeable" would be a fair characterization of the overall permeability of tested coarse graded pavement sections. Air voids and permeability results did not show obvious trends across the different sections. Comparing the air voids and permeability coefficient values with the APA rut results, the lower the air voids or permeability coefficient in a section the higher the rutting under the APA was observed. This observation is consistent with the conventional wisdom regarding the relationship between air voids and rutting.

REFERENCES

Asphalt Institute 1996. *Superpave Mix Design.* Superpave Series No. 2 (SP-2). Lexington: Asphalt Institute.

Buchanan, M.S., and L. Allen Cooley, Jr. 2002. *Case Studies of the Tender Zone in Coarse-Graded Superpave Mixtures.* National Center for Asphalt Technology (NCAT) Report No. 02-01. Auburn: NCAT.

Choubane, B., G.C. Page, and J.A. Musselman 1998. Investigation of Water Permeability of Coarse Graded Superpave Pavements. Association of Asphalt Paving Technologists, Volume 67.

Cooley, L. Allen, Jr., E.R. Brown, and S. Maghasoodloo 2001. Developing Critical Field Permeability and Pavement Density Values for Coarse-Graded Superpave Pavements. In Transportation Research Record: Journal of the Transportation Research Board, No. 1761: pp. 41–49. Washington, D.C.: TRB.

Kandhal, P., and L. Allen Cooley, Jr. 2002. Coarse Versus Fine-Graded Superpave Mixtures: Comparative Evaluation of Resistance to Rutting. National Center for Asphalt Technology (NCAT) Report No. 02-02. Auburn: NCAT.

Kandhal, P., and R. Mallick 1999. Evaluation of Asphalt Pavement Analyzer (APA) For HMA Mix Design. National Center for Asphalt Technology (NCAT) Report No. 99-4. Auburn: NCAT.

NCHRP 2002. *Recommended Performance-Related Specifications for Hot-Mix Asphalt Construction: Results of the WesTrack Project.* Washington, D.C.: Transportation Research Board.

NDDOT 2002. *Standard Specifications for Road and Bridge Construction.* Volume 1 of 2. Bismarck: North Dakota Department of Transportation.

Roberts, F.L., P.S. Kandhal, E.R. Brown, D.Y. Lee, and T.W., Kennedy 1996. Hot Mix Asphalt Materials, Mixture Design, and Construction. Lanham: NAPA Education Foundation.

Westerman, J.R. 1998. AHTD's Experience with Superpave Pavement Permeability. http://www.utexas.edu/research/superpave/articles/jrw10a.html)

Efficient Transportation and Pavement Systems – Al-Qadi, Sayed, Alnuaimi & Masad (eds)
 ISBN 978-0-415-48979-9

Analytical investigation of the correctness of formulas used in bending beam tests

A.C. Pronk
Delft University of Technology, Delft, The Netherlands

ABSTRACT: Stiffness and fatigue characteristics for asphalt mixes are usually measured in bending beam tests like the four point bending beam test (4PB). In these tests the applied load and the deflection response are measured. Based on the theory for the bending of a slender beam the occurring strains and the complex stiffness modulus are calculated. This is fine in the case of stiffness measurements in which the beam will not be fatigued. However, in fatigue tests the stiffness will not stay constant over the beam. Nevertheless, for the interpretation, formulas are used which are based on an assumed constant stiffness modulus throughout the beam. In the paper as an example three distributions for the stiffness moduli are considered for a beam in a 4PB test. In the mid span the stiffness modulus decreases but stays constant over the mid span. In the two outer spans, the stiffness modulus decreases also but at the outer supports (no strain) the original stiffness modulus remains. Next to the direct influence on the determination of the stiffness modulus from the deflection, these examples are also important for test devices in which the deflection is measured in a relative way. The deflection is measured by using reference supports which rest on the beam itself. By choosing appropriate distances, the measured value will be half the absolute center deflection. But this is only true in the start of a fatigue test when the stiffness is still constant over the whole beam. This paper examines what the error in the determination of deflections will be for these devices.

1 INTRODUCTION

In pavement design, the four point bending (4PB) test is often used for the determination of the stiffness and fatigue characteristics of asphalt mixes. A scheme of the test is given in Figure 1. The length L between the outer supports is in the order of 0.4 m. The distance A between the inner and outer supports is in most 4PB equipments equal to L/3 (1:1:1). The applied load F0 is equally divided over the two inner supports which can move freely up and downwards. In this way a constant bending moment is created in the mid span of the beam. The centre deflection V at L/2 can be measured absolutely or relatively by placing the reference supports on the beam. The deflection V consists out of two parts: a deflection V_B due to pure bending and a deflection V_S due to shear forces. The later one is for a slender beam very small and is often neglected in the processing and interpretation of the data.

Due to the higher strains and stresses the decrease in the stiffness modulus will be faster in the mid span of the beam ($A < x < L-A$) than in the outer sections ($0 < x < A$ & $L-A < x < L$).

So, after a while the stiffness modulus will vary over the beam in a fatigue test. Still, for the interpretation of the measured deflections the bending beam theory is used with the assumption of a constant stiffness modulus throughout the beam. In this note the influence is calculated for three arbitrarily chosen stiffness modulus variations. The variation in stiffness modulus over the height of the beam is not taken into account. It is assumed that for each location along the beam a

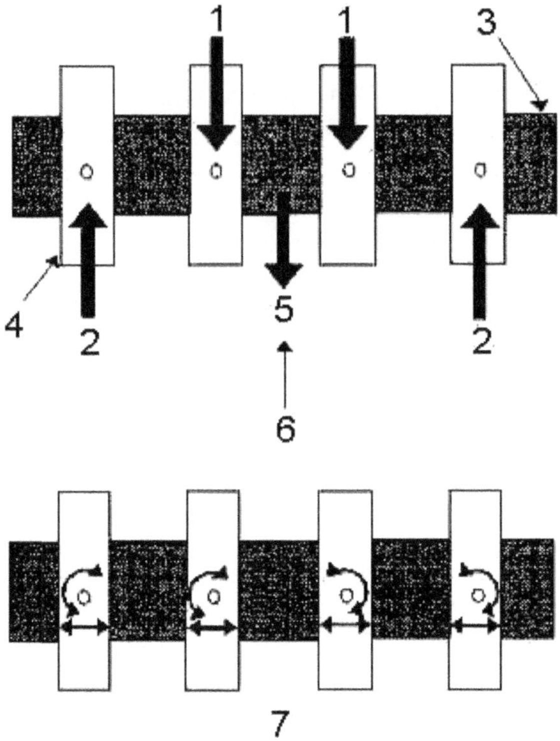

1. Applied load
2. Reaction
3. Specimen
4. Specimen clamp
5. Deflection
6. Return to original position
7. Free horizontal translation and rotation

Figure 1. Basic principals of 4-point bending.

mean value for the stiffness modulus can be taken which represents the weighed stiffness for that cross section in bending.

2 THEORY

The differential equation for the deflection V_B of a slender beam is given by Equation 1 in which I represents the stiffness moment of the beam, Q the applied force distribution and E the stiffness modulus which depends on the distance x.

$$I\frac{\partial^2}{\partial x^2}\left\{ E\{x\}\frac{\partial^2}{\partial x^2}\{V_B\{x\}\}\right\} = Q\{x\} \tag{1}$$

Instead of the derivation symbol d/dx the symbol $\partial/\partial x$ is used because in cyclic testing the deflection and load are functions in x and the time t. Integration with respect to x for the interval $0 < x < A$ (at $x = 0$ the moment is 0) leads to Equation 2 in which F is the applied force at the inner supports:

$$I\,E\{x\}\frac{\partial^2}{\partial x^2}\{V_B\{x\}\} = -\frac{F}{2}x \tag{2}$$

And for the interval A < x < L-A to the well known formula:

$$I\,E\{x\}\frac{\partial^2}{\partial x^2}\{V_B\{x\}\} = -\frac{F}{2}A \tag{3}$$

The deflection due to shear forces is given by Equation 4 in which G represents the shear modulus, μ is the Poisson constant, B is the width of the beam and H is the height of the beam. The factor k is a correction factor because the real distribution of the shear force over a cross section is not known. In this paper a value of 0,85 is adopted based on finite element calculations. More information is given in Annex I.

$$\frac{\partial}{\partial x}V_S\{x\} = \left\langle \begin{array}{ll} \dfrac{F}{2k\,B\,H\,G} = \dfrac{(1+\mu)F}{k\,B\,H}\dfrac{1}{E}; & 0 \le x \le A \\ 0 \Rightarrow V_S\{x\} = V_S\{A\}; & A \le x \le L-A \end{array} \right\rangle \tag{4}$$

3 CALCULATIONS

Three cases will be dealt with. Case I represents a situation in which the stiffness of the beam between the inner and outer supports decreases nearly with the same rate as the stiffness of the beam between the two inner supports. Only at the outer supports the stiffness will stay equal to the stiffness of the virgin material (no strains). In case II a more or less linear decrease in the stiffness modulus of the beam between the outer and inner support is adopted. Finally case III represents a non-realistic situation in which the stiffness of the beam between the outer and inner supports will keep the start value. In the author's view, based on the high exponent of the fatigue relationship between strain and fatigue life, the actual situation will be between case II and III.

3.1 *Case I*

On the interval 0 < x < A the stiffness modulus as function of the distance x is given by:

$$E = \frac{E_0}{1+2\lambda\dfrac{x}{A} - \lambda\left(\frac{x}{A}\right)^2} \qquad x = 0 \Rightarrow E = E_0; \quad x = A \Rightarrow E = \frac{E_0}{1+\lambda} \tag{5}$$

On the interval A < x < L-A the stiffness modulus is constant and equals the value at x = A.

The parameter λ represents the evolution in time. At $\lambda = 0$ no fatigue damage has occurred ($t = 0$). At a certain time bending of the beam will have created so much fatigue damage that the weighed stiffness modulus value in the mid section of the beam is decreased by 50% ($\lambda = 1$).

In Figure 2 the ratio of the (weighed mean) stiffness modulus at a certain time and the original stiffness modulus is given for four values of the parameter λ. At x = A the ratio becomes equal to a ratio $1/(1+\lambda)$ which stays constant in the mid span of the 4PB test.

The solutions for the deflections V_B and V_S for A equals L/3 (ASTM configuration) and the locations x = L/6 and x = L/2 are given by Equations 6, 7, 8 and 9.

$$V_B\{L/6\} = \frac{F\,L^3}{48\,E_0\,I}\left(\frac{23}{54}\right)\left(1+\frac{863}{920}\lambda\right) \tag{6}$$

$$V_B\{L/2\} = \frac{F\,L^3}{48\,E_0\,I}\left(\frac{23}{27}\right)\left(1+\frac{111}{115}\lambda\right) \tag{7}$$

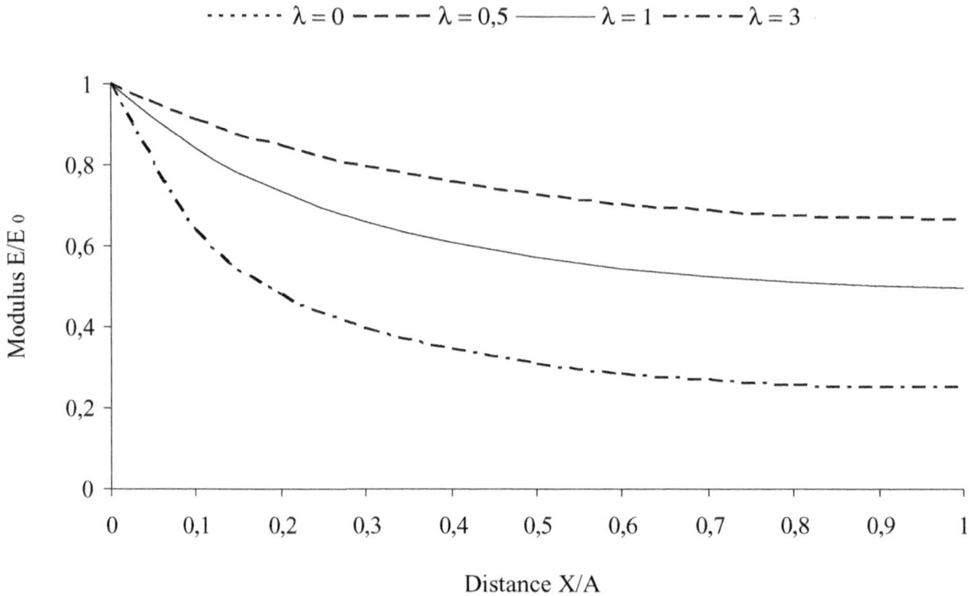

Figure 2. Stiffness modulus (Case I) for the beam between the inner and outer supports.

$$V_S\{L/6\} = \frac{(1+\mu)F}{6k\,B\,H\,E_0} L\left(1+\frac{5}{12}\lambda\right) \tag{8}$$

$$V_S\{L/2\} = \frac{(1+\mu)F}{6k\,B\,H\,E_0} L\left(1+\frac{2}{3}\lambda\right) \tag{9}$$

3.2 *Case II*

On the interval $0 < x < A$ the stiffness modulus is given by:

$$E = \frac{E_0}{1+\lambda(\frac{x}{A})} \qquad x = 0 \Rightarrow E = E_0; \quad x = A \Rightarrow E = \frac{E_0}{1+\lambda} \tag{10}$$

On the interval $A < x < L-A$ the stiffness modulus is constant and equals the value at $x = A$. If $A = L/3$ than the deflections V_B and V_S at $x = L/6$ and $x = L/2$ are:

$$V_B\{L/6\} = \frac{F\,L^3}{48\,E_0\,I}\left(\frac{23}{54}\right)\left(1+\frac{79}{92}\lambda\right) \tag{11}$$

$$V_B\{L/2\} = \frac{F\,L^3}{48\,E_0\,I}\left(\frac{23}{27}\right)\left(1+\frac{21}{23}\lambda\right) \tag{12}$$

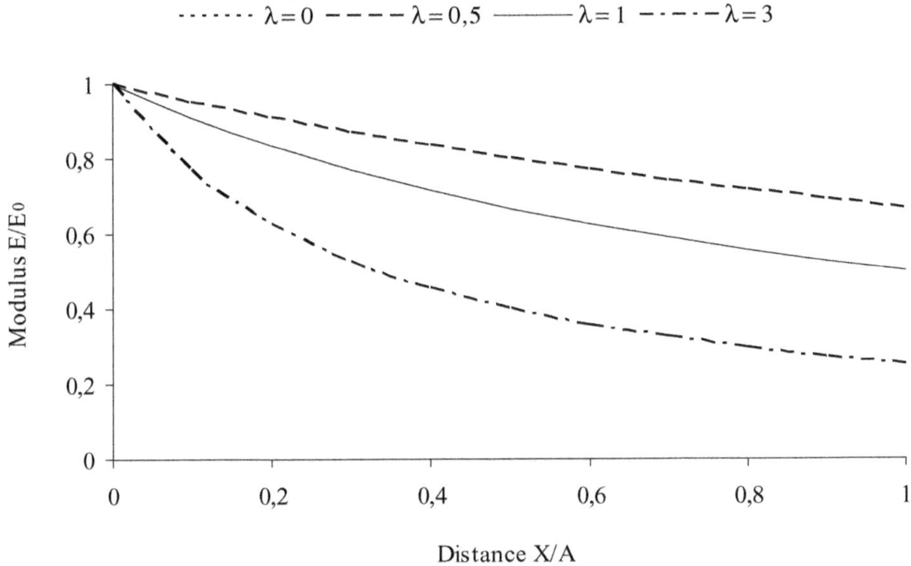

Figure 3. Stiffness modulus (Case II) for the beam between the inner and outer supports.

$$V_S\{L/6\} = \frac{(1+\mu)F}{6k\,B\,H\,E_0}L\left(1+\frac{\lambda}{4}\right) \tag{13}$$

$$V_S\{L/2\} = \frac{(1+\mu)\;F}{3k\,B\,H\;E_0}L\left(1+\frac{\lambda}{2}\right) \tag{14}$$

3.3 *Case III*

In case III the stiffness modulus on the interval $0 < x < A$ will not decrease. This is of course not realistic but will demonstrate the extreme case. For $A = L/3$ than the deflections at $x = L/6$ and $x = L/2$ are equal given by:

$$V_B\{L/6\} = \frac{F\,L^3}{48\,E_0\,I}\left(\frac{23}{54}\right)\left(1+\frac{18}{23}\lambda\right) \tag{15}$$

$$V_B\{L/2\} = \frac{F\,L^3}{48\,E_0\,I}\left(\frac{23}{27}\right)\left(1+\frac{15}{23}\lambda\right) \tag{16}$$

$$V_S\{L/6\} = \frac{(1+\mu)F}{6k\,B\,H\,E_0}L \tag{17}$$

$$V_S\{L/2\} = \frac{(1+\mu)\;F}{3k\,B\,H\;E_0}L \tag{18}$$

3.1 *Reference case*

In case of a constant stiffness $E = E_0/(1+\lambda)$ over the whole range from 0 to L the well known deflection formulas are obtained:

$$V_B\{0 < x < A\} = \frac{F L^3}{12 E_0 I} \frac{x}{L}\left(3\frac{A}{L} - 3\frac{A^2}{L^2} - \frac{x^2}{L^2}\right)(1+\lambda) \tag{19}$$

$$V_B\{A < x < (L-A)\} = \frac{F L^3}{12 E_0 I} \frac{A}{L}\left(3\frac{x}{L} - 3\frac{x^2}{L^2} - \frac{A^2}{L^2}\right)(1+\lambda) \tag{20}$$

$$V_S\{0 < x < A\} = \frac{(1+\mu)F}{k B H E_0} x (1+\lambda) \tag{21}$$

$$V_S\{A < x < (L-A)\} = V_S\{A\} = \frac{(1+\mu)F}{k B H E_0} A (1+\lambda) \tag{22}$$

$$V_B\{L/2\} = \frac{F A}{48 E_0 I}(3L^2 - 4A^2)(1+\lambda) \tag{23}$$

It should be noted that during a fatigue test in controlled force mode the deflection will increase with a factor $1 + \lambda$ and in controlled deflection mode the force will decrease with a factor $1 + \lambda$. For a better understanding the force is assumed to be constant during the test in this exercise. For $A = L/3$ the deflections at the locations $x = L/2$ and $x = L/6$ are:

$$V_B\{L/2\} = \frac{F L^3}{48 E_0 I} \frac{23}{27}(1+\lambda) \tag{24}$$

$$V_S\{L/2\} = \frac{(1+\mu)F}{3k B H E_0} L (1+\lambda) \tag{25}$$

$$V_B\{L/6\} = \frac{F L^3}{48 E_0 I} \frac{23}{54}(1+\lambda) \tag{26}$$

$$V_S\{L/6\} = \frac{(1+\mu)F}{6k B H E_0} L (1+\lambda) \tag{27}$$

4 ABSOLUTE DEFLECTION MEASUREMENTS

The absolute deflection measurement at $x = L/2$ for the three cases is compared with the deflection in the reference case (start of a fatigue test) in figure 4.

Due to the stiffer reaction of the beam in the two outer sections ($0 < x < A$ and $L-A < x < L$) the real measured deflection is lower than the 'expected' value on the assumption of a homogenous decrease in stiffness for the whole beam. This has consequences for the back calculation of the (weighed) beam stiffness modulus and the calculation of the strain in the mid span of the beam.

5 RELATIVE DEFLECTION MEASUREMENT

In some 4PB equipments the deflection is measured as twice the difference between the deflection at x = L/2 and x = L/6. This is possible for a homogenous elastic beam because than at the location x = L/6 both the deflections due to pure bending and due to shear are exactly equal to half the corresponding values at x = L/2. Due to the difference in stiffness decrease along the beam this relationship is not true anymore in a fatigue test. A comparison with the centre deflection of the reference system in the form of a ratio is given in Figure 5. Although the applied relationship does not hold anymore the error is less than in Figure 4. This is due to the fact that by using the deflection at x = L/6 (outer section) the effect of a stiffer beam reaction in the outer section is used for the calculation.

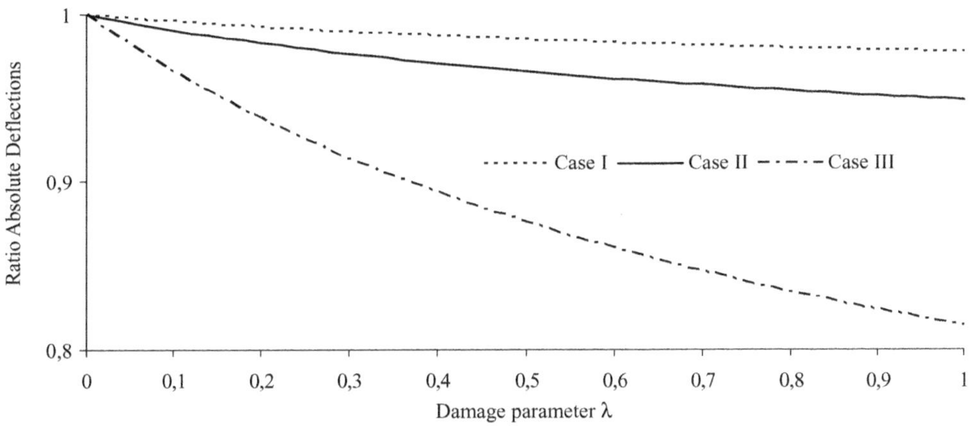

Figure 4. Ratio of the absolute deflections at x = L/2 and the centre deflection in the reference system as a function of the parameter λ for three cases.

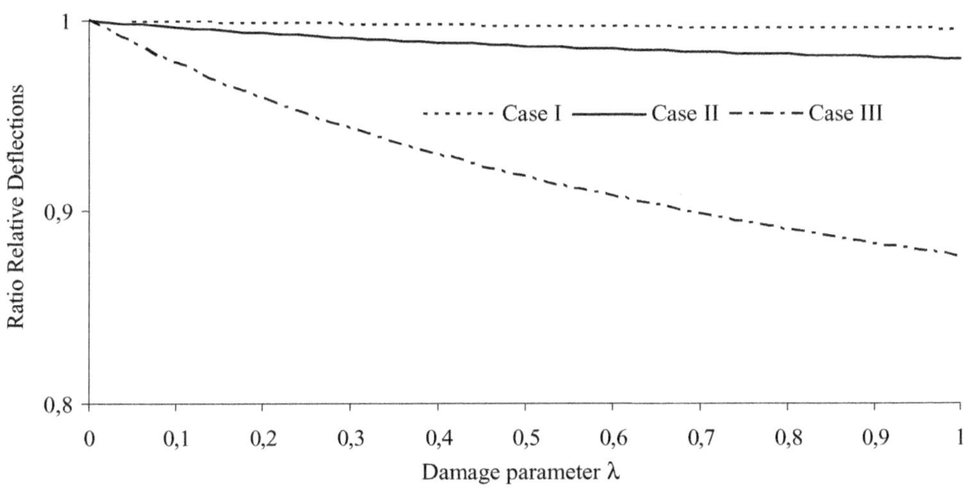

Figure 5. Ratio of the deflection in the centre for three cases calculated with the aid of the deflections at x = L/6 and x = L/2 and the centre deflection in the reference system as a function of the parameter λ.

6 STIFFNESS BACK CALCULATION

To illustrate the consequences for the back calculation of the stiffness modulus the error is calculated for $\lambda = 2/3$. This means that the stiffness modulus in the mid section of the beam is decreased to 60% of its initial value. If the stiffness in the outer sections had decreased in the same order this value would have been back calculated in the processing (provided that the deflection due to shear is taken into account). One would expect in controlled force test a deflection that is 1,67 times higher. However, due to the lesser decrease in stiffness for the outer sections a lower deflection will occur in the midsection. Thus, the back calculated stiffness modulus is overestimated. In Cases I and II the errors are small, but for Case III the error will be 14%. Case III is not realistic but in the author's view the real case will lie somewhere between Case II and III.

7 STRAIN DETERMINATION

In all cases the maximum strain in the mid section of the beam is given by Equation 28.

$$\varepsilon = \frac{H}{2}\frac{\partial^2}{\partial x^2}V_B = \frac{F_0 H L}{12 E_0 I}(1+\lambda) \tag{28}$$

This equation is based on the second derivate of the deflection due to bending and in spite of the differences in deflections this derivate is the same in all cases. However, in many data processes the strain is calculated on basis of the measured deflection and assuming a reference deflection profile. Moreover, no correction is applied for the deflection due to shear. This last deflection will be around 3 to 5% of the total measured deflection depending on the Poisson ratio and geometrical dimension of the beam (see Annex I). The relationship used in the data processing for calculating the strain is given by Equation 29 in which V_B {L/2} should be used. However, often the total measured or calculated deflection V at the location x = L/2 is taken in stead.

$$\varepsilon = \frac{108}{23}\frac{H}{L^2}V\{L/2\} \tag{29}$$

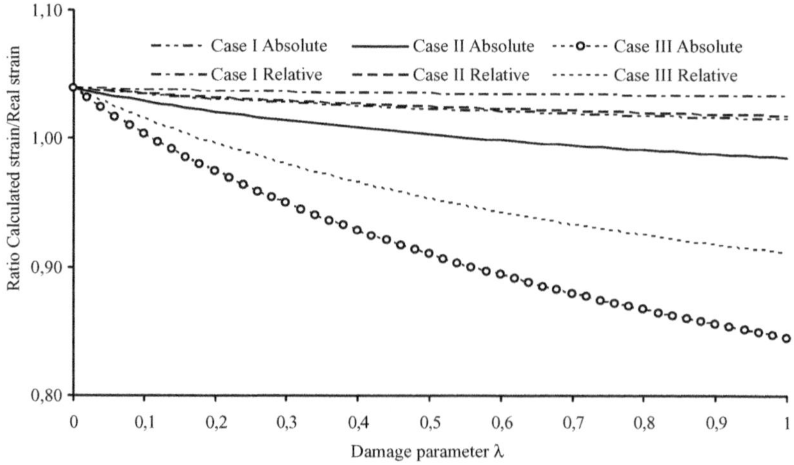

Figure 6. The ratio of the strain calculated using the measured total centre deflection and the real strain as a function of the damage parameter λ for three cases using controlled force mode.

The neglected contribution of the shear deflection will lead to a (initial) overestimation of the real strain in the mid span of the 4PB equipment. However, the total deflections in the three cases become soon lower than the total deflection in the reference system. This is due to the stiffer reaction of the beam in the two outer sections. Assuming that no correction is applied for the deflection due to shear the over- and underestimation of the calculated strain for the three cases expressed as a ratio of the correct strain value is given in figure 6 as a function of the damage parameter λ. As can be seen in Figure 6 the real occurring strain becomes much higher in case II and case III during the test than the strain calculated on the measured deflection.

In deflection controlled tests the applied force will decrease in time. Because the occurring strain is related to the second derivate of the deflection V_B due to bending, this value will stay constant only if the force decreases with a factor $1+\lambda$ (Equation 28). However, in the data processing the measured deflection is used with the wrong assumption that the deflection curve is similar to the curve in the reference case (Equation 29). But the strain is related to the second derivate of the deflection. Although the total deflection is kept constant the same error is induced as in the controlled force mode.

8 CONCLUSIONS

- It is clearly shown that errors are induced in the data processing of the 4PB measurements if the beam stiffness in the outer sections does not decrease in the same way as the beam stiffness in the mid span of the equipment.
- Using a relative deflection measurement the induced errors are smaller but the obtained deflection response is also a factor two smaller.
- Because the real difference for the decreases in the beam stiffness is not known, it is recommended to perform fatigue tests with strain gages in the mid span and the outer sections of the beam for several types of asphalt. In this way it might be possible to get an idea of the decrease function for the beam stiffness in the outer section and to establish correction formulas.

REFERENCES

Cowper G.R. 1966, The shear coefficient in Timoshenko's beam theory, *J. Appl. Mech.*, 33, p. 335–340.

Hutchinson J.R. 2001, Shear coefficient for Timoshenko beam theory, *ASME JAM* 68, p. 87–92.

Kaneko T. 1975, On Timoshenko's correction for shear in vibrating beams, *J. Phys. D: Appl. Phys.*, Vol. 8, p. 1927–1936.

Olsson R.G. 1935, *Z. angew. Math. Mech.*, 15, p. 245.

Pai P.F. *et al* 1999, Shear correction factors and an energy-consistent beam theory, *Int. J. Sol. & Str.*, 36, p. 1523–1540.

Pickett G. 1945, Equations for Computing Elastic Constants from Flexural and Torsional Resonant Frequencies of Vibration of Prisms and Cylinders, *Proceedings ASTM*, 45 846–865 (1945).

Pronk A.C.1996, Theory of the four point dynamic bending test- Part 1, Report P-DWW-96-008, Delft, The Netherlands.

Pronk A.C. 2007a, Theory of the four point dynamic bending test- Part IV: Pure bending & Shear deformation, Report Delft University of Technology, Road and Railway Engineering.

Pronk A.C. 2007b, Deflection in bending tests- Shear force—what do we neglect? presentation held at the 1st European 4PB Workshop, http://www.civil.uminho.pt/4pb/index.htm, Delft.

Sun C.T. 1972, J. *Appl. Mech.*, 39, p. 282–285.

Tanji Y. *et al* 1972, *J. Jap. Inst. Metals*, 36, p. 368–373.

Timoshenko S.P. 1921, On the correction for shear of the differential equation for transverse vibrations of prismatic bars, *Phil. Mag.*, 41, p. 744–746.

Timoshenko S.P. 1922, On the transverse vibrations of bars of uniform cross section, *Phil. Mag.*, 43, p. 125–131.

Uflyand, Ya. S. 1948, The propagation of waves in the transverse vibrations of bar and plates, Akad. Nauk SSSR Prikl. Mat. i Meh., 12, p. 287–300.

Yildirim V. 2007, "Vibration behavior of composite beams with rectangular sections considering the different shear corrections factors", *Vibrations Problems ICOVP 2005*, p. 531–536.

ANNEX I THE SHEAR CORRECTION FACTOR K

Because the shear stress is not uniformly distributed over the cross section of the beam a correction factor k is used in all bending theories. Several assumptions are made in the past for this distribution (Table A1). Nice overviews are given by Kaneko (1975) and Yildirim (2007). In many 4PB software a value of 2/3 is used, which was originally stated by Uflyand (1948) but this value for plate dynamics is very different from all other expressions. In case of a rectangular cross section the expressions given in Table A1 can be found in the literature.

Two other "impressive" expressions are given by Pai (1999) and Hutchinson (2001). The original expression by Pai (Equation A1) is rewritten in order to show the similarity with the expression of Hutchinson (Equation A2). Next to the dependency on the Poisson ratio μ the value for the correction coefficient k also depends on the ratio H/B of the beam in these two equations. This is in contrast with all other relationships in Table A1.

$$k_{\text{Pai}} = \frac{5}{6 + \frac{\mu^2}{(1+\mu)^2}\left(\frac{H}{B}\right)^4\left[1 - \frac{90}{\pi^4}\sum_{n=1}^{\infty}\left(\frac{\text{Tanh}\{n\pi B/H\}}{n^5(\pi B/H)}\right)\right]} \tag{A1}$$

$$k_{\text{Hutchinson}} = \frac{5(1+\mu)}{6 + 5\mu - \frac{\mu^2}{(1+\mu)^2}\left(\frac{H}{B}\right)^4\left[1 - \frac{90}{\pi^4}\sum_{n=1}^{\infty}\left(\frac{\text{Tanh}\{n\pi B/H\}}{n^5(\pi B/H)}\right)\right]} \tag{A2}$$

If the ratio H/B tends to zero the k factors become equal to the expressions given by Timoshenko (1921 & 1922). In Table A2 an overview is given for three Poisson ratios μ of the values for the k factor based on the formulas given in Table A1.

For a square cross section (H/B = 1) the k factors according to Pai *et al* and Hutchinson are for a Poisson ratio of 0,35 equal to 0,8269 and 0,8763 respectively.

Based on finite element calculations with the program ABAQUS a value of 0,85 is adopted for the correction coefficient k in this paper (Pronk, 2007b).

Table A1. Overview of k factor formulas.

Author	Year	K factor
Sun	1972	$\pi^2/12$
Cowper	1966	$(10 + 10\,\mu)/(12 + 11\,\mu)$
Timoshenko	1921	5/6
Timoshenko	1922	$(5 + 5\,\mu)/(6 + 5\,\mu)$
Olsson	1935	$(20 + 20\,\mu)/(24 + 15\mu)$
Pickett	1945	$24{,}612(1 + \mu)/(29{,}538 + 5{,}942\,\mu + 64{,}077\,\mu^2)$
Tanji *et al*	1972	$\approx(6 + 12\,\mu + 6\,\mu^2)/(7 + 12\,\mu + 4\,\mu^2)$

Table A2. K factors according to Table A1 for three Poisson ratios.

μ	Sun	Cowper	Timoshenko	Timoshenko	Ollson	Pickett	Tanji
0,30	0,8225	0,8497	0,8333	0,8667	0,9123	0,8627	0,9252
0,35	0,8225	0,8517	0,8333	0,8710	0,9231	0,8419	0,9354
0,40	0,8225	0,8537	0,8333	0,8750	0,9333	0,8171	0,9453

Efficient Transportation and Pavement Systems – Al-Qadi, Sayed, Alnuaimi & Masad (eds)
© 2009 Taylor & Francis Group, London, ISBN 978-0-415-48979-9

Assessment of rutting potential of asphalt mixture based on mixture resistivity

T.K. Pellinen
Helsinki University of Technology, Finland

G. Huber
Heritage Research Group, USA

S. Xiao
Resource International, Inc., USA

ABSTRACT: In this case study, the aim was to assess a new rutting model to be used as a simplified modeling tool in the mix design and construction stages of paving work. Rutting of selected intestate mixtures was predicted using a resistivity model developed by Christensen and Bonaquist for the National Cooperative Highway Research Program Project 9-25. The resistivity model estimates the aggregate structure's resistance to the asphalt binder flow. One essential input parameter is the age-hardened binder viscosity. The aging time is related to the time at which one wants to know the rutting in practice. Predictions were compared to the actual measured rutting. Research showed that predictions were promising and the developed models were able to identify inferior mixtures. However, the predictions were extremely sensitive for the binder viscosity values and aging time used in the analysis. Therefore, the selection of an appropriate aging time to obtain age-hardened viscosity that relates to the in service stiffening of asphalt pavement is of utmost importance. For practitioners, the advantage of the resistivity model is that, in principle, it is simple to use and with some enhancements could be employed as a simplified mix design tool.

1 INTRODUCTION

Christensen and Bonaquist (2005) developed a rational new model for estimating rutting based on compaction level, volumetric composition and binder properties for the National Cooperative Highway Research Program Project 9-25. A concept of mixture resistivity was introduced to estimate the given aggregate structure's resistance to the asphalt binder flow, which is expressed in Eq. (1). Equation utilizes Cox-Merz rule, which states that the dynamic viscosity (or modulus of complex viscosity) $|\eta^*|$ can be obtained from the modulus of dynamic (complex) stiffness using the relationship of $|G^*|/\omega$. Surface area, S_a, is calculated as the sum of the percent passing the 75, 150 and 300 micron sieves, divided by 5.0

$$P = \frac{|\eta^*| \; S_a^2 \; G_{sb}^2}{4.9 \; VMA^3} \tag{1}$$

where:
- P = resistivity, s/nm (seconds per nano-meter) (capital letter for Greek Rho)
- $|\eta^*|$ = binder dynamic viscosity, Pa-s
- S_a = aggregate surface area, m^2/kg

G_s = aggregate bulk specific gravity
VMA = Design Voids in the Mineral Aggregate, % by volume

Rut depth was considered to be proportional to the cubic root of the traffic loading expressed in 80 kN Equivalent Single Axle Loads (ESALs). The regression model using resistivity, number of design gyrations N_{design}, and relative density as predictors was developed to give the depth of rutting for the desired ESALs. The R^2 of the regression model was 89.1%, adjusted for degree of freedom. The regression model is given in Eq. (2), and the rut depth is calculated using Eq. (3):

$$RR = 224P^{-1.08}N^{-0.650}(RD)^{-18.6} \tag{2}$$

$$Rut\ Depth = RR \cdot h \cdot ESALs^{1/3} \tag{3}$$

where:
RR = Rutting rate, mm/m thickness/$ESALs^{1/3}$
P = resistivity, s/nm (seconds per nano-meter)
N = N_{design} or design Marshall hammer blows
RD = relative density, as-constructed density/as-designed density (decimal number)
ESALs = Equivalent 80 kN Single Axle Load
h = thickness of pavement layer (m)

According to the model developers, most of the rutting occurred in the top 100 mm of the pavement; this is based largely on trenches at MN/Road test pavements in Minnesota, which indicated that rutting in most cases was limited to this portion of the pavement. The assumption of a 100-mm thickness for the rutting rate was kept constant in the development work, which required the model to be calibrated for 100-mm thickness. According to the model developers, this method of calculation will significantly affect rutting rate calculations on thin overlays, where the actual thickness of the overlay should be used if less than 100 mm.

The viscosity used in the equations is the age-hardened viscosity, and the model developers suggested using the aging time given by the following equation, Eq. (4), where t_{total} is the time of the test duration or the age of the pavement in service for which rutting is predicted.

$$t_{aging} = \left[\frac{t_{total}^{1/3}}{2}\right]^3 = \frac{t_{total}}{8} \tag{4}$$

The rutting of selected intestate mixtures was predicted using the models presented above, and the predictions were then compared to the actual measured rutting. The aim was to assess the developed rutting model to be used as a simplified modeling tool in the mix design and construction stages of paving work. The mixture information, such as aggregate structure, binder content, voids in mineral aggregate, and air voids content were obtained from construction quality control (QC) testing. For some mixtures, forensic testing obtained after pavement failure was also used. Binder viscosity was estimated from testing of recovered binder and Tank binder during construction. Binder measurements were conducted using Dynamic Shear Rheometer (DSR) and rotational viscometer. Conventional binder testing was also undertaken, including penetration and softening point measurements.

2 ESTIMATION OF BINDER VISCOSITY

As discussed above, one essential parameter for the resistivity model is the age-hardened binder viscosity. The viscosity value must be obtained at a critical pavement temperature for rutting. The resistivity model developers recommended using pavement temperature at 50 mm deep in the pavement, which is a critical location for high shear loads caused by heavy truck traffic. As the binder dynamic and rotational viscosity measurements are conducted at ≥60°C temperatures, a conversion must be used to obtain the viscosity at desired pavement temperature. Also, if the DSR testing is used to measure binder stiffness, a conversion must be used between the binder shear stiffness and viscosity.

Bitumen is thermoplastic; it becomes harder i.e., stiffer (more viscous) with decrease in temperature and softer (less viscous) with increase in temperature. Temperature susceptibility varies from one source of crude oil to another. The binder temperature susceptibility can be expressed using Eq. (5), where A is the intercept and VTS is the slope of the regression line.

$$\log(\log(\eta)) = A + VTS \log(T_R) \tag{5}$$

where:

η = bitumen viscosity, cP
T_R = temperature in Rankine (T_R=T_F+459,7, T_F = temperature, Fahrenheit), °R
A = regression intercept
VTS = regression slope of viscosity temperature susceptibility

There also exists a relationship between the measured penetration for 100 g, 5 sec loading and viscosity in Poise. The penetration data is valid only for temperatures 10 and 25°C. The viscosity-penetration relation is considered valid between a penetration range of 3 to 300 and is more accurate between values of 3 and 15 (Mirza and Witczak, 1995). The statistical model relating penetration to viscosity is given by Eq. (6). The viscosity corresponding to the softening point and penetration 800, is equal to 13,000 Poise for most unmodified binders.

$$\log\eta = 10.5012 - 2.2601\log(Pen) + 0.00389(\log(Pen))^2 \tag{6}$$

Research shows that the VTS is related to the aging of binders and Mirza and Witczak (1995) developed a Global aging system (GAS) for asphalt mixtures where they expressed A and VTS for tank and mix lay-down conditions and then for long term aging, see Figure 1. The Global aging system predicts change in asphalt viscosity in pavement with time and depth based on statistical analysis of data from 40 field projects. The viscosity limit for the binder at cold temperatures as bitumen is approaching elastic stiffness is assumed to be $3 \cdot 10^9$ N/m^2 or $2{,}7 \cdot 10^{10}$ Poise.

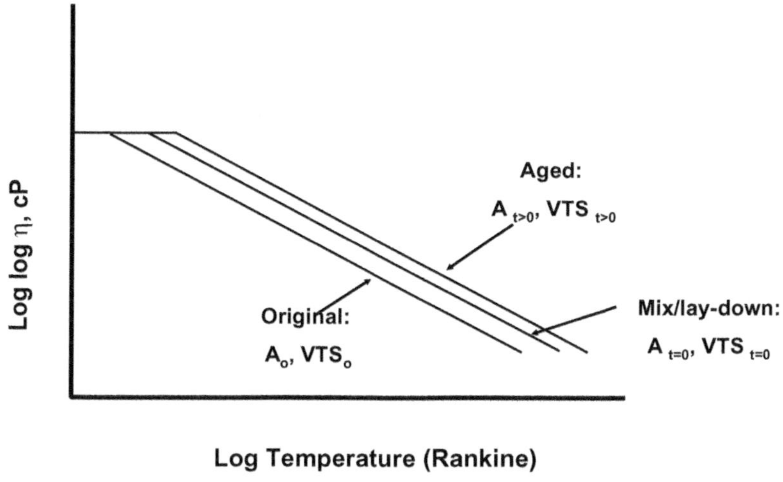

Figure 1. Global aging system for binder viscosity.

Table 1. Typical hardening resistance and code values for bitumen.

Mix/lay-down hardening resistance	Expected hardening resistance values	Code value
Excellent to good	HR ≤ 1,030	−1
Average	1,030 < HR ≤ 1,075	0
Fair	1,075 < HR ≤ 1,100	1
Poor to very poor	HR > 1,100	2

The GAS comprises four models: Short-term aging model Eq. (7), which predicts mix/lay-down viscosity as function of original viscosity and hardening code. Long-term aging model Eq. (8), which predicts aged viscosity at assumed depth of 6,3 mm as function of mix/lay-down viscosity, time, mean annual air temperature (MAAT), and temperature with optional adjustments for air voids and change in air voids with time. Depth model Eq. (9) predicts aged viscosity with depth as function of aged viscosity, MAAT, and depth. Table 1 gives the hardening resistance values used in Eq. (7). The code value is defined by the hardening ratio (HR) of log-log mix/lay-down viscosity (Rolling Thin Film Oven Test, RTFO) to the log-log original viscosity.

Original to mix/lay-down model:

$$\begin{aligned} \log\log \eta_{t=0} &= a_0 \cdot \log\log \eta_{\text{orig}} \\ a_0 &= (0.054405 + 0.004082 \cdot code) \\ a_1 &= (0.972305 + 0.010886 \cdot code) \end{aligned} \qquad (7)$$

where:
$\eta_{t=0}$ = mix/lay-down viscosity (cP) at temperature T_R (Rankine),
η_{orig} = original viscosity (cP) at temperature T_R (Rankine),
code = hardening resistance (code = 0 for average)

Surface aging from 0 to 6,3 mm from surface:

$$\begin{aligned} \log\log \eta_{\text{aged}} &= \frac{\log\log \eta_{t=0} + A \cdot t}{1 + B \cdot t} \\ A &= -0.004166 + 1.41213 \cdot C + C \cdot \log(MAAT) + D \cdot \log\log \eta_{t=0} \\ B &= 0.197725 + 0.068384 \cdot \log C \\ \log C &= 274.4946 - 193.831 \cdot \log(T_R) + 33.9366 \cdot (\log(T_R))^2 \\ D &= -14.5521 + 10.47662 \cdot \log(T_R) - 1.88161 \cdot (\log(T_R))^2 \end{aligned} \qquad (8)$$

where:
η_{aged} = aged viscosity (centipoises, cP)
$\eta_{t=0}$ = viscosity at mix/lay-down (centipoises, cP)
MAAT = mean annual air temperature (°F)
T_R = temperature in Rankine, °R
t = time in months

Viscosity depth relationship:

$$\begin{aligned} \eta_{t,z} &= \frac{\eta_t \cdot (4 + E) - E \cdot \eta_{t=0} \cdot (1 - 4 \cdot z)}{4 \cdot (1 + E \cdot z)} \\ E &= 23.82 \cdot e^{-0.0308 \cdot MAAT} \end{aligned} \qquad (9)$$

where:

z = depth (in),

η_t = surface aged viscosity,

$\eta_{t,z}$ = aged viscosity at any depth (z)

The relationship of the steady shear and dynamic rheological data can be approximated using the empirical Cox-Merz rule. The Cox-Merz rule supports the superposition of the shear rate dependence of the steady shear viscosity η and the frequency dependence of the complex viscosity $|\eta^*|$ at equal values of angular velocity ω and shear rate $\dot{\gamma}$ as follows.

$$|\eta^*(\omega)| = \eta(\dot{\gamma})_{\omega=\dot{\gamma}} \tag{10}$$

This rule applies for unfilled, structurally simple (linear) materials. For shear rate independent Newtonian fluids ($\delta = 90°$) the relationship between the zero shears viscosity η_0 and the shear complex stiffness $|G^*|$ can be approximated based on the Cox-Merz rule as follows. If the stiffness is given in Pa, the unit for viscosity is Pa·s as the unit for ω is rad/s.

$$\eta_0 = |\eta^*| = \frac{|G^*|}{\omega} \tag{11}$$

Based on the Cox-Merz rule, Bonaquist, Pellinen and Witczak (1998) developed an empirical model that can be applied for converting the binder shear stiffness $|G^*|$ to steady shear viscosity η which approximates the apparent Newtonian viscosity when the phase angle deviates from 90°:

$$\eta = \frac{|G^*|}{\omega}\left(\frac{1}{\sin\delta}\right)^{a_0 + a_1\omega + a_2\omega^2} \tag{12}$$

where:

η = apparent Newtonian viscosity, in Pas;

$|G^*|$ = binder complex shear modulus, in Pa;

ω = angular frequency, in radians/sec;

δ = phase angle in radians; a_0, a_1, a_2 = fitting parameters (for combined dataset of unmodified and modified binders $a_0 = 3.639216$, $a_1 = 0.131373$, $a_2 = -0.000901$).

Steps to calculate the age-hardened binder viscosity from the original Tank conditions for the resistivity model are listed below:

1. Determine pavement temperature at 50 mm deep in the pavement using Federal Highway Administration's LTTPBind software.
2. Convert DSR test results to viscosity using Eq. (11) or Eq. (12) and/or convert conventional penetration test results to viscosity using Eq. (6) For Superpave binder testing, ω is 10 rad/s.
3. Calculate binder viscosity at determined pavement temperature using Eq. (5)
4. Determine the aging time to be used in GAS using Eq. (4)
5. Calculate age-hardened binder viscosity using GAS models Eq. (7), Eq. (8) and Eq. (9)

3 STUDIED CASES

Table 2 lists the studied cases and gives the same basic information about the sites. All these cases are reported elsewhere and only a short introduction to each case is given below.

Table 2. Description of studied cases.

Case	Description	No. of mixtures	Research conducted	Reported by
I	Forensic study of rutting failure on I-74 near Indianapolis in Indiana	2	1999, 2002–2003	Andrewski 2001
II	Study on I-70 in Indiana to validate Superpave binder specification	6	1998–2004	Shah 2004
III	Indiana surface mixtures, study of mixtures stiffness at different stages of construction	11	2002–2005	Pellinen and Xiao, 2006

Figure 2. Trench studies in I-74 near Indianapolis, Indiana. Courtesy of Heritage Research Group.

3.1 *Case I, Interstate I-74*

This section of Interstate I-74 located near Indianapolis, Indiana, suffered rutting failure in the eastbound lanes in summer 1999. A study team was assembled by Indiana Department of Transportation (INDOT) in fall 1999 to investigate the failure. The forensic study conducted by the team included taking cores from the failed and surviving section of the road in fall 1999. The cores were obtained and tested by Heritage Research Group and the study team submitted its report in July 2001, "*I-74 Pavement Investigation, Contract: RS-23371.*" A set of additional cores were taken in spring 2002 for the Purdue University investigation before the road was resurfaced in the same spring.

Based on the gathered information, Interstate I-74 was constructed in 1998 in two phases. The interstate rehabilitation work consisted of a 2 × 76 mm of intermediate 19-mm Superpave base mixture placed in two lifts followed by 25 mm of 12.5-mm Superpave surface mixture. Both mixtures had a PG64-28 binder. The surface mixture contained blast furnace slag, dolomite, dolomite sand, and natural sand; and the intermediate mixture was fabricated using crushed stone, stone sand, and 15% reclaimed asphalt pavement (RAP). The westbound (WB) lanes and the eastbound (EB) lanes were constructed by the same paving crew using the same equipment. There were no changes in the Job Mix Formula (JMF) of the mixtures between the EB and WB construction projects.

During the summer of 1999, the EB lanes developed 12 to 15 mm of rutting, while the WB lanes had only 1 to 3 mm of rutting. Transverse slabs cut from the pavement confirmed that the rutting occurred in the surface layer and in the first lift of the intermediate base layer, shown in Figure 2. The WB lanes were constructed first followed by the EB lanes, and during EB construction all interstate traffic was directed to the WB lanes.

The study team's report concluded that the failure of the EB lanes was due to the soft mixture caused by the soft binder and slightly higher binder and finer material contents compared to the WB lanes. The binder stiffness measured from the extracted binder in the EB lanes was approximately one binder grade lower than that of the surviving WB lanes. The DSR tests from the recovered binder showed that the surviving lane, in fact, had a PG70-28 binder instead of the designed PG64-28 binder.

3.2 *Case II, Interstate I-70*

As a part of an on-going program to validate the Superpave binder selection, Shah (2004) studied the performance of six binders in six test sections constructed on interstate I-70 east of Indianapolis in 1998. The specification's recommended binder grade was PG58-16 at 50% reliability and 58-28 at 98% reliability. The selected grade was increased to PG64-28 due to the expected traffic of 30 to 100 million ESALs for 20 years of service life. The study concluded that the standard Superpave grade PG64-28 was sufficient for this site.

Pavement temperature measurements with thermocouples imbedded in the asphalt mixture showed that the 7-day average pavement temperature at 20 mm deep ranged from 38,0 to 43,0°C, while the air temperature ranged from 31,5 to 36,5°C. The test sites were monitored for 4 years and the rut depths were measured using a dipstick after 1,5 and 4 years of traffic loading. Surface mixtures were 9,5 mm nominal aggregate size mixtures (NMAS) and the intermediate mixtures were 19 mm NMAS mixtures.

3.3 *Case III, Stiffness study*

This research was initiated from a need to evaluate marginal mixtures that do not meet the Superpave mix design criteria for the air voids content requirements for the construction of hot-mix asphalt (HMA) pavements (Pellinen and Xiao, 2006). Therefore, this research investigated the possibility of using asphalt mixture stiffness and strength as performance indicators to assess the rutting resistance of marginal mixtures, and the major objective was to develop performance criteria or threshold values to be used as a QC/QA tool in future HMA construction projects. The research concentrated on the asphalt surface mixtures used in Indiana. All mixtures had 9,5 mm NMAS; and four mixtures were Stone mastic asphalt (SMA) mixtures and the remaining 11 mixtures were dense-graded mixtures. The mixtures were constructed using dolomite, sand stone, steel slag, and blast furnace slag. The binder grades used were PG64-22, PG70-22 and PG76-22.

4 RUTTING PREDICTIONS

4.1 *Case I, Interstate I-74*

Rutting predictions were conducted using the equations given above. Input data for the models are given in Table 3, which also shows the calculated age-hardened binder viscosity and resistivity values for each mixture. The VMA, G_{sb}, N_{des} and S_a were obtained from the QC data.

The pavement was in service for 12 months before failure, which gives t_{aged} of 1,5 months based on Eq. (4). The pavement temperature at 50 mm deep was estimated using FHWA's LTTPBind software. Table 4 gives the temperature susceptibility coefficients for the Tank (original) conditions and for the recovered binder. The DSR tests for the recovered binder were conducted at 58, 64 and 70°C temperature. The DSR test results were first converted to the viscosity using Eq. (12) Using these viscosity values, A and VTS coefficients were developed using Eq. (5). For the Tank binder, the temperature susceptibility coefficients were back-calculated by employing Global aging system. Finally, the age–hardened binder viscosity was obtained by interpolating t_{aged} = 1,5 months from the Tank and recovered binder data. For the EB lane, an aging code of 1 (low aging) was used, and for the WB lane an aging code of 0 (normal aging) was used based on the forensic study data of binder and mixture properties.

The predictions shown in Table 5 were done for 100-mm thick pavement layers as the analysis determined that using the actual surface layer thickness would yield drastically too small rutting for that layer. The relative density RD is the ratio of in-place and design air voids contents where the air voids content is expressed as % maximum density G_{mm}. Then 4,2% air voids is 95,8% G_{mm} and 4,4% is 95,6% G_{mm} and RD is 0,998. The predicted rutting for the surface and intermediate mixtures for the EB lanes was 10 and 7,8 mm respectively, which agreed quite well with the measured total rutting of 15 mm. For the WB lanes the predicted rutting was 6, and 2,2 mm, which was almost three times more than the measured total rutting of 3 mm. For the WB lanes, the use of actual surface layer thickness would have yielded better predictions. Table 3 shows that the rutted EB lanes had considerably lower resistivity values compared to the surviving WB lanes. The resistivity values reflect the binder viscosity values, as expected. Both the EB and WB mixtures had very similar mixture consistency, and the major difference was the in-situ binder stiffness, which was approximately one PG grade higher for the surviving WB lane.

4.2 *Case II, Interstate I-70*

Table 6 gives input data for the rutting predictions for the I-70 mixtures. Predictions were done for two cases: rutting after 1,5 and 4 years of traffic loading. The binder viscosity was estimated at t_{aged} = 2,25 and 6 months of aging time. The age-hardened viscosity for this site was estimated

Table 3. Mixture input data and calculated resitivity for Case I.

I-74 mixtures	VMA (%)	S_a (kg/m²)	G_{sb}	N_{des}	Pavement temperature (°C)	Aged η $t_{aged=1,5}$ (Pa-s)	P (s/nm)
Surface EB	15,5	4,69	2,596	109	46,0	1232	10,0
Surface WB	14,6	4,22	2,596	109	46,0	4109	32,3
Intermediate EB	14,0	4,74	2,615	109	46,0	1455	16,6
Intermediate WB	13,4	4,59	2,615	109	46,0	6575	80,3

Table 4. Binder temperature susceptibility and calculated viscosity values for Case I.

I-74 mixtures	Tank A_0	Tank VTS_0	Tank η (Pa-s)	Recov'd $A_{t=12\ mo}$	Recov'd $VTS_{t=12\ mo}$	Recov'd η $_{t=12\ mo}$ (Pa-s)
Surface EB	10,516	–3,5291	991	10,276	–3,4304	2914
Surface WB	11,774	–3,9721	3253	11,499	–3,8610	10106
Intermediate EB	10,537	–3,5351	1169	10,297	–3,4362	3461
Intermediate WB	11,397	–3,8308	5171	11,133	–3,7237	16403

Table 5. Predicted and measured rutting for Case I.

I-74 mixtures	Design VTM (%)	In-place VTM (%)	RD	*h* (m)	ESALs (10^6)	Predicted rut depth (mm) 1 Year	Meas'd rut depth (mm) 1 Year
Surface EB	4,2	4,4	0,998	0,1	30	10,5	10
Surface WB	4,2	8,4	0,956	0,1	30	6,5	3
Intermediate EB	4,0	5,5	0,984	0,1	30	7,8	5
Intermediate WB	3,9	7,5	0,963	0,1	30	2,2	0

from recovered binder viscosity measurements conducted at 2, 18, and 48 months time intervals during the research. Viscosity measurements included dynamic and rotational viscosity at 60 and 135°C, respectively. Also penetration was measured at 5 and 25°C. Table 6 shows that the resisitivity values varied between the mixtures, and clearly binder stiffness was the major contributor. For mixtures S2, S4, and S6, the same high temperature PG grade binder yielded large differences in the mixture resistivity, which ranged from 36,4 to 51 s/nm.

Table 7 gives developed temperature susceptibility coefficients A and VTS for 2 weeks, 18 months and 48 months of in service aging time from which the age-hardened binder viscosity at 2 months and 6 months were interpolated. Table 8 gives the predicted rutting and the measured rutting for layer thickness $h = 100$ mm. The model seems to under-predict and over-predict rutting. Also, the rutting predicted for 4 years of service-life is lower than that for 1,5 years. This is due to changes in binder stiffness; the increased binder stiffness and thus mixture resistivity at 6 months

Table 6. Input data for Case II.

I-70 Mixes	VMA (%)	S_a (kg/m²)	G_{sb}	N_{des}	Pavement temp. (°C)	Aged η $t_{aged=2,25}$ (Pa-s)	P $t_{aged=2,25}$ (s/nm)	Aged η $t_{aged=6}$ (Pa-s)	P $t_{aged=6}$ (s/nm)
S1	14,8	4,86	2,536	126	46,2	11006	105,4	18097	173,2
S2	14,4	5,33	2,573	126	46,2	2829	36,4	3181	40,9
S3	13,9	5,33	2,573	126	46,2	2421	34,5	3113	44,4
S4[1]	13,5	5,45	2,577	126	46,2	6337	103,5	9659	157,8
S5	15,0	4,98	2,573	126	46,2	15203	151,0	32732	325,1
S6	14,5	5,45	2,573	126	46,2	10780	141,8	29324	385,7

[1] 15% reclaimed asphalt pavement (RAP).

Table 7. Binder temperature susceptibility and calculated viscosity values for Case II.

I-70 Mixes	$A_{t=2\,mo}$	$VTS_{t=2\,mo}$	$\eta_{2\,mo}$ (Pa-s)	$A_{t=18\,mo}$	$VTS_{t=18\,mo}$	$\eta_{18\,mo}$ (Pa-s)	$A_{t=48\,mo}$	$VTS_{t=48\,mo}$	$\eta_{48\,mo}$ (Pa-s)
S1	11,0253	–3,6898	9315	11,2971	–3,7798	22472	11,4041	–3,8146	34929
S2	10,7772	–3,6127	2650	10,9731	–3,6793	4041	10,9651	–3,6760	4199
S3	10,8681	–3,6479	2142	10,9370	–3,6655	4307	11,0048	–3,6886	4972
S4	10,9544	–3,6687	5859	11,1271	–3,7264	9579	11,3243	–3,7923	16944
S5	10,3075	–3,4270	12269	10,5532	–3,5062	35092	10,5779	–3,5088	71951
S6	11,7786	–3,9645	7785	11,6136	–3,8915	31078	11,9694	–4,0132	70607

Table 8. Predicted and measured rutting for Case II.

I-70 Mixes	Binder type	Design VTM (%)	In-place VTM (%)	RD	Design ESALs (10^6)	Predicted rut depth (mm) 1,5 Years	Predicted rut depth (mm) 4 Years	Measured rut depth (mm) 1,5 Years	Measured rut depth (mm) 4 Years
S1	AC-20	7,6	7,7	0,999	92	1,2	1,0	3,0	3,2
S2	PG64-28	3,0	4,3	0,987	92	4,9	6,0	3,2	5,2
S3	PG58-28	4,3	6,5	0,977	92	6,2	6,5	4,2	4,9
S4	PG64-28	4,4	7,0	0,973	92	2,0	1,8	3,9	4,9
S5	PG70-28	5,1	7,9	0,970	92	1,4	0,9	3,0	4,4
S6	PG64-16	5,5	9,0	0,963	92	1,8	0,8	2,5	4,2

aging time reduces calculated rutting compared to 2,25 months of aging time. Thus, the proper selection of age-hardened binder viscosity is extremely important for correct predictions.

3.3 *Case III, Stiffness study mixtures*

Tables 9 and 10 give input data for rutting predictions for the Case III study mixtures. Predictions were conducted using measured Tank binder data, which was age-hardened employing the global aging model to obtain binder viscosity for $t_{aged} = 3$ months. Aging code of 0 was used in the analysis. The Tank binder measurements included penetration at 25°C, ring and ball softening point and rotational viscosity measurements at 90, 120, 135 and 175°C. Predictions were done for two years of in service traffic loading. Table 9 shows that there is some variation in mixture resistivity based on the mixture composition and design criteria.

Table 11 gives the predicted and measured rut depths for layer thickness $h = 100$ mm. The rutting predicted using 3 months aged binder stiffness is slightly on the high side for the mixtures with soft binders. However, based on the predictions, all mixtures should perform adequately, which they did, based on measured rutting.

Table 9. Input data for Case III.

Road no.	VMA (%)	S_a (kg/m^2)	G_{sb}	N_{des}	Pavement temp. (°C)	Aged η $t_{aged=3}$ (Pa-s)	P (s/nm)
SR15	15,9	4,72	2,682	75	46,4	8360	68,0
US24	16,1	4,76	2,607	100	46,5	9717	73,2
SR161	15,3	4,31	2,627	75	48,3	9513	69,5
SR56	15,4	3,82	2,830	100	48,3	17985	117,5
US30	15,9	4,17	2,620	125	46,5	26374	159,8
SR66	15,1	4,82	2,620	100	49,1	18052	170,6
SR135	15,4	4,75	2,583	100	49,1	15808	133,0
US31	17,8	5,95	3,314	100	47,5	29832	419,8
I74	17,7	6,36	3,476	100	47,0	44928	808,1
SR64	17,9	5,30	3,275	100	49,5	30938	331,7
I65	17,8	6,05	3,313	100	48,9	32659	474,8

[1] Estimated 3 months age-harmed binder.

Table 10. Binder temperature susceptibility and calculated viscosity values for Case III.

Road no.	Binder type	Tank A_0	Tank VTS_0	Tank η (Pa-s)
SR15	64-22	10,503	–3,5165	2207
US24	64-22	10,467	–3,5016	2581
SR161	64-22	10,738	–3,5971	2556
SR56	70-22	10,589	–3,5363	4921
US30	70-22	9,7503	–3,2312	7223
SR66	70-22	10,555	–3,5227	4958
SR135	70-22	10,351	–3,4512	4567
US31	76-22	9,0354	–2,9689	8699
I74	76-22	9,6311	–3,1815	13152
SR64	76-22	9,6311	–3,1815	9083
I65	76-22	9,5909	–3,1672	9580

Table 11. Predicted and measured rutting for Case III.

Road no.	Design VTM (%)	In-place VTM (%)	RD	Design ESALs (10^6)	Predicted rut depth (mm) 2 Years	Meas'd rut depth (mm) 2 Years
SR15	4,0	11,2	0,925	2,4	3,0	0,0
US24	4,0	9,5	0,943	9,6	2,6	0,6
SR161	4,0	10,8	0,929	0,5	1,6	0,0
SR56	4,0	9,6	0,942	7,5	1,4	0,0
US30	4,0	6,5	0,974	50,2	0,9	0,9
SR66	4,0	12,6	0,910	6,8	1,7	1,3
SR135	4,0	8,6	0,952	20	0,7	1,6
US31	4,0	8,3	0,955	21,4	0,4	0,0
I74	4,0	7,9	0,959	33	0,2	0,0
SR64	4,0	4,7	0,993	21,1	0,2	0,0
I65	4,0	8,9	0,949	155	0,8	0,0

Table 12. Minimum resistivity based on 12,5 mm rut depth.

Minimum Resistivity (s/nm)				
ESALs (10^6)				
<3	3–10	10–30	30–50	50–100
24,2	29,5	41,3	42,3	52,4

5 MINIMUM RESISTIVITY

By using Eq. (2) and Eq. (3) it is possible to back-calculate the needed resistivity and thus establish the resistivity criteria for different traffic levels by assuming a maximum rut depth of 12,5 mm, layer thickness of 100 mm, and the cubic root relationship between rut depth and ESALs. A general form of resistivity criteria are expressed by Eq. (13):

$$P_{\min} = \left[\left(\frac{12{,}5\,(mm)}{0{,}1(m)\cdot(ESALs)^{\frac{1}{3}}} \right) \cdot \left(\frac{1}{224} \right) \cdot N^{0.65} \cdot (RD)^{18.6} \right]^{-\frac{1}{1.08}} \tag{13}$$

The calculated resistivity criteria using Eq. (5) are given in Table 12, for a design air voids content of 4%, for an as-constructed air voids content of 7.5%, and for a layer thickness of 100-mm.

Table 13 summarizes the calculated resistivity values for different binder viscosity values obtained from the Tank and from the recovered binder measurements for Cases I and II. Based on Table 12, the minimum resistivity value needed for I-70 is 52,4 s/nm and for I-74, 42,3 s/nm in order that the rutting does not exceed 12,5 mm after 20 years of traffic loading. Only the I-74 WB mix with a 6 months aged binder meets the resistivity criteria.

The table shows that although the mixtures had the same binder grade, the resistivity values were quite different. The relative increase in resistivity was much higher for the S3 mix with a PG58-28 binder than that of the other binders. This indicates more rapid binder aging for softer binders and, therefore, increased relative rutting resistance.

Table 13. Comparison of Resistivity for Case I and II.

			P (s/nm)			
ESALs (10^6)	Road	PG Grade	Tank	1.5 mo.	2.25 mo.	6 mo.
30	Case I: I-74 Surface EB	PG64–28	8,1	10,0	11,0	15,9
30	Case I: I-74 Surface WB	PG70–28[1]	25,6	32,3	35,7	52,6
92	Case II: I-70 S2	PG64–28	21,9	35,6	36,4	40,9
92	Case II: 1-70 S3	PG58–28	10,9	33,2	34,5	44,4

[1]After RTFO criteria original grade PG64-28.

6 DISCUSSION

As a summary Figure 3 gives the comparison between predicted and measured rutting. The analysis excludes Case II mixtures with 4 years of in service time, which improves correlation slightly. As the figure shows, the developed models are able to distinguish the inferior mixture.

The introduced resistivity concept is an intriguing way of estimating asphalt mixture's resistance to rutting. The amount of permanent deformation or rutting is dependent on properties of asphalt mixture but it also depends upon the local climate, the traffic level, and the quality of construction.

Model developers discovered during analysis of repeated shear at constant height (RSCH) data collected during NCHRP Project 9-25 that there appeared to be a very good relationship between rut resistance, as indicated by this test, and aggregate permeability; "not permeability to water, but permeability of the aggregate to the particular asphalt binder at the selected test temperature". Three factors affect the permeability of the aggregate structure to flow of the binder: voids in mineral aggregate (*VMA)*, aggregate surface area, and binder viscosity. Permeability increases with increasing *VMA*, decreasing aggregate surface area, and decreasing binder viscosity at the temperature of interest. According to model developers "these factors also are among those most often associated with rut resistance, their effect on rut resistance being opposite to their effect on permeability".

Pavement engineers generally use the term permeability to refer to the permeability of an asphalt mixture to water; a different terminology was needed to describe this concept. To address this problem, the inverse of permeability was used, and the term "resistivity" was selected to describe this parameter. Resistivity in this case means the resistance to asphalt binder flow provided by a given aggregate and it is estimated based on Winterkorn's permeability equation" (Winterkorn and Fang, 1975).

This analysis confirms that resistivity and, therefore, mixture rutting resistance, is driven by binder stiffness. However, it is evident that the use of a Tank binder in the rutting predictions may lead to erroneous results, and the age-hardened binder viscosity must be used. Analysis also suggests that the layer thickness and location in the pavement may pose a problem as for some sites the default layer thickness of 100 mm was adequate and for some other sites the actual in situ layer thickness would yield better predictions. As the models give discrete rutting predictions and do not use an incremental way of calculating cumulative rutting, inconsistent results may occur when the time of aging is varied in the analysis. This suggests that the model may be better suited for a go/no-go type of analysis rather than trying to predict rutting after varying in service times (see Figure 3).

Therefore, a simple analysis to verify the mixture performance could be done by calculating the mixture resistivity at the mix design stage using Superpave binder grading test results with RTFOT-aged binder stiffness in the analysis. Table 14 compares the calculated resistivity values for the Tank and the RTFOT-aged binder stiffness obtained from DSR testing at 64°C temperature.

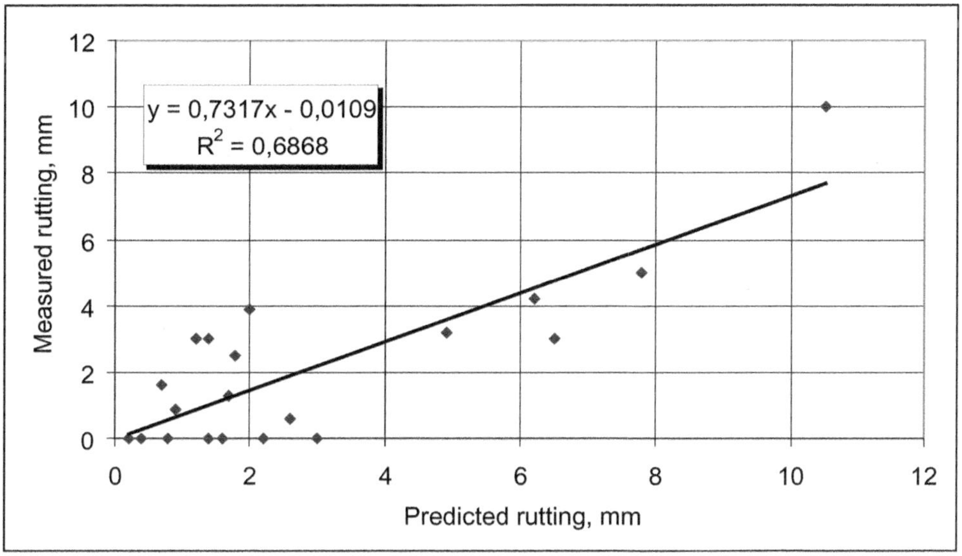

Figure 3. Correlation between predicted and measured rutting.

Table 14. Comparison of mixture properties for PG 64-xx binder.

		Tank		RTFOT		Recovered	
Mixture	PG (°C)	G*/sinδ (kPa)	P (s/nm)	G*/sinδ (kPa)	P (s/nm)	G*/sinδ (kPa)	P (s/nm)
Case I: I-74 EB	64	1,21	1,0	2,24	1,8	2,13	1,7
Case I: I-74 WB	64	1,61	1,3	2,92	2,3	4,34	3,4
Case II: I-70 S2	64	1,73	2,2	2,34	3,0	–	–
Case II: 1-70 S3	64	0,53	0,8	1,07	1,5	–	–

The binder viscosity η for the resistivity model was obtained by using Eq. (6). By knowing the field performance of the studied mixtures, it seems that resistivity P > 2 s/nm with the RTFOT-aged binder tested at 64°C is enough for a good mixture performance; while a mixture with resistivity P < 2 s/nm is vulnerable to failure. However, the failure depends on how the binder ages in-situ compared to RTFOT-aging in the laboratory. The grading temperature of 64°C used in the calculations represents the average pavement temperature at 25 mm deep in the pavement in Indiana for the adjusted traffic using 98% reliability. The binder stiffness variations and especially the variation in the age-hardened binder properties seem to be the key factor for identifying good and poor performing mixtures among those mixtures, which are designed with a low factor of safety against failure.

7 CONCLUDING REMARKS

In this case study, rutting of selected intestate mixtures was predicted using a resistivity model developed by Christensen and Bonaquist for the National Cooperative Highway Research Program Project 9-25. They introduced a concept of mixture resistivity to estimate the given aggregate structure's resistance to the asphalt binder flow. Research showed that the predictions were promising and the model was able to identify inferior mixtures. However, the predictions were extremely sensitive to the age-hardened binder viscosity and, subsequently, stiffness. Therefore, the selection of an appropriate aging time to obtain age-hardened viscosity is of utmost importance.

For the practitioner, the advantage of the resistivity model is that it is simple to use. However, some modifications should be introduced to the developed models that they can be applied to actual pavement problems. As the models gave discrete rutting predictions and do not use an incremental way of calculating cumulative rutting, inconsistent results may occur when the time of aging is varied in the analysis. This suggests that the models may be better suited for a go/no-go type of analysis rather than trying to predict rutting after varying in service times.

By rearranging equations, it is possible to back-calculate the required resistivity and thus establish rational resistivity criteria for different traffic levels by assuming a certain maximum rut depth and the cubic root relationship between the rut depth and traffic. Consequently, revised versions of the models are in development in a NCHRP contract, which will use binder stiffness instead of binder viscosity as input data for the resistivity model.

DISCLAIMER AND ACKNOWLEDGEMENT

This work was supported by the Joint Transportation Research Program (JTRP) administered by the Indiana Department of Transportation and Purdue University. The contents of this paper reflect the views of the authors, who are responsible for the facts and the accuracy of the data presented herein, and do not necessarily reflect the official views or policies of the Federal Highway Administration and the Indiana Department of Transportation, nor do the contents constitute a standard, specification, or regulation.

The authors wish to acknowledge Dr. Donald W. Christensen for his valuable help in conducting this research. Authors also wish to thank Mr. Dave Andrewski and Mr. Lee Gallivan for serving as JTRP Study Advisory Committee members and for providing their help and advice during the research.

REFERENCES

Andrewski, D. 2001. I-74 Pavement Investigation, Contract: RS-23371" Indiana Department of Transportation.

Bonaquist, R., Pellinen, T. and Witczak, M.W. 1998. Development of Relationship between Binder Viscosity and Stiffness. Superpave Support and Performance Models Management. Submitted to U.S. Department of Transportation, Federal Highway Administration. 77 p.

Christensen, D.W. and Bonaquist, R. 2005. Rut Resistance and Volumetric Composition of Asphalt Concrete Mixtures. Association of Asphalt Paving Technologists-Proceedings of the Technical Sessions, Vol. 74, pp. 410–439.

Pellinen, T.K. and Xiao, S. 2006. Development of Hot-mix asphalt performance Criteria for Indiana. Journal of the Association of Asphalt Paving Technologists, Vol. 75, pp. 1–40.

Pellinen, T., Zofka, A., Marasteanu, M. and Funk, N. 2007. The Use of Asphalt Mixture stiffness Predictive Models, Journal of the Association of Asphalt Paving Technologists, Vol. 73.

Mirza, M.W. and Witczak, M.W. 1995. Development of a Global Aging System for Short and Long Term Aging of Asphalt Cements, Journal of the Association of Asphalt Paving Technologists, Vol. 64, pp. 393–431.

Shah, A. 2004. Influence of Binder and Mixture Properties on the Performance of AC Pavements. Doctoral Dissertation. Purdue University, West Lafayette, IN.

Winterkorn, H.F. and Fang, H.Y. (1975) Soil Technology and Engineering Properties of Soils, in *Foundation Engineering Handbook*, H. F. Winterkorn and H. Y. Fang, Eds., New York: Van Nostrand Reinhold Co., 106 p.

Efficient Transportation and Pavement Systems – Al-Qadi, Sayed, Alnuaimi & Masad (eds)
© 2009 Taylor & Francis Group, London, ISBN 978-0-415-48979-9

Quality control of cold-in place recycled asphalt pavement using non-destructive tests

A. Loizos
National Technical University of Athens (NTUA), Athens, Greece

V. Papavasiliou & C. Plati
Laboratory of Highway Engineering (NTUA), Athens, Greece

ABSTRACT: The cold-in place recycling technique (CIR) using foamed asphalt and cement was used for the pavement rehabilitation of a heavily trafficked highway. In order to investigate the quality of the rehabilitated pavement during its early life, the Laboratory of Highway Engineering of the National Technical University of Athens (NTUA) undertook a field experiment. In-situ non-destructive testing (NDT) and a data analysis research study was conducted on a trial section using both the Falling Weight Deflectometer (FWD) and Ground Penetrating Radar (GPR). According to the data analysis results, GPR is a useful tool for backanalysis during early life of the pavement, when in-situ thickness data may not be available. Furthermore, the composite modulus of the pavement layers proved to be a practical indicator that gives important information about the condition of the newly constructed layers in terms of quality control of the rehabilitated pavement.

1 INTRODUCTION

Cold in-place recycling (CIR) offers an attractive alternative to other pavement rehabilitation options, as the cost of the hot mixed asphalt mixtures (HMA) continuously increases and the availability of good materials is limited (PIARC 2001). CIR is an advantageous rehabilitation technique since it is eminently suited for the reworking of the upper layers of distressed pavements to depths up to 30 cm. The foamed asphalt technique (Asphalt Academy 2002) is a promising strategy for its efficient use of salvage construction material.

With the growing need to construct new roads for rural access and to rehabilitate existing roads, the aforementioned technique offers multiple advantages, including:

- The ability to re-open the rehabilitated road to traffic shortly after construction, thereby minimizing traffic disruptions.
- Lowered construction costs in comparison to other standard methods of rehabilitation.
- The ability to effectively use marginal aggregates in pavement layers, when treated with foamed asphalt.

The use of foamed asphalt technology in the rehabilitation of road pavements, as well as new pavement construction was given a boost in the 1990s through the incorporation of the technology on advanced cold recycling machines (Wirtgen 2004). There are many reasons why the in-situ recycling process has gained wide acceptance in the road pavements industry, including economic, technical and environmental advantages over alternative road rehabilitation options. Benefits of cold recycling have been widely publicized, among others by Jenkins et al. (1995).

Information about foaming procedures in the laboratory, as well as about related construction aspects, although not standardized, are satisfactorily documented in international literature. However, not enough information is available concerning quality control issues and acceptance of rehabilitated pavements using the foamed asphalt CIR technique and when published it has been limited to low to medium volume roads. For this reason, a field experiment on a typical flexible pavement of a heavily trafficked Greek highway was undertaken by the Laboratory of Highway Engineering of the NTUA in order to investigate possible quality control options of the recycled material stabilized with foamed asphalt and cement, using mainly non-destructive tests (NDT).

2 BACKGROUND

A major heavy duty north-south highway, the Athens-Lamia link, a part of the Trans European Network (TEN), was upgraded to a six-lane dual carriageway facility. Traffic volume is in excess of 10^6 equivalent standard axle loads per year. The new flexible pavement some years after construction suffered from severe defects i.e. localized crocodile cracking and deformations (Fig. 1). The Road Authority decided to deal with these severe distresses by implementing an experimental study on a trial highway section, using the CIR with foamed asphalt stabilization technique (Fig. 2). Prior to the implementation of cold recycling, a thorough testing program was performed on the selected trial section as documented by Loizos et al. (2004).

Following an extensive pre-investigation, two components were used for the formulation of the pavement design (material or mix design and structural design) as presented briefly below.

Foamed asphalt mix design was undertaken to establish the application rates for foamed asphalt and active filler (cement), to achieve optimal strengths and to determine the strength characteristics for use in the structural design exercise on several different blends of material recovered from the test pits. These blends were treated with foamed asphalt using the appropriate laboratory unit and several briquettes were manufactured for testing purposes to determine the indirect tensile strength (ITS), the unconfined compressive strength (UCS), the cohesion (c) and the angle of internal friction (Φ), as well as the determination of the indirect tensile stiffness modulus (ITSM) (ASTM 2004). Details of the mix design blends can be found in (Loizos et al. 2004). A standard

Figure 1. Typical distressed section of the existing pavement.

Figure 2. Implementation of the CIR technique using foamed asphalt.

mix design of 2.5% foamed asphalt and 1% cement was adopted for the flexible pavement. The decision to introduce 1% cement was based on improvement in the achieved soaked strengths.

An analytical rehabilitation design approach was used based both on national and international experience to estimate the structural capacity of each pavement configuration. According to the analytical design with a structural capacity requirement in excess of 10 million 13-ton axle-loads, the pavement structure is described by an asphalt concrete (AC) layer (90 mm thick) and a cold in-place recycled and stabilized with foamed asphalt (FA) layer (250 mm thick). The AC layer was constructed of two courses, a 50 mm binder course and a 40 mm final semi-open graded surface course using polymer modified asphalt. Figure 3 shows the pavement cross-section after the rehabilitation.

For the analytical pavement design the stiffness modulus of the AC layers was considered to be 3500 MPa. The design of the pavement was dictated by the structural capacity requirements based on the ITS using 100 mm briquettes and UCS results for 150 mm diameter (120 mm high) briquettes and the underlying support conditions. A conservative modulus value of 1500 MPa was considered for the foamed asphalt stabilized recycled material. This was supplemented by using the stress ratio limit method based on the results of the tri-axial tests, as defined in reference (Jenkins et al. 2004).

Construction operations were carried out in four steps, requiring two "recycling" operations. No pre-milling was done before the pavement was pulverized, primarily to include as much RAP as possible in the recycled material. The sequence may be summarized as follows (Loizos et al. 2004):

Step 1: The first "recycling" operation called for the existing pavement to be pulverized by recycling with a new-generation Wirtgen WR2500S to a predetermined depth. This operation required careful control to ensure that the milling drum remained at least 50 mm above the horizon of the bottom of the new recycled/stabilized layer.

Step 2: The pulverized material was then compacted and shaped in accordance with the new design levels. This resulted in some areas having surplus material that was removed to stockpile. Other areas deficient were leveled up with material imported from stockpile.

Figure 3. Recycled pavement structure.

Step 3: The second "recycling" operation called for the material to be stabilized with a combination of foamed asphalt and cement. The Wirtgen WR2500S recycler drew stabilizing agents from the Wirtgen WM1000 cement slurry unit and a bulk bitumen tanker pushed ahead. The treated material was compacted to refusal density using a 19 ton vibrating padfoot roller before final profiling and finishing of the completed layer.

Step 4: The application of the AC binder layer was delayed until the moisture content in the upper 100 mm portion of the new recycled layer had reduced to below 50% of optimum moisture content, normally a period of between 2 and 4 days.

3 IN-SITU DATA COLLECTION AND ANALYSIS

3.1 *FWD testing*

During and after completion of the recycling works at the site of the field experiment a comprehensive FWD and GPR survey was undertaken, which was comprised of several monitoring levels in chronological order. The different pavement monitoring levels were comprised of in-situ measurements on the surface of the recycled layer after the laying of the binder course and finally on the surface course 3 weeks after the rehabilitation work. The post-construction monitoring levels were comprised of measurements on the related paths on the surface course approximately 16 and 19 months after construction. The trial section was approximately 550 m long. In-situ non-destructive tests (NDTs) were performed on two paths of the heavily trafficked lane: the outer traffic wheel path (OWP) as well as the respective path in between (BWP). The average temperature of the asphalt layers at the different monitoring levels ranged between 16 and 31°C. These temperature values were used for correction of the analysis results to a reference temperature.

3.2 *Ground penetrating radar*

The GPR system (Fig. 4) of the Laboratory of Highway Engineering of the NTUA (GSS 2002) was used for the analysis, with the aim to assess pavement layer thicknesses. This data is useful for backanalysis procedures. The system used, is appropriate for the evaluation of the upper part of the pavement structure, since it produces reliable information to an approximate 0.7 m penetration depth (Loulizi et al. 2003). The system follows the principles of (ASTM 2005) and is supported by the appropriate software (RoadScanners 2001). As far as the GPR antenna frequency is concerned, two different antennas were used (a 1 GHz air-coupled antenna and a 400 MHz ground couple antenna) in order to increase the accuracy of the thickness estimation. More specifically the horn

Table 1 Monitoring levels.

Monitoring level (ML)	Measurements on	Time since construction	Coring	GPR
1a	Recycled layer	1 day		×
1b	Recycled layer	2 days		×
2	Binder course	4 days		×
3	Surface course	3 weeks		×
4	Surface course	16 months	×	×
5	Surface course	19 months	×	×

Figure 4. Ground penetrating radar (GPR) measurements.

antenna gives more accurate thickness results for the upper pavement layers, while the ground couple provides with more accurate thicknesses for the deeper layers.

3.3 *Coring*

Several cores from the asphalt layer as well as from the foamed asphalt recycled layer were drilled at five specific FWD test locations, in accordance to a predetermined coring schedule set out at the beginning of the monitoring program. Cores were drilled along both the OWP and the BWP for ML4 and ML5. Before curing of the recycled material it was not possible to extract cores of the FA layer during the construction stages ML1a to ML3. The difference between estimated GPR thickness and measured core thickness was ranged from –28% to 14% for the AC layer, while for the recycled layer was ranged from –23% to 17%

3.4 *Backanalysis*

A thorough field data analysis was performed including a backanalysis with the aim to verify the robustness of the newly built trial recycled pavement. The backanalysis was undertaken using the latest metric version of the MODCOMP software (Irwin 2002). Considering the level of the subgrade at the bottom of the drainage layer (Fig. 3), the backanalysis model consisted of four layers. The backcalculation was performed at the five test points, where detailed AC and FA core information was available. GPR analysis enabled the estimation of the underneath the recycled layer remaining sublayer thickness. Thus a more reliable estimation of the FA modulus is achieved.

4 DATA ANALYSIS RESULTS

4.1 *Deflection indicators*

The overall pavement condition of the recycled pavement was determined using internationally accepted indicators (COST-336 1998), which are relevant to the measured elastic deflections. For this purpose, the center (maximum) deflection was taken into account, which represents the overall pavement performance at the time of the investigation (Hakim et al. 2002). The in-situ measured deflections were normalized to the reference temperature of 20°C (Van Gurp 1995).

The results (average values) are presented graphically in Figure 5. This figure clearly indicates a decrease of the average maximum deflections with time. The comparison of the results on the outer wheel path (OWP) and between wheel paths (BWP) showed no significant differences, with exception of measurements during ML4 (16 months), where the average center deflection on BWP was 19% increased.

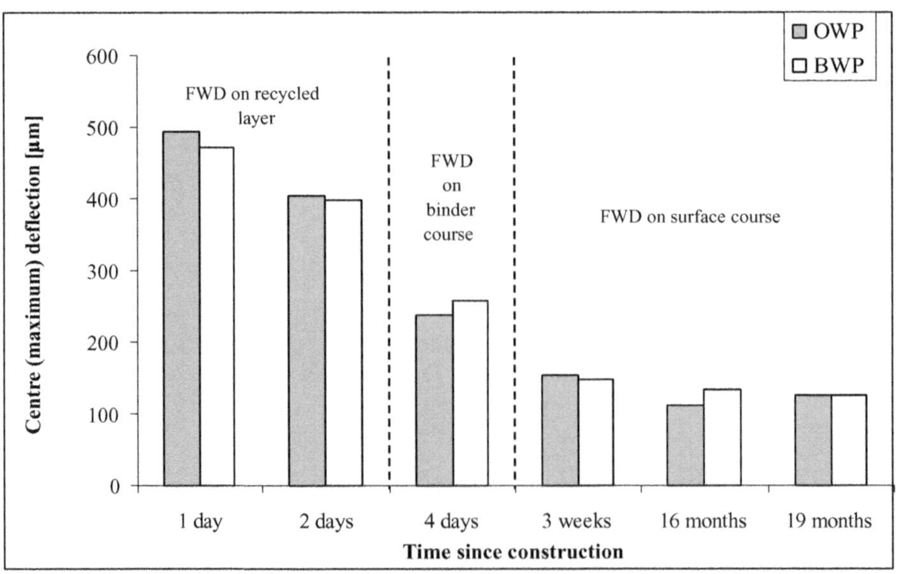

Figure 5. Average maximum deflection.

The level of homogeneity was determined using the coefficient of variation (CV). This parameter is defined as the ratio of the standard deviation over the mean value per section. The CV of the center deflection for the different monitoring levels ranged from 20% to 30%, indicating moderate homogeneity of the overall condition of the recycled pavement (COST-336 1998).

4.2 *Recycled layer modulus results*

The measured deflections were backanalyzed in order to estimate the modulus of the foamed asphalt recycled material and to compare with the modulus that was taken into account during the pavement design. As far as the thickness input is concerned, backcalculations were made by using GPR thickness estimations referred as E (GPR) and by using core thickness measurements referred as E (cores). In both cases the underneath the recycled layer remaining unbound sub-layer thickness was based on the GPR thickness.

The moduli obtained based on GPR thickness data were not significantly different, compared to the relative ones obtained from cores as shown in Table 3 and graphically in Figures 6–7. Three weeks after construction (ML3), the average backcalculated modulus of the

Table 3. GPR vs cores backanalysis results (recycled material).

Monitoring level (ML)	Time since construction	E(GPR)/E(cores) [OWP]			E(GPR)/E(cores) [BWP]		
		Avg	Min	Max	**Avg**	Min	Max
1a	1 day	**1.00**	0.94	1.09	**1.05**	0.96	1.12
1b	2 days	**0.95**	0.89	0.98	**1.00**	0.97	1.05
2	4 days	**1.06**	0.88	1.15	**0.94**	0.88	1.01
3	3 weeks	**0.96**	0.89	1.08	**0.97**	0.86	1.12
4	16 months	**1.00**	0.89	1.09	**1.01**	0.84	1.14
5	19 months	**0.90**	0.84	0.94	**0.86**	0.79	1.18

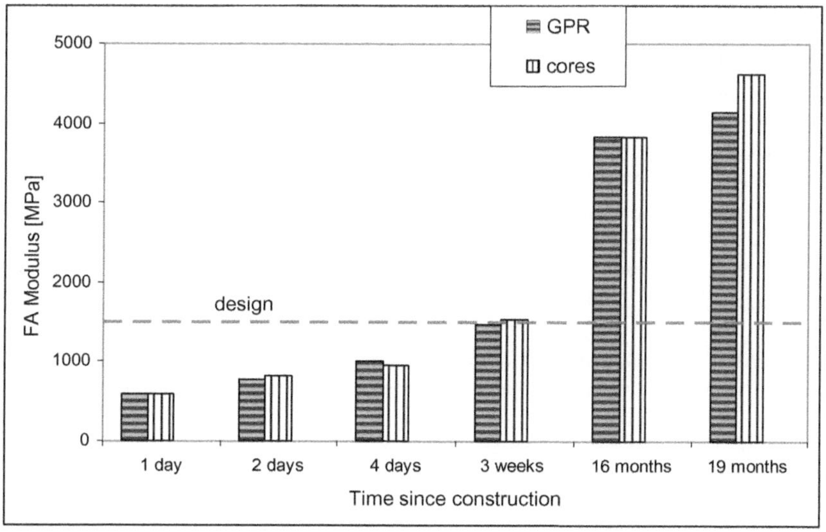

Figure 6. Backanalysis results (OWP).

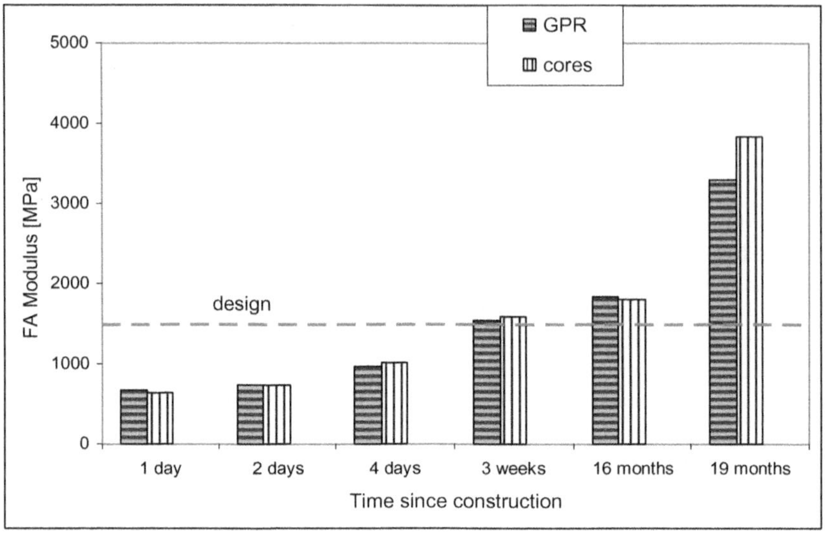

Figure 7. Backanalysis results (BWP).

recycled material has reached the design value (1500 MPa). However this is not observed for all of the respective modulus values, indicating that the recycled material has not fully reached its structural adequacy during this stage of measurements. An interesting observation is that during the ML4 (16 months) the average FA modulus on the OWP was higher than the relative one on the BWP. This fact was also observed in (Loizos & Papavasiliou 2007) during laboratory tests on FA cores from another pavement section. This difference seems to be the result of different loading along the two paths, i.e. the absence of post compaction between the wheel paths.

4.3 *Composite modulus*

The contribution of the asphalt layers to the structural condition of the recycled pavement is very important, especially during the first days of the pavement's life, where no complete curing of the stabilized with foamed asphalt material had occurred (Loizos & Papavasiliou 2006). A simplified approach, which takes into account the influence of the asphalt and the recycled layers as well, is the estimation of the composite modulus i.e. the combined asphalt layers and recycled layer modulus (Milton & Earland 1999). It can also be used for quality control of the rehabilitated pavement layer. Composite moduli were backcalculated based on core thickness values at ML3 (3 weeks), ML4 (16 months) and ML5 (19 months). The remaining unbound layer thickness was estimated using the GPR thickness. The backcalculated asphalt layer moduli were corrected to 25°C according to Baltzer & Jansen (1994), as this temperature was considered as representative of the in-situ measured AC temperatures. Figure 8 shows the estimated composite moduli as well as the equivalent to the design values composite modulus which was estimated using the SHRP (1993). Composite moduli were also backcalculated based on GPR thickness values, resulting in similar values.

Comparing the results at ML3 (3 weeks) presented in Figure 8 (composite modulus) with the results presented in Figures 6–7 (recycled material modulus), it can be concluded, that unlike the modulus of the recycled material, the composite modulus was higher than the calculated design value. This is an indication that the rehabilitated pavement layers have an adequate structural condition.

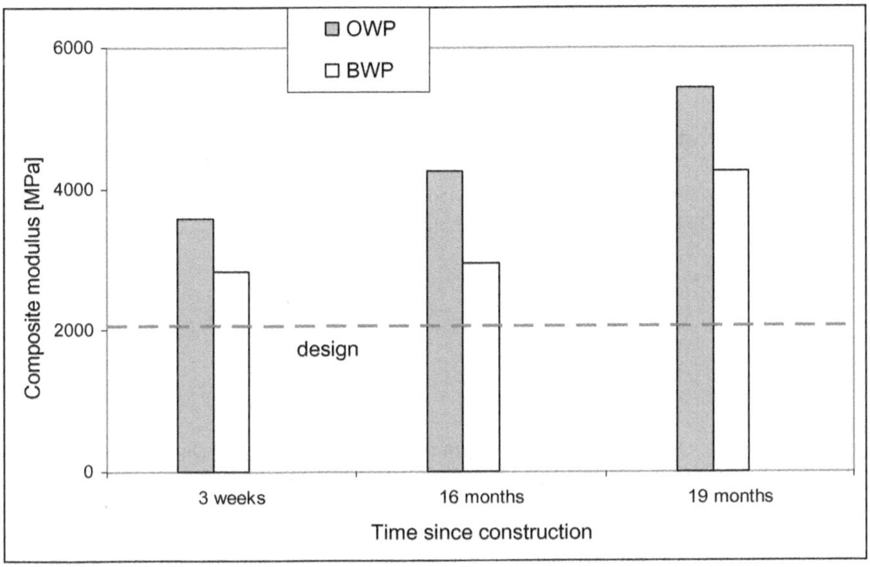

Figure 8. Composite modulus (OWP & BWP).

5 CONCLUSIONS

In the present research study, an effort was made to investigate possible quality control options of the cold in-place recycling (CIR) technique, using mainly non-destructive tests (NDT). The major findings and discussion points are the following:

The quality control of the CIR pavement taking into account the FWD deflection indicator (center deflection) presented an improvement to the overall pavement structural condition during the first 3 weeks of the life of the pavement and a tendency of stabilization thereafter.

The moduli of the foamed asphalt recycled material (FA) obtained from GPR results analysis thickness results were not significantly different, compared to the relative ones obtained from core thicknesses. GPR proved to be a useful tool for backanalysis during early life of the pavement, when in-situ thickness data may not be available.

The determination of the in-situ FA modulus is a very important quality control option for the CIR technique. In case that the in-situ obtained modulus is not always higher than the design value, the recycled material may not have reached its expected structural accuracy. This issue must be further investigated taking into account in the analysis process the presence of the asphalt concrete (AC) layers as well.

During the early life (3 weeks) of the recycled pavement, no significant differences between average recycled material moduli on the outer wheel path (OWP) and between wheel paths (BWP) were observed. The average modulus on the OWP was much higher than the relative one of the BWP. This difference reflects the impact of the traffic loads (post compaction) on the recycled material.

The composite modulus (combined AC and FA) seems to be a practical simplified indicator that gives important information about the condition of newly constructed layers, because it takes into account the influence of both the asphalt and recycled layers. Therefore it can be used as an option for the acceptance of the rehabilitated pavement.

The Laboratory of Highway Engineering of the NTUA continues to undertake further research on the subject, including continuous monitoring of pavement performance and characterization of the FA recycled material through detailed analysis. Due to the importance of the Trans European

Highway under rehabilitation, the investigated trial section is considered, both for the National and the European database, as a long term pavement performance (LTPP) site.

ACKNOWLEDGMENTS

The authors would like to thank the Greek Ministry of Public Work and the Responsible Road Authority (EYDE/PATHE) and the involved contractors for supporting the research work of this study.

REFERENCES

Asphalt Academy, 2002. The Design and Use of Foamed Bitumen Treated Materials, *Interim Technical Guideline 2*. 1st Ed., Pretoria.

ASTM D4123-82, 2004. Standard Test Method for Indirect Tension Test for Resilient Modulus of Bituminous Mixtures, *American Society of Testing and Materials*, Pennsylvania.

ASTM D4748, 2005. Standard Test Method for Determining the Thickness of Bound Pavement Layers Using Short-Pulse Radar, *Non-destructive Testing of Pavement Structures, American Society of Testing and Materials*. Pennsylvania.

Baltzer, S. & Jansen, J.M. 1994. Temperature Correction of Asphalt-Moduli for FWD-Measurements. *Proceedings, 4th International Conference on the Bearing Capacity of Roads and Airfields,*: 7-25, University of Minneapolis, Minnesota.

COST- 336. 1998. Guidelines for Evaluation of Flexible Pavements at Project Level Using Falling Weight Deflectometer. *Final Report, European Commission.* Brussels.

GSSI, 2002. RADAN for Windows NT (Version 4.0), *User's Manual*, Geophysical Survey Systems Inc., North Salem, New Hampshire.

Hakim, B.A., Brown, S.F. & Armitage, R.J. 2002. Pavement Evaluation and Strengthening Design: Sixteen Years Experience. *Proceedings, 9th International Conference on Asphalt Pavements, ISAP*. 3: 1–12. Copenhagen.

Irwin, L.H. 2002. Back-Calculation Analysis. *Tutorial, 9th International Conference on Asphalt Pavements,* International Society for Asphalt Pavements (ISAP), Copenhagen.

Jenkins, K.J., Lindsay, R.L., & Rossmann, D.R. 1995. The Deep In Situ Stabiliza-tion Process: Case Study. *Annual Traffic Convention (ATC.) Pavement Engineerng,* I 3 A: 1–13, Pretoria.

Jenkins, K.J., Van de Ven, M.F.C., Molennar, A.A.A. & De Groot J.L.A. 2002. Collins, D. & Jenkins, K. 2004. Performance Prediction of Cold Foamed Bitumen Mixes. *Proceedings, 9th International Conference on Asphalt Pavements (ISAP)* 1:4-4. Copenhagen.

Loizos, A., Collins, D. & Jenkins, K. 2004. Rehabilitation of a Major Greek Highway by Recycling/Stabilizing with Foamed Bitumen. *Proceedings, 8th Conference on Asphalt Pavements for Southern Africa (CAPSA 04,)*: 119–126. South Africa.

Loizos, A. & Papavasiliou, V. 2006. Evaluation of Foamed Asphalt Cold in-place Pavement Recycling Using Nondestructive Techniques. *Journal of Transportation Engineering.* Vol. 132, No. 12, Dec. 2006: 970–978.

Loizos, A. & Papavasiliou, V. 2007. In situ Characterization of Pavement Materials Stabilized with Foamed Asphalt and Cement. *Proceedings, International Conference on Advanced Characterization of Pavement and Soil Engineering Materials*: 875–882. Athens, Greece.

Loulizi, A., Al-Qadi, L.L. & Lahouar, S. 2003. Optimization of Ground Penetrating Radar to Predict Layer Thicknesses in Flexible Pavements. *Journal of Transportation Engineering*, Vol. 129, No. 1: 93–99.

Milton, L.J., & Earland, M.G. 1999. Design Guide and Specification for Structural Maintenance of Highway Pavements by Cold In-Situ Recycling. *Transportation Research Laboratory (TRL).* Report 386.

PIARC, 2001. Recycling of Existing Flexible Pavements. *Technical Committee on Road Pavements*, C7/8, World Road Association.

RoadScanners, 2001. Use of Ground Penetrating Radar in Relation with FWD. *Road Doctor Software, Version 1.1 User's Guide,* Rovaniemi, Finland.

SHRP's Layer Moduli Backcalculation Procedure. 1993. *Strategic Highway Research Program (SHRP), National Research Council.* Washington, DC.

Wirtgen, 2004. Cold Recycling Manual. *Wirtgen GmbH*, Honner Strabe 2, D-53578, Windhagen.

Efficient Transportation and Pavement Systems – Al-Qadi, Sayed, Alnuaimi & Masad (eds)
© 2009 Taylor & Francis Group, London, ISBN 978-0-415-48979-9

Investigation of sulfur modified asphalt concrete mixes for road construction in the gulf

Mirza G. Baig, Hamad I. Al-Abdul Wahhab & Isam A. Mahmoud
Department of Civil Engineering, King Fahd University of Petroleum and Minerals, Dhahran, Saudi Arabia

Mohammed Al-Mehthel, Saleh F. Al-Idi & Jonathan J. Grosch
Consulting Services Department, Saudi Aramco, Dhahran, Saudi Arabia

ABSTRACT: A study has been funded by Saudi Aramco and conducted at KFUPM on the feasibility of using sulfur as an additive for local asphalt concrete mixtures. The research work covered many aspects of utilizing sulfur modified asphalt in road construction including the field trial at Khursaniyah and the concerns related to air pollution due to sulfur containing gases. This study on sulfur-asphalt concrete consists of testing local sulfur, Shell Canada sulfur-extended asphalt modifier (SEAM™), with local asphalt-concrete mixes to assess the effect of sulfur and modified sulfur materials by comparing the performance of these paving mixes. Results from laboratory and field trials indicated that SEAM™ and sulfur modified asphalt concrete can be produced, hauled, placed and compacted easily with conventional methods and equipment. There will be no constructability problem with the use of sulfur or SEAM™ binder. Use of SEAM™ or sulfur material at 30% replacement of asphalt could be more economical as compared to regular asphalt as the amount of required asphalt will be reduced in proportion to the SEAM™/sulfur percentages added. The field tests on assessing the environmental impact of the sulfur-asphalt technology show that there is no long-term hazard for mixes as indicated by acceptable values of emission of hazardous gases such as H_2S and SO_2 (<1 PPM at 76°C). However, precautions must be taken during preparation and laying of mixes at 145°C.

Keywords: Sulfur, SEAM™, Sulfur-Asphalt Technology, Sulfur Modified Asphalt Mixes, Air Quality, Gas Emission

1 BACKGROUND

During the last decade in the Gulf region, particularly in Saudi Arabia the availability of sulfur as a by product has increased considerably. This is mainly due to the implementation of strict environmental restrictions on petroleum and gas refining processes which limit the maximum quantity of sulfur present at combustibles. The need to investigate the applications for sulfur has thus become fundamental in the region.

Asphalt binder is a thermoplastic material that behaves as an elastic solid at low service temperatures or during rapid loading and behaves as a viscous liquid at high temperatures or slow loading. This double behavior creates a need to improve the performance of an asphalt binder in order to minimize stress cracking, which occurs at low temperatures, and permanent deformation, which occurs at high in-service temperatures. Daily and seasonal temperature variations plus the growth in truck traffic volume, tire pressure and loading have increased stresses on asphalt pavements. Local asphalt pavement temperatures range between −10°C in winter and 73°C in summer (Al-Abdul Wahhab et al. 1997). This has led to an increased demand to modify asphalt binders. Different methods have been used to upgrade the properties of asphalt binders. One of the most commonly used procedures is the modification of asphalt binders by the addition of modifiers.

Sulfur's ability to modify and enhance the properties of construction materials has been extensively researched and exploited over the past four decades. Most of the research work has come to a halt as a result of the historic market collapse about two decades ago. Since then, the amount of sulfur produced from oil and gas has increased resulting in a drop of market prices. For the last five years, world sulfur production has exceeded consumption by one to two million tons per year. Some forecasts have predicted continued overproduction ranging from one to three million tons per year through 2010 (Weber and McBee 2000). In the last decade, mores sulfur quantities has become available. This is caused primarily by environmental restrictions in regard to the petroleum and gas refining processes. These processes limit the maximum quantity of sulfur present at combustibles. As a result, extremely large quantities of sulfur are produced as a by-product of these processes and the development of new methods for the use of sulfur becomes important (http://congress.cimne.upc.es/rilem04/admin/Files/FilePaper/p257.pdf) Sulfur asphalt used for paving roads is one of the prominent among its applications. Using sulfur to enhance or rejuvenate asphalt could consume significant amount of sulfur. Even if sulfur captured only a conservative 5% of the current asphalt market, it would represent a market for nearly one million tons of sulfur annually and can help alleviate the oncoming sulfur surplus. Sulfur-asphalt concrete has a relatively simple composition and manufacture, and very interesting characteristics and properties. Its extremely high corrosion resistance, mechanical strength and fast hardening make it a high performance material suitable for several applications, especially the ones in which other materials may not suffice (Gracia and Vazquez 2003).

In 1978, the Metrology, Standards and Materials Division of the Research Institute at King Fahd University of Petroleum and Minerals (KFUPM/RI) launched an in-house research study on sulfur-asphalt pavement development. Among the various available techniques of substituting asphalt with sulfur, the sulfur-extended asphalt (SEA) paving technology developed by Gulf Canada was considered to be the closest to practical applications. Three SEA test roads were laid in the Eastern Province in cooperation with Gulf Canada and the Ministry of Communications (MOC), Saudi Arabia as part of the ongoing road development program of MOC. A sulfur/asphalt ratio of 30/70 by weight was used in Test Road 1 (Kuwait Diversion) and Test Road 3 (KFUPM), whereas a higher percentage of 45/55 was used in Test Road 2 (Abu-Hadriyah Expressway). Performance of the test roads was monitored from time to time. For each test road, the control section using normal asphalt concrete showed better performance than the SEA sections (Arora et al. 1994). The SEA sections developed mostly longitudinal/transverse cracking in Test Road 1; alligator and block cracking in Test Road 2; and block and longitudinal/transverse cracking in Test Road 3. On the control AC sections, the most predominant distress types were found to be longitudinal/transverse cracking and polishing of aggregate. The inherently stiffer SEA mix of Test Road 2, where the sulfur/asphalt ratio was 45/55, resulted in earlier cracking of SEA pavement, particularly in a thinner section where the thickness was intentionally reduced by 20 percent.

Fatani and Sultan (1982) conducted a study to determine the feasibility of using dune sand in asphalt-concrete pavement in hot, desert like climates through the use of one-size crushed aggregates. Dense-graded aggregate and powdered sulfur were used in the sand-asphalt mixes. Engineering properties, including Marshall design parameters, compressive strength, tensile strength, modulus of rupture, and dynamic modulus of elasticity were evaluated. Results indicated that a mixture of dune sand and asphalt is weak, unstable, easily deformed under light loads, and therefore unacceptable for pavement construction in desert like environment. The use of powdered sulfur and sand-asphalt mixes reduces the optimum asphalt content, increases considerably the qualities of the mix even under severe environmental conditions, and reduces the pavement thickness.

Arora and Abdul-Rahman (1985) have explored the use of sulfur as a rejuvenation agent in recycling reclaimed asphalt pavement from a typical failed segment of Dammam-Abu-Hadriyah Expressway. They indicated that the addition of sulfur, at mixing temperature, would lower the viscosity of the aged asphalt. Upon cooling, recrystallization of sulfur is known to occur, which improves the strength of the mix. Properties like Marshall stability, resilient modulus and fatigue behavior of sulfur-recycled mix are compared with those of the conventional asphalt-concrete mix. The addition of sulfur results in higher Marshall stability without significant loss in flow values, higher retained strength index, and higher M_R and tensile strength, indicating superior engineering properties of the recycled mixture over the conventional asphalt hot mix. The above properties are particularly advantageous to the hot region of the Arab World since they provide adequate resistance to wheel track rutting otherwise associated with conventional asphalt-concrete mixtures.

Akili (1985) carried out an extensive laboratory testing program designed to measure improvements in engineering properties of sulfur-asphalt-sand (SAS) mixes attributable to the presence of sulfur in the mix, considering locally available sands and prevailing environmental conditions in eastern Saudi Arabia. The laboratory characterization data include Marshall design parameters, resilient moduli, and permanent strain characteristics under repeated triaxial loading. The results, in general, showed improvements in Marshall stability, resilient modulus values and reduced permanent deformation of SAS mixes in comparison to conventional sand-asphalt mixes. From Spring 2001 through February 2002, about 42 lane miles of roads containing sulfur were built in the southwest United States. These projects incorporated a formed, solid sulfur product that was added directly to existing hot mix plant equipment. Following mixing, the sulfur asphalt was hauled to the project location using conventional dump trucks, road paving, and compaction equipment. An additional 104 lane miles of roads containing sulfur are planned in the southwest U.S., and other road projects incorporating sulfur are also being considered in China, Kazakhstan, and Egypt. The use of a formed, solid material and the direct mixing method minimizes hot asphalt mix plant modifications and associated costs. Also, solid sulfur can be shipped freely without regulation; whereas, liquid sulfur requires special shipping considerations (Weber 2002).

The use of sulfur as an additive to extend or replace asphalt has been demonstrated successfully in both laboratory tests and actual construction. The current availability and low cost of sulfur offer the potential to reduce paving material costs by as much as 21 percent. Binder cost reductions as high as 32 percent are feasible (Weber and McBee 2000).

Sulfur asphalt is enjoying a resurgence of interest worldwide. The Sulfur Institute (TSI) has been actively working with the US Federal Highway Administration and through other channels to promote the utilization of sulfur asphalt, and at least one world-class oil and gas company is seriously interested in this outlet for its recovered sulfur. At the same time, Devco Company has already moved to penetrate the market. Using proprietary innovations to produce easily handled forms of pre-blended sulfur asphalt, the company has been constructing roads in Nevada and California for the last two to three years, along with parking lots and other paving projects.

Sulfur, a by-product of oil and gas production, is produced by Saudi Aramco at a rate of approximately 6,000 tons/day. The rate of production is expected to increase to 10,000 tons/day in a few years time. Although sulfur is a vital raw material to manufacture a myriad of products, its abundance has reduced its price worldwide. The storage of the sulfur will pose an environmental hazard. Other venues such as constructional usage should be explored to utilize this abundant sulfur in a useful, economical, and environmentally friendly way. This research funded by Saudi Aramco was carried out to study the effect of sulfur and sulfur-extended asphalt modifier (SEAM) modified asphalt mixes and verify their adequacies for safe use locally through laboratory and field studies.

2 RESEARCH APPROACH

2.1 *Material selection*

The locally available aggregate in the Eastern Province was selected for this study. The Saudi Aramco specification for Surface course (SC) and Base course (BC) gradation was followed in all mix designs. Asphalt cement of grade 60/70 was obtained from Ras Tanurah refinery. The materials obtained were tested for gradation and physical characteristics to assure their conformity to Aramco standards.

The elemental sulfur in pellet form, used in the study was obtained from Saudi Aramco. Sulfur-extended asphalt modifier (SEAM™), which was manufactured by Shell Co., was obtained from Shell (Canada). This material was selected to compare the improvements in engineering properties obtained by using it over the elemental sulfur-asphalt mixes, since SEAM is claimed to provide improved engineering properties with reduced gas emissions when used in asphalt concrete mixes.

2.2 *Sulfur-asphalt and SEAM-asphalt alends*

Three blends of modified asphalts were prepared by mixing SEAM™ or elemental sulfur with the local asphalt. Two blends were prepared using SEAM with ratios of 30/70 and 40/60 by weight of binder (as recommended by Shell Canada). Another blend was prepared using elemental sulfur with a ratio of 30/70 by weight of binder. Each blend was prepared by heating the plain asphalt to 135°C and slowly adding the required amount of modifier (sulfur or SEAM) as per the weight ratios using a shear-blender. The mixing was carried out for about 5 minutes in order to get a well-homogenized blend.

2.3 *Mix design*

Six different asphalt-concrete mixes were designed using the standard Marshall mix design method as per ASTM D 1559 test method with 75 blows, and the optimum asphalt content (OAC) was determined. Three mixes each for surface course and base course gradation were prepared using Sulfur 30/70, SEAM 30/70 and SEAM 40/60 blends. All the mixes were compacted at 140°C, and care has been taken not to exceed 145°C temperature limits for sulfur/SEAM modified mixes.

2.4 *Test section*

The access road for the Khursaniyah oil field under construction in eastern Saudi Arabia was selected to lay the test sections in March 2006. A 450 m long and two-lane width test section has been constructed to include SEAM™ 30/70, SEAM™ 40/70 and elemental sulfur 30/70 asphalt mixes, each with approximate length of 150 m. Base course (BC) and surface course (SC) mixes have both been prepared and laid using the same binder type. Each section consists of 25 cm of granular base course, 8 cm of asphalt base course, and 6 cm of asphalt surface (wearing) course as shown in Figure 1.

The construction quality control, environmental pollution monitoring, traffic analysis and condition surveys were all conducted to evaluate and monitor the test sections performance in comparison to conventional asphalt concrete.

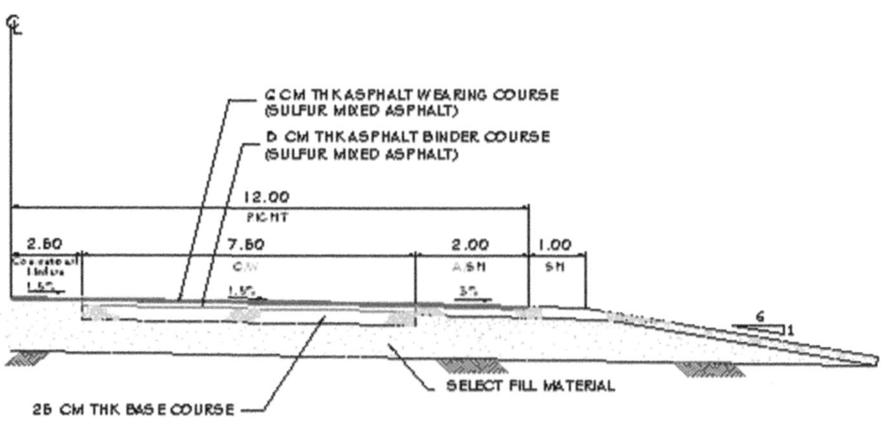

Figure 1. Typical Khursaniyah access road pavement cross-section.

3 RESULTS

3.1 *Optimum binder content*

The optimum binder content (OBC) of each mix was obtained as per the standard Marshall mix design procedure and after satisfying the Saudi Aramco specifications. The results obtained are shown in Table 1 for each of the six mixes. The results show that the optimum binder content required in both SC and BC mixes increases as the ratio of SEAM™ added as modifier is increased. The table lists all the necessary mix design properties for each mix along with the specification requirements.

3.2 *Monitoring during construction*

Construction monitoring at Khursaniyah access road was carried out under the supervision of KFUPM and Aramco to properly control the quality of constructed sections and to monitor the air quality both at the asphalt plant and construction site.

3.2.1 *Construction quality control*

The 30/70 sulfur-asphalt concrete and SEAM modified asphalt concrete mixtures were collected from behind the pavers during the construction of the test sections at Khursaniyah access road for both SC and BC layers. Each mix was sampled and used to assess the construction quality. For each mix type, at least three cores were taken and analyzed for thickness and bulk specific gravity (G_{mb}).

Comparing the results of both base course (BC) and wearing course (WC) layers for the three types of test sections laid (i.e. SEAM™ 30/70, SEAM™ 40/60 and Sulfur 30/70) with respect to the mix designs, it was found that the binder content in each mix was within the JMF limits. The gradation analysis indicated that the two sieve sizes, ¾″ and #40, tend towards the upper limit of the specified envelope for both mix types. The average percent compaction for both BC and WC layers achieved in the field was found to be slightly high.

3.2.2 *Air quality monitoring*

The investigation was intended to monitor the air quality of a road paved with 30/70 sulfur-asphalt concrete, 30/70 and 40/60 SEAM™ asphalt concrete at Khursaniyah in March 2006. The specific objective was the monitoring of gaseous emissions, particularly sulfur dioxide and hydrogen

Table 1. Summary of Marshall mix design results for SC and BC mixes, Khursaniyah access road.

	Surface Course (SC) Mix				Base Course (BC) Mix			
Mix property	Specification	SEAM 30/70 mix	SEAM 40/60 mix	Sulfur 30/70 mix	Specification	SEAM 30/70 mix	SEAM 40/60 mix	Sulfur 30/70 mix
Optimum Binder Content, % by wt of mix	Optimum +/–0.3%	5.4	5.6	5.5	Optimum +/–0.3%	4.8	5.0	5.0
Stability, kg	800 min.	1750	1826	1628	700 min.	1602	1768	1650
% Air Voids	3.0–5.0	4.1	4.0	4.05	3.0–7.0	4.2	4.15	4.2
Flow, mm	2.0–4.0	3.3	3.6	3.3	2.0–5.0	3.2	3.5	3.5
% Voids Filled w/Asphalt (VFA)	70–80	73.5	73.7	71.35	60–75	70.8	70.2	69.6
Stability loss, %	25 max.	18.1	18.9	19.0	25 max.	17.7	18.3	16.3
Voids in Mineral Agg. (VMA)	14 min.	14.3	14.3	14.23	13 min.	13.4	13.5	13.32

Table 2. Concentration of gases at construction site.

SO_2 (ppm)			H_2S (ppm)			
Maximum	Mean	Minimum	Maximum	Mean	Minimum	Remarks
30/70 Sulfur-Asphalt Mix						
3.118	0.56	0.156	3.17	2.00	0.26	C450 H_2S/SO_2 analyzer
8.0	1.89	0.0	–	–	–	S710 analyzer
40/60 SEAM™ Asphalt Mix						
3.399	0.724	0.134	2.11	0.512	0.0	C450 H_2S/SO_2 analyzer
12.0	5.469	0.0	–	–	–	S710 analyzer
30/70 SEAM™ Asphalt Mix						
2.804	0.119	0.356	1.252	0.688	0.297	C450 H_2S/SO_2 analyzer
8.0	3.524	0.0	–	–	–	S710 analyzer
30/70 Sulfur-Asphalt Mix						
SO_2 (ppm)			H_2S (ppm)			Probe position
0.39			0.47			At operator (driver) level (2.5 m)
0.404			0.51			At foreman level (1.8 m)
40/60 SEAM™ Asphalt Mix						
0.38			0.51			At operator (driver) level (2.5 m)
0.40			0.53			At foreman level (1.8 m)
30/70 SEAM™ Asphalt Mix						
0.342			0.468			At operator (driver) level (2.5 m)
0.307			0.464			At foreman level (1.8 m)

sulfide. The model C450 and S710 gas analyzers were used for ambient level and high concentration level of gases, respectively. Table 2 shows the concentration of gases obtained at the construction site.

Results indicated that the maximum values of SO_2 concentrations ranged from 0.0 to 12 ppm when measured close to the source (20–40 cm above the auger). The maximum measured values were 8, 12 and 8 ppm for 30/70 sulfur, 40/60 SEAM™ and 30/70 SEAM™ mixes, respectively. The measured values of SO_2 were found to be within the acceptable limits when measured at the foreman and driver levels ranging from 0.3 to 0.4. The measured values of hydrogen sulfide ranged from 0.0 to 3.17 ppm close to the source and 0.47 to 0.51 at the foreman and driver levels. It was also noted that the variation of temperature (124–145°C) did not significantly affect the concentration of fumes.

Apart from the measured SO_2 emission close to the source above the auger screw which continuously agitates the asphalt-concrete mix and releases trapped fumes, all measured gases concentrations are within the acceptable limits of OSHAA standards. Workers should not stand in downwind direction of the pavers close to the auger; otherwise special safety precautions should be taken.

3.3 *Traffic monitoring and analysis*

Khursaniyah access road has been reconstructed by Saudi Aramco to serve the traffic moving materials and personnel for Khursaniyah gas plant construction. Electronic counters were installed on all lanes shown in Figure 2 and traffic count was taken for five weeks in 2006 and another five weeks in 2007. Detailed traffic counts were obtained, a typical summary of the traffic counts for the right lane (test road) is shown in Table 3. It was noted that the total annual traffic for the test road was 519,395 vehicles in 2006 and 558,815 in 2007. The design lane (outer lane) has been

Figure 2. Installation of traffic counters on Khursaniyah access road.

Table 3. Traffic summary, Khursaniyah test road.

	Sept. 2006	July 2007
Average yearly volume	519,395	558,815
Average yearly ESAL	2,249,830	2,384,910
Design lane ratio	56%	60.7%

subjected to a total equivalent standard axle load of **2.7 million**, which indicates high volume of loaded trucks.

3.4 *Condition survey*

Khursaniyah access road was opened to traffic in March 2006. Similar to Dhahran-Jubail expressway, condition survey was carried out at regular intervals, usually after winter and summer, to evaluate pavement performance. In particular, the pavement surface was evaluated for distresses like cracks, raveling and rutting. The condition survey is summarized in Table 4.

Condition survey was carried out in September 2006, June 2007 and October 2007. Pavement condition is generally excellent with pavement condition index of about 98, as shown in Figure 3. No signs of distresses were observed other than minor roughness due to construction process. Up to date, no wheel path depressions or cracking were observed. Moreover, the elemental sulfur section has performance similar to the performance of SEAM sections.

Table 4. Summary of Khursaniyah access road test sections conditions.

Pavement section	Base course	Wearing course	Length meter	Condition PCI*	Remarks
–	Asphalt concrete	Asphalt concrete	Main road	98	
1	30/70 SEAM	30/70 SEAM	130 m	98	Excellent with no sign of any distress other than minor construction roughness.
2	40/60 SEAM	40/60 SEAM	130 m	98	
3	30/70 Sulfur Asphalt	30/70 Sulfur Asphalt	130 m	98	

*Initial PCI after construction is 100.

Figure 3. Pavement condition of Khursaniyah access road.

4 CONCLUSIONS

The Khursaniyah access road test sections performance has provided an insight towards the possible use of sulfur-extended asphalt mixes in the local paving industry. Based on the two-year monitoring and performance, the following conclusions can be made:

1. Sulfur asphalt technology can be used successfully with current road construction technology and expertise. Use of palletized sulfur and SEAM has resulted in convenient handling of sulfur.
2. The same construction equipment and procedures are used for sulfur asphalt and regular asphalt. No equipment modification is required. Minimal modification is needed to the batch plant to allow the addition of Sulfur/SEAM directly into the pug mill.

3. There are no major safety concerns with regard to careful monitoring of mixing and handling of hot sulfur asphalt mix.
4. Constructed test sections have met the construction quality requirements.
5. Performance of test sections, up to date, is found to be excellent even after almost two years of heavy traffic.
6. Elemental sulfur section performance is similar to the performance of SEAM sections.

ACKNOWLEDGMENTS

The authors would like to acknowledge the support provided by King Fahd University of Petroleum & Minerals (KFUPM) and Saudi Aramco for the execution of this research.

REFERENCES

Akili, W. (1985) Sulphur-Asphalt-Sands for Pavement Applications in Eastern Saudi Arabia, *Proceedings of the 2nd Arab Regional Conference on Sulphur and Its Usage,* III: 55–71.

Al-Abdul Wahhab, H.I., Asi, I.M., Al-Dubabe, I.A., and Ali, M.F. (1997) Development of Performance-Based Bitumen Specifications for the Gulf Countries, *Construction and Building Materials Journal*, 11(1): 15–22.

Arora, M.G. and Abdul-Rahman, K.M. (1985) Sulphur in Recyling Old Asphalt Pavements, *Proceedings of the 2nd Arab Regional Conference on Sulphur and Its Usage,* III: 37–53.

Arora, M.G., Al-Mana, A.I., Al-Tayyib, A.J., Ramadhan, R.H., and Khan, Z.A. (1994) "Long-Term Pavement Performance History of Sulfur-extended Asphalt Test Roads in Eastern Province of Saudi Arabia, *Transportation Research Record 14315,* Transportation Research Board, pp. 77–85.

Fatani, M.N. and Sultan, H.A. (1982) Dune Sand-Aggregate Mixes and Dune Sand-Sulfur Mixes for Asphalt Concrete Pavements, *Transportation Research Record No. 843,* National Academy of Science, Washington, D.C., pp. 72–80.

Gracia, V. and Vazquez, E. (2003) Utilization of By-product Sulfur for the Manufacture of Unmodified Sulfur Concrete, *Technical Report*, Department of Construction Engineering, University of Tecnica Federico, Santa Maria, Spain.

Weber, H.H. (2002) Market Opportunities for Sulphur Asphalt Road Paving Materials, presented at *The Sulfur Institute's Eighth International Symposium 'Sulphur Markets—Today and Tomorrow',* Amsterdam, The Netherlands, March 10–12, 2002.

Weber, H.H. and McBee, W.C. (2000) New Market Opportunities for Sulfur Asphalt, presented at *The Sulfur Institute's Seventh International Symposium 'Sulfur Markets—Today and Tomorrow',* Washington, D.C., March 28, 2000.

Efficient Transportation and Pavement Systems – Al-Qadi, Sayed, Alnuaimi & Masad (eds)
© 2009 Taylor & Francis Group, London, ISBN 978-0-415-48979-9

Discrete element analysis of aggregate resistance to fracture in asphalt mixtures

E. Masad
Texas A&M University at Qatar, Doha, Qatar

E. Mahmoud
Texas A&M University, College Station, Texas, USA

ABSTRACT: Aggregate fracture in hot mix asphalt is affected by strength, gradation, and shape of the aggregate particles. This paper presents an approach for accounting for all these variables within asphalt mix and its effect on the mix performance. This approach combines the discrete element modeling and image processing techniques, it's developed through calibrating both aggregate and asphalt mixes experimental results. This approach showed promising results that can be used to investigate the internal force distribution with different mixes, and how different aggregates behave within one mix, and so this method can be used to select the more appropriate aggregate type for a given mix type.

1 BACKGROUND AND RESEARCH OBJECTIVES

Asphalt mix is a combination of aggregates and asphalt binder where approximately 85% of the total volume is aggregate. Consequently, characteristics such as gradation, stiffness and strength are important factors that affect the performance of asphalt pavements. Most asphalt mixtures develop strength mainly from stone-on-stone contact of the coarse aggregate. However, evidence indicates that construction operations, particularly compaction of thin layers, plus subsequent traffic loadings can contribute to crushing and/or grinding of coarse aggregates at the contact points, which can significantly alter the original design gradation and create uncoated aggregate faces. Broken binder films provide inlets for water which, in concert with traffic loads, can exacerbate stripping. Therefore, strength properties of coarse aggregates are clearly important for asphalt mixtures to resist applied loads.

In this study, a discrete element model of asphalt mixture internal structure was developed in order to evaluate resistance of aggregates to fracture. The model inputs include aggregate stiffness, binder stiffness, aggregate-aggregate bond strength, and aggregate-binder bond strength, and image of the asphalt mixture internal structure. The results showed that the model is able to accurately capture the influence of aggregate properties and mixture designs on aggregate resistance to fracture. The model was used to select the appropriate aggregate type given a certain mix design in order to improve the mix resistance to failure under applied loads.

2 BACKGROUND ON DISCRETE ELEMENT METHOD

The discrete element method (DEM) is a finite difference scheme, used to study the interaction among assemblies of discrete particles. DEM was introduced by Cundall (1971) and later in 1979 this method was proposed by Cundall and Strack for the simulation of two-dimensional non-continuous materials (Cundall and Strack 1979).

Cundall and Hart (1992) summarized the advancements in discrete element codes. It was proposed that the name discrete element method (DEM) should only apply to codes that allow finite displacements and rotations of discrete objects, including full detachment, and recognize new contacts automatically as the calculation progresses. The DEM has been mainly utilized as a research tool in many studies in the last thirty years. In this study a commercially available DEM code called Particle Flow Code in 2-Dimensions (PFC2D Version 3.1), developed by Itasca Consulting Group is used. This code includes a user-friendly graphical interface, linear and non-linear contact models, linear and curvilinear boundary conditions.

The DEM concept is simple in principle; it is based on successively solving law of motion (Newton's second law) and the force-displacement law for each particle. An explicit time-stepping scheme is employed to integrate Newton's second law for each particle, given a set of contact forces acing on the particle, which results in the updated particles' positions and velocities. Based on the new positions, the relative displacements of each particle are calculated, and used to calculate the contact forces. The DEM is based upon the idea that the time step chosen is sufficiently small so that during a single time step, disturbances cannot propagate from any particle further than its immediate neighbors. Therefore, at all times the forces acting on any particle are determined exclusively by its interaction with particles that it is in contact with.

In PFC2D, particles are circular (balls). They are allowed to overlap at the contact points, which occur over a very small area (i.e., at a point). The amount of overlap is related to the contact force via the force-displacement law. All overlaps are assumed to be small in relation to particle sizes. Bonds can be added to the contacts between the particles, to either increase the stiffness of the contact and/or to include a strength parameter above which the bond breaks; PFC2D allows different types of bonds to be assigned. In the absence of bonding, particles slide over each other once the shear force exceeds the friction.

3 MATERIALS AND MIXES

Five different aggregates were selected: granite, hard limestone, soft limestone, gravel, and sandstone in preparing asphalt mixtures with different gradations as shown in Table 1.

The gradation curve for each mix type, is illustrated in Figure 1, these gradations differ from one another to provide different aggregate structure. The PFC is a coarse, almost uniform-graded mixture with a high percentage by weight of coarse aggregates. It is composed of 89% aggregates larger than a No. 8 sieve. In contrast, Superpave-C is a more fine-graded mixture. It consists of 35% coarse aggregates and 65% fine aggregates. The CMHB-C mix is a coarse-graded mixture that is composed of 63% coarse aggregates and 37% fine aggregates. The Type-D mix demonstrates a well-graded gradation with 40% coarse aggregates and 60% fine aggregates. The same bituminous binder (PG 76-22) was used for all the mixes to minimize the impact of binder grade on the result.

Table 1. Selection of aggregates and mixtures.

Aggregate type	Superpave-C	CMHB-C	PFC	Type-D
Granite	×	×	×	
Hard limestone	×	×	×	
Soft limestone	×	×	×	
Gravel	×	×	×	×
Sandstone	×	×	×	×

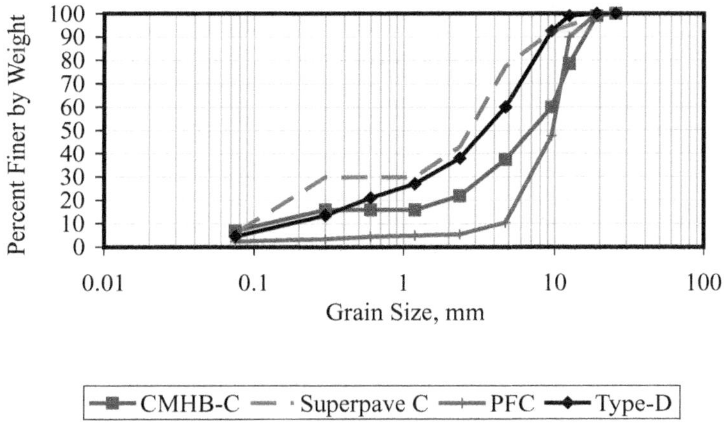

Figure 1. Aggregate gradation used in this study.

4 AGGREGATE TESTS

The splitting tensile tests on cores from rock masses retrieved from quarries were carried out to determine the potential tensile crushing strength of the aggregates. In this study, cylindrical rock specimens were tested; split tensile strength is a measure of a material's ability to resist a diametric compressive force.

Rock core samples were first extracted from bulk rocks and cut to 2.3-in. diameter by 2-in. height. Each sample was then placed and centered with its axis placed horizontally between the platens of a compression-testing machine. Once in place, a continuously increasing compressive load was applied to the test specimen until splitting or rupture occurred. The load was applied at a nominal constant rate of 250 psi per minute.

Rock cores similar to those for the indirect tensile tests extracted from bulk rocks were used determine the unconfined compressive crushing strength of the aggregates. Each sample was then placed and centered in the compression-testing machine. Once in place, the load was then continuously increased on the specimen until crushing failure occurred.

Modulus of aggregate rocks were obtained using nondestructive testing technique based on ultrasonic testing (V-meter).The V-meter is an ultrasonic device that measures the travel time of compressive waves by means of electric impulses. In this device, a transmitting transducer is securely placed on the top face of the specimen. The transducer is connected to the built-in high-voltage electrical pulse generator of the device. The electric pulse transformed to mechanical vibration is coupled to the specimen. A receiving transducer is then placed on the bottom face of the specimen, opposite the transmitting transducer. The receiving transducer, which senses the propagating waves, is connected to an internal clock of the device. The clock automatically displays the travel time of the compression wave. By dividing the length of the specimen by the travel time, the compression wave velocity and as such modulus of the material is determined. In this case, the two opposite faces of each rock mass was made smooth using a band saw. Table 2 summarizes the experimental results of the different aggregate tests.

5 ASPHALT MIXES TEST

For the asphalt mixes the indirect tensile test is conducted by applying a compressive load to a cylindrical specimen through two diametrically opposed, arc shaped rigid platens The IDT device

Table 2. Experimental results of the aggregates.

Material	Compressive strength, psi	Tensile strength, psi	Modulus, ksi
Hard limestone	10427	1412	10328
Granite	14034	1062	6686
Soft limestone	6970	682	5473
Sandstone	13952	1677	8659
Gravel	Not feasible		

Table 3. Experimental results of the asphalt mixes.

Material	Mix type	Tensile strength at failure, psi
Hard limestone	CMHB-C	106
	Superpave-C	120
	PFC	66
Granite	CMHB-C	83
	Superpave-C	116
	PFC	61
Soft limestone	CMHB-C	94
	Superpave-C	125
	PFC	50
Sandstone	CMHB-C	207
	Superpave-C	226
	PFC	78
Gravel	CMHB-C	204
	Superpave-C	183
	PFC	58

has four components: the testing apparatus, the test control unit and data acquisition system, load measuring device, and the environmental control chamber. The specimens are nominally 4 inch in diameter and 2 inch thick and are loaded at rate of 2 in/min. the testing temperature is 77 F, Table 3 summarizes the asphalt mixes results.

6 DISCRETE ELEMENT MODELING OF AGGREGATES

The DEM is used to model the modulus test, compressive strength test and indirect tensile strength of rock samples representing the five aggregates used in this study. A bonding model, stiffness model, and slip model are included in the constitutive representation of contact points between the elementary particles. The simple contact-bond model used in this model can be envisioned as a couple of elastic springs at the contact point. This bond present the maximum shear and normal forces the contact can carry before breaking. The bond breaks if either the shear force or the normal tension exceeds its limit. In the linear contact-stiffness model an effective normal and shear contact stiffness is calculated from the particles' stiffness's assuming that they act in series,

$$K^n = \frac{k_n^{[A]} k_n^{[B]}}{k_n^{[A]} + k_n^{[B]}}, \quad K^s = \frac{k_s^{[A]} k_s^{[B]}}{k_s^{[A]} + k_s^{[B]}} \tag{1}$$

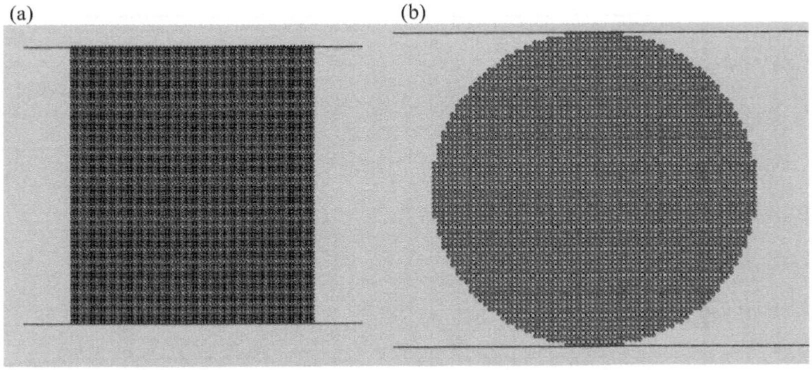

Figure 2. Models used in simulating: a) Compressive test, b) Indirect tensile test.

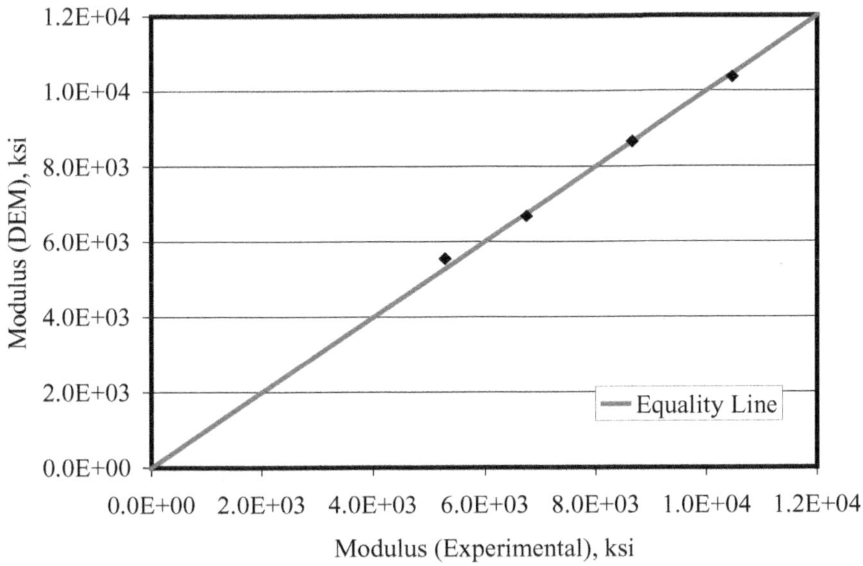

Figure 3. Comparison of modeling and experimental results of aggregate modulus.

where k_s: shear stiffness, k_n: normal stiffness, K^n: effective normal stiffness, K^s: effective shear stiffness, and A & B: ball designation

Aggregate samples of a diameter of about 2 in. and a height of about 2 in. were tested under compression and indirect tension loading. The model consisted of particles or balls with a diameter of 36 mm and a density of 160 pcf (Figure 2a & b). Following the work that was conducted by McDowell and Harireche (2002) and Cheng (2004), the simulation was conducted using a value of unity for the ratio normal stiffness to shear stiffness. The aggregate contact stiffness and strength in the model were determined such that the model results matched the experimental measurements.

Starting with very high bond strength, the contact stiffness among the model balls was varied until the aggregate modulus of the model matched the experimental modulus measurements. The next step is to vary the normal and shear bond strengths until the strength measurements (compressive and indirect tensile) of the model matched the experimental strength measurements.

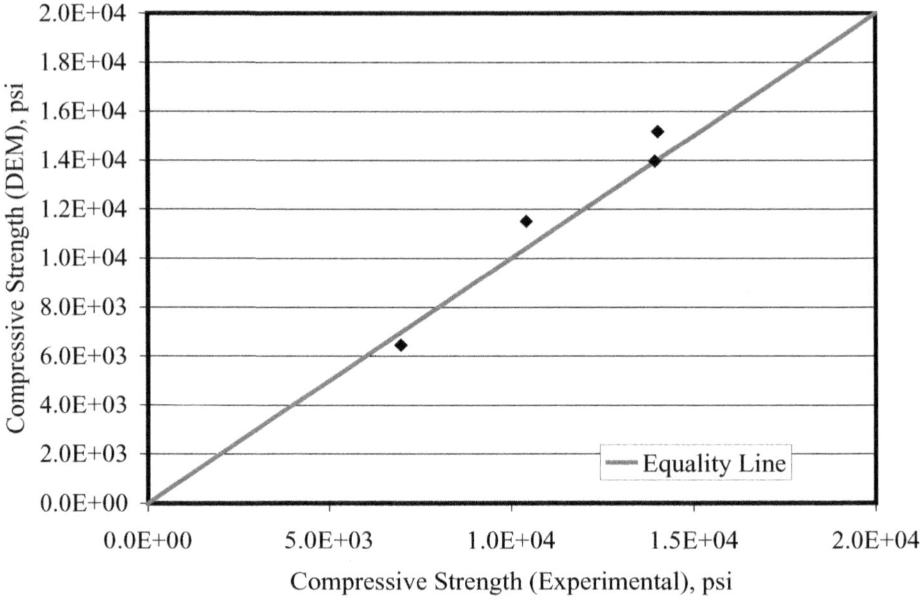

Figure 4. Comparison of modeling and experimental results of aggregate compressive strength.

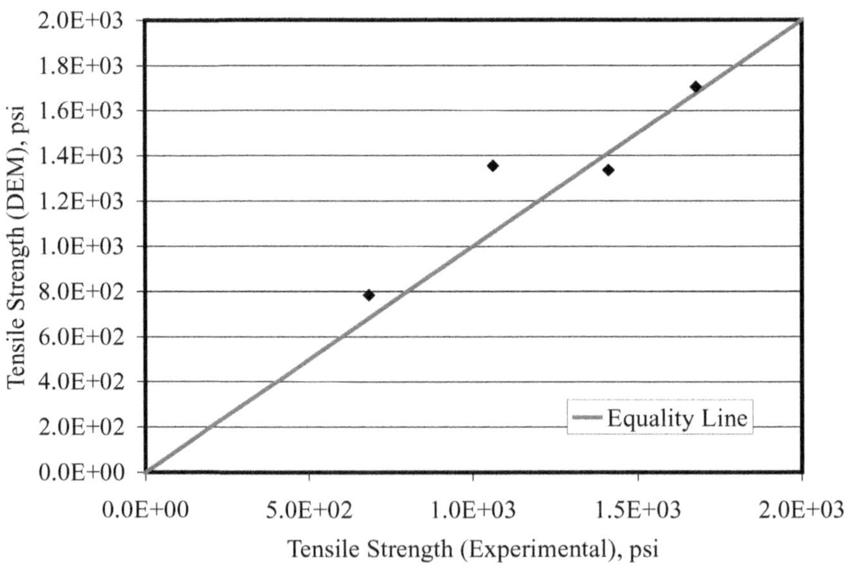

Figure 5. Comparison of modeling and experimental results of aggregate tensile strength.

This required conducting iterative analysis to determine the parameters that had the best match with both tests. The coefficient of friction between the model elements was set to a low value such that sliding occurs after the bond breaks. The friction between the loading walls and the model elements was set to 0.5 as recommended by Cheng (2004). A comparison between the experimental and modeling results are shown in Figures 3 through 5.

7 DISCRETE ELEMENT MODELING OF MIXES

Indirect tensile test results of the asphalt mixes were modeled as well; however, such modeling cannot be accomplished using simple geometry assignment because of the irregular shape of aggregates and the fact that there are two major phases (mastic and aggregates) in the mix. The asphalt mix model geometry is captured using X-ray computed tomography (CT) and then transferred to PFC2D. The image is transferred to a binary format (i.e., 0 for mastic and 255 for aggregate). The image pixels become the particles in the PFC2D model. The *Image-Pro Plus* (IPP) image analysis package is used to identify the outline pixels of each aggregate particle, and a FORTRAN code is used to group the elements of each aggregate particle in one group, and so is the case for the mastic phase. The FORTRAN code checks in all four directions (up, down, right, and left) whether the adjacent pixel is aggregate or mastic (Abbas 2004). Figure 6 shows the discrete element model after differentiating between mastic and aggregate. Each of the model phases can be assigned specific properties, such as bond strength and type, friction coefficient, and density.

The model developed by Abbas (2004) was used in this study; however the input parameters for the model treated differently as the aggregate phase parameters were already decided from the

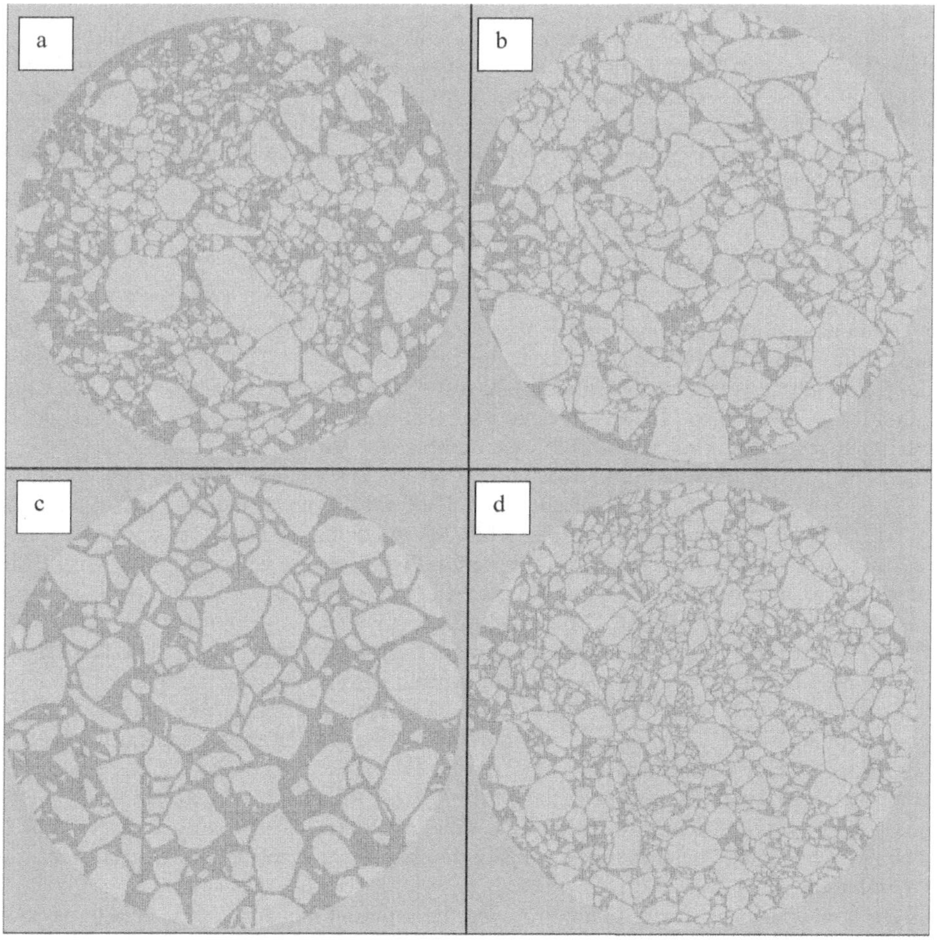

Figure 6. DEM of internal structure of asphalt mix (a) Superpave, (b) CMHB-C, (c) PFC, and (d) Type-D.

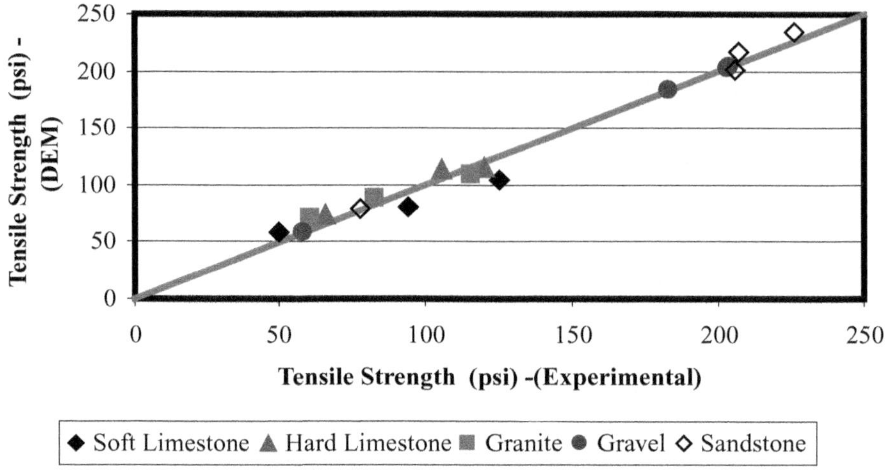

Figure 7. Comparison of modeling and experimental results of asphalt mixes tensile strength.

aggregate tests calibration, the matrix phase parameters were found such that each mix match the experimental results of the IDT from the laboratory (i.e. peak force, and the stiffness). The force displacement curve for each mix was the main output of the model, the model also were designed to track the internal forces developed among the discrete particles with different loading stages, these forces are shear and normal forces (compression and tension). Figure 7 shows the calibration results for the different mixes and aggregates.

8 ANALYSIS AND RESULTS

The distribution of internal forces in the mix models was studied. These forces were evaluated at three points (cases) in the force-displacement curve, case I is at the highest force applied (just before failure), case II is at 50% of the failure force, and case III is at a specific force which is common for all mixes. Case III was selected such that minimal cracking and bond loss occurred within the mixes. A force value of 450 lb was used for case III.

A general look at the distribution of internal forces was evaluated; these forces are shown graphically in Figure 8 for the gravel aggregate within the Type-D and PFC mix, in which the black color represents compression forces, while the red color represents tension forces. Higher forces are represented by thicker lines in these plots. The plots are generated at the case III loading stage. Same plots were generated for all the different aggregates and different mixes, the plots indicated that there is less uniform distribution of forces within the PFC mixes compared with the other mixes for all the aggregates. This is evident in the thicker black lines (higher forces).

Table 4 presents the maximum internal force within each of the mixes for the different cases. The maximum force values given in Table 4 indicate that the PFC mixes had, in general, the highest maximum internal forces among all mixtures. This indicates that aggregates in PFC experience higher forces than the other mixes.

Analysis of the data in Table 4 reveals that the ratio of the maximum internal force in PFC to the maximum internal force in the other mixes ranges from 1.1 to 2.0 with an average of 1.36. This comparison would be helpful in specifying minimum requirements for aggregates to be used in PFC mixes.

All the above analysis focused on comparing the different mixes within each aggregate type. The following analysis will evaluate the response of different aggregates within each mix type. In order to do so, force distributions are plotted for all aggregates within each mix. Figure 9 presents

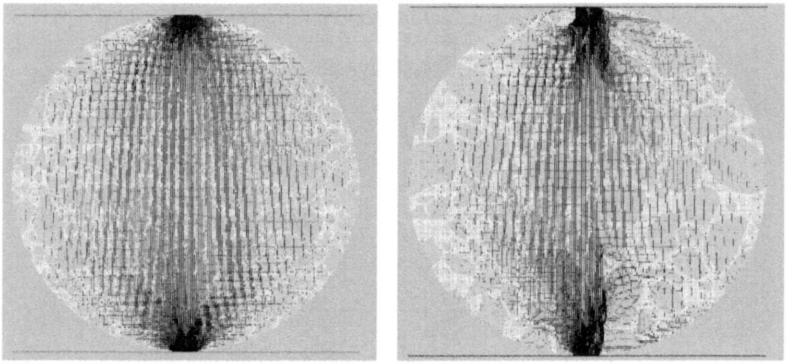

Figure 8. Comparison of internal force distribution between type-d mix and PFC mix.

Table 4. Maximum Internal force at different loading stages (lb).

		Loading stage		
Aggregate	Mix	450 lb	Case II	Case I
Hard limestone	CMHB	51.8	78.5	100.3
	PFC	113.1	113.4	114.6
	Superpave	67.4	104.1	123.9
Granite	CMHB	45.3	95.3	134.9
	PFC	66.8	62.6	159.5
	Superpave	57.5	84.7	194.2
Soft limestone	CMHB	44.0	46.8	66.6
	PFC	64.6	48.9	67.4
	Superpave	58.5	60.2	88.4
Gravel	Type D	43.3	135.1	562.0
	CMHB	41.4	147.8	327.3
	PFC	51.3	39.5	89.0
	Superpave	49.3	132.6	303.7
Sandstone	Type D	46.4	128.2	228.6
	CMHB	42.5	147.8	219.5
	PFC	59.7	67.2	119.0
	Superpave	54.3	115.6	177.0

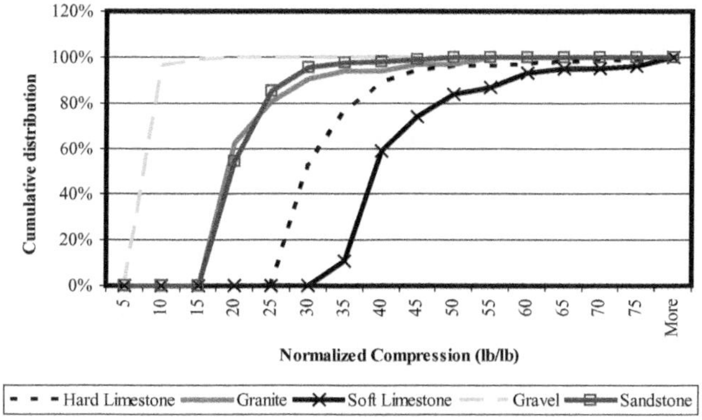

Figure 9. Internal compression forces distribution within different aggregates (CMHB-C).

Table 5. Aggregate ranking for the different mixes.

	Mix design		
Aggregate	CMHB-C	Superpave	PFC
Hard limestone	4	4	4
Soft limestone	5	5	5
Granite	2	2	2
Gravel	1	1	1
Sandstone	2	2	2

the compressive force distributions within the different aggregate types for the CMHB-C mix. Similar plots are generated for the other mixes; all the different plots followed the same trend as in Figure 9 with the soft limestone experiencing the highest internal forces compared with the other aggregates. On the other hand, the gravel had the least internal forces for all the different cases as well. Hard limestone had higher internal forces than the granite and the sandstone, but still below the soft limestone. Finally the sandstone and granite had similar internal forces. Based on these results, aggregates can be ranked for the different mixes as in Table 5.

9 CONCLUSIONS

The discrete element model was powerful in modeling the aggregate and mixture tests, the model allowed evaluating the internal forces in asphalt mixtures, which cannot be accomplished by conventional experimental methods, also the model was successful to large extent in representing the variability in aggregate properties and the influence of this variability on mixture response. The analysis of internal forces revealed that PFC mixtures experienced higher stresses than all other mixes. Based on the results, it is recommended that aggregate strength in PFC should be about 25% more than aggregate used in the other mixtures. However, With the exception of PFC, internal forces were comparable for all other mixtures for a given aggregate type. The soft limestone experienced the highest internal forces compared with the other aggregates. Aggregates were ranked based on the internal force values, this ranking can be used to select the appropriate aggregate type given a mixture design.

REFERENCES

Abbas, A.R. 2004. Simulation of the micromechanical behavior of asphaltic mixtures using discrete element method. PhD dissertation Washington State University.

Cheng, Yi P.H. 2004. Micromechanical investigation of soil plasticity. PhD dissertation, Churchill College, University of Cambridge.

Cundall, P.A. 1979. A computer model for simulating progressive large scale movements in blocky rock systems. In Proceedings for the symposium of the international society of rock mechanics (Nancy, France) Volume 1 paper no. II-8.

Cundall, P.A. and Hart, R. 1992. Numerical modeling of discontinua. In Journal of Engineering Computations, no 9, p. 101–113.

Cundall, P.A. and Stark O. 1979. A discrete numerical model for granular assemblies. Geotechnique 29(1), 47–65.

McDowell, G.R. and Harireche, O. 2002. Discrete element modeling of soil particle fracture. Geotechnique 52, No 2, pp: 131–135.

Efficient Transportation and Pavement Systems – Al-Qadi, Sayed, Alnuaimi & Masad (eds)
© 2009 Taylor & Francis Group, London, ISBN 978-0-415-48979-9

Field focused long term performance evaluation of asphalt concrete pavements

M. Solaimanian
Penn State University, University Park, PA, USA

ABSTRACT: A study was undertaken by Pennsylvania Transportation Institute and sponsored by Pennsylvania Department of Transportation to evaluate performance of several asphalt concrete pavements through close field monitoring. These pavements have been in service for several years. This paper deals with two of the nine constructed projects. Materials were collected from field and materials characterization testing was performed in the laboratory. Pavement distress survey and performance measurements of all the project pavement sections were conducted. Laboratory material characterization testing was also conducted on material from pavement cores drilled from several of these projects. The data was analyzed to evaluate the relationship between the laboratory characterization results and the actual measured performance of these projects in the field. Laboratory testing included evaluation of rut resistance properties and evaluation of resistance of thermal cracking. Field investigation included distress mapping, quantifying crack length, and measuring magnitude of rutting.

1 INTRODUCTION

Design of asphalt concrete is a first stage in a series of activities needed to achieve a high performing pavement. We rely on laboratory tests to determine the quality of the mixes we have designed. It is now more than a decade since hot mix asphalt concrete mixes have been designed and constructed under the Superpave system. There has been a strong need to evaluate performance quality of such mixes through close field monitoring of constructed pavements. This paper provides the results of such evaluation.

2 OBJECTIVES AND APPROACH

The goal of this study has been to evaluate performance of Superpave mixes after several years in service and to determine how well this performance relates to laboratory test results. The approach to address this question has been construction of several pavements, obtaining materials at the time of construction, testing the materials in the laboratory, and evaluating performance of these mixes in the pavements after several years in service.

All laboratory shear tests were conducted on specimens compacted using Superpave gyratory compactor from field obtained loose mixtures. Low temperature creep and strength tests were conducted on cores obtained from the field.

3 EXPERIMENTAL PROGRAM

This study presents the results for two of the nine projects constructed through the Commonwealth of Pennsylvania. These projects are located in the Central and the Northwest regions of Pennsylvania. These will be respectively referred to as Snyder and Clearfield, which identifies their respective counties. The study is focused on the wearing course of these projects. Table 1 indicates characteristics associated with each.

Loose mixtures were obtained form these projects at the time of construction. Materials were placed in the Superpave Gyratory compactor. The tests conducted on the specimens included the repeated shear constant height, simple shear, strain-controlled frequency sweep, and indirect tensile creep tests. Not all tests were conducted for all projects.

3.1 *Repeated Shear at Constant Height (RSCH)*

The testing system was a closed-loop feedback, servo hydraulic system. Repeated shear tests at constant height were conducted to determine rutting resistance of the mixes.

In this test, repeated synchronized haversine shear and axial load pulses were applied to the specimen for 5000 cycles. A load cycle consists of 0.7-second, which is comprised of 0.1-second load application followed by 0.6-second rest period (Figure 1). Tests were conducted at 52°C to simulate summer pavement temperatures in Pennsylvania.

Sousa and others have proposed a quantitative method of predicting rutting from SST/RSCH data, which appears to give reasonable predictions, although it has not been validated for a wide

Table 1. Characteristics of the projects considered under this study.

Project	Nominal Max. Agg. Size, mm	Performance grade binders	Design life	Traffic (ESAL's)*	Construction year
Clearfield	12.5	PG 64-22 & PG 64-28	10	917,000	1997
Snyder	12.5	PG 64–22 on travel lane & PG 76-22 on passing lane	10	7,304,000	1998

* Equivalent Single Axle Loads.

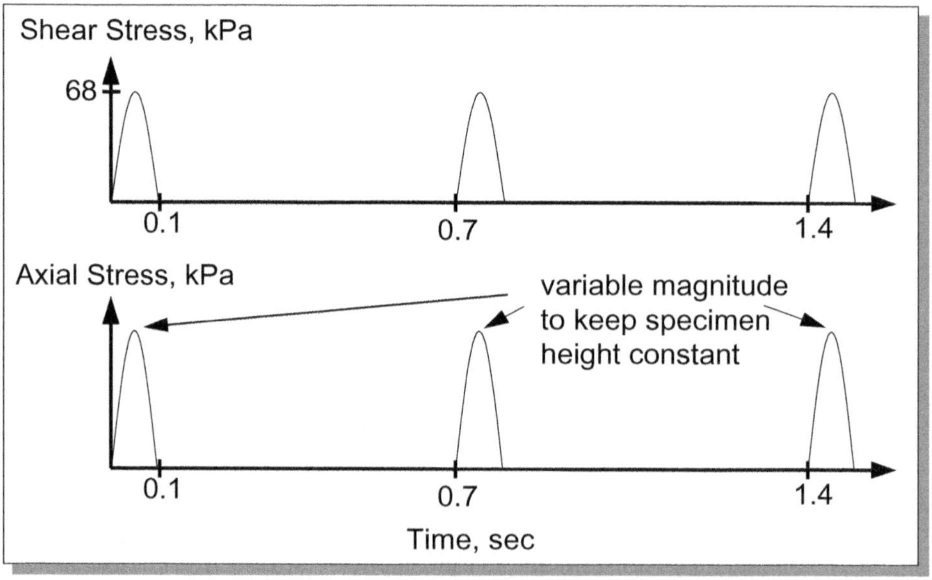

Figure 1. Stress pulses in repeated shear test at constant height.

range of mixtures, climates, and traffic levels (Sousa and Solaimanian, 1994; Sousa, et al., 1994). Other researchers have also looked at prediction potential of the repeated shear test for rutting (Williams et al. 1998 and Tayebali et al., 1999). Some have also used power law regression to develop the relationship between permanent shear strain and loading (Anderson et al., 2001).

3.2 *Simple shear test at constant height*

In this test, a controlled shearing stress is applied to the specimen. The test is performed at different stress levels and temperatures. Temperature used for this study included 4, 20, and 40°C. Example of stress levels of 20°C testing is shown in Figure 2.

3.3 *Frequency sweep test at constant height*

In this test, a repeated sinusoidal shearing load is applied to the specimen to achieve a controlled shearing strain of 0.005 percent (Figure 3). The following loading frequencies were used: 10, 5, 2, 1, 0.5, 0.2, 0.1, 0.05, 0.02, and 0.01 Hz. Testing temperatures included 4, 20, and 40°C. Testing at

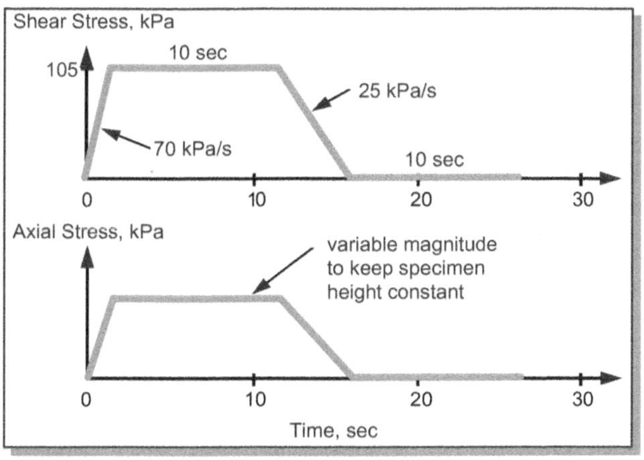

Figure 2. Stress applications in simple shear test at 20°C.

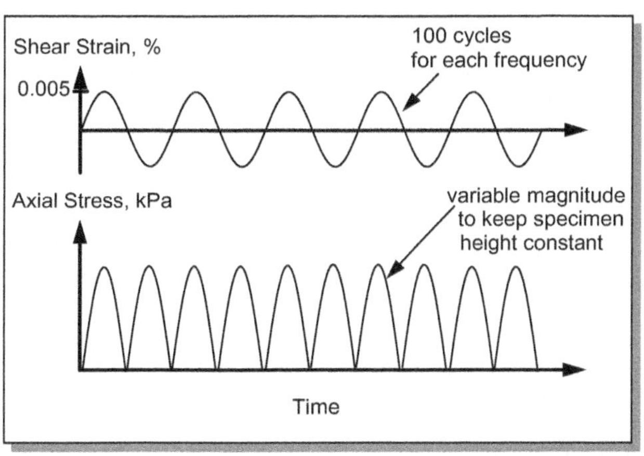

Figure 3. Shear strain and axial stress applications in frequency sweep test at constant height.

different temperatures and frequencies were all conducted on the same specimen, starting at the highest frequency and lowest temperature to minimize damage to the specimen during the test.

3.4 *Indirect tensile test*

The creep compliance and strength of mixes from Clearfield project were determined using an indirect tensile tester. The Indirect Tensile Test (IDT) was conducted on six of the cores of the wearing layers. The tests were conducted to provide data needed to evaluate thermal cracking sensitivity of the mixes. Three cores from the PG 68-22 section and three from the PG 68-28 were used.

The testing protocol was in accordance with AASHTO T322. In this test tensile creep and tensile strength are determined on the same specimen for thermal cracking analysis. The tensile creep is determined by applying a static load of fixed magnitude along the diametric axis of a specimen for 100s at temperatures of 0, −10,–20°C. The horizontal and vertical deformations measured across a gauge length of 38 mm near the center of the specimen are used to calculate creep compliance as a function of time. Loads are selected to keep horizontal strains in the linear viscoelastic range (typically below a horizontal strain of 500 micro strains) during the creep test. After the fixed load is applied the tensile strength will be determined by applying a load to the specimen at the rate of 12.5 mm per minute (vertical ram movement) to failure at −10°C. Typically the maximum strength of the specimen occurs at a point prior to failure and hence the vertical and horizontal LVDTs measurements are used to accurately determine the value of tensile strength (AASHTO T322).

4 FIELD PAVEMENT DISTRESS EVALUATION

Field work included transverse profiling to determine the rutting level and distress mapping to determine the level of cracking. Survey of sites for distress evaluation was conducted during summer 2007.

5 RESULTS AND DISCUSSION

Table 2 presents results for strain controlled frequency sweep test at 4, 20, and 40°C. The results in Table 2 are reported for 10 Hz frequency. For Snyder, not a significant difference is observed in dynamic modulus results between PG 64-22 and PG 76-22 sections. Overall, dynamic shear modulus over 350 MPa at 40°C represents good resistance of the mix against rutting (Anderson et al., 2003). It is unclear why the Clearfield 40°C modulus is similar to the modulus at 20°C. The 40°C modulus is expected to be significantly lower.

Permanent shear strain values reported in Table 3 are for 5000 cycles of repeated shear testing. All permanent strain values are less than 1% representing rut resistant mixes. Criteria proposed

Table 2. Results from strain controlled shear frequency sweep tests at constant height.

Project	Pavement course	PG Grade	Nominal maximum aggregate size (mm)	Dynamic shear modulus (MPa) @ 10 Hz 4°C	20°C	40°C
Snyder	SuperPave Wearing Course	64–22	12.5	6,425	3,164	480
	SuperPave Wearing Course	76–22	12.5	7,780	3,164	497
Clearfield	SuperPave Wearing Course	64–28	12.5	NA	1,491	1,391*

* This is unexpected as significantly lower result is expected.

Table 3. Results from stress controlled constant height repeated shear tests at 52C.

			Specimen information		Shear deformation (mm)		Permanent shear strain (%)	
Project	Pavement course	Air voids (%)	No.	Thickness (mm)	After 1 Cycle	After 5000 cycles	Max.	Avg
Snyder	12.5 mm	3	1	50.8	0.032	0.418	0.76	0.69
	PG 64-22	3	2	48.7	0.035	0.340	0.63	
	12.5 mm	3	1	47.3	0.026	0.303	0.58	0.56
	PG 76-22	3	2	48.1	0.026	0.288	0.54	
Clearfield	12.5 mm	3	1	50.7	0.015	0.111	0.19	0.41
	PG 64-28	3	2	51.1	0.017	0.342	0.64	
		7	1	50.4	0.196	0.250	0.11	0.11

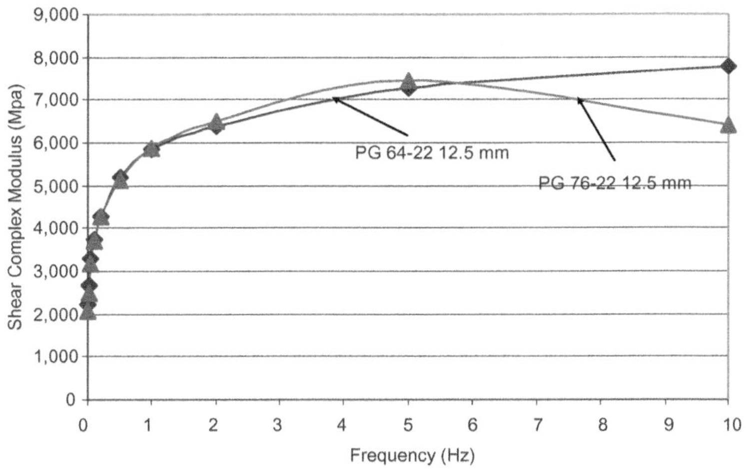

Figure 4. Shear complex modulus from frequency sweep test for snyder for different binder grades at 4°C.

by Asphalt Institute (Bukowski and Harman, 1997) identify mixes with less than 1% permanent strain as excellent mixes. However, as discussed later, rutting levels in some cases were close to 10 mm or even exceeding this value, suggesting that maybe strain levels less than 1 percent could suggest the mix to be in 'good category" rather than excellent. The results also show the impact of the binder stiffness on permanent deformation from laboratory shear tests. The PG 76-22 binder of Snyder showed less permanent deformation compared to the PG 64-22 (Figures 7 through 14).

Figures 15 and 16 present levels of rutting for the Snyder project after 10 years in service. What is noticeable from these graphs is higher rutting level in the travel lane (PG 64-22) compared to the passing lane (PG 76-22). The rutting level in the right wheel path for the passing lane is in the range of 4 to 6 mm and in the left wheel path it is in the range of 7 to 8 mm. However, for the travel lane, these two levels are, respectively, in the range of 10 to 12 mm, and 10 to 16 mm. This difference could be attributed to the use of stiffer binder in the passing lane. However, part of higher rutting in the travel lane could be due to higher traffic level and lower vehicle speed on this lane compared to the passing lane. It is also noticed that, as shown in Figure 15, as we approach the intersection rutting level increases and as we move further, the rutting decreases.

For the Clearfield project, no definite conclusion could be drawn in regard to rutting level of the PG 64-22 section compared to the PG 64-28 section. As shown in Figures 17 and 18 , the former has higher rutting in the right wheel path and lower rutting in the left wheel path compared with the latter. However, in regard to cracking, some difference is observed between the PG 64-22 section and the PG 64-28 section. The former exhibited much higher level of transverse and longitudinal cracking compared to the latter. Figure 19 indicates the results form laboratory measurements with indirect tensile creep tests −20°C, −10°C, and 0°C. The figure exhibits higher compliance of the PG 64-28 section (lower stiffness) which is consistent with observed performance in terms of cracking (Figure 20).

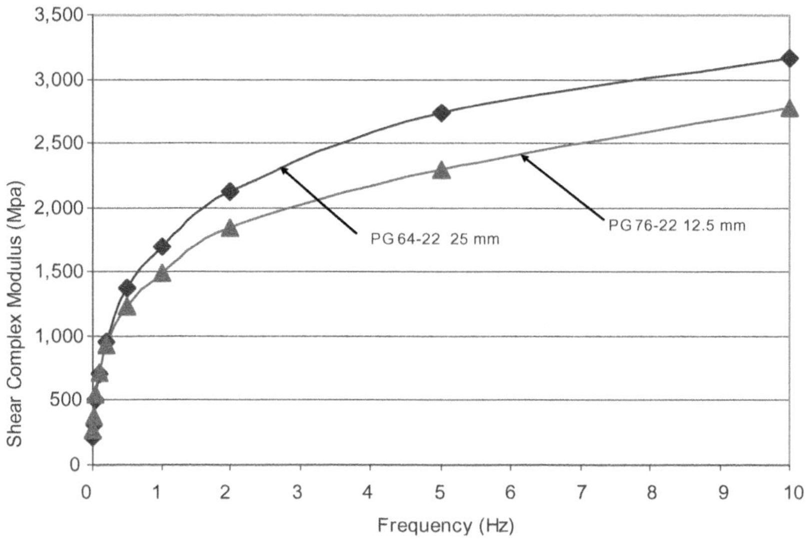

Figure 5. Shear complex modulus from frequency sweep test for snyder for different binder grades 20°C.

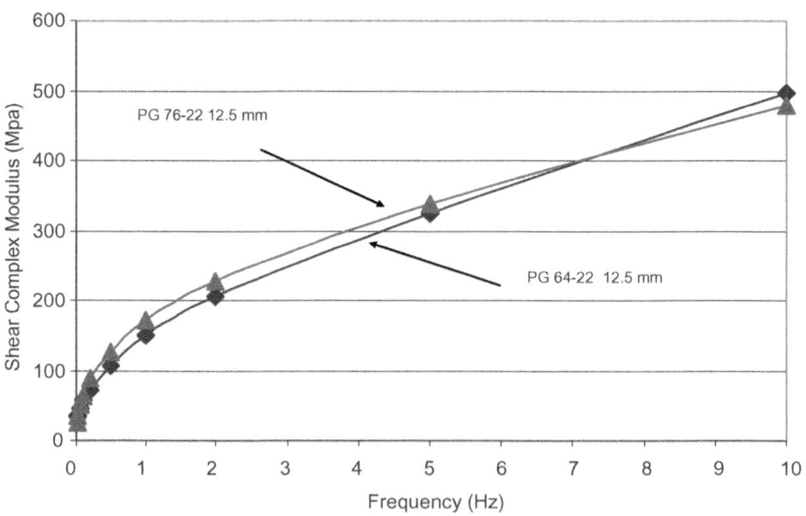

Figure 6. Shear complex modulus from frequency sweep test for snyder for different binder grades 40°C.

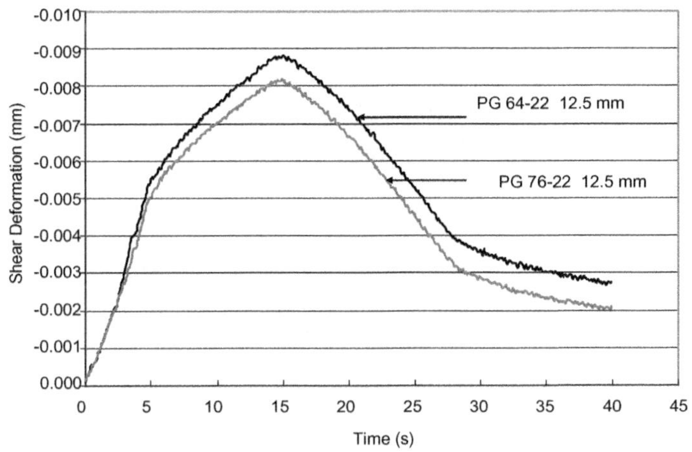

Figure 7. Shear deformation with time from simple shear test for snyder for different binder grades 2°C.

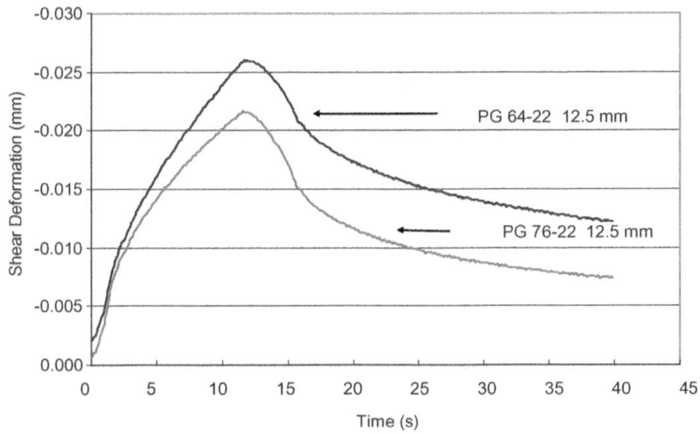

Figure 8. Shear deformation with time from simple shear test for snyder for different binder grades 20°C.

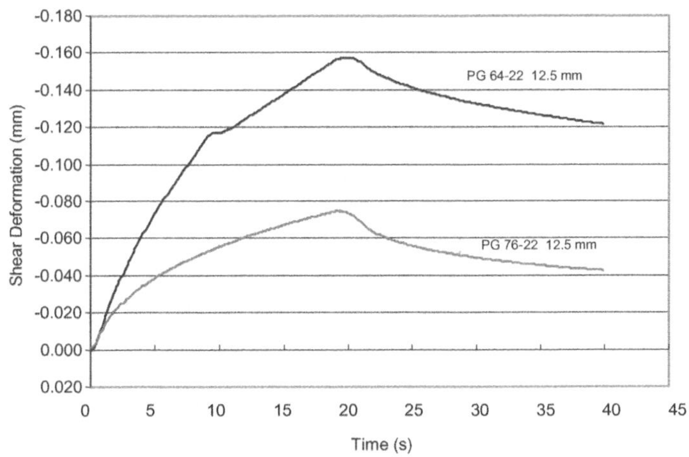

Figure 9. Shear deformation with time from simple shear test for snyder for different binder grades 40°C.

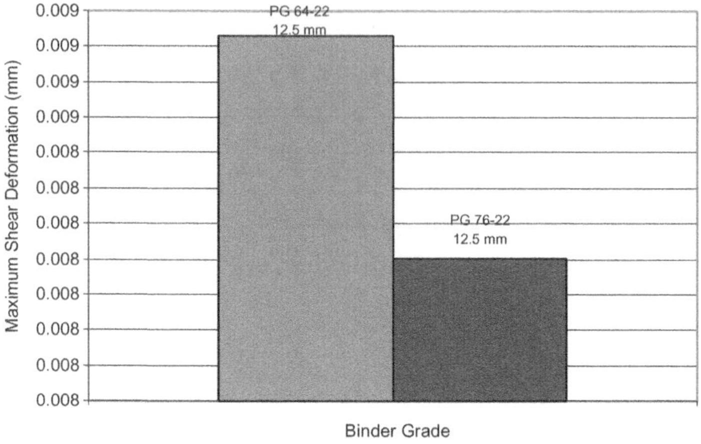

Figure 10. Maximum shear deformation from simple shear test for snyder for different binder grades at 4°C.

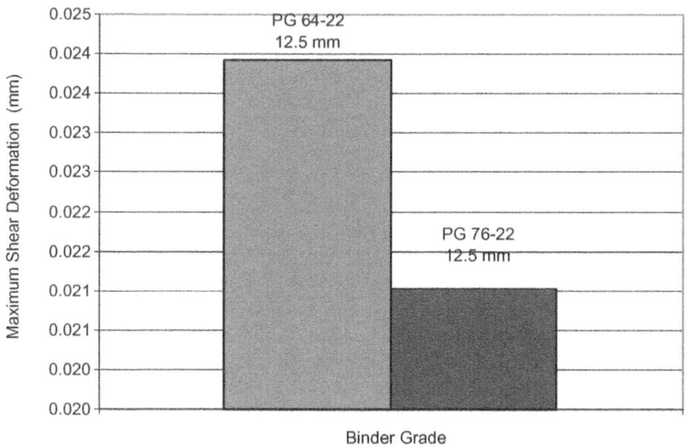

Figure 11. Maximum shear deformation from simple shear test for snyder for different binder grades at 20°C.

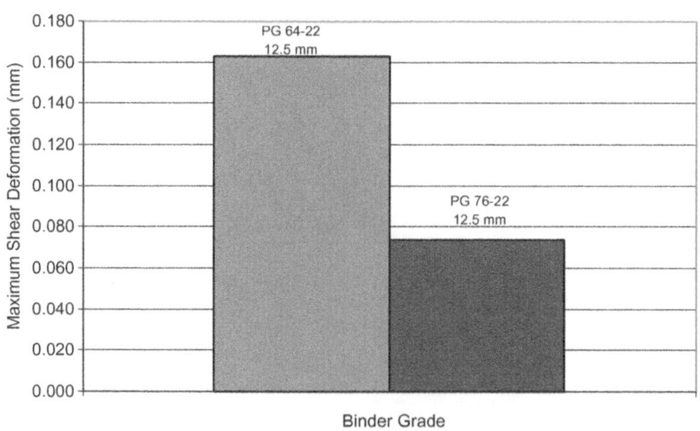

Figure 12. Maximum shear deformation from simple shear test for snyder for different binder grades at 40°C.

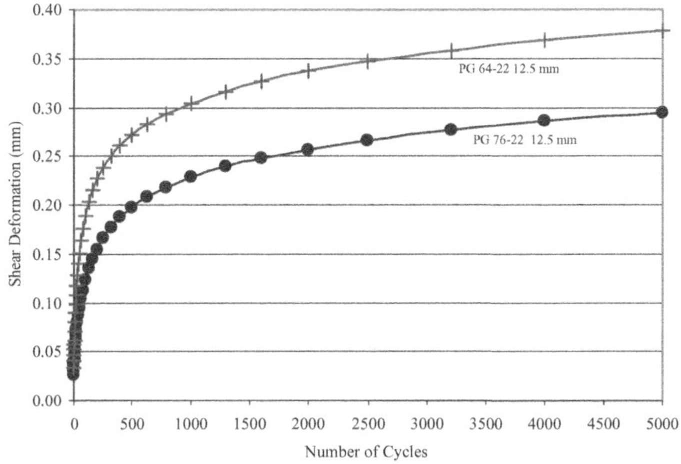

Figure 13. Shear deformation from RSCH test for snyder for different binder grades at 52°C.

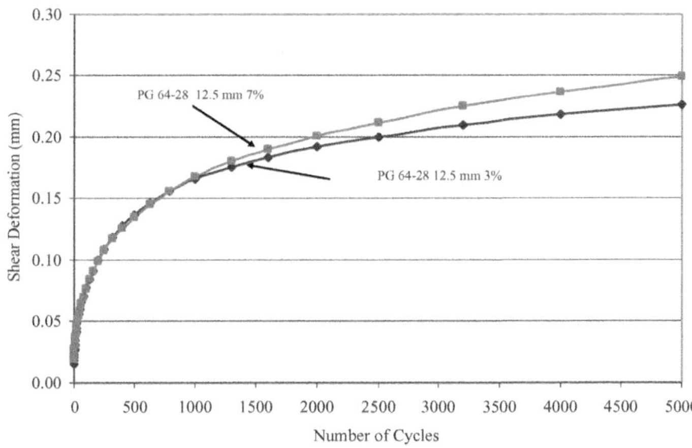

Figure 14. Shear deformation from RSCH test for clearfield for different binder grades at 52°C.

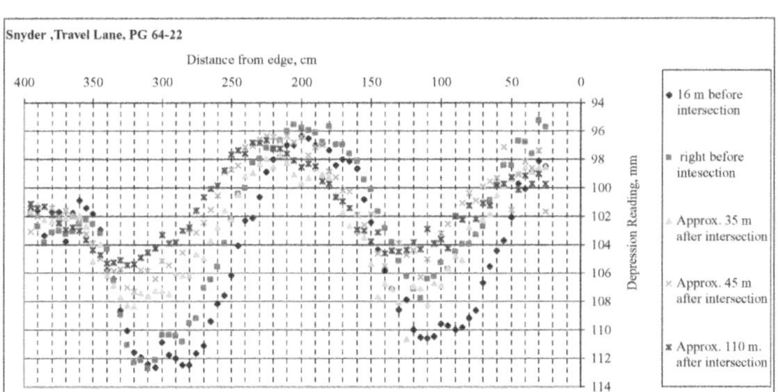

Figure 15. Rutting on travel lane of snyder project.

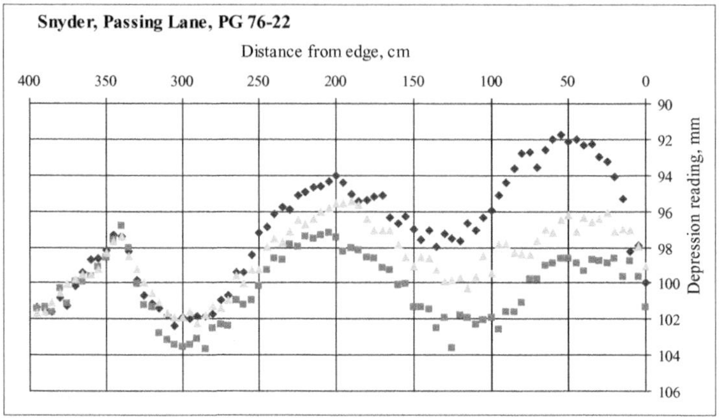

Figure 16. Rutting on passing lane of snyder project.

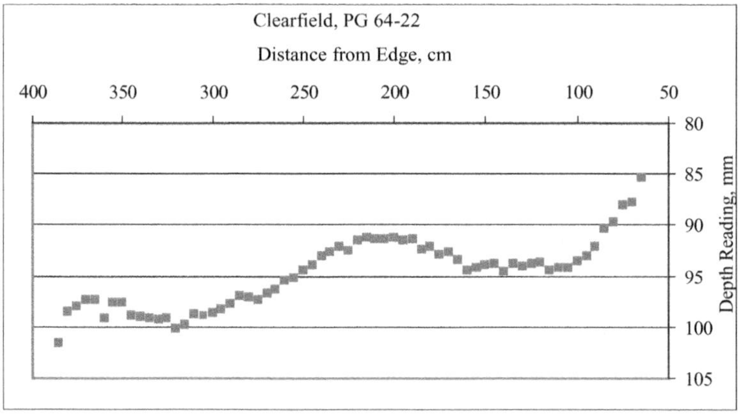

Figure 17. Rutting on PG 64-22 section of clearfield project.

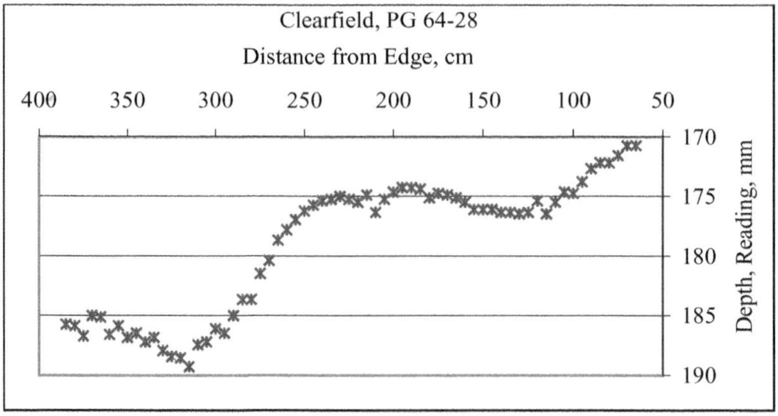

Figure 18. Rutting on PG 64-28 section of Clearfield Project.

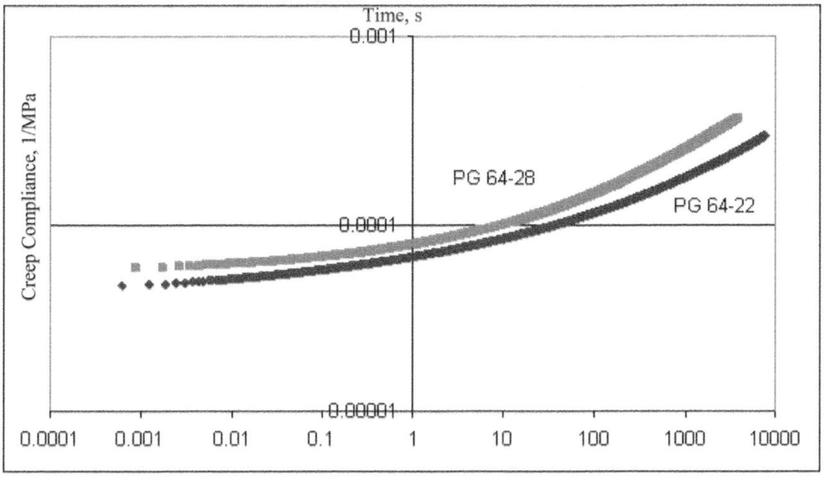

Figure 19. Comparison of master curves of both locations at cleafield on log-log domain.

Figure 20. Transverse and longitudinal cracks on PG 64-22 section of the clearfield project.

6 SUMMARY AND CONCLUSIONS

A study was undertaken to evaluate long term performance of several asphalt concrete pavements based on the Superpave system. Two of those projects are discussed in this paper. The work included pavement distress evaluation after 10 years of services as well as a series of laboratory tests to determine the rutting and cracking potential of the mixes.

One project included PG 76-22 and PG 64-22 binders in the wearing course and the other included PG 64-22 and PG 64-28 binders.

Repeated shear constant height tests at 52C indicated all mixes to be rut resistant. The level of rutting measured in the field after 10 years varies in the range of 4 to 16 mm. In general, the PG 76-22 section indicated less rutting than the PG 64-22 section. In addition, measured rutting was higher at points closer to the intersection indicating the adverse effect of low speed moving vehicles.

The indirect tensile creep tests at low temperature on cores from the PG 64-24 and PG 64-28 sections indicated higher compliance of the PG 64-28 section consistent with field observation of cracking, as PG 64-22 sections exhibited higher cracking levels compared with the PG 64-28 sections.

ACKNOWLEDGEMENT

The financial support of Pennsylvania Department of Transportation is greatly acknowledged.

REFERENCES

Anderson, R.M., R.B. McGennis, W. Tam, and T.W. Kennedy. "Sensitivity of Mixture Performance Properties to Changes in the Laboratory Compaction Using the Superpave Gyratory Compactor," *Proceeding of the Association of Asphalt Paving Technologists*, Volume 69, 2001.

Anderson, D.A., M. Solaimanian, D. Christensen, M. Marasteanu, and D. Hunter, "Superpave Validation Studies," *Pennsylvania Transportation Institute*, Report PTI 2003–14, January 2003.

Bukowski, J.R., and T. Harman. "Minutes of the Superpave Mixture Report Task Group," *Meeting of September 1997.*

Sousa J.B and M. Solaimanian, "An Abridged Procedure to Predict Permanent Deformation of Asphalt Concrete Pavements Using the Constant Height Repeated Shear Test,", *Transportation Research Record 1448*, TRB, National Research Council, Washington, D.C., January 1994, pp. 25–33.

Sousa, J.B., M. Solaimanian, and S.L. Weissman. "Development and Use of the Repeated Shear Test (Constant Height): An Optional SuperPave Mix Design Tool," Asphalt Research Program, *Strategic Highway Research Program, National Research Council*, Report SHRP-A-698, August 1994.

Tayebali, A.A., N. Khosla, G. Malpass, and F. Waller. "Evaluation of Superpave Repeated Shear at Constant Height Test to Predict Rutting Potential of Mixtures—Performance of Three Pavement Sections in North Carolina," Transportation Research Board, *Transportation Research Record 1681*, 1999.

Williams, R.C., P. Romero, and K. Stuart. *"Comparison of Superpave Shear Test Results to Westrack Performance," Symposium on Highway and Runway Technology*. The 1998, International Conference of Computational Science (ICES 98), 1998.

Flexible pavement design and assessment

Efficient Transportation and Pavement Systems – Al-Qadi, Sayed, Alnuaimi & Masad (eds)

Designing asphalt mixtures to withstand slow loading

P.E. Sebaaly & E.Y. Hajj
University of Nevada, Reno, Nevada, USA

ABSTRACT: Most asphalt pavements are designed to withstand highway traffic loading without any appreciable failures in rutting and fatigue cracking. However, as the asphalt mixtures are placed on city streets where traffic loads are slower and stop at intersection, these mixtures tend to experience severe rutting and shoving. These types of failures are caused by the complex stress conditions that are generated as heavy traffic loads move slowly and the weak behavior of the asphalt mixtures as they are subjected to the low rate of loading. Therefore, asphalt mixtures for highways must be carefully re-designed to withstand the loading conditions on urban facilities. It has been a common practice to increase the grade of the asphalt binder in order to compensate for the higher and more complex stresses under slower/stopped traffic. However, this approach did not lead to satisfactory results in most cases because adjusting the properties of the asphalt binder alone is not enough to resist these types of failures. This paper presents a fundamental approach to analyze the state of stresses generated under the slower/stopped heavy vehicles and their impact on the behavior of the asphalt mixtures. A process to modify the failure criteria that are currently used for the design of asphalt mixtures under highway traffic as they are placed under slow/stop traffic is recommended. In addition, laboratory data generated from the evaluation of asphalt mixtures for slow/stop traffic are presented.

1 INTRODUCTION

During the last decades there have been dramatic changes in traffic volumes, traffic weights, and tire pressures, resulting in a significant increase in permanent deformation of the hot mix asphalt (HMA) pavements. Permanent deformation is a major mode of failure in flexible pavements consisting of both rutting and shoving. Typically a rutting failure alone occurs under traffic loads moving at highway speed, while both rutting and shoving failure may occurs under traffic loads at intersections. In many instances, the same HMA mixtures that have a history of good performance in rutting did not perform well at intersections. Permanent deformation of HMA is a common problem at intersections as the HMA mixture exhibits lower stiffness as it is subjected to slow moving or stopped vehicle loads. This problem is more advanced in hot climates where the stiffness of the HMA is further decreased with the increase of the pavement temperature. Additionally, the different behavior of HMA pavement at intersection is related to the more complex stresses imposed at the pavement surface layer by the braking and turning movements of heavy loaded trucks. Therefore, a different type of HMA mixtures may be required at intersections from the one used on the main lane.

1.1 *Permanent deformation*

Rutting may be described as a longitudinal permanent deformation that occurs in the pavement's wheel path under the action of repeated traffic loads. Rutting of HMA pavements can result from different mechanisms: consolidation, plastic flow, and mechanical deformation (Roberts et al., 1996).

Consolidation can occur when there is insufficient compaction during the construction of the pavement. An asphalt mixture with insufficient density is prone to further compaction under traffic, especially in hot weather and at intersections where the loads are slow moving or static.

Plastic flow can result when HMA mixtures exhibit insufficient stability. Some of the more common reasons for mixture instability are: excessive amount of asphalt binder and insufficient air voids, too much rounded aggregate, or too high of the minus #200 material.

Lastly, mechanical deformation results from insufficient structural capacity of the pavement system and can occur when the strength and/or thickness of the pavement layers are insufficient to support the design traffic on the existing subgrade. A rut resulting from this type of action will generally be accompanied by longitudinal and/or alligator cracking.

In contrast to rutting, shoving is a common type of distress particularly confined to intersections and braking areas where traffic comes to a stop. It is defined as the horizontal displacement of an asphalt mixture and is caused either by shear flow of the mixture or slippage at the interface of the HMA lifts. The combination of excess asphalt binder and low air-voids leads to an unstable HMA mixture that fails under the high horizontal shear stresses. The slippage type of failure usually shows up at the surface of the pavement as a slippage cracking (AI 1999, Kandhal 1996).

The severity of permanent deformation is directly correlated to the in-service temperature of the HMA mixture and the speed of traffic loading. Asphalt pavements located in hot environments are expected to experience higher permanent deformations than those located in cold environment. On the other hand, asphalt pavements subjected to fast traffic loading are expected to experience lower permanent deformation than those subjected to slow traffic loading. When combining both environmental and rate of loading impacts, it is expected that asphalt pavements will experience the highest permanent deformations under hot weather and slow moving traffic loads.

2 RESEARCH PROGRAM

The objectives of this research effort are: 1) to develop guidelines to design HMA mixtures that can withstand the combined actions of hot environment and slow moving traffic loads and 2) to assess the performance of some HMA mixtures using the developed system. These conditions are typically encountered on urban streets where traffic loads travel at lower speeds and stop at intersections. In order to achieve the objectives of the research, the following tasks were completed:

- Examine the performance of HMA mixtures at urban intersections.
- Identify the loading conditions at urban intersections.
- Revise the permanent deformation criteria for urban intersections.
- Evaluate the performance of multiple HMA mixtures at urban intersections.

2.1 *Performance of HMA mixtures at urban intersections*

This investigation evaluated and compared the volumetric and gradation properties of mixtures from rutted and non-rutted intersections. Several in-service HMA intersections were identified in southern Nevada (i.e. Las Vegas), some with severe rutting and some without any rutting failures. Southern Nevada is characterized as having hot climate, therefore, urban intersections in this region are subjected to the severe combination of elevated temperatures and slow traffic loading.

Core samples were obtained from between wheel-path away from the intersection and from the wheel-path at the intersection. The idea behind the sampling plan was that cores from between wheel-path away from intersection provide an HMA mixture that has not been damaged due to traffic loading while the wheel-path cores at the intersection represent the damaged mixture. It should be noted that these cores were subjected to the same environmental conditions and therefore influences related to environment are absent when the performance of these samples are compared.

To evaluate and compare the volumetric and gradations properties of the various mixtures, the cores were tested for bulk specific gravity, asphalt content, and gradation. Table 1 summarizes the volumetric and gradation properties of the cores. For the projects experiencing severe rutting, the asphalt binder content from extraction was higher at the intersection when compared to the one at 500 ft away from the intersection. Additionally, the air-voids and voids in mineral aggregates (VMA) values are lower at the intersections when compared to the ones away from the intersection, except for the Lake Mead-Las Vegas Blvd intersection where both parameters had very close values. It should be noted that the lower asphalt content at the rutted intersections is only apparent which caused by the lower air-voids at these locations.

In general, if no excessive shoving is present, the air-voids and VMA values on rutted intersections are expected to be lower due to extra compactive effort by traffic. During the service life of the pavement, it is desirable to maintain the air-voids in the vicinity of 3 to 4% after initial traffic densification, but not less than 2%. The data in Table 1 show that the cores obtained from the intersection locations have air-voids less than 2.5% for the projects experiencing severe rutting, whereas the cores from projects with no significant rutting have air-voids greater than 2.5%. Field observations coupled with the data in Table 1 indicate that the intersections failed in the rutting mode caused by the excessive densifications of the HMA mixture under slow/stopped traffic.

Figure 1a shows the extracted gradation of the cores away from the intersections of the thirteen projects. By grouping the extracted aggregate gradations of mixtures that experienced rutting and mixtures that experienced no rutting, two gradation zones were identified as shown in Figure 1b. In other words, the no rut mixtures zone covers all the extracted aggregate gradations of the mixtures that experienced no rutting. In contrast, the rut mixtures zone covers all the extracted aggregate gradations of the mixtures that experienced rutting. Furthermore, the no rut mixtures zone

Table 1. Summary of field cores test data.

ID	Project	Rutting severity level	Core location	% Asphalt content	% Air voids	% VMA	% $P_{\#4}$	% $P_{\#200}$
1	SR593–Tropicana &		At intersection	7.7	1.3	15.2		
	Decatur	Severe	150 m away	5.8	3.0	15.3	61	8.3
2	SR593–Tropicana &		At intersection	6.8	0.2	12.9		
	Koval	Severe	150 m away	5.4	3.3	14.5	60	9.5
3	Blue Diamond &		At intersection	4.8	2.8	12.3		
	Industrial	No	150 m away	4.7	6.3	15.3	60	8.0
4	Lake Mead &		At intersection	5.7	1.6	13.4		
	Las Vegas Blvd	Severe	150 m away	5.4	1.4	13.0	65	8.5
5	Lake Mead &		At intersection	7.5	0.8	13.6		
	M. Luther King	Severe	150 m away	4.9	10.6	20.1	54	8.2
6	Cheyenne Ave. &		At intersection	4.8	3.5	13.6		
	North Fifth	No	150 m away	5.0	3.1	13.5	54	8.0
7	Cheyenne Ave. &		At intersection	6.9	2.6	15.0		
	Losee	Severe	150 m away	5.4	4.3	15.4	58	8.5
8	Tropicana & Valley		At intersection	6.3	0.0	12.5		
	View	Severe	150 m away	5.9	2.4	14.4	49	6.6
9	Spring MTN &		At intersection	5.8	4.3	15.1		
	Wynn	No	150 m away	5.3	6.1	16.3	53	7.0
10	Spring MTN &		At intersection	9.3	0.8	14.0		
	Polaris	Severe	150 m away	5.3	5.9	15.2	49	8.0
11	Industrial &		At intersection	5.4	2.5	12.8		
	Desert Inn	No	150 m away	4.4	4.9	14.2	38	5.7
12	Industrial &		At intersection	6.4	4.9	16.8		
	Warm Springs	No	150 m away	5.6	3.7	15.2	54	7.6
13	Buffalo & Desert		At intersection	4.4	8.9	17.2		
	Inn-SEAM 70/30	No	150 m away	6.0	2.1	12.4	42	4.4

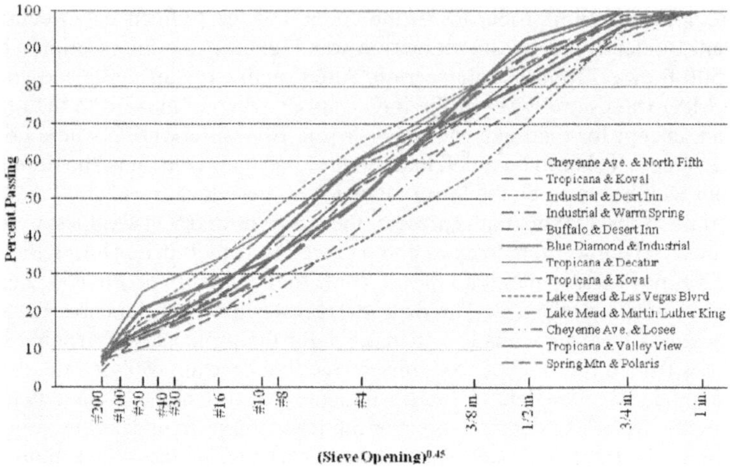

(a)

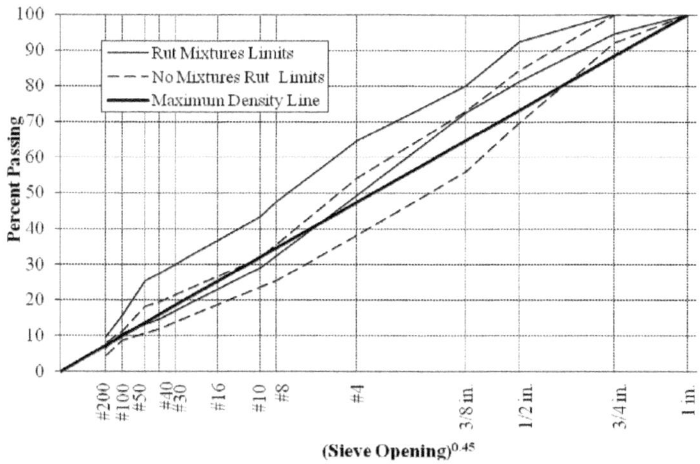

(b)

Figure 1. (a) Extracted aggregate gradation curves for selected intersections mixtures (b) Extracted aggregate gradation limits for Rut and No Rut Mixtures obtained away from intersections.

covers denser gradations (closer to the maximum density line) and coarser gradations than the rut mixtures zone. It should be stated that a narrow common region between the two zones exists.

In summary, this evaluation indicated that the gradation of the HMA mix plays the most significant role in the resistance of HMA mixtures to permanent deformation at urban intersections.

2.2 *Identify the loading conditions at urban intersections*

The objective of this effort is to identify the load distributions on a heavy tractor- semitrailer truck as it travels on urban streets and it stops at intersections. This effort consisted of determining the load distributions on the various axles of tractor-semitrailer as it travels on any given slope (i.e. downhill, uphill, or level) with the impact of braking as it approaches the intersection. In comparison with a two-axle vehicle, the braking characteristics of a tractor-semitrailer are significantly

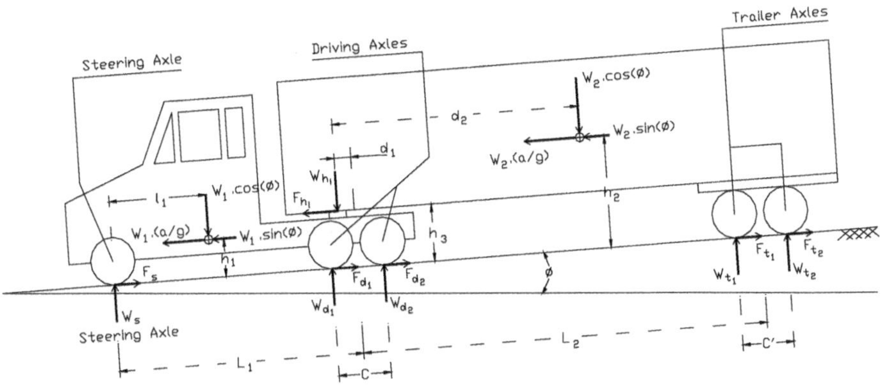

Figure 2. Forces acting on a tractor-semitrailer during braking on a downward slope.

more complex. For a given two-axle vehicle, the load transfer is only a function of the deceleration rate, whereas for a tractor-semitrailer, the load transfer during braking is dependent not only on the deceleration rate, but also on the braking force of the semitrailer.

Braking decelerates the vehicle, which causes load to transfer to the front of the vehicle. The resulting axle load can be higher or lower than the initial static load, depending on the location of the axle. Figure 2 shows the major forces acting on an eighteen-wheel tractor-semitrailer during braking on a sloping pavement. Since Brakes are the primary source of braking deceleration, the aerodynamic drag and rolling resistance are neglected in this study. The various axles include: the tractor steering axle, the tractor tandem axle (i.e. driving axle), and the semitrailer tandem axle (i.e. trailer axle). The tandem axles of the tractor and the semitrailer are considered without equalization, implying that an inter-axle load transfer will take place between the rear and the front axle of the tandem group during the braking period of the truck.

W_{hi} and F_{hi} are respectively the vertical and horizontal loads at the tractor-semitrailer articulation; "a" is the linear deceleration of the truck along the longitudinal axis; g is the deceleration due to gravity; W_1 and W_2 are respectively the tractor and semitrailer total weights; and θ is the angle of the slope with the horizontal (zero for level, positive for downhill and negative for uphill). W_s, W_{d1}, W_{d2}, W_{t1} and W_{t2} are the tires normal loads. F_s, F_{d1}, F_{d2}, F_{t1} and F_{t2} are the braking forces that originate from the brake system and developed on the tire-road interface.

In order to calculate the normal load on each axle, the tractor and the semitrailer are considered as free bodies separately and combined. The vertical, horizontal, and moment equilibrium equations for the tractor, semitrailer unit, and tractor-semitrailer combination can be written as a function of truck loads and geometry resulting in a total of eleven equilibrium equations, three characteristic equations, and fourteen unknowns. Full listing of the fourteen equations is provided by Hajj (2005).

Solving the fourteen equations and fourteen unknowns, the normal loads on the various axles can be expressed as follows:

Tractor front axle (steering axle):

$$\begin{aligned} W_s = {} & W_2\frac{[(a/g+\sin\theta)(h_2d_1+h_3L_2)+d_1\cos\theta(L_2+d_1-d_2)]}{L_1(L_2+d_1)} \\ & + W_1\frac{[(a/g+\sin\theta)h_1+(L_1-l_1)\cos\theta]}{L_1} - (F_{t_1}+F_{t_2})\frac{h_3L_2}{L_1(L_2+d_1)} \\ & - (W_{t_1}-W_{t_2})\frac{c'd_1}{2L_1(L_2+d_1)} - (W_{d_1}-W_{d_2})\frac{c}{2L_1} \end{aligned} \tag{1}$$

Tractor rear axles (driving axles):

$$\begin{aligned} W_d &= W_{d_1} + W_{d_2} \\ &= -W_2 \frac{\left[(a/g+\sin\theta)(h_2(d_1-L_1)+h_3(L_1+L_2))+(d_1-L_1)(L_2+d_1-d_2)\cos\theta\right]}{L_1(L_2+d_1)} \\ &\quad - W_1 \frac{\left[(a/g+\sin\theta)h_1 - l_1\cos\theta\right]}{L_1} + (F_{t_1}+F_{t_2})\frac{h_3(L_1+L_2)}{L_1(L_2+d_1)} \\ &\quad - (W_{t_1}-W_{t_2})\frac{c'(L_1-d_1)}{2L_1(L_2+d_1)} + (W_{d_1}-W_{d_2})\frac{c}{2L_1} \end{aligned} \tag{2}$$

$$\begin{aligned} W_{d_1} &= W_1 \frac{[-(a/g+\sin\theta)(h_3-h_1)+(L_1-l_1-d_1)\cos\theta]}{c} + (F_s+F_{d_1}+F_{d_2})\frac{h_3}{c} \\ &\quad - W_s\frac{(L_1-d_1)}{c} + W_d\frac{(d_1+c/2)}{c} \end{aligned} \tag{3}$$

$$\begin{aligned} W_{d_2} &= W_1 \frac{[(a/g+\sin\theta)(h_3-h_1)-(L_1-l_1-d_1)\cos\theta]}{c} - (F_s+F_{d_1}+F_{d_2})\frac{h_3}{c} \\ &\quad + W_s\frac{(L_1-d_1)}{c} + W_d\frac{(c/2-d_1)}{c} \end{aligned} \tag{4}$$

Semitrailer axles (trailer axles):

$$\begin{aligned} W_t = W_{t_1} + W_{t_2} &= W_2 \frac{[-(h_2-h_3)(a/g+\sin\theta)+d_2\cos\theta]}{(L_2+d_1)} \\ &\quad - (F_{t_1}+F_{t_2})\frac{h_3}{(L_2+d_1)} + (W_{t_1}-W_{t_2})\frac{c'}{2(L_2+d_1)} \end{aligned} \tag{5}$$

$$W_{t_1} = W_2 \frac{\left[(h_2-h_3)(a/g+\sin\theta)-d_2\cos\theta\right]}{c'} + W_t\frac{(L_2+d_1+c'/2)}{c'} + (F_{t_1}+F_{t_2})\frac{h_3}{c'} \tag{6}$$

$$W_{t_2} = W_2 \frac{\left[-(h_2-h_3)(a/g+\sin\theta)+d_2\cos\theta\right]}{c'} - W_t\frac{(L_2+d_1-c'/2)}{c'} - (F_{t_1}+F_{t_2})\frac{h_3}{c'} \tag{7}$$

The calculation of the normal loads on the various axles of the downhill braking 18-wheeler truck requires that the following characteristic properties be specified:

- Application (treadle) versus actuation (chamber) pressure at each axle: the application pressure is defined as the pressure produced at the output of the treadle valve, whereas the actuation pressure is the pressure experienced at the brake chamber. In the case where some sort of proportioning valve is used, these two pressures will differ significantly (Gillespie 1987, 1992).
- Brake force versus actuation pressure for the brakes on each axle: the braking force developed at the tire-road interface is determined by the actuation pressure applied to each brake and the gain of each (Gillespie 1987, 1992).

The braking force on individual wheels can be described by the following equation:

$$F_b = \frac{T_b}{r} = G\frac{P_{ac}}{r} \tag{8}$$

where, F_b = Brake force (lb), T_b = Brake torque (in-lb), r = Tire rolling radius (inch), G = Brake gain (in-lb/psi), and P_{ac} = Actuation pressure (psi).

The braking system properties of a standard United State (U.S.) 18-wheeler truck are taken from a National Highway Traffic Safety Administration (NHTSA) report that was conducted by T.D. Gillespie et al. (1987). The NHTSA study considered a linear brake system (i.e. linear relationship between the application and the actuation pressure). Table 2 summarizes the brake system properties of the U.S. eighteen-wheel.

The extent to which vertical load is transferred during braking from the rear tandem tires to the front tandem tires is called the dynamic load transfer coefficient (α) and is defined as follows (6).

$$\alpha = \frac{W_{t_1} - W_{t_2}}{2(F_{t_1} + F_{t_2})} = \frac{W_{d_1} - W_{d_2}}{2(F_{d_1} + F_{d_2})} \tag{9}$$

The load transfer coefficient has a value of 0.0 when the loads on the front and rear tandem tires are equal (i.e., $W_{t1} = W_{t2}$; $W_{d1} = W_{d2}$).

By solving the system of equilibrium equations, the normal loads on the various axles during braking on a slope can be expressed as a function of W_1, W_2, a, θ, braking forces, and truck geometry. Since an intersection located on a downward sloped pavement represents the worst case scenario, the tire loads distributions were determined for an 18-wheel truck braking on a 6% downward sloping pavement. Table 3 summarizes the load distributions on the various axles of the 18-wheel tractor-semitrailer combination under the normal highway traffic (without braking) and at intersections (with braking). The data in Table 3 show that there is a significant shift of loads to the front axles of the truck as it brakes to a stop at the intersection. Also the horizontal component of the tire load becomes significant due to the braking action. The combination of the load redistribution among axles and the horizontal load component is expected to play a major role in the rutting and shoving of HMA mixtures at urban intersections.

2.3 *Revise the permanent deformation criteria for urban intersections*

The objective of this effort is to revise the permanent deformation criteria used on highway pavements for HMA pavements at urban intersections. Currently, there are two commonly used permanent deformation criteria: one based on the repeated stress constant height (RSCH) test and one based on the repeated load triaxial (RLT) test. The RSCH criterion calls for a maximum of 5% permanent shear strain under 5,000 load cycles and the RLT criterion calls for a maximum of 3% permanent axial strain under 12,000 load cycles.

Table 2. Summary of a U.S. tractor-semitrailer braking system properties.

Truck unit	Axle	Torque gain* (in-lb/psi)	Pushout pressure (psi)
Tractor	Steering	1322.5	13.5
	Tandem leading driving	3280.0	5.8
	Tandem trailing driving	3280.0	5.8
Semitrailer	Tandem leading trailer	2818.8	5.5
	Tandem trailing trailer	2818.8	5.5

* For a loaded truck-semitrailer.

Table 3. Load distribution on various axles of the 18-wheel tractor-trailer.

Speed/braking action	Axle	Vertical load/tire (lbs)	Horizontal load/tire (lbs)
40 mph/no braking	Steering	6000	0
	Front driving	4250	0
	Rear driving	4250	0
	Front trailer	4250	0
	Rear trailer	4250	0
2 mph/braking	Steering	8474	2161
	Front driving	5122	2968
	Rear driving	3341	2968
	Front trailer	4394	2561
	Rear trailer	2858	2561

The process of revising the permanent deformation criteria for HMA pavement at urban intersections consisted of determining and comparing the permanent deformations under the tractor-semitrailer on both highway pavements and urban intersections. The ratio of the permanent deformation at intersections over the permanent deformation on highway pavements would provide the adjustment factor for revising the permanent deformation criteria. Determining the permanent deformations of HMA pavements on highways and at intersections necessitates the development of permanent deformation models in the laboratory followed by a mechanistic analysis to determine the pavement responses in the HMA layer, which in turn are the critical input to the permanent deformation models.

Hajj (2005) developed permanent deformation models for a typical HMA mixture manufactured with a PG76-22 binder based on responses measured in the laboratory under a variety of testing conditions. The permanent shear strain (γ_p) after 5,000 cycles measured from the RSCH test was correlated to the resilient shear strain (γ_r) as shown in equation 10. Similarly, the permanent axial strain (ε_p) after 12,000 cycles measured from the RLT test was correlated to the applied resilient axial strain (ε_r) as shown in equation 11. The resilient shear and axial strains were measured in the laboratory under the RSCH and RLT tests, respectively.

$$\text{RSCH: } \gamma_p = 15.0 \times \gamma_r^{1.136} \quad (R^2 = 89\%) \tag{10}$$

$$\text{RLT: } \varepsilon_p = 394.2 \times \varepsilon_r^{1.296} \quad (R^2 = 72\%) \tag{11}$$

It should be mentioned that the above are statistical relationships based on the laboratory analysis of the asphalt mixture and therefore a shift/adjustment factors are required to provide reasonable estimates of the permanent deformation in the field. The shift factors are outside the scope of this research and the laboratory-based models presented here were only used for a relative comparison.

Predicting permanent deformation in HMA pavement using equations 10 and 11 requires the determination of the resilient shear strain (γ_r) and the resilient axial strain (ε_r) under the loading of an 18-wheel truck at normal highway speed and at intersection. A mechanistic analysis was conducted to calculate the responses of the same mix in a thin (4 inch HMA) and a thick (8 inch HMA) pavement under the loading of an 18-wheel truck at normal highway speed (no braking) and at intersection (during braking). The computer code 3D-Moving Load Analysis (3D-MOVE) developed by Siddharthan et al. (1998, 2002) was used to estimate the pavement resilient responses.

Table 4 summarizes the properties of the pavement structures used in the 3D-MOVE analyses. The frequency-dependent viscoelastic properties of the HMA layer were used to characterize the

Table 4. Properties of pavement layers.

Layer	Thickness (inch)	Unit weight (pci)	Shear modulus (psi)	Damping ratio	Poisson's ratio
HMA	4.0 & 8.0	0.0876	Variable*	Variable*	0.40
Base	8.0	0.0663	12500	5.0%	0.40
Subgrade	236	0.0626	3570	5.0%	0.40

* Determined using the Shear Frequency Sweep at Constant Height (FSCH) test.

Table 5. Predicted permanent shear strains under 18-wheel truck for the typical HMA mixture.

HMA layer thickness (inch)	Traffic	Max. Resilient Shear Strain, $(\gamma_r)_{max}$ (%)	Max. Permanent Shear Strain, γ_p* (%)	Max. Resilient Axial Strain, $(\varepsilon_r)_{max}$ (%)	Max. Permanent Axial Strain, ε_{1p}^{+} (%)
4.0	Highway traffic	0.2363	1.56	0.1590	9.30
	At intersection	0.5602	4.15	0.2265	14.72
8.0	Highway traffic	0.2283	1.50	0.1298	7.16
	At intersection	0.5517	4.08	0.2136	13.64

* Calculated using Equation 10.
$^{+}$ Calculated using Equation 11.

rate dependent behavior of the HMA mixture under the moving truck. The shear frequency sweep at constant height (FSCH) test was used to measure the mixture's stiffness and internal damping as a function of the frequency. The damping ratio represents a measure of the ability of the materials to dissipate energy and dampens dynamic loads. The tires loads summarized in Table 3 were used in this analysis.

Two vehicle speeds were considered: 40 mph (away from intersection) and 2 mph (at intersection). The speed of 40 mph is more typical of a truck traveling on highway pavements while the 2 mph represents the truck as it approaches the intersection. Table 5 summarizes the maximum resilient shear strains $(\gamma_r)_{max}$ developed within the thin and thick HMA layer under the 18-wheel truck and the maximum resilient vertical strain $(\varepsilon_r)_{max}$ at the middle of the HMA layer.

The dynamic resilient responses $(\gamma_r)_{max}$ and $(\varepsilon_r)_{max}$, determined from the 3D-MOVE analysis were used to estimate the permanent deformation potential of the typical HMA mixture under 18-wheel truck at highway speed and at intersection. The permanent strains under the free rolling and braking 18-wheel truck are predicted by substituting the computed $(\gamma_r)_{max}$ and $(\varepsilon_r)_{max}$ in the HMA layer into the permanent deformation models (equations 10 and 11). Table 5 summarizes the computed maximum responses in the HMA layers along with the corresponding permanent strains.

Table 5 shows that, in both the 4-inch and 8-inch HMA layers the maximum permanent shear strain (γp) at intersection is as much as 2.7 times the response under normal highway traffic. In other words, the braking of the 18-wheel tractor-trailer combination at slow speed is expected to result in permanent deformation in the HMA layer that is 2.7 times higher than under the truck rolling at 40 mph and without braking. Therefore, it is suggested to reduce the allowable measured shear strain in the RSCH test by a factor of 2.7, when used as a permanent deformation criterion for intersections. This means, the 5% maximum permanent shear strain limit that is used in current specification for HMA mixtures under normal highway traffic should be reduced to 2% when the mixture is used on intersections.

Additionally, the permanent axial strain in the HMA layer at intersection was 1.5 times the permanent axial strain under normal highway traffic (Table 5). Therefore, for HMA mixtures at intersections, the maximum 3% allowable permanent axial strain in the RLT test should be reduced to 2%.

2.4 *Evaluate the performance of multiple HMA mixtures at urban intersections*

The objective of this task was to identify an HMA mixture that can withstand the high and complex state of stresses generated at intersections in southern Nevada. NDOT currently specifies a PG76-22 binder for all HMA mixtures in southern Nevada. Therefore, it is critical to identify an aggregate gradation that when designed with the PG76-22 binder provides a rut resistant mixture that can perform well at intersections.

Based on the evaluation of the field performance of various intersections in southern Nevada as summarized in section 2.1, the following three aggregate gradations were selected for evaluation:

- No Rut Mixture gradation, designated as NRM
- NDOT Type 2C gradation, designated as T2C
- Rut Mixture gradation, designated as RM.

The NRM gradation fits within the range of the mixtures that showed excellent rutting resistance at intersections as shown in Figure 1. The RM gradation fits within the range of mixtures that experienced rutting at intersections as shown in Figure 1. The T2C gradation represents the typical HMA mix that Nevada DOT uses on highway pavements and that has shown excellent resistance to rutting under normal highway traffic. Figure 3 shows the three gradations.

A Nevada DOT Hveem mix design was conducted for each of the three mixtures. All mixtures were treated with 1.5% hydrated lime by dry weight of aggregate following NDOT's specifications. The rutting resistances of the three mixtures were evaluated using the Repeated Shear at Constant Height (RSCH), and the repeated load triaxial test (RLT). Three specimens were compacted for each mixture to an air void content of 7±1% using the SGC. The 7±1% air voids was selected to represent the HMA layer during the first 2–3 years of the pavement life.

The RSCH test was run at 158°F (one grade lower than the binder maximum performance temperature) on 6-inch diameter by 2-inch height specimens (AASHTO 2004). The RSCH test simulates the resistance of the HMA mixtures to rutting under elevated temperature. During the test, the sample is maintained at a constant height while a repetitive haversine shear stress is applied. The magnitude of the shear stress is 10 psi with a duration of 0.1 seconds and a rest

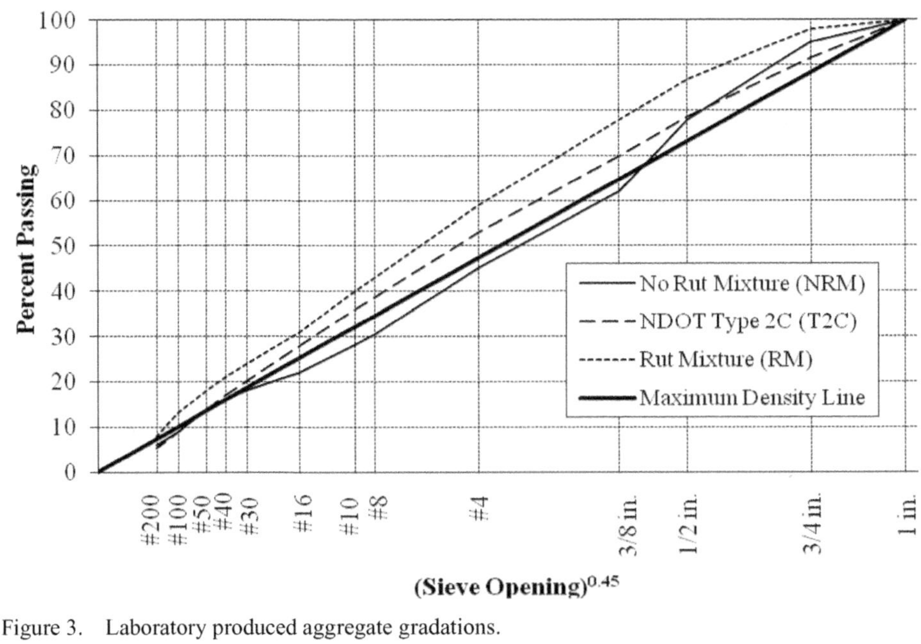

Figure 3. Laboratory produced aggregate gradations.

Table 6. Rutting resistance of the laboratory produced mixtures.

Tests	Property	No rut mixture, NRM	NDOT type 2C mixture, T2C	Rut mixture, RM
RSCH* at 158°F (5,000 Cycles)	Number of Replicates	3	3	3
	Air voids (%)	6.9	6.4	7.7
	Permanent shear strain (%)	1.3	1.8	4.0
	Criteria for Intersections (%)	2.0	2.0	2.0
	Expected Performance at Intersections	Excellent	Good	Poor
RLT* at 158°F (12,000 Cycles)	Number of Replicates	3	3	3
	Air voids (%)	6.4	6.5	7.7
	Permanent axial strain (%)	0.8	2.0	3.0
	Criteria for Intersections (%)	2.0	2.0	2.0
	Expected Performance at Intersections	Excellent	Good	Poor

* RSCH denotes "Repeated Shear at Constant Height".
RLT denotes "Repeated Load Triaxial".

period of 0.6 seconds. The permanent strain developed in the specimen is measured as a function of the number of load cycles. The test is run for a total of 12,000 cycles, and the magnitude of the permanent shear strain at the end of the test period provides an indication of the mixture's resistance to rutting.

The RLT test measures the axial permanent deformation in the HMA mixture as it is subjected to triaxial stress conditions. The test specimen preparation was similar to the preparation of samples for the triaxial compression test. The triaxial condition was achieved by applying a static radial confining pressure of 30 psi using compressed air and a repeated deviator vertical stress of 45 psi. The repeated deviator stress was applied for 0.1 seconds followed by a 0.6 seconds of rest period. The test was conducted on three replicates from each mixture at a temperature of 158°F for a total of 12,000 cycles with continuous measurements of the vertical deformation along the middle 4 inches of the sample. This enabled the evaluation of the permanent vertical strain in the sample as a function of load cycles.

Table 6 summarizes the performance of the three mixtures under all the two performance tests. The laboratory performance test results yielded a ranking of mixtures from best to worst as follows: NRM, T2C, and RM. The last step was to apply the rutting criteria that were revised for intersection loading under section 2.3 of this paper. Table 6 compares the performance of the three mixtures with the rutting criteria for intersections. The data in Table 6 shows that both the NRM and T2C mixtures are expected to perform well at urban intersections while the RM mixture is expected to experience rutting at urban intersections.

3 SUMMARY AND CONCLUSIONS

The objectives of this research effort were to define a system to evaluate the potential performance of HMA mixtures at urban intersections and to assess the performance of some HMA mixtures using the developed system. To accomplish these objectives the following work elements were carried out.

- Examine the performance of HMA mixtures at urban intersections: this effort led to the identification of poor and good performing mixtures at urban intersections.
- Identify the loading conditions at urban intersections: this effort evaluated the load distributions on the tires of an 18-wheeler truck as it approaches a urban intersection.

- Revise the permanent deformation criteria for urban intersections: this effort revised the rutting criteria used for highway pavements to be applicable at urban intersections.
- Evaluate the performance of multiple HMA mixtures at urban intersections: this effort evaluated the performance of three HMA mixtures against the revised rutting criteria.

The analyses of the laboratory data generated in this research effort led to the conclusion that the HMA mixtures that have performed well at intersections in southern Nevada (i.e. the no rut mixtures NRM) compare very well with the revised rutting criteria for the RSCH and RLT tests. On the other hand, the HMA mixtures that have performed poorly at intersections in southern Nevada (i.e. the rut mixtures RM) compare very poorly with the revised rutting criteria for the RSCH and RLT tests. This confirms that the revised rutting criteria for the RSCH and RLT are effective in determining the potential performance of HMA mixtures at urban intersections.

Finally, applying the revised rutting criteria for the RSCH and the RLT tests that have been validated on the NRM and RM mixtures to the Nevada DOT typical HMA mixture (T2C) with the PG76-22 polymer modified binder indicated that this mix will experience good performance at urban intersections. It should be noted that when the Nevada DOT typical mixture (T2C) was used with the AC-30 asphalt binder did not perform well at urban intersections. Therefore, it is anticipated, based on this research, that when the Nevada DOT T2C gradation is coupled with the PG76-22 polymer modified binder, it will perform well at urban intersections.

REFERENCES

AASHTO Standard Specifications for Transportation Materials and Methods of Sampling and Testing 2004. *Standard Method of Test for Determining the Permanent Shear and Stiffness of Asphalt Mixtures using the Superpave Shear Tester (SST)*. American Association of State Highway and Transportation Officials, T320, 24th edition.

Asphalt Institute. High performance hot mix asphalt intersection strategy. *Executive Offices and Research Center*, Lexington, Kentucky.

Gillespie, T.D. & Balderas, L. 1987. An analytical comparison of the dynamic performance of a european heavy vehicle and a generic U.S. heavy vehicle. The University of Michigan Transportation Research Institute, *Report No. UMTRI-87-17.*

Gillespie, T.D. 1992. *Fundamentals of vehicle dynamics*. ISBN 1-56091-199-9.

Hajj, E.Y. 2005. Hot Mix Asphalt Mixtures for Nevada's Intersections. PH.D. Dissertation, University of Nevada Reno, USA.

Kandhal, P.S., Mallick, R.B. & Brown, E.R. 1998. Hot mix asphalt for intersections in hot climates. National Center for Asphalt Technology, *NATC Report No. 98–6,* Auburn University, Alabama, USA.

Roberts, F.L., Kandhal, P.S., Brown, E.R., Lee, D.Y. & Kennedy T.W. 1996. Hot Mix Asphalt Materials, *Mixture Design, and Construction, Second Edition.* NAPA Education Foundation, Lanham, Maryland.

Siddharthan, R.V., Yao, J. & Sebaaly P.E. 1998. Pavement Strain from Moving Dynamic 3D Load Distribution. *American Society of Civil Engineers: Journal of Transportation Engineering, 124(6),* ASCE, pp. 557–566.

Siddharthan, R.V., Krishnamenon, N., El-Mously, M. & Sebaaly P.E. 2002. Validation of a Pavement Response Model Using Full-Scale Field Tests. *The International Journal of Pavement Engineering,* Vol. 3 (2), pp. 85–93.

Efficient Transportation and Pavement Systems – Al-Qadi, Sayed, Alnuaimi & Masad (eds)
© 2009 Taylor & Francis Group, London, ISBN 978-0-415-48979-9

Impact of increasing legal axle loads on truck factor in Egypt

Omar Osman, Mostafa El Ghazolly & Ragab M. Mousa
Department of Engineering, Cairo University, Cairo, Egypt

ABSTRACT: In 2005, the Ministry of Transport in Egypt issued new regulations for increasing legal truck axle loads. This paper presents an assessment of tuck factors (TF) under old and new regulations using actual truck loads and traffic data collected for three main roads in Egypt. Although axle load limits have increased by 17–36%, the TF increased by about 200%. The TF was used to determine the equivalent single axle load (ESAL) needed for pavement design and maintenance works. ESALs were found to increase by 75–136%. This impact was converted into additional thickness of asphalt layers, which ranged from 2.1–4.6 cm depending on restrictions on overloading and scenarios of freight volumes.

1 INTRODUCTION

The force transmitted to a road by a vehicle depends on its total weight and how that weight is spread via the axles to the road surface. It is the weight on each axle, rather than the total weight of the vehicle, that determines the amount of road damage. Road damage is approximately proportional to the weight of an axle raised to its 4th power (WSDOT 2006). Thus, 11-ton axle causes about 45% more damage than a 10- ton axle. Road damage is reduced by spreading loads evenly and avoiding very heavy single axles. Consequently, national regulations on the weights and dimensions of vehicles specify not only maximum gross weights but also maximum axle weights and spacing between axles.

Increasing the legal axle load is a challenging decision because it increases the truck factor (TF) of different truck types. The truck equivalence factor (TF) for any truck is the sum of load equivalency factors for the front and rear axle loads. A research project sponsored jointly by the Egyptian Ministry of Transportation and National Institute of Transportation was conducted in 1990 to study the effect of increasing axle load on the Egyptian road network. The study assessed six loading scenarios (10 to 15 ton axle load limits), three enforcement levels (poor, medium, and full), and two maintenance budget levels (limited and unlimited). The study covered all roads supervised by the General Authority for Roads, Bridges and Land Transport (GARBLAT). The main conclusion of the study was to implement 12 ton maximum axle load limit with medium enforcement level and increase maintenance budget by about 25% (MOT 1990).

In 2005, the Ministry of Transport in Egypt issued a new regulation for increasing the legal axle load limits from 10, 16, and 22 tons to 13, 20, and 30 tons for single, tandem, and triandem axles, respectively. The main objective of this research is to study the impact of such increase on the truck factors (TF) for different truck types operating on the Egyptian road network. The study considered different scenarios for the level of enforcement and the volume freight transported by the truck fleet.

Determining the TF for each truck type is essential for calculating the equivalent single axle load (ESAL), which is a basic input in different design procedures such as designing the thickness of new pavement layers or asphalt overlay on existing pavement. It is also an essential input in different strategic studies such as life cycle analysis considering road and agency user costs.

In order to achieve the research objectives, the first step is to select and group the types of vehicles that represent the running fleet on the selected roads. The second step is to analyze the available traffic and truck data and determine TF values under different scenarios of enforcement levels and freight volume.

2 APPLIED RESEARCH METHODOLOGY

The applied methodology in executing this research can be outlined as follows:

1. Data collection and Analysis: Data collection includes obtaining available data on vehicle types, traffic composition and volume, and axle loads and configurations (before and after new axle load limit regulations) for the three roads to be studied.
2. Identification of the representative vehicle groups: This step is to categorize the types of vehicles found in the running fleet into representative groups or types. This was done by comparing the types of vehicles used in GARBLT traffic data reports and those used in previous studies in Egypt (JICA 1993 & JICA 2002).
3. Truck Axle Load Analysis: Carrying out analysis for individual truck groups under old and new regulations to determine the truck factor under different loading conditions (empty, loaded, overloaded, etc.).
4. Identification of analysis scenarios: This step is to identify possible alternatives for the change in total freight weight transported by trucks under the new axle load limit and "cases" of law enforcement levels to be studied in this research.
5. Determination of TF and ESAL: This step is to determine average TF of each truck type and determine the ESAL under different analysis scenarios.
6. Comparison of TF and ESAL: This step is to establish a comparison of TF and ESAL under different analysis scenarios identified for this research.

3 SELECTED ROADS AND TRAFFIC DATA

Truck and traffic data were collected for three main rural roads of the national road network in Egypt, namely, Cairo–Ismailia, Cairo–Suez, and Cairo–Dommiata Roads. The selected roads were chosen so that they represent the GARBLT road network covering all road types (agricultural/desert roads and medium/high traffic volumes). The collected data for this study can be summarized as follows:

- Traffic data including volumes, composition, directional distribution, and growth rates.
- Truck data including truck type and axle configuration.
- Axle weights for randomly selected trucks.

4 ANALYSIS OF VEHICLE FLEET DATA

4.1 *Traffic volumes and composition*

Table 1 gives a summary of the collected traffic volumes (AADT) for the selected three roads. The obtained traffic data cover several years between 1993 and 2004. Two main items were extracted from the collected traffic data, namely, traffic composition and annual traffic growth for the selected roads. All available traffic count reports obtained from GARBLT indicated fixed traffic composition over the study period. The composition gives percentages of different types of vehicles represented in the vehicle fleet. In particular, the combined percentage of all truck types on the three roads was found to be 6%, 22.5%, and 20.7%,

Table 1. Total daily traffic volumes on the selected roads.

Year	Cairo-Ismailia road (vpd)	Cairo-Suez road (vpd)	Cairo-Dommiata road (vpd)
1993	8,716	4,696	10,083
1996	13,857	6,325	13,045
1999	28,791	10,044	17,420
2000	20,290	10,961	17,827
2001	35,027	12,170	18,848
2002	39,146	13,483	18,752
2003	41,451	14,409	18,973
2004	43,582	15,342	19,335

respectively. Average Traffic growth rates estimated for the same roads were 8.7%, 8.8%, and 2.1%, respectively.

4.2 *Identification of representative truck vehicles*

Reviewing the vehicle classification used in GARBLT traffic counts and other relevant reports revealed that grouping was based only on the number of axles regardless of the type of the vehicle itself. Such grouping criteria resulted in vehicles with different operational characteristics falling under the same class. For example, all semi-trailer trucks with 5 axles were grouped in the same class with truck-trailer trucks having the same number of axles. Since the vehicle operation cost (VOC) is a function of the vehicle type and its operational characteristics and performance, which significantly vary from one vehicle to another, it was necessary to subdivide the same truck group to differentiate between different vehicle types having the same number of axels. The results of this process are shown in Table 2. As shown in this table, GARBLT classified trucks into 9 groups based on the number of axles in each group. Three of these groups were subdivide into 6 sub-groups when the vehicle type was considered to make the total number of truck groups equal to 12. A code of two letters and two digits is given to these groups identifying the vehicle type and total weight under new law. Figure 1 shows an example of subgrouping 5-axles trucks.

4.3 *Axle loads under old and new regulations*

Six sets of sample axle load data were collected from weighing stations located on the selected three roads. Three of the collected data sets (one for each road) represent truck loading under the old legal axle load limits. The remaining three data sets represent truck loading under the new regulation. Table 3 summarizes the results of this axle load survey. Since only axle loads for overloaded trucks are recorded at weighing stations in Egypt, the obtained data represent only overloading conditions. Each of the collected data sets consists of actual measured axle load for randomly selected 100 axles of each type (front gear, dual, dual tandem, and triandem). Average axle loads for overloaded trucks under old and new regulations are shown in Table 4 for each axle type. As shown, average axle loads show less overloading (violation) under new law compared with old law where overloading in triandem axles exceeded 50% of the legal axle load. Results also show that overloaded truck axles vary between 5% and 32% under new regulations.

Based on these results, it could be reasonably assumed that overloaded trucks carry about 25% more than their legal load. Additional information regarding legally loaded trucks was collected via interviews with the officers in charge at the weighing stations. Obtained information was used in determining the percentages of empty, half loaded, and fully legally loaded trucks. Based on the outcome of these interviews, the percentages of empty, half loaded, fully loaded, and overloaded trucks can reasonably be assumed as 30%, 15%, 30%, and 25%. According to the interviewees, these shares did not change much after implementing the new axle load regulation.

Table 2. Classification of trucks representing the vehicle fleet.

Truck groups as per GARBLT classification	Adopted truck groups with subdivision	Total weight (tons)		Truck code
		Under old law	Under new law	
2 axles–6 tires	Light truck	16	20	LT20
3 axles (singles)	Medium truck	26	33	MT33
4 axles (singles)	Medium truck	36	46	MT46
<5 axles (single/double)	Medium truck	22	27	MT27
	Semi-trailer	32	40	ST40
5 axles (single/double)	Truck trailer	42	53	TT53
	Semi-trailer	38	47	ST47
>5 axles (single/double)	Truck trailer	48	60	TT60
	Semi-trailer	52	66	ST66
<6 axles (single/triple)	Semi-trailer	38	50	ST50
6 axles (single/double/triple)	Semi-trailer	44	57	ST57
>6 axles (single/triple)	Semi-trailer	50	67	ST67

Figure 1. Subdividing the 5-axle trucks into two subgroups.

Table 3. Average axle loads for overloaded trucks under old and new laws.

Axle type	Old axle load limit (ton)		Old axle load limit (ton)	
	Mean	St. dev	Mean	St. dev
Front axles	7.1	0.44	7.4	0.79
Dual axles	14.5	2.30	16.7	3.20
Dual tandem axles	21.3	4.92	26.4	6.34
Triandem axles	33.9	4.51	34.3	6.74

Table 4. Average axle loads for overloaded trucks under old and new laws.

Axle type	Legal axle load limit (ton)		Average actual axle load (ton)		Ratio of actual and legal axle load (%)	
	Old	New	Old	New	Old	New
Front axles	6	7	7.1	7.4	1.19	1.05
Dual axles	10	13	14.5	16.7	1.45	1.28
Dual tandem axles	16	20	21.3	26.4	1.33	1.32
Triandem axles	22	30	33.9	34.3	1.54	1.14

5 ANALYSIS SCENARIOS FOR TRUCK AXLE LOADS

In order to assess the expected effect of the increased axle load limits on truck factors (TF) and consequently on the total ESAls carried on each road, four different scenarios were considered in analysis based on possible axle load enforcement levels and expected change in total freight weight transported by trucks. These four analysis cases (or scenarios) are described below.

Case 1: old axle load limits with "poor enforcement level" (base scenario).
Case 2: new axle load limits with "poor enforcement level".
Case 3: new axle load limits with "medium enforcement level".
Case 4: new axle load limits with "high enforcement level".

In addition, two alternatives were considered for the expected change in freight weight under the new axle load limits as follows:

- Alternative 1: Assuming that the truck traffic volume does not change after adopting the new axle load limit. This means that the total freight weight transported by the truck fleet has increased after increasing the legal axle load. This alternative is referred to as "Same Traffic Volume" or "STV".
- Alternative 2: Assuming that freight weight does not change after adopting the new axle load limit. Accordingly, traffic volume and composition is adjusted to maintain the same fright weight (i.e., less truck traffic volume). This alternative is referred to as "Same Freight Weight" or "SFW".

Based on a symbolic questionnaire conducted with a sample of truck drivers after implementing the new axle load regulation, it could be concluded that increasing the legal axle load limit did not actually result in a reduction in truck trips. Accordingly, it could be fairly assumed that the first alternative "STV" is more realistic in the case of Egypt. This is particularly true since the capacity of the truck fleet in Egypt is known to be limited in comparison with market demand. However, the analysis is performed on both alternatives.

6 DETERMINATION OF TRUCK FACTORS

Based on the analysis in Section 4.3, the percentages of empty, half loaded, fully loaded, and overloaded trucks were estimated at 30%, 15%, 30%, and 25% of the total truck volumes, respectively. These percentages did not change much after implementing the new axle load limit. Based on the study of overloaded trucks under old and new regulations, the 25% overloaded trucks can be further divided into 15% "12.5% overloaded trucks" and 10% "25% overloaded trucks". These percentages are used to determine the TF values for the analysis scenarios under Case 1 (old axle load limits with "poor enforcement level") and Case 2 (new axle load limits with "poor enforcement level"). In Case 3 (new axle load limits with "medium enforcement level") and Case 4 (new axle load limits with "high enforcement level"), the percentages of overloaded trucks are reduced to reflect the stringent enforcement levels assumed in these cases. The truck loading distributions for each of the four cases are shown Table 5.

Table 5. Truck loading distribution for different loading condition.

Analysis Scenario	Empty	½ Loaded	Fully loaded	12.5% overloaded	25% overloaded
Case (1)	30.0%	15.0%	30.0%	15.0%	10.0%
Case (2)	30.0%	15.0%	30.0%	15.0%	10.0%
Case (3)	30.0%	15.0%	42.5%	7.5%	5.0%
Case (4)	30.0%	15.0%	55.0%	0.0%	0.0%

Table 6 gives truck factors (TF) for different truck types and loading conditions. These values are also shown in Figure 2. Table 7 gives an example for determining the TF for TT60 truck type.

Similarly, the truck factors (TF) obtained for new axle limits are presented in Table 8 and depicted in Figure 3. Comparing TF values of old and new axle loads indicates that TF under new loads is about 2 to 3 times the TF under old loads, especially for fully loaded and overloaded conditions.

The average truck factor for each truck type was then determined for different analysis cases (scenarios) using the percentages in Table 5 and truck factors in Tables 6 and 8. This average truck factor is shown in Table 9. Table 10 presents an example of how the average truck factor (TF) is determined for the TT60 truck type under old axle loads (Case 1). This factor is shown as 3.83 in Tables 9 and 10.

Table 6. Truck factors for different loading conditions under old axle limits.

Truck code	Empty	1/2 Loaded	Fully loaded	12.5% overloaded	25% overloaded
LT20	0.28	0.41	2.44	3.38	4.82
MT33	0.29	0.54	4.61	6.48	9.11
MT46	0.29	0.67	6.77	9.58	13.41
MT27	0.28	0.35	1.53	2.26	3.50
ST40	0.28	0.48	3.69	5.36	7.80
TT53	0.29	0.61	5.85	8.47	12.09
ST47	0.28	0.42	2.77	4.25	6.48
TT60	0.29	0.55	4.94	7.35	10.77
ST66	0.29	0.74	8.02	11.57	16.38
ST50	0.28	0.47	3.50	4.88	7.21
ST57	0.28	0.41	2.58	3.77	5.89
ST67	0.28	0.40	2.40	3.28	5.31

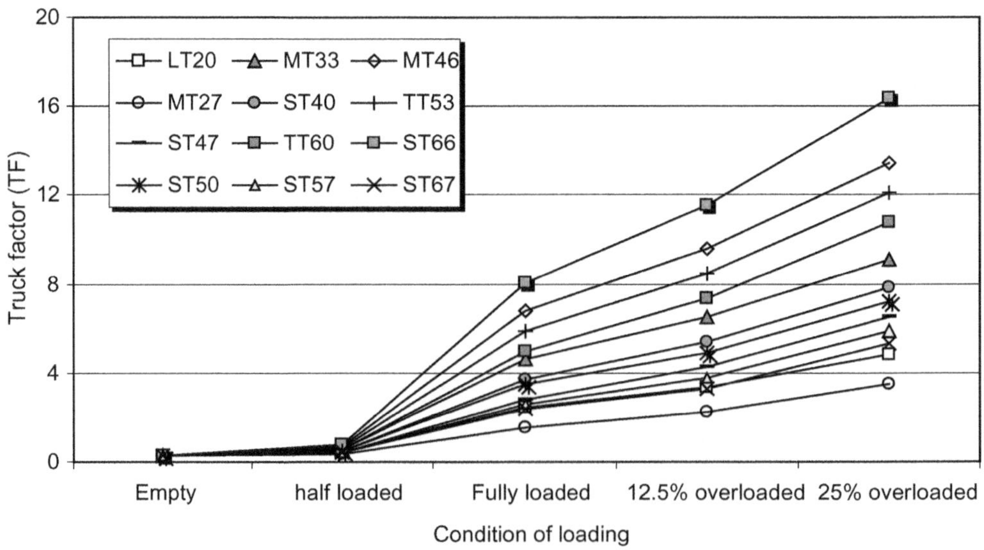

Figure 2. TF for different truck types and loading conditions under old axle load limits.

Table 7. Example of determining TF for TT60 truck under different loading conditions of old axle load limits.

Loading condition	Equivalency factor				Truck factor
	Front axle	Axle 2	Axle 3	Axle 4	
Empty	0.279	0.001	0.003	0.001	0.285
1/2 Loaded	0.279	0.070	0.130	0.070	0.549
Fully loaded	0.279	1.247	2.164	1.247	4.936
12.5% Overloaded	0.279	1.984	3.102	1.984	7.348
25.0% Overloaded	0.528	2.976	4.293	2.976	10.774

Table 8. Truck factors for different loading conditions under new axle limits.

Truck code	Empty	1/2 Loaded	Fully loaded	12.5% overloaded	25% overloaded
LT20	0.28	0.56	6.31	8.16	10.80
MT33	0.29	0.84	12.10	15.79	20.70
MT46	0.29	1.12	17.88	23.42	30.59
MT27	0.28	0.46	3.50	4.79	6.81
ST40	0.28	0.74	9.29	12.42	16.71
TT53	0.29	1.02	15.07	20.05	26.60
ST47	0.28	0.64	6.48	9.06	12.72
TT60	0.29	0.92	12.26	16.69	22.61
ST66	0.29	1.30	20.86	27.68	36.50
ST50	0.28	0.77	9.90	12.74	17.21
ST57	0.28	0.67	7.09	9.37	13.22
ST67	0.28	0.71	7.70	9.69	13.72

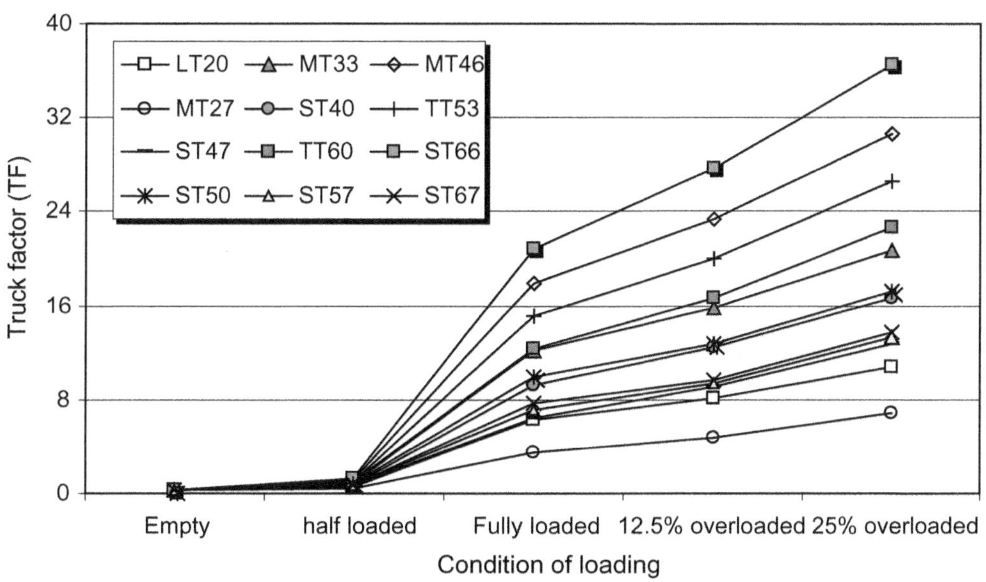

Figure 3. TF for different truck types and loading conditions under new axle load limits.

Table 9. Average truck factors under different loading cases.

Truck type (Code)	Average truck factor			
	Case (1)	Case (2)	Case (3)	Case (4)
LT20	1.87	4.37	4.00	3.64
MT33	3.43	8.28	7.57	6.86
MT46	5.00	12.19	11.14	10.09
MT27	1.28	2.60	2.34	2.08
ST40	2.85	6.52	5.91	5.30
TT53	4.41	10.43	9.48	8.53
ST47	2.26	4.76	4.25	3.74
TT60	3.83	8.67	7.82	6.97
ST66	5.98	14.34	13.05	11.75
ST50	2.66	6.80	6.22	5.64
ST57	2.07	5.04	4.56	4.09
ST67	1.89	5.33	4.88	4.43

Table 10. Example of Determining the Overall TF for TT60 Truck under Old Axle Loads.

Truck loading condition	TF for each loading condition	Weight of loading condition in truck volume	Weight of loading condition in average TF
Empty	0.285	30%	0.086
1/2 Loaded	0.549	15%	0.082
Fully loaded	4.936	30%	1.481
12.5% Overloaded	7.348	15%	1.102
25% Overloaded	10.773	10%	1.077
Average TF for TT60 (weighted average of loading conditions)			3.828

7 ESTIMATED ESALS FOR THE ANALYSIS PERIOD

The truck factors (TF) calculated for different loading conditions were utilized to estimate the total ESALs in year 2005 (base year) for STV alternative (Alternative 1) as shown in Table 11. The table also shows the cumulative ESAL for analysis periods of 15 and 25 years. Similar estimates of the ESAL for SFW alternative (Alternative 2) are given in Table 12.

Values in Table 11 indicate substantial increase in cumulative ESALs for all roads as a result of increasing the axle load limit. In particular, for Alternative 1 and Case 2 (no enforcement), increasing legal axle load limit is expected to increase the ESAL by about 136% that of the base scenario (Case 1-old axle load limit with poor enforcement). This increase becomes 94% when implementing a stringent enforcement level (Case 3). With intermediate enforcement level (Case 4), this factor becomes 115%. When Alternative 2 (SFW) is considered (Table 12), the corresponding increase in the ESAL, compared with the base case (Case 1), becomes 76%, 55%, and 66% for Cases 2, 3, and 4, respectively. The less impact associated with Alternative 2 is expected since the number of truck vehicles considered with this alternative is less than that considered with Alternative 1.

8 IMPACT ON ADDITIONAL ASPHALT THICKNESS

The ESAL values obtained earlier from truck factors were utilized to assess the impact of increasing the axle load and enforcement level in terms of the additional required asphalt thickness. For this purpose, the AASHTO procedure (AASHTO 1993) was utilized to determine the additional pavement structural number for Cases 2, 3, and 4 compared with Case 1, assuming a 15-year design period and initial effective structural number of 3.0. The additional structural number values were expressed in terms of additional asphalt thickness using reasonable asphalt layer coefficient of 0.145 per cm. Table 13 shows the additional asphalt overlay thicknesses for 15-year analysis period. As shown, the additional asphalt overlay needed for 15-year analysis period ranges between 3.3 cm and 4.6 cm in case of Alternative 1. This thickness ranges between 2.1 cm and 3.1 cm in case of Alternative 2.

Table 11. Total ESAL for STV alternative (Alt. 1) under different analysis cases.

		ESAL (millions per lane)			
Target analysis year (period)	Road	Case (1)	Case (2)	Case (3)	Case (4)
2005 (base year)	Cairo/Ismailia	0.55	1.29	1.07	1.18
	Cairo/Suez	0.86	2.01	1.65	1.83
	Cairo/Dommiata	1.13	2.70	2.22	2.46
2020 (15 years)	Cairo/Ismailia	15.74	36.98	30.56	33.77
	Cairo/Suez	24.43	57.37	47.13	52.25
	Cairo/Dommiata	32.35	77.01	63.52	70.27
2030 (25 years)	Cairo/Ismailia	44.31	104.12	86.04	95.08
	Cairo/Suez	68.79	161.51	132.69	147.10
	Cairo/Dommiata	91.09	216.83	178.85	197.84

Table 12. Total ESAL for SFW alternative (Alt. 2) under different analysis cases.

		ESAL (millions per lane)			
Target analysis year (period)	Road	Case (1)	Case (2)	Case (3)	Case (4)
2005 (base year)	Cairo/Ismailia	0.55	0.95	0.84	0.90
	Cairo/Suez	0.86	1.51	1.33	1.42
	Cairo/Dommiata	1.13	2.03	1.79	1.91
2020 (15 years)	Cairo/Ismailia	15.74	27.00	24.11	25.61
	Cairo/Suez	24.43	43.15	37.98	40.65
	Cairo/Dommiata	32.35	57.91	51.13	54.63
2030 (25 years)	Cairo/Ismailia	44.31	76.03	67.87	72.10
	Cairo/Suez	68.79	121.48	106.94	114.46
	Cairo/Dommiata	91.09	163.03	143.96	153.82

Table 13. Additional asphalt thickness (cm) due to increasing axle load limits.

	Alternative 1 (STV)			Alternative 2 (SFW)		
Road	Case (2)	Case (3)	Case (4)	Case (2)	Case (3)	Case (4)
Cairo/Ismailia	4.3	3.3	3.8	2.7	2.1	2.4
Cairo/Suez	4.4	3.4	3.9	2.9	2.2	2.6
Cairo/Dommiata	4.6	3.6	4.1	3.1	2.4	2.7

9 CONCLUSIONS AND RECOMMENDATION

This research presented an assessment of the truck factors on Egyptian roads under two axle load limits, three levels of enforcement, and two scenarios of transported freight weight. The analysis is based on actual truck axle loads collected at the weigh stations on three main roads of the national road network in Egypt. Although the increase in axle load limits ranged from 17% to 36%, the truck factor (TF) has increased to 2–3 times the TF under old axle load limits. The Equivalent single axle loads (ESAL) have also increased by about he same magnitude, especially under Alternative 1 where truck volume was maintained the same. Under Alternative 2 where truck vehicles were reduced to maintain same freight weight, the ESAL increased by about 75% only. These impacts were expressed in terms of additional asphalt layer thicknesses needed for 15-year analysis period, which ranges between 3.3 cm and 4.6 cm in case of Alternative 1 and between 2.1 cm and 3.1 cm in case of Alternative 2. It should be pointed out that the more thicknesses needed for Alt. 1 correspond to the higher freight weight transported under that alternative. The issue is now whether the benefits from transporting the additional freight weight can offset the maintenance cost of the additional asphalt thickness. This issue was investigated in a research conducted by the same authors (Mousa et. al 2008) and proved economic feasibility.

ACKNOWLEDGEMENTS

The authors would like to thank the Maintenance and Information Departments of GARBLT for their cooperation and providing the data utilized in this research. The authors would also like to thank Professor Essam Sharaf, the former Minister of Transport, and Professor of Highways and Traffic Engineering (Cairo University) for his valuable discussions concerning the ministry policy for the enfacement of old and new axle loads on the national road network in Egypt.

REFERENCES

AASHTO (1993). AASHTO Guide for Design of Pavement Structures", American Association of State Highway and Transportation Officials, Washington D.C.

El-Ghazolly, M.A. (2006). Assessment of the Impact of Increasing the Legal Truck Axle Loads on the Egyptian Roads, M.Sc. Thesis, Cairo University, Egypt.

Japan International Cooperation Agency, JICA, (1993). "A Study on the Transportation System and the National Road Transportation." Master Plan Main Report, Volumes I, II, III", Ministry of Transport, Cairo, Egypt.

Japan International Cooperation Agency, JICA, (2002), "Transport Master Plan." Phase I, Final Report, Volume III, Ministry of Transport, Cairo, Egypt.

Ministry of Transportation, (1990). "Effect of axle load increasing on the Egyptian network", Ministry of Transportation, National Institute of Transportation, Cairo, Egypt, 1990.

Mousa, R., El Ghazolly, M., and Osman, O (2008). "The Impact of Increasing Legal Truck Axle Loads on the Egyptian Roads" Proceedings of the 10th AATT 2008 Conference, Athens, May 27–31, 2008.

Washington State Department of Transportation, WSDOT, (2006). "Legal Load Limits and Overweight Loads on Pavements and Bridges." Website:www.wsdot.wa.gov.

Efficient Transportation and Pavement Systems – Al-Qadi, Sayed, Alnuaimi & Masad (eds)
© 2009 Taylor & Francis Group, London, ISBN 978-0-415-48979-9

Use of ANNs in the mechanistic-based overlay design of flexible pavements

E.A. Sharaf
Cairo University, Cairo, Egypt

M.A. Abo-Hashema
Fayoum University, Cairo, Egypt

E. Abdelbari
Arab Academy for Science and Technology, Egypt

ABSTRACT: This study presents an endeavor to apply Artificial Neural Networks (ANN) to recommend pavement overlay thickness, based on learning from Mechanistic-Empirical (M-E) overlay design cases. The design process of the M-E procedures includes three stages: Stage 1 is the development of input values for the analysis, which include material characterization, traffic input data, and environmental effects (temperature and moisture). Stage 2 of the design process is the structural/performance analysis. Stage 3 of the process includes those activities required to evaluate the structurally viable alternatives. The computational demands of the M-E design process associated with the differential equations and finite element matrix solutions employed by the various analysis models are very complicated and need high-speed computers. The ANN approach can be used to overcome this difficulty. Results indicate that the ANN approach is suitable for determining flexible overlay thickness with high accuracy.

1 INTRODUCTION

Repeated application of traffic loads, weather conditions and other factors related to materials and construction decrease the serviceability of the flexible pavements. In other words, the comfort decreases while the user costs and the operation cost increase. For this reason, a maintenance program should be set up to decide what, when and where to carry out maintenance activities. It is important that the maintenance activities be done at the right time and right places (Saltan et al. 2002). Designing an overlay thickness for flexible pavements is considered one of the most crucial maintenance activities. It necessitates a vast number of conditions that must be observed and planned for. The design process becomes convoluted when considering the environmental effects on material properties. Many design procedures have been developed, particularly the effort done by the American Association of State Highway and Transportation Officials (AASHTO), in the direction for the AASHTO 2002 design guide through NCHRP Project 1-37 A. The objective of the AASHTO 2002 is to develop and deliver the 2002 Guide for design of new and rehabilitated pavement structures based on Mechanistic-Empirical (M-E) principals as well as development of the necessary computational software (AASHTO 1993, NCHRP 2004).

The M-E design procedures were not practical until the advent of high-speed computers because of the computational demands associated with the differential equations and finite element matrix solutions employed by the various analysis models. The design process of the M-E procedures includes three stages: Stage 1 is the development of input values for the analysis, which include material characterization, traffic input data, and environmental effects (temperature for asphalt layers and moisture for unbound layers). Stage 2 of the design process is the structural/performance analysis. Stage 3 includes those activities required to evaluate the structurally viable alternatives. These activities include an engineering analysis and Life-Cycle Cost (LCC) analysis of the alternatives.

Use of Artificial Neural Networks (ANNs) in pavement systems has significantly increased in the past five years. Moreover, their successful applications in other fields of decision-making sciences and in computer and electrical engineering is expected to lead to further-increased interest and confidence in their applications in all fields of civil engineering. Despite the fact that ANNs have already proved to outperform traditional modeling counterparts in solving various complex engineering problems, their practical use in transportation or pavements particularly is still limited. The primary obstacles to advantageous implementation of ANNs in transportation systems are lack of understanding and current skepticism. Most of the reported ANN-based studies, even though successful, have not been implemented in practice since practicing engineers are still doubtful of their use. These obstacles can be overcome if the practicing engineers are provided with sources of necessary background information and involved in specifically-oriented ANN workshops and tutorials (A2K05 1999).

This paper can be considered as a step toward achieving the goal of overcoming these obstacles. The objective of this paper is to reveal the results of implementing ANNs to recommend appropriate pavement overlay thickness based on learning from M-E overlay design cases.

2 OVERVIEW OF M-E OVERLAY DESIGN PROCEDURE

2.1 *Concept*

In general, a mechanistic design procedure requires the calculation of critical stress, strain, or deflection in the pavement by some mechanistic method and prediction of resulting damages by some empirical failure criteria. The pavement layer thicknesses are calculated so that the damages in either the existing pavement or the new overlay will be within the allowable limits. This procedure considers many design factors to calculate the required thickness. These factors include the axle loading, environmental factors, and pavement material properties. It takes into consideration the effect of seasonal variation on the change of pavement layer properties and accounts for non-linearity of material properties. The main features of the method include: determination of design inputs, modeling pavement response, and damage analysis. Due to the complexity of such a design system, its procedure should be implemented in a computer program code to facilitate stress and strain calculations, which are essential in the calculation of the pavement service life (Bayomy & Abo-Hashema 2001, Bayomy et al. 2002, Bayomy et al. 1996, Fahmy 2004). This paper is not intended to describe in details the design process for flexible pavements, while it presents a brief description of the design process. Figure 1 shows the overall design process for flexible pavements (NCHRP 2004).

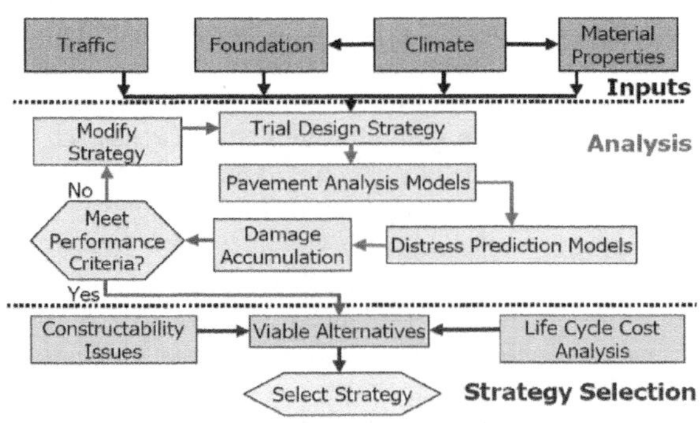

Figure 1. M-E overall design process for flexible pavements (NCHRP 2004).

Generally, the main steps in the design process include the following (NCHRP 2004):

1. Assemble a trial design for specific site conditions—define subgrade support, asphalt concrete and other paving material properties, traffic loads, climate, pavement type and design and construction features.
2. Establish criteria for acceptable pavement performance at the end of deign period.
3. Select the desired level of reliability for each of the applicable performance indicators.
4. Process input to obtain monthly values of traffic inputs and seasonal variations of material and climatic inputs needed in the design evaluations for the entire design period.
5. Compute structural responses (stress and strains) using multilayer elastic theory or finite element based pavement response models for each axle type and load and for each damage-calculation increment throughout the design period.
6. Calculate accumulated distress and/or damage at the end of each analysis period for the entire design period.
7. Predict key distresses (Rutting and Fatigue) at the end of each analysis period throughout the design life using available performance models.
8. Evaluate the expected performance of the trial design at the given reliability level.
9. If the trial design does not meet the performance criteria, modify the design and repeat the steps 4 through 9 until the design does meet the criteria.

2.2 *Design inputs*

A layered elastic model requires a minimum number of inputs to adequately characterize a pavement structure and its response to loading. These inputs are as follow (NCHRP 2004, WSDOT 2003, Huang 1993, Bayomy et al. 1993, Bayomy & Abo-Hashema 2000):

1. Material properties of each layer
   a. Modulus of elasticity
   b. Poisson's ratio
   c. Pavement layer thicknesses.
2. Traffic and Loading Conditions.
3. Environmental Effects.

2.3 *Pavement response models*

The purpose of the flexible pavement response model is to determine the structural response of the pavement system due to traffic loads and environmental influences. Environmental influences may be direct (e.g., strains due to thermal expansion and/or contraction) or indirect via effects on material properties (e.g., changes in stiffness due to temperature and/or moisture effects) (NCHRP 2004, Hall et al. 1992).

The outputs from the pavement response model are the stresses, strains, and displacements within the pavement layers. Of particular interests are the critical response variables required as inputs to the pavement distress models in the M-E design procedure. Example of critical pavement response variables include: tensile horizontal strain at the bottom of the asphalt layers, compressive vertical stresses/strains within the asphalt layer (for asphalt rutting), compressive vertical stresses/strains within the base/subbase layers (for rutting of unbound layers), and compressive vertical sresses/strains at the top of the subgrade (for subgrade rutting).

Each pavement response variable must be evaluated at the critical location within the pavement layer where the parameter is at its most extreme value. For a single wheel loading, the critical location can usually be determined by inspection. For example, the critical location for the tensile horizontal strain at the bottom of the asphalt layer under a single wheel load is directly beneath the center of the wheel. For multiple wheel and/or axles, the critical location will be a function of the wheel load configuration and the pavement structure. Mixed traffic conditions further complicates the problem, as the critical location within the pavement structure will not generally be the same over all vehicle types. The pavement response model must search for the critical location for each response parameter in these cases (NCHRP 2004).

2.4 *Damage analysis*

The design procedure calculates the damage due to fatigue and rutting using the Minor's law of linear cumulative damage concept (Huang 1993). Pavement is considered to have failed when the total damage has reached 100%, whether it is due to fatigue or rutting or other distress combinations.

2.5 *Overlay design computer applications*

Many Mechanistic-based overlay design computer applications for flexible pavements have been developed in the USA and Europe. Recently, the American Association of State Highway and Transportation Officials have also developed the AASHTO 2002 design guide through NCHRP Project 1-37 A. The objective of the AASHTO 2002 is to develop and deliver the 2002 Guide for design of new and rehabilitated pavement structures based on M-E principals as well as development of the necessary computational software (NCHRP 2004).

Following the current efforts, a mechanistic-empirical overlay design program for flexible pavements has been developed recently at Cairo University named OLFLEX (Fahmy 2004). OLFLEX has been developed based on ten Egypt pavement climatic zones, which have been also developed through the same research. The OLFLEX program has been selected in this study to create the design cases. More details about the OLFLEX program is out of scope of this paper, while it is presented in references (Fahmy 2004).

3 OVERVIEW OF ARTIFICIAL NEURAL NETWORKS

Imitating the biological nervous system, artificial neural networks are information processing computational tools capable of solving nonlinear relations in a specific problem. Like humans, they have the flexibility to learn from examples by means of interconnected elements, namely neurons. Neural network architectures, arranged in layers, involve synaptic connections amid neurons that receive signals and transmit them to the others via activation functions. Each connection has its own weight and learning is the process of adjusting the weight between neurons to minimize error between the calculated and predicted values. During the learning process, node biases are also adjusted, similar to the connection weights. Since interconnected neurons have the flexibility to adjust the weights, neural networks have powerful capacities for analyzing complex problems. Artificial neural networks, motivated by the neuronal architecture and operation of the brain, contribute to our understanding of several complexes, nonlinear pavement engineering problems with various pavement and soil variables. In Figure 2, a typical structure of ANNs that consists of a number of

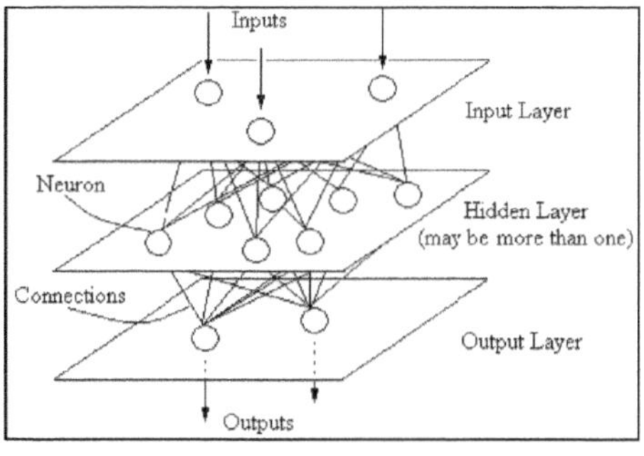

Figure 2. General schematic view of anns (Bayrak et al. 2005).

neurons that are usually arranged in layers, which are the input layer, hidden layers, and output layers. A comprehensive description of ANNs is beyond the scope of this paper (Bayrak et al. 2005).

4 ARTIFICIAL NEURAL NETWORKS IN CALCULATING OVERLAY THICKNESS FOR FLEXIBLE PAVEMENTS

4.1 *Creating design cases*

A comprehensive overlay design analysis, using the OLFLEX program, was performed using three typical flexible pavement cross sections to create design cases or database. The three cross sections are as follow: 3-layer, 4-layer, and 5-layer. For each cross section, seven cases to control the design have been selected creating 21 design cases. The seven cases consist of three conditions of controlling fatigue failure at the bottom of the AC layers with/without rutting failure on the top of the subgrade layer, which creates six cases, and the last case is controlling only rutting failure. The three condition of controlling fatigue failure mode are as follow:

- The first condition of fatigue failure is selected so that the design is controlled by considering fatigue failure in the old AC layer and the new overlay layer.
- The second fatigue failure mode is selected so that considering the fatigue failure in the new overlay layer only controls the design.
- The last condition is chosen with the intention of considering the fatigue failure in the old AC layer only.

A range of input material and traffic data has been suggested for each design case. The material data inputs for each layer consist of elastic modulus (E), layer thickness (H), and Poisson rations (ν). Because the Poisson's ratio has a relatively small effect on the pavement response (Huang 1993), it is assumed to be constant. However, other input data are assumed to be constant. As a result, 1200 runs can be performed using the following data range (Abdelbari 2006):

- Material properties of each layer
  - Modulus of elasticity
    - New overlay (E_{ov} = 2000 to 6000 Mpa)
    - Old asphalt (E_{old} = 650 to 3500 Mpa)
    - Base course (E_b = 345 to 2760 Mpa)
    - Subbase course (E_{sb} = 175 to 1750 Mpa)
    - Subgrade layer (E_{sg}= 20 to 275 Mpa).
  - Pavement layer thicknesses
    - Old AC layer (H_{old} = 25.4 to 254 mm)
    - Base layer (H_b = 50.8 to 254 mm)
    - Subbase layer (H_{sb} = 50.8 to 254 mm).
- Traffic and Loading Conditions (ESAL = 500,000 to 10,000,000).

4.2 *Training process*

Training data sets can then be selected from the developed database. NeuroSolutions software, ANN software version 4.2, was used in the training and testing processes (NeuroSolutions 2003). The networks were trained with 25%, 50%, 75%, and 100% of the size of training data. In addition, two transfer functions were tested: Sigmoid or Tanh function. Furthermore, one/two hidden layers and a range between 2 and 121 hidden nodes were used to get minimum error in getting the optimal overlay thickness. A multilayer ANN architecture was developed for each cross section as illustrated in Figure 2. The variation in the number of hidden layers, the number of hidden nodes, and the transfer function changes the size and construction of the built neural network. On the other hand, the variation in training database affects on the size of the training set employed (NeuroSolution 2003).

5 EVALUATION OF ANN IN PREDICTING THE OVERLAY THICKNESS

The objective is to express the fitness of neural networks as an effective way in calculating overlay thickness using M-E design approach with the most achievable accuracy that can be obtained for the most economical cost. The neural networks were influenced to several parameters that can guarantee the greatest achievable accuracy such as size of training data, type of transfer function, number of nodes, and hidden layers. Figure 3 shows the analysis results for the 4-layer pavement cross section, as an example, with different parameters.

5.1 *Size of training data*

In general, it can be concluded that using 100% of training data gives the most accurate result in the three pavement cross sections. Conversely, using 25% of training data does not reflect the facts. For example, using 25%, 50%, 75%, and 100% of training data in the 4-layer cross section, with two hidden layers, two nodes and Sigmoid function, provides 94.55%, 95.75%, 96.20% and 96.30% accuracy, respectively. As a result, it is not reliable to use size of training data less than 50%. In addition, using 50%, 75% and 100% of training data achieves the main target of ANN (Abdelbari 2006).

5.2 *Type of transfer function*

Changing the transfer function has a noticeable affect on the accuracy, where using the Tanh function is much better than using the Sigmoid function. Furthermore, the Tanh function is much interacting with number of nodes and the size of data than Sigmoid function. For example, using the Tanh function achieved on average 1.48% better accuracy than the Sigmoid function in case of using the 4-layer cross section (Abdelbari 2006).

5.3 *Number of nodes*

The number of nodes has an effect on the accuracy, where using more number of nodes gives high accuracy. It can be observed that number of nodes should be more than 25 nodes to achieve high accuracy (Abdelbari 2006).

5.4 *Number of hidden layers*

It has been noticed that using two hidden layers gives accuracy slightly higher than one hidden layer with an average 0.35 for the 4-layer cross section, as an example. Although, using of one

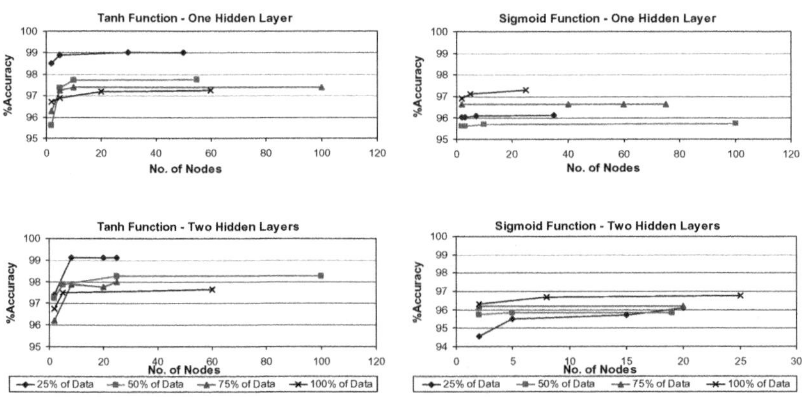

Figure 3. Accuracy results of ann for the 4-layer pavement cross section.

hidden layer gives good results, but it is recommended to use two hidden layers because the Neurosolutions predicts much better with the two hidden layer (Abdelbari 2006).

6 VALIDATION PROCESS

The validation process was conducted to verify the reliability of the trained network. Validating the network was achieved using testing data sets. The trained network should not be exposed to these data sets before. The outputs using these data sets should be compared with the actual calculated ones using OLFLEX to come up with the accuracy rate or reliability. If the accuracy rate is low, then the network is not properly trained and other training sets should be generated to retrain the network, otherwise, the network is considered to be reliable and ready for implementation.

To check the accuracy rate, testing data sets have been selected from the developed database different than the training data sets. The OLFLEX program has been used for the all selected cases to determine the calculated overlay thickness for the 3-layer, 4-layer and 5-layer cross section. The trained network has also been used for the same cases to determine the predicted overlay thickness. The calculated and predicted values were compared to come up with the accuracy rate. Table 1 shows example of the calculated (OLFLEX) and predicted (ANN) overlay thickness for the seven failure conditions of the 4-layer pavement cross section. It can be noticed that the trained network for overlay thickness gave a quite close approximation to the calculated values for the three pavement cross sections.

A regression analysis was carried out to see the approximation between the calculated and predicted overlay thickness, as shown in Figure 4 for the 4-layer cross section, as an example. The coefficient of multiple determination (R^2) is found to be 0.851, 0.913 and 0.953 for the 3-layer, 4-layer and 5-layer pavement cross section, respectively. These values are considered acceptable for the highway industry. Finally, these results indicate that ANN can be effectively used to determine the overlay thickness based on M-E design procedure.

7 CONCLUSIONS

Despite exponential advances in computational speed, pavement structural models still expend considerable amounts of computing time. Depending on application, the slowest Artificial Neural Networks (ANNs) can be two or three orders of magnitude faster than elastic layer programs and

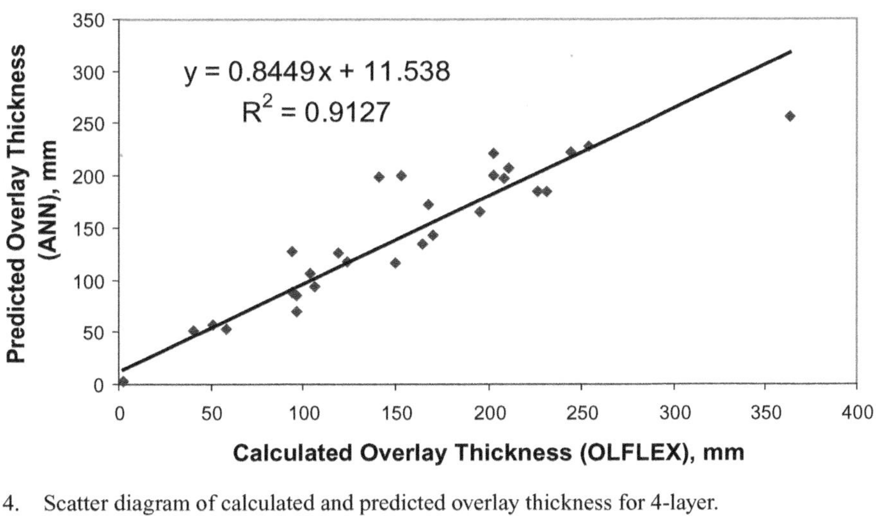

Figure 4. Scatter diagram of calculated and predicted overlay thickness for 4-layer.

Table 1. Calculated and predicted overlay thickness for seven failure conditions of the 4-layer pavement cross section.

	H_{ov} (mm) Calculated (OLFLEX)	H_{ov} (mm) Predicted (ANN)	Accuracy, %
Control the Design by Considering Fatigue Failure in New Overlay	40.64	51.94	78.24
	2.54	2.54	100.00
	96.52	69.35	71.85
	2.54	2.54	100.00
	2.54	2.54	100.00
Control the Design by Considering Fatigue Failure in Old Pavement	195.58	165.35	84.54
	231.14	185.44	80.23
	153.57	200.66	76.53
	226.06	184.53	81.63
	141.19	198.12	71.26
Control the Design by Considering Fatigue Failure in New Overlay & Old Pavement	363.22	255.53	70.35
	165.1	134.77	81.63
	58.42	53.5	91.58
	96.52	85.82	88.91
	170.18	142.62	83.81
Control the Design by Considering both Rutting & Fatigue Failure in New Overlay	243.84	223.22	91.54
	104.14	105.99	98.25
	119.38	126.29	94.53
	149.86	115.62	77.15
	167.64	171.96	97.49
Control the Design by Considering both Rutting & Fatigue Failure in Old Pavement	124.46	118.08	94.87
	203.2	199.6	98.23
	2.53	2.53	100.00
	50.8	57.6	88.19
	93.98	87.69	93.31
Control the Design by Considering both Rutting & Fatigue Failure in New Overlay & Old Pavement	210.82	207.42	98.39
	208.28	197.97	95.05
	2.54	2.54	100.00
	93.98	127.49	73.72
	2.54	2.21	87.01
Control the Design by Considering Rutting on the Subgrade Layer	254	228.6	90.00
	106.68	94.33	88.42
	2.54	2.62	96.95
	2.54	2.65	95.85
	203.2	220.97	91.96

several more orders of magnitude faster than the most sophisticated finite element programs. Therefore, this paper presents an additional step towards application of ANNs in the field of pavement design. In this paper, an ANN-based pavement overlay design tool has been developed. Several network architectures were trained using data sets developed by the Mechanistic-Empirical (M-E) overlay design program OLFLEX. Validation of the trained network was achieved using testing data sets to determine the predicted overlay thickness. Testing data sets were also used to determine the calculated overlay thickness using OLFLEX. The calculated and predicted overlay thicknesses have been compared to come up with the accuracy rate. The results indicate that the ANN technology can be used to determine the pavement overlay thickness with high accuracy based on M-E procedure. With an ANN, it is possible to obtain results within 5–17% error, which is considered acceptable for the highway industry. An important point of caution must be stated, which is comparison to measured field performance is needed prior to implementation or consideration of the proposed model.

REFERENCES

AASHTO 1993. *AASHTO Guide for Design of Pavement Structures*, American Association of State Highway and Transportation Officials, Washington D.C.

Abdalbari E. 2006. *Application of Artificial Neural Network to Pavement Overlay Design*, Master of Science Thesis, Master of Construction Management, Faculty of Engineering, Arab Academy For Science & Technology.

A2K05(3) 1999. *Use of Artificial Neural Networks in Geomechanical and Pavement Systems*, Subcommittee on Neural Nets and Other Computational Intelligence–Based Modeling Systems, Transportation Research Circular, TRB, Number E-C012, December 1999, ISSN 0097–8515.

Bayomy, F.M. & Abo-Hashema, M.A. 2001. *WINFLEX 2000: Mechanistic-Empirical Overlay Design System for Flexible Pavements, Program Documentation and User Guide,* Two Final Report, National Institute for Advanced Transportation Technology, University of Idaho, Moscow, 2001.

Bayomy, F.M. & Abo-Hashema, M.A. 2000. Inclusion of the Environmental Effects in the Empirical-Mechanistic Pavement Overlay Design System, *Sound Infrastructure Management, Infra 2000 International Conference*, November 13–15, 2000, Laval Convention Center, Laval Quebec, CANADA.

Bayomy, F.M., Abo-Hashema, M.A., Nassar, W. & Al-Kandari, F. 2002. WINFLEX 2000—A Mechanistic-Empirical Overlay Design System for Asphalt Pavements, *First Gulf Conference on Roads*, Paper Number GCR1E202, Vol. (2) Design, Kuwait, 11–13 March 2002, pp. 265–282.

Bayomy, F.M., Al-Kandari, F. & Smith, R. 1996. Mechanistic Based Overlay Design System for Idaho, *Transportation Research Record 1543, TRB*, National Research Council, Washington D.C., pp. 10–19.

Bayomy, F.M. & Shah, Z. 1993. *Development of Recommendations and Guidelines for Pavement Rehabilitation Design Procedure for the State of Idaho*, Final Report, Research Project No. 92-35-112, University of Idaho, October 1993.

Bayrak M.B., Guclu A. & Ceylan H. 2005. Rapid Pavement Backcalculation Technique for Evaluating Flexible Pavement Systems, *Proceedings of the 2005 Mid-Continent Transportation Research Symposium*, Ames, Iowa, August 2005, Iowa State University.

Fahmy, K.M. 2004. *Development of a Mechanistic-Based Overlay Design System for Flexible Pavements in Egypt*, Master of Science Thesis, Cairo University.

Huang, Y.H. 1993. *Pavement Analysis and Design,* Prentice Hall, Englewood Cliffs, New Jersey.

Hall, T.K., Darter, M.I. & Elliott, R.P. 1992. Revision of AASHTO Pavement Overlay Design Procedures, *Transportation Research Record 1374, TRB*, National Research Council, Washington D.C.

NCHRP 2004. *Guide for mechanistic-empirical design for a new and rehabilitated pavement structures*, NCHRP 1-37 A, Transportation Research Board, National Research Council.

NeuroSolutions-4.2 2003. Neuro Dimension, Inc., *NeuroSolutions 4.2*, www.nd.com, www.nerosolutions.com.

Saltan M., Tigdemir M. & Karasahin M. 2002. Artificial Neural Network Application for Flexible Pavement Thickness Modeling, *Turkish Journal Engineering Environment Science*, Number 26, pp. 243–248.

WSDOT 2003. *WSDOT Pavement Guide for Hot Mix Asphalt,* Washington Department of Transportation, Online: www.ce.washington.edu.

Efficient Transportation and Pavement Systems – Al-Qadi, Sayed, Alnuaimi & Masad (eds)
© 2009 Taylor & Francis Group, London, ISBN 978-0-415-48979-9

Evaluation of pavement cracking performance based on limiting critical strain value

Vivek Jha
Rowan University, New Jersey, USA

Yusuf Mehta
Rowan University, New Jersey, USA

Michael Byrne
Materials Department, Rhode Island Department of Transportation, Rhode Island, USA

Francis Manning
Research and Development Department, Rhode Island Department of Transportation, Rhode Island, USA

Kevin O'Brien
Rowan University, New Jersey, USA

Edward J. Saridaki III
Rowan University, New Jersey, USA

ABSTRACT: A large portion of the infrastructure budget of the Rhode Island Department of Transportation (RIDOT) is spent on resurfacing of existing roads. The standard method adopted by RIDOT involves removing the top 2.5–5 centimeters by milling & replacing it with new dense graded course. However this might not be the most cost effective way to rehabilitate pavement. Just replacing the surface course will not solve the problem in the long run if the problem lies in the substructure.

An objective procedure is needed for determining the optimum cost effective design. Currently, the RIDOT has no such standard procedure in place to make this determination. The focus of this study is to establish a relationship between the pavement performance and the mechanical response, such as the critical strain in the pavement under the surface layer, and to determine a range of critical strain values at which the pavement performs satisfactorily.

Thirteen sections were selected across the state of Rhode Island consisting of multiple severity of cracking. Falling weight deflectometer (FWD) data was measured on each of these sections and thickness was determined from the core data collected at the sections. The modulus values were determined using the backcalculation software, BAKFAA. All the modulus values for the top layer were normalized at a constant temperature of 7.8°C. The tensile strain value under the surface layer was then calculated using the forward calculation software, KENLAYER. A critical strain criterion was established by correlating it to the measured distress such as uncrack, transverse crack, and longitudinal crack. The critical strain value depends upon the modulus values of the top two layers as well as the thickness of the top layer. The research team will provide guidelines in the form of a decision tree to avoid exceeding the critical strain and hence in the process provide a long-lasting, cost-effective pavement. This mechanistic approach will save the RIDOT millions of dollars on future projects.

1 PROBLEM STATEMENT

A large portion of the infrastructure budget of the Rhode Island Department of Transportation (RIDOT) is spent on the resurfacing of existing roads. The standard method adopted by RIDOT involves removing the top 2.5–5 centimeters by milling & replacing it with a new dense graded course. However this might not be the most cost effective way to rehabilitate pavement. Just replacing the surface course will not solve the problem in the long run if the problem lies within the substructure. An objective procedure is needed for determining the optimum cost effective design.

The focus of this study is to develop tools in the form of a decision tree that can be utilized by the RIDOT to develop a cost effective pavement design procedure. The step towards making this tool is to identify a parameter(s) that correlates to pavement performance. Using the parameter(s) establish a decision tree in the form of pavement layer stiffness and thickness.

2 STUDY OBJECTIVES

The objectives can be divided into three major categories:

1. Analyze the data provided by the RIDOT by using various backcalculation and pavement evaluation software's to find the relationship between pavement condition and its performance.
2. Determine design criteria when no cracking occurs with respect to the construction of new pavement. Rehabilitation of existing pavement was not included in this study.
3. Develop a decision tree to select the appropriate thickness and stiffness of surface layer that can be utilized for flexible pavement design.

3 RESEARCH APPROACH

The following approaches were taken to accomplish each objective of this study:

3.1 *Objective 1*

1. Backcalculate layer moduli of the failed and unfailed pavement sections by utilizing BAKFAA software program.
2. Check for reasonableness of backcalculated layer moduli by comparing it with previous studies (Lee et al. 2003).
3. Calculate mechanical response beneath the surface layer of the pavement utilizing forward calculation software KENPAVE, after all the layer modulus values were calculated.

3.2 *Objective 2*

1. Compare pavement parameters like moduli, thickness, and tensile strain below the surface layer to various pavement cracking conditions to establish design criteria.
2. Calculate the limiting value of the criteria for new pavement construction to ensure no cracking.

3.3 *Objective 3*

1. Develop a decision tree to determine the thickness of the surface layer for new construction based on the criteria established in this study.

4 LITERATURE REVIEW

There was an extensive study done at the University of Rhode Island to develop flexible pavement parameters (Lee et al. 2003) like resilient modulus, drainage and layer coefficients, initial structural number and equivalent single axle loads (ESAL's) for the state of Rhode Island based on the 1993 AASHTO flexible pavement design procedures. The study recorded the resilient modulus of various layers for a range of temperature values for route 2 and route 146 in the state of Rhode Island and calculated the effective resilient modulus for those two routes as shown in Table 1 below. These resilient modulus values were used as the seed values for the backcalculation process in this study and also used as a check for the final modulus value determined from backcalculation. The moduli values calculated by backcalculation process were on the higher side as the conditions are different in the field as compared to the condition in the lab especially with respect to confinement. The confining stresses are far greater in the field than those applied in the lab; this will lead to higher modulus values in the field as compared to that measured in the lab. The study also calculated the resilient modulus of subbase for all the months to incorporate the change in weather condition as shown in Table 2 below.

The study done at Korea Institute of Construction Technology (Park et al., 2005) mainly focused on determining the thickness of Asphalt concrete (AC) layer for long life of the pavement. The major conclusion that came out of the study was thickness of the AC layer was the most influential factor for controlling the critical pavement responses. The study established a limiting value of critical tensile strain value below the AC layer (which was composed of a wearing course, intermediate course and a base course) and critical compressive strain above the subgrade to limit rutting. However the researchers noticed that if the strain value beneath the AC layer was kept below the limiting value of 65 microstrain then the strain above subgrade is well within the limit, thus

Table 1. Effective modulus of various layers for route 2 and route 146 (Lee et al. 2003).

Site	Effective resilient Modulus kN/m^2(10^4)			
	E_1	E_2	E_3	E_4
Route 2	224	372	331	9.4
Route 146	298	301	–	8.0

Table 2. Resilient modulus (M_{R}) for route 2 and route 146 for different months (Lee et al. 2003).

	Route 2		Route 146	
Months	Temperature °C	M_R kN/m^2(10^4)	Temperature °C	M_R kN/m^2(10^4)
January	1.5	6.9	–1.83	5.0
February	1.27	4.6	2.17	4.9
March	1.78	4.8	5.94	5.3
April	11.27	6.2	6.67	5.7
May	15.33	7.0	11.78	6.5
June	19.22	7.5	19.05	7.4
July	20.5	7.7	22.26	7.6
August	21.39	8.2	20.61	7.5
September	11.83	7.3	15.56	6.7
October	9.22	6.4	10.5	6.4
November	5.94	7.2	6.278	6.2
December	1.33	7.3	3.89	5.9

Table 3. FWD and core data provided by the RIDOT with varying degree of pavement distress.

Road	Location	Number of sections Total	Uncracked	Longitudinal cracked	Transverse Cracked	"Other" cracked
Child Street	Warren	9	2	2	1	4
Chestnut	Bristol	9	2	4	3	–
Route 3	Coventry	3	1	–	–	2
Route 3	Westerly	7	3	–	–	4
Route 12	Scituate	9	–	5	1	3
Route 112	Charlestown/ Richmond	5	–	–	–	5
Route 138	North Kingstown	5	–	–	–	5
Route 403		8	–	–	–	8
Total		55	8	11	5	31

Table 4. Thickness of various layers and seed value of moduli for backcalculation.

Route	Thickness cm	Seed value kN/m^2(10^4)
Chestnut Street		
Asphalt	6.4–10.8	207
Base	11.5–18	138
Gravel	30.5	14
Subgrade	508	8
Child Street		
Asphalt	5–21.6	207
Base	0–20.3	275
Gravel	30.5	14
Subgrade	508	10
Route 3 Coventry		
Asphalt	5–21.6	344
Gravel	30.5	14
Subgrade	508	8
Route 3 Westerly		
Asphalt	5	344
Base	15.2	70
Gravel	30.5	14
Subgrade	508	10
Route 12		
Asphalt	10.2–16.58	344
Base	34.3–40.6	69
Gravel	30.5	14
Subgrade	508	10
Route 112		
Asphalt	9–13.3	275
Gravel	30.5	70
Subgrade	508	70

(*continued*)

Table 4. (*continued*).

Route	Thickness cm	Seed value kN/m²(10^4)
Route 138		
Asphalt	9–9.5	206
Base	10.8–12.7	34
Gravel	30.5	14
Subgrade	508	14
Route 403		
Asphalt	7.6	206
Gravel	30.5	14
Subgrade	508	14

critical tensile strain value is the controlling factor to determine the thickness of AC layer for long life of the pavement.

5 DATA

The Table 3 below shows the sections evaluated in this study. The table also lists the number of stations along with pavement condition at those stations.

A total of 55 sections across the state of Rhode Island were selected in this study. All the sections analyzed in the study were taken on lanes and not on shoulders. Table 4 below shows the thickness of various layers at each section and the initial value of modulus assumed for backcalculation. The initial seed values assumed for backcalculation were selected from Tables 1 and 2.

6 RAW DATA ANALYSIS

Raw data analysis mainly focuses on using the deflection data obtained from the FWD testing and establishing some initial trends about the pavement condition prior to backcalculating modulus values. A typical deflection vs. station is shown in Figure 1. From the figure it is possible to establish an overall correlation (trends) between performance data and FWD deflection data.

From Figure 1 it can be seen that the difference between D_1 and D_7 (D_1–D_7) values for some stations are relatively very high. Higher D_1–D_7 values show cracking in the pavement and very low values for D_1–D_7 show that the station is in good condition compared to other stations. The above figure does not establish a clear relationship between various stations and the pavement conditions at those stations, but forms a baseline for future analysis with regards to the backcalculation as it establishes the range for modulus values for different layers as some stations can be identified as broken and some as good.

For the purpose of backcalculation, a large thickness of subgrade was assumed and very high modulus beneath the subgrade of the order of 68,94757 kN/m² was input into BAKFAA. The high thickness of the subgrade along with high modulus beneath it, provides stability to the pavement system similar to that observed in the field.

7 BACKCALCULATION AND FORWARDCALCULATION

The modulus values for different layers were obtained from the backcalculation software BAKFAA. BAKFAA was developed by the Federal Aviation Administration's Airport Technology

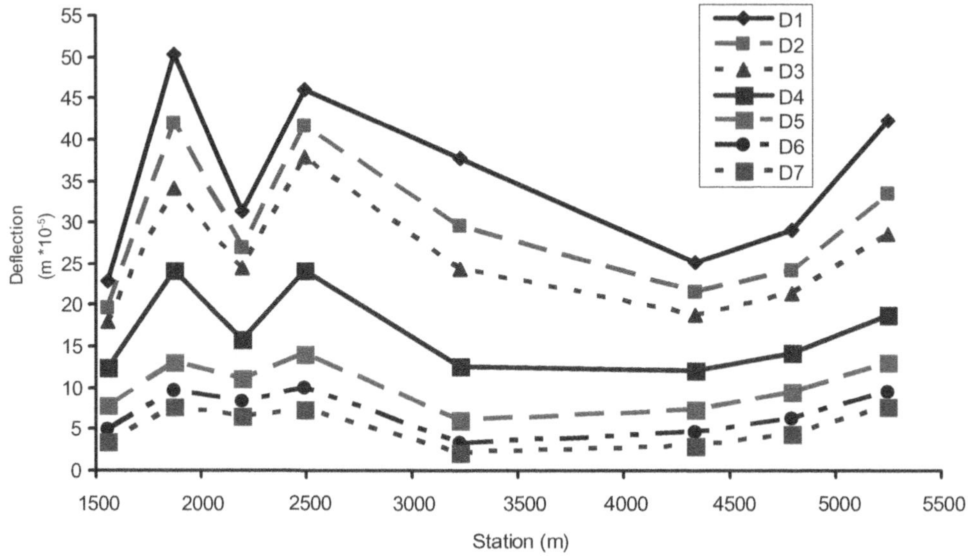

Figure 1. Station vs. deflection for Child Street.

Research and Development. The program uses a layered elastic analysis and a downhill multidimensional simplex minimization method (FAA, 2004). Backcalculation is the process of using iteration and optimization schemes to determine the modulus of pavement layers in a pavement system given the layer thickness, Poisson ratio and deflection data from the FWD test. It's an iterative method and is continued until the deflections observed in the field and calculated deflections match. The corresponding moduli are noted. Backcalculation is a useful tool but does have some disadvantages. The process can produce several different combinations of moduli for the same deflection profile. The initial seed values for backcalculation were selected from the previous studies (Lee et al. 2003). All the modulus values were calculated for a load range of 52.5–58 kN because all the FWD data had loads within that range of values. The backcalculated modulus value obtained for the surface layer were then normalized to a constant temperature of 7.8°C to establish a consistency in the modulus value as the FWD data was taken over a wide range of temperature. The final values selected were checked for reasonableness with both the past studies as well as with the deflection profile.

The modulus values calculated were then used to calculate the mechanical response of the pavement using forward calculation software KENPAVE. It was developed at the University of Kentucky by Dr. Yang H Huang. (Huang, 2004). It calculates the mechanical responses at any point in the pavement system. For uniformity all the responses were calculated within the pressure range of 758–827 kN/m^2. The pressure of 758–827 kN/m^2 was utilized because all the modulus values were calculated for the falling load of 52.5–58 kN.

8 ESTABLISHING PARAMETERS FOR DEVELOPING DECISION TREE

A typical section which showed a wide range of distress was selected as a typical example to demonstrate how the parameter(s) were established which differentiated uncracked stations from cracked stations. Once an appropriate parameter was identified that correlated reasonably well with field cracking performance the reliability of that parameter was established by plotting it for the other stations in this study against the various distress modes.

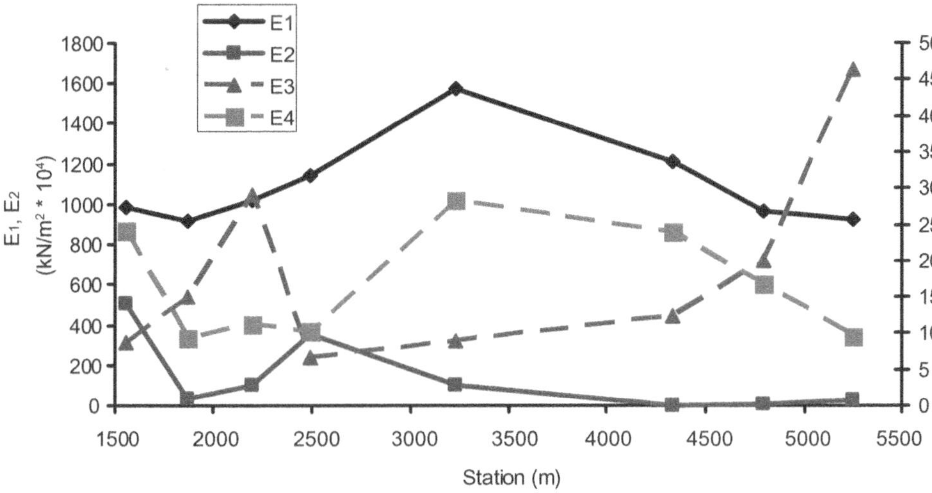

Figure 2. Moduli of various layers vs. Station for Child Street.

In this paper, Child Street was selected as a typical section as it showed wide range of pavement conditions to demonstrate the analysis conducted, similar analysis was conducted for all the sections. The modulus values calculated from the backcalculation process were compared to that of the FWD data to establish the reasonableness of the value obtained. The modulus value for Child Street were plotted against the stations as seen in Figure 2 and was compared with corresponding deflection values in Figure 1.

The surface modulus values and the deflection profile, D_1 show a clear inverse relationship for most of the stations except for two stations. At station 2498 meter there is an increase in the surface moduli profile at the same time there was an increase in the deflection D_1 profile as compared to the previous station. The values for that particular section were looked into more detail. The description of the sections showed that there was a crack in close proximity of the station however the crack was sealed according to the data provided by the RIDOT. It is possible that the crack was sealed recently thus a high modulus value due to the presence of new material. The higher deflection might be due to the fact that it was a considerably new construction during the time FWD data was taken, hence has not bonded well with the other layer as in the case of other stations.

However, an opposite trend was observed at station 4337 meter. There is a drop in the deflection D_1 and also a drop in the surface modulus profile as compared to the previous station. On further analysis it was seen that the station does not have a base as in other sections and is directly laid on the gravel bed and the thickness of the top surface is 19 cm. The high thickness of the surface layer explains the decrease in the deflection values as compared to previous station. The decrease in the modulus value as compared to previous station can be explained due to the absence of a base layer and thus the surface layer does not have a firm base to rest on thus the lower modulus value. But on close observation it can be seen that the E_1 value if compared to others is on the higher side which is true as the depth of the surface layer is very high.

One of the important conclusions that can be drawn from Figure 2 is that the modulus of surface layer E_1 does not show a strong correlation between the cracked and the uncracked section. The E_1 value at station 3231 meter in Figure 2 is highest however from the description provided with the FWD data states that "cracking around test area".

The cracking that was reported in the FWD data was mainly surface cracking or cracking in the base layer. E_1/E_2 was plotted along the stations to establish a relationship between the moduli value of surface and base layers to the pavement condition as shown in Figure 3.

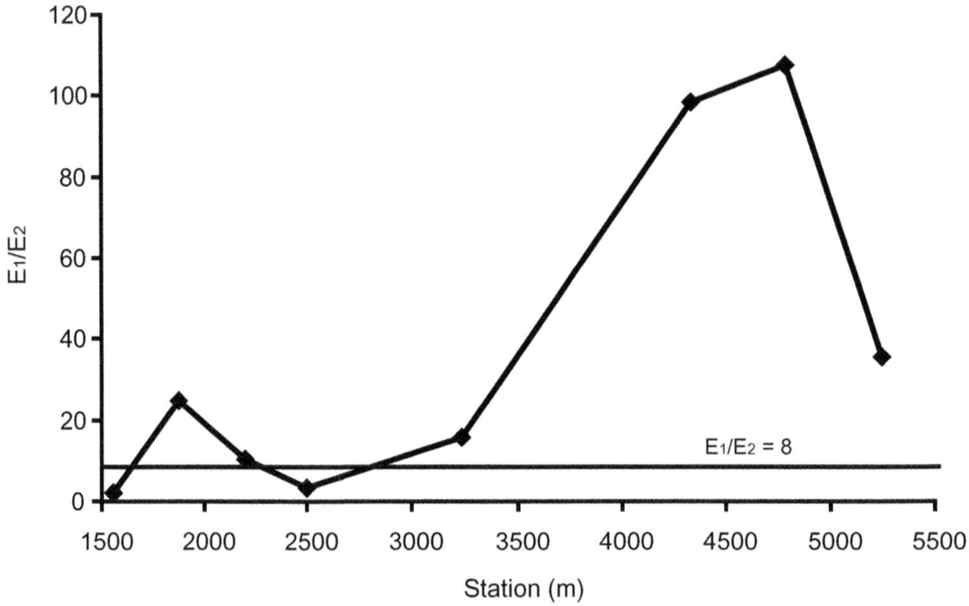

Figure 3. E_1/E_2 vs. Station for Child Street.

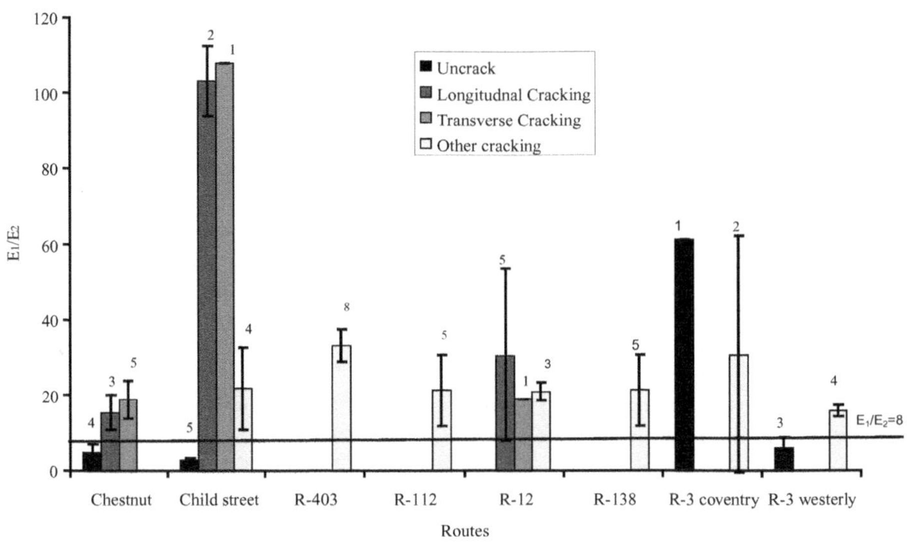

Figure 4. Stiffness ratio (E_1/E_2) for all routes.

From Figure 3 it can be observed that good sections showed no cracking (points below the line at $E_1/E_2 = 8$) in the field had a very low E_1/E_2 as compared to other sections. It also shows that to have no cracking in the pavement the value of base should not be very low as it will result in very high E_1/E_2. It can also be observed from Figure 3 that E_1/E_2 does not show a strong correlation between different pavement conditions.

The weak correlation between E_1/E_2 and the pavement conditions which was observed in Figure 3 was checked for different routes by plotting E_1/E_2 for all routes at different pavement condition as shown in Figure 4 below.

In the above figure, E_1/E_2 was plotted against various routes for varying degrees of pavement distress observed in the field. The number above each bar shows the number of stations at each section with that particular distress mode. Each bar in Figure 4 represents the average value for that specific distress mode for that particular route. The values were plotted along with their 95 percent confidence intervals to show the variablity of those values from their average value. A clear relationship cannot be made between various pavement conditions and the stiffness ratio.

The only relationship that can be obtained from the above graph is a limiting value of E_1/E_2. The problem associated with using the above condition is that, a limiting value of E_1 will lead to rutting and very high thicknesses of surface layers. As no substantial relationship was established between the modulus value and the pavement distress, the pavement mechanical response was calculated using forward calculation. The mechanical response in terms of tensile strain beneath the surface layer was calculated to obtain a relationship between the pavement response and pavement condition. Critical tensile strain beneath the surface layer was then plotted against various stations to check whether it showed a relationship between cracked and uncracked sections as shown in Figure 5 below.

From the above graph it can be seen that critical tensile strain changes considerably with change in the pavement condition. On comparing Figure 5 with Figure 1 it appears to show that there exist a strong correlation between the tensile strain and deflection. For the lowest deflection value in Figure 1, i.e. station at 1558 meter, the critical strain value is also the least which is confirmed by the description provided in the FWD data that the closest crack is about 0.762 meter away from the station. From Figure 5 it can also be seen that critical strain compares resonably well with field conditions. The relationship established in Figure 5 was confirmed for other routes by plotting the critical tensile strain for various routes for various distress modes as shown below in Figure 6.

In the above Figure the critical tensile strain is plotted for various routes. The number above the bar shows the number of station in the particular route with the respective pavement condition. From the above Figure it can be seen that to have no cracking in the pavement the critical strain value has to be limited to 90 microstrains. The tensile strain below surface layer is given by (Huang, 2004)

$$\varepsilon = \frac{q}{E_1} F_e \tag{1}$$

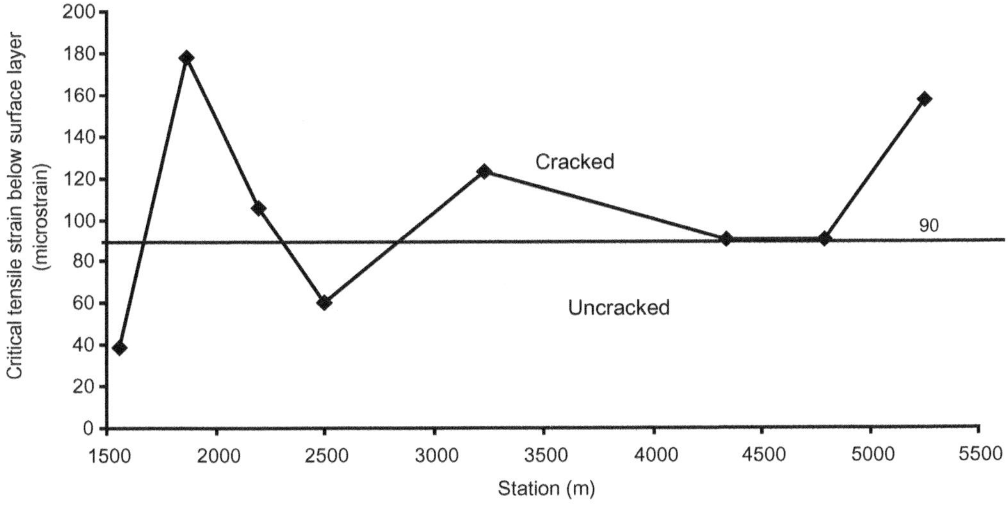

Figure 5. Critical tensile strain beneath the surface layer vs. Station for Child Street.

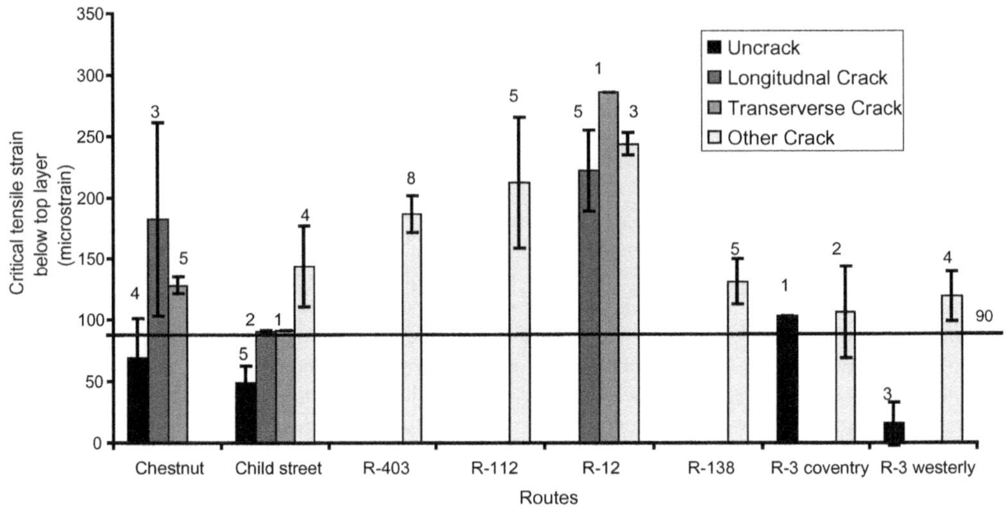

Figure 6. Critical tensile strain beneath surface layer for all routes.

where ε = critical tensile strain (micro strain); q = pressure (kN/m^2); E_1=Modulus of surface layer (kN/m^2); F_e = strain factor.

The tensile strain, as calculated from the strain factor depends on the pressure applied by the load and the stiffness ratio E_1/E_2 and h_1/a where "h_1" is the thickness of surface layer in centimeter and "a" is the contact radius of the load in centimeter.

9 DECISION TREE

The decision tree constructed was based on limiting the tensile strain beneath the surface layer to 90 microstrains. Flow chart depicting the process of developing decision tree is shown in Figure 7 below. The focus of the process was to provide reasonable thickness for the surface layer for a range of moduli values as it was the only parameter that can be changed easily in the field. The typical output that can be expected from the catalog are shown below in Table 5 below.

The fixed moduli input values for subbase and subgrade were based on the average value calculated form all the routes analyzed. The moduli values of these layers were on the higher side. However, it is not uncommon for in-situ moduli values to be slightly higher than the laboratory values. Lee et al. 2003 obtained values within a similar range from backcalculation. but even when the moduli values were reduced and the strain value was calculated it was observed that the effect of moduli value of subbase and subgrade on the tensile strain beneath the surface layer was minimal. The thickness of the subbase and the subgrade were selected as 30.5 cm and 508 cm respectively, as all the values from the field data were same as them or around that value. The range of values for E_1, E_2 and H_2 used in the decision tree incorporates all the values for the repective paramteres that was observed during this study. The typical output that can be expected from the decision tree can be seen in the Table 5 below.

All the thicknesses of H_1 in the above Table are for critical strain values of less than 90 microstrains beneath the surface layer as shown in the flowchart in Figure 7. In case 1 in Table 5 the thickness of the surface layer which was 66 cm was very high for practical purposes. In cases like case 1 the thickness of the surface layer can be reduced to a more practical value for the same set of values of E_3, E_4, H_3, H_4, and H_2 by changing the stiffness of the surface or the base layer which are the variable input as shown in the flowchart in Figure 7. The thickness of the surface layer for

Table 5. Typical output from the decision tree.

Input								Output
Case	E_4 kN/m^2 (10^4)	E_3 kN/m^2 (10^4)	E_2 kN/m^2 (10^4)	E_1 kN/m^2 (10^4)	H_4 cm	H_3 cm	H_2 cm	H_1 Cm
1	18.6	19.3	9	69	508	30.5	6.4	66
2	18.6	19.3	9	1103	508	30.5	6.4	20.3
3	18.6	19.3	414	1103	508	30.5	24	1.3
4	18.6	19.3	552	1103	508	30.5	10.2	2.54

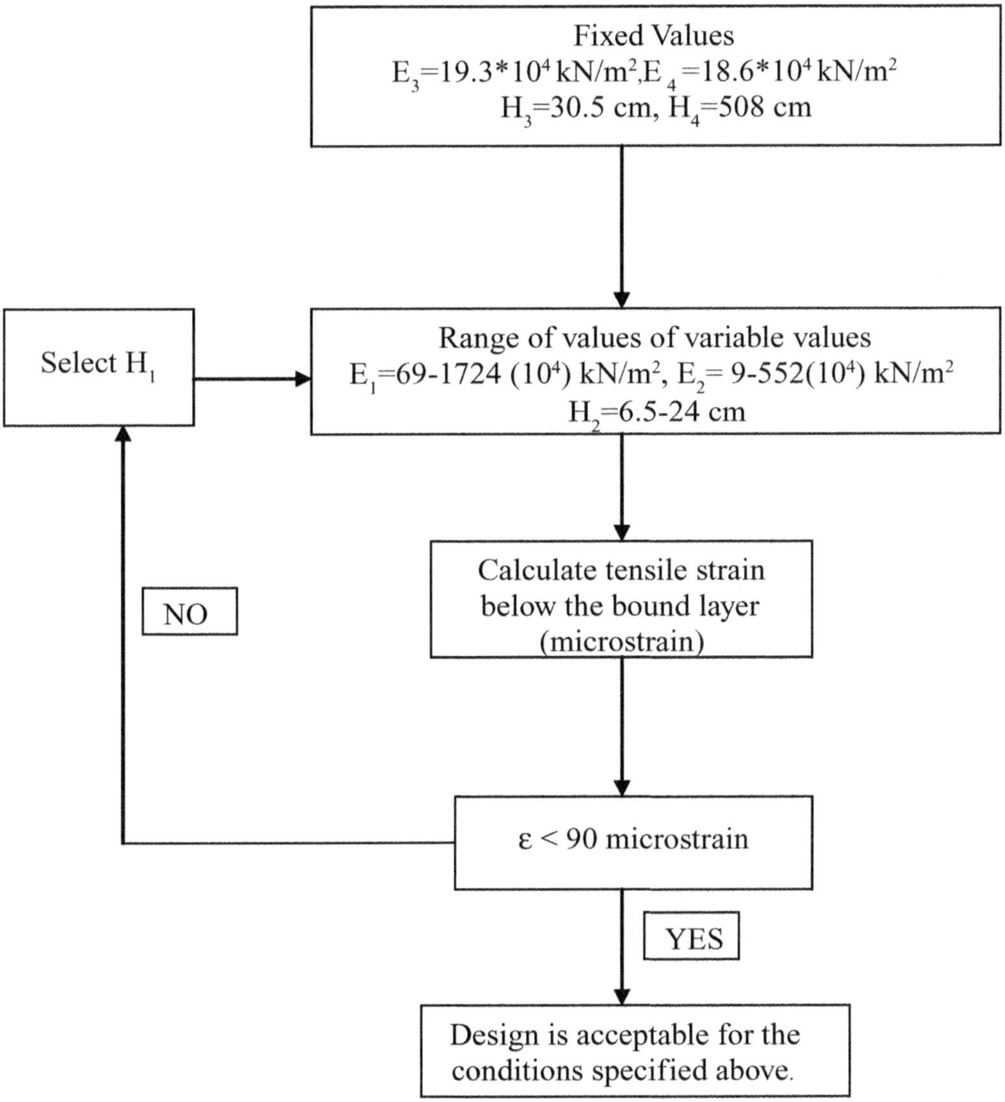

Figure 7. Flowchart depicting the process of developing a decision tree.

case 1 was reduced from 66 cm to 20.3 cm by increasing the stiffness of the surface layer which can be seen in case 2. Cases 3 and 4 depict that thickness of surface layer can also be reduced by increasing the thickness and the stiffness of the second layer

10 SUMMARY OF FINDINGS

The summary of findings based on the analysis conducted is as follows:

1. If the tensile strain beneath the bound surface layer was lower than 90 microstrains (at vertical stress of 782 kN/m^2) then no cracking was observed.
2. The decision tree was developed based on satisfying the condition to limit the tensile strain below the surface layer to 90 microstrains thus ensuring no cracking occurs below the bound surface layer.
3. The tensile strain value calculated for no cracking by researchers at Korea Institute of Construction Technology (Park et al., 2005) was 65 microstrain which are considerably lower than 90 micro strains as calculated in this study. The tensile strain calculated in that study was below the AC layer which composed of wearing course, intermediate course and a base course while the strain value calculated in this study was below the surface layer. The difference of thickness plays an important role in determination of the strain value which accounts for the higher strain value calculated in this study.
4. The decision tree developed in this study helps to find the thickness of the surface layer for a range of modulus value for surface layer and base layer. For uneconomical thickness the stiffness of the surface or the base layer can be modified to reduce the thickness of the surface layer. For example, it can be observed from the Table 5 that thickness for the surface layer in some cases were too high which might turn out to be uneconomical. The thickness problem can be solved by increasing the stiffness of the surface layer or the base layer which can be seen by comparing case 1 and case 2.

11 CONCLUSION AND RECOMMENDATIONS

The conclusions based on the summary of findings are as follows:

1. A mechanical response, tensile strain below the bound surface later was identified that correlated reasonably well with pavement performance. Based on this, a limiting value of 90 microstrain was established.
2. A decision tree that can be utilized as a design tool was developed for the state of Rhode Island based on limiting the tensile strain below the bound surface layer.

Since most of the construction are pavement rehabiliation and pavement preservation projects, it is recommended that a similar limiting value needs to be estalished below the surface layer (not necessarily the entire bound layer) and a new decision tree needs to be established.

REFERENCES

Lee, K.W., Marcus, A.S., Mooney, k., Vajjhala, S., Kraus, E., Park, K. 2003. Development of flexible pavement design parameters for use with the 1993 AASHTO pavement design procedures.
Huang, Yang H. 2004 Pavement analysis and design. Kentucky: Prentice Hall.
Park, H., Kim, J., Lee, H. 2005. Determination of the layer thickness for long life of asphalt pavements.

Efficient Transportation and Pavement Systems – Al-Qadi, Sayed, Alnuaimi & Masad (eds)
© 2009 Taylor & Francis Group, London, ISBN 978-0-415-48979-9

Geogrid-reinforced low-volume flexible pavement response to loadings by various tire configurations

I.L. Al-Qadi
Civil and Environmental Engineering Department, University of Illinois at Urbana-Champaign, Urbana, IL

S.H. Dessouky
Civil and Environmental Engineering, University of Texas at San Antonio, Texas, USA

E. Tutumluer
Civil and Environmental Engineering Department, University of Illinois at Urbana-Champaign, Urbana, IL

ABSTRACT: Nine fully instrumented low-volume flexible pavement sections were tested at the University of Illinois to evaluate the effect of three types of tire configurations on geogrid-reinforced pavements. The nine sections were divided into three cells comprised of 76 mm hot-mix asphalt (HMA) placed on 203, 305, and 457 mm aggregate layers, respectively. Geogrid-reinforcement was incorporated in at least one section in each cell. The pavement was instrumented during construction with pressure cells, strain gauges, LVDTs, thermocouples, piezometers, and time domain reflectometers. The test sections were exposed to accelerated loading. The tires used for loading were dual-tire assembly and 425 and 455 wide-base tires at five loading levels and two speeds. The study found that the 455 wide-base tire caused greater transverse strain in the thin HMA layer than the conventional dual-tire assembly when relatively thin granular layer is used on weak subgrade. However, as the supporting granular base layer stiffness increases, the difference in strain diminishes. It was evident that the new generation of 455 wide-base tire caused less damage to the low-volume flexible pavements than the old generation 425 wide-base tires. The measured longitudinal deflection in the base layer and vertical deflection in the subgrade were significantly lower for the 455 wide-base tire and dual-tire assembly when compared to the 425 wide-base tire. As the supporting base layer stiffness increases, the wide-base 455 tire's and the dual-tire assembly's overall effects on the pavement become similar for different loading conditions.

1 INTRODUCTION

In recent years, the trucking industry has introduced new tire structure and suspensions to improve the efficiency of the transportation network. The wide-base tires were introduced to replace the conventional dual-tire assembly. Compared to the conventional dual-tire assembly, it is reported that wide-base tires improve fuel efficiency, reduce emission, increase payload, exhibit superior braking and comfort, and reduce repair and maintenance cost (Al-Qadi et al. 2004). In contrast, state departments of transportation (DOTs) and pavement agencies have been concerned about the damage wide-base tires could cause to flexible pavements. Their concerns arose from the lower total contact area of the wide-base tires on the pavement surface compared to the dual-tire assembly, which may result in increased applied contact stresses. Furthermore, driving at relatively higher inflation pressures could increase surface deterioration. Although designed in accordance with current pavement regulations, such as "inch-width" laws, earlier studies indicated that the first generation of wide-base tires (385 and 425) clearly resulted in an increase in pavement damage compared to dual-tire assemblies (Bonaquist 1992; Sebaaly 1992). Investigation into how to obtain a balance between the economic advantages to the industry and the impact on pavement is needed.

Several projects were conducted to investigate the aforementioned concern. Sebaaly and Tabatabaee (1989) reported that measured tensile strains at the bottom of the hot-mix asphalt (HMA) layer under dual radial tires were the lowest for all axle loads and HMA thicknesses measured in their study. Metcalf (1996) also found that the wide-base tire 14.00/80R20 caused twice the measured rutting compared to that resulting from standard dual radial 10.00R20. FHWA (2006) reported that the 425 wide-base tires caused significantly more fatigue and rutting damages than conventional dual-tire assembly under the same loading conditions. Huhtala et al. (1989) suggested that wide-base tires were on average 2.3 to 4.0 times more damaging to the pavement than dual-tire assemblies. Other studies using numerical tools also reached the same conclusion (Siddharthan et al. 1998).

The majority of the studies comparing conventional wide-base tires to conventional dual-tire assembly reported that wide-base tires were more damaging to pavement systems than conventional dual-tire assembly. The conventional wide-base tires were designed with relatively thin tire shoulder and narrow rib width to enhance cooling running temperature. Although it sustained high operating inflation pressure to minimize tire deflection, high vertical contact stresses resulted at the pavement surface. The new generation of wide-base tires has improved engineering characteristics, which warrant a completely new look at their impact on pavements.

In addition to offering safety and cost savings, the structural characteristics of the new generation of wide-base tires include a wider and flatter transverse profile for a uniform stress distribution, larger load carrying capacity, and lower tire pressure to 690 kPa. The new generation of wide-base tires utilizes a new crown architecture design that allows for a lower aspect ratio geometry (the ratio between tire height to total width), and therefore, a wider tire. As a result, COST 334 has suggested the use of wide-base tires on trucks including the steering axles. Many studies, including Al-Qadi et al. (2002), Pascale et al. (2004), and Priest et al. (2005) investigated the new generation of wide-base tires and dual-tire assembly and compared the resulting data. The majority of these studies have investigated the effect of wide-base tires on primary roads.

This paper recounts a study to evaluate low-volume flexible pavement responses to various tire configurations. Accelerated pavement testing was conducted on thin, low-volume, geogrid-reinforced pavement sections. Full-scale accelerated loading is a viable option for testing because it compresses years of road loading into weeks of testing. A conventional dual-tire assembly, 11R22.5, and wide-base tires: 425/65R22.5 and 455/55R22 were used. The experimental testing included the following variable parameters: load level, speed, inflation tire pressure, tire configuration, and tire offset.

The pavement sections were originally built to evaluate the effectiveness of geogrids in low volume road performance when constructed on low strength (below 4% CBR—California Bearing Ratio) subgrade. The pavement consists of nine instrumented full-scale flexible test sections designed and constructed to measure pavement responses, monitor pavement performances, and quantify the effectiveness of geogrid-reinforced flexible pavements. Test section variables examined in the full-scale testing study include HMA and granular base layer thicknesses and type and location of geogrid within the base course. The effectiveness of geogrid reinforcement appears to be more pronounced when used in roads designed for low to moderate traffic volumes; especially when the pavement structure consists of a thin HMA layer on top of a granular base/subbase layer. These benefits are often presented as extending pavement service life or reducing base course thickness. This study used the constructed pavement sections to evaluate the performance of geogrid-reinforced pavement structure and to investigate the effect of different tire configurations on low-volume roads.

2 PAVEMENT LAYOUT

2.1 *Layers structure*

Nine instrumented flexible pavement test sections, a total of three pavement cells each 3.65 m wide, were constructed at the University of Illinois as shown in Figure 1. Three different granular

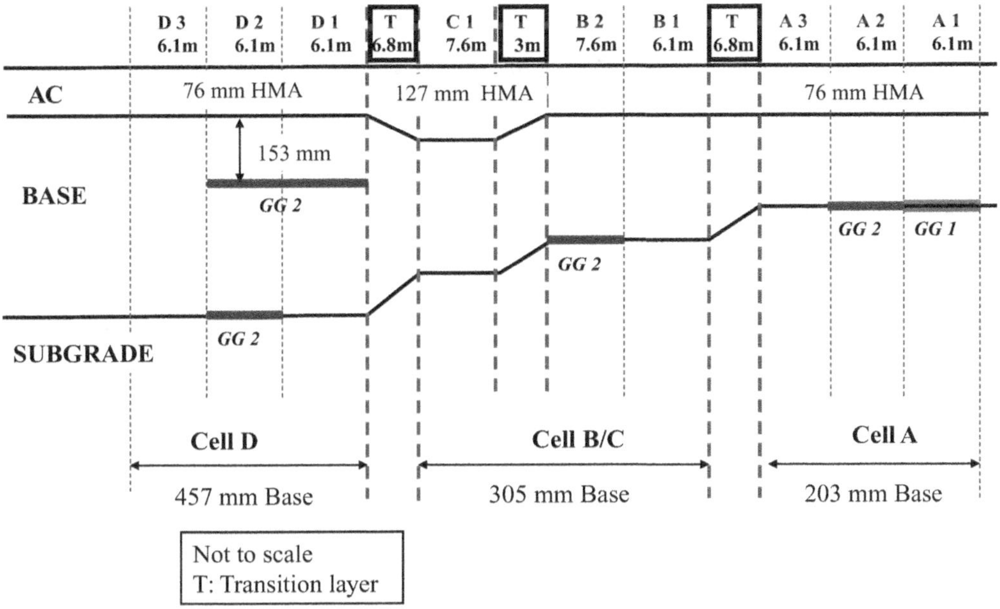

Figure 1. Pavement test sections layout.

base layer thicknesses of 203, 305, and 457 mm and two HMA layer thicknesses of 76 and 127 mm were used in pavement cells A, B/C, and D. Each cell, comprised of three pavement sections, was designed to be tested under the same loading and environmental conditions. Transitional sections were constructed between each cell to accommodate design changes during construction.

Pavement cell A was designed to allow for "thin" pavement response and performance comparisons between two different types of geogrids and a control section. Pavement cell B/C was designed to quantify the effectiveness of geogrid base reinforcement for different HMA thicknesses, while pavement cell D was to address the question of optimum geogrid location in a thicker granular base layer. The three cells collectively can be used to investigate the effect of base layer thickness on reinforced/unreinforced sections. Information on the types and properties of the geogrids used in this study can be obtained from Al-Qadi et al. (2007).

2.2 *Instrumentation*

A total of 173 instruments were installed in the pavement sections during construction. Two categories of instruments were used: those to measure environmental properties such as temperature, moisture, and pore-water pressure and those used to measure responses to loading such as pressure, strain, and deflection. Environmental instruments were installed at a 0.9 m offset from the centerline of the pavement (e.g. thermocouple, time domain reflectometer, piezometers); while load associated instruments were installed at the centerline (e.g. pressure cells, LVDT, strain gauges).

2.3 *Materials testing*

A hand auger was used to sample soils in the construction site down to a depth of 1.5 m. The subgrade was generally classified as ML, CL, or a combination of ML and CL using the dual classification in accordance with Unified Soil Classification System. The soil specific gravity is 2.72. Atterberg limit tests (ASTM D 4318) indicated that the subgrade has liquid limit (LL) of 21%

and an average plasticity limit (PL) of 16% and an average plasticity index (PI) of 5%. Several laboratory experiments were conducted to extract material properties such as unconfined compressive strength tests, Q_u (ASTM D 2166), California Bearing Ratio (CBR) tests (ASTM D 1883 and AASHTO T-193-81), moisture-density relationship using standard proctor tests (AASHTO D698), and dynamic cone penetrometer (DCP). The DCP evaluation of the test strip soil suggested that the CBR values ranged from 5 to 28% before the top subgrade soils were prepared for pavement construction and reduced it to 4%. More details on laboratory experiments can be found elsewhere (Al-Qadi et al. 2007).

A crushed limestone, designated as densely graded CA-6 aggregate by Illinois DOT, was used as the base course material. Modified proctor (AASHTO T180) moisture-density tests conducted on the CA-6 material gave a maximum dry density of 22.5 kN/m^3 corresponding to an optimum moisture content of 6.5%. Repeated load triaxial tests were conducted on the CA-6 aggregate material to determine its resilient moduli in accordance with AASHTO T307. The specimens were compacted at the field moisture content and dry density of 5.03% and 21.6 kN/m^3, respectively.

Two HMA were used in the test sections: wearing surface (SM-9.5) and base mix (BM-25.0). Ignition oven tests were performed in accordance with AASHTO T-308 on HMA specimens collected during construction. After the ignition oven tests, washed gradation analysis was performed on each specimen in accordance with Illinois DOT modified AASHTO T-30. Maximum specific gravity (G_{mm}) tests were performed in accordance with the Illinois DOT Modified AASHTO T-209. An average maximum specific gravity G_{mm} of 2.461 and 2.498 were obtained for the SM-9.5 and BM-25.0, respectively. The HMA air voids were 3.1% and 3.8% for SM-9.5 and BM-25.0, respectively. Complex modulus tests were conducted for each HMA lift placed at each cell. A total of 12 150-mm-diameter test specimens were prepared from field mixes using the gyratory compactor. Tests were conducted at −10, 4, and 20°C and frequencies of 0.01, 0.1, 1.0, 10, and 25 Hz. A master curve, as shown in Figure 2, was developed for each test section in accordance with the complex modulus protocol established by NCHRP projects 1-37 A and 9-19.

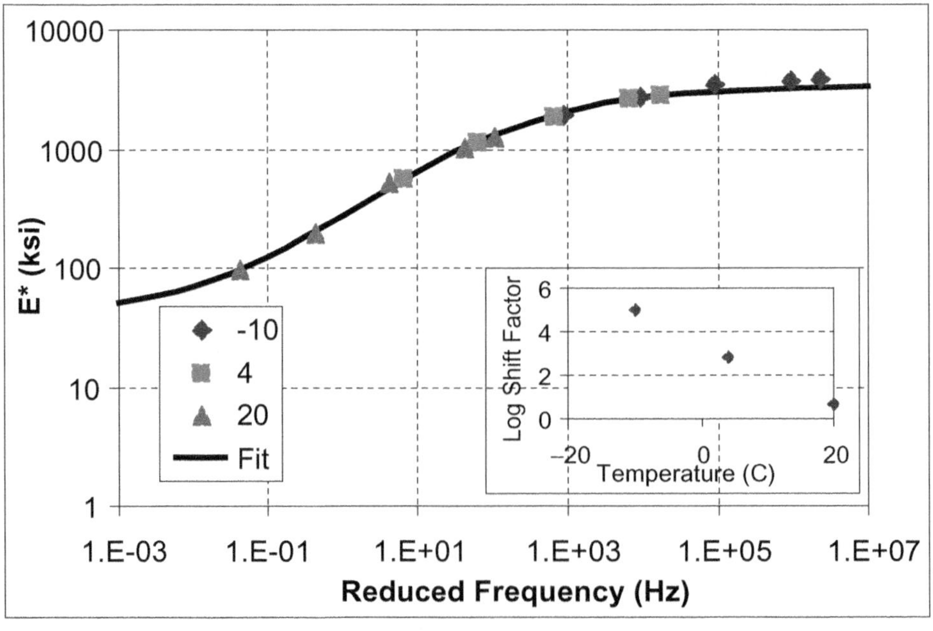

Figure 2. Complex modulus master curve for the SM-9.5 HMA (1 ksi = 6.89 MPa).

Table 1. Summary of the accelerated testing matrix.

Tire load (kN)	Tire configuration	Tire inflation pressure (kPa)	Tire Speed (km/h)	Offset from tire center (mm)	Number of passes per case
Cell A					
26	Dual (11R22.5)	551	8	Zero	
35				152	10
44	WB 455	689	16	305	
				457	
Cells B/C and D					
26	Dual (11R2.5)	550	8	Zero	
35	WB 425	689		152	
44			16	228	5
53	WB 455	758		305	
62					

2.4 *Accelerated testing program*

Using the mobile accelerated facility at University of Illinois, the testing program was conducted using various loading criteria including tire load, tire configuration, tire pressure, and speed. Each combination was conducted at 5 to 10 load cycle repetitions. With the exception of cell A, response testing parameters included two tire speeds 8 and 16 km/h, three inflation pressures 551, 689, and 758 kPa, and five load levels, 26, 35, 44, 53, and 62 kN using dual and wide-base tire configurations. However, due to its weak structure, cell A was tested using only two tire inflation pressures, i.e., 551 and 689 kPa, and three load levels, 26, 35, and 44 kN. A summary of the loading matrix is presented in Table 1. The 11R22.5 dual-tire along with two wide-base tires 425 and 455 were used in cells A, B/C, and D; with the exception of wide-base tire 425 that was not used on cell A.

3 RESULTS AND DATA ANALYSIS

A data acquisition system was developed using in-house custom software. The system consists of five modules to collect environmental and loading associated data. To quantify the effects of tire loading parameters on pavement responses, all data were shifted to a reference temperature of 25°C. The measured and shifted stresses and strains under the HMA layer were used as the basis for comparison. For example, this was accomplished by plotting the measured strain values with temperatures collected during testing. An exponential regression curve was used to best fit the data. The regression equation was then used to determine the response at the reference temperature of 25°C as shown in Figure 3. A correction factor, CF, was obtained according to the formula given in Equation 1. The original recorded response data were then multiplied by the CF in order to compare responses for the tested tires (Al-Qadi et al. 2002).

$$\text{Correction Factor} = \frac{\text{Response at reference temperature } (25^{\circ}\text{C})}{\text{Response at testing temperature}} \quad (1)$$

With exception to the environmental data, all responses were reduced by subtracting the peak value from the initial response measured at the initiation of each tire pass. This would reduce each response to an absolute value which was used in the data analysis.

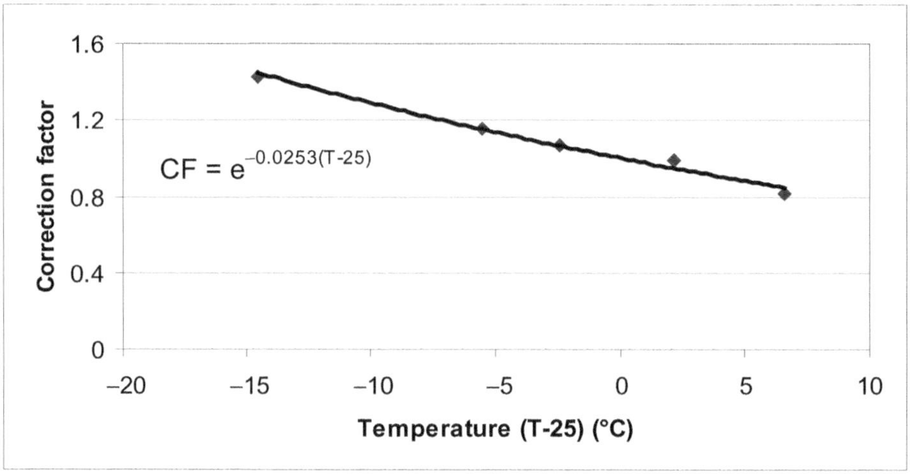

Figure 3. Correction factor applied to HMA strain at various pavement temperatures.

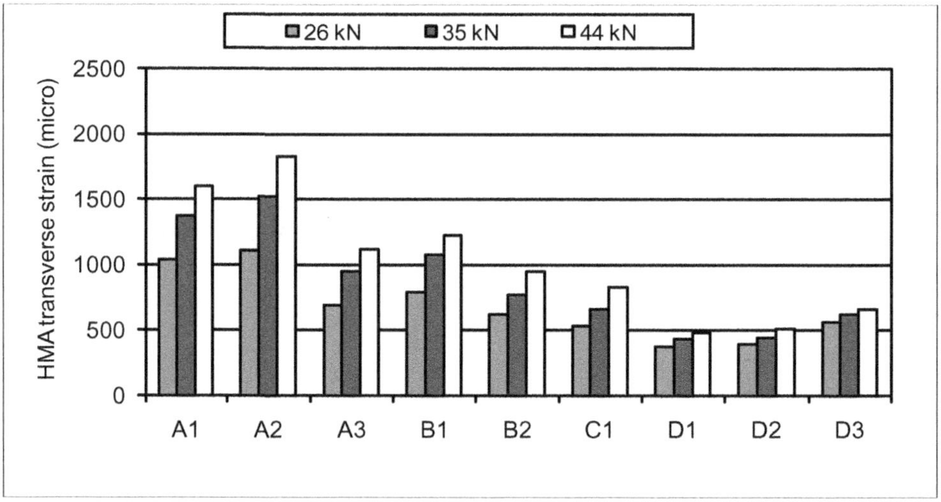

Figure 4. Effect of tire loads on HMA transverse strain.

3.1 *Effect of base layer thickness*

As expected, increasing the base layer thickness by 50% improves the pavement performance significantly by lowering the pressure, subgrade deflection, and HMA transverse strain. The reinforced sections have the lowest responses when compared to their counterpart unreinforced sections in each cell (Figure 4). Strain at the bottom of HMA was also affected by the thickness of HMA layer. The thicker the HMA layer was; the smaller the transverse strain was. However, the thicker base layer underneath the gauge could have also contributed to decreasing the HMA strain because of the increase in the supporting layer stiffness. Reinforcement could also reduce HMA strains, as shown in D3 compared to D1 and D2 and in B1 compared to B2. The trends shown in Figure 4 were similarly observed for the three tire configurations.

3.2 *Effect of speed*

The higher the tire speed, the lower the pavement response is. Recorded response data implied that vertical deflection, and HMA strain were lower at 16 km/h compared to 8 km/h for all tire configurations. As the speed decreased, the loading time over the sensor increased; and hence, the strain response values became higher. The effect of speed was more pronounced in weak pavement sections, i.e., cell A, compared to thicker ones as indicated in Figure 5. These results were observed for the three tire configurations.

3.3 *Effect of tire configurations*

As shown in Figure 5, it is clear that the wide-base 455 resulted in the lowest transverse HMA strain; while the 425 wide-base tire resulted in the highest strain. The difference between dual-tire assembly and wide-base 455 responses diminish as the stiffness of supporting layer increases. However, for a thin structure such as A2, the dual tire was the least damaging to the pavement compared to wide-base.

As shown in Figures 6 and 7, subgrade deflection and HMA transverse strain were the highest in cell A and the lowest in cell D for control and reinforced sections. The lowest pressure was in the sections having double-reinforcement (D2) and thicker HMA (C1). Subgrade vertical deflections were also in agreement with the layer thicknesses. Note that cell D pavements showed no significant differences in the subgrade vertical deflection measurements. However, it appears that geogrids have a potential benefit in reducing vertical deformations in the subgrade and the base layer. This effect is more pronounced in cells B/C and D because geogrid tends to reduce shear in the base layer (see Figures 6a and 7a). This is evident as geogrid reduces the overall longitudinal deformation of pavement structure, Figures 6c and 7c. Also, strain measurements indicated that increasing tire loading would increase the transverse strain, Figures 6b and 7b.

Operated at the design tire inflation pressure, the strain difference between the wide-base tires, 455 and 425, increases as the pavement structure thickness decreases (Figures 6 and 7). However, with respect to the vertical and longitudinal deformations, the 425 tire shows a slightly higher response compared to the 455 tire.

Figure 7c implies that the wide-base 455 tire induced less strain than the dual-tire assembly when operated at various tire loads and inflation pressures, except for transverse strain at cell A.

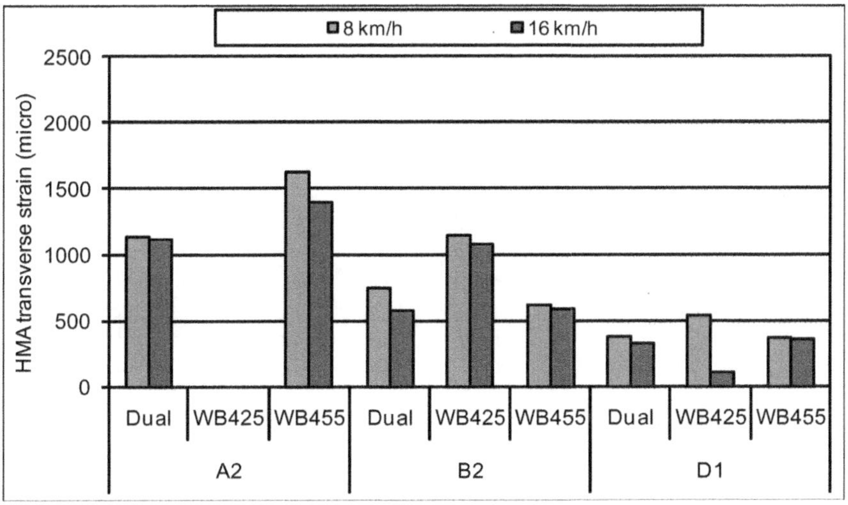

Figure 5. Effect of speed and tire configuration on HMA transverse strain.

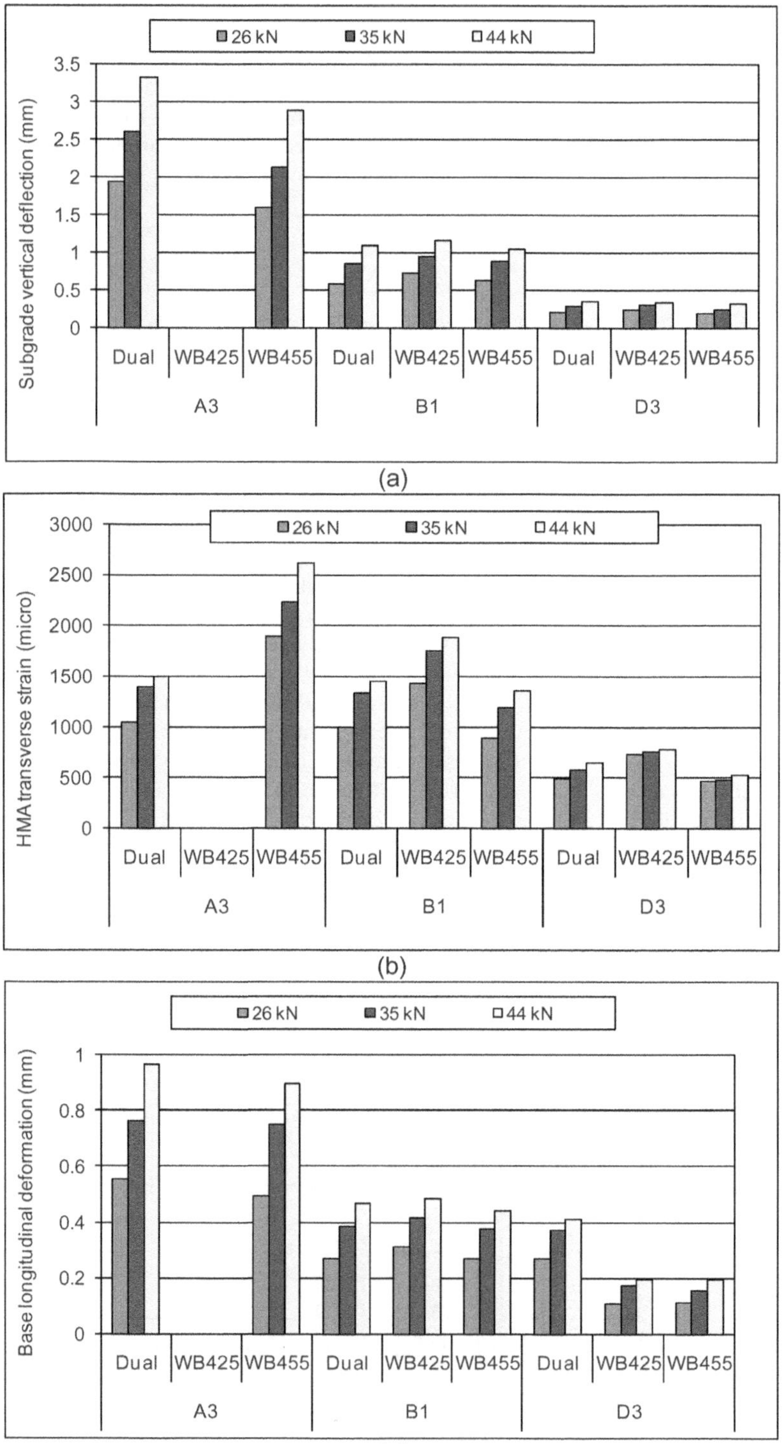

(c)

Figure 6. Control sections responses: a) subgrade deflection, b) HMA transverse strain and c) base longitudinal deformation.

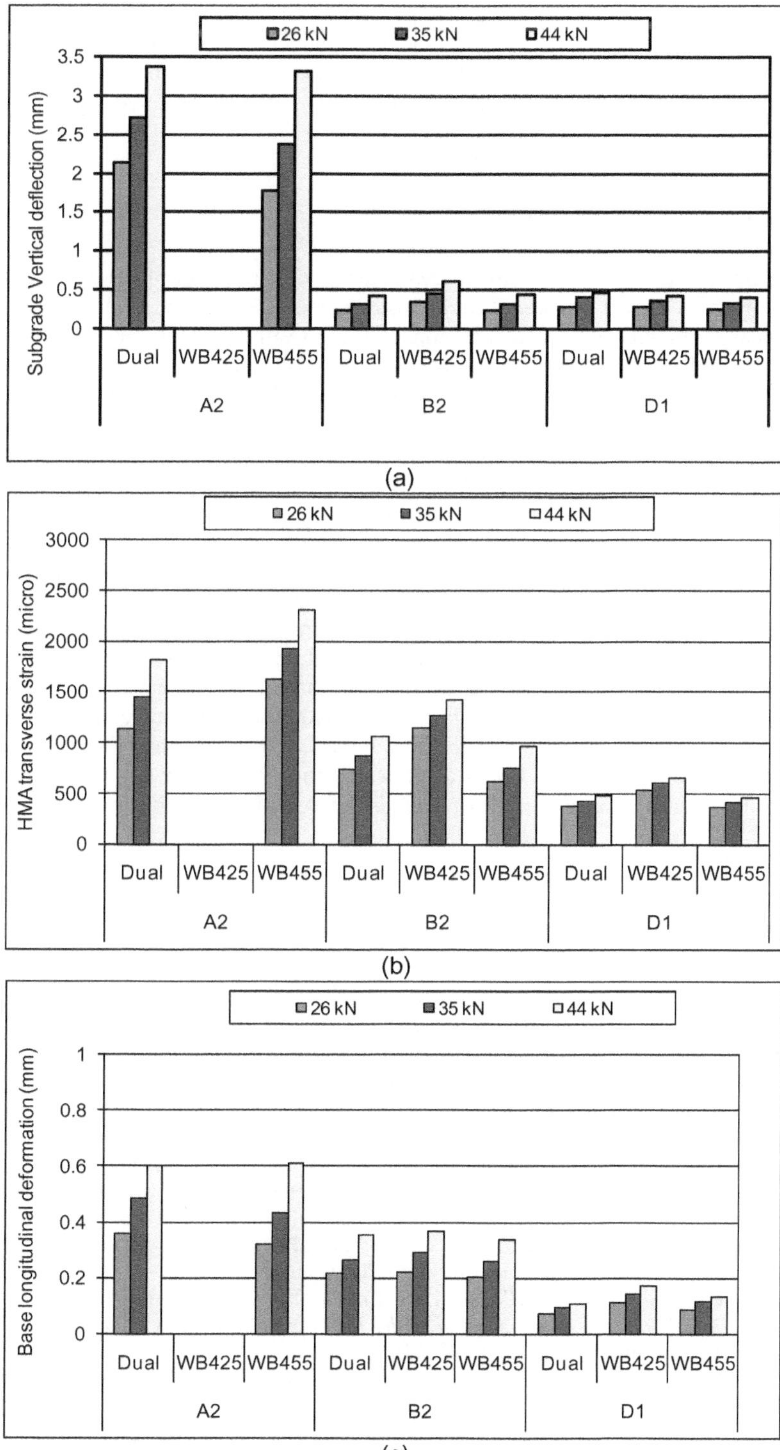

Figure 7. Reinforced sections responses include: a) subgrade deflection, b) HMA transverse strain and c) base longitudinal deformation.

In addition, it has been reported that the new generation of wide-base tire has 35% less dynamic loading than the conventional dual-tire assembly (Al-Qadi et al. 2004). This could not be captured in accelerated testing because suspensions were not used and tire loading was applied by pulling. In conclusion, in all pavement sections, the wide-base 425 was the most damaging among all tested tires.

4 SUMMARY AND CONCLUSION

Nine pavement test sections were built to evaluate the effectiveness of geogrids in low-volume flexible pavements. The sections were heavily instrumented with pressure cells, LVDTs, and strain gauges to measure moving wheel load response during testing and thermocouples, time domain reflectometers, and piezometers to capture environmental changes. The testing was conducted using the mobile Accelerated Testing Loading Assembly (ATLAS) housed at University of Illinois at Urbana-Champaign.

Instrument response analysis showed that the control sections had a greater degree of transverse strain, vertical deflection, and pressure compared to reinforced sections. The influence of tire load and speed was studied for several tire configurations such as dual-tire and wide-base 425 and 455 tires.

The study found that the 455 wide-base tire caused greater transverse strain in the thin HMA layer than the conventional dual-tire assembly when supported by relatively thin granular base layer built on weak subgrade. However, as the supporting base layer stiffness increases, the difference in strain diminishes. The measured longitudinal deflection in the base layer and vertical deflection in the subgrade were significantly lower for the 455 wide-base tire and dual-tire assembly when compared to the 425 wide-base tire. It was evident that the new generation of 455 wide-base tire caused less damage to the low-volume flexible pavements than the old generation 425 wide-base tires. With sufficient stiffness of the supporting base layer, the wide-base 455 would result in similar overall pavement performance as the dual-tire assembly when thin HMA layers are used.

REFERENCES

Al-Qadi, I.L., Loulizi, A., & Janajreh, I. 2002. Pavement response to dual and new wide-base tires at the same tire pressure. *Transportation Research Record: Journal of the Transportation Research Board* No. 1806, National Research Council, Washington D.C.: pp. 38–47.

Al-Qadi, I.L., Tutumluer, E., Kwon, J., & Dessouky, S. 2007. Accelerated full-scale testing of geogrid-reinforced flexible pavements. Paper No. 07-2317, Presented at Transportation Research Board 86th Annual Meeting, Washington, D.C.

Al-Qadi, I.L., Elseifi, M.A., & Yoo, P.J. 2004. Pavement damage due to different tires and vehicle configurations. Final Report Submitted To Michelin Americas Research and Development Corporation, Urbana, IL.

Bonaquist, R. 1992. An assessment of the increased damage potential of wide base single tires. *Proc, of the 7th International Conference on Asphalt Pavements*, Nottingham, UK: pp. 1–16.

COST 334. 2001. Effects of wide single tires and dual tires. Final Report of the Action (Version 29), European Cooperation in the field of Scientific and Technical Research.

Huhtala, M., Pihlajamäki, J., & Pienimäki, M. 1989. Effects of tires and tire pressures on road pavements. In *Transportation Research Record: Journal of the Transportation Research Board* No. 1227, National Research Council, Washington, D.C.: pp. 107–114.

Metcalf, J.B. 1996. Application of full scale accelerated pavement testing. National Cooperative Highway Research Program, Synthesis 235, Washington, D.C.

Pascale, P., Doré, G., & Prophète, F. 2004. Characterization of tire impact on the pavement behaviour. *Canadian Journal of Civil Engineering*, Vol. 31: pp. 860–869.

Priest, A.L., Timm, D.H., & Barrett, W.E. 2005. Mechanistic comparison of wide-base single vs. standard dual tire configurations. Draft Report, National Center of Asphalt Technology. University of Auburn, AL.

Sebaaly, P.E., 1992. Pavement damage as related to tires, pressures, axle loads, and configurations. Vehicle, tire, pavement interface. ASTM STP 1164, J.J. Henry and J.C. Wambold (eds), American Society for Testing and Materials, Philadelphia: pp. 54–68.

Sebaaly, P.E., & Tabatabaee, N. 1989. Effect of tire pressure and type on response of flexible pavement. *Transportation Research Record: Journal of the Transportation Research Board* No. 1227, National Research Council, Washington D.C.: pp. 115–127.

Siddharthan, R.V., Yao, J., & Sebaaly, P.E. 1998. Pavement strain from moving dynamic 3D load distribution. *Journal of Transportation Engineering,* American Society of Civil Engineers: pp. 557–566.

Turner Fairbank Highway Research Center (TFHRC). Pavement Testing Facility, Federal Highway Administration (FHWA), Accelerated Loading Facility (ALF). http://www.tfhrc.gov/pavement/utwweb/facilit.htm.

Efficient Transportation and Pavement Systems – Al-Qadi, Sayed, Alnuaimi & Masad (eds)

Characterizing the behavior of geogrid-reinforced crushed limestone base materials using monotonic and cyclic load tests

M. Abu-Farsakh, M. Nazzal & L. Mohammad
Louisiana Transportation Research Center, Louisiana State University, Baton Rouge, LA, USA

ABSTRACT: A series of monotonic triaxial compression tests and cyclic triaxial tests were conducted on unreinforced and geogrid-reinforced crushed limestone samples to evaluate the effects of geogrid type/stiffness, location and number of layers on the strength, resilient modulus and permanent deformations of these samples. Five different geogrid types were used in this study. For each geogrid type, four different reinforcement arrangements were investigated. The results indicated that the geogrid inclusion within crushed limestone samples significantly increased their elastic modulus and ultimate shear strength, and reduced their permanent deformations. This improvement was due to the geogrid stiffness and arrangement. However, the effect of the geogrid stiffness was found to be less significant than the effect of the geogrid arrangement. Finally, the results demonstrated that the geogrid did not have a significant effect on the resilient behavior of the crushed limestone samples.

1 INTRODUCTION

Over the last two decades, geogrids have been studied and used as reinforcement in the base course layer of flexible pavements. The attraction of this application lies in the possibility of reducing the thickness of the base course layer and/or improving the performance and extending the service life of the pavement structure. Many studies conducted tests on large-scale reinforced pavement sections showed that geogrids were able to mechanically improve the overall strength and stability of reinforced pavement structures and reduce their accumulated permanent deformation when placed within the base course layer (e.g. Al-Qadi et al., 1994; Collin et al., 1996; Berg et al., 1996; and Perkins, 2002), and investigate their benefits. These studies showed that geosynthetic reinforcements were able to mechanically improve the overall strength and stability of reinforced pavement structures and reduce the accumulated permanent deformation.

Few studies, however, were conducted to characterize the behavior of geogrid reinforced base course materials using monotonic and Repeated Load Triaxial tests (RLT) (e.g. Moghaddas-Nejad and Small 2003; Perkins et al., 2004; Nazzal et al., 2006; Abu-Farsakh et al., 2007). In general, these studies showed that the geogrid inclusion had an appreciable effect on the permanent deformation properties of unbound granular base material, while it did not have a significant effect on its resilient modulus properties. Moghaddas-Nejad and Small (2003) indicated that for a particular confining stress, the effect of a geogrid on the reduction in permanent deformation increases rapidly with an increase in the deviator stress, until a peak is reached, then decreases gradually. While Perkins et al. (2004) reported that the effect of the geogrids on permanent deformation was not significant until a mobilized friction angle of approximately 30 degrees was reached.

This paper aims at evaluating the influence of the geogrid type, location, and number of layers on the strength properties as well as permanent and resilient behavior of crushed limestone samples under monotonic and cyclic loading.

2 TESTING PROGRAM

2.1 *Materials properties*

2.1.1 *Crushed limestone*

The tested crushed limestone material has a maximum size of 19 mm, and a D_{10} and D_{60} sizes of 0.18 mm and 6 mm, respectively, giving it a uniformity coefficient of 30. In addition, it is classified as A-1-a and GW-GC according to the American Association of State Highway and Transportation (AASHTO) classification system and the Unified Soil Classification System (USCS), respectively. The tested material has a maximum dry unit weight of 17.2 kN/m^3 and an optimum moisture content of 7.0%, as measured by the standard proctor test. Finally, the material has an internal friction angle of 48° and a specific gravity value of 2.72.

2.1.2 *Geogrid*

The reinforcement material used in this study included five different types of biaxial geogrids, namely BX-6100, BX-1100, BX-6200, BX-1200, and BX-1500. Where BX-6100 represents the lowest stiffness geogrid and BX-1500 represents the stiffest geogrid. These geogrids are typically used to reinforce base course layer in pavement structures. The physical and mechanical properties of these products as reported by the manufacturer are presented in Table 1. Type I, II, III, IV, and V will be used thereafter to refer to BX-6100, BX-1100, BX-6200, BX-1200, and BX-1500, respectively.

2.2 *Sample preparation*

All crushed limestone samples were prepared following the AASHTO-T307 resilient modulus test procedure for untreated granular base materials. All samples were prepared with 150 mm diameter and 300 mm height using a split mold. The material were first mixed with water and then placed within the split mold and compacted using a vibratory compaction device to achieve the maximum dry density measured in the standard Proctor test. To achieve a uniform compaction throughout the thickness, samples were compacted in six-50 mm layers. The achieved dry densities of the prepared samples were within ±1 percent of the target value. Samples were enclosed in two latex membranes with a thickness of 0.3 mm. For reinforced samples, the geogrid was placed

Table 1. Physical and mechanical properties of geogrids.

		Tensile stiffness @ (strain %)[a]		Aperture dimension		Flexural stiffness[b]
Geogrid		MD[c] (kN/m)	CMD[d] (kN/m)	MD (mm)	CMD (mm)	(g-cm)
Type I	BX-6100	250 (2%) 550 (5%)	380 (2%) 720 (5%)	33	33	250
Type II	BX-1100	280 (2%) 580 (5%)	450 (2%) 920 (5%)	25	33	750
Type III	BX-6200	380 (2%) 720 (5%)	510 (2%) 1000 (5%)	33	33	250
Type IV	BX-1200	410 (2%) 810 (5%)	650 (2%) 1340 (5%)	25	33	750
Type V	BX-1500	580 (2%) 1200 (5%)	690 (2%) 1370 (5%)	25	30.5	2000

[a] Measured in accordance with the procedure of ASTM D6637.
[b] Measured in accordance with the cantilever test of ASTM D-5732-95.
[c] MD: machine direction.
[d] CMD: cross-machine direction.

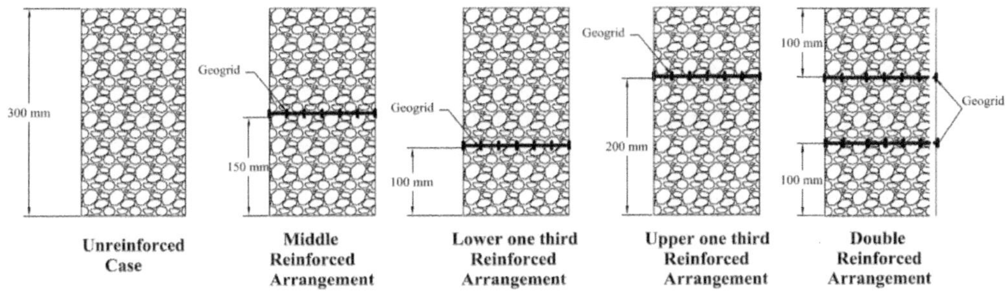

Figure 1. Reinforcement arrangements investigated in this study.

horizontally between layers at the desired locations. Four different arrangements of reinforcement were investigated. A sketch describing the four reinforcement arrangements investigated in this study is shown in Figure 1.

2.3 *Triaxial compression tests*

Drained triaxial compression tests were conducted on both unreinforced and reinforced samples. The tests were conducted with a strain rate less than 10% strain per hour to ensure that no excess pore water will develop during testing. In each test, the sample was loaded to a strain level of 1%, unloaded, and then reloaded to failure. Drained triaxial compression tests were first performed on unreinforced samples at three different confining pressures (21, 48, and 69 kPa) to obtain the strength properties of the crushed limestone. They were then performed at 21 kPa confinement pressure on reinforced samples. Three replicate samples were tested for each case to ensure repeatability, for a total of 69 triaxial compression tests.

2.4 *Cyclic triaxial tests*

2.4.1 *Resilient modulus tests*

Cyclic triaxial tests were performed to determine the resilient modulus of crushed limestone material. In this test, a series of steps consisting of different levels of cyclic deviatoric stress are followed such that the resilient modulus is measured at varying normal and shear stress levels. The cyclic loading consists of repeated cycles of a haversine shaped load-pulse. These load pulses have a 0.1 sec load duration and 0.9 sec rest period. Cyclic triaxial tests were conducted on unreinforced and reinforced samples. Three different arrangements were investigated for each reinforcement type (middle, upper one third, and double arrangement). Three replicate samples were tested for each case to ensure repeatability. A total of 48 resilient modulus tests were conducted.

2.4.2 *Permanent deformation tests*

Repeated Load Triaxial (RLT) tests were performed to determine the permanent deformations of unreinforced and reinforced crushed limestone samples at different number of load cycles. The tests consisted of applying 10,000 load cycles at a constant confining pressure (21 kPa) and peak cyclic stress (230 kPa). Each cycle consisted of the same load pulse used in resilient modulus tests. All samples were conditioned before the test by applying 1,000 repetitions of a specified cyclic deviatoric stress. The RLT tests were conducted on unreinforced samples and samples reinforced with geogrid types II, IV, and V. Three different arrangements, similar to resilient modulus tests, were investigated for each geogrid type, and three replicate samples were tested for each case.

3 DISCUSSION OF TEST RESULTS

3.1 *Triaxial compression test results*

The resulted stress-strain curves obtained from the triaxial compression tests conducted on unreinforced crushed limestone samples and samples reinforced with geogrid types I and V are presented in Figures 2a and 2b respectively. It is clear that the inclusion of geogrid reinforcement layer/s improved the strength and stiffness of the crushed limestone material substantially. This improvement was more pronounced at strain levels greater than 1%. This improvement can be attributed to the increase in the lateral confinement due to the presence of geogrid reinforcement. The placement of a geogrid layer(s) within the sample allows the development of shear interaction and interlocking between the aggregate and the geogrid. The shear stress is then transmitted from the aggregate to the geogrid, which places the geogrid in tension. The improvement of geogrid reinforcement under monotonic loading was evaluated at two response parameters, the secant elastic modulus at 2% strain level ($E_{s2\%}$) that corresponds to the average reinforcement strain in pavements, and the ultimate shear strength (USS). Improvement factors IM-$E_{s2\%}$, and IM-USS were then determined using the following equation:

$$\mathrm{IM} = \frac{\text{parameter from reinforced sample}}{\text{parameter from unreinforced sample}} \tag{1}$$

Figures 3a and 3b present the average improvement factor and the standard deviation values for each reinforced case. The figures show that the improvement depended on the type, location, and number of geogrid layers. It can be seen clearly that the improvement increased with increasing the geogrid stiffness. In addition, the double arrangement had the maximum improvement. The figures also show that the improvement was more appreciable in USS than $E_{s2\%}$.

3.2 *Resilient modulus test results*

The resilient modulus (M_r) for the different unreinforced and reinforced samples were calculated at a confining stress of 21 kPa, and three deviatoric stresses: 80 kPa, 160, and 250 kPa. Figures 4a and 4b present the M_r values obtained for unreinforced samples and samples reinforced with geogrid types I and V, respectively. It is clear that the resilient modulus values increased with

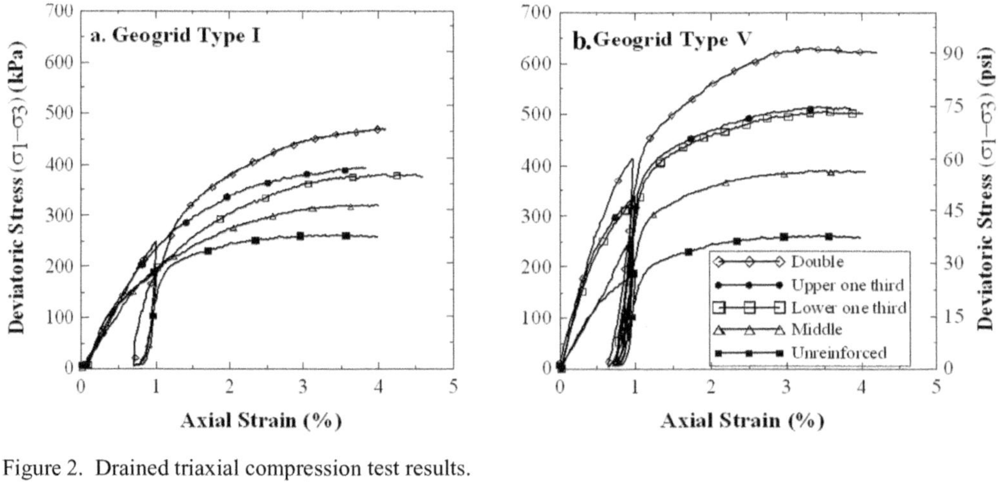

Figure 2. Drained triaxial compression test results.

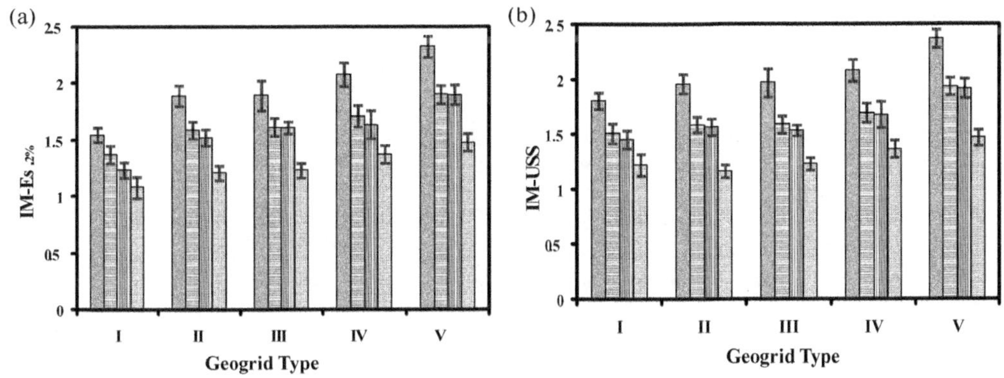

Figure 3. Improvement factors: a) IM-$E_{s2\%}$, and b) IM-USS.

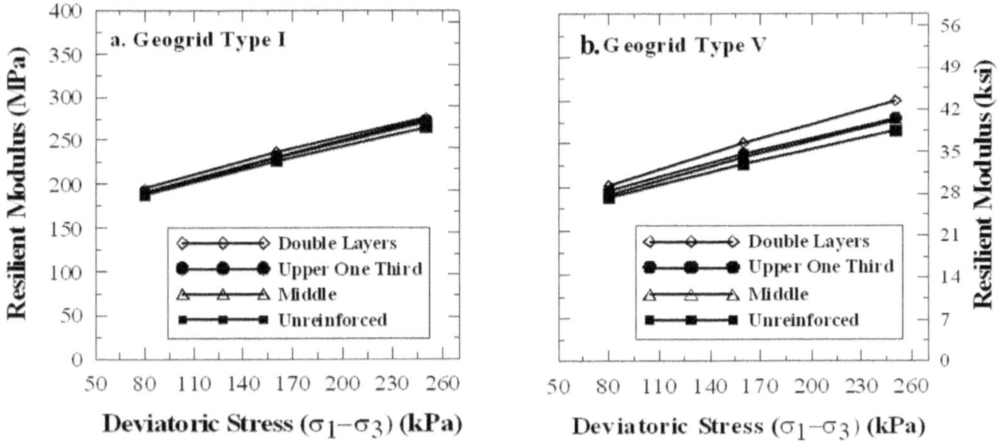

Figure 4. Resilient modulus values at different deviatoric stresses.

increasing the deviatoric stress. A slight improvement in resilient modulus was detected only for samples reinforced with two layers of geogrid Type V at high deviatoric stresses. However, no conclusion can be drawn since this improvement lies within the margin of error of the calculated values.

3.3 *Permanent deformation test results*

The curves of the average permanent strain value versus number of load cycles for unreinforced samples and samples reinforced with geogrid types II and V are shown in Figures 5a and 5b, respectively. It can be seen that the geogrid reinforcement resulted in a reduction in the permanent strain. This reduction was more significant at larger number of load cycles, which are associated with higher strain levels.

Based on the results of the permanent deformation tests, the reduction in the permanent strain at 10,000 load cycles ($RPS_{10,000}$) due to geogrid reinforcement was determined as follows:

$$RPS_{10,000}\,(\%) = \frac{\text{permanent strain without geogrid - permanent strain with geogrid}}{\text{permanent strain without geogrid}} \times 100\% \qquad (2)$$

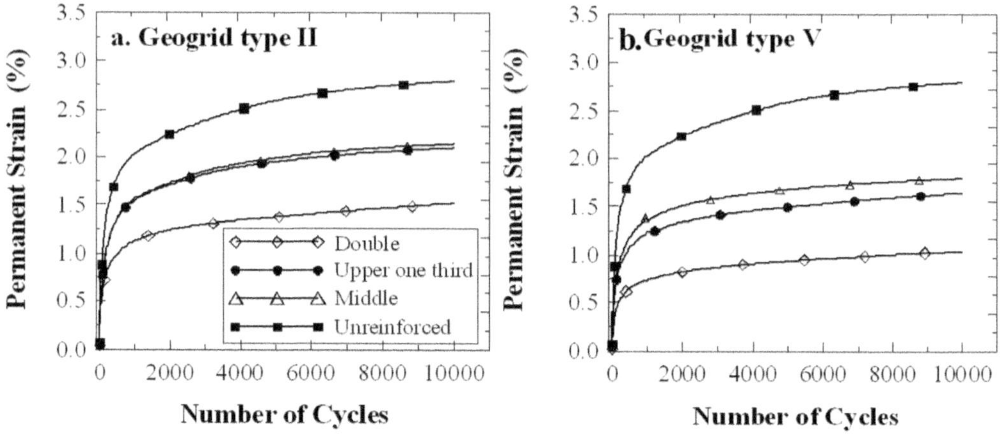

Figure 5. Permanent deformation curves.

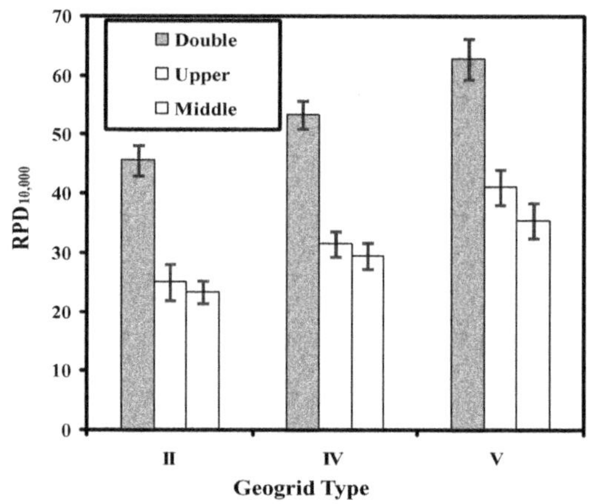

Figure 6. Reduction in permanent deformation.

The average $RPS_{10,000}$ and the standard deviation values for the reinforced cases investigated are presented in Figure 6. The inclusion of the geogrid reinforcement resulted in a reduction of the permanent strain up to 65%. Similar to the triaxial compression test results, the geogrid improvement in permanent deformation test depended on the geogrid type, location, and the number of reinforcement layers; such that stiffer geogrids exhibited a higher reduction in permanent strains than the ones with the least stiffness, as can be seen for type V geogrids compared to type II geogrids. The figure shows that the upper one third location had a better improvement than the middle location, while the maximum improvement was observed when using two geogrid layers.

4 CONCLUSIONS

A series of monotonic and cyclic triaxial tests were conducted on unreinforced and geogrid reinforced crushed limestone samples to evaluate the effects of geogrid location, number of layers, and stiffness on the strength properties and stress-strain response of reinforced samples. Based on this study, the following conclusions can be drawn:

- The test results showed that the inclusion of geogrid reinforcement layer/s significantly improves the compressive strength and stiffness and reduces the permanent deformation of crushed limestone material. The geogrid improvement was more pronounced at higher strain levels.
- The improvement was found to be a function of the geogrid location, type, and number of layers. At certain geogrid location, stiffer geogrids exhibited greater benefits. For a specific geogrid type, the highest improvement was achieved when using two geogrid layers placed at the upper and lower third of the sample height; while the lowest improvement was found for samples reinforced with a single geogrid layer placed at the mid-height.
- The results of the cyclic triaxial tests showed that the geogrid reinforcement did not have significant effect on the resilient behavior of crushed limestone samples.

ACKNOWLEDGMENTS

This research project is funded by the Louisiana Transportation Research Center (LTRC Project No. 05-5GT) and Louisiana Department of Transportation and Development (State Project No. 736-99-1312). The authors gratefully acknowledge the help and advice of Zhongjie Zhang, Pavement and Geotechnical Administrator at LTRC.

REFERENCES

Abu-Farsakh, M.Y., Nazzal M., and Mohammad, L. "Effect of Reinforcement on Resilient and Permanent Deformations of Base Course Material," *Journal of the Tranportation Research Board.*

Al-Qadi, I.L., Brandon, T.L., Valentine, R.J., Lacina, B.A., and Smith, T.E. (1994). "Laboratory evaluation of geosynthetic reinforced pavement sections." *Transportation Research Record, No. 1188*, 25–31.

Collin, J.G., Kinney, T.C., and Fu, X. (1996). "Full scale highway load test of flexible pavement systems with geogrid reinforced base courses." *Geosynthetics International*, Vol. 3, No. 4, 537–549.

Moghaddas-Nejad, F.J.C., and Small (2003). "Resilient and permanent characteristics of reinforced granular materials by repeated load triaxial tests." *ASTM Geotechnical Testing Journal,* Vol. 26, No. 2, 152–166.

Nazzal, M.D., Abu-Farsakh M.Y., and Mohammad, N.L. (2006). "Laboratory Characterization of Reinforced Crushed Limestone Material," *ASCE Journal of Materials* (In press).

Perkins, S.W. (2002). "*Evaluation of geosynthetic reinforced flexible pavement systems using two pavement test facilities*." Report No. FHWA/MT-02-008/20040, Department of Transportation, Federal Highway Administration, Washington, DC.

Perkins, S.W., Christopher, B., Cuelho, E.L., Eiksund, G.R., Hoff, I., Schwartz, C.W., Svano, G., and Want, A. (2004). "Development of design methods for geosynthetic reinforced flexible pavements." A report prepared for the U.S. Department of transportation, Federal Highway Administration, FHWA/DTFH61-01-X-00068, p. 263.

Efficient Transportation and Pavement Systems – Al-Qadi, Sayed, Alnuaimi & Masad (eds)
© 2009 Taylor & Francis Group, London, ISBN 978-0-415-48979-9

Model footing tests on geosynthetic reinforced soils

Q. Chen & M. Abu-Farsakh
Louisiana Transportation Research Center, Louisiana State University, Baton Rouge, LA, USA

R. Sharma
Louisiana State University, Baton Rouge, LA, USA

ABSTRACT: A series of laboratory model footing tests were conducted to investigate the behavior of foundations on geosynthetic reinforced sand and embankment silty clay soils for application to solve the bridge approach slab problem. The parameters investigated in this study included the top layer spacing, the number of reinforcement layers, vertical spacing between layers, tensile modulus and type of reinforcement. The effect of reinforcement on the vertical stress distribution in the soil and the strain distribution along the reinforcement was also investigated. The test results showed that the inclusion of reinforcement can appreciably improve the soil's bearing capacity and reduce the footing settlement. The geogrids with higher tensile moudulus performed better than geogrids with lower tensile modulus. The test results also showed that the inclusion of reinforcement can redistribute the applied load to a wider area, thus reducing the stress concentration and having more uniform stress distribution, which will result in reducing the consolidation settlement. The test results also showed that the effective length of reinforcement is about six times the footing width.

1 INTRODUCTION

Geosynthetic reinforcements have been used in many engineering applications, such as MSE walls, slopes, embankments, and pavements. Recently, there has been an increased concern toward the use of gosynthetics to reinforce soil foundations. One potential implementation is the use of reinforced soil foundation (RSF) in the design of approach slab for highway engineering applications to minimize the resulting differential settlements. The Louisiana Quality Initiative study recently recommended changing the design of approach slabs by increasing its rigidity. As a result, the slab and traffic loads will be carried by the two ends of the slab rather than distributed over the length of the slab. Accordingly, a footing will be needed at the far end of the approach slab away from the bridge. To increase the soil's bearing capacity and minimize settlement due to the concentrated load, the soil underneath the footing needs to be reinforced or replaced with granular fill in combination with geosynthetics reinforcement.

In the past thirty years, a significant amount of research effort has been made to investigate the behavior of the RSF for different soil types. Several experimental studies were conducted to evaluate the bearing capacity of footings on reinforced sandy soil (e.g., Guido et al., 1986; Huang and Tatsuoka, 1990; Yetimoglu et al., 1994; Adams and Collin, 1997) and clayey soil (e.g., Saki and Das, 1987; Ramaswamy and Purushothaman, 1992; Shine et al., 1993; Das et al., 1994). All these works indicated that the use of reinforcements can significantly increase the soil's bearing capacity and reduce the settlement of footing. Binquet and Lee (1975) identified three possible failure modes depending on the configuration and tensile strength of reinforcement. These failure modes are: shear failure above the reinforcement, pullout of reinforcement and tie break of reinforcement. They also developed a design method for a strip footing on reinforced sand based on the concept of tension membrane effect. Huang and Tatsuoka (1990) presented two mechanisms that can describe the increase in bearing capacity of RSF: deep footing mechanism and wide-slab mechanism. They substantiated the strain restraining effect (confinement effect) by successfully using short reinforcement with a length (l) equal to the footing width (B) to reinforce sand.

The main objective of this paper is to investigate the behavior of footings on reinforced sand as well as embankment silty clay. For this purpose, extensive laboratory model tests were conducted. The parameters investigated in the model tests include the top layer spacing, the number of reinforcement layers, vertical spacing between layers, tensile modulus and type of reinforcement. The experimental study also includes investigating the stress distribution in sand and silty clay, and the strain distribution along the reinforcement.

2 MODEL LOADING TESTS

2.1 *Material properties and model foundation*

A uniform sand with a mean particle size (D_{50}) of 0.45 mm was used in the present study. The uniformity coefficient (C_u) and coefficient of curvature (C_c) were 2.07 and 1.25, respectively. This sand was classified as SP according to the Unified Soil Classification System (USCS). Large scale (304.8 mm × 304.8 mm × 130.9 mm) direct shear tests on this sand at densities ranging from 1,686 to 1,764 kg/m^3 revealed internal friction angles of 44° to 48°. The embankment clayey soil used for the model tests has a liquid limit of 31 and a plastic index of 15. The maximum dry density of the silty clay is 1,670 kg/m^3 with an optimum moisture content of 18.75% as determined by Standard Proctor test. This clayey soil was classified as CL-ML according to the USCS. Large scale direct shear tests on this clayey soil at densities ranging from 1,525 kg/m^3 to 1,763 kg/m^3 revealed internal friction angles between 24.1° and 26.0° and cohesion intercepts between 5.1 kPa and 24.6 kPa.

The model footing used in the tests was a steel plate with dimensions of 152 mm × 152 mm (B × L) and 25.4 mm thickness. The model tests were conducted in a 1.5 m long, 0.91 m wide, and 0.91 m deep steel test box. The testing procedure was performed according to the ASTM D 1196-93 (ASTM 1997), where the load increments were applied and maintained until the rate of settlement was less than 0.03 mm/min over three consecutive minutes. The load and the corresponding footing settlement were measured by a ring load cell and two dial extension gauges, respectively.

Four types of geogrids: GG1, GG2, and GG3, and one type of geotextile, GT1, were used as reinforcement in the tests. A composite, GGT1, which is a combination of GG2 geogrid and GT1 geotextile (i.e. GT1 geotextile is placed directly on the top of GG2 geogrid to form a new reinforcement) was also used in the present study. The physical and mechanical properties of these reinforcements as provided by the manufacturers are listed in Table 1.

Table 1. Properties of geosynthetics.

Reinforcement	Polymer Type	E[a], kN/m		Aperture Size, mm
		MD[b]	CD[c]	
GG1 geogrid	Polyester	365	365	25.4 × 25.4
GG2 geogrid	Polypropylene	182	255	33 × 33
GG3 geogrid	Polypropylene	274	372	33 × 33
GT1 geotextile	Polypropylene	700	965	≈0

[a] Tensile Modulus (at 2% strain) per unit width,
[b] Machine Direction,
[c] Cross machine direction.

2.2 *Section preparation and compaction control*

The soil was placed and compacted in lifts inside the steel test box. The thickness of each lift varied from 25 mm to 102 mm depending on reinforcement spacing. The test samples were prepared by hand mixing the soil and water. The amount of soil needed for each lift was first calculated; then, the soil was poured into the box, raked level, and compacted using a 203 mm × 203 mm plate adapted to a vibratory jack hammer to the predetermined height. A nuclear density gauge was used to measure the density for each lift for construction control. The dry densities for sand with/without reinforcement varied from 1,690 to 1,763 kg/m^3 with moisture contents ranging from 4.5 to 5%. Similarly, the dry densities varied from 1,601 to 1,709 kg/m^3 for reinforced clay with moisture contents ranging from 18 to 18.5%.

3 TEST RESULTS AND ANALYSIS

3.1 *Effect of reinforcement's top spacing*

The optimum location of the first reinforcement layer (top layer spacing, u) was investigated for both sand and silty clay. Figures 1a and 1b show the variation in the bearing capacity ratio (BCR) values of the loads corresponding to different settlement ratios as a function of the top layer spacing ratio (u/B) for sand and silty clay, respectively. Here BCR is defined as the ratio of the bearing capacity of the RSF to that of the unreinforced soil foundation; and top layer spacing ratio (u/B) is defined as the ratio of the top layer spacing (u) to the footing width (B). The optimum top layer spacing ratio can be estimated to be about 0.33B for both sandy and clayey soil.

3.2 *Effect of number of reinforcement layers*

The variations of BCRs obtained at different settlement ratios for different numbers of reinforcement layers (N) and reinforcement depth ratios (d/B) are shown in Figures 2a and 2b for reinforced sand and silty clay, respectively. The reinforcement depth ratio is defined as the ratio of the total depth of reinforcement (d) to footing width (B). It can be seen from these Figures that the BCRs increase with N and d/B, and appear to become almost constant after $N = 3$ and $N = 4$ which are located at depths of 1.00B and 1.33B for reinforced sand and reinforced silty clay, respectively. Accordingly, the influence depth can be estimated to be 1.25B for reinforced sand and 1.5B for reinforced silty clay.

3.3 *Effect of vertical spacing of reinforcement layers*

Figures 3a and 3b depict the variations in the BCR values of the loads corresponding to different settlement ratios as a function of the vertical spacing ratio (h/B), which is defined as the ratio

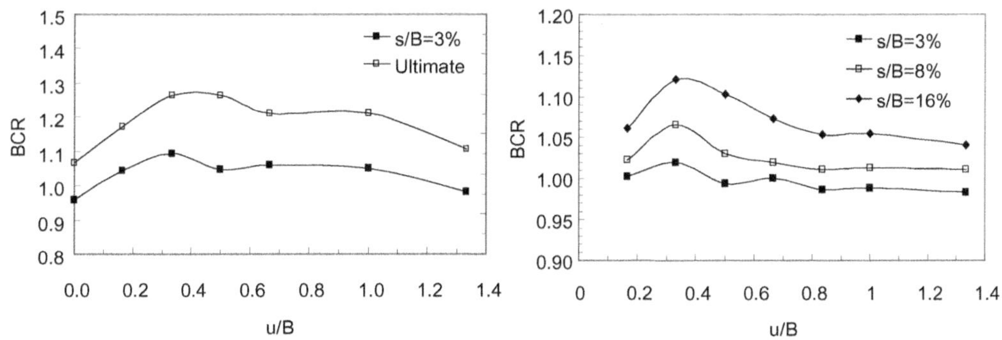

Figure 1. BCR versus u/B for one layer of reinforcement.

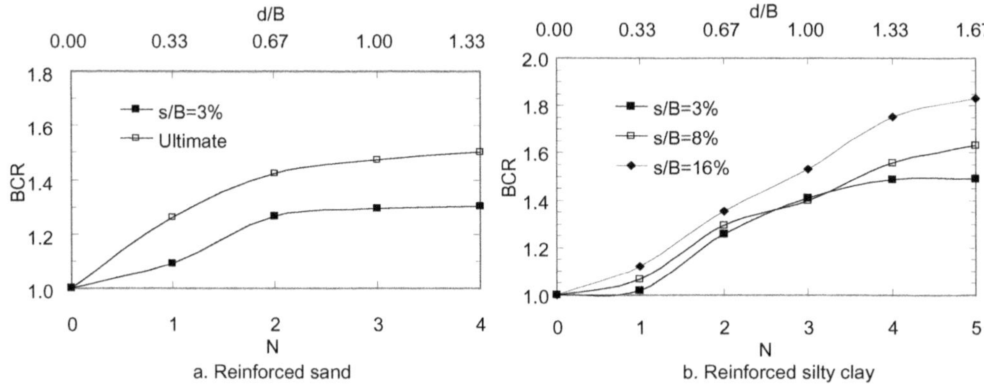

Figure 2. BCR versus *N* and *d/B* at different settlement ratio (*s/B*).

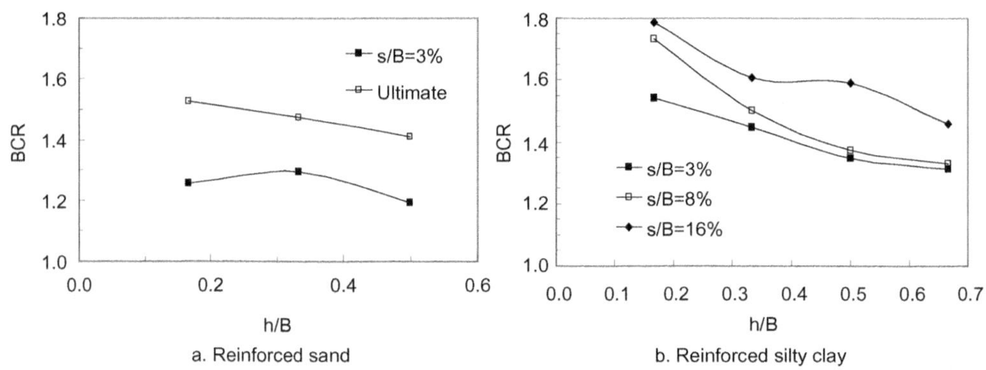

Figure 3. BCR versus *h/B* at different settlement ratio (*s/B*).

of the vertical spacing of reinforcement layers (h) to the footing width (B). It is obvious that the BCR values decrease with increasing vertical spacing of reinforcement layers with maximum BCR at $h = 0.167B$ for both reinforced sand and silty clay. No optimum vertical spacing was obtained. However, for the reinforced sand and reinforced silty clay tested in this study, one can realize that the smaller the spacing, the higher the BCR. For design purpose, engineers need to balance between reducing spacing and increasing geogrid modulus. The authors believe a value of $h/B = 0.33$ can be a reasonable value for use in the design of RSFs.

3.4 *Effect of tensile modulus and type of reinforcement*

Figures 4a to 4b compare the pressure-settlement curves obtained for different types of reinforcements on model footing tests conducted with multiple reinforcement layers placed at top layer spacing and a vertical spacing of 51 mm. As shown in these figures, this study demonstrates that the performance of reinforced sandy and clayey soil improves with increasing geogrid tensile modulus, but the effect of the tensile modulus seems a function of settlement. These figures also show that the sand and silty clay reinforced by geogrids performs better than those reinforced by geotextiles at relatively low settlement. However, the response of geotextile is better than geogrids after reaching a certain amount of settlement. This behavior can be attributed to the slack of woven

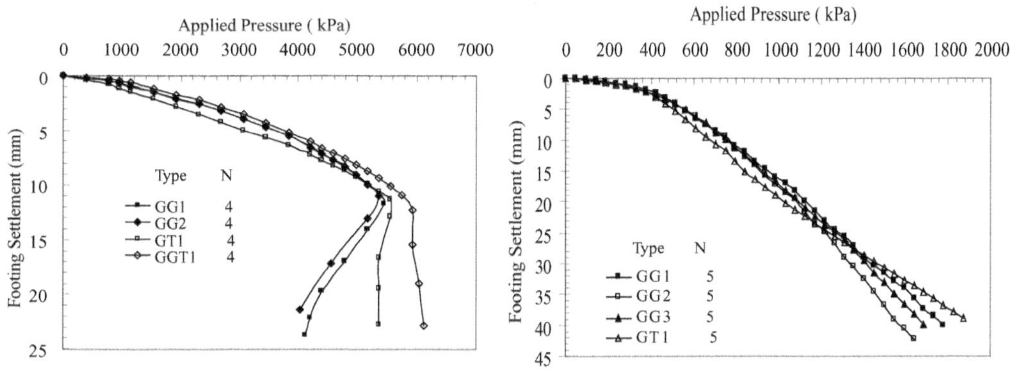

a. Reinforced sand

b. Reifnorced silty clay

Figure 4. Pressure-settlement curves for model footing tests.

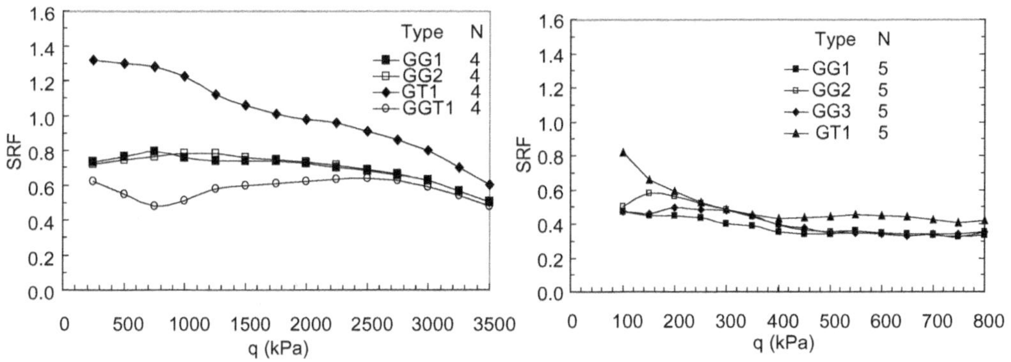

a. Reinforced sand

b. Reinforced silty sand

Figure 5. SRF versus applied footing pressure (q).

geotexitle. The slack of woven geotextile can be caused by stretching of woven, test setup, or both. Figure 4a also indicates that the sand reinforced by GGT1 composite performed better than that reinforced by either geogrid or geotextile alone.

The settlement reduction factors (SRF) at different footing pressure (q) for the model tests with different types of reinforcement are presented in Figures 5a and 5b for reinforced sand and silty clay, respectively. The SRF is defined here as the ratio of the immediate settlement of the footing on a reinforced soil to that on an unreinforced soil at a specified footing pressure. It is obvious that the inclusion of reinforcement would reduce the immediate settlement significantly. With three or more layers of reinforcement, the settlement in silty clay soil can even be reduced by 50% at a relatively medium footing pressure of 400 kPa. With two or more layers of geogrid, the settlement in sand can be reduced by 20% at all pressure levels.

3.5 *Stress distribution in soil*

Several laboratory model tests were conducted to evaluate the stress distribution in sandy and clayey soil with and without reinforcement inclusion. Pressure cells were placed at specified locations/depths for this purpose. The measured stress distributions along the center line of the footing

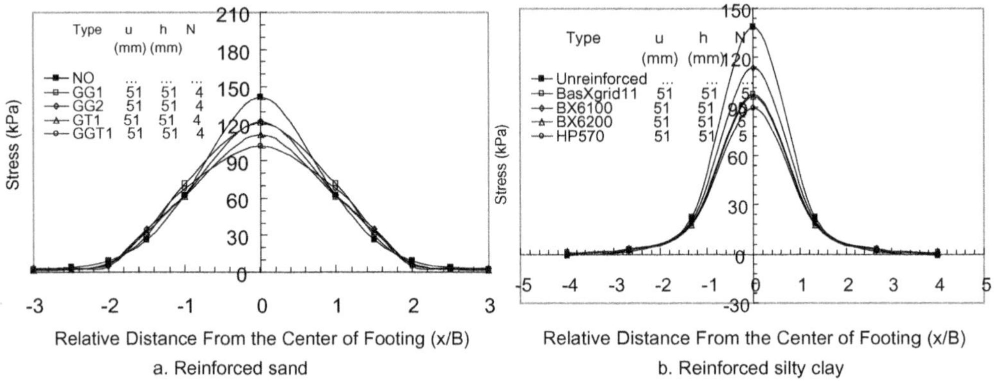

Figure 6. Stress distributions along the center line of footing at a depth of 1.67*B* below the footing.

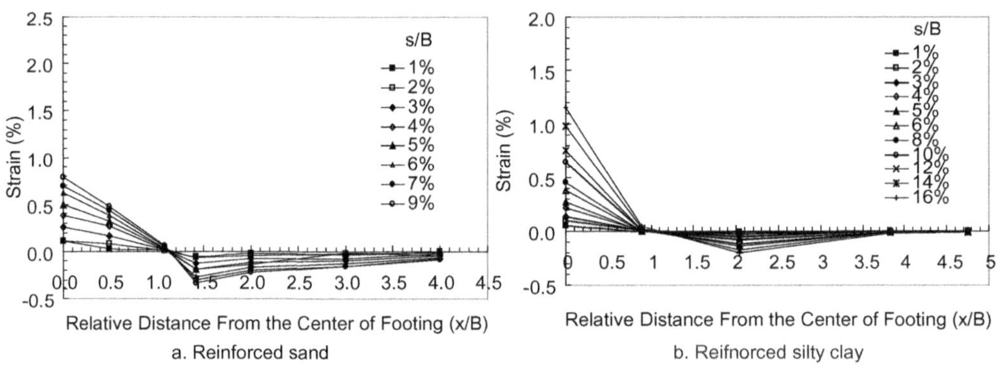

Figure 7. Strain distributions along the center line of geogrid.

at a depth of 254 mm (1.67B) below the footing are shown in Figures 6a and 6b for reinforced sand and silty clay soil, respectively. It should be noted that the stresses measured here by the pressure cells are the total vertical stresses induced by the applied load, not including the stresses induced by the weight of soil. As can be seen from these Figures, the induced stresses beneath the center of the footing in both reinforced sand and reinforced silty clay are appreciably reduced compared to those without reinforcement. This reduction in stress distribution below influence depth is expected to decrease the consolidation settlement of the footing, which is directly related to the induced stresses.

3.6 *Strain distribution along the reinforcement*

The variations of strains along the centerline of the geogrid at different settlement ratios(*s/B*) are shown in Figure 7a for reinforced sand and Figure 7b for reinforced silty clay. The measured tensile strain was maximum at the point beneath the center of the footing and becomes almost negligible at about 3.0*B* from the center of footing. It also indicates that the geogrid beyond the effective length of $l_e = 6.0B$ results in insignificant mobilized tensile strength, and thus provides negligible effects on the improved performance of reinforced sandy and clayey soil foundation.

4 CONCLUSIONS

Based on the results of model footing tests on reinforced sand and silty clay soils, the following conclusions can be drawn:

- The optimum depth of the top layer was found to be 0.33*B* for the square footing on reinforced sand and reinforced silty clay soils.
- The bearing capacity increases with increasing the number of reinforcement layers. The significance of adding a new reinforcement layer decreases with increasing number of layers, which becomes negligible below the influence depth. The influence depth was obtained at approximately 1.25*B* for reinforced sand and 1.5*B* for reinforced silty clay.
- The bearing capacity increases with the decrease of the vertical spacing of reinforcement layers. The authors believe that a spacing ratio $h/B = 0.33$ is a reasonable value for the design of RSFs
- The performance of RSF improves with increasing the tensile modulus of geosynthetics, which is also a function of footing settlement.
- Immediate settlement can be reduced significantly with the inclusion of reinforcement.
- The inclusion of reinforcement can significantly improve the vertical stress distributions below the reinforced zone. This will be resulted in reducing the soil's consolidation settlement.
- The effective length of geogrid was found to be about 6.0*B* for both reinforced sand and reinforced silty clay soils.

ACKNOWLEDGEMENTS

This research is funded by the Louisiana Transportation Research Center (LTRC Project No. 04-2GT) and Louisiana Department of Transportation and Development (State Project No. 736-99-1242). The authors would like to express their thanks to Mark Morvant, Zhongjie Zhang, and Gavin Gautreau of LTRC for providing valuable help and support in this study.

REFERENCES

Binquet, J. and Lee, K.L., 1975. "Bearing capacity analysis on reinforced earth slabs." *Journal of Geotechnical Engineering Division*, ASCE, Vol. 101, No. GT12, pp. 1257–1276.

Guido, V.A., Chang, D.K. and Sweeny, M.A., 1986. "Comparison of geogrid and geotextile reinforced slabs." *Canadian Geotechnical Journal*, Vol. 20, pp. 435–440.

Huang, C.C. and Tatsuoka, F., 1990. "Bearing capacity reinforced horizontal sandy ground." *Geotextiles and Geomembranes*, Vol. 9, pp. 51–82.

Yetimoglu, T., Wu, J.T.H. and Saglamer, A., 1994. "Bearing capacity of rectangular footings on geogrid-reinforced sand." *Journal of Geotechnical Engineering*, ASCE, Vol. 120, No.12, pp. 2083–2099.

Adams, M.T. and Collin, J.G., 1997. "Large model spread footing load tests on geosynthetic reinforced soil foundations." *Journal of Geotechnical and Geoenvironmental Engineering*, ASCE, Vol. 123, No.1, pp. 66–72.

Ramaswamy, S.D. and Puroshothama, P., 1992. "Model footings of geogrid reinforced clay." *Proceedings of the Indian Geotechnical Conference on Geotechnique Today*, Vol. 1, pp. 183–186.

Das, B.M., Shin, E.C. and Omar, M.T., 1994. "The bearing capacity of surface strip foundations on geogrid reinforced sand and clay—a comparative study." *Geotechnical and Geological Engineering*, Vol. 12, No. 1, pp. 1–14.

Sakti, J. and Das, B.M., 1987. "Model tests for strip foundation on clay reinforced with geotextile layers." *Transportation Research Record No. 1153*, National Academy of Sciences, Washington, D.C., pp. 40–45.

ASTM, 1993. "Standard test Method for Nonrepetitive Static Plate Load Tests of Soils and Flexible Pavement Components, for Use in Evaluation and Design of Airport and Highway Pavements". Reapproved 1997. pp. 112–113.

Shin, E.C., Das, B.M., Puri, V.K., Yen, S.C. and Cook, E.E., 1993. "Bearing capacity of strip foundation on geogrid-reinforced clay." *Geotechnical Testing Journal*, ASTM, Vol. 16, No. 4, pp. 534–541.

Efficient Transportation and Pavement Systems – Al-Qadi, Sayed, Alnuaimi & Masad (eds)
© 2009 Taylor & Francis Group, London, ISBN 978-0-415-48979-9

Numerical modeling and laboratory testing of geogrid-reinforced flexible pavements

X. Tang, M.S. Medeiros & G.R. Chehab
Department of Civil and Environmental Engineering, Pennsylvania State University, University Park, PA, USA

ABSTRACT: This paper presents a follow-up study to earlier work on the evaluation of geogrids for subgrade stabilization in flexible pavements. The previous work related index properties of geogrids and bench scale test results of four geogrids to performance of reinforced pavement sections under accelerated traffic loading using the one-third scale Model Mobile Load Simulator (MMLS3). The promising results of that study serve as a motivation to conduct further research which is presented herein. The paper presents an overview of instrumentation of additional model sections with the geogrid-stabilized subgrade. Information on selection of instrumentation, installation techniques, and calibration/validation methodologies are presented. The instrumentation allows for measurement of stresses and deformation of the subgrade, and the strains in the geogrid. The instrumentation study is performed on a model section using a model box with subgrade and aggregate layers. In parallel, a three-dimensional (3D) finite element (FE) model is developed and for analysis using the conditions and properties of the instrumented box and layers. Comparison of measured responses and those predicted using finite element analysis (FEA) yields very promising results. Findings from this work provide guidelines for further research on instrumented and full reinforced pavement sections under accelerated trafficking.

1 INTRODUCTION

1.1 *Background*

One of the applications of geogrids in pavements is stabilization of the subgrade by placing a geogrid at the interface between the subgrade and aggregate base course. The benefits of this application for flexible pavements have been demonstrated through several laboratory tests using static and cyclic plate loading ((1), (2), and (3)). Full-scale field experiments have also demonstrated that the performance of pavements on weak subgrade can be improved by including a geogrid at the subgrade-aggregate base layer interface ((4), (5), and (6)).

Using the one-third scale Model Mobile Load Simulator (MMLS3), accelerated testing is conducted on pavement sections with and without geogrids to investigate the effectiveness of geogrids integration. The MMLS3 is an accelerated pavement testing device that applies unidirectional trafficking to the pavement in a controlled laboratory environment. Materials used for the pavement construction are in accordance with the relevant specifications of the Pennsylvania Department of Transportation (PennDOT).

To provide additional means to quantify the benefits of geogrids in stabilizing weak subgrade, an instrumentation system composed of load and displacement measurements, in addition to data acquisition setup is presented. In order to examine the system and identify important aspects in terms of sensors selection and installation, data logging and refinement, a pilot test is conducted on a model scale instrumented box under static loading.

A 3D FE model using ABAQUS® is developed to numerically determine the responses of the geogrid-reinforced pavement layers and predict distresses under accelerated trafficking. The model which is undergoing staged development is applied to the model scale instrumented box. Comparison between FEA-predicted responses to measured yields promising results. Efforts for more accurate and enhanced instrumentation and development of more advanced material and structural models for the 3D FEA will ensue based on the findings and lessons learned from this study.

2 ACCELERATED TRAFFICKING ON MODEL PAVEMENTS

In order to investigate the benefits of geogrids for stabilizing weak subgrade, accelerated pavement testing is conducted to assess the induced rutting under wheel loading. The testing is performed on four pavement sections, one control section with no geogrid, and three others, each of which is reinforced with a specific type of geogrid.

2.1 *Overview of accelerated testing facility*

The pavement slabs are constructed in a test pit with reinforced concrete walls and foundations. The pit is backfilled with aggregate base and compacted to serve as pavement bedrock. Figure 1 shows the layout of the constructed pavement sections, labeled as P1, P2, P3, and P4. P1 is a control section, i.e. no geogrid, while sections P2, P3, and P4 are reinforced with three different geogrid products, designated as Grid A, Grid B, and Grid C, respectively.

The accelerated testing is performed using the one-third scale Model Mobile Load Simulator (MMLS3). The MMLS3 has four tires, each with diameter of 30.5 cm and width of 7.6 cm. The actual wheel path generated by the MMLS3 is 137 cm long. A rutted wheel path of a pavement section can be seen next to the MMLS3 in Figure 2. The load exerted by each wheel is 2.7 kN with a corresponding tire contact pressure of 690 kPa. The trafficking speed is set to 7, 200 wheel applications per hour. A contact-type profilometer is used to measure the rutting profile at regular trafficking intervals, and an environmental chamber is fitted on the MMLS3 to maintain the specified temperature.

2.2 *Materials*

The subgrade material used is a silty sand/SW-SM. Table 1 summarizes the tested properties of the subgrade soil. The soil grain size distribution is shown in Figure 3. Dense-graded crushed stone is used for the pavement aggregate base layer. The grain size analysis (Figure 3) shows that the base course aggregate meets the Pennsylvania Department of Transportation (PennDOT) 2 A

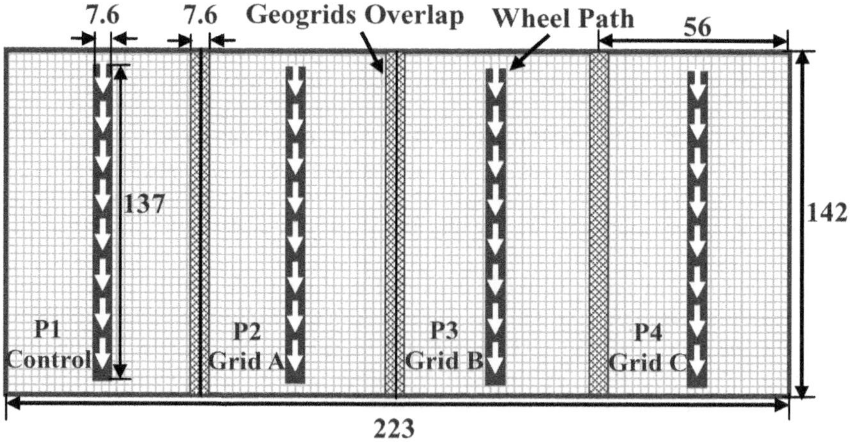

Figure 1. Sketch of the layout of the pavement sections;, units are in cm.

Figure 2. MMLS3 mounted on top of a pavement slab; rutted wheel path on another section seen on the right.

Table 1. Subgrade soil properties.

Property	Measurement methods	Result
Classification	AASHTO M145	A-2-4(0)
	ASTM D2487	SW-SM
Percent passing #200 sieve (%)	ASTM D422	6.2
Liquid limit (%)	ASTM D4318	17.6
Plastic limit (%)	ASTM D4318	14.7
Optimum moisture content (%)	ASTM D698 (standard compaction test)	10.0
Maximum dry density (kg/m^3)	ASTM D698 (standard compaction test)	2066
Friction angle ϕ (deg)	ASTM D3080 (direct shear test)	31.8

grading requirements. A standard Proctor test for the aggregates yields an optimum moisture content of 3.9% and maximum dry density of 2329 kg/m^3. A 12.5 mm Superpave mix is used for the surface asphalt layer.

Three biaxial geogrid products are used in this study: Grid A made of woven polypropylene (PP) yarns, Grid B is made of extruded PP sheets, and Grids C composed of high tenacity polyester (PET) multifilament yarns and coated with polyvinyl chloride (PVC) coating. Grids A and C are classified as flexible geogrids and Grid B as a stiff geogrid based on measured flexural rigidity per ASTM D 1388. Physical and mechanical properties of the geogrids were tested as shown in Table 2. Comprehensive evaluation of the geogrids' material properties is documented by Chehab et al. 2007 (7).

2.3 *Construction of pavement sections*

The same materials are used in constructing the four pavement sections. The subgrade is placed on top of a waterproof membrane to avoid moisture loss to the aggregate bedrock layer underneath.

Table 2. Tested index properties of the geogrids.

		Geogrid					
		Grid A		Grid B		Grid C	
Index property	Test method	MD*	TD**	MD	TD	MD	TD
Aperture size (mm)	Calipers	35.05	41.15	25.65	36.58	25.65	26.42
Rib thickness (mm)	Calipers	1.98	1.09	0.76	1.07	1.42	2.03
Junction thickness (mm)	ASTM D 5199	2.29		3.94		1.55	
Mass per unit area (g/m²)	ASTM D 5261	252.26		319.06		350.93	
Tensile strength at 2% strain (kN/m)	ASTM D 6637	14.8	15.0	9.8	15.6	10.3	11.2
Tensile strength at 5% strain (kN/m)	ASTM D 6637	30.1	30.0	16.8	29.2	18.1	17.4
Ultimate tensile strength (kN/m)	ASTM D 6637	36.5	35.7	23.9	32.9	39.5	52.8
Elongation at break (%)	ASTM D 6637	7.1	6.7	20.6	10.9	10.5	12.0
Junction Strength (kN/m)	GRI GG2	10.2	4.3	17.7	28.1	7.4	7.1
Flexural rigidity (mg-cm)	ASTM D 1388, mod.	271509		1429355		452671	
Torsional stiffness (cm-kg/degree)	COE/GRI GG9	3.97		7.50		3.43	

* MD: machine direction; ** TD: transverse/cross-machine direction.

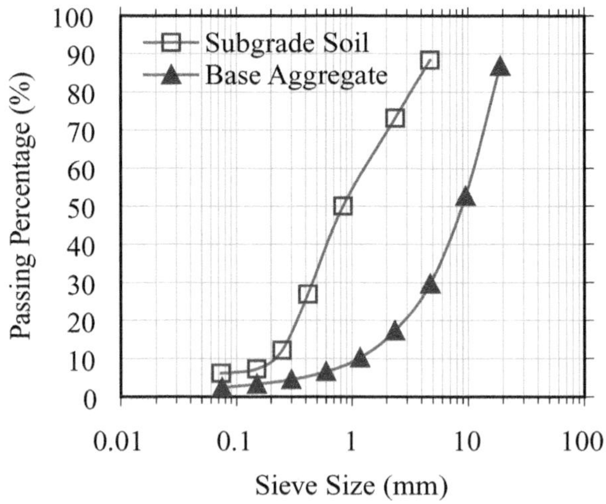

Figure 3. Grain size distribution of subgrade soil and base course aggregates.

The pavement layers are compacted using a vibratory plate compactor. The soil is compacted at the desired moisture content of 14.8% to achieve CBR of 1.5% (details can be found in (7)). The compactor travel direction ensures consistency of the soil density throughout the pit. In-situ soil density and moisture content are measured by means of the sand cone method. The as-built thicknesses of subgrade, base course, and asphalt layer are 15.2 cm, 6.6 cm, and 3.8 cm, respectively.

The geogrids are placed directly on the subgrade layer. The grids are carefully unrolled to avoid folds and wrinkles. Geogrids of each section overlap with those from adjacent sections by 7.6 cm (Figure 1). The ends of the geogrids are folded upwards against the pit walls to obtain necessary anchorage and slight pre-tensioning, as well as to prevent shifting of the geogrids out of position while trafficking. The geogrid is then overlaid with a compacted aggregate base course layer. A relatively low air void (AV) asphalt concrete (AC) layer is placed and compacted on top of the aggregate base. The low AV content minimizes the densification and rutting of the asphalt layer under loading. The AC layer density is measured along the wheel path for each section using a Pavement Quality Indicator™ (PQI) Model 301 A prior to MMLS3 trafficking.

2.4 *Results and discussion*

MMLS3 trafficking is applied to each section until rutting failure occurs. The control section P1 and section P2 exhibit extensive deformation after 40,000 and 70,000 wheel applications respectively. Trafficking is stopped at 100,000 wheel applications for sections P3 and P4.

Figure 4-a shows the rutted wheel paths for the four sections. The extensive rutting and resulting cracking at the edge of the wheel path for the control section P1 and section P2 is clearly visible. The measured rutting profile for section P3 is also shown in Figure 4-b.

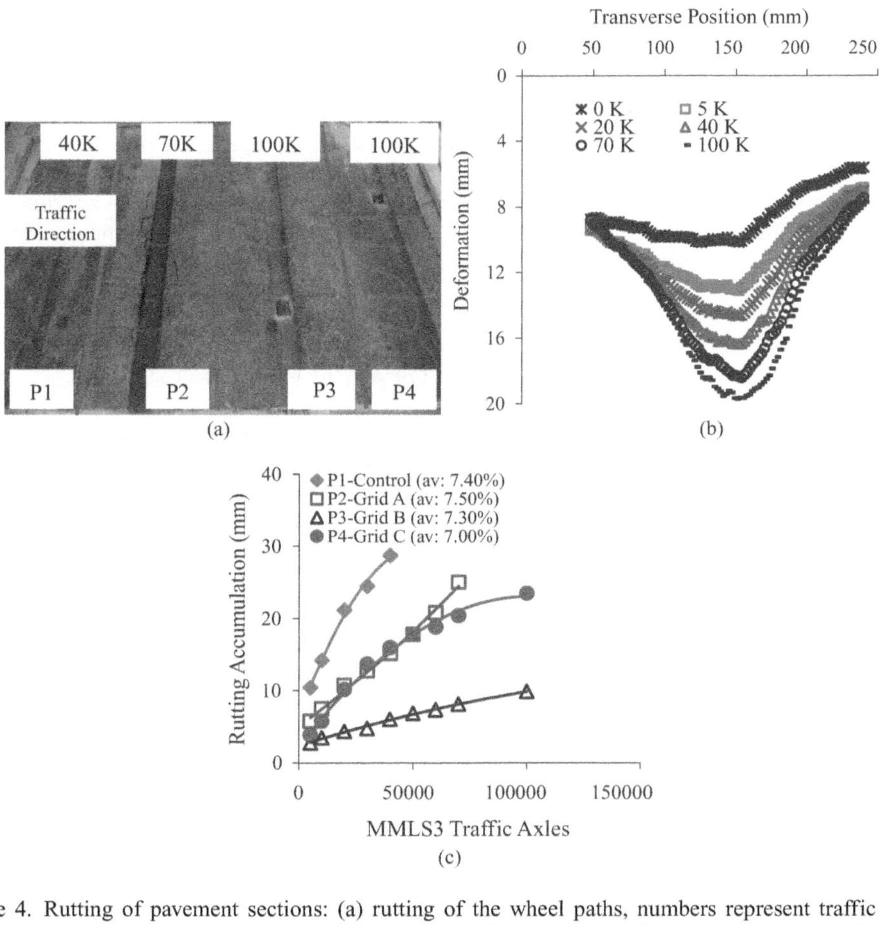

Figure 4. Rutting of pavement sections: (a) rutting of the wheel paths, numbers represent traffic cycles; (b) measured rutting profile at various stages of trafficking for section P3; (c) rutting accumulation for locations with similar AC air voids.

In order to minimize the effect of variation in air void (AV) content of the asphalt layer, the rutting accumulation at locations with similar compacted asphalt air voids is presented in Figure 1-c. Control section P1 exhibits significantly higher rutting which accumulated relatively quickly (failure occurs at 40 K cycles) compared to the reinforced sections (failure between 70 K and 100 K cycles). This illustrates the ability of geogrids to stabilize weak subgrade and the potential to minimize pavement deformation under traffic load. Section P2 (Grid A) has the second highest cumulative rutting among the four test sections. The rate of rutting accumulation for sections P1 and P2 is slightly higher than for sections P3 and P4. Significantly higher rutting is observed for section P4 (Grid C) when compared to that of P3 (Grid B). Overall, Grid B, the stiffest geogrid, in the study, shows the best performance with respect to the rutting resistance for the tested pavement on *this specific weak* subgrade.

The results of the APT testing confirm the ability of the geogrid in stabilizing the subgrade and minimizing its deformation under sustained trafficking. More importantly, however, is that the performance of the geogrids varies from one type to the other to the difference in their properties and interaction with the soil and aggregates. Previous studies confirmed that various properties or attributes can contribute to the performance difference as well as compatibility with the subgrade properties ((8) and (9)). In order to accurately quantify the reinforcement effectiveness for different geogrids and identify the optimal properties for given subgrade conditions, it is necessary to measure the stresses and strains prevalent at the interface in addition to the nature and value of the strain felt by the geogrid ribs.

Using instrumentation for making such measurements also allows for understanding and characterization of the mechanisms taking place at the subgrade-geogrid-subbase interface. Thus instrumentation becomes the keystone in developing mechanistic-empirical (M-E) based design method for geogrid-reinforced pavements. The challenge of instrumenting the geogrids and adjacent layers has been documented by many ((10), (11), and (12)). However, the opportunity of being able to correlate measured responses in the geogrid and sub layers to measured deformation at the surface due to APT trafficking is worth pursuing. Such experiments are needed to calibrate and validate any M-E design method for reinforced pavements. Presented next is an exploratory study on instrumentation of a subgrade-geogrid-aggregate base interface at model scale.

3 A PILOT TEST ON INSTRUMENTATION FOR ACCELERATED TESTING

An instrumentation system consisting of sensing devices, data acquisition hardware and software, is devised to complement the APT study and aid in further evaluating the behavior and mechanisms of geogrids in reinforcing the soil subgrade. Measurement of responses in the geogrids and surrounding unbound materials are needed for calibrating and validating mechanistic-empirical analytical models and numerical models used for performance prediction of subgrade-stabilized pavements.

An exploratory study is conducted to calibrate the various sensors used, including LVDTs, strain gages, load cells, identify proper installation techniques, and validate the measurements obtained under loading. It is cost and time efficient to evaluate the instrumentation system as a whole before the sensors are buried in the pavement sections. The exploratory evaluation study is conducted on the sensors themselves before being incorporated in any medium for calibration, and then evaluated within a small scale medium consisting of a subgrade and aggregate base layers in a customized test box.

3.1 *Instrumentation system*

A National Instruments® data acquisition (DAQ) hardware and software is used for the measurement instrumentation system (Figure 5). The DAQ chassis has modules for analog and digital inputs to accommodate the various measurement sensors. The sensor types selected are capable of

Figure 5. Data acquisition system hardware, the insertion shows the chassis and modules.

measuring responses under dynamic loading conditions resulting from accelerated trafficking. A sample of sensors and their intended use is listed below:

- Dynatest PAST-II-AC H-type strain gages installed in the asphalt layer to measure the horizontal transverse strains.
- Macro Sensors GHSE-750-1000 LVDTs for measurement of subgrade deformation.
- Geokon 3500 pressure cell for measurement of stresses at the interface between subgrade and base course.
- High-elongation Vishay EP-08-230DS-120 gages with narrow geometry for measuring strains of geogrids in both transverse and longitudinal directions.

Under the small scale controllable conditions as compared to those experienced in field full-scale pavement testing, the survivability rate of strain gages and other sensors is expected to be relatively high.

3.2 *Construction of model pavement and instruments installation*

The small-scale model structure used in the exploratory study consists of compacted subgrade, a geogrid layer, and compacted aggregate base constructed in a customized 51 cm × 51 cm × 33 cm poly-wood box. The same subgrade soil and aggregate base used previously in the APT tests were used for this model-scale study. Thicknesses of soil and aggregates layers are 15.2 cm and 5.1 cm, respectively separated with geogrid B.

The LVDT used for subgrade deformation measurement is housed in a steel tube with its base fixed to a concrete base at the bottom of the box. A thin yet rigid disk is installed onto the contact tip of the spring-loaded LVDT to provide sufficient contact area with the soil-geogrid interface. The LVDT is totally immersed in the soil with its contact disk flush with the soil surface but underneath the geogrid.

The pressure cell is also positioned flush with the top of the soil layer. The soil underneath the pressure cell is compacted to provide adequate support. Fine soil particles are then used to surround the load cell in order to avoid stress concentration caused by individual-aggregate contact. Strain gages are installed on geogrid ribs in both machine and cross-machine directions. Strains measured in each direction are expected to be different due to differential tensile properties in the two directions (13). To protect the strain gages from damage without affecting their functionality, a waterproof epoxy is applied. After curing, the gages are wrapped with a thin tape and then covered with a thin layer of fine aggregates. Strain gage wires are enclosed by thin flexible tubes for protection against damage caused by penetrating aggregates.

Following installation of the sensors, the geogrid is placed on top of the soil layer with edges attached to the box walls to ensure sufficient anchorage. Figure 6 is a picture showing the installed instrumentation. Aggregates are then placed and compacted with proper care to avoid damaging the strain gages.

3.3 *Results and discussion*

A 21 kg (0.206 kN) steel plate, 15.1 × 15.1 cm, is placed at four positions on the aggregates layer surface. The responses are measured at each of those positions. The surcharge is first placed at the center of the box from which the sensors are offset. Then, the load is placed on top of each of the LVDT, strain gages, and load cell. The first load location is referred to as "*center*", while the location above a particular sensor is referred to as "*top*" for that sensor (Figure 7).

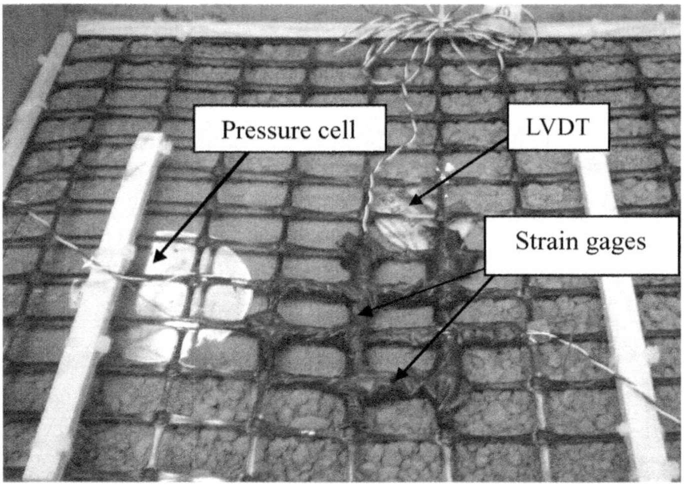

Figure 6. Instrumentation at the interface between subgrade and aggregates layers.

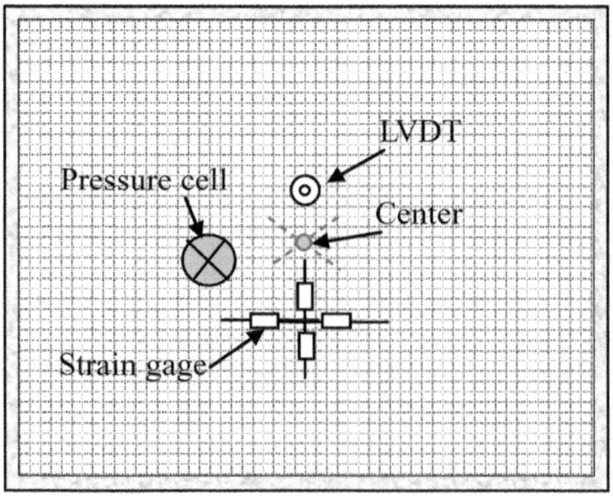

Figure 7. Schematic showing the positions of the sensors, and the offset from the center.

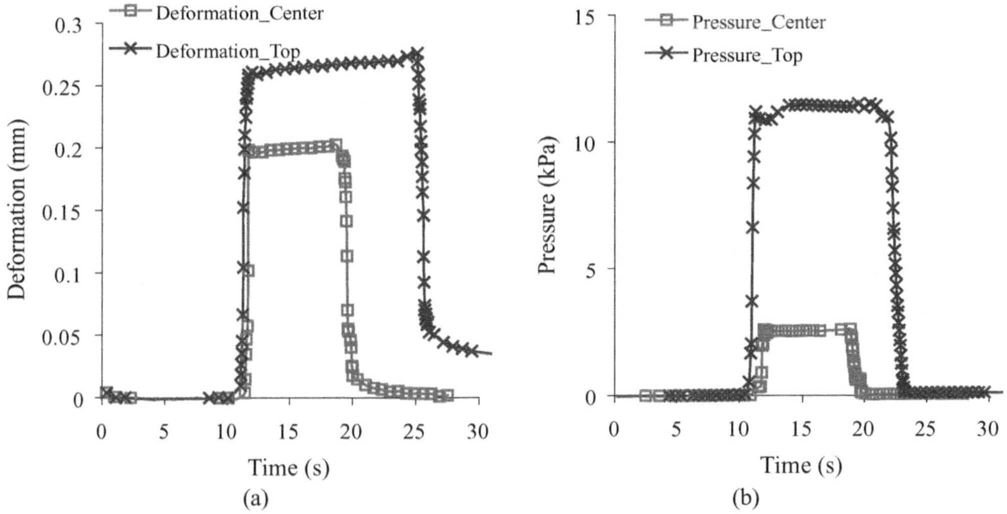

Figure 8. (a) Vertical deformation, and (b) stress response measured at the soil layer interface for two different loading positions.

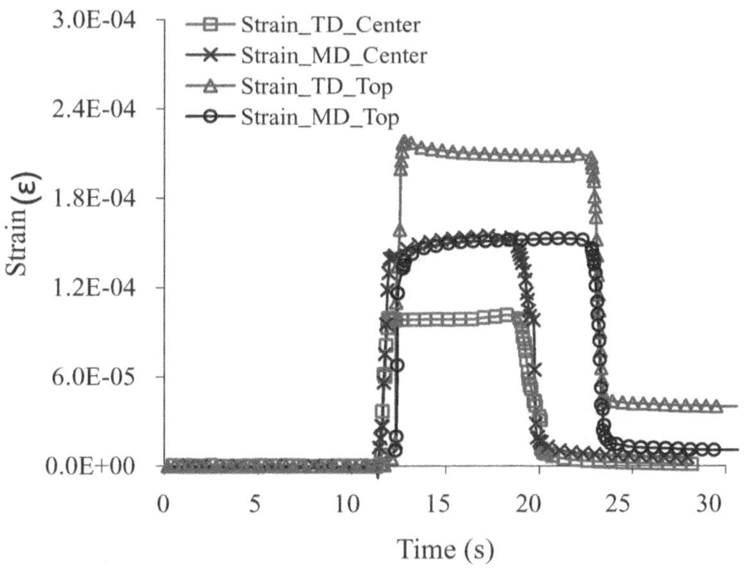

Figure 9. Strain responses of geogrids to loading at different locations.

Figure 8 shows the deformation of the interface under the load for two locations. As expected, the deformation is greater when the load is at the top of the LVDT than that when placed at the center. The difference in deformation measured is 0.07 mm. Similarly, the stress at the interface differs by 8.9 kPa between the two loading conditions. Strain measured along the machine direction (MD) is greater than that for transverse direction (TD) (Figure 9). This is expected because the ribs in the TD are stiffer than those in the MD. The observations made regarding trends of the measurements indicate proper functionality of the sensors after installation and construction.

4 FINITE ELEMENT MODELING

This section provides an overview of the 3D FE analysis that is undergoing continuous enhancements to achieve accurate determination of pavement responses and performance predictions. Such analysis requires both proper modeling of the pavement structure and realistic, and accurate characterization models for the materials comprising the various layers in the structure. While the use of FE analysis for geogrid-reinforced pavement is not new; however, few studies have implemented accurate material models, particularly characterization of the asphalt layers with growing damage, and characterization of the subgrade-geogrid-aggregate interaction under dynamic loading. Developing and incorporating characterization models for both in FEM is a subject of ongoing research, but has proved to be challenging and will require staged implementation.

4.1 *FEM model*

Similar to the approach presented earlier for implementation of measurement instrumentation, a pilot FE analysis is to be conducted first on a model scale section with controlled conditions and limited variables. In this study it is convenient to perform the FEM study on the instrumented box described earlier. The results of the analysis can be compared to the measured responses and observed distresses. The FEM study, conducted using ABAQUS, is at its first stage, aimed at verifying the measurements from the sensors installed in the instrumented box. The stress and strain responses are low; thus, any resulting errors in assuming the layers to be linear elastic are small.

The structure modeled in this work matches the dimensions of the instrumented box used earlier. It consists of two layers: 5.1 cm aggregate base and a 15.2 cm subgrade. The geogrid is placed at the interface between the base and subgrade layers. The model is constrained at the bottom and the sides. A 21 kg load is distributed over a 15.1 × 15.1 cm steel plate at the center of the surface layer. The geometry can be seen on the Figure 10.

The base and subgrade layers are meshed using C3D20R which is a reduced integration second-order element consisting of 20 nodes. Despite the fact that such quadratic elements require more computational time; they provide higher accuracy than linear elements for problems that do not involve complex contact conditions, impact, or severe element distortions (14). As the thickness of each component in geogrid layer is significantly smaller than the other layers, an eight-node continuum shell element (SC8R) is employed in this study, this type of element have been used to model the geogrid behavior by several authors (15, 16, 17). The layers are considered as fully bonded, and no slippage is allowed between the surfaces. Table 3 presents the properties used for the analysis.

The resilient modulus (M_r) was used to represent the elastic property of the aggregate and subgrade layers The values were obtained using the correlation between the resilient modulus and the

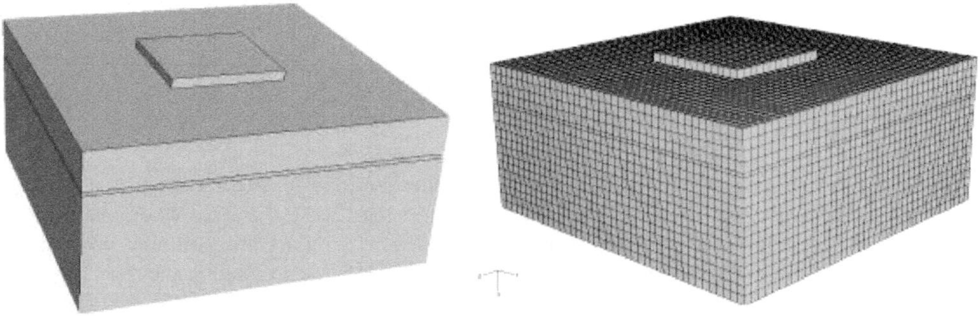

Figure 10. Model geometry and selected mesh.

CBR defined in the AASHTO design guide by the Equation 1. The elastic modulus of the geogrid used in the numerical model (geogrid B) was the true initial stiffness on the machine direction as reported by the manufacturer.

$$M_r = 2555 \times CBR^{0.64} \tag{1}$$

4.2 *FEM results*

Graphical results of vertical compressive strain and stress at the top of the subgrade are shown in Figure 11 and Figure 12 respectively. The responses shown are along a vertical plane passing through the center of the box. One can observe from the figures that discontinuities in strain and stress occur at the geogrid interface, suggesting that the geogrid, through its interaction with its surrounding material, is acting as a flexural member that is absorbing the compressive stresses and redistributing them into tensile stresses along its surface, thus minimizing the stresses transferred to the bottom of the geogrid. This results in a decrease of compressive strains on top of the subgrade layer. Figure 13 shows the stress and strain profiles along the depth. The discontinuity at the interface between the geogrid and the top and bottom layers can be seen in both profiles.

The maximum and average stresses from FEM correspond to the stress at the position where the load cell is located. The average stress in the area corresponding to a 150 mm circular area

Table 3. Young's modulus and Poisson's ratio of each material.

Material properties		
Layer	Young's modulus (PSI)	Poisson's ratio
Load plate	30,000,000	0.2
Aggregate	34,000	0.35
Geogrid	40,610	0.5
Subgrade	6000	0.40

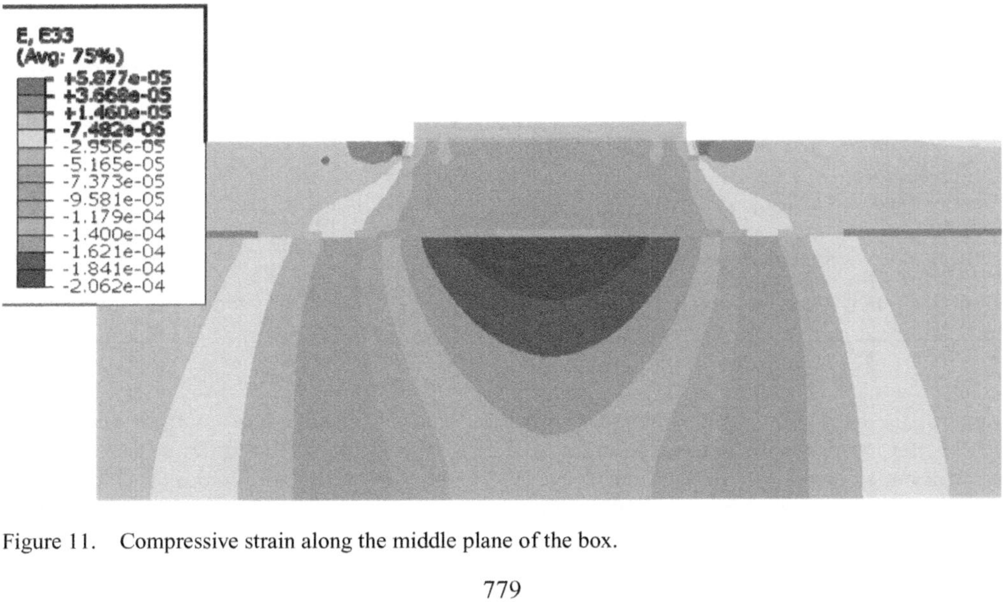

Figure 11. Compressive strain along the middle plane of the box.

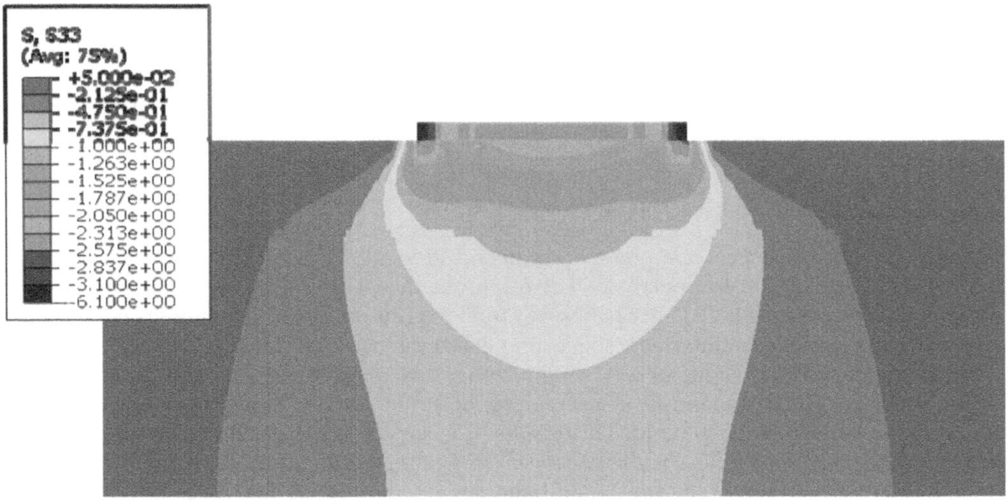

Figure 12. Vertical stress distribution along the middle plane of the box.

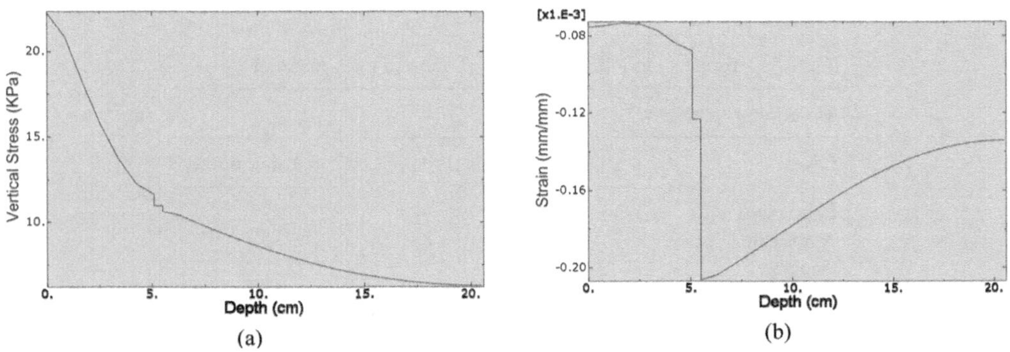

Figure 13. (a) vertical stress and (b) vertical strain along the depth.

Table 4. Comparison between measured and FEM-predicted responses (compression is positive).

Response	Measured	FEM
Deformation (top of) subgrade	0.025 mm	0.023 mm
Vertical stress (top of) subgrade	11.4 kPa	Max: 10.55 Avg: 9.55
Horizontal stress at interface	N/A	–2.940 kPa
Horizontal strain at interface	MD: 15E-05 TD: 12E-05	6.045E-05

(area of the load cell) is calculated. The measured and predicted responses from FEM match very well, except for the geogrid strain, as can be seen in Table 4. That difference is attributed to an assumed depth of the geogrid in case of the FEM. That depth varies from pre to post compaction of the aggregate base. The difference is also due to the bending effect of the geogrid, which leads to a slight error in strain gage measurement. Ideally, two strain gages should be installed, one on the top of the geogrid rib, and another at the bottom in order to counter the bending effects.

5 CONCLUSIONS

A previous study had been conducted on evaluation of geogrid-reinforced subgrade in flexible pavements. Different types of geogrids were tested in accelerated trafficking tests of pavement sections using the MMLS3. The study reveals valuable findings regarding the performance of the geogrids in stabilizing the subgrade. However, for further understanding of the reinforcing mechanisms of the geogrids, there is a need for instrumentation of the pavement structure and geogrids.

A pilot test for an instrumented geogrid-reinforced pavement model is conducted on a model scale for the purpose of examining the instrumentation system while embedded under pavement layers. The measured responses of pressure, deformation at the aggregates-subgrade interface, and strains of geogrid were compared to those predicted by a 3D FE model using ABAQUS. The model is in its staged development. The finalized model will be able to accurately characterize the viscoelastic asphalt layer behavior, as well as the interaction of the subgrade-geogrid-aggregate base composite layer. Findings from this study will aid in refining the instrumentation setup and developing a comprehensive 3D FEM model to be applied to pavement sections subjected to accelerated trafficking.

ACKNOWLEDGEMENTS

Financial support from the Pennsylvania Department of Transportation is gratefully acknowledged. The authors would like to thank Mr. Dan Fura for his technical support.

REFERENCES

1. Hass R., Wall, J., and Carroll, R.G. (1988). "Geogrid reinforcement of granular bases in flexible pavements." *Transportation Research Record 1188*, Washington D.C.:19–27.
2. Al-Qadi, I.L, Brandon, T.L., Valentine, R.J., Lacina, B.A., and Smith, T.E. (1994). "Laboratory Evaluation of Geosynthetic-reinforced Pavement Sections." *Transportation Research Record 1439*, Washington D.C.: 25-31.
3. Perkins, S.S. (1999). "Geosynthetic Reinforcement of Flexible Pavements: Laboratory Based Pavement Test Sections." *Report No. FHWA/MT-99-001/8138*, U.S. Department of Transportation, Federal Highway Administration, Washington, D.C.
4. Barksdale, R.D., Brown, S.F., and Chan, F. (1989). "Potential Benefits of Geosynthetics in Flexible Pavement Systems." *National Cooperative Highway Research Program (NCHRP) Report No.315*, Transportation Research Board, National Research Council, Washington D.C.
5. Hufenus, R., Rueegger R., Banjac, R., Mayor, P., Springman, S.M., and Bronimann, R. (2006). "Full-scale Field Tests on Geosynthetic Reinforced Unpaved Roads on Soft Subgrade." *Geotextiles and Geomembranes*, Vol. 24(1):21-37.
6. Al-Qadi, Tutumluer, E., Kwon, J., and Dessouky, S.H. (2007). "Accelerated Full-scale Testing of Geogrid-reinforced Flexible Pavements." *TRB 2007 Annual Meeting (CD-ROM)*, Transportation Research Board, National Research Council, Washington D.C.
7. Chehab, G.R., Palomino, A.M., and Tang, X. (2007). Laboratory evaluation and specification development for geogrids for highway engineering applications, *Report No. FHWA-PA-2007-009-050110*, Pennsylvania Department of Transportation, Harrisburg, PA, USA.
8. Tang, X., Chehab, G.R., and Palomino, A.M (2008). "Accelerated testing of geogrid-reinforced subgrade in flexible pavements." *Proceedings of GeoCongress 2008: Geosustainability and Geohazard Mitigation*, ASCE, Geotechnical Special Publication 178, pp. 1049–1056.
9. Tang, X., Chehab, G.R., Palomino, A.M, Allen, S., and Sprague J. (2008). "Effects of geogrids properties on subgrade stabilization of flexible pavements." *Proceedings of GeoCongress 2008: Geosustainability and Geohazard Mitigation*, ASCE, Geotechnical Special Publication 178, pp. 1089–1096.

10. Brandon, T.L., Al-Qadi, I.L., Lacina, B.A., and Bhutta, S.A. (1996). "Construction and instrumentation of geosynthetically stabilized secondary road test sections." *Transportation Research Record 1534*, pp. 50–57.
11. Maxwell, S., Kim, W.H., Edil, T.B., and Benson, C.H. (2005). "Geosynthetics in stabilizing soft subgrade with breaker run." *Report No. 0092-45-15*, Wisconsin Department of Transportation, Madison, WI, USA.
12. Warren, K., Brooks, J., and Howard, I. (2005). "Survivability of foil strain gages mounted on geosynthetics under full-scale construction loads." *ASCE, Geotechnical Special Publication No. 130–142*, pp. 4085–4090.
13. Tang, X., Palomino, A.M., and Chehab, G.R. (2008). "Laboratory evaluation of geogrids for flexible pavement reinforcement." Conference Proceedings, *International Geosynthetics Society (IGS) GeoAmericas*, Cancun, Mexico, accepted for publication.
14. Hibbitt, K., (2002), ABAQUS Theory Manual, Karlsson and Sorensen Inc, Pawtucket.
15. Peila D., Ogerri C., and Castiglia C. (2007) "Ground reinforced embankments for rockfall protection: design and evaluation of full scale tests" Landslides, Volume 4, Issue 3, pp. 255–265.
16. Huang J., Han J., and Collins J.G. (2005) "Geogrid-Reinforced Pile-Supported Railway Embankments" Transportation Research Record 1936, Washington D.C.:221–229.
17. Mazursky L. Ann (2006) "Three-Dimensional Analysis of Geosynthetic Reinforcement Used in Column-Supported Embankments" Master Thesis, Virginia Polytechnic Institute and State University, Blasburg, VA.

Pavement distresses and maintenance, and infrastructure management

Efficient Transportation and Pavement Systems – Al-Qadi, Sayed, Alnuaimi & Masad (eds)
 ISBN 978-0-415-48979-9

Automated in-vehicle data collection and treatment for existing roadway alignment

S. Cafiso & A. Di Graziano
Department of Civil and Environmental Engineering, University of Catania, Italy

ABSTRACT: Given the lack of suitable documentation regarding highway design data, the first problem to be solved in all situations related to the management of existing roads is the necessity to rapidly and effectively acquire information relating to the geometrical and functional characteristics of the road network. To provide a low cost solution to this problem, the paper presents a low cost hardware and software systems that uses a suitably-equipped vehicle for the high speed survey together with purposely developed computer procedures to elaborate the acquired data. The survey is performed using a GPS kinematical-differential method to locate both the vehicle position along the road axis and other data acquired with in-vehicle instruments. The algorithm developed is able to process the data acquired from the GPS and inertial systems in order to define the geometric horizontal and vertical features of the road.

Keywords: Roadway alignment, GPS, Spline, High-speed survey

1 INTRODUCTION

Efficient management of a road network greatly depends on the availability of adequate and reliable information that allows the state of the road infrastructure to be controlled. Then, a partial or total lack of such information represents the first obstacle that must be overcome for an optimized road management. It is, therefore, necessary both to characterize the parameters which affect the performance offered by the road in terms of safety and comfort and to set up the procedures for data acquisition and elaboration of these parameters. For some time now, high-performance acquisition procedures have been used to collect all the parameters connected to the use of the road network (Drakopoulos and Ornek 2000, Harkey et al. 2004, Wul 2006).

In this automated approach using a van, one of the main data elements which can now be collected is the horizontal alignment of the roadway. However, while the vehicle path can be reconstructed from the survey data with relative ease, the segmentation of the surveyed highway into elements of defined horizontal and vertical geometry is a much more challenging task.

Mathematical tools have been quite widely used to define highway alignment in computer applications whose objective is alignment optimization (Chew et al. 1989), or the definition of vehicle paths (Guarino et al. 2000). In addition, they have been used to graphically represent an highway alignment previously calculated by means of traditional procedures (Walton and Meek 2005). Also the use of cubic splines to estimate the highway alignment starting from a GPS survey was investigated by different authors (Bojkov et al. 1997, Ben-Arieh et al. 2004, Castro et al. 2006). In this context the added value of the present work consists in an integrated use of data obtained from low cost equipments with an evaluation procedure matching more smoothing splines algorithm with the road alignment geometric parameters (curvature radius, tangent length and spiral curve parameter).

The paper organization consists of two main parts:

- the first presents the hardware and software architecture of the mobile laboratory developed by the Department of Civil and Environmental Engineering (DICA) for data acquisition;
- the second reports the algorithm that was defined for the elaboration of survey data in order to determine the geometrical elements of the horizontal alignment.

The results and considerations arrived at are based on experience acquired in the field during the surveying of approximately 400 km of two-lane rural roads.

2 THE VEHICLE DATA COLLECTION

2.1 *DICA mobile laboratory*

At the Department of Civil and Environmental Engineering (DICA) of the University of Catania, a multi-function Mobile Laboratory for high-performance surveys of the road infrastructure has been developed (Cafiso et al. 2003) in order to satisfy needs connected to the considerable lack of information regarding highway design data. At present, the Mobile Laboratory (Figure 1) is equipped with:

1. a dual-frequency GPS operating in kinematical mode;
2. a WSS2 odometer with a digital output (128 pulses/tire rotation);
3. ES 261 Servoinclinometer in longitudinal direction with a micron/meter resolution;
4. a computer system (2 P4 Pc laptop 1.8 GHz 256 MB Ram);
5. a DAQ data acquisition card (PCI bus, channels 16SE/2DI);
6. a triaxial PCB accelerometer (sensitivity: 1000 mV/g; frequency: 0–150 Hz; precision: 30 μg);

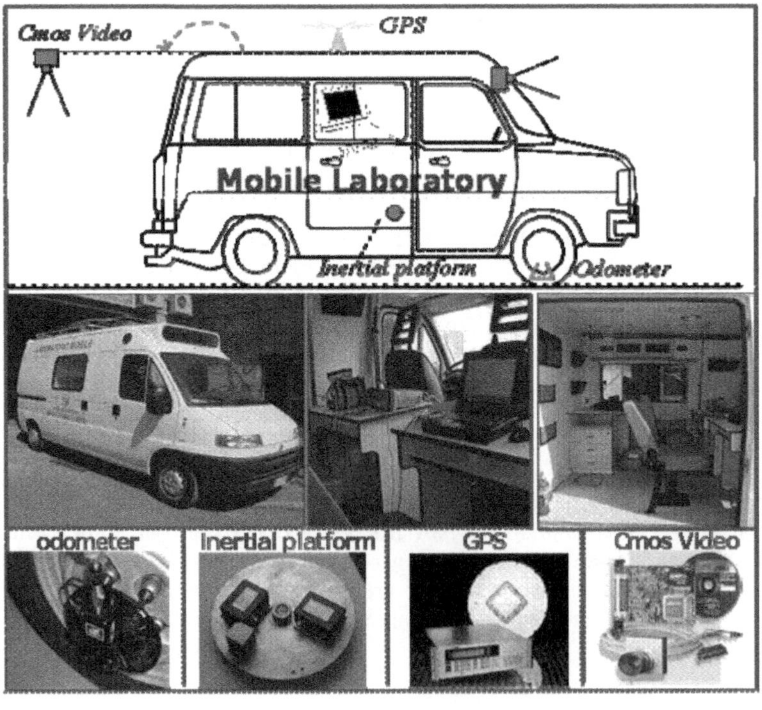

Figure 1. DICA mobile laboratory.

7. ES 261 Servoinclinometer in transversal direction with a micron/meter resolution;
8. two CMOS digital cameras (rear 1288 × 1032; front 640 × 480).

The procedure presented in the paper make use of equipments listed between 1 and 5 with the aim of achieve an effectiveness result in roadway alignment reconstruction using low cost instruments.

The Laboratory can be used to collect useful data for horizontal and vertical road alignment reconstruction, as well as other fundamental information such as section width, road markings, pavement condition, etc. The hardware system of the mobile laboratory is managed by a software for the synchronous acquisition and post-elaboration of both analogue and digital information coming from the various instruments (Figure 2) so that any data acquired has a common temporal reference (ID).

The acquisition of analogue and digital signals is carried out by means of a multi-function card (NI DAQ pci 622), using a personal computer equipped with a suitable interface (IEEE1394, RS-232, PCMCIA). The card has 16 analogue channels, 24 I/O digital lines and two counters for the differentiated acquisition of the signals emitted by the instruments on-board the vehicle. Data acquisition takes place with a maximum frequency of 100 Hz but registration is performed at a frequency of 2 Hz using suitable noise-reduction filters.

At the present time, the card channels have been programmed to acquire the distance covered by the odometer, the transversal and longitudinal slope, as well as the vertical, transversal and longitudinal acceleration from the inertial platform. The GPS spatial coordinates and the photographic images are acquired directly by the PC at regular time intervals (0.2 ÷ 1.0 sec) synchronized by the DAQ card (figure 2). The Street Invaders 2.0® software was implemented to manage the multifunction card, making it possible to define the signal acquisition methods according to the sampling frequency (Hz) and the event period (ms) or rather, using external PPS interrupt signals emitted by the satellite receiver at pre-established 1 s intervals (Figure 3).

2.2 *GPS data acquisition*

The acquisition of high quality GPS signals is fundamental for the geo-referencing of all the data acquired and also to carry out the geometric road alignment.

The GPS survey was performed using a kinematical-differential method (DGPS). The method for performing a GPS differential survey (DGPS) consists in positioning a fixed station, called the master, and using a mobile station, called the rover. The master is positioned at a fixed point in completely static surveying mode for the whole of the survey period, acquiring information relating to the position of the satellites. The rover is moved to perform the programmed static, 'stop and go' and kinematical surveys (Grewal et al. 2001). The survey was performed acquiring, with the rover receiver, one point a second at a speed of about 60 km/h, so as to obtain a precision of ±2 cm + 1 ppm of the distance from the master receiver.

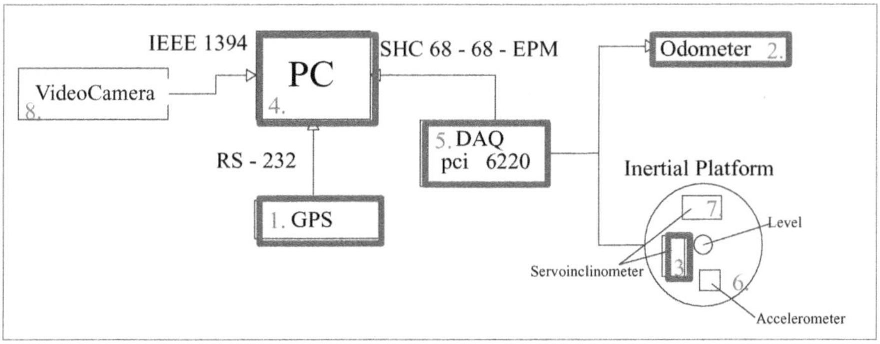

Figure 2. Instrument connection layout.

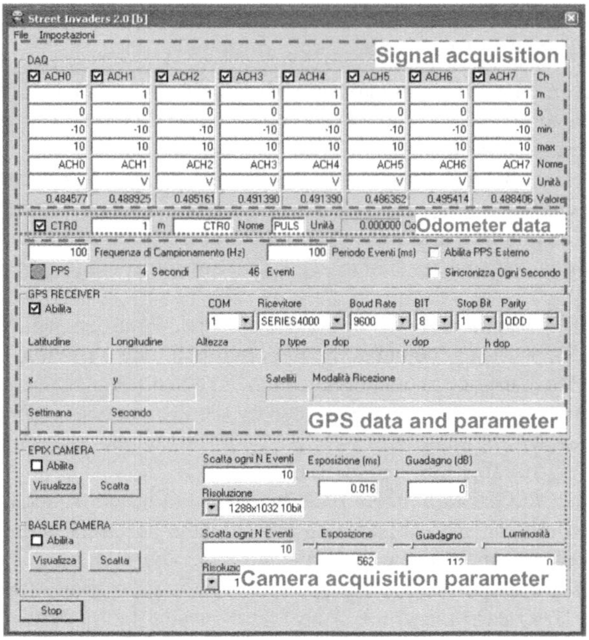

Figure 3. Layout of the street invaders 2.0® software.

The use of the post—processing differential method made it possible to eliminate, in the post-processing phase, systematic error effects. The use of real time correction, although improving precision, is not indispensable to achieve the aims of the survey and can lead to reception problems as regards the communication signal between the rover and master and therefore, ultimately compromise the results.

When using the DGPS cinematic mode, a horizontal precision of ±10 cm is sufficient for most of road survey applications and specifically for road alignment reconstruction. Vertically, the precision that it is possible to obtain is considerably lower in the region of ±50 cm. Moreover, precision is conditioned by the relative position of the two receivers. During trials, it was shown that it is adequate to maintain a maximum distance of 100 km between the two receivers to guarantee the necessary precision without excessively conditioning the functionality of the procedure. In fact, in the case of a continuous survey involving a large number of kilometers of road network it could be necessary to continually move the master station. Today, with a network of fixed stations spread across the territory it can be easier to acquire data for differential corrections.

Furthermore, the accuracy of the survey is a function of the parameters within which the receivers are set up, the most important being the PDOP (Position Dilution of Precision). The PDOP is linked to the volume of the ideal solid which has the satellite and the rover aerial as its vertex. The lower the value of the PDOP the greater the volume and, consequently, the greater the precision of the survey. After various trials the application was performed setting the PDOP < 7 so as to guarantee both the precision and continuity of the survey. In fact, excessively reduced values of PDOP would provide excessive precision as compared to the requirements of the survey which could lead, moreover, to a greater probability of the signal being interrupted as a result of insufficient satellite positioning.

Apart from these parameters, the sampling time was also defined, being set at 1 second so as to obtain an adequate density of points (about one every 17 meters). The DGPS is restricted by the minimum initialization time necessary for the survey to be carried out (generally 200 seconds) when visible by at least 5 satellites. Subsequently, the survey can also be performed with a lower number of satellites (3 or 4) but only for brief intervals which must, in any case, follow long enough periods in which at least 5 satellites are present. Insufficient signals over long stretches result in survey errors.

As a consequence, the main limitation to the efficient use of the DGPS method of road surveying is the presence of obstacles which block satellite reception (tunnels, trenches, lateral walls and trees, buildings and so on). If reliable inertial data are available, appropriate algorithms can be used to reconstruct the missing GPS information (Toledo Moreo et al 2007). However, in our research different procedures were set up and verified as will be shown in the following paragraphs.

3 ROAD GEOMETRY

3.1 *Horizontal alignment*

The GPS survey allows a series of points to be located along the trajectory of the vehicle. Although this information makes it possible to produce an efficient representation of the stretch, it is not enough for the geometric parameters of the alignment design to be identified.

Therefore, a post-elaboration process is necessary in order to identify axis elements (tangents, circular curves, clothoids) and determine their parameters (length, radius, angular extension). The main challenge in extracting plan view information from data collected at discrete points is the correct automated identification of curvature and points of tangency between different geometric elements.

The algorithm developed is able to process the data acquired from the GPS survey to define the horizontal alignment of the road starting from spline curves.

Splines are mathematical tools that have been used since the 1960's to interpolate or to approximate functions as an alternative to the classic definition of road geometry in horizontal and vertical alignment. In actual fact, they have been used to mathematically define highway alignment in computer applications whose objective is alignment optimization (Chew et al. 1989), or the definition of vehicle paths (Guarino et al. 2000). In addition, spline curves can be used to graphically represent a highway alignment previously calculated by means of traditional procedures (Walton and Meek 2005; Castro et al. 2006).

As splines provide an high degree of freedom resulting in the possibility of over-fitting the GPS data, a specific procedure was developed to improve the output of alignment parameters.

The proposed model is based on the use of regression splines with the aid of suitable smoothing factors to correct the inevitable axis survey errors caused by the actual trajectory followed by the vehicle or to measurement inaccuracies (Cafiso et al., 2002, 2007). In particular, smoothing functions are used on the spline to correct the curvature locally by means of a P parameter whose value varies between 0 and 1:

- P = 0, the spline represents a minimum squared linear regression;
- P = 1, the spline represents an interpolation spline passing exactly through the points.

Operationally, the parameter P is defined limiting the smoothing error to 1 m, or rather, the spline performed (figure 4) is a maximum of 1 m from the surveyed GPS points.

The alignment is composed of several cubic splines each one related to four successive points, while between two successive splines there is continuity in end point coordinates, tangent and curvature. The coefficients of the cubic spline "i" being defined as $f_i(x) = a_i x^3 + b_i x^2 + c_i x + d_i$, the curvilinear extension (s) of the spline from the point of origin of the co-ordinate (x0, f(x0)) to the point of the co-ordinate (x, f(x)) is calculated according to the expression:

$$s = \sum_i \int_{x_0}^{x} \sqrt{1 + f_i^{'}(x)dx} \qquad (1)$$

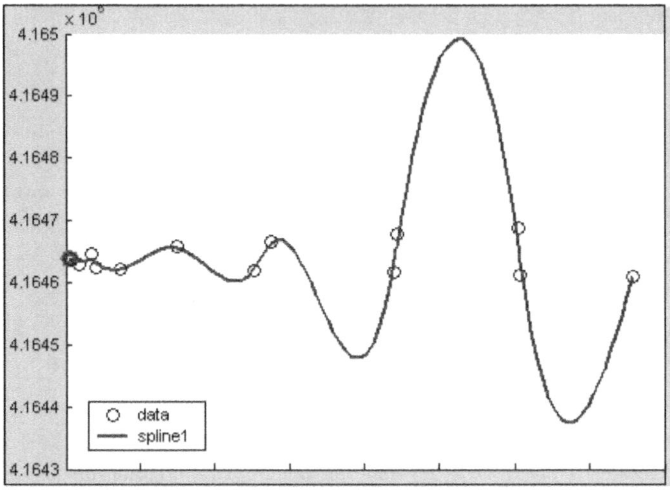

Figure 4. Spline performed by the smoothing routine. The circles are centered on the GPS points with a radius of 1 m (acceptable error).

where $f_i'(x)$ represents the first derivative of the spline function $f_i(x)$.

While, using the first and second spline derivative $f''(x)$ it is possible to calculate the $1/\rho$ curvature by means of the expression:

$$\frac{1}{\rho} = \frac{f''(x)}{(1 + f'(x)^2)^{3/2}} \quad (2)$$

The diagram of curvature vs. curvilinear extension of spline (s, $1/\rho$) highlights an oscillation in the values as compared to the actual values of the axis due to errors in the GPS data and to the inevitable shifting of the vehicle trajectory from the lane axis. Therefore, the former step of the procedure was defined in order to identify deviations from the straight trajectory ($\Delta\alpha = 0$) caused by measurement errors. The method is based on the definition of a band of tolerance having a constant width ($\pm 1/R$) within which the curvature values are returned to a nil value (figure 5). As far as the kind of road analyzed is concerned, a band width referring to a radius of 1500 m was assumed.

The latter step of the procedure concerns the definition of the curvature and extension of curvilinear elements of the horizontal alignment. Based on the curvature function (s, $1/\rho$), a definition of the actual geometrical elements of the alignment is obtained by looking for the succession of defined curvilinear elements (clothoids, circular curves) that guarantee the "*same angular and linear extension of the spline curvature*".

Having imposed equality on the angular development, the previous sentence is equivalent to impose "*equality between the areas subtended by the spline curvature and by the curvature of the alignment geometric element*". Therefore, the numerical values of the radius R of the curve and the parameter A of the clothoid are calculated imposing equality between the area subtended by the curve function of the spline and that of the trapezium, between the beginning and end points of the element under consideration. The best fitting solution is obtained establishing the minimum mean square variation of the error (figure 6).

In this way it is possible to carry out the reconstruction of the alignment geometry (tangents, clothoids and circular curves) of the stretch as shown in figure 7.

For old roads constructed without the use of clothoids, constant curve elements (rectangular shape) are identified having the same angular development and maximum curvature of the reference trapezium (figure 8).

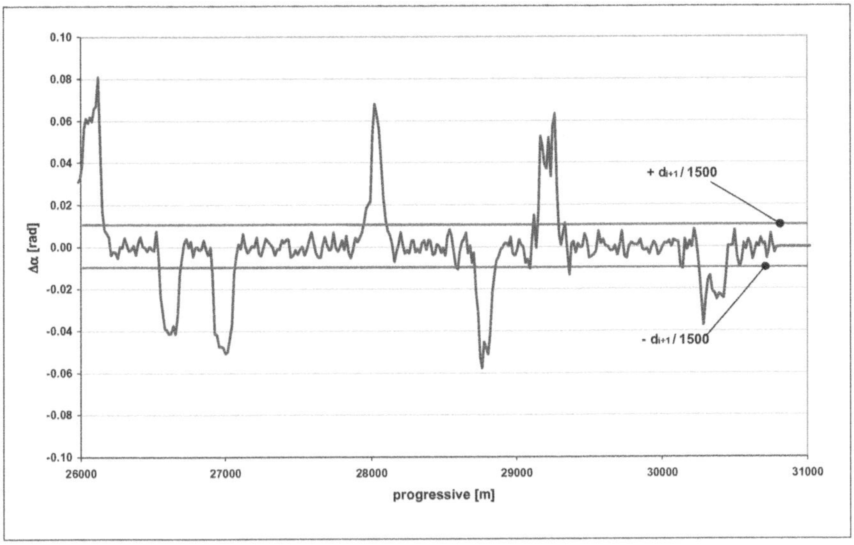

Figure 5. Data filtering by the way of a band of tolerance.

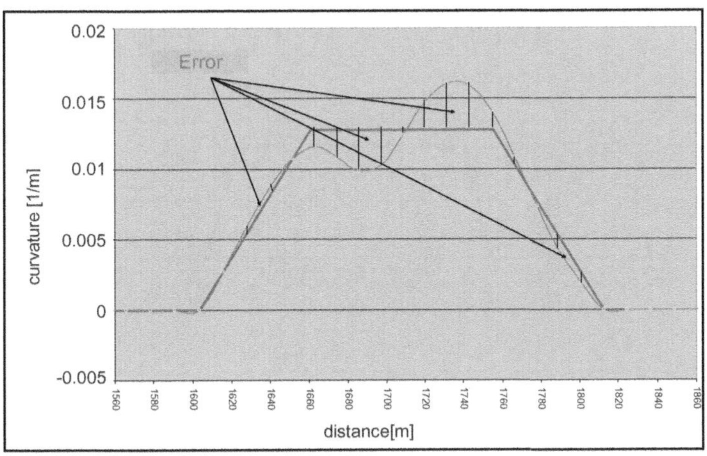

Figure 6. Error between the imposed curvature and that of the spline.

The horizontal alignment, obtained using the above method (figure 9), is congruent with the actual road design with respect to both the length and angular deviation, moreover, it corresponds well to the survey points (figure 10) highlighting the correct identification of the geometrical elements forming the stretch.

Data from inertial platform and odometer usually can be implemented (Grewal, 2001) in order to cover the parts of the road where it is not possible to acquire a GPS signal. Anyway, as mentioned above, a procedure was developed for the direct integration from maps of the points missing from the GPS survey with the aim to reduce the cost of the necessary equipment. The points are localized at constant steps (15 meters on straight stretches and every 5 meters on curves) on 1:2,000—1:10,000 scale maps using a CAD software. The points are then elaborated with the same procedure as previously described.

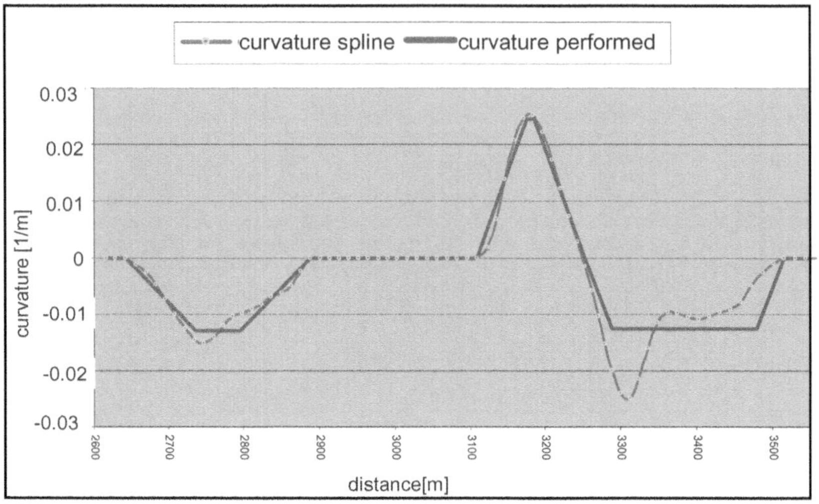

Figure 7. Comparison between curvature of spline and geometric elements.

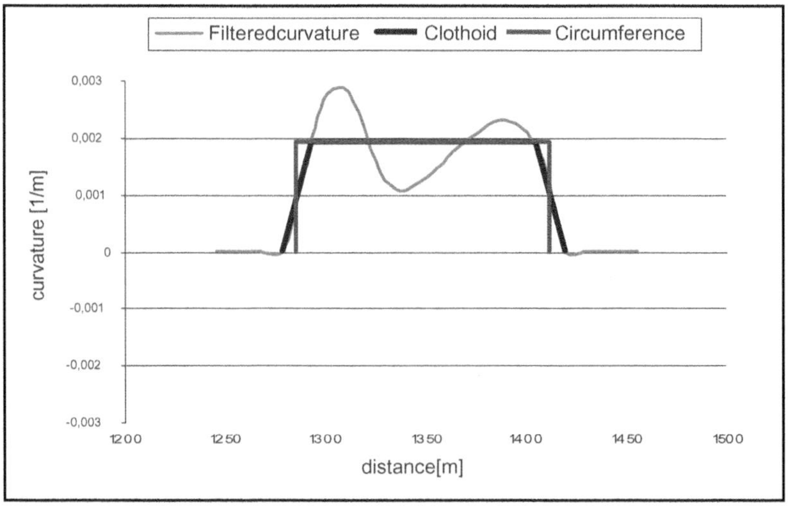

Figure 8. Example of spline, clothoids and circular elements with equivalent curvature development.

In order to verify the congruence of the two applications (GPS survey and map survey) the procedure was implemented on points acquired with both the survey methods.

As regards the comparison with GPS data (Figure 11), the manual integration procedure is able to detect all the curves shown on the maps. Curve length and degree-of-curve information is more reliable for longer curves (L > 300 m) and detection accuracy improves as the degree-of-curve on the curve increases. The maximum difference in the value of curve radius is about 2%, the difference in the length of the straight stretch is about 4% while the most significant difference is in the short curve length where a maximum error of 35% was found.

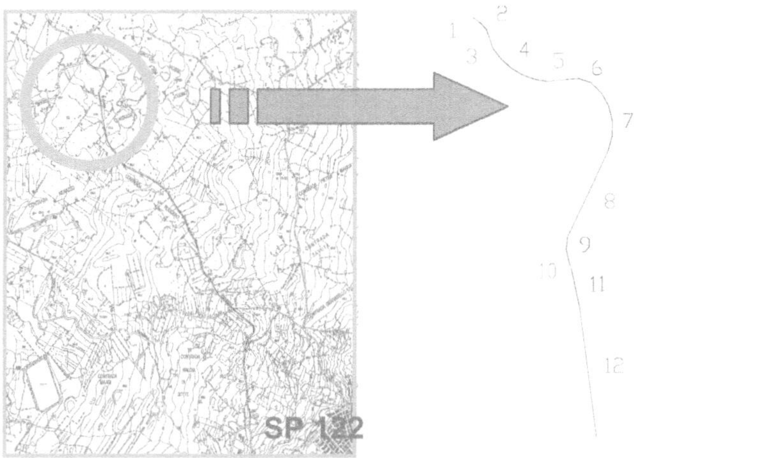

ID Road	ID Elements	FROM	TO	ELEMENT	Length[m]	Radius[m]
SP122	1	0,00	9,10	TANGENT		
SP122	2	9,10	29,50	CURVE	20,40	24,00
SP122	3	29,50	41,30	TANGENT	11,80	
SP122	4	41,30	113,20	CURVE	71,90	81,00
SP122	5	113,20	135,60	TANGENT	22,40	
SP122	6	135,60	164,50	CURVE	28,90	45,00
SP122	7	164,50	244,80	CURVE	80,30	57,00
SP122	8	244,80	345,30	TANGENT	100,50	
SP122	9	345,30	371,50	CURVE	26,20	36,00
SP122	10	371,50	380,80	TANGENT	9,30	
SP122	12	421,20	556,40	TANGENT	135,20	

Figure 9. Example of horizontal alignment data obtained from the procedure.

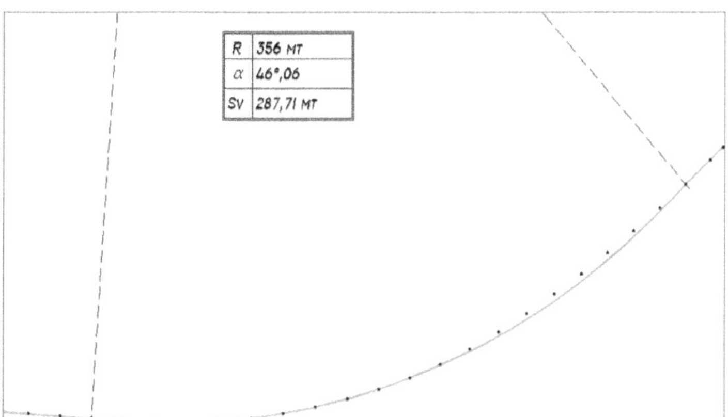

Figure 10. GPS points and horizontal alignment reconstruction.

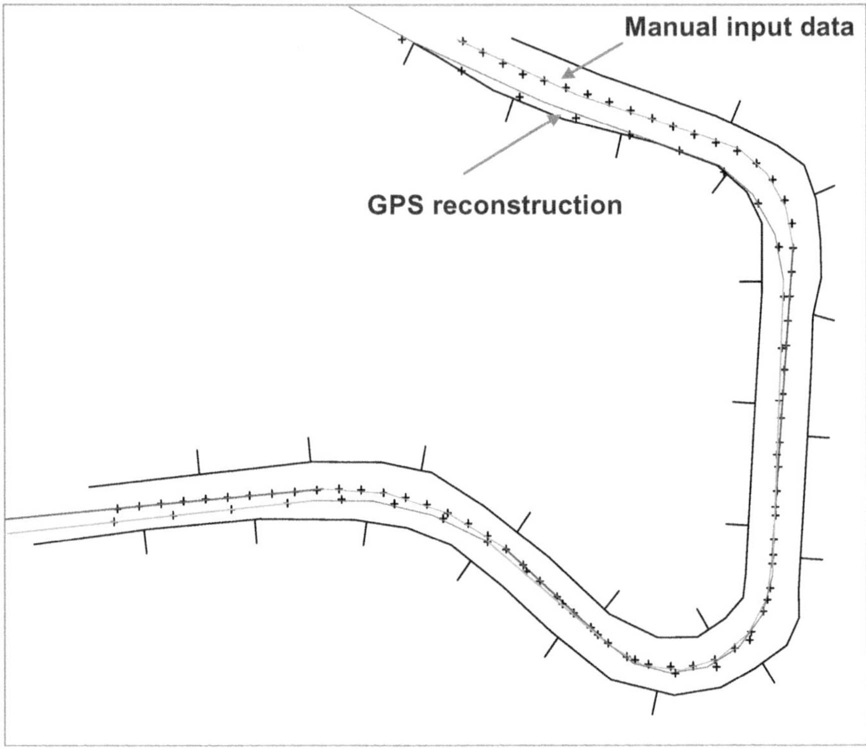

Figure 11. Comparison between CAD and GPS detected values.

3.2 *Vertical alignment*

The GPS survey of the coordinates (X, Y, Z) of the stretch can be used to define the vertical alignment. Due to the characteristics of vertical alignment composed above all by long stretch with constant slope and high value of vertical curve radius and due to the lower quota precision, the above mentioned spline method in not effective. Therefore the procedure uses linear regression lines of the GPS survey points in a Cartesian system in which the ordinate represents the quota (Z) of the point and the abscissa the progressive distance along the road axis (s):

$$Z = a \cdot s + b \text{ (equation of linear regression)} \qquad (3)$$

The constant slope stretches are identified from segments of the regression lines so as to guarantee a pre-established maximum error (err max) between the actual quota of the point and that on the equation. When the GPS survey is obtained with good precision the procedure supplies acceptable results (Figure 12).

As mentioned in the previous paragraph, when the signal quality is not good enough or there is a total loss of signal, the GPS survey is not effective enough to perform the vertical alignment with acceptable accuracy. In this case, the procedure is can be completed by the acquisition of complimentary geo-referenced data characterizing the road alignment (Cafiso et al. 2005). By using the output data of the inertial platform installed in the mobile laboratory it is possible to complete the geometrical reconstruction of the vertical alignment. In particular, the inclinometer placed along the vehicle axis is able to measure the slope of the longitudinal profile of the axis (Figure 13).

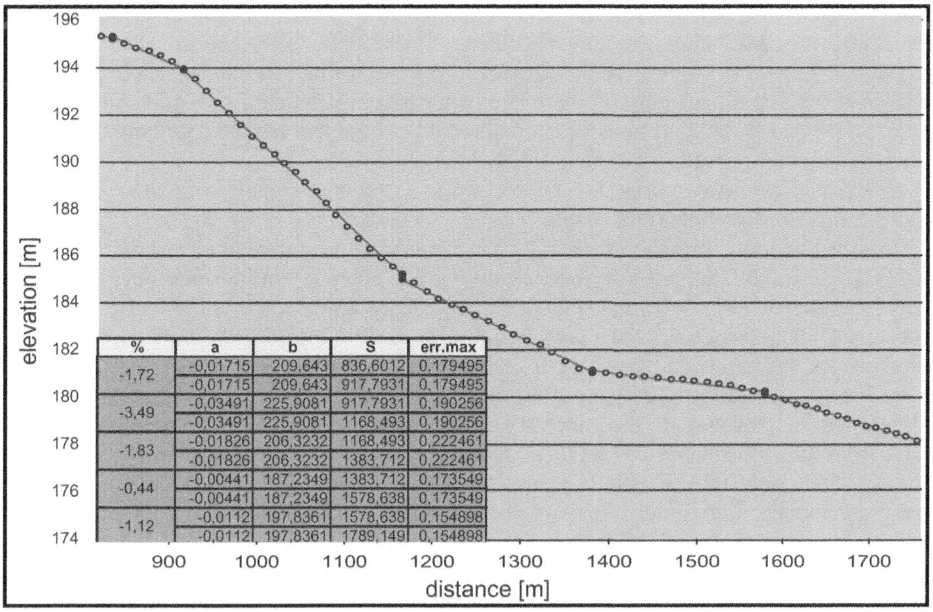

%	a	b	S	err.max
-1,72	-0,01715	209,643	836,6012	0,179495
	-0,01715	209,643	917,7931	0,179495
-3,49	-0,03491	225,9081	917,7931	0,190256
	-0,03491	225,9081	1168,493	0,190256
-1,83	-0,01826	206,3232	1168,493	0,222461
	-0,01826	206,3232	1383,712	0,222461
-0,44	-0,00441	187,2349	1383,712	0,173549
	-0,00441	187,2349	1578,638	0,173549
-1,12	-0,0112	197,8361	1578,638	0,154898
	-0,0112	197,8361	1789,149	0,154898

Figure 12. Example of vertical alignment output obtained using the linear regression procedure.

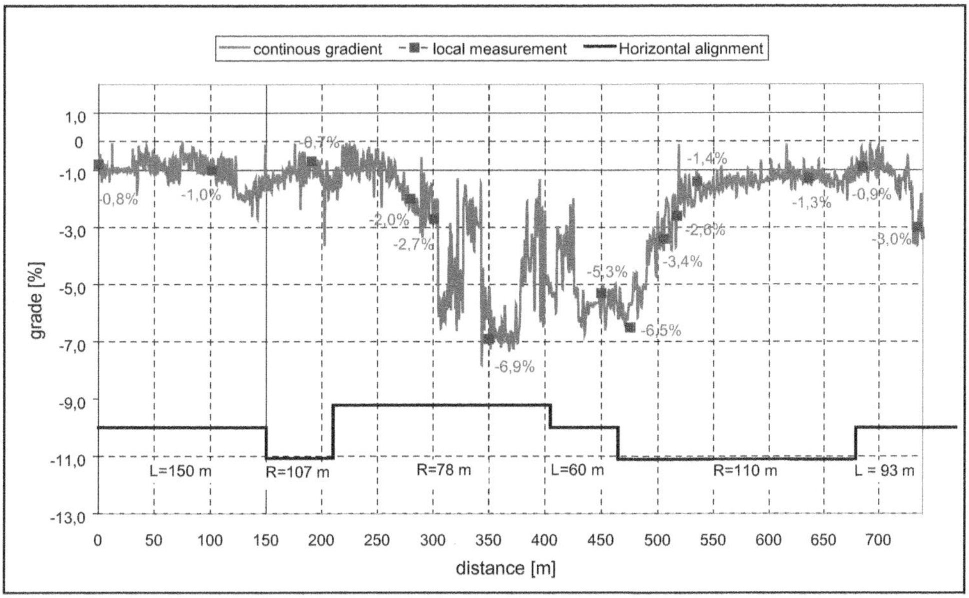

Figure 13. Example of longitudinal slope data acquired during the survey.

4 CONCLUSIONS

The necessity to rapidly and effectively acquire information relating to the geometrical and functional characteristics of the existing road network is widely felt, above all given the lack of suitable documentation available to management agencies.

In this context, the horizontal alignment of the roadway must be considered the fundamental basis for all future road management procedures.

To provide a low cost solution to this lack of information which is indispensable for the setting up of infrastructure management projects, the paper presents the hardware and software systems that were designed and experimented using a suitably-equipped vehicle together with purposely developed computer procedures to elaborate the acquired data.

The results and considerations are based on the experience acquired in the field during an extensive survey of two-lane rural roads.

As regards the surveying, the cinematic DGPS mode is accurate enough while the use of real time (RTK) correction, although further improving precision, is not indispensable to achieve the scope of the survey. The most important problem in GPS road survey is the loss of a signal that can occur for the presence of obstacles which block satellite reception (tunnels, trenches, lateral walls and trees, buildings and so on). If reliable inertial data are available and the missing GPS data are limited to short stretches, appropriate algorithms could be used to reconstruct missing GPS information. However, in the paper, a process of direct integration from maps of the points missing from the GPS survey was set up and verified.

The algorithm developed is able to process the data acquired from the GPS survey in order to define the geometric horizontal fetures of the road. The model is based on the use of regression splines with the aid of suitable smoothing factors to correct the inevitable axis survey errors due to the actual trajectory followed by the vehicle or to measurement inaccuracies.

The horizontal alignment obtained is congruent with the actual road design as regards length and angular deviation and, moreover, it corresponds well to the survey points thus highlighting the correct identification of the geometrical elements forming the stretch.

The GPS survey of the coordinates (X, Y, Z) of the stretch can be used to define the vertical alignment, too. The proposed procedure uses linear regression lines of the GPS survey points. On the other hand, when the GPS precision obtained in vertical coordinates is not effective enough to measure the vertical alignment with acceptable accuracy, it is possible to complete the geometrical reconstruction of the stretch if other information are collected during the survey. In particular, an inclinometer placed along the vehicle axis can be used to measure the slope of the longitudinal profile of the axis.

REFERENCES

Ben-Arieh, D., Chang, S., Rys, M. & Zhang, G. 2004. Geometric modeling of highways using global positioning system data and B-spline approximation. Journal of Transportation Engineering 130, pp. 632–636.

Bojkov, V. & Urmanov, I. 1997. Use of parametric spline smoothing in highways reconstruction. Proceedings of the XIIIth World Meeting of International Road Federation. Toronto, Canada.

Cafiso, S., Di Graziano, A. & Di pasquale, G. 2002. Procedure di rilievo e ricostruzione dei tracciati stradali per la realizzazione di un GIS per il catasto delle strade (IT). Proceedings of XII SIIV Annual Meeting. Parma. Italia.

Cafiso, S., Di Graziano, A., La Cava, G. & Calabrò, S. 2003. Progettazione e manutenzione di un sistema per il rilievo ad alto rendimento della viabilità in esercizio (IT). Proceedings of XIII SIIV Annual Meeting. Padova. Italia.

Cafiso, S., Di Graziano, A. & La Cava, G. 2005. Actual Driving Data Analysis for Design Consistency Evaluation. Transportation Research Record 1912, TRB Washington D.C. USA.

Cafiso, S., La Cava, G., Montella, A., Di Graziano, A., Pappalardo, G. & Taormina, S. 2007. Final Research Report "Identification of Hazard Location and Ranking of Measures to Improve Safety on Local Rural Roads: I.A.S.P.)" in website http://ec.europa.eu/transport/roadsafety/publications/projectfiles/iasp_en.htm

Castro, M., Iglesias, L., Rodríguez-Solano, R. & Sánchez, J.A. 2006. Geometric modelling of highways using global positioning system (GPS) data and spline approximation. Transportation Research Part C: Emerging Technologies Volume 14, Issue 4.

Chew, E.P., Goh, C.J. & Fwa, T.F. 1989. Simultaneous optimization of horizontal and vertical alignments for highways. Transportation Research part B: Methodological.

Drakopoulos, A. & Ornek, E. 2000. Use of vehicle-collected data to calculate existing roadway geometry. ASCE Journal of Transportation Engineering Vol. 126 March/April.
Grewal, M.S., Weill, L.R. & Andrews, A.P. 2001. Global Positioning Systems, Inertial Navigation and integration. John Wiley 6 Sons Inc. Publication. ISBN 0471-35032-X.
Guarino Lo Bianco, C. & Piazzi, A. 2000. Optimal trajectory planning with quintic G2-splines. Proceedings of the IEEE Intelligent Vehicles Symposium, Dearborn, USA.
Harkey, D.L., Yi Chang & Feagenes, J. 2004. Evaluation and validation of an automated in-vehicle data collection system for developing roadway alignments. Proceedings of 83rd TRB Annual Meeting. Washington D.C. USA.
Toledo-Moreo, R., Zamora-Izquierdo, M.A., Ubeda-Miarro, B. & Gomez-Skarmeta, A.F. 2007. High-Integrity IMM-EKF-Based Road Vehicle Navigation With Low-Cost GPS/SBAS/INS. Intelligent Transportation Systems, IEEE vol. 8, Issue: 3.
Walton, D.J. & Meek, D.S. 2005. A controlled clothoid spline. Computers and Graphics 29, pp. 353–363.
Wul, J. & Tsai l, Y. 2006. Enhanced Roadway Geometry Data Collection Using an Effective Video Log Image-Processing Algorithm TRR 1972 2006.

Efficient Transportation and Pavement Systems – Al-Qadi, Sayed, Alnuaimi & Masad (eds)
© 2009 Taylor & Francis Group, London, ISBN 978-0-415-48979-9

A framework for pavement RL prediction using KDD

E.A. Sharaf
Cairo University, Egypt

M.A. Abo-Hashema
Fayoum University, Egypt

M.M.S. El-Din El-Hawwary
AAW- UAE

ABSTRACT: In Pavement Management Systems (PMS), data analysis is considered crucial step due to the huge amount of data collected. Consequently, many studies have been conducted to incorporate technology within the analysis process. Recently, Knowledge Discovery in Databases (KDD) technology has gained special interest not only in analysis but also for knowledge extraction. The objective of this paper is to introduce KDD and its powerful way of handling huge amount of data, through predicting pavement Remaining Life (RL). As a matter of fact, discovering new relationships between RL and the different pavement characteristics isn't of major interest. This attempt is considered an initiative for incorporating KDD in PMS.

1 INTRODUCTION

The need for improved methods of decision making regarding the most cost-effective means of allocating limited resources, setting priorities, measuring performance, and budgeting for pavement maintenance and rehabilitation has led to the development of a variety of management systems as an aid for achieving this task. Such systems dated back to the mid to late 1960's in North America, the United Kingdom, and other European countries.

In these countries, they have been identified under a confusing variety of labels such as Road Maintenance Management Systems (RMMS), Pavement Maintenance Management Systems (PMMS), and Pavement Management Systems (PMS) (Haas & Hudson 1978, Haas, Hudson & Zaniewski 1994). While throughout this research, the label PMS will be used.

The main functions of a PMS are to assist highway authorities in finding optimum strategies for maintaining pavements at an acceptable level of serviceability over their intended design life at minimum cost. It is noteworthy that pavement, constructed poorly with substandard materials, will perform poorly regardless of the use of a PMS. However, while it cannot make decisions or eliminate financial deficiencies, it can minimize the impact of such deficiencies and assist the responsible authorities in arriving at the appropriate decisions in an objective and consistent manner (Pinard 1987).

Collecting data is considered the first step in developing a PMS, where the PMS Database becomes the heart of the system. There are three main types of data that should be fed into the PMS database: road inventory, pavement condition survey, and various repair strategies and options, as shown in Figure 1 (Fahmy 2003).

Several data items should be also collected taking into consideration the network identification system. Many of the existing systems focus on collecting data items related to the condition and performance of pavement structures with minimal attention given to other data items.

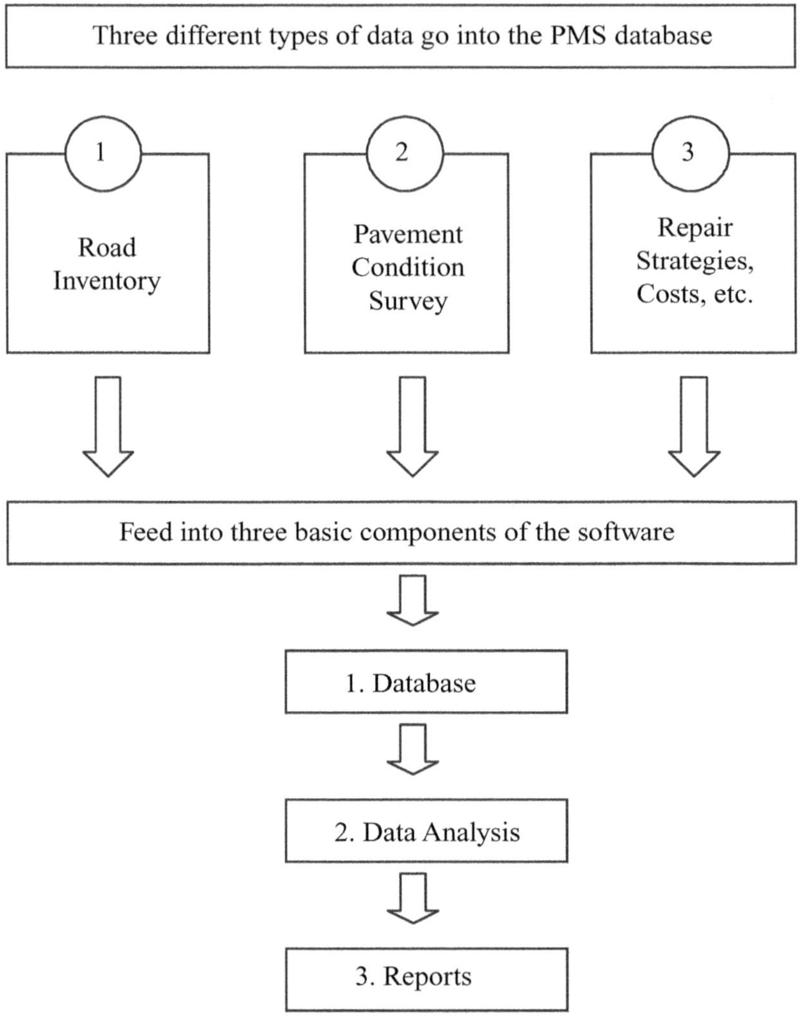

Figure 1. PMS software components.

Although, the pavement condition data is the key element of any PMS; other data play a vital role as well. The classes of data needed, other than pavement condition and performance, include the following (Haas 1991):

1. Historic data
2. Policy-related data
3. Geometric data
4. Environmental data
5. Costal data

The amount of data collected throughout developing the PMS database is recognized to be huge enough to require computer assistance for proper interpretation. Accordingly, many studies have been involved to include the automated bay for the PMS, from primitive computer programs to Expert Systems (ES).

Database scientists have been working hard developing new techniques and tools to pave the gap that is expanding between the analysis method used and the amount of data available. Through the past ten years, Knowledge Discovery in Databases (KDD) has gained a special interest for its powerful way of handling huge amount of data not only in analysis but also for knowledge extraction.

Many countries have expressed interest in the use of the KDD technology through academic research, though with fewer attempts for practical use. On the other hand, in developing countries such as Egypt, shortages in resources and updated technologies generated a barrier towards applying the new technique. This shortage is due the need for a real investment in developing a practical running project or even an academic one.

The objective of this paper is to implement the KDD techniques on the PMS database through a case study for predicting the Remaining Life (RL) of a pavement. As a matter of fact, discovering new relationships between RL and the different pavement characteristics is of major interest. This attempt is considered influential for enhancement of PMS.

2 KNOWLEDGE DISCOVERY IN DATABASES OVERVIEW

2.1 *General*

The advances in scientific data collection tools have led to a huge and continuously increasing amount of data in numerous forms and many types. This explosive growth of databases has created an urgent need for new techniques and tools that support human mind in transforming data into useful information and knowledge (Fayyad, Piatetsky-Shapiro & Smyth FPS 96).

Since 1960s, database technology has been evolving systematically from primitive file processing systems to powerful and sophisticated database systems. The research in database systems since 1970s has progressed from early hierarchy and network database systems to development of relational database systems (Han & Kamber 2001).

Through the same period, tools were introduced to build an environment for the users to easily access the data through a Graphical User Interface (GUI) and simple query languages. Since 1980, the concept of relational technology has been dominating even over the development of new techniques.

Currently, data can be stored in many different types of databases. An example of the recently emerged database architecture is the data warehouse. Data warehouse can be defined as a repository of multiple heterogeneous data sources organized under a unified schema at a single site in order to facilitate management decision-making (Han & Kamber 2001).

While these techniques were recorded as powerful tools and were strongly and widely used, problems exploded rapidly. Statement by John Naisbett, one of the KDD pioneers, stated that people are drowning in information but starving for knowledge, which clearly summarizes the problem, El-Toni, T. B. 1999).

2.2 *KDD definition*

In general, KDD is considered to be "the non-trivial extraction of implicit, previously unknown, and potentially useful information from database" (Frawley "AI Magazine" 1992).

A more descriptive definition, proposed by Usama Fayyad, explore KDD as " the non-trivial process of identifying valid, novel, potentially useful, and ultimately understandable patterns in data", (Fayyad, Piatetsky-Shapiro & Smyth FPS 96). In simple words, KDD is the process of using facts (data), in a process of discovering unexpected (novel), valid on new data, understandable and useful to us expression (pattern).

The KDD process is interactive and iterative involving numerous steps with many decisions being made by the user, (Fayyad, Piatetsky-Shapiro & Smyth FPS 96). Figure 2 propagate the steps of leading from raw data onto some form of knowledge.

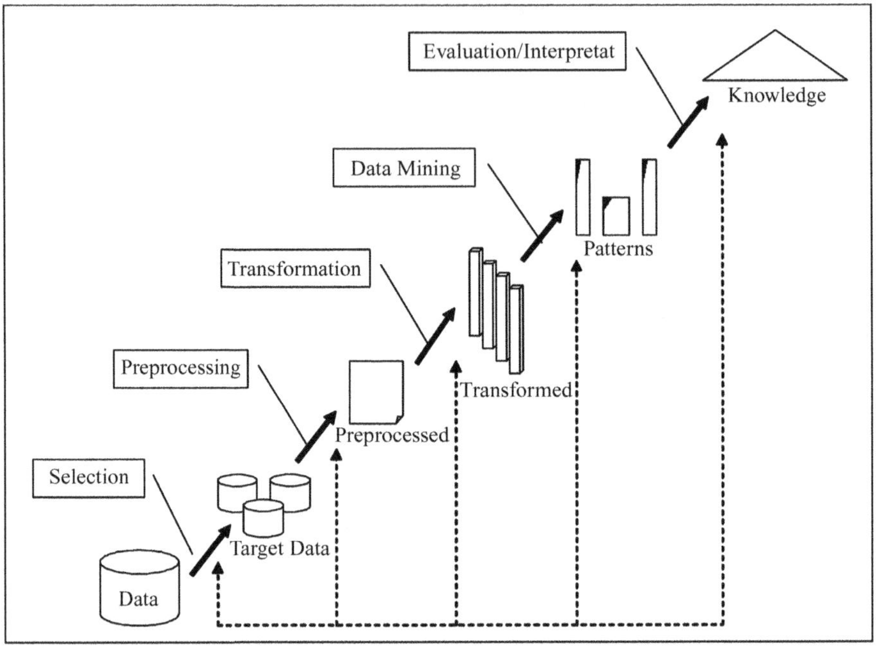

Figure 2. Overview of steps comprising KDD processes (Fayyad, Piatetsky-Shapiro & Smyth FPS 96).

The iterative process consists of the following steps (Osmar & Zaïane 1999):

1. Data cleaning: also known as data cleansing, it is a phase in which noise data and irrelevant data are removed from the collection.
2. Data integration: at this stage, multiple data sources, often heterogeneous, may be combined in a common source.
3. Data selection: at this step, the data relevant to the analysis is decided on and retrieved from the data collection.
4. Data transformation: also known as data consolidation, it is a phase in which the selected data is transformed into forms appropriate for the mining procedure.
5. Data mining: it is the crucial step in which clever techniques are applied to extract patterns potentially useful.
6. Pattern evaluation: in this step, strictly interesting patterns representing knowledge are identified based on given measures. This step is described briefly in the next section.
7. Knowledge representation: is the final phase in which the discovered knowledge is visually represented to the user. This essential step uses visualization techniques to help users understand and interpret the data mining results.

It is a common act merging steps as shown in Figure 2, data cleaning and data integration steps were merged to form one preprocessing phase.

3 DATA MINING

3.1 *General*

Data mining derives its name from the similarities between searching for valuable information in a large database and mining rocks for a vein of valuable ore (Osmar & Zaïane 1999). In industry, in media, and in the database milieu, the term data mining is becoming more popular than the long Knowledge Discovery in Databases (Han & Kamber 2001). However, in this research the term data mining is used as an essential step through the process of KDD.

The two "high-level" primary goals of data mining in practice tend to be prediction and description. Prediction involves using some variables or fields in the database to predict unknown or future values of other variables of interest. Description focuses on finding human-interpretable patterns describing the data (Osmar & Zaïane 1999).

In principle, data mining is applicable to any kind of data repository starting from simple databases repositories like flat files and moving through the paradigm of relational databases and the advanced databases systems. Following is a list of some repositories types:

- Flat files
- Relational Databases
- Multimedia Databases & Spatial Databases
- Temporal Databases
- World Wide Web

3.2 *Mining tasks*

The kinds of patterns that can be discovered depend upon the data mining tasks, which can be defined as the process of artificially analyzing and organizing data. One of the most popular tasks is classification. Classification analysis is the organization of data in given classes (Shekhar et al. 2000), usually through learning from set of data with known classes. A model is then generated for classification of new data through a prediction process.

3.3 *Mining methods*

To achieve the data mining tasks previously mentioned an algorithm should be constructed (Osmar & Zaïane 1999). Much effort has been expended in the statistics, computer science, artificial intelligence, and engineering communities to overcome the limitations of this basic model (Osmar & Zaïane 1999). Following is a list of some common mining methods:

- Neural networks
- Decision trees
- Rule induction
- K-nearest neighbor and Memory-Based Reasoning (MBR)
- Genetic algorithms

4 RESEARCH METHODOLOGY

The objective of this research is to apply the KDD process to discover new relationships between the RL of a pavement and the different pavement characteristics. To achieve this objective, a framework has been constructed as shown in Figure 3. The framework of this study directs into a mainstream of sequential steps, which consists of four modules. The KDD process is considered the core of the mainstream. However, adding or skipping steps through the process will be some how essential to achieve the collaborative approach of KDD to the PMS discipline with respect to the scope of the research.

Figure 3 shows an essentially needed feedback from the decision support system task to the problem identification so as to ensure orientation of results. However, some non-research related feedback paths were excluded.

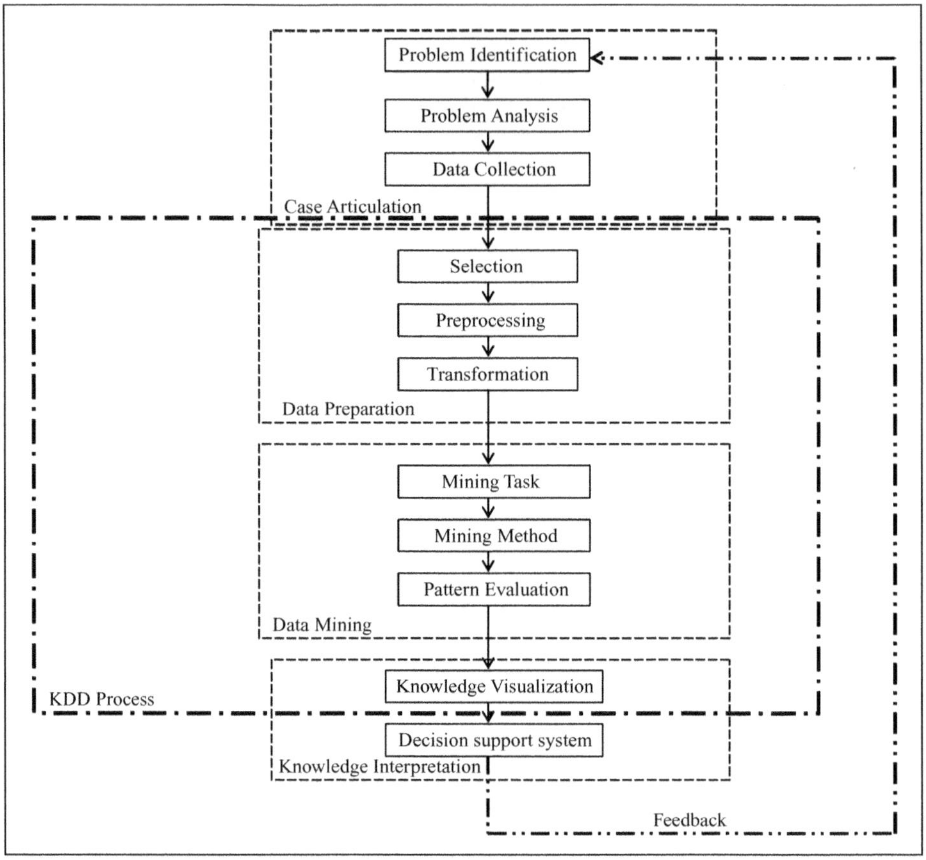

Figure 3. Research mainstream with core KDD process.

5 APPROACHING PAVEMENT REMAINING LIFE

5.1 *General*

The following sections describe the four modules of the research mainstream to approach pavement remaining life.

5.2 Case articulation

The case articulation includes three tasks: problem identification, problem analysis, and data collection. The problem identification task involves statementizing the problem in a clear and descriptive way usually in a form of a question.

The problem may call for phenomena description or trend prediction as previously stated. In this study, the question is calling for the relation description between the different pavement characteristics and RL as well as prediction of the RL of a pavement.

The statements questioning both goals are as follows:

- What is the relation between RL and pavement characteristics?
- What is the remaining life of a pavement section?

The problem analysis task provides a wider view of the problem definition, type of data needed and tools used. Remaining life of a pavement depends on known factors such as past traffic, pavement condition, and quality of existing materials. Describing relations between pavement characteristics and RL necessitates blending known related-factors and unexpected related-factors to discover novel relations. To achieve this goal, tools with both linguistic and numerical capabilities should be used. Accordingly, the GeneMiner-2001 software was selected.

The data collection procedure has direct relation with the problem definition, i.e. case uniqueness property. Therefore, it was involved in the module of case articulation and hasn't been introduced along with the other modules. Following is a brief description of the data source.

Since 1987, the General Authority for Roads and Bridges and Land Transport (GARBLT), of the Ministry of Transport in Egypt, performed maintenance and rehabilitation project for the Egyptian road network. The project has aimed at developing the maintenance and rehabilitation program through a computer database. The component of the database program consisted of road inventory related data, structural assessment related data, and maintenance cost related data. In 1993, some modifications have been performed to the database program. It is noteworthy that periodically updating of the database has been accomplished in a yearly basis. In 2000, it has been decided to develop a comprehensive PMS using Geographic Information System (GIS) that suits the GARBLT specifications (Shekhar et al. 2000).

Throughout the GARBLT study, a remarkable effort was done in data collection, which involved office and field data. The field data deliberated pavement surface condition data by visual inspection and Digital Video Mapping system. The available data contains 3580 recorded sections over the network under study. Each section has more than 200-parameter some of which were measured from field (Example: Section width) and others were post processed and calculated (Example: Remaining Life), (Shekhar et al. 2000).

5.3 *Data preparation*

The data preparation module consists of three tasks: data selection, data preprocessing and data transformation. Figure 3 reveals the boundaries of KDD coinciding with the data preparation module to formulating the formal beginning of knowledge discovery.

The data selection task is the step in which the data relevant to the analysis is decided on and retrieved from the data collected. In other words, data selection is considered as the focusing on a subset of variables or data sample on which discovery is to be performed, (Fayyad, Piatetsky-Shapiro & Smyth FPS 96).

Based on the problem definition, case articulation module, the question is calling for the prediction of the pavement remaining life. The RL of a pavement is affected by pavement surface condition, past and future traffic loading, and quality of the existing materials used in the different pavement layers. Therefore, the relevant data was considered to be the data associated with the pavement condition and that of the general identification for the highways in question as well as traffic loading.

It is noteworthy that the non-relevant and non-essential parameters were excluded through the available 3580 sections. Eventually due to software limitations, seven variables out of 200 variables for each section have been selected to represent the subset of variables on which discovery is to be performed. Impact of variables selection can be recorded as future plans. **Error! Reference source not found.** illustrates the selected attributes for the analysis including the examined parameter, which is the remaining life. The analysis was done on 1740 section only for knowledge extraction and the rest, 1840 section, were used for prediction and validation.

As shown in **Error! Reference source not found.**, the pavement surface condition is represented by Pavement Condition Index (PCI) value. The PCI value was calculated based on the 19 pavement surface distresses listed in the PCI procedure system (Shahin & Kohn 1981).

The data preprocessing task involves removal of some sections, to ensure data consistency, that have at least one of the following:

- Noise data such as the usual manual errors and outliers,
- Missing data fields, and
- Zero fields.

Table 1. Selected attributes for analysis including the examined parameter.

Data code	Description
Serial	Number of section
PCI	Pavement Condition Index
Rem_L	Remaining Life
Width	Section width
Per-Trk	Percent of Trucks
AADT	Average Annual Daily Traffic
TGF	Traffic Growth Factor

Ultimately, the selected 1740 section for the analysis were summed up into 1500 section. It is noteworthy that there was only one data source so integration of multiple data sources was out of needs and the data was totally homogeneous. The task of data transformation is also known as data consolidation. It is a phase in which the selected data is transformed into forms appropriate for the mining procedure. In this research, data transformation was applied to convert the selected data from a Microsoft Excel format into Text file format then the tables were combined to generate a single.

5.4 *Data mining*

This module fairly divides into three basic tasks: Mining Task, Mining Method, and Pattern Evaluation. The point here is applying the KDD process on the selected data-set to investigate two different issues: prediction of pavement remaining life and finding the relations governing it. Both trails were applied using the GeneMiner-2001 software.

GeneMiner is a tool developed by Kdiscovery and released on Feb 2001. The software includes a built-in MS Excel-compatible spreadsheet for easy data manipulation, user-friendly interface to simplify the KDD process in a parable way with a dynamic, high degree of automation, and a ready to deploy solutions state. The mining Method used by the GeneMiner is a composite system called the GenoFuzzy Hybrid System, which combines the advantages of fuzzy systems and Genetic algorithms. Fuzzy systems deal with explicit knowledge that can be explained and understood. On the other hand, Genetic algorithms deal with implicit knowledge that can be acquired by genetic search providing a good way to adjust the expert's knowledge and automatically generate additional fuzzy rules to meet certain specifications and reduce design time and costs (GenoFuzzy Help 2001).

Moreover, fuzzy logic enhances the generalization capability of a genetic system by providing more reliable output when extrapolation is needed beyond the limits of the training data (GenoFuzzy Help 2001).

Fuzzy logic makes it simple to represent concepts that are more linguistic than mathematical. Fuzzy models are fairly straightforward translations of the linguistic statements to a group of rules. The model begins to function roughly as soon as two or three rules are stated, and is easily refined by tuning up the sets or by addition of more rules (www.arts.ucsc.edu 2002). Figure 4 investigates a typical fuzzy logic system when applied for classification task.

Finally, after running the GeneMiner it was found that 1500 case were the total dataset evaluated and the success was 99.93%, which is usually called the learning success. It was noticed that the best tree is obtained after 60 generations.

The knowledge or the extracted pattern indicates that the most important factor affecting RL is the section width, which achieved 99.93% success; while the AADT is not of that much importance with only 32.60% success. This could be understood as in such a case of narrow sections where heavy truck tends to move very slowly causing more deterioration effect on the pavement layers, while for wider sections probably truck-driver tends to move faster causing less deterioration effect on the pavement layers. The PCI value and percent of trucks move side by side

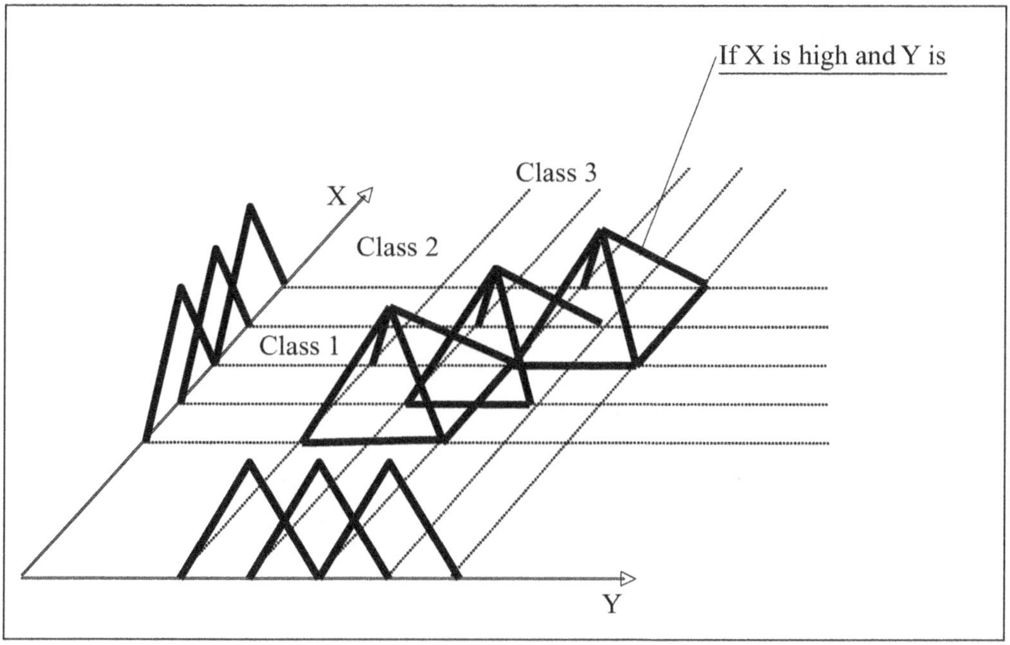

Figure 4. A typical fuzzy logic system applied for classification (El-Toni, T.B. 1999).

achieving 76.80% success. Other than that the traffic growth factor was excluded due to its limited effect shown in the provided dataset. Finally, the five parameters affecting RL can be ranked based on highest-first as follow: section width, PCI value, percent of trucks, AADT and TGF.

5.5 *Knowledge interpretation*

The Knowledge Interpretation Module: Consists of two steps; Knowledge Visualization and Decision Support System (DSS) implementation. Knowledge visualization step involves the visualization of the discovered patterns in a multisided environment using the proposed tool or an additional tool. Types of visualization include graphs, tables, trees etc.

On the other hand, a DSS was developed by visual basic programming language to predict the remaining life of a pavement section and results were validated as described below.

5.6 *Validation process*

The prediction and validation process was performed using the rest of dataset provided, which is the 1840 section. The predicted RL of the 1840 section was calculated using the DSS program compared to the corresponding available dataset. **Error! Reference source not found.** provides a sample of the outputs of the predicted RL compared to the provided RL.

Figure 5 shows the relationships between the predicted and the provided remaining life. It indicates that the DSS program predicts the pavement remaining life, which is given in ranges, around the equality line.

It can be noticed that some points are slightly above the equality line. The important thing to notice here is that the error between the predicted and the provided remaining life is of an acceptable range.

Statistical measurements were obtained to compare between the predicted and the provided remaining life. The correlation analysis was made between the predicted (average between upper and lower limits) and the provided remaining life with a test of the significant 2-tailed.

Table 2. Sample data information and predicted results.

Sample number	Per-Truck %	AADT V/Day	TGF	Width, M	PCI %	RL, Years	Pre-RL Years
1	56.3	7545	3	7.5	95.7	3	2.6–3.8
2	30	5000	5	7.5	99.6	4	2.6–3.8
3	10	500	3	7.5	69.5	1	1–1.4
4	64.9	1052	3	6	89.4	2	1.4–2.6
5	64.9	1052	3	6	70.8	1	1–1.4
6	49	2560	8	7.5	100	5	2.6–3.8

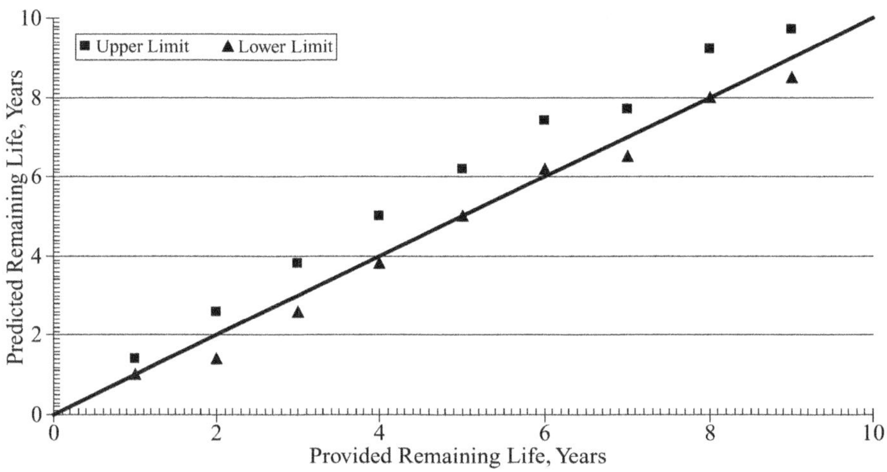

Figure 5. Comparison between predicted and provided remaining life.

A two-tailed significance is the probability of obtaining results as extreme as the one observed and in either direction when the null hypothesis is true. It tests a null hypothesis in which the direction of an effect is not specified in advance. This correlation analysis is to show how strong the relationship between the predicted and the provided remaining life is. The correlation coefficient is 0.975 with the Standard Error of Estimate (SEE) 0.75 Year.

6 CONCLUSIONS

Highway Management faces a problem of huge data availability outpacing the human ability to interpret and analysis this data especially in pavement deterioration studies. This research formulates a general framework for using the rapidly emerging field of Knowledge Discovery in Database to predict the reaming life of a pavement section. Were the results has shown an acceptable success of KDD use. The outcome of this effort is not considered to be a new theory, rather than an understanding of existing technology; uses, importance, and procedures.

REFERENCES

El-Toni, T. B. 1999, "*Knowledge Discovery in Databases and Decision Support systems*", Master of Science Thesis, Nottingham University.

Fahmy, M.M. 2003, "*Comparison between Ranking and Optimization Models in Setting Pavement Maintenance Priorities*", Master of Science Thesis, Department of Civil Engineering, Cairo University, Egypt.

FPS 96 Fayyad U.M., Piatetsky-Shapiro G., Smyth P.: "*From Data Mining to Knowledge Discovery: An Overview*".

Frawley, W, "*AI Magazine*" 1992.

GenoFuzzy Help 2001, *Computing Technical Report TR-2000-408*. Copyright ©, KDiscovery.com. All rights reserved.

Haas. R. and Hudson W.R. 1978, "*Pavement Management Systems*", McGraw-Hill.

Haas, R. 1991, "*Generically Based data Needs and Priorities for Pavement Management*", ASTM STP,

Haas, R., Hudson, W.R. and Zaniewski 1994, "*Modern Pavement Management*", Krieger Publishing Company, Malabar, Florida.

Han J., Kamber M. 2001 "*Data Mining: Concepts and Techniques*", Simon Fraser University.

Pinard, M.I., 1987, "*Factors Affecting the Development and Implementation of Pavement Management Systems in Developing Countries*", Second North American Conference on Managing Pavements, Proceedings, Volume 1 Toronto, Ontario, Canada, November 2–6, pp. 300–310.

Osmar R. Zaïane 1999, "*Principles of Knowledge Discovery in Databases*" Department of Computing Science, University of Alberta.

Shekhar S., Lu C.T., Chawla S., Zhang P. 2000, Computer science department, University of Minnesota August 2, (*http://www.cs.umn.edu/Research/shahi-group*).

Sharaf, E.A., Fahmy, M. M., Abo-Hashema, M.A. 2003, "*Pavement Maintenance Priorities Using Ranking and Optimization Models*", The 2nd International Operation & Maintenance Conference in the Arab Countries OMAINTEC 2003, 30 June–3 July, Phoenicia Inter-Continental, Beirut, Lebanon.

Shahin, M.Y. and Kohn, S.D. 1981, "*Pavement Maintenance Management for Roads and Parking Lots*", United States Army Corps of Engineers, Technical Report M-294. http://*arts.ucsc.edu*/EMS/Music/research/FuzzyLogicTutor/FuzzyTut.html#fuzzy (2002).

Efficient Transportation and Pavement Systems – Al-Qadi, Sayed, Alnuaimi & Masad (eds)
© 2009 Taylor & Francis Group, London, ISBN 978-0-415-48979-9

21st Century pavement design and maintenance of the English road network

B.W. Ferne, D. Gershkoff & M.A. Wright
Transport Research Laboratory, Wokingham, Berkshire, England

ABSTRACT: For a number of years the UK Transport Research Laboratory has been working for the English Highways Agency and Department for Transport in developing ever more sophisticated methods for designing and evaluating the 300,000 km road network in the United Kingdom and using this information to design the most economic maintenance treatments. With ever heavier traffic, the emphasis is on techniques that minimise the disruption to traffic and the whole life costs in a sustainable manner. The development of long-life pavement designs, that concentrate on pavements that do not deteriorate structurally, is core to this but designing surface layers that provide the properties required by the road user is also essential. When evaluating such pavements a range of assessment techniques is now available in the UK, from the use of comprehensive condition information measured at traffic speed covering the whole network annually to in-depth assessment techniques targeting specific schemes.

1 INTRODUCTION

For many years the UK Transport Research Laboratory (TRL) has been carrying out research for the UK Highways Agency and the Department for Transport in order to develop improved methods of designing, evaluating and maintaining their road networks, which cover over 300,000 km. As traffic has grown so has the demand for designs with longer lives that make better use of the raw materials. Careful study of the structural behaviour of road pavements over many years has lead to long-life designs for the structural elements of a pavement and more recently the potential for long-life surfacings. Although long-life pavements are inherently sustainable, the re-use of existing pavement materials, particularly for surfacings, is also an outcome of this research programme. Despite these long-life designs maintenance is still necessary and therefore evaluation of pavement condition is vital. When evaluating such pavements a range of assessment techniques are now available in the UK, from the use of comprehensive surface condition information provided by multi-function traffic-speed survey vehicles covering the whole network annually, or in some cases more frequently, to in-depth assessment techniques more generally targeted at specific schemes. Such in-depth techniques include Ground Penetrating Radar (GPR) to identify construction thicknesses and material deterioration, both at slow speed for detailed investigations and at traffic speed to provide network information as well as the use and interpretation of deflection measurements for assessing the structural condition of selected schemes and for use at network level.

Such techniques lead to large quantities of useful information that need careful management if the full benefit of the information is to be realised particularly in terms of the economic assessment of the whole life of the network. The following sections describe in more detail how each of these issues has been developed for the design and maintenance of the English road network in the 21st century.

2 PAVEMENT DESIGN

2.1 *Structural design*

Road pavement design in the UK has been developed at TRL over many years through studying the performance of a wide range of experimental pavements which formed part of the core English road network. In the 1980's the structural performance of these roads was interpreted with the help of theoretical design concepts. A design life of 40 years was advocated for flexible pavements, which was achieved by strengthening the road after about 20 years. Calculation of the costs of roads over 40 years, taking into account variability of pavement performance, cost of traffic delays and other costs associated with strengthening, showed this to be the optimum design strategy. Since this method was introduced, traffic levels and the consequent disruption at roadworks have continued to increase. Economic considerations indicate that it is now cost effective to increase the design life of very heavily trafficked routes to at least 40 years, without the requirement for structural strengthening, in order to reduce future maintenance and the associated traffic delay costs, as reported by Powell et al (1984).

In addition more knowledge has become available, mainly during the 1990's, on the performance of heavily trafficked roads. This has indicated that deterioration, as either cracking or deformation, is far more likely to be found in the surfacing than deeper in the pavement structure, as illustrated by Figures 1 and 2; this evidence is in conflict with conventional theory. Also, it was found that the great majority of the thick pavements examined in the UK have maintained their strength or become stronger over time, as reported by Nunn and Ferne (2001), rather than gradually weakening with trafficking as assumed in the current pavement assessment method based on deflection measurements.

For fully flexible asphalt pavements in the mid-90's the performance of full-scale experimental pavements, studies of deterioration mechanisms on the road network, long-term deflection monitoring of motorways and condition assessment reports prepared to aid the design of structural maintenance were analysed to produce a design method and strategy for condition assessment for roads expected to last at least 40 years without the need for structural maintenance, as described by Nunn et al (1997). These roads are described as long-life road pavements.

One of the main deterioration mechanisms in these long-life pavements is that of top-down cracking. TRL's International work had demonstrated this behaviour in tropical environments as long ago as the 1970's, Rolt et al (1986). Other examples of this behaviour were published in the France, Netherlands and South Africa in the early 80's and it is now widely accepted that this phenomena can occur in temperate as well as high temperature environments. In addition, further evidence has been found to show that relatively thick, well-constructed pavements deform very little in the lower layers, the rutting being formed largely in the upper surfacing course.

As a consequence some countries have formed the opinion that there is a threshold design thickness beyond which there is no benefit in constructing, provided the methods of construction are of the highest quality. For example, in the United Kingdom where TRL originated this concept, the design thicknesses for 80 million (80 kN) standard axles is thought to suffice for any traffic level over this. This is illustrated in the UK design chart for fully flexible pavements shown in Figure 3 and a similar chart exists for semi-rigid pavements. In developing such a design theory it has been realised that many parts of an existing road network already meet the criteria to be defined as a 'long-life' pavement. In other words they have shown no signs of structural deterioration over a period of time and traffic much exceeding their original design parameters, albeit they may have still required timely surface treatments to preserve the essential properties of these layers both in terms of safety, e.g. skid resistance, and waterproofing to protect the lower layers.

In the UK there is now a new emphasis on best value and sustainability in road construction. The use of recycled materials, secondary binders, secondary aggregates, engineered foundations and innovative designs is much more prevalent, with pavement designs falling between conventional fully flexible and flexible composite designs. The main limitations of the conventional mechanistic-empirical pavement design method are that it involves two distinctly different

Figure 1. Core from motorway showing top-down cracking.

Figure 2. Cross section of motorway showing rutting in surfacing only.

design methods, it can only deal with a limited number of materials and design configurations, no credit is given for stronger foundations, slow curing materials are not accommodated and long life design is reached at different traffic levels for flexible and flexible composite pavements, Nunn (2004). Modifications to the current design method have been proposed including a single design criterion that limits the flexural stress or strain to the underside of the base to a permissible level to achieve the required pavement life.

Pavement design can be carried out in two stages, firstly the foundation and secondly the properties of the constructed foundation can be used in an analytical model to design the pavement structure, as reported by Chaddock and Roberts (2006). This design procedure has been incorporated into the UK Design Manual for Roads and Bridges along with standard designs that will be available for smaller schemes. It is considered that the adoption of this new design procedure encourages innovation.

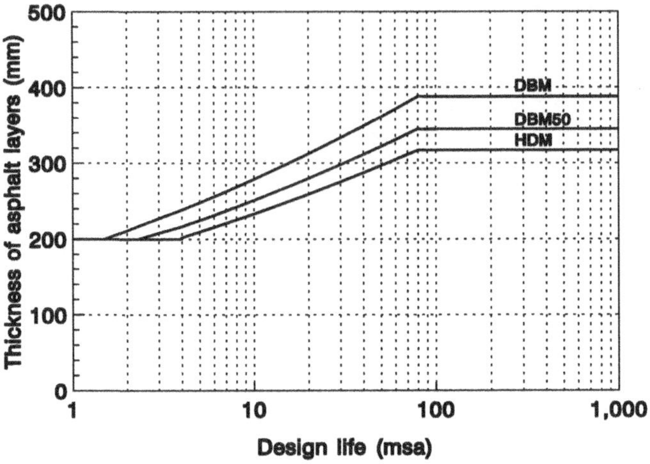

Figure 3. Example of UK design chart for fully flexible pavements incorporating long-life designs.

2.2 *Surfacing design*

In the UK, for many years traditional asphalt surface course layers for major roads, mainly based on hot rolled asphalt covered with coated chippings, have generally been laid at least 40 mm thick. Thinner surface course materials have been available, but were considered to be technically inferior and they were only used on roads carrying low traffic levels within the local authority road networks. However, during the 1990's, various categories of thin surfacing that have beneficial properties, in particular lower tyre/road noise emission, were introduced into the UK, mostly from continental Europe. Since then, thin asphalt surfacing systems have gained a major share of the surface course market on all parts of the network.

2.2.1 *Recycling surfacing*

Although thin surfacings have gained a major share of the surfacing market in the UK, there is concern about their use since they require the use of relatively scarce aggregates with high skid-resistance properties throughout the layer. The noisier alternative hot rolled asphalt wearing course only requires the use of such scare resource in the surface chipping layer. Therefore TRL has investigated, through laboratory investigations and site trials the viability of recycling thin surfacing systems. The trials were on the access road to an asphalt plant and on two heavily trafficked sites on the UK Highways Agency road network, and included the use of polymer modified binders (PMB) and up to 30% reclaimed asphalt (RA) in the mixed asphalt. The trials demonstrate that 10% RA can be easily added to new materials without processing the RA. The paver temperature monitoring for these trials is illustrated in Figure 4. As the proportion of RA increases up to 30%, greater care needs to be taken on assessing grading compatibility and how to treat the residual binder present in the RA as a proportion of the "active" binder content in the recycled surface course layer. The use of reclaimed asphalt in the surfacing has also been undertaken on two major UK motorway contracts, as reported by Carswell et al. (2005).

2.2.2 *Long-life surfacings*

In a more forward looking development, TRL has been exploring, as part of an OECD/ECMT (Organisation for Economic Co-operation and Development/European Conference of Ministers of Transport) joint research project on Economic Evaluation of Long Life Pavements, the potential for creating long-life wearing courses which would complement the current long-life designs for the underlying structure of the pavement, as reported by the OECD (2005). The first stage of the project comprised an economic appraisal and this has shown that there are likely to be economic

Figure 4. Temperature monitoring at the paver during recycling trial.

benefits from the development of road surfacing materials with a service life in excess of 30 years (Long Life Surfacing). This target service life is at least twice as long as that typically expected from surfacings generally used in road pavements and would involve higher initial expense, but would aim to avoid major maintenance costs over the lifetime of the pavement and thus generate overall savings and be environmentally more sustainable. Further work in the project included laboratory performance-based and pilot scale assessments of two advanced surfacing materials for high-traffic roads: (i) a widely used European asphalt surfacing material, modified with an epoxy binder, and; (ii) an entirely new high performance cementitious material (HPCM) surfacing alternative, suitable for application on an asphalt substrate. The laboratory testing comprised a comprehensive assessment of the mechanical properties of both materials using fundamental and simulative techniques, followed by accelerated loading at pilot scale in a pavement test facility (PTF). The success of these initial trials has lead to forthcoming road trials of the designs.

3 PAVEMENT MAINTENANCE

3.1 *Condition assessment*

In order to evaluate when, where and what maintenance is needed on a road network it is essential to have regular and reliable measures of the condition of the pavement in terms of its functional surface condition as well as its structural condition. Additionally, what could be considered part of its functionality is the safety condition of the surface.

3.1.1 *Safety condition*

The most obvious measure of pavement condition that relates to safety is skid resistance and much effort has and still is devoted to measuring this reliably and regularly. However there are several other condition measures that relate to safety e.g.

- profile—both longitudinal and transverse
- crossfall
- gradient
- curvature

- spray
- visibility
  - surface reflectance
  - road marking condition

Several of these have a direct effect on skid resistance. For example, water ponding on the pavement will clearly affect high-speed skid resistance in wet conditions but such ponding, in turn, is affected by the transverse profile, the gradient and the crossfall. However some of these measures can also affect the safety of the user more directly, for example a poor transverse profile can affect the handling of a vehicle.

For some years now the English road network has been monitored for skid resistance using the TRL-developed Sideway-force Coefficient Routine Investigation Machine (SCRIM), shown in Figure 5. This provides a measure of the wet road skid resistance at a slip speed of 20 km/h whilst travelling around 50 to 80 km/h. This property is largely provided by the fine scale microtexture on the surface of aggregate particles. However the skid resistance is also dependent on the texture depth of the pavement surface. This combines with the microtexture to influence the skid resistance at high speeds and is monitored at traffic speed by the use of non-contact laser sensors. The UK Highways Agency use the above measurements, together information on the nature of the site and its accident history, to guide the engineer in determining whether remedial works are required in conjunction with an economic assessment of the costs and benefits, as described by Parry and Viner (2005).

3.1.2 *Functional condition*

To the user of a road network, the most important feature is whether or not the road is serviceable i.e. does it function correctly? The user's primary concern is for the ride quality, i.e. longitudinal profile and transverse profile in extreme cases. However, as improved road surfaces have become available, e.g. porous asphalt, the user has become sensitive to other required functional properties, for example:

- in-car road noise
- spray
- glare
- rolling resistance
- tyre wear

Figure 5. Sideway-force coefficient routine investigation machine (SCRIM).

Many of these parameters cannot be measured directly in a practical routine way but the effects can be indirectly assessed by alternative measures. For example, tyre/road noise is significantly affected by the megatexture features of a road surface which can be relatively easily measured at traffic speed as can the profile features that have influence on rolling resistance and tyre wear.

Following the development by TRL of a test vehicle, HARRIS, as shown in Figure 6 and as described by Ferne et al (2003), the Highways Agency has since 2000, let a commercial contract, TRACS, to survey the surface condition of the core road network including the measurement of transverse and longitudinal profile, texture profile, road geometry and cracking, as described by Christie et al (2000). In addition the vital location referencing information is collected by inertial-supplemented high quality GPS. This contract now covers around 40,000 lane km every year with some parts covered every six months.

For the local authority network the Department for Transport has initiated a similar survey which runs under the name of SCANNER currently covering around 100,000 lane km of their 300,000 km network. Greater coverage is given to the principal roads with no coverage on the least trafficked roads.

With such large quantities of condition information being collected every year it is vital that this data is robust and reliable. Therefore the HA and the DfT have commissioned the TRL to provide a quality assurance system for such surveys which includes comprehensive accreditation of the survey vehicles and random checks on the data quality, as shown in Figure 7 and as described for the SCANNER surveys by Thomas et al (2007).

TRL have continued to develop and assess the latest technology for such surveys together with major improvements in the interpretation of such data. The latest version of HARRIS provides extremely detailed three dimensional plots of the road surface together with high resolution colour images of the road surface, which can for example be used in the detection of road edge deterioration, as reported by Watson and Wright (2006) and as shown in Figure 8.

Despite the relatively high capital cost of such equipment, surveys can still be extremely cost effective in view of the speed of coverage, approximately 80 km/h, and the lack of traffic management and disruption with this type of survey, as illustrated by its widespread application on even fairly minor roads on the local authority network.

3.1.3 *Structural condition*

To minimise the cost of maintaining a pavement we not only need to know about the safety and functional condition of the pavement but, more importantly, we need to know the structural condition.

Figure 6. The UK highways agency road research information system (HARRIS).

Figure 7. SCANNER survey vehicles taking part in accreditation trials at TRL.

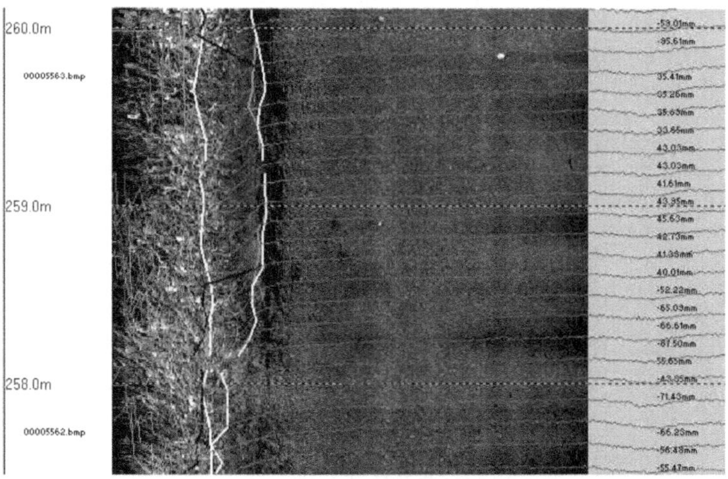

Figure 8. Example of combining 3D view of road surface with road surface image.

The assessment of structural condition generally involves the assessment of condition in-depth, either indirectly by measuring its response under load, i.e. deflection measurement or more directly by measuring material type, composition, thickness and condition (such as cracking, stripping or delamination).

For many years in the UK we have been using the slow speed UK Deflectograph, see Ferne and Roberts (1991), to measure pavement deflection response supplemented by the Falling Weight Deflectometer when more detailed investigations are required, as discussed by Ferne (2002). However, with the current high levels of traffic congestion, we have been investigating the options for measuring deflection response at traffic-speed. Consequently we are currently evaluating and developing for routine use, on behalf of the UK Highways Agency, a Danish-built Traffic Speed Deflectometer, see Figure 9, that can measure the deflection response of a pavement at up to 80 km/h, as reported by FORMAT (2002). As a first stage, the results are being related to the current UK Deflectograph interpretation method which provides an indication of whether the pavement is or is not likely to be a long-life pavement and to provide an estimate of the residual structural life and strengthening requirements for the latter pavement type. In due course we expect to develop a more fundamental interpretation of the response of this device.

Any interpretation of deflection response normally also requires information on the construction materials and thicknesses of the layers of a pavement. Although this can be provided from cores taken from the pavement, this is slow, disruptive and only provides limited samples. TRL has carried out extensive assessments of the capability of Ground Penetrating Radar to provide such information at traffic speed with limited coring, as reported by Forest et al (2004) and illustrated in Figure 10, and is currently working to provide the quality assurance for such surveys that will ensure that the UK Highways Agency can commission such surveys and obtain robust reliable data to store in their Pavement Management System.

Figure 9. UK highways agency TSD surveying deflection response of motorway at 80 km/h.

Figure 10. UK ground penetrating radar equipment taking part in comparative trials.

3.2 *Maintenance selection and design*

A number of UK studies developing the concepts, for long life pavements, have shown that in the vast majority of cases on the English trunk road network the deterioration mechanisms encountered concern defects originating and propagating from the surface of the pavement downwards. This has lead to developing a more appropriate and reliable pavement assessment regime by always first considering surface condition before structural condition. This new approach integrates existing procedures for assessing surface and structural maintenance and assumes that no maintenance of any kind is required unless there is evidence of wear on the road surface. Much more reliance is placed on the results of surface condition surveys that are now mostly carried out at traffic speed with little traffic disruption. A simple overview of the overall selection and design procedures is shown in Figure 11.

3.2.1 *Scheme level assessment*

The initial aim of scheme or project level assessment is to determine how far the surface deterioration extends into the pavement and, in particular, whether it is only in the surfacing. Details of the exact location and extent of rutting and pattern of cracking are available from routine survey data. Cores taken on the cracks or at crack ends determine the depth, direction and propagation of cracks. These cores should penetrate at least half-way into the base on fully flexible pavements to ensure that the full depth of cracking can be recorded. They also provide evidence of which layers are affected by rutting and any loss of integrity of the materials, such as stripping of the binder.

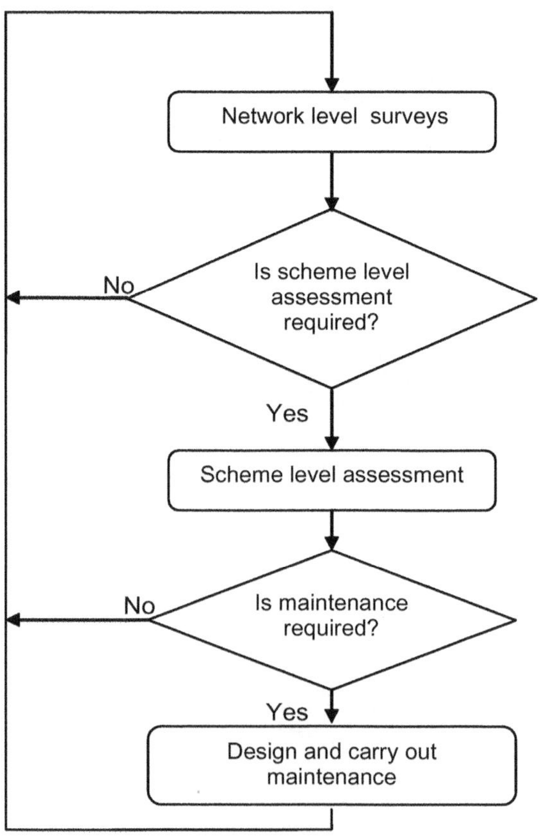

Figure 11. Simple overview of maintenance assessment, selection and design.

In future, on the basis of earlier TRL research, Ground Penetrating Radar is now being used to assess crack depth penetration over a much larger sample of the scheme than is possible with cores, as presented by Forest and Utsi (2004). If the cracking is found to penetrate into the structural layers of the pavement then a conventional thorough structural investigation will be necessary.

3.2.2 *Treatment selection*

Acknowledgement of the existence of long life pavements will of course affect the selection of suitable maintenance treatments. As was stated earlier, long life pavements will not be of infinite life. However, so long as the pavement has been well constructed and the foundations remain sound, no structural treatment such as overlays or reconstruction should be necessary. The surface of a long-life pavement will deteriorate as for determinate life pavements. Accumulated deformation will cause rutting which will need to be treated by replacement or by thin overlay before it becomes a safety hazard. Cracking may initiate at the surface which will need treating by replacement before it penetrates into the structural components of the pavement. Skid resistance will deteriorate and initiate the necessary remedial surface treatment. If structural deterioration has occurred, on a determinate life pavement, strengthening may need to be provided by overlays or reconstruction. However it should be noted that it may be that only thin overlays are necessary to convert a determinate life pavement into a long life equivalent.

The UK Highways Agency is responsible for building and maintaining the motorway and all-purpose trunk road network in England. Individual areas within the network are managed on behalf of HA by specialist consultants known as Managing Agents or Managing Agent/Contractors. However the final allocation of maintenance funds is determined by the Highways Agency which needs to ensure that maintenance is targeted at schemes that are likely to yield maximum benefits to the road user, minimise disruption and offer the best value for money. To facilitate the prioritisation process, a Value Management system was developed and implemented by the Highways Agency, with support from TRL engineers, in 1999. Under Value Management, all road maintenance schemes valued over £100,000 (circa $200,000 US) are assessed and scored against key criteria. The Value Management system has continually evolved encompassing whole-life cost, long-life pavements and many other concepts. The approach has also been adopted to cover other types of maintenance schemes including structures, drainage, geotechnical and lighting schemes. This Value Management approach has developed and evolved into a very effective and practical tool for prioritising maintenance.

4 ASSET MANAGEMENT

4.1 *Asset management plans*

One of the major challenges for governments and asset owners in a thriving economy is to ensure that assets are maintained in such a way as to mitigate the risk to future prosperity and that investment is carefully planned to ensure continuity. TRL has been involved in the delivery of asset management strategies for over 70 years, working for central and local governments both in the UK and internationally; our knowledge is extensive. An Asset Management Plan for Highways obviously needs to encompass all the highway assets including pavements, structures, drainage, above and below ground services, signs, street furniture, signalling and street lighting. Consideration also needs to be given to utilities, such as communication systems, water and electricity, that may run within or cross the highway boundaries.

One of the first questions that any Local Authority (LA) considering the preparation of an Asset Management Plan needs to address is whether the plan will incorporate the wider issues of transport within its plan or initially only consider highway issues. An asset management plan which includes transport will be significantly more complex in its preparation and implementation than a highways asset management plan. A review of government requirements showed that transportation matters should be considered in the wider aspects of asset management and therefore will need to be considered within the Asset Management Plan. In the preparation of a highways asset management plan it is essential to recognise that highway or transport asset management is not just the management and improvement of the physical asset based on lifecycle cost considerations, but also, the use and availability of the asset. Therefore careful assessment of appropriate service levels, risk and safety are an intrinsic part of the process. The key steps based on the template method proposed by TRL are:

- Define strategic objectives, performance measures and performance targets
- Determine current performance levels
- Determine asset attributes, inventory and condition
- Determine the asset value
- Determine depreciated replacement cost
- Determine a range of appropriate overall levels of service
- Determine goals for levels of service
- Determine lifecycle options for Provide—Maintain—Operate
- Determine current lifecycle plan for Provide—Maintain—Operate
- Determine benefits, risks and costs of options
- Determine the option priorities

Pavements constitute a large asset for the maintaining agency. The Current UK Highways Agency asset alone is worth over £60 billion making it the UK Government's largest single asset. Maintaining and operating pavements on a large network typically involves complex decisions on how and when to resurface or apply other treatments to keep the system performing and ensure operating costs remain at an acceptable level. Conventional methods result in decisions being made by a maintenance engineer who would select treatments based on his extensive knowledge and experience. More often than not, this has lead to the pavements in the worst state being treated first. This method is still widely practised and may provide an acceptable solution in lightly trafficked areas, or where repair/restoration funds are unlimited. However, in the majority of situations, this is not the case.

There are a number of problems with the traditional method. Firstly, there are seldom enough funds available to complete all identified road repairs. Secondly, where there are high levels of traffic the traffic management options severely limit the availability of closures or lane restrictions. Thirdly, a high level of traffic significantly increases user costs. In a heavily trafficked location the user costs contribute to over half the total costs involved. The traffic control proportion also increases significantly.

To overcome these problems a whole life cost analysis of maintenance must be considered. That is a consideration of the costs not only in the present but also future consequential costs as a result of maintenance actions taken.

One of the main advantages of preventative maintenance is that it delays the time to the need for reactive maintenance. It is also much cheaper and quicker to apply than maintenance needed when the pavement is in poorer condition. Reactive maintenance can cost up to 5 times more to achieve the same degree of restoration as preventative maintenance. Studies have shown that this approach not only leads to reduced maintenance expenditure but in time the overall network condition improves.

The main problem is that this is easy to consider for the situation at a single location, but a complete network has pavements at different stages of their lives. When funds are restricted, it may actually be better to allow one location to deteriorate and treat several others at an earlier stage of deterioration. Determining which sections of a network to treat and which to leave becomes a complex process that can only be effectively achieved by the use of a Pavement Management System (PMS).

4.2 *Other highway asset management issues*

The earlier sections have described some of the developments that have occurred in reaching the system currently used to manage the English road network. However there are a number of ancillary issues that affect the effectiveness of such a system, particularly to try and meet objectives set by Governments.

4.2.1 *Minimising traffic disruption*
Many governments are starting to realise that the transport network, and in particular the road network, is a vital part of the country's economy and maintaining traffic flows is part of the service they should be providing as a network operator. Thus one of their top priorities is to construct and maintain their network in a manner that minimises disruption to traffic. For some years this has been TRL's main objective in developing the techniques for design, construction and maintenance described in this paper. These range from long-life structural designs that only require surface treatment to condition monitoring techniques that operate within the normal traffic flow.

4.2.2 *Optimising safety*
Another high level objective for the operation of a national network is almost certainly to optimise the safe operation of that network. In terms of design, construction and maintenance this involves choosing appropriate surfacing materials, routinely assessing the potential of the surface for safe operation and investigating the need for improvement as appropriate, as has been discussed in earlier sections.

4.2.3 *Minimising whole life value*
In the UK it is now Government Treasury Procurement Policy that all construction procurement must be made based on best value for money in terms of the optimum combination of whole life costs and quality and must also address key considerations such as time, environment, social and sustainable development issues. In other words, the ultimate success and future implications of any investment decision will depend on:

- How well the decision makers take account of not just construction costs, but whole life costs;
- How well the decision makers address other stakeholders requirements, such as time, quality, environmental and sustainability issues.

4.2.4 *Managing the effect of climate change and sustainability*
TRL has recently carried out a study for the UK Department for Transport on how climate change will affect highway maintenance. This work had clearly identified that the network is being affected today and that climate change will exacerbate these effects. It should come as no surprise that poorly constructed and/or maintained networks and historically evolved roads are the most vulnerable. Perhaps more surprising is the finding that every network is unique and the fact that there is a lack of recorded detail about the impact of extreme weather events in the past.

Crucially, TRL's findings point to climate change being primarily a risk management issue and should be managed in broadly the same way as other risks. Highway Authorities need to record data on weather impacts and share this with colleagues from across their authority, particularly with those responsible for property. The cumulative effects of climate change matter, so it is important to know your network to understand where you have vulnerable soils, to know where your drainage is and what condition it is in. Finally, the report advises not to make climate change special, but to incorporate it as another element in the overall management strategy and to involve suppliers in identifying solutions.

Sustainability has been a major theme in UK government policy for a number of years and the choice of materials and methods used by local authorities for highways maintenance can have a major impact on the sustainability of the works. TRL has developed a series of milestones for sustainability in highway maintenance. These cover applications from surface course to subgrade; and include suggestions for appropriate Key Performance Indicators. The milestones are based on the application of currently available techniques and relate to materials, methods and waste. These can be used to set targets to demonstrate that the local authority is increasing the sustainability of its operations.

5 SUMMARY

This paper has shown how targeted research can provide an efficient and effective means of managing a large road network carrying a significant number of vehicles without major disruption from the essential improvement and maintenance of such a network whilst taking account of the need for consideration of whole life value and sustainability. Although these developments were made specifically for the UK road network many aspects have applicability to other road networks whatever their size and location. For example, the concept of long-life pavements and their associated maintenance strategy have found world-wide application. In fact, one of the key features of this concept, top down cracking, was first discovered in a high temperature climate. At the other end of the spectrum, techniques developed for the efficient management of large road networks have found applicability to very different networks in terms of both size and environment.

ACKNOWLEDGEMENTS

REFERENCES

Carswell, I. Nicholls, J.C. Elliott, R.C. Harris, J. & Strickland, D. 2005. Feasibility of recycling thin surfacing back into thin surfacing systems. *TRL Report TRL645,* TRL Limited, Wokingham.

Chaddock, B. & Roberts, C. 2006. Road to foundation design for major highways. *Published Project Report PPR127.* Transport Research Laboratory, Wokingham.

Christie, C. Ferne, B.W. Kennedy, C.K. & McQueen, S. 2000. Comprehensive pavement condition monitoring—from research to practice. *Surface Transport 2000 Conference*, Crowthorne, June 2000.

Ferne, B.W. & Roberts, P.K. 1991. Deflectograph—the way forward for structural assessment. *Highways and Transportation,* October 1991, pp5–6.

Ferne, B.W. 2002. Quality control using the FWD. *FWD/Backcalculation Workshop, Cascais, 21 June 2002, 6th International Conference on the Bearing Capacity of Roads, Railways and Airfields*, Lisbon.

Ferne, B.W. Wright, M.A. & Pynn, J. 2003. The development of HARRIS—a system for road surface condition monitoring at traffic speed. *TRL Annual Research Review 2002*, Crowthorne, TRL 2003.

Forest, R. Pynn, J. Alani, A. & Ferne, B.W. 2004a. Ground Penetrating radar for monitoring of English Roads. *International Journal of Pavement Engineering & Asphalt Technology*. ISSN 1464–8164.

Forest, R. & Utsi, V. 2004b. Non destructive crack depth measurements with Ground Penetrating Radar. *Tenth International Conference on Ground Penetrating Radar*, 21–24 June, Delft, The Netherlands.

FORMAT. 2002. Fully Optimised Road Maintenance, Assessment of high speed monitoring equipment, *Deliverable Report D12, Contract no GRD1/2000/25255 S12.3119665 FORMAT*, Project funded by the European Community under the competitive and sustainable Growth Programme.

Nunn, M.E. Brown, A. Weston, D. & Nicholls, J.C. 1997. Design of long-life flexible pavements for heavy traffic. *TRL Report 250.* Crowthorne: Transport and Road Research Laboratory.

Nunn, M. & Ferne. B.W. 2001. Design and assessment of long-life pavements : United Kingdom experience. *80th Annual Meeting Transportation Research Board*, Washington, D.C., January 2001.

Nunn, M.E. 2004. Development of a more versatile approach to flexible and flexible composite pavement design. *TRL Report TRL615*. Transport Research Laboratory, Wokingham.

OECD. 2005. Economic evaluation of long-life pavements: Phase 1. *OECD Document ISDN 926400856X.* Paris.

Parry, A.R. & Viner, H.E. 2005. Accidents and the skidding resistance standard for strategic roads in England. *TRL report TRL622*. TRL Limited, Wokingham, Berkshire, UK.

Powell, W.D. Potter, J.F. Mayhew, H. & Nunn, M.E. 1984. Design of bituminous roads. *TRRL Report LR1132*. Crowthorne: Transport and Road Research Laboratory.

Rolt, J. Smith, H.R. & Jones, C.R. 1986. The design and performance of bituminous overlays in tropical environments. *Proc. Second Int. Conf. on the Bearing Capacity of Roads and Airfields*. Trondheim, Norway.
Thomas, C. Werro, P. & Wright, M.A. 2007. SCANNER accredited surveys on local roads in England—Accreditation, QA and Audit testing—Annual Report 2005–06. *Published Project Report PPR229*. TRL Limited, Wokingham, Berkshire, UK.
Watson, P. & Wright, M.A. 2006. New pattern recognition methods and the detection of edge deterioration. *Published Project Report PPR141*. Transport Research Laboratory, Wokingham.

Efficient Transportation and Pavement Systems – Al-Qadi, Sayed, Alnuaimi & Masad (eds)
© 2009 Taylor & Francis Group, London, ISBN 978-0-415-48979-9

Photocatalytic road pavements: An analysis of structural and functional performances

M. Crispino, S. Lambrugo & M. Bacchi
Politecnico di Milano, DIIAR Sezione Infrastrutture Viarie, Italy

ABSTRACT: Air-quality related problems represent at the present time some of the most critical issues for urban areas. Further on the traditional pollution mitigation approaches, the photocatalytic materials represent a further method for air pollution control, active in pollutants degradation. The chance to apply photocatalytic materials to an increasing number of urban surfaces leads to focus the attention on the design of a photocatalytic road pavement.

In this paper, the Authors present the results of an analysis carried out to characterize a peculiar photocatalytic pavement combining asphalt concrete and a special photocatalytic mortar. The first phase of the experimental analysis allowed the Authors to evaluate the influence of titanium dioxide on the materials features, considering viscosity, compression resistance and indirect tensile stress resistance. The second part of the laboratory investigation aimed instead at characterizing the effectiveness of the pavement in the NOx concentration reduction, testing both traditional and carbon doped anatase.

1 INTRODUCTION

The chemical phenomenon of photocatalysis is based on the characteristic of the photocatalyst to accelerate, in presence of sunlight and air, the natural oxidation reactions turning the air pollutants into inert substances [Fujishima A. 1999].

The photocatalyst is a semiconductor that, because of its electronic configuration, when sunlight irradiated produces an excited state able to favour and accelerate chemical pollutants oxidation. The best known photocatalyst is titanium dioxide, in the anatase form, that, irradiated with light of wavelength lower than 388 nm, produces a very steady excited state, maximizing the effectiveness of the photocatalytic reactions [Fujishima A. 2000].

Referring to nitrogen oxides, some of the worse air pollutants for both human health and environment, the oxidative power of the photocatalyst is particularly meaningful, according to the equation:

$$NO_x + \mathrm{rad}(\lambda < 388\ \mathrm{nm}) + H_2O \rightarrow NO_3^-$$

The photocatalyst, irradiated with a light of wavelength lower than 388 nm, turns itself into its excited state, characterized by two conduction electrons and two positive vacancies. The water adsorbed on the surface generates OH radicals and the atmospherical oxygen is reduced in hydroperoxide ions. The two high reactive elements react with NOx producing nitric acid that, in presence of an alkaline sublayer, such as cement, is turned into nitrate ions [Quian Chunxiang et al. 2005].

Nowadays photocatalytic materials are used in several different indoor and outdoor contexts, such as hospitals, building facades and so on. Referring to the urban area, road pavements represent one of the most extended surfaces, directly in contact with traffic, one of the primary pollution sources, for these reasons the chance to design a photocatalytic pavement is a current key issue. At the present time the experimented photocatalytic road pavements are mainly three: cement blocks, bituminous pavements superficially treated with photocatalytic resins or mortars and open graded asphalt concrete layers partly o fully filled up with cement mortars.

The Authors present the results of analysis concerning a road pavement made up of an asphalt concrete layer completely filled up with a special photocatalytic cement mortar.

The analysis concerns on one hand the evaluation of the effects of the photocatalyst on the mortar and pavement performance, on the other the characterization of the effectiveness of the pavement in NOx reducing. In the following paragraphs the laboratory analysis is described and the results discussed.

2 THE ANALYSED PHOTOCATALYTIC PAVEMENT

The analysed photocatalytic pavement represents the evolution of a peculiar pavement typology combing asphalt and cement, suitable for high and slow traffic loads, characterized by the addiction of the photocatalyst to the hyper fluids cement mortar constituting the layer. The technology of pavements combining asphalt and cement had been already analysed and characterized by the Authors, regarding applications in both road and airport contexts [Crispino et al. 2007]. The pavement presents structural performance comparable to a concrete pavement [Da Rios et al. 2007]. The photocatalytic pavement combining asphalt and cement is made up of an open graded asphalt concrete layer, characterized by a high rate of voids, fill up with a hyper fluid cement mortar added with titanium dioxide.

The open graded asphalt concrete layer is designed in order to present a rate and a distribution of voids proper to allow the penetration of the cement mortar and the full obstruction of voids. The asphalt concrete presents mono-granular features, according to the curve reported in Figure 1, and a bitumen content of 4.5%. The penetration grade of the bitumen is chosen depending on the climatic conditions, generally 50/70 pen.

The photocatalytic cement mortar is obtained adding the photocatalyst, anatase, to a hyper fluid cement mortar, characterized by high fluidity and high final performance.

The construction of the photocatalytic layer is made up of two main phases: the first regarding the laying of the asphalt concrete layer, the second the application of the photoactive mortar. The asphalt concrete layer is laid with traditional vibro-finishing machines and then compacted in order to obtain a rate of voids of about 25–30%. When the surface reaches a temperature lower the 30°C the photocatalytic mortar is applied, then the surface is scoured by a rubber scraper machine to allow the full penetration and to remove the exceeding mortar. An additional operation of shot-

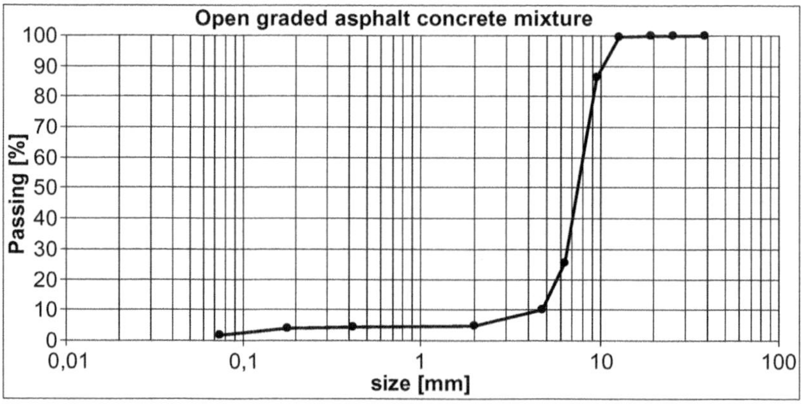

Figure 1. Open graded asphalt concrete mixture.

blasting can be carried out in order to improve the surface features of the pavement [Crispino & Lambrugo, 2007].

3 EXPERIMENTAL ANALYSIS

The experimental analysis carried out by the Authors regards the characterization of four photocatalytic pavements, made through different photocatalytic mortar mixtures.

The four analysed mortar mixtures are characterized by the use of different rates of two different typologies of photocatalyst: a traditional anatase and a carbon doped titanium dioxide.

The photocatalytic properties of the traditional titanium dioxide are based on the absorption of the ultra violet radiation, according to a wavelength lower than 388 nm, doping the anatase with carbon it is possible to shift the cut-off wavelength to 535 nm, including the visible light [Blöß & Elfenthal, 2007; Irie et al. 2003].

The four analysed mixtures are shortly described in Table 1.

The optimal water content of each mixture was determined through preliminary fluidity test [Crispino & Lambrugo, 2007; Crispino & Lambrugo 2008].

The main goals of this analysis are the evaluation of the alterations in materials characteristics due to the presence of titanium dioxide and the determination of the environmental performance of each one of the pavements through laboratory tests.

The evaluation of the mechanical alterations in materials behaviour due to titanium dioxide was carried out considering viscosity, hardening time, compression and indirect tensile stress resistance, taking as a reference a high performance cement mortar used for road applications. Moreover, a preliminary analysis of the two different titanium dioxide types, traditional and doped, allowed to observe that the physical properties of the two elements are almost the same, for this reason it seemed appropriate to consider the two TiO_2s as identical in the fist experimental phase. The two photocatalysts were considered as different only in the second phase of the analysis, evaluating the effectiveness of the pavements in the NOx concentration reduction, because of the basically chemical difference between the two TiO_2s, traditional and carbon doped.

3.1 *Viscosity test*

The first step of the experimental analysis regarded the evaluation of the viscosity of the photocatalytic mortar mixtures.

The laboratory tests were performed by the Brookfield Viscosimeter and focused at defining the workability time of the mortar before the beginning of the hardening phase [Toraldo & Lambrugo, 2007].

Three different mortar mixtures were analyzed:

- Mixture 0: reference high performance mortar, $TiO_2 = 0\%$ and $w = 16.8\ \%$;
- Mixture 1: photocatalytic mortar, $TiO_2 = 5\%$ and $w = 17.0\%$;
- Mixture 2: photocatalytic mortar, $TiO_2 = 7.5\%$ and $w = 17.2\%$

Table 1. Analysed mortar mixtures.

	TiO_2		H_2O
Mixture	Type	%	%
1	A	5	17
2	A	7,5	17,2
3	B	5	17
4	B	7,5	17,2

The laboratory tests were carried out at a temperature of 30°C, regarded as representative of the real application conditions on the asphalt surface. Each mixture was made up into a laboratory mechanical mixer, according to a mixing time of 5 minutes, set through preliminary tests as optimal for obtaining a homogeneous mortar, with no clots of powder.

A sample of 10 gr, the right quantity of mortar to be test into the Brookfield Viscosimeter, was taken from the mixer and tested. In Figure 2 an example of the viscosity behaviour is given.

The results showed a common behaviour for all the mixtures, characterized by two different phases: the first phase represents the good workability time of the mortar, with low viscosities, the second phase determines the hardening, the crossing point between the two phases points out the sudden loss in workability of the mixture and it can be considered as the starting point of the mortar setting.

The typical curve representative of each one of the two photocatalytic mixtures showed a strong reduction of the workability times, as illustrated in Figure 3.

Considering the obtained results it seemed necessary to verify the adequacy of the photocatalytic mixtures to the full penetration into the 5 cm asphalt concrete layer. An analytic model of infiltration was used to determine an average value for the time required by the mortars to completely penetrate into the layer.

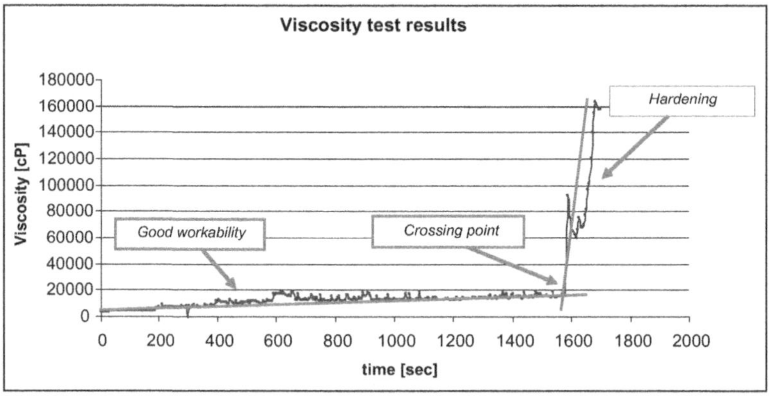

Figure 2. Example of viscosity test results.

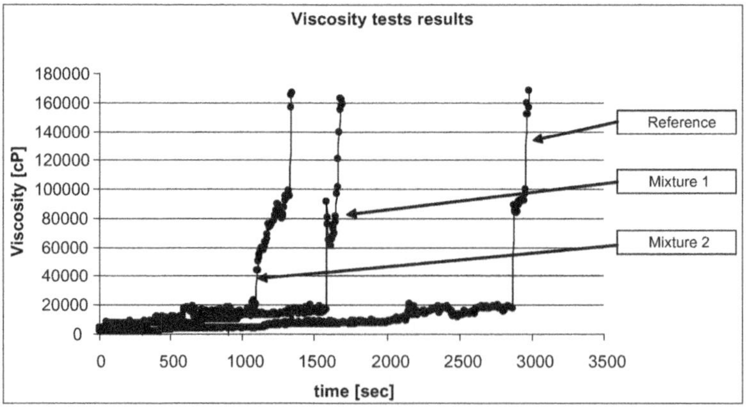

Figure 3. Comparison between the typical viscosity curves of the tested mortars.

The analytic model is based on the hypothesis that the mortar moves inside the open graded asphalt concrete layer in accordance with the laws of motion in a porous unsatured material. Considering a mono-dimensional geometry the description of the penetration of the mortar into the asphalt layer is derived from the Darcy and the Continuity equations, according to the three hypotheses of Green and Ampt [Haverkamp et al. 1999].

The motion of the mortar into the asphalt concrete, considering a unit pavement area, is described by the following explicit equation, depending on time t, expressed in the only variable z, representing the advancing front depth:

$$t = \frac{z^2(t)}{h + z(t)} \cdot \frac{\mu}{\rho \cdot k \cdot g}$$

The time t represents the time necessary to a mortar, characterized by a ρ density and a μ viscosity, to penetrate into the asphalt layer, of permeability k, as far as the depth z, being h the hydraulic load.

The asphalt concrete geometric factor (respect to permeability) k was derived from the hydraulic conductivity definition:

$$k = K \cdot \frac{\mu}{\rho \cdot g}$$

applied to water (density ρ as 1000 kg/m^3 and a viscosity at 20° of 0.001 kg/ms). In accordance with the assumed hydraulic conductivity K to water of 0.3 cm/s [Cooley, 2002] k was determined as 3.06E-10 m^2.

The parameters related to the mortars were determinate through experimental tests, the mortar density was fixed at 2,5 g/cm^3 and it was considered an average initial viscosity of about 100 Poise.

The hydraulic loads h was determined as the sum of two component, the first related to the thickness of the initial mortar layer applied over the surface (1 cm), the second derived from the loads applied on the surface by the rubber scraper machine (200 kg).

According to this hypothesis the time required by the mortar to completely fill the layer returned by the model resulted to be equal to 15 minutes. From the Brookfield Viscosimeter analysis the workability time of the described mortars resulted to be longer than the estimated required penetration time. As reported in Table 2 all the mortars resulted able to assure a full penetration into the layer.

3.2 *Essential mechanical test*

The second step of the experimental analysis concerned the evaluation of possible alterations to the mechanical characteristics of the mortar and of the composite material due to the presence of the titanium dioxide. Essential mechanical tests were chosen in order to quickly point out any relevant anomalous behaviour: compression resistance tests for the mortar and indirect tensile tests for the composite material.

Table 2. Workability times evaluated through viscosity tests.

Mixture	Workability time [Min]
0	48
1	25
2	20

3.2.1 *Compression resistance tests*

The compression resistance of the two different mortar mixtures defined in the previous paragraph was determinate and compared with the reference one.

An expressive number of 10 cm side cube-shapes samples were made up with the two photocatalytic mixtures and with the reference one, after a mixing time of 5 minutes.

The samples were subjected to different aging, 24 hours, 7 and 28 days, at the temperature of 20°C, and then tested.

The compression resistance test results are reported in Figure 4.

The photocatalytic mortars showed a reduced resistance in comparison with the reference one, but the final resistance measured value resulted to be always higher than 70 MPa.

3.2.2 *Indirect tensile stress*

The indirect tensile stress resistance was evaluated as well, referring to the two photocatalytic mortars and to the reference one.

In order to test the combined material asphalt concrete samples of pavement were made up and then filled up with the different mortars.

An expressive number of open graded asphalt concrete samples were made: the asphalt concrete mixture was heated at the temperature of 150°C and then compacted with gyratory compactor, in order to obtain a rate of voids of about 25–30%. The average rate of voids value of the samples was 27%, with a sqm of 0,8.

Each sample was then laterally confined and completely filled up with mortar, according to the three different mixtures. The samples were subjected to different aging, 18, 24 hours and 28 days, and then tested. The results are reported in Figure 5.

The results showed a reduction of the indirect tensile resistance due both to the presence of Titanium Dioxide and to the increase of the water rate. The final resistance was always higher than 1 MPa.

3.3 *Photocatalytic activity test*

The final step of the analysis regarded the evaluation of the effectiveness of the different photocatalytic pavements in the NOx reduction.

Four different photocatalytic mixtures were tested:

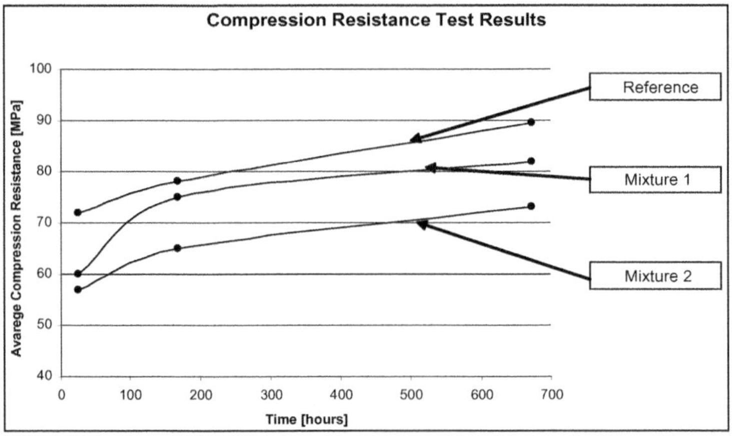

Figure 4. Average compression resistance test results.

– Mixture 1: photocatalytic mortar, traditional TiO_2 = 5% and w = 17%;
– Mixture 2: photocatalytic mortar, traditional TiO_2 = 7.5% and w = 17.2%;
– Mixture 3: photocatalytic mortar, carbon doped TiO_2 = 5% and w = 17%;
– Mixture 4: photocatalytic mortar, carbon doped TiO_2 = 7.5% and w = 17.2%.

Samples of photocatalytic pavement were realized, according to the four different mixtures, laterally confined and tested.

The test system, summed up in Figure 6, is chiefly made up of a NOx source, a test chamber and a NOx analyser. The pavement sample is put into the test chamber, a glass box irradiated with special lamps assuring into the chamber a radiant flux, integrated from 300 to 400 nm, of about 20 w/m². Into the test chamber both temperature and humidity are steadily monitored. The NOx, mixed with humidified air simulating polluted air, are blown into the test chamber, and a second flux is constantly taken from the chamber by the NOx analyser that returns the NOx concentrations.

The test samples were put into the chamber with the light off and the analyser monitored the air into the glass box in order to define the steady concentration in dark conditions, then light was turned on and the NOx concentrations continuously monitored. The test finished when the NOx concentration assumed a new steady value. The results of tests are shown in Figure 7.

The final NOx reduction of each analysed pavement is reported in Table 3.

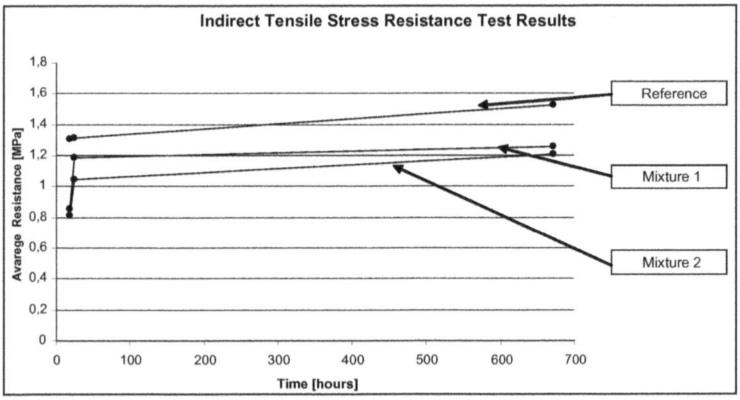

Figure 5. Average indirect tensile stress resistance test results.

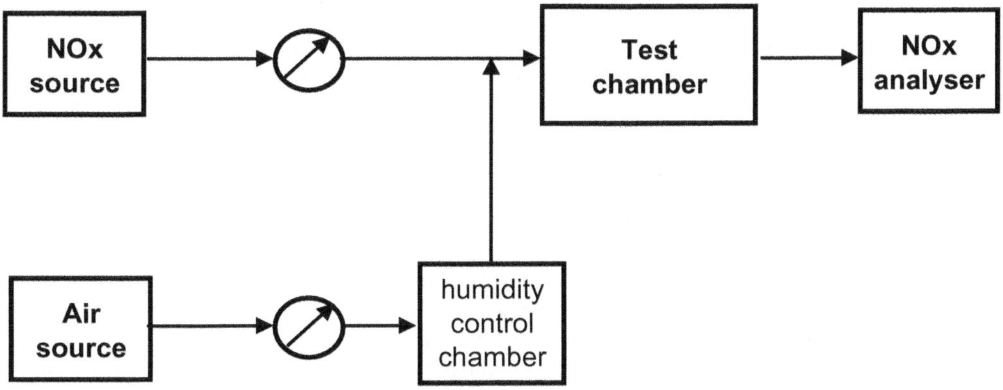

Figure 6. Scheme of the NOx reduction test apparatus.

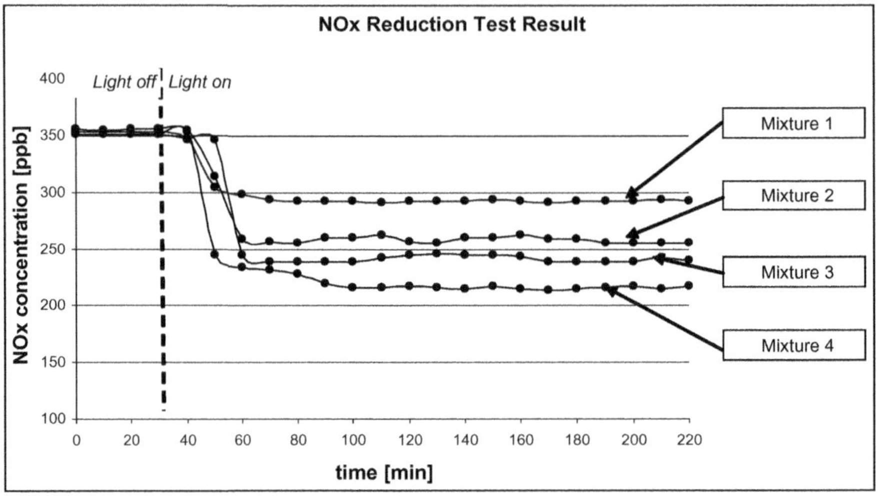

Figure 7. Nox reduction test results.

Table 3. Average final NOx reductions.

Mixture	NOx average reduction %
1	17
2	28
3	32
4	39

10 CONCLUSIONS

The experimental analysis aimed at evaluating the possible interferences of the photocatalyst in the pavement material characteristics. The viscosity tests showed a reduction of the mortar workability time due to the presence of titanium dioxide into the mixture, anyway the analytic models pointed out that the viscosity of the mixtures is still compatible with the full penetration into the 5 cm asphalt concrete layer.

The mortars were also subjected to compression resistance tests, at different aging time. The results showed a slower increasing of resistance of the two photocatalytic mortars in comparison with the reference one, however the average compression resistance measured after 28 days was, for each one of the considered mixtures, higher than 70 MPa, adequate for an urban road pavement.

Finally indirect tensile resistance tests pointed out a sensible reduction of the average resistances of the photocatalytic pavement samples if compared with the reference ones. The indirect tensile stress resistance measured after an aging of 28 days resulted to be higher than 1 MPa for all the samples, fully compatible with the requirement of an urban pavement.

Considering the results of the described analysis it seems reasonable to conclude that the presence of titanium dioxide affected the pavement materials characteristics, but the measured alterations are fully compatible with an effective urban road pavement.

The final part of the experimental analysis concerned the evaluation of the NOx reduction referred to each one of the photocatalytic pavement, considering both traditional and carbon doped anatase. The results showed a higher NOx concentration reduction for the mixtures containing carbon doped anatase, with a maximum reduction of about 40%.

In conclusion the experimental analysis pointed out an actual chance to use this peculiar kind of pavement in urban context, assuring a significant NOx concentration reduction even with low TiO_2 content (7.5%), especially in relation to the use of carbon doped anatase.

REFERENCES

Cooley L.A., 2002, "Issue pertaining to the permeability characteristics of course-graded superpave mixes", NCAT Report 2002–06, Auburn, Alabama.

Crispino M. & Lambrugo S., 2007, "An experimental characterization of a photocatalytic mortar for road bituminous pavement", International Rilem Symposium on Photocatalysis, Firenze, Italy.

Crispino M. & Lambrugo S., 2007, "*Surface characteristics and environmental performance of a photocatalytic innovative pavement*", RSS, Roma.

Crispino M. & Lambrugo S., 2008, "Effectiveness of a photocatalytic wearing course through experimental analysis", ISAP, Zurich.

Crispino M. & Toraldo E., 2007, "An innovative rehabilitation technique for airport concrete pavement", Advanced characterization of pavement and soil engineering materials, Athens.

Da Rios G., Agostinacchio M. & Fiori F., 2007, "Performance and durability of grouted open grade asphalt concretes", 5th MAIREPAV International Conference, ParkCity, Utah, USA.

Fujishima A., Hashimoto K. & Watanabe T., 1999, "*TiO2 photocatalysis: fundamentals and applications*", Tokyo: BKC, Inc.

Fujishima A., Rao T. & Tryk D.A., 2000, "Titanium dioxide photocatalysis", J. Photochem. Photobiol. C: Photochem. Reviews, 1.

Haverkamp R., Bouraoui F., Zammit C. & Angulo-Jaramillo R., 1999, "*Soil properties and moisture movement in the unsaturated zone*", CRC Press LLC.

Irie H., Watanabe Y. e Hashimoto K., 2003, "*Carbon doped Anatase TiO2 Powders as a visible light photocatalyst*", Chemistry letters 32.

Quian Chunxiang, Zhao Lianfang, Fu Dafang, April 2005, "Photocatalytic oxidation of nitrogen oxides bi NANO-TiO2 immobilized on road surface materials", Kuei Suan Jen Hsueh Pao/journal of the Chinese Ceramic Society.

Stephan P., Blöß. & Lothar Elfenthal., 2007, "*Doped titanium dioxide as a photocatalyst for UV and Visibile Light*", Rilem Symposium on photocatalysis, Florence, Italy.

Toraldo E. & Lambrugo S., 2007, "The optimization of photocatalytic mortars for road pavements", 4th International Conference Bituminous mixtures and pavements, Thessaloniki, Greece.

Efficient Transportation and Pavement Systems – Al-Qadi, Sayed, Alnuaimi & Masad (eds)
© 2009 Taylor & Francis Group, London, ISBN 978-0-415-48979-9

Factors affecting distress density behavior on urban streets

M. Mubaraki
Nottingham University, School of Civil Engineering, Nottingham, UK

ABSTRACT: An analysis of distress in flexible pavements for the urban street network at Riyadh city over 4 years is described. Pavement sections were categorized according to distress data, namely distress type, distress density, and distress severity as well as factors that can affect the behavior analysis such as traffic level, pavement condition, street class and time. From distress types found on the network, only the more common ones have been selected. Before proceeding to analyze the data, it was necessary to build a database to cover all possible classifications and to remove any irrelevant data. After building the required database, several statistical tests were implemented to ensure the reliability of the data collected. An analysis of normality and homogeneity was conducted to determine whether the distress density data was normally distributed or not. An analysis of variance was carried out to find the effect of different factors on distress density.

1 INTRODUCTION

Pavement distresses are visible imperfections on the surface of pavements. They are symptoms of the deterioration of pavement structures. Most, if not all, agencies that have implemented a Pavement Management System (PMS) to collect periodic surface distress information on their pavements through distress surveys [Haas et al 1994]. Information on distress type, severity, density, and sometimes location is usually gathered.

Pavement engineers have long recognized the importance of distress information in quantifying the quality of pavements. This information has been used to document present pavement condition, chart past performance history, and predict future pavement performance [Shahin M. 1994].

Pavement distress information is needed to assess maintenance requirements and plan rehabilitation. For immediate maintenance requirements, it is necessary that the details of individual distress types, severity, and density are known [Hajek et al 1987].

Surface distress evaluation was one such technique used to evaluate pavement condition in order to build the Pavement Maintenance Management System (PMMS) for the Riyadh Street Network.

The main objective of the study is to understand the behavior of common types of pavement distress on the Riyadh Street Network. These common types of distress are among those selected for the Riyadh Pavement Maintenance Management System (RPMMS). To achieve this objective the study will include the following:

- Identifying the most common types of pavement distress on the Riyadh Street Network.
- Studying the behavior of common types of pavement distress.

2 METHODOLOGY

There is a need to build a database for this study due to the enormous number of streets in the network. The database will be formatted to cover all possible cases. The possible cases depend on parameters that are under investigation in the study. The parameters are highway class (Main

Street or Secondary Street), traffic level (high, low), pavement condition rating (excellent, good, fair, and poor), distress type, distress severity (low, medium, high) and distress density (percentage). A normalization function for the data was required to reduce the variance of the data and to increase the homogeneity. The function was applied to transform all distress density values by taking the logarithm of the density value. Based on the developed database, the behavior of the common distresses will be investigated in terms of changing and progression of the distress density over time.

3 FACTORS AFFECTING DISTRESS DENSITY BEHAVIOR

In general, distress density initially propagates very slowly and accelerates more at a later stage, but the propagation rate varies according to the parameters under study which are: distress type, distress density, pavement condition, time, traffic level, and highway class.

4 FACTORS ANALYSIS

Two main tests were performed in factor analysis, namely analysis of variance (ANOVA) and the correlation test.

4.1 *Variance analysis*

An analysis of variance (ANOVA) was carried out to check the effect on distress density behavior of propagation.

4.1.1 *Analysis of variance for individual common distress types on main streets*

Each distress type has its own behavior progress over time, so this will cause different distress densities. Therefore, analysis of variance for each distress was needed. Seven common distress types were checked in this analysis. The results are presented in Table 1. Where, for example, the 'P-value' is 0.05 or less, that means there is a 95% or more chance that the parameter under study is influencing distress. Table 1 indicates that all parameters are probably related to all distress types, except for the case of traffic level, which seems not to relate to pothole density.

4.1.2 *Analysis of variance for individual common distresses on secondary streets*

Four common distress types were checked in this analysis. Table 2 shows the results. In this case, there are zero or near zero P-values (i.e. high chance of relationship) throughout, with a rather high value (less certain relationship) between severity level and pothole density.

5 CORRELATION ANALYSIS

Correlations between the parameters were investigated. Correlation coefficient provides a measure of the association between any two variables. It has a value between +1 and −1. Perfect statistically independent variables have a correlation coefficient value of zero.

5.1 *Analysis of correlation for all common distresses on main street*

Table 3 shows the correlation matrix among the considered factors. The matrix shows that the correlation varies from factor to factor. Distress density correlations as percentages were 34.6%, 36.7%, and −7% with pavement condition, distress severity, and traffic level. In fact, it is clear that traffic level has no clear correlation with any other variable.

Table 1. Analysis of variance for distress density for individual distresses on main streets data.

Distress name	Source	P-value
Long. & Trans. cracks (D3)	Pavement condition	.000
	Traffic level	.000
	Severity level	.000
	Time	.000
Patching (D4)	Pavement condition	.000
	Traffic level	.000
	Time	.000
Weath. & Raveling (D11)	Pavement condition	.001
	Traffic level	.000
	Severity level	.001
	Time	.000
Potholes (D5)	Pavement condition	.000
	Traffic level	.624
	Severity level	.005
	Time	.000
Patching cracks (D12)	Pavement condition	.000
	Traffic level	.030
	Time	0.00
Block cracks (D2)	Pavement condition	.000
	Traffic level	.000
	Severity level	.000
	Time	.000
Depression (D6)	Pavement condition	.000
	Traffic level	.076
	Severity level	.163
	Time	.000

Table 2. Analysis of variance for distress density for individual distresses on secondary streets.

Distress code	Source	P-value
Weathering and Raveling (D11)	Pavement condition	.000
	Severity level	.022
	Time	.000
Patching (D4)	Pavement condition	.000
	Time	.000
Long. & Trans. cracks (D3)	Pavement condition	.000
	Severity level	.006
	Time	.000
Potholes (D5)	Pavement condition	.002
	Severity level	.138
	Time	.000

5.2 *Analysis of correlation for all common distresses on secondary streets*

Similarly, factors relating to secondary streets were correlated with each other. The correlation matrix is presented in Table 4. In this case, there is a moderate correlation between distress density and pavement condition (24%), but no significant correlation of either with distress severity.

Table 3. Correlation matrix for main streets data.

	Distress density	Pavement condition	Distress severity	Traffic level
Pavement condition	.346	1		
Distress severity	.367	.112	1	
Traffic level	–.070	.03	.026	1

Table 4. Correlation matrix for secondary streets data.

	Distress density	Pavement condition	Distress severity
Pavement condition	.237	1	
Distress severity	–.014	.059	1

Table 5. Probability values between distresses on main streets.

	D3	D4	D11	D5	D12	D2	D6
D3 Long. & Trans. cracks							
D4 Patching	.535						
D11 Weath. & Raveling	.013	.521					
D5 Potholes	.024	.558	.037				
D12 Block cracks	.019	.515	.006	.043			
D2 Block cracks	.032	.567	.046	.009	.052		
D6 Depression	.058	.593	.072	.034	.077	.026	

6 THE PROBABILITY ANALYSIS BETWEEN THE COMMON DISTRESSES

Table 5 shows the probability for main streets that any two distress types occur together on the same section and Table 6 shows the same for secondary streets. Table 5 shows that patching (D4) has a relatively high probability of occurring together with other distress. For example, there is 52% probability that patching (D4) and weathering and raveling (D11) occur together; the figure is 60% between patching (D4) and depression (D6). Other distress types are approximately independent of each other, with the lowest probability 0.6% between weathering and raveling (D11) and patching cracks (D12).

7 DISTRESS BEHAVIOR

The database for common distress types on the Riyadh street network, on both main and secondary streets, is used in studying the behavior of these distresses.

7.1 *Behavior of common distress types on main streets*

The common distress types on main streets were monitored. Each distress will be investigated individually to study its behavior. Studying the trend of data will help to understand the propagation of these distresses. The distress density data was plotted as a percentage against time (in years) for each distress type. Observing the percentage of distress density on the network will also assist in understanding the distress behaviors and which types of distress increase the fastest. The percentage distress density was calculated at the second and the fourth year.

Figure 1 shows the general trend for propagation behavior of longitudinal and transverse cracks. The figure indicates that the average distress density reached 20% after 4 years. That means 20% of the total area of pavement network was subject to longitudinal and transverse cracks within 4 years. The distress density increased by 50% in the first two years. After four years, it had increased by 80%. And more or less the propagation for the others.

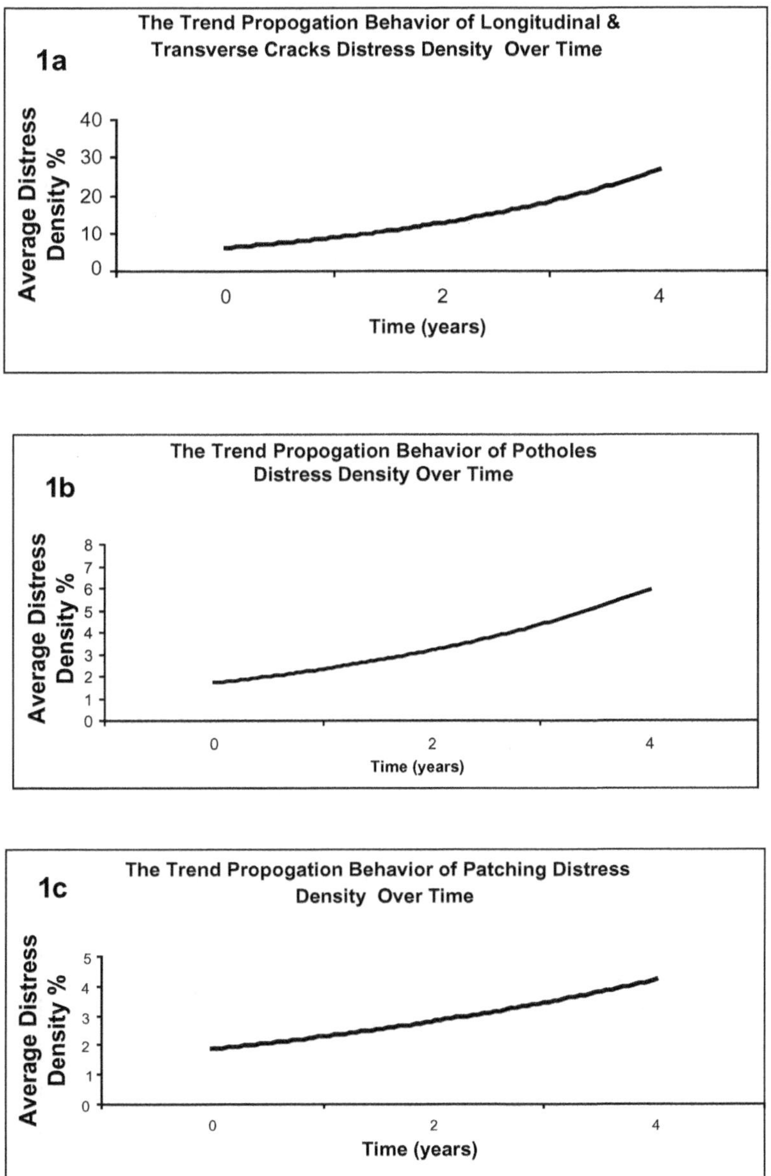

Figure 1a–1g. (Continued)

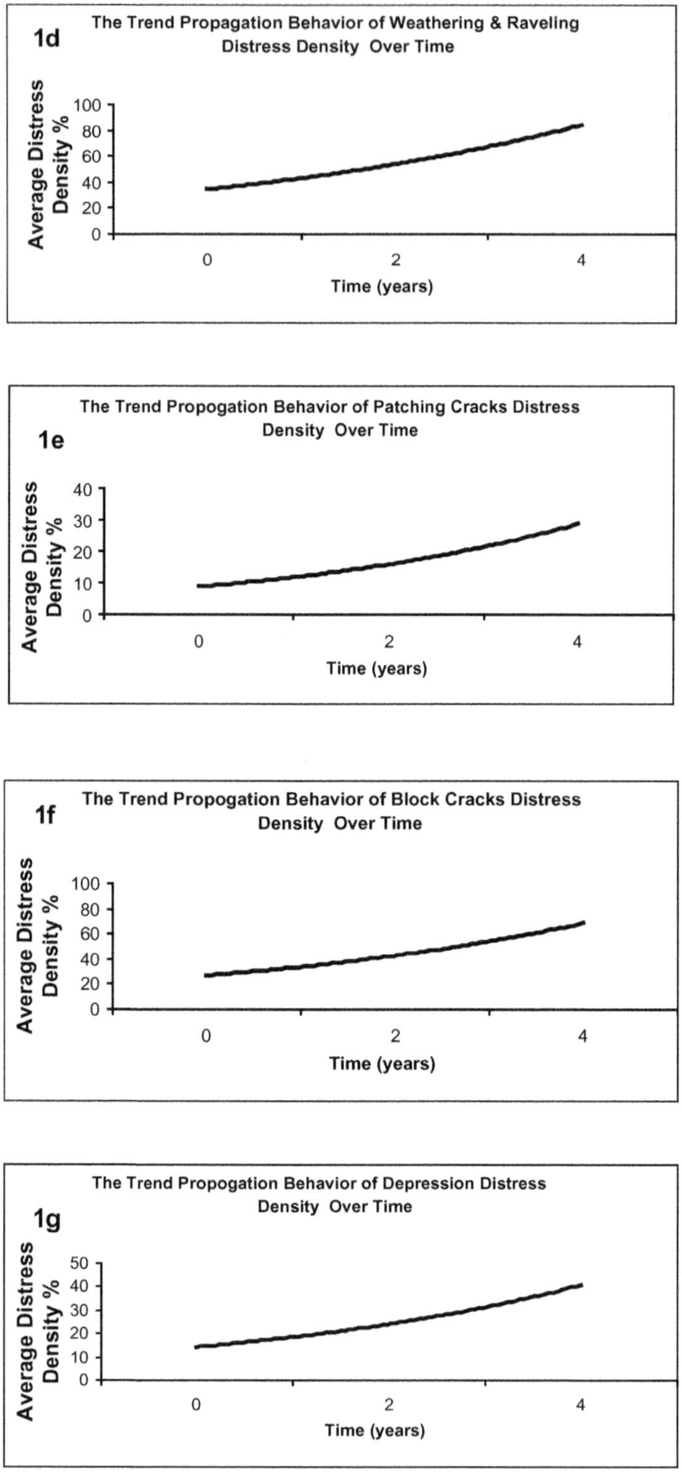

Figures 1a–1g. Propagation trends for distress on main roads.

7.2 *Behavior of common distress types on secondary streets*

Weathering and raveling, patching, longitudinal and transverse cracks and potholes are the common distress types on secondary streets. Development of each distress type is shown in Figure 2. The deterioration rate for weathering and raveling was noticed to be the fastest of all distress types.

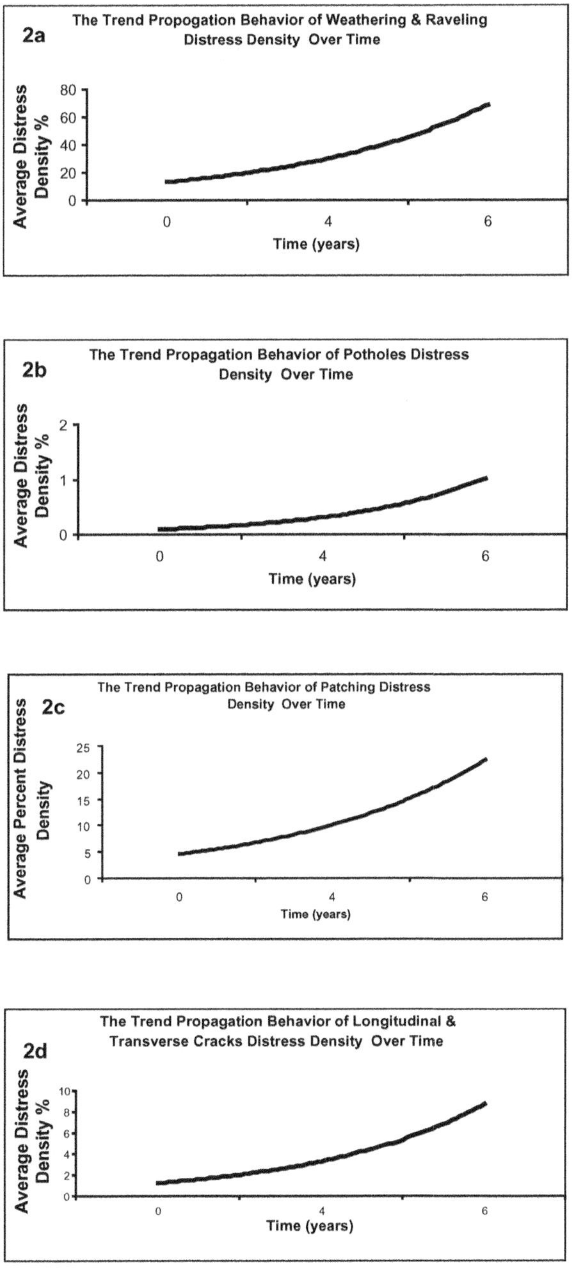

Figure 2a–2d. Propagation trends for distress on secondary streets.

8 RESULTS SUMMARY

This paper has presented data from Riyadh streets over a 4-year period. It was found that, for almost all distress types, distress density was related to time, traffic, overall pavement condition and the severity level for the relevant distress type. There were a few exceptions, notably the lack of relationship between pothole density and traffic on main streets. In fact, traffic level generally failed to correlate clearly with pavement condition. In terms of inter-relationship between distress types, patching was clearly associated with every other distress mode; otherwise distress types were practically independent of each other. This information, together with the data on the progression of each distress type, should prove useful to the maintenance engineer in determining the effective timing of maintenance work.

REFERENCES

Haas, R., Hudson, R. and Zaniewski, J., "Modern Pavement Management", Krieger Publishing Company, Malabor, Florida, 1994.

Hajek, J., and Hass, C., "Factors Analysis of Pavement Distresses for Surface Condition Prediction", Transportation Research Record (TRR), No. 1117, 1987, PP. 125–133.

Shahin, M., "Pavement for Airports, Roads, and Parking Lots", Chapman & Hall, New York, 1994.

Efficient Transportation and Pavement Systems – Al-Qadi, Sayed, Alnuaimi & Masad (eds)
© 2009 Taylor & Francis Group, London, ISBN 978-0-415-48979-9

Decision support tools/project programming: How to communicate the results?

Omar Smadi
Research Scientist, CTRE/Iowa State University, USA

ABSTRACT: Decision support tools (DST) like pavement, bridge, and other systems to manage assets are very critical to providing information to decision makers so that objective, consistent, and cost effective decisions are made regarding resource allocation and project selection.

This paper will focus on the importance of communicating the results from these DST to upper management, governing bodies, and the public to steer them in the direction of investing in the transportation infrastructure. Often times, agencies have very good management systems (in a recent survey of state DOTs by FHWA, all 50 states said they have a pavement management system), but the decisions made by the top level management don't utilize this crucial information. In some of these cases, it is lack of proper communication between the technical staff and the decision makers.

This paper shows 3 examples on how better communication can result in more use of the results from the DST using a variety of assets focusing on pavement management systems with the cities of Des Moines and Muscatine in the state of Iowa. The final example is for the Minnesota DOT pavement marking management system.

1 INTRODUCTION

Decision support tools (DST) like pavement, bridge, safety, signs, pavement markings, and other systems to manage assets are very critical to providing information to decision makers so that objective, consistent, and cost effective decisions are made regarding resource allocation and project selection in a comprehensive transportation asset management perspective.

Asset management is defined as

"Transportation Asset Management is a strategic and systematic process of operating, maintaining, upgrading and expanding physical assets effectively throughout their lifecycle. It focuses on business and engineering practices for resource allocation and utilization, with the objective of better decision-making based upon quality information and well-defined objectives."(1)

Asset management is not a new concept per say, but it still eludes full implementation by a majority of transportation agencies around the world. Managing the transportation infrastructure in a comprehensive manner allowing for the analysis of condition of the different assets a transportation agency manages (highways, bridges, signs, markings, drainage structures, signals, lighting, etc…), determining needs, resource allocation and trade-off analysis, and finally determining funding requests, is a crucial function in maintaining those assets for the next generation of users. Transportation infrastructure has a huge impact on the economy and society, so maintaining that infrastructure in good working order is critical to keeping the economy moving and growing. In a speech by the US President, Clinton, in 2000 titled "Technology for America's Economic Growth, A New Direction to Build Economic Strength", he stressed that "*a competitive, growing economy requires a transportation system that can move people, goods, and services quickly and efficiently.*

To meet this challenge, each transport sector must work effectively both by itself and as part of a larger, interconnected whole. Technologies that increase the speed, reliability, and cost-effectiveness of the transportation sector also will increase the economy's competitiveness and ability to create jobs. Today, one of the greatest challenges we face is to rehabilitate and properly maintain the huge stock of infrastructure facilities already in place. Providing a world-class transportation sector will require the Nation to meet the challenges posed both by increased congestion in many parts of the transportation system and by the need to rebuild and maintain a public capital stock valued at more than $2.4 trillion."(2)

Considering all of these factors, having an asset management system is clearly needed to help decision makers make informed, consistent, and objectives decisions regarding investment in the transportation infrastructure. Decision support tools (DST) are an important component in making asset management successful. Without effective DSTs, decisions such as project selection and resource allocation for individual assets becomes very difficult to make in an efficient manner which might result in improper use of public resources. Transportation agencies around the world have implemented these decision support tools in the form of management systems. It is very common for a transportation agency to have the following DSTs:

1. Pavement management system (PMS),
2. Bridge management system (BMS),
3. Safety management system (SMS), and
4. Maintenance management system (MMS).

The measure of effectiveness though should not be the fact that the agency has a system, but whether that system results are used in making decisions at the different levels within the organization that operates these systems. The best, most advanced and most expensive PMS does not provide any tangible benefits if we are not using its project selection and resource allocation recommendations is developing, for example, our 5-year Transportation Improvement Program (TIP).

The last component to the success of an asset management system is communication. Communication is very important at so many different levels within the organization to ensure effective implementation and operation of the different management systems. Communication starts early on when the system is being developed to the later stages when the system has been implemented and being operated. Proper communication of the system's capabilities, input and output data, results format, tabular and/or graphical representations are only few of the items that need to be addressed by the agency. Lack of communication can result in not utilizing the results for decision making which is detrimental to the management process.

This paper will focus on the importance of communicating the results from these DST to upper management, governing bodies, and the public to steer them in the direction of investing in the transportation infrastructure. Often times, agencies have very good management systems (in a recent survey of state DOTs by FHWA, all 50 states said they have a pavement management system), but the decisions made by the top level management don't utilize this crucial information. In some of these cases, it is lack of proper communication between the technical staff and the decision makers. The next sections will address decision support tools and how they function within a transportation agencies, examples of how some DSTs have been used to communicate needs resulting in increased funding and/or in the development of additional management systems for different assets, and finally conclude with some recommendations on how to address the discussed issues.

2 DECISION SUPPORT TOOLS

Decision support tools are management systems consisting of models (prioritization, economic analysis, and optimization) that support decision making in an organization at different levels covering field, technical, management, and administrative functions. Decision making can be regarded as an outcome of mental processes (cognitive process) leading to the selection of a course of action among several alternatives. Every decision making process produces a final choice.(3) Decisions made could impact any one of the following functions that a transportation agency deals with when it comes to the transportation infrastructure:

1. Allocate resources to different assets within the same mode or across modes ($X to highways—both pavements and bridges, $Y to transit, $Z to aviation, etc…),
2. Select projects for individual assets (bridges to maintain, rehabilitate, and/or replace),
3. Determine overall system condition considering existing funding levels,
4. Support the development of performance measures and provide data for continuous monitoring and evaluation,
5. Determine funding needs for both short (1 to 5 years) and long (up to 20 years) terms to achieve desired system performance, and
6. Develop funding requests for top management to increase investment in the transportation infrastructure.

The following is a description of the major types of decision support tools used by different transportation agencies around the world. Each DST will be described based on the analysis level (project vs. network), analysis period (short vs. long term), and finally on the efficiency of the results.

2.1 *Prioritization models*

Prioritization models are one of the simplest DSTs that an agency can use to support the development of management systems. Simply put, an agency would prioritize based on a simple criterion (mostly worst first—fix assets in the worst category) or multiple criteria. Prioritization techniques are used for project level decisions and mostly used for short term analysis. Even though prioritization models are simple to implement and operate, the results produced are far from being optimal especially when the worst-first approach is utilized. Figure 1 demonstrates how inefficient the worst-first approach when compared to other DST schemes. The data in figures one through three are based on a hypothetical network of roads (1000 miles long) and the only variable is how we allocate resources and select projects. The three criteria evaluated are worst-first, percentage based (spend a % of the available budget on each condition category-excellent/good, fair, and

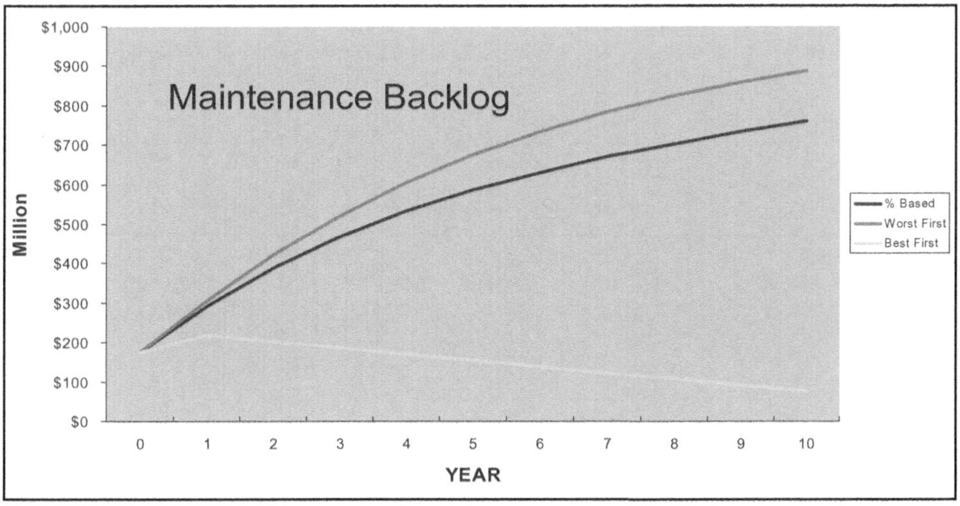

Figure 1. Maintenance backlog for different prioritization schemes (4).

poor- similar to the % of the condition category in terms of miles), and finally best-first (maintain the assets that are in good condition at that level and then use the remaining budget to fix the fair and poor). Figure 1 shows how the maintenance backlog compares between the three spending criteria. The worst-first approach resulted in almost 900 miles of backlog compared to 100 miles for the best-first.

Figures 2 and 3 show the distribution of the miles in the three different condition categories for the worst-first and best-first respectively. As illustrated in the two figures, there are over 500 more miles in the poor category utilizing the worst-first approach of distributing resources while the best-first resulted in 600 more miles in the excellent/good category.

Prioritization DSTs should be the last resort for an agency when it comes to management systems because of the inefficiencies in producing cost-effective results and the variability in the results produced.

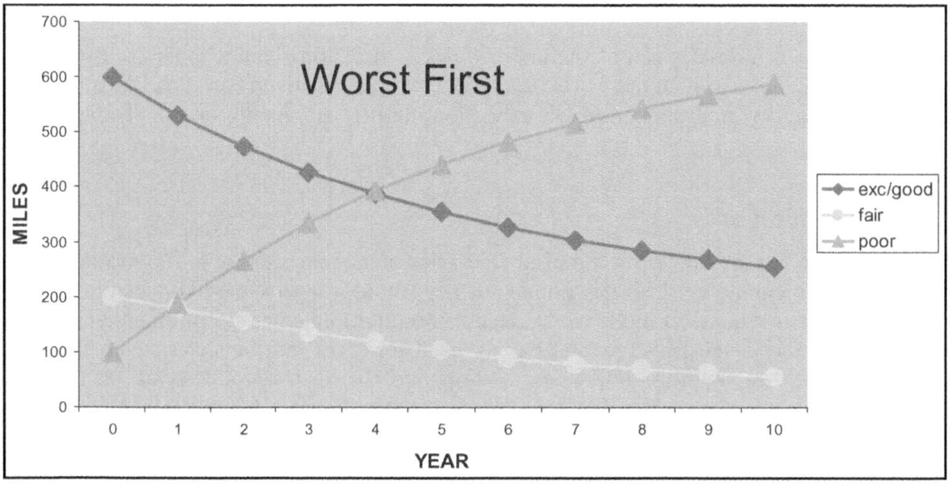

Figure 2. Worst-first prioritization scheme results (4).

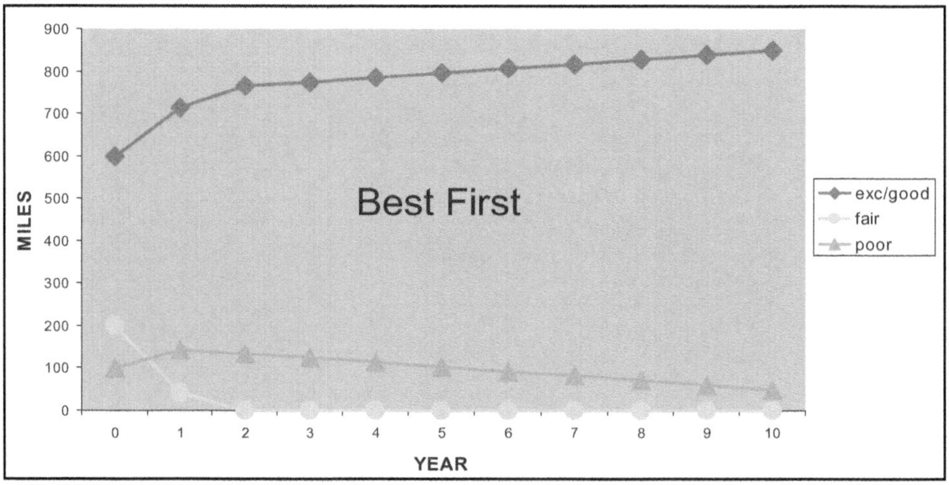

Figure 3. Best-first prioritization scheme results (4).

2.2 *Economic analysis models*

Economic analysis models such as incremental benefit cost (IBC), cost effectiveness, marginal cost effectiveness, and benefit cost ration (BC) are few of the techniques that can be used by DST to provide for resource allocation and project selection. These tools are used at both project and network levels and either short or long term analysis periods. Economic tools are computationally efficient, easily understood, and produce near optimal results. When compared to results from traditional pavement management systems, economic models prove to be effective and produce better asset condition using the same funding levels. In a study to quantify the benefits of pavement management systems using the Iowa Department of Transportation (DOT) Interstate system as a case study, the researchers were able to show that results from the IBC analysis produced a $5 million in benefits over a 5 year period.(5)

The analysis considered before and after PMS implementation results. Figures 4 and 5 show a comparison between what the Iowa DOT accomplished before and after PMS implementation and

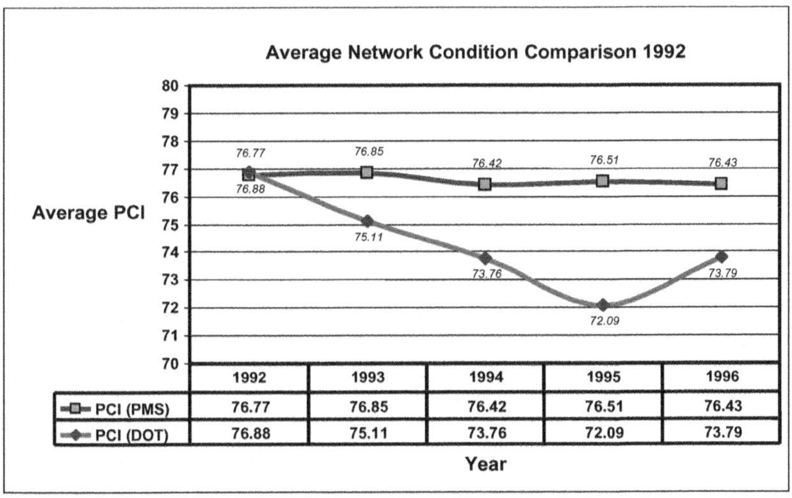

Figure 4. Before PMS average condition (PCI) comparison (5).

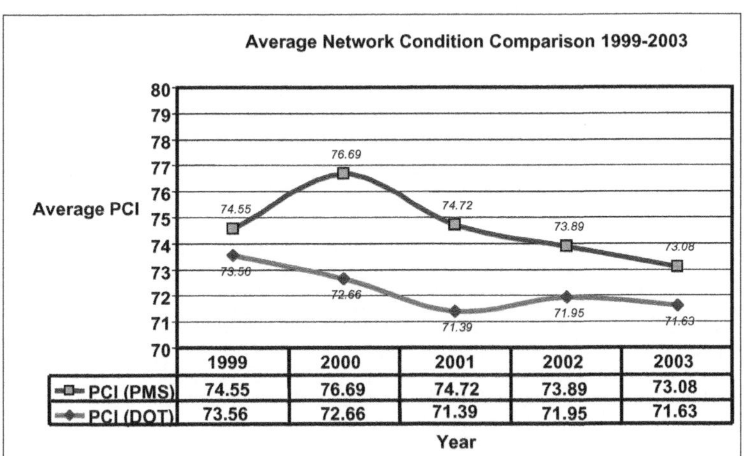

Figure 5. After PMS average condition (PCI) comparison (5).

the results from using an IBC model. The difference between the two scenarios was determined to be the benefits due to implementing a pavement management system. The before period covered the period from1992 to 1996 while the after period included the years 1999 to 2003.

2.3 *Optimization models*

Mathematical programming techniques (true optimization) are tools used to analyze large complex systems for resource allocation and project selection. Optimization models such as linear programming, integer programming, dynamic programming, and no-linear programming are examples of the mathematical algorithms that can be utilized to optimize the decision making progress. The components of an optimization model are the followings:

1. An objective function to minimize or maximize an event:
   a. Minimize agency cost
   b. Maximize benefits (asset condition)
   c. Minimize user cost
2. A set of constraints:
   a. Budget
   b. Performance standards
   c. Resources (human power, working days, machinery, etc…)
3. Decision variables.

Optimization models are computer intensive and require knowledge of specific software capabilities and systems. The results are more efficient than economic models but the additional effort computationally, technically, and the complexity renders some of that gain in efficiency as not necessary. In a study to compare the results from a multi-year prioritization (MYP) utilizing IBC analysis compared to optimization results from a dynamic program with knowledge-based expert system, the optimization program produced a higher pavement condition measure as the pavement condition index (PCI). Figure 6 shows the comparison.

The decision support tools discussed are critical in terms of communicating the results from the management systems to the end users. The same figures referenced in this paper (Figures 1 through 6) are examples of how results were communicate to influence management decisions. In the case of Figures 1, 2 and 3, the case was made against implementing a prioritization tool as a

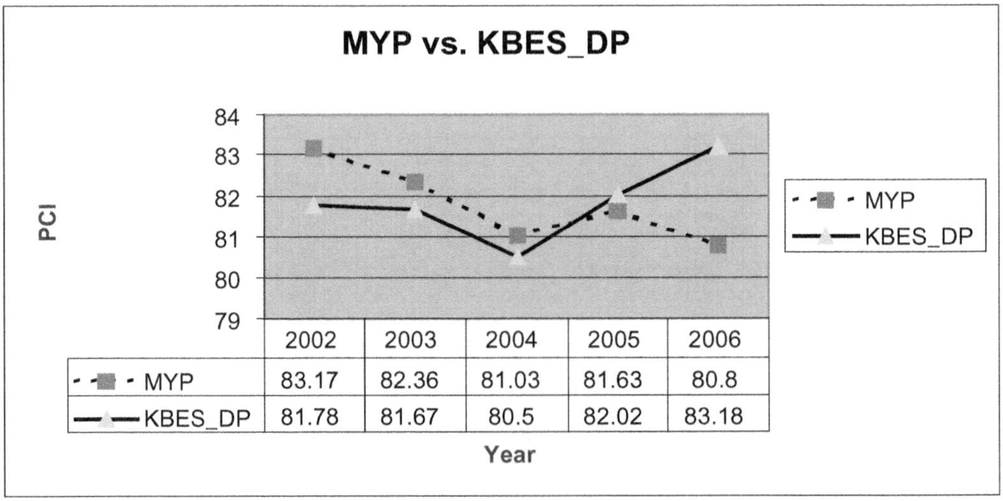

	2002	2003	2004	2005	2006
MYP	83.17	82.36	81.03	81.63	80.8
KBES_DP	81.78	81.67	80.5	82.02	83.18

Figure 6. PCI differences between optimization and IBC (6).

management system for a local agency. Figures 4 and 5 were used to encourage the development and implementation of a pavement management system for an additional 9,000 center-line miles that make the primary system in the state of Iowa. It also showed the decision makers that additional investment in collecting data and dedicating staff to the PMS process will pay off in terms of improved efficiency of allocating resources and improved conditions.

3 EXAMPLE CASE STUDIES

This section of the paper discusses how some transportation agencies utilized the results from the decision support tools to enhance their transportation infrastructure funding and convince their decision makers to support more asset management systems development.

3.1 *City of Des Moines, Iowa*

The city of Des Moines is the capital city of the state of Iowa. Des Moines has a population of 200, 000 residents and maintain a street network of about 800 center-line miles. In 2003, the city started their pavement management system and in 2005 worked with their city council to increase funding for street maintenance from $4 million to $6 million annually. The city staff was able to show their decision makers how their system condition is declining and that the current budget level is not enough to achieve the stated performance targets. The city though developed a plan to use the extra funding to address specific needs and work on achieving performance targets in a 5 year time frame (by 2011).

3.2 *City of Muscatine, Iowa*

The city of Muscatine has a population of 23,000 people and the public works department manages close to 175 center-line miles. The city's council was concerned about the condition of the street network and its impact on customer service and the ability of the city to attract new business to increase employment. The city staff, the public works department, decided that pavement management was the key to show the council the condition of the network and how additional funding is needed to improve the street network condition. The city was able to show the council that with additional funding ($1 million for 2007, $2 million for 2008 and 2009, and then $750,000 annually after that) will improve the street condition and ensure that condition for an extended period of time. In 2005 and 2006, the public works department had only $500,000 for street maintenance. The city staff spent a great deal of time educating their decision makers on pavement management tools, the benefits from implementing such a system, and how these tools can guide the allocation of the additional funding so that system improvement is evident. Figure 7 shows the impact of the different funding scenarios on the overall pavement condition index from 2007 to 2016 (a 10-year analysis).

3.3 *Minnesota department of transportation (DOT)*

The Office of Traffic, Security and Operations Pavement Marking, Work Zone and Product at the Minnesota DOT wanted to implement a pavement marking management system to manage the decisions made regarding product selection, highway miles painted, and pavement marking retroreflectivity monitoring. The Office of Traffic staff worked with their top level management, showed them a prototype application based on a system that was developed for the Iowa DOT, explained why the system is needed, the benefits from developing and implementing such a system. The Office of Traffic received funding to contract for the system development and they are in the final stages of developing a web-based application to allow all users (at different levels) almost immediate access to the pavement marking retroreflectivity data and paint information. Figure 8 shows a screen print from the prototype application.

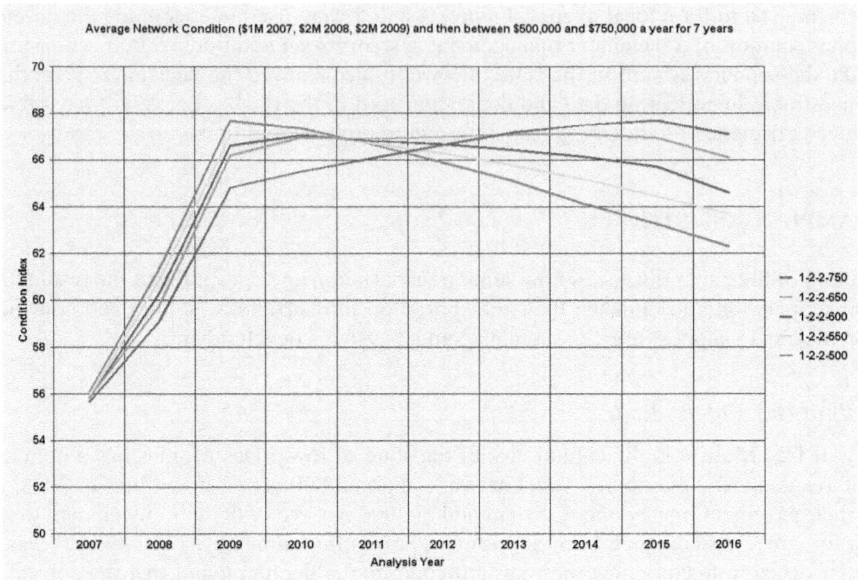

Figure 7. City of Muscatine PCI trends over a 10-year period.

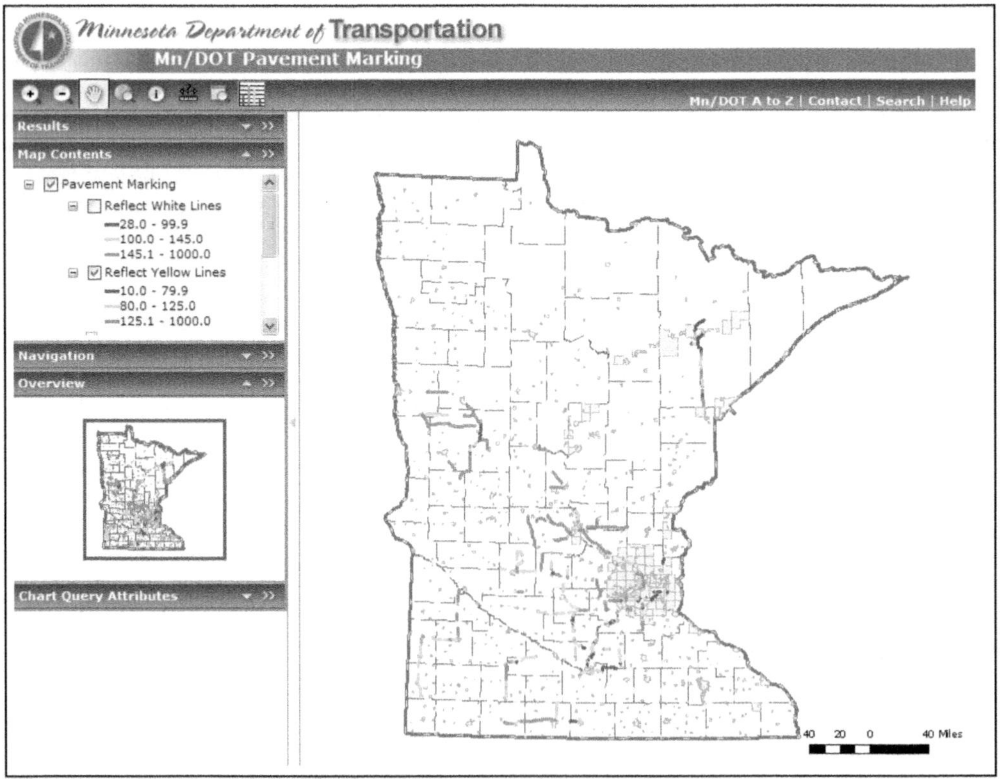

Figure 8. MN DOT pavement marking management system (7).

4 CONCLUDING REMARKS

The paper discussed how decision support tools are used to communicate with decision makers and influence the decision making process when it comes to project selection, resource allocation, infrastructure funding needs, and funding requests. The examples presented showed how some agencies utilize the results from their DSTs to either implement different management systems or secure additional funding.

The examples also illustrates how output from DSTs can be used to develop charts and figures to relay information to decision makers, the public, and the elected officials. Output from DSTs need to be tailored to fit the needs of the groups where the results are considered.

REFERENCES

1. AASHTO Subcommittee on Asset Management. American Association of State Highway and Transportation Officials, 2006.
2. Web document (http://www.ibiblio.org/darlene/tech/report9.html) accessed March 30th, 2008.
3. James Reason (1990). Human Error. Ashgate. ISBN 1840141042
4. Omar Smadi. Adapted from a presentation at the "Key Issues in Transportation Programming Conference," Seattle, November 2006.
5. Omar Smadi. "Quantifying the Benefits of Pavement Management." Proceedings of the 6th International Conference on Managing Pavements, Brisbane, Australia, October 2004.
6. Omar Ghaleb Smadi. Knowledge based expert system pavement management optimization. Ph.D. dissertation, Iowa State University, Ames, Iowa, 2000
7. Minnesota DOT Pavement Marking Management prototype, accesses April, 5th, 2008.

Efficient Transportation and Pavement Systems – Al-Qadi, Sayed, Alnuaimi & Masad (eds)
© 2009 Taylor & Francis Group, London, ISBN 978-0-415-48979-9

Quantifying the life-cycle value of optimal infrastructure condition sampling

Rabi G. Mishalani
The Ohio State University, Columbus, OH, USA

Liying Gong
Capital One, Glen Allen, VA, USA

ABSTRACT: Several methods have been developed to generate optimal infrastructure inspection, maintenance, and rehabilitation (IM&R) policies progressively taking into account various uncertainties associated with measurement and forecasting. Moreover, new non-destructive technologies are being applied in collecting and processing large quantities of raw condition data to produce useful input to decision-making. As a result, the uncertainty associated with condition sampling has been quantified and the sample size has been introduced as a decision variable in IM&R decision-making. After an overview of these advances is presented, a methodology developed to assess their value is discussed. An infrastructure facility reflecting realistic condition and cost characteristics is then considered. The magnitude of the value of the condition sampling advancements of interest is found to be appreciable in terms of reduced expected total life-cycle cost.

1 INTRODUCTION

Pavement management is a process consisting of three main elements: collecting infrastructure condition data, assessing and forecasting infrastructure condition, and making decisions on inspection, maintenance, and rehabilitation (IM&R) activities aimed at minimizing expected total life-cycle cost. Several new non-destructive technologies have been developed and are being applied in collecting raw condition data and processing them to produce useful condition input to IM&R decision-making (McGhee 2004). Such developments make it possible to estimate facilities' condition using large quantities of data. The quality of measurements, the sample size, and the nature of correlation between condition variables at different locations determine the accuracy of condition estimates, and more accurate estimates have the potential to lead to more effective IM&R decisions. Consequently, the expected combined user costs and IM&R costs are reduced over the planning horizon. However, more accurate information requires more resources such as increased inspection frequency, advanced inspection technologies and data processing methods, larger sample sizes, or less correlated observations.

Therefore, in response to the developments in inspection technologies, decision-making methods evolved whereby the optimum combination of inspection decisions on the one hand and maintenance and rehabilitation decisions on the other are determined based on an economic evaluation that captures the long-term costs and benefits. The comprehensive set of IM&R decision variables the optimal values of which are determined through such optimization processes includes the following: whether to inspect or not during a particular year, inspection technology to use for condition measurement (e.g., visual inspection, manual measurements, automated techniques, etc.) if inspection is to take place, sample size if inspection is to take place, and maintenance actions to apply (e.g., routine maintenance such as crack filling and rehabilitation such as applying overlays).

Pavement facility inspection deals with the gathering of data on the extent of facility damage expressed by variables such as cracking and rutting in the case of roadway pavements. The average of collected damage measurements over a facility is an estimate of the current condition of that facility, and, in turn, is one primary input to forecasting future condition and, consequently, life-cycle-based IM&R decision-making. A facility is defined as a section of infrastructure where condition behaves homogeneously over space and time (Mishalani & Koutsopoulos 1995, 2002).

In the inspection data collection and utilization process, two sources of error are considered: measurement and spatial sampling. A continuous distress variable x is distributed along a specific facility (defined to be a homogenous section) with mean μ_x, variance σ_x^2, and spatial correlation ρ_x. The values of distress variable x at n sample locations $\{x_1, x_2, \ldots, x_n\}$ vary around μ_x due to inherent spatial processes discussed by Mishalani and Koutsopoulos (1995, 2002). Measured distress values $\{\hat{x}_{1,r}, \hat{x}_{2,r}, \ldots, \hat{x}_{n,r}\}$ collected using inspection technology r are observations of $\{x_1, x_2, \ldots, x_n\}$, which reflect an additional degree of uncertainty due to technological limitations and environmental effects (Humplick 1992). The sample mean of measured distress $\bar{\hat{x}}_{n,r}$ is calculated from these observations as follows:

$$\bar{\hat{x}}_{n,r} = \frac{1}{n}\sum_{l=1}^{n}\hat{x}_{l,r} \tag{1}$$

Therefore, the difference between the condition estimate $\bar{\hat{x}}_{n,r}$ and the true condition μ_x reflects both measurement and spatial sampling uncertainty, where the latter is due to the spatial variation in the condition variables x and positive correlation among them. Together with forecasting uncertainty, which is introduced in predicting future condition, the combination of these three sources of uncertainty could affect the resulting IM&R decisions through their influence on the quality of condition estimation and forecasting.

Humplick (1992) developed a measurement error model whereby the difference between the true value of a distress variable at a given location and its measured value is explained in terms of systematic biases and a random error. Madanat (1991, 1993), Ben-Akiva et al. (1993), Madanat and Ben-Akiva (1994), and Ellis et al. (1995) extended the traditional infrastructure management framework known as the Markov Decision Process (MDP) (Carnahan et al. 1987), which captures the forecasting uncertainty, to the Latent Markov Decision Process (LMDP) by incorporating the measurement error resulting uncertainty into the decision-making framework. The LMDP framework differs from the MDP framework in several aspects including the absence of the assumption that condition measurements are error-free. That is, the LMDP framework assumes that the decision-maker's condition observations are only probabilistically related to the true condition of the facility. With this relationship, the LMDP framework therefore captures the measurement uncertainty in addition to forecasting uncertainty.

In the LMDP framework the uncertainty associated with condition sampling over the length of a facility has not been recognized or quantified. As a result, this framework has been recently extended to capture inspection sampling uncertainty and to include sample size as a decision variable resulting in the new Latent Markov Decision Process in the presence of Spatial Sampling (LMDPSS) framework. The total true condition assessment uncertainty captured by this framework is represented by the variance of the sample mean condition of a facility, which is derived by Mishalani and Gong (2007, forthcoming) and is given by the following:

$$\mathrm{Var}(\bar{\hat{x}}_{n,r}) = \frac{\sigma_r^2}{n} + \frac{\sigma_x^2}{n} + \frac{2\sigma_x^2 \sum_{i=1}^{n-1}(n-i)\cdot\rho(i\cdot h)}{n^2} \tag{2}$$

where, n = sample size; σ_r^2 = variance of the measured condition using inspection technology r after it has been corrected for the systematic bias captured by the measurement error model (Humplick 1992, Ben-Akiva et al. 1993); σ_x^2 = true variance of the condition variable over the length of the facility; and $\rho(s)$ = correlation between condition variables at two points within the facility that are separated from one another by a distance s.

Equation (2) combines both the measurement and sampling uncertainties through the variance of the corrected condition measurement, σ_r^2, variance of the condition variable, σ_x^2, spatial correlation among observations, $\rho(s)$, and sample size, n. The larger the sample size is, the smaller the contributions of the variances of the condition variable and its measurement are to the overall variance. And, the smaller the correlation among observations is, which is captured by the last additive component of equation (2), the more accurate the estimate of facility condition is.

In this paper, the quantification of the value of the condition sampling related advances reflected in the LMDPSS framework in improving decision-making is presented. In section 2 the evaluation framework is presented. Section 3 consists of a numerical analysis where the evaluation framework is applied to a realistic infrastructure facility leading to the quantification of interest. Section 4 concludes the paper with a summary and suggested directions for future research.

2 EVALUATION FRAMEWORK

2.1 *Introduction*

In order to quantify the benefits derived from each of the two main elements of the LMDPSS framework—namely the incorporation of sampling uncertainty and the introduction of sampling as a decision variable—with respect to the LMDP framework, it is necessary to establish a common frame of reference to conduct the various comparisons. The methodology presented in this paper focuses on quantifying the value of each of the above mentioned elements. Mishalani and Gong (2008) take the analysis further by examining the value of capturing sampling uncertainty due to each of the spatial variability of condition and the spatial correlation among the observations.

The basic idea behind comparing any two frameworks is to use each to produce optimal policies based on the specific assumptions they reflect, and then simulate these optimal policies within the framework considered to reflect the "truth" as far as capturing the most realistic assumptions. The former frameworks are referred to as the optimization systems and the latter framework is referred to as the simulation system. In what follows the various systems are described and the simulation process is discussed.

2.2 *Optimization and simulation systems*

The optimization systems are the ones based on which the "optimum" IM&R decision policies are determined. It is important to emphasize that the decisions are so called optimum by the decision-maker and they are dependent on the assumptions behind each of the optimization systems. The simulation system is the one in which the "optimum" decisions obtained from an optimization system are simulated and reflect more relaxed and realistic set of assumptions. Therefore, an optimization system is used to produce an IM&R decision policy, while the simulation system is used to simulate the resulting current and future conditions of the facility after the determined IM&R policy is applied. There are three frameworks used in this analysis ranging from the one that does not address the uncertainty associated with condition sampling to the one that does and includes sample size as part of the decision-making process. The three frameworks are described in what follows.

2.2.1 *Adjusted LMDP (ALMDP) framework*

In the traditional LMDP framework—capturing both measurement and forecasting uncertainties—a unique observation is taken as an estimate of the condition of the infrastructure facility. Therefore, the variance of the measurement random error, σ_r^2, represents the uncertainty in this case, since only a single observation is considered to represent the condition of a facility. Based on statistical theory, if multiple observations are averaged to represent the facility condition, the variance of that mean condition would be given by σ_r^2/n. In order to relax the assumption of a single observation in the original LMDP framework, the framework is revised to reflect multiple observations through the replacement of σ_r^2 with σ_r^2/n. Doing so renders the adjusted LMDP (ALMDP) framework consistent with the other frameworks where multiple observations over the length of a facility are made. Notice that the variance of the mean condition of the ALMDP framework is the first additive component of the true variance of the sample mean of the LMDPSS framework shown in Equation (2). This framework represents the most basic optimization system in this analysis.

2.2.2 *LMDPSS with predetermined sample size (SPU) framework*

There are two main contributions associated with the complete LMDPSS framework: (a) capturing the uncertainty associated with condition sampling, and (b) incorporating the sample size as a decision variable. In order to study the effects of these advances separately, the LMDPSS framework is adopted with the complete representation of uncertainty reflected in Equation (2), however, the sample size is pre-determined (i.e., it is not a decision variable). That is, the only difference between this framework, referred to as the SPU optimization system, and the ALMDP framework is the modeling of the complete condition sampling uncertainty in the case of the SPU framework.

2.2.3 *LMDPSS with the sample size as a decision variable (SVU) framework*

The most general and complete LMDPSS framework not only captures the true uncertainty associated with spatial sampling of Equation (2), but also incorporates the sample size as a decision variable. Being complete, this framework, referred to as SVU, serves as the most advanced optimization system. It also serves as the simulation system, in that it represents the truth as far as capturing all aspects of uncertainty.

2.3 *Scenarios of expected total life-cycle costs*

Scenarios of total expected life-cycle costs associated with simulating each optimization system in the SVU simulation system can be obtained by applying the IM&R decision policy provided by an optimization system in the SVU simulation system. More specifically, the optimal policy obtained from an optimization system consists of a series of inspection, maintenance, and rehabilitation activities given the state of the facility at a given point in time. Such a policy is simulated in the SVU system by applying the recommended activities as a function of the assessed condition state over the planning horizon, whereby the true representation of uncertainty is considered in calculating the total expected life-cycle cost.

The true expected total life-cycle cost associated with each of the ALMDP, SPU, and SVU frameworks are denoted by C_1, C_2, and C_3, respectively. The costs C_1 and C_2 are naturally different from the expected total life-cycle costs computed when determining each of the optimal policies (under the restricted assumptions inherent to the corresponding optimization system), and reflect the true costs associated with each of the optimization systems. Note that determining the cost C_3 does not require any simulation as the optimization itself reflects the true simulation system of interest.

2.4 *Evaluation measures*

It is important to compare the expected total life-cycle costs across the optimization systems in evaluating the condition sampling advances of interest. The cost C_2 represents the expected total life-cycle cost when the optimum policy is based on the SPU framework, whereby the sample size is predetermined but, otherwise, the true complete uncertainty is captured. The cost C_3 is the expected total life-cycle cost when the sample size is a decision variable. Therefore, the difference between C_2 and C_3 represents the value of incorporating sample size as a decision variable in the

LMDPSS framework. This value is quantified by the relative difference as a percentage, denoted by V_D, which is mathematically given by the following:

$$V_D = \frac{C_2 - C_3}{C_3} \times 100 \tag{3}$$

The cost C_1 represents the life-cycle cost when the optimum policy is based on the ALMDP framework, which ignores the uncertainty associated with spatial sampling due to the condition variability within a field and the correlation among the observations. Therefore, the difference between C_1 and C_2 represents the value of capturing the uncertainty associated with condition sampling. This value is quantified by the relative difference as a percentage, denoted by V_U, which is given by the following:

$$V_U = \frac{C_1 - C_2}{C_3} \times 100 \tag{4}$$

Hence, the total value of the complete LMDPSS (SVU) framework, denoted by V_T, whereby condition sampling uncertainty is represented and the sample size is included as a decision variable, is computed as a percentage by the following:

$$V_T = \frac{C_1 - C_3}{C_3} \times 100 \tag{5}$$

Clearly, V_T is the sum of the values V_D and V_U.

Other evaluation measures relating to agency cost and the accuracy associated with measurement and sampling are considered in Mishalani and Gong (2008, forthcoming).

3 NUMERICAL ANALYSIS

In this section, a realistic example pavement facility is introduced to quantify the value of the condition sampling advances as captured by the LMDPSS framework. The evaluation methodology is then applied to this example facility and the results are subsequently discussed.

3.1 *Realistic example facility*

A 1 km long, 3.66 m wide homogeneous pavement section consisting of the right lane of a two-lane interstate highway is assumed. Gong (2006) and Mishalani and Gong (forthcoming) motivated, developed, and presented in detail this realistic example facility. A brief summary is provided here for the purpose of presenting the results in a meaningful manner.

The section of interest is divided into 100 subsections with equal area. Each subsection is considered a potential observation. Therefore, the maximum sample size is 100. The planning horizon *T* is set at 10 years and the interest rate at 5%. There are 11 combinations of inspection activities: two types of inspection technologies and five levels of sample sizes (100%, 80%, 60%, 40% and 20%), in addition to the "no inspection" option.

The condition scale is described by the Pavement Condition Index (PCI), which takes a value ranging from 0 to 100. This range is equally divided into eight condition states (state 7 is the best and state 0 is the worst). The initial condition state is set at 7 with a probability of one. The evolution of facility condition from year to year is captured by transition probabilities adopted from chapter 5 in Madanat (1991) for the routine maintenance and rehabilitation activities. Ten sets of transition probabilities each corresponding to the ten years of the planning horizon are specified for the case of routine maintenance. That is, the deterioration rate varies over time. It is also assumed that rehabilitation brings the condition of the facility back to the best possible state. This assumption has also been made by Carnahan et al. (1987).

The variance of the condition variable, σ_x^2, is derived from results presented by Mishalani and Koutsopoulos (2002) as a function of the mean condition level based on actual roadway pavement observations. The spatial correlation function, $\rho(s)$, is derived from results presented in Mishalani (1993), again based on the same set of observations. The two types of inspection technologies considered in this example are manual mapping and a typical automatic data collecting method based on photographic or video technologies. More details on the inspection technologies are provided in Humplick (1992). Representative variances associated with the inspection technologies, σ_r^2, are derived from results presented by Humplick (1992), which are also based on actual roadway observations (though different from those mentioned above).

3.2 *Cost components*

The costs of interest considered in this study include costs related to maintenance and rehabilitation activities, inspection, users, and terminal condition of the facility in question. Each of these components is discussed in what follows. A more detailed presentation is provided in Mishalani and Gong (forthcoming).

3.2.1 *Maintenance and rehabilitation cost*

The cost of maintenance and rehabilitation activities depends on their type and extent. The activity type can range from basic routine maintenance to major rehabilitation. The extent of an activity is a function of the true condition state of the facility and has been addressed in earlier studies (Carnahan et al. 1987, Sharaf et al. 1985). The costs developed and used in these studies are adopted in this analysis.

3.2.2 *Inspection cost*

Inspection cost, incurred at the beginning of each year if inspection is scheduled to take place, includes two components: (i) fixed cost consisting of data collection equipment cost, labor safety training and insurance cost, data processing equipment cost, and logistics cost of transporting equipment and labor from the depot to the inspection sites; and (ii) variable cost incurred whenever an observation is made (i.e., a sample is taken), which is determined by the inspection time and the labor hourly wage including an economies of scale parameter. The quantification of each component is discussed by Gong (2006).

3.3.3 *User cost*

One of the components of user cost captures the impact of deterioration on the users as a function of the condition of the facility. The reduction in the user cost resulting from condition improvements is the benefit that the users derive from the application of maintenance and rehabilitation activities. Ben-Akiva and Gopinath (1995) developed a model that specifies the user cost as a function of condition and a set of independent variables including highway geometry, traffic composition, and traffic volume. This state-of-the-art user cost model is adopted in the evaluation analysis.

In addition, maintenance and rehabilitation results in further user cost in the form of delays at work zones. Similarly, inspection results in traffic disruption cost. Vadakpat et al. (2000) developed a model to estimate the additional user cost due to pavement work zones, which is a function of a set of variables including lane location, truck percentage, and the traffic volume for which the delay is to be computed. This model is used to quantify the delay and disruption costs of interest. Note that user cost relating to delays is a function of network effects as well. However, since this analysis focuses on decision-making at the facility level, such effects are not modeled.

3.3.4 *Terminal cost*

In this study the terminal cost is defined as the hypothetical expense incurred at the end of the planning horizon for bringing the facility condition back to the best condition level in order to equalize the service life from that point onward. It is a function of the true state of the facility at the end of the planning horizon. If the facility is in good condition, it can continue to supply high level of service resulting in a relatively low terminal cost. However, if the condition is poor, the facility's potential to provide a reasonable level of service is low resulting in a relatively high terminal cost. These costs are based on those presented by Carnahan et al. (1987).

3.3 *Evaluation results*

The total life-cycle cost in this analysis includes all the discounted cost components discussed above. The evaluation methodology is applied to the above example facility over the 10-year planning horizon whereby the costs C_1, C_2, and C_3 are computed in $/m². The results are shown in Figure 1. The sample size axis is pertinent to costs C_1 and C_2 as each depends on the predetermined sample size. The cost C_3 corresponds to the case where optimal sample sizes are employed over the planning horizon and, therefore, could vary by facility condition state and year. The cost C_3 is, therefore, presented in the same figure for the purpose of comparison only. Clearly, each advancement towards a more complete representation of uncertainty provides a progressively lower expected total cost, and as expected, the case where the sample size is determined optimally produces the IM&R policy resulting in the least cost possible.

The associated relative percentage savings in total cost with respect to C_3 given by Equations (3) and (4) along with the total percentage savings given by Equation (5) are computed. The application of the complete LMDPSS framework including all the condition sampling advances (i.e., the SVU optimization system) results in total life-cycle cost savings, V_T, ranging from 2.64% to 4.59% for this particular example facility. Clearly, each of the two assessed advancements contributes towards these total savings. However, for this example facility most of the contribution is due to capturing the condition sampling uncertainty with life-cycle cost savings, V_U, ranging from 2.31% to 4.35%, while the contribution due to including the sample size as a decision variable is

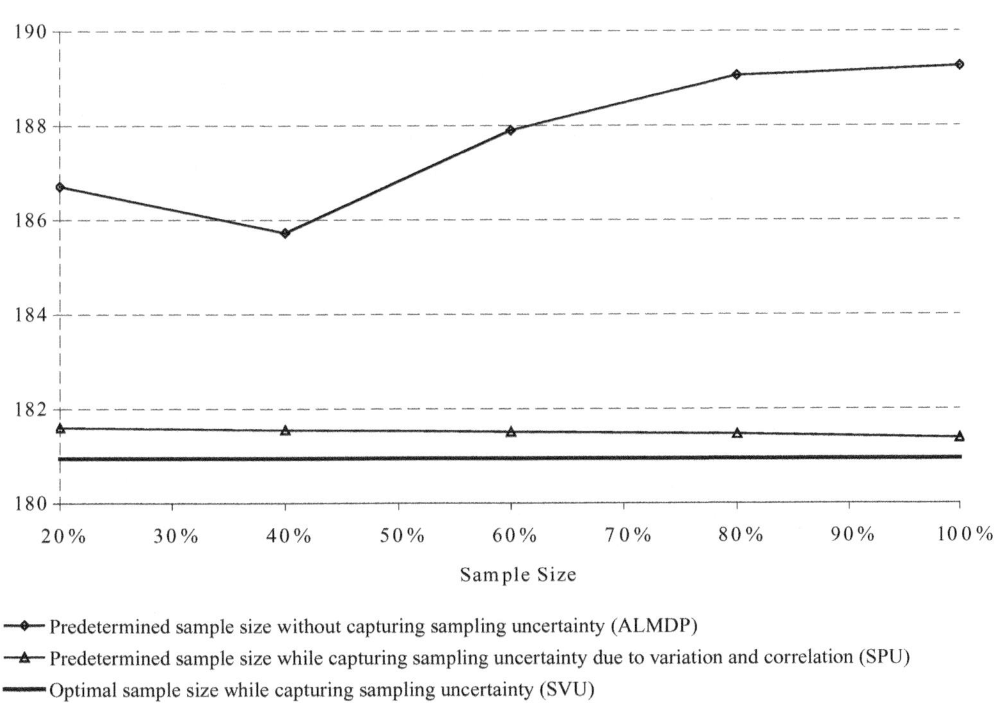

Figure 1. Total minimum expected life-cycle costs for optimization systems.

relatively small with savings, V_D, ranging from 0.24% to 0.36%. Nevertheless, the examination of a suite of additional 323 example facilities reflecting a spectrum of realistic characteristics and conditions—as specified and discussed by Gong (2006) and Mishalani and Gong (2007)—suggests that in general each of the condition sampling advances including the incorporation of the sample size as a decision variable is valuable. Moreover, while in percentage terms the savings seem modest, since the order of magnitude of the costs is large, such savings still represent substantial improvements. As an illustration, in maintaining the 209,107 lane-miles of interstate highways in the United States (FHWA 2003) over a 10-year planning horizon, the total savings of 4.59% would be in the order of $9.8 billion based on the condition, deterioration, and costs assumed for the example facility.

4 SUMMARY AND FUTURE RESEARCH

The development and application of several new non-destructive inspection technologies for the collection of infrastructure condition data have been producing large amounts of data spurring the need to use these data more effectively in IM&R decision-making. Such efforts first involved the quantification of measurement errors and capturing their effects in the decision-making process. Subsequently, the spatial variation of condition over the length of a facility has been quantified and the uncertainty associated with condition sampling has been incorporated in the decision-making process. In this study, a methodology for quantifying the value of such advances regarding improved decision-making is developed. The methodology consists of comparing different optimization systems with varying degrees of realism and completeness regarding condition sampling using a simulation system that captures the true condition sampling uncertainty.

The methodology is applied to a facility specified based on extensive material derived from the literature to ensure its realism. The results indicate that both modeling the uncertainty due to condition sampling and including sample size as a decision variable contribute to improved decision-making. Moreover, while the expected total life-cycle cost savings in percentage terms seem fairly modest, the corresponding magnitudes are appreciable.

Given the identified value of taking into account condition sampling in IM&R decision-making on the basis of examining a total of 324 example facilities, one of which is discussed in detail in the previous section, it is imperative that such demonstrated value be quantified in the context of a small set of actual facilities with condition and cost characteristics derived from field data and decision-making environments of the pavement management agencies responsible for these facilities. Such further confirmation should set the stage for agencies to take measures towards a comprehensive treatment of the condition sampling considerations discussed in this paper.

Another research direction worthwhile to pursue is the study of the sensitivity of IM&R decision outcomes and life-cycle cost to certain input factors. Doing so allows for identifying the factors that are important to be quantified accurately by agencies, and the factors that require improved understanding through research. Example factors of interest include inspection cost, user cost, and the nature of the spatial correlation of condition variables.

Finally, given the importance of condition sampling in infrastructure IM&R decision-making argued in this paper, it is essential that the treatment of condition sampling be extended to decision-making for a network of facilities where budget constraints and network effects can be addressed. Simlowitz and Madanat (2000) utilized the LMDP framework to address the network-level problem. However, condition sampling has not been considered. Therefore, extending the LMDPSS framework to address the network-level problem is expected to be fruitful.

ACKNOWLEDGEMENT

This study was funded by the US National Science Foundation through grant 0093452.

REFERENCES

Ben-Akiva, M., Humplick, F., Madanat, S., & Ramaswamy, R. 1993. Decision-Making under Uncertainty in Infrastructure Management: The Latent Performance Approach. *Journal of Transportation Engineering*, 119(1): 43–58.

Ben-Akiva, M., & Gopinath, D. 1995. Modeling Infrastructure Performance and User Costs. *Journal of infrastructure systems*, 1(1): 33–43.

Carnahan, J.V., Davis, E.J., Shahin, M.Y., Keane, P.L., & Wu, M.I. 1987. Optimal Maintenance Decisions for Pavement Management. *Journal of Transportation Engineering*, 113(5): 554–572.

Ellis, H., Jiang, M., & Corotis, R.B. 1995. Inspection, Maintenance, and Repair with Partial Observability. *Journal of Infrastructure Systems*, 1(2): 92–99.

Federal Highway Administration. 2003. U.S. Department of Transportation, Office of Policy Homepage. http://www.fhwa.dot.gov/policy/ohim/hs03/xls/ps1.xls.

Gong, L. 2006. Optimal Spatial Sampling of Infrastructure Condition: A Life-cycle-based Approach under Uncertainty. Ph.D. dissertation, The Ohio State University, Columbus, OH.

Humplick, F. 1992. Highway Pavement Distress Evaluation: Modeling Measurement Error. *Transportation Research Part B*, 26(2): 135–154.

Madanat, S.M. 1991. Optimizing Sequential Decisions under Measurement and Forecasting Uncertainty: Application to Infrastructure Inspection, Maintenance and Rehabilitation. Ph.D. Dissertation, Department of Civil Engineering, Massachusetts Institute of Technology, Cambridge, MA.

Madanat, S.M. 1993. Optimal Infrastructure Management Decisions under Uncertainty. *Transportation Research Part C*, 1(1): 77–88.

Madanat, S.M., & Ben-Akiva, M. 1994. Optimal Inspection and Repair Policies for Infrastructure Facilities. *Transportation Science*, 28(1): 55–62.

McGhee, K.H. 2004. Automated Pavement Distress Collection Techniques: A Synthesis of Highway Practice. *NCHRP SYNTHESIS 334*. Washington, D.C.

Mishalani, R.G. 1993. Extracting Spatial Information on Infrastructure Distress Using a Stochastic Modeling Approach: Application to Highway Maintenance Decision Making. Ph.D. dissertation, Massachusetts Institute of Technology, Cambridge, MA.

Mishalani, R.G., & Koutsopoulos. H.N. 1995. Uniform Infrastructure Fields: Definition and Identification. *Journal of Infrastructure Systems*, 1(1): 44–55.

Mishalani, R.G., & Koutsopoulos, H.N. 2002. Modeling the Spatial Behavior of Infrastructure Condition. *Transportation Research Part B*, 36(2): 171–194.

Mishalani, R.G., & Gong, L. 2007. Spatial Sampling of Infrastructure Condition: Motivation and Numerical Analysis. *Proceedings of the 11th World Conference on Transportation Research*, Berkeley, CA.

Mishalani, R.G., & Gong, L. 2008. Evaluating the Impact of Pavement Condition Sampling Advances on Life-cycle Management. *Transportation Research Record*.

Mishalani, R.G., & Gong, L. forthcoming. Optimal Infrastructure Condition Sampling over Space and Time for Maintenance Decision-making under Uncertainty. *Transportation Research Part B*.

Sharaf, E.A., Reichelt, E., Shahin, M.Y., & Sinha, K.C. 1985. Development of a Methodology to Estimate Pavement Maintenance and Repair Costs for Different Range of Pavement Condition Index. *Transportation Research Record: Journal of the Transportation Research Board*, 1123: 30–39.

Smilowitz, K., & Madanat. S. 2000. Optimal Inspection and Maintenance Policies for Infrastructure Networks. *Computer-Aided Civil and Infrastructure Engineering*, 15(1): 5–13.

Vadakpat, G., Stoffels, S., & Dixon, K. 2000. Road User Cost Models for Network-Level Pavement Management. *Transportation Research Record*, 1699: 49–57.

Efficient Transportation and Pavement Systems – Al-Qadi, Sayed, Alnuaimi & Masad (eds)

Development of pavement Distress Identification Manual for Motorway (M-2) Pakistan

K.M. Khan
Department of Civil Engineering, University of Engineering & Technology, Taxila, Pakistan

ABSTRACT: In the last several years, the volume of vehicles in terms of traffic as well as axle load has increased rapidly. This increase in number of vehicles has reduced the carrying capacity of roads, while the uncontrolled axle loads have adversely affected the quality of riding surface and pavement structure. Same is the case with M-2 Lahore to Islamabad Motorway. (M-2). The Distress Identification Manual for flexible Pavement Performance Studies enables agencies to assess the type and severity of pavement distress accurately and uniformly. This manual provides a method for obtaining consistent, repeatable, and quantifiable information. Also it is a basic tool that highway agencies can adopt easily for use with state-of-art pavement management systems. The Manual covers the analysis and evaluation of survey conducted during 2006 which shows the trend of deterioration of Lahore–Islamabad Motorway Pakistan. The manual also provides quantifiable information regarding extent of repairs.

Keywords: pavement, distress, manual

1 INTRODUCTION

Pavement surface distress is a key performance measure used by asset management systems for maintaining the current conditions of pavement networks and predicting long-term conditions. It is also one of the primary pavement indicators for network level pavement rehabilitation programming and can be used as a diagnostic tool at the project level to gain insight as to the possible causes of pavement deterioration. The pavement condition evaluation manual is designed to indicate the extent and type of surface distresses on a roadway and indicate the rating of severity for the defects. The manual prepared for Islamabad Lahore Motorway (M-2) is based on the survey data obtained from National Highway Authority (NHA). This surface condition survey was conducted during year 2006.

1.1 *Pavement condition survey legend*

1.1.1 *Rutting*

1. *Severity (depth in mm)*
   a. 6–12
   b. 12–25
   c. 25–40
   d. >40
2. *Extent (Length affected in meters)*
   a. <50
   b. 50–100
   c. 100–250
   d. 500–1000

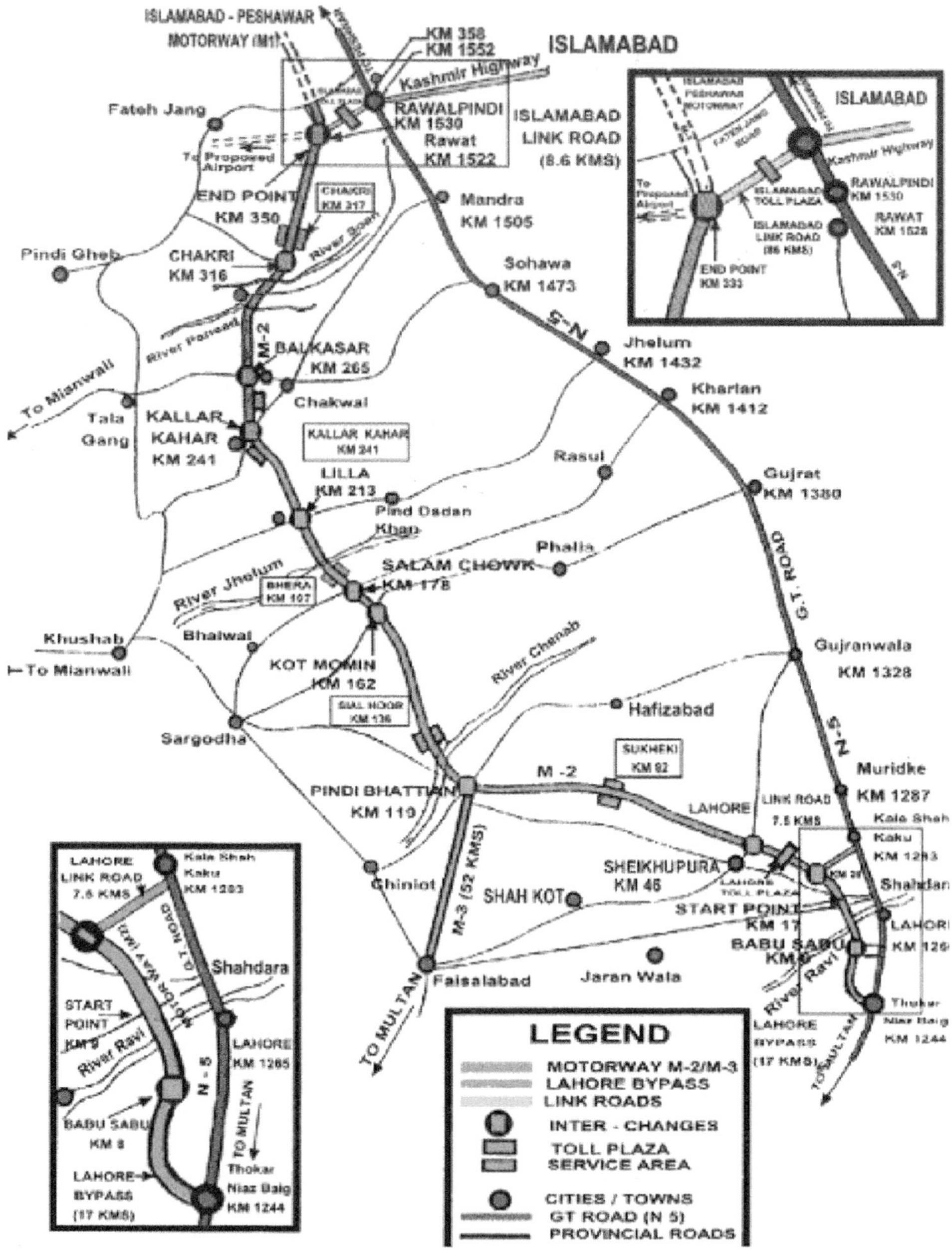

Figure 1. Layout of M-2 Motorway Pakistan.

1.1.2 *Potholes*

1. Severity (Area in 1 km
   a. <20
   b. 20–30
   c. 30–40
   d. >40

2. *Extent (Length affected in meters)*
    a. <2 SM
    b. 2–4 SM
    c. 4–6 SM
    d. 6–8 SM
    e. >8 SM

1.1.3 *Cracking*

1. *Severity (Width in mm)*
    a. <2
    b. 2–5
    c. 5–10
    d. >40
2. *Extent (% Area Effected)*
    a. <5
    b. 5–10
    c. 10–25
    d. 25–50
    e. 50–10

1.1.4 *Reveling*

1. *Severity (depth in mm)*
    a. AC/binder started to wear
    b. Surface texture is lightly rough and pitted
    c. Surface is moderately rough and pitted
    d. Surface texture is heavily rough and pitted
2. *Extent (Length Effected)*
    a. < 50
    b. 5–10
    c. 10–25

2 DATA ANALYSIS

Data collected has been analyzed and charted below:

2.1 *Rutting*

Table 1. Length affected due to rutting.

Affected length	North bound	South bound
a. Total surveyed length	342 Km	342 Km
b. Road width	11 M	11 M
c. Un-rutted length	252 Km	216 Km
d. Length having ruts, <6 mm depth	46 Km	56 Km
e. Length having ruts, 6–12 mm depth	29 Km	42 Km
f. Length having ruts, 12–25 mm depth	11 Km	18 Km
g. Length having ruts, >25 mm depth	4 Km	10 Km

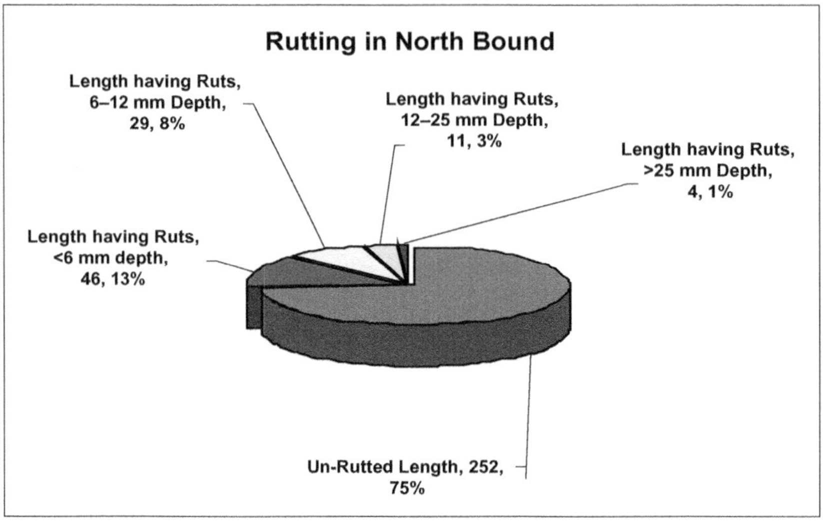

Figure 2. Representation of rutting length and depth (North bound).

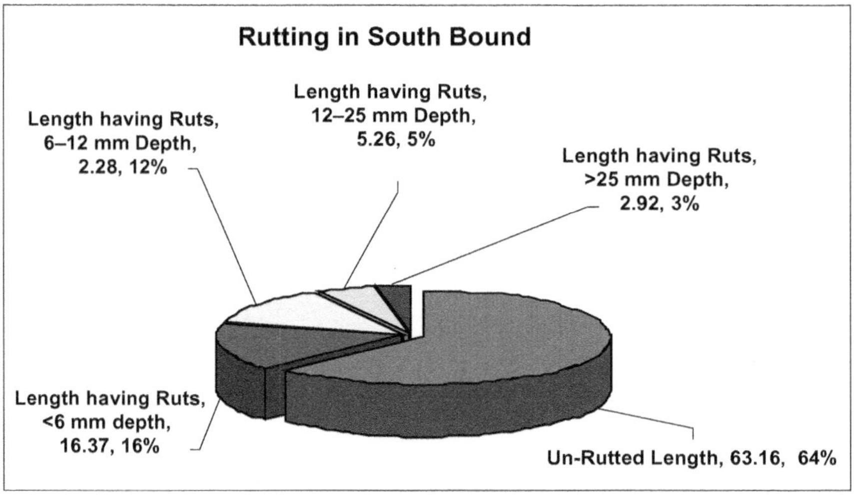

Figure 3. Representation of rutting length and depth (South bound).

2.2 *Cracking (all types)*

Table 2. Length affected due to cracking.

Affected length	North bound	South bound
a. Total surveyed length	342 Km	342 Km
b. Road width	11 M	11 M
c. Un-cracked length	8 Km	22 Km
d. Length having cracks, <2 mm depth	234 Km	76 Km
e. Length having cracks, 2–6 mm depth	77 Km	137 Km
f. Length having cracks, 6–10 mm depth	20 Km	70 Km
g. Length having cracks, >10 mmdepth	3 Km	35 Km

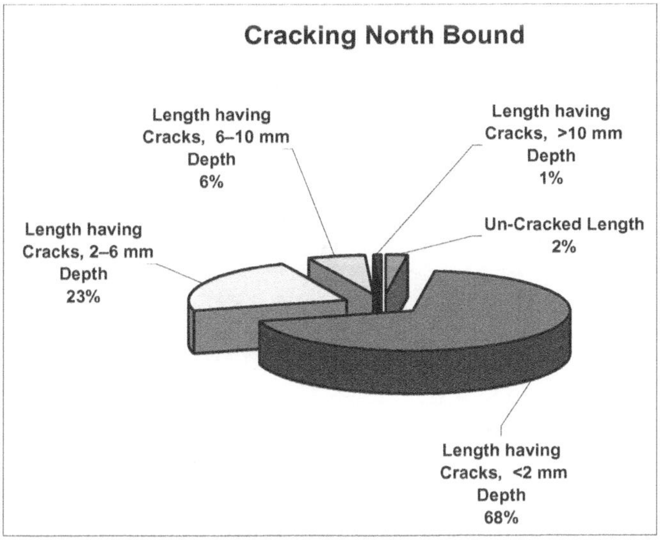

Figure 4. Representation of cracked length and depth (North bound).

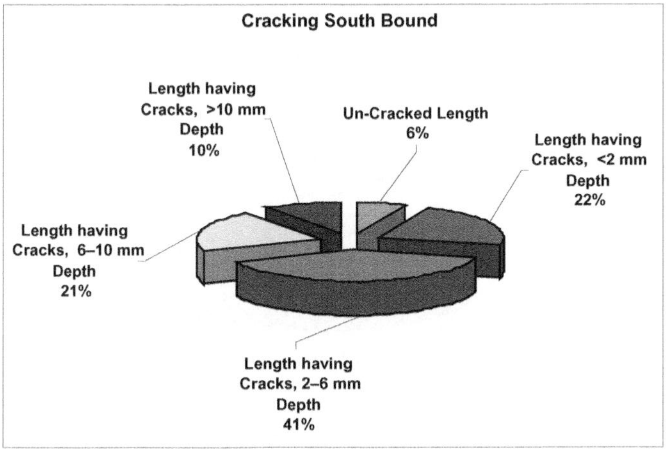

Figure 5. Representation of cracked length and depth (South bound).

2.3 *Potholes*

Data regarding potholes has been tabulated in table 3 below.

Table 3. Length affected due to potholes.

Affected length	North bound	South bound
a. Total surveyed length	342 Km	342 Km
b. Road width	11 M	11 M
c. Length having no potholes	337 Km	341 Km
d. Length having potholes of depth <20 mm	1 Km	1 Km
e. Length having potholes of depth 30–40 mm	3 Km	0 Km
f. Length having potholes of depth >40 mm	1 Km	0 Km

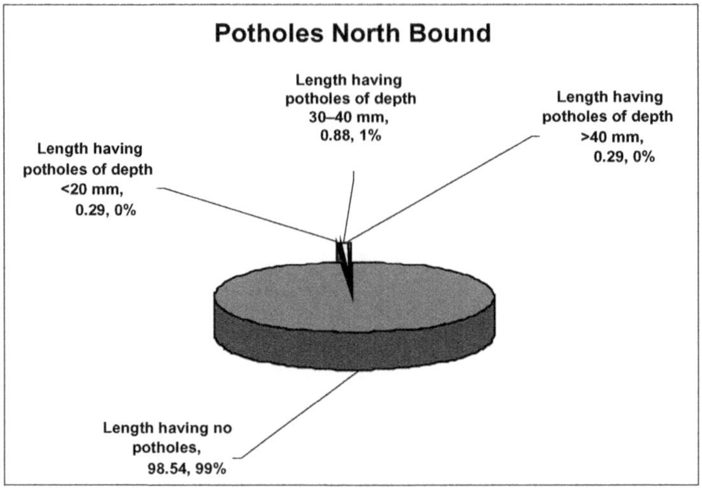

Figure 6. Representation of potholes with depth (North bound).

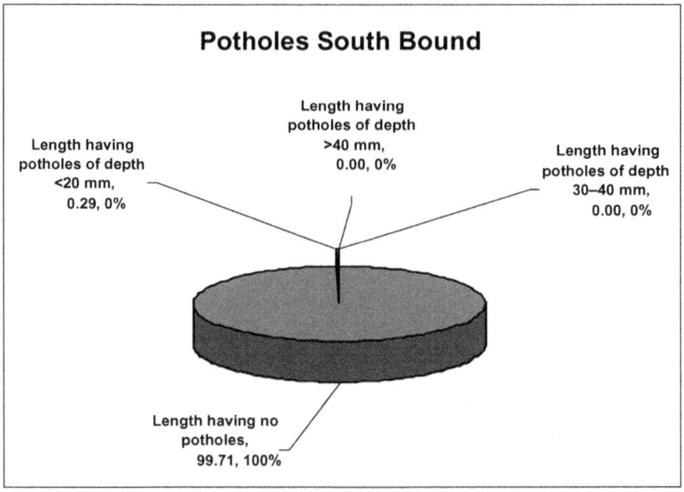

Figure 7. Representation of potholes with depth (South bound).

2.4 *Roughness*

Table 4. Length affected due to roughness.

Affected length	North bound	South bound
a. Total surveyed length	342 Km	342 Km
b. Road Width	11 M	11 M
c. Length having roughness >2000 mm per km	299 Km	308 Km
d. Length having roughness between 2000 to 3000 mm per km	41 Km	32 Km
e. Length having roughness between 3000 to 4000 mm per km	2 Km	2 Km

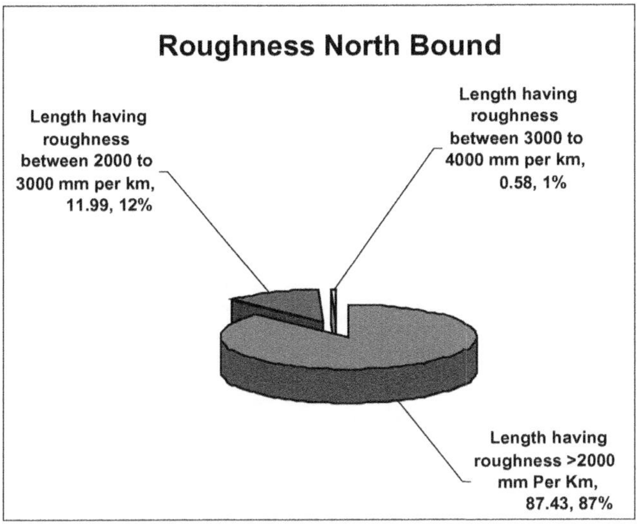

Figure 8. Representation of roughness (North bound).

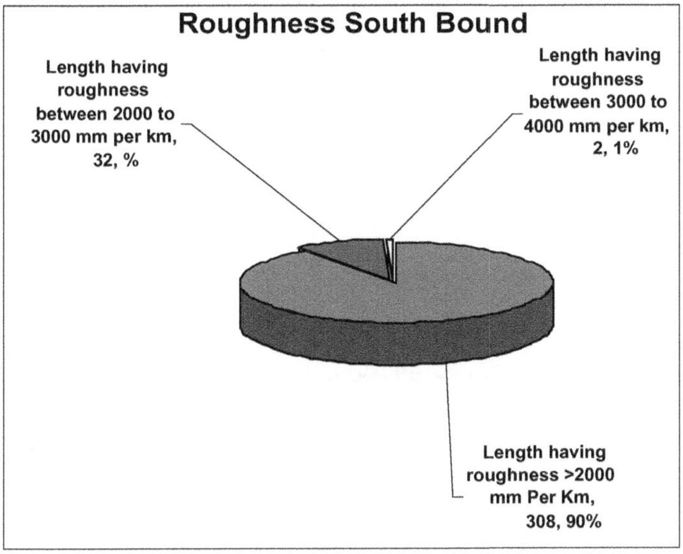

Figure 9. Representation of roughness (South bound).

2.5 *Remaining service life (Rsl)*

Remaining Service Life (RSL) is a measure of the existing pavement condition, as well as anticipated number of years before it will require reconstruction or rehabilitation. After evaluation of M-2, it was determined that the North bound area had a greater remaining service life as compared to the South bound as shown in table 5 and has been explained figures 10 and 11 below.

Table 5. Remaining service life (RSL) of M-2.

Affected length	North bound	South bound
a. Total surveyed length	342 Km	342 Km
b. Road width	11 M	11 M
c. Length having no life	6 km	8 km
d. Length having life > = 1 years	4 km	10 Km
e. Length having life > = 2 years	8 Km	18 Km
f. Length having life > = 3 years	19 Km	32 Km
g. Length having life > = 4 years	30 Km	63 Km
h. Length having life > = 5 years	86 Km	118 Km
j. Length having life > = 6 years	189 Km	93 Km

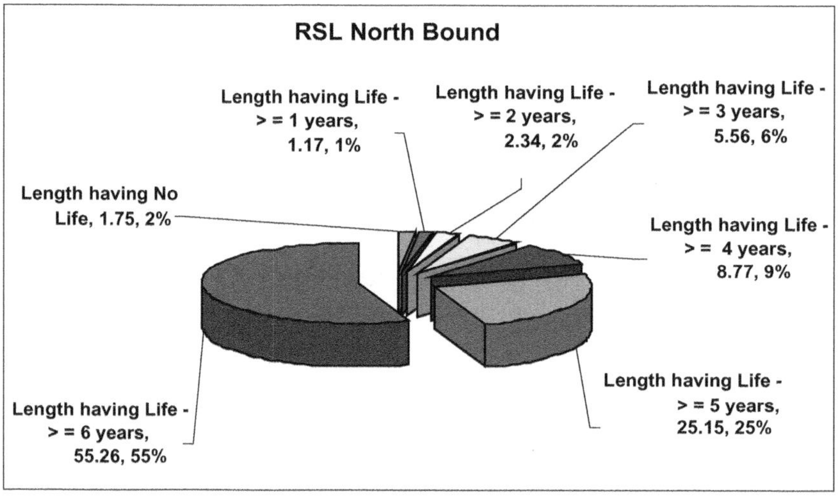

Figure 10. Representation of RSL (North Bound).

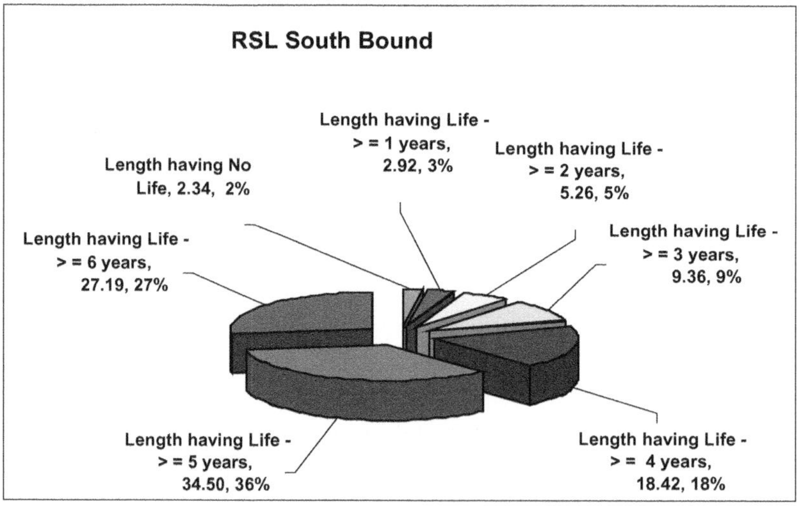

Figure 11. Representation of RSL (South Bound).

3 CONCLUSIONS

Based on the analysis of the data available, it is evident that National Highway Authority Pakistan (NHA) has to plan the distress management in terms of both:

- Roughness control
- Measures to enhance the service life of Lahore – Islamabad Motorway (M2 Section).

4 RECOMMENDATIONS

Keeping in view the data collected & analysis carried out, it is recommended to plan and develop an extensive schedule & methodology for maintenance of all the distresses identified in this manual.

REFERENCES

National Highway Authority (Road & Asset Management Department) Islamabad.
National Highway Authority (Maintenance Section Motorway M-2) Balkasar.
National Highway Authority Pakistan. (www.**nha**.gov.pk)
Federal Highway Authority USA. (www.fhwa.dot.gov)

Research, development, and standardization

Efficient Transportation and Pavement Systems – Al-Qadi, Sayed, Alnuaimi & Masad (eds)
© 2009 Taylor & Francis Group, London, ISBN 978-0-415-48979-9

Academic curriculum and research development in road safety, supported with ICT

Ghazwan Al-Haji, Malin Granath & Kenneth Asp
Linkoping University, Sweden

ABSTRACT: Today, there exists a large body of knowledge and research in different EU universities about issues related to road safety but there is a vast lack of academic knowledge and experiences in traffic safety at many less-developed countries. The paper is based on the SafeIT[1] project, which aims to develop and complement the existing master's curriculum and research in road safety at consortium partner universities in Sweden, Poland, Russia, Jordan and Syria. The developed joint curriculum will benefit from several courses applied currently at EU countries and will be adapted to the problems, needs and characteristics of traffic in each partner country. Additionally, it will be supported with new teaching methods and ICT knowledge and technologies. The project is structured into three phases: current curriculum analysis, development of a new curriculum, and realisation of the new curriculum and testing at Partner University. This paper will provide an overview of the structure and objectives of the project including specific results achieved so far. This paper concludes that a modern curriculum and research in the field of road safety in the targeted universities would also in the long run contribute to a better traffic safety situation in the whole country.

1 INTRODUCTION

The World Health Organisation (WHO, 2004) statistics have shown that more than 1.2 million people are killed in road accidents each year worldwide and an additional 40 million people are estimated injured. The majority of road deaths and injuries occur in Less Developed Countries (LDCs). The annual cost of road accidents is too high and it is estimated to be between one and three percent of a country's Gross National Product (GNP).

Research (e.g. Al-Haji & Asp, 2006b) shows that road safety problem is a key and urgent issue in LDCs in terms of its impact on public health, economic and social development. The efforts made in these countries in order to reduce number of accidents do not often meet the requirements of a modern scientific knowledge of road safety and its recent basic and applied research.

Improving road safety is a major priority for both European and LDCs. Today, there exists a large body of knowledge and research in the different European universities about issues related to road safety but there is a vast lack of knowledge and experiences in this field in many LDCs. Students seldom receive more than an introductory part of road safety during their study at university. Even at postgraduate level, students in transportation engineering learn very little about road safety. Road safety is in these cases often a little part of other course e.g. Traffic Engineering. Additionally, the differences in road safety curricula between the different taught

[1] See http://www.tempus-safeit.eu/ The SafeIT project started on November 2007 and sponsored by the European Commission. The consortium of the project are: Linköping University (Sweden); Gdansk University of Technology (Poland); University of Aleppo (Syria); Al-Baath University (Syria); Ministry of Transport (Syria); Al-Balqa Applied University (Jordan); and the Automobile & Road Technical University- MADI (Russia). The Project management and leadership is Linköping University, Sweden.

courses cause differences in knowledge and understanding not just among teachers but also among students.

In general, road safety is not yet a prioritised academic area at many universities in LDCs and it is not yet given as a separated course. However, there is an increased demand in these countries to reform education and research regarding road safety. One example to develop advanced curriculum at universities in LDCs is the SafeIT project. This project introduces, develops, tests and implements a joint modern master's curriculum in road safety in a LDC with the help of experienced countries within the European Union (EU). The LDC partner universities will be benefiting from several courses currently applied in EU universities. An advanced curriculum in road safety at specific universities should cover all knowledge, research and skills that can positively influencing traffic safety performance in their country.

This paper gives an overview of the structure and objectives of the SafeIT project including specific results and experiences achieved so far from previous project activities.

2 THE SAFEIT PROJECT

2.1 *Overview and consortium*

The SafeIT is a 2 years project sponsored by the European Commission. The consortium of the project are Linköping University (Sweden), Gdansk University of Technology (Poland), University of Aleppo (Syria), Al-Baath University (Syria), the Ministry of Transport (Syria), Al-Balqa Applied University (Jordan) and the Automobile & Road Technical University-MADI (Russia).

According to official statistics from the international databases (Globesafe, 2008; IRF, 2005), in year 2004, there were annually 29 deaths per 10,000 vehicles in Syria, 12.9 deaths per 10,000 in Jordan, 12 deaths per 10,000 in Russia, which were very high figure compared with 1.5 deaths according to the same measures in Sweden and 3.4 deaths in Poland. These figures clearly show that road safety is an important problem and that there are challenges to address in these countries. Therefore, in responding to the challenges and road accident problems in the partner countries, the universities, who produce future engineers, should be prepared for taking leading roles to tackle these problems.

2.2 *Project objectives*

The overall aim of the SafeIT project is to encourage sharing experiences and good practices between consortium partners as means to develop a modern curriculum and guidelines in road safety that are adapted to the problems, needs and characteristics of traffic in each Partner Country.

Within this overall aim the following four specific outcomes shall be achieved during the project implementation:

2.2.1 *Curriculum development*

The Curriculum development aims at developing and integrating new curricula in Road Safety at master and postgraduate level in transportation education programs in Partner Universities. The development of new curricula will be based on already existing road safety courses in the EU partner universities and these courses will be refined and adapted to the Partner's Universities' teaching practices and country traffic situation. According to Al-Haji & Asp (2006a) and Al-Haji & Lindskog (2005) western standards are developed to fit the needs of highly developed and motorised countries, these principles are often too costly or require an excessive maintenance and they do not take into account the high level of vulnerable road users VRU's flows, which characterise many LDCs countries. The developed curriculum will enhance understanding of the topic in maintaining economic and social development in the country (Komarov, 1968). Additionally, the multinational curriculum promotes cross-cultural interaction between staff, individual experts and students from different countries (OECD, 2002).

2.2.2 *Information and Communication Technologies (ICT)*

In order to improve the road safety curriculum at the Partner universities the European partners will provide educational web-based tools in order to allow students of the project to browse information, learn, conduct assignments and communicate over the web. An integrated and blended set of course materials and electronic library with the support of ICT will be developed and implemented to be suited the purpose of the Partner universities. As the advantage of integrating ICT into the curriculum is the flexibility in time, place and pace for it is important that the course materials (lectures, demonstrations and scenarios) will be made available over the Internet (Al-Haji *et al*, 2006).

A clear advantage in this project is that Linkoping University in Sweden has great previous experience in working with and integrating web based tools for communication and information exchange in their courses and research (e.g. the ASNet[2] project and TechTrans project[3]). This work was proven to be efficient for the development of higher education in general and networking between the partner countries (Al-Haji & Lindskog, 2005; Al-Haji & Asp, 2004, 2007).

2.2.3 *The pedagogical approach*

The pedagogical approach to be used is crucial in order to achieve good results of the education. New teaching methods, for example problem-based learning, are therefore provided and applied in the new curriculum of the project. Education in many LDCs is today characterised by traditional teaching methods such as rote learning which often fail to offer good results of training, practices and skills to students. However, road safety is such a complex area that it demands more than memorisation of facts. Proposed teaching method (PBL) encourages students to seek and share knowledge and practices with each other under the guidance of course tutors. It simply stimulates students' creative thinking (Elis, 1998).

2.2.4 *Improving the linkage between practice and research in road safety*

In many EU countries, road safety research is usually carried out between the universities and different road safety stakeholders. Pedagogical research (EC, 2995) show that a strong connection between real life situations/cases and studies lead to better results, that is, in other words, a strong connection between theory and practise will benefit the whole society. Therefore, educators and professionals must come together to create a shared vision of a new curriculum that will provide students with skills and real-world applications. This link will also broaden and deepen student's knowledge and understanding of the relationship between curriculum and real world applications. For instance, the Public Establishment for Road Communications (PERC) and the National Road Safety Council (NRSC) at the Ministry of transport in Syria can support the project implementation in Syria with road accident data and applications. This gives opportunities to create teaching PBL scenarios close to real traffic situations in Syria. The road safety situation and traffic nature in Syria can be seen in a similar way to other neighbouring countries in the region (Al-Haji & Asp, 2002).

3 METHODOLOGY

To correspond to the main aims mentioned above, the project is structured in three phases:

1. *Current curriculum analysis*:

The partner institutions will gather all necessary information and prepare a comprehensive survey of the present situation and analysis of existing curricula at the partner universities and practical experiences in road safety in their country.

---

[2] The ASEAN Road Safety Network (ASNet) is an internet based regional network designed to strengthen the regional cooperation between decision makers/professional in ASEAN countries.

[3] Techtrans project is designed to develope an internet based network and applications in Road Safety to Russian Universities.

2. *Development of a new curriculum integrating ICT and new teaching methods:*
Based on the analysis of the current state the main issues for the new curriculum will be defined. A course book (Traffic Safety Management) with an electronic library will be developed in two languages (English and Arabic) by the institutions and individual experts who have been the participants of the project. The editors of the book are Linkoping University- Sweden.

3. *Realisation of the new curriculum:*
At beneficiary universities in target countries, the new curriculum will be realised and accredited from the university board and the Ministry of Higher Education (MoHE).

In order to achieve the objectives and carry out the three phases several measures are taken. First, teachers, staff and students at the different partner universities are addressed as the particular target groups. Extensive training on information and communication technologies in combination with PBL will be carried out. Second, workshops and management meeting will be carried out within the consortium. Third, mobility of students, teachers and staff will be undertaken between partner institutions.

4 THE OUTCOMES ACHIEVED SO FAR FROM PREVIOUS PROJECT ACTIVITIES

The project is still in the first phase, i.e. current curriculum analysis. Hence, the results presented below are based on the performances and achievements during this particular time. To begin with;

- A checklist for analysing the existing curriculum in road safety at partner university has been developed and tested.
- A survey among students, who will be potentially taking the developed new course in road safety, has been made so that knowing their expectations.
- Project partners have submitted several teaching, course materials and reports related to road traffic safety. All material is under revision by EU partners. In principle the structure of the new curriculum will take nearly the following structure:
  - o Road traffic safety in a global perspective and partner countries.
  - o Accident theories and modelling.
  - o Statistical methods for road safety.
  - o Road user behaviour.
  - o Road safety problems and countermeasures.
  - o Road safety strategies and national action plan.
- A project ICT platform (closed with login) has been established, written in English and Arabic, and includes the new curriculum, knowledge database and communication tools.
- Dissemination outsides the consortium has been made to other stakeholders, universities and authorities. The outcomes of the project will be also disseminated through different publications, brochures, posters, articles in special journals or university publications, in international conferences, Websites, etc.

5 SUMMARY AND CONCLUSIONS

In brief the problems in LDCs that are stated in the SafeIT project are:

1. There is no clear and advanced academic curriculum in road safety at universities in LDCs.
2. There is no curriculum in road safety adapted to the nature, problems, needs and characteristics of traffic in the participant country of the project.
3. There is no effective link between the ministries and universities, between practice and research.

4. There are no teaching methods encouraging students to search knowledge independently and stimulating creative thinking
5. There is no efficient use of ICT tools.

An academic curriculum that corresponds to the above requirements in combination with delivery methods strongly linking practice with theory will form a good basis for future engineers to understand the complexity of road safety problems and development.

This paper can conclude that a modern separated curriculum in road safety is necessary at universities in LDCs and it would also in the long run contribute to a better traffic safety situation in the whole country.

The expected results of the project are also prepared for wider dissemination to other universities nationally and regionally.

ACKNOWLEDGMENT

The authors would like to express their appreciation to the partners and coordinators of the SafeIT project. We also acknowledge the grant and support of the European Commission through the project.

REFERENCES

Al-Haji, G. and Asp, K., (2007). *Traffic accidents reduction strategy, best practices from European States.* The International Conference in Traffic & Its Contemporary Issues, Kuwait, May 12–14, 2007.

Al-Haji, G., Asp, K., Lindskog, P., Eriksson, M., Wismadi A. and Silyanov, V., (2006). *Distance education and collaborative virtual environments—examples using ICT in global higher education*, EAIE conference (European Association for International Education), 13–16 Sep, 2006, Basel, Switzerland.

Al-Haji, G. and Asp, K., (2006a). *New Tools for Assessing and Monitoring National Road Safety Development.* The 2nd International Road Safety Conference, Dubai—United Arab Emirates, pp. 31.

Al-Haji, G. and Asp, K., (2006b). *The Evolution of International Road Safety Benchmarking Models: Towards a Road Safety Development Index (RSDI).* The International Journal Science & Technology for Highways, Vol.3, pp. 3–9.

Al-Haji, G. and Lindskog, P., (2005). *Road Safety in Southeast Asia- Factors Affecting Motorcycle Safety.* The ICTCT international workshop on measures to assess risk in traffic, Campo Grande, Brazil.

Al-Haji, G. and Asp, K., (2004). *Applying Road Safety Development Index (RSDI) for Big Cities.* The 6th International Conference: Traffic Safety Management for Big Cities, St. Petersburg, Russia, pp. 218–222.

Al-Haji, G., (2003). *Developing Road Safety Development Index*, The ICTCT international conference on improving safety by linking research with safety policies and management, Soesterberg, the Netherlands.

Al-Haji, G. and Asp, K., (2002). *Road Safety Perspective in Arab Countries- Comparative Study and Analysis of Progress*, SORIC' 02 Conference (Safety on Roads), Bahrain, pp. 116–121.

Ellis A. (chair) (1998). *Resources, Tools, and Techniques for Problem Based Learning in Computing.* ITiCSE, Dublin, Ireland.

European Commission (1995). *The European Report: European Pilot Project for Evaluating Quality in Higher Education.*

Globsafe database, (2008). A *database for road safety analysis.* Linkoping University, Sweden. Some parts of this database are available on-line: http://www.globesafe.org/ (Last visited: 2008–01–28).

IRF 'International Road Federation', (2005). *World Road Statistics*, Geneva, Switzerland, and Washington DC, United States.

Jacobs, B. and F. van der Ploeg (2006). *Guide to reform of higher education: A European perspective, Economic Policy*, 47, 535–592.

Komarov, V.E. (1968). *The relationship between economic development and the development of education.* In, UNESCO, Readings in the Economics of Education. Paris: UNESCO, pp. 85–92.

OECD publications- (2002), *Education Policy Analysis: The growth of cross-border education*, Centre for Educational Research and Innovation, vol. no.9, pp. 112–145, 2002.

WHO, World Health Organisation, (2004). *World Report on Road Traffic Injury Prevention*, Geneva.

Efficient Transportation and Pavement Systems – Al-Qadi, Sayed, Alnuaimi & Masad (eds)
© 2009 Taylor & Francis Group, London, ISBN 978-0-415-48979-9

Towards standardizing construction specifications and design criteria of agricultural roads in the GCC countries

Abdulaziz Al-Kulaib
Ministry of Public Works, Kuwait

Fadel Al-Ajmi
Maintenance Division, Ministry of Public Works, Kuwait

Sayed Metwali
Maintenance Division, Ministry of Public Works, Kuwait

Ahmad H. Aljassar
Civil Engineering Department, Kuwait University, Ministry of Public Works, Kuwait

ABSTRACT: The Gulf Cooperation Council (GCC) countries are among others that construct roads to serve agricultural areas. These roads are of simplified designs suitable for the expected low volumes of traffic on such roads. These low-volume roads are usually referred to by different names such as "Agricultural Roads" or "Temporary Roads". One main concern when constructing these roads is minimizing costs. The Ministry of Public Works has been building such type of roads in Kuwait for many years, and therefore has accumulated considerable experience in this field. As a gesture and a step towards standardizing specifications with other GCC countries, this paper presents a summary of the design criteria in addition to construction specifications for these roads in Kuwait.

Keywords: Agricultural roads, temporary roads, low-volume roads, LVR, farm roads, pavement design, Kuwait.

1 INTRODUCTION

In order to sustain agricultural development in Kuwait, the Ministry of Public Works (MPW) has been constructing high number of farm-to-market type roads for many years. Locally these roads are known as agricultural roads (AR). The main purpose of these roads is to provide all-weather accessibility to farming areas in the North and the South of the country. Initially, traffic was light, consisting mainly of light pickup trucks and occasional heavy trucks. Such traffic did not require thick pavements. Therefore, design engineers specified these pavements consisting of 5 cm of an asphalt wearing course on top of 15 cm of prepared subgrade. However, due to rapid developments in agricultural areas in recent years, large numbers of heavy trucks started to commute on a daily basis between farms and markets. This led to rapid deterioration of many of these roads. Therefore, design engineers recognized the need to use stronger pavements to sustain the increased traffic loads.

In recent years, MPW has tendered six contracts for the construction and maintenance of agricultural roads in the Northern and Southern agricultural areas, as well as border areas. The total value of these contracts is KD 12 million (1 Kuwaiti Dinar, KD ≈ 13 Qatar Riyals, QAR).

At present, construction cost of AR is about KD 30,000 per kilometer.

2 GEOMETRIC CHARACTERISTICS OF AR

2.1 *Cross section elements*

2.1.1 *Lane widths*

Figure 1 shows a typical cross-section of an AR. It can be noted that the lane width is 4 m. Although AASHTO standards (AASHTO, 2001) consider that 3.6 m-wide lanes are adequate for two-lane, two-way rural highways carrying commercial vehicles, wider lanes are used in Kuwait in order to increase safety and comfort of driving. The literature review revealed that lane width is a major contributor to safety on two-lane low-volume roads (Zeqear, 1994).

2.1.2 *Cross slope*

In order to facilitate drainage, the traveled way is crowned. The cross slope is usually taken at 2%.

2.1.3 *Shoulders*

ARs are provided with graded earth shoulders 3 m in width. The shoulder is sloped at 3% to drain away from the traveled way.

2.1.4 *Side slopes*

Because the general topography in Kuwait is flat, lower embankments are used for LVR. This allows designers to provide flat side slopes to enhance safety at a reasonable cost. Usually 6H:1V slide slopes are used for ARs in Kuwait. Based on experience, such flat side slopes provide a reasonable opportunity for recovering of an out-of-control vehicle.

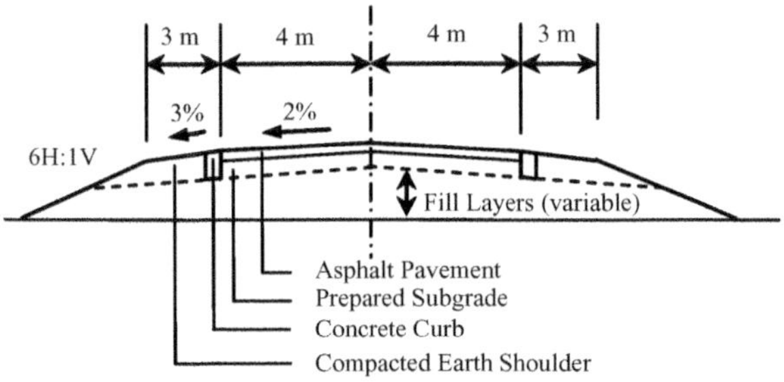

Figure 1. Typical cross section of a low-volume road.

Figure 2. An AR in Kuwait.

2.1.5 *Edge protection*

To provide lateral support to the paved traveled way and protect pavement edge, a sunken cast in-situ concrete curb is provided as shown in Figure 1.

2.2 *Road profile*

As previously mentioned, the terrain in farming areas in Kuwait is generally flat. Consequently, one or two layers of fill are used to construct the required embankment. Hence, the road profile follows the natural terrain creating an aesthetically pleasing appearance of the road. This type of design is known as "touch-grade". Such design reduces construction cost and provides improved safety (see Figure 2).

3 DRAINAGE

No drainage facilities are needed for LVR in Kuwait. Surface runoff is drained through the crowned pavement and down the side slopes to the roadside. Usually, during winter vegetation grows along the roadside which helps to protect the shoulder and side slopes against erosion.

4 EARTHWORKS

4.1 *Material properties*

4.1.1 *Soils*

In order to sustain the heavy loads using LVR, high quality materials are used in the construction. Soils used for fill and subgrade must be A-2-4 or better (AI, 1978). In addition, the maximum dry density for subgrade is set at 1922 kg/m^3 as a minimum when compacted to 100% of AASHTO T-180. Subgrade material must have a minimum CBR of 15 when compacted to 95% of AASHTO T-180 and a maximum swell of 1%. Material used in fill is required to have a density not less than 1840 kg/m^3.

MPW specifications (MPW 95) require that the thickness of the subgrade layer should not be less than 30 cm constructed in two layers each not more than 15 cm of compacted thickness.

Native soils in the Northern region of Kuwait are generally good and meet the requirements for road construction. In the Southern region, however, soils tend to be sandy and at some locations road alignment may pass through *sabkha* areas (*sabkha* is a local name in the Gulf region of a soil that is essentially a salt-contaminated deposit). Table 1 shows an example of the properties of *sabkha* soils. It can be noted that such soils exhibit very low CBR, high swell, high plasticity index and high liquid limit.

A range of options exists for improving the structural capacity of the roadway in areas of poor subgrades. These commonly include: (1) adding material of higher strength and quality, such as reclaimed asphalt pavement (RAP) (Aljassar et al., 2005, Scholz, 1991), (2) removing the poor soil and replacing it with a high quality granular material, and (3) bridging over the soft soil (such as *sabkha*) with a thick layer of large pieces of rock or salvaged concrete.

The selection of the most appropriate improvement option depends mainly on its cost-effectiveness as compared to the other available options as well as existing soil conditions.

4.1.2 *Water*

Use of fresh, sea, and brackish water is allowed for AR earthworks.

5 PAVEMENTS

5.1 *Design*

Early pavement designs of AR required an asphalt wearing course of 5 cm in thickness on top of one layer of subgrade as shown in Figure 1. However, with the increase in heavy commercial

Table 1. Typical properties of *sabkha* soil.

Sieve Size	Percent Passing by weight
3/4"	100
1/2"	99
3/8"	98
# 4	95
# 10	90
# 30	77
# 40	68
# 50	55
# 100	37
# 200	30
Liquid limit	42
Plasticity index	13
Soaked CBR	2
Swell (%)	5
Optimum moisture content (%)	16.5
Maximum dry density (kg/m^3)	1780
Chlorides (ppm)	757
Sulphates (ppm)	533
Organic impurities (%)	1.49

truck traffic thin pavements proved inadequate to sustain such loads and started to fail prematurely.

Pavement engineers started to use thicker pavements consisting of two layers: (1) a bituminous base course 8–10 cm in thickness, and (2) a bituminous wearing course 4–5 cm thick. This led to a considerable improvement in pavement performance.

A sunken concrete curb, usually 20 cm in width and 20 cm in depth, is used to protect pavement edges as shown in Figure 3. The presence of this curb, however, creates difficulties during maintenance when an overlay is needed. A longitudinal reflection crack develops at the pavement-curb line soon after the overlay. Although the construction of the concrete curb is costly and prolongs construction period, the curb provides excellent protection to pavement edge and reduces maintenance costs in the long run. Care should be exercised during the compaction of the asphalt pavement to avoid damaging the curb as the steel rollers have to operate very close to it to ensure proper compaction of pavement edge. Other alternatives to this concrete curb are available. One of these alternatives is to provide asphalted shoulders to provide lateral support to pavement and protect its edge.

5.2 *Materials properties*

The following paragraphs present the salient requirements for materials used in the construction of bituminous pavements for AR in Kuwait.

5.2.1 *Aggregates*

Table 2 provides a summary of main required aggregate properties.

5.2.2 *Bituminous mixes*

Table 3 shows the composition of bituminous mixes and mix requirements used for the construction of AR in Kuwait.

Figure 3. Construction of sunken concrete curb.

Table 2. Summary of main properties required for aggregates in bituminous mixes.

Material type	Size	Material description	Main properties
Fine Aggregate	Passing # 8 sieve	Natural Sand and crushed sand	Sand equivalent: 30 min. for crushed sand 45 min. for natural sand—soundness loss: 15% max. (AASHTO T-104)
Coarse Aggregate	Retained on # 8 sieve	Crushed natural stone and gravel	Cubic in shape. Degree of crushing 80% min. Has at least one fractured face. Flakiness: 30% max. (BS 812). Soft and friable particles: 3% max. for base course, 2% max. for binder and surface courses. Sodium sulfate soundness loss: 12% max. Abrasion loss: 40% max for surface course, 50% max for base and binder courses (AASHTO T-96).
Mineral Filler	100% passing #30 Min. 75% passing #200	Limestone, cement, or hydrated lime	Free Moisture 30% max. Lime Index: 50% min. Unhydrated lime: 9% max. Carbon Dioxide: 8% max.

5.3 *Pavement construction*

5.3.1 *Prime coat*

Prior to paving operations a tack coat is applied by means of an asphalt distributor at a rate of 0.45–1.5 kg/m^2 of a medium-curing cut back (MC-70). Application temperature must be within 60 °C to 85 °C. Ambient air temperature must be at least 13 °C. Priming is never allowed when there is free water present on the surface.

Table 3. Composition and mix requirements of bituminous mixes for LVR in Kuwait.

Sieve	% passing by weight		Acceptable tolerances
	Base course	Surface course	
1 1/2"	100	–	±5%
1"	72–100	–	
3/4"	60–89	100	
1/2"	46–76	66–95	
3/8"	40–67	54–88	
# 4	30–54	37–70	
# 8	22–43	26–52	±4%
# 16	15–36	18–40	
# 30	10–28	13–30	
# 50	6–22	8–23	
# 100	4–14	6–16	
# 200	2–8	4–10	
Bitumen (60/70)% by weight of total aggregate	3.5–5	4.5–6.5	±0.4%
Marshall Stability	1600 kg min.	1800 kg min.	
Index of Retained Strength (AASHTO T-165),	90% min.		
OR			
Net Retained Strength	14 kg/cm^2 min.		

5.3.2 *Asphalt concrete*

For the construction of AR pavements in Kuwait, plant-mixed asphalt concrete mixes are spread using mechanical pavers equipped with electronic screeds, capable of maintaining a constant level of material along the full length of the screed, as well as automatic grade and slope controls.

Production and spreading of asphalt concrete mixes is not permitted when ambient temperature is less than 8 °C nor during rain, fog or sand-storms. When emptied from the mixer, MPW specifications require that mix temperature does not exceed 171 °C. When delivered to the paver, mix temperature shall be between 120 °C and 160 °C.

Compaction is done in 3 stages: (1) breakdown rolling using steel-wheeled rollers, (2) intermediate rolling using pneumatic-tire rollers, and (3) finish rolling with steel-wheeled rollers. MPW specifications require that mix temperature not to be allowed to drop below 107 °C prior to breakdown rolling.

5.3.3 *Tack coat*

A slow-setting cationic emulsified asphalt (CSS-1h) diluted with water (1:1) is used as a tack coat. Application rate is between 0.15–0.75 kg/m^2 and application temperature should be between 10 °C–60 °C. Ambient temperature should not be less than 13 °C at the time of application.

6 CONCLUSION

ARs in Kuwait have become an important element in the national highway system. They are mainly used in the farming areas to provide farmers with an all-weather access to the markets. This paper presented the special features of designs of AR in Kuwait. The paper also presented the material properties required for AR. The traditional thin pavement design (5 cm wearing course over prepared subgrade) showed satisfactory performance in the past. However, with recent increases in traffic loads using AR, thin pavements started to show rapid deterioration. Pavement designers responded with new designs adopting a thicker pavement structure which is showing satisfactory performance.

ACKNOWLEDGEMENT

The authors would like to acknowledge with great appreciation the help and support of the following people: Engineer Ayedh N. Ghoneim, Director, Engineer Ayed Habbas, Supervising Engineer, Faisal Al-Ghanayem, Supervising Engineer, all from the Operations Department, MD, MPW for providing valuable information to this study.

REFERENCES

Aljassar, A.H., Metwali, S. 2005, "Using Reclaimed Asphalt Pavements to Improve Subgrade Soil Properties in Kuwait", 84th Annual Meeting of Transportation Research Board (TRB), Washington, D.C., USA.

American Association of State Highway and Transportation Officials (AASHTO), 2001, "A Policy on Geometric Design of Highways and Streets", USA.

Asphalt Institute (AI), 1978, "Soils Manual for the Design of Asphalt Pavement Structures", Manual Series No. 10 (MS-10), 2nd Edition, College Park, Maryland, USA.

Ministry of Public Works (MPW), 1995, "General Specifications for Kuwait Motorway/Expressway System", Kuwait.

Scholz, T., Hicks, R., Rogge, D., Allen, D. 1991, "Use of Cold In-Place Recycling on Low-Volume Roads", TRR No. 1291, Volume 2, TRB, Fifth International Conference on Low-Volume Roads.

Zeqear, C.V., Stewart, R., Council, F. 1994, "Roadway Widths for Low-Traffic-Volume Roads", NCHRP Report 362, TRB, National Research Council, Washington, D.C., USA.

Efficient Transportation and Pavement Systems – Al-Qadi, Sayed, Alnuaimi & Masad (eds)
© 2009 Taylor & Francis Group, London, ISBN 978-0-415-48979-9

Advancements in research and development

Amir Omar Abd El Halim
Stantec Consulting, Waterloo, Ontario, Canada

Abd El Halim Omar Abd El Halim
Carleton University, Ottawa, Ontario, Canada

Said M. Easa
Ryerson University, Toronto, Ontario, Canada

ABSTRACT: The past 100 years have witnessed several innovative and creative improvements and developments in the field of highways and pavements. Improved safety, efficient operations and high quality roads have been the mark of the past Century. However, in spite of investing time and resources to build paved roads that will meet their design objectives, early failures and premature deteriorated surfaces have been commonly observed worldwide. Several major research initiatives such as USA SHRP were lunched to address old problems that are still unsolved today such as reflection cracking, rutting, stripping and other similar pavement problems. Unfortunately, after more than 10 years since the start of the SHRP initiative same problems are still observed on many highways in the world today.

One of the main deficiencies in many of the current research approaches is their assumption that "*newly constructed asphalt pavements are structurally sound*". Subsequently, all attempts to treat and correct any problems or failures observed on the roads are related to causes and factors that may exist only after completion of the construction of the road. In other words, the solutions are mainly materials based and therefore modifications of the designs of the asphalt mixes as well as selection of different materials have been the main and only available course of action. Clearly, this approach neglects the fact that the construction process contributes to the main assumption of the current analytical approach. If the construction of asphalt pavements is shown to be flawed then the stated assumption may be questionable.

This paper gives a brief background of historical problems associated with asphalt pavements; discusses its traditional approaches and remedies. Also, a new innovative approach is presented which focuses on the interaction between the newly laid soft asphalt layer and the stiff steel drum during the compaction of the asphalt mat. The new approach shows that defects are created during compaction which causes the premature deterioration and early failure of the newly constructed roads. A new roller (AMIR) has been developed and tested in several countries including Egypt, Canada, Sweden and Australia. The results of the field tests and evaluation of the implementation of this new technology is presented and discussed.

1 INTRODUCTION

Over the past 100 years, the asphalt industry has become a multi billion-dollar business and has achieved significant technological advances. However, it has also been experiencing a number of problems. One such major problem is a perception, and arguably a reality, that asphalt pavements are not being designed and constructed to be sufficiently long lasting and cost effective. At the same time, there is a growing backlog of needs due to road deterioration. This deterioration includes such problems as early cracking, rutting, moisture damage, and roughness.

The result is that at the turn of the century, despite major efforts, better economic and technical solutions for these problems have not been achieved. On a positive note, however, the pressure to address these problems and develop better solutions has led to a number of key initiatives that are shaping the asphalt industry today. These initiatives include the Strategic Highway Research Program (SHRP), which deals in part with the characterization of asphalt materials and their performance under a wide range of in-service conditions, and the adoption of gyratory compaction as the main laboratory device for designing asphalt mixes. Also the use of advanced materials such as non-metallic polymeric grid and/or polymer modified asphalt binders to reinforce asphalt layers and/or improve their tensile and shear strengths and the introduction of end-result specifications and even long-term performance based specifications have contributed to an improvement in pavement performance.

These and other initiatives should have a significant impact on the performance of asphalt pavements both in the short and long term. There is no doubt that most, if not all have significant technical and economic potential. However, without addressing the initial cause of asphalt pavement deterioration, the impact of these initiatives and advancement is expected to be significantly reduced.

Current compaction equipment and techniques date back to the early twentieth century and have not undergone any significant change or improvement. This is despite the fact that compaction has been recognized by many experts as one of the most important factors affecting asphalt pavement performance. Traditionally, compaction problems such as construction cracks, also known as "checking", were attributed to deficiencies in base stability, operator error, or the asphalt mix itself. These cracks were either assumed to be "probably more unsightly than physically detrimental" (Gellar 1992), or assumed to be corrected with rubber-tired rollers.

The need for adequate asphalt compaction has been recognized since the beginning of the asphalt construction. In 1972, the chief Engineer of the Asphalt Institute made the following statement at the meeting of the Association of the Asphalt Paving Technologists (Marker 1972): "*The compaction and densification of asphalt mixtures are the most important construction operations with regard to the ultimate performance of the completed pavement, regardless of the thickness of the course being placed.*" It has been conclusively proven that construction cracks are caused by conventional cylindrical steel drum rollers and that such cracks are detrimental to the long-term pavement performance by reducing fatigue life by more than 50 percent (Abd El Halim 1993, 1994). This research work further concluded that pneumatic-rubber tired rollers do not eliminate the cracks. This is further illustrated by the fact that as more and more SuperPave mixtures go to the field, compaction problems are occurring (Brown 1998). It has been reported that the SuperPave mixes are either "too stiff" or that "tender zones" are occurring.

The overall purpose of this paper is to present an overview of the crack initiation due to the traditional compaction practices, the effect of construction cracks on pavement performance. The paper also highlights the success of the new AMIR compaction technology in improving the compaction process of asphalt concrete pavements.

2 CRACKING IN ASPHALT CONCRETE SURFACES

The most widely observed cracking in asphalt concrete pavements are fatigue cracking, thermal cracking, and what has been traditionally referred to as reflective cracking. Fatigue cracking is attributed to repeated loading due to traffic. Fatigue has been defined in pavements as the phenomenon of fracture under repeated fluctuating traffic loading. Himeno and Watnabe (1987) introduced a new criterion of fatigue failure, which states "*Fatigue failure can initiate at the top of the mix slab, when the mix stiffness modulus is low.*" Low stiffness could result in poor compaction and inadequate methods of compaction.

Low temperature cracking in asphalt is generally induced by a drop in the temperature. Basic mechanism of low temperature cracking has been investigated by Christison (1972). Under decreasing temperature, a micro-crack may develop at the pavement surface and then propagate through the full depth of the asphalt layer with successive thermal cycles. Although the excessively low temperature can cause the initiation of micro-cracks, the presence of these cracks due to other reasons, such as compaction, can significantly accelerate the full development of these cracks.

Reflective cracking have been traditionally thought to develop at the bottom of the asphalt concrete mat at the same location of the previously existing cracks in the under layers and then these cracks propagate to the surface. However, recent research has shown that this type of cracking is initiated at the surface and then propagates to the bottom of the mat near the existing cracks.

Based on the previous discussion, it can be concluded that the presence of surface cracking can significantly affect the life of the asphalt concrete mat. Although the presence of surface cracks developed during construction is not the only reason for the development of the different types of cracking in asphalt pavements, these surface cracks serve as a catalyst to accelerate of the development of cracking and ultimately the failure of the pavements.

3 MECHANICS OF ASPHALT COMPACTION

The mechanics of compaction involves three factors: compressive force of the roller, the force within the mix that resists the force of the roller, and the supporting forces provided by the stable surface beneath the asphalt concrete mat. For compaction to occur, the compressive force of the roller comes from either the weight of the roller or the combination of the weight and the dynamic energy provided by the rollers.

The supporting forces in the subgrade result from the subgrade stability. Firmness and resisting forces are the result of the inter-particle friction between the aggregate particles, in addition to the viscosity of the asphalt binder. When the mix density and the temperature reach a point where the resisting force equal the compressive force of the rollers and the resistant forces of the subgrade, equilibrium is reached and the compaction process is complete. In other words, compaction of an asphalt mix requires achieving a permanent deformation in the asphalt layer thickness. This requires displacing the aggregate particles in relation to each other to achieve minimal voids, in addition to decreasing the volume air, moisture, and bitumen vapour in the mix. Decreasing vapour volume can be achieved through cooling the vapour, thus decreasing its volume and increasing the pressure through rolling.

4 CURRENT PRACTICES AND CRACK DEVELOPMENT

The construction of asphalt pavements is traditionally carried out by placing the hot mix asphalt over a base course or an existing road surface, and then compaction is carried out with one or more of three different types of compactors. The first operation in the compaction procedure uses heavy vibratory steel rollers to obtain the desired density. Smoothing out the surface is accomplished with a multi wheeled rubber roller followed by a light steel roller. The finished product is assumed to be structurally sound and free of defects. The addition of vibratory units to the heavy steel rollers has been considered one of the major achievements in the compaction industry in the last 50 years.

However, the problem of compaction still consists of an asphalt layer to be compacted and a roller to induce the compaction energy to the asphalt mix, which has been placed. As a result, the interface between the surface of the compacted asphalt mix and the surface of the compacting device is a key element in the quality of the compacted product.

For a typical steel roller, the contact between the asphalt mix and the roller drum is a rectangular area with a length of 7–15 cm and a width equal to the width of the drum. Although this area is supposedly rectangular in shape, it takes the shape of the cylindrical drum reflecting the shape of the roller drum. Due to this small curved contact area, the applied force has both radial

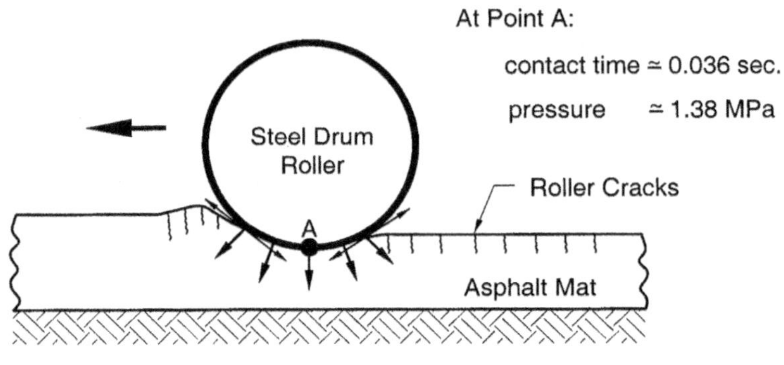

Figure 1. Schematic of conventional steel drum rolling with very short contact time.

and tangential components as shown in Figure 1. At a normal rolling speed of 10 km/hr, the contact between the roller and the mix is about 0.036 seconds. The small contact area and load time combine to apply an intense pressure impulse, typically 1.38 MPa, to the asphalt surface. Hence conventional equipment presents forces that are both rapidly applied and held for merely a fraction of a second.

These characteristics present two major negative effects concerning asphalt compaction. First, loads applied rapidly for a short duration cause the asphalt to respond with a high elastic stiffness and relatively small plastic deformation. To overcome the increased asphalt stiffness, conventional equipment seeks to increase the applied load or to use vibration, which may cause breakage of the aggregate particles. Second, the small contact area causes the application of horizontal (shoving) forces to the asphalt layer. As shown in Figure 1, the radial and tangential forces applied by the steel roller both contain horizontal components. As the roller travels, the horizontal forces push and pull the asphalt in front of and behind the drum, respectively, causing surface cracking with an interrelated dissipation of a significant portion of the applied compaction energy. The induction of surface cracking at this stage of construction reduces the strength and fatigue resistance of the asphalt layer, and facilitates the development of cracking, which can ultimately cause the premature failure of the pavement.

5 INNOVATIVE AMIR COMPACTOR TECHNOLOGY

The reason for roller checking has been attributed to the incompatibility between the geometry and relative rigidity of the soft, flat asphalt layer and the hard, cylindrical steel drums of conventional rollers (Abd EL Halim and Bauer 1986). To overcome these incompatibilities, the Asphalt Multi-Integrated Roller (AMIR) prototype, shown in Figure 2, was designed and fabricated by Carleton University and the National Research Council of Canada in 1989. The AMIR uses a multi-layered belt composed of special rubber compounds to create a single flat contact plate of approximately 3 m^2 for compaction. In addition to the flat contact area, the rubber belt is flexible, providing a closer match in rigidity to the asphalt surface. Due to the large contact area, the stress applied to the asphalt mat is a relatively small 41.6 KPa compared with stresses of 1.38 MPa applied by typical steel rollers. However, for the same rolling speed, the AMIR load duration is 30 times longer than that of conventional steel rollers. The AMIR prototype is well described in the literature, see for example Abd El Halim et al. (1993).

The HIPAC compactor, shown in Figure 3, was designed and fabricated in early 1998 by Pioneer Road Services in Australia based on AMIR technology. Significant operational modifications have been introduced in the HIPAC device since it represents an effort to produce a regular production, commercially available compaction device. As shown, the HIPAC is a double-belted device that provides greater manoeurvrebality than that of the AMIR compactor. The HIPAC

Figure 2. First full-scale (AMIR) prototype compactor.

Figure 3. New generation successor to AMIR, the HIPAC compactor.

development program is presented in greater depth in Svec and Abd El Halim (1991) and Abd El Halim and Haas (2004).

The initial stiffness response of the asphalt is kept low by the low compaction pressure applied gradually over a long duration. As shown in Figure 4, the applied stress from the compactor is more efficiently utilized because it is not "fighting" against a large initial damped elastic stiffness response that occurs with rapid, short duration loading. The long load duration also compensates for the low applied pressure by allowing visco-plastic flow of the asphalt, providing efficient particle contact and expulsion of entrained air. Furthermore, the large contact area minimizes the horizontal forces applied to the asphalt mat and provides a high degree of confinement during compaction. This, in turn, eliminates the roller induced cracking, reduces surface permeability, and increases tensile strength and resistance to fatigue damage. The elimination of surface cracking also permits the full compaction energy to be applied to the pavement layer.

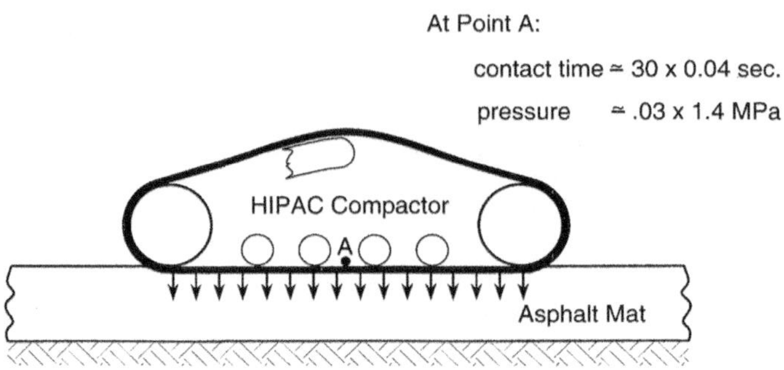

Figure 4. Schematic of flat plate type of compaction achieved with AMIR and HIPAC.

(a) Compacted Using AMIR Roller

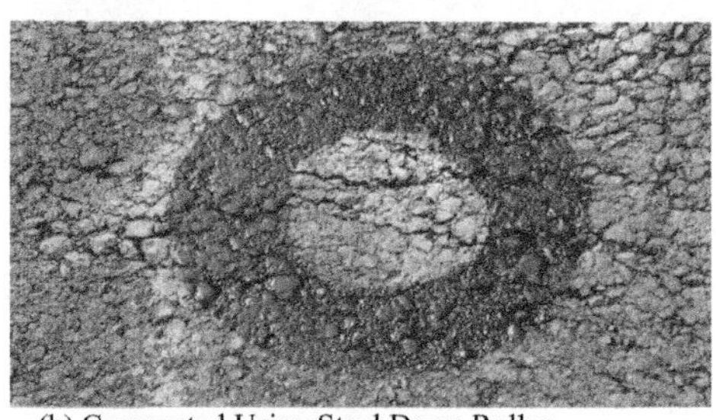

(b) Compacted Using Steel Drum Roller

Figure 5. Asphalt surface: (a) Compacted using AMIR roller (b) Compacted using steel rollers.

6 REVIEW OF AMIR FIELD TRIAL RESULTS

6.1 *Earlier field tests*

The AMIR (and subsequently the HIPAC) has been used in several field trials in Canada, during 1989–1993 and in Australia during 1997–2002. Detailed compaction and performance results observed with AMIR in Canada compared to conventional compaction equipment are found in numerous papers, including Abd El Halim (1993, 1994) and Rickards (1999). However, it is useful to present a brief review of the major conclusions of these field trials.

In general, it was found that the AMIR compacted sections showed a crack-free surface with tighter texture. Permeability tests competed in Australia indicated that the sections compacted using conventional equipment were between 2 and 4 times more permeable than the AMIR sections. Figure 5a shows an AMIR compacted asphalt surface, while Figure 5b shows a steel roller

Table 1. Permeability test results (Ottawa 2004).

Type	Initial height	Final height	Time (sec)	k
AMIR	67	51	600	$3.83e^{-5}$
AMIR	50	40	1800	$5.74e^{-5}$
AMIR	33.50	32.70	720	$3.84e^{-5}$
Steel	35	27	600	$4.94e^{-4}$
Steel	16.5	15.5	900	$3.47e^{-4}$
Steel	15.5	13.5	3600	$1.9e^{-4}$

Mean K value for AMIR = $4.47^{-5;}$ Mean K value for AMIR = $3.45^{-4;}$ Ratio of the K values of Steel and AMIR = 7.7.

compacted asphalt surface. As noted, the asphalt surface compacted using the AMIR is intact and crack free, while that compacted with the steel roller showed a rough surface with a large number of cracks.

Furthermore, AMIR achieved higher specific gravities and better uniformity of the density distribution than those achieved by the static steel roller. This was achieved the desired density after 6 passes while the combination of static steel and pneumatic rollers required 16 passes. In the filed tests carried out in Australia (Rickards et al. 1999), the HIPAC achieved a high compaction at the first pass, and the required compaction was achieved at a much lower number of passes.

6.2 *Ottawa field tests*

In 2004, due to new interest in the technology, a new rubber belt was installed and a couple of field tests were completed in Ottawa. One test included testing AMIR on compacting coarse sand while the second field test was performed on asphalt layer during the last week of September 2004. The asphalt layer was HL-3 provided by Lafarge Construction Materials in their site at Gloucester in Ottawa. The section consisted of 60 meter long by 3-meter width and 50 mm thickness. The section was compacted using AMIR side by side with the conventional steel roller. After the compaction, permeability tests were performed and the results are given in Table 1. As shown, the AMIR compaction provides much tighter surface which is confirmed by the measured K values and an average ratio of steel to AMIR of 7.7.

7 CONCLUDING REMARKS

Asphalt concrete pavements were known to have cracks and sometimes were found to significantly deteriorate over a few years following construction and to require rehabilitation at an age well below the economical age. One of the reasons for such rapid deterioration was found to be the construction cracks. Previous research showed that existing rollers are the main cause of construction cracks. These cracks, being already initiated at the construction stage, are then developed in the form of fatigue cracking, thermal cracking, and/or reflective cracking.

The construction of asphalt pavements is traditionally carried out by placing the hot mix asphalt over a base course or an existing road surface, and then compaction is carried out with one or more of three different types of compactors: heavy vibratory steel rollers, multi wheeled rubber roller, and light steel roller. This usually results in pushing a shoving of the hot mix ahead and behind the roller initiating cracking, which in turn caused premature failure of the asphalt pavements.

The development of an innovative approach for compacting asphalt concrete pavements has been introduced by describing a new compactor termed the Asphalt Multi Integrated Roller (AMIR). This approach has the advantage of producing crack-free asphalt surfaces at a much lower compacting effort and with less number of passes. Furthermore, AMIR achieved higher specific gravities and better uniformity of the density distribution than those achieved by the static steel roller.

The AMIR/HIPAC compaction technology has the capability of meeting current compaction standards and specifications as demonstrated in several field trials. In fact, the technology should also have major economic potential since only one machine can carry out the entire compaction process and it does not require highly trained operators.

ACKNOWLEDGEMENTS

The writers wish to acknowledge the cooperation and support of Mr. Ian Rickards of Pioneer Road Services of Australia, Professor R. Haas of University of Waterloo, late Dr. O. Svec, National Research Council of Canada (NRC), and Mr. J. Chander, formerly of the Industrial Research Assistance Program of NRC. Additional financial support provided by the Natural Sciences and Engineering Research Council of Canada (NSERC) and by the Ministry of Transportation of Ontario (MTO) is also gratefully acknowledged. The initial support of the Egyptian Ministry of Defence to build the first AMIR prototype is very much appreciated.

REFERENCES

Abd El Halim, A.O. and Bauer, G.E. (1986) "Premature Failure of Asphalt Overlays at Time of Construction". *Journal of Transportation Forum*, Road and Transportation Association of Canada, Vol. 3.2. Sept., 1986, pp. 52–58.

Abd El Halim, A.O., Haas, R. and Svec, O.J. (1994) "Improved Asphalt Pavement Performance Through a New Method of Compaction". Proc., *17th ARRB Conference,* Part. 3. pp. 175–191.

Abd El Halim, A.O., and R. Haas, (2004). "Process and Case Illustration of Construction Innovation: From Concept to Commercial Realization", *Journal of ASCE Construction Engineering and Management*, Volume 130, Issue 4, pp. 570–575.

Abd El Halim, A.O., Phang, W. and Haas, R.C. (1987). "Realizing Structural Design Objectives Through Minimizing Of Construction Induced Cracking", Proc., *Sixth International Conference on Structural Design of Asphalt Pavements*, Ann Arbor, U.S.A., July 13–16, Vol. I., pp. 965–970.

Abd El Halim, A.O., Phang, W. and Haas, R.C. (1993) "Unwanted Legacy of Asphalt Pavement Compaction". *Journal of Transportation Engineering*, Vol. 119, No. 6, November/December. pp. 914–932.

Brown, D. (1998) "Solving Compaction Problems with Superpave Mixes". *Engineering News Record*, March 23rd, pp. T21–T27.

Christison, T.J. (1972). "*The Response of Asphalt Concrete Pavements to Low Temperature*", Ph.D. Dissertation in Civil Engineering, Universtiy of Alberta, Canada.

Geller, M. (1982) "*Compaction Equipment for Asphalt Mixtures". Placement and Compaction of Asphalt Mixtures*, ASTM STP 829, American Society for Testing and Materials, Philadelphia, PA. p. 28.

Himeno, K. and Watnabe, T. (1987). "Design of Asphalt Pavements", Proc., *Six International Conference on Structural Design of Asphalt Pavements*, Ann-Arbour, Michigan.

Marker, V. "Construciton Methods—Symposium on Thick Lift Construction" Proc., *Association of Asphalt Pavement Technologists*, 1972, pp. 354.

Rickards, I., Goodman, S., Pagai, J., Abd El Halim, A.O. and Haas, R. (1999). "Practical Realization of a New Concept for Asphalt Compaction" *Transportation Research Record* No. 1654, pp. 27–35.

Svec, O.J. and Abd El Halim, A.O. (1991). "Field Verification of a New Asphalt Compactor, AMIR", *Canadian Journal of Civil Engineering*, Vol. 18, No. 3, 465–471.

Efficient Transportation and Pavement Systems – Al-Qadi, Sayed, Alnuaimi & Masad (eds)
© 2009 Taylor & Francis Group, London, ISBN 978-0-415-48979-9

Recycled waste materials as additives in cement-treated bases

Ashraf Rahim
Civil and Environmental Engineering, California Polytechnic State University, San Luis Obispo, CA

ABSTRACT: Four industrial byproducts were evaluated as additives in cement-treated aggregate bases. Additives included; Class F fly ash, Class C fly ash, Cement Kiln Dust (CKD), and Ground Granulated Blast Furnace Slag (GGBFS). Criteria based on the 7 and 28-day compressive strengths (350–450 psi @ 7 days) were adopted and optimal additive percentage was selected. Unconfined compressive strength along with splitting tensile strength and a modified version of ASTM D 559-03 tests were conducted to evaluate the different mixes. It was found that the four different byproduct additives could partially replace cement in mixtures while still meeting the strength criteria and maintaining durability.

Keywords: Fly Ash, CKD, GGBFS, Cement Treated Aggregate

1 INTRODUCTION

The American Concrete Institute defines soil-cement as a mixture of soil and measured amounts of portland cement and water compacted to a high density (ACI 1990). This definition can be further broadened to include other aggregates besides soil. As in concrete, aggregate particles in cement-treated bases (CTB) are bonded together by the cement paste. However, unlike the case of concrete individual particles are not completely covered by the cement paste.

Cement-treated base (CTB) is a compacted mixture of graded aggregate, cement and water. This mixture is used as a base course for the construction of highways, airport runways and taxiways. A CTB is used to provide a structurally sound base material over weak sub-grades such as expansive clay (http://gearsinc.com/prod02.htm, accessed 12/30/2007). Cement-treated aggregate is appealing as a low-cost construction material, and investigations have examined the use of marginal aggregates in combination with sand and cement, for use in low-cost pavement bases (Majumder et al. 1999). Other researchers have also studied the use of recycled concrete in CTA (Lim and Zollinger 2003).

Most problems with cement-stabilized base layers in pavement stem from the fact that current design practices are based only on strength (Guthrie et al. 2006). For example, many state department of transportation require sufficient cement to achieve a minimum unconfined compressive strength as high as 750 psi after seven days. While this level of cement results in a very stiff base layer, high dry shrinkage induced stresses could lead to cracking of the base layer. These cracks within heavily cement-stabilized base layers reflect into the surface asphalt concrete layer as transverse cracks.

As the cost of cement began to rise in the 1970s other additives/byproduct materials that possess cementitous properties are tried by partially replacing the cement in the concrete industry. Examples of these additives/byproduct materials include fly ash, cement kiln dust, granulated blast furnace slag among others.

Reducing the potential for cement treated bases to develop shrinkage cracks is crucial in order to improve the pavement serviceability and for a cost effective design. Recent studies suggest that crack-related degradation can be abated by adopting materials and/or methods that bring about a "desirable" crack pattern, where "desirable" being defined as numerous fine cracks at close spacing, which ensures adequate load transfer across the cracks (George 2002). It is not so much the number of cracks but the width of these cracks that has a significant influence on the long-term performance of the pavement since wider cracks have the tendency to reflect through the overlying pavement. Limiting/controlling drying shrinkage can effect the development of this "desirable" crack pattern in a stabilized layer. Several alternatives are available to control the drying shrinkage. These include: judiciously selecting the cement dosage, selecting a soil for stabilization having a limited fines content and plasticity, and the use of additives in conjunction with Portland cement, all of which promote development of a "desirable" crack pattern in a stabilized layer. While using byproduct additives to replace part of the cement may help reduce drying shrinkage strength and durability of the mixtures have to be evaluated. The main objective of this study is to investigate the effect of four byproduct additives on the strength characteristics of cement-treated bases.

2 EXPERIMENTAL WORK

2.1 *Materials*

Coarse, medium and fine aggregates were blended to produce a mix that meets Caltrans' gradation for CTB Class A (see Table 1). Portland cement type II was used in preparing all specimens tested in this study.

Four industrial byproducts were included in this research—fly ash class F, fly ash class C, Cement Kiln Dust (CKD) and Ground Granulated Blast Furnace Slag (GGBFS). Historically, class C fly ash has been the preferred supplemental material for concrete because of its self-cementitious properties, but class F fly ash is produced in larger quantities and finds fewer commercial applications. The source of fly ash is coal-fired electric power plants where fly ash is produced as a combustion byproduct formed as flue gasses cool. According to the American Coal Ash Association, as of 2001, 58% of the coal combustion products were Class F fly ash and 68% of this required disposal of for lack of use (Arora and Aydilek 2005). Fly ash class C and F used in this research were donated by Mineral Resource Technologies, Inc. Table 2 presents the chemical composition of the fly ash used in this study.

Cement Kiln Dust (CKD) is a byproduct of the cement production process. CKD is captured by air pollution control equipment installed on the chimneys of cement kilns. According to the

Table 1. Particle size distribution for aggregate used in the study.

Sieve size	Specification, %	% Passing
3-in	–	100
2 ½ -in	–	100
1-in	100	100
¾-in	90–100	94
No. 4	40–70	68
No. 30	12–40	33
No. 200	3–15	5

Table 2. Chemical composition of fly ash.

	SiO_2	Al_2O_3	Fe_2O_3	SO_3	CaO	MgO
FA class C	36.10	17.98	4.66	2.37	23.33	3.96
FA class F	47.87	21.30	5.08	1.24	13.83	2.84

Table 3. Chemical composition of CKD.

Component	SiO_2	Al_2O_3	Fe_2O_3	CaO	MgO	SO_3	Na_2O	K_2O	Ti_2O_5	P_2O_5	LOI*
%	12.86	4.46	0.91	51.11	2.39	2.52	0.40	1.72	0.07	0.20	22.2

* Loss on ignition.

EPA in 1990, the cement industry generated an estimated 12.7 million metric tons of CKD at 111 plants in 38 states of which 4 million metric tons were disposed of in piles, quarries, and landfills. The industry disposed of an estimated 3.3 million metric tons in 1995. There are currently 110 Portland cement plants operating in the United States and Puerto Rico with the highest production in California, Texas, Pennsylvania, and Michigan. CKD has cementitious properties which are dependent on the concentration of hydratable oxides such as free (un-reacted) lime or free Magnesia (MgO). But its fine powdery nature (0.3 mm max. size) and caustic properties (pH≈12) make CKD difficult material to handle in bulk. If moisture is added to control dust some of the cementitious properties are sacrificed (premature hydration) and the material becomes clumpy [Turner-Fairbanks Highway Research Center (TFHRC) 2002a]. It was included in the scope of this research because it is currently underutilized and has shown potential as a partial replacement for cement in mixtures (Udoeye and Hyee 2002). CKD for this study was donated by Colton Cement Plant, Colton, California. The chemical composition of CKD is shown in Table 3.

Ground Granulated Blast Furnace Slag (GGBFS) is a by product of the steel industry. Blast furnace slag is defined as "the non-metallic product consisting essentially of calcium silicates and other bases that is developed in a molten condition simultaneously with iron". In the production of iron, blast furnaces are loaded with iron ore, fluxing agents, and coke. When the iron ore, which is made up of iron oxides, silica, and alumina comes together with the fluxing agents, molten slag and iron are produced. The molten slag then goes through a particular process depending on what type of slag it will become. Air-cooled slag has a rough finish and larger surface area when compared to aggregates of that volume which allows it to bind well with Portland cements as well as asphalt mixtures. GGBFS is produced when molten slag is quenched rapidly using water jets, which produces a granular glassy aggregate (Cervantes and Roesler, 2007). The chemical composition for the GGBFS used in this study was not available and therefore, it was not included in this paper.

2.2 *Testing*

The optimal moisture content (OMC) required for compaction of the cement-aggregate mixture was determined based on the optimum Unconfined Compressive Strength (UCS) of 4.0-inch diameter by 4.6-inch height cement-aggregate specimens prepared at 5% cement and compacted employing standard proctor compaction and cured for 7 days. This is, however, the standard mold size used for both the moisture-density and the wetting and drying durability tests. Test Method A of ASTM 1633 uses this common mold size which has a height-to-diameter ratio (h/d) of 1.15. For comparison with samples molded with the ideal h/d of 2.0, a correction factor based on ASTM C 42 for concrete would use a strength reduction of 1/(1.1) (ACI 1990). As shown in Figure 1 the optimum moisture content (OMC) is approximately 6%.

For cement-aggregate mixtures with byproduct additives a portion of the 5% cement was replaced by an estimated equivalent portion of the byproduct additives based on the density of the respective byproducts. Several mixes with byproduct additives were tested for UCS and the mixtures that met the strength criterion were selected. This test was performed in general accordance with ASTM D 1633-00, a compressive strength test for molded cement-aggregate cylinders. Specimens from the same initial set of cylinders were also tested for their split tensile strength in accordance with ASTM C 496-04 designed for concrete. Specimens were cured at 95% humidity room for 3, 7 and 28 days before being tested. Table 4 lists the successful mixtures included in the study.

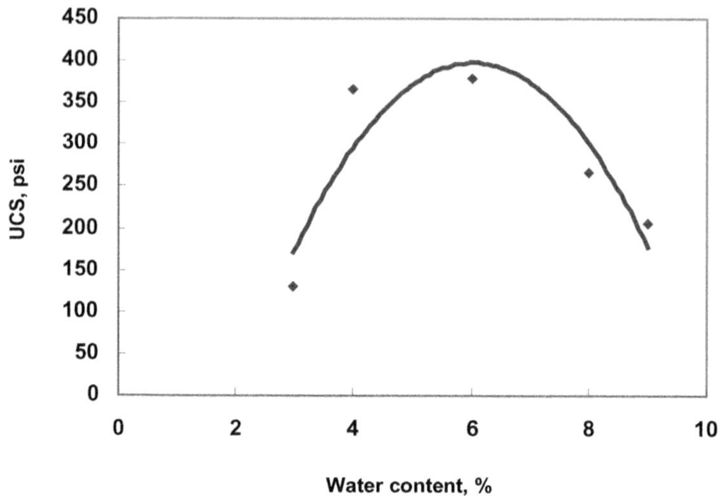

Figure 1. Water content *vs*. Ucs for 5%-cement mixture.

Table 4. Successful mixes with cement and byproduct proportions.

Symbol	Description	Cement, %	Byproduct, %	Water content,%
C	Cement-only mix	5	–	6.0
C/CKD	Mixes with cement and CKD byproduct	2	15	7.5
C/FAC	Mixes with cement and fly ash class C	4	6	6.5
C/FAF	Mixes with cement and fly ash class F	2	12	6.5
C/GGBFS	Mixes with cement and GGBFS	5	0.5	6.0

3 RESULTS AND DISCUSSION

3.1 *Unconfined compressive strength (ucs)*

Figures 2 presents the UCS test results. As noticed from Figure 2 all mixes with byproduct additives met the 7-day strength criterion adopted in this study (350–500) psi. This strength criterion was adopted following George's recommendation in a study conducted for the Portland Cement Association (PCA, 2002). By replacing 4.5% of the cement (expensive material) needed in the cement-only mix by 5% GGBFS (much cheaper byproduct) the aggregate mixture achieved approximately the same 7-day UCS. Mixes with fly ash class F exhibited the slowest strength gain trend among all mixes tested in this study. This trend could help in achieving the desirable crack pattern (several closely spaced minute cracks taking place early after construction) minimizing the possibility of reflection through the asphalt surface layer.

3.2 *Indirect Tensile Strength (ITS)*

The tensile strengths of selected mixture designs were evaluated using ASTM C 496-04—*Standard Test Method for Splitting Tensile Strength of Cylindrical Concrete Specimens*, a test designed for concrete which measures the tensile performance of a specimen indirectly.

In the test setup shown in Figure 3 the cement-aggregate cylinder is held on its side sandwiched in a jig. This produces loading which induces tensile stresses in the plane of applied loading. The test machine used for this procedure was also used for the unconfined compressive strength testing. Plywood bearing plates were added to distribute the load and prevent specimen movement.

It is seen in Figure 4 that Mixes with fly ash class F exhibited the lowest ITS gain during the first 2 weeks. This result suggests that drying shrinkage induced tensile stresses could overcome the tensile strength of this mix early after construction producing a desirable crack distribution (closely spaced minute cracks) rather than wide cracks at larger spacing which could reflect at a faster rate in the asphalt concrete surface. Narrow closely spaced cracks are less likely to translate through asphalt overlays built on cement-treated bases. Higher tensile strength allows greater stresses to develop before cracking is initiated. This leads to larger more widely spaced cracks which can reflect through pavement (George, 2002).

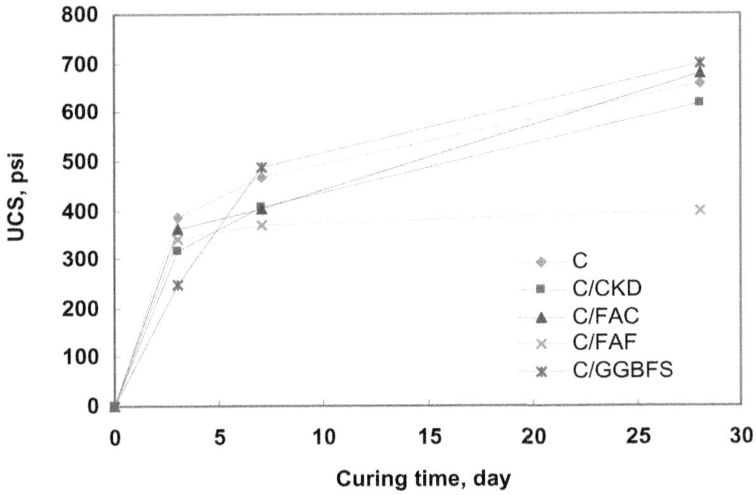

Figure 2. UCS for different mixes at different curing periods.

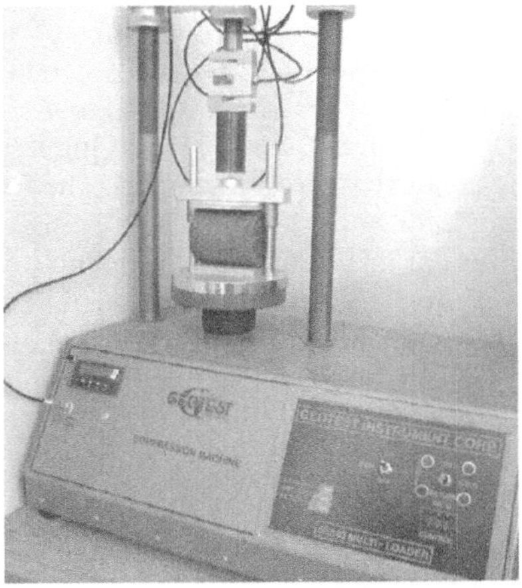

Figure 3. Machine used for ITS and UCS test.

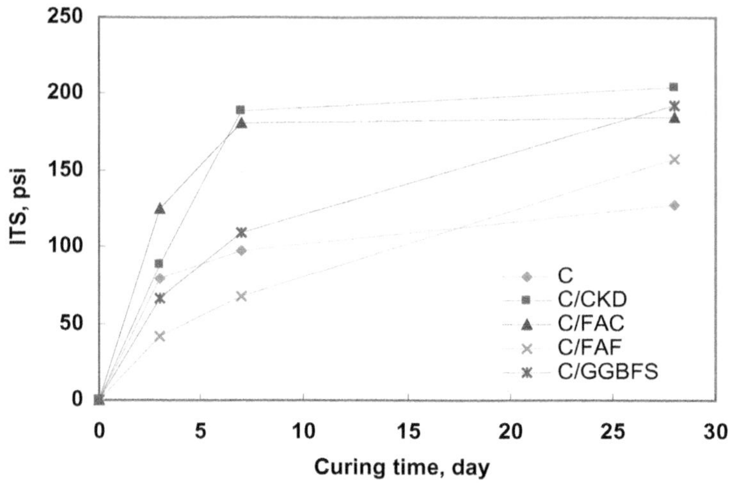

Figure 4. ITS for different mixes at different curing period.

Table 5. Loss in UCS after twelve wetting and drying cycles.

Mix	Base UCS[a], psi	WD-UCS[b], psi	% loss
C	656	523	20.0
C/CKD	618	351	43.0
C/FAC	680	578	15.0
C/FAF	398	323	19.0
C/GGBFS	698	327	53.0

[a] 28-day UCS without wetting drying
[b] UCS after wetting drying.

3.3 *Wetting and drying durability*

A modified ASTM D 559-03—*Standard Test Methods for Wetting and Drying Compacted Soil-Cement Mixtures* was performed to evaluate the durability of the different mixtures. In this modified version no samples were subjected to the manually conducted heavy brush which is considered subjective. However, the two samples were exposed to a series of 12 cycles of alternating immersion at room temperature for 5 hours followed by drying in an oven at 160°F for 42 hours. This is followed by UCS testing conducted on both samples. The UCSs determined from test were compared with the 28-day strength for same mixes without being exposed to wetting/drying. The loss in UCS was used as a measure of mixes durability (see Table 5).

As seen in Table 5, mixes with fly ash class C experienced the lowest loss in UCS following wetting and drying cycles. Mixes with fly ash class F and cement-only ranked second and third, respectively. Mixes containing GGBFS additive exhibited the poorest performance with regard to strength loss.

4 CONCLUSIONS

1. All compressive strengths for mixes with byproduct additives met the 7-day strength criterion set for this study.

2. The addition of GGBFS additive allowed a significant reduction in the amount of cement required to meet the compressive strength criterion established for this test program.
3. Mixes with CKD and fly ash class C exhibited the fastest rate of increase in strength gain while mix with fly ash class F experienced the slowest strength gain making it a candidate for treated base that could provide a desirable crack pattern.
4. Based on the loss of UCS after wetting-drying durability test, Mix with fly ash class C was ranked the best, followed by mix with fly ash class F and mix with cement only.
5. More research is needed to study the effect of these additives on shrinkage characteristics. Field trials will be needed to verify the findings of this study in-situ.

ACKNOWLEDGEMENT

This report includes the results of a study titled, "Recycled Waste Materials as Additives to Improve the Performance of Soil-cement—A Laboratory Investigation" sponsored by the Department of the Navy, Office of Naval Research, under Award # N00014-05-1-0855. The author wishes to thank Dr. Susan Opava for making this support available. The hard work of under-graduate students Alex Ubaldo and Bill Johnson who worked on this project and input from Jennifer Beck in preparing this report are also acknowledged. The author is greatly thankful to Mr. Kirk McDonald of Colton Cement Plant for donating the CKD and his help in obtaining the fly ash used in this study. Mineral Resource Technologies, Inc. donated fly ash class C and class F.

REFERENCES

American Concrete Institute (ACI) (1990), *State-of-the-Art Report on Soil-Cement*, ACI 230.1R-90, Committee 230, Farmington Hills, MI.

Arora, S. and Aydilek, A.H. (2005), "Class F Fly-Ash-Amended Soils as Highway Base Materials," *Journal of Materials in Civil Engineering*, Vol. 17, No. 6 November/December, pp. 640–649.

Cement Treated Base (2007), (http://gearsinc.com/prod02.htm, accessed 12/30/2007).

Cervantes, V. and Roesler, J. (2007), Ground Granulated Blast Furnace Slag, Technical Note No. 35 http://www.ceat.uiuc.edu/PUBLICATIONS/technotes, Accessed 12/30/2007.

George, K.P. (2002), "Minimizing Cracking in Cement Treated Materials for Improved Performance", *Research and Development Bulletin No. RD123*, S Portland Cement Association Skokie, Ill.

Guthrie, W.S., Sebesta, S. and Scullion, T. (2006), Improving Long-term Performance Of Cement-Treated Aggregate Base Materials. Texas Transportation Institute, *Report No. 7 4920-S.*

Lim, S. and Zollinger, D.G. (2003). Estimation of the Compressive Strength and Modulus of Elasticity of Cement-Treated Aggregate Base Materials. *Transportation Research Record* (1837), 30.

Majumder, B.K., Das, A. and Pandey, B.B. (1999). Cement Treated Marginal Aggregates for Roads. *Journal of Materials in Civil Engineering*, 11(3), 257–265.

Turner-Fairbanks Highway Research Center (TFHRC) (2002a), "Coal Fly Ash User Guideline—Stabilized Base," http://www.tfhrc.gov/hnr20/recycle/waste/cfa55.htm, Accessed on April 30, 2006.

Udoeyo, F.F. and Hyee, A. (2002), "Strengths of Cement Kiln Dust Concrete", *Journal of Materials in Civil Engineering*, Volume 14, Issue 6, pp. 524–526.

Efficient Transportation and Pavement Systems – Al-Qadi, Sayed, Alnuaimi & Masad (eds)
© 2009 Taylor & Francis Group, London, ISBN 978-0-415-48979-9

Statistical approach to analyze compressive strength of interlocking paving tiles in Kuwait

Sherida Al-Azmi, Sayed Metwali & Amal Merza
Maintenance Division, Ministry of Public Works, Kuwait

Ahmad H. Aljassar
Civil Engineering Department, Kuwait University, Kuwait
Maintenance Division, Ministry of Public Works, Kuwait

ABSTRACT: The Ministry of Public Works (MPW) has been using interlocking paving tiles for many years. The multicolor tiles were first used in highly trafficked pedestrian paths and walkways in commercial areas, especially in the commercial center of the Capital. The interlocking paving tiles have been widely used for aesthetic advantages and their high load-bearing capacity. These tiles are also used as a surface layer of asphalt pavements in many signalized intersections to control surface rutting. Their use was then expanded to include U-turns in many locations for the same reason. Due to the achieved success in using the interlocking paving tiles, they are also used as a surface layer of asphalt pavements in many locations on Kuwait road network. MPW specifications require special criteria for interlocking paving tiles, the most important of which is the compressive strength. This paper presents the newly-developed statistical approach to analyze compressive strength test results and acceptance criteria in accordance with Kuwait specifications.

1 INTRODUCTION

In the late 80s of the last century, pavement shoulders in Kuwait were mostly uncovered soil; except in some commercial and industrial areas where shoulders were paved with gray-color sand-cement square tiles. As a step towards improving public services, the Ministry of Public Works (MPW) was assigned the responsibility of paving road shoulders, medians, and other places prone to movement of surface soil. This was carried out on a large scale for its environmental importance, since cover-up of open surfaces prevents the movement of surface soils by rains and therefore eliminates the problem of its subsequent precipitation in storm-water gullies and pipes which is responsible for malfunctions of storm-water drainage networks and water collection in roads. The latter is a main cause of traffic problems and jeopardizes traffic safety. Paving open areas also improves aesthetics of pavement shoulders, medians, walkways and parking areas in commercial and residential areas in the urban areas.

In 1988, MPW started paving projects to pave road shoulders with colored paving tiles of different types. Mainly, there were two types of paving tiles used. The first was the square sand-cement type which is placed over a layer of cement paste. The other was the interlocking tiles in its various shapes which are laid over a layer of sand without using cement paste. The first type was mainly used in the main road shoulders where only occasional pedestrian traffic is expected. The interlocking tiles were used in commercial areas, especially in the city center of the Capital which witnesses a heavy pedestrian traffic.

The use of interlocking tiles was widely accepted by the public for its improvement of aesthetics in addition to its high bearing capacity and low-cost and ease of maintenance. This encouraged MPW to use it in large scales and to serve other purposes other than paving shoulders and walkways only. In 1989, MPW tested the use of interlocking tiles as paving material in parking lots in Arabian Gulf road in Salmiya district. The experience proved successful. These parking lots are still intact (after 18 years of service) although the asphalt pavement of the road witnessed many activities of maintenance during this same period because of the heavy traffic volumes it serves.

2 USING INTERLOCKING TILES TO RESIST RUTTING ON ROAD SURFACES

The pavement surface in U-turns is subjected to repeated applications of horizontal shear forces caused by the tires of the turning vehicles. This led to rutting of the pavement surface in all U-turns shortly after construction of the pavement. Therefore, maintenance activities at the U-turns were very frequent to reinstate the shape of the pavement surface. Hence, MPW's Maintenance Department of Hawalli Governorate started experimenting the use of interlocking tiles as a surface paving material in U-turns on the Third Ring Road. This was done in 1994. Since then the interlocking tiles have not needed any maintenance, although the material of the tiles itself is aging. The rutting phenomenon was almost eliminated.

The infrastructure in Kuwait is growing rapidly because of the rapid growth of the economy and the sudden increase in population. Therefore, the nation's road network is experiencing increases in traffic volumes at increasing rates. Moreover, axle loads are also increasing (No actual measurements are available to measure these axle loads accurately; there is a study underway to conduct such measurements). The climate in Kuwait is harsh. The weather is hot in most parts of the year, and the temperature variations are high between day and night and between summer and winter. In addition, weather humidity is common in some months of the year. Therefore, pavements in Kuwait are serving in severe conditions especially under the increasingly increasing and non-monitored traffic loads. These factors together are responsible for the pavement distresses that are common in Kuwait. One main distress is the surface rutting which occurs in wheel tracks on pavements. Rutting sometimes penetrates deeper than the surface layer in some locations, especially at stop-and-go spots during the hot weather season when high temperatures reduce the stiffness of the asphalt concrete pavement and therefore reduce its bearing capacity. This causes the need for very frequent maintenance activities to correct surface profiles at spots where rutting is occurring. Therefore, and because of the previous successes in using the interlocking tiles, the Maintenance Division of MPW was encouraged to enlarge the scale of using it as a pavement material for pavement surfaces at main intersections used by heavy vehicles such as busses and trucks.

Since then, the use of interlocking tiles as a paving material has become common. A complete road may now be surfaced with interlocking tiles for the two advantages of aesthetics and less frequent maintenance.

3 CONSTRUCTION OF INTERLOCKING TILES AS A PAVING COURSE

In general, the followed local practice of constructing interlocking tiles as a paving material for roads includes the construction of a 10 cm layer of Type II asphalt concrete (Type II is local name for this type of mix), then a 3 to 5 cm layer of washed sand is laid, on top of which 8 cm thick interlocking tiles are placed and compacted. Special attention is paid to the water table in the construction site to ensure that enough measures are applied to prevent the collapse of the pavement layers due to rising water tables.

The MPW specifications require certain properties for the interlocking tiles. The most important of which is the compressive strength requirement. It is well known that for Portland cement concrete, properties of the mix improve with increasing compressive strength. Therefore, the compressive strength is used as a quality control/quality assurance property for Portland cement concrete materials.

A system (supported by a software) was developed to analyze compressive strength data obtained from testing samples of interlocking paving tiles imported to sites where thy are to be laid and compacted. The system gives decision either to approve imported materials, reject them, or accept with appropriate discount. The following is a brief description of the system.

4 DISCOUNT SYSTEM ON INTERLOCKING TILES

The imported interlocking tiles imported to paving sites are tested for compressive strength. A system was developed to handle the results of the compressive strength tests. The system was coded as a software to be applied on interlocking tiles paving jobs executed by the Maintenance Division. The compressive strength results are entered in the system where they are statistically analyzed to determine the percentage of executed works which lie inside the discount region of the normal distribution curve.

The statistical discount system developed was programmed in two different computer software program formats. The first is programmed using Excel, and is intended for the supervising engineer to use. After receiving the results of the compressive strength test performed on interlocking tiles at the Government Center for Testing and Research, the supervising engineer enters them in his copy of the statistical system and determines whether the job is expected to be accepted, rejected, or accepted with a discount. If a discount was found to be applied to the job under consideration, he would inform the contractor of this penalty before executing the job. If the result was rejection, the contractor is informed to replace the imported interlocking tiles with another batch that is according to specifications, in which case the new material would go in the same checking cycle. If the result was acceptance, the contractor is allowed to execute the job. The developed program enables the supervising engineer (as well as the contractor) with the expected decision of the work before execution, and whether a discount is going to be applied by the Office of Planning and Follow Up after the job order has been completed. This certainly save time, effort, and money for all parties involved, and reduces the chances of removal and exchange of laid interlocking tiles which is the case when the compressive strength results were found out of specifications and cannot be accepted, in which case the executed interlocking tiles are removed and replaced.

The other computer software program format was developed in DOS environment using Basic programming language. This was done to integrate the program with the existing computer network of the Maintenance Division which is linked to KIMMS (Kuwait Integrated Infrastructure Maintenance Management System). This enables the Office of Planning and Follow Up to use the developed system and apply the appropriate discounts (if applicable) on executed job orders. Incorporating the developed system in the network also serves the purpose of archiving historical discount data on previous job orders.

5 BASIS OF THE DISCOUNT SYSTEM

The statistical discount system for analyzing strength results of interlocking tiles was developed based on scientific approach which applies discount rates that are calculated based on statistical principles. This was meant to develop a mechanism that enables the acceptance of interlocking tiles jobs which involve some deviations from specification limits but can be accepted with an appropriate discount rate that is comparable to the amount of deviation from specifications, but not resulting in jobs that are below the absolute acceptable level.

The compressive strength property was chosen as the basic property of which results are used in the developed system. The system is based on statistical values, focusing on the mean ($\bar{x}$) and the standard deviation (s). Since the tested samples of interlocking tiles are to be chosen randomly, the strength results of the tested samples are assumed to comply with the normal distribution where the frequency of a certain value of compressive strength is highest around the mean and reduces as the value is further from the mean. The standard deviation can be thought of as an index of the deviation of results of the measured property (compressive strength in this case) from the mean value. If more values deviate from the mean, the standard deviation will become

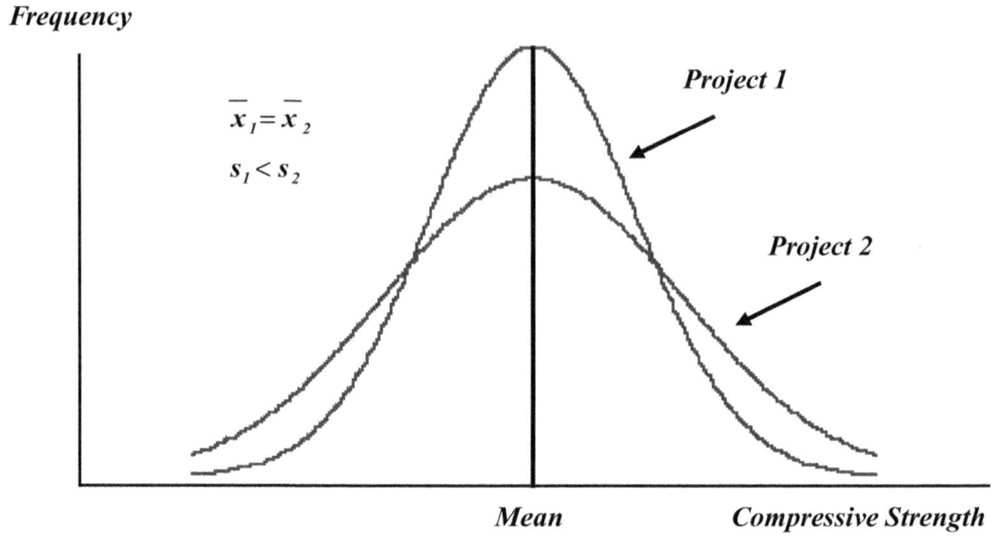

Figure 1. The normal distribution of compressive strength results of interlocking tiles for two different projects.

higher, and vice versa. In the old practice, a job would be accepted or rejected based on the mean value of strength results only. However this is not enough to expect how the strength results of the imported interlocking tiles are distributed around the mean. Interlocking tiles of two jobs may have the same mean (and both may be accepted in the old acceptance system) but one has more consistent results (less standard deviation) than the other (Figure 1). The new developed system incorporates the standard deviation and applies the known statistical analysis to determine the appropriate discount rate.

In order to verify the calculations of the developed program, data for different projects were entered into the algorithm and results were checked and found accurate. After approving the developed system in the Maintenance Division, it will be used to apply appropriate discounts on interlocking tiles jobs that the division executes all over Kuwait.

6 CONCLUSION

In this paper, an overview is presented for the newly-developed statistical approach to analyze the compressive strength test results of the interlocking paving tiles used by the Maintenance Division of the Ministry of Public Works in its various projects. These projects include pedestrian paths and walkways as well as paving of roads. The acceptance criteria of the interlocking paving tiles in such projects as per local specifications were presented.

ACKNOWLEDGEMENT

The authors would like to thanks Eng. Al-Numan Al-Shuar for the Office of Planning and Development, Maintenance Division, for writing the algorithms for the developed discount system on interlocking tiles, and Eng. Ahmed Al-Amriti for his valuable contribution in the team work. Thanks are also due to Eng. Fadel Al-Ajmi, Assistant Undersecretary for Maintenance, MPW, for his support to the team that was given the responsibility of developing the system described in this paper.

Efficient Transportation and Pavement Systems – Al-Qadi, Sayed, Alnuaimi & Masad (eds)
© 2009 Taylor & Francis Group, London, ISBN 978-0-415-48979-9

Author index

For Product Safety Concerns and Information please contact our EU representative GPSR@taylorandfrancis.com Taylor & Francis Verlag GmbH, Kaufingerstraße 24, 80331 München, Germany

Batch number: 10731482

Printed by Printforce, the Netherlands